# NFPA® 101®

# Life Safety Code®

## 2012 Edition

This edition of NFPA *101®*, *Life Safety Code*, was approved by NFPA at its June Association Technical Meeting held June 12–15, 2011, in Boston, MA. It was issued by the Standards Council on August 11, 2011, with an effective date of August 31, 2011, and supersedes all previous editions.

A tentative interim amendment (TIA) to Table 8.3.4.2 was issued on August 11, 2011. For further information on tentative interim amendments, see Section 5 of the NFPA Regulations Governing Committee Projects available at:

http://www.nfpa.org/assets/files/PDF/CodesStandards/TIAErrataFI/TIARegs.pdf

This edition of NFPA *101* was approved as an American National Standard on August 31, 2011.

## Origin and Development of NFPA 101

The *Life Safety Code* had its origin in the work of the Committee on Safety to Life of the National Fire Protection Association, which was appointed in 1913. In 1912, a pamphlet titled *Exit Drills in Factories, Schools, Department Stores and Theaters* was published following its presentation by the late Committee member R. H. Newbern at the 1911 Annual Meeting of the Association. Although the pamphlet's publication antedated the organization of the Committee, it was considered a Committee publication.

For the first few years of its existence, the Committee on Safety to Life devoted its attention to a study of the notable fires involving loss of life and to analyzing the causes of this loss of life. This work led to the preparation of standards for the construction of stairways, fire escapes, and other egress routes for fire drills in various occupancies, and for the construction and arrangement of exit facilities for factories, schools, and other occupancies. These reports were adopted by the National Fire Protection Association and published in pamphlet form as *Outside Stairs for Fire Exits* (1916) and *Safeguarding Factory Workers from Fire* (1918). These pamphlets served as a groundwork for the present *Code*. These pamphlets were widely circulated and put into general use.

In 1921, the Committee on Safety to Life was enlarged to include representatives of certain interested groups not previously participating in the standard's development. The Committee then began to further develop and integrate previous Committee publications to provide a comprehensive guide to exits and related features of life safety from fire in all classes of occupancy. Known as the *Building Exits Code*, various drafts were published, circulated, and discussed over a period of years, and the first edition of the *Building Exits Code* was published by the National Fire Protection Association in 1927. Thereafter, the Committee continued its deliberations, adding new material on features not originally covered and revising various details in the light of fire experience and practical experience in the use of the *Code*. New editions were published in 1929, 1934, 1936, 1938, 1939, 1942, and 1946 to incorporate the amendments adopted by the National Fire Protection Association.

National attention was focused on the importance of adequate exits and related fire safety features after the Cocoanut Grove Night Club fire in Boston in 1942 in which 492 lives were lost. Public attention to exit matters was further stimulated by the series of hotel fires in 1946 (LaSalle, Chicago — 61 dead; Canfield, Dubuque — 19 dead; and Winecoff, Atlanta — 119 dead). The *Building Exits Code*, thereafter, was used to an increasing extent for regulatory purposes. However, the *Code* was not written in language suitable for adoption into law, because it had been drafted as a reference document and contained advisory provisions that were useful to building designers but inappropriate for legal use. This led to a decision by the Committee to re-edit the entire *Code*, limiting the body of the text to requirements suitable for mandatory application and placing advisory and explanatory material in notes. The re-editing expanded *Code* provisions to cover additional occupancies and building features to produce a complete document. The *Code* expansion was carried on concurrently with development of the 1948, 1949, 1951, and 1952 editions.

The results were incorporated into the 1956 edition and further refined in subsequent editions dated 1957, 1958, 1959, 1960, 1961, and 1963.

In 1955, NFPA 101B, on nursing homes, and NFPA 101C, on interior finish, were published. NFPA 101C was revised in 1956. These publications have since been withdrawn.

In 1963, the Committee on Safety to Life was restructured to represent all interested factions and to include only those members with broad knowledge of fire matters. The Committee served as a review and correlating committee for seven sectional committees whose personnel included members having a special knowledge and interest in various portions of the *Code*.

Under the revised structure, the sectional committees, through the Committee on Safety to Life, prepared the 1966 edition of the *Code*, which was a complete revision of the 1963 edition. The *Code* title was changed from *Building Exits Code* to *Code for Safety to Life from Fire in Buildings and Structures*. The *Code* text was written in enforceable "code language," and all explanatory notes were placed in an appendix.

The *Code* was placed on a 3-year revision schedule, with new editions adopted in 1967, 1970, 1973, and 1976.

In 1977, the Committee on Safety to Life was reorganized as a technical committee, with an executive committee and standing subcommittees responsible for various chapters and sections. The 1981 edition contained major editorial changes, including reorganization within the occupancy chapters, to make them parallel to each other, and the splitting of requirements for new and existing buildings into separate chapters. Chapters on detention and correctional facilities were added, as well as new sections for ambulatory health care centers.

The 1985 edition contained a new Chapter 21 on residential board and care occupancies with related Appendix F and Appendix G, a new Appendix D on alternative calculations for stair width, and Appendix E, a fire safety evaluation system (FSES) for detention and correctional facilities.

The 1988 edition contained a major change in the method of determining egress capacity with the deletion of the traditional units of exit width and the substitution of a straight linear approach to calculating egress capacity. Appendix C through Appendix G were moved from NFPA *101* into a new document, NFPA 101M.

The 1991 edition contained numerous new requirements for mandatory sprinklers in new health care facilities, hotels, apartment buildings, lodging and room houses, and board and care facilities, as well as mandatory sprinkler requirements for existing high-rise hotels and apartment buildings. The requirements for board and care facilities were split into two chapters, Chapter 22 for new construction and Chapter 23 for existing buildings.

The 1994 edition contained new requirements for accessible means of egress, areas of refuge, and ramps, putting the *Code* in substantial agreement with the Americans with Disabilities Act Accessibility Guidelines (ADAAG).

The 1997 edition relocated the material on day-care occupancies from Chapters 10 and 11 for new and existing educational occupancies to new Chapters 30 and 31. The operating features requirements, previously contained in Chapter 31, were interspersed throughout the *Code*, as applicable.

The 2000 edition introduced a performance-based option via Section 4.4 and new Chapter 5. That edition also reformatted the *Code* for substantial compliance with the NFPA *Manual of Style:* (1) former Chapter 1, General, was split into Chapter 1, Administration, and Chapter 4, General; (2) the mandatory references list was moved from Chapter 33 to Chapter 2; (3) all definitions were moved into Chapter 3, and each defined term was numbered; (4) the paragraph numbering style that separated the chapter number from the section number using a hyphen was changed to the use of a decimal point as the separator; and (5) the appendixes were renamed annexes. Former Chapter 32 on special structures and high-rise buildings was moved to Chapter 11 to join the core chapters (i.e., the chapters that are not occupancy specific). The subject of interior finish, contents, and furnishings was moved from Section 6.5 into a separate new chapter, Chapter 10. The occupancy chapters, formerly Chapters 8 through 32, became Chapters 12 through 42, with some repositioning of chapters. For example, the day-care occupancies chapters were renumbered from Chapters 30/31 to Chapters 16/17, so as to be positioned immediately after the chapters for educational occupancies.

The 2003 edition reformatted all exceptions into numbered or lettered paragraphs. Some reformatting of paragraphs with multiple requirements was done for additional compliance with the NFPA *Manual of Style*.

The 2006 edition repositioned the inch-pound (U.S. Customary) units to appear first, followed by the metric equivalent (SI) units in parentheses. New Chapter 43, Building Rehabilitation, was added to promote the adaptive reuse of existing buildings without sacrificing needed life safety.

The 2009 edition added provisions to Chapter 7 for electrically controlled egress doors, horizontal-sliding doors serving an area with an occupant load of fewer than 10, elevator lobby access door locking, and door inspection and maintenance. The remoteness criteria of Chapter 7 were expanded to have applicability to all three portions of the means of egress — exit access, exit, and exit discharge. Extensive revisions were made throughout the *Code* to standardize the use of the terms *stories in height, finished ground level, grade plane, basement,* and *level of exit discharge*. Section 9.6 and the applicable occupancy chapters were revised to limit the use of public address systems for occupant alarm notification to large venue assembly occupancies and mercantile mall buildings, where the physical configuration, function, and human behavior present challenges with respect to effective occupant notification by standard means in accordance with NFPA 72®, *National Fire Alarm Code*®. A subsection was added to Chapter 11 for special provisions applicable to air traffic control towers. The criteria for assembly stage proscenium opening fire curtains were deleted from

Chapter 12 and replaced by a reference to the new fire curtain provisions of NFPA 80, *Standard for Fire Doors and Other Opening Protectives*. Provisions were added to Chapters 14 through 17 for the placement and use of alcohol-based hand-rub dispensers in educational and day-care occupancies. The provisions of Chapters 18 and 19 were expanded to address door locking where the needs of patients or clients require specialized protective measures for their safety and security in hospitals, nursing homes, and limited care facilities. Also, a limitation on common path of travel was added to Chapter 18 for new health care occupancies; the requirement for patient sleeping room windows was deleted for new and existing health care occupancies; and all existing high-rise health care occupancy buildings must be sprinklered within 12 years of the adoption of this edition of the *Code*. Numerous occupancy chapters were revised to require emergency plans in accordance with Section 4.8. Chapter 43 on building rehabilitation was revised to address issues not identified when the chapter was written for the 2003 edition and to delete redundancies. An adoptable annex was added for elevators for occupant evacuation prior to Phase I Emergency Recall Operations. Another adoptable annex was added for supplemental escape devices and systems.

The 2012 edition expanded what had been the definitions of *noncombustible material* and *limited-combustible material* and moved the material to new subsections in Chapter 4. The material addressing elevators for occupant controlled evacuation which had comprised Annex B was moved to Chapter 7. A new section was added to Chapter 7 to address normally unoccupied building service equipment support areas. The Chapter 8 table addressing minimum fire protection ratings for opening protectives was expanded. Provisions for carbon monoxide detection were added to Chapter 9. Requirements for carbon monoxide detection were added to some of the occupancy chapters. The health care occupancies provisions were modified to permit the health care setting to be made more homelike.

## To the User

The following comments are offered to assist in the use of the *Life Safety Code*. Additional help on using the *Life Safety Code* can be obtained by attending one of the seminars NFPA conducts on the *Life Safety Code* or by using the *Life Safety Code Handbook* available from NFPA. Further information on these seminars is available through the NFPA Division of Continuing Education.

Essentially, the *Code* is comprised of four major parts. The first part consists of Chapters 1 through 4, Chapters 6 through 11, and Chapter 43; these are often referred to as the base chapters or fundamental chapters. The second part is Chapter 5, which details the performance-based option. The next part consists of Chapters 12 through 42, which are the occupancy chapters. The fourth and last part consists of Annex A and Annex B, which contain useful additional information.

A thorough understanding of Chapters 1 through 4, Chapters 6 through 11, and Chapter 43 is necessary to use the *Code* effectively, because these chapters provide the building blocks on which the requirements of the occupancy chapters are based. Note that many of the provisions of Chapters 1 through 4 and Chapters 6 through 11 are mandatory for all occupancies. Some provisions are mandated only when referenced by a specific occupancy, while others are exempted for specific occupancies. Often, in one of the base chapters, especially in Chapter 7, the phrase "where permitted by Chapters 11 through 43" appears. In this case, that provision can be used only where specifically permitted by an occupancy chapter. For example, the provisions of 7.2.1.6.1 on delayed-egress locks are permitted only when permitted by Chapters 11 through 43. Permission to use the delayed-egress lock is normally found in the "____.2.2" subsection of each occupancy chapter. For example, 12.2.2.2.5 specifically permits the use of delayed-egress locks in new assembly occupancies. If this permission is not found in an occupancy chapter, the delayed-egress lock cannot be used. Similar types of restricted permission are found for such items as security grilles, double-cylinder locks, revolving doors, and so forth. In other locations in the base chapters, the wording "unless prohibited by Chapters 11 through 43" is used. In this case, the provision is permitted in all occupancies, unless specifically prohibited by an occupancy chapter.

Metric units of measurement in this *Code* are in accordance with the modernized metric system known as the International System of Units (SI). The unit liter, which is outside of but recognized by SI, is commonly used and is therefore used in this *Code*. In this *Code*, inch-pound units for measurements are followed by an equivalent in SI units, as noted in 1.5.2. The inch-pound value and the SI value are each acceptable for use as primary units for satisfying the requirements of this *Code*.

## Technical Correlating Committee on Safety to Life

**William E. Koffel,** *Chair*
Koffel Associates, Inc., MD [SE]

**Ron Coté,** *Administrative Secretary*
National Fire Protection Association, MA

**David S. Collins,** The Preview Group, Inc., OH [SE]
Rep. American Institute of Architects
**Wayne D. Holmes,** HSB Professional Loss Control, NC [I]
**Howard Hopper,** Underwriters Laboratories Inc., CA [RT]
**Kenneth E. Isman,** National Fire Sprinkler Association, Inc., NY [M]
**Thomas W. Jaeger,** Jaeger and Associates, LLC, VA [U]
Rep. American Health Care Association
**J. Edmund Kalie, Jr.,** Prince George's County Government, MD [E]
**David C. Lind,** North Shore Fire Department, WI [E]
Rep. International Fire Marshals Association

**George H. McCall,** Wade Hampton Fire Department, SC [U]
Rep. International Association of Fire Chiefs
**Jake Pauls,** Jake Pauls Consulting Services in Building Use & Safety, MD [C]
Rep. American Public Health Association
**James R. Quiter,** Arup, CA [SE]
**Todd C. Shearer,** Tyco/SimplexGrinnell, NJ [M]
Rep. National Electrical Manufacturers Association

### Alternates

**John F. Bender,** Underwriters Laboratories Inc., MD [RT]
(Alt. to H. Hopper)
**Sharon S. Gilyeat,** Koffel Associates, Inc., MD [SE]
(Alt. to W. E. Koffel)
**Stanley C. Harbuck,** School of Building Inspection, UT [C]
(Alt. to J. Pauls)

**Jeffrey M. Hugo,** National Fire Sprinkler Association, Inc., MI [M]
(Alt. to K. E. Isman)
**Isaac I. Papier,** Honeywell, Inc., IL [M]
(Alt. to T. C. Shearer)
**Martin H. Reiss,** The RJA Group, Inc., MA [SE]
(Alt. to J. R. Quiter)

### Nonvoting

**John A. Alderman,** Aon/Schirmer Engineering Corporation, TX [I]
Rep. TC on Industrial, Storage, & Miscellaneous Occupancies
**Chad E. Beebe,** American Society for Healthcare Engineering, WA [E]
Rep. TC on Board & Care Facilities
**Warren D. Bonisch,** Aon/Schirmer Engineering Corporation, TX [I]
Rep. TC on Residential Occupancies
**Kenneth E. Bush,** Maryland State Fire Marshals Office, MD [E]
Rep. TC on Mercantile & Business Occupancies
**Wayne G. Carson,** Carson Associates, Inc., VA [SE]
Rep. TC on Fundamentals
**Michael DiMascio,** Arup Fire, MA [SE]
Rep. TC on Detention & Correctional Occupancies
**Joseph M. Jardin,** Fire Department City of New York, NY [C]
Rep. TC on Building Service & Fire Protection Equipment
**David P. Klein,** U.S. Department of Veterans Affairs, DC [U]
Rep. TC on Health Care Occupancies

**James K. Lathrop,** Koffel Associates, Inc., CT [SE]
Rep. TC on Means of Egress
**Wayne D. Moore,** Hughes Associates, Inc., RI [SE]
Rep. Signaling Systems Correlating Committee
**Henry Paszczuk,** Connecticut Department of Public Safety, CT [E]
Rep. TC on Furnishings & Contents
**Eric R. Rosenbaum,** Hughes Associates, Inc., MD [SE]
Rep. TC on Fire Protection Features
**Aleksy L. Szachnowicz,** Anne Arundel County Public Schools, MD [U]
Rep. TC on Educational & Day-Care Occupancies
**Jeffrey S. Tubbs,** Arup, MA [SE]
Rep. TC on Assembly Occupancies & Membrane Structures
**Joseph H. Versteeg,** Versteeg Associates, CT [E]
Rep. TC on Alternative Approaches to Life Safety
**Shane M. Clary,** Bay Alarm Company, CA [IM]
(Alt. to W. D. Moore)
Rep. Signaling Systems Correlating Committee
**John L. Bryan,** Frederick, MD [SE]
(Member Emeritus)
**Harold E. Nelson,** Annandale, VA [SE]
(Member Emeritus)

**Ron Coté,** NFPA Staff Liaison

*This list represents the membership at the time the Committee was balloted on the final text of this edition. Since that time, changes in the membership may have occurred. A key to classifications is found at the back of the document.*

NOTE: Membership on a committee shall not in and of itself constitute an endorsement of the Association or any document developed by the committee on which the member serves.

**Committee Scope:** This Committee shall have primary responsibility for documents on the protection of human life from fire and other circumstances capable of producing similar consequences and for the non-emergency and emergency movement of people.

## Technical Committee on Assembly Occupancies and Membrane Structures

**Jeffrey S. Tubbs,** *Chair*
Arup, MA [SE]

**Ron Coté,** *Administrative Secretary*
National Fire Protection Association, MA

**Scott W. Adams,** Park City Fire Service District, UT [E]
    Rep. International Fire Marshals Association
**Raymond J. Battalora,** Aon/Schirmer Engineering Corporation, TX [I]
**David L. Bowman,** National Fire Sprinkler Association, Inc., FL [M]
**George D. Bushey,** Rosser International, GA [SE]
**William Conner,** Bill Conner Associates LLC, IL [SE]
    Rep. American Society of Theater Consultants
**Bhola Dhume,** City of New Orleans, LA [E]
**Ronald R. Farr,** Michigan Bureau of Fire Services, MI [E]
    Rep. Michigan Fire Inspectors Society
**Robert D. Fiedler,** City of Lincoln, NE [E]
**Daniel P. Finnegan,** Siemens Industry, Inc., NJ [M]
    Rep. National Electrical Manufacturers Association
**William E. Fitch,** Phyrefish Enterprises, Inc., FL [SE]
**Ralph D. Gerdes,** Ralph Gerdes Consultants, LLC, IN [SE]
**Harold C. Hansen,** International Association of Assembly Managers, IL [U]
**Wesley W. Hayes,** Polk County Fire Services Division, FL [E]
    Rep. International Fire Marshals Association
**Mike Hayward,** PlayPower LT Canada, Inc., Canada [M]
    Rep. International Play Equipment Manufacturers Assn.

**Jonathan Humble,** American Iron and Steel Institute, CT [M]
**John Lake,** City of Gainesville, FL [E]
    Rep. NE Florida Fire Prevention Association
**Vern L. Martindale,** Church of Jesus Christ of Latter-day Saints, UT [U]
**Gregory R. Miller,** Code Consultants, Inc., MO [U]
    Rep. National Association of Theatre Owners
**Jake Pauls,** Jake Pauls Consulting Services in Building Use & Safety, MD [SE]
**Steven W. Peavey,** Altamonte Springs Building Fire Safety Division, FL [E]
    Rep. Florida Fire Marshals & Inspectors Association
**Vincent Quinterno,** Rhode Island State Fire Marshal's Office, RI [E]
**Ed Roether,** Ed Roether Consulting LLC, KS [SE]
**Karl G. Ruling,** Entertainment Services & Technology Assn., NY [U]
    Rep. U.S. Institute for Theatre Technology
**Steven J. Scandaliato,** SDG, LLC, AZ [IM]
    Rep. American Fire Sprinkler Association
**Robert C. Schultz, Jr.,** University of Texas at Austin, TX [U]
**Philip R. Sherman,** Philip R. Sherman, PE, NH [SE]
**Stephen V. Skalko,** Portland Cement Association, GA [M]
**Paul L. Wertheimer,** Crowd Management Strategies, CA [SE]

### Alternates

**Farid Alfawakhiri,** American Iron and Steel Institute, IL [M]
    (Alt. to J. Humble)
**Gene Boecker,** Code Consultants, Inc., MO [U]
    (Alt. to G. R. Miller)
**David Cook,** Ralph Gerdes Consultants, LLC, IN [SE]
    (Alt. to R. D. Gerdes)
**Jerrold S. Gorrell,** Theatre Safety Programs, AZ [U]
    (Alt. to K. G. Ruling)
**Eugene Leitermann,** Theatre Projects Consultants, Inc., CT [SE]
    (Alt. to W. Conner)

**Vern T. Lewis,** Church of Jesus Christ of Latter-day Saints, UT [U]
    (Alt. to V. L. Martindale)
**Mark V. Smith,** Alachua County Fire Rescue, FL [E]
    (Alt. to J. Lake)
**Robert B. Treiber,** National Fire Sprinkler Association, Inc., OH [M]
    (Alt. to D. L. Bowman)
**Thomas G. Wellen,** American Fire Sprinkler Association, Inc., TX [IM]
    (Alt. to S. J. Scandaliato)

**Ron Coté,** NFPA Staff Liaison

*This list represents the membership at the time the Committee was balloted on the final text of this edition. Since that time, changes in the membership may have occurred. A key to classifications is found at the back of the document.*

NOTE: Membership on a committee shall not in and of itself constitute an endorsement of the Association or any document developed by the committee on which the member serves.

**Committee Scope:** This Committee shall have primary responsibility for documents on protection of human life and property from fire and other circumstances capable of producing similar consequences, and on the non-emergency and emergency movement of people in assembly occupancies, tents, and membrane structures.

## Technical Committee on Board and Care Facilities

**Chad E. Beebe,** *Chair*
American Society for Healthcare Engineering, WA [U]

**Gregory E. Harrington,** *Administrative Secretary*
National Fire Protection Association, MA

**Scott D. Allen,** LifeServices Management Corporation, PA [U]

**Gregory J. Austin,** Gentex Corporation, MI [M]
Rep. National Electrical Manufacturers Association

**Justin B. Biller,** Roanoke County Office of Building Safety, VA [E]

**Andrew Blum,** Exponent, Inc., MD [SE]

**Warren D. Bonisch,** Aon/Schirmer Engineering Corporation, TX [I]

**Harry L. Bradley,** Maryland State Fire Marshals Office, MD [E]
Rep. International Fire Marshals Association

**Richard T. Byrd,** Tennessee Department of Health, TN [E]

**Virgil Hall,** US Department of Veterans Affairs, IN [U]

**Kenneth E. Isman,** National Fire Sprinkler Association, Inc., NY [M]

**Thomas W. Jaeger,** Jaeger and Associates, LLC, VA [U]
Rep. American Health Care Association

**Philip R. Jose,** P. R. Jose & Associates, MI [SE]

**Henry Kowalenko,** Illinois Department of Public Health, IL [E]

**James K. Lathrop,** Koffel Associates, Inc., CT [SE]

**Cindy Mahan,** Friendship Community Care, Inc., AR [U]
Rep. American Network of Community Options & Resources

**Randy S. McDermott,** US Department of Health & Human Services, TX [E]

**Daniel E. Nichols,** New York State Department of State, NY [E]

**Carol Olson,** Alaska Department of Public Safety, AK [E]

**John A. Rickard,** Olicon Design, TX [SE]

**Terry Schultz,** Code Consultants, Inc., MO [SE]

**Jon Taluba,** Russell Phillips & Associates, LLC, NY [SE]

### Alternates

**Kerry M. Bell,** Underwriters Laboratories Inc., IL [RT]
(Voting Alt. to UL Rep.)

**Margaret R. Engwer,** U.S. Department of Veterans Affairs, PA [U]
(Alt. to V. Hall)

**Martin J. Farraher,** Siemens Industry, Inc., NJ [M]
(Alt. to G. J. Austin)

**Diana E. Hugue,** Koffel Associates, Inc., MD [SE]
(Alt. to J. K. Lathrop)

**Kaitlin McGillvray,** Code Consultants, Inc., NY [SE]
(Alt. to T. Schultz)

**David A. Seitz,** HCF Management, Inc., OH [U]
(Alt. to T. W. Jaeger)

### Nonvoting

**Harold E. Nelson,** Annandale, VA [SE]
(Member Emeritus)

**Gregory E. Harrington,** NFPA Staff Liaison

*This list represents the membership at the time the Committee was balloted on the final text of this edition. Since that time, changes in the membership may have occurred. A key to classifications is found at the back of the document.*

NOTE: Membership on a committee shall not in and of itself constitute an endorsement of the Association or any document developed by the committee on which the member serves.

**Committee Scope:** This Committee shall have primary responsibility for documents on protection of human life and property from fire and other circumstances capable of producing similar consequences, and on the emergency movement of people in residential board and care facilities.

# Technical Committee on Building Service and Fire Protection Equipment

**Joseph M. Jardin,** *Chair*
Fire Department City of New York, NY [C]
Rep. NFPA Fire Service Section

**Gregory E. Harrington,** *Administrative Secretary*
National Fire Protection Association, MA

**Keith A. Ball,** Tyco International, FL [M]
**Brian D. Black,** BDBlack Codes, Inc., NY [M]
Rep. National Elevator Industry Inc.
**Harry L. Bradley,** Maryland State Fire Marshals Office, MD [E]
Rep. International Fire Marshals Association
**Pat D. Brock,** Oklahoma State University, OK [SE]
**Phillip A. Brown,** American Fire Sprinkler Association, Inc., TX [IM]
**Paul M. Donga,** Boston Fire Department, MA [E]
**Raymond A. Grill,** Arup Fire, DC [SE]
**Thomas P. Hammerberg,** Automatic Fire Alarm Association, Inc., GA [M]
**Kenneth E. Isman,** National Fire Sprinkler Association, Inc., NY [M]
**Ignatius Kapalczynski,** Connecticut Department of Public Safety, CT [E]

**David A. Killian,** Walt Disney Parks & Resorts, CA [U]
**Roy C. Kimball,** Brooks Equipment Company, Inc., NC [M]
Rep. Fire Equipment Manufacturers' Association
**David L. Klepitch,** Whitman, Requardt & Associates, LLP, MD [SE]
**Richard L. Klinker,** Klinker & Associates, Inc., MD [SE]
**Peter A. Larrimer,** US Department of Veterans Affairs, PA [U]
**Martin H. Reiss,** The RJA Group, Inc., MA [SE]
**Robert A. Schmidt,** Combustion Science & Engineering, Inc., MD [SE]
**Lawrence J. Shudak,** Underwriters Laboratories Inc., IL [RT]
**Carl D. Wren,** Austin Fire Department, TX [E]
**David M. Wyatt,** Battelle/Pacific Northwest National Laboratory, WA [U]

## Alternates

**J. Robert Boyer,** Edwards, a UTC Company, NJ [M]
(Alt. to T. P. Hammerberg)
**James D. Brown,** Oklahoma State University, OK [SE]
(Alt. to P. D. Brock)
**Greg Gottlieb,** Hauppauge Fire District, NY [C]
(Alt. to J. M. Jardin)
**Claudia Hagood,** Klinker and Associates, Inc., MD [SE]
(Alt. to R. L. Klinker)
**Jeffrey M. Hugo,** National Fire Sprinkler Association, Inc., MI [M]
(Alt. to K. E. Isman)
**Michael Kellett,** Connecticut Department of Public Safety, CT [E]
(Alt. to I. Kapalczynski)

**Peter Leszczak,** U.S. Department of Veterans Affairs, CT [U]
(Alt. to P. A. Larrimer)
**Stephen M. Leyton,** Protection Design and Consulting, CA [IM]
(Alt. to P. A. Brown)
**John J. McSheffrey, Jr.,** en-Gauge Inc., MA [M]
(Alt. to R. C. Kimball)
**Gary L. Nuschler,** Otis Elevator Company, CT [M]
(Alt. to B. D. Black)
**Rodger Reiswig,** Tyco/SimplexGrinnell, FL [M]
(Alt. to K. A. Ball)
**Andrew M. Schneider,** Maryland State Fire Marshals Office, MD [E]
(Alt. to H. L. Bradley)

**Gregory E. Harrington,** NFPA Staff Liaison

*This list represents the membership at the time the Committee was balloted on the final text of this edition. Since that time, changes in the membership may have occurred. A key to classifications is found at the back of the document.*

NOTE: Membership on a committee shall not in and of itself constitute an endorsement of the Association or any document developed by the committee on which the member serves.

**Committee Scope:** This Committee shall have primary responsibility for documents on the application of fire protection systems including detection, alarm, and suppression, and the life safety impact of various building systems.

## Technical Committee on Detention and Correctional Occupancies

**Michael DiMascio,** *Chair*
Arup Fire, MA [SE]

**Ron Coté,** *Administrative Secretary*
National Fire Protection Association, MA

**James R. Ambrose,** Code Consultants, Inc., MO [SE]
**David L. Bondor,** Travelers Insurance Company, TX [I]
Rep. American Society of Safety Engineers
**Peter J. Collins,** U.S. Department of Justice, DC [U]
**Randy Gaw,** Correctional Service of Canada, Canada [E]
**Patrick G. Gordon,** Philadelphia Prison System, PA [U]
**Kenneth E. Isman,** National Fire Sprinkler Association, Inc., NY [M]
**Thomas W. Jaeger,** Jaeger and Associates, LLC, VA [SE]
**William E. Koffel,** Koffel Associates, Inc., MD [SE]
**Jack McNamara,** Bosch Security Systems, NY [M]
Rep. National Electrical Manufacturers Association
**E. Eugene Miller,** Washington, DC [SE]
**Robert R. Perry,** Robert Perry Associates Inc., IL [M]
Rep. Door and Hardware Institute

**Kurt A. Roeper,** Ingersoll-Rand Security Technologies, OH [M]
Rep. Builders Hardware Manufacturers Association
**Kenneth J. Schwartz,** Aon/Schirmer Engineering Corporation, IL [I]
**Wayne S. Smith,** Texas State Fire Marshal, TX [E]
Rep. International Fire Marshals Association
**David W. Spence,** Corrections Corporation of America, TN [U]
**James A. Stapleton, Jr.,** Habersham Metal Products Company, GA [M]
Rep. National Assn. of Architectural Metal Manufacturers

### Alternates

**Clay P. Aler,** Koffel Associates, Inc., MD [SE]
(Alt. to W. E. Koffel)
**A. Larry Iseminger, Jr.,** Maryland State Fire Marshals Office, MD [E]
(Alt. to W. S. Smith)
**Terry Schultz,** Code Consultants, Inc., MO [SE]
(Alt. to J. R. Ambrose)

**Doyle G. Sutton,** National Fire Sprinkler Association, Inc., CO [M]
(Alt. to K. E. Isman)
**John Younghusband,** Aon/Schirmer Engineering Corporation, CA [I]
(Alt. to K. J. Schwartz)

**Ron Coté,** NFPA Staff Liaison

*This list represents the membership at the time the Committee was balloted on the final text of this edition. Since that time, changes in the membership may have occurred. A key to classifications is found at the back of the document.*

NOTE: Membership on a committee shall not in and of itself constitute an endorsement of the Association or any document developed by the committee on which the member serves.

**Committee Scope:** This Committee shall have primary responsibility for documents on protection of human life and property from fire and other circumstances capable of producing similar consequences, and on the emergency movement of people in detention and correctional occupancies.

# Technical Committee on Educational and Day-Care Occupancies

**Aleksy L. Szachnowicz,** *Chair*
Anne Arundel County Public Schools, MD [U]

**Ron Coté,** *Administrative Secretary*
National Fire Protection Association, MA

**Steven D. Admire,** Communication Concepts, TX [IM]
**Thomas M. Beare,** Siemens Industry, TX [M]
  Rep. National Electrical Manufacturers Association
**Samuel S. Dannaway,** S. S. Dannaway Associates, Inc., HI [SE]
**Victor L. Dubrowski,** Code Consultants, Inc., MO [SE]
**Keith S. Frangiamore,** Fire Safety Consultants, Inc., IL [SE]
**Dominick G. Kasmauskas,** National Fire Sprinkler Association, Inc., NY [M]
**Alfred J. Longhitano,** Alfred J. Longhitano, P.E., LLC, NY [SE]
**Vern L. Martindale,** Church of Jesus Christ of Latter-day Saints, UT [U]

**Richard E. Merck,** Montgomery County Fire Rescue Service, MD [E]
**Kurt A. Roeper,** Ingersoll-Rand Security Technologies, OH [M]
  Rep. Steel Door Institute
**Michael L. Savage, Sr.,** Middle Department Inspection Agency, Inc., MD [E]
**Michael L. Sinsigalli,** West Hartford Fire Department, CT [E]
**Catherine L. Stashak,** Office of the Illinois State Fire Marshal, IL [E]
**Billy E. Upton,** Ballou Justice Upton Architects, VA [SE]

## Alternates

**Richard M. DiMisa,** Code Consultants, Inc., MO [SE]
  (Alt. to V. L. Dubrowski)
**Terry L. Phillips,** National Fire Sprinkler Association, Inc., WY [M]
  (Alt. to D. G. Kasmauskas)

**Kenneth Wood,** Office of the Illinois State Fire Marshal, IL [E]
  (Alt. to C. L. Stashak)

**Ron Coté,** NFPA Staff Liaison

*This list represents the membership at the time the Committee was balloted on the final text of this edition. Since that time, changes in the membership may have occurred. A key to classifications is found at the back of the document.*

NOTE: Membership on a committee shall not in and of itself constitute an endorsement of the Association or any document developed by the committee on which the member serves.

**Committee Scope:** This Committee shall have primary responsibility for documents on protection of human life and property from fire and other circumstances capable of producing similar consequences, and on the emergency movement of people in educational occupancies and day-care occupancies.

## Technical Committee on Fire Protection Features

**Eric R. Rosenbaum,** *Chair*
Hughes Associates, Inc., MD [SE]

**Kristin Collette,** *Administrative Secretary*
National Fire Protection Association, MA

**John F. Bender,** Underwriters Laboratories Inc., MD [RT]
**Gregory J. Cahanin,** Cahanin Fire & Code Consulting, FL [U]
    Rep. Louisiana State Firemen's Association
**Joseph A. Castellano,** The RJA Group, Inc., GA [SE]
**John F. Devlin,** Aon/Schirmer Engineering Corporation, MD [I]
**Sam W. Francis,** American Forest & Paper Association, PA [M]
**Ralph D. Gerdes,** Ralph Gerdes Consultants, LLC, IN [SE]
**Jack A. Gump,** Babcock & Wilcox Y-12, LLC, TN [U]
**Wayne D. Holmes,** HSB Professional Loss Control, NC [I]
**Jonathan Humble,** American Iron and Steel Institute, CT [M]
**Ignatius Kapalczynski,** Connecticut Department of Public Safety, CT [E]
**Marshall A. Klein,** Marshall A. Klein & Associates, Inc., MD [SE]

**William E. Koffel,** Koffel Associates, Inc., MD [M]
    Rep. Glazing Industry Code Committee
**David A. Lewis,** Code Consultants, Inc., MO [SE]
**Vickie J. Lovell,** InterCode Incorporated, FL [M]
    Rep. Air Movement & Control Association
**William J. McHugh, Jr.,** Firestop Contractors International Association, IL [IM]
**Jon W. Pasqualone,** Martin County Board of County Commissioners, FL [E]
    Rep. Florida Fire Marshals & Inspectors Association
**Kurt A. Roeper,** Ingersoll-Rand Security Technologies, OH [M]
    Rep. Steel Door Institute
**Andrew M. Schneider,** Maryland State Fire Marshals Office, MD [E]
    Rep. International Fire Marshals Association
**Stephen V. Skalko,** Portland Cement Association, GA [M]
**Kenneth Wood,** Office of the Illinois State Fire Marshal, IL [E]

### Alternates

**Farid Alfawakhiri,** American Iron and Steel Institute, IL [M]
    (Alt. to J. Humble)
**Richard C. Butcher,** Tarpon Springs Fire Rescue, FL [E]
    (Alt. to J. W. Pasqualone)
**David Cook,** Ralph Gerdes Consultants, LLC, IN [SE]
    (Alt. to R. D. Gerdes)
**Rick Glenn,** Aon/Schirmer Engineering Corporation, IL [I]
    (Alt. to J. F. Devlin)
**David M. Hammerman,** Marshall A. Klein and Associates, Inc., MD [SE]
    (Alt. to M. A. Klein)
**Howard Hopper,** Underwriters Laboratories Inc., CA [RT]
    (Alt. to J. F. Bender)

**Thomas R. Janicak,** Ceco Door Products, IL [M]
    (Alt. to K. A. Roeper)
**Robert LeClair, Jr.,** A. F. Underhill, Inc., MA [IM]
    (Alt. to W. J. McHugh, Jr.)
**Ronald B. Melucci,** The RJA Group, Inc., MA [SE]
    (Alt. to J. A. Castellano)
**Kevin D. Morin,** Code Consultants, Inc., NY [SE]
    (Alt. to D. A. Lewis)
**Timothy J. Orris,** AMCA International, Inc., IL [M]
    (Alt. to V. J. Lovell)
**Brian T. Rhodes,** Hughes Associates, Inc., MD [SE]
    (Alt. to E. R. Rosenbaum)
**Catherine L. Stashak,** Office of the Illinois State Fire Marshal, IL [E]
    (Alt. to K. Wood)

### Nonvoting

**Michael Earl Dillon,** Dillon Consulting Engineers, Inc., CA [SE]
    Rep. TC on Air Conditioning

**Kristin Collette,** NFPA Staff Liaison

*This list represents the membership at the time the Committee was balloted on the final text of this edition. Since that time, changes in the membership may have occurred. A key to classifications is found at the back of the document.*

NOTE: Membership on a committee shall not in and of itself constitute an endorsement of the Association or any document developed by the committee on which the member serves.

**Committee Scope:** This Committee shall have primary responsibility for documents on construction compartmentation, including the performance of assemblies, openings, and penetrations, as related to the protection of life and property from fire and other circumstances capable of producing similar consequences.

## Technical Committee on Fundamentals

**Wayne G. Carson,** *Chair*
Carson Associates, Inc., VA [SE]

**Ron Coté,** *Administrative Secretary*
National Fire Protection Association, MA

**Farid Alfawakhiri,** American Iron and Steel Institute, IL [M]

**Andrew Blum,** Exponent, Inc., MD [SE]

**Phillip A. Brown,** American Fire Sprinkler Association, Inc., TX [IM]

**Amy Y. Cheng,** Clark County Department of Development Services, NV [E]

**Salvatore DiCristina,** Rutgers, The State University of New Jersey, NJ [E]

**Joshua W. Elvove,** U.S. General Services Administration, CO [U]

**Robert J. Eugene,** Underwriters Laboratories Inc., WA [RT]

**Ralph D. Gerdes,** Ralph Gerdes Consultants, LLC, IN [SE]
 Rep. American Institute of Architects

**Ben Greene,** City of Englewood, CO [E]

**Norman E. Groner,** John Jay College of Criminal Justice, NY [SE]

**Morgan J. Hurley,** Society of Fire Protection Engineers, MD [U]

**David J. Jacoby,** Arup Fire, NY [SE]

**David P. Klein,** U.S. Department of Veterans Affairs, DC [U]

**Scott T. Laramee,** Aon/Schirmer Engineering Corporation, MD [I]

**James K. Lathrop,** Koffel Associates, Inc., CT [SE]

**David C. Lind,** North Shore Fire Department, WI [E]
 Rep. International Fire Marshals Association

**Jake Pauls,** Jake Pauls Consulting Services in Building Use & Safety, MD [C]
 Rep. American Public Health Association

**Dennis L. Pitts,** American Forest & Paper Association, TX [M]

**Milosh T. Puchovsky,** Worcester Polytechnic Institute, MA [SE]

**Rodger Reiswig,** Tyco/SimplexGrinnell, FL [M]
 Rep. Automatic Fire Alarm Association, Inc.

**Stephen V. Skalko,** Portland Cement Association, GA [M]

**Rick Thornberry,** The Code Consortium, Inc., CA [M]
 Rep. Alliance for Fire & Smoke Containment & Control, Inc.

**Victoria B. Valentine,** National Fire Sprinkler Association, Inc., NY [M]

### Alternates

**Michael W. Ashley,** Alliance for Fire & Smoke Containment & Control Inc., GA [M]
 (Alt. to R. Thornberry)

**David W. Frable,** U.S. General Services Administration, IL [U]
 (Alt. to J. W. Elvove)

**Thomas P. Hammerberg,** Automatic Fire Alarm Association, Inc., GA [M]
 (Alt. to R. Reiswig)

**Bonnie E. Manley,** American Iron and Steel Institute, MA [M]
 (Alt. to F. Alfawakhiri)

**Rodney A. McPhee,** Canadian Wood Council, Canada [M]
 (Alt. to D. L. Pitts)

### Nonvoting

**Pichaya Chantranuwat,** Fusion Consultants Co. Ltd/Thailand, Thailand [SE]

**Ron Coté,** NFPA Staff Liaison

*This list represents the membership at the time the Committee was balloted on the final text of this edition. Since that time, changes in the membership may have occurred. A key to classifications is found at the back of the document.*

NOTE: Membership on a committee shall not in and of itself constitute an endorsement of the Association or any document developed by the committee on which the member serves.

**Committee Scope:** This Committee shall have primary responsibility for documents on the basic goals, objectives, performance requirements, and definitions for protection of human life and property from fire, earthquake, flood, wind, and other circumstances capable of producing similar consequences, on the non-emergency and emergency movement of people, and on high-rise buildings.

## Technical Committee on Furnishings and Contents

**Henry Paszczuk,** *Chair*
Connecticut Department of Public Safety, CT [E]

**Kristin Collette,** *Administrative Secretary*
National Fire Protection Association, MA

**Vytenis Babrauskas,** Fire Science and Technology Inc.,
WA [SE]
**William E. Fitch,** Phyrefish Enterprises, Inc., FL [SE]
**Marcelo M. Hirschler,** GBH International, CA [SE]
**Alfred J. Hogan,** Winter Haven, FL [E]
     Rep. New England Association of Fire Marshals
**E. Ken McIntosh,** Carpet and Rug Institute, GA [M]
**C. Anthony Penaloza,** Intertek Testing Services, TX [RT]

**Shelley Siegel,** Universal Design and Education Network,
FL [U]
     Rep. American Society of Interior Designers
**Dwayne E. Sloan,** Underwriters Laboratories Inc.,
NC [RT]
**T. Hugh Talley,** Hugh Talley Company, TN [M]
     Rep. American Furniture Manufacturers Association

**Alternates**

**James K. Lathrop,** Koffel Associates, Inc., CT [M]
     (Alt. to E. K. McIntosh)
**Randall K. Laymon,** Underwriters Laboratories Inc.,
IL [RT]
     (Alt. to D. E. Sloan)

**Kristin Collette,** NFPA Staff Liaison

*This list represents the membership at the time the Committee was balloted on the final text of this edition. Since that time, changes in the membership may have occurred. A key to classifications is found at the back of the document.*

NOTE: Membership on a committee shall not in and of itself constitute an endorsement of the Association or any document developed by the committee on which the member serves.

**Committee Scope:** This Committee shall have primary responsibility for documents on limiting the impact of furnishings and building contents effect on protection of human life and property from fire and other circumstances capable of producing similar consequences, and on the emergency movement of people.

## Technical Committee on Health Care Occupancies

**David P. Klein,** *Chair*
U.S. Department of Veterans Affairs, DC [U]

**Ron Coté,** *Administrative Secretary*
National Fire Protection Association, MA

**James R. Ambrose,** Code Consultants, Inc., MO [SE]
**Kenneth E. Bush,** Maryland State Fire Marshals Office, MD [E]
Rep. International Fire Marshals Association
**Wayne G. Carson,** Carson Associates, Inc., VA [SE]
**Michael A. Crowley,** The RJA Group, Inc., TX [SE]
**Samuel S. Dannaway,** S. S. Dannaway Associates, Inc., HI [SE]
Rep. American Society of Safety Engineers
**Buddy Dewar,** National Fire Sprinkler Association, Inc., FL [M]
**Alice L. Epstein,** CNA Insurance, CO [I]
**Douglas S. Erickson,** American Society for Healthcare Engineering, VI [U]
**John E. Fishbeck,** The Joint Commission, IL [E]
**Gary Furdell,** State of Florida, FL [E]
**Robert J. Harmeyer,** MSKTD & Associates, IN [SE]
Rep. American Institute of Architects
**Donald W. Harris,** California Office of Health Planning & Development, CA [E]
**David R. Hood,** Russell Phillips & Associates, LLC, NY [U]
Rep. NFPA Health Care Section
**Richard M. Horeis,** HDR Architecture, Inc., NE [SE]

**Thomas W. Jaeger,** Jaeger and Associates, LLC, VA [U]
Rep. American Health Care Association
**Henry Kowalenko,** Illinois Department of Public Health, IL [E]
**James Merrill II,** U.S. Department of Health & Human Services, MD [E]
Rep. U.S. Dept. of Health & Human Services/CMS
**Daniel J. O'Connor,** Aon/Schirmer Engineering Corporation, IL [I]
**Ben Pethe,** Health Care Consultant, FL [SE]
**G. Brian Prediger,** U.S. Department of the Army, TX [U]
**John A. Rickard,** Olicon Design, TX [SE]
**Richard Jay Roberts,** Honeywell Life Safety, IL [M]
Rep. Automatic Fire Alarm Association, Inc.
**George F. Stevens,** U.S. Department of Health & Human Services, AZ [E]
Rep. U.S. Dept. of Health & Human Services/IHS
**Saundra J. Stevens,** Adams County Regional Medical Center, OH [U]
**Geza Szakats,** Arup North America Ltd., CA [SE]
**Peter W. Tately,** Siemens Building Technologies, MA [M]
**Michael D. Widdekind,** Zurich Services Corporation, MD [I]

### Alternates

**Doug Beardsley,** Care Providers of Minnesota, MN [U]
(Alt. to T. W. Jaeger)
**William M. Dorfler,** The RJA Group, Inc., IL [SE]
(Alt. to M. A. Crowley)
**A. Richard Fasano,** Russell Phillips & Associates Inc., CA [U]
(Alt. to D. R. Hood)
**Philip J. Hoge,** U.S. Army Corps of Engineers, VA [U]
(Alt. to G. Brian Prediger)

**William E. Koffel,** Koffel Associates, Inc., MD [U]
(Alt. to D. S. Erickson)
**Peter A. Larrimer,** U.S. Department of Veterans Affairs, PA [U]
(Alt. to D. P. Klein)
**Terry Schultz,** Code Consultants, Inc., MO [SE]
(Alt. to J. R. Ambrose)
**Kenneth Sun,** U.S. Public Health Service, CO [E]
(Alt. to J. Merrill II)

### Nonvoting

**Pichaya Chantranuwat,** Fusion Consultants Co. Ltd/Thailand, Thailand [SE]
**David M. Sine,** National Center for Patient Safety, MI [U]
Rep. National Association of Psychiatric Health Systems

**Ron Coté,** NFPA Staff Liaison

*This list represents the membership at the time the Committee was balloted on the final text of this edition. Since that time, changes in the membership may have occurred. A key to classifications is found at the back of the document.*

NOTE: Membership on a committee shall not in and of itself constitute an endorsement of the Association or any document developed by the committee on which the member serves.

**Committee Scope:** This Committee shall have primary responsibility for documents on protection of human life and property from fire and other circumstances capable of producing similar consequences, and on the emergency movement of people in health care occupancies.

# Technical Committee on Industrial, Storage, and Miscellaneous Occupancies

**John A. Alderman,** *Chair*
Aon/Schirmer Engineering Corporation, TX [I]
Rep. American Society of Safety Engineers

**Kristin Collette,** *Administrative Secretary*
National Fire Protection Association, MA

**Thomas L. Allison,** Savannah River Nuclear Solutions, SC [U]
**Raymond E. Arntson,** Rayden Research, LLC, WI [SE]
**Bradley B. Barnes,** UTC/Fire Detection & Alarm, OR [M]
　Rep. National Electrical Manufacturers Association
**Donald C. Birchler,** FP&C Consultants, Inc., MO [SE]
**Nicholas A. Dawe,** Cobb County Fire Marshal's Office, GA [E]
**Jeffry T. Dudley,** National Aeronautics & Space Administration, FL [U]
**John F. Farney, Jr.,** Sargent & Lundy Engineers, IL [SE]
**Larry L. Fluer,** Fluer, Inc., CA [M]
　Rep. Compressed Gas Association
**James E. Golinveaux,** Tyco Fire Suppression & Building Products, RI [IM]
　Rep. American Fire Sprinkler Association
**Wayne D. Holmes,** HSB Professional Loss Control, NC [I]
**Jonathan Humble,** American Iron and Steel Institute, CT [M]
**Marshall A. Klein,** Marshall A. Klein & Associates, Inc., MD [U]
　Rep. Automotive Oil Change Association

**Richard J. Kobelski,** Hanford Fire Department, WA [U]
**Neal W. Krantz, Sr.,** Krantz Systems & Associates, LLC, MI [M]
　Rep. Automatic Fire Alarm Association, Inc.
**Richard S. Kraus,** API/Petroleum Safety Consultants, VA [U]
　Rep. American Petroleum Institute
**Raymond W. Lonabaugh,** National Fire Sprinkler Association, Inc., PA [M]
**Patrick A. McLaughlin,** McLaughlin & Associates, RI [U]
　Rep. Semiconductor Industry Association
**Milton L. Norsworthy,** Cleveland, TN [M]
**Scot Pruett,** Black & Veatch Corporation, KS [SE]
**Roberto Lozano Rosales,** Delphi Corporation, TX [U]
　Rep. NFPA Industrial Fire Protection Section
**Stephen V. Skalko,** Portland Cement Association, GA [M]
**Cleveland B. Skinker,** Bechtel Power Corporation, MD [SE]
**Bruce J. Swiecicki,** National Propane Gas Association, IL [IM]
**David C. Tabar,** The Sherwin-Williams Company, OH [U]
**Carl D. Wren,** Austin Fire Department, TX [E]

## Alternates

**Farid Alfawakhiri,** American Iron and Steel Institute, IL [M]
　(Alt. to J. Humble)
**Kathryn Mead Allan,** Bechtel National, Inc., MD [SE]
　(Alt. to C. B. Skinker)
**Ryan J. Bierwerth,** Summit Fire Protection, MN [SE]
　(Alt. to R. E. Arntson)
**Daniel J. Gengler,** National Fire Sprinkler Association, Inc., WI [M]
　(Alt. to R. W. Lonabaugh)
**David M. Hammerman,** Marshall A. Klein and Associates, Inc., MD [U]
　(Alt. to M. A. Klein)
**Roland J. Huggins,** American Fire Sprinkler Association, Inc., TX [IM]
　(Alt. to J. E. Golinveaux)

**David W. Mertz,** Hanford Fire Department, WA [U]
　(Alt. to R. J. Kobelski)
**Jerald Pierrottie,** Arch Chemicals, Inc., TN [M]
　(Alt. to M. L. Norsworthy)
**Marko J. Saric, Jr.,** The Sherwin-Williams Company, OH [U]
　(Alt. to D. C. Tabar)
**Jeffrey A. Scott,** FP&C Consultants, Inc., MO [SE]
　(Alt. to D. C. Birchler)
**Roger A. Smith,** Compressed Gas Association, Inc., VA [M]
　(Alt. to L. L. Fluer)
**Bobbie L. Smith,** Micron Technology, Inc., ID [U]
　(Alt. to P. A. McLaughlin)
**Samuel Vanover,** Jefferson Parish Fire Department, LA [E]
　(Alt. to H. Key)

## Nonvoting

**Virginia G. Fitzner,** U.S. Department of Labor, DC [E]
　Rep. Occupational Safety & Health Administration

**Matthew I. Chibbaro,** U.S. Department of Labor, DC [E]
　Rep. Occupational Safety & Health Administration

**Kristin Collette,** NFPA Staff Liaison

*This list represents the membership at the time the Committee was balloted on the final text of this edition. Since that time, changes in the membership may have occurred. A key to classifications is found at the back of the document.*

NOTE: Membership on a committee shall not in and of itself constitute an endorsement of the Association or any document developed by the committee on which the member serves.

**Committee Scope:** This Committee shall have primary responsibility for documents on protection of human life and property from fire and other circumstances capable of producing similar consequences, and on the emergency movement of people in industrial and storage occupancies, special structures, and windowless and underground buildings.

# Technical Committee on Means of Egress

**James K. Lathrop,** *Chair*
Koffel Associates, Inc., CT [SE]

**Ron Coté,** *Administrative Secretary*
National Fire Protection Association, MA

**Jason D. Averill,** National Institute of Standards & Technology, MD [RT]
**Charles V. Barlow,** EverGlow NA, Inc., NC [M]
**Warren D. Bonisch,** Aon/Schirmer Engineering Corporation, TX [I]
**Kenneth E. Bush,** Maryland State Fire Marshals Office, MD [E]
Rep. International Fire Marshals Association
**David S. Collins,** The Preview Group, Inc., OH [SE]
Rep. American Institute of Architects
**David A. de Vries,** Firetech Engineering Inc., IL [SE]
**Joseph M. DeRosier,** U.S. Department of Veterans Affairs, MI [U]
**Steven Di Pilla,** ESIS Global Risk Control Services, NJ [I]
Rep. American Society of Safety Engineers
**David W. Frable,** U.S. General Services Administration, IL [U]
**Robert E. Goodwin, Jr.,** Kentucky State Fire Marshal's Office, KY [E]
**Rita C. Guest,** Carson Guest, Inc., GA [U]
Rep. American Society of Interior Designers
**Waymon Jackson,** University of Texas at Austin, TX [U]
**Robert L. Leon,** Life-Pack Technologies, Inc., PA [M]
Rep. The Safe Evacuation Coalition

**Christine McMahon,** Easter Seals NH/NY/VT/ME, NH [C]
**Gary L. Nuschler,** Otis Elevator Company, CT [M]
Rep. National Elevator Industry Inc.
**Steven Orlowski,** National Association of Home Builders, DC [U]
**Jake Pauls,** Jake Pauls Consulting Services in Building Use & Safety, MD [C]
Rep. American Public Health Association
**Robert R. Perry,** Robert Perry Associates Inc., IL [M]
Rep. Door and Hardware Institute
**Eric R. Rosenbaum,** Hughes Associates, Inc., MD [SE]
**Roy W. Schwarzenberg,** U.S. Central Intelligence Agency, DC [U]
**Michael S. Shulman,** Underwriters Laboratories Inc., CA [RT]
**Thomas Stoll,** Philips Emergency Lighting, TN [M]
Rep. National Electrical Manufacturers Association
**Leslie Strull,** The RJA Group, Inc., PA [SE]
**Phillip Z. Tapper,** U.S. Department of Defense, MD [U]
**Michael Tierney,** Builders Hardware Manufacturers Association, CT [M]
**Joseph H. Versteeg,** Versteeg Associates, CT [SE]

## Alternates

**Ryan Alles,** High Rise Escape Systems, Inc., FL [M]
(Alt. to R. L. Leon)
**Brian D. Black,** BDBlack Codes, Inc., NY [M]
(Alt. to G. L. Nuschler)
**Lawrence Brown,** National Association of Home Builders, DC [U]
(Alt. to S. Orlowski)
**Michael A. Crowley,** The RJA Group, Inc., TX [SE]
(Alt. to L. Strull)
**Joshua W. Elvove,** US General Services Administration, CO [U]
(Alt. to D. W. Frable)
**Steven D. Holmes,** Underwriters Laboratories Inc., CA [RT]
(Alt. to M. S. Shulman)
**William E. Koffel,** Koffel Associates, Inc., MD [SE]
(Alt. to J. K. Lathrop)

**R. T. Leicht,** State of Delaware, DE [E]
(Alt. to K. E. Bush)
**Denise L. Pappas,** Valcom, Inc., VA [M]
(Alt. to T. Stoll)
**Keith E. Pardoe,** Door and Hardware Institute, VA [M]
(Alt. to R. R. Perry)
**Richard D. Peacock,** National Institute of Standards & Technology, MD [RT]
(Alt. to J. D. Averill)
**Brian T. Rhodes,** Hughes Associates, Inc., MD [SE]
(Alt. to E. R. Rosenbaum)
**Kelly R. Tilton,** US Central Intelligence Agency, MD [U]
(Alt. to R. W. Schwarzenberg)
**John Woestman,** The Kellen Company, IA [M]
(Alt. to M. Tierney)

## Nonvoting

**Pichaya Chantranuwat,** Fusion Consultants Co. Ltd/Thailand, Thailand [SE]
**Matthew I. Chibbaro,** U.S. Department of Labor, DC [E]

**William R. Hamilton,** U.S. Department of Labor, DC [E]
**John L. Bryan,** Frederick, MD [SE]
(Member Emeritus)

**Ron Coté,** NFPA Staff Liaison

*This list represents the membership at the time the Committee was balloted on the final text of this edition. Since that time, changes in the membership may have occurred. A key to classifications is found at the back of the document.*

NOTE: Membership on a committee shall not in and of itself constitute an endorsement of the Association or any document developed by the committee on which the member serves.

**Committee Scope:** This Committee shall have primary responsibility for documents on the general requirements for safe egress for protection of human life from fire and other circumstances capable of producing similar consequences, and on the nonemergency and emergency movement of people.

# Technical Committee on Mercantile and Business Occupancies

**Kenneth E. Bush,** *Chair*
Maryland State Fire Marshals Office, MD [E]
Rep. International Fire Marshals Association

**Kristin Collette,** *Administrative Secretary*
National Fire Protection Association, MA

**Mark J. Aaby,** Koffel Associates, Inc., MD [SE]
**Mark T. Bedell,** The Taubman Company, MI [U]
**Tracey D. Bellamy,** Telgian Corporation, GA [U]
  Rep. The Home Depot
**William J. Burrus,** Aon/Schirmer Engineering Corporation, TX [I]
**Nicholas A. Dawe,** Cobb County Fire Marshal's Office, GA [E]
**David A. Dodge,** Safety and Forensic Consulting, ME [SE]
  Rep. American Society of Safety Engineers
**David W. Frable,** U.S. General Services Administration, IL [U]
**Sam W. Francis,** American Forest & Paper Association, PA [M]
**Douglas R. Freels,** UT-Batelle at Oak Ridge National Lab, TN [U]

**Daniel J. Gauvin,** Tyco/SimplexGrinnell, MA [M]
**Anthony C. Gumkowski,** Travelers Insurance Company, CT [I]
**Wayne D. Holmes,** HSB Professional Loss Control, NC [I]
**Jonathan Humble,** American Iron and Steel Institute, CT [M]
**Scott Jacobs,** ISC Electronic Systems, Inc., CA [IM]
**Jeff Martin,** Elite Fire Protection, Canada [IM]
  Rep. National Association of Fire Equipment Distributors
**Sarah A. Rice,** The Preview Group, Inc., OH [SE]
**Terry Schultz,** Code Consultants, Inc., MO [SE]
**David C. Tabar,** The Sherwin-Williams Company, OH [U]
**Rick Thornberry,** The Code Consortium, Inc., CA [SE]
**J. L. (Jim) Tidwell,** Tidwell Code Consulting, TX [M]
  Rep. Fire Equipment Manufacturers' Association

### Alternates

**Farid Alfawakhiri,** American Iron and Steel Institute, IL [M]
  (Alt. to J. Humble)
**Mark Budzinski,** Aon/Schirmer Engineering Corporation, CA [I]
  (Alt. to W. J. Burrus)
**Joshua W. Elvove,** U.S. General Services Administration, CO [U]
  (Alt. to D. W. Frable)
**William Hiotaky,** The Taubman Company, MI [U]
  (Alt. to M. T. Bedell)
**Raymond W. Lonabaugh,** National Fire Sprinkler Association, Inc., PA [M]
  (Voting Alt. to NFSA Rep.)

**Brian L. Marburger,** Travelers Insurance Company, CT [I]
  (Alt. to A. C. Gumkowski)
**Patrick A. McLaughlin,** McLaughlin & Associates, RI [U]
  (Alt. to D. C. Tabar)
**Amy J. Murdock,** Code Consultants, Inc., MO [SE]
  (Alt. to T. Schultz)
**Dennis L. Pitts,** American Forest & Paper Association, TX [M]
  (Alt. to S. W. Francis)
**William J. Tomes,** Telgian Corporation, CA [U]
  (Alt. to T. D. Bellamy)
**Jim Widmer,** Potter Roemer, LLC, GA [M]
  (Alt. to J. L. Tidwell)

**Kristin Collette,** NFPA Staff Liaison

*This list represents the membership at the time the Committee was balloted on the final text of this edition. Since that time, changes in the membership may have occurred. A key to classifications is found at the back of the document.*

NOTE: Membership on a committee shall not in and of itself constitute an endorsement of the Association or any document developed by the committee on which the member serves.

**Committee Scope:** This Committee shall have primary responsibility for documents on protection of human life and property from fire and other circumstances capable of producing similar consequences, and for the emergency movement of people in mercantile and business occupancies.

# Technical Committee on Residential Occupancies

**Warren D. Bonisch,** *Chair*
Aon/Schirmer Engineering Corporation, TX [I]

**Gregory E. Harrington,** *Administrative Secretary*
National Fire Protection Association, MA

**Gordon Bates,** Minneapolis Fire Department, MN [E]
**James R. Bell,** Marriott International, Inc., DC [U]
  Rep. American Hotel & Lodging Association
**H. Wayne Boyd,** U.S. Safety & Engineering Corporation, CA [M]
**Harry L. Bradley,** Maryland State Fire Marshals Office, MD [E]
  Rep. International Fire Marshals Association
**Phillip A. Brown,** American Fire Sprinkler Association, Inc., TX [IM]
**Daniel P. Finnegan,** Siemens Industry, Inc., NJ [M]
  Rep. Automatic Fire Alarm Association, Inc.
**Sam W. Francis,** American Forest & Paper Association, PA [M]
**Ralph D. Gerdes,** Ralph Gerdes Consultants, LLC, IN [SE]
**Stanley C. Harbuck,** School of Building Inspection, UT [C]
  Rep. American Public Health Association
**Kenneth E. Isman,** National Fire Sprinkler Association, Inc., NY [M]
**Marshall A. Klein,** Marshall A. Klein & Associates, Inc., MD [SE]

**James K. Lathrop,** Koffel Associates, Inc., CT [SE]
**Alfred J. Longhitano,** Alfred J. Longhitano, P.E., LLC, NY [SE]
**Eric N. Mayl,** Core Engineers Consulting Group, LLC, DC [SE]
**Ronald G. Nickson,** National Multi Housing Council, DC [U]
**Steven Orlowski,** National Association of Home Builders, DC [U]
**Henry Paszczuk,** Connecticut Department of Public Safety, CT [E]
**Peter Puhlick,** University of Connecticut, CT [U]
**Richard Jay Roberts,** Honeywell Life Safety, IL [M]
  Rep. National Electrical Manufacturers Association
**John A. Sharry,** Beakmann Properties, CA [U]
**Jeffrey L. Shearman,** Zurich Services Corporation, PA [U]
  Rep. NFPA Lodging Industry Section
**Stephen V. Skalko,** Portland Cement Association, GA [M]
**T. Hugh Talley,** Hugh Talley Company, TN [M]
  Rep. American Furniture Manufacturers Association
**Joseph H. Versteeg,** Versteeg Associates, CT [SE]
**Andrew F. Weisfield,** Michael Baker Jr. Corporation, PA [SE]

## Alternates

**Lawrence Brown,** National Association of Home Builders, DC [U]
  (Alt. to S. Orlowski)
**David Cook,** Ralph Gerdes Consultants, LLC, IN [SE]
  (Alt. to R. D. Gerdes)
**David M. Hammerman,** Marshall A. Klein and Associates, Inc., MD [SE]
  (Alt. to M. A. Klein)
**Michael F. Meehan,** VSC Fire & Security, VA [IM]
  (Alt. to P. A. Brown)

**Donald J. Pamplin,** National Fire Sprinkler Association, Inc., WA [M]
  (Alt. to K. E. Isman)
**Jake Pauls,** Jake Pauls Consulting Services in Building Use & Safety, MD [C]
  (Alt. to S. C. Harbuck)
**Dennis L. Pitts,** American Forest & Paper Association, TX [M]
  (Alt. to S. W. Francis)

**Gregory E. Harrington,** NFPA Staff Liaison

*This list represents the membership at the time the Committee was balloted on the final text of this edition. Since that time, changes in the membership may have occurred. A key to classifications is found at the back of the document.*

NOTE: Membership on a committee shall not in and of itself constitute an endorsement of the Association or any document developed by the committee on which the member serves.

**Committee Scope:** This Committee shall have primary responsibility for documents on protection of human life and property from fire and other circumstances capable of producing similar consequences, and on the emergency movement of people in hotels, dormitories, apartments, lodging and rooming houses, and one- and two-family dwellings.

# Contents

# NFPA 101®

# Life Safety Code®

### 2012 Edition

*IMPORTANT NOTE: This NFPA document is made available for use subject to important notices and legal disclaimers. These notices and disclaimers appear in all publications containing this document and may be found under the heading "Important Notices and Disclaimers Concerning NFPA Documents." They can also be obtained on request from NFPA or viewed at www.nfpa.org/disclaimers.*

NOTICE: An asterisk (*) following the number or letter designating a paragraph indicates that explanatory material on the paragraph can be found in Annex A.

Changes other than editorial are indicated by a vertical rule beside the paragraph, table, or figure in which the change occurred. These rules are included as an aid to the user in identifying changes from the previous edition. Where one or more complete paragraphs have been deleted, the deletion is indicated by a bullet (•) between the paragraphs that remain.

A reference in brackets [ ] following a section or paragraph indicates material that has been extracted from another NFPA document. As an aid to the user, the complete title and edition of the source documents for extracts in mandatory sections of the document are given in Chapter 2 and those for extracts in informational sections are given in Annex C. Extracted text may be edited for consistency and style and may include the revision of internal paragraph references and other references as appropriate. Requests for interpretations or revisions of extracted text shall be sent to the technical committee responsible for the source document.

Information on referenced publications can be found in Chapter 2 and Annex C.

## Chapter 1    Administration

**1.1\* Scope.**

**1.1.1 Title.** NFPA 101, *Life Safety Code,* shall be known as the *Life Safety Code®*, is cited as such, and shall be referred to herein as "this *Code*" or "the *Code.*"

**1.1.2 Danger to Life from Fire.** The *Code* addresses those construction, protection, and occupancy features necessary to minimize danger to life from the effects of fire, including smoke, heat, and toxic gases created during a fire.

**1.1.3 Egress Facilities.** The *Code* establishes minimum criteria for the design of egress facilities so as to allow prompt escape of occupants from buildings or, where desirable, into safe areas within buildings.

**1.1.4 Other Fire-Related Considerations.** The *Code* addresses other considerations that are essential to life safety in recognition of the fact that life safety is more than a matter of egress. The *Code* also addresses protective features and systems, building services, operating features, maintenance activities, and other provisions in recognition of the fact that achieving an acceptable degree of life safety depends on additional safeguards to provide adequate egress time or protection for people exposed to fire.

**1.1.5\* Considerations Not Related to Fire.** The *Code* also addresses other considerations that, while important in fire conditions, provide an ongoing benefit in other conditions of use, including non-fire emergencies.

**1.1.6 Areas Not Addressed.** The *Code* does not address the following:

(1) \*General fire prevention or building construction features that are normally a function of fire prevention codes and building codes
(2) Prevention of injury incurred by an individual due to that individual's failure to use reasonable care
(3) Preservation of property from loss by fire

**1.2\* Purpose.** The purpose of this *Code* is to provide minimum requirements, with due regard to function, for the design, operation, and maintenance of buildings and structures for safety to life from fire. Its provisions will also aid life safety in similar emergencies.

**1.3 Application.**

**1.3.1\* New and Existing Buildings and Structures.** The *Code* shall apply to both new construction and existing buildings and existing structures.

**1.3.2 Vehicles and Vessels.** The *Code* shall apply to vehicles, vessels, or other similar conveyances, as specified in Section 11.6, in which case such vehicles and vessels shall be treated as buildings.

**1.4\* Equivalency.** Nothing in this *Code* is intended to prevent the use of systems, methods, or devices of equivalent or superior quality, strength, fire resistance, effectiveness, durability, and safety over those prescribed by this *Code.*

**1.4.1 Technical Documentation.** Technical documentation shall be submitted to the authority having jurisdiction to demonstrate equivalency.

**1.4.2 Approval.** The system, method, or device shall be approved for the intended purpose by the authority having jurisdiction.

**1.4.3\* Equivalent Compliance.** Alternative systems, methods, or devices approved as equivalent by the authority having jurisdiction shall be recognized as being in compliance with this *Code.*

**1.5 Units and Formulas.**

**1.5.1 SI Units.** Metric units of measurement in this *Code* are in accordance with the modernized metric system known as the International System of Units (SI).

**1.5.2 Primary Values.** The inch-pound value for a measurement, and the SI value given in parentheses, shall each be acceptable for use as primary units for satisfying the requirements of this *Code.*

**1.6 Enforcement.** This *Code* shall be administered and enforced by the authority having jurisdiction designated by the governing authority.

## Chapter 2    Referenced Publications

**2.1 General.** The documents referenced in this chapter, or portions of such documents, are referenced within this *Code,* shall be considered part of the requirements of this *Code,* and the following shall also apply:

(1)\*Documents referenced in this chapter, or portion of such documents, shall only be applicable to the extent called for within other chapters of this *Code*.

(2) Where the requirements of a referenced code or standard differ from the requirements of this *Code*, the requirements of this *Code* shall govern.

(3)\*Existing buildings or installations that do not comply with the provisions of the codes or standards referenced in this chapter shall be permitted to be continued in service, provided that the lack of conformity with these documents does not present a serious hazard to the occupants as determined by the authority having jurisdiction.

**2.2\* NFPA Publications.** National Fire Protection Association, 1 Batterymarch Park, Quincy, MA 02169-7471.

NFPA 10, *Standard for Portable Fire Extinguishers*, 2010 edition.

NFPA 11, *Standard for Low-, Medium-, and High-Expansion Foam*, 2010 edition.

NFPA 12, *Standard on Carbon Dioxide Extinguishing Systems*, 2011 edition.

NFPA 12A, *Standard on Halon 1301 Fire Extinguishing Systems*, 2009 edition.

NFPA 13, *Standard for the Installation of Sprinkler Systems*, 2010 edition.

NFPA 13D, *Standard for the Installation of Sprinkler Systems in One- and Two-Family Dwellings and Manufactured Homes*, 2010 edition.

NFPA 13R, *Standard for the Installation of Sprinkler Systems in Residential Occupancies up to and Including Four Stories in Height*, 2010 edition.

NFPA 14, *Standard for the Installation of Standpipe and Hose Systems*, 2010 edition.

NFPA 15, *Standard for Water Spray Fixed Systems for Fire Protection*, 2012 edition.

NFPA 16, *Standard for the Installation of Foam-Water Sprinkler and Foam-Water Spray Systems*, 2011 edition.

NFPA 17, *Standard for Dry Chemical Extinguishing Systems*, 2009 edition.

NFPA 17A, *Standard for Wet Chemical Extinguishing Systems*, 2009 edition.

NFPA 25, *Standard for the Inspection, Testing, and Maintenance of Water-Based Fire Protection Systems*, 2011 edition.

NFPA 30, *Flammable and Combustible Liquids Code*, 2012 edition.

NFPA 30B, *Code for the Manufacture and Storage of Aerosol Products*, 2011 edition.

NFPA 31, *Standard for the Installation of Oil-Burning Equipment*, 2011 edition.

NFPA 40, *Standard for the Storage and Handling of Cellulose Nitrate Film*, 2011 edition.

NFPA 45, *Standard on Fire Protection for Laboratories Using Chemicals*, 2011 edition.

NFPA 54, *National Fuel Gas Code*, 2012 edition.

NFPA 58, *Liquefied Petroleum Gas Code*, 2011 edition.

NFPA 70®, *National Electrical Code®*, 2011 edition.

NFPA 72®, *National Fire Alarm and Signaling Code*, 2010 edition.

NFPA 80, *Standard for Fire Doors and Other Opening Protectives*, 2010 edition.

NFPA 82, *Standard on Incinerators and Waste and Linen Handling Systems and Equipment*, 2009 edition.

NFPA 88A, *Standard for Parking Structures*, 2011 edition.

NFPA 90A, *Standard for the Installation of Air-Conditioning and Ventilating Systems*, 2012 edition.

NFPA 90B, *Standard for the Installation of Warm Air Heating and Air-Conditioning Systems*, 2012 edition.

NFPA 91, *Standard for Exhaust Systems for Air Conveying of Vapors, Gases, Mists, and Noncombustible Particulate Solids*, 2010 edition.

NFPA 92, *Standard for Smoke Control Systems*, 2012 edition.

NFPA 96, *Standard for Ventilation Control and Fire Protection of Commercial Cooking Operations*, 2011 edition.

NFPA 99, *Health Care Facilities Code*, 2012 edition.

NFPA 101A, *Guide on Alternative Approaches to Life Safety*, 2010 edition.

NFPA 105, *Standard for Smoke Door Assemblies and Other Opening Protectives*, 2010 edition.

NFPA 110, *Standard for Emergency and Standby Power Systems*, 2010 edition.

NFPA 111, *Standard on Stored Electrical Energy Emergency and Standby Power Systems*, 2010 edition.

NFPA 160, *Standard for the Use of Flame Effects Before an Audience*, 2011 edition.

NFPA 170, *Standard for Fire Safety and Emergency Symbols*, 2009 edition.

NFPA 204, *Standard for Smoke and Heat Venting*, 2012 edition.

NFPA 211, *Standard for Chimneys, Fireplaces, Vents, and Solid Fuel–Burning Appliances*, 2010 edition.

NFPA 220, *Standard on Types of Building Construction*, 2012 edition.

NFPA 221, *Standard for High Challenge Fire Walls, Fire Walls, and Fire Barrier Walls*, 2012 edition.

NFPA 241, *Standard for Safeguarding Construction, Alteration, and Demolition Operations*, 2009 edition.

NFPA 251, *Standard Methods of Tests of Fire Resistance of Building Construction and Materials*, 2006 edition.

NFPA 252, *Standard Methods of Fire Tests of Door Assemblies*, 2008 edition.

NFPA 253, *Standard Method of Test for Critical Radiant Flux of Floor Covering Systems Using a Radiant Heat Energy Source*, 2011 edition.

NFPA 257, *Standard on Fire Test for Window and Glass Block Assemblies*, 2007 edition.

NFPA 259, *Standard Test Method for Potential Heat of Building Materials*, 2008 edition.

NFPA 260, *Standard Methods of Tests and Classification System for Cigarette Ignition Resistance of Components of Upholstered Furniture*, 2009 edition.

NFPA 261, *Standard Method of Test for Determining Resistance of Mock-Up Upholstered Furniture Material Assemblies to Ignition by Smoldering Cigarettes*, 2009 edition.

NFPA 265, *Standard Methods of Fire Tests for Evaluating Room Fire Growth Contribution of Textile or Expanded Vinyl Wall Coverings on Full Height Panels and Walls*, 2011 edition.

NFPA 271, *Standard Method of Test for Heat and Visible Smoke Release Rates for Materials and Products Using an Oxygen Consumption Calorimeter*, 2009 edition.

NFPA 286, *Standard Methods of Fire Tests for Evaluating Contribution of Wall and Ceiling Interior Finish to Room Fire Growth*, 2011 edition.

NFPA 288, *Standard Methods of Fire Tests of Floor Fire Door Assemblies Installed Horizontally in Fire Resistance–Rated Floor Systems*, 2007 edition.

NFPA 289, *Standard Method of Fire Test for Individual Fuel Packages*, 2009 edition.

NFPA 415, *Standard on Airport Terminal Buildings, Fueling Ramp Drainage, and Loading Walkways*, 2008 edition.

NFPA 418, *Standard for Heliports,* 2011 edition.

NFPA 701, *Standard Methods of Fire Tests for Flame Propagation of Textiles and Films,* 2010 edition.

NFPA 703, *Standard for Fire Retardant–Treated Wood and Fire-Retardant Coatings for Building Materials,* 2012 edition.

NFPA 720, *Standard for the Installation of Carbon Monoxide (CO) Detection and Warning Equipment,* 2012 edition.

NFPA 750, *Standard on Water Mist Fire Protection Systems,* 2010 edition.

NFPA 914, *Code for Fire Protection of Historic Structures,* 2010 edition.

NFPA 1124, *Code for the Manufacture, Transportation, Storage, and Retail Sales of Fireworks and Pyrotechnic Articles,* 2006 edition.

NFPA 1126, *Standard for the Use of Pyrotechnics Before a Proximate Audience,* 2011 edition.

NFPA 2001, *Standard on Clean Agent Fire Extinguishing Systems,* 2012 edition.

**2.3 Other Publications.**

**2.3.1 ACI Publications.** American Concrete Institute, P.O. Box 9094, Farmington Hills, MI 48333. www.concrete.org

ACI 216.1/TMS 0216.1, *Code Requirements for Determining Fire Resistance of Concrete and Masonry Construction Assemblies,* 2008.

**2.3.2 ANSI Publications.** American National Standards Institute, Inc., 25 West 43rd Street, 4th floor, New York, NY 10036.

ANSI A14.3, *Safety Requirements for Fixed Ladders,* 1992.

ICC/ANSI A117.1, *American National Standard for Accessible and Usable Buildings and Facilities,* 2009.

ANSI/BHMA A156.3 *Exit Devices,* 2008.

BHMA/ANSI A156.19, *American National Standard for Power Assist and Low Energy Power Operated Doors,* 2007.

ANSI Z22-3.1, *National Fuel Gas Code,* 2006.

**2.3.3 ASCE Publications.** American Society of Civil Engineers, 1801 Alexander Bell Drive, Reston, VA 20191-4400. www.asce.org

ASCE/SFPE 29, *Standard Calculation Methods for Structural Fire Protection,* 2005.

**2.3.4 ASME Publications.** American Society of Mechanical Engineers, Three Park Avenue, New York, NY 10016-5990. www.asme.org

ASME A17.1/CSA B44, *Safety Code for Elevators and Escalators,* 2007.

ASME A17.3, *Safety Code for Existing Elevators and Escalators,* 2008.

ASME A17.7/CSA B44.7, *Performance-Based Safety Code for Elevators and Escalators,* 2007.

**2.3.5 ASSE Publications.** American Society of Safety Engineers, 1800 East Oakton Street, Des Plaines, IL 60018.

ANSI/ASSE A1264.1, *Safety Requirements for Workplace Floor and Wall Openings, Stairs and Railing Systems,* 2007.

**2.3.6 ASTM Publications.** ASTM International, 100 Barr Harbor Drive, P.O. Box C700, West Conshohocken, PA 19428-2959. www.astm.org

ASTM C 1629/C 1629M, *Standard Classification for Abuse-Resistant Nondecorated Interior Gypsum Panel Products and Fiber-Reinforced Cement Panels,* 2006.

ASTM D 1929, *Standard Test Method for Determining Ignition Temperatures of Plastic,* 1996 (2001e1).

ASTM D 2859, *Standard Test Method for Ignition Characteristics of Finished Textile Floor Covering Materials,* 2006.

ASTM D 2898, *Standard Test Methods for Accelerated Weathering of Fire-Retardant-Treated Wood for Fire Testing,* 2010.

ASTM E 84, *Standard Test Method for Surface Burning Characteristics of Building Materials,* 2010.

ASTM E 108, *Standard Test Methods for Fire Tests of Roof Coverings,* 2010a.

ASTM E 119, *Standard Test Methods for Fire Tests of Building Construction and Materials,* 2010b.

ASTM E 136, *Standard Test Method for Behavior of Materials in a Vertical Tube Furnace at 750 Degrees C,* 2009b.

ASTM E 648, *Standard Test Method for Critical Radiant Flux of Floor Covering Systems Using a Radiant Heat Energy Source,* 2010.

ASTM E 814, *Standard Test Method for Fire Tests of Through-Penetration Fire Stops,* 2010.

ASTM E 1352, *Standard Test Method for Cigarette Ignition Resistance of Mock-Up Upholstered Furniture Assemblies,* 2008a.

ASTM E 1353, *Standard Test Methods for Cigarette Ignition Resistance of Components of Upholstered Furniture,* 2008a(e1).

ASTM E 1354, *Standard Test Method for Heat and Visible Smoke Release Rates for Materials and Products Using an Oxygen Consumption Calorimeter,* 2009.

ASTM E 1537, *Standard Test Method for Fire Testing of Upholstered Furniture,* 2007.

ASTM E 1590, *Standard Test Method for Fire Testing of Mattresses,* 2007.

ASTM E 1591, *Standard Guide for Obtaining Data for Deterministic Fire Models,* 2007.

ASTM E 1966, *Standard Test Method for Fire-Resistive Joint Systems,* 2007.

ASTM E 2072, *Standard Specification for Photoluminescent (Phosphorescent) Safety Markings,* 2010.

ASTM E 2074, *Standard Test Method for Fire Tests of Door Assemblies, Including Positive Pressure Testing of Side-Hinged and Pivoted Swinging Door Assemblies,* 2000, Revised 2004.

ASTM E 2307, *Standard Test Method for Determining Fire Resistance of Perimeter Fire Barrier Systems Using Intermediate-Scale, Multi-Story Test Apparatus,* 2010.

ASTM E 2404, *Standard Practice for Specimen Preparation and Mounting of Textile, Paper or Vinyl Wall or Ceiling Coverings to Assess Surface Burning Characteristics,* 2008.

ASTM E 2573, *Standard Practice for Specimen Preparation and Mounting of Site-Fabricated Stretch Systems to Assess Surface Burning Characteristics,* 2007a.

ASTM E 2599, *Standard Practice for Specimen Preparation and Mounting of Reflective Insulation Materials and Radiant Barrier Materials for Building Applications to Assess Surface Burning Characteristics,* 2009.

ASTM E 2652, *Standard Test Method for Behavior of Materials in a Tube Furnace with a Cone-shaped Airflow Stabilizer, at 750 Degrees C,* 2009a.

ASTM F 851, *Standard Test Method for Self-Rising Seat Mechanisms*, 1987 (2005).

ASTM F 1577, *Standard Test Methods for Detention Locks for Swinging Doors*, 2005.

ASTM G 155, *Standard Practice for Operating Xenon Arc Light Apparatus for Exposure of Non-Metallic Materials*, 2005a.

**2.3.7 FMGR Publications.** FM Global Research, FM Global, 1301 Atwood Avenue, P.O. Box 7500, Johnston, RI 02919. www.fmglobal.com

ANSI/FM 4880, *American National Standard for Evaluating Insulated Wall or Wall and Roof/Ceiling Assemblies, Plastic Interior Finish Materials, Plastic Exterior Building Panels, Wall/Ceiling Coating Systems, Interior or Exterior Finish Systems*, 2007.

FM Approval Standard 6921, *Containers for Combustible Waste*, 2004.

UL 300, *Standard for Fire Testing of Fire Extinguishing Systems for Protection of Commercial Cooking Equipment*, 2005.

**2.3.8 NEMA Publications.** National Electrical Manufacturers Association, 1300 North 17th Street, Suite 1847, Rosslyn, VA 22209.

NEMA SB 30, *Fire Service Annunciator and Interface*, 2005.

**2.3.9 UL Publications.** Underwriters Laboratories Inc., 333 Pfingsten Road, Northbrook, IL 60062-2096. www.ul.com

ANSI/UL 9, *Standard for Fire Tests of Window Assemblies*, 2009.

ANSI/UL 10B, *Standard for Fire Tests of Door Assemblies*, 2008, Revised 2009.

ANSI/UL 10C, *Standard for Positive Pressure Fire Tests of Door Assemblies*, 2009.

ANSI/UL 263, *Standard for Fire Tests of Building Construction and Materials*, 2007.

ANSI/UL 294, *Standard for Access Control System Units*, 1999, Revised 2010.

UL 300A, *Extinguishing System Units for Residential Range Top Cooking Surfaces*, 2006.

ANSI/UL 305, *Standard for Safety Panic Hardware*, 1997.

ANSI/UL 555, *Standard for Fire Dampers*, 2006, Revised 2010.

ANSI/UL 555S, *Standard for Smoke Dampers*, 1999, Revised 2010.

ANSI/UL 723, *Standard for Test for Surface Burning Characteristics of Building Materials*, 2008, Revised 2010.

ANSI/UL 790, *Test Methods for Fire Tests of Roof Coverings*, 2004, Revised 2008.

ANSI/UL 924, *Standard for Emergency Lighting and Power Equipment*, 2006, Revised 2009.

ANSI/UL 1040, *Standard for Fire Test of Insulated Wall Construction*, 1996, Revised 2007.

ANSI/UL 1315, *Standard for Safety for Metal Waste Paper Containers*, 2007.

ANSI/UL 1479, *Standard for Fire Tests of Through-Penetration Firestops*, 2003, Revised 2010.

ANSI/UL 1715, *Standard for Fire Test of Interior Finish Material*, 1997, Revised 2008.

ANSI/UL 1784, *Standard for Air Leakage Tests for Door Assemblies*, 2001, Revised 2009.

ANSI/UL 1975, *Standard for Fire Tests for Foamed Plastics Used for Decorative Purposes*, 2006.

ANSI/UL 1994, *Standard for Luminous Egress Path Marking Systems*, 2004, Revised 2010.

ANSI/UL 2079, *Standard for Tests for Fire Resistance of Building Joint Systems*, 2004, Revised 2008.

**2.3.10 U.S. Government Publications.** U.S. Government Printing Office, Washington, DC 20402. www.access.gpo.gov

Title 16, *Code of Federal Regulations*, Part 1500 and Part 1507.

Title 16, *Code of Federal Regulations*, Part 1632, "Standard for the Flammability of Mattresses and Mattress Pads" (FF 4-72).

**2.3.11 Other Publication.**

*Merriam-Webster's Collegiate Dictionary*, 11th edition, Merriam-Webster, Inc., Springfield, MA, 2003.

**2.4 References for Extracts in Mandatory Sections.**

NFPA 1, *Fire Code*, 2012 edition.
*NFPA 72®, National Fire Alarm and Signaling Code*, 2010 edition.
NFPA 80, *Standard for Fire Doors and Other Opening Protectives*, 2010 edition.
NFPA 88A, *Standard for Parking Structures*, 2011 edition.
NFPA 252, *Standard Methods of Fire Tests of Door Assemblies*, 2008 edition.
NFPA 288, *Standard Methods of Fire Tests of Floor Fire Door Assemblies Installed Horizontally in Fire Resistance–Rated Floor Systems*, 2007 edition.
NFPA 301, *Code for Safety to Life from Fire on Merchant Vessels*, 2008 edition.
NFPA 415, *Standard on Airport Terminal Buildings, Fueling Ramp Drainage, and Loading Walkways*, 2008 edition.
NFPA 914, *Code for Fire Protection of Historic Structures*, 2010 edition.
NFPA 921, *Guide for Fire and Explosion Investigations*, 2011 edition.

ASCE/SEI 7, *Minimum Design Loads for Buildings and Other Structures*, 2010.

## Chapter 3   Definitions

**3.1 General.** The definitions contained in this chapter shall apply to the terms used in this *Code*. Where terms are not defined in this chapter or within another chapter, they shall be defined using their ordinarily accepted meanings within the context in which they are used. *Merriam-Webster's Collegiate Dictionary*, 11th edition, shall be the source for the ordinarily accepted meaning.

**3.2 NFPA Official Definitions.**

**3.2.1\* Approved.** Acceptable to the authority having jurisdiction.

**3.2.2\* Authority Having Jurisdiction (AHJ).** An organization, office, or individual responsible for enforcing the requirements of a code or standard, or for approving equipment, materials, an installation, or a procedure.

**3.2.3\* Code.** A standard that is an extensive compilation of provisions covering broad subject matter or that is suitable for adoption into law independently of other codes and standards.

**3.2.4 Labeled.** Equipment or materials to which has been attached a label, symbol, or other identifying mark of an organization that is acceptable to the authority having jurisdiction and concerned with product evaluation, that maintains periodic inspection of production of labeled equipment or materials, and by whose labeling the manufacturer indicates compliance with appropriate standards or performance in a specified manner.

**3.2.5\* Listed.** Equipment, materials, or services included in a list published by an organization that is acceptable to the authority having jurisdiction and concerned with evaluation of products or services, that maintains periodic inspection of production of listed equipment or materials or periodic evaluation of services, and whose listing states that either the equipment, material, or service meets appropriate designated standards or has been tested and found suitable for a specified purpose.

**3.2.6 Shall.** Indicates a mandatory requirement.

**3.2.7 Should.** Indicates a recommendation or that which is advised but not required.

**3.3 General Definitions.**

**3.3.1 Accessible Area of Refuge.** See 3.3.22.1.

**3.3.2 Accessible Means of Egress.** See 3.3.170.1.

**3.3.3 Accessible Route.** A continuous unobstructed path that complies with this *Code* and ICC/ANSI A117.1, *American National Standard for Accessible and Usable Buildings and Facilities.*

**3.3.4\* Actuating Member or Bar.** The activating mechanism of a panic hardware or fire exit hardware device located on the egress side of a door.

**3.3.5 Addition.** An increase in building area, aggregate floor area, building height, or number of stories of a structure.

**3.3.6 Air Traffic Control Tower.** See 3.3.280.1.

**3.3.7 Aircraft Loading Walkway.** An aboveground device through which passengers move between a point in an airport terminal building and an aircraft. Included in this category are walkways that are essentially fixed and permanently placed, or walkways that are essentially mobile in nature and that fold, telescope, or pivot from a fixed point at the airport terminal building. [**415**, 2008]

**3.3.8 Air-Inflated Structure.** See 3.3.271.1.

**3.3.9 Airport Terminal Building.** See 3.3.36.1.

**3.3.10 Air-Supported Structure.** See 3.3.271.2.

**3.3.11\* Aisle Accessway.** The initial portion of an exit access that leads to an aisle.

**3.3.12 Aisle Ramp.** See 3.3.219.1.

**3.3.13 Aisle Stair.** See 3.3.263.1.

**3.3.14 Alarm.**

**3.3.14.1 *Single Station Alarm.*** A detector comprising an assembly that incorporates a sensor, control components, and an alarm notification appliance in one unit operated from a power source either located in the unit or obtained at the point of installation. [**72**, 2010]

**3.3.14.2 *Smoke Alarm.*** A single or multiple station alarm responsive to smoke. [**72**, 2010]

**3.3.15 Alternative Calculation Procedure.** A calculation procedure that differs from the procedure originally employed by the design team but that provides predictions for the same variables of interest.

**3.3.16 Ambulatory Health Care Occupancy.** See 3.3.188.1.

**3.3.17 Analysis.**

**3.3.17.1 *Sensitivity Analysis.*** An analysis performed to determine the degree to which a predicted output will vary given a specified change in an input parameter, usually in relation to models.

**3.3.17.2 *Uncertainty Analysis.*** An analysis performed to determine the degree to which a predicted value will vary.

**3.3.18 Anchor Building.** See 3.3.36.2.

**3.3.19 Apartment Building.** See 3.3.36.3.

**3.3.20 Approved Existing.** See 3.3.79.1.

**3.3.21 Area.**

**3.3.21.1 *Detention and Correctional Residential Housing Area.*** Sleeping areas and any contiguous day room, group activity space, or other common space for customary access of residents.

**3.3.21.2 *Floor Area.***

**3.3.21.2.1\* *Gross Floor Area.*** The floor area within the inside perimeter of the outside walls of the building under consideration with no deductions for hallways, stairs, closets, thickness of interior walls, columns, elevator and building services shafts, or other features.

**3.3.21.2.2 *Net Floor Area.*** The floor area within the inside perimeter of the outside walls, or the outside walls and fire walls of the building under consideration with deductions for hallways, stairs, closets, thickness of interior walls, columns, or other features.

**3.3.21.3 *Gross Leasable Area.*** Fifty percent of major tenant areas, and 100 percent of all other floor areas designated for tenant occupancy and exclusive use, including storage areas. The area of tenant occupancy is measured from the centerlines of joint partitions to the outside of the tenant walls.

**3.3.21.4\* *Hazardous Area.*** An area of a structure or building that poses a degree of hazard greater than that normal to the general occupancy of the building or structure.

**3.3.21.5 *Living Area.*** Any normally occupiable space in a residential occupancy, other than sleeping rooms or rooms that are intended for combination sleeping/living, bathrooms, toilet compartments, kitchens, closets, halls, storage or utility spaces, and similar areas.

**3.3.21.6\* *Normally Unoccupied Building Service Equipment Support Area.*** A building service equipment support area in which people are not expected to be present on a regular basis.

**3.3.21.7 *Occupiable Area.*** An area of a facility occupied by people on a regular basis.

**3.3.21.8** *Rehabilitation Work Area.* That portion of a building affected by any renovation, modification, or reconstruction work as initially intended by the owner, and indicated as such in the permit, but excluding other portions of the building where incidental work entailed by the intended work must be performed, and excluding portions of the building where work not initially intended by the owner is specifically required.

**3.3.22\*** **Area of Refuge.** An area that is either (1) a story in a building where the building is protected throughout by an approved, supervised automatic sprinkler system and has not less than two accessible rooms or spaces separated from each other by smoke-resisting partitions; or (2) a space located in a path of travel leading to a public way that is protected from the effects of fire, either by means of separation from other spaces in the same building or by virtue of location, thereby permitting a delay in egress travel from any level.

**3.3.22.1** *Accessible Area of Refuge.* An area of refuge that complies with the accessible route requirements of ICC/ANSI A117.1, *American National Standard for Accessible and Usable Buildings and Facilities.*

**3.3.23** **Assembly.**

**3.3.23.1** *Door Assembly.* Any combination of a door, frame, hardware, and other accessories that is placed in an opening in a wall that is intended primarily for access or for human entrance or exit. [252, 2008]

**3.3.23.1.1** *Fire Door Assembly.* Any combination of a fire door, a frame, hardware, and other accessories that together provide a specific degree of fire protection to the opening. [80, 2010]

**3.3.23.1.1.1** *Floor Fire Door Assembly.* A combination of a fire door, a frame, hardware, and other accessories installed in a horizontal plane, which together provide a specific degree of fire protection to a through-opening in a fire resistance–rated floor. [288, 2007]

**3.3.23.2** *Fire Window Assembly.* A window or glass block assembly having a fire protection rating. [80, 2010]

**3.3.24** **Assembly Occupancy.** See 3.3.188.2.

**3.3.25** **Assisted Mechanical Type Parking Structure.** See 3.3.271.7.1.

**3.3.26** **Atmosphere.**

**3.3.26.1** *Common Atmosphere.* The atmosphere that exists between rooms, spaces, or areas within a building that are not separated by an approved smoke barrier.

**3.3.26.2** *Separate Atmosphere.* The atmosphere that exists between rooms, spaces, or areas that are separated by an approved smoke barrier.

**3.3.27\*** **Atrium.** A large-volume space created by a floor opening or series of floor openings connecting two or more stories that is covered at the top of the series of openings and is used for purposes other than an enclosed stairway; an elevator hoistway; an escalator opening; or as a utility shaft used for plumbing, electrical, air-conditioning, or communications facilities.

**3.3.28\*** **Attic.** The space located between the ceiling of a story and the roof directly above that habitable story.

**3.3.29** **Automated Type Parking Structure.** See 3.3.271.7.2.

**3.3.30** **Automatic.** Capable of performing a function without the necessity of human intervention.

**3.3.31** **Barrier.**

**3.3.31.1\*** *Fire Barrier.* A continuous membrane or a membrane with discontinuities created by protected openings with a specified fire protection rating, where such membrane is designed and constructed with a specified fire resistance rating to limit the spread of fire, that also restricts the movement of smoke.

**3.3.31.2\*** *Smoke Barrier.* A continuous membrane, or a membrane with discontinuities created by protected openings, where such membrane is designed and constructed to restrict the movement of smoke.

**3.3.31.3\*** *Thermal Barrier.* A material that limits the average temperature rise of an unexposed surface to not more than 250°F (139°C) for a specified fire exposure complying with the standard time-temperature curve of ASTM E 119, *Standard Test Methods for Fire Tests of Building Construction and Materials,* or ANSI/UL 263, *Standard for Fire Tests of Building Construction and Materials.*

**3.3.32** **Basement.** Any story of a building wholly or partly below grade plane that is not considered the first story above grade plane. *(See also 3.3.124.1, First Story Above Grade Plane.)*

**3.3.33\*** **Birth Center.** A facility in which low-risk births are expected following normal, uncomplicated pregnancies, and in which professional midwifery care is provided to women during pregnancy, birth, and postpartum.

**3.3.34** **Bleachers.** A grandstand in which the seats are not provided with backrests.

**3.3.35** **Board and Care.** See 3.3.188.12, Residential Board and Care Occupancy.

**3.3.36\*** **Building.** Any structure used or intended for supporting or sheltering any use or occupancy.

**3.3.36.1** *Airport Terminal Building.* A structure used primarily for air passenger enplaning or deplaning, including ticket sales, flight information, baggage handling, and other necessary functions in connection with air transport operations. This term includes any extensions and satellite buildings used for passenger handling or aircraft flight service functions. Aircraft loading walkways and "mobile lounges" are excluded. [415, 2008]

**3.3.36.2** *Anchor Building.* A building housing any occupancy having low or ordinary hazard contents and having direct access to a mall building, but having all required means of egress independent of the mall.

**3.3.36.3\*** *Apartment Building.* A building or portion thereof containing three or more dwelling units with independent cooking and bathroom facilities.

**3.3.36.4** *Bulk Merchandising Retail Building.* A building in which the sales area includes the storage of combustible materials on pallets, in solid piles, or in racks in excess of 12 ft (3660 mm) in storage height.

**3.3.36.5\*** *Existing Building.* A building erected or officially authorized prior to the effective date of the adoption of this edition of the *Code* by the agency or jurisdiction.

**3.3.36.6\*** *Flexible Plan and Open Plan Educational or Day-Care Building.* A building or portion of a building designed for multiple teaching stations.

**3.3.36.7\*** *High-Rise Building.* A building where the floor of an occupiable story is greater than 75 ft (23 m) above the lowest level of fire department vehicle access.

**3.3.36.8\*** *Historic Building.* A building or facility deemed to have historical, architectural, or cultural significance by a local, regional, or national jurisdiction.

**3.3.36.9\*** *Mall Building.* A single building enclosing a number of tenants and occupancies wherein two or more tenants have a main entrance into one or more malls. For the purpose of this *Code*, anchor buildings shall not be considered as a part of the mall building.

**3.3.36.10\*** *Special Amusement Building.* A building that is temporary, permanent, or mobile and contains a device or system that conveys passengers or provides a walkway along, around, or over a course in any direction as a form of amusement arranged so that the egress path is not readily apparent due to visual or audio distractions or an intentionally confounded egress path, or is not readily available due to the mode of conveyance through the building or structure.

**3.3.37\* Building Code.** The building code enforced by the jurisdiction or agency enforcing this *Code*.

**3.3.38 Bulk Merchandising Retail Building.** See 3.3.36.4.

**3.3.39 Business Occupancy.** See 3.3.188.3.

**3.3.40 Categories of Rehabilitation Work.** The nature and extent of rehabilitation work undertaken in an existing building.

**3.3.41\* Cellular or Foamed Plastic.** A heterogeneous system comprised of not less than two phases, one of which is a continuous, polymeric, organic material, and the second of which is deliberately introduced for the purpose of distributing gas in voids throughout the material.

**3.3.42 Change of Occupancy Classification.** The change in the occupancy classification of a structure or portion of a structure.

**3.3.43 Change of Use.** A change in the purpose or level of activity within a structure that involves a change in application of the requirements of the *Code*.

**3.3.44 Combustible (Material).** See 3.3.169.1.

**3.3.45 Combustion.** A chemical process of oxidation that occurs at a rate fast enough to produce heat and usually light in the form of either a glow or flame.

**3.3.46 Common Atmosphere.** See 3.3.26.1.

**3.3.47\* Common Path of Travel.** The portion of exit access that must be traversed before two separate and distinct paths of travel to two exits are available.

**3.3.48 Compartment.**

**3.3.48.1\*** *Fire Compartment.* A space within a building that is enclosed by fire barriers on all sides, including the top and bottom.

**3.3.48.2\*** *Smoke Compartment.* A space within a building enclosed by smoke barriers on all sides, including the top and bottom.

**3.3.49\* Consumer Fireworks, 1.4G.** *(Formerly known as Class C, Common Fireworks.)* Any small fireworks device designed primarily to produce visible effects by combustion that complies with the construction, chemical composition, and labeling regulations of the U.S. Consumer Product Safety Commission, as set forth in 16 CFR, Parts 1500 and 1507. Some small devices designed to produce audible effects are included, such as whistling devices, ground devices containing 0.8 gr (50 mg) or less of explosive composition (salute powder), and aerial devices containing 2 gr (130 mg) or less of explosive composition (salute powder) per explosive unit.

**3.3.50 Contents and Furnishings.** Any movable objects in a building that normally are secured or otherwise put in place for functional reasons, excluding (1) parts of the internal structure of the building, and (2) any items meeting the definition of interior finish.

**3.3.51 Court.** An open, uncovered, unoccupied space, unobstructed to the sky, bounded on three or more sides by exterior building walls.

**3.3.51.1** *Enclosed Court.* A court bounded on all sides by the exterior walls of a building or by the exterior walls and lot lines on which walls are permitted.

**3.3.51.2** *Food Court.* A public seating area located in a mall that serves adjacent food preparation tenant spaces.

**3.3.52\* Critical Radiant Flux.** The level of incident radiant heat energy in units of $W/cm^2$ on a floor-covering system at the most distant flameout point. [**253**, 2010]

**3.3.53 Data Conversion.** The process of developing the input data set for the assessment method of choice.

**3.3.54 Day-Care Home.** See 3.3.140.1.

**3.3.55 Day-Care Occupancy.** See 3.3.188.4.

**3.3.56 Design Fire Scenario.** See 3.3.103.1.

**3.3.57 Design Specification.** See 3.3.260.1.

**3.3.58 Design Team.** A group of stakeholders including, but not limited to, representatives of the architect, client, and any pertinent engineers and other designers.

**3.3.59 Detention and Correctional Occupancy.** See 3.3.188.5.

**3.3.60 Detention and Correctional Residential Housing Area.** See 3.3.21.1.

**3.3.61 Device.**

**3.3.61.1** *Multiple Station Alarm Device.* Two or more single station alarm devices that can be interconnected so that actuation of one causes all integral or separate audible alarms to operate; or one single station alarm device having connections to other detectors or to a manual fire alarm box. [**72**, 2010]

**3.3.61.2\*** *Stair Descent Device.* A portable device, incorporating a means to control the rate of descent, used to transport a person with a severe mobility impairment downward on stairs during emergency egress.

**3.3.62 Door.**

**3.3.62.1** *Elevator Lobby Door.* A door between an elevator lobby and another building space other than the elevator shaft.

**3.3.62.2 *Fire Door.*** The door component of a fire door assembly.

**3.3.63 Door Assembly.** See 3.3.23.1.

**3.3.64\* Dormitory.** A building or a space in a building in which group sleeping accommodations are provided for more than 16 persons who are not members of the same family in one room, or a series of closely associated rooms, under joint occupancy and single management, with or without meals, but without individual cooking facilities.

**3.3.65 Draft Stop.** A continuous membrane used to subdivide a concealed space to resist the passage of smoke and heat.

**3.3.66\* Dwelling Unit.** One or more rooms arranged for complete, independent housekeeping purposes with space for eating, living, and sleeping; facilities for cooking; and provisions for sanitation.

**3.3.66.1\* *One- and Two-Family Dwelling Unit.*** A building that contains not more than two dwelling units with independent cooking and bathroom facilities.

**3.3.66.2 *One-Family Dwelling Unit.*** A building that consists solely of one dwelling unit with independent cooking and bathroom facilities.

**3.3.66.3 *Two-Family Dwelling Unit.*** A building that consists solely of two dwelling units with independent cooking and bathroom facilities.

**3.3.67 Educational Occupancy.** See 3.3.188.6.

**3.3.68\* Electroluminescent.** Refers to a light-emitting capacitor in which alternating current excites phosphor atoms placed between electrically conductive surfaces and produces light.

**3.3.69 Elevator Evacuation System.** See 3.3.273.1.

**3.3.70 Elevator Lobby.** A landing from which occupants directly enter an elevator car(s) and into which occupants directly enter upon leaving an elevator car(s).

**3.3.71 Elevator Lobby Door.** See 3.3.62.1.

**3.3.72 Enclosed Court.** See 3.3.51.1.

**3.3.73 Enclosed Parking Structure.** See 3.3.271.7.1

**3.3.74 Equipment or Fixture.** Any plumbing, heating, electrical, ventilating, air-conditioning, refrigerating, and fire protection equipment; and elevators, dumbwaiters, escalators, boilers, pressure vessels, or other mechanical facilities or installations that are related to building services.

**3.3.75 Equivalency.** An alternative means of providing an equal or greater degree of safety than that afforded by strict conformance to prescribed codes and standards.

**3.3.76\* Evacuation Capability.** The ability of occupants, residents, and staff as a group either to evacuate a building or to relocate from the point of occupancy to a point of safety.

**3.3.76.1 *Impractical Evacuation Capability.*** The inability of a group to reliably move to a point of safety in a timely manner.

**3.3.76.2 *Prompt Evacuation Capability.*** The ability of a group to move reliably to a point of safety in a timely manner that is equivalent to the capacity of a household in the general population.

**3.3.76.3 *Slow Evacuation Capability.*** The ability of a group to move reliably to a point of safety in a timely manner, but not as rapidly as members of a household in the general population.

**3.3.77 Exhibit.** A space or portable structure used for the display of products or services.

**3.3.78 Exhibitor.** An individual or entity engaged in the display of the products or services offered.

**3.3.79\* Existing.** That which is already in existence on the date this edition of the *Code* goes into effect.

**3.3.79.1 *Approved Existing.*** That which is already in existence on the date this edition of the *Code* goes into effect and is acceptable to the authority having jurisdiction.

**3.3.80 Existing Building.** See 3.3.36.5.

**3.3.81\* Exit.** That portion of a means of egress that is separated from all other spaces of a building or structure by construction or equipment as required to provide a protected way of travel to the exit discharge.

**3.3.81.1\* *Horizontal Exit.*** A way of passage from one building to an area of refuge in another building on approximately the same level, or a way of passage through or around a fire barrier to an area of refuge on approximately the same level in the same building that affords safety from fire and smoke originating from the area of incidence and areas communicating therewith.

**3.3.82 Exit Access.** That portion of a means of egress that leads to an exit.

**3.3.83 Exit Discharge.** That portion of a means of egress between the termination of an exit and a public way.

**3.3.83.1\* *Level of Exit Discharge.*** The story that is either (1) the lowest story from which not less than 50 percent of the required number of exits and not less than 50 percent of the required egress capacity from such a story discharge directly outside at the finished ground level; or (2) where no story meets the conditions of item (1), the story that is provided with one or more exits that discharge directly to the outside to the finished ground level via the smallest elevation change.

**3.3.84 Exposition.** An event in which the display of products or services is organized to bring together the provider and user of the products or services.

**3.3.85 Exposition Facility.** See 3.3.88.1.

**3.3.86\* Exposure Fire.** A fire that starts at a location that is remote from the area being protected and grows to expose that which is being protected.

**3.3.87 Externally Illuminated.** See 3.3.144.1.

**3.3.88 Facility.**

**3.3.88.1 *Exposition Facility.*** A convention center, hotel, or other building at which exposition events are held.

**3.3.88.2\* *Limited Care Facility.*** A building or portion of a building used on a 24-hour basis for the housing of four or more persons who are incapable of self-preservation because of age; physical limitations due to accident or illness; or limitations such as mental retardation/developmental disability, mental illness, or chemical dependency.

**3.3.89 Festival Seating.** See 3.3.237.1.

**3.3.90 Finish.**

   **3.3.90.1** *Interior Ceiling Finish.* The interior finish of ceilings.

   **3.3.90.2\*** *Interior Finish.* The exposed surfaces of walls, ceilings, and floors within buildings.

   **3.3.90.3\*** *Interior Floor Finish.* The interior finish of floors, ramps, stair treads and risers, and other walking surfaces.

   **3.3.90.4** *Interior Wall Finish.* The interior finish of columns, fixed or movable walls, and fixed or movable partitions.

**3.3.91 Finished Ground Level (Grade).** The level of the finished ground (earth or other surface on ground). *(See also 3.3.124, Grade Plane.)*

**3.3.92 Fire Barrier.** See 3.3.31.1.

**3.3.93 Fire Barrier Wall.** See 3.3.287.1.

**3.3.94\* Fire Code.** The fire code enforced by the jurisdiction or agency enforcing this *Code.*

**3.3.95 Fire Compartment.** See 3.3.48.1.

**3.3.96 Fire Door.** See 3.3.62.2.

**3.3.97 Fire Door Assembly.** See 3.3.23.1.1.

**3.3.98 Fire Exit Hardware.** See 3.3.133.1.

**3.3.99\* Fire Model.** A structured approach to predicting one or more effects of a fire.

**3.3.100 Fire Protection Rating.** See 3.3.221.1.

**3.3.101 Fire Resistance Rating.** See 3.3.221.2.

**3.3.102 Fire Safety Functions.** Building and fire control functions that are intended to increase the level of life safety for occupants or to control the spread of the harmful effects of fire. [*72*, 2010]

**3.3.103\* Fire Scenario.** A set of conditions that defines the development of fire, the spread of combustion products throughout a building or portion of a building, the reactions of people to fire, and the effects of combustion products.

   **3.3.103.1** *Design Fire Scenario.* A fire scenario selected for evaluation of a proposed design. [*914*, 2010]

**3.3.104 Fire Watch.** The assignment of a person or persons to an area for the express purpose of notifying the fire department, the building occupants, or both of an emergency; preventing a fire from occurring; extinguishing small fires; or protecting the public from fire or life safety dangers. [*1*, 2012]

**3.3.105 Fire Window Assembly.** See 3.3.23.2.

**3.3.106 Fire-Rated Glazing.** Glazing with either a fire protection rating or a fire resistance rating.

**3.3.107 Fire-Retardant-Treated Wood.** A wood product impregnated with chemical by a pressure process or other means during manufacture, which is tested in accordance with ASTM E 84, *Standard Test Method for Surface Burning Characteristics of Building Materials,* or ANSI/UL 723, *Standard for Test for Surface Burning Characteristics of Burning Materials,* has a listed flame spread index of 25 or less, and shows no evidence of significant progressive combustion when the test is continued for an additional 20-minute period; nor does the flame front progress more than 10.5 ft (3.2 m) beyond the centerline of the burners at any time during the test.

**3.3.108 First Story Above Grade Plane.** See 3.3.124.1.

**3.3.109 Fixed Seating.** See 3.3.237.2.

**3.3.110\* Flame Spread.** The propagation of flame over a surface.

**3.3.111 Flame Spread Index.** See 3.3.147.1.

**3.3.112 Flashover.** A stage in the development of a contained fire in which all exposed surfaces reach ignition temperature more or less simultaneously and fire spreads rapidly throughout the space.

**3.3.113 Flexible Plan and Open Plan Educational or Day-Care Building.** See 3.3.36.6.

**3.3.114 Floor Fire Door Assembly.** See 3.3.23.1.1.1.

**3.3.115 Flow Time.** A component of total evacuation time that is the time during which there is crowd flow past a point in the means of egress system.

**3.3.116 Fly Gallery.** A raised floor area above a stage from which the movement of scenery and operation of other stage effects are controlled.

**3.3.117 Foam Plastic Insulation.** See 3.3.150.1.

**3.3.118 Folding and Telescopic Seating.** See 3.3.237.3.

**3.3.119 Food Court.** See 3.3.51.2.

**3.3.120 Fuel Load.** See 3.3.162.1.

**3.3.121 General Industrial Occupancy.** See 3.3.188.8.1.

**3.3.122 Goal.** A nonspecific overall outcome to be achieved that is measured on a qualitative basis.

**3.3.123 Grade.** See 3.3.91, Finished Ground Level (Grade).

**3.3.124 Grade Plane.** A reference plane representing the average of the finished ground level adjoining the building at all exterior walls. When the finished ground level slopes down from the exterior walls, the grade plane is established by the lowest points within the area between the building and the lot line or, when the lot line is more than 6 ft (1830 mm) from the building, between the building and a point 6 ft (1830 mm) from the building.

   **3.3.124.1** *First Story Above Grade Plane.* Any story having its finished floor surface entirely above grade plane, except that a basement is to be considered as a first story above grade plane where the finished surface of the floor above the basement is (1) more than 6 ft (1830 mm) above grade plane or (2) more than 12 ft (3660 mm) above the finished ground level at any point.

**3.3.125\* Grandstand.** A structure that provides tiered or stepped seating.

**3.3.126 Gridiron.** The structural framing over a stage supporting equipment for hanging or flying scenery and other stage effects.

**3.3.127 Gross Floor Area.** See 3.3.21.2.1.

**3.3.128 Gross Leasable Area.** See 3.3.21.3.

**3.3.129 Guard.** A vertical protective barrier erected along exposed edges of stairways, balconies, and similar areas.

**3.3.130 Guest Room.** An accommodation combining living, sleeping, sanitary, and storage facilities within a compartment.

**3.3.131 Guest Suite.** See 3.3.272.1.

**3.3.132 Handrail.** A bar, pipe, or similar member designed to furnish persons with a handhold.

**3.3.133 Hardware.**

**3.3.133.1** *Fire Exit Hardware.* A door-latching assembly incorporating an actuating member or bar that releases the latch bolt upon the application of a force in the direction of egress travel and that additionally provides fire protection where used as part of a fire door assembly.

**3.3.133.2** *Panic Hardware.* A door-latching assembly incorporating an actuating member or bar that releases the latch bolt upon the application of a force in the direction of egress travel.

**3.3.134 Hazardous Area.** See 3.3.21.4.

**3.3.135 Health Care Occupancy.** See 3.3.188.7.

**3.3.136\* Heat Release Rate (HRR).** The rate at which heat energy is generated by burning. [**921**, 2011]

**3.3.137 High Hazard Industrial Occupancy.** See 3.3.188.8.2.

**3.3.138 High-Rise Building.** See 3.3.36.7.

**3.3.139 Historic Building.** See 3.3.36.8.

**3.3.140 Home.**

**3.3.140.1\*** *Day-Care Home.* A building or portion of a building in which more than 3 but not more than 12 clients receive care, maintenance, and supervision, by other than their relative(s) or legal guardians(s), for less than 24 hours per day.

**3.3.140.2** *Nursing Home.* A building or portion of a building used on a 24-hour basis for the housing and nursing care of four or more persons who, because of mental or physical incapacity, might be unable to provide for their own needs and safety without the assistance of another person.

**3.3.141 Horizontal Exit.** See 3.3.81.1.

**3.3.142 Hospital.** A building or portion thereof used on a 24-hour basis for the medical, psychiatric, obstetrical, or surgical care of four or more inpatients.

**3.3.143\* Hotel.** A building or groups of buildings under the same management in which there are sleeping accommodations for more than 16 persons and primarily used by transients for lodging with or without meals.

**3.3.144 Illuminated.**

**3.3.144.1\*** *Externally Illuminated.* Refers to an illumination source that is contained outside of the device or sign legend area that is to be illuminated.

**3.3.144.2\*** *Internally Illuminated.* Refers to an illumination source that is contained inside the device or legend that is illuminated.

**3.3.145 Impractical Evacuation Capability.** See 3.3.76.1.

**3.3.146 Incapacitation.** A condition under which humans do not function adequately and become unable to escape untenable conditions.

**3.3.147 Index.**

**3.3.147.1** *Flame Spread Index.* A comparative measure, expressed as a dimensionless number, derived from visual measurements of the spread of flame versus time for a material tested in accordance with ASTM E 84, *Standard Test Method for Surface Burning Characteristics of Building Materials,*

or ANSI/UL 723, *Standard for Test for Surface Burning Characteristics of Burning Materials.*

**3.3.147.2** *Smoke Developed Index.* A comparative measure, expressed as a dimensionless number, derived from measurements of smoke obscuration versus time for a material tested in accordance with ASTM E 84, *Standard Test Method for Surface Burning Characteristics of Building Materials,* or ANSI/UL 723, *Standard for Test for Surface Burning Characteristics of Burning Materials.*

**3.3.148 Industrial Occupancy.** See 3.3.188.8.

**3.3.149 Input Data Specification.** See 3.3.260.2.

**3.3.150 Insulation.**

**3.3.150.1** *Foam Plastic Insulation.* A cellular plastic, used for thermal insulating or acoustical applications, having a density of 20 lb/ft$^3$ (320 kg/m$^3$) or less, containing open or closed cells, and formed by a foaming agent.

**3.3.150.2** *Reflective Insulation.* Thermal insulation consisting of one or more low-emittance surfaces bounding one or more enclosed air spaces.

**3.3.151 Interior Ceiling Finish.** See 3.3.90.1.

**3.3.152 Interior Finish.** See 3.3.90.2

**3.3.153 Interior Floor Finish.** See 3.3.90.3.

**3.3.154 Interior Wall Finish.** See 3.3.90.4.

**3.3.155 Internally Illuminated.** See 3.3.144.2.

**3.3.156 Legitimate Stage.** See 3.3.262.1.

**3.3.157 Level of Exit Discharge.** See 3.3.83.1.

**3.3.158 Life Safety Evaluation.** A written review dealing with the adequacy of life safety features relative to fire, storm, collapse, crowd behavior, and other related safety considerations.

**3.3.159 Limited Access Structure.** See 3.3.271.3.

**3.3.160 Limited Care Facility.** See 3.3.88.2.

**3.3.161 Living Area.** See 3.3.21.5.

**3.3.162 Load.**

**3.3.162.1\*** *Fuel Load.* The total quantity of combustible contents of a building, space, or fire area.

**3.3.162.2** *Occupant Load.* The total number of persons that might occupy a building or portion thereof at any one time.

**3.3.163 Load-Bearing Element.** Any column, girder, beam, joist, truss, rafter, wall, floor, or roof sheathing that supports any vertical load in addition to its own weight, or any lateral load.

**3.3.164 Lock-Up.** An incidental use area in other than a detention and correctional occupancy where occupants are restrained and such occupants are mostly incapable of self-preservation because of security measures not under the occupants' control.

**3.3.165 Lodging or Rooming House.** A building or portion thereof that does not qualify as a one- or two-family dwelling, that provides sleeping accommodations for a total of 16 or fewer people on a transient or permanent basis, without personal care services, with or without meals, but without separate cooking facilities for individual occupants.

**3.3.166 Major Tenant.** A tenant space, in a mall building, with one or more main entrances from the exterior that also serve as exits and are independent of the mall.

**3.3.167 Mall.** A roofed or covered common pedestrian area within a mall building that serves as access for two or more tenants and does not exceed three levels that are open to each other.

**3.3.168 Mall Building.** See 3.3.36.9.

**3.3.169 Material.**

    **3.3.169.1 *Combustible (Material).*** A material that, in the form in which it is used and under the conditions anticipated, will ignite and burn; a material that does not meet the definition of noncombustible or limited-combustible.

    **3.3.169.2 *Limited-Combustible (Material).*** See 4.6.14.

    **3.3.169.3 *Metal Composite Material (MCM).*** A factory-manufactured panel consisting of metal skins bonded to both faces of a core made of any plastic other than foamed plastic insulation as defined in 3.3.150.1.

    **3.3.169.4 *Noncombustible (Material).*** See 4.6.13.

    **3.3.169.5 *Weathered-Membrane Material.*** Membrane material that has been subjected to a minimum of 3000 hours in a weatherometer in accordance with ASTM G 155, *Standard Practice for Operating Xenon Arc Light Apparatus for Exposure of Non-Metallic Materials,* or approved equivalent.

**3.3.170\* Means of Egress.** A continuous and unobstructed way of travel from any point in a building or structure to a public way consisting of three separate and distinct parts: (1) the exit access, (2) the exit, and (3) the exit discharge.

    **3.3.170.1 *Accessible Means of Egress.*** A means of egress that provides an accessible route to an area of refuge, a horizontal exit, or a public way.

**3.3.171 Means of Escape.** A way out of a building or structure that does not conform to the strict definition of means of egress but does provide an alternate way out.

**3.3.172\* Membrane.** A thin layer of construction material.

**3.3.173 Membrane Structure.** See 3.3.271.4.

**3.3.174 Mercantile Occupancy.** See 3.3.188.9.

**3.3.175 Metal Composite Material (MCM).** See 3.3.169.3.

**3.3.176 Mezzanine.** An intermediate level between the floor and the ceiling of any room or space.

**3.3.177 Mixed Occupancy.** See 3.3.188.10.

**3.3.178\* Modification.** The reconfiguration of any space; the addition or elimination of any door or window; the addition or elimination of load-bearing elements; the reconfiguration or extension of any system; or the installation of any additional equipment.

**3.3.179 Multilevel Play Structure.** See 3.3.271.5.

**3.3.180 Multiple Occupancy.** See 3.3.188.11.

**3.3.181 Multiple Station Alarm Device.** See 3.3.61.1.

**3.3.182 Multipurpose Assembly Occupancy.** See 3.3.188.2.1.

**3.3.183 Net Floor Area.** See 3.3.21.2.2.

**3.3.184 Non-Patient-Care Suite (Health Care Occupancies).** See 3.3.272.2.

**3.3.185 Normally Unoccupied Building Service Equipment Support Area.** See 3.3.21.6.

**3.3.186 Nursing Home.** See 3.3.140.2.

**3.3.187\* Objective.** A requirement that needs to be met to achieve a goal.

**3.3.188 Occupancy.** The purpose for which a building or other structure, or part thereof, is used or intended to be used. [**ASCE/SEI 7:**1.2]

    **3.3.188.1\* *Ambulatory Health Care Occupancy.*** An occupancy used to provide services or treatment simultaneously to four or more patients that provides, on an outpatient basis, one or more of the following: (1) treatment for patients that renders the patients incapable of taking action for self-preservation under emergency conditions without the assistance of others; (2) anesthesia that renders the patients incapable of taking action for self-preservation under emergency conditions without the assistance of others; (3) emergency or urgent care for patients who, due to the nature of their injury or illness, are incapable of taking action for self-preservation under emergency conditions without the assistance of others.

    **3.3.188.2\* *Assembly Occupancy.*** An occupancy (1) used for a gathering of 50 or more persons for deliberation, worship, entertainment, eating, drinking, amusement, awaiting transportation, or similar uses; or (2) used as a special amusement building, regardless of occupant load.

    **3.3.188.2.1 *Multipurpose Assembly Occupancy.*** An assembly room designed to accommodate temporarily any of several possible assembly uses.

    **3.3.188.3\* *Business Occupancy.*** An occupancy used for the transaction of business other than mercantile.

    **3.3.188.4\* *Day-Care Occupancy.*** An occupancy in which four or more clients receive care, maintenance, and supervision, by other than their relatives or legal guardians, for less than 24 hours per day.

    **3.3.188.5\* *Detention and Correctional Occupancy.*** An occupancy used to house one or more persons under varied degrees of restraint or security where such occupants are mostly incapable of self-preservation because of security measures not under the occupants' control.

    **3.3.188.6\* *Educational Occupancy.*** An occupancy used for educational purposes through the twelfth grade by six or more persons for 4 or more hours per day or more than 12 hours per week.

    **3.3.188.7\* *Health Care Occupancy.*** An occupancy used to provide medical or other treatment or care simultaneously to four or more patients on an inpatient basis, where such patients are mostly incapable of self-preservation due to age, physical or mental disability, or because of security measures not under the occupants' control.

    **3.3.188.8\* *Industrial Occupancy.*** An occupancy in which products are manufactured or in which processing, assembling, mixing, packaging, finishing, decorating, or repair operations are conducted.

    **3.3.188.8.1\* *General Industrial Occupancy.*** An industrial occupancy in which ordinary and low hazard industrial operations are conducted in buildings of conventional design suitable for various types of industrial processes.

**3.3.188.8.2\*** *High Hazard Industrial Occupancy.* An industrial occupancy in which industrial operations that include high hazard materials, processes, or contents are conducted.

**3.3.188.8.3** *Special-Purpose Industrial Occupancy.* An industrial occupancy in which ordinary and low hazard industrial operations are conducted in buildings designed for, and suitable only for, particular types of operations, characterized by a relatively low density of employee population, with much of the area occupied by machinery or equipment.

**3.3.188.9\*** *Mercantile Occupancy.* An occupancy used for the display and sale of merchandise.

**3.3.188.10** *Mixed Occupancy.* A multiple occupancy where the occupancies are intermingled.

**3.3.188.11** *Multiple Occupancy.* A building or structure in which two or more classes of occupancy exist.

**3.3.188.12\*** *Residential Board and Care Occupancy.* An occupancy used for lodging and boarding of four or more residents, not related by blood or marriage to the owners or operators, for the purpose of providing personal care services.

**3.3.188.13\*** *Residential Occupancy.* An occupancy that provides sleeping accommodations for purposes other than health care or detention and correctional.

**3.3.188.14** *Separated Occupancy.* A multiple occupancy where the occupancies are separated by fire resistance–rated assemblies.

**3.3.188.15\*** *Storage Occupancy.* An occupancy used primarily for the storage or sheltering of goods, merchandise, products, or vehicles.

**3.3.189 Occupant Characteristics.** The abilities or behaviors of people before and during a fire.

**3.3.190 Occupant Load.** See 3.3.162.2.

**3.3.191 Occupiable Area.** See 3.3.21.7.

**3.3.192 Occupiable Story.** See 3.3.268.1.

**3.3.193 One- and Two-Family Dwelling Unit.** See 3.3.66.1.

**3.3.194 One-Family Dwelling Unit.** See 3.3.66.2.

**3.3.195 Open Parking Structure.** See 3.3.271.7.4.

**3.3.196 Open Structure.** See 3.3.271.6.

**3.3.197 Open-Air Mercantile Operation.** An operation conducted outside of all structures, with the operations area devoid of all walls and roofs except for small, individual, weather canopies.

**3.3.198 Outside Stair.** See 3.3.263.2.

**3.3.199 Panic Hardware.** See 3.3.133.2.

**3.3.200 Parking Structure.** See 3.3.271.7.

**3.3.201 Patient Care Non-Sleeping Suite (Health Care Occupancies).** See 3.3.272.3.

**3.3.202 Patient Care Sleeping Suite (Health Care Occupancies).** See 3.3.272.4.

**3.3.203 Patient Care Suite (Health Care Occupancies).** See 3.3.272.5.

**3.3.204\* Performance Criteria.** Threshold values on measurement scales that are based on quantified performance objectives.

**3.3.205 Permanent Structure.** See 3.3.271.8.

**3.3.206\* Personal Care.** The care of residents who do not require chronic or convalescent medical or nursing care.

**3.3.207\* Photoluminescent.** Having the ability to store incident electromagnetic radiation typically from ambient light sources, and release it in the form of visible light. [**301,** 2008]

**3.3.208 Pinrail.** A rail on or above a stage through which belaying pins are inserted and to which lines are fastened.

**3.3.209\* Platform.** The raised area within a building used for the presentation of music, plays, or other entertainment.

> **3.3.209.1** *Temporary Platform.* A platform erected within an area for not more than 30 days.

**3.3.210 Plenum.** A compartment or chamber to which one or more air ducts are connected and that forms part of the air-distribution system.

**3.3.211 Point of Safety.** A location that (a) is exterior to and away from a building; or (b) is within a building of any construction type protected throughout by an approved automatic sprinkler system and that is either (1) within an exit enclosure meeting the requirements of this *Code*, or (2) within another portion of the building that is separated by smoke barriers in accordance with Section 8.5 having a minimum ½-hour fire resistance rating, and that portion of the building has access to a means of escape or exit that conforms to the requirements of this *Code* and does not necessitate return to the area of fire involvement; or (c) is within a building of Type I, Type II(222), Type II(111), Type III(211), Type IV, or Type V(111) construction *(see 8.2.1.2)* and is either (1) within an exit enclosure meeting the requirements of this *Code*, or (2) within another portion of the building that is separated by smoke barriers in accordance with Section 8.5 having a minimum ½-hour fire resistance rating, and that portion of the building has access to a means of escape or exit that conforms to the requirements of this *Code* and does not necessitate return to the area of fire involvement.

**3.3.212 Previously Approved.** That which was acceptable to the authority having jurisdiction prior to the date this edition of the *Code* went into effect.

**3.3.213 Private Party Tent.** See 3.3.278.1.

**3.3.214 Professional Engineer.** A person registered or licensed to practice engineering in a jurisdiction, subject to all laws and limitations imposed by the jurisdiction.

**3.3.215 Prompt Evacuation Capability.** See 3.3.76.2.

**3.3.216\* Proposed Design.** A design developed by a design team and submitted to the authority having jurisdiction for approval.

**3.3.217 Proscenium Wall.** See 3.3.287.2.

**3.3.218 Public Way.** A street, alley, or other similar parcel of land essentially open to the outside air deeded, dedicated, or otherwise permanently appropriated to the public for public use and having a clear width and height of not less than 10 ft (3050 mm).

**3.3.219\* Ramp.** A walking surface that has a slope steeper than 1 in 20.

**3.3.219.1 *Aisle Ramp*.** A ramp within a seating area of an assembly occupancy that directly serves rows of seating to the side of the ramp.

**3.3.220 Ramp Type Parking Structure.** See 3.3.271.7.5.

**3.3.221 Rating.**

**3.3.221.1 *Fire Protection Rating*.** The designation indicating the duration of the fire test exposure to which a fire door assembly or fire window assembly was exposed and for which it met all the acceptance criteria as determined in accordance with NFPA 252, *Standard Methods of Fire Tests of Door Assemblies*, or NFPA 257, *Standard on Fire Test for Window and Glass Block Assemblies*, respectively.

**3.3.221.2 *Fire Resistance Rating*.** The time, in minutes or hours, that materials or assemblies have withstood a fire exposure as determined by the tests, or methods based on tests, prescribed by this *Code*.

**3.3.222\* Reconstruction.** The reconfiguration of a space that affects an exit or a corridor shared by more than one occupant space; or the reconfiguration of a space such that the rehabilitation work area is not permitted to be occupied because existing means of egress and fire protection systems, or their equivalent, are not in place or continuously maintained.

**3.3.223 Reflective Insulation.** See 3.3.150.2.

**3.3.224 Registered Architect.** A person licensed to practice architecture in a jurisdiction, subject to all laws and limitations imposed by the jurisdiction.

**3.3.225 Registered Design Professional (RDP).** An individual who is registered or licensed to practice his/her respective design profession as defined by the statutory requirements of the professional registration laws of the state or jurisdiction in which the project is to be constructed.

**3.3.226 Regular Stage.** See 3.3.262.2.

**3.3.227 Rehabilitation Work Area.** See 3.3.21.8

**3.3.228 Renovation.** The replacement in kind, strengthening, or upgrading of building elements, materials, equipment, or fixtures that does not result in a reconfiguration of the building or spaces within.

**3.3.229 Repair.** The patching, restoration, or painting of materials, elements, equipment, or fixtures for the purpose of maintaining such materials, elements, equipment, or fixtures in good or sound condition.

**3.3.230 Residential Board and Care Occupancy.** See 3.3.188.12.

**3.3.231 Residential Board and Care Resident.** A person who receives personal care and resides in a residential board and care facility.

**3.3.232 Residential Occupancy.** See 3.3.188.13.

**3.3.233 Safe Location.** A location remote or separated from the effects of a fire so that such effects no longer pose a threat.

**3.3.234 Safety Factor.** A factor applied to a predicted value to ensure that a sufficient safety margin is maintained.

**3.3.235 Safety Margin.** The difference between a predicted value and the actual value where a fault condition is expected.

**3.3.236 Sally Port (Security Vestibule).** A compartment provided with two or more doors where the intended purpose is to prevent continuous and unobstructed passage by allowing the release of only one door at a time.

**3.3.237 Seating.**

**3.3.237.1\* *Festival Seating*.** A form of audience/spectator accommodation in which no seating, other than a floor or finished ground level, is provided for the audience/spectators gathered to observe a performance.

**3.3.237.2 *Fixed Seating*.** Seating that is secured to the building structure.

**3.3.237.3 *Folding and Telescopic Seating*.** A structure that is used for tiered seating of persons and whose overall shape and size can be reduced, without being dismantled, for purposes of moving or storing.

**3.3.237.4 *Smoke-Protected Assembly Seating*.** Seating served by means of egress that is not subject to smoke accumulation within or under the structure.

**3.3.238 Self-Closing.** Equipped with an approved device that ensures closing after opening.

**3.3.239\* Self-Luminous.** Illuminated by a self-contained power source and operated independently of external power sources.

**3.3.240\* Self-Preservation (Day-Care Occupancy).** The ability of a client to evacuate a day-care occupancy without direct intervention by a staff member.

**3.3.241 Sensitivity Analysis.** See 3.3.17.1.

**3.3.242 Separate Atmosphere.** See 3.3.26.2.

**3.3.243 Separated Occupancy.** See 3.3.188.14.

**3.3.244 Severe Mobility Impairment.** The ability to move to stairs but without the ability to use the stairs.

**3.3.245 Single Station Alarm.** See 3.3.14.1.

**3.3.246 Site-Fabricated Stretch System.** See 3.3.273.2.

**3.3.247\* Situation Awareness.** The perception of the elements in the environment within a volume of time and space, the comprehension of their meaning, and the projection of their status in the near future.

**3.3.248 Slow Evacuation Capability.** See 3.3.76.3.

**3.3.249 Smoke Alarm.** See 3.3.14.2.

**3.3.250 Smoke Barrier.** See 3.3.31.2.

**3.3.251 Smoke Compartment.** See 3.3.48.2.

**3.3.252 Smoke Detector.** A device that detects visible or invisible particles of combustion. [*72*, 2010]

**3.3.253 Smoke Developed Index.** See 3.3.147.2.

**3.3.254\* Smoke Partition.** A continuous membrane that is designed to form a barrier to limit the transfer of smoke.

**3.3.255\* Smokeproof Enclosure.** An enclosure designed to limit the movement of products of combustion produced by a fire.

**3.3.256 Smoke-Protected Assembly Seating.** See 3.3.237.4.

**3.3.257 Special Amusement Building.** See 3.3.36.10.

**3.3.258 Special Inspection.** Services provided by a qualified person, retained by the owner and approved by the authority having jurisdiction, who observes the installation and witnesses the pretesting and operation of the system or systems.

**3.3.259 Special-Purpose Industrial Occupancy.** See 3.3.188.8.3.

**3.3.260 Specification.**

**3.3.260.1\* *Design Specification.*** A building characteristic and other conditions that are under the control of the design team.

**3.3.260.2 *Input Data Specification.*** Information required by the verification method.

**3.3.261 Staff (Residential Board and Care).** Persons who provide personal care services, supervision, or assistance.

**3.3.262 Stage.** A space within a building used for entertainment and utilizing drops or scenery or other stage effects.

**3.3.262.1 *Legitimate Stage.*** A stage with a height greater than 50 ft (15 m) measured from the lowest point on the stage floor to the highest point of the roof or floor deck above.

**3.3.262.2 *Regular Stage.*** A stage with a height of 50 ft (15 m) or less measured from the lowest point on the stage floor to the highest point of the roof or floor deck above.

**3.3.263 Stair.**

**3.3.263.1 *Aisle Stair.*** A stair within a seating area of an assembly occupancy that directly serves rows of seating to the side of the stair.

**3.3.263.2\* *Outside Stair.*** A stair with not less than one side open to the outer air.

**3.3.264 Stair Descent Device.** See 3.3.61.2.

**3.3.265 Stakeholder.** An individual, or representative of same, having an interest in the successful completion of a project.

**3.3.266 Storage Occupancy.** See 3.3.188.15.

**3.3.267\* Stories in Height.** The story count starting with the level of exit discharge and ending with the highest occupiable story containing the occupancy considered.

**3.3.268 Story.** The portion of a building located between the upper surface of a floor and the upper surface of the floor or roof next above.

**3.3.268.1\* *Occupiable Story.*** A story occupied by people on a regular basis.

**3.3.269 Street.** A public thoroughfare that has been dedicated for vehicular use by the public and can be used for access by fire department vehicles.

**3.3.270\* Street Floor.** A story or floor level accessible from the street or from outside the building at the finished ground level, with the floor level at the main entrance located not more than three risers above or below the finished ground level, and arranged and utilized to qualify as the main floor.

**3.3.271\* Structure.** That which is built or constructed.

**3.3.271.1 *Air-Inflated Structure.*** A structure whose shape is maintained by air pressure in cells or tubes forming all or part of the enclosure of the usable area and in which the occupants are not within the pressurized area used to support the structure.

**3.3.271.2\* *Air-Supported Structure.*** A structure where shape is maintained by air pressure and in which occupants are within the elevated pressure area.

**3.3.271.3 *Limited Access Structure.*** A structure or portion of a structure lacking emergency openings.

**3.3.271.4 *Membrane Structure.*** A building or portion of a building incorporating an air-inflated, air-supported, tensioned-membrane structure; a membrane roof; or a membrane-covered rigid frame to protect habitable or usable space.

**3.3.271.5 *Multilevel Play Structure.*** A structure that consists of tubes, slides, crawling areas, and jumping areas that is located within a building and is used for climbing and entertainment, generally by children.

**3.3.271.6\* *Open Structure.*** A structure that supports equipment and operations not enclosed within building walls.

**3.3.271.7\* *Parking Structure.*** A building, structure, or portion thereof used for the parking, storage, or both, of motor vehicles. [**88A,** 2011]

**3.3.271.7.1 *Assisted Mechanical Type Parking Structure.*** A parking structure that uses lifts or other mechanical devices to transport vehicles to the floors of a parking structure, where the vehicles are then parked by a person. [**88A,** 2011]

**3.3.271.7.2 *Automated Type Parking Structure.*** A parking structure that uses computer controlled machines to store and retrieve vehicles, without drivers, in multi-level storage racks with no floors. [**88A,** 2011]

**3.3.271.7.3 *Enclosed Parking Structure.*** Any parking structure that is not an open parking structure. [**88A,** 2011]

**3.3.271.7.4 *Open Parking Structure.*** A parking structure that meets the requirements of 42.8.1.3.

**3.3.271.7.5 *Ramp Type Parking Structure.*** A parking structure that utilizes sloped floors for vertical vehicle circulation. [**88A,** 2011]

**3.3.271.8 *Permanent Structure.*** A building or structure that is intended to remain in place for a period of more than 180 days in any consecutive 12-month period.

**3.3.271.9 *Temporary Structure.*** A building or structure not meeting the definition of *permanent structure. (See also 3.3.271.8, Permanent Structure.)*

**3.3.271.10 *Tensioned-Membrane Structure.*** A membrane structure incorporating a membrane and a structural support system such as arches, columns and cables, or beams wherein the stresses developed in the tensioned membrane interact with those in the structural support so that the entire assembly acts together to resist the applied loads.

**3.3.271.11\* *Underground Structure.*** A structure or portions of a structure in which the floor level is below the level of exit discharge.

**3.3.271.12 *Water-Surrounded Structure.*** A structure fully surrounded by water.

**3.3.272 Suite.**

**3.3.272.1 *Guest Suite.*** An accommodation with two or more contiguous rooms comprising a compartment, with or without doors between such rooms, that provides living, sleeping, sanitary, and storage facilities.

**3.3.272.2** *Non-Patient-Care Suite (Heath Care Occupancies).* A suite within a health care occupancy that is not intended for sleeping or treating patients.

**3.3.272.3** *Patient Care Non-Sleeping Suite (Health Care Occupancies).* A suite for treating patients with or without patient beds not intended for overnight sleeping.

**3.3.272.4** *Patient Care Sleeping Suite (Health Care Occupancies).* A suite containing one or more patient beds intended for overnight sleeping.

**3.3.272.5** *Patient Care Suite (Health Care Occupancies).* A series of rooms or spaces or a subdivided room separated from the remainder of the building by walls and doors.

**3.3.273 System.**

**3.3.273.1** *Elevator Evacuation System.* A system, including a vertical series of elevator lobbies and associated elevator lobby doors, an elevator shaft(s), and a machine room(s), that provides protection from fire effects for elevator passengers, people waiting to use elevators, and elevator equipment so that elevators can be used safely for egress.

**3.3.273.2** *Site-Fabricated Stretch System.* A system, fabricated on-site, and intended for acoustical, tackable, or aesthetic purposes, that is comprised of three elements: (1) a frame (constructed of plastic, wood, metal, or other material) used to hold fabric in place, (2) a core material (infill, with the correct properties for the application), and (3) an outside layer, comprised of a textile, fabric, or vinyl, that is stretched taut and held in place by tension or mechanical fasteners via the frame.

**3.3.274 Technically Infeasible.** A change to a building that has little likelihood of being accomplished because the existing structural conditions require the removal or alteration of a load-bearing member that is an essential part of the structural frame, or because other existing physical or site constraints prohibit modification or addition of elements, spaces, or features that are in full and strict compliance with applicable requirements.

**3.3.275 Temporary Platform.** See 3.3.209.1.

**3.3.276 Temporary Structure.** See 3.3.271.9.

**3.3.277 Tensioned-Membrane Structure.** See 3.3.271.10.

**3.3.278* Tent.** A temporary structure, the covering of which is made of pliable material that achieves its support by mechanical means such as beams, columns, poles, or arches, or by rope or cables, or both.

**3.3.278.1** *Private Party Tent.* A tent erected in the yard of a private residence for entertainment, recreation, dining, a reception, or similar function.

**3.3.279 Thermal Barrier.** See 3.3.31.3.

**3.3.280 Tower.** An enclosed independent structure or portion of a building with elevated levels for support of equipment or occupied for observation, control, operation, signaling, or similar limited use.

**3.3.280.1** *Air Traffic Control Tower.* An enclosed structure or building at airports with elevated levels for support of equipment and occupied for observation, control, operation, and signaling of aircraft in flight and on the ground.

**3.3.281 Two-Family Dwelling Unit.** See 3.3.66.3.

**3.3.282 Uncertainty Analysis.** See 3.3.17.2.

**3.3.283 Underground Structure.** See 3.3.271.11.

**3.3.284 Verification Method.** A procedure or process used to demonstrate or confirm that the proposed design meets the specified criteria.

**3.3.285* Vertical Opening.** An opening through a floor or roof.

**3.3.286 Vomitory.** An entrance to a means of egress from an assembly seating area that pierces the seating rows.

**3.3.287 Wall.**

**3.3.287.1** *Fire Barrier Wall.* A wall, other than a fire wall, that has a fire resistance rating.

**3.3.287.2** *Proscenium Wall.* The wall that separates the stage from the auditorium or house.

**3.3.288* Wall or Ceiling Covering.** A textile-, paper-, or polymeric-based product designed to be attached to a wall or ceiling surface for decorative or acoustical purposes.

**3.3.289 Water-Surrounded Structure.** See 3.3.271.12.

**3.3.290 Weathered-Membrane Material.** See 3.3.169.5.

**3.3.291 Yard.** An open, unoccupied space other than a court, unobstructed from the finished ground level to the sky on the lot on which a building is situated.

## Chapter 4   General

**4.1* Goals.**

**4.1.1* Fire.** A goal of this *Code* is to provide an environment for the occupants that is reasonably safe from fire by the following means:

(1)*Protection of occupants not intimate with the initial fire development
(2) Improvement of the survivability of occupants intimate with the initial fire development

**4.1.2* Comparable Emergencies.** An additional goal is to provide life safety during emergencies that can be mitigated using methods comparable to those used in case of fire.

**4.1.3* Crowd Movement.** An additional goal is to provide for reasonably safe emergency crowd movement and, where required, reasonably safe nonemergency crowd movement.

**4.2 Objectives.**

**4.2.1 Occupant Protection.** A structure shall be designed, constructed, and maintained to protect occupants who are not intimate with the initial fire development for the time needed to evacuate, relocate, or defend in place.

**4.2.2 Structural Integrity.** Structural integrity shall be maintained for the time needed to evacuate, relocate, or defend in place occupants who are not intimate with the initial fire development.

**4.2.3 Systems Effectiveness.** Systems utilized to achieve the goals of Section 4.1 shall be effective in mitigating the hazard or condition for which they are being used, shall be reliable, shall be maintained to the level at which they were designed to operate, and shall remain operational.

**4.3\* Assumptions.**

**4.3.1\* General.** The protection methods of this *Code* are based on the hazards associated with fire and other events that have comparable impact on a building and its occupants.

**4.3.2 Single Fire Source.** The fire protection methods of this *Code* assume a single fire source.

**4.4 Life Safety Compliance Options.**

**4.4.1 Options.** Life safety meeting the goals and objectives of Sections 4.1 and 4.2 shall be provided in accordance with either of the following:

(1) Prescriptive-based provisions per 4.4.2
(2) Performance-based provisions per 4.4.3

**4.4.2 Prescriptive-Based Option.**

**4.4.2.1** A prescriptive-based life safety design shall be in accordance with Chapters 1 through 4, Chapters 6 through 11, Chapter 43, and the applicable occupancy chapter, Chapters 12 through 42.

**4.4.2.2** Prescriptive-based designs meeting the requirements of Chapters 1 through 3, Sections 4.5 through 4.8, and Chapters 6 through 43 of this *Code* shall be deemed to satisfy the provisions of Sections 4.1 and 4.2.

**4.4.2.3** Where specific requirements contained in Chapters 11 through 43 differ from general requirements contained in Chapters 1 through 4, and Chapters 6 through 10, the requirements of Chapters 11 through 43 shall govern.

**4.4.3 Performance-Based Option.** A performance-based life safety design shall be in accordance with Chapters 1 through 5.

**4.5 Fundamental Requirements.**

**4.5.1 Multiple Safeguards.** The design of every building or structure intended for human occupancy shall be such that reliance for safety to life does not depend solely on any single safeguard. An additional safeguard(s) shall be provided for life safety in case any single safeguard is ineffective due to inappropriate human actions or system failure.

**4.5.2 Appropriateness of Safeguards.** Every building or structure shall be provided with means of egress and other fire and life safety safeguards of the kinds, numbers, locations, and capacities appropriate to the individual building or structure, with due regard to the following:

(1) Character of the occupancy, including fire load
(2) Capabilities of the occupants
(3) Number of persons exposed
(4) Fire protection available
(5) Capabilities of response personnel
(6) Height and construction type of the building or structure
(7) Other factors necessary to provide occupants with a reasonable degree of safety

**4.5.3 Means of Egress.**

**4.5.3.1 Number of Means of Egress.** Two means of egress, as a minimum, shall be provided in every building or structure, section, and area where size, occupancy, and arrangement endanger occupants attempting to use a single means of egress that is blocked by fire or smoke. The two means of egress shall be arranged to minimize the possibility that both might be rendered impassable by the same emergency condition.

**4.5.3.2 Unobstructed Egress.** In every occupied building or structure, means of egress from all parts of the building shall be maintained free and unobstructed. Means of egress shall be accessible to the extent necessary to ensure reasonable safety for occupants having impaired mobility.

**4.5.3.3 Awareness of Egress System.** Every exit shall be clearly visible, or the route to reach every exit shall be conspicuously indicated. Each means of egress, in its entirety, shall be arranged or marked so that the way to a place of safety is indicated in a clear manner.

**4.5.3.4 Lighting.** Where artificial illumination is needed in a building or structure, egress facilities shall be included in the lighting design.

**4.5.4\* Occupant Notification.** In every building or structure of such size, arrangement, or occupancy that a fire itself might not provide adequate occupant warning, fire alarm systems shall be provided where necessary to warn occupants of the existence of fire.

**4.5.5\* Situation Awareness.** Systems used to achieve the goals of Section 4.1 shall be effective in facilitating and enhancing situation awareness, as appropriate, by building management, other occupants and emergency responders of the functionality or state of critical building systems, the conditions that might warrant emergency response, and the appropriate nature and timing of such responses.

**4.5.6 Vertical Openings.** Every vertical opening between the floors of a building shall be suitably enclosed or protected, as necessary, to afford reasonable safety to occupants while using the means of egress and to prevent the spread of fire, smoke, or fumes through vertical openings from floor to floor before occupants have entered exits.

**4.5.7 System Design/Installation.** Any fire protection system, building service equipment, feature of protection, or safeguard provided to achieve the goals of this *Code* shall be designed, installed, and approved in accordance with applicable NFPA standards.

**4.5.8 Maintenance.** Whenever or wherever any device, equipment, system, condition, arrangement, level of protection, or any other feature is required for compliance with the provisions of this *Code*, such device, equipment, system, condition, arrangement, level of protection, or other feature shall thereafter be maintained, unless the *Code* exempts such maintenance.

**4.6 General Requirements.**

**4.6.1 Authority Having Jurisdiction.**

**4.6.1.1** The authority having jurisdiction shall determine whether the provisions of this *Code* are met.

**4.6.1.2** Any requirements that are essential for the safety of building occupants and that are not specifically provided for by this *Code* shall be determined by the authority having jurisdiction.

**4.6.1.3** Where it is evident that a reasonable degree of safety is provided, any requirement shall be permitted to be modified if, in the judgment of the authority having jurisdiction, its application would be hazardous under normal occupancy conditions.

**4.6.1.4 Technical Assistance.**

**4.6.1.4.1** The authority having jurisdiction shall be permitted to require a review by an approved independent third party with expertise in the matter to be reviewed at the submitter's expense. [1:1.15.1]

**4.6.1.4.2** The independent reviewer shall provide an evaluation and recommend necessary changes of the proposed design, operation, process, or new technology to the authority having jurisdiction. [1:1.15.2]

**4.6.1.4.3** The authority having jurisdiction shall be authorized to require design submittals to bear the stamp of a registered design professional. [1:1.15.3]

**4.6.2 Previously Approved Features.** Where another provision of this *Code* exempts a previously approved feature from a requirement, the exemption shall be permitted, even where the following conditions exist:

(1) The area is being modernized, renovated, or otherwise altered.
(2) A change of occupancy has occurred, provided that the feature's continued use is approved by the authority having jurisdiction.

**4.6.3 Stories in Height.** Unless otherwise specified in another provision of this *Code*, the stories in height of a building shall be determined as follows:

(1) The stories in height shall be counted starting with the level of exit discharge and ending with the highest occupiable story containing the occupancy considered.
(2) Stories below the level of exit discharge shall not be counted as stories.
(3) Interstitial spaces used solely for building or process systems directly related to the level above or below shall not be considered a separate story.
(4) A mezzanine shall not be counted as a story for the purpose of determining the allowable stories in height.
(5) For purposes of application of the requirements for occupancies other than assembly, health care, detention and correctional, and ambulatory health care, where a maximum one-story abovegrade parking structure, enclosed, open, or a combination thereof, of Type I or Type II (222) construction or open Type IV construction, with grade entrance, is provided under a building, the number of stories shall be permitted to be measured from the floor above such a parking area.

**4.6.4 Historic Buildings.**

**4.6.4.1** Rehabilitation projects in historic buildings shall comply with Chapter 43.

**4.6.4.2\*** The provisions of this *Code* shall be permitted to be modified by the authority having jurisdiction for buildings or structures identified and classified as historic buildings or structures where it is evident that a reasonable degree of safety is provided.

**4.6.5\* Modification of Requirements for Existing Buildings.** Where it is evident that a reasonable degree of safety is provided, the requirements for existing buildings shall be permitted to be modified if their application would be impractical in the judgment of the authority having jurisdiction.

**4.6.6 Time Allowed for Compliance.** A limited but reasonable time, commensurate with the magnitude of expenditure, disruption of services, and degree of hazard, shall be allowed for compliance with any part of this *Code* for existing buildings.

**4.6.7 Building Rehabilitation.**

**4.6.7.1** Rehabilitation work on existing buildings shall be classified as one of the following work categories in accordance with 43.2.2.1:

(1) Repair
(2) Renovation
(3) Modification
(4) Reconstruction
(5) Change of use or occupancy classification
(6) Addition

**4.6.7.2** Rehabilitation work on existing buildings shall comply with Chapter 43.

**4.6.7.3** Except where another provision of this *Code* exempts a previously approved feature from a requirement, the resulting feature shall be not less than that required for existing buildings.

**4.6.7.4\*** Existing life safety features that exceed the requirements for new buildings shall be permitted to be decreased to those required for new buildings.

**4.6.7.5\*** Existing life safety features that do not meet the requirements for new buildings, but that exceed the requirements for existing buildings, shall not be further diminished.

**4.6.8 Provisions in Excess of *Code* Requirements.** Nothing in this *Code* shall be construed to prohibit a better building construction type, an additional means of egress, or an otherwise safer condition than that specified by the minimum requirements of this *Code*.

**4.6.9 Conditions for Occupancy.**

**4.6.9.1** No new construction or existing building shall be occupied in whole or in part in violation of the provisions of this *Code*, unless the following conditions exist:

(1) A plan of correction has been approved.
(2) The occupancy classification remains the same.
(3) No serious life safety hazard exists as judged by the authority having jurisdiction.

**4.6.9.2** Where compliance with this *Code* is effected by means of a performance-based design, the owner shall annually certify compliance with the conditions and limitations of the design by submitting a warrant of fitness acceptable to the authority having jurisdiction. The warrant of fitness shall attest that the building features, systems, and use have been inspected and confirmed to remain consistent with design specifications outlined in the documentation required by Section 5.8 and that such features, systems, and use continue to satisfy the goals and objectives specified in Sections 4.1 and 4.2. *(See Chapter 5.)*

**4.6.10 Construction, Repair, and Improvement Operations.**

**4.6.10.1\*** Buildings, or portions of buildings, shall be permitted to be occupied during construction, repair, alterations, or additions only where required means of egress and required fire protection features are in place and continuously maintained for the portion occupied or where alternative life safety measures acceptable to the authority having jurisdiction are in place.

**4.6.10.2*** In buildings under construction, adequate escape facilities shall be maintained at all times for the use of construction workers. Escape facilities shall consist of doors, walkways, stairs, ramps, fire escapes, ladders, or other approved means or devices arranged in accordance with the general principles of the *Code* insofar as they can reasonably be applied to buildings under construction.

**4.6.10.3** Flammable or explosive substances or equipment for repairs or alterations shall be permitted in a building while the building is occupied if the condition of use and safeguards provided do not create any additional danger or impediment to egress beyond the normally permissible conditions in the building.

**4.6.11 Change of Use or Occupancy Classification.** In any building or structure, whether or not a physical alteration is needed, a change from one use or occupancy classification to another shall comply with 4.6.7.

**4.6.12 Maintenance, Inspection, and Testing.**

**4.6.12.1** Whenever or wherever any device, equipment, system, condition, arrangement, level of protection, fire-resistive construction, or any other feature is required for compliance with the provisions of this *Code*, such device, equipment, system, condition, arrangement, level of protection, fire-resistive construction, or other feature shall thereafter be continuously maintained. Maintenance shall be provided in accordance with applicable NFPA requirements or requirements developed as part of a performance-based design, or as directed by the authority having jurisdiction.

**4.6.12.2** No existing life safety feature shall be removed or reduced where such feature is a requirement for new construction.

**4.6.12.3*** Existing life safety features obvious to the public, if not required by the *Code*, shall be either maintained or removed.

**4.6.12.4** Any device, equipment, system, condition, arrangement, level of protection, fire-resistive construction, or any other feature requiring periodic testing, inspection, or operation to ensure its maintenance shall be tested, inspected, or operated as specified elsewhere in this *Code* or as directed by the authority having jurisdiction.

**4.6.12.5** Maintenance, inspection, and testing shall be performed under the supervision of a responsible person who shall ensure that testing, inspection, and maintenance are made at specified intervals in accordance with applicable NFPA standards or as directed by the authority having jurisdiction.

**4.6.13* Noncombustible Material.**

**4.6.13.1** A material that complies with any of the following shall be considered a noncombustible material:

(1)*A material that, in the form in which it is used and under the conditions anticipated, will not ignite, burn, support combustion, or release flammable vapors when subjected to fire or heat

(2) A material that is reported as passing ASTM E 136, *Standard Test Method for Behavior of Materials in a Vertical Tube Furnace at 750 Degrees C*

(3) A material that is reported as complying with the pass/fail criteria of ASTM E 136 when tested in accordance with the test method and procedure in ASTM E 2652, *Standard Test Method for Behavior of Materials in a Tube Furnace with a Cone-shaped Airflow Stabilizer, at 750 Degrees C*

**4.6.13.2** Where the term *limited-combustible* is used in this *Code*, it shall also include the term *noncombustible*.

**4.6.14* Limited-Combustible Material.** A material shall be considered a limited-combustible material where all the conditions of 4.6.14.1 and 4.6.14.2, and the conditions of either 4.6.14.3 or 4.6.14.4, are met.

**4.6.14.1** The material shall not comply with the requirements for noncombustible material in accordance with 4.6.13.

**4.6.14.2** The material, in the form in which it is used, shall exhibit a potential heat value not exceeding 3500 Btu/lb (8141 kJ/kg) where tested in accordance with NFPA 259, *Standard Test Method for Potential Heat of Building Materials*.

**4.6.14.3** The material shall have the structural base of a noncombustible material with a surfacing not exceeding a thickness of ⅛ in. (3.2 mm) where the surfacing exhibits a flame spread index not greater than 50 when tested in accordance with ASTM E 84, *Standard Test Method for Surface Burning Characteristics of Building Materials*, or ANSI/UL 723, *Standard for Test for Surface Burning Characteristics of Building Materials*.

**4.6.14.4** The material shall be composed of materials that, in the form and thickness used, neither exhibit a flame spread index greater than 25 nor evidence of continued progressive combustion when tested in accordance with ASTM E 84, *Standard Test Method for Surface Burning Characteristics of Building Materials*, or ANSI/UL 723, *Standard for Test for Surface Burning Characteristics of Building Materials*, and shall be of such composition that all surfaces that would be exposed by cutting through the material on any plane would neither exhibit a flame spread index greater than 25 nor exhibit evidence of continued progressive combustion when tested in accordance with ASTM E 84 or ANSI/UL 723.

**4.6.14.5** Where the term *limited-combustible* is used in this *Code*, it shall also include the term *noncombustible*.

**4.7* Fire Drills.**

**4.7.1 Where Required.** Emergency egress and relocation drills conforming to the provisions of this *Code* shall be conducted as specified by the provisions of Chapters 11 through 43, or by appropriate action of the authority having jurisdiction. Drills shall be designed in cooperation with the local authorities.

**4.7.2* Drill Frequency.** Emergency egress and relocation drills, where required by Chapters 11 through 43 or the authority having jurisdiction, shall be held with sufficient frequency to familiarize occupants with the drill procedure and to establish conduct of the drill as a matter of routine. Drills shall include suitable procedures to ensure that all persons subject to the drill participate.

**4.7.3 Orderly Evacuation.** When conducting drills, emphasis shall be placed on orderly evacuation rather than on speed.

**4.7.4* Simulated Conditions.** Drills shall be held at expected and unexpected times and under varying conditions to simulate the unusual conditions that can occur in an actual emergency.

**4.7.5 Relocation Area.** Drill participants shall relocate to a predetermined location and remain at such location until a recall or dismissal signal is given.

**4.7.6\*** A written record of each drill shall be completed by the person responsible for conducting the drill and maintained in an approved manner.

**4.8 Emergency Plan.**

**4.8.1 Where Required.** Emergency plans shall be provided as follows:

(1) Where required by the provisions of Chapters 11 through 42
(2) Where required by action of the authority having jurisdiction

**4.8.2 Plan Requirements.**

**4.8.2.1\*** Emergency plans shall include the following:

(1) Procedures for reporting of emergencies
(2) Occupant and staff response to emergencies
(3) \*Evacuation procedures appropriate to the building, its occupancy, emergencies, and hazards *(see Section 4.3)*
(4) Appropriateness of the use of elevators
(5) Design and conduct of fire drills
(6) Type and coverage of building fire protection systems
(7) Other items required by the authority having jurisdiction

**4.8.2.2** Required emergency plans shall be submitted to the authority having jurisdiction for review.

**4.8.2.3** Emergency plans shall be reviewed and updated as required by the authority having jurisdiction.

## Chapter 5    Performance-Based Option

**5.1 General Requirements.**

**5.1.1\* Application.** The requirements of this chapter shall apply to life safety systems designed to the performance-based option permitted by 4.4.1 and 4.4.3.

**5.1.2 Goals and Objectives.** The performance-based design shall meet the goals and objectives of this *Code* in accordance with Sections 4.1 and 4.2.

**5.1.3 Qualifications.** The performance-based design shall be prepared by a registered design professional.

**5.1.4\* Independent Review.** The authority having jurisdiction shall be permitted to require an approved, independent third party to review the proposed design and provide an evaluation of the design to the authority having jurisdiction.

**5.1.5 Sources of Data.** Data sources shall be identified and documented for each input data requirement that must be met using a source other than a design fire scenario, an assumption, or a building design specification. The degree of conservatism reflected in such data shall be specified, and a justification for the source shall be provided.

**5.1.6\* Final Determination.** The authority having jurisdiction shall make the final determination as to whether the performance objectives have been met.

**5.1.7\* Maintenance of Design Features.** The design features required for the building to continue to meet the performance goals and objectives of this *Code* shall be maintained for the life of the building. Such performance goals and objectives shall include complying with all documented assumptions and design specifications. Any variations shall require the approval of the authority having jurisdiction prior to the actual change. *(See also 4.6.9.2.)*

**5.1.8 Definitions.**

**5.1.8.1 General.** For definitions, see Chapter 3, Definitions.

**5.1.8.2 Special Definitions.** A list of special terms used in this chapter follows:

(1) **Alternative Calculation Procedure.** See 3.3.15.
(2) **Data Conversion.** See 3.3.53.
(3) **Design Fire Scenario.** See 3.3.103.1.
(4) **Design Specification.** See 3.3.260.1.
(5) **Design Team.** See 3.3.58.
(6) **Exposure Fire.** See 3.3.86.
(7) **Fire Model.** See 3.3.99.
(8) **Fire Scenario.** See 3.3.103.
(9) **Fuel Load.** See 3.3.162.1.
(10) **Incapacitation.** See 3.3.146.
(11) **Input Data Specification.** See 3.3.260.2.
(12) **Occupant Characteristics.** See 3.3.189.
(13) **Performance Criteria.** See 3.3.204.
(14) **Proposed Design.** See 3.3.216.
(15) **Safe Location.** See 3.3.233.
(16) **Safety Factor.** See 3.3.234.
(17) **Safety Margin.** See 3.3.235.
(18) **Sensitivity Analysis.** See 3.3.17.1.
(19) **Stakeholder.** See 3.3.265.
(20) **Uncertainty Analysis.** See 3.3.17.2.
(21) **Verification Method.** See 3.3.284.

**5.2 Performance Criteria.**

**5.2.1 General.** A design shall meet the objectives specified in Section 4.2 if, for each design fire scenario, assumption, and design specification, the performance criterion in 5.2.2 is met.

**5.2.2\* Performance Criterion.** Any occupant who is not intimate with ignition shall not be exposed to instantaneous or cumulative untenable conditions.

**5.3 Retained Prescriptive Requirements.**

**5.3.1\* Systems and Features.** All fire protection systems and features of the building shall comply with applicable NFPA standards for those systems and features.

**5.3.2 Means of Egress.** The design shall comply with the following requirements in addition to the performance criteria of Section 5.2 and the methods of Sections 5.4 through 5.8:

(1) Changes in level in means of egress — 7.1.7
(2) Guards — 7.1.8
(3) Doors — 7.2.1
(4) Stairs — 7.2.2, excluding the provisions of 7.2.2.5.1, 7.2.2.5.2, 7.2.2.6.2, 7.2.2.6.3, and 7.2.2.6.4
(5) Ramps — 7.2.5, excluding the provisions of 7.2.5.3.1, 7.2.5.5, and 7.2.5.6.1
(6) Fire escape ladders — 7.2.9
(7) Alternating tread devices — 7.2.11

(8) Capacity of means of egress — Section 7.3, excluding the provisions of 7.3.3 and 7.3.4
(9) Impediments to egress — 7.5.2
(10) Illumination of means of egress — Section 7.8
(11) Emergency lighting — Section 7.9
(12) Marking of means of egress — Section 7.10

**5.3.3 Equivalency.** Equivalent designs for the features covered in the retained prescriptive requirements mandated by 5.3.2 shall be addressed in accordance with the equivalency provisions of Section 1.4.

**5.4 Design Specifications and Other Conditions.**

**5.4.1\* Clear Statement.** Design specifications and other conditions used in the performance-based design shall be clearly stated and shown to be realistic and sustainable.

**5.4.2 Assumptions and Design Specifications Data.**

**5.4.2.1** Each assumption and design specification used in the design shall be accurately translated into input data specifications, as appropriate for the method or model.

**5.4.2.2** Any assumption and design specifications that the design analyses do not explicitly address or incorporate and that are, therefore, omitted from input data specifications shall be identified, and a sensitivity analysis of the consequences of that omission shall be performed.

**5.4.2.3** Any assumption and design specifications modified in the input data specifications, because of limitations in test methods or other data-generation procedures, shall be identified, and a sensitivity analysis of the consequences of the modification shall be performed.

**5.4.3 Building Characteristics.** Characteristics of the building or its contents, equipment, or operations that are not inherent in the design specifications, but that affect occupant behavior or the rate of hazard development, shall be explicitly identified.

**5.4.4\* Operational Status and Effectiveness of Building Features and Systems.** The performance of fire protection systems, building features, and emergency procedures shall reflect the documented performance and reliability of the components of those systems or features, unless design specifications are incorporated to modify the expected performance.

**5.4.5 Occupant Characteristics.**

**5.4.5.1\* General.** The selection of occupant characteristics to be used in the design calculations shall be approved by the authority having jurisdiction and shall provide an accurate reflection of the expected population of building users. Occupant characteristics shall represent the normal occupant profile, unless design specifications are used to modify the expected occupant features. Occupant characteristics shall not vary across fire scenarios, except as authorized by the authority having jurisdiction.

**5.4.5.2\* Response Characteristics.** The basic response characteristics of sensibility, reactivity, mobility, and susceptibility shall be evaluated. Such evaluation shall include the expected distribution of characteristics of a population appropriate to the use of the building. The source of data for these characteristics shall be documented.

**5.4.5.3 Location.** It shall be assumed that, in every normally occupied room or area, at least one person shall be located at the most remote point from the exits.

**5.4.5.4\* Number of Occupants.** The design shall be based on the maximum number of people that every occupied room or area is expected to contain. Where the success or failure of the design is contingent on the number of occupants not exceeding a specified maximum, operational controls shall be used to ensure that the maximum number of occupants is not exceeded.

**5.4.5.5\* Staff Assistance.** The inclusion of trained employees as part of the fire safety system shall be identified and documented.

**5.4.6 Emergency Response Personnel.** Design characteristics or other conditions related to the availability, speed of response, effectiveness, roles, and other characteristics of emergency response personnel shall be specified, estimated, or characterized sufficiently for evaluation of the design.

**5.4.7\* Post-Construction Conditions.** Design characteristics or other conditions related to activities during the life of a building that affect the ability of the building to meet the stated goals and objectives shall be specified, estimated, or characterized sufficiently for evaluation of the design.

**5.4.8 Off-Site Conditions.** Design characteristics or other conditions related to resources or conditions outside the property being designed that affect the ability of the building to meet the stated goals and objectives shall be specified, estimated, or characterized sufficiently for evaluation of the design.

**5.4.9\* Consistency of Assumptions.** The design shall not include mutually inconsistent assumptions, specifications, or statements of conditions.

**5.4.10\* Special Provisions.** Additional provisions that are not covered by the design specifications, conditions, estimations, and assumptions provided in Section 5.4, but that are required for the design to comply with the performance objectives, shall be documented.

**5.5\* Design Fire Scenarios.**

**5.5.1 Approval of Parameters.** The authority having jurisdiction shall approve the parameters involved in design fire scenarios. The proposed design shall be considered to meet the goals and objectives if it achieves the performance criteria for each required design fire scenario. *(See 5.5.3.)*

**5.5.2\* Evaluation.** Design fire scenarios shall be evaluated using a method acceptable to the authority having jurisdiction and appropriate for the conditions. Each design fire scenario shall be as challenging as any that could occur in the building, but shall be realistic, with respect to at least one of the following scenario specifications:

(1) Initial fire location
(2) Early rate of growth in fire severity
(3) Smoke generation

**5.5.3\* Required Design Fire Scenarios.** Design fire scenarios shall comply with the following:

(1) Scenarios selected as design fire scenarios shall include, but shall not be limited to, those specified in 5.5.3.1 through 5.5.3.8.
(2) Design fire scenarios demonstrated by the design team to the satisfaction of the authority having jurisdiction as inappropriate for the building use and conditions shall not be required to be evaluated fully.

2012 Edition

**5.5.3.1\* Design Fire Scenario 1.** Design Fire Scenario 1 shall be described as follows:

(1) It is an occupancy-specific fire representative of a typical fire for the occupancy.
(2) It explicitly accounts for the following:
    (a) Occupant activities
    (b) Number and location of occupants
    (c) Room size
    (d) Contents and furnishings
    (e) Fuel properties and ignition sources
    (f) Ventilation conditions
    (g) Identification of the first item ignited and its location

**5.5.3.2\* Design Fire Scenario 2.** Design Fire Scenario 2 shall be described as follows:

(1) It is an ultrafast-developing fire, in the primary means of egress, with interior doors open at the start of the fire.
(2) It addresses the concern regarding a reduction in the number of available means of egress.

**5.5.3.3\* Design Fire Scenario 3.** Design Fire Scenario 3 shall be described as follows:

(1) It is a fire that starts in a normally unoccupied room, potentially endangering a large number of occupants in a large room or other area.
(2) It addresses the concern regarding a fire starting in a normally unoccupied room and migrating into the space that potentially holds the greatest number of occupants in the building.

**5.5.3.4\* Design Fire Scenario 4.** Design Fire Scenario 4 shall be described as follows:

(1) It is a fire that originates in a concealed wall or ceiling space adjacent to a large occupied room.
(2) It addresses the concern regarding a fire originating in a concealed space that does not have either a detection system or a suppression system and then spreading into the room within the building that potentially holds the greatest number of occupants.

**5.5.3.5\* Design Fire Scenario 5.** Design Fire Scenario 5 shall be described as follows:

(1) It is a slowly developing fire, shielded from fire protection systems, in close proximity to a high occupancy area.
(2) It addresses the concern regarding a relatively small ignition source causing a significant fire.

**5.5.3.6\* Design Fire Scenario 6.** Design Fire Scenario 6 shall be described as follows:

(1) It is the most severe fire resulting from the largest possible fuel load characteristic of the normal operation of the building.
(2) It addresses the concern regarding a rapidly developing fire with occupants present.

**5.5.3.7\* Design Fire Scenario 7.** Design Fire Scenario 7 shall be described as follows:

(1) It is an outside exposure fire.
(2) It addresses the concern regarding a fire starting at a location remote from the area of concern and either spreading into the area, blocking escape from the area, or developing untenable conditions within the area.

**5.5.3.8\* Design Fire Scenario 8.** Design Fire Scenario 8 shall be described as follows:

(1) It is a fire originating in ordinary combustibles in a room or area with each passive or active fire protection system independently rendered ineffective.
(2) It addresses concerns regarding the unreliability or unavailability of each fire protection system or fire protection feature, considered individually.
(3)\*It is not required to be applied to fire protection systems for which both the level of reliability and the design performance in the absence of the system are acceptable to the authority having jurisdiction.

**5.5.4 Design Fire Scenarios Data.**

**5.5.4.1** Each design fire scenario used in the performance-based design proposal shall be translated into input data specifications, as appropriate for the calculation method or model.

**5.5.4.2** Any design fire scenario specifications that the design analyses do not explicitly address or incorporate and that are, therefore, omitted from input data specifications shall be identified, and a sensitivity analysis of the consequences of that omission shall be performed.

**5.5.4.3** Any design fire scenario specifications modified in input data specifications, because of limitations in test methods or other data-generation procedures, shall be identified, and a sensitivity analysis of the consequences of the modification shall be performed.

**5.6\* Evaluation of Proposed Designs.**

**5.6.1 General.** A proposed design's performance shall be assessed relative to each performance objective in Section 4.2 and each applicable scenario in 5.5.3, with the assessment conducted through the use of appropriate calculation methods. The authority having jurisdiction shall approve the choice of assessment methods.

**5.6.2 Use.** The design professional shall use the assessment methods to demonstrate that the proposed design will achieve the goals and objectives, as measured by the performance criteria in light of the safety margins and uncertainty analysis, for each scenario, given the assumptions.

**5.6.3 Input Data.**

**5.6.3.1 Data.** Input data for computer fire models shall be obtained in accordance with ASTM E 1591, *Standard Guide for Obtaining Data for Deterministic Fire Models*. Data for use in analytical models that are not computer-based fire models shall be obtained using appropriate measurement, recording, and storage techniques to ensure the applicability of the data to the analytical method being used.

**5.6.3.2 Data Requirements.** A complete listing of input data requirements for all models, engineering methods, and other calculation or verification methods required or proposed as part of the performance-based design shall be provided.

**5.6.3.3\* Uncertainty and Conservatism of Data.** Uncertainty in input data shall be analyzed and, as determined appropriate by the authority having jurisdiction, addressed through the use of conservative values.

**5.6.4\* Output Data.** The assessment methods used shall accurately and appropriately produce the required output data from input data, based on the design specifications, assumptions, and scenarios.

**5.6.5 Validity.** Evidence shall be provided to confirm that the assessment methods are valid and appropriate for the proposed building, use, and conditions.

**5.7\* Safety Factors.** Approved safety factors shall be included in the design methods and calculations to reflect uncertainty in the assumptions, data, and other factors associated with the performance-based design.

**5.8 Documentation Requirements.**

**5.8.1\* General.** All aspects of the design, including those described in 5.8.2 through 5.8.14, shall be documented. The format and content of the documentation shall be acceptable to the authority having jurisdiction.

**5.8.2\* Technical References and Resources.** The authority having jurisdiction shall be provided with sufficient documentation to support the validity, accuracy, relevance, and precision of the proposed methods. The engineering standards, calculation methods, and other forms of scientific information provided shall be appropriate for the particular application and methodologies used.

**5.8.3 Building Design Specifications.** All details of the proposed building design that affect the ability of the building to meet the stated goals and objectives shall be documented.

**5.8.4 Performance Criteria.** Performance criteria, with sources, shall be documented.

**5.8.5 Occupant Characteristics.** Assumptions about occupant characteristics shall be documented.

**5.8.6 Design Fire Scenarios.** Descriptions of design fire scenarios shall be documented.

**5.8.7 Input Data.** Input data to models and assessment methods, including sensitivity analyses, shall be documented.

**5.8.8 Output Data.** Output data from models and assessment methods, including sensitivity analyses, shall be documented.

**5.8.9 Safety Factors.** The safety factors utilized shall be documented.

**5.8.10 Prescriptive Requirements.** Retained prescriptive requirements shall be documented.

**5.8.11\* Modeling Features.**

**5.8.11.1** Assumptions made by the model user, and descriptions of models and methods used, including known limitations, shall be documented.

**5.8.11.2** Documentation shall be provided to verify that the assessment methods have been used validly and appropriately to address the design specifications, assumptions, and scenarios.

**5.8.12 Evidence of Modeler Capability.** The design team's relevant experience with the models, test methods, databases, and other assessment methods used in the performance-based design proposal shall be documented.

**5.8.13 Performance Evaluation.** The performance evaluation summary shall be documented.

**5.8.14 Use of Performance-Based Design Option.** Design proposals shall include documentation that provides anyone involved in the ownership or management of the building with notification of the following:

(1) Approval of the building as a performance-based design with certain specified design criteria and assumptions

(2) Need for required re-evaluation and reapproval in cases of remodeling, modification, renovation, change in use, or change in established assumptions

# Chapter 6   Classification of Occupancy and Hazard of Contents

**6.1 Classification of Occupancy.**

**6.1.1 General.**

**6.1.1.1 Occupancy Classification.** The occupancy of a building or structure, or portion of a building or structure, shall be classified in accordance with 6.1.2 through 6.1.13. Occupancy classification shall be subject to the ruling of the authority having jurisdiction where there is a question of proper classification in any individual case.

**6.1.1.2 Special Structures.** Occupancies in special structures shall conform to the requirements of the specific occupancy chapter, Chapters 12 through 43, except as modified by Chapter 11.

**6.1.2 Assembly.** For requirements, see Chapters 12 and 13.

**6.1.2.1\* Definition — Assembly Occupancy.** An occupancy (1) used for a gathering of 50 or more persons for deliberation, worship, entertainment, eating, drinking, amusement, awaiting transportation, or similar uses; or (2) used as a special amusement building, regardless of occupant load.

**6.1.2.2 Other. (Reserved)**

**6.1.3 Educational.** For requirements, see Chapters 14 and 15.

**6.1.3.1\* Definition — Educational Occupancy.** An occupancy used for educational purposes through the twelfth grade by six or more persons for 4 or more hours per day or more than 12 hours per week.

**6.1.3.2 Other Occupancies.** Other occupancies associated with educational institutions shall be in accordance with the appropriate parts of this *Code.*

**6.1.3.3 Incidental Instruction.** In cases where instruction is incidental to some other occupancy, the section of this *Code* governing such other occupancy shall apply.

**6.1.4 Day Care.** For requirements, see Chapters 16 and 17.

**6.1.4.1\* Definition — Day-Care Occupancy.** An occupancy in which four or more clients receive care, maintenance, and supervision, by other than their relatives or legal guardians, for less than 24 hours per day.

**6.1.4.2 Other. (Reserved)**

**6.1.5 Health Care.** For requirements, see Chapters 18 and 19.

**6.1.5.1\* Definition — Health Care Occupancy.** An occupancy used to provide medical or other treatment or care simultaneously to four or more patients on an inpatient basis, where such patients are mostly incapable of self-preservation due to age, physical or mental disability, or because of security measures not under the occupants' control.

**6.1.5.2 Other. (Reserved)**

**6.1.6 Ambulatory Health Care.** For requirements, see Chapters 20 and 21.

**6.1.6.1\* Definition — Ambulatory Health Care Occupancy.** An occupancy used to provide services or treatment simultaneously to four or more patients that provides, on an outpatient basis, one or more of the following:

(1) Treatment for patients that renders the patients incapable of taking action for self-preservation under emergency conditions without the assistance of others
(2) Anesthesia that renders the patients incapable of taking action for self-preservation under emergency conditions without the assistance of others
(3) Emergency or urgent care for patients who, due to the nature of their injury or illness, are incapable of taking action for self-preservation under emergency conditions without the assistance of others

**6.1.6.2 Other. (Reserved)**

**6.1.7 Detention and Correctional.** For requirements, see Chapters 22 and 23.

**6.1.7.1\* Definition — Detention and Correctional Occupancy.** An occupancy used to house one or more persons under varied degrees of restraint or security where such occupants are mostly incapable of self-preservation because of security measures not under the occupants' control.

**6.1.7.2\* Nonresidential Uses.** Within detention and correctional facilities, uses other than residential housing shall be in accordance with the appropriate chapter of the *Code*. (See *22.1.3.3 and 23.1.3.3.*)

**6.1.8 Residential.** For requirements, see Chapters 24 through 31.

**6.1.8.1 Definition — Residential Occupancy.** An occupancy that provides sleeping accommodations for purposes other than health care or detention and correctional.

**6.1.8.1.1\* Definition — One- and Two-Family Dwelling Unit.** A building that contains not more than two dwelling units with independent cooking and bathroom facilities.

**6.1.8.1.2 Definition — Lodging or Rooming House.** A building or portion thereof that does not qualify as a one- or two-family dwelling, that provides sleeping accommodations for a total of 16 or fewer people on a transient or permanent basis, without personal care services, with or without meals, but without separate cooking facilities for individual occupants.

**6.1.8.1.3\* Definition — Hotel.** A building or groups of buildings under the same management in which there are sleeping accommodations for more than 16 persons and primarily used by transients for lodging with or without meals.

**6.1.8.1.4\* Definition — Dormitory.** A building or a space in a building in which group sleeping accommodations are provided for more than 16 persons who are not members of the same family in one room, or a series of closely associated rooms, under joint occupancy and single management, with or without meals, but without individual cooking facilities.

**6.1.8.1.5 Definition — Apartment Building.** A building or portion thereof containing three or more dwelling units with independent cooking and bathroom facilities.

**6.1.8.2 Other. (Reserved)**

**6.1.9 Residential Board and Care.** For requirements, see Chapters 32 and 33.

**6.1.9.1\* Definition — Residential Board and Care Occupancy.** An occupancy used for lodging and boarding of four or more residents, not related by blood or marriage to the owners or operators, for the purpose of providing personal care services.

**6.1.9.2 Other. (Reserved)**

**6.1.10 Mercantile.** For requirements, see Chapters 36 and 37.

**6.1.10.1\* Definition — Mercantile Occupancy.** An occupancy used for the display and sale of merchandise.

**6.1.10.2 Other. (Reserved)**

**6.1.11 Business.** For requirements, see Chapters 38 and 39.

**6.1.11.1\* Definition — Business Occupancy.** An occupancy used for the transaction of business other than mercantile.

**6.1.11.2 Other. (Reserved)**

**6.1.12 Industrial.** For requirements, see Chapter 40.

**6.1.12.1\* Definition — Industrial Occupancy.** An occupancy in which products are manufactured or in which processing, assembling, mixing, packaging, finishing, decorating, or repair operations are conducted.

**6.1.12.2 Other. (Reserved)**

**6.1.13 Storage.** For requirements, see Chapter 42.

**6.1.13.1\* Definition — Storage Occupancy.** An occupancy used primarily for the storage or sheltering of goods, merchandise, products, or vehicles.

**6.1.13.2 Other. (Reserved)**

**6.1.14 Multiple Occupancies.**

**6.1.14.1 General.**

**6.1.14.1.1** Multiple occupancies shall comply with the requirements of 6.1.14.1 and one of the following:

(1) Mixed occupancies — 6.1.14.3
(2) Separated occupancies — 6.1.14.4

**6.1.14.1.2** Where exit access from an occupancy traverses another occupancy, the multiple occupancy shall be treated as a mixed occupancy.

**6.1.14.1.3\*** Where incidental to another occupancy, areas used as follows shall be permitted to be considered part of the predominant occupancy and shall be subject to the provisions of the *Code* that apply to the predominant occupancy:

(1) Mercantile, business, industrial, or storage use
(2)\*Nonresidential use with an occupant load fewer than that established by Section 6.1 for the occupancy threshold

**6.1.14.2 Definitions.**

**6.1.14.2.1 Multiple Occupancy.** A building or structure in which two or more classes of occupancy exist.

**6.1.14.2.2 Mixed Occupancy.** A multiple occupancy where the occupancies are intermingled.

**6.1.14.2.3 Separated Occupancy.** A multiple occupancy where the occupancies are separated by fire resistance–rated assemblies.

**6.1.14.3 Mixed Occupancies.**

**6.1.14.3.1** Each portion of the building shall be classified as to its use in accordance with Section 6.1.

**6.1.14.3.2\*** The building shall comply with the most restrictive requirements of the occupancies involved, unless separate safeguards are approved.

**6.1.14.4 Separated Occupancies.**

**6.1.14.4.1** Where separated occupancies are provided, each part of the building comprising a distinct occupancy, as described in this chapter, shall be completely separated from other occupancies by fire-resistive assemblies, as specified in 6.1.14.4.2, 6.1.14.4.3, Table 6.1.14.4.1(a), and Table 6.1.14.4.1(b), unless separation is provided by approved existing separations.

**6.1.14.4.2** Occupancy separations shall be classified as 3-hour fire resistance–rated, 2-hour fire resistance–rated, or 1-hour fire resistance–rated and shall meet the requirements of Chapter 8.

**6.1.14.4.3** The minimum fire resistance rating specified in Table 6.1.14.4.1(a) and Table 6.1.14.4.1(b) shall be permitted to be reduced by 1 hour, but in no case shall it be reduced to less than 1 hour, where the building is protected throughout by an approved automatic sprinkler system in accordance with 9.7.1.1(1) and supervised in accordance with 9.7.2, unless prohibited by the double-dagger footnote entries in the tables.

**6.1.14.4.4** Occupancy separations shall be vertical, horizontal, or both or, when necessary, of such other form as required to provide complete separation between occupancy divisions in the building.

**6.2 Hazard of Contents.**

**6.2.1 General.**

**6.2.1.1** For the purpose of this *Code*, the hazard of contents shall be the relative danger of the start and spread of fire, the danger of smoke or gases generated, and the danger of explosion or other occurrence potentially endangering the lives and safety of the occupants of the building or structure.

**6.2.1.2** Hazard of contents shall be classified by the registered design professional (RDP) or owner and submitted to the authority having jurisdiction for review and approval on the basis of the character of the contents and the processes or operations conducted in the building or structure.

**6.2.1.3\*** For the purpose of this *Code*, where different degrees of hazard of contents exist in different parts of a building or structure, the most hazardous shall govern the classification, unless hazardous areas are separated or protected as specified in Section 8.7 and the applicable sections of Chapters 11 through 43.

**6.2.2 Classification of Hazard of Contents.**

**6.2.2.1\* General.** The hazard of contents of any building or structure shall be classified as low, ordinary, or high in accordance with 6.2.2.2, 6.2.2.3, and 6.2.2.4.

**6.2.2.2\* Low Hazard Contents.** Low hazard contents shall be classified as those of such low combustibility that no self-propagating fire therein can occur.

**6.2.2.3\* Ordinary Hazard Contents.** Ordinary hazard contents shall be classified as those that are likely to burn with moderate rapidity or to give off a considerable volume of smoke.

**6.2.2.4\* High Hazard Contents.** High hazard contents shall be classified as those that are likely to burn with extreme rapidity or from which explosions are likely. *(For means of egress requirements, see Section 7.11.)*

## Chapter 7  Means of Egress

**7.1 General.**

**7.1.1\* Application.** Means of egress for both new and existing buildings shall comply with this chapter. *(See also 5.5.3.)*

**7.1.2 Definitions.**

**7.1.2.1 General.** For definitions see Chapter 3 Definitions.

**7.1.2.2 Special Definitions.** A list of special terms used in this chapter follows:

(1)  **Accessible Area of Refuge.** See 3.3.22.1.
(2)  **Accessible Means of Egress.** See 3.3.170.1.
(3)  **Area of Refuge.** See 3.3.22.
(4)  **Common Path of Travel.** See 3.3.47.
(5)  **Electroluminescent.** See 3.3.68.
(6)  **Elevator Evacuation System.** See 3.3.69.
(7)  **Elevator Lobby.** See 3.3.70.
(8)  **Elevator Lobby Door.** See 3.3.62.1.
(9)  **Exit.** See 3.3.81.
(10)  **Exit Access.** See 3.3.82.
(11)  **Exit Discharge.** See 3.3.83.
(12)  **Externally Illuminated.** See 3.3.144.1.
(13)  **Horizontal Exit.** See 3.3.81.1.
(14)  **Internally Illuminated.** See 3.3.144.2.
(15)  **Means of Egress.** See 3.3.170.
(16)  **Photoluminescent.** See 3.3.207.
(17)  **Ramp.** See 3.3.219.
(18)  **Self-Luminous.** See 3.3.239.
(19)  **Severe Mobility Impairment.** See 3.3.244.
(20)  **Smokeproof Enclosure.** See 3.3.255.

**7.1.3 Separation of Means of Egress.** See also Section 8.2.

**7.1.3.1 Exit Access Corridors.** Corridors used as exit access and serving an area having an occupant load exceeding 30 shall be separated from other parts of the building by walls having not less than a 1-hour fire resistance rating in accordance with Section 8.3, unless otherwise permitted by one of the following:

(1)  This requirement shall not apply to existing buildings, provided that the occupancy classification does not change.
(2)  This requirement shall not apply where otherwise provided in Chapters 11 through 43.

## Table 6.1.14.4.1(a)  Required Separation of Occupancies (hours),[†] Part 1

| Occupancy | Assembly ≤300 | Assembly >300 to ≤1000 | Assembly >1000 | Educational | Day-Care >12 Clients | Day-Care Homes | Health Care | Ambulatory Health Care | Detention & Correctional | One- & Two Family Dwellings | Lodging or Rooming Houses | Hotels & Dormitories |
|---|---|---|---|---|---|---|---|---|---|---|---|---|
| Assembly ≤ 300 | — | 0 | 0 | 2 | 2 | 1 | 2‡ | 2 | 2‡ | 2 | 2 | 2 |
| Assembly >300 to ≤1000 | 0 | — | 0 | 2 | 2 | 2 | 2‡ | 2 | 2‡ | 2 | 2 | 2 |
| Assembly >1000 | 0 | 0 | — | 2 | 2 | 2 | 2‡ | 2 | 2‡ | 2 | 2 | 2 |
| Educational | 2 | 2 | 2 | — | 2 | 2 | 2‡ | 2 | 2‡ | 2 | 2 | 2 |
| Day-Care >12 Clients | 2 | 2 | 2 | 2 | — | 1 | 2‡ | 2 | 2‡ | 2 | 2 | 2 |
| Day-Care Homes | 1 | 2 | 2 | 2 | 1 | — | 2‡ | 2 | 2‡ | 2 | 2 | 2 |
| Health Care | 2‡ | 2‡ | 2‡ | 2‡ | 2‡ | 2‡ | — | 2‡ | 2‡ | 2‡ | 2‡ | 2‡ |
| Ambulatory Health Care | 2 | 2 | 2 | 2 | 2 | 2 | 2‡ | — | 2‡ | 2 | 2 | 2 |
| Detention & Correctional | 2‡ | 2‡ | 2‡ | 2‡ | 2‡ | 2‡ | 2‡ | 2‡ | — | 2‡ | 2‡ | 2‡ |
| One- & Two- Family Dwellings | 2 | 2 | 2 | 2 | 2 | 2 | 2‡ | 2 | 2‡ | — | 1 | 1 |
| Lodging or Rooming Houses | 2 | 2 | 2 | 2 | 2 | 2 | 2‡ | 2 | 2‡ | 1 | — | 1 |
| Hotels & Dormitories | 2 | 2 | 2 | 2 | 2 | 2 | 2‡ | 2 | 2‡ | 1 | 1 | — |
| Apartment Buildings | 2 | 2 | 2 | 2 | 2 | 2 | 2‡ | 2 | 2‡ | 1 | 1 | 1 |
| Board & Care, Small | 2 | 2 | 2 | 2 | 2 | 2 | 2‡ | 2 | 2‡ | 1 | 2 | 2 |
| Board & Care, Large | 2 | 2 | 2 | 2 | 2 | 2 | 2‡ | 2 | 2‡ | 2 | 2 | 2 |
| Mercantile | 2 | 2 | 2 | 2 | 2 | 2 | 2‡ | 2 | 2‡ | 2 | 2 | 2 |
| Mercantile, Mall | 2 | 2 | 2 | 2 | 2 | 2 | 2‡ | 2 | 2‡ | 2 | 2 | 2 |
| Mercantile, Bulk Retail | 3 | 3 | 3 | 3 | 3 | 3 | 2‡ | 2‡ | 2‡ | 3 | 3 | 3 |
| Business | 1 | 2 | 2 | 2 | 2 | 2 | 2‡ | 1 | 2‡ | 2 | 2 | 2 |
| Industrial, General Purpose | 2 | 2 | 3 | 3 | 3 | 3 | 2‡ | 2 | 2‡ | 2 | 2 | 2 |
| Industrial, Special-Purpose | 2 | 2 | 2 | 3 | 3 | 3 | 2‡ | 2 | 2‡ | 2 | 2 | 2 |
| Industrial, High Hazard | 3 | 3 | 3 | 3 | 3 | 3 | 2‡ | 2‡ | NP | 3 | 3 | 3 |
| Storage, Low & Ordinary Hazard | 2 | 2 | 3 | 3 | 3 | 2 | 2‡ | 2 | 2‡ | 2 | 2 | 2 |
| Storage, High Hazard | 3 | 3 | 3 | 3 | 3 | 3 | 2‡ | 2‡ | NP | 3 | 3 | 3 |

NP: Not permitted.

[†]*Minimum Fire Resistance Rating.* The fire resistance rating is permitted to be reduced by 1 hour, but in no case to less than 1 hour, where the building is protected throughout by an approved automatic sprinkler system in accordance with 9.7.1.1(1) and supervised in accordance with 9.7.2.

[‡]The 1-hour reduction due to the presence of sprinklers in accordance with the single-dagger footnote is not permitted.

## Table 6.1.14.4.1(b)  Required Separation of Occupancies (hours)[†], Part 2

| Occupancy | Apartment Buildings | Board & Care, Small | Board & Care, Large | Mercantile | Mercantile, Mall | Mercantile, Bulk Retail | Business | Industrial, General Purpose | Industrial, Special-Purpose | Industrial, High Hazard | Storage, Low & Ordinary Hazard | Storage, High Hazard |
|---|---|---|---|---|---|---|---|---|---|---|---|---|
| Assembly ≤ 300 | 2 | 2 | 2 | 2 | 2 | 3 | 1 | 2 | 2 | 3 | 2 | 3 |
| Assembly >300 to ≤1000 | 2 | 2 | 2 | 2 | 2 | 3 | 2 | 2 | 2 | 3 | 2 | 3 |
| Assembly >1000 | 2 | 2 | 2 | 2 | 2 | 3 | 2 | 3 | 2 | 3 | 3 | 3 |
| Educational | 2 | 2 | 2 | 2 | 2 | 3 | 2 | 3 | 3 | 3 | 3 | 3 |
| Day-Care >12 Clients | 2 | 2 | 2 | 2 | 2 | 3 | 2 | 3 | 3 | 3 | 3 | 3 |
| Day-Care Homes | 2 | 2 | 2 | 2 | 2 | 3 | 2 | 3 | 3 | 3 | 2 | 3 |
| Health Care | 2[‡] | 2[‡] | 2[‡] | 2[‡] | 2[‡] | 2[‡] | 2[‡] | 2[‡] | 2[‡] | 2[‡] | 2[‡] | 2[‡] |
| Ambulatory Health Care | 2 | 2 | 2 | 2 | 2 | 2[‡] | 1 | 2 | 2 | 2[‡] | 2 | 2[‡] |
| Detention & Correctional | 2[‡] | 2[‡] | 2[‡] | 2[‡] | 2[‡] | 2[‡] | 2[‡] | 2[‡] | 2[‡] | NP | 2[‡] | NP |
| One- & Two-Family Dwellings | 1 | 1 | 2 | 2 | 2 | 3 | 2 | 2 | 2 | 3 | 2 | 3 |
| Lodging or Rooming Houses | 1 | 2 | 2 | 2 | 2 | 3 | 2 | 2 | 2 | 3 | 2 | 3 |
| Hotels & Dormitories | 1 | 2 | 2 | 2 | 2 | 3 | 2 | 2 | 2 | 3 | 2 | 3 |
| Apartment Buildings | — | 2 | 2 | 2 | 2 | 3 | 2 | 2 | 2 | 3 | 2 | 3 |
| Board & Care, Small | 2 | — | 1 | 2 | 2 | 3 | 2 | 3 | 3 | 3 | 3 | 3 |
| Board & Care, Large | 2 | 1 | — | 2 | 2 | 3 | 2 | 3 | 3 | 3 | 3 | 3 |
| Mercantile | 2 | 2 | 2 | — | 0 | 3 | 2 | 2 | 2 | 3 | 2 | 3 |
| Mercantile, Mall | 2 | 2 | 2 | 0 | — | 3 | 2 | 3 | 3 | 3 | 2 | 3 |
| Mercantile, Bulk Retail | 3 | 3 | 3 | 3 | 3 | — | 2 | 2 | 2 | 3 | 2 | 2 |
| Business | 2 | 2 | 2 | 2 | 2 | 2 | — | 2 | 2 | 2 | 2 | 2 |
| Industrial, General Purpose | 2 | 3 | 3 | 2 | 3 | 2 | 2 | — | 1 | 1 | 1 | 1 |
| Industrial, Special-Purpose | 2 | 3 | 3 | 2 | 3 | 2 | 2 | 1 | — | 1 | 1 | 1 |
| Industrial, High Hazard | 3 | 3 | 3 | 3 | 3 | 3 | 2 | 1 | 1 | — | 1 | 1 |
| Storage, Low & Ordinary Hazard | 2 | 3 | 3 | 2 | 2 | 2 | 2 | 1 | 1 | 1 | — | 1 |
| Storage, High Hazard | 3 | 3 | 3 | 3 | 3 | 2 | 2 | 1 | 1 | 1 | 1 | — |

NP: Not permitted.

[†]*Minimum Fire Resistance Rating.* The fire resistance rating is permitted to be reduced by 1 hour, but in no case to less than 1 hour, where the building is protected throughout by an approved automatic sprinkler system in accordance with 9.7.1.1(1) and supervised in accordance with 9.7.2.

[‡]The 1-hour reduction due to the presence of sprinklers in accordance with the single-dagger footnote is not permitted.

2012 Edition

**7.1.3.2 Exits.**

**7.1.3.2.1** Where this *Code* requires an exit to be separated from other parts of the building, the separating construction shall meet the requirements of Section 8.2 and the following:

(1)*The separation shall have a minimum 1-hour fire resistance rating where the exit connects three or fewer stories.

(2) The separation specified in 7.1.3.2.1(1), other than an existing separation, shall be supported by construction having not less than a 1-hour fire resistance rating.

(3)*The separation shall have a minimum 2-hour fire resistance rating where the exit connects four or more stories, unless one of the following conditions exists:

(a) In existing non-high-rise buildings, existing exit stair enclosures shall have a minimum 1-hour fire resistance rating.

(b) In existing buildings protected throughout by an approved, supervised automatic sprinkler system in accordance with Section 9.7, existing exit stair enclosures shall have a minimum 1-hour fire resistance rating.

(c) The minimum 1-hour enclosures in accordance with 28.2.2.1.2, 29.2.2.1.2, 30.2.2.1.2, and 31.2.2.1.2 shall be permitted as an alternative to the requirement of 7.1.3.2.1(3).

(4) Reserved.

(5) The minimum 2-hour fire resistance–rated separation required by 7.1.3.2.1(3) shall be constructed of an assembly of noncombustible or limited-combustible materials and shall be supported by construction having a minimum 2-hour fire resistance rating, unless otherwise permitted by 7.1.3.2.1(7).

(6)*Structural elements, or portions thereof, that support exit components and either penetrate into a fire resistance–rated assembly or are installed within a fire resistance–rated wall assembly shall be protected, as a minimum, to the fire resistance rating required by 7.1.3.2.1(1) or (3).

(7) In Type III, Type IV, and Type V construction, as defined in NFPA 220, *Standard on Types of Building Construction (see 8.2.1.2)*, fire-retardant-treated wood enclosed in noncombustible or limited-combustible materials shall be permitted.

(8) Openings in the separation shall be protected by fire door assemblies equipped with door closers complying with 7.2.1.8.

(9)*Openings in exit enclosures shall be limited to door assemblies from normally occupied spaces and corridors and door assemblies for egress from the enclosure, unless one of the following conditions exists:

(a) Openings in exit passageways in mall buildings as provided in Chapters 36 and 37 shall be permitted.

(b) In buildings of Type I or Type II construction, as defined in NFPA 220, *Standard on Types of Building Construction (see 8.2.1.2)*, existing fire protection–rated door assemblies to interstitial spaces shall be permitted, provided that such spaces meet all of the following criteria:

i. The space is used solely for distribution of pipes, ducts, and conduits.

ii. The space contains no storage.

iii. The space is separated from the exit enclosure in accordance with Section 8.3.

(c) Existing openings to mechanical equipment spaces protected by approved existing fire protection–rated door assemblies shall be permitted, provided that the following criteria are met:

i. The space is used solely for non-fuel-fired mechanical equipment.

ii. The space contains no storage of combustible materials.

iii. The building is protected throughout by an approved, supervised automatic sprinkler system in accordance with Section 9.7.

(10) Penetrations into, and openings through, an exit enclosure assembly shall be limited to the following:

(a) Door assemblies permitted by 7.1.3.2.1(9)

(b)*Electrical conduit serving the exit enclosure

(c) Required exit door openings

(d) Ductwork and equipment necessary for independent stair pressurization

(e) Water or steam piping necessary for the heating or cooling of the exit enclosure

(f) Sprinkler piping

(g) Standpipes

(h) Existing penetrations protected in accordance with 8.3.5

(i) Penetrations for fire alarm circuits, where the circuits are installed in metal conduit and the penetrations are protected in accordance with 8.3.5

(11) Penetrations or communicating openings shall be prohibited between adjacent exit enclosures.

(12) Membrane penetrations shall be permitted on the exit access side of the exit enclosure and shall be protected in accordance with 8.3.5.6.

**7.1.3.2.2** An exit enclosure shall provide a continuous protected path of travel to an exit discharge.

**7.1.3.2.3\*** An exit enclosure shall not be used for any purpose that has the potential to interfere with its use as an exit and, if so designated, as an area of refuge. *(See also 7.2.2.5.3.)*

**7.1.4 Interior Finish in Exit Enclosures.**

**7.1.4.1\* Interior Wall and Ceiling Finish in Exit Enclosures.** Interior wall and ceiling finish shall be in accordance with Section 10.2. In exit enclosures, interior wall and ceiling finish materials complying with Section 10.2 shall be Class A or Class B.

**7.1.4.2\* Interior Floor Finish in Exit Enclosures.** New interior floor finish in exit enclosures, including stair treads and risers, shall be not less than Class II in accordance with Section 10.2.

**7.1.5\* Headroom.**

**7.1.5.1** Means of egress shall be designed and maintained to provide headroom in accordance with other sections of this *Code*, and such headroom shall be not less than 7 ft 6 in. (2285 mm), with projections from the ceiling not less than 6 ft 8 in. (2030 mm) with a tolerance of −¾ in. (−19 mm), above the finished floor, unless otherwise specified by any of the following:

(1) In existing buildings, the ceiling height shall be not less than 7 ft (2135 mm) from the floor, with projections from the ceiling not less than 6 ft 8 in. (2030 mm) nominal above the floor.

(2) Headroom in industrial equipment access areas as provided in 40.2.5.2 shall be permitted.

**7.1.5.2** The minimum ceiling height shall be maintained for not less than two-thirds of the ceiling area of any room or space, provided that the ceiling height of the remaining ceiling area is not less than 6 ft 8 in. (2030 mm).

**7.1.5.3** Headroom on stairs shall be not less than 6 ft 8 in. (2030 mm) and shall be measured vertically above a plane parallel to, and tangent with, the most forward projection of the stair tread.

**7.1.6 Walking Surfaces in the Means of Egress.**

**7.1.6.1 General.**

**7.1.6.1.1** Walking surfaces in the means of egress shall comply with 7.1.6.2 through 7.1.6.4.

**7.1.6.1.2** Approved existing walking surfaces shall be permitted.

**7.1.6.2 Changes in Elevation.** Abrupt changes in elevation of walking surfaces shall not exceed ¼ in. (6.3 mm). Changes in elevation exceeding ¼ in. (6.3 mm), but not exceeding ½ in. (13 mm), shall be beveled with a slope of 1 in 2. Changes in elevation exceeding ½ in. (13 mm) shall be considered a change in level and shall be subject to the requirements of 7.1.7.

**7.1.6.3 Level.** Walking surfaces shall comply with all of the following:

(1) Walking surfaces shall be nominally level.
(2) The slope of a walking surface in the direction of travel shall not exceed 1 in 20, unless the ramp requirements of 7.2.5 are met.
(3) The slope perpendicular to the direction of travel shall not exceed 1 in 48.

**7.1.6.4\* Slip Resistance.** Walking surfaces shall be slip resistant under foreseeable conditions. The walking surface of each element in the means of egress shall be uniformly slip resistant along the natural path of travel.

**7.1.7 Changes in Level in Means of Egress.**

**7.1.7.1** Changes in level in means of egress shall be achieved by an approved means of egress where the elevation difference exceeds 21 in. (535 mm).

**7.1.7.2\*** Changes in level in means of egress not in excess of 21 in. (535 mm) shall be achieved either by a ramp complying with the requirements of 7.2.5 or by a stair complying with the requirements of 7.2.2.

**7.1.7.2.1** Where a ramp is used, the presence and location of ramped portions of walkways shall be readily apparent.

**7.1.7.2.2** Where a stair is used, the tread depth of such stair shall be not less than 13 in. (330 mm).

**7.1.7.2.3** Tread depth in industrial equipment access areas as provided in 40.2.5.2 shall be permitted.

**7.1.7.2.4** The presence and location of each step shall be readily apparent.

**7.1.8\* Guards.** Guards in accordance with 7.2.2.4 shall be provided at the open sides of means of egress that exceed 30 in. (760 mm) above the floor or the finished ground level below.

**7.1.9 Impediments to Egress.** Any device or alarm installed to restrict the improper use of a means of egress shall be designed and installed so that it cannot, even in case of failure, impede or prevent emergency use of such means of egress, unless otherwise provided in 7.2.1.6 and Chapters 18, 19, 22, and 23.

**7.1.10 Means of Egress Reliability.**

**7.1.10.1\* General.** Means of egress shall be continuously maintained free of all obstructions or impediments to full instant use in the case of fire or other emergency.

**7.1.10.2 Furnishings and Decorations in Means of Egress.**

**7.1.10.2.1** No furnishings, decorations, or other objects shall obstruct exits or their access thereto, egress therefrom, or visibility thereof.

**7.1.10.2.2** No obstruction by railings, barriers, or gates shall divide the means of egress into sections appurtenant to individual rooms, apartments, or other occupied spaces. Where the authority having jurisdiction finds the required path of travel to be obstructed by furniture or other movable objects, the authority shall be permitted to require that such objects be secured out of the way or shall be permitted to require that railings or other permanent barriers be installed to protect the path of travel against encroachment.

**7.1.10.2.3** Mirrors shall not be placed on exit door leaves. Mirrors shall not be placed in or adjacent to any exit in such a manner as to confuse the direction of egress.

**7.1.11 Sprinkler System Installation.** Where another provision of this chapter requires an automatic sprinkler system, the sprinkler system shall be installed in accordance with the subparts of 9.7.1.1 permitted by the applicable occupancy chapters.

**7.2 Means of Egress Components.**

**7.2.1 Door Openings.**

**7.2.1.1 General.**

**7.2.1.1.1** A door assembly in a means of egress shall conform to the general requirements of Section 7.1 and to the special requirements of 7.2.1.

**7.2.1.1.2** Every door opening and every principal entrance that is required to serve as an exit shall be designed and constructed so that the path of egress travel is obvious and direct. Windows that, because of their physical configuration or design and the materials used in their construction, have the potential to be mistaken for door openings shall be made inaccessible to the occupants by barriers or railings.

**7.2.1.1.3 Occupied Building.**

**7.2.1.1.3.1** For the purposes of Section 7.2, a building shall be considered to be occupied at any time it meets any of the following criteria:

(1) It is open for general occupancy.
(2) It is open to the public.
(3) It is occupied by more than 10 persons.

**7.2.1.1.3.2** Where means of egress doors are locked in a building that is not considered occupied, occupants shall not be locked beyond their control in buildings or building spaces, except for lockups in accordance with 22.4.5 and 23.4.5, detention and correctional occupancies, and health care occupancies.

**7.2.1.2 Door Leaf Width.**

**7.2.1.2.1\* Measurement of Clear Width.**

**7.2.1.2.1.1 Swinging Door Assemblies.** For swinging door assemblies, clear width shall be measured as follows:

(1) The measurement shall be taken at the narrowest point in the door opening.

(2) The measurement shall be taken between the face of the door leaf and the stop of the frame.

(3) For new swinging door assemblies, the measurement shall be taken with the door leaf open 90 degrees.

(4) For any existing door assembly, the measurement shall be taken with the door leaf in the fully open position.

(5) Projections of not more than 4 in. (100 mm) into the door opening width on the hinge side shall not be considered reductions in clear width, provided that such projections are for purposes of accommodating panic hardware or fire exit hardware and are located not less than 34 in. (865 mm), and not more than 48 in. (1220 mm), above the floor.

(6) Projections exceeding 6 ft 8 in. (2030 mm) above the floor shall not be considered reductions in clear width.

**7.2.1.2.1.2 Other than Swinging Door Assemblies.** For other than swinging door assemblies, clear width shall be measured as follows:

(1) The measurement shall be taken at the narrowest point in the door opening.

(2) The measurement shall be taken as the door opening width when the door leaf is in the fully open position.

(3) Projections exceeding 6 ft 8 in. (2030 mm) above the floor shall not be considered reductions in clear width.

**7.2.1.2.2* Measurement of Egress Capacity Width.**

**7.2.1.2.2.1 Swinging Door Assemblies.** For swinging door assemblies, egress capacity width shall be measured as follows:

(1) The measurement shall be taken at the narrowest point in the door opening.

(2) The measurement shall be taken between the face of the door leaf and the stop of the frame.

(3) For new swinging doors assemblies, the measurement shall be taken with the door leaf open 90 degrees.

(4) For any existing door assembly, the measurement shall be taken with the door leaf in the fully open position.

(5) Projections not more than 3½ in. (90 mm) at each side of the door openings at a height of not more than 38 in. (965 mm) shall not be considered reductions in egress capacity width.

(6) Projections exceeding 6 ft 8 in. (2030 mm) above the floor shall not be considered reductions in egress capacity width.

**7.2.1.2.2.2 Other than Swinging Door Assemblies.** For other than swinging door assemblies, egress capacity width shall be measured as follows:

(1) The measurement shall be taken at the narrowest point in the door opening.

(2) The measurement shall be taken as the door opening width when the door leaf is in the fully open position.

(3) Projections not more than 3½ in. (90 mm) at each side of the door openings at a height of not more than 38 in. (965 mm) shall not be considered reductions in egress capacity width.

(4) Projections exceeding 6 ft 8 in. (2030 mm) above the floor shall not be considered reductions in egress capacity width.

**7.2.1.2.3 Minimum Door Leaf Width.**

**7.2.1.2.3.1** For purposes of determining minimum door opening width, the clear width in accordance with 7.2.1.2.1 shall be used, unless door leaf width is specified.

**7.2.1.2.3.2** Door openings in means of egress shall be not less than 32 in. (810 mm) in clear width, except under any of the following conditions:

(1) Where a pair of door leaves is provided, one door leaf shall provide not less than a 32 in. (810 mm) clear width opening.

(2) Exit access door assemblies serving a room not exceeding 70 ft² (6.5 m²) and not required to be accessible to persons with severe mobility impairments shall be not less than 24 in. (610 mm) in door leaf width.

(3) Door openings serving a building or portion thereof not required to be accessible to persons with severe mobility impairments shall be permitted to be 28 in. (710 mm) in door leaf width.

(4) In existing buildings, the existing door leaf width shall be not less than 28 in. (710 mm).

(5) Door openings in detention and correctional occupancies, as otherwise provided in Chapters 22 and 23, shall not be required to comply with 7.2.1.2.3.

(6) Interior door openings in dwelling units as otherwise provided in Chapter 24 shall not be required to comply with 7.2.1.2.3.

(7) A power-operated door leaf located within a two-leaf opening shall be exempt from the minimum 32 in. (810 mm) single-leaf requirement in accordance with 7.2.1.9.1.5.

(8) Revolving door assemblies, as provided in 7.2.1.10, shall be exempt from the minimum 32 in. (810 mm) width requirement.

(9)*Where a single door opening is provided for discharge from a stairway required to be a minimum of 56 in. (1420 mm) wide in accordance with 7.2.2.2.1.2(B), and such door assembly serves as the sole means of exit discharge from such stairway, the clear width of the door opening, measured in accordance with 7.2.1.2.2, shall be not less than two-thirds the required width of the stairway.

**7.2.1.3 Floor Level.**

**7.2.1.3.1** The elevation of the floor surfaces on both sides of a door opening shall not vary by more than ½ in. (13 mm), unless otherwise permitted by 7.2.1.3.5 or 7.2.1.3.6.

**7.2.1.3.2** The elevation of the floor surfaces required by 7.2.1.3.1 shall be maintained on both sides of the door openings for a distance not less than the width of the widest leaf.

**7.2.1.3.3** Thresholds at door openings shall not exceed ½ in. (13 mm) in height.

**7.2.1.3.4** Raised thresholds and floor level changes in excess of ¼ in. (6.3 mm) at door openings shall be beveled with a slope not steeper than 1 in 2.

**7.2.1.3.5** In existing buildings, where the door opening discharges to the outside or to an exterior balcony or exterior exit access, the floor level outside the door opening shall be permitted to be one step lower than that of the inside, but shall be not more than 8 in. (205 mm) lower.

**7.2.1.3.6** In existing buildings, a door assembly at the top of a stair shall be permitted to open directly at a stair, provided that the door leaf does not swing over the stair and that the door opening serves an area with an occupant load of fewer than 50 persons.

**7.2.1.4 Swing and Force to Open.**

**7.2.1.4.1* Swinging-Type Door Assembly Requirement.** Any door assembly in a means of egress shall be of the side-hinged or pivoted-swinging type, and shall be installed to be capable of swinging from any position to the full required width of the opening in which it is installed, unless otherwise specified as follows:

(1) Door assemblies in dwelling units, as provided in Chapter 24, shall be permitted.

(2) Door assemblies in residential board and care occupancies, as provided in Chapters 32 and 33, shall be permitted.

(3) Where permitted in Chapters 11 through 43, horizontal-sliding or vertical-rolling security grilles or door assemblies that are part of the required means of egress shall be permitted, provided that all of the following criteria are met:

(a) Such grilles or door assemblies shall remain secured in the fully open position during the period of occupancy by the general public.

(b) On or adjacent to the grille or door opening, there shall be a readily visible, durable sign in letters not less than 1 in. (25 mm) high on a contrasting background that reads as follows: THIS DOOR TO REMAIN OPEN WHEN THE BUILDING IS OCCUPIED.

(c) Door leaves or grilles shall not be brought to the closed position when the space is occupied.

(d) Door leaves or grilles shall be operable from within the space without the use of any special knowledge or effort.

(e) Where two or more means of egress are required, not more than half of the means of egress shall be equipped with horizontal-sliding or vertical-rolling grilles or door assemblies.

(4) Horizontal-sliding door assemblies shall be permitted under any of the following conditions:

(a) Horizontal-sliding door assemblies in detention and correctional occupancies, as provided in Chapters 22 and 23, shall be permitted.

(b) Horizontal-sliding door assemblies complying with 7.2.1.14 shall be permitted.

(c) Unless prohibited by Chapters 11 through 43, horizontal-sliding door assemblies serving a room or area with an occupant load of fewer than 10 shall be permitted, provided that all of the following criteria are met:

    i. The area served by the door assembly has no high hazard contents.

    ii. The door assembly is readily operable from either side without special knowledge or effort.

    iii. The force required to operate the door assembly in the direction of door leaf travel is not more than 30 lbf (133 N) to set the door leaf in motion and is not more than 15 lbf (67 N) to close the door assembly or open it to the minimum required width.

    iv. The door assembly complies with any required fire protection rating, and, where rated, is self-closing or automatic-closing by means of smoke detection in accordance with 7.2.1.8 and is installed in accordance with NFPA 80, *Standard for Fire Doors and Other Opening Protectives.*

    v. Corridor door assemblies required to be self-latching have a latch or other mechanism that ensures that the door leaf will not rebound into a partially open position if forcefully closed.

(d) Where private garages, business areas, industrial areas, and storage areas with an occupant load not exceeding 10 contain only low or ordinary hazard contents, door openings to such areas and private garages shall be permitted to be horizontal-sliding door assemblies.

(5) Where private garages, business areas, industrial areas, and storage areas with an occupant load not exceeding 10 contain only low or ordinary hazard contents, door openings to such areas and private garages shall be permitted to be vertical-rolling door assemblies.

(6) Revolving door assemblies complying with 7.2.1.10 shall be permitted.

(7) Existing fusible link–operated horizontal-sliding or vertical-rolling fire door assemblies shall be permitted to be used as provided in Chapters 39, 40, and 42.

**7.2.1.4.2 Door Leaf Swing Direction.** Door leaves required to be of the side-hinged or pivoted-swinging type shall swing in the direction of egress travel under any of the following conditions:

(1) Where serving a room or area with an occupant load of 50 or more, except under any of the following conditions:

(a) Door leaves in horizontal exits shall not be required to swing in the direction of egress travel where permitted by 7.2.4.3.8.1 or 7.2.4.3.8.2.

(b) Door leaves in smoke barriers shall not be required to swing in the direction of egress travel in existing health care occupancies, as provided in Chapter 19.

(2) Where the door assembly is used in an exit enclosure, unless the door opening serves an individual living unit that opens directly into an exit enclosure

(3) Where the door opening serves a high hazard contents area

**7.2.1.4.3 Door Leaf Encroachment.**

**7.2.1.4.3.1\*** During its swing, any door leaf in a means of egress shall leave not less than one-half of the required width of an aisle, a corridor, a passageway, or a landing unobstructed and shall project not more than 7 in. (180 mm) into the required width of an aisle, a corridor, a passageway, or a landing, when fully open, unless both of the following conditions are met:

(1) The door opening provides access to a stair in an existing building.

(2) The door opening meets the requirement that limits projection to not more than 7 in. (180 mm) into the required width of the stair landing when the door leaf is fully open.

**7.2.1.4.3.2** Surface-mounted latch release hardware on the door leaf shall be exempt from being included in the maximum 7 in. (180 mm) projection requirement of 7.2.1.4.3.1, provided that both of the following criteria are met:

(1) The hardware is mounted to the side of the door leaf that faces the aisle, corridor, passageway, or landing when the door leaf is in the open position.

(2) The hardware is mounted not less than 34 in. (865 mm), and not more than 48 in. (1220 mm), above the floor.

**7.2.1.4.4 Screen Door Assemblies and Storm Door Assemblies.** Screen door assemblies and storm door assemblies used in a means of egress shall be subject to the requirements for direction of swing that are applicable to other door assemblies used in a means of egress.

**7.2.1.4.5 Door Leaf Operating Forces.**

**7.2.1.4.5.1** The forces required to fully open any door leaf manually in a means of egress shall not exceed 15 lbf (67 N) to release the latch, 30 lbf (133 N) to set the leaf in motion, and 15 lbf (67 N) to open the leaf to the minimum required width, unless otherwise specified as follows:

(1) The opening forces for interior side-hinged or pivoted-swinging door leaves without closers shall not exceed 5 lbf (22 N).
(2) The opening forces for existing door leaves in existing buildings shall not exceed 50 lbf (222 N) applied to the latch stile.
(3) The opening forces for horizontal-sliding door leaves in detention and correctional occupancies shall be as provided in Chapters 22 and 23.
(4) The opening forces for power-operated door leaves shall be as provided in 7.2.1.9.

**7.2.1.4.5.2** The forces specified in 7.2.1.4.5 shall be applied to the latch stile.

**7.2.1.5 Locks, Latches, and Alarm Devices.**

**7.2.1.5.1** Door leaves shall be arranged to be opened readily from the egress side whenever the building is occupied.

**7.2.1.5.2\*** The requirement of 7.2.1.5.1 shall not apply to door leaves of listed fire door assemblies after exposure to elevated temperature in accordance with the listing, based on laboratory fire test procedures.

**7.2.1.5.3** Locks, if provided, shall not require the use of a key, a tool, or special knowledge or effort for operation from the egress side.

**7.2.1.5.4** The requirements of 7.2.1.5.1 and 7.2.1.5.3 shall not apply where otherwise provided in Chapters 18 through 23.

**7.2.1.5.5 Key-Operated Locks.**

**7.2.1.5.5.1** Exterior door assemblies shall be permitted to have key-operated locks from the egress side, provided that all of the following criteria are met:

(1) This alternative is permitted in Chapters 11 through 43 for the specific occupancy.
(2) A readily visible, durable sign in letters not less than 1 in. (25 mm) high on a contrasting background that reads as follows is located on or adjacent to the door leaf: THIS DOOR TO REMAIN UNLOCKED WHEN THE BUILDING IS OCCUPIED.
(3) The locking device is of a type that is readily distinguishable as locked.
(4) A key is immediately available to any occupant inside the building when it is locked.

**7.2.1.5.5.2** The alternative provisions of 7.2.1.5.5.1 shall be permitted to be revoked by the authority having jurisdiction for cause.

**7.2.1.5.6 Electrically Controlled Egress Door Assemblies.** Door assemblies in the means of egress shall be permitted to be electrically locked if equipped with approved, listed hardware, provided that all of the following conditions are met:

(1) The hardware for occupant release of the lock is affixed to the door leaf.
(2) The hardware has an obvious method of operation that is readily operated in the direction of egress.
(3) The hardware is capable of being operated with one hand in the direction of egress.

(4) Operation of the hardware interrupts the power supply directly to the electric lock and unlocks the door assembly in the direction of egress.
(5)\*Loss of power to the listed releasing hardware automatically unlocks the door assembly in the direction of egress.
(6) Hardware for new installations is listed in accordance with ANSI/UL 294, *Standard for Access Control System Units*.

**7.2.1.5.7** Where permitted in Chapters 11 through 43, key operation shall be permitted, provided that the key cannot be removed when the door leaf is locked from the side from which egress is to be made.

**7.2.1.5.8\*** Every door assembly in a stair enclosure serving more than four stories, unless permitted by 7.2.1.5.8.2, shall meet one of the following conditions:

(1) Re-entry from the stair enclosure to the interior of the building shall be provided.
(2) An automatic release that is actuated with the initiation of the building fire alarm system shall be provided to unlock all stair enclosure door assemblies to allow re-entry.
(3) Selected re-entry shall be provided in accordance with 7.2.1.5.8.1.

**7.2.1.5.8.1** Door assemblies on stair enclosures shall be permitted to be equipped with hardware that prevents re-entry into the interior of the building, provided that all of the following criteria are met:

(1) There shall be not less than two levels where it is possible to leave the stair enclosure to access another exit.
(2) There shall be not more than four stories intervening between stories where it is possible to leave the stair enclosure to access another exit.
(3) Re-entry shall be possible on the top story or next-to-top story served by the stair enclosure, and such story shall allow access to another exit.
(4) Door assemblies allowing re-entry shall be identified as such on the stair side of the door leaf.
(5) Door assemblies not allowing re-entry shall be provided with a sign on the stair side indicating the location of the nearest door opening, in each direction of travel, that allows re-entry or exit.

**7.2.1.5.8.2** The requirements of 7.2.1.5.8, except as provided in 7.2.1.5.8.3, shall not apply to the following:

(1) Existing installations in buildings that are not high-rise buildings as permitted in Chapters 11 through 43
(2) Existing installations in high-rise buildings as permitted in Chapters 11 through 43 where the occupancy is within a building protected throughout by an approved, supervised automatic sprinkler system in accordance with Section 9.7
(3) Existing approved stairwell re-entry installations as permitted by Chapters 11 through 43
(4) Stair enclosures serving a building permitted to have a single exit in accordance with Chapters 11 through 43
(5) Stair enclosures in health care occupancies where otherwise provided in Chapter 18
(6) Stair enclosures in detention and correctional occupancies where otherwise provided in Chapter 22

**7.2.1.5.8.3** When the provisions of 7.2.1.5.8.2 are used, signage on the stair door leaves shall be required as follows;

(1) Door assemblies allowing re-entry shall be identified as such on the stair side of the door leaf.

(2) Door assemblies not allowing re-entry shall be provided with a sign on the stair side indicating the location of the nearest door opening, in each direction of travel, that allows re-entry or exit.

**7.2.1.5.9** If a stair enclosure allows access to the roof of the building, the door assembly to the roof either shall be kept locked or shall allow re-entry from the roof.

**7.2.1.5.10*** A latch or other fastening device on a door leaf shall be provided with a releasing device that has an obvious method of operation and that is readily operated under all lighting conditions.

**7.2.1.5.10.1** The releasing mechanism for any latch shall be located as follows:

(1) Not less than 34 in. (865 mm) above the finished floor for other than existing installations
(2) Not more than 48 in. (1220 mm) above the finished floor

**7.2.1.5.10.2** The releasing mechanism shall open the door leaf with not more than one releasing operation, unless otherwise specified in 7.2.1.5.10.3, 7.2.1.5.10.4, or 7.2.1.5.10.6.

**7.2.1.5.10.3*** Egress door assemblies from individual living units and guest rooms of residential occupancies shall be permitted to be provided with devices, including automatic latching devices, that require not more than one additional releasing operation, provided that such device is operable from the inside without the use of a key or tool and is mounted at a height not exceeding 48 in. (1220 mm) above the finished floor.

**7.2.1.5.10.4** Existing security devices permitted by 7.2.1.5.10.3 shall be permitted to have two additional releasing operations.

**7.2.1.5.10.5** Existing security devices permitted by 7.2.1.5.10.3, other than automatic latching devices, shall be located not more than 60 in. (1525 mm) above the finished floor.

**7.2.1.5.10.6** Two releasing operations shall be permitted for existing hardware on a door leaf serving an area having an occupant load not exceeding three, provided that releasing does not require simultaneous operations.

**7.2.1.5.11** Where pairs of door leaves are required in a means of egress, one of the following criteria shall be met:

(1) Each leaf of the pair shall be provided with a releasing device that does not depend on the release of one leaf before the other.
(2) Approved automatic flush bolts shall be used and arranged such that both of the following criteria are met:
    (a) The door leaf equipped with the automatic flush bolts shall have no doorknob or surface-mounted hardware.
    (b) Unlatching of any leaf shall not require more than one operation.

**7.2.1.5.12*** Devices shall not be installed in connection with any door assembly on which panic hardware or fire exit hardware is required where such devices prevent or are intended to prevent the free use of the leaf for purposes of egress, unless otherwise provided in 7.2.1.6.

**7.2.1.6*** **Special Locking Arrangements.**

**7.2.1.6.1** **Delayed-Egress Locking Systems.**

**7.2.1.6.1.1** Approved, listed, delayed-egress locking systems shall be permitted to be installed on door assemblies serving low and ordinary hazard contents in buildings protected throughout by an approved, supervised automatic fire detection system in accordance with Section 9.6 or an approved, supervised automatic sprinkler system in accordance with Section 9.7, and where permitted in Chapters 11 through 43, provided that all of the following criteria are met:

(1) The door leaves shall unlock in the direction of egress upon actuation of one of the following:
    (a) Approved, supervised automatic sprinkler system in accordance with Section 9.7
    (b) Not more than one heat detector of an approved, supervised automatic fire detection system in accordance with Section 9.6
    (c) Not more than two smoke detectors of an approved, supervised automatic fire detection system in accordance with Section 9.6
(2) The door leaves shall unlock in the direction of egress upon loss of power controlling the lock or locking mechanism.
(3)*An irreversible process shall release the lock in the direction of egress within 15 seconds, or 30 seconds where approved by the authority having jurisdiction, upon application of a force to the release device required in 7.2.1.5.10 under all of the following conditions:
    (a) The force shall not be required to exceed 15 lbf (67 N).
    (b) The force shall not be required to be continuously applied for more than 3 seconds.
    (c) The initiation of the release process shall activate an audible signal in the vicinity of the door opening.
    (d) Once the lock has been released by the application of force to the releasing device, relocking shall be by manual means only.
(4)*A readily visible, durable sign in letters not less than 1 in. (25 mm) high and not less than ⅛ in. (3.2 mm) in stroke width on a contrasting background that reads as follows shall be located on the door leaf adjacent to the release device in the direction of egress:
PUSH UNTIL ALARM SOUNDS
DOOR CAN BE OPENED IN 15 SECONDS
(5) The egress side of doors equipped with delayed-egress locks shall be provided with emergency lighting in accordance with Section 7.9.

**7.2.1.6.1.2** The provisions of 7.2.1.6.2 for access-controlled egress door assemblies shall not apply to door assemblies with delayed-egress locking systems.

**7.2.1.6.2*** **Access-Controlled Egress Door Assemblies.** Where permitted in Chapters 11 through 43, door assemblies in the means of egress shall be permitted to be equipped with electrical lock hardware that prevents egress, provided that all of the following criteria are met:

(1) A sensor shall be provided on the egress side, arranged to unlock the door leaf in the direction of egress upon detection of an approaching occupant.
(2) Door leaves shall automatically unlock in the direction of egress upon loss of power to the sensor or to the part of the access control system that locks the door leaves.
(3) Door locks shall be arranged to unlock in the direction of egress from a manual release device complying with all of the following criteria:
    (a) The manual release device shall be located on the egress side, 40 in. to 48 in. (1015 mm to 1220 mm ) vertically above the floor, and within 60 in. (1525 mm) of the secured door openings.

(b) The manual release device shall be readily accessible and clearly identified by a sign that reads as follows: PUSH TO EXIT.

(c) When operated, the manual release device shall result in direct interruption of power to the lock — independent of the locking system electronics — and the lock shall remain unlocked for not less than 30 seconds.

(4) Activation of the building fire-protective signaling system, if provided, shall automatically unlock the door leaves in the direction of egress, and the door leaves shall remain unlocked until the fire-protective signaling system has been manually reset.

(5) The activation of manual fire alarm boxes that activate the building fire-protective signaling system specified in 7.2.1.6.2(4) shall not be required to unlock the door leaves.

(6) Activation of the building automatic sprinkler or fire detection system, if provided, shall automatically unlock the door leaves in the direction of egress, and the door leaves shall remain unlocked until the fire-protective signaling system has been manually reset.

(7) The egress side of access-controlled egress doors, other than existing access-controlled egress doors, shall be provided with emergency lighting in accordance with Section 7.9.

**7.2.1.6.3 Elevator Lobby Exit Access Door Assemblies Locking.** Where permitted in Chapters 11 through 43, door assemblies separating the elevator lobby from the exit access required by 7.4.1.6.1 shall be permitted to be electrically locked, provided that all the following criteria are met:

(1) The lock is listed in accordance with ANSI/UL 294, *Standard for Access Control System Units.*

(2) The building is protected throughout by a fire alarm system in accordance with Section 9.6.

(3) The building is protected throughout by an approved, supervised automatic sprinkler system in accordance with Section 9.7.

(4) Waterflow in the sprinkler system required by 7.2.1.6.3(3) is arranged to initiate the building fire alarm system.

(5) The elevator lobby is protected by an approved, supervised smoke detection system in accordance with Section 9.6.

(6) Detection of smoke by the detection system required by 7.2.1.6.3(5) is arranged to initiate the building fire alarm system and notify building occupants.

(7) Initiation of the building fire alarm system by other than manual fire alarm boxes unlocks the elevator lobby door assembly.

(8) Loss of power to the elevator lobby electronic lock system unlocks the elevator lobby door assemblies.

(9) Once unlocked, the elevator lobby door assemblies remain unlocked until the building fire alarm system has been manually reset.

(10) Where the elevator lobby door assemblies remain latched after being unlocked, latch-releasing hardware in accordance with 7.2.1.5.10 is affixed to the door leaves.

(11) A two-way communication system is provided for communication between the elevator lobby and a central control point that is constantly staffed.

(12) The central control point staff required by 7.2.1.6.3(11) is capable, trained, and authorized to provide emergency assistance.

(13) The provisions of 7.2.1.6.1 for delayed-egress locking systems are not applied to the elevator lobby door assemblies.

(14)*The provisions of 7.2.1.6.2 for access-controlled egress door assemblies are not applied to the elevator lobby door assemblies.

**7.2.1.7 Panic Hardware and Fire Exit Hardware.**

**7.2.1.7.1** Where a door assembly is required to be equipped with panic or fire exit hardware, such hardware shall meet all of the following criteria:

(1) It shall consist of a cross bar or a push pad, the actuating portion of which extends across not less than one-half of the width of the door leaf.

(2) It shall be mounted as follows:

(a) New installations shall be not less than 34 in. (865 mm), and not more than 48 in. (1220 mm), above the floor.

(b) Existing installations shall be not less than 30 in. (760 mm), and not more than 48 in. (1220 mm), above the floor.

(3) It shall be constructed so that a horizontal force not to exceed 15 lbf (66 N) actuates the cross bar or push pad and latches.

**7.2.1.7.2** Only approved panic hardware shall be used on door assemblies that are not fire-rated door assemblies. Only approved fire exit hardware shall be used on fire-rated door assemblies. New panic hardware and new fire exit hardware shall comply with ANSI/UL 305, *Standard for Safety Panic Hardware,* and ANSI/BHMA A156.3, *Exit Devices.*

**7.2.1.7.3** Required panic hardware and fire exit hardware, in other than detention and correctional occupancies as otherwise provided in Chapters 22 and 23, shall not be equipped with any locking device, set screw, or other arrangement that prevents the release of the latch when pressure is applied to the releasing device.

**7.2.1.7.4** Devices that hold the latch in the retracted position shall be prohibited on fire exit hardware, unless such devices are listed and approved for such a purpose.

**7.2.1.8 Self-Closing Devices.**

**7.2.1.8.1*** A door leaf normally required to be kept closed shall not be secured in the open position at any time and shall be self-closing or automatic-closing in accordance with 7.2.1.8.2, unless otherwise permitted by 7.2.1.8.3.

**7.2.1.8.2** In any building of low or ordinary hazard contents, as defined in 6.2.2.2 and 6.2.2.3, or where approved by the authority having jurisdiction, door leaves shall be permitted to be automatic-closing, provided that all of the following criteria are met:

(1) Upon release of the hold-open mechanism, the leaf becomes self-closing.

(2) The release device is designed so that the leaf instantly releases manually and, upon release, becomes self-closing, or the leaf can be readily closed.

(3) The automatic releasing mechanism or medium is activated by the operation of approved smoke detectors installed in accordance with the requirements for smoke detectors for door leaf release service in *NFPA 72, National Fire Alarm and Signaling Code.*

(4) Upon loss of power to the hold-open device, the hold-open mechanism is released and the door leaf becomes self-closing.

(5) The release by means of smoke detection of one door leaf in a stair enclosure results in closing all door leaves serving that stair.

**7.2.1.8.3** The elevator car doors, and the associated hoistway enclosure doors, at the floor level designated for recall in accordance with the requirements of 9.4.3 shall be permitted to remain open during Phase I Emergency Recall Operation.

**7.2.1.9\* Powered Door Leaf Operation.**

**7.2.1.9.1\* General.** Where means of egress door leaves are operated by power upon the approach of a person or are provided with power-assisted manual operation, the design shall be such that, in the event of power failure, the leaves open manually to allow egress travel or close when necessary to safeguard the means of egress.

**7.2.1.9.1.1** The forces required to manually open the door leaves specified in 7.2.1.9.1 shall not exceed those required in 7.2.1.4.5, except that the force required to set the leaf in motion shall not exceed 50 lbf (222 N).

**7.2.1.9.1.2** The door assembly shall be designed and installed so that, when a force is applied to the door leaf on the side from which egress is made, it shall be capable of swinging from any position to provide full use of the required width of the opening in which it is installed. *(See 7.2.1.4.)*

**7.2.1.9.1.3** A readily visible, durable sign in letters not less than 1 in. (25 mm) high on a contrasting background that reads as follows shall be located on the egress side of each door opening:

IN EMERGENCY, PUSH TO OPEN

**7.2.1.9.1.4** Sliding, power-operated door assemblies in an exit access serving an occupant load of fewer than 50 that manually open in the direction of door leaf travel, with forces not exceeding those required in 7.2.1.4.5, shall not be required to have the swing-out feature required by 7.2.1.9.1.2. The required sign shall be in letters not less than 1 in. (25 mm) high on a contrasting background and shall read as follows:

IN EMERGENCY, SLIDE TO OPEN

**7.2.1.9.1.5\*** In the emergency breakout mode, a door leaf located within a two-leaf opening shall be exempt from the minimum 32 in. (810 mm) single-leaf requirement of 7.2.1.2.3.2(1), provided that the clear width of the single leaf is not less than 30 in. (760 mm).

**7.2.1.9.1.6** For a biparting sliding door assembly in the emergency breakout mode, a door leaf located within a multiple-leaf opening shall be exempt from the minimum 32 in. (810 mm) single-leaf requirement of 7.2.1.2.3.2(1) if a clear opening of not less than 32 in. (810 mm) is provided by all leafs broken out.

**7.2.1.9.1.7** Door assemblies complying with 7.2.1.14 shall be permitted to be used.

**7.2.1.9.1.8** The requirements of 7.2.1.9.1 through 7.2.1.9.1.7 shall not apply in detention and correctional occupancies where otherwise provided in Chapters 22 and 23.

**7.2.1.9.2 Self-Closing or Self-Latching Door Leaf Operation.** Where door leaves are required to be self-closing or self-latching and are operated by power upon the approach of a person, or are provided with power-assisted manual operation,

they shall be permitted in the means of egress where they meet the following criteria:

(1) The door leaves can be opened manually in accordance with 7.2.1.9.1 to allow egress travel in the event of power failure.

(2) New door leaves remain in the closed position, unless actuated or opened manually.

(3) When actuated, new door leaves remain open for not more than 30 seconds.

(4) Door leaves held open for any period of time close — and the power-assist mechanism ceases to function — upon operation of approved smoke detectors installed in such a way as to detect smoke on either side of the door opening in accordance with the provisions of *NFPA 72, National Fire Alarm and Signaling Code.*

(5) Door leaves required to be self-latching are either self-latching or become self-latching upon operation of approved smoke detectors per 7.2.1.9.2(4).

(6) New power-assisted swinging door assemblies comply with BHMA/ANSI A156.19, *American National Standard for Power Assist and Low Energy Power Operated Doors.*

**7.2.1.10 Revolving Door Assemblies.**

**7.2.1.10.1** Revolving door assemblies, whether used or not used in the means of egress, shall comply with all of the following:

(1) Revolving door wings shall be capable of being collapsed into a book-fold position, unless they are existing revolving doors approved by the authority having jurisdiction.

(2) When revolving door wings are collapsed into the book-fold position, the parallel egress paths formed shall provide an aggregate width of 36 in. (915 mm), unless they are approved existing revolving door assemblies.

(3) Revolving door assemblies shall not be used within 10 ft (3050 mm) of the foot or the top of stairs or escalators.

(4) A dispersal area acceptable to the authority having jurisdiction shall be located between stairs or escalators and the revolving door assembly.

(5) The revolutions per minute (rpm) of revolving door wings shall not exceed the values in Table 7.2.1.10.1.

(6) Each revolving door assembly shall have a conforming side-hinged swinging door assembly in the same wall as the revolving door within 10 ft (3050 mm) of the revolving door, unless one of the following conditions applies:

   (a) Revolving door assemblies shall be permitted without adjacent swinging door assemblies, as required by 7.2.1.10.1(6), in street floor elevator lobbies, provided that no stairways or door openings from other parts of the building discharge through the lobby and the lobby has no occupancy other than as a means of travel between the elevators and street.

   (b) The requirement of 7.2.1.10.1(6) shall not apply to existing revolving door assemblies where the number of revolving door assemblies does not exceed the number of swinging door assemblies within 20 ft (6100 mm) of the revolving door assembly.

**7.2.1.10.2** Where permitted in Chapters 11 through 43, revolving door assemblies shall be permitted as a component in a means of egress, provided that all of the following criteria are met:

(1) Revolving door openings shall not be given credit for more than 50 percent of the required egress capacity.

**Table 7.2.1.10.1 Revolving Door Assembly Maximum Speed**

| Inside Diameter | | Power-Driven Speed Control (rpm) | Manual Speed Control (rpm) |
|---|---|---|---|
| **ft/in.** | **mm** | | |
| 6 ft 6 in. | 1980 | 11 | 12 |
| 7 ft | 2135 | 10 | 11 |
| 7 ft 6 in. | 2285 | 9 | 11 |
| 8 ft | 2440 | 9 | 10 |
| 8 ft 6 in. | 2590 | 8 | 9 |
| 9 ft | 2745 | 8 | 9 |
| 9 ft 6 in. | 2895 | 7 | 8 |
| 10 ft | 3050 | 7 | 8 |

(2) Each revolving door opening shall not be credited with more than a 50-person capacity or, if of not less than a 9 ft (2745 mm) diameter, a revolving door assembly shall be permitted egress capacity based on the clear opening width provided when collapsed into a book-fold position.

(3) Revolving door wings shall be capable of being collapsed into a book-fold position when a force not exceeding 130 lbf (580 N) is applied to the wings within 3 in. (75 mm) of the outer edge.

**7.2.1.10.3** Revolving door assemblies not used as a component of a means of egress shall have a collapsing force not exceeding 180 lbf (800 N) applied at a point 3 in. (75 mm) from the outer edge of the outer wing stile and 40 in. (1015 mm) above the floor.

**7.2.1.10.4** The requirement of 7.2.1.10.3 shall not apply to revolving door assemblies, provided that the collapsing force is reduced to a force not to exceed 130 lbf (580 N) under all of the following conditions:

(1) Power failure, or removal of power to the device holding the wings in position

(2) Actuation of the automatic sprinkler system, where such a system is provided

(3) Actuation of a smoke detection system that is installed to provide coverage in all areas within the building that are within 75 ft (23 m) of the revolving door assemblies

(4) Actuation of a clearly identified manual control switch in an approved location that reduces the holding force to a force not to exceed 130 lbf (580 N)

**7.2.1.11 Turnstiles.**

**7.2.1.11.1** Turnstiles or similar devices that restrict travel to one direction or are used to collect fares or admission charges shall not be placed so as to obstruct any required means of egress, unless otherwise specified in 7.2.1.11.1.1 and 7.2.1.11.1.2.

**7.2.1.11.1.1** Approved turnstiles not exceeding 39 in. (990 mm) in height that turn freely in the direction of egress travel shall be permitted where revolving door assemblies are permitted in Chapters 11 through 43.

**7.2.1.11.1.2** Where turnstiles are approved by the authority having jurisdiction and permitted in Chapters 11 through 43, each turnstile shall be credited for a capacity of 50 persons, provided that such turnstiles meet all of the following criteria:

(1) They freewheel in the egress direction when primary power is lost, and freewheel in the direction of egress travel upon manual release by an employee assigned in the area.

(2) They are not given credit for more than 50 percent of the required egress width.

(3) They are not in excess of 39 in. (990 mm) in height and have a clear width of not less than 16½ in. (420 mm).

**7.2.1.11.2** Turnstiles exceeding 39 in. (990 mm) in height shall meet the requirements for revolving door assemblies in 7.2.1.10.

**7.2.1.11.3** Turnstiles located in, or furnishing access to, required exits shall provide not less than 16½ in. (420 mm) clear width at and below a height of 39 in. (990 mm) and at least 22 in. (560 mm) clear width at heights above 39 in. (990 mm).

**7.2.1.12 Door Openings in Folding Partitions.** Where permanently mounted folding or movable partitions divide a room into smaller spaces, a swinging door leaf or open doorway shall be provided as an exit access from each such space, unless otherwise specified in 7.2.1.12.1 and 7.2.1.12.2.

**7.2.1.12.1** A door leaf or opening in the folding partition shall not be required, provided that all of the following criteria are met:

(1) The subdivided space is not used by more than 20 persons at any time.

(2) The use of the space is under adult supervision.

(3) The partitions are arranged so that they do not extend across any aisle or corridor used as an exit access to the required exits from the story.

(4) The partitions conform to the interior finish and other requirements of this *Code*.

(5) The partitions are of an approved type, have a simple method of release, and are capable of being opened quickly and easily by experienced persons in case of emergency.

**7.2.1.12.2** Where a subdivided space is provided with not less than two means of egress, the swinging door leaf in the folding partition specified in 7.2.1.12 shall not be required, and one such means of egress shall be permitted to be equipped with a horizontal-sliding door assembly complying with 7.2.1.14.

**7.2.1.13 Balanced Door Assemblies.** If panic hardware is installed on balanced door leaves, the panic hardware shall be of the push-pad type, and the pad shall not extend more than approximately one-half the width of the door leaf, measured from the latch stile. *[See 7.2.1.7.1(1).]*

**7.2.1.14 Horizontal-Sliding Door Assemblies.** Horizontal-sliding door assemblies shall be permitted in means of egress, provided that all of the following criteria are met:

(1) The door leaf is readily operable from either side without special knowledge or effort.

(2) The force that, when applied to the operating device in the direction of egress, is required to operate the door leaf is not more than 15 lbf (67 N).

(3) The force required to operate the door leaf in the direction of travel is not more than 30 lbf (133 N) to set the leaf in motion and is not more than 15 lbf (67 N) to close the leaf or open it to the minimum required width.

(4) The door leaf is operable using a force of not more than 50 lbf (222 N) when a force of 250 lbf (1100 N) is applied perpendicularly to the leaf adjacent to the operating device, unless the door opening is an existing horizontal-sliding exit access door assembly serving an area with an occupant load of fewer than 50.

(5) The door assembly complies with the fire protection rating, if required, and, where rated, is self-closing or automatic-closing by means of smoke detection in accordance with 7.2.1.8 and is installed in accordance with NFPA 80, *Standard for Fire Doors and Other Opening Protectives.*

**7.2.1.15 Inspection of Door Openings.**

**7.2.1.15.1*** Where required by Chapters 11 through 43, the following door assemblies shall be inspected and tested not less than annually in accordance with 7.2.1.15.2 through 7.2.1.15.8:

(1) Door leaves equipped with panic hardware or fire exit hardware in accordance with 7.2.1.7
(2) Door assemblies in exit enclosures
(3) Electrically controlled egress doors
(4) Door assemblies with special locking arrangements subject to 7.2.1.6

**7.2.1.15.2** Fire-rated door assemblies shall be inspected and tested in accordance with NFPA 80, *Standard for Fire Doors and Other Opening Protectives.* Smoke door assemblies shall be inspected and tested in accordance with NFPA 105, *Standard for Smoke Door Assemblies and Other Opening Protectives.*

**7.2.1.15.3** The inspection and testing interval for fire-rated and nonrated door assemblies shall be permitted to exceed 12 months under a written performance-based program in accordance with 5.2.2 of NFPA 80, *Standard for Fire Doors and Other Opening Protectives.*

**7.2.1.15.4** A written record of the inspections and testing shall be signed and kept for inspection by the authority having jurisdiction.

**7.2.1.15.5** Functional testing of door assemblies shall be performed by individuals who can demonstrate knowledge and understanding of the operating components of the type of door being subjected to testing.

**7.2.1.15.6** Door assemblies shall be visually inspected from both sides of the opening to assess the overall condition of the assembly.

**7.2.1.15.7** As a minimum, the following items shall be verified:

(1) Floor space on both sides of the openings is clear of obstructions, and door leaves open fully and close freely.
(2) Forces required to set door leaves in motion and move to the fully open position do not exceed the requirements in 7.2.1.4.5.
(3) Latching and locking devices comply with 7.2.1.5.
(4) Releasing hardware devices are installed in accordance with 7.2.1.5.10.1.
(5) Door leaves of paired openings are installed in accordance with 7.2.1.5.11.
(6) Door closers are adjusted properly to control the closing speed of door leaves in accordance with accessibility requirements.
(7) Projection of door leaves into the path of egress does not exceed the encroachment permitted by 7.2.1.4.3.
(8) Powered door openings operate in accordance with 7.2.1.9.
(9) Signage required by 7.2.1.4.1(3), 7.2.1.5.5, 7.2.1.6, and 7.2.1.9 is intact and legible.
(10) Door openings with special locking arrangements function in accordance with 7.2.1.6
(11) Security devices that impede egress are not installed on openings, as required by 7.2.1.5.12.

**7.2.1.15.8** Door openings not in proper operating condition shall be repaired or replaced without delay.

**7.2.2 Stairs.**

**7.2.2.1 General.**

**7.2.2.1.1** Stairs used as a component in the means of egress shall conform to the general requirements of Section 7.1 and to the special requirements of 7.2.2, unless otherwise specified in 7.2.2.1.2.

**7.2.2.1.2** The requirement of 7.2.2.1.1 shall not apply to the following:

(1) Aisle stairs in assembly occupancies, as provided in Chapters 12 and 13
(2) Approved existing noncomplying stairs

**7.2.2.2 Dimensional Criteria.**

**7.2.2.2.1 Standard Stairs.**

**7.2.2.2.1.1** Stairs shall meet the following criteria:

(1) New stairs shall be in accordance with Table 7.2.2.2.1.1(a) and 7.2.2.2.1.2.
(2)*Existing stairs shall be permitted to remain in use, provided that they meet the requirements for existing stairs shown in Table 7.2.2.2.1.1(b).
(3) Approved existing stairs shall be permitted to be rebuilt in accordance with the following:
    (a) Dimensional criteria of Table 7.2.2.2.1.1(b)
    (b) Other stair requirements of 7.2.2
(4) The requirements for new and existing stairs shall not apply to stairs located in industrial equipment access areas where otherwise provided in 40.2.5.2.

**7.2.2.2.1.2 Minimum New Stair Width.**

**(A)** Where the total occupant load of all stories served by the stair is fewer than 50, the minimum width clear of all obstructions, except projections not more than 4½ in. (114 mm) at or below handrail height on each side, shall be 36 in. (915 mm).

**(B)*** Where stairs serve occupant loads exceeding that permitted by 7.2.2.2.1.2(A), the minimum width clear of all obstructions, except projections not more than 4½ in. (114 mm) at or below handrail height on each side, shall be in accordance with Table 7.2.2.2.1.2(B) and the requirements of 7.2.2.2.1.2(C), 7.2.2.2.1.2(D), 7.2.2.2.1.2(E), and 7.2.2.2.1.2(F).

**Table 7.2.2.2.1.1(a) New Stairs**

| Feature | Dimensional Criteria | |
| --- | --- | --- |
| | ft/in. | mm |
| Minimum width | See 7.2.2.2.1.2. | |
| Maximum height of risers | 7 in. | 180 |
| Minimum height of risers | 4 in. | 100 |
| Minimum tread depth | 11 in. | 280 |
| Minimum headroom | 6 ft 8 in. | 2030 |
| Maximum height between landings | 12 ft | 3660 |
| Landing | See 7.2.1.3, 7.2.1.4.3.1, and 7.2.2.3.2. | |

**Table 7.2.2.2.1.1(b)  Existing Stairs**

| Feature | Dimensional Criteria | |
| --- | --- | --- |
| | ft/in. | mm |
| Minimum width clear of all obstructions, except projections not more than 4½ in. (114 mm) at or below handrail height on each side | 36 in. | 915 |
| Maximum height of risers | 8 in. | 205 |
| Minimum tread depth | 9 in. | 230 |
| Minimum headroom | 6 ft 8 in. | 2030 |
| Maximum height between landings | 12 ft | 3660 |
| Landing | See 7.2.1.3 and 7.2.1.4.3.1. | |

**Table 7.2.2.2.1.2(B)  New Stair Width**

| Total Cumulative Occupant Load Assigned to the Stair | Width | |
| --- | --- | --- |
| | in. | mm |
| <2000 persons | 44 | 1120 |
| ≥2000 persons | 56 | 1420 |

**(C)**  The total cumulative occupant load assigned to a particular stair shall be that stair's prorated share of the total occupant load, as stipulated in 7.2.2.2.1.2(D) and 7.2.2.2.1.2(E), calculated in proportion to the stair width.

**(D)**  For downward egress travel, stair width shall be based on the total number of occupants from stories above the level where the width is measured.

**(E)**  For upward egress travel, stair width shall be based on the total number of occupants from stories below the level where the width is measured.

**(F)**  The clear width of door openings discharging from stairways required to be a minimum of 56 in. (1420 mm) wide in accordance with 7.2.2.2.1.2(B) shall be in accordance with 7.2.1.2.3.2(9).

**7.2.2.2.2 Curved Stairs.**

**7.2.2.2.2.1**  New curved stairs shall be permitted as a component in a means of egress, provided that the depth of tread is not less than 11 in. (280 mm) at a point 12 in. (305 mm) from the narrower end of the tread and the smallest radius is not less than twice the stair width.

**7.2.2.2.2.2**  Existing curved stairs shall be permitted as a component in a means of egress, provided that the depth of tread is not less than 10 in. (255 mm) at a point 12 in. (305 mm) from the narrower end of the tread and the smallest radius is not less than twice the stair width.

**7.2.2.2.3 Spiral Stairs.**

**7.2.2.2.3.1**  Where specifically permitted for individual occupancies by Chapters 11 through 43, spiral stairs shall be permitted as a component in a means of egress in accordance with 7.2.2.2.3.2 through 7.2.2.2.3.4.

**7.2.2.2.3.2**  Spiral stairs shall be permitted, provided that all of the following criteria are met:

(1)  Riser heights shall not exceed 7 in. (180 mm).
(2)  The stairway shall have a tread depth of not less than 11 in. (280 mm) for a portion of the stairway width sufficient to provide egress capacity for the occupant load served in accordance with 7.3.3.1.
(3)  At the outer side of the stairway, an additional 10½ in. (265 mm) of width shall be provided clear to the other handrail, and this width shall not be included as part of the required egress capacity.
(4)  Handrails complying with 7.2.2.4 shall be provided on both sides of the spiral stairway.
(5)  The inner handrail shall be located within 24 in. (610 mm), measured horizontally, of the point where a tread depth of not less than 11 in. (280 mm) is provided.
(6)  The turn of the stairway shall be such that the outer handrail is at the right side of descending users.

**7.2.2.2.3.3**  Where the occupant load served does not exceed three, spiral stairs shall be permitted, provided that all of the following criteria are met:

(1)  The clear width of the stairs shall be not less than 26 in. (660 mm).
(2)  The height of risers shall not exceed 9½ in. (240 mm).
(3)  The headroom shall be not less than 6 ft 6 in. (1980 mm).
(4)  Treads shall have a depth not less than 7½ in. (190 mm) at a point 12 in. (305 mm) from the narrower edge.
(5)  All treads shall be identical.
(6)  Handrails shall be provided on both sides of the stairway.

**7.2.2.2.3.4**  Where the occupant load served does not exceed five, existing spiral stairs shall be permitted, provided that the requirements of 7.2.2.2.3.3(1) through (5) are met.

**7.2.2.2.4* Winders.**

**7.2.2.2.4.1**  Where specified in Chapters 11 through 43, winders shall be permitted in stairs, provided that they meet the requirements of 7.2.2.2.4.2 and 7.2.2.2.4.3.

**7.2.2.2.4.2**  New winders shall have a tread depth of not less than 6 in. (150 mm) and a tread depth of not less than 11 in. (280 mm) at a point 12 in. (305 mm) from the narrowest edge.

**7.2.2.2.4.3**  Existing winders shall be permitted to be continued in use, provided that they have a tread depth of not less than 6 in. (150 mm) and a tread depth of not less than 9 in. (230 mm) at a point 12 in. (305 mm) from the narrowest edge.

**7.2.2.3 Stair Details.**

**7.2.2.3.1 Construction.**

**7.2.2.3.1.1**  All stairs serving as required means of egress shall be of permanent fixed construction, unless they are stairs serving seating that is designed to be repositioned in accordance with Chapters 12 and 13.

**7.2.2.3.1.2**  Each stair, platform, and landing, not including handrails and existing stairs, in buildings required in this *Code* to be of Type I or Type II construction shall be of noncombustible material throughout.

**7.2.2.3.2 Landings.**

**7.2.2.3.2.1**  Stairs shall have landings at door openings, except as permitted in 7.2.2.3.2.5.

**7.2.2.3.2.2** Stairs and intermediate landings shall continue with no decrease in width along the direction of egress travel.

**7.2.2.3.2.3** In new buildings, every landing shall have a dimension, measured in the direction of travel, that is not less than the width of the stair.

**7.2.2.3.2.4** Landings shall not be required to exceed 48 in. (1220 mm) in the direction of travel, provided that the stair has a straight run.

**7.2.2.3.2.5** In existing buildings, a door assembly at the top of a stair shall be permitted to open directly to the stair, provided that the door leaf does not swing over the stair and the door opening serves an area with an occupant load of fewer than 50 persons.

**7.2.2.3.3 Tread and Landing Surfaces.**

**7.2.2.3.3.1** Stair treads and landings shall be solid, without perforations, unless otherwise permitted in 7.2.2.3.3.4.

**7.2.2.3.3.2\*** Stair treads and landings shall be free of projections or lips that could trip stair users.

**7.2.2.3.3.3** If not vertical, risers on other than existing stairs shall be permitted to slope under the tread at an angle not to exceed 30 degrees from vertical, provided that the projection of the nosing does not exceed 1½ in. (38 mm).

**7.2.2.3.3.4** The requirement of 7.2.2.3.3.1 shall not apply to noncombustible grated stair treads and landings in the following occupancies:

(1) Assembly occupancies as otherwise provided in Chapters 12 and 13
(2) Detention and correctional occupancies as otherwise provided in Chapters 22 and 23
(3) Industrial occupancies as otherwise provided in Chapter 40
(4) Storage occupancies as otherwise provided in Chapter 42

**7.2.2.3.4\* Tread and Landing Slope.** The tread and landing slope shall not exceed ¼ in./ft (21 mm/m) (a slope of 1 in 48).

**7.2.2.3.5\* Riser Height and Tread Depth.** Riser height shall be measured as the vertical distance between tread nosings. Tread depth shall be measured horizontally, between the vertical planes of the foremost projection of adjacent treads and at a right angle to the tread's leading edge, but shall not include beveled or rounded tread surfaces that slope more than 20 degrees (a slope of 1 in 2.75). At tread nosings, such beveling or rounding shall not exceed ½ in. (13 mm) in horizontal dimension.

**7.2.2.3.6\* Dimensional Uniformity.**

**7.2.2.3.6.1** Variation in excess of 3⁄16 in. (4.8 mm) in the sizes of adjacent tread depths or in the height of adjacent risers shall be prohibited, unless otherwise permitted in 7.2.2.3.6.3.

**7.2.2.3.6.2** The variation between the sizes of the largest and smallest riser or between the largest and smallest tread depths shall not exceed 3⁄8 in. (9.5 mm) in any flight.

**7.2.2.3.6.3** Where the bottom or top riser adjoins a sloping public way, walk, or driveway having an established finished ground level and serves as a landing, the bottom or top riser shall be permitted to have a variation in height of not more than 1 in. in every 12 in. (25 mm in every 305 mm) of stairway width.

**7.2.2.3.6.4** The size of the variations addressed by 7.2.2.3.6.1, 7.2.2.3.6.2, and 7.2.2.3.6.3 shall be based on the nosing-to-nosing dimensions of the tread depths and riser heights, consistent with the measurement details set out in 7.2.2.3.5.

**7.2.2.3.6.5\*** All tread nosings of stairs utilizing the provision of 7.2.2.3.6.3 shall be marked in accordance with 7.2.2.5.4.3. Those portions of the marking stripe at locations where the riser height below the nosing is inconsistent by more than 3⁄16 in. (4.8 mm), relative to other risers in the stair flight, shall be distinctively colored or patterned, incorporating safety yellow, to warn descending users of the inconsistent geometry relative to other steps in the flight.

**7.2.2.3.6.6** The variation in the horizontal projection of all nosings, including the projection of the landing nosing, shall not exceed 3⁄8 in. (9.5 mm) within each stair flight and, for other than existing nosings, shall not exceed 3⁄16 in. (4.8 mm) between adjacent nosings.

**7.2.2.4 Guards and Handrails.**

**7.2.2.4.1 Handrails.**

**7.2.2.4.1.1** Stairs and ramps shall have handrails on both sides, unless otherwise permitted in 7.2.2.4.1.5 or 7.2.2.4.1.6.

**7.2.2.4.1.2** In addition to the handrails required at the sides of stairs by 7.2.2.4.1.1, both of the following provisions shall apply:

(1) For new stairs, handrails shall be provided within 30 in. (760 mm) of all portions of the required egress width.
(2) For existing stairs, handrails shall meet the following criteria:
  (a) They shall be provided within 44 in. (1120 mm) of all portions of the required egress width.
  (b) Such stairs shall not have their egress capacity adjusted to a higher occupant load than permitted by the capacity factor in Table 7.3.3.1 if the stair's clear width between handrails exceeds 60 in. (1525 mm).

**7.2.2.4.1.3** Where new intermediate handrails are provided in accordance with 7.2.2.4.1.2, the minimum clear width between handrails shall be 20 in. (510 mm).

**7.2.2.4.1.4\*** The required egress width shall be provided along the natural path of travel.

**7.2.2.4.1.5** If a single step or a ramp is part of a curb that separates a sidewalk from a vehicular way, it shall not be required to have a handrail.

**7.2.2.4.1.6** Existing stairs, existing ramps, stairs within dwelling units and within guest rooms, and ramps within dwelling units and guest rooms shall be permitted to have a handrail on one side only.

**7.2.2.4.2 Continuity.** Required guards and handrails shall continue for the full length of each flight of stairs. At turns of new stairs, inside handrails shall be continuous between flights at landings.

**7.2.2.4.3 Projections.** The design of guards and handrails and the hardware for attaching handrails to guards, balusters, or walls shall be such that there are no projections that might engage loose clothing. Openings in guards shall be designed to prevent loose clothing from becoming wedged in such openings.

**7.2.2.4.4\* Handrail Details.**

**7.2.2.4.4.1** New handrails on stairs shall be not less than 34 in. (865 mm), and not more than 38 in. (965 mm), above the surface of the tread, measured vertically to the top of the rail from the leading edge of the tread.

**7.2.2.4.4.2** Existing required handrails shall be not less than 30 in. (760 mm), and not more than 38 in. (965 mm), above the surface of the tread, measured vertically to the top of the rail from the leading edge of the tread.

**7.2.2.4.4.3** The height of required handrails that form part of a guard shall be permitted to exceed 38 in. (965 mm), but shall not exceed 42 in. (1065 mm), measured vertically to the top of the rail from the leading edge of the tread.

**7.2.2.4.4.4\*** Additional handrails that are lower or higher than the main handrail shall be permitted.

**7.2.2.4.4.5** New handrails shall be installed to provide a clearance of not less than 2¼ in. (57 mm) between the handrail and the wall to which it is fastened.

**7.2.2.4.4.6** Handrails shall include one of the following features:

(1) Circular cross section with an outside diameter of not less than 1¼ in. (32 mm) and not more than 2 in. (51 mm)
(2)\*Shape that is other than circular with a perimeter dimension of not less than 4 in. (100 mm), but not more than 6¼ in. (160 mm), and with the largest cross-sectional dimension not more than 2¼ in. (57 mm), provided that graspable edges are rounded so as to provide a radius of not less than ⅛ in. (3.2 mm)

**7.2.2.4.4.7** New handrails shall be continuously graspable along their entire length.

**7.2.2.4.4.8** Handrail brackets or balusters attached to the bottom surface of the handrail shall not be considered to be obstructions to graspability, provided that both of the following criteria are met:

(1) They do not project horizontally beyond the sides of the handrail within 1½ in. (38 mm) of the bottom of the handrail and provided that, for each additional ½ in. (13 mm) of handrail perimeter dimension greater than 4 in. (100 mm), the vertical clearance dimension of 1½ in. (38 mm) is reduced by ⅛ in. (3.2 mm).
(2) They have edges with a radius of not less than 0.01 in. (0.25 mm).

**7.2.2.4.4.9** New handrail ends shall be returned to the wall or floor or shall terminate at newel posts.

**7.2.2.4.4.10** In other than dwelling units, new handrails that are not continuous between flights shall extend horizontally, at the required height, not less than 12 in. (305 mm) beyond the top riser and continue to slope for a depth of one tread beyond the bottom riser.

**7.2.2.4.4.11** Within dwelling units, handrails shall extend, at the required height, to at least those points that are directly above the top and bottom risers.

**7.2.2.4.5 Guard Details.** See 7.1.8 for guard requirements.

**7.2.2.4.5.1** The height of guards required in 7.1.8 shall be measured vertically to the top of the guard from the surface adjacent thereto.

**7.2.2.4.5.2** Guards shall be not less than 42 in. (1065 mm) high, except as permitted by one of the following:

(1) Existing guards within dwelling units shall be permitted to be not less than 36 in. (915 mm) high.

(2) The requirement of 7.2.2.4.5.2 shall not apply in assembly occupancies where otherwise provided in Chapters 12 and 13.
(3)\*Existing guards on existing stairs shall be permitted to be not less than 30 in. (760 mm) high.

**7.2.2.4.5.3\*** Open guards, other than approved existing open guards, shall have intermediate rails or an ornamental pattern such that a sphere 4 in. (100 mm) in diameter is not able to pass through any opening up to a height of 34 in. (865 mm), and the following also shall apply:

(1) The triangular openings formed by the riser, tread, and bottom element of a guardrail at the open side of a stair shall be of such size that a sphere 6 in. (150 mm) in diameter is not able to pass through the triangular opening.
(2) In detention and correctional occupancies, in industrial occupancies, and in storage occupancies, the clear distance between intermediate rails, measured at right angles to the rails, shall not exceed 21 in. (535 mm).

**7.2.2.5 Enclosure and Protection of Stairs.**

**7.2.2.5.1 Enclosures.**

**7.2.2.5.1.1** All inside stairs serving as an exit or exit component shall be enclosed in accordance with 7.1.3.2.

**7.2.2.5.1.2** Inside stairs, other than those serving as an exit or exit component, shall be protected in accordance with Section 8.6.

**7.2.2.5.1.3** In existing buildings, where a two-story exit enclosure connects the story of exit discharge with an adjacent story, the exit shall be permitted to be enclosed only on the story of exit discharge, provided that not less than 50 percent of the number and capacity of exits on the story of exit discharge are independent of such enclosures.

**7.2.2.5.2\* Exposures.**

**7.2.2.5.2.1** Where nonrated walls or unprotected openings enclose the exterior of a stairway, other than an existing stairway, and the walls or openings are exposed by other parts of the building at an angle of less than 180 degrees, the building enclosure walls within 10 ft (3050 mm) horizontally of the nonrated wall or unprotected opening shall be constructed as required for stairway enclosures, including opening protectives.

**7.2.2.5.2.2** Construction shall extend vertically from the finished ground level to a point 10 ft (3050 mm) above the topmost landing of the stairs or to the roofline, whichever is lower.

**7.2.2.5.2.3** The fire resistance rating of the separation extending 10 ft (3050 mm) from the stairs shall not be required to exceed 1 hour where openings have a minimum ¾-hour fire protection rating.

**7.2.2.5.3\* Usable Space.** Enclosed, usable spaces within exit enclosures shall be prohibited, including under stairs, unless otherwise permitted by 7.2.2.5.3.2.

**7.2.2.5.3.1** Open space within the exit enclosure shall not be used for any purpose that has the potential to interfere with egress.

**7.2.2.5.3.2** Enclosed, usable space shall be permitted under stairs, provided that both of the following criteria are met:

(1) The space shall be separated from the stair enclosure by the same fire resistance as the exit enclosure.
(2) Entrance to the enclosed, usable space shall not be from within the stair enclosure. *(See also 7.1.3.2.3.)*

**7.2.2.5.4\* Stairway Identification.**

**7.2.2.5.4.1** New enclosed stairs serving three or more stories and existing enclosed stairs serving five or more stories shall comply with 7.2.2.5.4.1(A) through 7.2.2.5.4.1(M).

**(A)** The stairs shall be provided with special signage within the enclosure at each floor landing.

**(B)** The signage shall indicate the floor level.

**(C)** The signage shall indicate the terminus of the top and bottom of the stair enclosure.

**(D)** The signage shall indicate the identification of the stair enclosure.

**(E)** The signage shall indicate the floor level of, and the direction to, exit discharge.

**(F)** The signage shall be located inside the enclosure approximately 60 in. (1525 mm) above the floor landing in a position that is visible when the door is in the open or closed position.

**(G)** The signage shall comply with 7.10.8.1 and 7.10.8.2 of this *Code*.

**(H)** The floor level designation shall also be tactile in accordance with ICC/ANSI A117.1, *American National Standard for Accessible and Usable Buildings and Facilities.*

**(I)** The signage shall be painted or stenciled on the wall or on a separate sign securely attached to the wall.

**(J)** The stairway identification letter shall be located at the top of the sign in minimum 1 in. (25 mm) high lettering and shall be in accordance with 7.10.8.2.

**(K)\*** Signage that reads NO ROOF ACCESS and is located under the stairway identification letter shall designate stairways that do not provide roof access. Lettering shall be a minimum of 1 in. (25 mm) high and shall be in accordance with 7.10.8.2.

**(L)** The floor level number shall be located in the middle of the sign in minimum 5 in. (125 mm) high numbers and shall be in accordance with 7.10.8.2. Mezzanine levels shall have the letter "M" or other appropriate identification letter preceding the floor number, while basement levels shall have the letter "B" or other appropriate identification letter preceding the floor level number.

**(M)** Identification of the lower and upper terminus of the stairway shall be located at the bottom of the sign in minimum 1 in. (25 mm) high letters or numbers and shall be in accordance with 7.10.8.2.

**7.2.2.5.4.2** Wherever an enclosed stair requires travel in an upward direction to reach the level of exit discharge, special signs with directional indicators showing the direction to the level of exit discharge shall be provided at each floor level landing from which upward direction of travel is required, unless otherwise provided in 7.2.2.5.4.2(A) and 7.2.2.5.4.2(B), and both of the following also shall apply:

(1) Such signage shall comply with 7.10.8.1 and 7.10.8.2.
(2) Such signage shall be visible when the door leaf is in the open or closed position.

**(A)** The requirement of 7.2.2.5.4.2 shall not apply where signs required by 7.2.2.5.4.1 are provided.

**(B)** The requirement of 7.2.2.5.4.2 shall not apply to stairs extending not more than one story below the level of exit discharge where the exit discharge is clearly obvious.

**7.2.2.5.4.3\* Stairway Tread Marking.** Where new contrasting marking is applied to stairs, such marking shall comply with all of the following:

(1) The marking shall include a continuous strip as a coating on, or as a material integral with, the full width of the leading edge of each tread.
(2) The marking shall include a continuous strip as a coating on, or as a material integral with, the full width of the leading edge of each landing nosing.
(3) The marking strip width, measured horizontally from the leading vertical edge of the nosing, shall be consistent at all nosings.
(4) The marking strip width shall be 1 in. to 2 in. (25 mm to 51 mm).

**7.2.2.5.4.4\*** Where new contrast marking is provided for stairway handrails, it shall be applied to, or be part of, at least the upper surface of the handrail; have a minimum width of ½ in. (13 mm); and extend the full length of each handrail. After marking, the handrail shall comply with 7.2.2.4.4. Where handrails or handrail extensions bend or turn corners, the stripe shall be permitted to have a gap of not more than 4 in. (100 mm).

**7.2.2.5.5 Exit Stair Path Markings.** Where exit stair path markings are required in Chapters 11 through 43, such markings shall be installed in accordance with 7.2.2.5.5.1 through 7.2.2.5.5.11.

**7.2.2.5.5.1 Exit Stair Treads.** Exit stair treads shall incorporate a marking stripe that is applied as a paint/coating or be a material that is integral with the nosing of each step. The marking stripe shall be installed along the horizontal leading edge of the step and shall extend the full width of the step. The marking stripe shall also meet all of the following requirements:

(1) The marking stripe shall be not more than ½ in. (13 mm) from the leading edge of each step and shall not overlap the leading edge of the step by more than ½ in. (13 mm) down the vertical face of the step.
(2) The marking stripe shall have a minimum horizontal width of 1 in. (25 mm) and a maximum width of 2 in. (51 mm).
(3) The dimensions and placement of the marking stripe shall be uniform and consistent on each step throughout the exit enclosure.
(4) Surface-applied marking stripes using adhesive-backed tapes shall not be used.

**7.2.2.5.5.2 Exit Stair Landings.** The leading edge of exit stair landings shall be marked with a solid and continuous marking stripe consistent with the dimensional requirements for stair treads and shall be the same length as, and consistent with, the stripes on the steps.

**7.2.2.5.5.3 Exit Stair Handrails.** All handrails and handrail extensions shall be marked with a solid and continuous marking stripe and meet all of the following requirements:

(1) The marking stripe shall be applied to the upper surface of the handrail or be a material integral with the upper surface of the handrail for the entire length of the handrail, including extensions.

(2) Where handrails or handrail extensions bend or turn corners, the marking stripe shall be permitted to have a gap of not more than 4 in. (100 mm).

(3) The marking stripe shall have a minimum horizontal width of 1 in. (25 mm), which shall not apply to outlining stripes listed in accordance with UL 1994, *Standard for Luminous Egress Path Marking Systems*.

(4) The dimensions and placement of the marking stripe shall be uniform and consistent on each handrail throughout the exit enclosure.

**7.2.2.5.5.4 Perimeter Demarcation Marking.** Stair landings, exit passageways, and other parts of the floor areas within the exit enclosure shall be provided with a solid and continuous perimeter demarcation marking stripe on the floor or on the walls or a combination of both. The marking stripe shall also meet all of the following requirements:

(1) The marking stripe shall have a minimum horizontal width of 1 in. (25 mm) and a maximum width of 2 in. (51 mm), with interruptions not exceeding 4 in. (100 mm).

(2) The minimum marking stripe width of 1 in. (25 mm) shall not apply to outlining stripes listed in accordance with UL 1994, *Standard for Luminous Egress Path Marking Systems*.

(3) The dimensions and placement of the perimeter demarcation marking stripe shall be uniform and consistent throughout the exit enclosure.

(4) Surface-applied marking stripes using adhesive-backed tapes shall not be used.

**(A)** Perimeter floor demarcation lines shall comply with all of the following:

(1) They shall be placed within 4 in. (100 mm) of the wall and extend to within 2 in. (51 mm) of the markings on the leading edge of landings.

(2) They shall continue across the floor in front of all doors.

(3) They shall not extend in front of exit doors leading out of an exit enclosure and through which occupants must travel to complete the egress path.

**(B)** Perimeter wall demarcation lines shall comply with all of the following:

(1) They shall be placed on the wall with the bottom edge of the stripe not more than 4 in. (100 mm) above the finished floor.

(2) At the top or bottom of the stairs, they shall drop vertically to the floor within 2 in. (51 mm) of the step or landing edge.

(3) They shall transition vertically to the floor and then extend across the floor where a line on the floor is the only practical method of outlining the path.

(4) Where the wall line is broken by a door, they shall continue across the face of the door or transition to the floor and extend across the floor in front of such door.

(5) They shall not extend in front of doors leading out of an exit enclosure and through which occupants must travel to complete the egress path.

(6) Where a wall-mounted demarcation line transitions to a floor-mounted demarcation line, or vice versa, the wall-mounted demarcation line shall drop vertically to the floor to meet a complementary extension of the floor-mounted demarcation line, thus forming a continuous marking.

**7.2.2.5.5.5\* Obstacles.** Obstacles that are in the exit enclosure at or below 6 ft 6 in. (1980 mm) in height, and that project more than 4 in. (100 mm) into the egress path, shall be identified with markings not less than 1 in. (25 mm) in horizontal width comprised of a pattern of alternating equal bands of luminescent material and black; and with the alternating bands not more than 2 in. (51 mm) in horizontal width and angled at 45 degrees.

**7.2.2.5.5.6 Doors Serving Exit Enclosure.** All doors serving the exit enclosure that swing out from the enclosure in the direction of egress travel shall be provided with a marking stripe on the top and sides of the door(s) frame(s). The marking stripe shall also meet all of the following requirements:

(1) The marking stripe shall have a minimum horizontal width of 1 in. (25 mm) and a maximum width of 2 in. (51 mm).

(2) Gaps shall be permitted in the continuity of door frame markings where a line is fitted into a corner or bend, but shall be as small as practicable, and in no case shall gaps be greater than 1 in. (25 mm).

(3) Where the door molding does not provide enough flat surface on which to locate the marking stripe, the marking stripe shall be located on the wall surrounding the frame.

(4) The dimensions and placement of the marking stripe shall be uniform and consistent on all doors in the exit enclosure.

**7.2.2.5.5.7 Door Hardware Marking.** The door hardware for the doors serving the exit enclosure that swing out from the enclosure in the direction of egress travel shall be provided with a marking stripe. The marking stripe shall also meet the following requirements:

(1)\*The door hardware necessary to release the latch shall be outlined with an approved marking stripe having a minimum width of 1 in. (25 mm).

(2) Where panic hardware is installed, both of the following criteria shall be met:

    (a) The marking stripe shall have a minimum horizontal width of 1 in. (25 mm) and be applied to the entire length of the actuating bar or touch pad.

    (b) The placement of the marking stripe shall not interfere with viewing of any instructions on the actuating bar or touch pad.

**7.2.2.5.5.8 Emergency Exit Symbol.** An emergency exit symbol with a luminescent background shall be applied on all doors serving the exit enclosure that swing out from the enclosure in the direction of egress travel. The emergency exit symbol shall also meet both of the following requirements:

(1) The emergency exit symbol shall meet the requirements of NFPA 170, *Standard for Fire Safety and Emergency Symbols*.

(2) The emergency exit symbol applied on the door shall be a minimum of 4 in. (100 mm) in height and shall be applied on the door, centered horizontally, with the top of the symbol not higher than 18 in. (455 mm) above the finished floor.

**7.2.2.5.5.9 Uniformity.** Placement and dimensions of the marking stripes shall be consistent and uniform throughout the same exit enclosure.

**7.2.2.5.5.10 Materials.** Exit stair path markings shall be made of any material, including paint, provided that an electrical charge is not required to maintain the required luminescence. Such materials shall include, but shall not be limited to, self-luminous materials and photoluminescent materials. Materials shall comply with one of the following:

(1) ASTM E 2072, *Standard Specification for Photoluminescent (Phosphorescent) Safety Markings*, with the following exceptions:
  (a) The charging source shall be 1 ft-candle (10.8 lux) of fluorescent illumination for 60 minutes.
  (b) The minimum luminance shall be 5 millicandelas/m² after 90 minutes.

(2) ANSI/UL 1994, *Standard for Luminous Egress Path Marking Systems*

**7.2.2.5.5.11 Exit Stair Illumination.** Exit enclosures where photoluminescent materials are installed shall comply with all of the following:

(1) The exit enclosure shall be continuously illuminated for at least 60 minutes prior to periods when the building is occupied.
(2) The illumination shall remain on when the building is occupied.
(3) Lighting control devices provided for illumination within the exit enclosure shall meet all of the following requirements:
  (a) Lighting control devices that automatically turn exit enclosure lighting on and off, based on occupancy, shall be permitted, provided that they turn on illumination for charging photoluminescent materials for at least 60 minutes prior to periods when the building is occupied.
  (b) Lighting used to charge photoluminescent materials shall not be controlled by motion sensors.
  (c) Lighting control devices that dim the lighting levels within the exit enclosure shall not be installed unless they provide a minimum of 1 ft-candle (10.8 lux) of illumination within the exit enclosure measured at the walking surface.

**7.2.2.6 Special Provisions for Outside Stairs.**

**7.2.2.6.1 Access.** Where approved by the authority having jurisdiction, outside stairs shall be permitted to lead to roofs of other sections of a building or an adjoining building where the construction is fire resistive and there is a continuous and safe means of egress from the roof. *(See also 7.7.6.)*

**7.2.2.6.2\* Visual Protection.** Outside stairs shall be arranged to avoid any impediments to their use by persons having a fear of high places. Outside stairs more than 36 ft (11 m) above the finished ground level, other than previously approved existing stairs, shall be provided with an opaque visual obstruction not less than 48. in. (1220 mm) in height.

**7.2.2.6.3 Separation and Protection of Outside Stairs.**

**7.2.2.6.3.1\*** Outside stairs shall be separated from the interior of the building by construction with the fire resistance rating required for enclosed stairs with fixed or self-closing opening protectives, except as follows:

(1) Outside stairs serving an exterior exit access balcony that has two remote outside stairways or ramps shall be permitted to be unprotected.
(2) Outside stairs serving two or fewer adjacent stories, including the story where the exit discharges, shall be permitted to be unprotected where there is a remotely located second exit.
(3) In existing buildings, existing outside stairs serving three or fewer adjacent stories, including the story where the exit discharges, shall be permitted to be unprotected where there is a remotely located second exit.

(4) The fire resistance rating of a separation extending 10 ft (3050 mm) from the stairs shall not be required to exceed 1 hour where openings have a minimum ¾-hour fire protection rating.
(5) Outside stairs in existing buildings protected throughout by an approved, supervised automatic sprinkler system in accordance with Section 9.7 shall be permitted to be unprotected.

**7.2.2.6.3.2** Wall construction required by 7.2.2.6.3.1 shall extend as follows:

(1) Vertically from the finished ground level to a point 10 ft (3050 mm) above the topmost landing of the stairs or to the roofline, whichever is lower
(2) Horizontally for not less than 10 ft (3050 mm)

**7.2.2.6.3.3** Roof construction required by 7.2.2.6.3.1 shall meet both of the following criteria:

(1) It shall provide protection beneath the stairs.
(2) It shall extend horizontally to each side of the stair for not less than 10 ft (3050 mm).

**7.2.2.6.4 Protection of Openings.** All openings below an outside stair shall be protected with an assembly having a minimum ¾-hour fire protection rating as follows:

(1) Where located in an enclosed court *(see 3.3.51.1)*, the smallest dimension of which does not exceed one-third its height
(2) Where located in an alcove having a width that does not exceed one-third its height and a depth that does not exceed one-fourth its height

**7.2.2.6.5\* Water Accumulation.** Outside stairs and landings, other than existing outside stairs and landings, shall be designed to minimize water accumulation on their surfaces.

**7.2.2.6.6 Openness.** Outside stairs, other than existing outside stairs, shall be not less than 50 percent open on one side. Outside stairs shall be arranged to restrict the accumulation of smoke.

**7.2.3 Smokeproof Enclosures.**

**7.2.3.1 General.** Where smokeproof enclosures are required in other sections of this *Code*, they shall comply with 7.2.3, unless they are approved existing smokeproof enclosures.

**7.2.3.2 Performance Design.** An appropriate design method shall be used to provide a system that meets the definition of *smokeproof enclosure (see 3.3.255)*. The smokeproof enclosure shall be permitted to be created by using natural ventilation, by using mechanical ventilation incorporating a vestibule, or by pressurizing the stair enclosure.

**7.2.3.3 Enclosure.**

**7.2.3.3.1** A smokeproof enclosure shall be continuously enclosed by barriers having a 2-hour fire resistance rating from the highest point to the level of exit discharge, except as otherwise permitted in 7.2.3.3.3.

**7.2.3.3.2** Where a vestibule is used, it shall be within the 2-hour-rated enclosure and shall be considered part of the smokeproof enclosure.

**7.2.3.3.3** A smokeproof enclosure comprised of an enclosed stair and serving floors below the level of exit discharge shall not be required to comply with 7.2.3.3.1 where the portion of the stairway below is separated from the stairway enclosure at the level of exit discharge by barriers with a 1-hour fire resistance rating.

**7.2.3.4 Vestibule.** Where a vestibule is provided, the door opening into the vestibule shall be protected with an approved fire door assembly having a minimum 1½-hour fire protection rating, and the fire door assembly from the vestibule to the smokeproof enclosure shall have a minimum 20-minute fire protection rating. Door leaves shall be designed to minimize air leakage and shall be self-closing or shall be automatic-closing by actuation of a smoke detector within 10 ft (3050 mm) of the vestibule door opening. New door assemblies shall be installed in accordance with NFPA 105, *Standard for Smoke Door Assemblies and Other Opening Protectives.*

**7.2.3.5 Discharge.**

**7.2.3.5.1** Every smokeproof enclosure shall discharge into a public way, into a yard or court having direct access to a public way, or into an exit passageway. Such exit passageways shall be without openings, other than the entrance to the smokeproof enclosure and the door opening to the outside yard, court, or public way. The exit passageway shall be separated from the remainder of the building by a 2-hour fire resistance rating.

**7.2.3.5.2** The smokeproof enclosure shall be permitted to discharge through interior building areas, provided that all of the following criteria are met:

(1) The building shall be protected throughout by an approved, supervised automatic sprinkler system in accordance with Section 9.7.
(2) The discharge from the smokeproof enclosure shall lead to a free and unobstructed way to an exterior exit, and such way shall be readily visible and identifiable from the point of discharge from the smokeproof enclosure.
(3) Not more than 50 percent of the required number and capacity of exits comprised of smokeproof enclosures shall discharge through interior building areas in accordance with 7.7.2.

**7.2.3.6 Access.** For smokeproof enclosures other than those consisting of a pressurized enclosure complying with 7.2.3.9, access to the smokeproof enclosure shall be by way of a vestibule or by way of an exterior balcony.

**7.2.3.7 Natural Ventilation.** Smokeproof enclosures using natural ventilation shall comply with 7.2.3.3 and all of the following:

(1) Where access to the enclosure is by means of an open exterior balcony, the door assembly to the enclosure shall have a minimum 1½-hour fire protection rating and shall be self-closing or shall be automatic-closing by actuation of a smoke detector.
(2) Openings adjacent to the exterior balcony specified in 7.2.3.7(1) shall be protected in accordance with 7.2.2.6.4.
(3) Every vestibule shall have a net area of not less than 16 ft² (1.5 m²) of opening in an exterior wall facing an exterior court, yard, or public space not less than 20 ft (6100 mm) in width.
(4) Every vestibule shall have a minimum dimension of not less than the required width of the corridor leading to it and a dimension of not less than 6 ft (1830 mm) in the direction of travel.

**7.2.3.8 Mechanical Ventilation.** Smokeproof enclosures using mechanical ventilation shall comply with 7.2.3.3 and the requirements of 7.2.3.8.1 through 7.2.3.8.4.

**7.2.3.8.1** Vestibules shall have a dimension of not less than 44 in. (1120 mm) in width and not less than 6 ft (1830 mm) in the direction of travel.

**7.2.3.8.2** The vestibule shall be provided with not less than one air change per minute, and the exhaust shall be 150 percent of the supply. Supply air shall enter and exhaust air shall discharge from the vestibule through separate tightly constructed ducts used only for such purposes. Supply air shall enter the vestibule within 6 in. (150 mm) of the floor level. The top of the exhaust register shall be located not more than 6 in. (150 mm) below the top of the trap and shall be entirely within the smoke trap area. Door leaves, when in the open position, shall not obstruct duct openings. Controlling dampers shall be permitted in duct openings if needed to meet the design requirements.

**7.2.3.8.3** To serve as a smoke and heat trap and to provide an upward-moving air column, the vestibule ceiling shall be not less than 20 in. (510 mm) higher than the door opening into the vestibule. The height shall be permitted to be decreased where justified by engineering design and field testing.

**7.2.3.8.4** The stair shall be provided with a dampered relief opening at the top and supplied mechanically with sufficient air to discharge at least 2500 ft³/min (70.8 m³/min) through the relief opening while maintaining a positive pressure of not less than 0.10 in. water column (25 N/m²) in the stair, relative to the vestibule with all door leaves closed.

**7.2.3.9 Enclosure Pressurization.**

**7.2.3.9.1\*** Smokeproof enclosures using pressurization shall use an approved engineered system with a design pressure difference across the barrier of not less than 0.05 in. water column (12.5 N/m²) in sprinklered buildings, or 0.10 in. water column (25 N/m²) in nonsprinklered buildings, and shall be capable of maintaining these pressure differences under likely conditions of stack effect or wind. The pressure difference across door openings shall not exceed that which allows the door leaves to begin to be opened by a force of 30 lbf (133 N) in accordance with 7.2.1.4.5.

**7.2.3.9.2** Equipment and ductwork for pressurization shall be located in accordance with one of the following specifications:

(1) Exterior to the building and directly connected to the enclosure by ductwork enclosed in noncombustible construction
(2) Within the enclosure with intake and exhaust air vented directly to the outside or through ductwork enclosed by a 2-hour fire-resistive rating
(3) Within the building under the following conditions:
    (a) Where the equipment and ductwork are separated from the remainder of the building, including other mechanical equipment, by a 2-hour fire-resistive rating
    (b) Where the building, including the enclosure, is protected throughout by an approved, supervised automatic sprinkler system installed in accordance with Section 9.7, and the equipment and ductwork are separated from the remainder of the building, including other mechanical equipment, by not less than a 1-hour fire-resistive rating

**7.2.3.9.3** In all cases specified by 7.2.3.9.2(1) through (3), openings into the required fire resistance–rated construction shall be limited to those needed for maintenance and operation and shall be protected by self-closing fire protection–rated devices in accordance with 8.3.4.

**7.2.3.10 Activation of Mechanical Ventilation and Pressurized Enclosure Systems.**

**7.2.3.10.1** For both mechanical ventilation and pressurized enclosure systems, the activation of the systems shall be initiated by a smoke detector installed in an approved location within 10 ft (3050 mm) of each entrance to the smokeproof enclosure.

**7.2.3.10.2** The required mechanical system shall operate upon the activation of the smoke detectors specified in 7.2.3.10.1 and by manual controls accessible to the fire department. The required system also shall be initiated by the following, if provided:

(1) Waterflow signal from a complete automatic sprinkler system
(2) General evacuation alarm signal *(see 9.6.3.6)*

**7.2.3.11 Door Leaf Closers.** The activation of an automatic-closing device on any door leaf in the smokeproof enclosure shall activate all other automatic-closing devices on door leaves in the smokeproof enclosure.

**7.2.3.12 Emergency Power Supply System (EPSS).** Power shall be provided as follows:

(1) A Type 60, Class 2, Level 2 EPSS for new mechanical ventilation equipment shall be provided in accordance with NFPA 110, *Standard for Emergency and Standby Power Systems.*
(2) A previously approved existing standby power generator installation with a fuel supply adequate to operate the equipment for 2 hours shall be permitted in lieu of 7.2.3.12(1).
(3) The generator shall be located in a room separated from the remainder of the building by fire barriers having a minimum 1-hour fire resistance rating.

**7.2.3.13 Testing.** Before the mechanical equipment is accepted by the authority having jurisdiction, it shall be tested to confirm that it is operating in compliance with the requirements of 7.2.3. All operating parts of the system shall be tested semiannually by approved personnel, and a log shall be kept of the results.

**7.2.4 Horizontal Exits.**

**7.2.4.1 General.**

**7.2.4.1.1** Where horizontal exits are used in the means of egress, they shall conform to the general requirements of Section 7.1 and the special requirements of 7.2.4.

**7.2.4.1.2\*** Horizontal exits shall be permitted to be substituted for other exits where the total egress capacity and the total number of the other exits (stairs, ramps, door openings leading outside the building) is not less than half that required for the entire area of the building or connected buildings, and provided that none of the other exits is a horizontal exit, unless otherwise permitted by 7.2.4.1.3.

**7.2.4.1.3** The requirement of 7.2.4.1.2 shall not apply to the following:

(1) Health care occupancies as otherwise provided in Chapters 18 and 19
(2) Detention and correctional occupancies as otherwise provided in Chapters 22 and 23

**7.2.4.2 Fire Compartments.**

**7.2.4.2.1** Every fire compartment for which credit is permitted in connection with a horizontal exit(s) also shall have at least one additional exit, but not less than 50 percent of the required number and capacity of exits, that is not a horizontal exit, unless otherwise provided in 7.2.4.2.1.2.

**7.2.4.2.1.1** Any fire compartment not having an exit leading outside shall be considered as part of an adjoining compartment with an exit leading to the outside.

**7.2.4.2.1.2** The requirement of 7.2.4.2.1 shall not apply to the following:

(1) Health care occupancies as otherwise provided in Chapters 18 and 19
(2) Detention and correctional occupancies as otherwise provided in Chapters 22 and 23

**7.2.4.2.2** Every horizontal exit for which credit is permitted shall be arranged so that there are continuously available paths of travel leading from each side of the exit to stairways or other means of egress leading to outside the building.

**7.2.4.2.3** Wherever either side of a horizontal exit is occupied, the door leaves used in connection with the horizontal exit shall be unlocked from the egress side, unless otherwise permitted for the following:

(1) Health care occupancies as provided in Chapters 18 and 19
(2) Detention and correctional occupancies as provided in Chapters 22 and 23

**7.2.4.2.4** The floor area on either side of a horizontal exit shall be sufficient to hold the occupants of both floor areas and shall provide at least 3 ft$^2$ (0.28 m$^2$) clear floor area per person, unless otherwise permitted for the following:

(1) Health care occupancies as provided in Chapters 18 and 19
(2) Detention and correctional occupancies as provided in Chapters 22 and 23

**7.2.4.3 Fire Barriers.**

**7.2.4.3.1** Fire barriers separating buildings or areas between which there are horizontal exits shall have a minimum 2-hour fire resistance rating, unless otherwise provided in 7.2.4.4.1, and shall provide a separation that is continuous to the finished ground level. *(See also Section 8.3.)*

**7.2.4.3.2 Reserved.**

**7.2.4.3.3** Where a fire barrier provides a horizontal exit in any story of a building, such fire barrier shall not be required on other stories, provided that all of the following criteria are met:

(1) The stories on which the fire barrier is omitted are separated from the story with the horizontal exit by construction having a fire resistance rating at least equal to that of the horizontal exit fire barrier.
(2) Vertical openings between the story with the horizontal exit and the open fire area story are enclosed with construction having a fire resistance rating at least equal to that of the horizontal exit fire barrier.
(3) All required exits, other than horizontal exits, discharge directly to the outside.

**7.2.4.3.4** Where fire barriers serving horizontal exits, other than existing horizontal exits, terminate at outside walls, and the outside walls are at an angle of less than 180 degrees for a distance of 10 ft (3050 mm) on each side of the horizontal exit, the outside walls shall have a minimum 1-hour fire resistance rating, with opening protectives having a minimum ¾-hour fire protection rating, for a distance of 10 ft (3050 mm) on each side of the horizontal exit.

**7.2.4.3.5** Fire barriers forming horizontal exits shall not be penetrated by ducts, unless one of the following criteria is met:

(1) The ducts are existing penetrations protected by approved and listed fire dampers.
(2) The building is protected throughout by an approved, supervised automatic sprinkler system in accordance with Section 9.7.
(3) The duct penetrations are those permitted in detention and correctional occupancies as otherwise provided in Chapters 22 and 23 and are protected by combination fire dampers/smoke leakage–rated dampers that meet the smoke damper actuation requirements of 8.5.5.

**7.2.4.3.6** Any opening in the fire barriers specified in 7.2.4.3.5 shall be protected as provided in 8.3.4.

**7.2.4.3.7** Door assemblies in horizontal exits shall comply with 7.2.1.4, unless they are sliding door assemblies in industrial or storage occupancies as otherwise provided in Chapters 40 and 42.

**7.2.4.3.8** Unless otherwise specified in 7.2.4.3.8.1 and 7.2.4.3.8.2, swinging fire door assemblies shall be permitted in horizontal exits, provided that the criteria of both 7.2.4.3.8(1) and (2), or the criteria of both 7.2.4.3.8(1) and (3), are met as follows:

(1) The door leaves shall swing in the direction of egress travel.
(2) In other than sleeping room areas in detention and correctional occupancies, where a horizontal exit serves areas on both sides of a fire barrier, adjacent openings with swinging door leaves that open in opposite directions shall be provided, with signs on each side of the fire barrier identifying the door leaf that swings with the travel from that side.
(3) The door assemblies shall be of any other approved arrangement, provided that the door leaves always swing with any possible egress travel.

**7.2.4.3.8.1** The requirements of 7.2.4.3.8 shall not apply to horizontal exit door leaf swing as provided in Chapters 19 and 23.

**7.2.4.3.8.2** The requirements of 7.2.4.3.8 shall not apply to horizontal exit door assemblies in corridors not more than 6 ft (1830 mm) wide in existing buildings.

**7.2.4.3.9** Door leaves in horizontal exits shall be designed and installed to minimize air leakage. New door assemblies in horizontal exits shall be installed in accordance with NFPA 105, *Standard for Smoke Door Assemblies and Other Opening Protectives*.

**7.2.4.3.10\*** All fire door assemblies in horizontal exits shall be self-closing or automatic-closing in accordance with 7.2.1.8.

**7.2.4.3.11** Horizontal exit door assemblies located across a corridor, other than approved existing door assemblies, shall be automatic-closing in accordance with 7.2.1.8.2.

**7.2.4.4 Bridges Serving Horizontal Exits Between Buildings.** The provisions of 7.2.4.4 shall apply to bridges serving horizontal exits between buildings and to the associated horizontal exit fire barrier.

**7.2.4.4.1** The minimum 2-hour fire resistance–rated barrier required by 7.2.4.3.1 shall extend as follows:

(1) Vertically from the ground to a point 10 ft (3050 mm) above the bridge or to the roofline, whichever is lower
(2) Horizontally for not less than 10 ft (3050 mm) to each side of the bridge

**7.2.4.4.2** Any opening in the fire barrier addressed in 7.2.4.4.1 shall be protected with fire door assemblies or fixed fire window assemblies having a ¾-hour fire protection rating, unless otherwise provided in 7.2.4.4.3.

**7.2.4.4.3** The requirement of 7.2.4.4.2 shall not apply to approved existing bridges.

**7.2.4.4.4** Where the bridge serves as a horizontal exit in one direction, the horizontal exit door leaf shall be required to swing only in the direction of egress travel, unless the door leaf complies with the swing requirements for the following:

(1) Existing health care occupancies in Chapter 19
(2) Existing detention and correctional occupancies in Chapter 23

**7.2.4.4.5** Where the bridge serves as a horizontal exit in both directions, door leaves shall be provided in pairs that swing in opposite directions, with only the door leaf swinging in the direction of egress travel included when determining egress capacity, unless otherwise provided in 7.2.4.4.5.1 through 7.2.4.4.5.3.

**7.2.4.4.5.1** Approved existing door assemblies on both ends of the bridge shall be permitted to swing out from the building.

**7.2.4.4.5.2** The requirement of 7.2.4.4.5 shall not apply to existing bridges if the bridge has sufficient floor area to accommodate the occupant load of either connected building or fire area based on 3 ft² (0.28 m²) per person.

**7.2.4.4.5.3** The requirement of 7.2.4.4.5 shall not apply to horizontal exit door leaf swing as provided for the following:

(1) Existing health care occupancies in Chapter 19
(2) Existing detention and correctional occupancies in Chapter 23

**7.2.4.4.6** Every bridge shall be not less than the width of the door opening to which it leads and shall be not less than 44 in. (1120 mm) wide for new construction.

**7.2.4.4.7** In climates subject to the accumulation of snow and ice, the bridge floor shall be protected to prevent the accumulation of snow and ice.

**7.2.4.4.8** In existing buildings, one step not exceeding 8 in. (205 mm) shall be permitted below the level of the inside floor.

**7.2.5 Ramps.**

**7.2.5.1 General.** Every ramp used as a component in a means of egress shall conform to the general requirements of Section 7.1 and to the special requirements of 7.2.5.

**7.2.5.2 Dimensional Criteria.** The following dimensional criteria shall apply to ramps:

(1) New ramps shall be in accordance with Table 7.2.5.2(a), unless otherwise permitted by the following:
   (a) Table 7.2.5.2(a) shall not apply to industrial equipment access areas as provided in 40.2.5.2.
   (b) The maximum slope requirement shall not apply to ramps in assembly occupancies as provided in Chapter 12.
   (c) The maximum slope or maximum rise for a single ramp run shall not apply to ramps providing access to vehicles, vessels, mobile structures, and aircraft.
(2) Existing ramps shall be permitted to remain in use or be rebuilt, provided that they meet the requirements shown

in Table 7.2.5.2(b), unless otherwise permitted by any of the following:

(a) The requirements of Table 7.2.5.2(b) shall not apply to industrial equipment access areas as provided in 40.2.5.2.

(b) The maximum slope or maximum height between landings for a single ramp run shall not apply to ramps providing access to vehicles, vessels, mobile structures, and aircraft.

(c) Approved existing ramps with slopes not steeper than 1 in 6 shall be permitted to remain in use.

(d) Existing ramps with slopes not steeper than 1 in 10 shall not be required to be provided with landings.

**Table 7.2.5.2(a) New Ramps**

| Feature | Dimensional Criteria | |
| --- | --- | --- |
| | in. | mm |
| Minimum width clear of all obstructions, except projections not more than 4½ in. (114 mm) at or below handrail height on each side | 44 | 1120 |
| Maximum slope | 1 in 12 | |
| Maximum cross slope | 1 in 48 | |
| Maximum rise for a single ramp run | 30 | 760 |

**Table 7.2.5.2(b) Existing Ramps**

| Feature | Dimensional Criteria | |
| --- | --- | --- |
| | ft/in. | mm |
| Minimum width | 30 in. | 760 |
| Maximum slope | 1 in 8 | |
| Maximum height between landings | 12 ft | 3660 |

**7.2.5.3 Ramp Details.**

**7.2.5.3.1 Construction.** Ramp construction shall be as follows:

(1) All ramps serving as required means of egress shall be of permanent fixed construction.

(2) Each ramp in buildings required by this *Code* to be of Type I or Type II construction shall be any combination of noncombustible or limited-combustible material or fire-retardant-treated wood.

(3) Ramps constructed with fire-retardant-treated wood shall be not more than 30 in. (760 mm) high, shall have an area of not more than 3000 ft² (277 m²), and shall not occupy more than 50 percent of the room area.

(4) The ramp floor and landings shall be solid and without perforations.

**7.2.5.3.2 Landings.** Ramp landings shall be as follows:

(1) Ramps shall have landings located at the top, at the bottom, and at door leaves opening onto the ramp.

(2) The slope of the landing shall be not steeper than 1 in 48.

(3) Every landing shall have a width not less than the width of the ramp.

(4) Every landing, except as otherwise provided in 7.2.5.3.2(5), shall be not less than 60 in. (1525 mm) long in the direction of travel, unless the landing is an approved existing landing.

(5) Where the ramp is not part of an accessible route, the ramp landings shall not be required to exceed 48 in. (1220 mm) in the direction of travel, provided that the ramp has a straight run.

(6) Any changes in travel direction shall be made only at landings, unless the ramp is an existing ramp.

(7) Ramps and intermediate landings shall continue with no decrease in width along the direction of egress travel.

**7.2.5.3.3 Drop-Offs.** Ramps and landings with drop-offs shall have curbs, walls, railings, or projecting surfaces that prevent people from traveling off the edge of the ramp. Curbs or barriers shall be not less than 4 in. (100 mm) in height.

**7.2.5.4 Guards and Handrails.**

**7.2.5.4.1** Guards complying with 7.2.2.4 shall be provided for ramps, unless otherwise provided in 7.2.5.4.4.

**7.2.5.4.2** Handrails complying with 7.2.2.4 shall be provided along both sides of a ramp run with a rise greater than 6 in. (150 mm), unless otherwise provided in 7.2.5.4.4.

**7.2.5.4.3** The height of handrails and guards shall be measured vertically to the top of the guard or rail from the walking surface adjacent thereto.

**7.2.5.4.4** The requirements of 7.2.5.4.1 and 7.2.5.4.2 shall not apply to guards and handrails provided for ramped aisles in assembly occupancies as otherwise provided in Chapters 12 and 13.

**7.2.5.5 Enclosure and Protection of Ramps.** Ramps in a required means of egress shall be enclosed or protected as a stair in accordance with 7.2.2.5 and 7.2.2.6.

**7.2.5.6 Special Provisions for Outside Ramps.**

**7.2.5.6.1\* Visual Protection.** Outside ramps shall be arranged to avoid any impediments to their use by persons having a fear of high places. Outside ramps more than 36 ft (11 m) above the finished ground level shall be provided with an opaque visual obstruction not less than 48. in. (1220 mm) in height.

**7.2.5.6.2\* Water Accumulation.** Outside ramps and landings shall be designed to minimize water accumulation on their surfaces.

**7.2.6\* Exit Passageways.**

**7.2.6.1\* General.** Exit passageways used as exit components shall conform to the general requirements of Section 7.1 and to the special requirements of 7.2.6.

**7.2.6.2 Enclosure.** An exit passageway shall be separated from other parts of the building as specified in 7.1.3.2, and the following alternatives shall be permitted:

(1) Fire windows in accordance with 8.3.3 shall be permitted to be installed in the separation in a building protected throughout by an approved, supervised automatic sprinkler system in accordance with Section 9.7.

(2) Existing fixed wired glass panels in steel sash shall be permitted to be continued in use in the separation in buildings protected throughout by an approved, supervised automatic sprinkler system in accordance with Section 9.7.

**7.2.6.3 Stair Discharge.** An exit passageway that serves as a discharge from a stair enclosure shall have not less than the same fire resistance rating and opening protective fire protection rating as those required for the stair enclosure.

**7.2.6.4 Width.**

**7.2.6.4.1** The width of an exit passageway shall be sized to accommodate the aggregate required capacity of all exits that discharge through it, unless one of the following conditions applies:

(1)*Where an exit passageway serves occupants of the level of exit discharge as well as other stories, the capacity shall not be required to be aggregated.
(2) As provided in Chapters 36 and 37, an exit passageway in a mall building shall be permitted to accommodate occupant loads independently from the mall and the tenant spaces. *(See 36.2.2.7.2 and 37.2.2.7.2.)*

**7.2.6.4.2** In new construction, the minimum width of any exit passageway into which an exit stair discharges, or that serves as a horizontal transfer within an exit stair system, shall meet the following criteria:

(1) The minimum width of the exit passageway shall be not less than two-thirds of the width of the exit stair.
(2) Where stairs are credited with egress capacity in accordance with 7.3.3.2, the exit passageway width shall be sized to accommodate the same capacity as the stair, with such capacity determined by use of the capacity factors in Table 7.3.3.1.

**7.2.6.5 Floor.** The floor shall be solid and without perforations.

**7.2.7 Escalators and Moving Walks.** Escalators and moving walks shall not constitute a part of the required means of egress, unless they are previously approved existing escalators and moving walks.

**7.2.8 Fire Escape Stairs.**

**7.2.8.1 General.**

**7.2.8.1.1** Fire escape stairs shall comply with the provisions of 7.2.8, unless they are approved existing fire escape stairs.

**7.2.8.1.2** Fire escape stairs shall not constitute any of the required means of egress, unless otherwise provided in 7.2.8.1.2.1 and 7.2.8.1.2.2.

**7.2.8.1.2.1** Fire escape stairs shall be permitted on existing buildings as provided in Chapters 11 through 43 but shall not constitute more than 50 percent of the required means of egress.

**7.2.8.1.2.2** New fire escape stairs shall be permitted to be erected on existing buildings only where the authority having jurisdiction has determined that outside stairs are impractical. *(See 7.2.2.)*

**7.2.8.1.2.3** New fire escape stairs permitted by 7.2.8.1.2.2 shall not incorporate ladders or access windows, regardless of occupancy classification or occupant load served.

**7.2.8.1.3** Fire escape stairs of the return-platform type with superimposed runs, or of the straight-run type with a platform that continues in the same direction, shall be permitted. Either type shall be permitted to be parallel to, or at right angles to, buildings. Either type shall be permitted to be attached to buildings or erected independently of buildings and connected by walkways.

**7.2.8.2 Protection of Openings.** Fire escape stairs shall be exposed to the smallest possible number of window and door openings, and each opening shall be protected with approved fire door or fire window assemblies where the opening or any portion of the opening is located as follows:

(1) Horizontally, within 15 ft (4570 mm) of any balcony, platform, or stairway constituting a component of the fire escape stair
(2) Below, within three stories or 36 ft (11 m) of any balcony, platform, walkway, or stairway constituting a component of the fire escape stair, or within two stories or 24 ft (7320 mm) of a platform or walkway leading from any story to the fire escape stair
(3) Above, within 10 ft (3050 mm) of any balcony, platform, or walkway, as measured vertically, or within 10 ft (3050 mm) of any stair tread surface, as measured vertically
(4) Facing a court served by a fire escape stair, where the least dimension of the court does not exceed one-third of the height to the uppermost platform of the fire escape stair, measured from the finished ground level
(5) Facing an alcove served by a fire escape stair, where the width of the alcove does not exceed one-third, or the depth of the alcove does not exceed one-fourth, of the height to the uppermost platform of the fire escape stair, measured from the finished ground level

**7.2.8.2.1** The requirements of 7.2.8.2 shall not apply to openings located on the top story where stairs do not lead to the roof.

**7.2.8.2.2** The requirements of 7.2.8.2 shall be permitted to be modified by the authority having jurisdiction where automatic sprinkler protection is provided, where the occupancy is limited to low hazard contents, or where other special conditions exist.

**7.2.8.2.3** The requirements of 7.2.8.2 for the protection of window openings shall not apply where such window openings are necessary for access to existing fire escape stairs.

**7.2.8.3 Access.**

**7.2.8.3.1** Access to fire escape stairs shall be in accordance with 7.2.8.4 and 7.5.1.1.1 through 7.5.1.2.2.

**7.2.8.3.2** Where access is permitted by way of windows, the windows shall be arranged and maintained so as to be easily opened. Screening or storm windows that restrict free access to the fire escape stair shall be prohibited.

**7.2.8.3.3** Fire escape stairs shall extend to the roof in all cases where the roof is subject to occupancy or provides an area of safe refuge, unless otherwise provided in 7.2.8.3.4.

**7.2.8.3.4** Where a roof has a pitch that does not exceed 1 to 6, fire escape ladders in accordance with 7.2.9 or alternating tread devices in accordance with 7.2.11 shall be permitted to provide access to the roof.

**7.2.8.3.5** Access to a fire escape stair shall be directly to a balcony, landing, or platform; shall not exceed the floor or windowsill level; and shall not be more than 8 in. (205 mm) below the floor level or 18 in. (455 mm) below the windowsill level.

**7.2.8.4 Stair Details.**

**7.2.8.4.1 General.** Fire escape stairs shall comply with the requirements of Table 7.2.8.4.1(a). Replacement of fire escape stairs shall comply with the requirements of Table 7.2.8.4.1(b).

**Table 7.2.8.4.1(a) Fire Escape Stairs**

| Feature | Serving More Than 10 Occupants | Serving 10 or Fewer Occupants |
|---|---|---|
| Minimum widths | 22 in. (560 mm) clear between rails | 18 in. (455 mm) clear between rails |
| Minimum horizontal dimension of any landing or platform | 22 in. (560 mm) clear | 18 in. (455 mm) clear |
| Maximum riser height | 9 in. (230 mm) | 12 in. (305 mm) |
| Minimum tread, exclusive of nosing | 9 in. (230 mm) | 6 in. (150 mm) |
| Minimum nosing or projection | 1 in. (25 mm) | No requirement |
| Tread construction | Solid ½ in. (13 mm) diameter perforations permitted | Flat metal bars on edge or square bars secured against turning, spaced 1¼ in. (32 mm) maximum on centers |
| Winders | None | Permitted subject to capacity penalty |
| Risers | None | No requirement |
| Spiral | None | Permitted subject to capacity penalty |
| Maximum height between landings | 12 ft (3660 mm) | No requirement |
| Minimum headroom | 6 ft 8 in. (2030 mm) | 6 ft 8 in. (2030 mm) |
| Access to escape | Door or casement windows, 24 in. × 6 ft 8 in. (610 mm × 1980 mm); or double-hung windows, 30 in. × 36 in. (760 mm × 915 mm) clear opening | Windows providing a clear opening of at least 20 in. (510 mm) in width, 24 in. (610 mm) in height, and 5.7 ft² (0.53 m²) in area |
| Level of access opening | Not over 12 in. (305 mm) above floor; steps if higher | Not over 12 in. (305 mm) above floor; steps if higher |
| Discharge to the finished ground level | Swinging stair section permitted if approved by authority having jurisdiction | Swinging stair, or ladder if approved by authority having jurisdiction |
| Capacity | ½ in. (13 mm) per person, if access by door; 1 in. (25 mm) per person, if access by climbing over windowsill | 10 persons; if winders or ladder from bottom balcony, 5 persons; if both, 1 person |

**7.2.8.4.2 Slip Resistance.** Stair treads and landings of new or replacement fire escape stairs shall have slip-resistant surfaces.

**7.2.8.5 Guards, Handrails, and Visual Enclosures.**

**7.2.8.5.1** All fire escape stairs shall have walls or guards and handrails on both sides in accordance with 7.2.2.4.

**7.2.8.5.2** Replacement fire escape stairs in occupancies serving more than 10 occupants shall have visual enclosures to avoid any impediments to their use by persons having a fear of high places. Fire escape stairs more than 36 ft (11 m) above the finished ground level shall be provided with an opaque visual obstruction not less than 48 in. (1220 mm) in height.

**7.2.8.6 Materials and Strength.**

**7.2.8.6.1** Noncombustible materials shall be used for the construction of all components of fire escape stairs.

**7.2.8.6.2** The authority having jurisdiction shall be permitted to approve any existing fire escape stair that has been shown by load test or other satisfactory evidence to have adequate strength.

**7.2.8.7\* Swinging Stairs.**

**7.2.8.7.1** A single swinging stair section shall be permitted to terminate fire escape stairs over sidewalks, alleys, or driveways where it is impractical to make the termination with fire escape stairs.

**7.2.8.7.2** Swinging stair sections shall not be located over doors, over the path of travel from any other exit, or in any locations where there are likely to be obstructions.

**7.2.8.7.3** The width of swinging stair sections shall be at least that of the fire escape stairs above.

**7.2.8.7.4** The pitch of swinging stair sections shall not exceed the pitch of the fire escape stairs above.

**Table 7.2.8.4.1(b)  Replacement Fire Escape Stairs**

| Feature | Serving More Than 10 Occupants | Serving 10 or Fewer Occupants |
|---|---|---|
| Minimum widths | 22 in. (560 mm) clear between rails | 22 in. (560 mm) clear between rails |
| Minimum horizontal dimension of any landing or platform | 22 in. (560 mm) | 22 in. (560 mm) |
| Maximum riser height | 9 in. (230 mm) | 9 in. (230 mm) |
| Minimum tread, exclusive of nosing | 10 in. (255 mm) | 10 in. (255 mm) |
| Tread construction | Solid, ½ in. (13 mm) diameter perforations permitted | Solid, ½ in. (13 mm) diameter perforations permitted |
| Winders | None | Permitted subject to 7.2.2.2.4 |
| Spiral | None | Permitted subject to 7.2.2.2.3 |
| Risers | None | None |
| Maximum height between landings | 12 ft (3660 mm) | 12 ft (3660 mm) |
| Minimum headroom | 6 ft 8 in. (2030 mm) | 6 ft 8 in. (2030 mm) |
| Access to escape | Door or casement windows, 24 in. × 6 ft 8 in. (610 mm × 1980 mm); or double-hung windows, 30 in. × 36 in. (760 mm × 915 mm) clear opening | Windows providing a clear opening of at least 20 in. (510 mm) in width, 24 in. (610 mm) in height, and 5.7 ft$^2$ (0.53 m$^2$) in area |
| Level of access opening | Not over 12 in. (305 mm) above floor; steps if higher | Not over 12 in. (305 mm) above floor; steps if higher |
| Discharge to the finished ground level | Swinging stair section permitted if approved by authority having jurisdiction | Swinging stair section permitted if approved by authority having jurisdiction |
| Capacity | ½ in. (13 mm) per person, if access by door; 1 in. (25 mm) per person, if access by climbing over windowsill | 10 persons |

**7.2.8.7.5** Guards and handrails shall be provided in accordance with 7.2.2.4 and shall be similar in height and construction to those used with the fire escape stairs above. Guards and handrails shall be designed to prevent any possibility of injury to persons where stairs swing downward. The clearance between moving sections and any other portion of the stair system where hands have the potential to be caught shall be not less than 4 in. (100 mm).

**7.2.8.7.6** If the distance from the lowest platform to the finished ground level is not less than 12 ft (3660 mm), an intermediate balcony not more than 12 ft (3660 mm) from the finished ground level and not less than 7 ft (2135 mm) in the clear underneath shall be provided, with width not less than that of the stairs and length not less than 48 in. (1220 mm).

**7.2.8.7.7** Swinging stairs shall be counterbalanced about a pivot, and cables shall not be used. A weight of 150 lb (68 kg) located one step from the pivot shall not cause the stairs to swing downward, and a weight of 150 lb (68 kg) located one-quarter of the length of the swinging stairs from the pivot shall cause the stairs to swing down.

**7.2.8.7.8** The pivot for swinging stairs shall be of a corrosion-resistant assembly or shall have clearances to prevent sticking due to corrosion.

**7.2.8.7.9\*** Devices shall not be installed to lock a swinging stair section in the up position.

**7.2.8.8 Intervening Spaces.**

**7.2.8.8.1** Where approved by the authority having jurisdiction, fire escape stairs shall be permitted to lead to an adjoining roof that is crossed before continuing downward travel. The direction of travel shall be clearly marked, and walkways with guards and handrails complying with 7.2.2.4 shall be provided.

**7.2.8.8.2** Where approved by the authority having jurisdiction, fire escape stairs shall be permitted to be used in combination with inside or outside stairs complying with 7.2.2, provided that a continuous safe path of travel is maintained.

**7.2.9 Fire Escape Ladders.**

**7.2.9.1 General.** Fire escape ladders complying with 7.2.9.2 and 7.2.9.3 shall be permitted in the means of egress only where providing one of the following:

(1) Access to unoccupied roof spaces as permitted in 7.2.8.3.4
(2) Second means of egress from storage elevators as permitted in Chapter 42
(3) Means of egress from towers and elevated platforms around machinery or similar spaces subject to occupancy not to exceed three persons who are all capable of using the ladder
(4) Secondary means of egress from boiler rooms or similar spaces subject to occupancy not to exceed three persons who are all capable of using the ladder
(5) Access to the finished ground level from the lowest balcony or landing of a fire escape stair for small buildings as permitted in 7.2.8.4 where approved by the authority having jurisdiction

**7.2.9.2 Construction and Installation.**

**7.2.9.2.1** Fire escape ladders shall comply with ANSI A14.3, *Safety Requirements for Fixed Ladders*, unless one of the following criteria is met:

(1) Approved existing ladders complying with the edition of this *Code* that was in effect when the ladders were installed shall be permitted.
(2) Industrial stairs complying with the minimum requirements for fixed stairs of ANSI/ASSE A1264.1, *Safety Requirements for Workplace Floor and Wall Openings, Stairs and Railing Systems*, shall be permitted where fire escape ladders are permitted in accordance with Chapter 40.

**7.2.9.2.2** Ladders shall be installed with a pitch that exceeds 75 degrees.

**7.2.9.3 Access.** The lowest rung of any ladder shall be not more than 12 in. (305 mm) above the level of the surface beneath it.

**7.2.10 Slide Escapes.**

**7.2.10.1 General.**

**7.2.10.1.1** A slide escape shall be permitted as a component in a means of egress where permitted in Chapters 11 through 43.

**7.2.10.1.2** Each slide escape shall be of an approved type.

**7.2.10.2 Capacity.**

**7.2.10.2.1** Slide escapes, where permitted as a required means of egress, shall be rated at a capacity of 60 persons.

**7.2.10.2.2** Slide escapes shall not constitute more than 25 percent of the required egress capacity from any building or structure or any individual story thereof, unless otherwise provided for industrial occupancies in Chapter 40.

**7.2.11* Alternating Tread Devices.**

**7.2.11.1** Alternating tread devices complying with 7.2.11.2 shall be permitted in the means of egress only where providing one of the following:

(1) Access to unoccupied roof spaces as permitted in 7.2.8.3.4
(2) Second means of egress from storage elevators as permitted in Chapter 42
(3) Means of egress from towers and elevated platforms around machinery or similar spaces subject to occupancy not to exceed three persons who are all capable of using the alternating tread device
(4) Secondary means of egress from boiler rooms or similar spaces subject to occupancy not to exceed three persons who are all capable of using the alternating tread device

**7.2.11.2** Alternating tread devices shall comply with all of the following:

(1) Handrails shall be provided on both sides of alternating tread devices in accordance with 7.2.2.4.4, except as provided in 7.2.11.3.
(2) The clear width between handrails shall be not less than 17 in. (430 mm) and not more than 24 in. (610 mm).
(3) Headroom shall be not less than 6 ft 8 in. (2030 mm).
(4) The angle of the device shall be between 50 degrees and 68 degrees to horizontal.
(5) The height of the riser shall not exceed 9½ in. (240 mm).
(6) Treads shall have a projected tread depth of not less than 5⅘ in. (145 mm), measured in accordance with 7.2.2, with each tread providing 9½ in. (240 mm) of depth, including tread overlap.
(7) A distance of not less than 6 in. (150 mm) shall be provided between the alternating tread device handrail and any other object.
(8) The initial tread of the alternating tread device shall begin at the same elevation as the platform, landing, or floor surface.
(9) The alternating treads shall not be laterally separated by a distance of more than 2 in. (51 mm).
(10) The occupant load served shall not exceed three.

**7.2.11.3** Handrails of alternating tread devices shall comply with the following:

(1) The handrail height of alternating tread devices, measured above tread nosings, shall be uniform, not less than 30 in. (760 mm), and not more than 34 in. (865 mm).
(2) Handrails for alternating tread devices shall be permitted to terminate at a location vertically above the top and bottom risers.
(3) Handrails for alternating tread devices shall not be required to be continuous between flights or to extend beyond the top or bottom risers.
(4) Alternating tread device guards, with a top rail that also serves as a handrail, shall have a height of not less than 30 in. (760 mm), and not more than 34 in. (865 mm), measured vertically from the leading edge of the device tread nosing.
(5) Open guards of alternating tread devices shall have rails such that a sphere 21 in. (535 mm) in diameter is not able to pass through any opening.

**7.2.12 Areas of Refuge.**

**7.2.12.1 General.**

**7.2.12.1.1** An area of refuge used as part of a required accessible means of egress in accordance with 7.5.4; consisting of a story in a building that is protected throughout by an approved, supervised automatic sprinkler system in accordance with Section 9.7; and having an accessible story that is one or more stories above or below a story of exit discharge shall meet the following criteria:

(1) Each elevator landing shall be provided with a two-way communication system for communication between the elevator landing and the fire command center or a central control point approved by the authority having jurisdiction.
(2) Directions for the use of the two-way communication system, instructions for summoning assistance via the two-way communication system, and written identification of the location shall be posted adjacent to the two-way communication system.
(3) The two-way communication system shall include both audible and visible signals.

**7.2.12.1.2** An area of refuge used as part of a required accessible means of egress in accordance with 7.5.4 in other than a building that is protected throughout by an approved, supervised automatic sprinkler system in accordance with Section 9.7 shall meet both of the following criteria:

(1) The area of refuge shall meet the general requirements of Section 7.1.
(2) The area of refuge shall meet the requirements of 7.2.12.2 and 7.2.12.3.

**7.2.12.2 Accessibility.**

**7.2.12.2.1** Required portions of an area of refuge shall be accessible from the space they serve by an accessible means of egress.

**7.2.12.2.2** Required portions of an area of refuge shall have access to a public way via an exit or an elevator without requiring return to the building spaces through which travel to the area of refuge occurred.

**7.2.12.2.3*** Where the exit providing egress from an area of refuge to a public way that is in accordance with 7.2.12.2.2 includes stairs, the clear width of landings and stair flights, measured between handrails and at all points below handrail height, shall be not less than 48 in. (1220 mm), unless otherwise permitted by the following:

(1) The minimum 48 in. (1220 mm) clear width shall not be required where the area of refuge is separated from the remainder of the story by a horizontal exit meeting the requirements of 7.2.4. *(See also 7.2.12.3.4.)*
(2) Existing stairs and landings that provide a clear width of not less than 37 in. (940 mm), measured at and below handrail height, shall be permitted.

**7.2.12.2.4*** Where an elevator provides access from an area of refuge to a public way that is in accordance with 7.2.12.2.2, all of the following criteria shall be met:

(1) The elevator shall be approved for fire fighters' emergency operations as provided in ASME A17.1/CSA B44, *Safety Code for Elevators and Escalators.*
(2) The power supply shall be protected against interruption from fire occurring within the building but outside the area of refuge.
(3) The elevator shall be located in a shaft system meeting the requirements for smokeproof enclosures in accordance with 7.2.3, unless otherwise provided in 7.2.12.2.4.1 and 7.2.12.2.4.2.

**7.2.12.2.4.1** The smokeproof enclosure specified in 7.2.12.2.4(3) shall not be required for areas of refuge that are more than 1000 ft$^2$ (93 m$^2$) and that are created by a horizontal exit meeting the requirements of 7.2.4.

**7.2.12.2.4.2** The smokeproof enclosure specified in 7.2.12.2.4(3) shall not be required for elevators complying with 7.2.13.

**7.2.12.2.5** The area of refuge shall be provided with a two-way communication system for communication between the area of refuge and a central control point. The door opening to the stair enclosure or the elevator door and the associated portion of the area of refuge that the stair enclosure door opening or elevator door serves shall be identified by signage. *(See 7.2.12.3.5.)*

**7.2.12.2.6*** Instructions for summoning assistance, via the two-way communication system, and written identification of the area of refuge location shall be posted adjacent to the two-way communication system.

**7.2.12.3 Details.**

**7.2.12.3.1*** Each area of refuge shall be sized to accommodate one wheelchair space of 30 in. × 48 in. (760 mm × 1220 mm) for every 200 occupants, or portion thereof, based on the occupant load served by the area of refuge. Such wheelchair spaces shall maintain the width of a means of egress to not less than that required for the occupant load served and to not less than 36 in. (915 mm).

**7.2.12.3.2*** For any area of refuge that does not exceed 1000 ft$^2$ (93 m$^2$), it shall be demonstrated by calculation or test that tenable conditions are maintained within the area of refuge for a period of 15 minutes when the exposing space on the other side of the separation creating the area of refuge is subjected to the maximum expected fire conditions.

**7.2.12.3.3** Access to any designated wheelchair space in an area of refuge shall not pass through more than one adjoining wheelchair space.

**7.2.12.3.4*** Each area of refuge shall be separated from the remainder of the story by a barrier having a minimum 1-hour fire resistance rating, unless one of the following criteria applies:

(1) A greater rating is required in other provisions of this *Code.*
(2) The barrier is an existing barrier with a minimum 30-minute fire resistance rating.

**7.2.12.3.4.1** New fire door assemblies serving an area of refuge shall be smoke leakage–rated in accordance with 8.2.2.4.

**7.2.12.3.4.2** The barriers specified in 7.2.12.3.4, and any openings in them, shall minimize air leakage and resist the passage of smoke.

**7.2.12.3.4.3** Door assemblies in the barriers specified in 7.2.12.3.4 shall have not less than a 20-minute fire protection rating, unless a greater rating is required in other provisions of this *Code,* and shall be either self-closing or automatic-closing in accordance with 7.2.1.8.

**7.2.12.3.4.4** Ducts shall be permitted to penetrate the barrier specified in 7.2.12.3.4, unless prohibited in other provisions of this *Code,* and shall be provided with smoke-actuated dampers or other approved means to resist the transfer of smoke into the area of refuge.

**7.2.12.3.5** Each area of refuge shall be identified by a sign that reads as the follows:

### AREA OF REFUGE

**7.2.12.3.5.1** The sign required by 7.2.12.3.5 shall conform to the requirements of ICC/ANSI A117.1, *American National Standard for Accessible and Usable Buildings and Facilities,* for such signage and shall display the international symbol of accessibility. Signs also shall be located as follows:

(1) At each door opening providing access to the area of refuge
(2) At all exits not providing an accessible means of egress, as defined in 3.3.170.1
(3) Where necessary to indicate clearly the direction to an area of refuge

**7.2.12.3.5.2** Signs required by 7.2.12.3.5 shall be illuminated as required for exit signs where exit sign illumination is required.

**7.2.12.3.6** Tactile signage complying with ICC/ANSI A117.1, *American National Standard for Accessible and Usable Buildings and Facilities*, shall be located at each door opening to an area of refuge.

### 7.2.13 Elevators.

**7.2.13.1\* General.** An elevator complying with the requirements of Section 9.4 and 7.2.13 shall be permitted to be used as a second means of egress from a tower, as defined in 3.3.280, provided that all of the following criteria are met:

(1) The tower and any attached structure shall be protected throughout by an approved, supervised automatic sprinkler system in accordance with Section 9.7.
(2) The tower shall be subject to occupancy not to exceed 90 persons.
(3) Primary egress discharges shall be directly to the outside.
(4) No high hazard content areas shall exist in the tower or attached structure.
(5) One hundred percent of the egress capacity shall be provided independent of the elevators.
(6) An evacuation plan that specifically includes the elevator shall be implemented, and staff personnel shall be trained in operations and procedures for elevator emergency use in normal operating mode prior to fire fighter recall.
(7) The tower shall not be used by the general public.

**7.2.13.2 Elevator Evacuation System Capacity.**

**7.2.13.2.1** The elevator car shall have a capacity of not less than eight persons.

**7.2.13.2.2** The elevator lobby shall have a capacity of not less than 50 percent of the occupant load of the area served by the lobby. The capacity shall be calculated based on 3 ft² (0.28 m²) per person and shall also include one wheelchair space of 30 in. × 48 in. (760 mm × 1220 mm) for every 50 persons, or portion thereof, of the total occupant load served by that lobby.

**7.2.13.3 Elevator Lobby.** Every floor served by the elevator shall have an elevator lobby. Barriers forming the elevator lobby shall have a minimum 1-hour fire resistance rating and shall be arranged as a smoke barrier in accordance with Section 8.5.

**7.2.13.4 Elevator Lobby Door Assemblies.** Elevator lobby door assemblies shall have a minimum 1-hour fire protection rating. The transmitted temperature end point shall not exceed 450°F Δ (250°C Δ) above ambient at the end of 30 minutes of the fire exposure specified in the test method referenced in 8.3.3.2. Elevator lobby door leaves shall be self-closing or automatic-closing in accordance with 7.2.1.8.

**7.2.13.5 Door Leaf Activation.** The elevator lobby door leaves shall close in response to a signal from a smoke detector located directly outside the elevator lobby adjacent to or on each door opening. Elevator lobby door leaves shall be permitted to close in response to a signal from the building fire alarm system. Where one elevator lobby door leaf closes by means of a smoke detector or a signal from the building fire alarm system, all elevator lobby door leaves serving that elevator evacuation system shall close.

**7.2.13.6\* Water Protection.** Building elements shall be used to restrict water exposure of elevator equipment.

**7.2.13.7\* Power and Control Wiring.** Elevator equipment, elevator communications, elevator machine room cooling, and elevator controller cooling shall be supplied by both normal and standby power. Wiring for power and control shall be located and properly protected to ensure a minimum 1 hour of operation in the event of a fire.

**7.2.13.8\* Communications.** Two-way communication systems shall be provided between elevator lobbies and a central control point and between elevator cars and a central control point. Communications wiring shall be protected to ensure a minimum 1 hour of operation in the event of fire.

**7.2.13.9\* Elevator Operation.** Elevators shall be provided with fire fighters' emergency operations in accordance with ASME A17.1/CSA B44, *Safety Code for Elevators and Escalators*.

**7.2.13.10 Maintenance.** Where an elevator lobby is served by only one elevator car, the elevator evacuation system shall have a program of scheduled maintenance during times of building shutdown or low building activity. Repairs shall be performed within 24 hours of breakdown.

**7.2.13.11 Earthquake Protection.** Elevators shall have the capability of orderly shutdowns during earthquakes at locations where such shutdowns are an option of ASME A17.1/CSA B44, *Safety Code for Elevators and Escalators*.

**7.2.13.12 Signage.** Signage shall comply with 7.10.8.4.

### 7.3 Capacity of Means of Egress.

### 7.3.1 Occupant Load.

**7.3.1.1 Sufficient Capacity.**

**7.3.1.1.1** The total capacity of the means of egress for any story, balcony, tier, or other occupied space shall be sufficient for the occupant load thereof.

**7.3.1.1.2** For other than existing means of egress, where more than one means of egress is required, the means of egress shall be of such width and capacity that the loss of any one means of egress leaves available not less than 50 percent of the required capacity.

**7.3.1.2\* Occupant Load Factor.** The occupant load in any building or portion thereof shall be not less than the number of persons determined by dividing the floor area assigned to that use by the occupant load factor for that use as specified in Table 7.3.1.2, Figure 7.3.1.2(a), and Figure 7.3.1.2(b). Where both gross and net area figures are given for the same occupancy, calculations shall be made by applying the gross area figure to the gross area of the portion of the building devoted to the use for which the gross area figure is specified and by applying the net area figure to the net area of the portion of the building devoted to the use for which the net area figure is specified.

**7.3.1.3 Occupant Load Increases.**

**7.3.1.3.1** The occupant load in any building or portion thereof shall be permitted to be increased from the occupant load established for the given use in accordance with 7.3.1.2 where all other requirements of this *Code* are also met, based on such increased occupant load.

**7.3.1.3.2** The authority having jurisdiction shall be permitted to require an approved aisle, seating, or fixed equipment diagram to substantiate any increase in occupant load and shall be permitted to require that such a diagram be posted in an approved location.

**Table 7.3.1.2 Occupant Load Factor**

| Use | (ft²/person)ᵃ | (m²/person)ᵃ |
|---|---|---|
| **Assembly Use** | | |
| Concentrated use, without fixed seating | 7 net | 0.65 net |
| Less concentrated use, without fixed seating | 15 net | 1.4 net |
| Bench-type seating | 1 person/ 18 linear in. | 1 person/ 455 linear mm |
| Fixed seating | Use number of fixed seats | Use number of fixed seats |
| Waiting spaces | See 12.1.7.2 and 13.1.7.2. | See 12.1.7.2 and 13.1.7.2. |
| Kitchens | 100 | 9.3 |
| Library stack areas | 100 | 9.3 |
| Library reading rooms | 50 net | 4.6 net |
| Swimming pools | 50 (water surface) | 4.6 (water surface) |
| Swimming pool decks | 30 | 2.8 |
| Exercise rooms with equipment | 50 | 4.6 |
| Exercise rooms without equipment | 15 | 1.4 |
| Stages | 15 net | 1.4 net |
| Lighting and access catwalks, galleries, gridirons | 100 net | 9.3 net |
| Casinos and similar gaming areas | 11 | 1 |
| Skating rinks | 50 | 4.6 |
| **Educational Use** | | |
| Classrooms | 20 net | 1.9 net |
| Shops, laboratories, vocational rooms | 50 net | 4.6 net |
| **Day-Care Use** | 35 net | 3.3 net |
| **Health Care Use** | | |
| Inpatient treatment departments | 240 | 22.3 |
| Sleeping departments | 120 | 11.1 |
| Ambulatory health care | 100 | 9.3 |
| **Detention and Correctional Use** | 120 | 11.1 |
| **Residential Use** | | |
| Hotels and dormitories | 200 | 18.6 |
| Apartment buildings | 200 | 18.6 |
| Board and care, large | 200 | 18.6 |
| **Industrial Use** | | |
| General and high hazard industrial | 100 | 9.3 |
| Special-purpose industrial | NA | NA |
| **Business Use (other than below)** | 100 | 9.3 |
| Air traffic control tower observation levels | 40 | 3.7 |

**Table 7.3.1.2  *Continued***

| Use | (ft²/person)ᵃ | (m²/person)ᵃ |
|---|---|---|
| **Storage Use** | | |
| In storage occupancies | NA | NA |
| In mercantile occupancies | 300 | 27.9 |
| In other than storage and mercantile occupancies | 500 | 46.5 |
| **Mercantile Use** | | |
| Sales area on street floorᵇ,ᶜ | 30 | 2.8 |
| Sales area on two or more street floorsᶜ | 40 | 3.7 |
| Sales area on floor below street floorᶜ | 30 | 2.8 |
| Sales area on floors above street floorᶜ | 60 | 5.6 |
| Floors or portions of floors used only for offices | See business use. | See business use. |
| Floors or portions of floors used only for storage, receiving, and shipping, and not open to general public | 300 | 27.9 |
| Mall buildingsᵈ | Per factors applicable to use of spaceᵉ | |

NA: Not applicable. The occupant load is the maximum probable number of occupants present at any time.

ᵃAll factors are expressed in gross area unless marked "net."

ᵇFor the purpose of determining occupant load in mercantile occupancies where, due to differences in the finished ground level of streets on different sides, two or more floors directly accessible from streets (not including alleys or similar back streets) exist, each such floor is permitted to be considered a street floor. The occupant load factor is one person for each 40 ft² (3.7 m²) of gross floor area of sales space.

ᶜFor the purpose of determining occupant load in mercantile occupancies with no street floor, as defined in 3.3.253, but with access directly from the street by stairs or escalators, the floor at the point of entrance to the mercantile occupancy is considered the street floor.

ᵈFor any food court or other assembly use areas located in the mall that are not included as a portion of the gross leasable area of the mall building, the occupant load is calculated based on the occupant load factor for that use as specified in Table 7.3.1.2. The remaining mall area is not required to be assigned an occupant load.

ᵉThe portions of the mall that are considered a pedestrian way and not used as gross leasable area are not required to be assessed an occupant load based on Table 7.3.1.2. However, means of egress from a mall pedestrian way are required to be provided for an occupant load determined by dividing the gross leasable area of the mall building (not including anchor stores) by the appropriate lowest whole number occupant load factor from Figure 7.3.1.2(a) or Figure 7.3.1.2(b).

Each individual tenant space is required to have means of egress to the outside or to the mall based on occupant loads calculated by using the appropriate occupant load factor from Table 7.3.1.2.

Each individual anchor store is required to have means of egress independent of the mall.

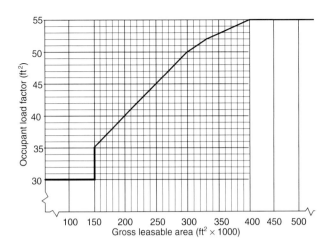

**FIGURE 7.3.1.2(a)   Mall Building Occupant Load Factors (U.S. Customary Units).**

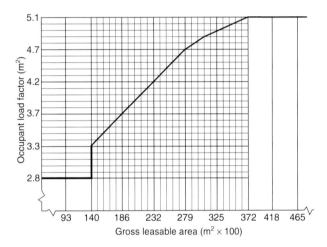

**FIGURE 7.3.1.2(b)   Mall Building Occupant Load Factors (SI Units).**

**7.3.1.4 Exits Serving More than One Story.** Where an exit serves more than one story, only the occupant load of each story considered individually shall be used in computing the required capacity of the exit at that story, provided that the required egress capacity of the exit is not decreased in the direction of egress travel.

**7.3.1.5 Capacity from a Point of Convergence.** Where means of egress from a story above and a story below converge at an intermediate story, the capacity of the means of egress from the point of convergence shall be not less than the sum of the required capacity of the two means of egress.

**7.3.1.6 Egress Capacity from Balconies and Mezzanines.** Where any required egress capacity from a balcony or mezzanine passes through the room below, that required capacity shall be added to the required egress capacity of the room below.

**7.3.2 Measurement of Means of Egress.**

**7.3.2.1** The width of means of egress shall be measured in the clear at the narrowest point of the egress component under consideration, unless otherwise provided in 7.3.2.2 or 7.3.2.3.

**7.3.2.2** Projections within the means of egress of not more than 4½ in. (114 mm) on each side shall be permitted at a height of 38 in. (965 mm) and below. In the case of stair and landing handrails forming part of a guard, in accordance with 7.2.2.4.4.3, such projections shall be permitted at a height of 42 in. (1065 mm) and below.

**7.3.2.3** In health care and ambulatory health care occupancies, projections shall be permitted in corridors in accordance with Chapters 18 through 21.

**7.3.3\* Egress Capacity.**

**7.3.3.1** Egress capacity for approved components of means of egress shall be based on the capacity factors shown in Table 7.3.3.1, unless otherwise provided in 7.3.3.2 .

**Table 7.3.3.1   Capacity Factors**

| Area | Stairways (width/person) | | Level Components and Ramps (width/person) | |
|---|---|---|---|---|
| | in. | mm | in. | mm |
| Board and care | 0.4 | 10 | 0.2 | 5 |
| Health care, sprinklered | 0.3 | 7.6 | 0.2 | 5 |
| Health care, nonsprinklered | 0.6 | 15 | 0.5 | 13 |
| High hazard contents | 0.7 | 18 | 0.4 | 10 |
| All others | 0.3 | 7.6 | 0.2 | 5 |

**7.3.3.2\*** For stairways wider than 44 in. (1120 mm) and subject to the 0.3 in. (7.6 mm) width per person capacity factor, the capacity shall be permitted to be increased using the following equation:

$$C = 146.7 + \left( \frac{Wn - 44}{0.218} \right)$$

where:
$C$ = capacity, in persons, rounded to the nearest integer
$Wn$ = nominal width of the stair as permitted by 7.3.2.2 (in.)

**7.3.3.3** The required capacity of a corridor shall be the occupant load that utilizes the corridor for exit access divided by the required number of exits to which the corridor connects, but the corridor capacity shall be not less than the required capacity of the exit to which the corridor leads.

**7.3.4 Minimum Width.**

**7.3.4.1** The width of any means of egress, unless otherwise provided in 7.3.4.1.1 through 7.3.4.1.3, shall be as follows:

(1) Not less than that required for a given egress component in this chapter or Chapters 11 through 43

(2) Not less than 36 in. (915 mm) where another part of this chapter and Chapters 11 through 43 do not specify a minimum width

**7.3.4.1.1\*** The width of exit access that is formed by furniture and movable partitions, that serves not more than six people, and that has a length not exceeding 50 ft (15 m) shall meet both of the following criteria:

(1) The width shall be not less than 18 in. (455 mm), at and below a height of 38 in. (965 mm), and not less than 28 in. (710 mm) above a height of 38 in. (965 mm).
(2) A width of not less than 36 in. (915 mm) for new exit access, and not less than 28 in. (710 mm) for existing exit access, shall be capable of being provided without moving permanent walls.

**7.3.4.1.2** In existing buildings, the width of exit access shall be permitted to be not less than 28 in. (710 mm).

**7.3.4.1.3** The requirement of 7.3.4.1 shall not apply to the following:

(1) Doors as otherwise provided for in 7.2.1.2
(2) Aisles and aisle accessways in assembly occupancies as otherwise provided in Chapters 12 and 13
(3) Industrial equipment access as otherwise provided in 40.2.5.2

**7.3.4.2** Where a single exit access leads to an exit, its capacity in terms of width shall be not less than the required capacity of the exit to which it leads.

**7.3.4.3** Where more than one exit access leads to an exit, each shall have a width adequate for the number of persons it accommodates.

**7.4\* Number of Means of Egress.**

**7.4.1 General.**

**7.4.1.1** The number of means of egress from any balcony, mezzanine, story, or portion thereof shall be not less than two, except under one of the following conditions:

(1) A single means of egress shall be permitted where permitted in Chapters 11 through 43.
(2) A single means of egress shall be permitted for a mezzanine or balcony where the common path of travel limitations of Chapters 11 through 43 are met.

**7.4.1.2** The number of means of egress from any story or portion thereof, other than for existing buildings as permitted in Chapters 11 through 43, shall be as follows:

(1) Occupant load more than 500 but not more than 1000 — not less than 3
(2) Occupant load more than 1000 — not less than 4

**7.4.1.3** Accessible means of egress in accordance with 7.5.4 that do not utilize elevators shall be permitted to serve as any or all of the required minimum number of means of egress.

**7.4.1.4** The occupant load of each story considered individually shall be required to be used in computing the number of means of egress at each story, provided that the required number of means of egress is not decreased in the direction of egress travel.

**7.4.1.5** Doors other than the hoistway door; the elevator car door; and doors that are readily openable from the car side without a key, a tool, special knowledge, or special effort shall be prohibited at the point of access to an elevator car.

**7.4.1.6 Elevator Landing and Lobby Exit Access.**

**7.4.1.6.1** Each elevator landing and lobby shall have access to at least one exit.

**7.4.1.6.2** The elevator landing and lobby exit access required by 7.4.1.6.1 shall not require the use of a key, a tool, special knowledge, or special effort, unless permitted by 7.4.1.6.3.

**7.4.1.6.3** Doors separating the elevator lobby from the exit access required by 7.4.1.6.1 shall be permitted to be electronically locked in accordance with 7.2.1.6.3.

**7.4.2 Spaces About Electrical Equipment.**

**7.4.2.1 600 Volts, Nominal, or Less.** The minimum number of means of egress for working space about electrical equipment, other than existing electrical equipment, shall be in accordance with *NFPA 70, National Electrical Code*, Section 110.26(C).

**7.4.2.2 Over 600 Volts, Nominal.** The minimum number of means of egress for working space about electrical equipment, other than existing electrical equipment, shall be in accordance with *NFPA 70, National Electrical Code*, Section 110.33(A).

**7.5 Arrangement of Means of Egress.**

**7.5.1 General.**

**7.5.1.1** Exits shall be located, and exit access shall be arranged, so that exits are readily accessible at all times.

**7.5.1.1.1\*** Where exits are not immediately accessible from an open floor area, continuous passageways, aisles, or corridors leading directly to every exit shall be maintained and shall be arranged to provide access for each occupant to not less than two exits by separate ways of travel, unless otherwise provided in 7.5.1.1.3 and 7.5.1.1.4.

**7.5.1.1.2** Exit access corridors shall provide access to not less than two approved exits, unless otherwise provided in 7.5.1.1.3 and 7.5.1.1.4.

**7.5.1.1.3** The requirements of 7.5.1.1.1 and 7.5.1.1.2 shall not apply where a single exit is permitted in Chapters 11 through 43

**7.5.1.1.4** Where common paths of travel are permitted for an occupancy in Chapters 11 through 43, such common paths of travel shall be permitted but shall not exceed the limit specified.

**7.5.1.2** Corridors shall provide exit access without passing through any intervening rooms other than corridors, lobbies, and other spaces permitted to be open to the corridor, unless otherwise provided in 7.5.1.2.1 and 7.5.1.2.2.

**7.5.1.2.1** Approved existing corridors that require passage through a room to access an exit shall be permitted to continue to be used, provided that all of the following criteria are met:

(1) The path of travel is marked in accordance with Section 7.10.
(2) Doors to such rooms comply with 7.2.1.
(3) Such arrangement is not prohibited by the applicable occupancy chapter.

**7.5.1.2.2** Corridors that are not required to be fire resistance rated shall be permitted to discharge into open floor plan areas.

**7.5.1.3** Remoteness shall be provided in accordance with 7.5.1.3.1 through 7.5.1.3.7.

**7.5.1.3.1** Where more than one exit, exit access, or exit discharge is required from a building or portion thereof, such exits, exit accesses, or exit discharges shall be remotely located from each other and be arranged to minimize the possibility that more than one has the potential to be blocked by any one fire or other emergency condition.

**7.5.1.3.2\*** Where two exits, exit accesses, or exit discharges are required, they shall be located at a distance from one another not less than one-half the length of the maximum overall diagonal dimension of the building or area to be served, measured in a straight line between the nearest edge of the exits, exit accesses, or exit discharges, unless otherwise provided in 7.5.1.3.3 through 7.5.1.3.5.

**7.5.1.3.3** In buildings protected throughout by an approved, supervised automatic sprinkler system in accordance with Section 9.7, the minimum separation distance between two exits, exit accesses, or exit discharges, measured in accordance with 7.5.1.3.2, shall be not less than one-third the length of the maximum overall diagonal dimension of the building or area to be served.

**7.5.1.3.4\*** In other than high-rise buildings, where exit enclosures are provided as the required exits specified in 7.5.1.3.2 or 7.5.1.3.3 and are interconnected by not less than a 1-hour fire resistance–rated corridor, exit separation shall be measured along the shortest line of travel within the corridor.

**7.5.1.3.5** In existing buildings, where more than one exit, exit access, or exit discharge is required, such exits, exit accesses, or exit discharges shall be exempt from the diagonal measurement separation distance criteria of 7.5.1.3.2 and 7.5.1.3.3, provided that such exits, exit accesses, or exit discharges are remotely located in accordance with 7.5.1.3.1.

**7.5.1.3.6** In other than existing buildings, where more than two exits, exit accesses, or exit discharges are required, at least two of the required exits, exit accesses, or exit discharges shall be arranged to comply with the minimum separation distance requirement.

**7.5.1.3.7** The balance of the exits, exit accesses, or exit discharges specified in 7.5.1.3.6 shall be located so that, if one becomes blocked, the others are available.

**7.5.1.4** Interlocking or scissor stairs shall comply with 7.5.1.4.1 and 7.5.1.4.2.

**7.5.1.4.1** New interlocking or scissor stairs shall be permitted to be considered only as a single exit.

**7.5.1.4.2\*** Existing interlocking or scissor stairs shall be permitted to be considered separate exits, provided that they meet all of the following criteria:

(1) They are enclosed in accordance with 7.1.3.2.
(2) They are separated from each other by 2-hour fire resistance–rated noncombustible construction.
(3) No protected or unprotected penetrations or communicating openings exist between the stair enclosures.

**7.5.1.5\*** Exit access shall be arranged so that there are no dead ends in corridors, unless permitted by, and limited to the lengths specified in, Chapters 11 through 43.

**7.5.1.6** Exit access from rooms or spaces shall be permitted to be through adjoining or intervening rooms or areas, provided that such rooms or areas are accessory to the area served. Foyers, lobbies, and reception rooms constructed as required for corridors shall not be construed as intervening rooms. Exit access shall be arranged so that it is not necessary to pass through any area identified under Protection from Hazards in Chapters 11 through 43.

**7.5.2 Impediments to Egress.** See also 7.1.9 and 7.2.1.5.

**7.5.2.1\*** Access to an exit shall not be through kitchens, storerooms other than as provided in Chapters 36 and 37, restrooms, workrooms, closets, bedrooms or similar spaces, or other rooms or spaces subject to locking, unless passage through such rooms or spaces is permitted for the occupancy by Chapter 18, 19, 22, or 23.

**7.5.2.2\*** Exit access and exit doors shall be designed and arranged to be clearly recognizable.

**7.5.2.2.1** Hangings or draperies shall not be placed over exit doors or located so that they conceal or obscure any exit, unless otherwise provided in 7.5.2.2.2.

**7.5.2.2.2** Curtains shall be permitted across means of egress openings in tent walls, provided that all of the following criteria are met:

(1) They are distinctly marked in contrast to the tent wall so as to be recognizable as means of egress.
(2) They are installed across an opening that is at least 6 ft (1830 mm) in width.
(3) They are hung from slide rings or equivalent hardware so as to be readily moved to the side to create an unobstructed opening in the tent wall that is of the minimum width required for door openings.

**7.5.3 Exterior Ways of Exit Access.**

**7.5.3.1** Exit access shall be permitted to be by means of any exterior balcony, porch, gallery, or roof that conforms to the requirements of this chapter.

**7.5.3.2** The long side of the balcony, porch, gallery, or similar space shall be at least 50 percent open and shall be arranged to restrict the accumulation of smoke.

**7.5.3.3** Exterior exit access balconies shall be separated from the interior of the building by walls and opening protectives as required for corridors, unless the exterior exit access balcony is served by at least two remote stairs that can be accessed without any occupant traveling past an unprotected opening to reach one of the stairs, or unless dead ends on the exterior exit access do not exceed 20 ft (6100 mm).

**7.5.3.4** Exterior exit access shall be arranged so that there are no dead ends in excess of those permitted for dead-end corridors in Chapters 11 through 43.

**7.5.4 Accessible Means of Egress.**

**7.5.4.1\*** Areas accessible to people with severe mobility impairment, other than in existing buildings, shall have not less than two accessible means of egress, unless otherwise provided in 7.5.4.1.2 through 7.5.4.1.4.

**7.5.4.1.1** Access within the allowable travel distance shall be provided to not less than one accessible area of refuge or one accessible exit providing an accessible route to an exit discharge.

**7.5.4.1.2** A single accessible means of egress shall be permitted from buildings or areas of buildings permitted to have a single exit.

**7.5.4.1.3** Accessible means of egress shall not be required in health care occupancies protected throughout by an approved, supervised automatic sprinkler system in accordance with Section 9.7.

**7.5.4.1.4** Exit access travel along the accessible means of egress shall be permitted to be common for the distances permitted as common paths of travel.

**7.5.4.2** Where two accessible means of egress are required, the exits serving such means of egress shall be located at a distance from one another not less than one-half the length of the maximum overall diagonal dimension of the building or area to be served. This distance shall be measured in a straight line between the nearest edge of the exit doors or exit access doors, unless otherwise provided in 7.5.4.2.1 through 7.5.4.2.3.

**7.5.4.2.1** Where exit enclosures are provided as the required exits specified in 7.5.4.2 and are interconnected by not less than a 1-hour fire resistance–rated corridor, exit separation shall be permitted to be measured along the line of travel within the corridor.

**7.5.4.2.2** The requirement of 7.5.4.2 shall not apply to buildings protected throughout by an approved, supervised automatic sprinkler system in accordance with Section 9.7.

**7.5.4.2.3** The requirement of 7.5.4.2 shall not apply where the physical arrangement of means of egress prevents the possibility that access to both accessible means of egress will be blocked by any one fire or other emergency condition as approved by the authority having jurisdiction.

**7.5.4.3** Each required accessible means of egress shall be continuous from each accessible occupied area to a public way or area of refuge in accordance with 7.2.12.2.2.

**7.5.4.4** Where an exit stair is used in an accessible means of egress, it shall comply with 7.2.12 and either shall incorporate an area of refuge within an enlarged story-level landing or shall be accessed from an area of refuge.

**7.5.4.5** To be considered part of an accessible means of egress, an elevator shall be in accordance with 7.2.12.2.4.

**7.5.4.6** To be considered part of an accessible means of egress, a smoke barrier in accordance with Section 8.5 with not less than a 1-hour fire resistance rating, or a horizontal exit in accordance with 7.2.4, shall discharge to an area of refuge in accordance with 7.2.12.

**7.5.4.7** Accessible stories that are four or more stories above or below a story of exit discharge shall have not less than one elevator complying with 7.5.4.5, except as modified in 7.5.4.8.

**7.5.4.8** Where elevators are required by 7.5.4.7, the smokeproof enclosure required by 7.2.12.2.4 shall not be required in buildings protected throughout by an approved, supervised automatic sprinkler system in accordance with 9.7.1.1(1).

**7.5.4.9** An area of refuge used as part of a required accessible means of egress shall be in accordance with 7.2.12.

**7.6\* Measurement of Travel Distance to Exits.**

**7.6.1\*** The travel distance to an exit shall be measured on the floor or other walking surface as follows:

(1) Along the centerline of the natural path of travel, starting from the most remote point subject to occupancy
(2) Curving around any corners or obstructions, with a 12 in. (305 mm) clearance therefrom
(3) Terminating at one of the following:

(a) Center of the doorway
(b) Other point at which the exit begins
(c) Smoke barrier in an existing detention and correctional occupancy as provided in Chapter 23

**7.6.2** Where outside stairs that are not separated from the building are permitted as required exits, the travel distance shall be measured from the most remote point subject to occupancy to the leading nosing of the stair landing at the floor level under consideration.

**7.6.3\*** Where open stairways or ramps are permitted as a path of travel to required exits, the distance shall include the travel on the stairway or ramp and the travel from the end of the stairway or ramp to an outside door or other exit in addition to the distance traveled to reach the stairway or ramp.

**7.6.4** Where any part of an exterior exit is within 10 ft (3050 mm) of horizontal distance of any unprotected building opening, as permitted by 7.2.2.6.3 for outside stairs, the travel distance to the exit shall include the length of travel to the finished ground level.

**7.6.5** Where measurement includes stairs, the measurement shall be taken in the plane of the tread nosing.

**7.6.6** The travel distance in any occupied space to not less than one exit, measured in accordance with 7.6.1 through 7.6.5, shall not exceed the limits specified in this *Code*. *(See 7.6.7.)*

**7.6.7** Travel distance limitations shall be as provided in Chapters 11 through 43 and, for high hazard areas, shall be in accordance with Section 7.11.

**7.7 Discharge from Exits.**

**7.7.1\* Exit Termination.** Exits shall terminate directly, at a public way or at an exterior exit discharge, unless otherwise provided in 7.7.1.2 through 7.7.1.4.

**7.7.1.1** Yards, courts, open spaces, or other portions of the exit discharge shall be of the required width and size to provide all occupants with a safe access to a public way.

**7.7.1.2** The requirement of 7.7.1 shall not apply to interior exit discharge as otherwise provided in 7.7.2.

**7.7.1.3** The requirement of 7.7.1 shall not apply to rooftop exit discharge as otherwise provided in 7.7.6.

**7.7.1.4** Means of egress shall be permitted to terminate in an exterior area for detention and correctional occupancies as otherwise provided in Chapters 22 and 23.

**7.7.2 Exit Discharge Through Interior Building Areas.** Exits shall be permitted to discharge through interior building areas, provided that all of the following are met:

(1) Not more than 50 percent of the required number of exits, and not more than 50 percent of the required egress capacity, shall discharge through areas on any level of discharge, except as otherwise permitted by one of the following:
   (a) One hundred percent of the exits shall be permitted to discharge through areas on any level of discharge in detention and correctional occupancies as otherwise provided in Chapters 22 and 23.
   (b) In existing buildings, the 50 percent limit on egress capacity shall not apply if the 50 percent limit on the required number of exits is met.

(2) Each level of discharge shall discharge directly outside at the finished ground level or discharge directly outside and provide access to the finished ground level by outside stairs or outside ramps.

(3) The interior exit discharge shall lead to a free and unobstructed way to the exterior of the building, and such way shall be readily visible and identifiable from the point of discharge from the exit.

(4) The interior exit discharge shall be protected by one of the following methods:

(a) The level of discharge shall be protected throughout by an approved automatic sprinkler system in accordance with Section 9.7, or the portion of the level of discharge used for interior exit discharge shall be protected by an approved automatic sprinkler system in accordance with Section 9.7 and shall be separated from the nonsprinklered portion of the floor by fire barriers with a fire resistance rating meeting the requirements for the enclosure of exits. *(See 7.1.3.2.1.)*

(b) The interior exit discharge area shall be in a vestibule or foyer that meets all of the following criteria:

i. The depth from the exterior of the building shall be not more than 10 ft (3050 mm), and the length shall be not more than 30 ft (9.1 m).

ii. The foyer shall be separated from the remainder of the level of discharge by construction providing protection not less than the equivalent of wired glass in steel frames or 45 minutes fire-resistive construction.

iii. The foyer shall serve only as means of egress and shall include an exit directly to the outside.

(5) The entire area on the level of discharge shall be separated from areas below by construction having a fire resistance rating not less than that required for the exit enclosure, unless otherwise provided in 7.7.2(6).

(6) Levels below the level of discharge in an atrium shall be permitted to be open to the level of discharge where such level of discharge is protected in accordance with 8.6.7.

**7.7.3 Arrangement and Marking of Exit Discharge.**

**7.7.3.1** Where more than one exit discharge is required, exit discharges shall be arranged to meet the remoteness criteria of 7.5.1.3.

**7.7.3.2** The exit discharge shall be arranged and marked to make clear the direction of egress travel from the exit discharge to a public way.

**7.7.3.3** Stairs and ramps shall be arranged so as to make clear the direction of egress travel from the exit discharge to a public way.

**7.7.3.4\*** Stairs and ramps that continue more than one-half story beyond the level of discharge shall be provided with an approved means to prevent or dissuade occupants from traveling past the level of discharge during emergency building evacuation.

**7.7.4 Components of Exit Discharge.** Doors, stairs, ramps, corridors, exit passageways, bridges, balconies, escalators, moving walks, and other components of an exit discharge shall comply with the detailed requirements of this chapter for such components.

**7.7.5 Signs.** See 7.2.2.5.4.

**7.7.6 Discharge to Roofs.** Where approved by the authority having jurisdiction, exits shall be permitted to discharge to roofs or other sections of the building or an adjoining building where all of the following criteria are met:

(1) The roof/ceiling assembly construction has a fire resistance rating not less than that required for the exit enclosure.

(2) A continuous and safe means of egress from the roof is available.

**7.8 Illumination of Means of Egress.**

**7.8.1 General.**

**7.8.1.1\*** Illumination of means of egress shall be provided in accordance with Section 7.8 for every building and structure where required in Chapters 11 through 43. For the purposes of this requirement, exit access shall include only designated stairs, aisles, corridors, ramps, escalators, and passageways leading to an exit. For the purposes of this requirement, exit discharge shall include only designated stairs, aisles, corridors, ramps, escalators, walkways, and exit passageways leading to a public way.

**7.8.1.2** Illumination of means of egress shall be continuous during the time that the conditions of occupancy require that the means of egress be available for use, unless otherwise provided in 7.8.1.2.2.

**7.8.1.2.1** Artificial lighting shall be employed at such locations and for such periods of time as are necessary to maintain the illumination to the minimum criteria values herein specified.

**7.8.1.2.2** Unless prohibited by Chapters 11 through 43, automatic, motion sensor–type lighting switches shall be permitted within the means of egress, provided that the switch controllers comply with all of the following:

(1) The switch controllers are listed.

(2) The switch controllers are equipped for fail-safe operation and evaluated for this purpose.

(3) The illumination timers are set for a minimum 15-minute duration.

(4) The motion sensor is activated by any occupant movement in the area served by the lighting units.

(5) The switch controller is activated by activation of the building fire alarm system, if provided.

**7.8.1.2.3\*** Energy-saving sensors, switches, timers, or controllers shall be approved and shall not compromise the continuity of illumination of the means of egress required by 7.8.1.2.

**7.8.1.3\*** The floors and other walking surfaces within an exit and within the portions of the exit access and exit discharge designated in 7.8.1.1 shall be illuminated as follows:

(1) During conditions of stair use, the minimum illumination for new stairs shall be at least 10 ft-candle (108 lux), measured at the walking surfaces.

(2) The minimum illumination for floors and walking surfaces, other than new stairs during conditions of stair use, shall be to values of at least 1 ft-candle (10.8 lux), measured at the floor.

(3) In assembly occupancies, the illumination of the walking surfaces of exit access shall be at least 0.2 ft-candle (2.2 lux) during periods of performances or projections involving directed light.

(4)\*The minimum illumination requirements shall not apply where operations or processes require low lighting levels.

**7.8.1.4\*** Required illumination shall be arranged so that the failure of any single lighting unit does not result in an illumination level of less than 0.2 ft-candle (2.2 lux) in any designated area.

**7.8.1.5** The equipment or units installed to meet the requirements of Section 7.10 also shall be permitted to serve the function of illumination of means of egress, provided that all requirements of Section 7.8 for such illumination are met.

**7.8.2 Sources of Illumination.**

**7.8.2.1\*** Illumination of means of egress shall be from a source considered reliable by the authority having jurisdiction.

**7.8.2.2** Battery-operated electric lights and other types of portable lamps or lanterns shall not be used for primary illumination of means of egress. Battery-operated electric lights shall be permitted to be used as an emergency source to the extent permitted under Section 7.9.

**7.9 Emergency Lighting.**

**7.9.1 General.**

**7.9.1.1\*** Emergency lighting facilities for means of egress shall be provided in accordance with Section 7.9 for the following:

(1) Buildings or structures where required in Chapters 11 through 43
(2) Underground and limited access structures as addressed in Section 11.7
(3) High-rise buildings as required by other sections of this *Code*
(4) Doors equipped with delayed-egress locks
(5) Stair shafts and vestibules of smokeproof enclosures, for which the following also apply:
   (a) The stair shaft and vestibule shall be permitted to include a standby generator that is installed for the smokeproof enclosure mechanical ventilation equipment.
   (b) The standby generator shall be permitted to be used for the stair shaft and vestibule emergency lighting power supply.
(6) New access-controlled egress doors in accordance with 7.2.1.6.2

**7.9.1.2** For the purposes of 7.9.1.1, exit access shall include only designated stairs, aisles, corridors, ramps, escalators, and passageways leading to an exit. For the purposes of 7.9.1.1, exit discharge shall include only designated stairs, ramps, aisles, walkways, and escalators leading to a public way.

**7.9.1.3** Where maintenance of illumination depends on changing from one energy source to another, a delay of not more than 10 seconds shall be permitted.

**7.9.2 Performance of System.**

**7.9.2.1\*** Emergency illumination shall be provided for a minimum of 1½ hours in the event of failure of normal lighting. Emergency lighting facilities shall be arranged to provide initial illumination that is not less than an average of 1 ft-candle (10.8 lux) and, at any point, not less than 0.1 ft-candle (1.1 lux), measured along the path of egress at floor level. Illumination levels shall be permitted to decline to not less than an average of 0.6 ft-candle (6.5 lux) and, at any point, not less than 0.06 ft-candle (0.65 lux) at the end of 1½ hours. A maximum-to-minimum illumination uniformity ratio of 40 to 1 shall not be exceeded.

**7.9.2.2** New emergency power systems for emergency lighting shall be at least Type 10, Class 1.5, Level 1, in accordance with NFPA 110, *Standard for Emergency and Standby Power Systems*.

**7.9.2.3\*** The emergency lighting system shall be arranged to provide the required illumination automatically in the event of any interruption of normal lighting due to any of the following:

(1) Failure of a public utility or other outside electrical power supply
(2) Opening of a circuit breaker or fuse
(3) Manual act(s), including accidental opening of a switch controlling normal lighting facilities

**7.9.2.4** Emergency generators providing power to emergency lighting systems shall be installed, tested, and maintained in accordance with NFPA 110, *Standard for Emergency and Standby Power Systems*. Stored electrical energy systems, where required in this *Code*, other than battery systems for emergency luminaires in accordance with 7.9.2.5, shall be installed and tested in accordance with NFPA 111, *Standard on Stored Electrical Energy Emergency and Standby Power Systems*.

**7.9.2.5** Unit equipment and battery systems for emergency luminaires shall be listed to ANSI/UL 924, *Standard for Emergency Lighting and Power Equipment*.

**7.9.2.6\*** Existing battery-operated emergency lights shall use only reliable types of rechargeable batteries provided with suitable facilities for maintaining them in properly charged condition. Batteries used in such lights or units shall be approved for their intended use and shall comply with *NFPA 70, National Electrical Code*.

**7.9.2.7** The emergency lighting system shall be either continuously in operation or shall be capable of repeated automatic operation without manual intervention.

**7.9.3 Periodic Testing of Emergency Lighting Equipment.**

**7.9.3.1** Required emergency lighting systems shall be tested in accordance with one of the three options offered by 7.9.3.1.1, 7.9.3.1.2, or 7.9.3.1.3.

**7.9.3.1.1** Testing of required emergency lighting systems shall be permitted to be conducted as follows:

(1) Functional testing shall be conducted monthly, with a minimum of 3 weeks and a maximum of 5 weeks between tests, for not less than 30 seconds, except as otherwise permitted by 7.9.3.1.1(2).
(2)\*The test interval shall be permitted to be extended beyond 30 days with the approval of the authority having jurisdiction.
(3) Functional testing shall be conducted annually for a minimum of 1½ hours if the emergency lighting system is battery powered.
(4) The emergency lighting equipment shall be fully operational for the duration of the tests required by 7.9.3.1.1(1) and (3).
(5) Written records of visual inspections and tests shall be kept by the owner for inspection by the authority having jurisdiction.

**7.9.3.1.2** Testing of required emergency lighting systems shall be permitted to be conducted as follows:

(1) Self-testing/self-diagnostic battery-operated emergency lighting equipment shall be provided.
(2) Not less than once every 30 days, self-testing/self-diagnostic battery-operated emergency lighting equipment shall automatically perform a test with a duration of a minimum of 30 seconds and a diagnostic routine.

(3) Self-testing/self-diagnostic battery-operated emergency lighting equipment shall indicate failures by a status indicator.

(4) A visual inspection shall be performed at intervals not exceeding 30 days.

(5) Functional testing shall be conducted annually for a minimum of 1½ hours.

(6) Self-testing/self-diagnostic battery-operated emergency lighting equipment shall be fully operational for the duration of the 1½-hour test.

(7) Written records of visual inspections and tests shall be kept by the owner for inspection by the authority having jurisdiction.

**7.9.3.1.3** Testing of required emergency lighting systems shall be permitted to be conducted as follows:

(1) Computer-based, self-testing/self-diagnostic battery-operated emergency lighting equipment shall be provided.

(2) Not less than once every 30 days, emergency lighting equipment shall automatically perform a test with a duration of a minimum of 30 seconds and a diagnostic routine.

(3) The emergency lighting equipment shall automatically perform annually a test for a minimum of 1½ hours.

(4) The emergency lighting equipment shall be fully operational for the duration of the tests required by 7.9.3.1.3(2) and (3).

(5) The computer-based system shall be capable of providing a report of the history of tests and failures at all times.

**7.10 Marking of Means of Egress.**

**7.10.1 General.**

**7.10.1.1 Where Required.** Means of egress shall be marked in accordance with Section 7.10 where required in Chapters 11 through 43.

**7.10.1.2 Exits.**

**7.10.1.2.1\*** Exits, other than main exterior exit doors that obviously and clearly are identifiable as exits, shall be marked by an approved sign that is readily visible from any direction of exit access.

**7.10.1.2.2\*** Horizontal components of the egress path within an exit enclosure shall be marked by approved exit or directional exit signs where the continuation of the egress path is not obvious.

**7.10.1.3 Exit Door Tactile Signage.** Tactile signage shall be provided to meet all of the following criteria, unless otherwise provided in 7.10.1.4:

(1) Tactile signage shall be located at each exit door requiring an exit sign.

(2) Tactile signage shall read as follows: EXIT.

(3) Tactile signage shall comply with ICC/ANSI A117.1, *American National Standard for Accessible and Usable Buildings and Facilities.*

**7.10.1.4 Existing Exemption.** The requirements of 7.10.1.3 shall not apply to existing buildings, provided that the occupancy classification does not change.

**7.10.1.5 Exit Access.**

**7.10.1.5.1** Access to exits shall be marked by approved, readily visible signs in all cases where the exit or way to reach the exit is not readily apparent to the occupants.

**7.10.1.5.2\*** New sign placement shall be such that no point in an exit access corridor is in excess of the rated viewing distance or 100 ft (30 m), whichever is less, from the nearest sign.

**7.10.1.6\* Floor Proximity Exit Signs.** Where floor proximity exit signs are required in Chapters 11 through 43, such signs shall comply with 7.10.3, 7.10.4, 7.10.5, and 7.10.6 for externally illuminated signs and 7.10.7 for internally illuminated signs. Such signs shall be located near the floor level in addition to those signs required for doors or corridors. The bottom of the sign shall be not less than 6 in. (150 mm), but not more than 18 in. (455 mm), above the floor. For exit doors, the sign shall be mounted on the door or adjacent to the door, with the nearest edge of the sign within 4 in. (100 mm) of the door frame.

**7.10.1.7\* Floor Proximity Egress Path Marking.** Where floor proximity egress path marking is required in Chapters 11 through 43, an approved floor proximity egress path marking system that is internally illuminated shall be installed within 18 in. (455 mm) of the floor. Floor proximity egress path marking systems shall be listed in accordance with ANSI/UL 1994, *Standard for Luminous Egress Path Marking Systems.* The system shall provide a visible delineation of the path of travel along the designated exit access and shall be essentially continuous, except as interrupted by doorways, hallways, corridors, or other such architectural features. The system shall operate continuously or at any time the building fire alarm system is activated. The activation, duration, and continuity of operation of the system shall be in accordance with 7.9.2. The system shall be maintained in accordance with the product manufacturing listing.

**7.10.1.8\* Visibility.** Every sign required in Section 7.10 shall be located and of such size, distinctive color, and design that it is readily visible and shall provide contrast with decorations, interior finish, or other signs. No decorations, furnishings, or equipment that impairs visibility of a sign shall be permitted. No brightly illuminated sign (for other than exit purposes), display, or object in or near the line of vision of the required exit sign that could detract attention from the exit sign shall be permitted.

**7.10.1.9 Mounting Location.** The bottom of new egress markings shall be located at a vertical distance of not more than 6 ft 8 in. (2030 mm) above the top edge of the egress opening intended for designation by that marking. Egress markings shall be located at a horizontal distance of not more than the required width of the egress opening, as measured from the edge of the egress opening intended for designation by that marking to the nearest edge of the marking.

**7.10.2 Directional Signs.**

**7.10.2.1\*** A sign complying with 7.10.3, with a directional indicator showing the direction of travel, shall be placed in every location where the direction of travel to reach the nearest exit is not apparent.

**7.10.2.2** Directional exit signs shall be provided within horizontal components of the egress path within exit enclosures as required by 7.10.1.2.2.

**7.10.3\* Sign Legend.**

**7.10.3.1** Signs required by 7.10.1 and 7.10.2 shall read as follows in plainly legible letters, or other appropriate wording shall be used:

EXIT

**7.10.3.2\*** Where approved by the authority having jurisdiction, pictograms in compliance with NFPA 170, *Standard for Fire Safety and Emergency Symbols*, shall be permitted.

**7.10.4\* Power Source.** Where emergency lighting facilities are required by the applicable provisions of Chapters 11 through 43 for individual occupancies, the signs, other than approved self-luminous signs and listed photoluminescent signs in accordance with 7.10.7.2, shall be illuminated by the emergency lighting facilities. The level of illumination of the signs shall be in accordance with 7.10.6.3 or 7.10.7 for the required emergency lighting duration as specified in 7.9.2.1. However, the level of illumination shall be permitted to decline to 60 percent at the end of the emergency lighting duration.

**7.10.5 Illumination of Signs.**

**7.10.5.1\* General.** Every sign required by 7.10.1.2, 7.10.1.5, or 7.10.8.1, other than where operations or processes require low lighting levels, shall be suitably illuminated by a reliable light source. Externally and internally illuminated signs shall be legible in both the normal and emergency lighting mode.

**7.10.5.2\* Continuous Illumination.**

**7.10.5.2.1** Every sign required to be illuminated by 7.10.6.3, 7.10.7, and 7.10.8.1 shall be continuously illuminated as required under the provisions of Section 7.8, unless otherwise provided in 7.10.5.2.2.

**7.10.5.2.2\*** Illumination for signs shall be permitted to flash on and off upon activation of the fire alarm system.

**7.10.6 Externally Illuminated Signs.**

**7.10.6.1\* Size of Signs.**

**7.10.6.1.1** Externally illuminated signs required by 7.10.1 and 7.10.2, other than approved existing signs, unless otherwise provided in 7.10.6.1.2, shall read EXIT or shall use other appropriate wording in plainly legible letters sized as follows:

(1) For new signs, the letters shall be not less than 6 in. (150 mm) high, with the principal strokes of letters not less than ¾ in. (19 mm) wide.
(2) For existing signs, the required wording shall be permitted to be in plainly legible letters not less than 4 in. (100 mm) high.
(3) The word EXIT shall be in letters of a width not less than 2 in. (51 mm), except the letter I, and the minimum spacing between letters shall be not less than ⅜ in. (9.5 mm).
(4) Sign legend elements larger than the minimum established in 7.10.6.1.1(1) through (3) shall use letter widths, strokes, and spacing in proportion to their height.

**7.10.6.1.2** The requirements of 7.10.6.1.1 shall not apply to marking required by 7.10.1.3 and 7.10.1.7.

**7.10.6.2\* Size and Location of Directional Indicator.**

**7.10.6.2.1** Directional indicators, unless otherwise provided in 7.10.6.2.2, shall comply with all of the following:

(1) The directional indicator shall be located outside of the EXIT legend, not less than ⅜ in. (9.5 mm) from any letter.
(2) The directional indicator shall be of a chevron type, as shown in Figure 7.10.6.2.1.

(3) The directional indicator shall be identifiable as a directional indicator at a distance of 40 ft (12 m).
(4) A directional indicator larger than the minimum established for compliance with 7.10.6.2.1(3) shall be proportionately increased in height, width, and stroke.
(5) The directional indicator shall be located at the end of the sign for the direction indicated.

**FIGURE 7.10.6.2.1 Chevron-Type Indicator.**

**7.10.6.2.2** The requirements of 7.10.6.2.1 shall not apply to approved existing signs.

**7.10.6.3\* Level of Illumination.** Externally illuminated signs shall be illuminated by not less than 5 ft-candles (54 lux) at the illuminated surface and shall have a contrast ratio of not less than 0.5.

**7.10.7 Internally Illuminated Signs.**

**7.10.7.1 Listing.** Internally illuminated signs shall be listed in accordance with ANSI/UL 924, *Standard for Emergency Lighting and Power Equipment*, unless they meet one of the following criteria:

(1) They are approved existing signs.
(2) They are existing signs having the required wording in legible letters not less than 4 in. (100 mm) high.
(3) They are signs that are in accordance with 7.10.1.3 and 7.10.1.6.

**7.10.7.2\* Photoluminescent Signs.** The face of a photoluminescent sign shall be continually illuminated while the building is occupied. The illumination levels on the face of the photoluminescent sign shall be in accordance with its listing. The charging illumination shall be a reliable light source, as determined by the authority having jurisdiction. The charging light source, shall be of a type specified in the product markings.

**7.10.8 Special Signs.**

**7.10.8.1 Sign Illumination.**

**7.10.8.1.1** Where required by other provisions of this *Code*, special signs shall be illuminated in accordance with 7.10.5, 7.10.6.3, and 7.10.7.

**7.10.8.1.2** Where emergency lighting facilities are required by the applicable provisions of Chapters 11 through 43, the required illumination of special signs shall additionally be provided under emergency lighting conditions.

**7.10.8.2 Characters.** Special signs, where required by other provisions of this *Code*, shall comply with the visual character requirements of ICC/ANSI A117.1, *American National Standard for Accessible and Usable Buildings and Facilities*.

**7.10.8.3\* No Exit.**

**7.10.8.3.1** Any door, passage, or stairway that is neither an exit nor a way of exit access and that is located or arranged so that it is likely to be mistaken for an exit shall be identified by a sign that reads as follows:

# NO

### EXIT

**7.10.8.3.2** The NO EXIT sign shall have the word NO in letters 2 in. (51 mm) high, with a stroke width of ⅜ in. (9.5 mm), and the word EXIT in letters 1 in. (25 mm) high, with the word EXIT below the word NO, unless such sign is an approved existing sign.

**7.10.8.4 Elevator Signs.** Elevators that are a part of a means of egress (*see 7.2.13.1*) shall have both of the following signs with a minimum letter height of ⅝ in. (16 mm) posted in every elevator lobby:

(1)\*Signs that indicate that the elevator can be used for egress, including any restrictions on use
(2)\*Signs that indicate the operational status of elevators

**7.10.8.5\* Evacuation Diagram.** Where a posted floor evacuation diagram is required in Chapters 11 through 43, floor evacuation diagrams reflecting the actual floor arrangement and exit locations shall be posted and oriented in a location and manner acceptable to the authority having jurisdiction.

**7.10.9 Testing and Maintenance.**

**7.10.9.1 Inspection.** Exit signs shall be visually inspected for operation of the illumination sources at intervals not to exceed 30 days or shall be periodically monitored in accordance with 7.9.3.1.3.

**7.10.9.2 Testing.** Exit signs connected to, or provided with, a battery-operated emergency illumination source, where required in 7.10.4, shall be tested and maintained in accordance with 7.9.3.

**7.11 Special Provisions for Occupancies with High Hazard Contents.** See Section 6.2.

**7.11.1\*** Where the contents are classified as high hazard, exits shall be provided and arranged to allow all occupants to escape from the building or structure, or from the hazardous area thereof, to the outside or to a place of safety with a travel distance of not more than 75 ft (23 m), measured as required in 7.6.1, unless otherwise provided in 7.11.2.

**7.11.2** The requirement of 7.11.1 shall not apply to storage occupancies as otherwise provided in Chapter 42.

**7.11.3** Egress capacity for high hazard contents areas shall be based on 0.7 in./person (18 mm/person) for stairs or 0.4 in./person (10 mm/person) for level components and ramps in accordance with 7.3.3.1.

**7.11.4** Not less than two means of egress shall be provided from each building or hazardous area thereof, unless all of the following criteria are met:

(1) Rooms or spaces do not exceed 200 ft² (18.6 m²).
(2) Rooms or spaces have an occupant load not exceeding three persons.
(3) Rooms or spaces have a travel distance to the room door not exceeding 25 ft (7620 mm).

**7.11.5** Means of egress, for rooms or spaces other than those that meet the criteria of 7.11.4(1) through (3), shall be arranged so that there are no dead ends in corridors.

**7.11.6** Doors serving high hazard contents areas with occupant loads in excess of five shall be permitted to be provided with a latch or lock only if the latch or lock is panic hardware or fire exit hardware complying with 7.2.1.7.

**7.12 Mechanical Equipment Rooms, Boiler Rooms, and Furnace Rooms.**

**7.12.1** Mechanical equipment rooms, boiler rooms, furnace rooms, and similar spaces shall be arranged to limit common path of travel to a distance not exceeding 50 ft (15 m), unless otherwise permitted by the following:

(1) A common path of travel not exceeding 100 ft (30 m) shall be permitted in the following locations:
   (a) In buildings protected throughout by an approved, supervised automatic sprinkler system in accordance with Section 9.7
   (b) In mechanical equipment rooms with no fuel-fired equipment
   (c) In existing buildings
(2) In an existing building, a common path of travel not exceeding 150 ft (46 m) shall be permitted, provided that all of the following criteria are met:
   (a) The building is protected throughout by an approved, supervised automatic sprinkler system installed in accordance with Section 9.7.
   (b) No fuel-fired equipment is within the space.
   (c) The egress path is readily identifiable.
(3) The requirement of 7.12.1 shall not apply to rooms or spaces in existing health care occupancies complying with the arrangement of means of egress provisions of 19.2.5 and the travel distance limits of 19.2.6.

**7.12.2** Stories used exclusively for mechanical equipment, furnaces, or boilers shall be permitted to have a single means of egress where the travel distance to an exit on that story is not in excess of the common path of travel limitations of 7.12.1.

**7.13 Normally Unoccupied Building Service Equipment Support Areas.**

**7.13.1\* Hazard of Contents.**

**7.13.1.1** Unless prohibited by Chapters 11 through 43, the provisions of Section 7.13 shall apply, in lieu of the provisions of Sections 7.1 through 7.12, to normally unoccupied building service equipment support areas where such areas do not contain high hazard contents or operations.

**7.13.1.2** Building service equipment support areas shall not contain fuel-fired equipment or be used for the storage of combustibles.

**7.13.2 Egress Doors.**

**7.13.2.1\*** Egress from normally unoccupied building service equipment support areas shall be provided by doors complying with 7.2.1 where the normally unoccupied building service equipment support area exceeds 45,000 ft² (4180 m²) in buildings not protected throughout by an approved, supervised automatic sprinkler system in accordance with 9.7.1.1(1).

**7.13.2.2** Egress from normally unoccupied building service equipment support areas shall be provided by doors complying with 7.2.1 where the normally unoccupied building service equipment support area exceeds 90,000 ft² (8370 m²) in buildings protected throughout by an approved, supervised automatic sprinkler system in accordance with 9.7.1.1(1).

**7.13.2.3** The absence of sprinklers in the normally unoccupied building service equipment support area, as permitted by an exemption of NFPA 13, *Standard for the Installation of Sprinkler Systems*, shall not cause a building to be classified as nonsprinklered for purposes of applying the provisions of 7.13.2.2.

### 7.13.3 Means of Egress Path.

**7.13.3.1** A designated means of egress path shall be provided within the normally unoccupied building service equipment support area where the normally unoccupied area exceeds 45,000 ft² (4180 m²) in buildings not protected throughout by an approved, supervised automatic sprinkler system in accordance with 9.7.1.1(1).

**7.13.3.2** A designated means of egress path shall be provided within the normally unoccupied building service equipment support area where the normally unoccupied area exceeds 90,000 ft² (8370 m²) in buildings protected throughout by an approved, supervised automatic sprinkler system in accordance with 9.7.1.1(1).

**7.13.3.3** The absence of sprinklers in the normally unoccupied building service equipment support area, as permitted by an exemption of NFPA 13, *Standard for the Installation of Sprinkler Systems*, shall not cause a building to be classified as nonsprinklered for purposes of applying the provisions of 7.13.3.2.

**7.13.3.4** Where a means of egress path is required, the path shall be a minimum of 28 in. (710 mm) clear width.

**7.13.3.5** Where a means of egress path is required, minimum headroom shall be 6 ft 8 in. (2030 mm) along the entire designated means of egress path.

**7.13.3.6** Exit signage shall not be required along the means of egress path within normally unoccupied building service equipment support areas.

**7.13.3.7** Where two means of egress are required, the means of egress path shall connect the two required means of egress.

**7.13.3.8** The designated means of egress path shall be within 25 ft (7.6 m) of any portion of the space where the only available access requires crossing over or under obstructions, unless the space is completely inaccessible.

### 7.13.4 Illumination.

**7.13.4.1** The minimum illumination of means of egress along the required means of egress path shall be 0.2 ft-candle (2.2 lux), except as otherwise provided in 7.13.4.2.

**7.13.4.2** Illumination of means of egress shall not be required in normally unoccupied building service equipment support areas where illumination of means of egress is not required by the applicable occupancy chapter for the remainder of the building.

### 7.13.5 Number of Means of Egress.

**7.13.5.1** Two remotely located means of egress shall be provided within the normally unoccupied building service equipment suport area where the normally unoccupied area exceeds 45,000 ft² (4180 m²) in buildings not protected throughout by an approved, supervised automatic sprinkler system in accordance with 9.7.1.1(1).

**7.13.5.2** Two remotely located means of egress shall be provided within the normally unoccupied building service equipment support area where the normally unoccupied area exceeds 90,000 ft² (8370 m²) in buildings protected throughout by an approved, supervised automatic sprinkler system in accordance with 9.7.1.1(1).

**7.13.5.3** The absence of sprinklers in the normally unoccupied building service equipment support area, as permitted by an exemption of NFPA 13, *Standard for the Installation of Sprinkler Systems*, shall not cause a building to be classified as nonsprinklered for purposes of applying the provisions of 7.13.5.2.

## 7.14 Elevators for Occupant-Controlled Evacuation Prior to Phase I Emergency Recall Operations.

### 7.14.1 General.

**7.14.1.1\*** Where passenger elevators for general public use are permitted to be used for occupant-controlled evacuation prior to Phase I Emergency Recall Operation mandated by the firefighters' emergency operation provisions of ASME A17.1/CSA B44, *Safety Code for Elevators and Escalators*, the elevator system shall also comply with this section, except as otherwise permitted by 7.14.1.2.

**7.14.1.2** The provisions of Section 7.14 shall not apply where the limited or supervised use of elevators for evacuation is part of a formal or informal evacuation strategy, including the relocation or evacuation of patients in health care occupancies and the relocation or evacuation of occupants with disabilities in other occupancies.

**7.14.1.3** Occupant evacuation elevators in accordance with Section 7.14 shall not be permitted to satisfy requirements of this *Code* applicable to the following:

(1) Number of means of egress
(2) Capacity of means of egress
(3) Arrangement of means of egress

### 7.14.2 Occupant Information Features.

**7.14.2.1\*** An emergency plan approved by the authority having jurisdiction shall be implemented, specifically including the procedures for occupant evacuation using the exit stairs and the occupant evacuation elevators.

**7.14.2.2** Occupant evacuation elevators shall be marked with signage indicating the elevators are suitable for use by building occupants for evacuation during fires.

### 7.14.2.3 Conditions for Safe Continued Operation.

**7.14.2.3.1** Conditions necessary for the continued safe operation of the occupant evacuation elevators and the associated elevator lobbies and elevator machine rooms shall be continuously monitored and displayed at the building emergency command center by a standard emergency service interface system meeting the requirements of *NFPA 72, National Fire Alarm and Signaling Code*, and NEMA SB 30, *Fire Service Annunciator and Interface*.

**7.14.2.3.2** The monitoring and display required by 7.14.2.3.1 shall include all of the following:

(1) Floor location of each elevator car
(2) Direction of travel of each elevator car
(3) Status of each elevator car with respect to whether it is occupied

(4) Status of normal power to the elevator equipment, elevator controller cooling equipment, and elevator machine room ventilation and cooling equipment

(5) Status of standby or emergency power system that provides backup power to the elevator equipment, elevator controller cooling equipment, and elevator machine room ventilation and cooling equipment

(6) Activation of any fire alarm–initiating device in any elevator lobby, elevator machine room or machine space, or elevator hoistway

**7.14.2.4** The building emergency command center location specified in 7.14.2.3 shall be provided with a means to override normal elevator operation and to initiate manually a Phase I Emergency Recall Operation of the occupant-controlled elevators in accordance with ASME A17.1/CSA B44, *Safety Code for Elevators and Escalators.*

**7.14.2.5** Occupant evacuation elevator lobbies shall be equipped with a status indicator arranged to display the following:

(1) Illuminated green light and the message "Elevators available for occupant evacuation" while the elevators are operating under emergency conditions but before Phase I Emergency Recall Operation in accordance with the fire fighters' emergency operation requirements of ASME A17.1/CSA B44, *Safety Code for Elevators and Escalators*

(2) Illuminated red light and the message "Elevators out of service, use exit stairs" once the elevators are under Phase I Emergency Recall Operation

(3) No illuminated light but the message "Elevators are operating normally" while the elevators are operating under nonemergency conditions

**7.14.3 Fire Detection, Alarm, and Communication.**

**7.14.3.1** The building shall be protected throughout by an approved fire alarm system in accordance with Section 9.6.

**7.14.3.2\*** The fire alarm system shall include an emergency voice/alarm communication system in accordance with *NFPA 72, National Fire Alarm and Signaling Code,* with the ability to provide voice directions on a selective basis to any building floor.

**7.14.3.3\*** The emergency voice/alarm communication system shall be arranged so that intelligible voice instructions are audible in the elevator lobbies under conditions where the elevator lobby doors are in the closed position.

**7.14.3.4 Two-way Communication System.** A two-way communication system shall be provided in each occupant evacuation elevator lobby for the purpose of initiating communication with the emergency command center or an alternative location approved by the fire department.

**7.14.3.4.1 Design and Installation.** The two-way communication system shall include audible and visible signals and shall be designed and installed in accordance with the requirements of ICC/ANSI A117.1, *American National Standard for Accessible and Usable Buildings and Facilities.*

**7.14.3.4.2 Instructions.** Instructions for the use of the two-way communication system, along with the location of the station, shall be permanently located adjacent to each station. Signage shall comply with the requirements of ICC/ANSI A117.1,

*American National Standard for Accessible and Usable Buildings and Facilities,* for visual characters.

**7.14.4 Sprinklers.**

**7.14.4.1** The building shall be protected throughout by an approved, supervised automatic sprinkler system in accordance with 9.7.1.1(1), except as otherwise specified in 7.14.4.2.

**7.14.4.1.1** A sprinkler control valve and a waterflow device shall be provided for each floor.

**7.14.4.1.2** The sprinkler control valves and waterflow devices required by 7.14.4.1.1 shall be monitored by the building fire alarm system.

**7.14.4.2\*** Sprinklers shall not be installed in elevator machine rooms serving occupant evacuation elevators, and such prohibition shall not cause an otherwise fully sprinklered building to be classified as nonsprinklered.

**7.14.4.3\*** Where a hoistway serves occupant evacuation elevators, sprinklers shall not be installed at the top of the elevator hoistway or at other points in the hoistway more than 24 in. (610 mm) above the pit floor, and such prohibition shall not cause the building to be classified as nonsprinklered.

**7.14.5 Elevator Installation.**

**7.14.5.1** Except as modified by 7.14.5.2, occupant evacuation elevators shall be installed in accordance with ASME A17.1/CSA B44, *Safety Code for Elevators and Escalators.*

**7.14.5.2\*** Shunt breakers shall not be installed on elevator systems used for occupant evacuation.

**7.14.5.3** Occupant evacuation elevators shall be limited to passenger elevators that are located in noncombustible hoistways and for which the car enclosure materials meet the requirements of ASME A17.1/CSA B44, *Safety Code for Elevators and Escalators.*

**7.14.6 Elevator Machine Rooms.**

**7.14.6.1\*** Elevator machine rooms associated with occupant evacuation elevators shall be separated from all building areas, other than elevator hoistways, by minimum 2-hour fire resistance–rated construction.

**7.14.6.2\*** Elevator machine rooms associated with occupant evacuation elevators shall be used for no purpose other than elevator machine rooms.

**7.14.7 Electrical Power and Control Wiring.**

**7.14.7.1** The following features associated with occupant evacuation elevators shall be supplied by both normal power and Type 60, Class 2, Level 1 standby power:

(1) Elevator equipment
(2) Elevator machine room ventilation and cooling equipment
(3) Elevator controller cooling equipment

**7.14.7.2\*** Wiring for power of the elevators shall meet one of the following criteria:

(1) The wiring shall utilize Type CI cable with a minimum 1-hour fire resistance rating.
(2) The wiring shall be enclosed in a minimum 1-hour fire resistance construction.

**7.14.8 Occupant Evacuation Shaft System.**

**7.14.8.1** Occupant evacuation elevators shall be provided with an occupant evacuation shaft system consisting of all of the following:

(1) Elevator hoistway
(2) Enclosed elevator lobby outside the bank or group of hoistway doors on each floor served by the elevators, with the exception that elevator lobbies not be required to be enclosed where located either on the street floor or level of exit discharge
(3) Enclosed exit stair with doors to all floors, at and above grade level, served by the elevators

**7.14.8.2 Elevator Lobby Size**

**7.14.8.2.1\*** Occupant evacuation elevator lobbies shall have minimum floor area, except as otherwise provided in 7.14.8.2.2, as follows:

(1) The elevator lobby floor area shall accommodate, at 3 ft$^2$ (0.28 m$^2$) per person, a minimum of 25 percent of the occupant load of the floor area served by the lobby.
(2) The elevator lobby floor area also shall accommodate one wheelchair space of 30 in. × 48 in. (760 mm × 1220 mm) for each 50 persons, or portion thereof, of the occupant load of the floor area served by the lobby.

**7.14.8.2.2** The size of lobbies serving multiple banks of elevators shall be exempt from the requirement of 7.14.8.2.1(1), provided that the area of such lobbies is approved on an individual basis and is consistent with the building's emergency plan.

**7.14.8.3** Access to the exit stair required by 7.14.8.1(3) shall be directly from the enclosed elevator lobby on each floor.

**7.14.8.4** The occupant evacuation shaft system shall be enclosed and separated from the remainder of the building by walls complying with the following:

(1) The shaft system walls shall be smoke barriers in accordance with Section 8.5.
(2) The shaft system walls separating the elevator lobby from the remainder of the building shall have a minimum 1-hour fire resistance rating and minimum ¾-hour fire protection–rated opening protectives.
(3) The shaft system walls separating the elevator hoistway from the remainder of the building shall have a minimum 2-hour fire resistance rating and minimum 1½-hour fire protection–rated opening protectives.
(4) The shaft system walls separating the enclosed exit stair from the remainder of the building shall have a minimum 2-hour fire resistance rating and minimum 1½-hour fire protection–rated opening protectives.

**7.14.8.5** Occupant evacuation shaft system enclosures shall be constructed to provide a minimum of classification Level 2 in accordance with ASTM C 1629/C 1629M, *Standard Classification for Abuse-Resistant Nondecorated Interior Gypsum Panel Products and Fiber-Reinforced Cement Panels.*

**7.14.8.6\*** An approved method to prevent water from infiltrating into the hoistway enclosure from the operation of the automatic sprinkler system outside the enclosed occupant evacuation elevator lobby shall be provided.

**7.14.8.7\*** Occupant evacuation shaft system elevator lobby doors shall have all of the following features:

(1) The doors shall have a fire protection rating of not less than ¾ hour.
(2) The doors shall be smoke leakage–rated assemblies in accordance with NFPA 105, *Standard for Smoke Door Assemblies and Other Opening Protectives.*
(3) The doors shall have an automatic positioning bottom seal to resist the passage of water at floor level from outside the shaft system.

**7.14.8.8** Occupant evacuation shaft system elevator lobby doors shall have the following features:

(1) Each door shall be automatic-closing in accordance with 7.2.1.8.2, as modified by 7.14.8.8(2).
(2) In addition to the automatic-closing means addressed by 7.2.1.8.2, the elevator lobby door on any floor shall also close in response to any alarm signal initiated on that floor.
(3) Each door shall be provided with a vision panel arranged to allow people within the lobby to view conditions on the other side of the door.

**7.14.8.9** Each occupant evacuation shaft system exit stair enclosure door shall be provided with a vision panel arranged to allow people on either side of the door to view conditions on the other side of the door.

## Chapter 8    Features of Fire Protection

**8.1 General.**

**8.1.1 Application.** The features of fire protection set forth in this chapter shall apply to both new construction and existing buildings.

**8.1.2 Automatic Sprinkler Systems.** Where another provision of this chapter requires an automatic sprinkler system, the automatic sprinkler system shall be installed in accordance with the subparts of 9.7.1.1, as permitted by the applicable occupancy chapter.

**8.2 Construction and Compartmentation.**

**8.2.1 Construction.**

**8.2.1.1** Buildings or structures occupied or used in accordance with the individual occupancy chapters, Chapters 11 through 43, shall meet the minimum construction requirements of those chapters.

**8.2.1.2\*** NFPA 220, *Standard on Types of Building Construction,* shall be used to determine the requirements for the construction classification.

**8.2.1.3** Where the building or facility includes additions or connected structures of different construction types, the rating and classification of the structure shall be based on one of the following:

(1) Separate buildings, if a 2-hour or greater vertically aligned fire barrier wall in accordance with NFPA 221, *Standard for High Challenge Fire Walls, Fire Walls, and Fire Barrier Walls,* exists between the portions of the building
(2) Separate buildings, if provided with previously approved separations
(3) Least fire-resistive construction type of the connected portions, if separation as specified in 8.2.1.3(1) or (2) is not provided

### 8.2.2 General.

**8.2.2.1** Where required by other chapters of this *Code*, every building shall be divided into compartments to limit the spread of fire and restrict the movement of smoke.

**8.2.2.2** Fire compartments shall be formed with fire barriers that comply with Section 8.3.

**8.2.2.3** Fire compartments shall be formed by fire barriers complying with 8.3.1.2.

**8.2.2.4** Where door assemblies are required elsewhere in this *Code* to be smoke leakage–rated in accordance with 8.2.2.4, door assemblies shall comply with all of the following:

(1) They shall be tested in accordance with ANSI/UL 1784, *Standard for Air Leakage Tests for Door Assemblies*.
(2) The maximum air leakage rate of the door assembly shall be 3.0 ft$^3$/min/ft$^2$ (0.9 m$^3$/min/m$^2$) of door opening at 0.10 in. water column (25 N/m$^2$) for both the ambient and elevated temperature tests.
(3) Door assemblies shall be installed in accordance with NFPA 105, *Standard for Smoke Door Assemblies and Other Opening Protectives*.
(4) Door assemblies shall be inspected in accordance with 7.2.1.15.

### 8.2.3 Fire Resistance–Rated Construction.

**8.2.3.1\*** The fire resistance of structural elements and building assemblies shall be determined in accordance with test procedures set forth in ASTM E 119, *Standard Test Methods for Fire Tests of Building Construction and Materials*, or ANSI/UL 263, *Standard for Fire Tests of Building Construction and Materials*; other approved test methods; or analytical methods approved by the authority having jurisdiction.

**8.2.3.1.1** Materials used to construct fire resistance–rated elements and assemblies shall be limited to those permitted in this *Code*.

**8.2.3.1.2** In new construction, end-jointed lumber used in an assembly required to have a fire resistance rating shall have the designation "Heat Resistant Adhesive" or "HRA" included in its grade mark.

**8.2.3.2** Fire resistance–rated floor and roof assemblies shall be classified as restrained or unrestrained in accordance with ASTM E 119, *Standard Test Methods for Fire Tests of Building Construction and Materials*, or ANSI/UL 263, *Standard for Fire Tests of Building Construction and Materials*; or other approved test methods. The construction shall be considered restrained only where a registered design professional has furnished the authority having jurisdiction with satisfactory documentation verifying that the construction is restrained. The classification of fire resistance–rated floor and roof construction shall be identified on the plans as restrained or unrestrained.

**8.2.3.3** Structural elements that support fire barriers shall be permitted to have only the fire resistance rating required for the construction classification of the building, provided that both of the following criteria are met:

(1) Such structural elements support nonbearing wall or partition assemblies that have a required 1-hour fire resistance rating or less.
(2) Such structural elements do not serve as exit enclosures or protection for vertical openings.

**8.2.3.4** The requirement of 8.2.3.3 shall not apply to health care occupancy structural elements supporting floor assemblies in accordance with the provisions of 18.1.6 and 19.1.6.

### 8.2.4 Analytical Methods.

**8.2.4.1** Analytical methods utilized to determine the fire resistance of building assemblies shall comply with 8.2.4.2 or 8.2.4.3.

**8.2.4.2\*** Where calculations are used to establish the fire resistance rating of structural elements or assemblies, they shall be permitted to be performed in accordance with ASCE/SFPE 29, *Standard Calculation Methods for Structural Fire Protection*. Where calculations are used to establish the fire resistance rating of concrete or masonry elements or assemblies, the provisions of ACI 216.1/TMS 0216.1, *Standard Method for Determining Fire Resistance of Concrete and Masonry Construction Assemblies*, shall be permitted to be used.

**8.2.4.3** Except for the methods specified in 8.2.4.2, analytical methods used to calculate the fire resistance of building assemblies or structural elements shall be approved. Where an approved analytical method is utilized to establish the fire resistance rating of a structural element or building assembly, the calculations shall be based upon the fire exposure and acceptance criteria specified in ASTM E 119, *Standard Test Methods for Fire Tests of Building Construction and Materials*, or ANSI/UL 263, *Standard for Fire Tests of Building Construction and Materials*.

## 8.3 Fire Barriers.

### 8.3.1 General.

**8.3.1.1** Fire barriers used to provide enclosure, subdivision, or protection under this *Code* shall be classified in accordance with one of the following fire resistance ratings:

(1) 3-hour fire resistance rating
(2) 2-hour fire resistance rating
(3) 1-hour fire resistance rating
(4)\*½-hour fire resistance rating

**8.3.1.2\*** Fire barriers shall comply with one of the following:

(1) The fire barriers are continuous from outside wall to outside wall or from one fire barrier to another, or a combination thereof, including continuity through all concealed spaces, such as those found above a ceiling, including interstitial spaces.
(2) The fire barriers are continuous from outside wall to outside wall or from one fire barrier to another, and from the floor to the bottom of the interstitial space, provided that the construction assembly forming the bottom of the interstitial space has a fire resistance rating not less than that of the fire barrier.

**8.3.1.3** Walls used as fire barriers shall comply with Chapter 7 of NFPA 221, *Standard for High Challenge Fire Walls, Fire Walls, and Fire Barrier Walls*. The NFPA 221 limitation on percentage width of openings shall not apply.

### 8.3.2 Walls.

**8.3.2.1** The fire-resistive materials, assemblies, and systems used shall be limited to those permitted in this *Code* and this chapter.

**8.3.2.1.1\*** Fire resistance–rated glazing tested in accordance with ASTM E 119, *Standard Test Methods for Fire Tests of Building Construction and Materials*, or ANSI/UL 263, *Standard for Fire Tests of Building Construction and Materials*, shall be permitted.

**8.3.2.1.2** New fire resistance–rated glazing shall bear the identifier "W-XXX" where "XXX" is the fire resistance rating in minutes. Such identification shall be permanently affixed.

**8.3.2.2** The construction materials and details for fire-resistive assemblies and systems for walls described shall comply with all other provisions of this *Code*, except as modified herein.

**8.3.2.3** Interior walls and partitions of nonsymmetrical construction shall be evaluated from both directions and assigned a fire resistance rating based on the shorter duration obtained in accordance with ASTM E 119, *Standard Test Methods for Fire Tests of Building Construction and Materials*, or ANSI/UL 263, *Standard for Fire Tests of Building Construction and Materials*. When the wall is tested with the least fire-resistive side exposed to the furnace, the wall shall not be required to be subjected to tests from the opposite side.

### 8.3.3 Fire Doors and Windows.

**8.3.3.1** Openings required to have a fire protection rating by Table 8.3.4.2 shall be protected by approved, listed, labeled fire door assemblies and fire window assemblies and their accompanying hardware, including all frames, closing devices, anchorage, and sills in accordance with the requirements of NFPA 80, *Standard for Fire Doors and Other Opening Protectives*, except as otherwise specified in this *Code*.

**8.3.3.1.1** Fire resistance–rated glazing tested in accordance with ASTM E 119, *Standard Test Methods for Fire Tests of Building Construction and Materials*, or ANSI/UL 263, *Standard for Fire Tests of Building Construction and Materials*, shall be permitted in fire door assemblies and fire window assemblies where tested and installed in accordance with their listings.

**8.3.3.1.2** New fire resistance–rated glazing shall be marked in accordance with Table 8.3.3.12 and Table 8.3.4.2. Such marking shall be permanently affixed.

**8.3.3.2\*** Fire protection ratings for products required to comply with 8.3.3 shall be as determined and reported by a nationally recognized testing agency in accordance with NFPA 252, *Standard Methods of Fire Tests of Door Assemblies;* ANSI/UL 10B, *Standard for Fire Tests of Door Assemblies;* ANSI/UL 10C, *Standard for Positive Pressure Fire Tests of Door Assemblies;* NFPA 257, *Standard on Fire Test for Window and Glass Block Assemblies;* or ANSI/UL 9, *Standard for Fire Tests of Window Assemblies.*

**8.3.3.2.1** Fire protection–rated glazing shall be evaluated under positive pressure in accordance with NFPA 257, *Standard on Fire Test for Window and Glass Block Assemblies.*

**8.3.3.2.2** All products required to comply with 8.3.3.2 shall bear an approved label.

**8.3.3.2.3\*** Labels on fire door assemblies shall be maintained in a legible condition.

**8.3.3.3** Unless otherwise specified, fire doors shall be self-closing or automatic-closing in accordance with 7.2.1.8.

**8.3.3.4** Floor fire door assemblies shall be tested in accordance with NFPA 288, *Standard Methods of Fire Tests of Floor Fire Door Assemblies Installed Horizontally in Fire Resistance–Rated Floor Systems*, and shall achieve a fire resistance rating not less than the assembly being penetrated. Floor fire door assemblies shall be listed and labeled.

**8.3.3.5** Fire protection–rated glazing shall be permitted in fire barriers having a required fire resistance rating of 1 hour or less and shall be of an approved type with the appropriate fire protection rating for the location in which the barriers are installed.

**8.3.3.6\*** Glazing in fire window assemblies, other than in existing fire window installations of wired glass and other fire-rated glazing material, shall be of a design that has been tested to meet the conditions of acceptance of NFPA 257, *Standard on Fire Test for Window and Glass Block Assemblies*, or ANSI/UL 9, *Standard for Fire Tests of Window Assemblies*. Fire protection–rated glazing in fire door assemblies, other than in existing fire-rated door assemblies, shall be of a design that has been tested to meet the conditions of acceptance of NFPA 252, *Standard Methods of Fire Tests of Door Assemblies;* ANSI/UL 10B, *Standard for Fire Tests of Door Assemblies;* or ANSI/UL 10C, *Standard for Positive Pressure Fire Tests of Door Assemblies.*

**8.3.3.7** Fire resistance–rated glazing complying with 8.3.2.1.1 shall be permitted in fire doors and fire window assemblies in accordance with their listings.

**8.3.3.8** Glazing materials that have been tested, listed, and labeled to indicate the type of opening to be protected for fire protection purposes shall be permitted to be used in approved opening protectives in accordance with Table 8.3.4.2 and in sizes in accordance with NFPA 80, *Standard for Fire Doors and Other Opening Protectives.*

**8.3.3.9** Existing installations of wired glass of ¼ in. (6.3 mm) thickness and labeled for fire protection purposes shall be permitted to be used in approved opening protectives, provided that the maximum size specified by the listing is not exceeded.

**8.3.3.10** Nonsymmetrical fire protection–rated glazing systems shall be tested with each face exposed to the furnace, and the assigned fire protection rating shall be that of the shortest duration obtained from the two tests conducted in compliance with NFPA 257, *Standard on Fire Test for Window and Glass Block Assemblies*, or ANSI/UL 9, *Standard for Fire Tests of Window Assemblies.*

**8.3.3.11** The total combined area of glazing in fire-rated window assemblies and fire-rated door assemblies used in fire barriers shall not exceed 25 percent of the area of the fire barrier that is common with any room, unless the installation meets one of the following criteria:

(1) The installation is an existing fire window installation of wired glass and other fire-rated glazing materials in approved frames.

(2) The fire protection–rated glazing material is installed in approved existing frames.

**8.3.3.12** New fire protection-rated glazing shall be marked in accordance with Table 8.3.3.12 and Table 8.3.4.2, and such marking shall be permanently affixed.

### 8.3.4 Opening Protectives.

**8.3.4.1** Every opening in a fire barrier shall be protected to limit the spread of fire and restrict the movement of smoke from one side of the fire barrier to the other.

**8.3.4.2\*** The fire protection rating for opening protectives in fire barriers, fire-rated smoke barriers, and fire-rated smoke partitions shall be in accordance with Table 8.3.4.2, except as otherwise permitted in 8.3.4.3 or 8.3.4.4.

Table 8.3.3.12 Marking Fire-Rated Glazing Assemblies

| Fire Test Standard | Marking | Definition of Marking |
|---|---|---|
| ASTM E119, or ANSI/UL 263[a] | W | Meets wall assembly criteria |
| NFPA 257 | OH | Meets fire window assembly criteria, including the hose stream test |
| NFPA 252 | D | Meets fire door assembly criteria |
| | H | Meets fire door assembly hose stream test |
| | T | Meets 450° F (232°C) temperature rise criteria for 30 minutes |
| | XXX | The time, in minutes, of fire resistance or fire protection rating of the glazing assembly |

[a]ASTM E 119, *Standard Test Methods for Fire Tests of Building Construction and Materials* and ANSI/UL 263, *Standard for Fire Tests of Building Construction and Materials.*

---

Table 8.3.4.2 was revised by a tentative interim amendment (TIA). See page 1.

---

Table 8.3.4.2 Minimum Fire Protection Ratings for Opening Protectives in Fire Resistance–Rated Assemblies and Fire-Rated Glazing Markings

| Component | Walls and Partitions (hr) | Fire Door Assemblies (hr) | Door Vision Panel Maximum Size (in²)[a] | Fire-Rated Glazing Marking Door Vision Panel | Minimum Side Light/Transom Assembly Rating (hr) — Fire Protection | Fire Resistance | Fire-Rated Glazing Marking Side Light/Transom Panel — Fire Protection | Fire Resistance | Fire Window Assemblies[b, c] (hr) | Fire-Rated Glazing Marking Window |
|---|---|---|---|---|---|---|---|---|---|---|
| Elevator hoistways | 2 | 1½ | 155 in.²  d | D-H-90 or D-H-W-90 | NP | 2 | NP | D-H-W-120 | NP | W-120 |
| | 1 | 1 | 155 in.²  d | D-H-60 or D-H-W-60 | NP | 1 | NP | D-H-W-60 | NP | W-60 |
| | ½ | ⅓ | 85 in.²  e | D-20 or D-W-20 | ⅓ | ⅓ | D-H-20 | D-W-20 | NP | W-30 |
| Elevator lobby (per 7.2.13.4) | 1 | 1 | 100 in.²  b | ≤100 in.², D-H-T-60 or D-H-W-60[a] >100 in.², D-H-W-60[a] | NP | 1 | NP | D-H-W-60 | NP | W-60 |
| Vertical shafts, including stairways, exits, and refuse chutes | 2 | 1½ | Maximum size tested | D-H-90 or D-H-W-90 | NP | 2 | NP | D-H-W-120 | NP | W-120 |
| | 1 | 1 | Maximum size tested | D-H-60 or D-H-W-60 | NP | 1 | NP | D-H-W-60 | NP | W-60 |
| Replacement panels in existing vertical shafts | ½ | ⅓ | Maximum size tested | D-20 or D-W-20 | ⅓ | ⅓ | D-H-20 | D-W-20 | NP | W-30 |

*(continues)*

**Table 8.3.4.2** *Continued*

| Component | Walls and Partitions (hr) | Fire Door Assemblies (hr) | Door Vision Panel Maximum Size (in²)[a] | Fire-Rated Glazing Marking Door Vision Panel | Minimum Side Light/Transom Assembly Rating (hr) Fire Protection | Fire Resistance | Fire-Rated Glazing Marking Side Light/Transom Panel Fire Protection | Fire Resistance | Fire Window Assemblies[b,c] (hr) | Fire-Rated Glazing Marking Window |
|---|---|---|---|---|---|---|---|---|---|---|
| Fire barriers | 3 | 3 | 100 in.²  [b] | ≤100 in.², D-H-180 or D-H-W-180[h] | NP | 3 | NP | D-H-W-180 | NP | W-180 |
|  |  |  |  | >100 in.², D-H-W-180[h] |  |  |  |  |  |  |
|  | 2 | 1½ | Maximum size tested | D-H-90 or D-H-W-90 | NP | 2 | NP | D-H-W-120 | NP | W-120 |
|  | 1 | ¾ | Maximum size tested[f] | D-H-45 or D-H-W-45 | ¾[f] | ¾[f] | D-H-45 | D-H-W-45 | ¾ | OH-45 or W-60 |
|  | ½ | ⅓ | Maximum size tested | D-20 or D-W-20 | ⅓ | ⅓ | D-H-20 | D-W-20 | ⅓ | OH-20 or W-30 |
| Horizontal exits | 2 | 1½ | Maximum size tested | D-H-90 or D-H-W-90 | NP | 2 | NP | D-H-W-120 | NP | W-120 |
| Horizontal exits served by bridges between buildings | 2 | ¾ | Maximum size tested[f] | D-H-45 or D-H-W-45 | ¾[f] | ¾[f] | D-H-45 | D-H-W-45 | ¾ | OH-45 or W-120 |
| Exit access corridors[g] | 1 | ⅓ | Maximum size tested | D-20 or D-W-20 | ¾ | ¾ | D-H-45 | D-H-W-20 | ¾ | OH-45 or W-60 |
|  | ½ | ⅓ | Maximum size tested | D-20 or D-W-20 | ⅓ | ⅓ | D-H-20 | D-H-W-20 | ⅓ | OH-20 or W-30 |
| Smoke barriers[a] | 1 | ⅓ | Maximum size tested | D-20 or D-W-20 | ¾ | ¾ | D-H-45 | D-H-W-20 | ¾ | OH-45 or W-60 |
| Smoke partitions[g, h] | ½ | ⅓ | Maximum size tested | D-20 or D-W-20 | ⅓ | ⅓ | D-H-20 | D-H-W-20 | ⅓ | OH-20 or W-30 |

NP: Not permitted.

[a]Note: 1 inch² = .00064516 m².

[b]Fire resistance–rated glazing tested to ASTM E 119, *Standard Test Methods for Fire Tests of Building Construction and Materials*, or ANSI/UL 263, *Standard for Fire Tests of Building Construction and Materials*, shall be permitted in the maximum size tested. *(See 8.3.3.7.)*

[c]Fire-rated glazing in exterior windows shall be marked in accordance with Table 8.3.3.12.

[d]See ASME A17.1, *Safety Code for Elevators and Escalators*, for additional information.

[e]See ASTM A17.3, *Safety Code for Existing Elevators and Escalators*, for additional information.

[f]Maximum area of individual exposed lights shall be 1296 in.² (0.84 m²) with no dimension exceeding 54 in. (1.37 m) unless otherwise tested. [**80:** Table 4.4.5, Note b, and 80:4.4.5.1]

[g]Fire doors are not required to have a hose stream test per NFPA 252, *Standard Methods of Fire Tests of Door Assemblies*; ASTM E 2074, *Standard Test Method for Fire Tests of Door Assemblies, Including Positive Pressure Testing of Side-Hinged and Pivoted Swinging Door Assemblies*; ANSI/UL 10B, *Standard for Fire Tests of Door Assemblies*; or ANSI/UL 10C, *Standard for Positive Pressure Fire Tests of Door Assemblies*.

[h]For residential board and care, see 32.2.3.1 and 33.2.3.1.

**8.3.4.2.1** Fire-rated glazing assemblies marked as complying with hose stream requirements (H) shall be permitted in applications that do not require compliance with hose stream requirements. Fire-rated glazing assemblies marked as complying with temperature rise requirements (T) shall be permitted in applications that do not require compliance with temperature rise requirements. Fire-rated glazing assemblies marked with ratings that exceed the ratings required by this *Code* (XXX) shall be permitted.

**8.3.4.3** Existing fire door assemblies having a minimum ¾-hour fire protection rating shall be permitted to continue to be used in vertical openings and in exit enclosures in lieu of the minimum 1-hour fire protection rating required by Table 8.3.4.2.

**8.3.4.4** Where a 20-minute fire protection–rated door is required in existing buildings, an existing 1¾ in. (44 mm) solid-bonded wood-core door, an existing steel-clad (tin-clad) wood door, or an existing solid-core steel door with positive latch and closer shall be permitted, unless otherwise specified by Chapters 11 through 43.

**8.3.5 Penetrations.** The provisions of 8.3.5 shall govern the materials and methods of construction used to protect through-penetrations and membrane penetrations in fire walls, fire barrier walls, and fire resistance–rated horizontal assemblies. The provisions of 8.3.5 shall not apply to approved existing materials and methods of construction used to protect existing through-penetrations and existing membrane penetrations in fire walls, fire barrier walls, or fire resistance–rated horizontal assemblies, unless otherwise required by Chapters 11 through 43.

**8.3.5.1\* Firestop Systems and Devices Required.** Penetrations for cables, cable trays, conduits, pipes, tubes, combustion vents and exhaust vents, wires, and similar items to accommodate electrical, mechanical, plumbing, and communications systems that pass through a wall, floor, or floor/ceiling assembly constructed as a fire barrier shall be protected by a firestop system or device. The firestop system or device shall be tested in accordance with ASTM E 814, *Standard Test Method for Fire Tests of Through Penetration Fire Stops*, or ANSI/UL 1479, *Standard for Fire Tests of Through-Penetration Firestops*, at a minimum positive pressure differential of 0.01 in. water column (2.5 N/m²) between the exposed and the unexposed surface of the test assembly.

**8.3.5.1.1** The requirements of 8.3.5.1 shall not apply where otherwise permitted by any one of the following:

(1) Where penetrations are tested and installed as part of an assembly tested and rated in accordance with ASTM E 119, *Standard Test Methods for Fire Tests of Building Construction and Materials*, or ANSI/UL 263, *Standard for Fire Tests of Building Construction and Materials*
(2) Where penetrations through floors are enclosed in a shaft enclosure designed as a fire barrier
(3) Where concrete, grout, or mortar has been used to fill the annular spaces around cast-iron, copper, or steel piping that penetrates one or more concrete or masonry fire resistance–rated assemblies and both of the following criteria are also met:
  (a) The nominal diameter of each penetrating item shall not exceed 6 in. (150 mm), and the opening size shall not exceed 1 ft² (0.09 m²).
  (b) The thickness of the concrete, grout, or mortar shall be the full thickness of the assembly.

(4) Where firestopping materials are used with the following penetrating items, the penetration is limited to one floor, and the firestopping material is capable of preventing the passage of flame and hot gases sufficient to ignite cotton waste when subjected to the time–temperature fire conditions of ASTM E119 under a minimum positive pressure differential of 0.01 in. water column (2.5 Pa) at the location of the penetration for the time period equivalent to the required fire resistance rating of the assembly penetrated:

  (a) Steel, ferrous, or copper cables
  (b) Cable or wire with steel jackets
  (c) Cast-iron, steel, or copper pipes
  (d) Steel conduit or tubing

**8.3.5.1.2** The maximum nominal diameter of the penetrating item, as indicated in 8.3.5.1.1(4)(a) through (d), shall not be greater than 4 in. (100 mm) and shall not exceed an aggregate 100 in.² (64,520 mm²) opening in any 100 ft² (9.3 m²) of floor or wall area.

**8.3.5.1.3** Firestop systems and devices shall have a minimum 1-hour F rating, but not less than the required fire resistance rating of the fire barrier penetrated.

**8.3.5.1.4** Penetrations in fire-rated horizontal assemblies shall have a minimum 1-hour T rating, but not less than the fire resistance rating of the horizontal assembly. Rated penetrations shall not be required for either of the following:

(1) Floor penetrations contained within the cavity of a wall assembly
(2) Penetrations through floors or floor assemblies where the penetration is not in direct contact with combustible material

**8.3.5.2 Sleeves.** Where the penetrating item uses a sleeve to penetrate the wall or floor, the sleeve shall be securely set in the wall or floor, and the space between the item and the sleeve shall be filled with a material that complies with 8.3.5.1.

**8.3.5.3 Insulation and Coverings.** Insulation and coverings for penetrating items shall not pass through the wall or floor unless the insulation or covering has been tested as part of the firestop system or device.

**8.3.5.4 Transmission of Vibrations.** Where designs take transmission of vibrations into consideration, any vibration isolation shall meet one of the following conditions:

(1) It shall be provided on either side of the wall or floor.
(2) It shall be designed for the specific purpose.

**8.3.5.5 Transitions.**

**8.3.5.5.1** Where piping penetrates a fire resistance–rated wall or floor assembly, combustible piping shall not connect to noncombustible piping within 36 in. (915 mm) of the firestop system or device without demonstration that the transition will not reduce the fire resistance rating, except in the case of previously approved installations.

**8.3.5.5.2** Unshielded couplings shall not be used to connect noncombustible piping to combustible piping unless it can be demonstrated that the transition complies with the fire-resistive requirements of 8.3.5.1.

**8.3.5.6 Membrane Penetrations.**

**8.3.5.6.1** Membrane penetrations for cables, cable trays, conduits, pipes, tubes, combustion vents and exhaust vents, wires, and similar items to accommodate electrical, mechanical, plumbing, and communications systems that pass through a membrane of a wall, floor, or floor/ceiling assembly constructed as a fire barrier shall be protected by a firestop system or device and shall comply with 8.3.5.1 through 8.3.5.5.2.

**8.3.5.6.2** The firestop system or device shall be tested in accordance with ASTM E 814, *Standard Test Method for Fire Tests of Through Penetration Fire Stops*, or ANSI/UL 1479, *Standard for Fire Tests of Through-Penetration Firestops*, at a minimum positive pressure differential of 0.01 in. water column ($2.5 \text{ N/m}^2$) between the exposed and the unexposed surface of the test assembly, unless one of the following applies:

(1) Membrane penetrations of ceilings that are not an integral part of a fire resistance–rated floor/ceiling or roof/ceiling assembly shall be permitted.
(2) Membrane penetrations of steel, ferrous, or copper conduits, and pipes, tubes, or combustion vents or exhaust vents, shall be permitted where the annular space is protected with an approved material and the aggregate area of the openings does not exceed 0.7 $\text{ft}^2$ ($0.06 \text{ m}^2$) in any 100 $\text{ft}^2$ ($9.3 \text{ m}^2$) of ceiling area.
(3) Electrical outlet boxes and fittings shall be permitted, provided that such devices are listed for use in fire resistance–rated assemblies and are installed in accordance with their listing.
(4) The annular space created by the membrane penetration of a fire sprinkler shall be permitted, provided that the space is covered by a metal escutcheon plate.

**8.3.5.6.3** Where walls or partitions are required to have a minimum 1-hour fire resistance rating, recessed fixtures shall be installed in the wall or partition in such a manner that the required fire resistance is not reduced, unless one of the following is met:

(1) Any steel electrical box not exceeding 0.1 $\text{ft}^2$ ($0.01 \text{ m}^2$) shall be permitted where the aggregate area of the openings provided for the boxes does not exceed 0.7 $\text{ft}^2$ ($0.06 \text{ m}^2$) in any 100 $\text{ft}^2$ ($9.3 \text{ m}^2$) of wall area, and, where outlet boxes are installed on opposite sides of the wall, the boxes shall be separated by one of the following:
    (a) Horizontal distance of not less than 24 in. (610 mm)
    (b) Horizontal distance of not less than the depth of the wall cavity, where the wall cavity is filled with cellulose loose-fill, rock wool, or slag wool insulation
    (c)*Solid fireblocking
    (d) Other listed materials and methods
(2) Membrane penetrations for any listed electrical outlet box made of any material shall be permitted, provided that such boxes have been tested for use in fire resistance–rated assemblies and are installed in accordance with the instructions included in the listing.
(3) The annular space created by the membrane penetration of a fire sprinkler shall be permitted, provided that the space is covered by a metal escutcheon plate.

**8.3.5.7 Openings for Air-Handling Ductwork.** Openings in fire barriers for air-handling ductwork or air movement shall be protected in accordance with 9.2.1.

**8.3.6 Joints.**

**8.3.6.1** The provisions of 8.3.6 shall govern the materials and methods of construction used to protect joints in between and at the perimeter of fire barriers or, where fire barriers meet other fire barriers, the floor or roof deck above, or the outside walls. The provisions of 8.3.6 shall not apply to approved existing materials and methods of construction used to protect existing joints in fire barriers, unless otherwise required by Chapters 11 through 43.

**8.3.6.2** Joints made within or at the perimeter of fire barriers shall be protected with a joint system that is capable of limiting the transfer of smoke.

**8.3.6.3** Joints made within or between fire barriers shall be protected with a smoke-tight joint system that is capable of limiting the transfer of smoke.

**8.3.6.4** Testing of the joint system in a fire barrier shall be representative of the actual installation suitable for the required engineering demand without compromising the fire resistance rating of the assembly or the structural integrity of the assembly.

**8.3.6.5\*** Joints made within or between fire resistance–rated assemblies shall be protected with a joint system that is designed and tested to prevent the spread of fire for a time period equal to that of the assembly in which the joint is located. Such materials, systems, or devices shall be tested as part of the assembly in accordance with the requirements of ASTM E 1966, *Standard Test Method for Fire-Resistive Joint Systems*, or ANSI/UL 2079, *Standard for Tests for Fire Resistance of Building Joint Systems*.

**8.3.6.6** All joint systems shall be tested at their maximum joint width in accordance with the requirements of ASTM E 1966, *Standard Test Method for Fire-Resistive Joint Systems*, or ANSI/UL 2079, *Standard for Tests for Fire Resistance of Building Joint Systems*, under a minimum positive pressure differential of 0.01 in. water column ($2.5 \text{ N/m}^2$) for a time period equal to that of the assembly. All test specimens shall comply with the minimum height or length required by the standard. Wall assemblies shall be subjected to a hose stream test in accordance with ASTM E 119, *Standard Test Methods for Fire Tests of Building Construction and Materials*, or ANSI/UL 263, *Standard for Fire Tests of Building Construction and Materials*.

**8.3.6.7\* Exterior Curtain Walls and Perimeter Joints.**

**8.3.6.7.1** Voids created between the fire resistance–rated floor assembly and the exterior curtain wall shall be protected with a perimeter joint system that is designed and tested in accordance with ASTM E 2307, *Standard Test Method for Fire Resistance of Perimeter Fire Barriers Using Intermediate-Scale, Multi-story Apparatus*.

**8.3.6.7.2** The perimeter joint system shall have an F rating equal to the fire resistance rating of the floor assembly.

**8.4 Smoke Partitions.**

**8.4.1\* General.** Where required elsewhere in this *Code*, smoke partitions shall be provided to limit the transfer of smoke.

**8.4.2 Continuity.** Smoke partitions shall comply with the following:

(1) They shall extend from the floor to the underside of the floor or roof deck above, through any concealed spaces, such as those above suspended ceilings, and through interstitial structural and mechanical spaces.
(2)*They shall be permitted to extend from the floor to the underside of a monolithic or suspended ceiling system where all of the following conditions are met:

(a) The ceiling system forms a continuous membrane.
(b) A smoke-tight joint is provided between the top of the smoke partition and the bottom of the suspended ceiling.
(c) The space above the ceiling is not used as a plenum.

(3) Smoke partitions enclosing hazardous areas shall be permitted to terminate at the underside of a monolithic or suspended ceiling system where all of the following conditions are met:

(a) The ceiling system forms a continuous membrane.
(b) A smoke-tight joint is provided between the top of the smoke partition and the bottom of the suspended ceiling.
(c) Where the space above the ceiling is used as a plenum, return grilles from the hazardous area into the plenums are not permitted.

**8.4.3 Opening Protectives.**

**8.4.3.1** Doors in smoke partitions shall comply with 8.4.3.2 through 8.4.3.5.

**8.4.3.2** Doors shall comply with the provisions of 7.2.1.

**8.4.3.3** Doors shall not include louvers.

**8.4.3.4\*** Door clearances shall be in accordance with NFPA 80, *Standard for Fire Doors and Other Opening Protectives.*

**8.4.3.5** Doors shall be self-closing or automatic-closing in accordance with 7.2.1.8.

**8.4.4 Penetrations.** The provisions of 8.4.4 shall govern the materials and methods of construction used to protect through-penetrations and membrane penetrations of smoke partitions.

**8.4.4.1** Penetrations for cables, cable trays, conduits, pipes, tubes, vents, wires, and similar items to accommodate electrical, mechanical, plumbing, and communications systems that pass through a smoke partition shall be protected by a system or material that is capable of limiting the transfer of smoke.

**8.4.4.2** Where designs take transmission of vibrations into consideration, any vibration isolation shall meet one of the following conditions:

(1) It shall be provided on either side of the smoke partition.
(2) It shall be designed for the specific purpose.

**8.4.5 Joints.**

**8.4.5.1** The provisions of 8.4.5 shall govern the materials and methods of construction used to protect joints in between and at the perimeter of smoke partitions or, where smoke partitions meet other smoke partitions, the floor or roof deck above, or the outside walls. The provisions of 8.4.5 shall not apply to approved existing materials and methods of construction used to protect existing joints in smoke partitions, unless otherwise required by Chapters 11 through 43.

**8.4.5.2** Joints made within or at the perimeter of smoke partitions shall be protected with a joint system that is capable of limiting the transfer of smoke.

**8.4.6 Air-Transfer Openings.**

**8.4.6.1 General.** The provisions of 8.4.6 shall govern the materials and methods of construction used to protect air-transfer openings in smoke partitions.

**8.4.6.2\* Smoke Dampers.** Air-transfer openings in smoke partitions shall be provided with approved smoke dampers designed and tested in accordance with the requirements of ANSI/UL 555S, *Standard for Smoke Dampers,* to limit the transfer of smoke.

**8.4.6.3 Smoke Damper Ratings.** Smoke damper leakage ratings shall be not less than Class II. Elevated temperature ratings shall be not less than 250°F (140°C).

**8.4.6.4 Smoke Detectors.** Dampers in air-transfer openings shall close upon detection of smoke by approved smoke detectors installed in accordance with *NFPA 72, National Fire Alarm and Signaling Code.*

**8.5 Smoke Barriers.**

**8.5.1\* General.** Where required by Chapters 11 through 43, smoke barriers shall be provided to subdivide building spaces for the purpose of restricting the movement of smoke.

**8.5.2\* Continuity.**

**8.5.2.1** Smoke barriers required by this *Code* shall be continuous from an outside wall to an outside wall, from a floor to a floor, or from a smoke barrier to a smoke barrier, or by use of a combination thereof.

**8.5.2.2** Smoke barriers shall be continuous through all concealed spaces, such as those found above a ceiling, including interstitial spaces.

**8.5.2.3** A smoke barrier required for an occupied space below an interstitial space shall not be required to extend through the interstitial space, provided that the construction assembly forming the bottom of the interstitial space provides resistance to the passage of smoke equal to that provided by the smoke barrier.

**8.5.3 Fire Barrier Used as Smoke Barrier.** A fire barrier shall be permitted to be used as a smoke barrier, provided that it meets the requirements of Section 8.5.

**8.5.4 Opening Protectives.**

**8.5.4.1\*** Doors in smoke barriers shall close the opening, leaving only the minimum clearance necessary for proper operation, and shall be without louvers or grilles. The clearance under the bottom of a new door shall be a maximum of ¾ in. (19 mm).

**8.5.4.2** Where required by Chapters 11 through 43, doors in smoke barriers that are required to be smoke leakage–rated shall comply with the requirements of 8.2.2.4.

**8.5.4.3** Latching hardware shall be required on doors in smoke barriers, unless specifically exempted by Chapters 11 through 43.

**8.5.4.4\*** Doors in smoke barriers shall be self-closing or automatic-closing in accordance with 7.2.1.8 and shall comply with the provisions of 7.2.1.

**8.5.4.5** Fire window assemblies shall comply with 8.3.3.

**8.5.5 Ducts and Air-Transfer Openings.**

**8.5.5.1 General.** The provisions of 8.5.5 shall govern the materials and methods of construction used to protect ducts and air-transfer openings in smoke barriers.

**8.5.5.2 Smoke Dampers.** Where a smoke barrier is penetrated by a duct or air-transfer opening, a smoke damper designed and tested in accordance with the requirements of

ANSI/UL 555S, *Standard for Smoke Dampers*, shall be installed. Where a smoke barrier is also constructed as a fire barrier, a combination fire/smoke damper designed and tested in accordance with the requirements of ANSI/UL 555, *Standard for Fire Dampers*, and ANSI/UL 555S, *Standard for Smoke Dampers*, shall be installed.

**8.5.5.3 Smoke Damper Exemptions.** Smoke dampers shall not be required under any of the following conditions:

(1) Where specifically exempted by provisions in Chapters 11 through 43
(2) Where ducts or air-transfer openings are part of an engineered smoke control system
(3) Where the air in ducts continues to move and the air-handling system installed is arranged to prevent recirculation of exhaust or return air under fire emergency conditions
(4) Where the air inlet or outlet openings in ducts are limited to a single smoke compartment
(5) Where ducts penetrate floors that serve as smoke barriers
(6) Where ducts penetrate smoke barriers forming a communicating space separation in accordance with 8.6.6(4)(a).

**8.5.5.4 Installation, Testing, and Maintenance.**

**8.5.5.4.1** Air-conditioning, heating, ventilating ductwork, and related equipment, including smoke dampers and combination fire and smoke dampers, shall be installed in accordance with NFPA 90A, *Standard for the Installation of Air-Conditioning and Ventilating Systems*, and NFPA 105, *Standard for Smoke Door Assemblies and Other Opening Protectives*.

**8.5.5.4.2** Smoke dampers and combination fire and smoke dampers required by this *Code* shall be inspected, tested, and maintained in accordance with NFPA 105, *Standard for Smoke Door Assemblies and Other Opening Protectives*.

**8.5.5.4.3** The equipment specified in 8.5.5.4.1 shall be installed in accordance with the requirements of 8.5.5, the manufacturer's installation instructions, and the equipment listing.

**8.5.5.5 Access and Identification.** Access to the dampers shall be provided for inspection, testing, and maintenance. The access openings shall not reduce the fire resistance rating of the fire barrier assembly.

**8.5.5.6 Smoke Damper Ratings.** Smoke damper leakage ratings shall be not less than Class II. Elevated temperature ratings shall be not less than 250°F (140°C).

**8.5.5.7 Smoke Detectors.**

**8.5.5.7.1** Required smoke dampers in ducts penetrating smoke barriers shall close upon detection of smoke by approved smoke detectors in accordance with *NFPA 72, National Fire Alarm and Signaling Code*, unless one of the following conditions exists:

(1) The ducts penetrate smoke barriers above the smoke barrier doors, and the door release detector actuates the damper.
(2) Approved smoke detector installations are located within the ducts in existing installations.

**8.5.5.7.2** Where a duct is provided on one side of the smoke barrier, the smoke detectors on the duct side shall be in accordance with 8.5.5.7.1.

**8.5.5.7.3** Required smoke dampers in air-transfer openings shall close upon detection of smoke by approved smoke detectors in accordance with *NFPA 72, National Fire Alarm and Signaling Code*.

**8.5.6 Penetrations.**

**8.5.6.1** The provisions of 8.5.6 shall govern the materials and methods of construction used to protect through-penetrations and membrane penetrations of smoke barriers.

**8.5.6.2** Penetrations for cables, cable trays, conduits, pipes, tubes, vents, wires, and similar items to accommodate electrical, mechanical, plumbing, and communications systems that pass through a wall, floor, or floor/ceiling assembly constructed as a smoke barrier, or through the ceiling membrane of the roof/ceiling of a smoke barrier assembly, shall be protected by a system or material capable of restricting the transfer of smoke.

**8.5.6.3** Where a smoke barrier is also constructed as a fire barrier, the penetrations shall be protected in accordance with the requirements of 8.3.5 to limit the spread of fire for a time period equal to the fire resistance rating of the assembly and 8.5.6 to restrict the transfer of smoke, unless the requirements of 8.5.6.4 are met.

**8.5.6.4** Where sprinklers penetrate a single membrane of a fire resistance–rated assembly in buildings equipped throughout with an approved automatic fire sprinkler system, noncombustible escutcheon plates shall be permitted, provided that the space around each sprinkler penetration does not exceed ½ in. (13 mm), measured between the edge of the membrane and the sprinkler.

**8.5.6.5** Where the penetrating item uses a sleeve to penetrate the smoke barrier, the sleeve shall be securely set in the smoke barrier, and the space between the item and the sleeve shall be filled with a material capable of restricting the transfer of smoke.

**8.5.6.6** Where designs take transmission of vibrations into consideration, any vibration isolation shall meet one of the following conditions:

(1) It shall be provided on either side of the smoke barrier.
(2) It shall be designed for the specific purpose.

**8.5.7 Joints.**

**8.5.7.1** The provisions of 8.5.7 shall govern the materials and methods of construction used to protect joints in between and at the perimeter of smoke barriers or, where smoke barriers meet other smoke barriers, the floor or roof deck above, or the outside walls. The provisions of 8.5.7 shall not apply to approved existing materials and methods of construction used to protect existing joints in smoke barriers, unless otherwise required by Chapters 11 through 43.

**8.5.7.2** Joints made within or at the perimeter of smoke barriers shall be protected with a joint system that is capable of limiting the transfer of smoke.

**8.5.7.3** Joints made within or between smoke barriers shall be protected with a smoke-tight joint system that is capable of limiting the transfer of smoke.

**8.5.7.4** Smoke barriers that are also constructed as fire barriers shall be protected with a joint system that is designed and tested to resist the spread of fire for a time period equal to the required fire resistance rating of the assembly and restrict the transfer of smoke.

**8.5.7.5** Testing of the joint system in a smoke barrier that also serves as fire barrier shall be representative of the actual installation suitable for the required engineering demand without compromising the fire resistance rating of the assembly or the structural integrity of the assembly.

**8.6 Vertical Openings.**

**8.6.1 Floor Smoke Barriers.** Every floor that separates stories in a building shall meet the following criteria:

(1) It shall be constructed as a smoke barrier in accordance with Section 8.5.
(2) It shall be permitted to have openings as described by 8.6.6, 8.6.7, 8.6.8, 8.6.9, or Chapters 11 through 43.

**8.6.2\* Continuity.** Openings through floors shall be enclosed with fire barrier walls, shall be continuous from floor to floor, or floor to roof, and shall be protected as appropriate for the fire resistance rating of the barrier.

**8.6.3 Continuity Exemptions.** The requirements of 8.6.2 shall not apply where otherwise permitted by any of the following:

(1) Where penetrations for cables, cable trays, conduits, pipes, tubes, combustion vents and exhaust vents, wires, pneumatic tube conveyors, and similar items to accommodate electrical, mechanical, plumbing, and communications systems are protected in accordance with 8.3.5.1 and 8.5.6
(2) Where specified by 8.6.6, 8.6.7, 8.6.8, 8.6.9.1, 8.6.9.2, 8.6.9.3, or Chapters 11 through 43
(3) Where escalators and moving walks are protected in accordance with 8.6.9.6 or 8.6.9.7
(4) Where expansion or seismic joints are designed to prevent the penetration of fire and are shown to have a fire resistance rating of not less than that required for the floor when tested in accordance with ANSI/UL 2079, *Standard for Tests for Fire Resistance of Building Joint Systems*
(5) Where existing mail chutes meet one of the following criteria:
    (a) The cross-sectional area does not exceed 0.1 ft$^2$ (0.01 m$^2$).
    (b) The building is protected throughout by an approved automatic sprinkler system in accordance with Section 9.7.

**8.6.4 Shafts.** Shafts that do not extend to the bottom or the top of the building or structure shall comply with either 8.6.4.1, 8.6.4.2, or 8.6.4.3.

**8.6.4.1** Shafts shall be enclosed at the lowest or highest level of the shaft, respectively, with construction in accordance with 8.6.5.

**8.6.4.2** Shafts shall be permitted to terminate in a room or space having a use related to the purpose of the shaft, provided that the room or space is separated from the remainder of the building by construction having a fire resistance rating and opening protectives in accordance with 8.6.5 and 8.3.4.

**8.6.4.3** Shafts that do not extend to the bottom or top of the building or structure shall be permitted to be protected by approved fire dampers installed in accordance with their listing at the lowest or highest floor level, as applicable, within the shaft enclosure.

**8.6.5\* Required Fire Resistance Rating.** The minimum fire resistance rating for the enclosure of floor openings shall be as follows *(see 7.1.3.2.1 for enclosure of exits):*

(1) Enclosures connecting four or more stories in new construction — 2-hour fire barriers
(2) Other enclosures in new construction — 1-hour fire barriers

(3) Existing enclosures in existing buildings — ½-hour fire barriers
(4) Enclosures for lodging and rooming houses — as specified in Chapter 26
(5) Enclosures for new hotels — as specified in Chapter 28
(6) Enclosures for new apartment buildings — as specified in Chapter 30

**8.6.6 Communicating Space.** Unless prohibited by Chapters 11 through 43, unenclosed floor openings forming a communicating space between floor levels shall be permitted, provided that the following conditions are met:

(1) The communicating space does not connect more than three contiguous stories.
(2) The lowest or next-to-lowest story within the communicating space is a street floor.
(3) The entire floor area of the communicating space is open and unobstructed, such that a fire in any part of the space will be readily obvious to the occupants of the space prior to the time it becomes an occupant hazard.
(4) The communicating space is separated from the remainder of the building by fire barriers with not less than a 1-hour fire resistance rating, unless one of the following is met:
    (a) In buildings protected throughout by an approved automatic sprinkler system in accordance with Section 9.7, a smoke barrier in accordance with Section 8.5 shall be permitted to serve as the separation required by 8.6.6(4).
    (b) The requirement of 8.6.6(4) shall not apply to fully sprinklered residential housing units of detention and correctional occupancies in accordance with 22.3.1(2) and 23.3.1.1(2).
(5) The communicating space has ordinary hazard contents protected throughout by an approved automatic sprinkler system in accordance with Section 9.7 or has only low hazard contents. *(See 6.2.2.)*
(6) Egress capacity is sufficient to allow all the occupants of all levels within the communicating space to simultaneously egress the communicating space by considering it as a single floor area in determining the required egress capacity.
(7)\*Each occupant within the communicating space has access to not less than one exit without having to traverse another story within the communicating space.
(8) Each occupant not in the communicating space has access to not less than one exit without having to enter the communicating space.

**8.6.7\* Atriums.** Unless prohibited by Chapters 11 through 43, an atrium shall be permitted, provided that the following conditions are met:

(1) The atrium is separated from the adjacent spaces by fire barriers with not less than a 1-hour fire resistance rating, with opening protectives for corridor walls, unless one of the following is met:
    (a) The requirement of 8.6.7(1) shall not apply to existing, previously approved atriums.
    (b) Any number of levels of the building shall be permitted to open directly to the atrium without enclosure, based on the results of the engineering analysis required in 8.6.7(5).

(c)*Glass walls and inoperable windows shall be permitted in lieu of the fire barriers where all the following are met:

    i. Automatic sprinklers are spaced along both sides of the glass wall and the inoperable windows at intervals not to exceed 6 ft (1830 mm).

    ii. The automatic sprinklers specified in 8.6.7(1)(c)(i) are located at a distance from the glass wall not to exceed 12 in. (305 mm) and arranged so that the entire surface of the glass is wet upon operation of the sprinklers.

    iii. The glass wall is of tempered, wired, or laminated glass held in place by a gasket system that allows the glass framing system to deflect without breaking (loading) the glass before the sprinklers operate.

    iv. The automatic sprinklers required by 8.6.7(1)(c)(i) are not required on the atrium side of the glass wall and the inoperable window where there is no walkway or other floor area on the atrium side above the main floor level.

    v. Doors in the glass walls are of glass or other material that resists the passage of smoke.

    vi. Doors in the glass walls are self-closing or automatic-closing upon detection of smoke.

    vii. The glass is continuous vertically, without horizontal mullions, window treatments, or other obstructions that would interfere with the wetting of the entire glass surface.

(2) Access to exits is permitted to be within the atrium, and exit discharge in accordance with 7.7.2 is permitted to be within the atrium.

(3) The occupancy within the atrium meets the specifications for classification as low or ordinary hazard contents. *(See 6.2.2.)*

(4) The entire building is protected throughout by an approved, supervised automatic sprinkler system in accordance with Section 9.7.

(5)*For other than existing, previously approved atriums, an engineering analysis is performed that demonstrates that the building is designed to keep the smoke layer interface above the highest unprotected opening to adjoining spaces, or 6 ft (1830 mm) above the highest floor level of exit access open to the atrium, for a period equal to 1.5 times the calculated egress time or 20 minutes, whichever is greater.

(6)*In other than existing, previously approved atriums, where an engineered smoke control system is installed to meet the requirements of 8.6.7(5), the system is independently activated by each of the following:

    (a) Required automatic sprinkler system

    (b) Manual controls that are readily accessible to the fire department

**8.6.8 Two-Story Openings with Partial Enclosure.** A vertical opening serving as other than an exit enclosure, connecting only two adjacent stories and piercing only one floor, shall be permitted to be open to one of the two stories.

**8.6.9 Convenience Openings.**

**8.6.9.1** Where permitted by Chapters 11 through 43, unenclosed vertical openings not concealed within the building construction shall be permitted as follows:

(1) Such openings shall connect not more than two adjacent stories (one floor pierced only).

(2) Such openings shall be separated from unprotected vertical openings serving other floors by a barrier complying with 8.6.5.

(3) Such openings shall be separated from corridors.

(4)*In other than approved, existing convenience openings, such openings shall be separated from other fire or smoke compartments on the same floor.

(5) In new construction, the convenience opening shall be separated from the corridor referenced in 8.6.9.1(3) by a smoke partition, unless Chapters 11 through 43 require the corridor to have a fire resistance rating.

(6)*Such openings shall not serve as a required means of egress.

**8.6.9.2** Where permitted by Chapters 11 through 43, unenclosed vertical openings created by convenience stairways shall be permitted as follows:

(1) The convenience stair openings shall not serve as required means of egress.

(2) The building shall be protected throughout by an approved, supervised automatic sprinkler system in accordance with Section 9.7.

(3) The convenience stair openings shall be protected in accordance with the method detailed for the protection of vertical openings in NFPA 13, *Standard for the Installation of Sprinkler Systems.*

**8.6.9.3** Convenience stairs shall be permitted to be unenclosed in large open areas such as atriums and shopping malls.

**8.6.9.4** For other than existing hoistways in existing buildings, elevator cars located within a building shall be enclosed as follows:

(1) Where there are three or fewer elevator cars in the building, they shall be permitted to be located within the same hoistway enclosure.

(2) Where there are four elevator cars in the building, they shall be divided in such a manner that not less than two separate hoistway enclosures are provided.

(3) Where there are more than four elevator cars in the building, the number of elevator cars located within a single hoistway enclosure shall not exceed four.

**8.6.9.5** Service openings for conveyors, elevators, and dumbwaiters, where required to be open on more than one story at the same time for purposes of operation, shall be provided with closing devices in accordance with 7.2.1.8.

**8.6.9.6** Any escalators and moving walks serving as a required exit in existing buildings shall be enclosed in the same manner as exit stairways. *(See 7.1.3.2.)*

**8.6.9.7** Any escalators and moving walks not constituting an exit shall have their floor openings enclosed or protected as required for other vertical openings, unless otherwise permitted by one of the following:

(1) The requirement of 8.6.9.7 shall not apply to escalators in large open areas, such as atriums and enclosed shopping malls.

(2)*In buildings protected throughout by an approved automatic sprinkler system in accordance with Section 9.7, escalator and moving walk openings shall be permitted to be protected in accordance with the method detailed in NFPA 13, *Standard for the Installation of Sprinkler Systems,* or in accordance with a method approved by the authority having jurisdiction.

(3) In buildings protected throughout by an approved automatic sprinkler system in accordance with Section 9.7, escalator and moving walk openings shall be permitted to be protected by rolling steel shutters appropriate for the

fire resistance rating of the vertical opening and complying with all of the following:

(a) The shutters shall close automatically and independently of each other upon smoke detection and sprinkler operation.

(b) A manual means of operating and testing the operation of the shutters shall be provided.

(c) The shutters shall be operated not less than once a week to ensure that they remain in proper operating condition.

(d) The shutters shall operate at a speed not to exceed 30 ft/min (0.15 m/s) and shall be equipped with a sensitive leading edge.

(e) The leading edge shall arrest the progress of a moving shutter and cause it to retract a distance of approximately 6 in. (150 mm) upon the application of a force not exceeding 20 lbf (90 N) applied to the surface of the leading edge.

(f) The shutter, following the retraction specified in 8.6.9.7(3)(e), shall continue to close.

(g) The operating mechanism for the rolling shutter shall be provided with standby power complying with the provisions of *NFPA 70, National Electrical Code.*

### 8.6.10 Mezzanines.

**8.6.10.1 General.** Multilevel residential housing areas in detention and correctional occupancies in accordance with Chapters 22 and 23 shall be exempt from the provisions of 8.6.10.2 and 8.6.10.3.

### 8.6.10.2 Area Limitations.

**8.6.10.2.1** The aggregate area of mezzanines located within a room, other than those located in special-purpose industrial occupancies, shall not exceed one-third the open area of the room in which the mezzanines are located. Enclosed space shall not be included in a determination of the size of the room in which the mezzanine is located.

**8.6.10.2.2** No limit on the number of mezzanines in a room shall be required.

**8.6.10.2.3** For purposes of determining the allowable mezzanine area, the aggregate area of the mezzanines shall not be included in the area of the room.

**8.6.10.3 Openness.** The openness of mezzanines shall be in accordance with 8.6.10.3.1 or 8.6.10.3.2.

**8.6.10.3.1** All portions of a mezzanine, other than walls not more than 42 in. (1065 mm) high, columns, and posts, shall be open to and unobstructed from the room in which the mezzanine is located, unless the occupant load of the aggregate area of the enclosed space does not exceed 10.

**8.6.10.3.2** A mezzanine having two or more means of egress shall not be required to open into the room in which it is located if not less than one of the means of egress provides direct access from the enclosed area to an exit at the mezzanine level.

### 8.6.11 Concealed Spaces and Draftstops.

**8.6.11.1** Any concealed combustible space in which building materials having a flame spread index greater than Class A are exposed shall be draftstopped as follows:

(1) Every exterior and interior wall and partition shall be firestopped at each floor level, at the top story ceiling level, and at the level of support for roofs.

(2) Every unoccupied attic space shall be subdivided by draftstops into areas not to exceed 3000 ft² (280 m²).

(3) Any concealed space between the ceiling and the floor or roof above shall be draftstopped for the full depth of the space along the line of support for the floor or roof structural members and, if necessary, at other locations to form areas not to exceed 1000 ft² (93 m²) for any space between the ceiling and floor, and 3000 ft² (280 m²) for any space between the ceiling and roof.

**8.6.11.2** The requirements of 8.6.11.1 shall not apply where any of the following conditions are met:

(1) Where the space is protected throughout by an approved automatic sprinkler system in accordance with Section 9.7

(2)*Where concealed spaces serve as plenums

(3) Where the installation is an existing installation

**8.6.11.3** Draftstopping materials shall be not less than ½ in. (13 mm) thick gypsum board, ¹⁵/₃₂ in. (12 mm) thick plywood, or other approved materials that are adequately supported.

**8.6.11.4** The integrity of all draftstops shall be maintained.

**8.6.11.5** In existing buildings, firestopping and draftstopping shall be provided as required by Chapters 11 through 43.

## 8.7 Special Hazard Protection.

### 8.7.1 General.

**8.7.1.1\*** Protection from any area having a degree of hazard greater than that normal to the general occupancy of the building or structure shall be provided by one of the following means:

(1) Enclosing the area with a fire barrier without windows that has a 1-hour fire resistance rating in accordance with Section 8.3

(2) Protecting the area with automatic extinguishing systems in accordance with Section 9.7

(3) Applying both 8.7.1.1(1) and (2) where the hazard is severe or where otherwise specified by Chapters 11 through 43

**8.7.1.2** In new construction, where protection is provided with automatic extinguishing systems without fire-resistive separation, the space protected shall be enclosed with smoke partitions in accordance with Section 8.4, unless otherwise permitted by one of the following conditions:

(1) Where mercantile occupancy general storage areas and stockrooms are protected by automatic sprinklers in accordance with Section 9.7

(2) Where hazardous areas in industrial occupancies are protected by automatic extinguishing systems in accordance with 40.3.2

(3) Where hazardous areas in detention and correctional occupancies are protected by automatic sprinklers in accordance with 22.3.2

**8.7.1.3** Doors in barriers required to have a fire resistance rating shall have a minimum ¾-hour fire protection rating and shall be self-closing or automatic-closing in accordance with 7.2.1.8.

**8.7.2\* Explosion Protection.** Where hazardous processes or storage is of such a character as to introduce an explosion potential, an explosion venting system or an explosion suppression system specifically designed for the hazard involved shall be provided.

**8.7.3 Flammable Liquids and Gases.**

**8.7.3.1** The storage and handling of flammable liquids or gases shall be in accordance with the following applicable standards:

(1) NFPA 30, *Flammable and Combustible Liquids Code*
(2) NFPA 54, *National Fuel Gas Code*
(3) NFPA 58, *Liquefied Petroleum Gas Code*

**8.7.3.2\*** No storage or handling of flammable liquids or gases shall be permitted in any location where such storage would jeopardize egress from the structure, unless otherwise permitted by 8.7.3.1.

**8.7.4 Laboratories.**

**8.7.4.1** Laboratories that use chemicals shall comply with NFPA 45, *Standard on Fire Protection for Laboratories Using Chemicals*, unless otherwise modified by other provisions of this *Code*.

**8.7.4.2** Laboratories in health care occupancies and medical and dental offices shall comply with NFPA 99, *Health Care Facilities Code*.

**8.7.5\* Hyperbaric Facilities.** All occupancies containing hyperbaric facilities shall comply with NFPA 99, *Health Care Facilities Code*, Chapter 20, unless otherwise modified by other provisions of this *Code*.

## Chapter 9   Building Service and Fire Protection Equipment

**9.1 Utilities.**

**9.1.1 Gas.** Equipment using gas and related gas piping shall be in accordance with NFPA 54, *National Fuel Gas Code*, or NFPA 58, *Liquefied Petroleum Gas Code*, unless such installations are approved existing installations, which shall be permitted to be continued in service.

**9.1.2 Electrical Systems.** Electrical wiring and equipment shall be in accordance with *NFPA 70, National Electrical Code*, unless such installations are approved existing installations, which shall be permitted to be continued in service.

**9.1.3 Emergency Generators and Standby Power Systems.** Where required for compliance with this *Code*, emergency generators and standby power systems shall comply with 9.1.3.1 and 9.1.3.2.

**9.1.3.1** Emergency generators and standby power systems shall be installed, tested, and maintained in accordance with NFPA 110, *Standard for Emergency and Standby Power Systems*.

**9.1.3.2** New generator controllers shall be monitored by the fire alarm system, where provided, or at an attended location, for the following conditions:

(1) Generator running
(2) Generator fault
(3) Generator switch in nonautomatic position

**9.1.4 Stored Electrical Energy Systems.** Stored electrical energy systems shall be installed, tested, and maintained in accordance with NFPA 111, *Standard on Stored Electrical Energy Emergency and Standby Power Systems*.

**9.2 Heating, Ventilating, and Air-Conditioning.**

**9.2.1 Air-Conditioning, Heating, Ventilating Ductwork, and Related Equipment.** Air-conditioning, heating, ventilating ductwork, and related equipment shall be in accordance with NFPA 90A, *Standard for the Installation of Air-Conditioning and Ventilating Systems*, or NFPA 90B, *Standard for the Installation of Warm Air Heating and Air-Conditioning Systems*, as applicable, unless such installations are approved existing installations, which shall be permitted to be continued in service.

**9.2.2 Ventilating or Heat-Producing Equipment.** Ventilating or heat-producing equipment shall be in accordance with NFPA 91, *Standard for Exhaust Systems for Air Conveying of Vapors, Gases, Mists, and Noncombustible Particulate Solids*; NFPA 211, *Standard for Chimneys, Fireplaces, Vents, and Solid Fuel–Burning Appliances*; NFPA 31, *Standard for the Installation of Oil-Burning Equipment*; NFPA 54, *National Fuel Gas Code*; or *NFPA 70, National Electrical Code*, as applicable, unless such installations are approved existing installations, which shall be permitted to be continued in service.

**9.2.3 Commercial Cooking Equipment.** Commercial cooking equipment shall be in accordance with NFPA 96, *Standard for Ventilation Control and Fire Protection of Commercial Cooking Operations*, unless such installations are approved existing installations, which shall be permitted to be continued in service.

**9.2.4 Ventilating Systems in Laboratories Using Chemicals.** Ventilating systems in laboratories using chemicals shall be in accordance with NFPA 45, *Standard on Fire Protection for Laboratories Using Chemicals*, or NFPA 99, *Health Care Facilities Code*, as appropriate.

**9.3 Smoke Control.**

**9.3.1** Where required by the provisions of another section of this *Code*, smoke control systems shall be installed, inspected, tested, and maintained in accordance with NFPA 92, *Standard for Smoke Control Systems*; NFPA 204, *Standard for Smoke and Heat Venting*; or nationally recognized standards, engineering guides, or recommended practices, as approved by the authority having jurisdiction.

**9.3.2** The engineer of record shall clearly identify the intent of the system, the design method used, the appropriateness of the method used, and the required means of inspecting, testing, and maintaining the system.

**9.3.3** Acceptance testing shall be performed by a special inspector in accordance with Section 9.9.

**9.3.4 Smoke Control System Operation.**

**9.3.4.1** Floor- or zone-dependent smoke control systems shall be automatically activated by sprinkler waterflow or smoke detection systems.

**9.3.4.2** Means for manual operation of smoke control systems shall be provided at an approved location.

**9.4 Elevators, Escalators, and Conveyors.**

**9.4.1\* General.** An elevator, other than an elevator in accordance with 7.2.13, shall not be considered a component in a required means of egress but shall be permitted as a component in an accessible means of egress.

**9.4.2 Code Compliance.**

**9.4.2.1** Except as modified herein, new elevators, escalators, dumbwaiters, and moving walks shall be in accordance with the requirements of ASME A17.1/CSA B44, *Safety Code for Elevators and Escalators.*

**9.4.2.2** Except as modified herein, existing elevators, escalators, dumbwaiters, and moving walks shall be in accordance with the requirements of ASME A17.3, *Safety Code for Existing Elevators and Escalators.*

**9.4.2.3** Elevators in accordance with ASME A17.7/CSA B44.7, *Performance-Based Safety Code for Elevators and Escalators,* shall be deemed to comply with ASME A17.1/CSA B44, *Safety Code for Elevators and Escalators,* or ASME A17.3, *Safety Code for Existing Elevators and Escalators.*

**9.4.2.4** For other than elevators used for occupant-controlled evacuation in accordance with Section 7.14 and other than existing elevators, the elevator corridor call station pictograph specified in 2.27.9 of ASME A17.1/CSA B44, *Safety Code for Elevators and Escalators,* shall be provided at each elevator landing.

### 9.4.3 Fire Fighters' Emergency Operations.

**9.4.3.1** All new elevators shall conform to the fire fighters' emergency operations requirements of ASME A17.1/CSA B44, *Safety Code for Elevators and Escalators.*

**9.4.3.2** All existing elevators having a travel distance of 25 ft (7620 mm) or more above or below the level that best serves the needs of emergency personnel for fire-fighting or rescue purposes shall conform to the fire fighters' emergency operations requirements of ASME A17.3, *Safety Code for Existing Elevators and Escalators.*

**9.4.4 Number of Cars.** The number of elevator cars permitted in a hoistway shall be in accordance with 8.6.9.4.

**9.4.5\* Elevator Machine Rooms.** Elevator machine rooms that contain solid-state equipment for elevators, other than existing elevators, having a travel distance exceeding 50 ft (15 m) above the level of exit discharge, or exceeding 30 ft (9.1 m) below the level of exit discharge, shall be provided with independent ventilation or air-conditioning systems to maintain temperature during fire fighters' emergency operations for elevator operation *(see 9.4.3).* The operating temperature shall be established by the elevator equipment manufacturer's specifications. When standby power is connected to the elevator, the machine room ventilation or air-conditioning shall be connected to standby power.

### 9.4.6 Elevator Testing.

**9.4.6.1** Elevators shall be subject to periodic inspections and tests as specified in ASME A17.1/CSA B44, *Safety Code for Elevators and Escalators.*

**9.4.6.2** All elevators equipped with fire fighters' emergency operations in accordance with 9.4.3 shall be subject to a monthly operation with a written record of the findings made and kept on the premises as required by ASME A17.1/CSA B44, *Safety Code for Elevators and Escalators.*

**9.4.6.3** The elevator inspections and tests required by 9.4.6.1 shall be performed at frequencies complying with one of the following:

(1) Inspection and test frequencies specified in Appendix N of ASME A17.1/CSA B44, *Safety Code for Elevators and Escalators*
(2) Inspection and test frequencies specified by the authority having jurisdiction

**9.4.7 Openings to Exit Enclosures.** Conveyors, elevators, dumbwaiters, and pneumatic conveyors serving various stories of a building shall not open to an exit enclosure.

### 9.5 Rubbish Chutes, Incinerators, and Laundry Chutes.

### 9.5.1 Enclosure.

**9.5.1.1** Rubbish chutes and laundry chutes shall be separately enclosed by walls or partitions in accordance with the provisions of Section 8.3.

**9.5.1.2** Inlet openings serving chutes shall be protected in accordance with Section 8.3.

**9.5.1.3** The doors of chutes specified in 9.5.1.2 shall open only to a room that is designed and used exclusively for accessing the chute opening.

**9.5.1.4** The room used for accessing the chute opening shall be separated from other spaces in accordance with Section 8.7.

**9.5.1.5** The requirements of 9.5.1.1 through 9.5.1.4 shall not apply where otherwise permitted by the following:

(1) Existing installations having properly enclosed service chutes and properly installed and maintained service openings shall be permitted to have inlets open to a corridor or normally occupied space.
(2) Rubbish chutes and laundry chutes shall be permitted to open into rooms not exceeding 400 ft² (37 m²) that are used for storage, provided that the room is protected by automatic sprinklers.

**9.5.2 Installation and Maintenance.** Rubbish chutes, laundry chutes, and incinerators shall be installed and maintained in accordance with NFPA 82, *Standard on Incinerators and Waste and Linen Handling Systems and Equipment,* unless such installations are approved existing installations, which shall be permitted to be continued in service.

### 9.6 Fire Detection, Alarm, and Communications Systems.

### 9.6.1\* General.

**9.6.1.1** The provisions of Section 9.6 shall apply only where specifically required by another section of this *Code.*

**9.6.1.2** Fire detection, alarm, and communications systems installed to make use of an alternative permitted by this *Code* shall be considered required systems and shall meet the provisions of this *Code* applicable to required systems.

**9.6.1.3** A fire alarm system required for life safety shall be installed, tested, and maintained in accordance with the applicable requirements of *NFPA 70, National Electrical Code,* and *NFPA 72, National Fire Alarm and Signaling Code,* unless it is an approved existing installation, which shall be permitted to be continued in use.

**9.6.1.4** All systems and components shall be approved for the purpose for which they are installed.

**9.6.1.5\*** To ensure operational integrity, the fire alarm system shall have an approved maintenance and testing program complying with the applicable requirements of *NFPA 70, National Electrical Code,* and *NFPA 72, National Fire Alarm and Signaling Code.*

**9.6.1.6\*** Where a required fire alarm system is out of service for more than 4 hours in a 24-hour period, the authority having jurisdiction shall be notified, and the building shall be evacuated, or an approved fire watch shall be provided for all parties left unprotected by the shutdown until the fire alarm system has been returned to service.

**9.6.1.7** For the purposes of this *Code*, a complete fire alarm system shall provide functions for initiation, notification, and control, which shall perform as follows:

(1) The initiation function provides the input signal to the system.
(2) The notification function is the means by which the system advises that human action is required in response to a particular condition.
(3) The control function provides outputs to control building equipment to enhance protection of life.

**9.6.1.8 Protection of Fire Alarm System.**

**9.6.1.8.1\*** In areas that are not continuously occupied, and unless otherwise permitted by 9.6.1.8.1.1 or 9.6.1.8.1.2, automatic smoke detection shall be installed to provide notification of fire at the following locations:

(1) Each fire alarm control unit
(2) Notification appliance circuit power extenders
(3) Supervising station transmitting equipment

**9.6.1.8.1.1** The provisions of 9.6.1.8.1(2) and (3) shall not apply to existing alarm systems.

**9.6.1.8.1.2** Where ambient conditions prohibit installation of a smoke detector, a heat detector shall be used.

**9.6.2 Signal Initiation.**

**9.6.2.1** Where required by other sections of this *Code*, actuation of the complete fire alarm system shall be initiated by, but shall not be limited to, any or all of the following means:

(1) Manual fire alarm initiation
(2) Automatic detection
(3) Extinguishing system operation

**9.6.2.2** Manual fire alarm boxes shall be used only for fire-protective signaling purposes. Combination fire alarm and guard's tour stations shall be acceptable.

**9.6.2.3** A manual fire alarm box shall be provided as follows, unless modified by another section of this *Code*:

(1) For new alarm system installations, the manual fire alarm box shall be located within 60 in. (1525 mm) of exit doorways.
(2) For existing alarm system installations, the manual fire alarm box either shall be provided in the natural exit access path near each required exit or within 60 in. (1525 mm) of exit doorways.

**9.6.2.4** Manual fire alarm boxes shall be mounted on both sides of grouped openings over 40 ft (12.2 m) in width, and within 60 in. (1525 mm) of each side of the opening.

**9.6.2.5\*** Additional manual fire alarm boxes shall be located so that, on any given floor in any part of the building, no horizontal distance on that floor exceeding 200 ft (61 m) shall need to be traversed to reach a manual fire alarm box.

**9.6.2.6\*** For fire alarm systems using automatic fire detection or waterflow detection devices to initiate the fire alarm system in accordance with Chapters 11 through 43, not less than one manual fire alarm box shall be provided to initiate a fire alarm signal. The manual fire alarm box shall be located where required by the authority having jurisdiction.

**9.6.2.7\*** Each manual fire alarm box on a system shall be accessible, unobstructed, and visible.

**9.6.2.8** Where a sprinkler system provides automatic detection and alarm system initiation, it shall be provided with an approved alarm initiation device that operates when the flow of water is equal to or greater than that from a single automatic sprinkler.

**9.6.2.9** Where a total (complete) coverage smoke detection system is required by another section of this *Code*, automatic detection of smoke in accordance with *NFPA 72, National Fire Alarm and Signaling Code*, shall be provided in all occupiable areas in environments that are suitable for proper smoke detector operation.

**9.6.2.10 Smoke Alarms.**

**9.6.2.10.1 General.**

**9.6.2.10.1.1** Where required by another section of this *Code*, single-station and multiple-station smoke alarms shall be in accordance with *NFPA 72, National Fire Alarm and Signaling Code*, unless otherwise provided in 9.6.2.10.1.2, 9.6.2.10.1.3, or 9.6.2.10.1.4.

**9.6.2.10.1.2** The installation of smoke alarms in sleeping rooms shall be required where required by Chapters 11 through 43.

**9.6.2.10.1.3\*** The interconnection of smoke alarms shall apply only to new construction as provided in 9.6.2.10.3.

**9.6.2.10.1.4** System smoke detectors in accordance with *NFPA 72, National Fire Alarm and Signaling Code*, and arranged to function in the same manner as single-station or multiple-station smoke alarms shall be permitted in lieu of smoke alarms.

**9.6.2.10.2** Smoke alarms, other than existing battery-operated smoke alarms as permitted by other sections of this *Code*, shall be powered in accordance with the requirements of *NFPA 72, National Fire Alarm and Signaling Code*.

**9.6.2.10.3\*** In new construction, where two or more smoke alarms are required within a dwelling unit, suite of rooms, or similar area, they shall be arranged so that operation of any smoke alarm shall cause the alarm in all smoke alarms within the dwelling unit, suite of rooms, or similar area to sound, unless otherwise permitted by the following:

(1) The requirement of 9.6.2.10.3 shall not apply where permitted by another section of this *Code*.
(2) The requirement of 9.6.2.10.3 shall not apply to configurations that provide equivalent distribution of the alarm signal.

**9.6.2.10.4** The alarms shall sound only within an individual dwelling unit, suite of rooms, or similar area and shall not actuate the building fire alarm system, unless otherwise permitted by the authority having jurisdiction. Remote annunciation shall be permitted.

**9.6.2.11** Where required by Chapters 11 through 43, an automatic fire detection system shall be provided in hazardous areas for initiation of the signaling system.

**9.6.3 Occupant Notification.**

**9.6.3.1** Occupant notification shall be provided to alert occupants of a fire or other emergency where required by other sections of this *Code*.

**9.6.3.2** Occupant notification shall be in accordance with 9.6.3.3 through 9.6.3.10.2, unless otherwise provided in 9.6.3.2.1 through 9.6.3.2.4.

**9.6.3.2.1\*** Elevator lobby, hoistway, and associated machine room smoke detectors used solely for elevator recall, and heat detectors used solely for elevator power shutdown, shall not be required to activate the building evacuation alarm if the power supply and installation wiring to such detectors are monitored by the building fire alarm system, and if the activation of such detectors initiates a supervisory signal at a constantly attended location.

**9.6.3.2.2\*** Smoke detectors used solely for closing dampers or heating, ventilating, and air-conditioning system shutdown shall not be required to activate the building evacuation alarm, provided that the power supply and installation wiring to the detectors are monitored by the building fire alarm system, and the activation of the detectors initiates a supervisory signal at a constantly attended location.

**9.6.3.2.3\*** Smoke detectors located at doors for the exclusive operation of automatic door release shall not be required to activate the building evacuation alarm, provided that the power supply and installation wiring to the detectors are monitored by the building fire alarm system, and the activation of the detectors initiates a supervisory signal at a constantly attended location.

**9.6.3.2.4** Detectors in accordance with 22.3.4.3.1(2) and 23.3.4.3.1(2) shall not be required to activate the building evacuation alarm.

**9.6.3.3** Where permitted by Chapters 11 through 43, a presignal system shall be permitted where the initial fire alarm signal is automatically transmitted without delay to a municipal fire department, to a fire brigade (if provided), and to an on-site staff person trained to respond to a fire emergency.

**9.6.3.4** Where permitted by Chapters 11 through 43, a positive alarm sequence shall be permitted, provided that it is in accordance with *NFPA 72, National Fire Alarm and Signaling Code.*

**9.6.3.5** Unless otherwise provided in 9.6.3.5.1 through 9.6.3.5.8, notification signals for occupants to evacuate shall be audible, and visible signals in accordance with *NFPA 72, National Fire Alarm and Signaling Code,* and ICC/ANSI A117.1, *American National Standard for Accessible and Usable Buildings and Facilities,* or other means of notification acceptable to the authority having jurisdiction shall be provided.

**9.6.3.5.1** Areas not subject to occupancy by persons who are hearing impaired shall not be required to comply with the provisions for visible signals.

**9.6.3.5.2** Visible-only signals shall be provided where specifically permitted in health care occupancies in accordance with the provisions of Chapters 18 and 19.

**9.6.3.5.3** Existing alarm systems shall not be required to comply with the provision for visible signals.

**9.6.3.5.4** Visible signals shall not be required in lodging or rooming houses in accordance with the provisions of Chapter 26.

**9.6.3.5.5** Visible signals shall not be required in exit stair enclosures.

**9.6.3.5.6** Visible signals shall not be required in elevator cars.

**9.6.3.5.7\*** Public mode visual notification appliances in accordance with *NFPA 72, National Fire Alarm and Signaling Code,* shall not be required in designated areas as permitted by Chapters 11 through 43, provided that they are replaced with approved alternative visible means.

**9.6.3.5.8\*** Where visible signals are not required, as permitted by 9.6.3.5.7, documentation of such omission shall be maintained in accordance with 9.7.7.

**9.6.3.6** The general evacuation alarm signal shall operate in accordance with one of the methods prescribed by 9.6.3.6.1 through 9.6.3.6.3.

**9.6.3.6.1** The general evacuation alarm signal shall operate throughout the entire building.

**9.6.3.6.2\*** Where total evacuation of occupants is impractical due to building configuration, only the occupants in the affected zones shall be notified initially. Provisions shall be made to selectively notify occupants in other zones to afford orderly evacuation of the entire building.

**9.6.3.6.3** Where occupants are incapable of evacuating themselves because of age, physical or mental disabilities, or physical restraint, the private operating mode, as described in *NFPA 72, National Fire Alarm and Signaling Code,* shall be permitted to be used. Only the attendants and other personnel required to evacuate occupants from a zone, area, floor, or building shall be required to be notified. The notification shall include means to readily identify the zone, area, floor, or building in need of evacuation.

**9.6.3.6.4** The general evacuation signal shall not be required in exit stair enclosures.

**9.6.3.6.5** The general evacuation signal shall not be required in elevator cars.

**9.6.3.7** Audible alarm notification appliances shall be of such character and so distributed as to be effectively heard above the average ambient sound level that exists under normal conditions of occupancy.

**9.6.3.8** Audible alarm notification appliances shall produce signals that are distinctive from audible signals used for other purposes in a given building.

**9.6.3.9** Automatically transmitted or live voice evacuation or relocation instructions shall be permitted to be used to notify occupants and shall comply with either 9.6.3.9.1 or 9.6.3.9.2.

**9.6.3.9.1** Automatically transmitted or live voice evacuation or relocation instructions shall be in accordance with *NFPA 72, National Fire Alarm and Signaling Code.*

**9.6.3.9.2\*** Where permitted by Chapters 11 through 43, automatically transmitted or live voice announcements shall be permitted to be made via a voice communication or public address system that complies with the following:

(1) Occupant notification, either live or recorded, shall be initiated at a constantly attended receiving station by personnel trained to respond to an emergency.

(2) An approved secondary power supply shall be provided for other than existing, previously approved systems.

(3) The system shall be audible above the expected ambient noise level.

(4) Emergency announcements shall take precedence over any other use.

**9.6.3.10** Unless otherwise permitted by another section of this *Code,* audible and visible fire alarm notification appliances shall comply with either 9.6.3.10.1 or 9.6.3.10.2.

**9.6.3.10.1** Audible and visible fire alarm notification appliances shall be used only for fire alarm system or other emergency purposes.

**9.6.3.10.2** Emergency voice/alarm communication systems shall be permitted to be used for other purposes, subject to the approval of the authority having jurisdiction, if the fire alarm system takes precedence over all other signals, with the exception of mass notification inputs.

### 9.6.4 Emergency Forces Notification.

**9.6.4.1** Where required by another section of this *Code*, emergency forces notification shall be provided to alert the municipal fire department and fire brigade (if provided) of fire or other emergency.

**9.6.4.2** Where fire department notification is required by another section of this *Code*, the fire alarm system shall be arranged to transmit the alarm automatically via any of the following means acceptable to the authority having jurisdiction and shall be in accordance with *NFPA 72, National Fire Alarm and Signaling Code*:

(1) Auxiliary fire alarm system
(2) Central station fire alarm system
(3) Proprietary supervising station fire alarm system
(4) Remote supervising station fire alarm system

**9.6.4.3** For existing installations where none of the means of notification specified in 9.6.4.2(1) through (4) are available, an approved plan for notification of the municipal fire department shall be permitted.

### 9.6.5 Fire Safety Functions.

**9.6.5.1** Fire safety functions shall be installed in accordance with the requirements of *NFPA 72, National Fire Alarm and Signaling Code*.

**9.6.5.2** Where required by another section of this *Code*, the following functions shall be actuated:

(1) Release of hold-open devices for doors or other opening protectives
(2) Stairwell or elevator shaft pressurization
(3) Smoke management or smoke control systems
(4) Unlocking of doors
(5) Elevator recall and shutdown
(6) HVAC shutdown

**9.6.6 Location of Controls.** Operator controls, alarm indicators, and manual communications capability shall be installed at a convenient location acceptable to the authority having jurisdiction.

### 9.6.7 Annunciation.

**9.6.7.1** Where alarm annunciation is required by another section of this *Code*, it shall comply with 9.6.7.2 through 9.6.7.7.

**9.6.7.2** Alarm annunciation at the control center shall be by means of audible and visible indicators.

**9.6.7.3** For the purposes of alarm annunciation, each floor of the building, other than floors of existing buildings, shall be considered as not less than one zone, unless otherwise permitted by 9.6.7.4.3, 9.6.7.4.4, 9.6.7.4.5, or another section of this *Code*.

**9.6.7.4** If a floor area exceeds 22,500 ft$^2$ (2090 m$^2$), additional fire alarm zoning shall be provided, and the length of any single fire alarm zone shall not exceed 300 ft (91 m) in any

direction, except as provided in 9.6.7.4.1 through 9.6.7.4.5 or as otherwise modified by another section of this *Code*.

**9.6.7.4.1** Where permitted by another section of this *Code*, fire alarm zones shall be permitted to exceed 22,500 ft$^2$ (2090 m$^2$), and the length of a zone shall be permitted to exceed 300 ft (91 m) in any direction.

**9.6.7.4.2** Where the building is protected by an automatic sprinkler system in accordance with 9.7.1.1(1), the area of the fire alarm zone shall be permitted to coincide with the allowable area of the sprinkler system.

**9.6.7.4.3** Unless otherwise prohibited elsewhere in this *Code*, where a building not exceeding four stories in height is protected by an automatic sprinkler system in accordance with 9.7.1.1(1), the sprinkler system shall be permitted to be annunciated on the fire alarm system as a single zone.

**9.6.7.4.4** Where the building is protected by an automatic sprinkler system in accordance with 9.7.1.1(2), the sprinkler system shall be permitted to be annunciated on the fire alarm system as a single zone.

**9.6.7.4.5** Where the building is protected by an automatic sprinkler system in accordance with 9.7.1.1(3), the sprinkler system shall be permitted to be annunciated on the fire alarm system as a single zone.

**9.6.7.5** A system trouble signal shall be annunciated at the control center by means of audible and visible indicators.

**9.6.7.6** A system supervisory signal shall be annunciated at the control center by means of audible and visible indicators.

**9.6.7.7** Where the system serves more than one building, each building shall be annunciated separately.

### 9.7 Automatic Sprinklers and Other Extinguishing Equipment.

### 9.7.1 Automatic Sprinklers.

**9.7.1.1\*** Each automatic sprinkler system required by another section of this *Code* shall be in accordance with one of the following:

(1) NFPA 13, *Standard for the Installation of Sprinkler Systems*
(2) NFPA 13D, *Standard for the Installation of Sprinkler Systems in One- and Two-Family Dwellings and Manufactured Homes*
(3) NFPA 13R, *Standard for the Installation of Sprinkler Systems in Residential Occupancies up to and Including Four Stories in Height*

**9.7.1.2** Sprinkler piping serving not more than six sprinklers for any isolated hazardous area shall be permitted to be connected directly to a domestic water supply system having a capacity sufficient to provide 0.15 gpm/ft$^2$ (6.1 mm/min) throughout the entire enclosed area. An indicating shutoff valve, supervised in accordance with 9.7.2 or NFPA 13, *Standard for the Installation of Sprinkler Systems*, shall be installed in an accessible, visible location between the sprinklers and the connection to the domestic water supply.

**9.7.1.3\*** In areas protected by automatic sprinklers, automatic heat-detection devices required by other sections of this *Code* shall not be required.

**9.7.1.4** Automatic sprinkler systems installed to make use of an alternative permitted by this *Code* shall be considered required systems and shall meet the provisions of this *Code* that apply to required systems.

## 9.7.2 Supervision.

**9.7.2.1\* Supervisory Signals.** Where supervised automatic sprinkler systems are required by another section of this *Code*, supervisory attachments shall be installed and monitored for integrity in accordance with *NFPA 72, National Fire Alarm and Signaling Code*, and a distinctive supervisory signal shall be provided to indicate a condition that would impair the satisfactory operation of the sprinkler system. Supervisory signals shall sound and shall be displayed either at a location within the protected building that is constantly attended by qualified personnel or at an approved, remotely located receiving facility.

**9.7.2.2 Alarm Signal Transmission.** Where supervision of automatic sprinkler systems is provided in accordance with another provision of this *Code*, waterflow alarms shall be transmitted to an approved, proprietary alarm-receiving facility, a remote station, a central station, or the fire department. Such connection shall be in accordance with 9.6.1.3.

## 9.7.3 Other Automatic Extinguishing Equipment.

**9.7.3.1** In any occupancy where the character of the fuel for fire is such that extinguishment or control of fire is accomplished by a type of automatic extinguishing system in lieu of an automatic sprinkler system, such system shall be installed in accordance with the appropriate standard, as determined in accordance with Table 9.7.3.1.

**Table 9.7.3.1 Fire Suppression System Installation Standards**

| Fire Suppression System | Installation Standard |
| --- | --- |
| Low-, medium-, and high-expansion foam systems | NFPA 11, *Standard for Low-, Medium-, and High-Expansion Foam* |
| Carbon dioxide systems | NFPA 12, *Standard on Carbon Dioxide Extinguishing Systems* |
| Halon 1301 systems | NFPA 12A, *Standard on Halon 1301 Fire Extinguishing Systems* |
| Water spray fixed systems | NFPA 15, *Standard for Water Spray Fixed Systems for Fire Protection* |
| Deluge foam-water sprinkler systems | NFPA 16, *Standard for the Installation of Foam-Water Sprinkler and Foam-Water Spray Systems* |
| Dry chemical systems | NFPA 17, *Standard for Dry Chemical Extinguishing Systems* |
| Wet chemical systems | NFPA 17A, *Standard for Wet Chemical Extinguishing Systems* |
| Water mist systems | NFPA 750, *Standard on Water Mist Fire Protection Systems* |
| Clean agent extinguishing systems | NFPA 2001, *Standard on Clean Agent Fire Extinguishing Systems* |

**9.7.3.2** If the extinguishing system is installed in lieu of a required, supervised automatic sprinkler system, the activation of the extinguishing system shall activate the building fire alarm system, where provided. The actuation of an extinguishing system that is not installed in lieu of a required, supervised automatic sprinkler system shall be indicated at the building fire alarm system, where provided.

## 9.7.4 Manual Extinguishing Equipment.

**9.7.4.1\*** Where required by the provisions of another section of this *Code*, portable fire extinguishers shall be selected, installed, inspected, and maintained in accordance with NFPA 10, *Standard for Portable Fire Extinguishers*.

**9.7.4.2** Where required by the provisions of another section of this *Code*, standpipe and hose systems shall be provided in accordance with NFPA 14, *Standard for the Installation of Standpipe and Hose Systems*. Where standpipe and hose systems are installed in combination with automatic sprinkler systems, installation shall be in accordance with the appropriate provisions established by NFPA 13, *Standard for the Installation of Sprinkler Systems*, and NFPA 14, *Standard for the Installation of Standpipe and Hose Systems*.

**9.7.5 Maintenance and Testing.** All automatic sprinkler and standpipe systems required by this *Code* shall be inspected, tested, and maintained in accordance with NFPA 25, *Standard for the Inspection, Testing, and Maintenance of Water-Based Fire Protection Systems*.

**9.7.6 Sprinkler System Impairments.** Sprinkler impairment procedures shall comply with NFPA 25, *Standard for the Inspection, Testing, and Maintenance of Water-Based Fire Protection Systems*.

**9.7.7 Documentation.** All required documentation regarding the design of the fire protection system and the procedures for maintenance, inspection, and testing of the fire protection system shall be maintained at an approved, secured location for the life of the fire protection system.

**9.7.8 Record Keeping.** Testing and maintenance records required by NFPA 25, *Standard for the Inspection, Testing, and Maintenance of Water-Based Fire Protection Systems*, shall be maintained at an approved, secured location.

**9.8 Carbon Monoxide (CO) Detection and Warning Equipment.** Where required by another section of this *Code*, carbon monoxide (CO) detection and warning equipment shall be provided in accordance with NFPA 720, *Standard for the Installation of Carbon Monoxide (CO) Detection and Warning Equipment*.

## 9.9 Special Inspections and Tests.

**9.9.1** Where required by another section of this *Code*, special inspections and tests shall be performed to verify the operation of the fire protection system in its final condition for acceptance by the authority having jurisdiction.

**9.9.2** The special inspector's relevant experience in the design, installation, and testing of the fire protection systems being tested shall be documented.

**9.9.3** The design documents shall provide the procedures and methods to be used and items subject to special inspections and tests.

**9.9.4** The special inspector shall submit an inspection and test report to the authority having jurisdiction and registered design professional in responsible charge.

## Chapter 10 Interior Finish, Contents, and Furnishings

## 10.1 General.

**10.1.1 Application.** The interior finish, contents, and furnishings provisions set forth in this chapter shall apply to new construction and existing buildings.

**10.1.2 Automatic Sprinkler Systems.** Where another provision of this chapter requires an automatic sprinkler system, the automatic sprinkler system shall be installed in accordance with the subparts of 9.7.1.1 as permitted by the applicable occupancy chapter.

**10.1.3 Definitions.**

**10.1.3.1 General.** For definitions see Chapter 3 Definitions.

**10.1.3.2 Special Definitions.** A list of special terms used in this chapter follows:

(1) **Contents and Furnishings.** See 3.3.50.
(2) **Flashover.** See 3.3.112.
(3) **Interior Finish.** See 3.3.90.2.
(4) **Interior Ceiling Finish.** See 3.3.90.1.
(5) **Interior Floor Finish.** See 3.3.90.3.
(6) **Interior Wall Finish.** See 3.3.90.4.

**10.2* Interior Finish.**

**10.2.1* General.**

**10.2.1.1** Classification of interior finish materials shall be in accordance with tests made under conditions simulating actual installations, provided that the authority having jurisdiction is permitted to establish the classification of any material on which classification by a standard test is not available, unless otherwise provided in 10.2.1.2 or 10.2.1.3.

**10.2.1.2** Materials applied directly to the surface of walls and ceilings in a total thickness of less than ⅟₂₈ in. (0.9 mm) shall not be considered interior finish and shall be exempt from tests simulating actual installation if they meet the requirements of Class A interior wall or ceiling finish when tested in accordance with 10.2.3 using fiber cement board as the substrate material.

**10.2.1.3** Approved existing installations of materials applied directly to the surface of walls and ceilings in a total thickness of less than ⅟₂₈ in. (0.9 mm) shall be permitted to remain in use, and the provisions of 10.2.2 through 10.2.3.7.2 shall not apply.

**10.2.1.4*** Fixed or movable walls and partitions, paneling, wall pads, and crash pads applied structurally or for decoration, acoustical correction, surface insulation, or other purposes shall be considered interior finish and shall not be considered decorations or furnishings.

**10.2.1.5** Lockers constructed of combustible materials shall be considered interior finish.

**10.2.2* Use of Interior Finishes.**

**10.2.2.1** Requirements for interior wall and ceiling finish shall apply as follows:

(1) Where specified elsewhere in this *Code* for specific occupancies *(see Chapter 7 and Chapters 11 through 43)*
(2) As specified in 10.2.4

**10.2.2.2*** Requirements for interior floor finish shall apply under any of the following conditions:

(1) Where floor finish requirements are specified elsewhere in the *Code*
(2)*Where carpet or carpetlike material not meeting the requirements of ASTM D 2859, *Standard Test Method for Ignition Characteristics of Finished Textile Floor Covering Materials*, is used

(3) Where the fire performance of the floor finish cannot be demonstrated to be equivalent to floor finishes with a critical radiant flux of at least 0.1 W/cm²
(4) Where the fire performance of the floor finish is unknown

**10.2.3* Interior Wall or Ceiling Finish Testing and Classification.** Interior wall or ceiling finish that is required elsewhere in this *Code* to be Class A, Class B, or Class C shall be classified based on test results from ASTM E 84, *Standard Test Method for Surface Burning Characteristics of Building Materials*, or ANSI/UL 723, *Standard for Test for Surface Burning Characteristics of Building Materials*, except as indicated in 10.2.3.1 or 10.2.3.2.

**10.2.3.1** Exposed portions of structural members complying with the requirements for Type IV(2HH) construction in accordance with NFPA 220, *Standard on Types of Building Construction*, or with the building code shall be exempt from testing and classification in accordance with ASTM E 84, *Standard Test Method for Surface Burning Characteristics of Building Materials*, or ANSI/UL 723, *Standard for Test for Surface Burning Characteristics of Building Materials*.

**10.2.3.2** Interior wall and ceiling finish tested in accordance with NFPA 286, *Standard Methods of Fire Tests for Evaluating Contribution of Wall and Ceiling Interior Finish to Room Fire Growth*, and meeting the conditions of 10.2.3.7.2 shall be permitted to be used where interior wall and ceiling finish is required to be Class A in accordance with ASTM E 84, *Standard Test Method for Surface Burning Characteristics of Building Materials*, or ANSI/UL 723, *Standard for Test for Surface Burning Characteristics of Building Materials*.

**10.2.3.3** For fire-retardant coatings, see 10.2.6.

**10.2.3.4*** Products required to be tested in accordance with ASTM E 84, *Standard Test Method for Surface Burning Characteristics of Building Materials*, or ANSI/UL 723, *Standard for Test for Surface Burning Characteristics of Building Materials*, shall be classified as follows in accordance with their flame spread index and smoke developed index, except as indicated in 10.2.3.4(4):

(1) Class A interior wall and ceiling finish shall be characterized by the following:
   (a) Flame spread index, 0–25
   (b) Smoke developed index, 0–450
(2) Class B interior wall and ceiling finish shall be characterized by the following:
   (a) Flame spread index, 26–75
   (b) Smoke developed index, 0–450
(3) Class C interior wall and ceiling finish shall be characterized by the following:
   (a) Flame spread index, 76–200
   (b) Smoke developed index, 0–450
(4) Existing interior finish shall be exempt from the smoke developed index criteria of 10.2.3.4(1)(b), (2)(b), and (3)(b).

**10.2.3.5** The classification of interior finish specified in 10.2.3.4 shall be that of the basic material used by itself or in combination with other materials.

**10.2.3.6** Wherever the use of Class C interior wall and ceiling finish is required, Class A or Class B shall be permitted. Where Class B interior wall and ceiling finish is required, Class A shall be permitted.

**10.2.3.7\*** Products tested in accordance with NFPA 265, *Standard Methods of Fire Tests for Evaluating Room Fire Growth Contribution of Textile or Expanded Vinyl Wall Coverings on Full Height Panels and Walls*, shall comply with the criteria of 10.2.3.7.1. Products tested in accordance with NFPA 286, *Standard Methods of Fire Tests for Evaluating Contribution of Wall and Ceiling Interior Finish to Room Fire Growth*, shall comply with the criteria of 10.2.3.7.2.

**10.2.3.7.1** The interior finish shall comply with all of the following when tested using method B of the test protocol of NFPA 265, *Standard Methods of Fire Tests for Evaluating Room Fire Growth Contribution of Textile or Expanded Vinyl Wall Coverings on Full Height Panels and Walls*:

(1) During the 40 kW exposure, flames shall not spread to the ceiling.
(2) The flame shall not spread to the outer extremities of the samples on the 8 ft × 12 ft (2440 mm × 3660 mm) walls.
(3) Flashover, as described in NFPA 265, shall not occur.
(4) For new installations, the total smoke released throughout the test shall not exceed 1000 m$^2$.

**10.2.3.7.2** The interior finish shall comply with all of the following when tested using the test protocol of NFPA 286, *Standard Methods of Fire Tests for Evaluating Contribution of Wall and Ceiling Interior Finish to Room Fire Growth*:

(1) During the 40 kW exposure, flames shall not spread to the ceiling.
(2) The flame shall not spread to the outer extremity of the sample on any wall or ceiling.
(3) Flashover, as described in NFPA 286, shall not occur.
(4) The peak heat release rate throughout the test shall not exceed 800 kW.
(5) For new installations, the total smoke released throughout the test shall not exceed 1000 m$^2$.

**10.2.4\* Specific Materials.**

**10.2.4.1\* Textile Wall and Textile Ceiling Materials.** The use of textile materials on walls or ceilings shall comply with one of the following conditions:

(1) Textile materials meeting the requirements of Class A when tested in accordance with ASTM E 84, *Standard Test Method for Surface Burning Characteristics of Building Materials*, or ANSI/UL 723, *Standard for Test for Surface Burning Characteristics of Building Materials*, using the specimen preparation and mounting method of ASTM E 2404, *Standard Practice for Specimen Preparation and Mounting of Textile, Paper or Vinyl Wall or Ceiling Coverings to Assess Surface Burning Characteristics (see 10.2.3.4)*, shall be permitted on the walls or ceilings of rooms or areas protected by an approved automatic sprinkler system.
(2) Textile materials meeting the requirements of Class A when tested in accordance with ASTM E 84 or ANSI/UL 723, using the specimen preparation and mounting method of ASTM E 2404 *(see 10.2.3.4)*, shall be permitted on partitions that do not exceed three-quarters of the floor-to-ceiling height or do not exceed 8 ft (2440 mm) in height, whichever is less.
(3) Textile materials meeting the requirements of Class A when tested in accordance with ASTM E 84 or ANSI/UL 723, using the specimen preparation and mounting method of ASTM E 2404 *(see 10.2.3.4)*, shall be permitted to extend not more than 48 in. (1220 mm) above the finished floor on ceiling-height walls and ceiling-height partitions.

(4) Previously approved existing installations of textile material meeting the requirements of Class A when tested in accordance with ASTM E 84 or ANSI/UL 723 *(see 10.2.3.4)* shall be permitted to be continued to be used.
(5) Textile materials shall be permitted on walls and partitions where tested in accordance with NFPA 265, *Standard Methods of Fire Tests for Evaluating Room Fire Growth Contribution of Textile or Expanded Vinyl Wall Coverings on Full Height Panels and Walls. (See 10.2.3.7.)*
(6) Textile materials shall be permitted on walls, partitions, and ceilings where tested in accordance with NFPA 286, *Standard Methods of Fire Tests for Evaluating Contribution of Wall and Ceiling Interior Finish to Room Fire Growth. (See 10.2.3.7.)*

**10.2.4.2\* Expanded Vinyl Wall and Expanded Vinyl Ceiling Materials.** The use of expanded vinyl wall or expanded vinyl ceiling materials shall comply with one of the following conditions:

(1) Materials meeting the requirements of Class A when tested in accordance with ASTM E 84, *Standard Test Method for Surface Burning Characteristics of Building Materials*, or ANSI/UL 723, *Standard for Test for Surface Burning Characteristics of Building Materials*, using the specimen preparation and mounting method of ASTM E 2404, *Standard Practice for Specimen Preparation and Mounting of Textile, Paper or Vinyl Wall or Ceiling Coverings to Assess Surface Burning Characteristics (see 10.2.3.4)*, shall be permitted on the walls or ceilings of rooms or areas protected by an approved automatic sprinkler system.
(2) Materials meeting the requirements of Class A when tested in accordance with ASTM E 84 or ANSI/UL 723, using the specimen preparation and mounting method of ASTM E 2404 *(see 10.2.3.4)*, shall be permitted on partitions that do not exceed three-quarters of the floor-to-ceiling height or do not exceed 8 ft (2440 mm) in height, whichever is less.
(3) Materials meeting the requirements of Class A when tested in accordance with ASTM E 84 or ANSI/UL 723, using the specimen preparation and mounting method of ASTM E 2404 *(see 10.2.3.4)*, shall be permitted to extend not more than 48 in. (1220 mm) above the finished floor on ceiling-height walls and ceiling-height partitions.
(4) Previously approved existing installations of materials meeting the requirements for the occupancy involved, when tested in accordance with ASTM E 84 or ANSI/UL 723 *(see 10.2.3.4)*, shall be permitted to be continued to be used.
(5) Materials shall be permitted on walls and partitions where tested in accordance with NFPA 265, *Standard Methods of Fire Tests for Evaluating Room Fire Growth Contribution of Textile or Expanded Vinyl Wall Coverings on Full Height Panels and Walls. (See 10.2.3.7.)*
(6) Textile materials shall be permitted on walls, partitions, and ceilings where tested in accordance with NFPA 286, *Standard Methods of Fire Tests for Evaluating Contribution of Wall and Ceiling Interior Finish to Room Fire Growth. (See 10.2.3.7.)*

**10.2.4.3 Cellular or Foamed Plastic.** Cellular or foamed plastic materials shall not be used as interior wall and ceiling finish unless specifically permitted by 10.2.4.3.1 or 10.2.4.3.2. The requirements of 10.2.4.3 through 10.2.4.3.2 shall apply both to exposed foamed plastics and to foamed plastics used in conjunction with a textile or vinyl facing or cover.

**10.2.4.3.1\*** Cellular or foamed plastic materials shall be permitted where subjected to large-scale fire tests that substantiate their combustibility and smoke release characteristics for the use intended under actual fire conditions. The tests shall be performed on a finished foamed plastic assembly related to the actual end-use configuration, including any cover or facing, and at the maximum thickness intended for use. Suitable large-scale fire tests shall include those shown in 10.2.4.3.1.1.

**10.2.4.3.1.1** One of the following fire tests shall be used for assessing the combustibility of cellular or foamed plastic materials as interior finish:

(1) NFPA 286, *Standard Methods of Fire Tests for Evaluating Contribution of Wall and Ceiling Interior Finish to Room Fire Growth*, with the acceptance criteria of 10.2.3.7.2
(2) ANSI/UL 1715, *Standard for Fire Test of Interior Finish Material* (including smoke measurements, with total smoke release not to exceed 1000 m$^2$)
(3) ANSI/UL 1040, *Standard for Fire Test of Insulated Wall Construction*
(4) ANSI/FM 4880, *Approval Standard for Class 1 Insulated Wall or Wall and Roof/Ceiling Panels; Plastic Interior Finish Materials; Plastic Exterior Building Panels; Wall/Ceiling Coating Systems; Interior or Exterior Finish Systems*

**10.2.4.3.1.2\*** New installations of cellular or foamed plastic materials tested in accordance with ANSI/UL 1040, *Standard for Fire Test of Insulated Wall Construction*, or ANSI/FM 4880, *Approval Standard for Class 1 Insulated Wall or Wall and Roof/Ceiling Panels; Plastic Interior Finish Materials; Plastic Exterior Building Panels; Wall/Ceiling Coating Systems; Interior or Exterior Finish Systems*, shall also be tested for smoke release using NFPA 286, *Standard Methods of Fire Tests for Evaluating Contribution of Wall and Ceiling Interior Finish to Room Fire Growth*, with the acceptance criterion of 10.2.3.7.2(4).

**10.2.4.3.2** Cellular or foamed plastic shall be permitted for trim not in excess of 10 percent of the wall or ceiling area, provided that it is not less than 20 lb/ft$^3$ (320 kg/m$^3$) in density, is limited to ½ in. (13 mm) in thickness and 4 in. (100 mm) in width, and complies with the requirements for Class A or Class B interior wall and ceiling finish as described in 10.2.3.4; however, the smoke developed index shall not be limited.

**10.2.4.4\* Light-Transmitting Plastics.** Light-transmitting plastics shall be permitted to be used as interior wall and ceiling finish if approved by the authority having jurisdiction.

**10.2.4.5 Decorations and Furnishings.** Decorations and furnishings that do not meet the definition of interior finish, as defined in 3.3.90.2, shall be regulated by the provisions of Section 10.3.

**10.2.4.6 Metal Ceiling and Wall Panels.** Listed factory finished metal ceiling and wall panels meeting the requirements of Class A when tested in accordance with ASTM E 84, *Standard Test Method for Surface Burning Characteristics of Building Materials*, or ANSI/UL 723, *Standard for Test for Surface Burning Characteristics of Building Materials (see 10.2.3.4)*, shall be permitted to be finished with one additional application of paint. Such painted panels shall be permitted for use in areas where Class A interior finishes are required. The total paint thickness shall not exceed ⅛ in. (0.9 mm).

**10.2.4.7 Polypropylene (PP) and High-Density Polyethylene (HDPE).** Polypropylene and high-density polyethylene materials shall not be permitted as interior wall or ceiling finish unless the material complies with the requirements of 10.2.3.7.2. The tests shall be performed on a finished assembly and on the maximum thickness intended for use.

**10.2.4.8 Site-Fabricated Stretch Systems.** For new installations, site-fabricated stretch systems containing all three components described in the definition in Chapter 3 shall be tested in the manner intended for use and shall comply with the requirements of 10.2.3 or 10.2.3.2. If the materials are tested in accordance with ASTM E 84, *Standard Test Method for Surface Burning Characteristics of Building Materials*, or ANSI/UL 723, *Standard for Test for Surface Burning Characteristics of Building Materials*, specimen preparation and mounting shall be in accordance with ASTM E 2573, *Standard Practice for Specimen Preparation and Mounting of Site-Fabricated Stretch Systems to Assess Surface Burning Characteristics*.

**10.2.4.9 Reflective Insulation Materials.** Reflective insulation materials shall be tested in the manner intended for use and shall comply with the requirements of 10.2.3. If the materials are tested in accordance with ASTM E 84, *Standard Test Method for Surface Burning Characteristics of Building Materials*, or ANSI/UL 723, *Standard for Test for Surface Burning Characteristics of Building Materials*, specimen preparation and mounting shall be in accordance with ASTM E 2599, *Standard Practice for Specimen Preparation and Mounting of Reflective Insulation Materials and Radiant Barrier Materials for Building Applications to Assess Surface Burning Characteristics*.

**10.2.5 Trim and Incidental Finish.**

**10.2.5.1 General.** Interior wall and ceiling trim and incidental finish, other than wall base in accordance with 10.2.5.2 and bulletin boards, posters, and paper in accordance with 10.2.5.3, not in excess of 10 percent of the aggregate wall and ceiling areas of any room or space shall be permitted to be Class C materials in occupancies where interior wall and ceiling finish of Class A or Class B is required.

**10.2.5.2 Wall Base.** Interior floor trim material used at the junction of the wall and the floor to provide a functional or decorative border, and not exceeding 6 in. (150 mm) in height, shall meet the requirements for interior wall finish for its location or the requirements for Class II interior floor finish as described in 10.2.7.4 using the test described in 10.2.7.3. If a Class I floor finish is required, the interior floor trim shall be Class I.

**10.2.5.3 Bulletin Boards, Posters, and Paper.**

**10.2.5.3.1** Bulletin boards, posters, and paper attached directly to the wall shall not exceed 20 percent of the aggregate wall area to which they are applied.

**10.2.5.3.2** The provision of 10.2.5.3.1 shall not apply to artwork and teaching materials in sprinklered educational or day-care occupancies in accordance with 14.7.4.3(2), 15.7.4.3(2), 16.7.4.3(2), or 17.7.4.3(2).

**10.2.6\* Fire-Retardant Coatings.**

**10.2.6.1\*** The required flame spread index or smoke developed index of existing surfaces of walls, partitions, columns, and ceilings shall be permitted to be secured by applying approved fire-retardant coatings to surfaces having higher flame spread index values than permitted. Such treatments shall be tested, or shall be listed and labeled for application to the material to which they are applied, and shall comply with the requirements of NFPA 703, *Standard for Fire Retardant–Treated Wood and Fire-Retardant Coatings for Building Materials*.

**10.2.6.2** In new construction, surfaces of walls, partitions, columns, and ceilings shall be permitted to be finished with factory-applied fire-retardant coated assemblies that have been listed and labeled to demonstrate compliance with the following: (a) a flame spread index of 25 or less, when tested in accordance with ASTM E 84, *Standard Test Method of Surface Burning Characteristics of Building Materials,* or ANSI/UL 723, *Standard for Test for Surface Burning Characteristics of Building Materials,* (b) show no evidence of significant progressive combustion when the test is continued for an additional 20-minute period, and (c) result in a flame front that does not progress more than 10 ft 6 in. (3.2 m) beyond the centerline of the burners at any time during the test.

**10.2.6.3** Fire-retardant coatings or factory-applied fire-retardant coated assemblies shall possess the desired degree of permanency and shall be maintained so as to retain the effectiveness of the treatment under the service conditions encountered in actual use.

**10.2.7 Interior Floor Finish Testing and Classification.**

**10.2.7.1** Carpet and carpetlike interior floor finishes shall comply with ASTM D 2859, *Standard Test Method for Ignition Characteristics of Finished Textile Floor Covering Materials.*

**10.2.7.2\*** Floor coverings, other than carpet for which 10.2.2.2 establishes requirements for fire performance, shall have a minimum critical radiant flux of 0.1 W/cm$^2$.

**10.2.7.3\*** Interior floor finishes shall be classified in accordance with 10.2.7.4, based on test results from NFPA 253, *Standard Method of Test of Critical Radiant Flux of Floor Covering Systems Using a Radiant Heat Energy Source,* or ASTM E 648, *Standard Test Method for Critical Radiant Flux of Floor Covering Systems Using a Radiant Heat Energy Source.*

**10.2.7.4** Interior floor finishes shall be classified as follows in accordance with their critical radiant flux values:

(1) Class I interior floor finish shall be characterized by a critical radiant flux not less than 0.45 W/cm$^2$, as determined by the test described in 10.2.7.3.
(2) Class II interior floor finish shall be characterized by a critical radiant flux not less than 0.22 W/cm$^2$ but less than 0.45 W/cm$^2$, as determined by the test described in 10.2.7.3.

**10.2.7.5** Wherever the use of Class II interior floor finish is required, Class I interior floor finish shall be permitted.

**10.2.8 Automatic Sprinklers.**

**10.2.8.1** Other than as required in 10.2.4, where an approved automatic sprinkler system is installed in accordance with Section 9.7, Class C interior wall and ceiling finish materials shall be permitted in any location where Class B is required, and Class B interior wall and ceiling finish materials shall be permitted in any location where Class A is required.

**10.2.8.2** Where an approved automatic sprinkler system is installed in accordance with Section 9.7, Class II interior floor finish shall be permitted in any location where Class I interior floor finish is required, and where Class II is required, the provisions of 10.2.7.2 shall apply.

**10.3 Contents and Furnishings.**

**10.3.1\*** Where required by the applicable provisions of this *Code,* draperies, curtains, and other similar loosely hanging furnishings and decorations shall meet the flame propagation performance criteria contained in NFPA 701, *Standard Methods of Fire Tests for Flame Propagation of Textiles and Films.*

**10.3.2 Smoldering Ignition of Upholstered Furniture and Mattresses.**

**10.3.2.1\* Upholstered Furniture.** Newly introduced upholstered furniture, except as otherwise permitted by Chapters 11 through 43, shall be resistant to a cigarette ignition (i.e., smoldering) in accordance with one of the following:

(1) The components of the upholstered furniture shall meet the requirements for Class I when tested in accordance with NFPA 260, *Standard Methods of Tests and Classification System for Cigarette Ignition Resistance of Components of Upholstered Furniture,* or with ASTM E 1353, *Standard Test Methods for Cigarette Ignition Resistance of Components of Upholstered Furniture.*
(2) Mocked-up composites of the upholstered furniture shall have a char length not exceeding 1½ in. (38 mm) when tested in accordance with NFPA 261, *Standard Method of Test for Determining Resistance of Mock-Up Upholstered Furniture Material Assemblies to Ignition by Smoldering Cigarettes,* or with ASTM E 1352, *Standard Test Method for Cigarette Ignition Resistance of Mock-Up Upholstered Furniture Assemblies.*

**10.3.2.2\* Mattresses.** Newly introduced mattresses, except as otherwise permitted by Chapters 11 through 43, shall have a char length not exceeding 2 in. (51 mm) when tested in accordance with 16 CFR 1632, "Standard for the Flammability of Mattresses and Mattress Pads" (FF 4-72).

**10.3.3\*** Where required by the applicable provisions of this *Code,* upholstered furniture, unless the furniture is located in a building protected throughout by an approved automatic sprinkler system, shall have limited rates of heat release when tested in accordance with ASTM E 1537, *Standard Test Method for Fire Testing of Upholstered Furniture,* as follows:

(1) The peak rate of heat release for the single upholstered furniture item shall not exceed 80 kW.
(2) The total heat released by the single upholstered furniture item during the first 10 minutes of the test shall not exceed 25 MJ.

**10.3.4\*** Where required by the applicable provisions of this *Code,* mattresses, unless the mattress is located in a building protected throughout by an approved automatic sprinkler system, shall have limited rates of heat release when tested in accordance with ASTM E 1590, *Standard Test Method for Fire Testing of Mattresses,* as follows:

(1) The peak rate of heat release for the mattress shall not exceed 100 kW.
(2) The total heat released by the mattress during the first 10 minutes of the test shall not exceed 25 MJ.

**10.3.5\*** Furnishings or decorations of an explosive or highly flammable character shall not be used.

**10.3.6** Fire-retardant coatings shall be maintained to retain the effectiveness of the treatment under service conditions encountered in actual use.

**10.3.7\*** Where required by the applicable provisions of this *Code,* furnishings and contents made with foamed plastic materials that are unprotected from ignition shall have a heat release rate not exceeding 100 kW when tested in accordance with ANSI/UL 1975, *Standard for Fire Tests for Foamed Plastics Used for Decorative Purposes,* or when tested in accordance with NFPA 289, *Standard Method of Fire Test for Individual Fuel Packages,* using the 20 kW ignition source.

**10.3.8 Lockers.**

**10.3.8.1 Combustible Lockers.** Where lockers constructed of combustible materials other than wood are used, the lockers shall be considered interior finish and shall comply with Section 10.2, except as permitted by 10.3.8.2.

**10.3.8.2 Wood Lockers.** Lockers constructed entirely of wood and of noncombustible materials shall be permitted to be used in any location where interior finish materials are required to meet a Class C classification in accordance with 10.2.3.

**10.3.9 Containers for Rubbish, Waste, or Linen.**

**10.3.9.1** Where required by Chapters 11 through 43, newly introduced containers for rubbish, waste, or linen, with a capacity of 20 gal (75.7 L) or more, shall meet both of the following:

(1) Such containers shall be provided with lids.
(2) Such containers and their lids shall be constructed of noncombustible materials or of materials that meet a peak rate of heat release not exceeding 300 kW/m$^2$ when tested, at an incident heat flux of 50 kW/m$^2$ in the horizontal orientation, and at a thickness as used in the container but not less than ¼ in. (6.3 mm), in accordance with ASTM E 1354, *Test Method for Heat and Visible Smoke Release Rates for Materials and Products Using an Oxygen Consumption Calorimeter,* or NFPA 271, *Standard Method of Test for Heat and Visible Smoke Release Rates for Materials and Products Using an Oxygen Consumption Calorimeter.*

**10.3.9.2** Where required by Chapters 11 through 43, newly introduced metal wastebaskets and other metal rubbish, waste, or linen containers with a capacity of 20 gal (75.7 L) or more shall be listed in accordance with ANSI/UL 1315, *Standard for Safety for Metal Waste Paper Containers,* and shall be provided with a noncombustible lid.

## Chapter 11  Special Structures and High-Rise Buildings

**11.1 General Requirements.**

**11.1.1 Application.** The requirements of Sections 11.1 through 11.11 shall apply to occupancies regulated by Chapters 12 through 42 that are in a special structure. The applicable provisions of Chapters 12 through 42 shall apply, except as modified by this chapter. Section 11.8 shall apply to high-rise buildings only where specifically required by Chapters 12 through 42.

**11.1.2 Multiple Occupancies.** See 6.1.14.

**11.1.3 Definitions.**

**11.1.3.1 General.** For definitions see Chapter 3 Definitions.

**11.1.3.2 Special Definitions.** Special terms used in this chapter are located within each special structure section.

**11.1.4 Classification of Occupancy.** Occupancies regulated by Chapters 12 through 42 that are in special structures shall meet the requirements of those chapters, except as modified by this chapter.

**11.1.5 Classification of Hazard of Contents.** Classification of hazard of contents shall be in accordance with Section 6.2.

**11.1.6 Minimum Construction Requirements.** Minimum construction requirements shall be in accordance with the applicable occupancy chapter.

**11.1.7 Occupant Load.** The occupant load of special structures shall be based on the use of the structure as regulated by Chapters 12 through 42.

**11.1.8 Automatic Sprinkler Systems.** Where another provision of this chapter requires an automatic sprinkler system, the automatic sprinkler system shall be installed in accordance with the subparts of 9.7.1.1 as permitted by the applicable occupancy chapter.

**11.2 Open Structures.**

**11.2.1 Application.**

**11.2.1.1 General.** The provisions of Section 11.1 shall apply.

**11.2.1.2 Definition — Open Structure.** See 3.3.271.6.

**11.2.2\* Means of Egress.**

**11.2.2.1 General.** The means of egress provisions of the applicable occupancy chapter, Chapters 12 through 42, shall apply, except as modified by 11.2.2.2 through 11.2.2.10.

**11.2.2.2 Means of Egress Components.**

**11.2.2.2.1 Fire Escape Ladders.** Open structures that are designed for occupancy by not more than three persons shall be permitted to be served by fire escape ladders complying with 7.2.9.

**11.2.2.2.2 Reserved.**

**11.2.2.3 Capacity of Means of Egress.** Open structures shall be exempt from the requirements for capacity of means of egress.

**11.2.2.4 Number of Means of Egress.**

**11.2.2.4.1\*** Open structures at the finished ground level are exempt from the requirements for number of means of egress.

**11.2.2.4.2** Open structures occupied by not more than three persons, with travel distance of not more than 200 ft (61 m), shall be permitted to have a single exit.

**11.2.2.5 Arrangement of Means of Egress.** (No modifications.)

**11.2.2.6 Travel Distance to Exits.** Open structures shall be exempt from travel distance limitations.

**11.2.2.7 Discharge from Exits.** Open structures permitted to have a single exit per 11.2.2.4 shall be permitted to have 100 percent of the exit discharge through areas on the level of exit discharge.

**11.2.2.8 Illumination of Means of Egress.** Open structures shall be exempt from illumination of means of egress requirements.

**11.2.2.9 Emergency Lighting.** Open structures shall be exempt from emergency lighting requirements.

**11.2.2.10 Marking of Means of Egress.** Open structures shall be exempt from marking of means of egress requirements.

**11.2.3 Protection.**

**11.2.3.1 Protection of Vertical Openings.** Open structures shall be exempt from protection of vertical opening requirements.

**11.2.3.2 Protection from Hazards.** Every open structure, other than those structures with only occasional occupancy, shall have automatic, manual, or other protection that is appropriate to the particular hazard and that is designed to minimize danger to occupants in case of fire or other emergency before they have time to use the means of egress.

**11.2.3.3 Interior Finish.** (No modifications.)

**11.2.3.4 Detection, Alarm, and Communications Systems.** Open structures shall be exempt from requirements for detection, alarm, and communications systems.

**11.2.3.5 Extinguishing Requirements.** (No modifications.)

**11.3 Towers.**

**11.3.1 Application.**

**11.3.1.1 General.** The provisions of Section 11.1 shall apply.

**11.3.1.2 Definition — Tower.** See 3.3.280.

**11.3.1.3 Use of Accessory Levels.**

**11.3.1.3.1 Sprinklered Towers.** In towers protected throughout by an automatic sprinkler system in accordance with Section 9.7, the levels located below the observation level shall be permitted to be occupied only for the following uses that support tower operations:

(1) Use as electrical and mechanical equipment rooms, including emergency power, radar, communications, and electronics rooms
(2)*Incidental accessory uses

**11.3.1.3.2 Nonsprinklered Towers.** The levels located within a tower below the observation level and the equipment room for that level in nonsprinklered towers shall not be occupied.

**11.3.2 Means of Egress.**

**11.3.2.1 General.** The means of egress provisions of the applicable occupancy chapter, Chapters 12 through 42, shall apply, except as modified by 11.3.2.2 through 11.3.2.10.

**11.3.2.2 Means of Egress Components.**

**11.3.2.2.1 Fire Escape Ladders.** Towers, such as forest fire observation or railroad signal towers, that are designed for occupancy by not more than three persons shall be permitted to be served by fire escape ladders complying with 7.2.9.

**11.3.2.2.2 Elevators.** Towers subject to occupancy by not more than 90 persons shall be permitted to use elevators in the means of egress in accordance with 7.2.13.

**11.3.2.3 Capacity of Means of Egress.**

**11.3.2.3.1** Means of egress for towers shall be provided for the number of persons expected to occupy the space.

**11.3.2.3.2** Spaces not subject to human occupancy because of machinery or equipment shall be excluded from consideration.

**11.3.2.4\* Number of Means of Egress.**

**11.3.2.4.1** Towers shall be permitted to have a single exit, provided that the following conditions are met:

(1) The tower shall be subject to occupancy by fewer than 25 persons.
(2) The tower shall not be used for living or sleeping purposes.

(3) The tower shall be of Type I, Type II, or Type IV construction. *(See 8.2.1.)*
(4) The tower interior wall and ceiling finish shall be Class A or Class B.
(5) No combustible materials shall be located within the tower, under the tower, or within the immediate vicinity of the tower, except necessary furniture.
(6) No high hazard occupancies shall be located within the tower or within its immediate vicinity.
(7) Where the tower is located above a building, the single exit from the tower shall be provided by one of the following:

  (a) Exit enclosure separated from the building with no door openings to or from the building
  (b) Exit enclosure leading directly to an exit enclosure serving the building, with walls and door separating the exit enclosures from each other, and another door allowing access to the top floor of the building that provides access to a second exit serving that floor

**11.3.2.4.2** Towers with 360-degree line-of-sight requirements shall be permitted to have a single means of egress for a distance of travel not exceeding 75 ft (23 m), or 100 ft (30 m) if the tower is protected throughout by an approved, supervised automatic sprinkler system in accordance with Section 9.7.

**11.3.2.5 Arrangement of Means of Egress.** (No modifications.)

**11.3.2.6 Travel Distance to Exits.** Towers where ladders are permitted by 11.3.2.2.1 shall be exempt from travel distance limitations.

**11.3.2.7 Discharge from Exits.** Towers permitted to have a single exit per 11.3.2.4 shall be permitted to have 100 percent of the exit discharge through areas on the level of exit discharge.

**11.3.2.8 Illumination of Means of Egress.** Towers where ladders are permitted by 11.3.2.2.1 shall be exempt from illumination of means of egress requirements.

**11.3.2.9 Emergency Lighting.**

**11.3.2.9.1** Towers where ladders are permitted by 11.3.2.2.1 shall be exempt from emergency lighting requirements.

**11.3.2.9.2** Locations not routinely inhabited by humans shall be exempt from emergency lighting requirements.

**11.3.2.9.3** Structures occupied only during daylight hours, with windows arranged to provide the required level of illumination of all portions of the means of egress during such hours, shall be exempt from emergency lighting requirements where approved by the authority having jurisdiction.

**11.3.2.10 Marking of Means of Egress.**

**11.3.2.10.1** Towers where ladders are permitted by 11.3.2.2.1 shall be exempt from marking of means of egress requirements.

**11.3.2.10.2** Locations not routinely inhabited by humans shall be exempt from marking of means of egress requirements.

**11.3.3 Protection.**

**11.3.3.1 Protection of Vertical Openings.**

**11.3.3.1.1** Towers where ladders are permitted by 11.3.2.2.1 shall be exempt from protection of vertical opening requirements.

**11.3.3.1.2** In towers where the support structure is open and there is no occupancy below the top floor level, stairs shall be permitted to be open with no enclosure required, or fire escape stairs shall be permitted.

**11.3.3.2 Protection from Hazards.** Every tower, other than structures with only occasional occupancy, shall have automatic, manual, or other protection that is appropriate to the particular hazard and that is designed to minimize danger to occupants in case of fire or other emergency before they have time to use the means of egress.

**11.3.3.3 Interior Finish.** (No modifications.)

**11.3.3.4 Detection, Alarm, and Communications Systems.** Towers designed for occupancy by not more than three persons shall be exempt from requirements for detection, alarm, and communications systems.

**11.3.3.5 Extinguishing Requirements.** (No modifications.)

**11.3.3.6 Corridors.** (No modifications.)

**11.3.4 Additional Requirements for Air Traffic Control Towers.**

**11.3.4.1 Definition — Air Traffic Control Tower.** See 3.3.280.1.

**11.3.4.2 Use of Accessory Levels.** The levels located below the observation level shall be permitted to be occupied only for the following uses that support tower operations:

(1) Use as electrical and mechanical equipment rooms, including emergency and standby power, radar, communications, and electronics rooms
(2)*Incidental accessory uses

**11.3.4.3 Minimum Construction Requirements.** New air traffic control towers shall be of Type I or Type II construction. *(See 8.2.1.)*

**11.3.4.4 Means of Egress.**

**11.3.4.4.1 Number of Means of Egress.** Air traffic control towers shall be permitted to have a single exit, provided that the following conditions are met in addition to the requirements of 11.3.2.4:

(1) Each level of new air traffic control towers, served by a single exit, shall be subject to a calculated occupant load of 15 or fewer persons.
(2) The requirements of 11.3.2.4.1(1) shall not apply to existing air traffic control towers.
(3) Smoke detection shall be provided throughout air traffic control towers to meet the requirements of partial coverage, as defined in 5.5.2.2 of *NFPA 72, National Fire Alarm and Signaling Code*, and shall include coverage of all of the following:
   (a) Occupiable areas
   (b) Common areas
   (c) Work spaces
   (d) Equipment areas
   (e) Means of egress
   (f) Accessible utility shafts
(4) The requirements of 11.3.2.4.1(5) shall not apply.
(5) Rooms or spaces used for the storage, processing, or use of combustible supplies shall be permitted in quantities deemed acceptable by the authority having jurisdiction.

**11.3.4.4.2 Egress for Occupant Load.** Means of egress for air traffic control towers shall be provided for the occupant load, as determined in accordance with 7.3.1.

**11.3.4.4.3 Areas Excluded from Occupant Load.** Shafts, stairs, and spaces and floors not subject to human occupancy shall be excluded from consideration in determining the total calculated occupant load of the tower as required by 11.3.2.4.1(1) and 11.3.4.4.1(1).

**11.3.4.4.4 Single Means of Egress.** A single means of egress shall be permitted from the observation level of an air traffic control tower, as permitted by 11.3.2.4.2.

**11.3.4.4.5 Smokeproof Enclosures.** For other than existing, previously approved air traffic control towers, smokeproof exit enclosures complying with 7.2.3 shall be provided for all air traffic control tower exit stair enclosures.

**11.3.4.4.6 Discharge from Exits.**

**11.3.4.4.6.1** Air traffic control towers shall comply with the requirements of 7.7.2, except as permitted by 11.3.4.4.6.2.

**11.3.4.4.6.2** Existing, single-exit air traffic control towers shall be permitted to have discharge of the exit comply with one of the following:

(1) Discharge of the exit in a previously approved, single-exit air traffic control tower is permitted to a vestibule or foyer complying with the requirements of 7.7.2(4)(b).
(2)*Discharge of the exit in a single-exit air traffic control tower is permitted within the building to a location where two means of egress are available and are arranged to allow travel in independent directions after leaving the exit enclosure, so that both means of egress do not become compromised by the same fire or similar emergency.

**11.3.4.5 Protection.**

**11.3.4.5.1 Detection, Alarm, and Communications Systems.** For other than existing, previously approved air traffic control towers, air traffic control towers shall be provided with a fire alarm system in accordance with Section 9.6. Smoke detection shall be provided throughout the air traffic control tower to meet the requirements for selective coverage, as defined in 17.5.3.2 of *NFPA 72, National Fire Alarm and Signaling Code*, and shall include coverage of all of the following:

(1) At equipment areas
(2) Outside each opening into exit enclosures
(3) Along the single means of egress permitted from observation levels in 11.3.2.4.2
(4) Outside each opening into the single means of egress permitted from observation levels in 11.3.2.4.2

**11.3.4.5.2 Extinguishing Requirements.** New air traffic control towers shall be protected throughout by an approved, supervised automatic sprinkler system in accordance with Section 9.7.

**11.3.4.5.3 Standpipe Requirements.** New air traffic control towers where the floor of the cab is greater than 30 ft (9.1 m) above the lowest level of fire department vehicle access shall be protected throughout with a Class I standpipe system in accordance with Section 9.7. Class I standpipes shall be manual standpipes, as defined in NFPA 14, *Standard for the Installation of Standpipe and Hose Systems*, where permitted by the authority having jurisdiction.

**11.3.4.6 Contents and Furnishings.** Contents and furnishings in air traffic control towers shall comply with 10.3.1, 10.3.2, 10.3.6, and 10.3.7.

**11.3.4.7 Uses.** Sleeping areas shall be prohibited in air traffic control towers.

**11.4 Water-Surrounded Structures.**

**11.4.1 Application.**

**11.4.1.1 General.** The provisions of Sections 11.1 and 11.4 shall apply to those structures that are not under the jurisdiction of the U.S. Coast Guard and not designed and arranged in accordance with U.S. Coast Guard regulations.

**11.4.1.2 Definition — Water-Surrounded Structure.** See 3.3.271.12.

**11.4.2 Means of Egress.**

**11.4.2.1 General.** The means of egress provisions of the applicable occupancy chapter, Chapters 12 through 42, shall apply, except as modified by 11.4.2.2 through 11.4.2.10.

**11.4.2.2 Means of Egress Components.** (No modifications.)

**11.4.2.3 Capacity of Means of Egress.** Spaces in water-surrounded structures that are not subject to human occupancy because of machinery or equipment shall be exempt from the requirements for capacity of means of egress.

**11.4.2.4 Number of Means of Egress.** (No modifications.)

**11.4.2.5 Arrangement of Means of Egress.** (No modifications.)

**11.4.2.6 Travel Distance to Exits.** (No modifications.)

**11.4.2.7 Discharge from Exits.** Structures permitted to have a single exit per the applicable occupancy chapter shall be permitted to have 100 percent of the exit discharge through areas on the level of exit discharge.

**11.4.2.8 Illumination of Means of Egress.** (No modifications.)

**11.4.2.9 Emergency Lighting.**

**11.4.2.9.1** Locations not routinely inhabited by humans are exempt from emergency lighting requirements.

**11.4.2.9.2** Structures occupied only during daylight hours, with windows arranged to provide the required level of illumination of all portions of the means of egress during such hours, shall be exempt from emergency lighting requirements where approved by the authority having jurisdiction.

**11.4.2.10 Marking of Means of Egress.** Locations not routinely inhabited by humans shall be exempt from marking of means of egress requirements.

**11.4.3 Protection.**

**11.4.3.1 Protection of Vertical Openings.** (No modifications.)

**11.4.3.2 Protection from Hazards.** Every water-surrounded structure, other than structures with only occasional occupancy, shall have automatic, manual, or other protection that is appropriate to the particular hazard and that is designed to minimize danger to occupants in case of fire or other emergency before they have time to use the means of egress.

**11.4.3.3 Interior Finish.** (No modifications.)

**11.4.3.4 Detection, Alarm, and Communications Systems.** (No modifications.)

**11.4.3.5 Extinguishing Requirements.** (No modifications.)

**11.4.3.6 Corridors.** (No modifications.)

**11.5\* Piers.**

**11.5.1 Application.** The provisions of Section 11.1 shall apply.

**11.5.2 Number of Means of Egress.**

**11.5.2.1** Piers used exclusively to moor cargo vessels and to store material shall be exempt from number of means of egress requirements where provided with proper means of egress from structures thereon to the pier and a single means of access to the mainland, as appropriate to the pier's arrangement.

**11.5.2.2** Buildings on piers not meeting the requirements of 11.5.2.1 and occupied for other than cargo handling and storage shall be in accordance with both of the following:

(1) Means of egress shall be arranged in accordance with Chapters 12 through 43.
(2) One of the following measures shall be provided on piers extending over 150 ft (46 m) from shore to minimize the possibility that fire under or on the pier blocks the escape of occupants to shore:
   (a) The pier shall be arranged to provide two separate ways to travel to shore, such as by two well-separated walkways or independent structures.
   (b) The pier deck shall be open, fire resistive, and set on noncombustible supports.
   (c) The pier shall be open, unobstructed, and not less than 50 ft (15 m) in width if less than 500 ft (150 m) long, or its width shall be not less than 10 percent of its length if more than 500 ft (150 m) long.
   (d) The pier deck shall be provided with an approved automatic sprinkler system in accordance with Section 9.7 for combustible substructures and all superstructures.
   (e) The sprinkler system specified in 11.5.2.2(2)(d) shall be supervised where required by the applicable occupancy chapter, Chapters 12 through 42.

**11.6\* Vehicles and Vessels.**

**11.6.1 Vehicles.** Where immobile, attached to a building, or permanently fixed to a foundation, and where subject to human occupancy, the following vehicles shall comply with the requirements of this *Code* that are appropriate to buildings of similar occupancy:

(1) Trailers
(2) Railroad cars
(3) Streetcars
(4) Buses
(5) Conveyances similar to those in 11.6.1(1) through (4)

**11.6.2 Vessels.** Any ship, barge, or other vessel permanently fixed to a foundation or mooring, or unable to get underway by means of its own power, and occupied for purposes other than navigation shall be subject to the requirements of this *Code* that apply to buildings of similar occupancy.

**11.7 Underground and Limited Access Structures.**

**11.7.1 Application.** The provisions of Section 11.1 shall apply.

**11.7.2 Special Definitions.** A list of special terms used in Section 11.7 follows:

(1) **Limited Access Structure.** See 3.3.271.3.
(2) **Underground Structure.** See 3.3.271.11.

**11.7.3 Special Provisions for Underground and Limited Access Structures.**

**11.7.3.1** A structure or portion of a structure that does not have openings in compliance with 11.7.3.1.1 and 11.7.3.1.2

shall be designated as a limited access structure and shall comply with 11.7.3.4 and 11.7.3.5.

**11.7.3.1.1 One-Story Structures.** One-story structures shall have finished ground level doors or emergency access openings in accordance with 11.7.3.2 on two sides of the building, spaced not more than 125 ft (38 m) apart on the exterior walls.

**11.7.3.1.2 Multiple-Story Structures.** Multiple-story structures shall comply with the following:

(1) The story at the finished ground level shall comply with 11.7.3.1.1.
(2) Other stories shall be provided with emergency access openings in accordance with 11.7.3.2 on two sides of the building, spaced not more than 30 ft (9.1 m) apart.

**11.7.3.2\*** Emergency access openings shall consist of a window, panel, or similar opening that complies with all of the following:

(1) The opening shall have dimensions of not less than 22 in. (560 mm) in width and 24 in. (610 mm) in height and shall be unobstructed to allow for ventilation and rescue operations from the exterior.
(2) The bottom of the opening shall be not more than 44 in. (1120 mm) above the floor.
(3) The opening shall be readily identifiable from both the exterior and interior.
(4) The opening shall be readily openable from both the exterior and interior.

**11.7.3.3** A structure or portion of a structure shall not be considered an underground structure if the story is provided, on not less than two sides, with not less than 20 ft² (1.9 m²) of emergency access opening located entirely above the adjoining finished ground level in each 50 lineal ft (15 lineal m) of exterior enclosing wall area.

**11.7.3.4** Underground and limited access structures, and all areas and floor levels traversed in traveling to the exit discharge, shall be protected by an approved, supervised automatic sprinkler system in accordance with Section 9.7, unless such structures meet one of the following criteria:

(1) They have an occupant load of 50 or fewer persons in new underground or limited access portions of the structure.
(2) They have an occupant load of 100 or fewer persons in existing underground or limited access portions of the structure.
(3) The structure is a one-story underground or limited access structure that is permitted to have a single exit per Chapters 12 through 43, with a common path of travel not greater than 50 ft (15 m).

**11.7.3.5** Underground or limited access portions of structures and all areas traversed in traveling to the exit discharge, other than in one- and two-family dwellings, shall be provided with emergency lighting in accordance with Section 7.9.

**11.7.4 Additional Provisions for Underground Structures.**

**11.7.4.1** A structure or portion of a structure shall not be considered an underground structure if the story is provided, on not less than two sides, with not less than 20 ft² (1.9 m²) of emergency access opening located entirely above the adjoining finished ground level in each 50 lineal ft (15 lineal m) of exterior enclosing wall area.

**11.7.4.2** The requirements of 11.7.3 shall apply.

**11.7.4.3** Exits from underground structures with an occupant load of more than 100 persons in the underground portions of the structure and having a floor used for human occupancy located more than 30 ft (9.1 m) below the lowest level of exit discharge, or having more than one level located below the lowest level of exit discharge, shall be provided with outside smoke-venting facilities or other means to prevent the exits from becoming charged with smoke from any fire in the areas served by the exits.

**11.7.4.4** The underground portions of an underground structure, other than an existing underground structure, shall be provided with approved automatic smoke venting in accordance with Section 9.3 where the underground structure has the following features:

(1) Occupant load of more than 100 persons in the underground portions of the structure
(2) Floor level used for human occupancy located more than 30 ft (9.1 m) below the lowest level of exit discharge, or more than one level located below the lowest level of exit discharge
(3) Combustible contents, combustible interior finish, or combustible construction

**11.7.4.5** Exit stair enclosures in underground structures having a floor level used for human occupancy located more than 30 ft (9.1 m) below the lowest level of exit discharge, or having more than one level located below the lowest level of exit discharge, shall be provided with signage in accordance with 7.2.2.5.4 at each floor level landing traversed in traveling to the exit discharge. The signs shall include a chevron-shaped indicator to show direction to the exit discharge.

**11.8 High-Rise Buildings.**

**11.8.1 General.**

**11.8.1.1** The provisions of Section 11.8 shall apply to the following:

(1) New high-rise buildings, as defined in 3.3.36.7
(2) Existing high-rise buildings as required by Chapters 13, 15, 17, 19, 21, 23, 26, 29, 31, 33, 37, 39, 40, 41, or 43

**11.8.1.2** In addition to the requirements of Section 11.8, compliance with all other applicable provisions of this *Code* shall be required.

**11.8.2 Means of Egress Requirements.**

**11.8.2.1 Reserved.**

**11.8.2.2 Elevator Lobby Exit Access Door Locking.** In other than newly constructed high-rise buildings, locks in accordance with 7.2.1.6.3 shall be permitted.

**11.8.3 Extinguishing Requirements.**

**11.8.3.1\*** High-rise buildings shall be protected throughout by an approved, supervised automatic sprinkler system in accordance with Section 9.7. A sprinkler control valve and a waterflow device shall be provided for each floor.

**11.8.3.2** High-rise buildings shall be protected throughout by a Class I standpipe system in accordance with Section 9.7.

**11.8.4 Detection, Alarm, and Communications Systems.**

**11.8.4.1\*** A fire alarm system using an approved emergency voice/alarm communication system shall be installed in accordance with Section 9.6.

**11.8.4.2** Two-way telephone service shall be in accordance with 11.8.4.2.1 and 11.8.4.2.2.

**11.8.4.2.1** Two-way telephone communication service shall be provided for fire department use. This system shall be in accordance with *NFPA 72, National Fire Alarm and Signaling Code.* The communications system shall operate between the emergency command center and every elevator car, every elevator lobby, and each floor level of exit stairs.

**11.8.4.2.2** The requirement of 11.8.4.2.1 shall not apply where the fire department radio system is approved as an equivalent system.

**11.8.5 Emergency Lighting and Standby Power.**

**11.8.5.1** Emergency lighting in accordance with Section 7.9 shall be provided.

**11.8.5.2** Requirements for standby power shall be as specified in 11.8.5.2.1 through 11.8.5.2.4.

**11.8.5.2.1** Type 60, Class 1, Level 1, standby power in accordance with Article 701 of *NFPA 70, National Electrical Code,* and NFPA 110, *Standard for Emergency and Standby Power Systems,* shall be provided.

**11.8.5.2.2** The standby power system shall have a capacity and rating sufficient to supply all equipment required to be connected by 11.8.5.2.4.

**11.8.5.2.3** Selective load pickup and load shedding shall be permitted in accordance with *NFPA 70, National Electrical Code.*

**11.8.5.2.4** The standby power system shall be connected to the following:

(1) Electric fire pump
(2) Jockey pump, except as otherwise provided in 40.4.2 for special-purpose industrial occupancies
(3) Air compressor serving dry-pipe and pre-action systems, except as otherwise provided in 40.4.2 for special-purpose industrial occupancies
(4) Emergency command center equipment and lighting
(5) Not less than one elevator serving all floors, with standby power transferable to any elevator
(6) Mechanical equipment for smokeproof enclosures
(7) Mechanical equipment required to conform with the requirements of Section 9.3

**11.8.5.3** Power for detection, alarm, and communications systems shall be in accordance with *NFPA 72, National Fire Alarm and Signaling Code.*

**11.8.6\* Emergency Command Center.**

**11.8.6.1** An emergency command center shall be provided in a location approved by the fire department.

**11.8.6.2** The emergency command center shall contain the following:

(1) Voice fire alarm system panels and controls
(2) Fire department two-way telephone communication service panels and controls where required by another section of this *Code*
(3) Fire detection and fire alarm system annunciation panels
(4) Elevator floor location and operation annunciators
(5) Elevator fire recall switch in accordance with ASME A17.1/CSA B44, *Safety Code for Elevators and Escalators*
(6) Elevator emergency power selector switch(es) where provided in accordance with ASME A17.1/CSA B44

(7) Sprinkler valve and waterflow annunciators
(8) Emergency generator status indicators
(9) Controls for any automatic stairway door unlocking system
(10) Fire pump status indicators
(11) Telephone for fire department use with controlled access to the public telephone system

**11.8.7 Emergency Plans.** Emergency plans shall be provided in accordance with 4.8.2.

**11.9 Permanent Membrane Structures.**

**11.9.1 Application.**

**11.9.1.1 General.** The provisions of Section 11.1 shall apply.

**11.9.1.2 Use of Membrane Roofs.** Membrane roofs shall be used in accordance with the following:

(1) Membrane materials shall not be used where fire resistance ratings are required for walls or roofs.
(2) Where every part of the roof, including the roof membrane, is not less than 20 ft (6100 mm) above any floor, balcony, or gallery, a noncombustible or limited-combustible membrane shall be permitted to be used as the roof in any construction type.
(3) With approval of the authority having jurisdiction, membrane materials shall be permitted to be used where every part of the roof membrane is sufficiently above every significant fire potential, such that the imposed temperature cannot exceed the capability of the membrane, including seams, to maintain its structural integrity.

**11.9.1.3 Testing.** Testing of membrane materials for compliance with the requirements of Section 11.9 for use of the categories of noncombustible and limited-combustible materials shall be performed on weathered-membrane material, as defined in 3.3.169.5.

**11.9.1.4 Flame Spread Index.** The flame spread index of all membrane materials exposed within the structure shall be Class A in accordance with Section 10.2.

**11.9.1.5 Roof Covering Classification.** Roof membranes shall have a roof covering classification, as required by the applicable building codes, when tested in accordance with ASTM E 108, *Standard Test Methods for Fire Tests of Roof Coverings,* or ANSI/UL 790, *Test Methods for Fire Tests of Roof Coverings.*

**11.9.1.6 Flame Propagation Performance.**

**11.9.1.6.1** All membrane structure fabric shall meet the flame propagation performance criteria contained in NFPA 701, *Standard Methods of Fire Tests for Flame Propagation of Textiles and Films.*

**11.9.1.6.2** One of the following shall serve as evidence that the fabric materials have the required flame propagation performance:

(1) The authority having jurisdiction shall require a certificate or other evidence of acceptance by an organization acceptable to the authority having jurisdiction.
(2) The authority having jurisdiction shall require a report of tests made by other inspection authorities or organizations acceptable to the authority having jurisdiction.

**11.9.1.6.3** Where required by the authority having jurisdiction, confirmatory field tests shall be conducted using test specimens from the original material, which shall have been affixed at the time of manufacture to the exterior of the structure.

## 11.9.2 Tensioned-Membrane Structures.

**11.9.2.1** The design, materials, and construction of the building shall be based on plans and specifications prepared by a licensed architect or engineer knowledgeable in tensioned-membrane construction.

**11.9.2.2** Material loads and strength shall be based on physical properties of the materials verified and certified by an approved testing laboratory.

**11.9.2.3** The membrane roof for structures in climates subject to freezing temperatures and ice buildup shall be composed of two layers separated by an air space through which heated air can be moved to guard against ice accumulation. As an alternative to the two layers, other approved methods that protect against ice accumulation shall be permitted.

**11.9.2.4** Roof drains shall be equipped with electrical elements to protect against ice buildup that can prevent the drains from functioning. Such heating elements shall be served by on-site standby electrical power in addition to the normal public service. As an alternative to such electrical elements, other approved methods that protect against ice accumulation shall be permitted.

## 11.9.3 Air-Supported and Air-Inflated Structures.

**11.9.3.1 General.** In addition to the general provisions of 11.9.1, the requirements of 11.9.3 shall apply to air-supported and air-inflated structures.

**11.9.3.2 Pressurization (Inflation) System.** The pressurization system shall consist of one or more operating blower units. The system shall include automatic control of auxiliary blower units to maintain the required operating pressure. Such equipment shall meet the following requirements:

(1) Blowers shall be powered by continuous-rated motors at the maximum power required.
(2) Blowers shall have personnel protection, such as inlet screens and belt guards.
(3) Blower systems shall be weather protected.
(4) Blower systems shall be equipped with backdraft check dampers.
(5) Not less than two blower units shall be provided, each of which has capacity to maintain full inflation pressure with normal leakage.
(6) Blowers shall be designed to be incapable of overpressurization.
(7) The auxiliary blower unit(s) shall operate automatically if there is any loss of internal pressure or if an operating blower unit becomes inoperative.
(8) The design inflation pressure and the capacity of each blower system shall be certified by a professional engineer.

## 11.9.3.3 Standby Power System.

**11.9.3.3.1\*** A fully automatic standby power system shall be provided. The system shall be either an auxiliary engine generator set capable of running the blower system or a supplementary blower unit that is sized for 1 times the normal operating capacity and is powered by an internal combustion engine.

**11.9.3.3.2** The standby power system shall be fully automatic to ensure continuous inflation in the event of any failure of the primary power. The system shall be capable of operating continuously for a minimum of 4 hours.

**11.9.3.3.3** The sizing and capacity of the standby power system shall be certified by a professional engineer.

## 11.9.4 Maintenance and Operation.

**11.9.4.1** Instructions in both operation and maintenance shall be transmitted to the owner by the manufacturer of the tensioned-membrane, air-supported, or air-inflated structure.

**11.9.4.2** Annual inspection and required maintenance of each structure shall be performed to ensure safety conditions. At least biennially, the inspection shall be performed by a professional engineer, registered architect, or individual certified by the manufacturer.

## 11.9.5 Services.

### 11.9.5.1 Fired Heaters.

**11.9.5.1.1** Only labeled heating devices shall be used.

**11.9.5.1.2** Fuel-fired heaters and their installation shall be approved by the authority having jurisdiction.

**11.9.5.1.3** Containers for liquefied petroleum gases shall be installed not less than 60 in. (1525 mm) from any temporary membrane structure and shall be in accordance with the provisions of NFPA 58, *Liquefied Petroleum Gas Code*.

**11.9.5.1.4** Tanks shall be secured in the upright position and protected from vehicular traffic.

### 11.9.5.2 Electric Heaters.

**11.9.5.2.1** Only labeled heaters shall be permitted.

**11.9.5.2.2** Electric heaters, their placement, and their installation shall be approved by the authority having jurisdiction.

**11.9.5.2.3** Heaters shall be connected to electricity by electric cable that is suitable for outside use and is of sufficient size to handle the electrical load.

## 11.10 Temporary Membrane Structures.

### 11.10.1 Application.

**11.10.1.1 General.** The provisions of Section 11.1 shall apply.

**11.10.1.2 Required Approval.** Membrane structures designed to meet all the requirements of Section 11.10 shall be permitted to be used as temporary buildings subject to the approval of the authority having jurisdiction.

**11.10.1.3 Alternative Requirements.** Temporary tensioned-membrane structures shall be permitted to comply with Section 11.11 instead of Section 11.10.

**11.10.1.4 Roof Covering Classification.** Roof membranes shall have a roof covering classification, as required by the applicable building codes, when tested in accordance with ASTM E 108, *Standard Test Methods for Fire Tests of Roof Coverings*, or ANSI/UL 790, *Test Methods for Fire Tests of Roof Coverings*.

### 11.10.1.5 Flame Propagation Performance.

**11.10.1.5.1** All membrane structure fabric shall meet the flame propagation performance criteria contained in NFPA 701, *Standard Methods of Fire Tests for Flame Propagation of Textiles and Films*.

**11.10.1.5.2** One of the following shall serve as evidence that the fabric materials have the required flame propagation performance:

(1) The authority having jurisdiction shall require a certificate or other evidence of acceptance by an organization acceptable to the authority having jurisdiction.

(2) The authority having jurisdiction shall require a report of tests made by other inspection authorities or organizations acceptable to the authority having jurisdiction.

**11.10.1.5.3** Where required by the authority having jurisdiction, confirmatory field tests shall be conducted using test specimens from the original material, which shall have been affixed at the time of manufacture to the exterior of the structure.

### 11.10.2 Fire Hazards.

**11.10.2.1** The finished ground level enclosed by any temporary membrane structure, and the finished ground level for a reasonable distance but for not less than 10 ft (3050 mm) outside of such a structure, shall be cleared of all flammable or combustible material or vegetation that is not used for necessary support equipment. The clearing work shall be accomplished to the satisfaction of the authority having jurisdiction prior to the erection of such a structure. The premises shall be kept free from such flammable or combustible materials during the period for which the premises are used by the public.

**11.10.2.2** Where prohibited by the authority having jurisdiction, smoking shall not be permitted in any temporary membrane structure.

**11.10.3 Fire-Extinguishing Equipment.** Portable fire-extinguishing equipment of approved types shall be furnished and maintained in temporary membrane structures in such quantity and in such locations as directed by the authority having jurisdiction.

### 11.10.4 Tensioned-Membrane Structures.

**11.10.4.1** The design, materials, and construction of the building shall be based on plans and specifications prepared by a licensed architect or engineer knowledgeable in tensioned-membrane construction.

**11.10.4.2** Material loads and strength shall be based on physical properties of the materials verified and certified by an approved testing laboratory.

**11.10.4.3** The membrane roof for structures in climates subject to freezing temperatures and ice buildup shall be composed of two layers separated by an air space through which heated air can be moved to guard against ice accumulation. As an alternative to the two layers, other approved methods that protect against ice accumulation shall be permitted.

**11.10.4.4** Roof drains shall be equipped with electrical elements to protect against ice buildup that can prevent the drains from functioning. Such heating elements shall be served by on-site standby electrical power in addition to the normal public service. As an alternative to such electrical elements, other approved methods that protect against ice accumulation shall be permitted.

### 11.10.5 Air-Supported and Air-Inflated Structures.

**11.10.5.1 General.** In addition to the general provisions of 11.10.1, the requirements of 11.10.5 shall apply to air-supported and air-inflated structures.

**11.10.5.2 Pressurization (Inflation) System.** The pressurization system shall consist of one or more operating blower units. The system shall include automatic control of auxiliary blower units to maintain the required operating pressure. Such equipment shall meet the following requirements:

(1) Blowers shall be powered by continuous-rated motors at the maximum power required.

(2) Blowers shall have personnel protection, such as inlet screens and belt guards.
(3) Blower systems shall be weather protected.
(4) Blower systems shall be equipped with backdraft check dampers.
(5) Not less than two blower units shall be provided, each of which has capacity to maintain full inflation pressure with normal leakage.
(6) Blowers shall be designed to be incapable of overpressurization.
(7) The auxiliary blower unit(s) shall operate automatically if there is any loss of internal pressure or if an operating blower unit becomes inoperative.
(8) The design inflation pressure and the capacity of each blower system shall be certified by a professional engineer.

### 11.10.5.3 Standby Power System.

**11.10.5.3.1** A fully automatic standby power system shall be provided. The system shall be either an auxiliary engine generator set capable of running the blower system or a supplementary blower unit that is sized for 1 times the normal operating capacity and is powered by an internal combustion engine.

**11.10.5.3.2** The standby power system shall be fully automatic to ensure continuous inflation in the event of any failure of the primary power. The system shall be capable of operating continuously for a minimum of 4 hours.

**11.10.5.3.3** The sizing and capacity of the standby power system shall be certified by a professional engineer.

### 11.10.6 Maintenance and Operation.

**11.10.6.1** Instructions in both operation and maintenance shall be transmitted to the owner by the manufacturer of the tensioned-membrane, air-supported, or air-inflated structure.

**11.10.6.2** Annual inspection and required maintenance of each structure shall be performed to ensure safety conditions. At least biennially, the inspection shall be performed by a professional engineer, registered architect, or individual certified by the manufacturer.

### 11.10.7 Services.

#### 11.10.7.1 Fired Heaters.

**11.10.7.1.1** Only labeled heating devices shall be used.

**11.10.7.1.2** Fuel-fired heaters and their installation shall be approved by the authority having jurisdiction.

**11.10.7.1.3** Containers for liquefied petroleum gases shall be installed not less than 60 in. (1525 mm) from any temporary membrane structure and shall be in accordance with the provisions of NFPA 58, *Liquefied Petroleum Gas Code*.

**11.10.7.1.4** Tanks shall be secured in the upright position and protected from vehicular traffic.

#### 11.10.7.2 Electric Heaters.

**11.10.7.2.1** Only labeled heaters shall be permitted.

**11.10.7.2.2** Heaters used inside a temporary membrane structure shall be approved.

**11.10.7.2.3** Heaters shall be connected to electricity by electric cable that is suitable for outside use and is of sufficient size to handle the electrical load.

## 11.11 Tents.

### 11.11.1 General.

**11.11.1.1** The provisions of Section 11.1 shall apply.

**11.11.1.2** Tents shall be permitted only on a temporary basis.

**11.11.1.3** Tents shall be erected to cover not more than 75 percent of the premises, unless otherwise approved by the authority having jurisdiction.

### 11.11.2 Flame Propagation Performance.

**11.11.2.1** All tent fabric shall meet the flame propagation performance criteria contained in NFPA 701, *Standard Methods of Fire Tests for Flame Propagation of Textiles and Films.*

**11.11.2.2** One of the following shall serve as evidence that the tent fabric materials have the required flame propagation performance:

(1) The authority having jurisdiction shall require a certificate or other evidence of acceptance by an organization acceptable to the authority having jurisdiction.
(2) The authority having jurisdiction shall require a report of tests made by other inspection authorities or organizations acceptable to the authority having jurisdiction.

**11.11.2.3** Where required by the authority having jurisdiction, confirmatory field tests shall be conducted using test specimens from the original material, which shall have been affixed at the time of manufacture to the exterior of the tent.

### 11.11.3 Location and Spacing.

**11.11.3.1** There shall be a minimum of 10 ft (3050 mm) between stake lines.

**11.11.3.2** Adjacent tents shall be spaced to provide an area to be used as a means of emergency egress. Where 10 ft (3050 mm) between stake lines does not meet the requirements for means of egress, the distance necessary for means of egress shall govern.

**11.11.3.3** Tents not occupied by the public and not used for the storage of combustible material shall be permitted to be erected less than 10 ft (3050 mm) from other structures where the authority having jurisdiction deems such close spacing to be safe from hazard to the public.

**11.11.3.4** Tents, each not exceeding 1200 ft$^2$ (112 m$^2$) in finished ground level area and located in fairgrounds or similar open spaces, shall not be required to be separated from each other, provided that safety precautions meet the approval of the authority having jurisdiction.

**11.11.3.5** The placement of tents relative to other structures shall be at the discretion of the authority having jurisdiction, with consideration given to occupancy, use, opening, exposure, and other similar factors.

### 11.11.4 Fire Hazards.

**11.11.4.1** The finished ground level enclosed by any tent, and the finished ground level for a reasonable distance, but for not less than 10 ft (3050 mm) outside of such a tent, shall be cleared of all flammable or combustible material or vegetation that is not used for necessary support equipment. The clearing work shall be accomplished to the satisfaction of the authority having jurisdiction prior to the erection of such a tent. The premises shall be kept free from such flammable or combustible materials during the period for which the premises are used by the public.

### 11.11.4.2 Smoking.

**11.11.4.2.1** Smoking shall not be permitted in any tent, unless approved by the authority having jurisdiction.

**11.11.4.2.2** In rooms or areas where smoking is prohibited, plainly visible signs shall be posted that read as follows:

NO SMOKING

**11.11.5 Fire-Extinguishing Equipment.** Portable fire-extinguishing equipment of approved types shall be furnished and maintained in tents in such quantity and in such locations as directed by the authority having jurisdiction.

### 11.11.6 Services.

### 11.11.6.1 Fired Heaters.

**11.11.6.1.1** Only labeled heating devices shall be used.

**11.11.6.1.2** Fuel-fired heaters and their installation shall be approved by the authority having jurisdiction.

**11.11.6.1.3** Containers for liquefied petroleum gases shall be installed not less than 60 in. (1525 mm) from any tent and shall be in accordance with the provisions of NFPA 58, *Liquefied Petroleum Gas Code.*

**11.11.6.1.4** Tanks shall be secured in the upright position and protected from vehicular traffic.

### 11.11.6.2 Electric Heaters.

**11.11.6.2.1** Only labeled heaters shall be permitted.

**11.11.6.2.2** Heaters used inside a tent shall be approved.

**11.11.6.2.3** Heaters shall be connected to electricity by electric cable that is suitable for outside use and is of sufficient size to handle the electrical load.

## Chapter 12   New Assembly Occupancies

### 12.1 General Requirements.

### 12.1.1 Application.

**12.1.1.1** The requirements of this chapter shall apply to new buildings or portions thereof used as an assembly occupancy. *(See 1.3.1.)*

**12.1.1.2 Administration.** The provisions of Chapter 1, Administration, shall apply.

**12.1.1.3 General.** The provisions of Chapter 4, General, shall apply.

**12.1.2\* Classification of Occupancy.** See 6.1.2.

### 12.1.3 Multiple Occupancies.

**12.1.3.1 General.** Multiple occupancies shall be in accordance with 6.1.14.

**12.1.3.2\* Simultaneous Occupancy.** Exits shall be sufficient for simultaneous occupancy of both the assembly occupancy and other parts of the building, except where the authority having jurisdiction determines that the conditions are such that simultaneous occupancy will not occur.

**12.1.3.3 Assembly and Mercantile Occupancies in Mall Buildings.**

**12.1.3.3.1** The provisions of Chapter 12 shall apply to the assembly occupancy tenant space.

**12.1.3.3.2** The provisions of 36.4.4 shall be permitted to be used outside the assembly occupancy tenant space.

**12.1.4 Definitions.**

**12.1.4.1 General.** For definitions, see Chapter 3, Definitions.

**12.1.4.2* Special Definitions.** A list of special terms used in this chapter follows:

(1) **Aisle Accessway.** See 3.3.11.
(2) **Exhibit.** See 3.3.77.
(3) **Exhibitor.** See 3.3.78.
(4) **Exposition.** See 3.3.84.
(5) **Exposition Facility.** See 3.3.88.1.
(6) **Festival Seating.** See 3.3.237.1.
(7) **Flow Time.** See 3.3.115.
(8) **Fly Gallery.** See 3.3.116.
(9) **Gridiron.** See 3.3.126.
(10) **Legitimate Stage.** See 3.3.262.1.
(11) **Life Safety Evaluation.** See 3.3.158.
(12) **Multilevel Play Structure.** See 3.3.271.5.
(13) **Multipurpose Assembly Occupancy.** See 3.3.188.2.1.
(14) **Pinrail.** See 3.3.208.
(15) **Platform.** See 3.3.209.
(16) **Proscenium Wall.** See 3.3.287.2.
(17) **Regular Stage.** See 3.3.262.2.
(18) **Smoke-Protected Assembly Seating.** See 3.3.237.4.
(19) **Special Amusement Building.** See 3.3.36.10.
(20) **Stage.** See 3.3.262.
(21) **Temporary Platform.** See 3.3.209.1.

**12.1.5 Classification of Hazard of Contents.** Contents of assembly occupancies shall be classified in accordance with the provisions of Section 6.2.

**12.1.6 Minimum Construction Requirements.** Assembly occupancies shall be limited to the building construction types specified in Table 12.1.6, based on the number of stories in height as defined in 4.6.3, unless otherwise permitted by the following *(see 8.2.1)*:

(1) This requirement shall not apply to outdoor grandstands of Type I or Type II construction.
(2) This requirement shall not apply to outdoor grandstands of Type III, Type IV, or Type V construction that meet the requirements of 12.4.8.
(3) This requirement shall not apply to grandstands of noncombustible construction supported by the floor in a building meeting the construction requirements of Table 12.1.6.
(4) This requirement shall not apply to assembly occupancies within mall buildings in accordance with 36.4.4.

**12.1.7 Occupant Load.**

**12.1.7.1* General.** The occupant load, in number of persons for whom means of egress and other provisions are required, shall be determined on the basis of the occupant load factors of Table 7.3.1.2 that are characteristic of the use of the space or shall be determined as the maximum probable population of the space under consideration, whichever is greater.

**12.1.7.1.1** In areas not in excess of 10,000 ft² (930 m²), the occupant load shall not exceed one person in 5 ft² (0.46 m²).

**12.1.7.1.2** In areas in excess of 10,000 ft² (930 m²), the occupant load shall not exceed one person in 7 ft² (0.65 m²).

**12.1.7.2 Waiting Spaces.** In theaters and other assembly occupancies where persons are admitted to the building at times when seats are not available, or when the permitted occupant load has been reached based on 12.1.7.1 and persons are allowed to wait in a lobby or similar space until seats or space is available, all of the following requirements shall apply:

(1) Such use of a lobby or similar space shall not encroach upon the required clear width of exits.
(2) The waiting spaces shall be restricted to areas other than the required means of egress.
(3) Exits shall be provided for the waiting spaces on the basis of one person for each 3 ft² (0.28 m²) of waiting space area.
(4) Exits for waiting spaces shall be in addition to the exits specified for the main auditorium area and shall conform in construction and arrangement to the general rules for exits given in this chapter.

**12.1.7.3 Life Safety Evaluation.** Where the occupant load of an assembly occupancy exceeds 6000, a life safety evaluation shall be performed in accordance with 12.4.1.

**12.1.7.4 Outdoor Facilities.** In outdoor facilities, where approved by the authority having jurisdiction, the number of occupants who are each provided with not less than 15 ft² (1.4 m²) of lawn surface shall be permitted to be excluded from the maximum occupant load of 6000 of 12.1.7.3 in determining the need for a life safety evaluation.

**12.2 Means of Egress Requirements.**

**12.2.1 General.** All means of egress shall be in accordance with Chapter 7 and this chapter.

**12.2.2 Means of Egress Components.**

**12.2.2.1 Components Permitted.** Components of means of egress shall be limited to the types described in 12.2.2.2 through 12.2.2.12.

**12.2.2.2 Doors.**

**12.2.2.2.1** Doors complying with 7.2.1 shall be permitted.

**12.2.2.2.2** Assembly occupancies with occupant loads of 300 or less in malls *(see 36.4.4.2.2)* shall be permitted to have horizontal or vertical security grilles or doors complying with 7.2.1.4.1(3) on the main entrance/exits.

**12.2.2.2.3** Any door in a required means of egress from an area having an occupant load of 100 or more persons shall be permitted to be provided with a latch or lock only if the latch or lock is panic hardware or fire exit hardware complying with 7.2.1.7, unless otherwise permitted by one of the following:

(1) This requirement shall not apply to delayed-egress locks as permitted in 12.2.2.2.5.
(2) This requirement shall not apply to access-controlled egress doors as permitted in 12.2.2.2.6.

**12.2.2.2.4** Locking devices complying with 7.2.1.5.5 shall be permitted to be used on a single door or a single pair of doors if both of the following conditions apply:

(1) The door or pair of doors serve as the main exit and the assembly occupancy has an occupant load not greater than 500.
(2) Any latching devices on such a door(s) from an assembly occupancy having an occupant load of 100 or more are released by panic hardware or fire exit hardware.

**Table 12.1.6　Construction Type Limitations**

| Construction Type | Sprinklered[a] | Stories Below | Stories in Height[b] | | | | |
|---|---|---|---|---|---|---|---|
| | | | 1 | 2 | 3 | 4 | ≥5 |
| I (442)[c, d, e] | Yes | X | X | X | X | X | X |
| | No | NP | X4 | X4 | X4 | X4 | X4 |
| I (332)[c, d, e] | Yes | X | X | X | X | X | X |
| | No | NP | X4 | X4 | X4 | X4 | X4 |
| II (222)[c, d, e] | Yes | X | X | X | X | X | X |
| | No | NP | X4 | X4 | X4 | X4 | X4 |
| II (111)[c, d, e] | Yes | X1 | X | X | X | X3 | NP |
| | No | NP | X4 | X4 | X4 | NP | NP |
| II (000) | Yes | X2 | X | X4 | NP | NP | NP |
| | No | NP | X4 | NP | NP | NP | NP |
| III (211)[d] | Yes | X1 | X | X | X | X3 | NP |
| | No | NP | X4 | X4 | X4 | NP | NP |
| III (200) | Yes | X2 | X3 | X4 | NP | NP | NP |
| | No | NP | X4 | NP | NP | NP | NP |
| IV (2HH) | Yes | X1 | X | X | X | X3 | NP |
| | No | NP | X4 | X4 | X4 | NP | NP |
| V (111) | Yes | X1 | X | X | X | X3 | NP |
| | No | NP | X4 | X4 | X4 | NP | NP |
| V (000) | Yes | X2 | X3 | X4 | NP | NP | NP |
| | No | NP | X4 | NP | NP | NP | NP |

X: Permitted for assembly of any occupant load.

X1: Permitted for assembly of any occupant load, but limited to one story below the level of exit discharge.

X2: Permitted for assembly limited to an occupant load of 1000 or less, and limited to one story below the level of exit discharge.

X3: Permitted for assembly limited to an occupant load of 1000 or less.

X4: Permitted for assembly limited to an occupant load of 300 or less.

NP: Not permitted.

[a]Protected by an approved, supervised automatic sprinkler system in accordance with Section 9.7 in the following locations:

(1) Throughout the story of the assembly occupancy

(2) Throughout all stories below the story of the assembly occupancy, including all stories below the level of exit discharge

(3) In the case of an assembly occupancy located below the level of exit discharge, throughout all stories intervening between the story of the assembly occupancy and the level of exit discharge, including the level of exit discharge

[b]See 4.6.3.

[c]Where every part of the structural framework of roofs in Type I or Type II construction is 20 ft (6100 mm) or more above the floor immediately below, omission of all fire protection of the structural members is permitted, including protection of trusses, roof framing, decking, and portions of columns above 20 ft (6100 mm).

[d]In open-air fixed seating facilities, including stadia, omission of fire protection of structural members exposed to the outside atmosphere is permitted where substantiated by an approved engineering analysis.

[e]Where seating treads and risers serve as floors, such seating treads and risers are permitted to be of 1-hour fire resistance–rated construction. Structural members supporting seating treads and risers are required to conform to the requirements of Table 12.1.6. Joints between seating tread and riser units are permitted to be unrated, provided that such joints do not involve separation from areas containing high hazard contents and the facility is constructed and operated in accordance with 12.4.2.

**12.2.2.2.5** Delayed-egress locks complying with 7.2.1.6.1 shall be permitted on doors other than main entrance/exit doors.

**12.2.2.2.6** Doors in the means of egress shall be permitted to be equipped with an approved access control system complying with 7.2.1.6.2, and such doors shall not be locked from the egress side when the assembly occupancy is occupied. *(See 7.2.1.1.3.)*

**12.2.2.2.7** Elevator lobby exit access door locking in accordance with 7.2.1.6.3 shall be permitted.

**12.2.2.2.8** Revolving doors complying with the requirements of 7.2.1.10 shall be permitted.

**12.2.2.2.9** The provisions of 7.2.1.11.1.1 to permit turnstiles where revolving doors are permitted shall not apply.

**12.2.2.2.10** No turnstiles or other devices that restrict the movement of persons shall be installed in any assembly occupancy in such a manner as to interfere with required means of egress facilities.

**12.2.2.3 Stairs.**

**12.2.2.3.1 General.** Stairs complying with 7.2.2 shall be permitted, unless one of the following criteria applies:

(1)*Stairs serving seating that is designed to be repositioned shall not be required to comply with 7.2.2.3.1.
(2) This requirement shall not apply to stages and platforms as permitted by 12.4.5.1.2.
(3) The stairs connecting only a stage or platform and the immediately adjacent assembly seating shall be permitted to have a handrail in the center only or on one side only.
(4) The stairs connecting only a stage or platform and the immediately adjacent assembly seating shall be permitted to omit the guards required by 7.1.8 where both of the following criteria are met:
    (a) The guard would restrict audience sight lines to the stage or platform.
    (b) The height between any part of the stair and the adjacent floor is not more than 42 in. (1065 mm).

**12.2.2.3.2 Catwalk, Gallery, and Gridiron Stairs.**

**12.2.2.3.2.1** Noncombustible grated stair treads and landing floors shall be permitted in means of egress from lighting and access catwalks, galleries, and gridirons.

**12.2.2.3.2.2** Spiral stairs complying with 7.2.2.2.3 shall be permitted in means of egress from lighting and access catwalks, galleries, and gridirons.

**12.2.2.4 Smokeproof Enclosures.** Smokeproof enclosures complying with 7.2.3 shall be permitted.

**12.2.2.5 Horizontal Exits.** Horizontal exits complying with 7.2.4 shall be permitted.

**12.2.2.6 Ramps.** Ramps complying with 7.2.5 shall be permitted, and the following alternatives shall also apply:

(1) Ramps not part of an accessible means of egress and serving only stages or nonpublic areas shall be permitted to have a slope not steeper than 1 in 8.
(2) Ramped aisles not part of an accessible means of egress shall be permitted to have a slope not steeper than 1 in 8.

**12.2.2.7 Exit Passageways.** Exit passageways complying with 7.2.6 shall be permitted.

**12.2.2.8 Reserved.**

**12.2.2.9 Reserved.**

**12.2.2.10 Fire Escape Ladders.**

**12.2.2.10.1** Fire escape ladders complying with 7.2.9 shall be permitted.

**12.2.2.10.2** For ladders serving catwalks, the three-person limitation in 7.2.9.1(3) shall be permitted to be increased to ten persons.

**12.2.2.11 Alternating Tread Devices.** Alternating tread devices complying with 7.2.11 shall be permitted.

**12.2.2.12 Areas of Refuge.** Areas of refuge complying with 7.2.12 shall be permitted.

**12.2.3 Capacity of Means of Egress.**

**12.2.3.1 General.** The capacity of means of egress shall be in accordance with one of the following:

(1) Section 7.3 for other than theater-type seating or smoke-protected assembly seating
(2) 12.2.3.2 for rooms with theater-type seating or similar seating arranged in rows
(3) 12.4.2 for smoke-protected assembly seating

**12.2.3.2\* Theater-Type Seating.** Minimum clear widths of aisles and other means of egress serving theater-type seating, or similar seating arranged in rows, shall be in accordance with Table 12.2.3.2.

**Table 12.2.3.2 Capacity Factors**

| No. of Seats | Clear Width per Seat Served | | | |
| | Stairs | | Passageways, Ramps, and Doorways | |
| | in. | mm | in. | mm |
|---|---|---|---|---|
| Unlimited | 0.3 AB | 7.6 AB | 0.22 C | 5.6 C |

**12.2.3.3 Width Modifications.** The minimum clear widths shown in Table 12.2.3.2 shall be modified in accordance with all of the following:

(1) If risers exceed 7 in. in height, the stair width in Table 12.2.3.2 shall be multiplied by factor $A$, where $A$ equals the following:

$$A = 1 + \frac{\text{riser height} - 7}{5}$$

(2) If risers exceed 178 mm in height, the stair width in Table 12.2.3.2 shall be multiplied by factor $A$, where $A$ equals the following:

$$A = 1 + \frac{\text{riser height} - 178}{125}$$

(3) Stairs not having a handrail within a 30 in. (760 mm) horizontal distance shall be 25 percent wider than otherwise calculated; that is, their width shall be multiplied by factor $B$, where $B$ equals the following:

$$B = 1.25$$

(4) Ramps steeper than 1 in 10 slope where used in ascent shall have their width increased by 10 percent; that is, their width shall be multiplied by factor $C$, where $C$ equals the following:

$$C = 1.10$$

**12.2.3.4 Lighting and Access Catwalks.** The requirements of 12.2.3.2 and 12.2.3.3 shall not apply to lighting and access catwalks as permitted by 12.4.5.9.

**12.2.3.5 Reserved.**

**12.2.3.6 Main Entrance/Exit.**

**12.2.3.6.1** Every assembly occupancy shall be provided with a main entrance/exit.

**12.2.3.6.2** The main entrance/exit width shall be as follows:

(1) The main entrance/exit shall be of a width that accommodates two-thirds of the total occupant load in the following assembly occupancies:
  (a) Dance halls
  (b) Discotheques
  (c) Nightclubs
  (d) Assembly occupancies with festival seating
(2) In assembly occupancies, other than those listed in 12.2.3.6.2(1), the main entrance/exit shall be of a width that accommodates one-half of the total occupant load.

**12.2.3.6.3** The main entrance/exit shall be at the level of exit discharge or shall connect to a stairway or ramp leading to a street.

**12.2.3.6.4** Access to the main entrance/exit shall be as follows:

(1) Each level of the assembly occupancy shall have access to the main entrance/exit, and such access shall have the capacity to accommodate two-thirds of the occupant load of such levels in the following assembly occupancies:
  (a) Bars with live entertainment
  (b) Dance halls
  (c) Discotheques
  (d) Nightclubs
  (e) Assembly occupancies with festival seating
(2) In assembly occupancies, other than those listed in 12.2.3.6.4(1), each level of the assembly occupancy shall have access to the main entrance/exit, and such access shall have the capacity to accommodate one-half of the occupant load of such levels.

**12.2.3.6.5** Where the main entrance/exit from an assembly occupancy is through a lobby or foyer, the aggregate capacity of all exits from the lobby or foyer shall be permitted to provide the required capacity of the main entrance/exit, regardless of whether all such exits serve as entrances to the building.

**12.2.3.6.6\*** In assembly occupancies where there is no well-defined main entrance/exit, exits shall be permitted to be distributed around the perimeter of the building, provided that the total exit width furnishes not less than 100 percent of the width needed to accommodate the permitted occupant load.

**12.2.3.7 Other Exits.** Each level of an assembly occupancy shall have access to the main entrance/exit and shall be provided with additional exits of a width to accommodate not less than one-half of the total occupant load served by that level.

**12.2.3.7.1** Additional exits shall discharge in accordance with 12.2.7.

**12.2.3.7.2** Additional exits shall be located as far apart as practicable and as far from the main entrance/exit as practicable.

**12.2.3.7.3** Additional exits shall be accessible from a cross aisle or a side aisle.

**12.2.3.7.4** In assembly occupancies where there is no well-defined main entrance/exit, exits shall be permitted to be distributed around the perimeter of the building, provided that the total exit width furnishes not less than 100 percent of the width required to accommodate the permitted occupant load.

**12.2.3.8 Minimum Corridor Width.** The width of any exit access corridor serving 50 or more persons shall be not less than 44 in. (1120 mm).

**12.2.4\* Number of Means of Egress.**

**12.2.4.1** The number of means of egress shall be in accordance with Section 7.4, other than exits for fenced outdoor assembly occupancies in accordance with 12.2.4.4.

**12.2.4.2 Reserved.**

**12.2.4.3 Reserved.**

**12.2.4.4** A fenced outdoor assembly occupancy shall have not less than two remote means of egress from the enclosure in accordance with 7.5.1.3, unless otherwise required by one of the following:

(1) If more than 6000 persons are to be served by such means of egress, there shall be not less than three means of egress.
(2) If more than 9000 persons are to be served by such means of egress, there shall be not less than four means of egress.

**12.2.4.5** Balconies or mezzanines having an occupant load not exceeding 50 shall be permitted to be served by a single means of egress, and such means of egress shall be permitted to lead to the floor below.

**12.2.4.6** Balconies or mezzanines having an occupant load exceeding 50, but not exceeding 100, shall have not less than two remote means of egress, but both such means of egress shall be permitted to lead to the floor below.

**12.2.4.7** Balconies or mezzanines having an occupant load exceeding 100 shall have means of egress as described in 7.4.1.

**12.2.4.8** A second means of egress shall not be required from lighting and access catwalks, galleries, and gridirons where a means of escape to a floor or a roof is provided. Ladders, alternating tread devices, or spiral stairs shall be permitted in such means of escape.

**12.2.5 Arrangement of Means of Egress.**

**12.2.5.1 General.**

**12.2.5.1.1** Means of egress shall be arranged in accordance with Section 7.5.

**12.2.5.1.2** A common path of travel shall be permitted for the first 20 ft (6100 mm) from any point where the common path serves any number of occupants, and for the first 75 ft (23 m) from any point where the common path serves not more than 50 occupants.

**12.2.5.1.3** Dead-end corridors shall not exceed 20 ft (6100 mm).

**12.2.5.2 Access Through Hazardous Areas.** Means of egress from a room or space for assembly purposes shall not be permitted through kitchens, storerooms, restrooms, closets, platforms, stages, projection rooms, or hazardous areas as described in 12.3.2.

**12.2.5.3 Auditorium and Area Floors.** Where the floor area of auditoriums and arenas is used for assembly occupancy activities/events, not less than 50 percent of the occupant load shall have means of egress provided without passing through adjacent fixed seating areas.

**12.2.5.4 General Requirements for Access and Egress Routes Within Assembly Areas.**

**12.2.5.4.1** Festival seating, as defined in 3.3.237.1, shall be prohibited within a building, unless otherwise permitted by one of the following:

(1) Festival seating shall be permitted in assembly occupancies having occupant loads of 250 or less.
(2) Festival seating shall be permitted in assembly occupancies where occupant loads exceed 250, provided that an approved life safety evaluation has been performed. (*See 12.4.1.*)

**12.2.5.4.2\*** Access and egress routes shall be maintained so that any individual is able to move without undue hindrance, on personal initiative and at any time, from an occupied position to the exits.

**12.2.5.4.3\*** Access and egress routes shall be maintained so that crowd management, security, and emergency medical personnel are able to reach any individual at any time, without undue hindrance.

**12.2.5.4.4\*** The width of aisle accessways and aisles shall provide sufficient egress capacity for the number of persons accommodated by the catchment area served by the aisle accessway or aisle in accordance with 12.2.3.2, or for smoke-protected assembly seating in accordance with 12.4.2.

**12.2.5.4.5** Where aisle accessways or aisles converge to form a single path of egress travel, the required egress capacity of that path shall be not less than the combined required capacity of the converging aisle accessways and aisles.

**12.2.5.4.6** Those portions of aisle accessways and aisles where egress is possible in either of two directions shall be uniform in required width, unless otherwise permitted by 12.2.5.4.7.

**12.2.5.4.7** The requirement of 12.2.5.4.6 shall not apply to those portions of aisle accessways where the required width, not including the seat space described by 12.2.5.7.3, does not exceed 12 in. (305 mm).

**12.2.5.4.8** In the case of side boundaries for aisle accessways or aisles, other than those for nonfixed seating at tables, the clear width shall be measured to boundary elements such as walls, guardrails, handrails, edges of seating, tables, and side edges of treads, and said measurement shall be made horizontally to the vertical projection of the elements, resulting in the smallest width measured perpendicularly to the line of travel.

**12.2.5.5\* Aisle Accessways Serving Seating Not at Tables.**

**12.2.5.5.1\*** The required clear width of aisle accessways between rows of seating shall be determined as follows:

(1) Horizontal measurements shall be made, between vertical planes, from the back of one seat to the front of the most forward projection of the seat immediately behind it.
(2) Where the entire row consists of automatic- or self-rising seats that comply with ASTM F 851, *Standard Test Method for Self-Rising Seat Mechanisms*, the measurement shall be permitted to be made with the seats in the up position.

**12.2.5.5.2** The aisle accessway between rows of seating shall have a clear width of not less than 12 in. (305 mm), and this minimum shall be increased as a function of row length in accordance with 12.2.5.5.4 and 12.2.5.5.5.

**12.2.5.5.3** If used by not more than four persons, no minimum clear width shall be required for the portion of an aisle accessway having a length not exceeding 6 ft (1830 mm), measured from the center of the seat farthest from the aisle.

**12.2.5.5.4\*** Rows of seating served by aisles or doorways at both ends shall not exceed 100 seats per row.

**12.2.5.5.4.1** The 12 in. (305 mm) minimum clear width of aisle accessway specified in 12.2.5.5.2 shall be increased by 0.3 in. (7.6 mm) for every seat over a total of 14 but shall not be required to exceed 22 in. (560 mm).

**12.2.5.5.4.2** The requirement of 12.2.5.5.4.1 shall not apply to smoke-protected assembly seating as permitted by 12.4.2.7.

**12.2.5.5.5** Rows of seating served by an aisle or doorway at one end only shall have a path of travel not exceeding 30 ft (9.1 m) in length from any seat to an aisle.

**12.2.5.5.5.1** The 12 in. (305 mm) minimum clear width of aisle accessway specified in 12.2.5.5.2 shall be increased by 0.6 in. (15 mm) for every seat over a total of seven.

**12.2.5.5.5.2** The requirements of 12.2.5.5.5 and 12.2.5.5.5.1 shall not apply to smoke-protected assembly seating as permitted by 12.4.2.8 and 12.4.2.9.

**12.2.5.5.6** Rows of seating using tablet-arm chairs shall be permitted only if the clear width of aisle accessways complies with the requirements of 12.2.5.5 when measured under one of the following conditions:

(1) The clear width is measured with the tablet arm in the usable position.
(2) The clear width is measured with the tablet arm in the stored position where the tablet arm automatically returns to the stored position when raised manually to a vertical position in one motion and falls to the stored position by force of gravity.

**12.2.5.5.7** The depth of seat boards shall be not less than 9 in. (230 mm) where the same level is not used for both seat boards and footboards.

**12.2.5.5.8** Footboards, independent of seats, shall be provided so that there is no horizontal opening that allows the passage of a ½ in. (13 mm) diameter sphere.

**12.2.5.6 Aisles Serving Seating Not at Tables.**

**12.2.5.6.1 General.**

**12.2.5.6.1.1** Aisles shall be provided so that the number of seats served by the nearest aisle is in accordance with 12.2.5.5.2 through 12.2.5.5.5, unless otherwise permitted by 12.2.5.6.1.2.

**12.2.5.6.1.2** Aisles shall not be required in bleachers, provided that all of the following conditions are met:

(1) Egress from the front row shall not be obstructed by a rail, a guard, or other obstruction.
(2) The row spacing shall be 28 in. (710 mm) or less.
(3) The rise per row, including the first row, shall be 6 in. (150 mm) or less.

(4) The number of rows shall not exceed 16.

(5) The seat spaces shall not be physically defined.

(6) Seat boards that are also used as stepping surfaces for descent shall provide a walking surface with a width not less than 12 in. (305 mm), and, where a depressed footboard exists, the gap between seat boards of adjacent rows shall not exceed 12 in. (305 mm), measured horizontally.

(7) The leading edges of seat boards used as stepping surfaces shall be provided with a contrasting marking stripe so that the location of the leading edge is readily apparent, particularly where viewed in descent, and the following shall also apply:

   (a) The marking stripe shall be not less than 1 in. (25 mm) wide and shall not exceed 2 in. (51 mm) in width.

   (b) The marking stripe shall not be required where bleacher surfaces and environmental conditions, under all conditions of use, are such that the location of each leading edge is readily apparent, particularly when viewed in descent.

**12.2.5.6.2 Dead-End Aisles.** Dead-end aisles shall not exceed 20 ft (6100 mm) in length, unless otherwise permitted by one of the following:

(1) A dead-end aisle shall be permitted to exceed 20 ft (6100 mm) in length where seats served by the dead-end aisle are not more than 24 seats from another aisle, measured along a row of seats having a clear width of not less than 12 in. (305 mm) plus 0.6 in. (15 mm) for each additional seat over a total of 7 in the row.

(2) A 16-row, dead-end aisle shall be permitted in folding and telescopic seating and grandstands.

(3) Aisle termination in accordance with 12.4.2.11 for smoke-protected assembly seating shall be permitted.

**12.2.5.6.3\* Minimum Aisle Width.** The minimum clear width of aisles shall be sufficient to provide egress capacity in accordance with 12.2.3.1 but shall be not less than the following:

(1) 48 in. (1220 mm) for stairs having seating on each side, or 36 in. (915 mm) where the aisle does not serve more than 50 seats

(2) 36 in. (915 mm) for stairs having seating on only one side

(3) 23 in. (585 mm) between a handrail and seating, or between a guardrail and seating where the aisle is subdivided by a handrail

(4) 42 in. (1065 mm) for level or ramped aisles having seating on both sides, or 36 in. (915 mm) where the aisle does not serve more than 50 seats

(5) 36 in. (915 mm) for level or ramped aisles having seating on only one side

(6) 23 in. (585 mm) between a handrail or a guardrail and seating where the aisle does not serve more than five rows on one side

**12.2.5.6.4 Aisle Stairs and Aisle Ramps.**

**12.2.5.6.4.1\*** The following shall apply to aisle stairs and aisle ramps:

(1) Aisles having a gradient steeper than 1 in 20, but not steeper than 1 in 8, shall consist of an aisle ramp.

(2) Aisles having a gradient steeper than 1 in 8 shall consist of an aisle stair.

**12.2.5.6.4.2** The limitation on height between landings in Table 7.2.2.2.1.1(a) and Table 7.2.2.2.1.1(b) shall not apply to aisle stairs and landings.

**12.2.5.6.4.3** The limitation on height between landings in Table 7.2.5.2(a) and Table 7.2.5.2(b) shall not apply to aisle ramps and landings.

**12.2.5.6.5 Aisle Stair Treads.** Aisle stair treads shall meet all of the following criteria:

(1) There shall be no variation in the depth of adjacent treads that exceeds ³⁄₁₆ in. (4.8 mm), unless otherwise permitted by 12.2.5.6.5(2).

(2) Construction-caused nonuniformities in tread depth shall be permitted, provided that both of the following criteria are met:

   (a) The nonuniformity does not exceed ³⁄₈ in. (10 mm).

   (b) The aisle tread depth is 22 in. (560 mm) or greater.

(3)\*Tread depth shall be not less than 11 in. (280 mm).

(4) All treads shall extend the full width of the aisle.

**12.2.5.6.6 Aisle Stair Risers.** Aisle stair risers shall meet all of the following criteria:

(1) Riser heights shall be not less than 4 in. (100 mm) in aisle stairs, unless aisle stairs are those in folding and telescopic seating.

(2) The riser height of aisle stairs in folding and telescopic seating shall be permitted to be not less than 3½ in. (90 mm).

(3) Riser heights shall not exceed 8 in. (205 mm), unless otherwise permitted by 12.2.5.6.6(4) or (5).

(4) The riser height of aisle stairs in folding and telescopic seating shall be permitted to be not more than 11 in. (280 mm).

(5) Where the gradient of an aisle is steeper than 8 in. (205 mm) in rise in 11 in. (280 mm) of run for the purpose of maintaining necessary sight lines in the adjoining seating area, the riser height shall be permitted to exceed 8 in. (205 mm) but shall not exceed 9 in. (230 mm).

(6) Riser heights shall be designed to be uniform in each aisle, and the construction-caused nonuniformities shall not exceed ³⁄₁₆ in. (4.8 mm) between adjacent risers, unless the conditions of 12.2.5.6.6(7) or (8) are met.

(7) Riser height shall be permitted to be nonuniform where both of the following criteria are met:

   (a) The nonuniformity shall be only for the purpose of accommodating changes in gradient necessary to maintain sight lines within a seating area, in which case the riser height shall be permitted to exceed ³⁄₁₆ in. (4.8 mm) in any flight.

   (b) Where nonuniformities exceed ³⁄₁₆ in. (4.8 mm) between adjacent risers, the exact location of such nonuniformities shall be indicated by a distinctive marking stripe on each tread at the nosing or leading edge adjacent to the nonuniform risers.

(8) Construction-caused nonuniformities in riser height shall be permitted to exceed ³⁄₁₆ in. (4.8 mm) where all of the following criteria are met:

   (a) The riser height shall be designed to be nonuniform.

   (b) The construction-caused nonuniformities shall not exceed ³⁄₈ in. (10 mm) where the aisle tread depth is less than 22 in. (560 mm).

   (c) The construction-caused nonuniformities shall not exceed ³⁄₄ in. (19 mm) where the aisle tread depth is 22 in. (560 mm) or greater.

(d) Where nonuniformities exceed ³⁄₁₆ in. (4.8 mm) between adjacent risers, the exact location of such nonuniformities shall be indicated by a distinctive marking stripe on each tread at the nosing or leading edge adjacent to the nonuniform risers.

**12.2.5.6.7 Aisle Stair Profile.** Aisle stairs shall comply with all of the following:

(1) Aisle risers shall be vertical or sloped under the tread projection at an angle not to exceed 30 degrees from vertical.
(2) Tread projection not exceeding 1½ in. (38 mm) shall be permitted.
(3) Tread projection shall be uniform in each aisle, except as otherwise permitted by 12.2.5.6.7(4).
(4) Construction-caused projection nonuniformities not exceeding ¼ in. (6.4 mm) shall be permitted.

**12.2.5.6.8* Aisle Handrails.**

**12.2.5.6.8.1** Ramped aisles having a gradient exceeding 1 in 20 and aisle stairs shall be provided with handrails at one side or along the centerline and shall also be in accordance with 7.2.2.4.4.1, 7.2.2.4.4.5, and 7.2.2.4.4.6.

**12.2.5.6.8.2** Where seating exists on both sides of the aisle, the handrails shall be noncontinuous with gaps or breaks at intervals not exceeding five rows to facilitate access to seating and to allow crossing from one side of the aisle to the other.

**12.2.5.6.8.3** The gaps or breaks permitted by 12.2.5.6.8.2 shall have a clear width of not less than 22 in. (560 mm) and shall not exceed 36 in. (915 mm), measured horizontally, and the handrail shall have rounded terminations or bends.

**12.2.5.6.8.4** Where handrails are provided in the middle of aisle stairs, an additional intermediate rail shall be located approximately 12 in. (305 mm) below the main handrail.

**12.2.5.6.8.5** Handrails shall not be required where otherwise permitted by one of the following:

(1) Handrails shall not be required for ramped aisles having a gradient not steeper than 1 in 8 and having seating on both sides where the aisle does not serve as an accessible route.
(2) The requirement for a handrail shall be satisfied by the use of a guard provided with a rail that complies with the graspability requirements for handrails and is located at a consistent height between 34 in. and 42 in. (865 mm and 1065 mm), measured as follows:
   (a) Vertically from the top of the rail to the leading edge (nosing) of stair treads
   (b) Vertically from the top of the rail to the adjacent walking surface in the case of a ramp

**12.2.5.6.9* Aisle Marking.**

**12.2.5.6.9.1** A contrasting marking stripe shall be provided on each tread at the nosing or leading edge so that the location of such tread is readily apparent, particularly when viewed in descent.

**12.2.5.6.9.2** The marking stripe shall be not less than 1 in. (25 mm) wide and shall not exceed 2 in. (51 mm) in width.

**12.2.5.6.9.3** The marking stripe shall not be required where tread surfaces and environmental conditions, under all conditions of use, are such that the location of each tread is readily apparent, particularly when viewed in descent.

**12.2.5.7* Aisle Accessways Serving Seating at Tables.**

**12.2.5.7.1** The required clear width of an aisle accessway shall be not less than 12 in. (305 mm) where measured in accordance with 12.2.5.7.3 and shall be increased as a function of length in accordance with 12.2.5.7.4, unless otherwise permitted by 12.2.5.7.2.

**12.2.5.7.2*** If used by not more than four persons, no minimum clear width shall be required for the portion of an aisle accessway having a length not exceeding 6 ft (1830 mm) and located farthest from an aisle.

**12.2.5.7.3*** Where nonfixed seating is located between a table and an aisle accessway or aisle, the measurement of required clear width of the aisle accessway or aisle shall be made to a line 19 in. (485 mm), measured perpendicularly to the edge of the table, away from the edge of said table.

**12.2.5.7.4*** The minimum required clear width of an aisle accessway, measured in accordance with 12.2.5.4.8 and 12.2.5.7.3, shall be increased beyond the 12 in. (305 mm) requirement of 12.2.5.7.1 by ½ in. (13 mm) for each additional 12 in. (305 mm) or fraction thereof beyond 12 ft (3660 mm) of aisle accessway length, where measured from the center of the seat farthest from an aisle.

**12.2.5.7.5** The path of travel along the aisle accessway shall not exceed 36 ft (11 m) from any seat to the closest aisle or egress doorway.

**12.2.5.8 Aisles Serving Seating at Tables.**

**12.2.5.8.1*** Aisles that contain steps or that are ramped, such as aisles serving dinner theater–style configurations, shall comply with the requirements of 12.2.5.6.

**12.2.5.8.2*** The width of aisles serving seating at tables shall be not less than 44 in. (1120 mm) where serving an occupant load exceeding 50, and 36 in. (915 mm) where serving an occupant load of 50 or fewer.

**12.2.5.8.3*** Where nonfixed seating is located between a table and an aisle, the measurement of required clear width of the aisle shall be made to a line 19 in. (485 mm), measured perpendicularly to the edge of the table, away from the edge of said table.

**12.2.5.9 Approval of Layouts.**

**12.2.5.9.1** Where required by the authority having jurisdiction, plans drawn to scale showing the arrangement of furnishings or equipment shall be submitted to the authority by the building owner, manager, or authorized agent to substantiate conformance with the provisions of 12.2.5.

**12.2.5.9.2** The layout plans shall constitute the only acceptable arrangement, unless one of the following criteria is met:

(1) The plans are revised.
(2) Additional plans are submitted and approved.
(3) Temporary deviations from the specifications of the approved plans are used, provided that the occupant load is not increased and the intent of 12.2.5.9 is maintained.

**12.2.6 Travel Distance to Exits.**

**12.2.6.1** Travel distance shall be measured in accordance with Section 7.6.

**12.2.6.2** Exits shall be arranged so that the total length of travel from any point to reach an exit shall not exceed 200 ft

(61 m) in any assembly occupancy, unless otherwise permitted by one of the following:

(1) The travel distance shall not exceed 250 ft (76 m) in assembly occupancies protected throughout by an approved, supervised automatic sprinkler system in accordance with Section 9.7.
(2) The travel distance requirement shall not apply to smoke-protected assembly seating as permitted by 12.4.2.12, 12.4.2.13, and 12.4.2.14.

**12.2.7 Discharge from Exits.**

**12.2.7.1** Exit discharge shall comply with Section 7.7.

**12.2.7.2** The level of exit discharge shall be measured at the point of principal entrance to the building.

**12.2.7.3** Where the principal entrance to an assembly occupancy is via a terrace, either raised or depressed, such terrace shall be permitted to be considered to be the first story in height for the purposes of Table 12.1.6 where all of the following criteria are met:

(1) The terrace is at least as long, measured parallel to the building, as the total width of the exit(s) it serves but not less than 60 in. (1525 mm) long.
(2) The terrace is at least as wide, measured perpendicularly to the building, as the exit(s) it serves but not less than 10 ft (3050 mm) wide.
(3) Required stairs leading from the terrace to the finished ground level are protected in accordance with 7.2.2.6.3 or are not less than 10 ft (3050 mm) from the building.

**12.2.8 Illumination of Means of Egress.** Means of egress, other than for private party tents not exceeding 1200 ft² (112 m²), shall be illuminated in accordance with Section 7.8.

**12.2.9 Emergency Lighting.**

**12.2.9.1** Emergency lighting shall be provided in accordance with Section 7.9.

**12.2.9.2** Private party tents not exceeding 1200 ft² (112 m²) shall not be required to have emergency lighting.

**12.2.10 Marking of Means of Egress.**

**12.2.10.1** Means of egress shall be provided with signs in accordance with Section 7.10.

**12.2.10.2** Exit markings shall not be required on the seating side of vomitories from seating areas where exit marking is provided in the concourse and where such marking is readily apparent from the vomitories.

**12.2.10.3** Evacuation diagrams in accordance with 7.10.8.5 shall be provided.

**12.2.11 Special Means of Egress Features.**

**12.2.11.1 Guards and Railings.**

**12.2.11.1.1\* Sight Line–Constrained Rail Heights.** Unless subject to the requirements of 12.2.11.1.2, a fasciae or railing system complying with the guard requirements of 7.2.2.4, and having a height of not less than 26 in. (660 mm), shall be provided where the floor or footboard elevation is more than 30 in. (760 mm) above the floor or the finished ground level below, and where the fasciae or railing system would otherwise interfere with the sight lines of immediately adjacent seating.

**12.2.11.1.2 At Foot of Aisles.**

**12.2.11.1.2.1** A fasciae or railing system complying with the guard requirements of 7.2.2.4 shall be provided for the full width of the aisle where the foot of the aisle is more than 30 in. (760 mm) above the floor or the finished ground level below.

**12.2.11.1.2.2** The fasciae or railing shall be not less than 36 in. (915 mm) high and shall provide not less than 42 in. (1065 mm), measured diagonally, between the top of the rail and the nosing of the nearest tread.

**12.2.11.1.3 At Cross Aisles.** Guards and railings at cross aisles shall meet the following criteria:

(1) Cross aisles located behind seating rows shall be provided with railings not less than 26 in. (660 mm) above the adjacent floor of the aisle.
(2) The requirement of 12.2.11.1.3(1) shall not apply where the backs of seats located at the front of the aisle project 24 in. (610 mm) or more above the adjacent floor of the aisle.
(3) Where cross aisles exceed 30 in. (760 mm) above the floor or the finished ground level below, guards shall be provided in accordance with 7.2.2.4.

**12.2.11.1.4 At Side and Back of Seating Areas.** Guards complying with the guard requirements of 7.2.2.4 shall be provided with a height not less than 42 in. (1065 mm) above the aisle, aisle accessway, or footboard where the floor elevation exceeds 30 in. (760 mm) above the floor or the finished ground level to the side or back of seating.

**12.2.11.1.5 Below Seating.** Openings between footboards and seat boards shall be provided with intermediate construction so that a 4 in. (100 mm) diameter sphere cannot pass through the opening.

**12.2.11.1.6 Locations Not Requiring Guards.** Guards shall not be required in the following locations:

(1) Guards shall not be required on the audience side of stages, of raised platforms, and of other raised floor areas such as runways, ramps, and side stages used for entertainment or presentations.
(2) Permanent guards shall not be required at vertical openings in the performance area of stages.
(3) Guards shall not be required where the side of an elevated walking surface is required to be open for the normal functioning of special lighting or for access and use of other special equipment.

**12.2.11.2 Lockups.** Lockups in assembly occupancies shall comply with the requirements of 22.4.5.

**12.3 Protection.**

**12.3.1 Protection of Vertical Openings.** Any vertical opening shall be enclosed or protected in accordance with Section 8.6, unless otherwise permitted by one of the following:

(1)\*Stairs or ramps shall be permitted to be unenclosed between balconies or mezzanines and main assembly areas located below, provided that the balcony or mezzanine is open to the main assembly area.
(2) Exit access stairs from lighting and access catwalks, galleries, and gridirons shall not be required to be enclosed.

(3) Assembly occupancies protected by an approved, supervised automatic sprinkler system in accordance with Section 9.7 shall be permitted to have unprotected vertical openings between any two adjacent floors, provided that such openings are separated from unprotected vertical openings serving other floors by a barrier complying with 8.6.5.

(4) Assembly occupancies protected by an approved, supervised automatic sprinkler system in accordance with Section 9.7 shall be permitted to have convenience stair openings in accordance with 8.6.9.2.

**12.3.2 Protection from Hazards.**

**12.3.2.1 Service Equipment, Hazardous Operations or Processes, and Storage Facilities.**

**12.3.2.1.1** Rooms containing high-pressure boilers, refrigerating machinery of other than the domestic refrigerator type, large transformers, or other service equipment subject to explosion shall meet both of the following requirements:

(1) Such rooms shall not be located directly under or abutting required exits.

(2) Such rooms shall be separated from other parts of the building by fire barriers in accordance with Section 8.3 having a minimum 1-hour fire resistance rating or shall be protected by automatic extinguishing systems in accordance with Section 8.7.

**12.3.2.1.2** Rooms or spaces for the storage, processing, or use of materials specified in 12.3.2.1.2(1) through (3) shall be protected in accordance with one of the following:

(1) Separation from the remainder of the building by fire barriers having a minimum 1-hour fire resistance rating or protection of such rooms by automatic extinguishing systems as specified in Section 8.7 in the following areas:

(a) Boiler and furnace rooms, unless otherwise permitted by one of the following:

 i. The requirement of 12.3.2.1.2(1)(a) shall not apply to rooms enclosing furnaces, heating and air-handling equipment, or compressor equipment with a total aggregate input rating less than 200,000 Btu (211 MJ), provided that such rooms are not used for storage.

 ii. The requirement of 12.3.2.1.2(1)(a) shall not apply to attic locations of the rooms addressed in 12.3.2.1.2(1)(a)(i), provided that such rooms comply with the draftstopping requirements of 8.6.11.

(b) Rooms or spaces used for the storage of combustible supplies in quantities deeme d hazardous by the authority having jurisdiction

(c) Rooms or spaces used for the storage of hazardous materials or flammable or combustible liquids in quantities deemed hazardous by recognized standards

(2) Separation from the remainder of the building by fire barriers having a minimum 1-hour fire resistance rating and protection of such rooms by automatic extinguishing systems as specified in Section 8.7 in the following areas:

(a) Laundries

(b) Maintenance shops, including woodworking and painting areas

(c) Rooms or spaces used for processing or use of combustible supplies deemed hazardous by the authority having jurisdiction

(d) Rooms or spaces used for processing or use of hazardous materials or flammable or combustible liquids in quantities deemed hazardous by recognized standards

(3) Protection as permitted in accordance with 9.7.1.2 where automatic extinguishing is used to meet the requirements of 12.3.2.1.2(1) or (2)

**12.3.2.2 Cooking Equipment.** Cooking equipment shall be protected in accordance with 9.2.3, unless the cooking equipment is one of the following types:

(1) Outdoor equipment

(2) Portable equipment not flue-connected

(3) Equipment used only for food warming

**12.3.3 Interior Finish.**

**12.3.3.1 General.** Interior finish shall be in accordance with Section 10.2.

**12.3.3.2 Corridors, Lobbies, and Enclosed Stairways.** Interior wall and ceiling finish materials complying with Section 10.2 shall be Class A or Class B in all corridors and lobbies and shall be Class A in enclosed stairways.

**12.3.3.3 Assembly Areas.** Interior wall and ceiling finish materials complying with Section 10.2 shall be Class A or Class B in general assembly areas having occupant loads of more than 300 and shall be Class A, Class B, or Class C in assembly areas having occupant loads of 300 or fewer.

**12.3.3.4 Screens.** Screens on which pictures are projected shall comply with requirements of Class A or Class B interior finish in accordance with Section 10.2.

**12.3.3.5 Interior Floor Finish.**

**12.3.3.5.1** Interior floor finish shall comply with Section 10.2.

**12.3.3.5.2** Interior floor finish in exit enclosures and exit access corridors and in spaces not separated from them by walls complying with 12.3.6 shall be not less than Class II.

**12.3.3.5.3** Interior floor finish shall comply with 10.2.7.1 or 10.2.7.2, as applicable.

**12.3.4 Detection, Alarm, and Communications Systems.**

**12.3.4.1 General.**

**12.3.4.1.1** Assembly occupancies with occupant loads of more than 300 and all theaters with more than one audience-viewing room shall be provided with an approved fire alarm system in accordance with 9.6.1 and 12.3.4, unless otherwise permitted by 12.3.4.1.2.

**12.3.4.1.2** Assembly occupancies that are a part of a multiple occupancy protected as a mixed occupancy *(see 6.1.14)* shall be permitted to be served by a common fire alarm system, provided that the individual requirements of each occupancy are met.

**12.3.4.2 Initiation.**

**12.3.4.2.1** Initiation of the required fire alarm system shall be by both of the following means:

(1) Manual means in accordance with 9.6.2.1(1), unless otherwise permitted by one of the following:

(a) The requirement of 12.3.4.2.1(1) shall not apply where initiation is by means of an approved automatic fire detection system in accordance with 9.6.2.1(2) that provides fire detection throughout the building.

(b) The requirement of 12.3.4.2.1(1) shall not apply where initiation is by means of an approved automatic sprinkler system in accordance with 9.6.2.1(3) that provides fire detection and protection throughout the building.

(2) Where automatic sprinklers are provided, initiation of the fire alarm system by sprinkler system waterflow, even where manual fire alarm boxes are provided in accordance with 12.3.4.2.1(1)

**12.3.4.2.2** The initiating device shall be capable of transmitting an alarm to a receiving station, located within the building, that is constantly attended when the assembly occupancy is occupied.

**12.3.4.2.3\*** In assembly occupancies with occupant loads of more than 300, automatic detection shall be provided in all hazardous areas that are not normally occupied, unless such areas are protected throughout by an approved, supervised automatic sprinkler system in accordance with Section 9.7.

**12.3.4.3 Notification.** The required fire alarm system shall activate an audible and visible alarm in a constantly attended receiving station within the building when occupied for purposes of initiating emergency action.

**12.3.4.3.1** Positive alarm sequence in accordance with 9.6.3.4 shall be permitted.

**12.3.4.3.2 Reserved.**

**12.3.4.3.3** Occupant notification shall be by means of voice announcements in accordance with 9.6.3.9, initiated by the person in the constantly attended receiving station.

**12.3.4.3.4** Occupant notification shall be by means of visible signals in accordance with 9.6.3.5, initiated by the person in the constantly attended receiving station, unless otherwise permitted by 12.3.4.3.5.

**12.3.4.3.5\*** Visible signals shall not be required in the assembly seating area, or the floor area used for the contest, performance, or entertainment, where the occupant load exceeds 1000 and an approved, alternative visible means of occupant notification is provided. (*See 9.6.3.5.7.*)

**12.3.4.3.6** The announcement shall be permitted to be made via a voice communication or public address system in accordance with 9.6.3.9.2.

**12.3.4.3.7** Where the authority having jurisdiction determines that a constantly attended receiving station is impractical, both of the following shall be provided:

(1) Automatically transmitted evacuation or relocation instructions shall be provided in accordance with *NFPA 72, National Fire Alarm and Signaling Code.*
(2) The system shall be monitored by a supervising station in accordance with *NFPA 72.*

**12.3.5 Extinguishment Requirements.**

**12.3.5.1** The following assembly occupancies shall be protected throughout by an approved, supervised automatic sprinkler system in accordance with 9.7.1.1(1):

(1) Dance halls
(2) Discotheques
(3) Nightclubs
(4) Assembly occupancies with festival seating

**12.3.5.2** Any building containing one or more assembly occupancies where the aggregate occupant load of the assembly occupancies exceeds 300 shall be protected by an approved, supervised automatic sprinkler system in accordance with Section 9.7 as follows (*see also 12.1.6, 12.2.6, 12.3.2, and 12.3.6*):

(1) Throughout the story containing the assembly occupancy
(2) Throughout all stories below the story containing the assembly occupancy
(3) In the case of an assembly occupancy located below the level of exit discharge, throughout all stories intervening between that story and the level of exit discharge, including the level of exit discharge

**12.3.5.3** The requirements of 12.3.5.2 shall not apply to the following:

(1)\*Assembly occupancies consisting of a single multipurpose room of less than 12,000 ft$^2$ (1115 m$^2$) that are not used for exhibition or display and are not part of a mixed occupancy
(2) Gymnasiums, skating rinks, and swimming pools used exclusively for participant sports with no audience facilities for more than 300 persons
(3) Locations in stadia and arenas as follows:
  (a) Over the floor areas used for contest, performance, or entertainment, provided that the roof construction is more than 50 ft (15 m) above the floor level, and use is restricted to low fire hazard uses
  (b) Over the seating areas, provided that use is restricted to low fire hazard uses
  (c) Over open-air concourses where an approved engineering analysis substantiates the ineffectiveness of the sprinkler protection due to building height and combustible loading
(4) Locations in unenclosed stadia and arenas as follows:
  (a) Press boxes of less than 1000 ft$^2$ (93 m$^2$)
  (b) Storage facilities of less than 1000 ft$^2$ (93 m$^2$) if enclosed with not less than 1-hour fire resistance–rated construction
  (c) Enclosed areas underneath grandstands that comply with 12.4.8.5

**12.3.5.4** Where another provision of this chapter requires an automatic sprinkler system, the sprinkler system shall be installed in accordance with 9.7.1.1(1).

**12.3.6 Corridors.** Interior corridors and lobbies shall be constructed in accordance with 7.1.3.1 and Section 8.3, unless otherwise permitted by one of the following:

(1) Corridor and lobby protection shall not be required where assembly rooms served by the corridor or lobby have at least 50 percent of their exit capacity discharging directly to the outside, independent of corridors and lobbies.
(2) Corridor and lobby protection shall not be required in buildings protected throughout by an approved, supervised automatic sprinkler system in accordance with Section 9.7.
(3) Lobbies serving only one assembly area that meet the requirements for intervening rooms (*see 7.5.1.6*) shall not be required to have a fire resistance rating.
(4) Where the corridor ceiling is an assembly having a 1-hour fire resistance rating where tested as a wall, the corridor walls shall be permitted to terminate at the corridor ceiling.
(5) Corridor and lobby protection shall not be required in buildings protected throughout by an approved, total (complete) coverage smoke detection system providing occupant notification and installed in accordance with Section 9.6.

## 12.4 Special Provisions.

### 12.4.1 Life Safety Evaluation.

**12.4.1.1\*** Where a life safety evaluation is required by other provisions of the *Code*, it shall comply with all of the following:

(1) The life safety evaluation shall be performed by persons acceptable to the authority having jurisdiction.
(2) The life safety evaluation shall include a written assessment of safety measures for conditions listed in 12.4.1.2.
(3) The life safety evaluation shall be approved annually by the authority having jurisdiction and shall be updated for special or unusual conditions.

**12.4.1.2** Life safety evaluations shall include an assessment of all of the following conditions and related appropriate safety measures:

(1) Nature of the events and the participants and attendees
(2) Access and egress movement, including crowd density problems
(3) Medical emergencies
(4) Fire hazards
(5) Permanent and temporary structural systems
(6) Severe weather conditions
(7) Earthquakes
(8) Civil or other disturbances
(9) Hazardous materials incidents within and near the facility
(10) Relationships among facility management, event participants, emergency response agencies, and others having a role in the events accommodated in the facility

**12.4.1.3\*** Life safety evaluations shall include assessments of both building systems and management features upon which reliance is placed for the safety of facility occupants, and such assessments shall consider scenarios appropriate to the facility.

### 12.4.2\* Smoke-Protected Assembly Seating.

**12.4.2.1** To be considered smoke protected, an assembly seating facility shall comply with both of the following:

(1) All enclosed areas with walls and ceilings in buildings or structures containing smoke-protected assembly seating shall be protected with an approved, supervised automatic sprinkler system in accordance with Section 9.7, unless otherwise permitted by one of the following:
  (a) The requirement of 12.4.2.1(1) shall not apply to the floor area used for contest, performance, or entertainment, provided that the roof construction is more than 50 ft (15 m) above the floor level and use is restricted to low fire hazard uses.
  (b)\*Sprinklers shall not be required to be located over the floor area used for contest, performance, or entertainment and over the seating areas where an approved engineering analysis substantiates the ineffectiveness of the sprinkler protection due to building height and combustible loading.
(2) All means of egress serving a smoke-protected assembly seating area shall be provided with smoke-actuated ventilation facilities or natural ventilation designed in accordance with both of the following criteria:
  (a) The ventilation system shall be designed to maintain the level of smoke at not less than 6 ft (1830 mm) above the floor of the means of egress.
  (b) The ventilation system shall be in accordance with NFPA 92, *Standard for Smoke Control Systems.*

**12.4.2.2** To use the provisions of smoke-protected assembly seating, a facility shall be subject to a life safety evaluation in accordance with 12.4.1.

**12.4.2.3** Minimum clear widths of aisles and other means of egress serving smoke-protected assembly seating shall be in accordance with Table 12.4.2.3.

**Table 12.4.2.3 Capacity Factors for Smoke-Protected Assembly Seating**

| No. of Seats | Clear Width per Seat Served | | | |
| | Stairs | | Passageways, Ramps, and Doorways | |
| | in. | mm | in. | mm |
|---|---|---|---|---|
| 2,000 | 0.300 *AB* | 7.6 *AB* | 0.220 *C* | 5.6 *C* |
| 5,000 | 0.200 *AB* | 5.1 *AB* | 0.150 *C* | 3.8 *C* |
| 10,000 | 0.130 *AB* | 3.3 *AB* | 0.100 *C* | 2.5 *C* |
| 15,000 | 0.096 *AB* | 2.4 *AB* | 0.070 *C* | 1.8 *C* |
| 20,000 | 0.076 *AB* | 1.9 *AB* | 0.056 *C* | 1.4 *C* |
| ≥25,000 | 0.060 *AB* | 1.5 *AB* | 0.044 *C* | 1.1 *C* |

#### 12.4.2.4 Outdoor Smoke-Protected Assembly Seating.

**12.4.2.4.1** Where smoke-protected assembly seating and its means of egress are located wholly outdoors, capacity shall be permitted to be provided in accordance with Table 12.4.2.4.1 and the provision of 12.4.2.4.2 shall apply.

**Table 12.4.2.4.1 Capacity Factors for Outdoor Smoke-Protected Assembly Seating**

| Feature | Clear Width per Seat Served | | | |
| | Stairs | | Passageways, Ramps, and Doorways | |
| | in. | mm | in. | mm |
|---|---|---|---|---|
| Outdoor smoke-protected assembly seating | 0.08 *AB* | 2.0 *AB* | 0.06 *C* | 1.5 *C* |

**12.4.2.4.2** Where the number of seats in outdoor smoke-protected assembly seating exceeds 20,000, the capacity factors of Table 12.4.2.3 shall be permitted to be used.

**12.4.2.5** Where using Table 12.4.2.3, the number of seats specified shall be within a single assembly space, and interpolation shall be permitted between the specific values shown. A single seating space shall be permitted to have multiple levels, floors, or mezzanines.

**12.4.2.6** The minimum clear widths shown in Table 12.4.2.3 and Table 12.4.2.4.1 shall be modified in accordance with all of the following:

(1) If risers exceed 7 in. in height, the stair width in Table 12.4.2.3 and Table 12.4.2.4.1 shall be multiplied by factor A, where A equals the following:

$$A = 1 + \frac{\text{riser height} - 7}{5}$$

(2) If risers exceed 178 mm in height, the stair width in Table 12.4.2.3 and Table 12.4.2.4.1 shall be multiplied by factor A, where A equals the following:

$$A = 1 + \frac{\text{riser height} - 178}{125}$$

(3) Stairs not having a handrail within a 30 in. (760 mm) horizontal distance shall be 25 percent wider than otherwise calculated; that is, their width shall be multiplied by factor B, where B equals the following:

$$B = 1.25$$

(4) Ramps steeper than 1 in 10 slope where used in ascent shall have their width increased by 10 percent; that is, their width shall be multiplied by factor C, where C equals the following:

$$C = 1.10$$

**12.4.2.7** Where smoke-protected assembly seating conforms to the requirements of 12.4.2, for rows of seats served by aisles or doorways at both ends, the number of seats per row shall not exceed 100, and the clear width of not less than 12 in. (305 mm) for aisle accessways shall be increased by 0.3 in. (7.6 mm) for every additional seat beyond the number stipulated in Table 12.4.2.7; however, the minimum clear width shall not be required to exceed 22 in. (560 mm).

**Table 12.4.2.7 Smoke-Protected Assembly Seating Aisle Accessways**

| | Number of Seats per Row Permitted to Have a Clear Width Aisle Accessway of Not Less than 12 in. (305 mm) | |
| Total Number of Seats in the Space | Aisle or Doorway at Both Ends of Row | Aisle or Doorway at One End of Row |
|---|---|---|
| <4,000 | 14 | 7 |
| 4,000–6,999 | 15 | 7 |
| 7,000–9,999 | 16 | 8 |
| 10,000–12,999 | 17 | 8 |
| 13,000–15,999 | 18 | 9 |
| 16,000–18,999 | 19 | 9 |
| 19,000–21,999 | 20 | 10 |
| ≥22,000 | 21 | 11 |

**12.4.2.8** Where smoke-protected assembly seating conforms to the requirements of 12.4.2, for rows of seats served by an aisle or doorway at one end only, the aisle accessway clear width of not less than 12 in. (305 mm) shall be increased by 0.6 in. (15 mm) for every additional seat beyond the number stipulated in Table 12.4.2.7; however, the minimum clear width shall not be required to exceed 22 in. (560 mm).

**12.4.2.9** Smoke-protected assembly seating conforming with the requirements of 12.4.2 shall be permitted to have a common path of travel of 50 ft (15 m) from any seat to a point where a person has a choice of two directions of egress travel.

**12.4.2.10** Aisle accessways shall be permitted to serve as one or both of the required exit accesses addressed in 12.4.2.9, provided that the aisle accessway has a minimum width of 12 in. (305 mm) plus 0.3 in. (7.6 mm) for every additional seat over a total of 7 in a row.

**12.4.2.11** Where smoke-protected assembly seating conforms to the requirements of 12.4.2, the dead ends in aisle stairs shall not exceed a distance of 21 rows, unless both of the following criteria are met:

(1) The seats served by the dead-end aisle are not more than 40 seats from another aisle.
(2) The 40-seat distance is measured along a row of seats having an aisle accessway with a clear width of not less than 12 in. (305 mm) plus 0.3 in. (7.6 mm) for each additional seat above 7 in the row.

**12.4.2.12** Where smoke-protected assembly seating conforms to the requirements of 12.4.2, the travel distance from each seat to the nearest entrance to an egress vomitory or egress concourse shall not exceed 400 ft (122 m).

**12.4.2.13** Where smoke-protected assembly seating conforms to the requirements of 12.4.2, the travel distance from the entrance to the vomitory or from the egress concourse to an approved egress stair, ramp, or walk at the building exterior shall not exceed 200 ft (61 m).

**12.4.2.14** The travel distance requirements of 12.4.2.12 and 12.4.2.13 shall not apply to outdoor assembly seating facilities of Type I or Type II construction where all portions of the means of egress are essentially open to the outside.

### 12.4.3 Limited Access or Underground Buildings.

**12.4.3.1** Limited access or underground buildings shall comply with 12.4.3 and Section 11.7.

**12.4.3.2** Underground buildings or portions of buildings having a floor level more than 30 ft (9.1 m) below the level of exit discharge shall comply with the requirements of 12.4.3.3 through 12.4.3.5, unless otherwise permitted by one of the following:

(1) This requirement shall not apply to areas within buildings used only for service to the building, such as boiler/heater rooms, cable vaults, and dead storage.
(2) This requirement shall not apply to auditoriums without intervening occupiable levels.

**12.4.3.3** Each level more than 30 ft (9.1 m) below the level of exit discharge shall be divided into not less than two smoke compartments by a smoke barrier complying with Section 8.5 and shall have a minimum 1-hour fire resistance rating.

**12.4.3.3.1** Smoke compartments shall comply with both of the following:

(1) Each smoke compartment shall have access to not less than one exit without passing through the other required compartment.
(2) Any doors connecting required compartments shall be tight-fitting, minimum 1-hour-rated fire door assemblies designed and installed to minimize smoke leakage and to close and latch automatically upon detection of smoke.

**12.4.3.3.2** Each smoke compartment shall be provided with a mechanical means of moving people vertically, such as an elevator or escalator.

**12.4.3.3.3** Each smoke compartment shall have an independent air supply and exhaust system capable of smoke control or smoke exhaust functions. The system shall be in accordance with NFPA 92, *Standard for Smoke Control Systems.*

**12.4.3.3.4** Throughout each smoke compartment shall be provided an automatic smoke detection system designed such that the activation of any two detectors causes the smoke control system to operate and the building voice alarm to sound.

**12.4.3.4** Any required smoke control or exhaust system shall be provided with a standby power system complying with Article 701 of *NFPA 70, National Electrical Code.*

**12.4.3.5** The building shall be provided with an approved, supervised voice alarm system, in accordance with Section 9.6, that complies with 9.6.3.9 and provides a prerecorded evacuation message.

**12.4.4 High-Rise Buildings.** High-rise assembly occupancy buildings and high-rise mixed occupancy buildings that house assembly occupancies in the high-rise portions of the building shall comply with Section 11.8.

**12.4.5 Stages and Platforms.** See 3.3.262 and 3.3.209.

**12.4.5.1 Materials and Design.**

**12.4.5.1.1** Materials used in the construction of platforms and stages shall conform to the applicable requirements of the local building code.

**12.4.5.1.2** Stage stairs shall be permitted to be of combustible materials, regardless of building construction type.

**12.4.5.2 Platform Construction.**

**12.4.5.2.1** Temporary platforms shall be permitted to be constructed of any materials.

**12.4.5.2.2** The space between the floor and the temporary platform above shall not be used for any purpose other than the electrical wiring to platform equipment.

**12.4.5.2.3** Permanent platforms shall be of the materials required for the building construction type in which the permanent platform is located, except that the finish floor shall be permitted to be of wood in all types of construction.

**12.4.5.2.4** Where the space beneath the permanent platform is used for storage or any purpose other than equipment wiring or plumbing, the floor construction shall not be less than 1-hour fire resistive.

**12.4.5.3 Stage Construction.**

**12.4.5.3.1** Regular stages shall be of the materials required for the building construction type in which they are located. In all cases, the finish floor shall be permitted to be of wood.

**12.4.5.3.2** Legitimate stages shall be constructed of materials required for Type I buildings, except that the area extending from the proscenium opening to the back wall of the stage, and for a distance of 6 ft (1830 mm) beyond the proscenium opening on each side, shall be permitted to be constructed of steel or heavy timber covered with a wood floor not less than 1½ in. (38 mm) in actual thickness.

**12.4.5.3.3** Openings through stage floors shall be equipped with tight-fitting traps with approved safety locks, and such traps shall comply with one of the following:

(1) The traps shall be of wood having an actual thickness of not less than 1½ in. (38 mm).
(2) The traps shall be of a material that provides fire and heat resistance at least equivalent to that provided by wood traps having an actual thickness of not less than 1½ in. (38 mm).

**12.4.5.4 Accessory Rooms.**

**12.4.5.4.1** Workshops, storerooms, permanent dressing rooms, and other accessory spaces contiguous to stages shall be separated from each other and other building areas by 1-hour fire resistance–rated construction and protected openings.

**12.4.5.4.2** The separation requirements of 12.4.5.4.1 shall not be required for stages having a floor area not exceeding 1000 ft² (93 m²).

**12.4.5.5 Ventilators.** Regular stages in excess of 1000 ft² (93 m²) and legitimate stages shall be provided with emergency ventilation to provide a means of removing smoke and combustion gases directly to the outside in the event of a fire, and such ventilation shall be achieved by one or a combination of the methods specified in 12.4.5.5.1 through 12.4.5.5.3.

**12.4.5.5.1 Smoke Control.**

**12.4.5.5.1.1** A means complying with Section 9.3 shall be provided to maintain the smoke level at not less than 6 ft (1830 mm) above the highest level of assembly seating or above the top of the proscenium opening where a proscenium wall and opening protection are provided.

**12.4.5.5.1.2** Smoke control systems used for compliance with 12.4.5.5.1.1 shall be in accordance with NFPA 92, *Standard for Smoke Control Systems.*

**12.4.5.5.1.3** The smoke control system shall be activated independently by each of the following:

(1) Activation of the sprinkler system in the stage area
(2) Activation of smoke detectors over the stage area
(3) Activation by manually operated switch at an approved location

**12.4.5.5.1.4** The emergency ventilation system shall be supplied by both normal and standby power.

**12.4.5.5.1.5** The fan(s) power wiring and ducts shall be located and properly protected to ensure a minimum of 20 minutes of operation in the event of activation.

**12.4.5.5.2 Roof Vents.**

**12.4.5.5.2.1** Two or more vents shall be located near the center of and above the highest part of the stage area.

**12.4.5.5.2.2** The vents shall be raised above the roof and shall provide a net free vent area equal to 5 percent of the stage area.

**12.4.5.5.2.3** Vents shall be constructed to open automatically by approved heat-activated devices, and supplemental means shall be provided for manual operation and periodic testing of the ventilator from the stage floor.

**12.4.5.5.2.4** Vents shall be labeled.

**12.4.5.5.3 Other Means.** Approved, alternate means of removing smoke and combustion gases shall be permitted.

**12.4.5.6 Proscenium Walls.** Legitimate stages shall be completely separated from the seating area by a proscenium wall of not less than 2-hour fire-resistive, noncombustible construction.

**12.4.5.6.1** The proscenium wall shall extend not less than 48 in. (1220 mm) above the roof of the auditorium in combustible construction.

**12.4.5.6.2** All openings in the proscenium wall of a legitimate stage shall be protected by a fire assembly having a minimum 1½-hour fire protection rating.

**12.4.5.6.3** The main proscenium opening used for viewing performances shall be provided with proscenium opening protection as described in 12.4.5.7.

**12.4.5.6.4** Proscenium walls shall not be required in smoke-protected assembly seating facilities constructed and operated in accordance with 12.4.2.

**12.4.5.7 Proscenium Opening Protection.**

**12.4.5.7.1** Where required by 12.4.5.6, the proscenium opening shall be protected by a listed, minimum 20-minute opening protective assembly, a fire curtain complying with NFPA 80, *Standard for Fire Doors and Other Opening Protectives*, or an approved water curtain complying with NFPA 13, *Standard for the Installation of Sprinkler Systems*.

**12.4.5.7.2** Proscenium opening protection provided by other than a fire curtain shall activate upon automatic detection of a fire and upon manual activation.

**12.4.5.8 Gridiron, Fly Galleries, and Pinrails.**

**12.4.5.8.1** Structural framing designed only for the attachment of portable or fixed theater equipment, gridirons, galleries, and catwalks shall be constructed of materials consistent with the building construction type, and a fire resistance rating shall not be required.

**12.4.5.8.2** Combustible materials shall be permitted to be used for the floors of galleries and catwalks of all construction types.

**12.4.5.9 Catwalks.** The clear width of lighting and access catwalks and the means of egress from galleries and gridirons shall be not less than 22 in. (560 mm).

**12.4.5.10 Fire Protection.** Every stage shall be protected by an approved, supervised automatic sprinkler system in compliance with Section 9.7.

**12.4.5.10.1** Protection shall be provided throughout the stage and in storerooms, workshops, permanent dressing rooms, and other accessory spaces contiguous to stages.

**12.4.5.10.2** Sprinklers shall not be required for stages 1000 ft$^2$ (93 m$^2$) or less in area and 50 ft (15 m) or less in height where both of the following criteria are met:

(1) Curtains, scenery, or other combustible hangings are not retractable vertically.
(2) Combustible hangings are limited to borders, legs, a single main curtain, and a single backdrop.

**12.4.5.10.3** Sprinklers shall not be required under stage areas less than 48 in. (1220 mm) in clear height that are used exclusively for chair or table storage and lined on the inside with ⅝ in. (16 mm) Type X gypsum wallboard or the approved equivalent.

**12.4.5.11 Flame-Retardant Requirements.**

**12.4.5.11.1** Combustible scenery of cloth, film, vegetation (dry), and similar materials shall comply with one of the following:

(1) They shall meet the flame propagation performance criteria contained in NFPA 701, *Standard Methods of Fire Tests for Flame Propagation of Textiles and Films*.
(2) They shall exhibit a heat release rate not exceeding 100 kW when tested in accordance with NFPA 289, *Standard Method of Fire Test for Individual Fuel Packages*, using the 20 kW ignition source.

**12.4.5.11.2** Foamed plastics *(see definition of cellular or foamed plastic in 3.3.41)* shall be permitted to be used if they exhibit a heat release rate not exceeding 100 kW when tested in accordance with NFPA 289, *Standard Method of Fire Test for Individual Fuel Packages*, using the 20 kW ignition source or by specific approval of the authority having jurisdiction.

**12.4.5.11.3** Scenery and stage properties not separated from the audience by proscenium opening protection shall be of noncombustible materials, limited-combustible materials, or fire-retardant-treated wood.

**12.4.5.11.4** In theaters, motion picture theaters, and television stage settings, with or without horizontal projections, and in simulated caves and caverns of foamed plastic, any single fuel package shall have a heat release rate not to exceed 100 kW where tested in accordance with one of the following:

(1) ANSI/UL 1975, *Standard for Fire Tests for Foamed Plastics Used for Decorative Purposes*
(2) NFPA 289, *Standard Method of Fire Test for Individual Fuel Packages*, using the 20 kW ignition source

**12.4.5.12\* Standpipes.**

**12.4.5.12.1** Regular stages over 1000 ft$^2$ (93 m$^2$) in area and all legitimate stages shall be equipped with 1½ in. (38 mm) hose lines for first aid fire fighting at each side of the stage.

**12.4.5.12.2** Hose connections shall be in accordance with NFPA 13, *Standard for the Installation of Sprinkler Systems*, unless Class II or Class III standpipes in accordance with NFPA 14, *Standard for the Installation of Standpipe and Hose Systems*, are used.

**12.4.6 Projection Rooms.**

**12.4.6.1** Projection rooms shall comply with 12.4.6.2 through 12.4.6.10.

**12.4.6.2** Where cellulose nitrate film is used, the projection room shall comply with NFPA 40, *Standard for the Storage and Handling of Cellulose Nitrate Film.*

**12.4.6.3** Film or video projectors or spotlights utilizing light sources that produce particulate matter or toxic gases, or light sources that produce hazardous radiation, without protective shielding shall be located within a projection room complying with 12.3.2.1.2.

**12.4.6.4** Every projection room shall be of permanent construction consistent with the building construction type in which the projection room is located and shall comply with the following:

(1) Openings shall not be required to be protected.
(2) The room shall have a floor area of not less than 80 ft² (7.4 m²) for a single machine and not less than 40 ft² (3.7 m²) for each additional machine.
(3) Each motion picture projector, floodlight, spotlight, or similar piece of equipment shall have a clear working space of not less than 30 in. (760 mm) on each side and at its rear, but only one such space shall be required between adjacent projectors.

**12.4.6.5** The projection room and the rooms appurtenant to it shall have a ceiling height of not less than 7 ft 6 in. (2285 mm).

**12.4.6.6** Each projection room for safety film shall have not less than one out-swinging, self-closing door not less than 30 in. (760 mm) wide and 6 ft 8 in. (2030 mm) high.

**12.4.6.7** The aggregate of ports and openings for projection equipment shall not exceed 25 percent of the area of the wall between the projection room and the auditorium, and all openings shall be provided with glass or other approved material so as to completely close the opening.

**12.4.6.8** Projection room ventilation shall comply with 12.4.6.8.1 and 12.4.6.8.2.

**12.4.6.8.1 Supply Air.**

**12.4.6.8.1.1** Each projection room shall be provided with adequate air supply inlets arranged to provide well-distributed air throughout the room.

**12.4.6.8.1.2** Air inlet ducts shall provide an amount of air equivalent to the amount of air being exhausted by projection equipment.

**12.4.6.8.1.3** Air shall be permitted to be taken from the outside; from adjacent spaces within the building, provided that the volume and infiltration rate is sufficient; or from the building air-conditioning system, provided that it is arranged to supply sufficient air whether or not other systems are in operation.

**12.4.6.8.2 Exhaust Air.**

**12.4.6.8.2.1** Projection booths shall be permitted to be exhausted through the lamp exhaust system.

**12.4.6.8.2.2** The lamp exhaust system shall be positively interconnected with the lamp so that the lamp cannot operate unless there is sufficient airflow required for the lamp.

**12.4.6.8.2.3** Exhaust air ducts shall terminate at the exterior of the building in such a location that the exhaust air cannot be readily recirculated into any air supply system.

**12.4.6.8.2.4** The projection room ventilation system shall be permitted also to serve appurtenant rooms, such as the generator room and the rewind room.

**12.4.6.9** Each projection machine shall be provided with an exhaust duct that draws air from each lamp and exhausts it directly to the outside of the building.

**12.4.6.9.1** The lamp exhaust shall be permitted to exhaust air from the projection room to provide room air circulation.

**12.4.6.9.2** Lamp exhaust ducts shall be of rigid materials, except for a flexible connector approved for the purpose.

**12.4.6.9.3** The projection lamp and projection room exhaust systems shall be permitted to be combined but shall not be interconnected with any other exhaust system or return-air system within the buildings.

**12.4.6.9.4** Specifications for electric arc and xenon projection equipment shall comply with 12.4.6.9.4.1 and 12.4.6.9.4.2.

**12.4.6.9.4.1 Electric Arc Projection Equipment.** The exhaust capacity shall be 200 ft³/min (0.09 m³/s) for each lamp connected to the lamp exhaust system, or as recommended by the equipment manufacturer, and auxiliary air shall be permitted to be introduced into the system through a screened opening to stabilize the arc.

**12.4.6.9.4.2 Xenon Projection Equipment.** The lamp exhaust system shall exhaust not less than 300 ft³/min (0.14 m³/s) per lamp, or not less than the exhaust volume required or recommended by the equipment manufacturer, whichever is greater.

**12.4.6.10** Miscellaneous equipment and storage shall be protected as follows:

(1) Each projection room shall be provided with rewind and film storage facilities.
(2) Flammable liquids containers shall be permitted in projection rooms, provided that all of the following criteria are met:
    (a) There are not more than four containers per projection room.
    (b) No container has a capacity exceeding 16 oz (0.5 L).
    (c) The containers are of a nonbreakable type.
(3) Appurtenant electrical equipment, such as rheostats, transformers, and generators, shall be permitted to be located within the booth or in a separate room of equivalent construction.

**12.4.7\* Special Amusement Buildings.**

**12.4.7.1\* General.** Special amusement buildings, regardless of occupant load, shall meet the requirements for assembly occupancies in addition to the requirements of 12.4.7, unless the special amusement building is a multilevel play structure that is not more than 10 ft (3050 mm) in height and has aggregate horizontal projections not exceeding 160 ft² (15 m²).

**12.4.7.2\* Automatic Sprinklers.** Every special amusement building, other than buildings or structures not exceeding 10 ft (3050 mm) in height and not exceeding 160 ft² (15 m²) in aggregate horizontal projection, shall be protected throughout by an approved, supervised automatic sprinkler system installed and maintained in accordance with Section 9.7.

**12.4.7.3 Temporary Water Supply.** Where the special amusement building required to be sprinklered by 12.4.7.2 is movable or portable, the sprinkler water supply shall be permitted to be provided by an approved temporary means.

**12.4.7.4 Smoke Detection.** Where the nature of the special amusement building is such that it operates in reduced lighting levels, the building shall be protected throughout by an approved automatic smoke detection system in accordance with Section 9.6.

**12.4.7.5 Alarm Initiation.** Actuation of any smoke detection system device shall sound an alarm at a constantly attended location on the premises.

**12.4.7.6 Illumination.** Actuation of the automatic sprinkler system, or any other suppression system, or actuation of a smoke detection system having an approved verification or cross-zoning operation capability shall provide for both of the following:

(1) Increase in illumination in the means of egress to that required by Section 7.8
(2) Termination of any conflicting or confusing sounds and visuals

**12.4.7.7 Exit Marking.**

**12.4.7.7.1** Exit marking shall be in accordance with Section 7.10.

**12.4.7.7.2** Floor proximity exit signs shall be provided in accordance with 7.10.1.6.

**12.4.7.7.3\*** In special amusement buildings where mazes, mirrors, or other designs are used to confound the egress path, approved directional exit marking that becomes apparent in an emergency shall be provided.

**12.4.7.8 Interior Finish.** Interior wall and ceiling finish materials complying with Section 10.2 shall be Class A throughout.

**12.4.8 Grandstands.**

**12.4.8.1 General.** Grandstands shall comply with the provisions of this chapter as modified by 12.4.8.

**12.4.8.2 Seating.**

**12.4.8.2.1** Where grandstand seating without backs is used indoors, rows of seats shall be spaced not less than 22 in. (560 mm) back-to-back.

**12.4.8.2.2** The depth of footboards and seat boards in grandstands shall be not less than 9 in. (230 mm); where the same level is not used for both seat foundations and footrests, footrests independent of seats shall be provided.

**12.4.8.2.3** Seats and footrests of grandstands shall be supported securely and fastened in such a manner that they cannot be displaced inadvertently.

**12.4.8.2.4** Individual seats or chairs shall be permitted only if secured in rows in an approved manner, unless seats do not exceed 16 in number and are located on level floors and within railed-in enclosures, such as boxes.

**12.4.8.2.5** The maximum number of seats permitted between the farthest seat and an aisle in grandstands and bleachers shall not exceed that shown in Table 12.4.8.2.5.

**Table 12.4.8.2.5 Maximum Number of Seats Between Farthest Seat and an Aisle**

| Application | Outdoors | Indoors |
|---|---|---|
| Grandstands | 11 | 6 |
| Bleachers *(See 12.2.5.6.1.2.)* | 20 | 9 |

**12.4.8.3 Special Requirements — Wood Grandstands.**

**12.4.8.3.1** An outdoor wood grandstand shall be erected within not less than two-thirds of its height, and, in no case, within not less than 10 ft (3050 mm), of a building, unless otherwise permitted by one of the following:

(1) The distance requirement shall not apply to buildings having minimum 1-hour fire resistance–rated construction with openings protected against the fire exposure hazard created by the grandstand.
(2) The distance requirement shall not apply where a wall having minimum 1-hour fire resistance–rated construction separates the grandstand from the building.

**12.4.8.3.2** An outdoor wood grandstand unit shall not exceed 10,000 ft² (929 m²) in finished ground level area or 200 ft (61 m) in length, and all of the following requirements also shall apply:

(1) Grandstand units of the maximum size shall be placed not less than 20 ft (6100 mm) apart or shall be separated by walls having a minimum 1-hour fire resistance rating.
(2) The number of grandstand units erected in any one group shall not exceed three.
(3) Each group of grandstand units shall be separated from any other group by a wall having minimum 2-hour fire resistance–rated construction extending 24 in. (610 mm) above the seat platforms or by an open space of not less than 50 ft (15 m).

**12.4.8.3.3** The finished ground level area or length required by 12.4.8.3.2 shall be permitted to be doubled where one of the following criteria is met:

(1) Where the grandstand is constructed entirely of labeled fire-retardant-treated wood that has passed the standard rain test, ASTM D 2898, *Standard Test Methods for Accelerated Weathering of Fire-Retardant-Treated Wood for Fire Testing*
(2) Where the grandstand is constructed of members conforming to dimensions for heavy timber construction [Type IV (2HH)]

**12.4.8.3.4** The highest level of seat platforms above the finished ground level or the surface at the front of any wood grandstand shall not exceed 20 ft (6100 mm).

**12.4.8.3.5** The highest level of seat platforms above the finished ground level, or the surface at the front of a portable grandstand within a tent or membrane structure, shall not exceed 12 ft (3660 mm).

**12.4.8.3.6** The height requirements specified in 12.4.8.3.4 and 12.4.8.3.5 shall be permitted to be doubled where constructed entirely of labeled fire-retardant-treated wood that has passed the standard rain test, ASTM D 2898, *Standard Test Methods for Accelerated Weathering of Fire-Retardant-Treated Wood for Fire Testing*, or where constructed of members conforming to dimensions for heavy timber construction [Type IV (2HH)].

**12.4.8.4 Special Requirements — Portable Grandstands.**

**12.4.8.4.1** Portable grandstands shall conform to the requirements of 12.4.8 for grandstands and the requirements of 12.4.8.4.2 through 12.4.8.4.7.

**12.4.8.4.2** Portable grandstands shall be self-contained and shall have within them all necessary parts to withstand and restrain all forces that might be developed during human occupancy.

**12.4.8.4.3** Portable grandstands shall be designed and manufactured so that, if any structural members essential to the strength and stability of the structure have been omitted during erection, the presence of unused connection fittings shall make the omissions self-evident.

**12.4.8.4.4** Portable grandstand construction shall be skillfully accomplished to produce the strength required by the design.

**12.4.8.4.5** Portable grandstands shall be provided with base plates, sills, floor runners, or sleepers of such area that the permitted bearing capacity of the supporting material is not exceeded.

**12.4.8.4.6** Where a portable grandstand rests directly on a base of such character that it is incapable of supporting the load without appreciable settlement, mud sills of suitable material, having sufficient area to prevent undue or dangerous settlement, shall be installed under base plates, runners, or sleepers.

**12.4.8.4.7** All bearing surfaces of portable grandstands shall be in contact with each other.

**12.4.8.5 Spaces Underneath Grandstands.** Spaces underneath a grandstand shall be kept free of flammable or combustible materials, unless protected by an approved, supervised automatic sprinkler system in accordance with Section 9.7 or unless otherwise permitted by one of the following:

(1) This requirement shall not apply to accessory uses of 300 ft$^2$ (28 m$^2$) or less, such as ticket booths, toilet facilities, or concession booths, where constructed of noncombustible or fire-resistive construction in otherwise nonsprinklered facilities.
(2) This requirement shall not apply to rooms that are enclosed in not less than 1-hour fire resistance–rated construction and are less than 1000 ft$^2$ (93 m$^2$) in otherwise nonsprinklered facilities.

**12.4.8.6 Guards and Railings.**

**12.4.8.6.1** Railings or guards not less than 42 in. (1065 mm) above the aisle surface or footrest or not less than 36 in. (915 mm) vertically above the center of the seat or seat board surface, whichever is adjacent, shall be provided along those portions of the backs and ends of all grandstands where the seats are more than 48 in. (1220 mm) above the floor or the finished ground level.

**12.4.8.6.2** The requirement of 12.4.8.6.1 shall not apply where an adjacent wall or fence affords equivalent safeguard.

**12.4.8.6.3** Where the front footrest of any grandstand is more than 24 in. (610 mm) above the floor, railings or guards not less than 33 in. (825 mm) above such footrests shall be provided.

**12.4.8.6.4** The railings required by 12.4.8.6.3 shall be permitted to be not less than 26 in. (660 mm) high in grandstands or where the front row of seats includes backrests.

**12.4.8.6.5** Cross aisles located within the seating area shall be provided with rails not less than 26 in. (660 mm) high along the front edge of the cross aisle.

**12.4.8.6.6** The railings specified by 12.4.8.6.5 shall not be required where the backs of the seats in front of the cross aisle project 24 in. (610 mm) or more above the surface of the cross aisle.

**12.4.8.6.7** Vertical openings between guardrails and footboards or seat boards shall be provided with intermediate construction so that a 4 in. (100 mm) diameter sphere cannot pass through the opening.

**12.4.8.6.8** An opening between the seat board and footboard located more than 30 in. (760 mm) above the finished ground level shall be provided with intermediate construction so that a 4 in. (100 mm) diameter sphere cannot pass through the opening.

**12.4.9 Folding and Telescopic Seating.**

**12.4.9.1 General.** Folding and telescopic seating shall comply with the provisions of this chapter as modified by 12.4.9.

**12.4.9.2 Seating.**

**12.4.9.2.1** The horizontal distance of seats, measured back-to-back, shall be not less than 22 in. (560 mm) for seats without backs, and all of the following requirements shall also apply:

(1) There shall be a space of not less than 12 in. (305 mm) between the back of each seat and the front of each seat immediately behind it.
(2) If seats are of the chair type, the 12 in. (305 mm) dimension shall be measured to the front edge of the rear seat in its normal unoccupied position.
(3) All measurements shall be taken between plumb lines.

**12.4.9.2.2** The depth of footboards (footrests) and seat boards in folding and telescopic seating shall be not less than 9 in. (230 mm).

**12.4.9.2.3** Where the same level is not used for both seat foundations and footrests, footrests independent of seats shall be provided.

**12.4.9.2.4** Individual chair-type seats shall be permitted in folding and telescopic seating only if firmly secured in groups of not less than three.

**12.4.9.2.5** The maximum number of seats permitted between the farthest seat in an aisle in folding and telescopic seating shall not exceed that shown in Table 12.4.8.2.5.

**12.4.9.3 Guards and Railings.**

**12.4.9.3.1** Railings or guards not less than 42 in. (1065 mm) above the aisle surface or footrest, or not less than 36 in. (915 mm) vertically above the center of the seat or seat board surface, whichever is adjacent, shall be provided along those portions of the backs and ends of all folding and telescopic seating where the seats are more than 48 in. (1220 mm) above the floor or the finished ground level.

**12.4.9.3.2** The requirement of 12.4.9.3.1 shall not apply where an adjacent wall or fence affords equivalent safeguard.

**12.4.9.3.3** Where the front footrest of folding or telescopic seating is more than 24 in. (610 mm) above the floor, railings or guards not less than 33 in. (825 mm) above such footrests shall be provided.

**12.4.9.3.4** The railings required by 12.4.9.3.3 shall be permitted to be not less than 26 in. (660 mm) high where the front row of seats includes backrests.

**12.4.9.3.5** Cross aisles located within the seating area shall be provided with rails not less than 26 in. (660 mm) high along the front edge of the cross aisle.

**12.4.9.3.6** The railings specified by 12.4.9.3.5 shall not be required where the backs of the seats in front of the cross aisle project 24 in. (610 mm) or more above the surface of the cross aisle.

**12.4.9.3.7** Vertical openings between guardrails and footboards or seat boards shall be provided with intermediate construction so that a 4 in. (100 mm) diameter sphere cannot pass through the opening.

**12.4.9.3.8** An opening between the seat board and footboard located more than 30 in. (760 mm) above the finished ground level shall be provided with intermediate construction so that a 4 in. (100 mm) diameter sphere cannot pass through the opening.

### 12.4.10 Airport Loading Walkways.

**12.4.10.1** Airport loading walkways shall conform to NFPA 415, *Standard on Airport Terminal Buildings, Fueling Ramp Drainage, and Loading Walkways*, and the provisions of 12.4.10.2 and 12.4.10.3.

**12.4.10.2** Doors in the egress path from the aircraft through the airport loading walkway into the airport terminal building shall meet both of the following criteria:

(1) They shall swing in the direction of egress from the aircraft.
(2)*They shall not be permitted to have delayed-egress locks.

**12.4.10.3** Exit access shall be unimpeded from the airport loading walkway to the nonsecured public areas of the airport terminal building.

### 12.5 Building Services.

**12.5.1 Utilities.** Utilities shall comply with the provisions of Section 9.1.

**12.5.2 Heating, Ventilating, and Air-Conditioning Equipment.** Heating, ventilating, and air-conditioning equipment shall comply with the provisions of Section 9.2.

**12.5.3 Elevators, Escalators, and Conveyors.** Elevators, escalators, and conveyors shall comply with the provisions of Section 9.4.

**12.5.4 Rubbish Chutes, Incinerators, and Laundry Chutes.** Rubbish chutes, incinerators, and laundry chutes shall comply with the provisions of Section 9.5.

### 12.6 Reserved.

### 12.7 Operating Features.

### 12.7.1 Means of Egress Inspection.

**12.7.1.1** The building owner or agent shall inspect the means of egress to ensure it is maintained free of obstructions, and correct any deficiencies found, prior to each opening of the building to the public.

**12.7.1.2** The building owner or agent shall prepare and maintain records of the date and time of each inspection on approved forms, listing any deficiencies found and actions taken to correct them.

**12.7.1.3 Inspection of Door Openings.** Door openings shall be inspected in accordance with 7.2.1.15.

### 12.7.2 Special Provisions for Food Service Operations.

**12.7.2.1** All devices in connection with the preparation of food shall be installed and operated to avoid hazard to the safety of occupants.

**12.7.2.2** All devices in connection with the preparation of food shall be of an approved type and shall be installed in an approved manner.

**12.7.2.3** Food preparation facilities shall be protected in accordance with 9.2.3 and shall not be required to have openings protected between food preparation areas and dining areas.

**12.7.2.4** Portable cooking equipment that is not flue-connected shall be permitted only as follows:

(1) Equipment fueled by small heat sources that can be readily extinguished by water, such as candles or alcohol-burning equipment, including solid alcohol, shall be permitted to be used, provided that precautions satisfactory to the authority having jurisdiction are taken to prevent ignition of any combustible materials.
(2) Candles shall be permitted to be used on tables used for food service where securely supported on substantial non-combustible bases located to avoid danger of ignition of combustible materials and only where approved by the authority having jurisdiction.
(3) Candle flames shall be protected.
(4) "Flaming sword" or other equipment involving open flames and flamed dishes, such as cherries jubilee or crêpes suzette, shall be permitted to be used, provided that precautions subject to the approval of the authority having jurisdiction are taken.
(5)*Listed and approved LP-Gas commercial food service appliances shall be permitted to be used where in accordance with NFPA 58, *Liquefied Petroleum Gas Code*.

**12.7.3 Open Flame Devices and Pyrotechnics.** No open flame devices or pyrotechnic devices shall be used in any assembly occupancy, unless otherwise permitted by one of the following:

(1) Pyrotechnic special effect devices shall be permitted to be used on stages before proximate audiences for ceremonial or religious purposes, as part of a demonstration in exhibits, or as part of a performance, provided that both of the following criteria are met:
    (a) Precautions satisfactory to the authority having jurisdiction are taken to prevent ignition of any combustible material.
    (b) Use of the pyrotechnic device complies with NFPA 1126, *Standard for the Use of Pyrotechnics Before a Proximate Audience*.
(2) Flame effects before an audience shall be permitted in accordance with NFPA 160, *Standard for the Use of Flame Effects Before an Audience*.
(3) Open flame devices shall be permitted to be used in the following situations, provided that precautions satisfactory to the authority having jurisdiction are taken to prevent ignition of any combustible material or injury to occupants:
    (a)*For ceremonial or religious purposes
    (b) On stages and platforms where part of a performance
    (c) Where candles on tables are securely supported on substantial noncombustible bases and candle flame is protected
(4) The requirement of 12.7.3 shall not apply to heat-producing equipment complying with 9.2.2.
(5) The requirement of 12.7.3 shall not apply to food service operations in accordance with 12.7.2.
(6) Gas lights shall be permitted to be used, provided that precautions are taken, subject to the approval of the authority having jurisdiction, to prevent ignition of any combustible materials.

### 12.7.4 Furnishings, Decorations, and Scenery.

**12.7.4.1** Fabrics and films used for decorative purposes, all draperies and curtains, and similar furnishings shall be in accordance with the provisions of 10.3.1.

**12.7.4.2** The authority having jurisdiction shall impose controls on the quantity and arrangement of combustible contents in assembly occupancies to provide an adequate level of safety to life from fire.

**12.7.4.3\*** Exposed foamed plastic materials and unprotected materials containing foamed plastic used for decorative purposes or stage scenery shall have a heat release rate not exceeding 100 kW where tested in accordance with one of the following:

(1) ANSI/UL 1975, *Standard for Fire Tests for Foamed Plastics Used for Decorative Purposes*
(2) NFPA 289, *Standard Method of Fire Test for Individual Fuel Packages*, using the 20 kW ignition source

**12.7.4.4** The requirement of 12.7.4.3 shall not apply to individual foamed plastic items and items containing foamed plastic where the foamed plastic does not exceed 1 lb (0.45 kg) in weight.

**12.7.5 Special Provisions for Exposition Facilities.**

**12.7.5.1 General.** No display or exhibit shall be installed or operated to interfere in any way with access to any required exit or with the visibility of any required exit or required exit sign; nor shall any display block access to fire-fighting equipment.

**12.7.5.2 Materials Not On Display.** A storage room having an enclosure consisting of a smoke barrier having a minimum 1-hour fire resistance rating and protected by an automatic extinguishing system shall be provided for combustible materials not on display, including combustible packing crates used to ship exhibitors' supplies and products.

**12.7.5.3 Exhibits.**

**12.7.5.3.1** Exhibits shall comply with 12.7.5.3.2 through 12.7.5.3.11.

**12.7.5.3.2** The travel distance within the exhibit booth or exhibit enclosure to an exit access aisle shall not exceed 50 ft (15 m).

**12.7.5.3.3** The upper deck of multilevel exhibits exceeding 300 ft² (28 m²) shall have not less than two remote means of egress.

**12.7.5.3.4** Exhibit booth construction materials shall be limited to the following:

(1) Noncombustible or limited-combustible materials
(2) Wood exceeding ¼ in. (6.3 mm) nominal thickness
(3) Wood that is pressure-treated, fire-retardant wood meeting the requirements of NFPA 703, *Standard for Fire Retardant–Treated Wood and Fire-Retardant Coatings for Building Materials*
(4) Flame-retardant materials complying with one of the following:
　(a) They shall meet the flame propagation performance criteria contained in NFPA 701, *Standard Methods of Fire Tests for Flame Propagation of Textiles and Films.*
　(b) They shall exhibit a heat release rate not exceeding 100 kW when tested in accordance with NFPA 289, *Standard Method of Fire Test for Individual Fuel Packages,* using the 20 kW ignition source.
(5) Textile wall coverings, such as carpeting and similar products used as wall or ceiling finishes, complying with the provisions of 10.2.2 and 10.2.4
(6) Plastics limited to those that comply with 12.3.3 and Section 10.2

(7) Foamed plastics and materials containing foamed plastics having a heat release rate for any single fuel package that does not exceed 100 kW where tested in accordance with one of the following:
　(a) ANSI/UL 1975, *Standard for Fire Tests for Foamed Plastics Used for Decorative Purposes*
　(b) NFPA 289, using the 20 kW ignition source
(8) Cardboard, honeycombed paper, and other combustible materials having a heat release rate for any single fuel package that does not exceed 150 kW where tested in accordance with one of the following:
　(a) ANSI/UL 1975
　(b) NFPA 289, using the 20 kW ignition source

**12.7.5.3.5** Curtains, drapes, and decorations shall comply with 10.3.1.

**12.7.5.3.6** Acoustical and decorative material including, but not limited to, cotton, hay, paper, straw, moss, split bamboo, and wood chips shall be flame-retardant treated to the satisfaction of the authority having jurisdiction.

**12.7.5.3.6.1** Materials that cannot be treated for flame retardancy shall not be used.

**12.7.5.3.6.2** Foamed plastics, and materials containing foamed plastics and used as decorative objects such as, but not limited to, mannequins, murals, and signs, shall have a heat release rate for any single fuel package that does not exceed 150 kW where tested in accordance with one of the following:

(1) ANSI/UL 1975, *Standard for Fire Tests for Foamed Plastics Used for Decorative Purposes*
(2) NFPA 289, *Standard Method of Fire Test for Individual Fuel Packages*, using the 20 kW ignition source

**12.7.5.3.6.3** Where the aggregate area of acoustical and decorative materials is less than 10 percent of the individual floor or wall area, such materials shall be permitted to be used subject to the approval of the authority having jurisdiction.

**12.7.5.3.7** The following shall be protected by automatic extinguishing systems:

(1) Single-level exhibit booths exceeding 300 ft² (28 m²) and covered with a ceiling
(2) Each level of multilevel exhibit booths, including the uppermost level where the uppermost level is covered with a ceiling

**12.7.5.3.7.1** The requirements of 12.7.5.3.7 shall not apply where otherwise permitted by the following:

(1) Ceilings that are constructed of open grate design or listed dropout ceilings in accordance with NFPA 13, *Standard for the Installation of Sprinkler Systems*, shall not be considered ceilings within the context of 12.7.5.3.7.
(2) Vehicles, boats, and similar exhibited products having over 100 ft² (9.3 m²) of roofed area shall be provided with smoke detectors acceptable to the authority having jurisdiction.
(3)\*The requirement of 12.7.5.3.7(2) shall not apply where fire protection of multilevel exhibit booths is consistent with the criteria developed through a life safety evaluation of the exhibition hall in accordance with 12.4.1, subject to approval of the authority having jurisdiction.

**12.7.5.3.7.2** A single exhibit or group of exhibits with ceilings that do not require sprinklers shall be separated by a distance of not less than 10 ft (3050 mm) where the aggregate ceiling exceeds 300 ft² (28 m²).

**12.7.5.3.7.3** The water supply and piping for the sprinkler system shall be permitted to be of an approved temporary means that is provided by a domestic water supply, a standpipe system, or a sprinkler system.

**12.7.5.3.8** Open flame devices within exhibit booths shall comply with 12.7.3.

**12.7.5.3.9** Cooking and food-warming devices in exhibit booths shall comply with 12.7.2 and all of the following:

(1) Gas-fired devices shall comply with the following:
  (a) Natural gas-fired devices shall comply with 9.1.1.
  (b) The requirement of 12.7.5.3.9(1)(a) shall not apply to compressed natural gas where permitted by the authority having jurisdiction.
  (c) The use of LP-Gas cylinders shall be prohibited.
  (d) Nonrefillable LP-Gas cylinders shall be approved for use where permitted by the authority having jurisdiction.
(2) The devices shall be isolated from the public by not less than 48 in. (1220 mm) or by a barrier between the devices and the public.
(3) Multi-well cooking equipment using combustible oils or solids shall comply with 9.2.3.
(4) Single-well cooking equipment using combustible oils or solids shall meet all of the following criteria:
  (a) The equipment shall have lids available for immediate use.
  (b) The equipment shall be limited to 2 ft² (0.2 m²) of cooking surface.
  (c) The equipment shall be placed on noncombustible surface materials.
  (d) The equipment shall be separated from each other by a horizontal distance of not less than 24 in. (610 mm).
  (e) The requirement of 12.7.5.3.9(4)(d) shall not apply to multiple single-well cooking equipment where the aggregate cooking surface area does not exceed 2 ft² (0.2 m²).
  (f) The equipment shall be kept at a horizontal distance of not less than 24 in. (610 mm) from any combustible material.
(5) A portable fire extinguisher in accordance with 9.7.4.1 shall be provided within the booth for each device, or an approved automatic extinguishing system shall be provided.

**12.7.5.3.10** Combustible materials within exhibit booths shall be limited to a one-day supply. Storage of combustible materials behind the booth shall be prohibited. *(See 12.7.4.2 and 12.7.5.2.)*

**12.7.5.3.11** Plans for the exposition, in an acceptable form, shall be submitted to the authority having jurisdiction for approval prior to setting up any exhibit.

**12.7.5.3.11.1** The plan shall show all details of the proposed exposition.

**12.7.5.3.11.2** No exposition shall occupy any exposition facility without approved plans.

**12.7.5.4 Vehicles.** Vehicles on display within an exposition facility shall comply with 12.7.5.4.1 through 12.7.5.4.5.

**12.7.5.4.1** All fuel tank openings shall be locked and sealed in an approved manner to prevent the escape of vapors; fuel tanks shall not contain in excess of one-half their capacity or contain in excess of 10 gal (38 L) of fuel, whichever is less.

**12.7.5.4.2** At least one battery cable shall be removed from the batteries used to start the vehicle engine, and the disconnected battery cable shall then be taped.

**12.7.5.4.3** Batteries used to power auxiliary equipment shall be permitted to be kept in service.

**12.7.5.4.4** Fueling or defueling of vehicles shall be prohibited.

**12.7.5.4.5** Vehicles shall not be moved during exhibit hours.

**12.7.5.5 Prohibited Materials.**

**12.7.5.5.1** The following items shall be prohibited within exhibit halls:

(1) Compressed flammable gases
(2) Flammable or combustible liquids
(3) Hazardous chemicals or materials
(4) Class II or greater lasers, blasting agents, and explosives

**12.7.5.5.2** The authority having jurisdiction shall be permitted to allow the limited use of any items specified in 12.7.5.5.1 under special circumstances.

**12.7.5.6 Alternatives.** See Section 1.4.

**12.7.6\* Crowd Managers.**

**12.7.6.1** Assembly occupancies shall be provided with a minimum of one trained crowd manager or crowd manager supervisor. Where the occupant load exceeds 250, additional trained crowd managers or crowd manager supervisors shall be provided at a ratio of one crowd manager or crowd manager supervisor for every 250 occupants, unless otherwise permitted by one of the following:

(1) This requirement shall not apply to assembly occupancies used exclusively for religious worship with an occupant load not exceeding 2000.
(2) The ratio of trained crowd managers to occupants shall be permitted to be reduced where, in the opinion of the authority having jurisdiction, the existence of an approved, supervised automatic sprinkler system and the nature of the event warrant.

**12.7.6.2** The crowd manager shall receive approved training in crowd management techniques.

**12.7.7\* Drills.**

**12.7.7.1** The employees or attendants of assembly occupancies shall be trained and drilled in the duties they are to perform in case of fire, panic, or other emergency to effect orderly exiting.

**12.7.7.2** Employees or attendants of assembly occupancies shall be instructed in the proper use of portable fire extinguishers and other manual fire suppression equipment where provided.

**12.7.7.3\*** In the following assembly occupancies, an audible announcement shall be made, or a projected image shall be shown, prior to the start of each program that notifies occupants of the location of the exits to be used in case of a fire or other emergency:

(1) Theaters
(2) Motion picture theaters

(3) Auditoriums

(4) Other similar assembly occupancies with occupant loads exceeding 300 where there are noncontinuous programs

**12.7.7.4** The requirement of 12.7.7.3 shall not apply to assembly occupancies in schools where used for nonpublic events.

### 12.7.8 Smoking.

**12.7.8.1** Smoking in assembly occupancies shall be regulated by the authority having jurisdiction.

**12.7.8.2** In rooms or areas where smoking is prohibited, plainly visible signs shall be posted that read as follows:

NO SMOKING

**12.7.8.3** No person shall smoke in prohibited areas that are so posted, unless permitted by the authority having jurisdiction under both of the following conditions:

(1) Smoking shall be permitted on a stage only where it is a necessary and rehearsed part of a performance.

(2) Smoking shall be permitted only where the smoker is a regular performing member of the cast.

**12.7.8.4** Where smoking is permitted, suitable ashtrays or receptacles shall be provided in convenient locations.

### 12.7.9 Seating.

#### 12.7.9.1 Secured Seating.

**12.7.9.1.1** Seats in assembly occupancies accommodating more than 200 persons shall be securely fastened to the floor, except where fastened together in groups of not less than three and as permitted by 12.7.9.1.2 and 12.7.9.2.

**12.7.9.1.2** Balcony and box seating areas that are separated from other areas by rails, guards, partial-height walls, or other physical barriers and have a maximum of 14 seats shall be exempt from the requirement of 12.7.9.1.1.

#### 12.7.9.2 Unsecured Seating.

**12.7.9.2.1** Seats not secured to the floor shall be permitted in restaurants, night clubs, and other occupancies where fastening seats to the floor might be impracticable.

**12.7.9.2.2** Unsecured seats shall be permitted, provided that, in the area used for seating, excluding such areas as dance floors and stages, there is not more than one seat for each 15 ft$^2$ (1.4 m$^2$) of net floor area, and adequate aisles to reach exits are maintained at all times.

**12.7.9.2.3** Seating diagrams shall be submitted for approval by the authority having jurisdiction to permit an increase in occupant load per 7.3.1.3.

#### 12.7.9.3 Occupant Load Posting.

**12.7.9.3.1** Every room constituting an assembly occupancy and not having fixed seats shall have the occupant load of the room posted in a conspicuous place near the main exit from the room.

**12.7.9.3.2** Approved signs shall be maintained in a legible manner by the owner or authorized agent.

**12.7.9.3.3** Signs shall be durable and shall indicate the number of occupants permitted for each room use.

### 12.7.10 Maintenance of Outdoor Grandstands.

**12.7.10.1** The owner shall provide for not less than annual inspection and required maintenance of each outdoor grandstand to ensure safe conditions.

**12.7.10.2** At least biennially, the inspection shall be performed by a professional engineer, registered architect, or individual certified by the manufacturer.

**12.7.10.3** Where required by the authority having jurisdiction, the owner shall provide a copy of the inspection report and certification that the inspection required by 12.7.10.2 has been performed.

### 12.7.11 Maintenance and Operation of Folding and Telescopic Seating.

**12.7.11.1** Instructions in both maintenance and operation shall be transmitted to the owner by the manufacturer of the seating or his or her representative.

**12.7.11.2** Maintenance and operation of folding and telescopic seating shall be the responsibility of the owner or his or her duly authorized representative and shall include all of the following:

(1) During operation of the folding and telescopic seats, the opening and closing shall be supervised by responsible personnel who shall ensure that the operation is in accordance with the manufacturer's instructions.

(2) Only attachments specifically approved by the manufacturer for the specific installation shall be attached to the seating.

(3) An annual inspection and required maintenance of each grandstand shall be performed to ensure safe conditions.

(4) At least biennially, the inspection shall be performed by a professional engineer, registered architect, or individual certified by the manufacturer.

**12.7.12 Clothing.** Clothing and personal effects shall not be stored in corridors, and spaces not separated from corridors, unless otherwise permitted by one of the following:

(1) This requirement shall not apply to corridors, and spaces not separated from corridors, that are protected by an approved, supervised automatic sprinkler system in accordance with Section 9.7.

(2) This requirement shall not apply to corridors, and spaces not separated from corridors, that are protected by a smoke detection system in accordance with Section 9.6.

(3) This requirement shall not apply to storage in metal lockers, provided that the required egress width is maintained.

### 12.7.13 Emergency Plans.

**12.7.13.1** Emergency plans shall be provided in accordance with Section 4.8.

**12.7.13.2** Where assembly occupancies are located in the high-rise portion of a building, the emergency plan shall include egress procedures, methods, and preferred evacuation routes for each event considered to be a life safety hazard that could impact the building, including the appropriateness of the use of elevators.

## Chapter 13   Existing Assembly Occupancies

### 13.1 General Requirements.

#### 13.1.1 Application.

**13.1.1.1** The requirements of this chapter shall apply to existing buildings or portions thereof currently occupied as assembly occupancies, unless otherwise specified by 13.1.1.4. (*See 3.3.188.2 for definition of assembly occupancy.*)

**13.1.1.2 Administration.** The provisions of Chapter 1, Administration, shall apply.

**13.1.1.3 General.** The provisions of Chapter 4, General, shall apply.

**13.1.1.4** An existing building housing an assembly occupancy established prior to the effective date of this *Code* shall be permitted to be approved for continued use if it conforms to, or is made to conform to, the provisions of this *Code* to the extent that, in the opinion of the authority having jurisdiction, reasonable life safety against the hazards of fire, explosion, and panic is provided and maintained.

**13.1.1.5** Additions to existing buildings shall conform to the requirements of 4.6.7.

**13.1.1.6** Existing portions of buildings shall be upgraded if the addition results in an increase in the required minimum number of separate means of egress in accordance with 7.4.1.2.

**13.1.1.7** Existing portions of the structure shall not be required to be modified, provided that both of the following criteria are met:

(1) The new construction has not diminished the fire safety features of the facility.
(2) The addition does not result in an increase in the required minimum number of separate means of egress in accordance with 7.4.1.2.

**13.1.1.8** An assembly occupancy in which an occupant load increase results in an increase in the required minimum number of separate means of egress, in accordance with 7.4.1.2, shall meet the requirements for new construction.

**13.1.2\* Classification of Occupancy.** See 6.1.2.

**13.1.3 Multiple Occupancies.**

**13.1.3.1 General.** Multiple occupancies shall be in accordance with 6.1.14.

**13.1.3.2\* Simultaneous Occupancy.** Exits shall be sufficient for simultaneous occupancy of both the assembly occupancy and other parts of the building, except where the authority having jurisdiction determines that the conditions are such that simultaneous occupancy will not occur.

**13.1.3.3 Assembly and Mercantile Occupancies in Mall Buildings.**

**13.1.3.3.1** The provisions of Chapter 13 shall apply to the assembly occupancy tenant space.

**13.1.3.3.2** The provisions of 37.4.4 shall be permitted to be used outside the assembly occupancy tenant space.

**13.1.4 Definitions.**

**13.1.4.1 General.** For definitions, see Chapter 3, Definitions.

**13.1.4.2\* Special Definitions.** A list of special terms used in this chapter follows:

(1) **Aisle Accessway.** See 3.3.11.
(2) **Exhibit.** See 3.3.77.
(3) **Exhibitor.** See 3.3.78.
(4) **Exposition.** See 3.3.84.
(5) **Exposition Facility.** See 3.3.88.1.
(6) **Festival Seating.** See 3.3.237.1.
(7) **Flow Time.** See 3.3.115.
(8) **Fly Gallery.** See 3.3.116.
(9) **Gridiron.** See 3.3.126.
(10) **Legitimate Stage.** See 3.3.262.1.
(11) **Life Safety Evaluation.** See 3.3.158.
(12) **Multilevel Play Structure.** See 3.3.271.5.
(13) **Pinrail.** See 3.3.208.
(14) **Platform.** See 3.3.209.
(15) **Proscenium Wall.** See 3.3.287.2.
(16) **Regular Stage.** See 3.3.262.2.
(17) **Smoke-Protected Assembly Seating.** See 3.3.237.4.
(18) **Special Amusement Building.** See 3.3.36.10.
(19) **Stage.** See 3.3.262.
(20) **Temporary Platform.** See 3.3.209.1.

**13.1.5 Classification of Hazard of Contents.** Contents of assembly occupancies shall be classified in accordance with the provisions of Section 6.2.

**13.1.6 Minimum Construction Requirements.** Assembly occupancies shall be limited to the building construction types specified in Table 13.1.6, based on the number of stories in height as defined in 4.6.3, unless otherwise permitted by the following *(see 8.2.1)*:

(1) This requirement shall not apply to outdoor grandstands of Type I or Type II construction.
(2) This requirement shall not apply to outdoor grandstands of Type III, Type IV, or Type V construction that meet the requirements of 13.4.8.
(3) This requirement shall not apply to grandstands of noncombustible construction supported by the floor in a building meeting the construction requirements of Table 13.1.6.
(4) This requirement shall not apply to assembly occupancies within mall buildings in accordance with 37.4.4.

**13.1.7 Occupant Load.**

**13.1.7.1\* General.** The occupant load, in number of persons for whom means of egress and other provisions are required, shall be determined on the basis of the occupant load factors of Table 7.3.1.2 that are characteristic of the use of the space or shall be determined as the maximum probable population of the space under consideration, whichever is greater.

**13.1.7.1.1** In areas not in excess of 10,000 ft$^2$ (930 m$^2$), the occupant load shall not exceed one person in 5 ft$^2$ (0.46 m$^2$).

**13.1.7.1.2** In areas in excess of 10,000 ft$^2$ (930 m$^2$), the occupant load shall not exceed one person in 7 ft$^2$ (0.65 m$^2$).

**13.1.7.1.3** The authority having jurisdiction shall be permitted to establish the occupant load as the number of persons for which the existing means of egress is adequate, provided that measures are established to prevent occupancy by a greater number of persons.

**13.1.7.2 Waiting Spaces.** In theaters and other assembly occupancies where persons are admitted to the building at times when seats are not available, or when the permitted occupant load has been reached based on 13.1.7.1 and persons are allowed to wait in a lobby or similar space until seats or space is available, all of the following requirements shall apply:

(1) Such use of a lobby or similar space shall not encroach upon the required clear width of exits.
(2) The waiting spaces shall be restricted to areas other than the required means of egress.
(3) Exits shall be provided for the waiting spaces on the basis of one person for each 3 ft$^2$ (0.28 m$^2$) of waiting space area.
(4) Exits for waiting spaces shall be in addition to the exits specified for the main auditorium area and shall conform in construction and arrangement to the general rules for exits given in this chapter.

**Table 13.1.6  Construction Type Limitations**

| Construction Type | Sprinklered[a] | Stories Below | Stories in Height[b] | | | | |
|---|---|---|---|---|---|---|---|
| | | | 1 | 2 | 3 | 4 | ≥5 |
| I (442)[c, d] | Yes | X | X | X | X | X | X |
| | No | NP | X | X | X | X | X3 |
| I (332)[c, d] | Yes | X | X | X | X | X | X |
| | No | NP | X | X | X | X | X3 |
| II (222)[c, d] | Yes | X | X | X | X | X | X |
| | No | NP | X | X | X | X | X3 |
| II (111)[c, d] | Yes | X1 | X | X | X | X3 | NP |
| | No | NP | X | X | X3 | NP | NP |
| II (000) | Yes | X2 | X | X4 | NP | NP | NP |
| | No | NP | X3 | NP | NP | NP | NP |
| III (211) | Yes | X1 | X | X | X | X3 | NP |
| | No | NP | X | X | X4 | NP | NP |
| III (200) | Yes | X2 | X | X4 | NP | NP | NP |
| | No | NP | X3 | NP | NP | NP | NP |
| IV (2HH) | Yes | X1 | X | X | X | X3 | NP |
| | No | NP | X | X | X4 | NP | NP |
| V (111) | Yes | X1 | X | X | X | X3 | NP |
| | No | NP | X | X | X4 | NP | NP |
| V (000) | Yes | X2 | X | X4 | NP | NP | NP |
| | No | NP | X3 | NP | NP | NP | NP |

X: Permitted for assembly of any occupant load.

X1: Permitted for assembly of any occupant load, but limited to one story below the level of exit discharge.

X2: Permitted for assembly limited to an occupant load of 1000 or less, and limited to one story below the level of exit discharge.

X3: Permitted for assembly limited to an occupant load of 1000 or less.

X4: Permitted for assembly limited to an occupant load of 300 or less.

NP: Not permitted.

[a]Protected by an approved automatic sprinkler system in accordance with Section 9.7 in the following locations:

(1) Throughout the story of the assembly occupancy

(2) Throughout all stories intervening between the story of the assembly occupancy and the level of exit discharge

(3) Throughout the level of exit discharge if there are any openings between the level of exit discharge and the exits serving the assembly occupancy

[b]See 4.6.3.

[c]Where every part of the structural framework of roofs in Type I or Type II construction is 20 ft (6100 mm) or more above the floor immediately below, omission of all fire protection of the structural members is permitted, including protection of trusses, roof framing, decking, and portions of columns above 20 ft (6100 mm).

[d]In open-air fixed seating facilities, including stadia, omission of fire protection of structural members exposed to the outside atmosphere is permitted where substantiated by an approved engineering analysis.

**13.1.7.3 Life Safety Evaluation.** Where the occupant load of an assembly occupancy exceeds 6000, a life safety evaluation shall be performed in accordance with 13.4.1.

**13.1.7.4 Outdoor Facilities.** In outdoor facilities, where approved by the authority having jurisdiction, the number of occupants who are each provided with not less than 15 ft² (1.4 m²) of lawn surface shall be permitted to be excluded from the maximum occupant load of 6000 of 13.1.7.3 in determining the need for a life safety evaluation.

**13.2 Means of Egress Requirements.**

**13.2.1 General.** All means of egress shall be in accordance with Chapter 7 and this chapter.

**13.2.2 Means of Egress Components.**

**13.2.2.1 Components Permitted.** Components of means of egress shall be limited to the types described in 13.2.2.2 through 13.2.2.12.

**13.2.2.2 Doors.**

**13.2.2.2.1** Doors complying with 7.2.1 shall be permitted.

**13.2.2.2.2** Assembly occupancies with occupant loads of 300 or less in malls (*see 37.4.4.2.2*) shall be permitted to have horizontal or vertical security grilles or doors complying with 7.2.1.4.1(3) on the main entrance/exits.

**13.2.2.2.3** Any door in a required means of egress from an area having an occupant load of 100 or more persons shall be permitted to be provided with a latch or lock only if the latch or lock is panic hardware or fire exit hardware complying with 7.2.1.7, unless otherwise permitted by one of the following:

(1) This requirement shall not apply to delayed-egress locks as permitted in 13.2.2.2.5.
(2) This requirement shall not apply to access-controlled egress doors as permitted in 13.2.2.2.6.

**13.2.2.2.4** Locking devices complying with 7.2.1.5.5 shall be permitted to be used on a single door or a single pair of doors if both of the following conditions apply:

(1) The door or pair of doors serve as the main exit from assembly occupancies having an occupant load not greater than 600.
(2) Any latching devices on such a door(s) from an assembly occupancy having an occupant load of 100 or more are released by panic hardware or fire exit hardware.

**13.2.2.2.5** Delayed-egress locks complying with 7.2.1.6.1 shall be permitted on doors other than main entrance/exit doors.

**13.2.2.2.6** Doors in the means of egress shall be permitted to be equipped with an approved access control system complying with 7.2.1.6.2, and such doors shall not be locked from the egress side when the assembly occupancy is occupied. *(See 7.2.1.1.3.)*

**13.2.2.2.7** Elevator lobby exit access door locking in accordance with 7.2.1.6.3 shall be permitted.

**13.2.2.2.8** Revolving doors complying with the requirements of 7.2.1.10 for new construction shall be permitted.

**13.2.2.2.9** The provisions of 7.2.1.11.1.1 to permit turnstiles where revolving doors are permitted shall not apply.

**13.2.2.2.10** No turnstiles or other devices that restrict the movement of persons shall be installed in any assembly occupancy in such a manner as to interfere with required means of egress facilities.

**13.2.2.3 Stairs.**

**13.2.2.3.1 General.** Stairs complying with 7.2.2 shall be permitted, unless one of the following criteria applies:

(1)*Stairs serving seating that is designed to be repositioned shall not be required to comply with 7.2.2.3.1.
(2) This requirement shall not apply to stages and platforms as permitted by 13.4.5.
(3) The stairs connecting only a stage or platform and the immediately adjacent assembly seating shall be permitted to have a handrail in the center only or on one side only.
(4) The stairs connecting only a stage or platform and the immediately adjacent assembly seating shall be permitted to omit the guards required by 7.1.8 where both of the following criteria are met:
   (a) The guard would restrict audience sight lines to the stage or platform.
   (b) The height between any part of the stair and the adjacent floor is not more than 42 in. (1065 mm).

**13.2.2.3.2 Catwalk, Gallery, and Gridiron Stairs.**

**13.2.2.3.2.1** Noncombustible grated stair treads and landing floors shall be permitted in means of egress from lighting and access catwalks, galleries, and gridirons.

**13.2.2.3.2.2** Spiral stairs complying with 7.2.2.2.3 shall be permitted in means of egress from lighting and access catwalks, galleries, and gridirons.

**13.2.2.4 Smokeproof Enclosures.** Smokeproof enclosures complying with 7.2.3 shall be permitted.

**13.2.2.5 Horizontal Exits.** Horizontal exits complying with 7.2.4 shall be permitted.

**13.2.2.6 Ramps.** Ramps complying with 7.2.5 shall be permitted.

**13.2.2.7 Exit Passageways.** Exit passageways complying with 7.2.6 shall be permitted.

**13.2.2.8 Escalators and Moving Walks.** Escalators and moving walks complying with 7.2.7 shall be permitted.

**13.2.2.9 Fire Escape Stairs.** Fire escape stairs complying with 7.2.8 shall be permitted.

**13.2.2.10 Fire Escape Ladders.**

**13.2.2.10.1** Fire escape ladders complying with 7.2.9 shall be permitted.

**13.2.2.10.2** For ladders serving catwalks, the three-person limitation in 7.2.9.1(3) shall be permitted to be increased to ten persons.

**13.2.2.11 Alternating Tread Devices.** Alternating tread devices complying with 7.2.11 shall be permitted.

**13.2.2.12 Areas of Refuge.** Areas of refuge complying with 7.2.12 shall be permitted.

**13.2.3 Capacity of Means of Egress.**

**13.2.3.1 General.** The capacity of means of egress shall be in accordance with one of the following:

(1) Section 7.3 for other than theater-type seating or smoke-protected assembly seating

(2) 13.2.3.2 for rooms with theater-type seating or similar seating arranged in rows

(3) 13.4.2 for smoke-protected assembly seating

**13.2.3.2\* Theater-Type Seating.** Minimum clear widths of aisles and other means of egress serving theater-type seating, or similar seating arranged in rows, shall be in accordance with Table 13.2.3.2.

**Table 13.2.3.2 Capacity Factors**

| | Clear Width per Seat Served | | | |
|---|---|---|---|---|
| | Stairs | | Passageways, Ramps, and Doorways | |
| No. of Seats | in. | mm | in. | mm |
| Unlimited | 0.3 *AB* | 7.6 *AB* | 0.22 *C* | 5.6 *C* |

**13.2.3.3 Width Modifications.** The minimum clear widths shown in Table 13.2.3.2 shall be modified in accordance with all of the following:

(1) If risers exceed 7 in. in height, the stair width in Table 13.2.3.2 shall be multiplied by factor *A*, where *A* equals the following:

$$A = 1 + \frac{\text{riser height} - 7}{5}$$

(2) If risers exceed 178 mm in height, the stair width in Table 13.2.3.2 shall be multiplied by factor *A*, where *A* equals the following:

$$A = 1 + \frac{\text{riser height} - 178}{125}$$

(3) Stairs not having a handrail within a 30 in. (760 mm) horizontal distance shall be 25 percent wider than otherwise calculated; that is, their width shall be multiplied by factor *B*, where *B* equals the following:

$$B = 1.25$$

(4) Ramps steeper than 1 in 10 slope where used in ascent shall have their width increased by 10 percent; that is, their width shall be multiplied by factor *C*, where *C* equals the following:

$$C = 1.10$$

**13.2.3.4 Lighting and Access Catwalks.** The requirements of 13.2.3.2 and 13.2.3.3 shall not apply to lighting and access catwalks as permitted by 13.4.5.9.

**13.2.3.5 Bleachers Aisles.** In seating composed entirely of bleachers for which the row-to-row dimension is 28 in. (710 mm) or less, and from which front egress is not limited, aisles shall not be required to exceed 66 in. (1675 mm) in width.

**13.2.3.6 Main Entrance/Exit.**

**13.2.3.6.1** Every assembly occupancy shall be provided with a main entrance/exit.

**13.2.3.6.2** The main entrance/exit shall be of a width that accommodates one-half of the total occupant load.

**13.2.3.6.3** The main entrance/exit shall be at the level of exit discharge or shall connect to a stairway or ramp leading to a street.

**13.2.3.6.4 Reserved.**

**13.2.3.6.5** Where the main entrance/exit from an assembly occupancy is through a lobby or foyer, the aggregate capacity of all exits from the lobby or foyer shall be permitted to provide the required capacity of the main entrance/exit, regardless of whether all such exits serve as entrances to the building.

**13.2.3.6.6\*** In assembly occupancies where there is no well-defined main entrance/exit, exits shall be permitted to be distributed around the perimeter of the building, provided that the total exit width furnishes not less than 100 percent of the width needed to accommodate the permitted occupant load.

**13.2.3.7 Other Exits.** Each level of an assembly occupancy shall have access to the main entrance/exit and shall be provided with additional exits of a width to accommodate not less than one-half of the total occupant load served by that level.

**13.2.3.7.1** Additional exits shall discharge in accordance with 13.2.7.

**13.2.3.7.2** Additional exits shall be located as far apart as practicable and as far from the main entrance/exit as practicable.

**13.2.3.7.3** Additional exits shall be accessible from a cross aisle or a side aisle.

**13.2.3.7.4** In assembly occupancies where there is no well-defined main entrance/exit, exits shall be permitted to be distributed around the perimeter of the building, provided that the total exit width furnishes not less than 100 percent of the width required to accommodate the permitted occupant load.

**13.2.4\* Number of Means of Egress.**

**13.2.4.1** The number of means of egress shall be in accordance with Section 7.4, other than fenced outdoor assembly occupancies in accordance with 13.2.4.4, unless otherwise permitted by 13.2.4.2 or 13.2.4.3.

**13.2.4.2** Assembly occupancies with occupant loads of 600 or fewer shall have two separate means of egress.

**13.2.4.3** Assembly occupancies with occupant loads greater than 600 but fewer than 1000 shall have three separate means of egress.

**13.2.4.4** A fenced outdoor assembly occupancy shall have not less than two widely separated means of egress from the enclosure, unless otherwise required by one of the following:

(1) If more than 6000 persons are to be served by such means of egress, there shall be not less than three means of egress.

(2) If more than 9000 persons are to be served by such means of egress, there shall be not less than four means of egress.

**13.2.4.5** Balconies or mezzanines having an occupant load not exceeding 50 shall be permitted to be served by a single means of egress, and such means of egress shall be permitted to lead to the floor below.

**13.2.4.6** Balconies or mezzanines having an occupant load exceeding 50, but not exceeding 100, shall have not less than two remote means of egress, but both such means of egress shall be permitted to lead to the floor below.

**13.2.4.7** Balconies or mezzanines having an occupant load exceeding 100 shall have means of egress as described in 7.4.1.

**13.2.4.8** A second means of egress shall not be required from lighting and access catwalks, galleries, and gridirons where a means of escape to a floor or a roof is provided. Ladders, alternating tread devices, or spiral stairs shall be permitted in such means of escape.

**13.2.5 Arrangement of Means of Egress.**

**13.2.5.1 General.**

**13.2.5.1.1** Means of egress shall be arranged in accordance with Section 7.5.

**13.2.5.1.2** A common path of travel shall be permitted for the first 20 ft (6100 mm) from any point where the common path serves any number of occupants, and for the first 75 ft (23 m) from any point where the common path serves not more than 50 occupants.

**13.2.5.1.3** Dead-end corridors shall not exceed 20 ft (6100 mm).

**13.2.5.2 Access Through Hazardous Areas.** Means of egress shall not be permitted through kitchens, storerooms, restrooms, closets, platforms, stages, or hazardous areas as described in 13.3.2.

**13.2.5.3 Reserved.**

**13.2.5.4 General Requirements for Access and Egress Routes Within Assembly Areas.**

**13.2.5.4.1** Festival seating, as defined in 3.3.237.1, shall be prohibited within a building, unless otherwise permitted by one of the following:

(1) Festival seating shall be permitted in assembly occupancies having occupant loads of 250 or less.
(2) Festival seating shall be permitted in assembly occupancies where occupant loads exceed 250, provided that an approved life safety evaluation has been performed. *(See 13.4.1.)*

**13.2.5.4.2\*** Access and egress routes shall be maintained so that any individual is able to move without undue hindrance, on personal initiative and at any time, from an occupied position to the exits.

**13.2.5.4.3\*** Access and egress routes shall be maintained so that crowd management, security, and emergency medical personnel are able to reach any individual at any time, without undue hindrance.

**13.2.5.4.4\*** The width of aisle accessways and aisles shall provide sufficient egress capacity for the number of persons accommodated by the catchment area served by the aisle accessway or aisle in accordance with 13.2.3.2, or for smoke-protected assembly seating in accordance with 13.4.2.

**13.2.5.4.5** Where aisle accessways or aisles converge to form a single path of egress travel, the required egress capacity of that path shall be not less than the combined required capacity of the converging aisle accessways and aisles.

**13.2.5.4.6** Those portions of aisle accessways and aisles where egress is possible in either of two directions shall be uniform in required width, unless otherwise permitted by 13.2.5.4.7.

**13.2.5.4.7** The requirement of 13.2.5.4.6 shall not apply to those portions of aisle accessways where the required width, not including the seat space described by 13.2.5.7.3, does not exceed 12 in. (305 mm).

**13.2.5.4.8** In the case of side boundaries for aisle accessways or aisles, other than those for nonfixed seating at tables, the clear width shall be measured to boundary elements such as walls, guardrails, handrails, edges of seating, tables, and side edges of treads, and said measurement shall be made horizontally to the vertical projection of the elements, resulting in the smallest width measured perpendicularly to the line of travel.

**13.2.5.5\* Aisle Accessways Serving Seating Not at Tables.**

**13.2.5.5.1\*** The required clear width of aisle accesses between rows of seating shall be determined as follows:

(1) Horizontal measurements shall be made, between vertical planes, from the back of one seat to the front of the most forward projection of the seat immediately behind it.
(2) Where the entire row consists of automatic- or self-rising seats that comply with ASTM F 851, *Standard Test Method for Self-Rising Seat Mechanisms*, the measurement shall be permitted to be made with the seats in the up position.

**13.2.5.5.2** The aisle accessway between rows of seating shall have a clear width of not less than 12 in. (305 mm), and this minimum shall be increased as a function of row length in accordance with 13.2.5.5.4 and 13.2.5.5.5.

**13.2.5.5.3** If used by not more than four persons, no minimum clear width shall be required for the portion of an aisle accessway having a length not exceeding 6 ft (1830 mm), measured from the center of the seat farthest from the aisle.

**13.2.5.5.4\*** Rows of seating served by aisles or doorways at both ends shall not exceed 100 seats per row.

**13.2.5.5.4.1** The 12 in. (305 mm) minimum clear width of aisle accessway specified in 13.2.5.5.2 shall be increased by 0.3 in. (7.6 mm) for every seat over a total of 14 but shall not be required to exceed 22 in. (560 mm).

**13.2.5.5.4.2** The requirement of 13.2.5.5.4.1 shall not apply to smoke-protected assembly seating as permitted by 13.4.2.7.

**13.2.5.5.5** Rows of seating served by an aisle or doorway at one end only shall have a path of travel not exceeding 30 ft (9.1 m) in length from any seat to an aisle.

**13.2.5.5.5.1** The 12 in. (305 mm) minimum clear width of aisle accessway specified in 13.2.5.5.2 shall be increased by 0.6 in. (15 mm) for every seat over a total of seven.

**13.2.5.5.5.2** The requirements of 13.2.5.5.5 and 13.2.5.5.5.1 shall not apply to smoke-protected assembly seating as permitted by 13.4.2.8 and 13.4.2.9.

**13.2.5.5.6** Rows of seating using tablet-arm chairs shall be permitted only if the clear width of aisle accessways complies with the requirements of 13.2.5.5 when measured under one of the following conditions:

(1) The clear width is measured with the tablet arm in the usable position.

(2) The clear width is measured with the tablet arm in the stored position where the tablet arm automatically returns to the stored position when raised manually to a vertical position in one motion and falls to the stored position by force of gravity.

**13.2.5.5.7** The depth of seat boards shall be not less than 9 in. (230 mm) where the same level is not used for both seat boards and footboards.

**13.2.5.5.8** Footboards, independent of seats, shall be provided so that there is no horizontal opening that allows the passage of a ½ in. (13 mm) diameter sphere.

**13.2.5.6 Aisles Serving Seating Not at Tables.**

**13.2.5.6.1 General.**

**13.2.5.6.1.1** Aisles shall be provided so that the number of seats served by the nearest aisle is in accordance with 13.2.5.5.2 through 13.2.5.5.5, unless otherwise permitted by 13.2.5.6.1.2.

**13.2.5.6.1.2** Aisles shall not be required in bleachers, provided that all of the following conditions are met:

(1) Egress from the front row shall not be obstructed by a rail, a guard, or other obstruction.
(2) The row spacing shall be 28 in. (710 mm) or less.
(3) The rise per row, including the first row, shall be 6 in. (150 mm) or less.
(4) The number of rows shall not exceed 16.
(5) The seat spaces shall not be physically defined.
(6) Seat boards that are also used as stepping surfaces for descent shall provide a walking surface with a width of not less than 12 in. (305 mm), and, where a depressed footboard exists, the gap between seat boards of adjacent rows shall not exceed 12 in. (305 mm), measured horizontally.
(7) The leading edges of seat boards used as stepping surfaces shall be provided with a contrasting marking stripe so that the location of the leading edge is readily apparent, particularly where viewed in descent, and the following shall also apply:
  (a) The marking stripe shall be not less than 1 in. (25 mm) wide and shall not exceed 2 in. (51 mm) in width.
  (b) The marking stripe shall not be required where bleacher surfaces and environmental conditions, under all conditions of use, are such that the location of each leading edge is readily apparent, particularly when viewed in descent.

**13.2.5.6.2 Dead-End Aisles.** Dead-end aisles shall not exceed 20 ft (6100 mm) in length, unless otherwise permitted by one of the following:

(1) A dead-end aisle shall be permitted to exceed 20 ft (6100 mm) in length where seats served by the dead-end aisle are not more than 24 seats from another aisle, measured along a row of seats having a clear width of not less than 12 in. (305 mm) plus 0.6 in. (15 mm) for each additional seat over a total of 7 in the row.
(2) A 16-row, dead-end aisle shall be permitted in folding and telescopic seating and grandstands.
(3) Aisle termination in accordance with 13.4.2.11 for smoke-protected assembly seating shall be permitted.
(4) Bleacher aisles in accordance with 13.2.3.5 shall not be considered as dead-end aisles.

**13.2.5.6.3\* Minimum Aisle Width.** The minimum clear width of aisles shall be sufficient to provide egress capacity in accordance with 13.2.3.1 but shall be not less than the following:

(1) 42 in. (1065 mm) for stairs having seating on each side, except that the minimum clear width shall be permitted to be not less than 30 in. (760 mm) for catchment areas having not more than 60 seats
(2) 36 in. (915 mm) for stairs having seating on only one side, or 30 in. (760 mm) for catchment areas having not more than 60 seats
(3) 20 in. (510 mm) between a handrail and seating or between a guardrail and seating where the aisle is subdivided by a handrail
(4) 42 in. (1065 mm) for level or ramped aisles having seating on both sides, except that the minimum clear width shall be not less than 30 in. (760 mm) for catchment areas having not more than 60 seats
(5) 36 in. (915 mm) for level or ramped aisles having seating on only one side, or 30 in. (760 mm) for catchment areas having not more than 60 seats
(6) 23 in. (585 mm) between a handrail or a guardrail and seating where the aisle does not serve more than five rows on one side

**13.2.5.6.4 Aisle Stairs and Aisle Ramps.**

**13.2.5.6.4.1\*** The following shall apply to aisle stairs and aisle ramps:

(1) Aisles having a gradient steeper than 1 in 20, but not steeper than 1 in 8, shall consist of an aisle ramp.
(2) Aisles having a gradient steeper than 1 in 8 shall consist of an aisle stair.

**13.2.5.6.4.2** The limitation on height between landings in Table 7.2.2.2.1.1(a) and Table 7.2.2.2.1.1(b) shall not apply to aisle stairs and landings.

**13.2.5.6.4.3** The limitation on height between landings in Table 7.2.5.2(a) and Table 7.2.5.2(b) shall not apply to aisle ramps and landings.

**13.2.5.6.5 Aisle Stair Treads.** Aisle stair treads shall meet all of the following criteria:

(1) There shall be no variation in the depth of adjacent treads that exceeds 3/16 in. (4.8 mm), unless otherwise permitted by 13.2.5.6.5(2), (5), or (6).
(2) Construction-caused nonuniformities in tread depth shall be permitted, provided that both of the following criteria are met:
  (a) The nonuniformity does not exceed 3/8 in. (10 mm).
  (b) The aisle tread depth is 22 in. (560 mm) or greater.
(3)\*Tread depth shall be not less than 11 in. (280 mm).
(4) All treads shall extend the full width of the aisle.
(5)\*In aisle stairs where a single intermediate tread is provided halfway between seating platforms, such intermediate treads shall be permitted to be of a relatively smaller but uniform depth but shall be not less than 13 in. (330 mm).
(6) All of the following shall apply to grandstands, bleachers, and folding and telescopic seating:
  (a) Steps shall not be required to be provided in aisles to overcome differences in level unless the gradient exceeds 1 unit of rise in 10 units of run.
  (b) Where the rise of the seating platform exceeds 11 in. (280 mm), an intermediate step shall be provided for the full width of the aisle and shall be proportioned to provide two steps of equal rise per platform.

(c) Where the rise of the seating platform exceeds 18 in. (455 mm), two intermediate steps for the full width of the aisle shall be provided and proportioned to provide three steps of equal rise per platform that are uniform and not less than 9 in. (230 mm).

(d) The full length of the nose of each step in the aisle, as required by 13.2.5.6.5(6)(c), shall be conspicuously marked.

**13.2.5.6.6 Aisle Stair Risers.** Aisle stair risers shall meet the following criteria:

(1) Riser heights shall be not less than 4 in. (100 mm) in aisle stairs, unless aisle stairs are those in folding and telescopic seating.

(2) The riser height of aisle stairs in folding and telescopic seating shall be permitted to be not less than 3½ in. (90 mm).

(3) Riser heights shall not exceed 8 in. (205 mm), unless otherwise permitted by 13.2.5.6.6(4) or (5).

(4) The riser height of aisle stairs in folding and telescopic seating shall be permitted to be not more than 11 in. (280 mm).

(5) Where the gradient of an aisle is steeper than 8 in. (205 mm) in rise in 11 in. (280 mm) of run for the purpose of maintaining necessary sight lines in the adjoining seating area, the riser height shall be permitted to exceed 8 in. (205 mm) but shall not exceed 11 in. (280 mm).

(6) Riser heights shall be designed to be uniform in each aisle, and the construction-caused nonuniformities shall not exceed 3/16 in. (4.8 mm) between adjacent risers, unless the conditions of 13.2.5.6.6(7) or (8) are met.

(7) Riser height shall be permitted to be nonuniform where both of the following criteria are met:

(a) The uniformity shall be only for the purpose of accommodating changes in gradient necessary to maintain sight lines within a seating area, in which case the riser height shall be permitted to exceed 3/16 in. (4.8 mm) in any flight.

(b) Where nonuniformities exceed 3/16 in. (4.8 mm) between adjacent risers, the exact location of such nonuniformities shall be indicated by a distinctive marking stripe on each tread at the nosing or leading edge adjacent to the nonuniform risers.

(8) Construction-caused nonuniformities in riser height shall be permitted to exceed 3/16 in. (4.8 mm) where all of the following criteria are met:

(a) The riser height shall be designed to be nonuniform.

(b) The construction-caused nonuniformities shall not exceed 3/8 in. (10 mm) where the aisle tread depth is less than 22 in. (560 mm).

(c) The construction-caused nonuniformities shall not exceed 3/4 in. (19 mm) where the aisle tread depth is 22 in. (560 mm) or greater.

(d) Where nonuniformities exceed 3/16 in. (4.8 mm) between adjacent risers, the exact location of such nonuniformities shall be indicated by a distinctive marking stripe on each tread at the nosing or leading edge adjacent to the nonuniform risers.

**13.2.5.6.7 Aisle Stair Profile.** Aisle stairs shall comply with all of the following:

(1) Aisle risers shall be vertical or sloped under the tread projection at an angle not to exceed 30 degrees from vertical.

(2) Tread projection not exceeding 1½ in. (38 mm) shall be permitted.

(3) Tread projection shall be uniform in each aisle, except as otherwise permitted by 13.2.5.6.7(4).

(4) Construction-caused projection nonuniformities not exceeding ¼ in. (6.4 mm) shall be permitted.

**13.2.5.6.8\* Aisle Handrails.**

**13.2.5.6.8.1** Ramped aisles having a gradient exceeding 1 in 12 and aisle stairs shall be provided with handrails at one side or along the centerline and shall also be in accordance with 7.2.2.4.4.1, 7.2.2.4.4.5, and 7.2.2.4.4.6.

**13.2.5.6.8.2** Where seating exists on both sides of the aisle, the handrails shall be noncontinuous with gaps or breaks at intervals not exceeding five rows to facilitate access to seating and to allow crossing from one side of the aisle to the other.

**13.2.5.6.8.3** The gaps or breaks permitted by 13.2.5.6.8.2 shall have a clear width of not less than 22 in. (560 mm) and shall not exceed 36 in. (915 mm), measured horizontally, and the handrail shall have rounded terminations or bends.

**13.2.5.6.8.4** Where handrails are provided in the middle of aisle stairs, an additional intermediate rail shall be located approximately 12 in. (305 mm) below the main handrail.

**13.2.5.6.8.5** Handrails shall not be required where otherwise permitted by one of the following:

(1) Handrails shall not be required for ramped aisles having a gradient not steeper than 1 in 8 and having seating on both sides.

(2) The requirement for a handrail shall be satisfied by the use of a guard provided with a rail that complies with the graspability requirements for handrails and is located at a consistent height between 34 in. and 42 in. (865 mm and 1065 mm), measured as follows:

(a) Vertically from the top of the rail to the leading edge (nosing) of stair treads

(b) Vertically from the top of the rail to the adjacent walking surface in the case of a ramp

(3) Handrails shall not be required where risers do not exceed 7 in. (180 mm) in height.

**13.2.5.6.9\* Aisle Marking.**

**13.2.5.6.9.1** A contrasting marking stripe shall be provided on each tread at the nosing or leading edge so that the location of such tread is readily apparent, particularly when viewed in descent.

**13.2.5.6.9.2** The marking stripe shall be not less than 1 in. (25 mm) wide and shall not exceed 2 in. (51 mm) in width.

**13.2.5.6.9.3** The marking stripe shall not be required where tread surfaces and environmental conditions, under all conditions of use, are such that the location of each tread is readily apparent, particularly when viewed in descent.

**13.2.5.7\* Aisle Accessways Serving Seating at Tables.**

**13.2.5.7.1** The required clear width of an aisle accessway shall be not less than 12 in. (305 mm) where measured in accordance with 13.2.5.7.3 and shall be increased as a function of length in accordance with 13.2.5.7.4, unless otherwise permitted by 13.2.5.7.2.

**13.2.5.7.2\*** If used by not more than four persons, no minimum clear width shall be required for the portion of an aisle accessway having a length not exceeding 6 ft (1830 mm) and located farthest from an aisle.

**13.2.5.7.3\*** Where nonfixed seating is located between a table and an aisle accessway or aisle, the measurement of required clear width of the aisle accessway or aisle shall be made to a line 19 in. (485 mm), measured perpendicularly to the edge of the table, away from the edge of said table.

**13.2.5.7.4\*** The minimum required clear width of an aisle accessway, measured in accordance with 13.2.5.4.8 and 13.2.5.7.3, shall be increased beyond the 12 in. (305 mm) requirement of 13.2.5.7.1 by ½ in. (13 mm) for each additional 12 in. (305 mm) or fraction thereof beyond 12 ft (3660 mm) of aisle accessway length, where measured from the center of the seat farthest from an aisle.

**13.2.5.7.5** The path of travel along the aisle accessway shall not exceed 36 ft (11 m) from any seat to the closest aisle or egress doorway.

**13.2.5.8 Aisles Serving Seating at Tables.**

**13.2.5.8.1\*** Aisles that contain steps or that are ramped, such as aisles serving dinner theater–style configurations, shall comply with the requirements of 13.2.5.6.

**13.2.5.8.2\*** The width of aisles serving seating at tables shall be not less than 44 in. (1120 mm) where serving an occupant load exceeding 50, and 36 in. (915 mm) where serving an occupant load of 50 or fewer.

**13.2.5.8.3\*** Where nonfixed seating is located between a table and an aisle, the measurement of required clear width of the aisle shall be made to a line 19 in. (485 mm), measured perpendicularly to the edge of the table, away from the edge of said table.

**13.2.5.9 Approval of Layouts.**

**13.2.5.9.1** Where required by the authority having jurisdiction, plans drawn to scale showing the arrangement of furnishings or equipment shall be submitted to the authority by the building owner, manager, or authorized agent to substantiate conformance with the provisions of 13.2.5.

**13.2.5.9.2** The layout plans shall constitute the only acceptable arrangement, unless one of the following criteria is met:

(1) The plans are revised.
(2) Additional plans are submitted and approved.
(3) Temporary deviations from the specifications of the approved plans are used, provided that the occupant load is not increased and the intent of 13.2.5.9 is maintained.

**13.2.6 Travel Distance to Exits.**

**13.2.6.1** Travel distance shall be measured in accordance with Section 7.6.

**13.2.6.2** Exits shall be arranged so that the total length of travel from any point to reach an exit shall not exceed 200 ft (61 m) in any assembly occupancy, unless otherwise permitted by one of the following:

(1) The travel distance shall not exceed 250 ft (76 m) in assembly occupancies protected throughout by an approved automatic sprinkler system in accordance with Section 9.7.
(2) The travel distance requirement shall not apply to smoke-protected assembly seating as permitted by 13.4.2.12, 13.4.2.13, and 13.4.2.14.

**13.2.7 Discharge from Exits.**

**13.2.7.1** Exit discharge shall comply with Section 7.7.

**13.2.7.2** The level of exit discharge shall be measured at the point of principal entrance to the building.

**13.2.7.3** Where the principal entrance to an assembly occupancy is via a terrace, either raised or depressed, such terrace shall be permitted to be considered to be the first story in height for the purposes of Table 13.1.6 where all of the following criteria are met:

(1) The terrace is at least as long, measured parallel to the building, as the total width of the exit(s) it serves but not less than 60 in. (1525 mm) long.
(2) The terrace is at least as wide, measured perpendicularly to the building, as the exit(s) it serves but not less than 60 in. (1525 mm) wide.
(3) Required stairs leading from the terrace to the finished ground level are protected in accordance with 7.2.2.6.3 or are not less than 10 ft (3050 mm) from the building.

**13.2.8 Illumination of Means of Egress.** Means of egress, other than for private party tents not exceeding 1200 ft² (112 m²), shall be illuminated in accordance with Section 7.8.

**13.2.9 Emergency Lighting.**

**13.2.9.1** Emergency lighting, other than that permitted by 13.2.9.3, shall be provided in accordance with Section 7.9.

**13.2.9.2** Private party tents not exceeding 1200 ft² (112 m²) shall not be required to have emergency lighting.

**13.2.9.3** Assembly occupancies with an occupant load not exceeding 300 and used exclusively for a place of worship shall not be required to have emergency lighting.

**13.2.10 Marking of Means of Egress.**

**13.2.10.1** Means of egress shall be provided with signs in accordance with Section 7.10.

**13.2.10.2** Exit markings shall not be required on the seating side of vomitories from seating areas where exit marking is provided in the concourse and where such marking is readily apparent from the vomitories.

**13.2.10.3** Evacuation diagrams in accordance with 7.10.8.5 shall be provided.

**13.2.11 Special Means of Egress Features.**

**13.2.11.1 Guards and Railings: Boxes, Balconies, and Galleries.** Boxes, balconies, and galleries shall meet the following criteria:

(1) The fasciae of boxes, balconies, and galleries shall rise not less than 26 in. (660 mm) above the adjacent floor or shall have substantial railings not less than 26 in. (660 mm) above the adjacent floor.
(2) The height of the rail above footrests on the adjacent floor immediately in front of a row of seats shall be not less than 26 in. (660 mm), and the following also shall apply:
    (a) Railings at the ends of aisles shall be not less than 36 in. (915 mm) high for the full width of the aisle.
    (b) Railings at the end of aisles shall be not less than 36 in. (915 mm) high at the ends of aisles where steps occur.
(3) Aisle accessways adjacent to orchestra pits and vomitories, and all cross aisles, shall be provided with railings not less than 26 in. (660 mm) above the adjacent floor.
(4) The requirement of 13.2.11.1(3) shall not apply where the backs of seats located at the front of the aisle project 24 in. (610 mm) or more above the adjacent floor of the aisle.

(5) Guardrails shall not be required on the audience side of stages, raised platforms, and other raised floor areas such as runways, ramps, and side stages used for entertainment or presentations.

(6) Permanent guardrails shall not be required at vertical openings in the performance area of stages.

(7) Guardrails shall not be required where the side of an elevated walking surface is required to be open for the normal functioning of special lighting or for access and use of other special equipment.

**13.2.11.2 Lockups.** Lockups in assembly occupancies, other than approved existing lockups, shall comply with the requirements of 23.4.5.

### 13.3 Protection.

**13.3.1 Protection of Vertical Openings.** Any vertical opening shall be enclosed or protected in accordance with Section 8.6, unless otherwise permitted by one of the following:

(1)*Stairs or ramps shall be permitted to be unenclosed between balconies or mezzanines and main assembly areas located below, provided that the balcony or mezzanine is open to the main assembly area.

(2) Exit access stairs from lighting and access catwalks, galleries, and gridirons shall not be required to be enclosed.

(3) Assembly occupancies protected by an approved, supervised automatic sprinkler system in accordance with Section 9.7 shall be permitted to have unprotected vertical openings between any two adjacent floors, provided that such openings are separated from unprotected vertical openings serving other floors by a barrier complying with 8.6.5.

(4) Assembly occupancies protected by an approved, supervised automatic sprinkler system in accordance with Section 9.7 shall be permitted to have convenience stair openings in accordance with 8.6.9.2.

(5) Use of the following alternative materials shall be permitted where assemblies constructed of such materials are in good repair and free of any condition that would diminish their original fire resistance characteristics:

(a) Existing wood lath and plaster

(b) Existing ½ in. (13 mm) gypsum wallboard

(c) Existing installations of ¼ in. (6.3 mm) thick wired glass that are, or are rendered, inoperative and fixed in the closed position

(d) Other existing materials having similar fire resistance capabilities

**13.3.2 Protection from Hazards.**

**13.3.2.1 Service Equipment, Hazardous Operations or Processes, and Storage Facilities.**

**13.3.2.1.1** Rooms containing high-pressure boilers, refrigerating machinery of other than the domestic refrigerator type, large transformers, or other service equipment subject to explosion shall meet both of the following requirements:

(1) Such rooms shall not be located directly under or abutting required exits.

(2) Such rooms shall be separated from other parts of the building by fire barriers in accordance with Section 8.3 that have a minimum 1-hour fire resistance rating or shall be protected by automatic extinguishing systems in accordance with Section 8.7.

**13.3.2.1.2** Rooms or spaces for the storage, processing, or use of materials specified in 13.3.2.1.2(1) through (3) shall be protected in accordance with the following:

(1) Separation from the remainder of the building by fire barriers having a minimum 1-hour fire resistance rating or protection of such rooms by automatic extinguishing systems as specified in Section 8.7 in the following areas:

(a) Boiler and furnace rooms, unless otherwise protected by one of the following:

i. The requirement of 13.3.2.1.2(1)(a) shall not apply to rooms enclosing furnaces, heating and air-handling equipment, or compressor equipment with a total aggregate input rating less than 200,000 Btu (211 MJ), provided that such rooms are not used for storage.

ii. The requirement of 13.3.2.1.2(1)(a) shall not apply to attic locations of the rooms addressed in 13.3.2.1.2(1)(a)i, provided that such rooms comply with the draftstopping requirements of 8.6.11.

(b) Rooms or spaces used for the storage of combustible supplies in quantities deemed hazardous by the authority having jurisdiction

(c) Rooms or spaces used for the storage of hazardous materials or flammable or combustible liquids in quantities deemed hazardous by recognized standards

(2) Separation from the remainder of the building by fire barriers having a minimum 1-hour fire resistance rating and protection of such rooms by automatic extinguishing systems as specified in Section 8.7 in the following areas:

(a) Laundries

(b) Maintenance shops, including woodworking and painting areas

(c) Rooms or spaces used for processing or use of combustible supplies deemed hazardous by the authority having jurisdiction

(d) Rooms or spaces used for processing or use of hazardous materials or flammable or combustible liquids in quantities deemed hazardous by recognized standards

(3) Protection as permitted in accordance with 9.7.1.2 where automatic extinguishing is used to meet the requirements of 13.3.2.1.2(1) or (2)

**13.3.2.2 Cooking Equipment.** Cooking equipment shall be protected in accordance with 9.2.3, unless the cooking equipment is one of the following types:

(1) Outdoor equipment

(2) Portable equipment not flue-connected

(3) Equipment used only for food warming

**13.3.3 Interior Finish.**

**13.3.3.1 General.** Interior finish shall be in accordance with Section 10.2.

**13.3.3.2 Corridors, Lobbies, and Enclosed Stairways.** Interior wall and ceiling finish materials complying with Section 10.2 shall be Class A or Class B in all corridors and lobbies and shall be Class A in enclosed stairways.

**13.3.3.3 Assembly Areas.** Interior wall and ceiling finish materials complying with Section 10.2 shall be Class A or Class B in general assembly areas having occupant loads of more than 300 and shall be Class A, Class B, or Class C in assembly areas having occupant loads of 300 or fewer.

**13.3.3.4 Screens.** Screens on which pictures are projected shall comply with requirements of Class A or Class B interior finish in accordance with Section 10.2.

**13.3.3.5 Interior Floor Finish.** (No requirements.)

**13.3.4 Detection, Alarm, and Communications Systems.**

**13.3.4.1 General.**

**13.3.4.1.1** Assembly occupancies with occupant loads of more than 300 and all theaters with more than one audience-viewing room shall be provided with an approved fire alarm system in accordance with 9.6.1 and 13.3.4, unless otherwise permitted by 13.3.4.1.2, 13.3.4.1.3, or 13.3.4.1.4.

**13.3.4.1.2** Assembly occupancies that are a part of a multiple occupancy protected as a mixed occupancy *(see 6.1.14)* shall be permitted to be served by a common fire alarm system, provided that the individual requirements of each occupancy are met.

**13.3.4.1.3** Voice communication or public address systems complying with 13.3.4.3.6 shall not be required to comply with 9.6.1.

**13.3.4.1.4** The requirement of 13.3.4.1.1 shall not apply to assembly occupancies where, in the judgment of the authority having jurisdiction, adequate alternative provisions exist or are provided for the discovery of a fire and for alerting the occupants promptly.

**13.3.4.2 Initiation.**

**13.3.4.2.1** Initiation of the required fire alarm system shall be by both of the following means, and the system shall be provided with an emergency power source:

(1) Manual means in accordance with 9.6.2.1(1), unless otherwise permitted by one of the following:
   (a) The requirement of 13.3.4.2.1(1) shall not apply where initiation is by means of an approved automatic fire detection system in accordance with 9.6.2.1(2) that provides fire detection throughout the building.
   (b) The requirement of 13.3.4.2.1(1) shall not apply where initiation is by means of an approved automatic sprinkler system in accordance with 9.6.2.1(3) that provides fire detection and protection throughout the building.
(2) Where automatic sprinklers are provided, initiation of the fire alarm system by sprinkler system waterflow, even where manual fire alarm boxes are provided in accordance with 13.3.4.2.1(1)

**13.3.4.2.2** The initiating device shall be capable of transmitting an alarm to a receiving station, located within the building, that is constantly attended when the assembly occupancy is occupied.

**13.3.4.2.3\*** In assembly occupancies with occupant loads of more than 300, automatic detection shall be provided in all hazardous areas that are not normally occupied, unless such areas are protected throughout by an approved automatic sprinkler system in accordance with Section 9.7.

**13.3.4.3 Notification.** The required fire alarm system shall activate an audible alarm in a constantly attended receiving station within the building when occupied for purposes of initiating emergency action.

**13.3.4.3.1** Positive alarm sequence in accordance with 9.6.3.4 shall be permitted.

**13.3.4.3.2** A presignal system in accordance with 9.6.3.3 shall be permitted.

**13.3.4.3.3** Occupant notification shall be by means of voice announcements in accordance with 9.6.3.9 initiated by the person in the constantly attended receiving station.

**13.3.4.3.4 Reserved.**

**13.3.4.3.5 Reserved.**

**13.3.4.3.6** The announcement shall be permitted to be made via a voice communication or public address system in accordance with 9.6.3.9.2.

**13.3.4.3.7** Where the authority having jurisdiction determines that a constantly attended receiving station is impractical, automatically transmitted evacuation or relocation instructions shall be provided in accordance with *NFPA 72, National Fire Alarm and Signaling Code.*

**13.3.5 Extinguishment Requirements.** See also 13.1.6, 13.2.6, and 13.3.2.

**13.3.5.1** Where the occupant load exceeds 100, the following assembly occupancies shall be protected throughout by an approved, supervised automatic sprinkler system in accordance with 9.7.1.1(1):

(1) Dance halls
(2) Discotheques
(3) Nightclubs
(4) Assembly occupancies with festival seating

**13.3.5.2** Any assembly occupancy used or capable of being used for exhibition or display purposes shall be protected throughout by an approved automatic sprinkler system in accordance with Section 9.7 where the exhibition or display area exceeds 15,000 ft$^2$ (1400 m$^2$).

**13.3.5.3** The sprinklers specified by 13.3.5.2 shall not be required where otherwise permitted in the following locations:

(1) Locations in stadia and arenas as follows:
   (a) Over the floor areas used for contest, performance, or entertainment
   (b) Over the seating areas
   (c) Over open-air concourses where an approved engineering analysis substantiates the ineffectiveness of the sprinkler protection due to building height and combustible loading
(2) Locations in unenclosed stadia and arenas as follows:
   (a) Press boxes of less than 1000 ft$^2$ (93 m$^2$)
   (b) Storage facilities of less than 1000 ft$^2$ (93 m$^2$) if enclosed with not less than 1-hour fire resistance–rated construction
   (c) Enclosed areas underneath grandstands that comply with 13.4.8.5

**13.3.5.4** Where another provision of this chapter requires an automatic sprinkler system, the sprinkler system shall be installed in accordance with 9.7.1.1(1).

**13.3.6 Corridors.** (No requirements.)

**13.4 Special Provisions.**

**13.4.1 Life Safety Evaluation.**

**13.4.1.1\*** Where a life safety evaluation is required by other provisions of the *Code,* it shall comply with all of the following:

(1) The life safety evaluation shall be performed by persons acceptable to the authority having jurisdiction.

(2) The life safety evaluation shall include a written assessment of safety measures for conditions listed in 13.4.1.2.

(3) The life safety evaluation shall be approved annually by the authority having jurisdiction and shall be updated for special or unusual conditions.

**13.4.1.2** Life safety evaluations shall include an assessment of all of the following conditions and the related appropriate safety measures:

(1) Nature of the events and the participants and attendees
(2) Access and egress movement, including crowd density problems
(3) Medical emergencies
(4) Fire hazards
(5) Permanent and temporary structural systems
(6) Severe weather conditions
(7) Earthquakes
(8) Civil or other disturbances
(9) Hazardous materials incidents within and near the facility
(10) Relationships among facility management, event participants, emergency response agencies, and others having a role in the events accommodated in the facility

**13.4.1.3\*** Life safety evaluations shall include assessments of both building systems and management features upon which reliance is placed for the safety of facility occupants, and such assessments shall consider scenarios appropriate to the facility.

**13.4.2\* Smoke-Protected Assembly Seating.**

**13.4.2.1** To be considered smoke protected, an assembly seating facility shall comply with both of the following:

(1) All enclosed areas with walls and ceilings in buildings or structures containing smoke-protected assembly seating shall be protected with an approved automatic sprinkler system in accordance with Section 9.7, unless otherwise permitted by one of the following:

  (a) The requirement of 13.4.2.1(1) shall not apply to the floor area used for contest, performance, or entertainment, provided that the roof construction is more than 50 ft (15 m) above the floor level and use is restricted to low fire hazard uses.

  (b)\*Sprinklers shall not be required to be located over the floor area used for contest, performance, or entertainment and over the seating areas where an approved engineering analysis substantiates the ineffectiveness of the sprinkler protection due to building height and combustible loading.

(2) All means of egress serving a smoke-protected assembly seating area shall be provided with smoke-actuated ventilation facilities or natural ventilation designed to maintain the level of smoke at not less than 6 ft (1830 mm) above the floor of the means of egress.

**13.4.2.2** To use the provisions of smoke-protected assembly seating, a facility shall be subject to a life safety evaluation in accordance with 13.4.1.

**13.4.2.3** Minimum clear widths of aisles and other means of egress serving smoke-protected assembly seating shall be in accordance with Table 13.4.2.3.

**Table 13.4.2.3 Capacity Factors for Smoke-Protected Assembly Seating**

| | Clear Width per Seat Served | | | |
| | Stairs | | Passageways, Ramps, and Doorways | |
| Number of Seats | in. | mm | in. | mm |
|---|---|---|---|---|
| 2,000 | 0.300 AB | 7.6 AB | 0.220 C | 5.6 C |
| 5,000 | 0.200 AB | 5.1 AB | 0.150 C | 3.8 C |
| 10,000 | 0.130 AB | 3.3 AB | 0.100 C | 2.5 C |
| 15,000 | 0.096 AB | 2.4 AB | 0.070 C | 1.8 C |
| 20,000 | 0.076 AB | 1.9 AB | 0.056 C | 1.4 C |
| ≥25,000 | 0.060 AB | 1.5 AB | 0.044 C | 1.1 C |

**13.4.2.4 Outdoor Smoke-Protected Assembly Seating.**

**13.4.2.4.1** Where smoke-protected assembly seating and its means of egress are located wholly outdoors, capacity shall be permitted to be provided in accordance with Table 13.4.2.4.1 and the provision of 13.4.2.4.2 shall apply.

**Table 13.4.2.4.1 Capacity Factors for Outdoor Smoke-Protected Assembly Seating**

| | Clear Width per Seat Served | | | |
| | Stairs | | Passageways, Ramps, and Doorways | |
| Feature | in. | mm | in. | mm |
|---|---|---|---|---|
| Outdoor smoke-protected assembly seating | 0.08 AB | 2.0 AB | 0.06 C | 1.5 C |

**13.4.2.4.2** Where the number of seats in outdoor smoke-protected assembly seating exceeds 20,000, the capacity factors of Table 13.4.2.3 shall be permitted to be used.

**13.4.2.5** Where using Table 13.4.2.3, the number of seats specified shall be within a single assembly space, and interpolation shall be permitted between the specific values shown. A single seating space shall be permitted to have multiple levels, floors, or mezzanines.

**13.4.2.6** The minimum clear widths shown in Table 13.4.2.3 and Table 13.4.2.4.1 shall be modified in accordance with all of the following:

(1) If risers exceed 7 in. in height, the stair width in Table 13.4.2.3 and Table 13.4.2.4.1 shall be multiplied by factor $A$, where $A$ equals the following:

$$A = 1 + \frac{\text{riser height} - 7}{5}$$

(2) If risers exceed 178 mm in height, the stair width in Table 13.4.2.3 and Table 13.4.2.4.1 shall be multiplied by factor $A$, where $A$ equals the following:

$$A = 1 + \frac{\text{riser height} - 178}{125}$$

(3) Stairs not having a handrail within a 30 in. (760 mm) horizontal distance shall be 25 percent wider than otherwise calculated; that is, their width shall be multiplied by factor B, where B equals the following:

$$B = 1.25$$

(4) Ramps steeper than 1 in 10 slope used in ascent shall have their width increased by 10 percent; that is, their width shall be multiplied by factor C, where C equals the following:

$$C = 1.10$$

**13.4.2.7** Where smoke-protected assembly seating conforms to the requirements of 13.4.2, for rows of seats served by aisles or doorways at both ends, the number of seats per row shall not exceed 100, and the clear width of not less than 12 in. (305 mm) for aisle accessways shall be increased by 0.3 in. (7.6 mm) for every additional seat beyond the number stipulated in Table 13.4.2.7; however, the minimum clear width shall not be required to exceed 22 in. (560 mm).

**Table 13.4.2.7 Smoke-Protected Assembly Seating Aisle Accessways**

| Total Number of Seats in the Space | Number of Seats per Row Permitted to Have a Clear Width Aisle Accessway of Not Less than 12 in. (305 mm) | |
|---|---|---|
| | Aisle or Doorway at Both Ends of Row | Aisle or Doorway at One End of Row |
| <4,000 | 14 | 7 |
| 4,000–6,999 | 15 | 7 |
| 7,000–9,999 | 16 | 8 |
| 10,000–12,999 | 17 | 8 |
| 13,000–15,999 | 18 | 9 |
| 16,000–18,999 | 19 | 9 |
| 19,000–21,999 | 20 | 10 |
| ≥22,000 | 21 | 11 |

**13.4.2.8** Where smoke-protected assembly seating conforms to the requirements of 13.4.2, for rows of seats served by an aisle or doorway at one end only, the aisle accessway clear width of not less than 12 in. (305 mm) shall be increased by 0.6 in. (15 mm) for every additional seat beyond the number stipulated in Table 13.4.2.7; however, the minimum clear width shall not be required to exceed 22 in. (560 mm).

**13.4.2.9** Smoke-protected assembly seating conforming with the requirements of 13.4.2 shall be permitted to have a common path of travel of 50 ft (15 m) from any seat to a point where a person has a choice of two directions of egress travel.

**13.4.2.10** Aisle accessways shall be permitted to serve as one or both of the required exit accesses addressed in 12.4.2.9, provided that the aisle accessway has a minimum width of 12 in. (305 mm) plus 0.3 in. (7.6 mm) for every additional seat over a total of 7 in a row.

**13.4.2.11** Where smoke-protected assembly seating conforms to the requirements of 13.4.2, the dead ends in aisle stairs shall not exceed a distance of 21 rows, unless both of the following criteria are met:

(1) The seats served by the dead-end aisle are not more than 40 seats from another aisle.
(2) The 40-seat distance is measured along a row of seats having an aisle accessway with a clear width of not less than 12 in. (305 mm) plus 0.3 in. (7.6 mm) for each additional seat above 7 in the row.

**13.4.2.12** Where smoke-protected assembly seating conforms to the requirements of 13.4.2, the travel distance from each seat to the nearest entrance to an egress vomitory or egress concourse shall not exceed 400 ft (122 m).

**13.4.2.13** Where smoke-protected assembly seating conforms to the requirements of 13.4.2, the travel distance from the entrance to the vomitory or from the egress concourse to an approved egress stair, ramp, or walk at the building exterior shall not exceed 200 ft (61 m).

**13.4.2.14** The travel distance requirements of 13.4.2.12 and 13.4.2.13 shall not apply to outdoor assembly seating facilities of Type I or Type II construction where all portions of the means of egress are essentially open to the outside.

**13.4.3 Limited Access or Underground Buildings.** Limited access or underground buildings shall comply with Section 11.7.

**13.4.4 High-Rise Buildings.** Existing high-rise buildings that house assembly occupancies in high-rise portions of the building shall have the highest level of the assembly occupancy and all levels below protected by an approved, supervised automatic sprinkler system in accordance with Section 9.7. *(See also 13.1.6.)*

**13.4.5 Stages and Platforms.** See 3.3.262 and 3.3.209.

**13.4.5.1 Materials and Design.**

**13.4.5.1.1 Reserved.**

**13.4.5.1.2** Stage stairs shall be permitted to be of combustible materials, regardless of building construction type.

**13.4.5.2 Platform Construction. (Reserved)**

**13.4.5.3 Stage Construction. (Reserved)**

**13.4.5.4 Accessory Rooms. (Reserved)**

**13.4.5.5 Ventilators.** Regular stages in excess of 1000 ft² (93 m²) and legitimate stages shall be provided with emergency ventilation to provide a means of removing smoke and combustion gases directly to the outside in the event of a fire, and such ventilation shall be achieved by one or a combination of the methods specified in 13.4.5.5.1 through 13.4.5.5.3.

**13.4.5.5.1 Smoke Control.**

**13.4.5.5.1.1** A means complying with Section 9.3 shall be provided to maintain the smoke level at not less than 6 ft (1830 mm) above the highest level of assembly seating or above the top of the proscenium opening where a proscenium wall and opening protection are provided.

**13.4.5.5.1.2 Reserved.**

**13.4.5.5.1.3** The smoke control system shall be activated independently by each of the following:

(1) Activation of the sprinkler system in the stage area
(2) Activation of smoke detectors over the stage area
(3) Activation by manually operated switch at an approved location

**13.4.5.5.1.4** The emergency ventilation system shall be supplied by both normal and standby power.

**13.4.5.5.1.5** The fan(s) power wiring and ducts shall be located and properly protected to ensure a minimum of 20 minutes of operation in the event of activation.

**13.4.5.5.2 Roof Vents.**

**13.4.5.5.2.1** Two or more vents shall be located near the center of and above the highest part of the stage area.

**13.4.5.5.2.2** The vents shall be raised above the roof and shall provide a net free vent area equal to 5 percent of the stage area.

**13.4.5.5.2.3** Vents shall be constructed to open automatically by approved heat-activated devices, and supplemental means shall be provided for manual operation and periodic testing of the ventilator from the stage floor.

**13.4.5.5.2.4** Vents shall be labeled.

**13.4.5.5.2.5** Existing roof vents that are not labeled shall be permitted where they open by spring action or force of gravity sufficient to overcome the effects of neglect, rust, dirt, frost, snow, or expansion by heat or warping of the framework, and the following requirements also shall apply:

(1) Glass, if used in vents, shall be protected against falling onto the stage.
(2) A wire screen, if used under the glass, shall be placed so that, if clogged, it does not reduce the required venting area, interfere with the operating mechanism, or obstruct the distribution of water from an automatic sprinkler.
(3) Vents shall be arranged to open automatically by the use of fusible links.
(4) The fusible links and operating cable shall hold each door closed against a minimum 30 lb (133 N) counterforce that shall be exerted on each door through its entire arc of travel and for not less than 115 degrees.
(5) Vents shall be provided with manual control.
(6) Springs, where employed to actuate vent doors, shall be capable of maintaining full required tension.
(7) Springs shall not be stressed more than 50 percent of their rated capacity and shall not be located directly in the airstream nor exposed to the outside.
(8) A fusible link shall be placed in the cable control system on the underside of the vent at or above the roofline, or as approved by the building official.
(9) The fusible link shall be located so as not to be affected by the operation of an automatic sprinkler system.
(10) Remote, manual, or electric controls shall provide for both opening and closing of the vent doors for periodic testing and shall be located at a point on stage designated by the authority having jurisdiction.
(11) Where remote control vents are electrical, power failure shall not affect instant operation of the vent in the event of fire.
(12) Hand winches shall be permitted to be employed to facilitate operation of manually controlled vents.

**13.4.5.5.3 Other Means.** Approved, alternate means of removing smoke and combustion gases shall be permitted.

**13.4.5.6 Proscenium Walls. (Reserved)**

**13.4.5.7 Proscenium Opening Protection.**

**13.4.5.7.1** On every legitimate stage, the main proscenium opening used for viewing performances shall be provided with proscenium opening protection as follows:

(1) The proscenium opening protection shall comply with 12.4.5.7.
(2) Asbestos shall be permitted in lieu of a listed fabric.
(3) Manual curtains of any size shall be permitted.

**13.4.5.7.2** In lieu of the protection required by 13.4.5.7.1(1), all the following shall be provided:

(1) A noncombustible opaque fabric curtain shall be arranged so that it closes automatically.
(2) An automatic, fixed waterspray deluge system shall be located on the auditorium side of the proscenium opening and shall be arranged so that the entire face of the curtain will be wetted, and all of the following requirements also shall apply:
  (a) The system shall be activated by a combination of rate-of-rise and fixed-temperature detectors located on the ceiling of the stage.
  (b) Detectors shall be spaced in accordance with their listing.
  (c) The water supply shall be controlled by a deluge valve and shall be sufficient to keep the curtain completely wet for 30 minutes or until the valve is closed by fire department personnel.
(3) The curtain shall be automatically operated in case of fire by a combination of rate-of-rise and fixed-temperature detectors that also activates the deluge spray system.
(4) Stage sprinklers and vents shall be automatically operated by fusible elements in case of fire.
(5) Operation of the stage sprinkler system or spray deluge valve shall automatically activate the emergency ventilating system and close the curtain.
(6) The curtain, vents, and spray deluge system valve shall also be capable of manual operation.

**13.4.5.7.3** Proscenium opening protection provided by other than a fire curtain in accordance with 12.4.5.7 [see 13.4.5.7.1(1)] shall activate upon automatic detection of a fire and upon manual activation.

**13.4.5.8 Gridirons, Fly Galleries, and Pinrails. (Reserved)**

**13.4.5.9 Catwalks.** The clear width of lighting and access catwalks and the means of egress from galleries and gridirons shall be not less than 22 in. (560 mm).

**13.4.5.10 Fire Protection.** Every stage shall be protected by an approved automatic sprinkler system in compliance with Section 9.7.

**13.4.5.10.1** Protection shall be provided throughout the stage and in storerooms, workshops, permanent dressing rooms, and other accessory spaces contiguous to stages.

**13.4.5.10.2** Sprinklers shall not be required for stages 1000 ft$^2$ (93 m$^2$) or less in area where both of the following criteria are met:

(1) Curtains, scenery, or other combustible hangings are not retractable vertically.
(2) Combustible hangings are limited to borders, legs, a single main curtain, and a single backdrop.

**13.4.5.10.3** Sprinklers shall not be required under stage areas less than 48 in. (1220 mm) in clear height that are used exclusively for chair or table storage and lined on the inside with ⅝ in. (16 mm) Type X gypsum wallboard or the approved equivalent.

### 13.4.5.11 Flame-Retardant Requirements.

**13.4.5.11.1** Combustible scenery of cloth, film, vegetation (dry), and similar materials shall comply with one of the following:

(1) They shall meet the flame propagation performance criteria contained in NFPA 701, *Standard Methods of Fire Tests for Flame Propagation of Textiles and Films*

(2) They shall exhibit a heat release rate not exceeding 100 kW when tested in accordance with NFPA 289, *Standard Method of Fire Test for Individual Fuel Packages*, using the 20 kW ignition source

**13.4.5.11.2** Foamed plastics *(see definition of cellular or foamed plastic in 3.3.41)* shall be permitted to be used if they exhibit a heat release rate not exceeding 100 kW when tested in accordance with NFPA 289, *Standard Method of Fire Test for Individual Fuel Packages*, using the 20 kW ignition source or by specific approval of the authority having jurisdiction.

**13.4.5.11.3** Scenery and stage properties not separated from the audience by proscenium opening protection shall be of noncombustible materials, limited-combustible materials, or fire-retardant-treated wood.

**13.4.5.11.4** In theaters, motion picture theaters, and television stage settings, with or without horizontal projections, and in simulated caves and caverns of foamed plastic, any single fuel package shall have a heat release rate not to exceed 100 kW where tested in accordance with one of the following:

(1) ANSI/UL 1975, *Standard for Fire Tests for Foamed Plastics Used for Decorative Purposes*

(2) NFPA 289, *Standard Method of Fire Test for Individual Fuel Packages*, using the 20 kW ignition source

### 13.4.5.12* Standpipes.

**13.4.5.12.1** Stages over 1000 ft² (93 m²) in area shall be equipped with 1½ in. (38 mm) hose lines for first aid fire fighting at each side of the stage.

**13.4.5.12.2** Hose connections shall be in accordance with NFPA 13, *Standard for the Installation of Sprinkler Systems*, unless Class II or Class III standpipes in accordance with NFPA 14, *Standard for the Installation of Standpipe and Hose Systems*, are used.

### 13.4.6 Projection Rooms.

**13.4.6.1** Projection rooms shall comply with 13.4.6.2 through 13.4.6.10.

**13.4.6.2** Where cellulose nitrate film is used, the projection room shall comply with NFPA 40, *Standard for the Storage and Handling of Cellulose Nitrate Film.*

**13.4.6.3** Film or video projectors or spotlights utilizing light sources that produce particulate matter or toxic gases, or light sources that produce hazardous radiation, without protective shielding shall be located within a projection room complying with 13.3.2.1.2.

**13.4.6.4** Every projection room shall be of permanent construction consistent with the building construction type in which the projection room is located and shall comply with the following:

(1) Openings shall not be required to be protected.

(2) The room shall have a floor area of not less than 80 ft² (7.4 m²) for a single machine and not less than 40 ft² (3.7 m²) for each additional machine.

(3) Each motion picture projector, floodlight, spotlight, or similar piece of equipment shall have a clear working space of not less than 30 in. (760 mm) on each side and at its rear, but only one such space shall be required between adjacent projectors.

**13.4.6.5** The projection room and the rooms appurtenant to it shall have a ceiling height of not less than 7 ft 6 in. (2285 mm).

**13.4.6.6** Each projection room for safety film shall have not less than one out-swinging, self-closing door not less than 30 in. (760 mm) wide and 6 ft 8 in. (2030 mm) high.

**13.4.6.7** The aggregate of ports and openings for projection equipment shall not exceed 25 percent of the area of the wall between the projection room and the auditorium, and all openings shall be provided with glass or other approved material so as to completely close the opening.

**13.4.6.8** Projection room ventilation shall comply with 13.4.6.8.1 and 13.4.6.8.2.

### 13.4.6.8.1 Supply Air.

**13.4.6.8.1.1** Each projection room shall be provided with adequate air supply inlets arranged to provide well-distributed air throughout the room.

**13.4.6.8.1.2** Air inlet ducts shall provide an amount of air equivalent to the amount of air being exhausted by projection equipment.

**13.4.6.8.1.3** Air shall be permitted to be taken from the outside; from adjacent spaces within the building, provided that the volume and infiltration rate is sufficient; or from the building air-conditioning system, provided that it is arranged to supply sufficient air whether or not other systems are in operation.

### 13.4.6.8.2 Exhaust Air.

**13.4.6.8.2.1** Projection booths shall be permitted to be exhausted through the lamp exhaust system.

**13.4.6.8.2.2** The lamp exhaust system shall be positively interconnected with the lamp so that the lamp cannot operate unless there is sufficient airflow required for the lamp.

**13.4.6.8.2.3** Exhaust air ducts shall terminate at the exterior of the building in such a location that the exhaust air cannot be readily recirculated into any air supply system.

**13.4.6.8.2.4** The projection room ventilation system shall be permitted also to serve appurtenant rooms, such as the generator room and the rewind room.

**13.4.6.9** Each projection machine shall be provided with an exhaust duct that draws air from each lamp and exhausts it directly to the outside of the building.

**13.4.6.9.1** The lamp exhaust shall be permitted to exhaust air from the projection room to provide room air circulation.

**13.4.6.9.2** Lamp exhaust ducts shall be of rigid materials, except for a flexible connector approved for the purpose.

**13.4.6.9.3** The projection lamp and projection room exhaust systems shall be permitted to be combined but shall not be interconnected with any other exhaust system or return-air system within the buildings.

**13.4.6.9.4** Specifications for electric arc and xenon projection equipment shall comply with 13.4.6.9.4.1 and 13.4.6.9.4.2.

**13.4.6.9.4.1 Electric Arc Projection Equipment.** The exhaust capacity shall be 200 ft$^3$/min (0.09 m$^3$/s) for each lamp connected to the lamp exhaust system or as recommended by the equipment manufacturer, and auxiliary air shall be permitted to be introduced into the system through a screened opening to stabilize the arc.

**13.4.6.9.4.2 Xenon Projection Equipment.** The lamp exhaust system shall exhaust not less than 300 ft$^3$/min (0.14 m$^3$/s) per lamp, or not less than the exhaust volume required or recommended by the equipment manufacturer, whichever is greater.

**13.4.6.10** Miscellaneous equipment and storage shall be protected as follows:

(1) Each projection room shall be provided with rewind and film storage facilities.
(2) Flammable liquids containers shall be permitted in projection rooms, provided that all of the following criteria are met:
    (a) There are not more than four containers per projection room.
    (b) No container has a capacity exceeding 16 oz (0.5 L).
    (c) The containers are of a nonbreakable type.
(3) Appurtenant electrical equipment, such as rheostats, transformers, and generators, shall be permitted to be located within the booth or in a separate room of equivalent construction.

**13.4.7\* Special Amusement Buildings.**

**13.4.7.1\* General.** Special amusement buildings, regardless of occupant load, shall meet the requirements for assembly occupancies in addition to the requirements of 13.4.7, unless the special amusement building is a multilevel play structure that is not more than 10 ft (3050 mm) in height and has aggregate horizontal projections not exceeding 160 ft$^2$ (15 m$^2$).

**13.4.7.2\* Automatic Sprinklers.** Every special amusement building, other than buildings or structures not exceeding 10 ft (3050 mm) in height and not exceeding 160 ft$^2$ (15 m$^2$) in aggregate horizontal projection, shall be protected throughout by an approved, supervised automatic sprinkler system installed and maintained in accordance with Section 9.7.

**13.4.7.3 Temporary Water Supply.** Where the special amusement building required to be sprinklered by 13.4.7.2 is movable or portable, the sprinkler water supply shall be permitted to be provided by an approved temporary means.

**13.4.7.4 Smoke Detection.** Where the nature of the special amusement building is such that it operates in reduced lighting levels, the building shall be protected throughout by an approved automatic smoke detection system in accordance with Section 9.6.

**13.4.7.5 Alarm Initiation.** Actuation of any smoke detection system device shall sound an alarm at a constantly attended location on the premises.

**13.4.7.6 Illumination.** Actuation of the automatic sprinkler system, or any other suppression system, or actuation of a smoke detection system having an approved verification or cross-zoning operation capability shall provide for both of the following:

(1) Increase in illumination in the means of egress to that required by Section 7.8
(2) Termination of any conflicting or confusing sounds and visuals

**13.4.7.7 Exit Marking.**

**13.4.7.7.1** Exit marking shall be in accordance with Section 7.10.

**13.4.7.7.2** Floor proximity exit signs shall be provided in accordance with 7.10.1.6.

**13.4.7.7.3\*** In special amusement buildings where mazes, mirrors, or other designs are used to confound the egress path, approved directional exit marking that becomes apparent in an emergency shall be provided.

**13.4.7.8 Interior Finish.** Interior wall and ceiling finish materials complying with Section 10.2 shall be Class A throughout.

**13.4.8 Grandstands.**

**13.4.8.1 General.**

**13.4.8.1.1** Grandstands shall comply with the provisions of this chapter as modified by 13.4.8.

**13.4.8.1.2** Approved existing grandstands shall be permitted to be continued to be used.

**13.4.8.2 Seating.**

**13.4.8.2.1** Where grandstand seating without backs is used indoors, rows of seats shall be spaced not less than 22 in. (560 mm) back-to-back.

**13.4.8.2.2** The depth of footboards and seat boards in grandstands shall be not less than 9 in. (230 mm); where the same level is not used for both seat foundations and footrests, footrests independent of seats shall be provided.

**13.4.8.2.3** Seats and footrests of grandstands shall be supported securely and fastened in such a manner that they cannot be displaced inadvertently.

**13.4.8.2.4** Individual seats or chairs shall be permitted only if secured firmly in rows in an approved manner, unless seats do not exceed 16 in number and are located on level floors and within railed-in enclosures, such as boxes.

**13.4.8.2.5** The maximum number of seats permitted between the farthest seat and an aisle in grandstands and bleachers shall not exceed that shown in Table 13.4.8.2.5.

**Table 13.4.8.2.5 Maximum Number of Seats Between Farthest Seat and an Aisle**

| Application | Outdoors | Indoors |
|---|---|---|
| Grandstands | 11 | 6 |
| Bleachers (See 13.2.5.6.1.2.) | 20 | 9 |

**13.4.8.3 Special Requirements — Wood Grandstands.**

**13.4.8.3.1** An outdoor wood grandstand shall be erected within not less than two-thirds of its height, and, in no case, within not less than 10 ft (3050 mm), of a building, unless otherwise permitted by one of the following:

(1) The distance requirement shall not apply to buildings having minimum 1-hour fire resistance–rated construction with openings protected against the fire exposure hazard created by the grandstand.

(2) The distance requirement shall not apply where a wall having minimum 1-hour fire resistance–rated construction separates the grandstand from the building.

**13.4.8.3.2** An outdoor wood grandstand unit shall not exceed 10,000 ft² (929 m²) in finished ground level area or 200 ft (61 m) in length, and all of the following requirements also shall apply:

(1) Grandstand units of the maximum size shall be placed not less than 20 ft (6100 mm) apart or shall be separated by walls having a minimum 1-hour fire resistance rating.

(2) The number of grandstand units erected in any one group shall not exceed three.

(3) Each group of grandstand units shall be separated from any other group by a wall having minimum 2-hour fire resistance–rated construction extending 24 in. (610 mm) above the seat platforms or by an open space of not less than 50 ft (15 m).

**13.4.8.3.3** The finished ground level area or length required by 13.4.8.3.2 shall be permitted to be doubled where one of the following criteria is met:

(1) Where the grandstand is constructed entirely of labeled fire-retardant-treated wood that has passed the standard rain test, ASTM D 2898, *Standard Test Methods for Accelerated Weathering of Fire-Retardant-Treated Wood for Fire Testing*

(2) Where the grandstand is constructed of members conforming to dimensions for heavy timber construction [Type IV (2HH)]

**13.4.8.3.4** The highest level of seat platforms above the finished ground level or the surface at the front of any wood grandstand shall not exceed 20 ft (6100 mm).

**13.4.8.3.5** The highest level of seat platforms above the finished ground level, or the surface at the front of a portable grandstand within a tent or membrane structure, shall not exceed 12 ft (3660 mm).

**13.4.8.3.6** The height requirements specified in 13.4.8.3.4 and 13.4.8.3.5 shall be permitted to be doubled where the grandstand is constructed entirely of labeled fire-retardant-treated wood that has passed the standard rain test, ASTM D 2898, *Standard Test Methods for Accelerated Weathering of Fire-Retardant-Treated Wood for Fire Testing*, or where constructed of members conforming to dimensions for heavy timber construction [Type IV (2HH)].

**13.4.8.4 Special Requirements — Portable Grandstands.**

**13.4.8.4.1** Portable grandstands shall conform to the requirements of 13.4.8 for grandstands and the requirements of 13.4.8.4.2 through 13.4.8.4.7.

**13.4.8.4.2** Portable grandstands shall be self-contained and shall have within them all necessary parts to withstand and restrain all forces that might be developed during human occupancy.

**13.4.8.4.3** Portable grandstands shall be designed and manufactured so that, if any structural members essential to the strength and stability of the structure have been omitted during erection, the presence of unused connection fittings shall make the omissions self-evident.

**13.4.8.4.4** Portable grandstand construction shall be skillfully accomplished to produce the strength required by the design.

**13.4.8.4.5** Portable grandstands shall be provided with base plates, sills, floor runners, or sleepers of such area that the permitted bearing capacity of the supporting material is not exceeded.

**13.4.8.4.6** Where a portable grandstand rests directly on a base of such character that it is incapable of supporting the load without appreciable settlement, mud sills of suitable material, having sufficient area to prevent undue or dangerous settlement, shall be installed under base plates, runners, or sleepers.

**13.4.8.4.7** All bearing surfaces shall be in contact with each other.

**13.4.8.5 Spaces Underneath Grandstands.** Spaces underneath a grandstand shall be kept free of flammable or combustible materials, unless protected by an approved, supervised automatic sprinkler system in accordance with Section 9.7 or unless otherwise permitted by one of the following:

(1) This requirement shall not apply to accessory uses of 300 ft² (28 m²) or less, such as ticket booths, toilet facilities, or concession booths, where constructed of noncombustible or fire-resistive construction in otherwise nonsprinklered facilities.

(2) This requirement shall not apply to rooms that are enclosed in not less than 1-hour fire resistance–rated construction and are less than 1000 ft² (93 m²) in otherwise nonsprinklered facilities.

**13.4.8.6 Guards and Railings.**

**13.4.8.6.1** Railings or guards not less than 42 in. (1065 mm) above the aisle surface or footrest or not less than 36 in. (915 mm) vertically above the center of the seat or seat board surface, whichever is adjacent, shall be provided along those portions of the backs and ends of all grandstands where the seats are in excess of 48 in. (1220 mm) above the floor or the finished ground level.

**13.4.8.6.2** The requirement of 13.4.8.6.1 shall not apply where an adjacent wall or fence affords equivalent safeguard.

**13.4.8.6.3** Where the front footrest of any grandstand is more than 24 in. (610 mm) above the floor, railings or guards not less than 33 in. (825 mm) above such footrests shall be provided.

**13.4.8.6.4** The railings required by 13.4.8.6.3 shall be permitted to be not less than 26 in. (660 mm) high in grandstands or where the front row of seats includes backrests.

**13.4.8.6.5** Cross aisles located within the seating area shall be provided with rails not less than 26 in. (660 mm) high along the front edge of the cross aisle.

**13.4.8.6.6** The railings specified by 13.4.8.6.5 shall not be required where the backs of the seats in front of the cross aisle project 24 in. (610 mm) or more above the surface of the cross aisle.

**13.4.8.6.7** Vertical openings between guardrails and footboards or seat boards shall be provided with intermediate construction so that a 4 in. (100 mm) diameter sphere cannot pass through the opening.

**13.4.8.6.8** An opening between the seat board and footboard located more than 30 in. (760 mm) above the finished ground level shall be provided with intermediate construction so that a 4 in. (100 mm) diameter sphere cannot pass through the opening.

### 13.4.9 Folding and Telescopic Seating.

#### 13.4.9.1 General.

**13.4.9.1.1** Folding and telescopic seating shall comply with the provisions of this chapter as modified by 13.4.9.

**13.4.9.1.2** Approved existing folding and telescopic seating shall be permitted to be continued to be used.

#### 13.4.9.2 Seating.

**13.4.9.2.1** The horizontal distance of seats, measured back-to-back, shall be not less than 22 in. (560 mm) for seats without backs, and all of the following requirements shall also apply:

(1) There shall be a space of not less than 12 in. (305 mm) between the back of each seat and the front of each seat immediately behind it.
(2) If seats are of the chair type, the 12 in. (305 mm) dimension shall be measured to the front edge of the rear seat in its normal unoccupied position.
(3) All measurements shall be taken between plumb lines.

**13.4.9.2.2** The depth of footboards (footrests) and seat boards in folding and telescopic seating shall be not less than 9 in. (230 mm).

**13.4.9.2.3** Where the same level is not used for both seat foundations and footrests, footrests independent of seats shall be provided.

**13.4.9.2.4** Individual chair-type seats shall be permitted in folding and telescopic seating only if firmly secured in groups of not less than three.

**13.4.9.2.5** The maximum number of seats permitted between the farthest seat in an aisle in folding and telescopic seating shall not exceed that shown in Table 13.4.8.2.5.

#### 13.4.9.3 Guards and Railings.

**13.4.9.3.1** Railings or guards not less than 42 in. (1065 mm) above the aisle surface or footrest, or not less than 36 in. (915 mm) vertically above the center of the seat or seat board surface, whichever is adjacent, shall be provided along those portions of the backs and ends of all folding and telescopic seating where the seats are more than 48 in. (1220 mm) above the floor or the finished ground level.

**13.4.9.3.2** The requirement of 13.4.9.3.1 shall not apply where an adjacent wall or fence affords equivalent safeguard.

**13.4.9.3.3** Where the front footrest of folding or telescopic seating is more than 24 in. (610 mm) above the floor, railings or guards not less than 33 in. (825 mm) above such footrests shall be provided.

**13.4.9.3.4** The railings required by 13.4.9.3.3 shall be permitted to be not less than 26 in. (660 mm) high where the front row of seats includes backrests.

**13.4.9.3.5** Cross aisles located within the seating area shall be provided with rails not less than 26 in. (660 mm) high along the front edge of the cross aisle.

**13.4.9.3.6** The railings specified by 13.4.9.3.5 shall not be required where the backs of the seats in front of the cross aisle project 24 in. (610 mm) or more above the surface of the cross aisle.

**13.4.9.3.7** Vertical openings between guardrails and footboards or seat boards shall be provided with intermediate construction so that a 4 in. (100 mm) diameter sphere cannot pass through the opening.

**13.4.9.3.8** An opening between the seat board and footboard located more than 30 in. (760 mm) above the finished ground level shall be provided with intermediate construction so that a 4 in. (100 mm) diameter sphere cannot pass through the opening.

### 13.4.10 Airport Loading Walkways.

**13.4.10.1** Airport loading walkways shall conform to NFPA 415, *Standard on Airport Terminal Buildings, Fueling Ramp Drainage, and Loading Walkways*, and the provisions of 13.4.10.2 and 13.4.10.3.

**13.4.10.2** Doors in the egress path from the aircraft through the airport loading walkway into the airport terminal building shall meet both of the following criteria:

(1) They shall swing in the direction of egress from the aircraft.
(2)*They shall not be permitted to have delayed-egress locks.

**13.4.10.3** Exit access shall be unimpeded from the airport loading walkway to the nonsecured public areas of the airport terminal building.

### 13.5 Building Services.

**13.5.1 Utilities.** Utilities shall comply with the provisions of Section 9.1.

**13.5.2 Heating, Ventilating, and Air-Conditioning Equipment.** Heating, ventilating, and air-conditioning equipment shall comply with the provisions of Section 9.2.

**13.5.3 Elevators, Escalators, and Conveyors.** Elevators, escalators, and conveyors shall comply with the provisions of Section 9.4.

**13.5.4 Rubbish Chutes, Incinerators, and Laundry Chutes.** Rubbish chutes, incinerators, and laundry chutes shall comply with the provisions of Section 9.5.

### 13.6 Reserved.

### 13.7 Operating Features.

#### 13.7.1 Means of Egress Inspection.

**13.7.1.1** The building owner or agent shall inspect the means of egress to ensure it is maintained free of obstructions, and correct any deficiencies found, prior to each opening of the building to the public.

**13.7.1.2** The building owner or agent shall prepare and maintain records of the date and time of each inspection on approved forms, listing any deficiencies found and actions taken to correct them.

**13.7.1.3 Inspection of Door Openings.** Door openings shall be inspected in accordance with 7.2.1.15.

#### 13.7.2 Special Provisions for Food Service Operations.

**13.7.2.1** All devices in connection with the preparation of food shall be installed and operated to avoid hazard to the safety of occupants.

**13.7.2.2** All devices in connection with the preparation of food shall be of an approved type and shall be installed in an approved manner.

**13.7.2.3** Food preparation facilities shall be protected in accordance with 9.2.3 and shall not be required to have openings protected between food preparation areas and dining areas.

**13.7.2.4** Portable cooking equipment that is not flue-connected shall be permitted only as follows:

(1) Equipment fueled by small heat sources that can be readily extinguished by water, such as candles or alcohol-burning equipment, including solid alcohol, shall be permitted to be used, provided that precautions satisfactory to the authority having jurisdiction are taken to prevent ignition of any combustible materials.

(2) Candles shall be permitted to be used on tables used for food service where securely supported on substantial non-combustible bases located to avoid danger of ignition of combustible materials and only where approved by the authority having jurisdiction.

(3) Candle flames shall be protected.

(4) "Flaming sword" or other equipment involving open flames and flamed dishes, such as cherries jubilee or crêpe suzette, shall be permitted to be used, provided that precautions subject to the approval of the authority having jurisdiction are taken.

(5)*Listed and approved LP-Gas commercial food service appliances shall be permitted to be used where in accordance with NFPA 58, *Liquefied Petroleum Gas Code.*

**13.7.3 Open Flame Devices and Pyrotechnics.** No open flame devices or pyrotechnic devices shall be used in any assembly occupancy, unless otherwise permitted by one of the following:

(1) Pyrotechnic special effect devices shall be permitted to be used on stages before proximate audiences for ceremonial or religious purposes, as part of a demonstration in exhibits, or as part of a performance, provided that both of the following criteria are met:

(a) Precautions satisfactory to the authority having jurisdiction are taken to prevent ignition of any combustible material.

(b) Use of the pyrotechnic device complies with NFPA 1126, *Standard for the Use of Pyrotechnics Before a Proximate Audience.*

(2) Flame effects before an audience shall be permitted in accordance with NFPA 160, *Standard for the Use of Flame Effects Before an Audience.*

(3) Open flame devices shall be permitted to be used in the following situations, provided that precautions satisfactory to the authority having jurisdiction are taken to prevent ignition of any combustible material or injury to occupants:

(a)*For ceremonial or religious purposes

(b) On stages and platforms where part of a performance

(c) Where candles on tables are securely supported on substantial noncombustible bases and candle flame is protected

(4) The requirement of 13.7.3 shall not apply to heat-producing equipment complying with 9.2.2.

(5) The requirement of 13.7.3 shall not apply to food service operations in accordance with 13.7.2.

(6) Gas lights shall be permitted to be used, provided that precautions are taken, subject to the approval of authority having jurisdiction, to prevent ignition of any combustible materials.

**13.7.4 Furnishings, Decorations, and Scenery.**

**13.7.4.1** Fabrics and films used for decorative purposes, all draperies and curtains, and similar furnishings shall be in accordance with the provisions of 10.3.1.

**13.7.4.2** The authority having jurisdiction shall impose controls on the quantity and arrangement of combustible contents in assembly occupancies to provide an adequate level of safety to life from fire.

**13.7.4.3\*** Exposed foamed plastic materials and unprotected materials containing foamed plastic used for decorative purposes or stage scenery shall have a heat release rate not exceeding 100 kW where tested in accordance with one of the following:

(1) ANSI/UL 1975, *Standard for Fire Tests for Foamed Plastics Used for Decorative Purposes*

(2) NFPA 289, *Standard Method of Fire Test for Individual Fuel Packages,* using the 20 kW ignition source

**13.7.4.4** The requirement of 13.7.4.3 shall not apply to individual foamed plastic items and items containing foamed plastic where the foamed plastic does not exceed 1 lb (0.45 kg) in weight.

**13.7.5 Special Provisions for Exposition Facilities.**

**13.7.5.1 General.** No display or exhibit shall be installed or operated to interfere in any way with access to any required exit or with the visibility of any required exit or required exit sign; nor shall any display block access to fire-fighting equipment.

**13.7.5.2 Materials Not on Display.** A storage room having an enclosure consisting of a smoke barrier having a minimum 1-hour fire resistance rating and protected by an automatic extinguishing system shall be provided for combustible materials not on display, including combustible packing crates used to ship exhibitors' supplies and products.

**13.7.5.3 Exhibits.**

**13.7.5.3.1** Exhibits shall comply with 13.7.5.3.2 through 13.7.5.3.11.

**13.7.5.3.2** The travel distance within the exhibit booth or exhibit enclosure to an exit access aisle shall not exceed 50 ft (15 m).

**13.7.5.3.3** The upper deck of multilevel exhibits exceeding 300 ft² (28 m²) shall have not less than two remote means of egress.

**13.7.5.3.4** Exhibit booth construction materials shall be limited to the following:

(1) Noncombustible or limited-combustible materials

(2) Wood exceeding ¼ in. (6.3 mm) nominal thickness

(3) Wood that is pressure-treated, fire-retardant wood meeting the requirements of NFPA 703, *Standard for Fire Retardant–Treated Wood and Fire-Retardant Coatings for Building Materials*

(4) Flame-retardant materials complying with one of the following:

(a) They shall meet the flame propagation performance criteria contained in NFPA 701, *Standard Methods of Fire Tests for Flame Propagation of Textiles and Films.*

(b) They shall exhibit a heat release rate not exceeding 100 kW when tested in accordance with NFPA 289, *Standard Method of Fire Test for Individual Fuel Packages,* using the 20 kW ignition source.

(5) Textile wall coverings, such as carpeting and similar products used as wall or ceiling finishes, complying with the provisions of 10.2.2 and 10.2.4

(6) Plastics limited to those that comply with 13.3.3 and Section 10.2

(7) Foamed plastics and materials containing foamed plastics having a heat release rate for any single fuel package that does not exceed 100 kW where tested in accordance with one of the following:

    (a) ANSI/UL 1975, *Standard for Fire Tests for Foamed Plastics Used for Decorative Purposes*

    (b) NFPA 289, using the 20 kW ignition source

(8) Cardboard, honeycombed paper, and other combustible materials having a heat release rate for any single fuel package that does not exceed 150 kW where tested in accordance with one of the following:

    (a) ANSI/UL 1975

    (b) NFPA 289, using the 20 kW ignition source

**13.7.5.3.5** Curtains, drapes, and decorations shall comply with 10.3.1.

**13.7.5.3.6** Acoustical and decorative material including, but not limited to, cotton, hay, paper, straw, moss, split bamboo, and wood chips shall be flame-retardant treated to the satisfaction of the authority having jurisdiction.

**13.7.5.3.6.1** Materials that cannot be treated for flame retardancy shall not be used.

**13.7.5.3.6.2** Foamed plastics, and materials containing foamed plastics and used as decorative objects such as, but not limited to, mannequins, murals, and signs shall have a heat release rate for any single fuel package that does not exceed 150 kW where tested in accordance with one of the following:

(1) ANSI/UL 1975, *Standard for Fire Tests for Foamed Plastics Used for Decorative Purposes*

(2) NFPA 289, *Standard Method of Fire Test for Individual Fuel Packages*, using the 20 kW ignition source

**13.7.5.3.6.3** Where the aggregate area of acoustical and decorative materials is less than 10 percent of the individual floor or wall area, such materials shall be permitted to be used subject to the approval of the authority having jurisdiction.

**13.7.5.3.7** The following shall be protected by automatic extinguishing systems:

(1) Single-level exhibit booths exceeding 300 ft$^2$ (28 m$^2$) and covered with a ceiling

(2) Each level of multilevel exhibit booths, including the uppermost level where the uppermost level is covered with a ceiling

**13.7.5.3.7.1** The requirements of 13.7.5.3.7 shall not apply where otherwise permitted by the following:

(1) Ceilings that are constructed of open grate design or listed dropout ceilings in accordance with NFPA 13, *Standard for the Installation of Sprinkler Systems*, shall not be considered ceilings within the context of 13.7.5.3.7.

(2) Vehicles, boats, and similar exhibited products having over 100 ft$^2$ (9.3 m$^2$) of roofed area shall be provided with smoke detectors acceptable to the authority having jurisdiction.

(3)*The requirement of 13.7.5.3.7(2) shall not apply where fire protection of multilevel exhibit booths is consistent with the criteria developed through a life safety evaluation of the exhibition hall in accordance with 13.4.1, subject to approval of the authority having jurisdiction.

**13.7.5.3.7.2** A single exhibit or group of exhibits with ceilings that do not require sprinklers shall be separated by a distance not less than 10 ft (3050 mm) where the aggregate ceiling exceeds 300 ft$^2$ (28 m$^2$).

**13.7.5.3.7.3** The water supply and piping for the sprinkler system shall be permitted to be of approved temporary means that is provided by a domestic water supply, a standpipe system, or a sprinkler system.

**13.7.5.3.8** Open flame devices within exhibit booths shall comply with 13.7.3.

**13.7.5.3.9** Cooking and food-warming devices in exhibit booths shall comply with 13.7.2 and all of the following:

(1) Gas-fired devices shall comply with all of the following:

    (a) Natural gas-fired devices shall comply with 9.1.1.

    (b) The requirement of 13.7.5.3.9(1)(a) shall not apply to compressed natural gas where permitted by the authority having jurisdiction.

    (c) The use of LP-Gas cylinders shall be prohibited.

    (d) Nonrefillable LP-Gas cylinders shall be approved for use where permitted by the authority having jurisdiction.

(2) The devices shall be isolated from the public by not less than 48 in. (1220 mm) or by a barrier between the devices and the public.

(3) Multi-well cooking equipment using combustible oils or solids shall comply with 9.2.3.

(4) Single-well cooking equipment using combustible oils or solids shall meet all of the following criteria:

    (a) The equipment shall have lids available for immediate use.

    (b) The equipment shall be limited to 2 ft$^2$ (0.2 m$^2$) of cooking surface.

    (c) The equipment shall be placed on noncombustible surface materials.

    (d) The equipment shall be separated from each other by a horizontal distance of not less than 24 in. (610 mm).

    (e) The requirement of 13.7.5.3.9(4)(d) shall not apply to multiple single-well cooking equipment where the aggregate cooking surface area does not exceed 2 ft$^2$ (0.2 m$^2$).

    (f) The equipment shall be kept at a horizontal distance of not less than 24 in. (610 mm) from any combustible material.

(5) A portable fire extinguisher in accordance with 9.7.4.1 shall be provided within the booth for each device, or an approved automatic extinguishing system shall be provided.

**13.7.5.3.10** Combustible materials within exhibit booths shall be limited to a one-day supply. Storage of combustible materials behind the booth shall be prohibited. *(See 13.7.4.2 and 13.7.5.2.)*

**13.7.5.3.11** Plans for the exposition, in an acceptable form, shall be submitted to the authority having jurisdiction for approval prior to setting up any exhibit.

**13.7.5.3.11.1** The plan shall show all details of the proposed exposition.

**13.7.5.3.11.2** No exposition shall occupy any exposition facility without approved plans.

**13.7.5.4 Vehicles.** Vehicles on display within an exposition facility shall comply with 13.7.5.4.1 through 13.7.5.4.5.

**13.7.5.4.1** All fuel tank openings shall be locked and sealed in an approved manner to prevent the escape of vapors; fuel tanks shall not contain in excess of one-half their capacity or contain in excess of 10 gal (38 L) of fuel, whichever is less.

**13.7.5.4.2** At least one battery cable shall be removed from the batteries used to start the vehicle engine, and the disconnected battery cable shall then be taped.

**13.7.5.4.3** Batteries used to power auxiliary equipment shall be permitted to be kept in service.

**13.7.5.4.4** Fueling or defueling of vehicles shall be prohibited.

**13.7.5.4.5** Vehicles shall not be moved during exhibit hours.

**13.7.5.5 Prohibited Materials.**

**13.7.5.5.1** The following items shall be prohibited within exhibit halls:

(1) Compressed flammable gases
(2) Flammable or combustible liquids
(3) Hazardous chemicals or materials
(4) Class II or greater lasers, blasting agents, and explosives

**13.7.5.5.2** The authority having jurisdiction shall be permitted to allow the limited use of any items specified in 13.7.5.5.1 under special circumstances.

**13.7.5.6 Alternatives.** See Section 1.4.

**13.7.6\* Crowd Managers.**

**13.7.6.1** Assembly occupancies shall be provided with a minimum of one trained crowd manager or crowd manager supervisor. Where the occupant load exceeds 250, additional trained crowd managers or crowd manager supervisors shall be provided at a ratio of one crowd manager or crowd manager supervisor for every 250 occupants, unless otherwise permitted by one of the following:

(1) This requirement shall not apply to assembly occupancies used exclusively for religious worship with an occupant load not exceeding 2000.
(2) The ratio of trained crowd managers to occupants shall be permitted to be reduced where, in the opinion of the authority having jurisdiction, the existence of an approved, supervised automatic sprinkler system and the nature of the event warrant.

**13.7.6.2** The crowd manager shall receive approved training in crowd management techniques.

**13.7.7\* Drills.**

**13.7.7.1** The employees or attendants of assembly occupancies shall be trained and drilled in the duties they are to perform in case of fire, panic, or other emergency to effect orderly exiting.

**13.7.7.2** Employees or attendants of assembly occupancies shall be instructed in the proper use of portable fire extinguishers and other manual fire suppression equipment where provided.

**13.7.7.3\*** In the following assembly occupancies, an audible announcement shall be made, or a projected image shall be shown, prior to the start of each program that notifies occupants of the location of the exits to be used in case of a fire or other emergency:

(1) Theaters
(2) Motion picture theaters
(3) Auditoriums

(4) Other similar assembly occupancies with occupant loads exceeding 300 where there are noncontinuous programs

**13.7.7.4** The requirement of 13.7.7.3 shall not apply to assembly occupancies in schools where used for nonpublic events.

**13.7.8 Smoking.**

**13.7.8.1** Smoking in assembly occupancies shall be regulated by the authority having jurisdiction.

**13.7.8.2** In rooms or areas where smoking is prohibited, plainly visible signs shall be posted that read as follows:

NO SMOKING

**13.7.8.3** No person shall smoke in prohibited areas that are so posted, unless permitted by the authority having jurisdiction under both of the following conditions:

(1) Smoking shall be permitted on a stage only where it is a necessary and rehearsed part of a performance.
(2) Smoking shall be permitted only where the smoker is a regular performing member of the cast.

**13.7.8.4** Where smoking is permitted, suitable ashtrays or receptacles shall be provided in convenient locations.

**13.7.9 Seating.**

**13.7.9.1 Secured Seating.**

**13.7.9.1.1** Seats in assembly occupancies accommodating more than 200 persons shall be securely fastened to the floor, except where fastened together in groups of not less than three and as permitted by 13.7.9.1.2 and 13.7.9.2.

**13.7.9.1.2** Balcony and box seating areas that are separated from other areas by rails, guards, partial-height walls, or other physical barriers and have a maximum of 14 seats shall be exempt from the requirement of 13.7.9.1.1.

**13.7.9.2 Unsecured Seating.**

**13.7.9.2.1** Seats not secured to the floor shall be permitted in restaurants, night clubs, and other occupancies where fastening seats to the floor might be impracticable.

**13.7.9.2.2** Unsecured seats shall be permitted, provided that, in the area used for seating, excluding such areas as dance floors and stages, there is not more than one seat for each 15 ft$^2$ (1.4 m$^2$) of net floor area, and adequate aisles to reach exits are maintained at all times.

**13.7.9.2.3** Seating diagrams shall be submitted for approval by the authority having jurisdiction to permit an increase in occupant load per 7.3.1.3.

**13.7.9.3 Occupant Load Posting.**

**13.7.9.3.1** Every room constituting an assembly occupancy and not having fixed seats shall have the occupant load of the room posted in a conspicuous place near the main exit from the room.

**13.7.9.3.2** Approved signs shall be maintained in a legible manner by the owner or authorized agent.

**13.7.9.3.3** Signs shall be durable and shall indicate the number of occupants permitted for each room use.

**13.7.10 Maintenance of Outdoor Grandstands.**

**13.7.10.1** The owner shall provide for not less than annual inspection and required maintenance of each outdoor grandstand to ensure safe conditions.

**13.7.10.2** At least biennially, the inspection shall be performed by a professional engineer, registered architect, or individual certified by the manufacturer.

**13.7.10.3** Where required by the authority having jurisdiction, the owner shall provide a copy of the inspection report and certification that the inspection required by 13.7.10.2 has been performed.

**13.7.11 Maintenance and Operation of Folding and Telescopic Seating.**

**13.7.11.1** Instructions in both maintenance and operation shall be transmitted to the owner by the manufacturer of the seating or his or her representative.

**13.7.11.2** Maintenance and operation of folding and telescopic seating shall be the responsibility of the owner or his or her duly authorized representative and shall include all of the following:

(1) During operation of the folding and telescopic seats, the opening and closing shall be supervised by responsible personnel who shall ensure that the operation is in accordance with the manufacturer's instructions.
(2) Only attachments specifically approved by the manufacturer for the specific installation shall be attached to the seating.
(3) An annual inspection and required maintenance of each grandstand shall be performed to ensure safe conditions.
(4) At least biennially, the inspection shall be performed by a professional engineer, registered architect, or individual certified by the manufacturer.

**13.7.12 Clothing.** Clothing and personal effects shall not be stored in corridors, and spaces not separated from corridors, unless otherwise permitted by one of the following:

(1) This requirement shall not apply to corridors, and spaces not separated from corridors, that are protected by an approved automatic sprinkler system in accordance with Section 9.7.
(2) This requirement shall not apply to corridors, and spaces not separated from corridors, that are protected by a smoke detection system in accordance with Section 9.6.
(3) This requirement shall not apply to storage in metal lockers, provided that the required egress width is maintained.

**13.7.13 Emergency Plans.**

**13.7.13.1** Emergency plans shall be provided in accordance with Section 4.8.

**13.7.13.2** Where assembly occupancies are located in the high-rise portion of a building, the emergency plan shall include egress procedures, methods, and preferred evacuation routes for each event considered to be a life safety hazard that could impact the building, including the appropriateness of the use of elevators.

## Chapter 14   New Educational Occupancies

**14.1 General Requirements.**

**14.1.1 Application.**

**14.1.1.1** The requirements of this chapter shall apply to new buildings or portions thereof used as educational occupancies. *(See 1.3.1.)*

**14.1.1.2 Administration.** The provisions of Chapter 1, Administration, shall apply.

**14.1.1.3 General.** The provisions of Chapter 4, General, shall apply.

**14.1.1.4** Educational facilities that do not meet the definition of an educational occupancy shall not be required to comply with this chapter but shall comply with the following requirements:

(1) Instructional building — business occupancy
(2) Classrooms under 50 persons — business occupancy
(3) Classrooms, 50 persons and over — assembly occupancy
(4) Laboratories, instructional — business occupancy
(5) Laboratories, noninstructional — industrial occupancy

**14.1.2 Classification of Occupancy.** See 6.1.3.

**14.1.2.1** Educational occupancies shall include all buildings used for educational purposes through the twelfth grade by six or more persons for 4 or more hours per day or more than 12 hours per week.

**14.1.2.2** Educational occupancies shall include part-day preschools, kindergartens, and other schools whose purpose is primarily educational, even though the children who attend such schools are of preschool age.

**14.1.2.3** In cases where instruction is incidental to some other occupancy, the section of this *Code* governing such other occupancy shall apply.

**14.1.2.4** Other occupancies associated with educational institutions shall be in accordance with the appropriate parts of this *Code. (See Chapters 18, 20, 26, 28, 30, 40, and 42 and 6.1.14.)*

**14.1.3 Multiple Occupancies.**

**14.1.3.1 General.** Multiple occupancies shall be in accordance with 6.1.14.

**14.1.3.2 Assembly and Educational.**

**14.1.3.2.1** Spaces subject to assembly occupancy shall comply with Chapter 12, including 12.1.3.2, which provides that, where auditorium and gymnasium egress lead through corridors or stairways also serving as egress for other parts of the building, the egress capacity shall be sufficient to allow simultaneous egress from auditorium and classroom sections.

**14.1.3.2.2** In the case of an assembly occupancy of a type suitable for use only by the school occupant load, and therefore not subject to simultaneous occupancy, the same egress capacity shall be permitted to serve both sections.

**14.1.3.3 Dormitory and Classrooms.**

**14.1.3.3.1** Any building used for both classroom and dormitory purposes shall comply with the applicable provisions of Chapter 28 in addition to complying with Chapter 14.

**14.1.3.3.2** Where classroom and dormitory sections are not subject to simultaneous occupancy, the same egress capacity shall be permitted to serve both sections.

**14.1.4 Definitions.**

**14.1.4.1 General.** For definitions, see Chapter 3, Definitions.

**14.1.4.2 Special Definitions.** A list of special terms used in this chapter follows:

(1) **Common Atmosphere.** See 3.3.26.1.

(2) **Flexible Plan and Open Plan Educational or Day-Care Building.** See 3.3.36.6.

(3) **Separate Atmosphere.** See 3.3.26.2.

**14.1.5 Classification of Hazard of Contents.** The contents of educational occupancies shall be classified in accordance with the provisions of Section 6.2.

**14.1.6 Minimum Construction Requirements.** (No requirements.)

**14.1.7 Occupant Load.**

**14.1.7.1** The occupant load, in number of persons for whom means of egress and other provisions are required, shall be determined on the basis of the occupant load factors of Table 7.3.1.2 that are characteristic of the use of the space or shall be determined as the maximum probable population of the space under consideration, whichever is greater.

**14.1.7.2** The occupant load of an educational occupancy, or a portion thereof, shall be permitted to be modified from that specified in 14.1.7.1 if the necessary aisles and exits are provided.

**14.1.7.3** An approved aisle or seating diagram shall be required by the authority having jurisdiction to substantiate the modification permitted in 14.1.7.2.

**14.2 Means of Egress Requirements.**

**14.2.1 General.**

**14.2.1.1** Means of egress shall be in accordance with Chapter 7 and Section 14.2.

**14.2.1.2** Rooms normally occupied by preschool, kindergarten, or first-grade students shall be located on a level of exit discharge, unless otherwise permitted by 14.2.1.4.

**14.2.1.3** Rooms normally occupied by second-grade students shall not be located more than one story above a level of exit discharge, unless otherwise permitted by 14.2.1.4.

**14.2.1.4** Rooms or areas located on floor levels other than as specified in 14.2.1.2 and 14.2.1.3 shall be permitted to be used where provided with independent means of egress dedicated for use by the preschool, kindergarten, first-grade, or second-grade students.

**14.2.2 Means of Egress Components.**

**14.2.2.1 Components Permitted.** Components of means of egress shall be limited to the types described in 14.2.2.2 through 14.2.2.10.

**14.2.2.2 Doors.**

**14.2.2.2.1** Doors complying with 7.2.1 shall be permitted.

**14.2.2.2.2** Any door in a required means of egress from an area having an occupant load of 100 or more persons shall be permitted to be provided with a latch or lock only if the latch or lock is panic hardware or fire exit hardware complying with 7.2.1.7.

**14.2.2.2.3 Special Locking.**

**14.2.2.2.3.1** Delayed-egress locking systems complying with 7.2.1.6.1 shall be permitted.

**14.2.2.2.3.2** Access-controlled egress door assemblies complying with 7.2.1.6.2 shall be permitted.

**14.2.2.2.3.3** Elevator lobby exit access door assemblies locking in accordance with 7.2.1.6.3 shall be permitted.

**14.2.2.3\* Stairs.** Stairs complying with 7.2.2 shall be permitted.

**14.2.2.4 Smokeproof Enclosures.** Smokeproof enclosures complying with 7.2.3 shall be permitted.

**14.2.2.5 Horizontal Exits.** Horizontal exits complying with 7.2.4 shall be permitted.

**14.2.2.6 Ramps.** Ramps complying with 7.2.5 shall be permitted.

**14.2.2.7 Exit Passageways.** Exit passageways complying with 7.2.6 shall be permitted.

**14.2.2.8 Fire Escape Ladders.** Fire escape ladders complying with 7.2.9 shall be permitted.

**14.2.2.9 Alternating Tread Devices.** Alternating tread devices complying with 7.2.11 shall be permitted.

**14.2.2.10 Areas of Refuge.** Areas of refuge complying with 7.2.12 shall be permitted.

**14.2.3 Capacity of Means of Egress.**

**14.2.3.1 General.** Capacity of means of egress shall be in accordance with Section 7.3.

**14.2.3.2 Minimum Corridor Width.** Exit access corridors shall have not less than 6 ft (1830 mm) of clear width.

**14.2.4 Number of Means of Egress.**

**14.2.4.1** The number of means of egress shall be in accordance with Section 7.4.

**14.2.4.2** Not less than two separate exits shall be in accordance with the following criteria:

(1) They shall be provided on every story.
(2) They shall be accessible from every part of every story and mezzanine; however, exit access travel shall be permitted to be common for the distance permitted as common path of travel by 14.2.5.3.

**14.2.5 Arrangement of Means of Egress.** See also Section 7.5.

**14.2.5.1** Means of egress shall be arranged in accordance with Section 7.5.

**14.2.5.2** No dead-end corridor shall exceed 20 ft (6100 mm), other than in buildings protected throughout by an approved, supervised automatic sprinkler system in accordance with Section 9.7, in which case dead-end corridors shall not exceed 50 ft (15 m).

**14.2.5.3** Limitations on common path of travel shall be in accordance with 14.2.5.3.1 and 14.2.5.3.2.

**14.2.5.3.1** Common path of travel shall not exceed 100 ft (30 m) in a building protected throughout by an approved, supervised automatic sprinkler system in accordance with Section 9.7.

**14.2.5.3.2** Common path of travel shall not exceed 75 ft (23 m) in a building not protected throughout by an approved, supervised automatic sprinkler system in accordance with Section 9.7.

**14.2.5.4** Every room or space larger than 1000 ft² (93 m²) or with an occupant load of more than 50 persons shall comply with the following:

(1) The room or space shall have a minimum of two exit access doors.
(2) The doors required by 14.2.5.4(1) shall provide access to separate exits.

(3) The doors required by 14.2.5.4(1) shall be permitted to open onto a common corridor, provided that such corridor leads to separate exits located in opposite directions.

**14.2.5.5** Every room that is normally subject to student occupancy shall have an exit access door leading directly to an exit access corridor or exit, unless otherwise permitted by one of the following:

(1) This requirement shall not apply where an exit door opens directly to the outside or to an exterior balcony or corridor as described in 14.2.5.9.
(2) One room shall be permitted to intervene between a normally occupied student room and an exit access corridor, provided that all of the following criteria are met:
    (a) The travel from a room served by an intervening room to the corridor door or exit shall not exceed 75 ft (23 m).
    (b) Clothing, personal effects, or other materials deemed hazardous by the authority having jurisdiction shall be stored in metal lockers, provided that they do not obstruct the exit access, or the intervening room shall be sprinklered in accordance with Section 9.7.
    (c) One of the following means of protection shall be provided:
        i. The intervening room shall have approved fire detection that activates the building alarm.
        ii. The building shall be protected by an approved, supervised automatic sprinkler system in accordance with Section 9.7.

**14.2.5.6** Doors that swing into an exit access corridor shall be arranged to prevent interference with corridor travel. *(See also 7.2.1.4.3.)*

**14.2.5.7** Aisles shall be not less than 30 in. (760 mm) wide.

**14.2.5.8** The space between parallel rows of seats shall not be subject to the minimum aisle width, provided that the number of seats that intervenes between any seat and an aisle does not exceed six.

**14.2.5.9\*** Exterior exit access shall comply with 7.5.3.

**14.2.6 Travel Distance to Exits.** Travel distance shall comply with 14.2.6.1 through 14.2.6.3.

**14.2.6.1** Travel distance shall be measured in accordance with Section 7.6.

**14.2.6.2** Travel distance to an exit shall not exceed 150 ft (46 m) from any point in a building, unless otherwise provided in 14.2.6.3. *(See also Section 7.6.)*

**14.2.6.3** Travel distance shall not exceed 200 ft (61 m) in educational occupancies protected throughout by an approved, supervised automatic sprinkler system in accordance with Section 9.7.

**14.2.7 Discharge from Exits.** Discharge from exits shall be arranged in accordance with Section 7.7.

**14.2.8 Illumination of Means of Egress.** Means of egress shall be illuminated in accordance with Section 7.8.

**14.2.9 Emergency Lighting.** Emergency lighting shall be provided in accordance with Section 7.9.

**14.2.10 Marking of Means of Egress.** Means of egress shall have signs in accordance with Section 7.10.

**14.2.11 Special Means of Egress Features.**

**14.2.11.1\* Windows for Rescue.**

**14.2.11.1.1** Every room or space greater than 250 ft$^2$ (23.2 m$^2$) and used for classroom or other educational purposes or normally subject to student occupancy shall have not less than one outside window for emergency rescue that complies with all of the following, unless otherwise permitted by 14.2.11.1.2:

(1) Such windows shall be openable from the inside without the use of tools and shall provide a clear opening of not less than 20 in. (510 mm) in width, 24 in. (610 mm) in height, and 5.7 ft$^2$ (0.5 m$^2$) in area.
(2) The bottom of the opening shall be not more than 44 in. (1120 mm) above the floor, and any latching device shall be capable of being operated from not more than 54 in. (1370 mm) above the finished floor.
(3) The clear opening shall allow a rectangular solid, with a width and height that provides not less than the required 5.7 ft$^2$ (0.5 m$^2$) opening and a depth of not less than 20 in. (510 mm), to pass fully through the opening.
(4) Such windows shall be accessible by the fire department and shall open into an area having access to a public way.

**14.2.11.1.2** The requirements of 14.2.11.1.1 shall not apply to any of the following:

(1) Buildings protected throughout by an approved, supervised automatic sprinkler system in accordance with Section 9.7
(2) Where the room or space has a door leading directly to an exit or directly to the outside of the building
(3) Reserved
(4) Rooms located four or more stories above the finished ground level

**14.2.11.2 Lockups.** Lockups in educational occupancies shall comply with the requirements of 22.4.5.

**14.3 Protection.**

**14.3.1 Protection of Vertical Openings.**

**14.3.1.1** Any vertical opening, other than unprotected vertical openings in accordance with 8.6.9.1, shall be enclosed or protected in accordance with Section 8.6.

**14.3.1.2** Where the provisions of 8.6.6 are used, the requirements of 14.3.5.4 shall be met.

**14.3.2 Protection from Hazards.**

**14.3.2.1** Rooms or spaces for the storage, processing, or use of materials shall be protected in accordance with the following:

(1) Such rooms or spaces shall be separated from the remainder of the building by fire barriers having a minimum 1-hour fire resistance rating or protected by automatic extinguishing systems as specified in Section 8.7 in the following areas:
    (a) Boiler and furnace rooms, unless such rooms enclose only air-handling equipment
    (b) Rooms or spaces used for the storage of combustible supplies in quantities deemed hazardous by the authority having jurisdiction
    (c) Rooms or spaces used for the storage of hazardous materials or flammable or combustible liquids in quantities deemed hazardous by recognized standards
    (d) Janitor closets *[see also 14.3.2.1(4)]*

(2) Such rooms or spaces shall be separated from the remainder of the building by fire barriers having a minimum 1-hour fire resistance rating and protected by automatic extinguishing systems as specified in Section 8.7 in the following areas:

(a) Laundries

(b) Maintenance shops, including woodworking and painting areas

(c) Rooms or spaces used for processing or use of combustible supplies deemed hazardous by the authority having jurisdiction

(d) Rooms or spaces used for processing or use of hazardous materials or flammable or combustible liquids in quantities deemed hazardous by recognized standards

(3) Where automatic extinguishing is used to meet the requirements of 14.3.2.1(1) or (2), the protection shall be permitted in accordance with 9.7.1.2.

(4) Where janitor closets addressed in 14.3.2.1(1)(d) are protected in accordance with the sprinkler option of 14.3.2.1(1), the janitor closet doors shall be permitted to have ventilating louvers.

**14.3.2.2** Cooking facilities shall be protected in accordance with 9.2.3. Openings shall not be required to be protected between food preparation areas and dining areas.

**14.3.2.3** Stages and platforms shall be protected in accordance with Chapter 12.

**14.3.2.4 Alcohol-Based Hand-Rub Dispensers.** Alcohol-based hand-rub dispensers shall be protected in accordance with 8.7.3.1, unless all of the following conditions are met:

(1) Dispensers shall be installed in rooms or spaces separated from corridors and exits.

(2) The maximum individual dispenser fluid capacity shall be as follows:

(a) 0.32 gal (1.2 L) for dispensers in rooms

(b) 0.53 gal (2.0 L) for dispensers in suites of rooms

(3) The dispensers shall be separated from each other by horizontal spacing of not less than 48 in. (1220 mm).

(4) Storage of quantities greater than 5 gal (18.9 L) in a single fire compartment shall meet the requirements of NFPA 30, *Flammable and Combustible Liquids Code.*

(5) The dispensers shall not be installed over or directly adjacent to an ignition source.

(6) Dispensers installed directly over carpeted floors shall be permitted only in sprinklered rooms or spaces.

**14.3.2.5** Educational occupancy laboratories using chemicals shall be in accordance with 8.7.4.

**14.3.3 Interior Finish.**

**14.3.3.1 General.** Interior finish shall be in accordance with Section 10.2.

**14.3.3.2\* Interior Wall and Ceiling Finish.** Interior wall and ceiling finish materials complying with Section 10.2 shall be permitted as follows:

(1) Exits — Class A

(2) Other than exits — Class A or Class B

(3) Low-height partitions not exceeding 60 in. (1525 mm) and used in locations other than exits — Class A, Class B, or Class C

**14.3.3.3 Interior Floor Finish.**

**14.3.3.3.1** Interior floor finish shall comply with Section 10.2.

**14.3.3.3.2** Interior floor finish in exit enclosures and exit access corridors and spaces not separated from them by walls complying with 14.3.6 shall be not less than Class II.

**14.3.3.3.3** Interior floor finish shall comply with 10.2.7.1 or 10.2.7.2, as applicable.

**14.3.4 Detection, Alarm, and Communications Systems.**

**14.3.4.1 General.**

**14.3.4.1.1** Educational occupancies shall be provided with a fire alarm system in accordance with Section 9.6.

**14.3.4.1.2** The requirement of 14.3.4.1.1 shall not apply to buildings meeting all of the following criteria:

(1) Buildings having an area not exceeding 1000 ft$^2$ (93 m$^2$)

(2) Buildings containing a single classroom

(3) Buildings located not less than 30 ft (9.1 m) from another building

**14.3.4.2 Initiation.**

**14.3.4.2.1 General.** Initiation of the required fire alarm system, other than as permitted by 14.3.4.2.3, shall be by manual means in accordance with 9.6.2.1(1).

**14.3.4.2.2 Automatic Initiation.** In buildings provided with automatic sprinkler protection, the operation of the sprinkler system shall automatically activate the fire alarm system in addition to the initiation means required in 14.3.4.2.1.

**14.3.4.2.3 Alternative Protection System.** Manual fire alarm boxes shall be permitted to be eliminated in accordance with 14.3.4.2.3.1 or 14.3.4.2.3.2.

**14.3.4.2.3.1\*** Manual fire alarm boxes shall be permitted to be eliminated where all of the following conditions apply:

(1) Interior corridors are protected by smoke detectors using an alarm verification system as described in *NFPA 72, National Fire Alarm and Signaling Code.*

(2) Auditoriums, cafeterias, and gymnasiums are protected by heat-detection devices or other approved detection devices.

(3) Shops and laboratories involving dusts or vapors are protected by heat-detection devices or other approved detection devices.

(4) Provision is made at a central point to manually activate the evacuation signal or to evacuate only affected areas.

**14.3.4.2.3.2\*** Manual fire alarm boxes shall be permitted to be eliminated where both of the following conditions apply:

(1) The building is protected throughout by an approved, supervised automatic sprinkler system in accordance with Section 9.7.

(2) Provision is made at a central point to manually activate the evacuation signal or to evacuate only affected areas.

**14.3.4.3 Notification.**

**14.3.4.3.1 Occupant Notification.**

**14.3.4.3.1.1\*** Occupant notification shall be accomplished automatically in accordance with 9.6.3.

**14.3.4.3.1.2** Positive alarm sequence shall be permitted in accordance with 9.6.3.4.

**14.3.4.3.1.3** Where installed and operated per *NFPA 72, National Fire Alarm and Signaling Code,* the fire alarm system shall be permitted to be used for other emergency signaling or for class changes.

**14.3.4.3.1.4**   To prevent students from being returned to a building that is burning, the recall signal shall be separate and distinct from any other signals, and such signal shall be permitted to be given by use of distinctively colored flags or banners.

**14.3.4.3.1.5**   If the recall signal required by 14.3.4.3.1.4 is electric, the push buttons or other controls shall be kept under lock, the key for which shall be in the possession of the principal or another designated person in order to prevent a recall at a time when there is an actual fire.

**14.3.4.3.1.6**   Regardless of the method of recall signal, the means of giving the recall signal shall be kept under lock.

**14.3.4.3.2 Emergency Forces Notification.** Emergency forces notification shall be accomplished in accordance with 9.6.4.

**14.3.5 Extinguishment Requirements.**

**14.3.5.1\*** Educational occupancy buildings exceeding 12,000 ft$^2$ (1120 m$^2$) shall be protected throughout by an approved, supervised automatic sprinkler system in accordance with Section 9.7.

**14.3.5.2**   Educational occupancy buildings four or more stories in height shall be protected throughout by an approved, supervised automatic sprinkler system in accordance with Section 9.7.

**14.3.5.3**   Every portion of educational buildings below the level of exit discharge shall be protected throughout by an approved, supervised automatic sprinkler system in accordance with Section 9.7.

**14.3.5.4**   Buildings with unprotected openings in accordance with 8.6.6 shall be protected throughout by an approved, supervised automatic sprinkler system in accordance with Section 9.7.

**14.3.5.5**   Where another provision of this chapter requires an automatic sprinkler system, the sprinkler system shall be installed in accordance with 9.7.1.1(1).

**14.3.6 Corridors.** Corridors shall be separated from other parts of the story by walls having a 1-hour fire resistance rating in accordance with Section 8.3, unless otherwise permitted by one of the following:

(1)  Corridor protection shall not be required where all spaces normally subject to student occupancy have not less than one door opening directly to the outside or to an exterior exit access balcony or corridor in accordance with 7.5.3.
(2)  The following shall apply to buildings protected throughout by an approved, supervised automatic sprinkler system in accordance with Section 9.7:
    (a)  Corridor walls shall not be required to be rated, provided that such walls form smoke partitions in accordance with Section 8.4.
    (b)  The provisions of 8.4.3.5 shall not apply to normally occupied classrooms.
(3)  Where the corridor ceiling is an assembly having a 1-hour fire resistance rating where tested as a wall, the corridor walls shall be permitted to terminate at the corridor ceiling.
(4)  Lavatories shall not be required to be separated from corridors, provided that they are separated from all other spaces by walls having not less than a 1-hour fire resistance rating in accordance with Section 8.3.
(5)  Lavatories shall not be required to be separated from corridors, provided that both of the following criteria are met:

(a)  The building is protected throughout by an approved, supervised automatic sprinkler system in accordance with Section 9.7.
(b)  The walls separating the lavatory from other rooms form smoke partitions in accordance with Section 8.4.

**14.3.7 Subdivision of Building Spaces.**

**14.3.7.1**   Educational occupancies shall be subdivided into compartments by smoke partitions having not less than a 1-hour fire resistance rating and complying with Section 8.4 where one or both of the following conditions exist:

(1)  The maximum floor area, including the aggregate area of all floors having a common atmosphere, exceeds 30,000 ft$^2$ (2800 m$^2$).
(2)  The length or width of the building exceeds 300 ft (91 m).

**14.3.7.2**   The requirement of 14.3.7.1 shall not apply to either of the following:

(1)  Where all spaces normally subject to student occupancy have not less than one door opening directly to the outside or to an exterior or exit access balcony or corridor in accordance with 7.5.3
(2)  Buildings protected throughout by an approved, supervised automatic sprinkler system in accordance with Section 9.7

**14.3.7.3**   The area of any smoke compartment required by 14.3.7.1 shall not exceed 30,000 ft$^2$ (2800 m$^2$), with no dimension exceeding 300 ft (91 m).

**14.4 Special Provisions.**

**14.4.1 Limited Access Buildings and Underground Buildings.** Limited access buildings and underground buildings shall comply with Section 11.7.

**14.4.2 High-Rise Buildings.** High-rise buildings shall comply with Section 11.8.

**14.4.3 Flexible Plan and Open Plan Buildings.**

**14.4.3.1**   Flexible plan and open plan buildings shall comply with the requirements of this chapter as modified by 14.4.3.2 through 14.4.3.5.

**14.4.3.2**   Each room occupied by more than 300 persons shall have two or more means of egress entering into separate atmospheres.

**14.4.3.3**   Where three or more means of egress are required, the number of means of egress permitted to enter into the same atmosphere shall not exceed two.

**14.4.3.4**   Flexible plan buildings shall be permitted to have walls and partitions rearranged periodically only if revised plans or diagrams have been approved by the authority having jurisdiction.

**14.4.3.5**   Flexible plan buildings shall be evaluated while all folding walls are extended and in use as well as when they are in the retracted position.

**14.5 Building Services.**

**14.5.1 Utilities.** Utilities shall comply with the provisions of Section 9.1.

**14.5.2 Heating, Ventilating, and Air-Conditioning Equipment.**

**14.5.2.1**   Heating, ventilating, and air-conditioning equipment shall comply with the provisions of Section 9.2.

**14.5.2.2** Unvented fuel-fired heating equipment, other than gas space heaters in compliance with NFPA 54/ANSI Z223.1, *National Fuel Gas Code*, shall be prohibited.

**14.5.3 Elevators, Escalators, and Conveyors.** Elevators, escalators, and conveyors shall comply with the provisions of Section 9.4.

**14.5.4 Rubbish Chutes, Incinerators, and Laundry Chutes.** Rubbish chutes, incinerators, and laundry chutes shall comply with the provisions of Section 9.5.

**14.6 Reserved.**

**14.7 Operating Features.**

**14.7.1 Emergency Plan.** Emergency plans shall be provided in accordance with Section 4.8.

**14.7.2 Emergency Egress Drills.**

**14.7.2.1\*** Emergency egress drills shall be conducted in accordance with Section 4.7 and the applicable provisions of 14.7.2.3 as otherwise provided in 14.7.2.2.

**14.7.2.2** Approved training programs designed for education and training and for the practice of emergency egress to familiarize occupants with the drill procedure, and to establish conduct of the emergency egress as a matter of routine, shall be permitted to receive credit on a one-for-one basis for not more than four of the emergency egress drills required by 14.7.2.3, provided that a minimum of four emergency egress drills are completed prior to the conduct of the first such training and practice program.

**14.7.2.3** Emergency egress drills shall be conducted as follows:

(1) Not less than one emergency egress drill shall be conducted every month the facility is in session, unless both of the following criteria are met:
  (a) In climates where the weather is severe, the monthly emergency egress drills shall be permitted to be deferred.
  (b) The required number of emergency egress drills shall be conducted, and not less than four shall be conducted before the drills are deferred.
(2) All occupants of the building shall participate in the drill.
(3) One additional emergency egress drill, other than for educational occupancies that are open on a year-round basis, shall be required within the first 30 days of operation.

**14.7.2.4** All emergency drill alarms shall be sounded on the fire alarm system.

**14.7.3 Inspection.**

**14.7.3.1\*** It shall be the duty of principals, teachers, or staff to inspect all exit facilities daily to ensure that all stairways, doors, and other exits are in proper condition.

**14.7.3.2** Open plan buildings shall require extra surveillance to ensure that exit paths are maintained clear of obstruction and are obvious.

**14.7.3.3 Inspection of Door Openings.** Door openings shall be inspected in accordance with 7.2.1.15.

**14.7.4 Furnishings and Decorations.**

**14.7.4.1** Draperies, curtains, and other similar furnishings and decorations in educational occupancies shall be in accordance with the provisions of 10.3.1.

**14.7.4.2** Clothing and personal effects shall not be stored in corridors, unless otherwise permitted by one of the following:

(1) This requirement shall not apply to corridors protected by an automatic sprinkler system in accordance with Section 9.7.
(2) This requirement shall not apply to corridor areas protected by a smoke detection system in accordance with Section 9.6.
(3) This requirement shall not apply to storage in metal lockers, provided that the required egress width is maintained.

**14.7.4.3** Artwork and teaching materials shall be permitted to be attached directly to the walls in accordance with the following:

(1) The artwork and teaching materials shall not exceed 20 percent of the wall area in a building that is not protected throughout by an approved, supervised automatic sprinkler system in accordance with Section 9.7.
(2) The artwork and teaching materials shall not exceed 50 percent of the wall area in a building that is protected throughout by an approved, supervised automatic sprinkler system in accordance with Section 9.7.

**14.7.5 Open Flames.** Approved open flames shall be permitted in laboratories and vocational/technical areas.

# Chapter 15 Existing Educational Occupancies

**15.1 General Requirements.**

**15.1.1 Application.**

**15.1.1.1** The requirements of this chapter shall apply to existing buildings or portions thereof currently occupied as educational occupancies.

**15.1.1.2 Administration.** The provisions of Chapter 1, Administration, shall apply.

**15.1.1.3 General.** The provisions of Chapter 4, General, shall apply.

**15.1.1.4** Educational facilities that do not meet the definition of an educational occupancy shall not be required to comply with this chapter but shall comply with the following requirements:

(1) Instructional building — business occupancy
(2) Classrooms under 50 persons — business occupancy
(3) Classrooms, 50 persons and over — assembly occupancy
(4) Laboratories, instructional — business occupancy
(5) Laboratories, noninstructional — industrial occupancy

**15.1.2 Classification of Occupancy.** See 6.1.3.

**15.1.2.1** Educational occupancies shall include all buildings used for educational purposes through the twelfth grade by six or more persons for 4 or more hours per day or more than 12 hours per week.

**15.1.2.2** Educational occupancies shall include part-day preschools, kindergartens, and other schools whose purpose is primarily educational, even though the children who attend such schools are of preschool age.

**15.1.2.3** In cases where instruction is incidental to some other occupancy, the section of this *Code* governing such other occupancy shall apply.

**15.1.2.4** Other occupancies associated with educational institutions shall be in accordance with the appropriate parts of this *Code.* (*See Chapters 19, 21, 26, 29, 31, 40, and 42 and 6.1.14.*)

### 15.1.3 Multiple Occupancies.

**15.1.3.1 General.** Multiple occupancies shall be in accordance with 6.1.14.

**15.1.3.2 Assembly and Educational.**

**15.1.3.2.1** Spaces subject to assembly occupancy shall comply with Chapter 13, including 13.1.3.2, which provides that, where auditorium and gymnasium egress lead through corridors or stairways also serving as egress for other parts of the building, the egress capacity shall be sufficient to allow simultaneous egress from auditorium and classroom sections.

**15.1.3.2.2** In the case of an assembly occupancy of a type suitable for use only by the school occupant load, and therefore not subject to simultaneous occupancy, the same egress capacity shall be permitted to serve both sections.

**15.1.3.3 Dormitory and Classrooms.**

**15.1.3.3.1** Any building used for both classroom and dormitory purposes shall comply with the applicable provisions of Chapter 29 in addition to complying with Chapter 15.

**15.1.3.3.2** Where classroom and dormitory sections are not subject to simultaneous occupancy, the same egress capacity shall be permitted to serve both sections.

### 15.1.4 Definitions.

**15.1.4.1 General.** For definitions, see Chapter 3, Definitions.

**15.1.4.2 Special Definitions.** A list of special terms used in this chapter follows:

(1) **Common Atmosphere.** See 3.3.26.1.
(2) **Flexible Plan and Open Plan Educational or Day-Care Building.** See 3.3.36.6.
(3) **Separate Atmosphere.** See 3.3.26.2.

**15.1.5 Classification of Hazard of Contents.** The contents of educational occupancies shall be classified in accordance with the provisions of Section 6.2.

**15.1.6 Minimum Construction Requirements.** (No requirements.)

### 15.1.7 Occupant Load.

**15.1.7.1** The occupant load, in number of persons for whom means of egress and other provisions are required, shall be determined on the basis of the occupant load factors of Table 7.3.1.2 that are characteristic of the use of the space or shall be determined as the maximum probable population of the space under consideration, whichever is greater.

**15.1.7.2** The occupant load of an educational occupancy, or a portion thereof, shall be permitted to be modified from that specified in 15.1.7.1 if the necessary aisles and exits are provided.

**15.1.7.3** An approved aisle or seating diagram shall be required by the authority having jurisdiction to substantiate the modification permitted in 15.1.7.2.

### 15.2 Means of Egress Requirements.

### 15.2.1 General.

**15.2.1.1** Means of egress shall be in accordance with Chapter 7 and Section 15.2.

**15.2.1.2** Rooms normally occupied by preschool, kindergarten, or first-grade students shall be located on a level of exit discharge, unless otherwise permitted by 15.2.1.4.

**15.2.1.3** Rooms normally occupied by second-grade students shall not be located more than one story above a level of exit discharge, unless otherwise permitted by 15.2.1.4.

**15.2.1.4** Rooms or areas located on floor levels other than as specified in 15.2.1.2 and 15.2.1.3 shall be permitted to be used where provided with independent means of egress dedicated for use by the preschool, kindergarten, first-grade, or second-grade students.

### 15.2.2 Means of Egress Components.

**15.2.2.1 Components Permitted.** Components of means of egress shall be limited to the types described in 15.2.2.2 through 15.2.2.10.

**15.2.2.2 Doors.**

**15.2.2.2.1** Doors complying with 7.2.1 shall be permitted.

**15.2.2.2.2** Any required exit door subject to use by 100 or more persons shall be permitted to be provided with a latch or lock only if the latch or lock is panic hardware or fire exit hardware complying with 7.2.1.7.

**15.2.2.2.3 Special Locking.**

**15.2.2.2.3.1** Delayed-egress locking systems complying with 7.2.1.6.1 shall be permitted.

**15.2.2.2.3.2** Access-controlled egress door assemblies complying with 7.2.1.6.2 shall be permitted.

**15.2.2.2.3.3** Elevator lobby exit access door assemblies locking in accordance with 7.2.1.6.3 shall be permitted.

**15.2.2.3\* Stairs.** Stairs complying with 7.2.2 shall be permitted.

**15.2.2.4 Smokeproof Enclosures.** Smokeproof enclosures complying with 7.2.3 shall be permitted.

**15.2.2.5 Horizontal Exits.** Horizontal exits complying with 7.2.4 shall be permitted.

**15.2.2.6 Ramps.** Ramps complying with 7.2.5 shall be permitted.

**15.2.2.7 Exit Passageways.** Exit passageways complying with 7.2.6 shall be permitted.

**15.2.2.8 Fire Escape Ladders.** Fire escape ladders complying with 7.2.9 shall be permitted.

**15.2.2.9 Alternating Tread Devices.** Alternating tread devices complying with 7.2.11 shall be permitted.

**15.2.2.10 Areas of Refuge.** Areas of refuge complying with 7.2.12 shall be permitted.

### 15.2.3 Capacity of Means of Egress.

**15.2.3.1 General.** Capacity of means of egress shall be in accordance with Section 7.3.

**15.2.3.2 Minimum Corridor Width.** Exit access corridors shall have not less than 6 ft (1830 mm) of clear width.

### 15.2.4 Number of Mean of Egress.

**15.2.4.1** The number of means of egress shall be in accordance with 7.4.1.1 and 7.4.1.3. through 7.4.1.6.

**15.2.4.2** Not less than two separate exits shall be in accordance with the following criteria:

(1) They shall be provided on every story.
(2) They shall be accessible from every part of every story and mezzanine; however, exit access travel shall be permitted to be common for the distance permitted as common path of travel by 15.2.5.3.

**15.2.5 Arrangement of Means of Egress.**

**15.2.5.1** Means of egress shall be arranged in accordance with Section 7.5.

**15.2.5.2** No dead-end corridor shall exceed 20 ft (6100 mm), other than in buildings protected throughout by an approved, supervised automatic sprinkler system in accordance with Section 9.7, in which case dead-end corridors shall not exceed 50 ft (15 m).

**15.2.5.3** Limitations on common path of travel shall be in accordance with 15.2.5.3.1 and 15.2.5.3.2.

**15.2.5.3.1** Common path of travel shall not exceed 100 ft (30 m) in a building protected throughout by an approved, supervised automatic sprinkler system in accordance with Section 9.7.

**15.2.5.3.2** Common path of travel shall not exceed 75 ft (23 m) in a building not protected throughout by an approved, supervised automatic sprinkler system in accordance with Section 9.7.

**15.2.5.4** Every room or space larger than 1000 ft$^2$ (93 m$^2$) or with an occupant load of more than 50 persons shall comply with the following:

(1) The room or space shall have a minimum of two exit access doors.
(2) The doors required by 15.2.5.4(1) shall provide access to separate exits.
(3) The doors required by 15.2.5.4(1) shall be permitted to open onto a common corridor, provided that such corridor leads to separate exits located in opposite directions.

**15.2.5.5** Every room that is normally subject to student occupancy shall have an exit access door leading directly to an exit access corridor or exit, unless otherwise permitted by one of the following:

(1) This requirement shall not apply where an exit door opens directly to the outside or to an exterior balcony or corridor as described in 15.2.5.9.
(2) One room shall be permitted to intervene between a normally occupied student room and an exit access corridor, provided that all of the following criteria are met:
   (a) The travel from a room served by an intervening room to the corridor door or exit shall not exceed 75 ft (23 m).
   (b) Clothing, personal effects, or other materials deemed hazardous by the authority having jurisdiction shall be stored in metal lockers, provided that they do not obstruct the exit access, or the intervening room shall be sprinklered in accordance with Section 9.7.
   (c) One of the following means of protection shall be provided:
      i. The intervening room shall have approved fire detection that activates the building alarm.
      ii. The building shall be protected by an approved automatic sprinkler system in accordance with Section 9.7.

(3) Approved existing arrangements shall be permitted to continue in use.

**15.2.5.6** Doors that swing into an exit access corridor shall be arranged to prevent interference with corridor travel. *(See also 7.2.1.4.3.)*

**15.2.5.7** Aisles shall be not less than 30 in. (760 mm) wide.

**15.2.5.8** The space between parallel rows of seats shall not be subject to the minimum aisle width, provided that the number of seats that intervenes between any seat and an aisle does not exceed six.

**15.2.5.9*** Exterior exit access shall comply with 7.5.3.

**15.2.6 Travel Distance to Exits.** Travel distance shall comply with 15.2.6.1 through 15.2.6.4.

**15.2.6.1** Travel distance shall be measured in accordance with Section 7.6.

**15.2.6.2** Travel distance to an exit shall not exceed 150 ft (46 m) from any point in a building, unless otherwise permitted by 15.2.6.3 or 15.2.6.4. *(See also Section 7.6.)*

**15.2.6.3** Travel distance shall not exceed 200 ft (61 m) in educational occupancies protected throughout by an approved automatic sprinkler system in accordance with Section 9.7.

**15.2.6.4** Approved existing travel distances shall be permitted to continue in use.

**15.2.7 Discharge from Exits.** Discharge from exits shall be arranged in accordance with Section 7.7.

**15.2.8 Illumination of Means of Egress.** Means of egress shall be illuminated in accordance with Section 7.8.

**15.2.9 Emergency Lighting.**

**15.2.9.1** Emergency lighting shall be provided in accordance with Section 7.9, unless otherwise permitted by 15.2.9.2.

**15.2.9.2** Approved existing emergency lighting installations shall be permitted to be continued in use.

**15.2.10 Marking of Means of Egress.** Means of egress shall have signs in accordance with Section 7.10.

**15.2.11 Special Means of Egress Features.**

**15.2.11.1* Windows for Rescue.**

**15.2.11.1.1** Every room or space greater than 250 ft$^2$ (23.2 m$^2$) and used for classroom or other educational purposes or normally subject to student occupancy shall have not less than one outside window for emergency rescue that complies with all of the following, unless otherwise permitted by 15.2.11.1.2:

(1) Such windows shall be openable from the inside without the use of tools and shall provide a clear opening of not less than 20 in. (510 mm) in width, 24 in. (610 mm) in height, and 5.7 ft$^2$ (0.5 m$^2$) in area.
(2) The bottom of the opening shall be not more than 44 in. (1120 mm) above the floor, and any latching device shall be capable of being operated from not more than 54 in. (1370 mm) above the finished floor.
(3) The clear opening shall allow a rectangular solid, with a width and height that provides not less than the required 5.7 ft$^2$ (0.5 m$^2$) opening and a depth of not less than 20 in. (510 mm), to pass fully through the opening.

**15.2.11.1.2** The requirements of 15.2.11.1.1 shall not apply to any of the following:

(1) Buildings protected throughout by an approved automatic sprinkler system in accordance with Section 9.7

(2) Where the room or space has a door leading directly to an exit or directly to the outside of the building

(3) Where the room has a door, in addition to the door that leads to the exit access corridor as required by 15.2.5.5, and such door leads directly to another corridor located in a compartment separated from the compartment housing the corridor addressed in 15.2.5.5 by smoke partitions in accordance with Section 8.4

(4) Rooms located four or more stories above the finished ground level

(5) Where awning-type or hopper-type windows that are hinged or subdivided to provide a clear opening of not less than 4 ft$^2$ (0.38 m$^2$) or any dimension of not less than 22 in. (560 mm) meet the following criteria:

    (a) Such windows shall be permitted to continue in use.

    (b) Screen walls or devices located in front of required windows shall not interfere with rescue requirements.

(6) Where the room or space complies with all of the following:

    (a) One door providing direct access to an adjacent classroom and a second door providing direct access to another adjacent classroom shall be provided.

    (b) The two classrooms to which exit access travel is made in accordance with 15.2.11.1.2(6)(a) shall each provide exit access in accordance with 15.2.11.1.2(2) or 15.2.11.1.2(3).

    (c) The corridor required by 15.2.5.5, and the corridor addressed by 15.2.11.1.2(3), if provided, shall be separated from the classrooms by a wall that resists the passage of smoke, and all doors between the classrooms and the corridor shall be self-closing or automatic-closing in accordance with 7.2.1.8.

    (d) The length of travel to exits along such paths shall not exceed 150 ft (46 m).

    (e) Each communicating door shall be marked in accordance with Section 7.10.

    (f) No locking device shall be permitted on the communicating doors.

**15.2.11.2 Lockups.** Lockups in educational occupancies, other than approved existing lockups, shall comply with the requirements of 23.4.5.

**15.3 Protection.**

**15.3.1 Protection of Vertical Openings.**

**15.3.1.1** Any vertical opening, other than unprotected vertical openings in accordance with 8.6.9.1, shall be enclosed or protected in accordance with Section 8.6.

**15.3.1.2** Where the provisions of 8.6.6 are used, the requirements of 15.3.5.4 shall be met.

**15.3.1.3** Stairway enclosures shall not be required where all of the following conditions are met:

(1) The stairway serves only one adjacent floor, other than a basement.

(2) The stairway is not connected with stairways serving other floors.

(3) The stairway is not connected with corridors serving other than the two floors involved.

**15.3.2 Protection from Hazards.**

**15.3.2.1** Rooms or spaces for the storage, processing, or use of materials shall be protected in accordance with the following:

(1) Such rooms or spaces shall be separated from the remainder of the building by fire barriers having a minimum 1-hour fire resistance rating or protected by automatic extinguishing systems as specified in Section 8.7 in the following areas:

    (a) Boiler and furnace rooms, unless such rooms enclose only air-handling equipment

    (b) Rooms or spaces used for the storage of combustible supplies in quantities deemed hazardous by the authority having jurisdiction

    (c) Rooms or spaces used for the storage of hazardous materials or flammable or combustible liquids in quantities deemed hazardous by recognized standards

    (d) Janitor closets *[see also 15.3.2.1(4)]*

(2) Such rooms or spaces shall be separated from the remainder of the building by fire barriers having a minimum 1-hour fire resistance rating and protected by automatic extinguishing systems as specified in Section 8.7 in the following areas:

    (a) Laundries

    (b) Maintenance shops, including woodworking and painting areas

    (c) Rooms or spaces used for processing or use of combustible supplies deemed hazardous by the authority having jurisdiction

    (d) Rooms or spaces used for processing or use of hazardous materials or flammable or combustible liquids in quantities deemed hazardous by recognized standards

(3) Where automatic extinguishing is used to meet the requirements of 15.3.2.1(1) or (2), the protection shall be permitted in accordance with 9.7.1.2.

(4) Where janitor closets addressed in 15.3.2.1(1)(d) are protected in accordance with the sprinkler option of 15.3.2.1(1), the janitor closet doors shall be permitted to have ventilating louvers.

**15.3.2.2** Cooking facilities shall be protected in accordance with 9.2.3. Openings shall not be required to be protected between food preparation areas and dining areas.

**15.3.2.3** Stages and platforms shall be protected in accordance with Chapter 13.

**15.3.2.4 Alcohol-Based Hand-Rub Dispensers.** Alcohol-based hand-rub dispensers shall be protected in accordance with 8.7.3.1, unless all of the following conditions are met:

(1) Dispensers shall be installed in rooms or spaces separated from corridors and exits.

(2) The maximum individual dispenser fluid capacity shall be as follows:

    (a) 0.32 gal (1.2 L) for dispensers in rooms

    (b) 0.53 gal (2.0 L) for dispensers in suites of rooms

(3) The dispensers shall be separated from each other by horizontal spacing of not less than 48 in. (1220 mm).

(4) Storage of quantities greater than 5 gal (18.9 L) in a single fire compartment shall meet the requirements of NFPA 30, *Flammable and Combustible Liquids Code.*

(5) The dispensers shall not be installed over or directly adjacent to an ignition source.

(6) Dispensers installed directly over carpeted floors shall be permitted only in sprinklered rooms or spaces.

**15.3.2.5** Educational occupancy laboratories using chemicals shall be in accordance with 8.7.4.

### 15.3.3 Interior Finish.

**15.3.3.1 General.** Interior finish shall be in accordance with Section 10.2.

**15.3.3.2 Interior Wall and Ceiling Finish.** Interior wall and ceiling finish materials complying with Section 10.2 shall be permitted as follows:

(1) Exits — Class A
(2) Corridors and lobbies — Class A or Class B
(3) Low-height partitions not exceeding 60 in. (1525 mm) and used in locations other than exits — Class A, Class B, or Class C

**15.3.3.3 Interior Floor Finish.** (No requirements.)

### 15.3.4 Detection, Alarm, and Communications Systems.

**15.3.4.1 General.**

**15.3.4.1.1** Educational occupancies shall be provided with a fire alarm system in accordance with Section 9.6.

**15.3.4.1.2** The requirement of 15.3.4.1.1 shall not apply to buildings meeting all of the following criteria:

(1) Buildings having an area not exceeding 1000 ft² (93 m²)
(2) Buildings containing a single classroom
(3) Buildings located not less than 30 ft (9.1 m) from another building

**15.3.4.2 Initiation.**

**15.3.4.2.1 General.** Initiation of the required fire alarm system shall be by manual means in accordance with 9.6.2.1(1), unless otherwise permitted by one of the following:

(1) Manual fire alarm boxes shall not be required where permitted by 15.3.4.2.3.
(2) In buildings where all normally occupied spaces are provided with a two-way communication system between such spaces and a constantly attended receiving station from where a general evacuation alarm can be sounded, the manual fire alarm boxes shall not be required, except in locations specifically designated by the authority having jurisdiction.

**15.3.4.2.2 Automatic Initiation.** In buildings provided with automatic sprinkler protection, the operation of the sprinkler system shall automatically activate the fire alarm system in addition to the initiation means required in 15.3.4.2.1.

**15.3.4.2.3 Alternative Protection System.** Manual fire alarm boxes shall be permitted to be eliminated in accordance with 15.3.4.2.3.1 or 15.3.4.2.3.2.

**15.3.4.2.3.1\*** Manual fire alarm boxes shall be permitted to be eliminated where all of the following conditions apply:

(1) Interior corridors are protected by smoke detectors using an alarm verification system as described in *NFPA 72, National Fire Alarm and Signaling Code.*
(2) Auditoriums, cafeterias, and gymnasiums are protected by heat-detection devices or other approved detection devices.
(3) Shops and laboratories involving dusts or vapors are protected by heat-detection devices or other approved detection devices.
(4) Provision is made at a central point to manually activate the evacuation signal or to evacuate only affected areas.

**15.3.4.2.3.2\*** Manual fire alarm boxes shall be permitted to be eliminated where both of the following conditions apply:

(1) The building is protected throughout by an approved, supervised automatic sprinkler system in accordance with Section 9.7.
(2) Provision is made at a central point to manually activate the evacuation signal or to evacuate only affected areas.

**15.3.4.3 Notification.**

**15.3.4.3.1 Occupant Notification.**

**15.3.4.3.1.1\*** Occupant notification shall be accomplished automatically in accordance with 9.6.3.

**15.3.4.3.1.2** Positive alarm sequence shall be permitted in accordance with 9.6.3.4.

**15.3.4.3.1.3** Where acceptable to the authority having jurisdiction, the fire alarm system shall be permitted to be used for other emergency signaling or for class changes, provided that the fire alarm is distinctive in signal and overrides all other use.

**15.3.4.3.1.4** To prevent students from being returned to a building that is burning, the recall signal shall be separate and distinct from any other signals, and such signal shall be permitted to be given by use of distinctively colored flags or banners.

**15.3.4.3.1.5** If the recall signal required by 15.3.4.3.1.4 is electric, the push buttons or other controls shall be kept under lock, the key for which shall be in the possession of the principal or another designated person in order to prevent a recall at a time when there is an actual fire.

**15.3.4.3.1.6** Regardless of the method of recall signal, the means of giving the recall signal shall be kept under lock.

**15.3.4.3.2 Emergency Forces Notification.**

**15.3.4.3.2.1** Wherever any of the school authorities determine that an actual fire exists, they shall immediately call the local fire department using the public fire alarm system or other available facilities.

**15.3.4.3.2.2** Emergency forces notification shall be accomplished in accordance with 9.6.4 where the existing fire alarm system is replaced.

**15.3.5 Extinguishment Requirements.**

**15.3.5.1** Where student occupancy exists below the level of exit discharge, every portion of such floor shall be protected throughout by an approved automatic sprinkler system in accordance with Section 9.7.

**15.3.5.2** Where student occupancy does not exist on floors below the level of exit discharge, such floors shall be separated from the rest of the building by 1-hour fire resistance–rated construction or shall be protected throughout by an approved automatic sprinkler system in accordance with Section 9.7.

**15.3.5.3** Automatic sprinkler protection shall not be required where student occupancy exists below the level of exit discharge, provided that both of the following criteria are met:

(1) The approval of the authority having jurisdiction shall be required.
(2) Windows for rescue and ventilation shall be provided in accordance with 15.2.11.1.

**15.3.5.4** Buildings with unprotected openings in accordance with 8.6.6 shall be protected throughout by an approved, supervised automatic sprinkler system in accordance with Section 9.7.

**15.3.5.5** Where another provision of this chapter requires an automatic sprinkler system, the sprinkler system shall be installed in accordance with 9.7.1.1(1).

**15.3.6 Corridors.** Corridors shall be separated from other parts of the story by walls having a minimum ½-hour fire resistance rating in accordance with Section 8.3, unless otherwise permitted by one of the following:

(1) Corridor protection shall not be required where all spaces normally subject to student occupancy have not less than one door opening directly to the outside or to an exterior exit access balcony or corridor in accordance with 7.5.3.
(2)*The following shall apply to buildings protected throughout by an approved automatic sprinkler system with valve supervision in accordance with Section 9.7:
   (a) Corridor walls shall not be required to be rated, provided that such walls form smoke partitions in accordance with Section 8.4.
   (b) The provisions of 8.4.3.5 shall not apply to normally occupied classrooms.
(3) Where the corridor ceiling is an assembly having a minimum ½-hour fire resistance rating where tested as a wall, the corridor wall shall be permitted to terminate at the corridor ceiling.
(4) Lavatories shall not be required to be separated from corridors, provided that they are separated from all other spaces by walls having a minimum ½-hour fire resistance rating in accordance with Section 8.3.
(5) Lavatories shall not be required to be separated from corridors, provided that both of the following criteria are met:
   (a) The building is protected throughout by an approved, supervised automatic sprinkler system in accordance with Section 9.7.
   (b) The walls separating the lavatory from other rooms form smoke partitions in accordance with Section 8.4.

**15.3.7 Subdivision of Building Spaces.**

**15.3.7.1** Educational occupancies shall be subdivided into compartments by smoke partitions having not less than a 1-hour fire resistance rating and complying with Section 8.4 where one or both of the following conditions exist:

(1) The maximum area of a compartment, including the aggregate area of all floors having a common atmosphere, exceeds 30,000 ft² (2800 m²).
(2) The length or width of the building exceeds 300 ft (91 m).

**15.3.7.2** The requirement of 15.3.7.1 shall not apply to either of the following:

(1) Where all classrooms have exterior exit access in accordance with 7.5.3
(2) Buildings protected throughout by an approved automatic sprinkler system in accordance with Section 9.7

**15.3.7.3** The area of any smoke compartment required by 15.3.7.1 shall not exceed 30,000 ft² (2800 m²), with no dimension exceeding 300 ft (91 m).

**15.4 Special Provisions.**

**15.4.1 Limited Access Buildings and Underground Buildings.** Limited access buildings and underground buildings shall comply with Section 11.7.

**15.4.2 High-Rise Buildings.** High-rise buildings shall comply with 11.8.3.1.

**15.4.3 Flexible Plan and Open Plan Buildings.**

**15.4.3.1** Flexible plan and open plan buildings shall comply with the requirements of this chapter as modified by 15.4.3.2 through 15.4.3.5.

**15.4.3.2** Each room occupied by more than 300 persons shall have two or more means of egress entering into separate atmospheres.

**15.4.3.3** Where three or more means of egress are required, the number of means of egress permitted to enter into the same atmosphere shall not exceed two.

**15.4.3.4** Flexible plan buildings shall be permitted to have walls and partitions rearranged periodically only if revised plans or diagrams have been approved by the authority having jurisdiction.

**15.4.3.5** Flexible plan buildings shall be evaluated while all folding walls are extended and in use as well as when they are in the retracted position.

**15.5 Building Services.**

**15.5.1 Utilities.** Utilities shall comply with the provisions of Section 9.1.

**15.5.2 Heating, Ventilating, and Air-Conditioning Equipment.**

**15.5.2.1** Heating, ventilating, and air-conditioning equipment shall comply with the provisions of Section 9.2.

**15.5.2.2** Unvented fuel-fired heating equipment, other than gas space heaters in compliance with NFPA 54/ANSI Z223.1, *National Fuel Gas Code*, shall be prohibited.

**15.5.3 Elevators, Escalators, and Conveyors.** Elevators, escalators, and conveyors shall comply with the provisions of Section 9.4.

**15.5.4 Rubbish Chutes, Incinerators, and Laundry Chutes.** Rubbish chutes, incinerators, and laundry chutes shall comply with the provisions of Section 9.5.

**15.6 Reserved.**

**15.7 Operating Features.**

**15.7.1 Emergency Plan.** Emergency plans shall be provided in accordance with Section 4.8.

**15.7.2 Emergency Egress Drills.**

**15.7.2.1\*** Emergency egress drills shall be conducted in accordance with Section 4.7 and the applicable provisions of 15.7.2.3 as otherwise provided by 15.7.2.2.

**15.7.2.2** Approved training programs designed for education and training and for the practice of emergency egress to familiarize occupants with the drill procedure, and to establish conduct of the emergency egress as a matter of routine, shall be permitted to receive credit on a one-for-one basis for not more than four of the emergency egress drills required by 15.7.2.3, provided that a minimum of four emergency egress drills are completed prior to the conduct of the first such training and practice program.

**15.7.2.3** Emergency egress drills shall be conducted as follows:

(1) Not less than one emergency egress drill shall be conducted every month the facility is in session, unless both of the following criteria are met:

(a) In climates where the weather is severe, the monthly emergency egress drills shall be permitted to be deferred.
(b) The required number of emergency egress drills shall be conducted, and not less than four shall be conducted before the drills are deferred.

(2) All occupants of the building shall participate in the drill.
(3) One additional emergency egress drill, other than for educational occupancies that are open on a year-round basis, shall be required within the first 30 days of operation.

**15.7.2.4** All emergency drill alarms shall be sounded on the fire alarm system.

### 15.7.3 Inspection.

**15.7.3.1\*** It shall be the duty of principals, teachers, or staff to inspect all exit facilities daily to ensure that all stairways, doors, and other exits are in proper condition.

**15.7.3.2** Open plan buildings shall require extra surveillance to ensure that exit paths are maintained clear of obstruction and are obvious.

**15.7.3.3 Inspection of Door Openings.** Door openings shall be inspected in accordance with 7.2.1.15.

### 15.7.4 Furnishings and Decorations.

**15.7.4.1** Draperies, curtains, and other similar furnishings and decorations in educational occupancies shall be in accordance with the provisions of 10.3.1.

**15.7.4.2** Clothing and personal effects shall not be stored in corridors, unless otherwise permitted by one of the following:

(1) This requirement shall not apply to corridors protected by an automatic sprinkler system in accordance with Section 9.7.
(2) This requirement shall not apply to corridor areas protected by a smoke detection system in accordance with Section 9.6.
(3) This requirement shall not apply to storage in metal lockers, provided that the required egress width is maintained.

**15.7.4.3** Artwork and teaching materials shall be permitted to be attached directly to the walls in accordance with the following:

(1) The artwork and teaching materials shall not exceed 20 percent of the wall area in a building that is not protected throughout by an approved automatic sprinkler system in accordance with Section 9.7.
(2) The artwork and teaching materials shall not exceed 50 percent of the wall area in a building that is protected throughout by an approved automatic sprinkler system in accordance with Section 9.7.

**15.7.5 Open Flames.** Approved open flames shall be permitted in laboratories and vocational/technical areas.

## Chapter 16   New Day-Care Occupancies

### 16.1 General Requirements.

#### 16.1.1\* Application.

**16.1.1.1** The requirements of this chapter shall apply to new buildings or portions thereof used as day-care occupancies. *(See 1.3.1.)*

**16.1.1.2 Administration.** The provisions of Chapter 1, Administration, shall apply.

**16.1.1.3 General.** The provisions of Chapter 4, General, shall apply.

**16.1.1.4** The requirements of Sections 16.1 through 16.5 and Section 16.7 shall apply to day-care occupancies in which more than 12 clients receive care, maintenance, and supervision by other than their relative(s) or legal guardian(s) for less than 24 hours per day.

**16.1.1.5** The requirements of Section 16.1 and Sections 16.4 through 16.7 shall apply to day-care homes as defined in 16.1.4.

**16.1.1.6** Where a facility houses more than one age group or self-preservation capability, the strictest requirements applicable to any group present shall apply throughout the day-care occupancy or building, as appropriate to a given area, unless the area housing such a group is maintained as a separate fire area.

**16.1.1.7** Places of religious worship shall not be required to meet the provisions of this chapter where providing day care while services are being held in the building.

**16.1.2 Classification of Occupancy.** See 6.1.4.

**16.1.2.1 General.** Occupancies that include part-day preschools, kindergartens, and other schools whose purpose is primarily educational, even though the children who attend such schools are of preschool age, shall comply with the provisions of Chapter 14.

#### 16.1.2.2 Adult Day-Care Occupancies.

**16.1.2.2.1** Adult day-care occupancies shall include any building or portion thereof used for less than 24 hours per day to house more than three adults requiring care, maintenance, and supervision by other than their relative(s).

**16.1.2.2.2** Clients in adult day-care occupancies shall be ambulatory or semiambulatory and shall not be bedridden.

**16.1.2.2.3** Clients in adult day-care occupancies shall not exhibit behavior that is harmful to themselves or to others.

**16.1.2.3\* Conversions.** A conversion from a day-care home to a day-care occupancy with more than 12 clients shall be permitted only if the day-care occupancy conforms to the requirements of this chapter for new day-care occupancies with more than 12 clients.

#### 16.1.3 Multiple Occupancies.

**16.1.3.1 General.** Multiple occupancies shall be in accordance with 6.1.14.

**16.1.3.2 Day-Care Occupancies in Apartment Buildings.** If the two exit accesses from a day-care occupancy enter the same corridor as an apartment occupancy, the exit accesses shall be separated in the corridor by a smoke partition complying with both of the following:

(1) It shall have not less than a 1-hour fire resistance rating and shall be constructed in accordance with Section 8.4.
(2) It shall be located so that it has an exit on each side.

#### 16.1.4 Definitions.

**16.1.4.1 General.** For definitions, see Chapter 3, Definitions.

**16.1.4.2 Special Definitions.** A list of special terms used in this chapter follows:

(1) **Day-Care Home.** See 3.3.140.1.
(2) **Flexible Plan and Open Plan Educational or Day-Care Building.** See 3.3.36.6.
(3) **Self-Preservation (Day-Care Occupancy).** See 3.3.240.
(4) **Separate Atmosphere.** See 3.3.26.2.

**16.1.5 Classification of Hazard of Contents.** The contents of day-care occupancies shall be classified as ordinary hazard in accordance with Section 6.2.

**16.1.6 Minimum Construction Requirements.**

**16.1.6.1** Day-care occupancies, other than day-care homes, shall be limited to the building construction types specified in Table 16.1.6.1 based on the number of stories in height as defined in 4.6.3. *(See 8.2.1.)*

**16.1.6.2** Where day-care occupancies, other than day-care homes, with clients who are 24 months or less in age or who are incapable of self-preservation, are located one or more stories above the level of exit discharge, or where day-care occupancies are located two or more stories above the level of exit discharge, smoke partitions shall be provided to divide such stories into not less than two compartments. The smoke partitions shall be constructed in accordance with Section 8.4 but shall not be required to have a fire resistance rating.

**16.1.7 Occupant Load.**

**16.1.7.1** The occupant load, in number of persons for whom means of egress and other provisions are required, either shall be determined on the basis of the occupant load factors of Table 7.3.1.2 that are characteristic of the use of the space or shall be determined as the maximum probable population of the space under consideration, whichever is greater.

**16.1.7.2** Where the occupant load is determined as the maximum probable population of the space in accordance with 16.1.7.1, an approved aisle, seating, and exiting diagram shall be required by the authority having jurisdiction to substantiate such a modification.

**16.2 Means of Egress Requirements.**

**16.2.1 General.** Means of egress shall be in accordance with Chapter 7 and Section 16.2.

**Table 16.1.6.1 Construction Type Limitations**

| Construction Type | Sprinklered[a] | Stories in Height[b] | | | | | |
|---|---|---|---|---|---|---|---|
| | | One Story Below[c] | 1 | 2 | 3–4 | >4 but Not High-Rise | High-Rise |
| I (442) | Yes | X | X | X | X | X | X |
| | No | NP | X | X | X | NP | NP |
| I (332) | Yes | X | X | X | X | X | X |
| | No | NP | X | X | X | NP | NP |
| II (222) | Yes | X | X | X | X | X | X |
| | No | NP | X | X | X | NP | NP |
| II (111) | Yes | X | X | X | X | X | NP |
| | No | NP | X | NP | NP | NP | NP |
| II (000) | Yes | X | X | X | X | NP | NP |
| | No | NP | X | NP | NP | NP | NP |
| III (211) | Yes | X | X | X | X | NP | NP |
| | No | NP | X | NP | NP | NP | NP |
| III (200) | Yes | NP | X | X | NP | NP | NP |
| | No | NP | X | NP | NP | NP | NP |
| IV (2HH) | Yes | X | X | X | NP | NP | NP |
| | No | NP | X | NP | NP | NP | NP |
| V (111) | Yes | X | X | X | X | NP | NP |
| | No | NP | X | NP | NP | NP | NP |
| V (000) | Yes | NP | X | X | NP | NP | NP |
| | No | NP | X | NP | NP | NP | NP |

X: Permitted. NP: Not Permitted.
[a]Sprinklered throughout by an approved, supervised automatic sprinkler system in accordance with Section 9.7.
[b]See 4.6.3.
[c]One story below the level of exit discharge.

**16.2.2 Means of Egress Components.**

**16.2.2.1 Components Permitted.** Components of means of egress shall be limited to the types described in 16.2.2.2 through 16.2.2.10.

**16.2.2.2 Doors.**

**16.2.2.2.1 General.** Doors complying with 7.2.1 shall be permitted.

**16.2.2.2.2 Panic Hardware or Fire Exit Hardware.** Any door in a required means of egress from an area having an occupant load of 100 or more persons shall be permitted to be provided with a latch or lock only if the latch or lock is panic hardware or fire exit hardware complying with 7.2.1.7.

**16.2.2.2.3 Special Locking Arrangements.**

**16.2.2.2.3.1** Delayed-egress locking systems complying with 7.2.1.6.1 shall be permitted.

**16.2.2.2.3.2** Access-controlled egress door assemblies complying with 7.2.1.6.2 shall be permitted.

**16.2.2.2.3.3** Elevator lobby exit access door assemblies locking in accordance with 7.2.1.6.3 shall be permitted.

**16.2.2.2.4\* Door Latches.** Every door latch to closets, storage areas, kitchens, and other similar spaces or areas shall be such that clients can open the door from inside the space or area.

**16.2.2.2.5 Bathroom Doors.** Every bathroom door lock shall be designed to allow opening of the locked door from the outside by an opening device that shall be readily accessible to the staff.

**16.2.2.3\* Stairs.** Stairs complying with 7.2.2 shall be permitted.

**16.2.2.4 Smokeproof Enclosures.** Smokeproof enclosures complying with 7.2.3 shall be permitted.

**16.2.2.5 Horizontal Exits.** Horizontal exits complying with 7.2.4 shall be permitted.

**16.2.2.6 Ramps.** Ramps complying with 7.2.5 shall be permitted.

**16.2.2.7 Exit Passageways.** Exit passageways complying with 7.2.6 shall be permitted.

**16.2.2.8 Fire Escape Ladders.** Fire escape ladders complying with 7.2.9 shall be permitted.

**16.2.2.9 Alternating Tread Devices.** Alternating tread devices complying with 7.2.11 shall be permitted.

**16.2.2.10 Areas of Refuge.** Areas of refuge complying with 7.2.12 shall be permitted.

**16.2.3 Capacity of Means of Egress.** Capacity of means of egress shall be in accordance with Section 7.3.

**16.2.4 Number of Means of Egress.**

**16.2.4.1** The number of means of egress shall be in accordance with Section 7.4.

**16.2.4.2** Not less than two separate exits shall be in accordance with both of the following criteria:

(1) They shall be provided on every story.
(2) They shall be accessible from every part of every story and mezzanine; however, exit access travel shall be permitted to be common for the distance permitted as common path of travel by 16.2.5.3.

**16.2.4.3 Reserved.**

**16.2.5 Arrangement of Means of Egress.** See also 16.1.6.2.

**16.2.5.1** Means of egress shall be arranged in accordance with Section 7.5.

**16.2.5.2** No dead-end corridor shall exceed 20 ft (6100 mm), other than in buildings protected throughout by an approved, supervised automatic sprinkler system in accordance with Section 9.7, in which case dead-end corridors shall not exceed 50 ft (15 m).

**16.2.5.3** Limitations on common path of travel shall be in accordance with 16.2.5.3.1 and 16.2.5.3.2.

**16.2.5.3.1** Common path of travel shall not exceed 100 ft (30 m) in a building protected throughout by an approved, supervised automatic sprinkler system in accordance with Section 9.7.

**16.2.5.3.2** Common path of travel shall not exceed 75 ft (23 m) in a building not protected throughout by an approved, supervised automatic sprinkler system in accordance with Section 9.7.

**16.2.6 Travel Distance to Exits.**

**16.2.6.1** Travel distance shall be measured in accordance with Section 7.6.

**16.2.6.2** Travel distance shall meet all of the following criteria, unless otherwise permitted by 16.2.6.3:

(1) The travel distance between any room door intended as an exit access and an exit shall not exceed 100 ft (30 m).
(2) The travel distance between any point in a room and an exit shall not exceed 150 ft (46 m).
(3) The travel distance between any point in a sleeping room and an exit access door in that room shall not exceed 50 ft (15 m).

**16.2.6.3** The travel distance required by 16.2.6.2(1) and (2) shall be permitted to be increased by 50 ft (15 m) in buildings protected throughout by an approved, supervised automatic sprinkler system in accordance with Section 9.7.

**16.2.7 Discharge from Exits.** Discharge from exits shall be arranged in accordance with Section 7.7.

**16.2.8 Illumination of Means of Egress.** Means of egress shall be illuminated in accordance with Section 7.8.

**16.2.9 Emergency Lighting.** Emergency lighting shall be provided in accordance with Section 7.9 in the following areas:

(1) Interior stairs and corridors
(2) Assembly use spaces
(3) Flexible and open plan buildings
(4) Interior or limited access portions of buildings
(5) Shops and laboratories

**16.2.10 Marking of Means of Egress.** Means of egress shall have signs in accordance with Section 7.10.

**16.2.11 Special Means of Egress Features.**

**16.2.11.1 Windows for Rescue.**

**16.2.11.1.1** Every room or space normally subject to client occupancy, other than bathrooms, shall have not less than one outside window for emergency rescue that complies with all of the following, unless otherwise permitted by 16.2.11.1.2:

(1) Such windows shall be openable from the inside without the use of tools and shall provide a clear opening of not less than 20 in. (510 mm) in width, 24 in. (610 mm) in height, and 5.7 ft$^2$ (0.5 m$^2$) in area.

(2) The bottom of the opening shall be not more than 44 in. (1120 mm) above the floor.

(3) The clear opening shall allow a rectangular solid, with a width and height that provides not less than the required 5.7 ft$^2$ (0.5 m$^2$) opening and a depth of not less than 20 in. (510 mm), to pass fully through the opening.

**16.2.11.1.2** The requirements of 16.2.11.1.1 shall not apply to either of the following:

(1) Buildings protected throughout by an approved, supervised automatic sprinkler system in accordance with Section 9.7

(2) Where the room or space has a door leading directly to an exit or directly to the outside of the building

**16.2.11.2 Lockups.** Lockups in day-care occupancies shall comply with the requirements of 22.4.5.

### 16.3 Protection.

**16.3.1 Protection of Vertical Openings.** Any vertical opening, other than unprotected vertical openings in accordance with 8.6.9.1, shall be enclosed or protected in accordance with Section 8.6.

**16.3.2 Protection from Hazards.**

**16.3.2.1** Rooms or spaces for the storage, processing, or use of materials specified in 16.3.2.1(1) through (3) shall be protected in accordance with the following:

(1) Separation from the remainder of the building by fire barriers having a minimum 1-hour fire resistance rating, or protection of such rooms by automatic extinguishing systems as specified in Section 8.7, in the following areas:
  (a) Boiler and furnace rooms, unless such rooms enclose only air-handling equipment
  (b) Rooms or spaces used for the storage of combustible supplies in quantities deemed hazardous by the authority having jurisdiction
  (c) Rooms or spaces used for the storage of hazardous materials or flammable or combustible liquids in quantities deemed hazardous by recognized standards
  (d) Janitor closets

(2) Separation from the remainder of the building by fire barriers having a minimum 1-hour fire resistance rating and protection of such rooms by automatic extinguishing systems as specified in Section 8.7 in the following areas:
  (a)*Laundries
  (b) Maintenance shops, including woodworking and painting areas
  (c) Rooms or spaces used for processing or use of combustible supplies deemed hazardous by the authority having jurisdiction
  (d) Rooms or spaces used for processing or use of hazardous materials or flammable or combustible liquids in quantities deemed hazardous by recognized standards

(3) Where automatic extinguishing is used to meet the requirements of 16.3.2.1(1) and (2), protection as permitted in accordance with 9.7.1.2

**16.3.2.2** Janitor closets protected in accordance with 16.3.2.1(1)(d) shall be permitted to have doors fitted with ventilating louvers where the space is protected by automatic sprinklers.

**16.3.2.3** Cooking facilities shall be protected in accordance with 9.2.3, unless otherwise permitted by 16.3.2.4 or 16.3.2.5.

**16.3.2.4** Openings shall not be required to be protected between food preparation areas and dining areas.

**16.3.2.5** Approved domestic cooking equipment used for food warming or limited cooking shall not be required to be protected.

**16.3.2.6 Alcohol-Based Hand-Rub Dispensers.** Alcohol-based hand-rub dispensers shall be protected in accordance with 8.7.3.1, unless all of the following conditions are met:

(1) Dispensers shall be installed in rooms or spaces separated from corridors and exits.

(2) The maximum individual dispenser fluid capacity shall be as follows:
  (a) 0.32 gal (1.2 L) for dispensers in rooms
  (b) 0.53 gal (2.0 L) for dispensers in suites of rooms

(3) Dispensers shall be separated from each other by horizontal spacing of not less than 48 in. (1220 mm).

(4) Storage of quantities greater than 5 gal (18.9 L) in a single fire compartment shall meet the requirements of NFPA 30, *Flammable and Combustible Liquids Code.*

(5) Dispensers shall not be installed over or directly adjacent to an ignition source.

(6) Dispensers installed directly over carpeted floors shall be permitted only in sprinklered rooms or spaces.

**16.3.3 Interior Finish.**

**16.3.3.1 General.** Interior finish shall be in accordance with Section 10.2.

**16.3.3.2 Interior Wall and Ceiling Finish.** Interior wall and ceiling finish materials complying with Section 10.2 shall be Class A in stairways, corridors, and lobbies; in all other occupied areas, interior wall and ceiling finish shall be Class A or Class B.

**16.3.3.3 Interior Floor Finish.**

**16.3.3.3.1** Interior floor finish shall comply with Section 10.2.

**16.3.3.3.2** Interior floor finish in exit enclosures and exit access corridors and spaces not separated from them by walls complying with 14.3.6 shall be not less than Class II.

**16.3.3.3.3** Interior floor finish shall comply with 10.2.7.1 or 10.2.7.2, as applicable.

**16.3.4 Detection, Alarm, and Communications Systems.**

**16.3.4.1 General.** Day-care occupancies, other than day-care occupancies housed in one room having at least one door opening directly to the outside at grade plane or to an exterior exit access balcony in accordance with 7.5.3, shall be provided with a fire alarm system in accordance with Section 9.6.

**16.3.4.2 Initiation.** Initiation of the required fire alarm system shall be by manual means and by operation of any required smoke detectors and required sprinkler systems. *(See 16.3.4.5.)*

**16.3.4.3 Occupant Notification.**

**16.3.4.3.1** Occupant notification shall be in accordance with 9.6.3.

**16.3.4.3.2** Positive alarm sequence shall be permitted in accordance with 9.6.3.4.

**16.3.4.3.3** Private operating mode in accordance with 9.6.3.6.3 shall be permitted.

**16.3.4.4 Emergency Forces Notification.** Emergency forces notification shall be accomplished in accordance with 9.6.4.

**16.3.4.5 Detection.** A smoke detection system in accordance with Section 9.6 shall be installed in day-care occupancies, other than those housed in one room having at least one door opening directly to the outside at grade plane or to an exterior exit access balcony in accordance with 7.5.3, and such system shall comply with both of the following:

(1) Detectors shall be installed on each story in front of the doors to the stairways and in the corridors of all floors occupied by the day-care occupancy.
(2) Detectors shall be installed in lounges, recreation areas, and sleeping rooms in the day-care occupancy.

**16.3.5 Extinguishment Requirements.**

**16.3.5.1** Any required sprinkler systems shall be in accordance with Section 9.7.

**16.3.5.2** Required sprinkler systems shall be installed in accordance with 9.7.1.1(1).

**16.3.5.3** Buildings with unprotected openings in accordance with 8.6.6 shall be protected throughout by an approved, supervised automatic sprinkler system in accordance with Section 9.7.

**16.3.6 Corridors.** Every interior corridor shall be constructed of walls having not less than a 1-hour fire resistance rating in accordance with Section 8.3, unless otherwise permitted by any of the following:

(1) Corridor protection shall not be required where all spaces normally subject to client occupancy have not less than one door opening directly to the outside or to an exterior exit access balcony or corridor in accordance with 7.5.3.
(2) In buildings protected throughout by an approved, supervised automatic sprinkler system in accordance with Section 9.7, corridor walls shall not be required to be rated, provided that such walls form smoke partitions in accordance with Section 8.4.
(3) Where the corridor ceiling is an assembly having a 1-hour fire resistance rating where tested as a wall, the corridor walls shall be permitted to terminate at the corridor ceiling.
(4) Lavatories shall not be required to be separated from corridors, provided that they are separated from all other spaces by walls having not less than a 1-hour fire resistance rating in accordance with Section 8.3.
(5) Lavatories shall not be required to be separated from corridors, provided that both of the following criteria are met:
  (a) The building is protected throughout by an approved, supervised automatic sprinkler system in accordance with Section 9.7.
  (b) The walls separating the lavatory from other rooms form smoke partitions in accordance with Section 8.4.

**16.4 Special Provisions.**

**16.4.1 Limited Access Buildings and Underground Buildings.** Limited access buildings and underground buildings shall comply with Section 11.7.

**16.4.2 High-Rise Buildings.** High-rise buildings that house day-care occupancies on floors more than 75 ft (23 m) above the lowest level of fire department vehicle access shall comply with Section 11.8.

**16.4.3 Flexible Plan and Open Plan Buildings.**

**16.4.3.1** Flexible plan and open plan buildings shall comply with the requirements of this chapter as modified by 16.4.3.2 through 16.4.3.5.

**16.4.3.2** Flexible plan buildings shall be permitted to have walls and partitions rearranged periodically only if revised plans or diagrams have been approved by the authority having jurisdiction.

**16.4.3.3** Flexible plan buildings shall be evaluated while all folding walls are extended and in use as well as when they are in the retracted position.

**16.4.3.4** Each room occupied by more than 300 persons shall have two or more means of egress entering into separate atmospheres.

**16.4.3.5** Where three or more means of egress are required from a single room, the number of means of egress permitted to enter into a common atmosphere shall not exceed two.

**16.5 Building Services.**

**16.5.1 Utilities.**

**16.5.1.1** Utilities shall comply with the provisions of Section 9.1.

**16.5.1.2** Special protective covers for all electrical receptacles shall be installed in all areas occupied by clients.

**16.5.2 Heating, Ventilating, and Air-Conditioning Equipment.**

**16.5.2.1** Heating, ventilating, and air-conditioning equipment shall be in accordance with Section 9.2.

**16.5.2.2** Unvented fuel-fired heating equipment, other than gas space heaters in compliance with NFPA 54/ANSI Z223.1, *National Fuel Gas Code*, shall be prohibited.

**16.5.2.3** Any heating equipment in spaces occupied by clients shall be provided with partitions, screens, or other means to protect clients from hot surfaces and open flames; if solid partitions are used to provide such protection, provisions shall be made to ensure adequate air for combustion and ventilation for the heating equipment.

**16.5.3 Elevators, Escalators, and Conveyors.** Elevators, escalators, and conveyors, other than those in day-care homes, shall comply with the provisions of Section 9.4.

**16.5.4 Rubbish Chutes, Incinerators, and Laundry Chutes.** Rubbish chutes, incinerators, and laundry chutes, other than those in day-care homes, shall comply with the provisions of Section 9.5.

**16.6 Day-Care Homes.**

**16.6.1 General Requirements.**

**16.6.1.1 Application.**

**16.6.1.1.1** The requirements of Section 16.6 shall apply to new buildings or portions thereof used as day-care homes. *(See 1.3.1.)*

**16.6.1.1.2** The requirements of Section 16.6 shall apply to day-care homes in which more than 3, but not more than 12, clients receive care, maintenance, and supervision by other than their relative(s) or legal guardian(s) for less than 24 hours per day, generally within a dwelling unit. *(See also 16.6.1.4.)*

**16.6.1.1.3** Where a facility houses more than one age group or one self-preservation capability, the strictest requirements applicable to any group present shall apply throughout the day-care home or building, as appropriate to a given area, unless the area housing such a group is maintained as a separate fire area.

**16.6.1.1.4** Facilities that supervise clients on a temporary basis with a parent or guardian in close proximity shall not be required to meet the provisions of Section 16.6.

**16.6.1.1.5** Places of religious worship shall not be required to meet the provisions of Section 16.6 where operating a day-care home while services are being held in the building.

**16.6.1.2 Multiple Occupancies.** See 16.1.3.

**16.6.1.3 Definitions.** See 16.1.4.

**16.6.1.4 Classification of Occupancy.**

**16.6.1.4.1 Subclassification of Day-Care Homes.** Subclassification of day-care homes shall comply with 16.6.1.4.1.1 and 16.6.1.4.1.2.

**16.6.1.4.1.1 Family Day-Care Home.** A family day-care home shall be a day-care home in which more than three, but fewer than seven, clients receive care, maintenance, and supervision by other than their relative(s) or legal guardian(s) for less than 24 hours per day, generally within a dwelling unit.

**16.6.1.4.1.2 Group Day-Care Home.** A group day-care home shall be a day-care home in which not less than 7, but not more than 12, clients receive care, maintenance, and supervision by other than their relative(s) or legal guardian(s) for less than 24 hours per day, generally within a dwelling unit.

**16.6.1.4.2\* Conversions.** A conversion from a day-care home to a day-care occupancy with more than 12 clients shall be permitted only if the day-care occupancy conforms to the requirements of Chapter 16 for new day-care occupancies with more than 12 clients.

**16.6.1.5 Classification of Hazard of Contents.** See 16.1.5.

**16.6.1.6 Location and Construction.** No day-care home shall be located more than one story below the level of exit discharge.

**16.6.1.7 Occupant Load.**

**16.6.1.7.1** In family day-care homes, both of the following shall apply:

(1) The minimum staff-to-client ratio shall be not less than one staff for up to six clients, including the caretaker's own children under age six.
(2) There shall be not more than two clients incapable of self-preservation.

**16.6.1.7.2** In group day-care homes, all of the following shall apply:

(1) The minimum staff-to-client ratio shall be not less than two staff for up to 12 clients.
(2) There shall be not more than 3 clients incapable of self-preservation.
(3) The staff-to-client ratio shall be permitted to be modified by the authority having jurisdiction where safeguards in addition to those specified by Section 16.6 are provided.

**16.6.2 Means of Escape Requirements.**

**16.6.2.1 General.** Means of escape shall comply with Section 24.2.

**16.6.2.2 Reserved.**

**16.6.2.3 Reserved.**

**16.6.2.4 Number and Type of Means of Escape.**

**16.6.2.4.1** The number and type of means of escape shall comply with Section 24.2 and 16.6.2.4.2 through 16.6.2.4.4.

**16.6.2.4.2** Every room used for sleeping, living, recreation, education, or dining purposes shall have the number and type of means of escape in accordance with Section 24.2.

**16.6.2.4.3** No room or space that is accessible only by a ladder or folding stairs or through a trap door shall be occupied by clients.

**16.6.2.4.4** In group day-care homes where spaces on the story above the level of exit discharge are used by clients, that story shall have not less than one means of escape complying with one of the following:

(1) Door leading directly to the outside with access to finished ground level
(2) Door leading directly to an outside stair to finished ground level
(3) Interior stair leading directly to the outside with access to finished ground level separated from other stories by a ½-hour fire barrier in accordance with Section 8.3

**16.6.2.4.5** Where clients occupy a story below the level of exit discharge, that story shall have not less than one means of escape complying with one of the following:

(1) Door leading directly to the outside with access to finished ground level
(2) Door leading directly to an outside stair going to finished ground level
(3) Bulkhead enclosure complying with 24.2.7
(4) Interior stair leading directly to the outside with access to finished ground level, separated from other stories by a ½-hour fire barrier in accordance with Section 8.3

**16.6.2.5 Arrangement of Means of Escape.**

**16.6.2.5.1** A story used above or below the level of exit discharge shall be in accordance with 16.6.2.4.3 and 16.6.2.4.4.

**16.6.2.5.2** For group day-care homes, means of escape shall be arranged in accordance with Section 7.5.

**16.6.2.5.3** No dead-end corridors shall exceed 20 ft (6100 mm).

**16.6.2.5.4** Doors in means of escape shall be protected from obstructions, including snow and ice.

**16.6.2.6 Travel Distance.** Travel distance shall comply with 16.6.2.6.1 through 16.6.2.6.3.

**16.6.2.6.1** Travel distance shall be measured in accordance with Section 7.6.

**16.6.2.6.2** Travel distance shall meet all of the following criteria, unless otherwise permitted by 16.6.2.6.3:

(1) The travel distance between any room door intended as an exit access and an exit shall not exceed 100 ft (30 m).
(2) The travel distance between any point in a room and an exit shall not exceed 150 ft (46 m).
(3) The travel distance between any point in a sleeping room and an exit access to that room shall not exceed 50 ft (15 m).

**16.6.2.6.3** The travel distance required by 16.6.2.6.2(1) and (2) shall be permitted to be increased by 50 ft (15 m) in buildings protected throughout by an approved, supervised automatic sprinkler system in accordance with Section 9.7.

**16.6.2.7 Discharge from Exits.** See 16.6.2.4.

**16.6.2.8 Illumination of Means of Egress.** Means of egress shall be illuminated in accordance with Section 7.8.

**16.6.2.9 Emergency Lighting.** (No requirements.)

**16.6.2.10 Marking of Means of Egress.** (No requirements.)

**16.6.3 Protection.**

**16.6.3.1 Protection of Vertical Openings.**

**16.6.3.1.1** For group day-care homes, the doorway between the level of exit discharge and any story below shall be equipped with a fire door assembly having a 20-minute fire protection rating.

**16.6.3.1.2** For group day-care homes where the story above the level of exit discharge is used for sleeping purposes, there shall be a fire door assembly having a 20-minute fire protection rating at the top or bottom of each stairway.

**16.6.3.2 Protection from Hazards.**

**16.6.3.2.1 Alcohol-Based Hand-Rub Dispensers.** Alcohol-based hand-rub dispensers shall be protected in accordance with 8.7.3.1, unless all of the following conditions are met:

(1) Dispensers shall be installed in rooms or spaces separated from corridors and exits.
(2) The maximum individual dispenser fluid capacity shall be as follows:
  (a) 0.32 gal (1.2 L) for dispensers in rooms
  (b) 0.53 gal (2.0 L) for dispensers in suites of rooms
(3) Dispensers shall be separated from each other by horizontal spacing of not less than 48 in. (1220 mm).
(4) Storage of quantities greater than 5 gal (18.9 L) in a single fire compartment shall meet the requirements of NFPA 30, *Flammable and Combustible Liquids Code.*
(5) Dispensers shall not be installed over or directly adjacent to an ignition source.
(6) Dispensers installed directly over carpeted floors shall be permitted only in sprinklered rooms or spaces.

**16.6.3.2.2 Reserved.**

**16.6.3.3 Interior Finish.**

**16.6.3.3.1 General.** Interior finish shall be in accordance with Section 10.2.

**16.6.3.3.2 Interior Wall and Ceiling Finish.**

**16.6.3.3.2.1** Interior wall and ceiling finish materials complying with Section 10.2 shall be Class A or Class B in corridors, stairways, lobbies, and exits. In the exits of family day-care homes, interior wall and ceiling finish materials in accordance with Section 10.2 shall be Class A or Class B.

**16.6.3.3.2.2** Interior wall and ceiling finish materials complying with Section 10.2 shall be Class A, Class B, or Class C in occupied spaces.

**16.6.3.3.3 Interior Floor Finish.**

**16.6.3.3.3.1** Interior floor finish shall comply with Section 10.2.

**16.6.3.3.3.2** Interior floor finish in exit enclosures shall be not less than Class II.

**16.6.3.3.3.3** Interior floor finish shall comply with 10.2.7.1 or 10.2.7.2, as applicable.

**16.6.3.4 Detection, Alarm, and Communications Systems.**

**16.6.3.4.1** Smoke alarms shall be installed within day-care homes in accordance with 9.6.2.10.

**16.6.3.4.2** Where a day-care home is located within a building of another occupancy, such as in an apartment building or office building, any corridors serving the day-care home shall be provided with a smoke detection system in accordance with Section 9.6.

**16.6.3.4.3** Single-station or multiple-station smoke alarms or smoke detectors shall be provided in all rooms used for sleeping in accordance with 9.6.2.10.

**16.6.3.4.4 Reserved.**

**16.6.3.4.5** Single-station or multiple-station carbon monoxide alarms or detectors shall be provided in accordance with Section 9.8 in day-care homes where client sleeping occurs and one or both of the following conditions exist:

(1) Fuel-fired equipment is present.
(2) An enclosed parking structure is attached to the day-care home.

**16.6.3.5 Extinguishment Requirements.** Any required sprinkler systems shall be in accordance with Section 9.7 and shall be installed in accordance with 9.7.1.1(1), (2), or (3), as appropriate with respect to the scope of the installation standard.

**16.7 Operating Features.**

**16.7.1\* Emergency Plans.** Emergency plans shall be provided in accordance with Section 4.8.

**16.7.2 Emergency Egress and Relocation Drills.**

**16.7.2.1\*** Emergency egress and relocation drills shall be conducted in accordance with Section 4.7 and the applicable provisions of 16.7.2.2.

**16.7.2.2** Emergency egress and relocation drills shall be conducted as follows:

(1) Not less than one emergency egress and relocation drill shall be conducted every month the facility is in session, unless both of the following criteria are met:
  (a) In climates where the weather is severe, the monthly emergency egress and relocation drills shall be permitted to be deferred.
  (b) The required number of emergency egress and relocation drills shall be conducted, and not less than four shall be conducted before the drills are deferred.
(2) All occupants of the building shall participate in the drill.
(3) One additional emergency egress and relocation drill, other than for day-care occupancies that are open on a year-round basis, shall be required within the first 30 days of operation.

**16.7.3 Inspections.**

**16.7.3.1** Fire prevention inspections shall be conducted monthly by a trained senior member of the staff, after which a copy of the latest inspection report shall be posted in a conspicuous place in the day-care facility.

**16.7.3.2\*** It shall be the duty of site administrators and staff members to inspect all exit facilities daily to ensure that all stairways, doors, and other exits are in proper condition.

**16.7.3.3** Open plan buildings shall require extra surveillance to ensure that exit paths are maintained clear of obstruction and are obvious.

**16.7.3.4 Inspection of Door Openings.** Door openings shall be inspected in accordance with 7.2.1.15.

**16.7.4 Furnishings and Decorations.**

**16.7.4.1** Draperies, curtains, and other similar furnishings and decorations in day-care occupancies, other than in day-care homes, shall be in accordance with the provisions of 10.3.1.

**16.7.4.2** Clothing and personal effects shall not be stored in corridors, unless otherwise permitted by one of the following:

(1) This requirement shall not apply to corridors protected by an automatic sprinkler system in accordance with Section 9.7.
(2) This requirement shall not apply to corridor areas protected by a smoke detection system in accordance with Section 9.6.
(3) This requirement shall not apply to storage in metal lockers, provided that the required egress width is maintained.

**16.7.4.3** Artwork and teaching materials shall be permitted to be attached directly to the walls in accordance with the following:

(1) The artwork and teaching materials shall not exceed 20 percent of the wall area in a building that is not protected throughout by an approved, supervised automatic sprinkler system in accordance with Section 9.7.
(2) The artwork and teaching materials shall not exceed 50 percent of the wall area in a building that is protected throughout by an approved, supervised automatic sprinkler system in accordance with Section 9.7.

**16.7.4.4** The provision of 10.3.2 for cigarette ignition resistance of newly introduced upholstered furniture and mattresses shall not apply to day-care homes.

**16.7.5\* Day-Care Staff.** Adequate adult staff shall be on duty in the facility and alert at all times where clients are present.

## Chapter 17   Existing Day-Care Occupancies

**17.1 General Requirements.**

**17.1.1\* Application.**

**17.1.1.1** The requirements of this chapter shall apply to existing buildings or portions thereof currently occupied as day-care occupancies.

**17.1.1.2 Administration.** The provisions of Chapter 1, Administration, shall apply.

**17.1.1.3 General.** The provisions of Chapter 4, General, shall apply.

**17.1.1.4** The requirements of Sections 17.1 through 17.5 and Section 17.7 shall apply to existing day-care occupancies in which more than 12 clients receive care, maintenance, and supervision by other than their relative(s) or legal guardian(s) for less than 24 hours per day. An existing day-care occupancy shall be permitted the option of meeting the requirements of Chapter 16 in lieu of Chapter 17. An existing day-care occupancy that meets the

requirements of Chapter 16 shall be judged as meeting the requirements of Chapter 17.

**17.1.1.5** The requirements of Section 17.1 and Sections 17.4 through 17.7 shall apply to existing day-care homes as defined in 17.1.4. An existing day-care home shall be permitted the option of meeting the requirements of Chapter 16 in lieu of Chapter 17. An existing day-care home that meets the requirements of Chapter 16 shall be judged as meeting the requirements of Chapter 17.

**17.1.1.6** Where a facility houses clients of more than one self-preservation capability, the strictest requirements applicable to any group present shall apply throughout the day-care occupancy or building, as appropriate to a given area, unless the area housing such a group is maintained as a separate fire area.

**17.1.1.7** Places of religious worship shall not be required to meet the provisions of this chapter where providing day care while services are being held in the building.

**17.1.2 Classification of Occupancy.** See 6.1.4.

**17.1.2.1 General.** Occupancies that include part-day preschools, kindergartens, and other schools whose purpose is primarily educational, even though the children who attend such schools are of preschool age, shall comply with the provisions of Chapter 15.

**17.1.2.2 Adult Day-Care Occupancies.**

**17.1.2.2.1** Adult day-care occupancies shall include any building or portion thereof used for less than 24 hours per day to house more than three adults requiring care, maintenance, and supervision by other than their relative(s).

**17.1.2.2.2** Clients in adult day-care occupancies shall be ambulatory or semiambulatory and shall not be bedridden.

**17.1.2.2.3** Clients in adult day-care occupancies shall not exhibit behavior that is harmful to themselves or to others.

**17.1.2.3\* Conversions.** A conversion from a day-care home to a day-care occupancy with more than 12 clients shall be permitted only if the day-care occupancy conforms to the requirements of Chapter 16 for new day-care occupancies with more than 12 clients.

**17.1.3 Multiple Occupancies.**

**17.1.3.1 General.** Multiple occupancies shall be in accordance with 6.1.14.

**17.1.3.2 Day-Care Occupancies in Apartment Buildings.** If the two exit accesses from a day-care occupancy enter the same corridor as an apartment occupancy, the exit accesses shall be separated in the corridor by a smoke partition complying with both of the following:

(1) It shall have not less than a 1-hour fire resistance rating and shall be constructed in accordance with Section 8.4.
(2) It shall be located so that it has an exit on each side.

**17.1.4 Definitions.**

**17.1.4.1 General.** For definitions, see Chapter 3, Definitions.

**17.1.4.2 Special Definitions.** A list of special terms used in this chapter follows:

(1) **Day-Care Home.** See 3.3.140.1.
(2) **Flexible Plan and Open Plan Educational or Day-Care Building.** See 3.3.36.6.

(3) **Self-Preservation (Day-Care Occupancy).** See 3.3.240.

(4) **Separate Atmosphere.** See 3.3.26.2.

**17.1.5 Classification of Hazard of Contents.** The contents of day-care occupancies shall be classified as ordinary hazard in accordance with Section 6.2.

**17.1.6 Minimum Construction Requirements.**

**17.1.6.1** Day-care occupancies, other than day-care homes, shall be limited to the building construction types specified in Table 17.1.6.1 based on the number of stories in height as defined in 4.6.3. *(See 8.2.1.)*

**17.1.6.2 Reserved.**

**17.1.7 Occupant Load.**

**17.1.7.1** The occupant load, in number of persons for whom means of egress and other provisions are required, either shall be determined on the basis of the occupant load factors of Table 7.3.1.2 that are characteristic of the use of the space or shall be determined as the maximum probable population of the space under consideration, whichever is greater.

**17.1.7.2** Where the occupant load is determined as the maximum probable population of the space in accordance with 17.1.7.1, an approved aisle, seating, and exiting diagram shall be required by the authority having jurisdiction to substantiate such a modification.

**17.2 Means of Egress Requirements.**

**17.2.1 General.** Means of egress shall be in accordance with Chapter 7 and Section 17.2.

**17.2.2 Means of Egress Components.**

**17.2.2.1 Components Permitted.** Components of means of egress shall be limited to the types described in 17.2.2.2 through 17.2.2.10.

**17.2.2.2 Doors.**

**17.2.2.2.1 General.** Doors complying with 7.2.1 shall be permitted.

**17.2.2.2.2 Panic Hardware or Fire Exit Hardware.** Any door in a required means of egress from an area having an occupant

**Table 17.1.6.1 Construction Type Limitations**

| Construction Type | Sprinklered[a] | Stories in Height[b] | | | | | |
|---|---|---|---|---|---|---|---|
| | | One Story Below[c] | 1 | 2 | 3–4 | >4 but Not High-Rise | High-Rise |
| I (442) | Yes | X | X | X | X | X | X |
| | No | X | X | X | X | X | NP |
| I (332) | Yes | X | X | X | X | X | X |
| | No | X | X | X | X | X | NP |
| II (222) | Yes | X | X | X | X | X | X |
| | No | X | X | X | X | X | NP |
| II (111) | Yes | X | X | X | X[d] | X[d] | NP |
| | No | X | X | X[d] | NP | NP | NP |
| II (000) | Yes | X | X | X | NP | NP | NP |
| | No | NP | X | X | NP | NP | NP |
| III (211) | Yes | X | X | X | X[d] | NP | NP |
| | No | X | X | X[d] | NP | NP | NP |
| III (200) | Yes | NP | X | X | NP | NP | NP |
| | No | NP | X | X | NP | NP | NP |
| IV (2HH) | Yes | X | X | X | NP | NP | NP |
| | No | X | X | X | NP | NP | NP |
| V (111) | Yes | X | X | X | X[d] | NP | NP |
| | No | X | X | X[d] | NP | NP | NP |
| V (000) | Yes | NP | X | X | NP | NP | NP |
| | No | NP | X | NP | NP | NP | NP |

X: Permitted. NP: Not Permitted.

[a]Sprinklered throughout by an approved, supervised automatic sprinkler system in accordance with Section 9.7. *(See 17.3.5.)*

[b]See 4.6.3.

[c]One story below the level of exit discharge.

[d] Permitted only if clients capable of self-preservation.

load of 100 or more persons shall be permitted to be provided with a latch or lock only if the latch or lock is panic hardware or fire exit hardware complying with 7.2.1.7.

**17.2.2.2.3 Special Locking Arrangements.**

**17.2.2.2.3.1** Delayed-egress locking systems complying with 7.2.1.6.1 shall be permitted.

**17.2.2.2.3.2** Access-controlled egress door assemblies complying with 7.2.1.6.2 shall be permitted.

**17.2.2.2.3.3** Elevator lobby exit access door assemblies locking in accordance with 7.2.1.6.3 shall be permitted.

**17.2.2.2.4\* Door Latches.** Every door latch to closets, storage areas, kitchens, and other similar spaces or areas shall be such that clients can open the door from inside the space or area.

**17.2.2.2.5 Bathroom Doors.** Every bathroom door lock shall be designed to allow opening of the locked door from the outside by an opening device that shall be readily accessible to the staff.

**17.2.2.3\* Stairs.** Stairs complying with 7.2.2 shall be permitted.

**17.2.2.4 Smokeproof Enclosures.** Smokeproof enclosures complying with 7.2.3 shall be permitted.

**17.2.2.5 Horizontal Exits.**

**17.2.2.5.1** Horizontal exits complying with 7.2.4 shall be permitted.

**17.2.2.5.2** Day-care occupancies located six or more stories above the level of exit discharge shall have horizontal exits to provide areas of refuge, unless the building meets one of the following criteria:

(1) The building is provided with smokeproof enclosures.
(2) The building is protected throughout by an approved, supervised automatic sprinkler system in accordance with Section 9.7.

**17.2.2.6 Ramps.** Ramps complying with 7.2.5 shall be permitted.

**17.2.2.7 Exit Passageways.** Exit passageways complying with 7.2.6 shall be permitted.

**17.2.2.8 Fire Escape Ladders.** Fire escape ladders complying with 7.2.9 shall be permitted.

**17.2.2.9 Alternating Tread Devices.** Alternating tread devices complying with 7.2.11 shall be permitted.

**17.2.2.10 Areas of Refuge.** Areas of refuge complying with 7.2.12 shall be permitted.

**17.2.3 Capacity of Means of Egress.** Capacity of means of egress shall be in accordance with Section 7.3.

**17.2.4 Number of Means of Egress.**

**17.2.4.1** The number of means of egress shall be in accordance with 7.4.1.1 and 7.4.1.3 through 7.4.1.6.

**17.2.4.2** Not less than two separate exits shall be in accordance with both of the following criteria:

(1) They shall be provided on every story.
(2) They shall be accessible from every part of every story and mezzanine; however, exit access travel shall be permitted to be common for the distance permitted as common path of travel by 17.2.5.3.

**17.2.4.3** Where the story below the level of exit discharge is occupied as a day-care occupancy, 17.2.4.3.1 and 17.2.4.3.2 shall apply.

**17.2.4.3.1** One means of egress shall be an outside or interior stair in accordance with 7.2.2. An interior stair, if used, shall serve only the story below the level of exit discharge. The interior stair shall be permitted to communicate with the level of exit discharge; however, the exit route from the level of exit discharge shall not pass through the stair enclosure.

**17.2.4.3.2** The second means of egress shall be permitted to be via an unenclosed stairway separated from the level of exit discharge in accordance with 8.6.5.

**17.2.4.3.3** The path of egress travel on the level of exit discharge shall be protected in accordance with 7.1.3.1, unless one of the following criteria is met:

(1) The path of egress on the level of exit discharge shall be permitted to be unprotected if the level of exit discharge and the level below the level of exit discharge are protected throughout by a smoke detection system.
(2) The path of egress on the level of exit discharge shall be permitted to be unprotected if the level of exit discharge and the level below the level of exit discharge are protected throughout by an approved automatic sprinkler system.

**17.2.5 Arrangement of Means of Egress.**

**17.2.5.1** Means of egress shall be arranged in accordance with Section 7.5.

**17.2.5.2** No dead-end corridor shall exceed 20 ft (6100 mm), other than in buildings protected throughout by an approved, supervised automatic sprinkler system in accordance with Section 9.7, in which case dead-end corridors shall not exceed 50 ft (15 m).

**17.2.5.3** Limitations on common path of travel shall be in accordance with 17.2.5.3.1 and 17.2.5.3.2.

**17.2.5.3.1** Common path of travel shall not exceed 100 ft (30 m) in a building protected throughout by an approved, supervised automatic sprinkler system in accordance with Section 9.7.

**17.2.5.3.2** Common path of travel shall not exceed 75 ft (23 m) in a building not protected throughout by an approved, supervised automatic sprinkler system in accordance with Section 9.7.

**17.2.5.4** The story used below the level of exit discharge shall be in accordance with 17.2.4.3.

**17.2.6 Travel Distance to Exits.**

**17.2.6.1** Travel distance shall be measured in accordance with Section 7.6.

**17.2.6.2** Travel distance shall meet all of the following criteria, unless otherwise permitted by 17.2.6.3:

(1) The travel distance between any room door intended as an exit access and an exit shall not exceed 100 ft (30 m).
(2) The travel distance between any point in a room and an exit shall not exceed 150 ft (46 m).
(3) The travel distance between any point in a sleeping room and an exit access door in that room shall not exceed 50 ft (15 m).

**17.2.6.3** The travel distance required by 17.2.6.2(1) and (2) shall be permitted to be increased by 50 ft (15 m) in buildings protected throughout by an approved automatic sprinkler system in accordance with Section 9.7.

**17.2.7 Discharge from Exits.** Discharge from exits shall be arranged in accordance with Section 7.7, unless otherwise provided in 17.2.4.3.

**17.2.8 Illumination of Means of Egress.** Means of egress shall be illuminated in accordance with Section 7.8.

**17.2.9 Emergency Lighting.** Emergency lighting shall be provided in accordance with Section 7.9 in the following areas:

(1) Interior stairs and corridors
(2) Assembly use spaces
(3) Flexible and open plan buildings
(4) Interior or limited access portions of buildings
(5) Shops and laboratories

**17.2.10 Marking of Means of Egress.** Means of egress shall have signs in accordance with Section 7.10.

**17.2.11 Special Means of Egress Features.**

**17.2.11.1 Windows for Rescue.**

**17.2.11.1.1** Every room or space greater than 250 ft$^2$ (23.2 m$^2$) and normally subject to client occupancy shall have not less than one outside window for emergency rescue that complies with all of the following, unless otherwise permitted by 17.2.11.1.2:

(1) Such windows shall be openable from the inside without the use of tools and shall provide a clear opening of not less than 20 in. (510 mm) in width, 24 in. (610 mm) in height, and 5.7 ft$^2$ (0.5 m$^2$) in area.
(2) The bottom of the opening shall be not more than 44 in. (1120 mm) above the floor
(3) The clear opening shall allow a rectangular solid, with a width and height that provides not less than the required 5.7 ft$^2$ (0.5 m$^2$) opening and a depth of not less than 20 in. (510 mm), to pass fully through the opening.

**17.2.11.1.2** The requirements of 17.2.11.1.1 shall not apply to any of the following:

(1) Buildings protected throughout by an approved, supervised automatic sprinkler system in accordance with Section 9.7
(2) Where the room or space has a door leading directly to an exit or directly to the outside of the building
(3) Where the room has a door, in addition to the door that leads to the exit access corridor and such door leads directly to an exit or directly to another corridor located in a compartment separated from the compartment housing the initial corridor by smoke partitions in accordance with Section 8.4
(4) Rooms located four or more stories above the finished ground level
(5) Where awning-type or hopper-type windows that are hinged or subdivided to provide a clear opening of not less than 4 ft$^2$ (0.38 m$^2$) or any dimension of not less than 22 in. (560 mm) meet the following criteria:
  (a) Such windows shall be permitted to continue in use.
  (b) Screen walls or devices in front of required windows shall not interfere with normal rescue requirements.
(6) Where the room or space complies with all of the following:
  (a) One door providing direct access to an adjacent room and a second door providing direct access to another adjacent room shall be provided.

(b) The two rooms to which exit access travel is made in accordance with 17.2.11.1.2(6)(a) shall each provide exit access in accordance with 17.2.11.1.2(2) or (3)
(c) The corridor required by 17.2.5.5 and the corridor addressed by 17.2.11.1.2(3), if provided, shall be separated from the rooms by a wall that resists the passage of smoke, and all doors between the rooms and the corridor shall be self-closing in accordance with 7.2.1.8.
(d) The length of travel to exits along such paths shall not exceed 150 ft (46 m).
(e) Each communicating door shall be marked in accordance with Section 7.10.
(f) No locking device shall be permitted on the communicating doors.

**17.2.11.2 Lockups.** Lockups in day-care occupancies, other than approved existing lockups, shall comply with the requirements of 23.4.5.

**17.3 Protection.**

**17.3.1 Protection of Vertical Openings.** Any vertical opening, other than unprotected vertical openings in accordance with 8.6.9.1, shall be enclosed or protected in accordance with Section 8.6.

**17.3.2 Protection from Hazards.**

**17.3.2.1** Rooms or spaces for the storage, processing, or use of materials specified in 17.3.2.1(1) through (3) shall be protected in accordance with the following:

(1) Separation from the remainder of the building by fire barriers having a minimum 1-hour fire resistance rating, or protection of such rooms by automatic extinguishing systems as specified in Section 8.7, in the following areas:
  (a) Boiler and furnace rooms, unless such rooms enclose only air-handling equipment
  (b) Rooms or spaces used for the storage of combustible supplies in quantities deemed hazardous by the authority having jurisdiction
  (c) Rooms or spaces used for the storage of hazardous materials or flammable or combustible liquids in quantities deemed hazardous by recognized standards
  (d) Janitor closets
(2) Separation from the remainder of the building by fire barriers having a minimum 1-hour fire resistance rating and protection of such rooms by automatic extinguishing systems as specified in Section 8.7 in the following areas:
  (a)*Laundries
  (b) Maintenance shops, including woodworking and painting areas
  (c) Rooms or spaces used for processing or use of combustible supplies deemed hazardous by the authority having jurisdiction
  (d) Rooms or spaces used for processing or use of hazardous materials or flammable or combustible liquids in quantities deemed hazardous by recognized standards
(3) Where automatic extinguishing is used to meet the requirements of 17.3.2.1(1) and (2), protection as permitted in accordance with 9.7.1.2

**17.3.2.2** Janitor closets protected in accordance with 17.3.2.1(1)(d) shall be permitted to have doors fitted with ventilating louvers where the space is protected by automatic sprinklers.

**17.3.2.3** Cooking facilities shall be protected in accordance with 9.2.3, unless otherwise permitted by 17.3.2.4 or 17.3.2.5.

**17.3.2.4** Openings shall not be required to be protected between food preparation areas and dining areas.

**17.3.2.5** Approved domestic cooking equipment used for food warming or limited cooking shall not be required to be protected.

**17.3.2.6 Alcohol-Based Hand-Rub Dispensers** Alcohol-based hand-rub dispensers shall be protected in accordance with 8.7.3.1, unless all of the following conditions are met:

(1) Dispensers shall be installed in rooms or spaces separated from corridors and exits.
(2) The maximum individual dispenser fluid capacity shall be as follows:
   (a) 0.32 gal (1.2 L) for dispensers in rooms
   (b) 0.53 gal (2.0 L) for dispensers in suites of rooms
(3) Dispensers shall be separated from each other by horizontal spacing of not less than 48 in. (1220 mm).
(4) Storage of quantities greater than 5 gal (18.9 L) in a single fire compartment shall meet the requirements of NFPA 30, *Flammable and Combustible Liquids Code*.
(5) Dispensers shall not be installed over or directly adjacent to an ignition source.
(6) Dispensers installed directly over carpeted floors shall be permitted only in sprinklered rooms or spaces.

**17.3.3 Interior Finish.**

**17.3.3.1 General.** Interior finish shall be in accordance with Section 10.2.

**17.3.3.2 Interior Wall and Ceiling Finish.** Interior wall and ceiling finish materials complying with Section 10.2 shall be Class A or Class B throughout.

**17.3.3.3 Interior Floor Finish.** (No requirements.)

**17.3.4 Detection, Alarm, and Communications Systems.**

**17.3.4.1 General.** Day-care occupancies, other than day-care occupancies housed in one room, shall be provided with a fire alarm system in accordance with Section 9.6.

**17.3.4.2 Initiation.** Initiation of the required fire alarm system shall be by manual means and by operation of any required smoke detectors and required sprinkler systems. *(See 17.3.4.5.)*

**17.3.4.3 Occupant Notification.**

**17.3.4.3.1** Occupant notification shall be in accordance with 9.6.3.

**17.3.4.3.2** Positive alarm sequence shall be permitted in accordance with 9.6.3.4.

**17.3.4.3.3** Private operating mode in accordance with 9.6.3.6.3 shall be permitted.

**17.3.4.4 Emergency Forces Notification.**

**17.3.4.4.1** Emergency forces notification, other than for day-care occupancies with not more than 100 clients, shall be accomplished in accordance with 9.6.4.

**17.3.4.4.2** Emergency forces notification shall be accomplished in accordance with 9.6.4 where the existing fire alarm system is replaced.

**17.3.4.5 Detection.** A smoke detection system in accordance with Section 9.6 shall be installed in day-care occupancies, other than those housed in one room or those housing clients capable of self-preservation where no sleeping facilities are provided, and such system shall comply with both of the following:

(1) Detectors shall be installed on each story in front of the doors to the stairways and in the corridors of all floors occupied by the day-care occupancy.
(2) Detectors shall be installed in lounges, recreation areas, and sleeping rooms in the day-care occupancy.

**17.3.5 Extinguishment Requirements.**

**17.3.5.1** Any required sprinkler system shall be in accordance with Section 9.7.

**17.3.5.2** Required sprinkler systems, other than approved existing systems, shall be installed in accordance with 9.7.1.1(1).

**17.3.5.3** Buildings with unprotected openings in accordance with 8.6.6 shall be protected throughout by an approved, supervised automatic sprinkler system in accordance with Section 9.7.

**17.3.6 Corridors.** Every interior corridor shall be constructed of walls having a minimum ½-hour fire resistance rating in accordance with Section 8.3, unless otherwise permitted by any of the following:

(1) Corridor protection shall not be required where all spaces normally subject to student occupancy have not less than one door opening directly to the outside or to an exterior exit access balcony or corridor in accordance with 7.5.3.
(2) In buildings protected throughout by an approved automatic sprinkler system with valve supervision in accordance with Section 9.7, corridor walls shall not be required to be rated, provided that such walls form smoke partitions in accordance with Section 8.4.
(3) Where the corridor ceiling is an assembly having a minimum ½-hour fire resistance rating where tested as a wall, the corridor walls shall be permitted to terminate at the corridor ceiling.
(4) Lavatories shall not be required to be separated from corridors, provided that they are separated from all other spaces by walls having a minimum ½-hour fire resistance rating in accordance with Section 8.3.
(5) Lavatories shall not be required to be separated from corridors, provided that both of the following criteria are met:
   (a) The building is protected throughout by an approved, supervised automatic sprinkler system in accordance with Section 9.7.
   (b) The walls separating the lavatory from other rooms form smoke partitions in accordance with Section 8.4.

**17.4 Special Provisions.**

**17.4.1 Limited Access Buildings and Underground Buildings.** Limited access buildings and underground buildings shall comply with Section 11.7.

**17.4.2 High-Rise Buildings.** High-rise buildings that house day-care occupancies on floors more than 75 ft (23 m) above the lowest level of fire department vehicle access shall comply with Section 11.8.

**17.4.3 Flexible Plan and Open Plan Buildings.**

**17.4.3.1** Flexible plan and open plan buildings shall comply with the requirements of this chapter as modified by 17.4.3.2 and 17.4.3.3.

**17.4.3.2** Flexible plan buildings shall be permitted to have walls and partitions rearranged periodically only if revised plans or diagrams have been approved by the authority having jurisdiction.

**17.4.3.3** Flexible plan buildings shall be evaluated while all folding walls are extended and in use as well as when they are in the retracted position.

## 17.5 Building Services.

### 17.5.1 Utilities.

**17.5.1.1** Utilities shall comply with the provisions of Section 9.1.

**17.5.1.2** Special protective covers for all electrical receptacles shall be installed in all areas occupied by clients.

### 17.5.2 Heating, Ventilating, and Air-Conditioning Equipment.

**17.5.2.1** Heating, ventilating, and air-conditioning equipment shall be in accordance with Section 9.2.

**17.5.2.2** Unvented fuel-fired heating equipment, other than gas space heaters in compliance with NFPA 54/ANSI Z 223.1, *National Fuel Gas Code*, shall be prohibited.

**17.5.2.3** Any heating equipment in spaces occupied by clients shall be provided with partitions, screens, or other means to protect clients from hot surfaces and open flames; if solid partitions are used to provide such protection, provisions shall be made to ensure adequate air for combustion and ventilation for the heating equipment.

### 17.5.3 Elevators, Escalators, and Conveyors.
Elevators, escalators, and conveyors, other than those in day-care homes, shall comply with the provisions of Section 9.4.

### 17.5.4 Rubbish Chutes, Incinerators, and Laundry Chutes.
Rubbish chutes, incinerators, and laundry chutes, other than those in day-care homes, shall comply with the provisions of Section 9.5.

## 17.6 Day-Care Homes.

### 17.6.1 General Requirements.

#### 17.6.1.1 Application.

**17.6.1.1.1 Reserved.**

**17.6.1.1.2\*** The requirements of Section 17.6 shall apply to existing day-care homes in which more than 3, but not more than 12, clients receive care, maintenance, and supervision by other than their relative(s) or legal guardian(s) for less than 24 hours per day, generally within a dwelling unit. An existing day-care home shall be permitted the option of meeting the requirements of Section 16.6 in lieu of Section 17.6. Any existing day-care home that meets the requirements of Chapter 16 shall be judged as meeting the requirements of this chapter. *(See also 17.6.1.4.)*

**17.6.1.1.3** Where a facility houses clients of more than one self-preservation capability, the strictest requirements applicable to any group present shall apply throughout the day-care home or building, as appropriate to a given area, unless the area housing such a group is maintained as a separate fire area.

**17.6.1.1.4** Facilities that supervise clients on a temporary basis with a parent or guardian in close proximity shall not be required to meet the provisions of Section 17.6.

**17.6.1.1.5** Places of religious worship shall not be required to meet the provisions of Section 17.6 where operating a day-care home while services are being held in the building.

**17.6.1.2 Multiple Occupancies.** See 17.1.3.

**17.6.1.3 Definitions.** See 17.1.4.

**17.6.1.4 Classification of Occupancy.**

**17.6.1.4.1 Subclassification of Day-Care Homes.** Subclassification of day-care homes shall comply with 17.6.1.4.1.1 and 17.6.1.4.1.2.

**17.6.1.4.1.1 Family Day-Care Home.** A family day-care home shall be a day-care home in which more than three, but fewer than seven, clients receive care, maintenance, and supervision by other than their relative(s) or legal guardian(s) for less than 24 hours per day, generally within a dwelling unit.

**17.6.1.4.1.2 Group Day-Care Home.** A group day-care home shall be a day-care home in which not less than 7, but not more than 12, clients receive care, maintenance, and supervision by other than their relative(s) or legal guardian(s) for less than 24 hours per day, generally within a dwelling unit.

**17.6.1.4.2\* Conversions.** A conversion from a day-care home to a day-care occupancy with more than 12 clients shall be permitted only if the day-care occupancy conforms to the requirements of Chapter 16 for new day-care occupancies with more than 12 clients.

**17.6.1.5 Classification of Hazard of Contents.** See 17.1.5.

**17.6.1.6 Location and Construction.** No day-care home shall be located more than one story below the level of exit discharge.

**17.6.1.7 Occupant Load.**

**17.6.1.7.1** In family day-care homes, both of the following shall apply:

(1) The minimum staff-to-client ratio shall be not less than one staff for up to six clients, including the caretaker's own children under age six.
(2) There shall be not more than two clients incapable of self-preservation.

**17.6.1.7.2** In group day-care homes, all of the following shall apply:

(1) The minimum staff-to-client ratio shall be not less than two staff for up to 12 clients.
(2) There shall be not more than 3 clients incapable of self-preservation.
(3) The staff-to-client ratio shall be permitted to be modified by the authority having jurisdiction where safeguards in addition to those specified by Section 17.6 are provided.

### 17.6.2 Means of Escape Requirements.

**17.6.2.1 General.** Means of escape shall comply with Section 24.2.

**17.6.2.2 Reserved.**

**17.6.2.3 Reserved.**

**17.6.2.4 Number and Type of Means of Escape.**

**17.6.2.4.1** The number and type of means of escape shall comply with Section 24.2 and 17.6.2.4.1 through 17.6.2.4.4.

**17.6.2.4.2** Every room used for sleeping, living, recreation, education, or dining purposes shall have the number and type of means of escape in accordance with Section 24.2.

**17.6.2.4.3** No room or space that is accessible only by a ladder or folding stairs or through a trap door shall be occupied by clients.

**17.6.2.4.4** In group day-care homes where spaces on the story above the level of exit discharge are used by clients, that story shall have not less than one means of escape complying with one of the following:

(1) Door leading directly to the outside with access to finished ground level
(2) Door leading directly to an outside stair to finished ground level
(3) Interior stair leading directly to the outside with access to finished ground level separated from other stories by a ½-hour fire barrier in accordance with Section 8.3
(4) Interior stair leading directly to the outside with access to finished ground level separated from other stories by a barrier that has been previously approved for use in a group day-care home

**17.6.2.4.5** Where clients occupy a story below the level of exit discharge, that story shall have not less than one means of escape complying with one of the following:

(1) Door leading directly to the outside with access to finished ground level
(2) Door leading directly to an outside stair to finished ground level
(3) Bulkhead enclosure complying with 24.2.7
(4) Interior stair leading directly to the outside with access to finished ground level separated from other stories by a ½-hour fire barrier in accordance with Section 8.3
(5) Interior stair leading directly to the outside with access to finished ground level separated from other stories by a barrier that has been previously approved for use in a group day-care home

**17.6.2.5 Arrangement of Means of Escape.**

**17.6.2.5.1** A story used above or below the level of exit discharge shall be in accordance with 17.6.2.4.3 or 17.6.2.4.4.

**17.6.2.5.2** For group day-care homes, means of escape shall be arranged in accordance with Section 7.5.

**17.6.2.5.3** No dead-end corridor shall exceed 20 ft (6100 mm), other than in buildings protected throughout by an approved, supervised automatic sprinkler system in accordance with Section 9.7, in which case dead-end corridors shall not exceed 50 ft (15 m).

**17.6.2.5.4** Doors in means of escape shall be protected from obstructions, including snow and ice.

**17.6.2.6 Travel Distance.** Travel distance shall comply with 17.6.2.6.1 through 17.6.2.6.3.

**17.6.2.6.1** Travel distance shall be measured in accordance with Section 7.6.

**17.6.2.6.2** Travel distance shall meet all of the following criteria, unless otherwise permitted by 17.6.2.6.3:

(1) The travel distance between any room door intended as an exit access and an exit shall not exceed 100 ft (30 m).
(2) The travel distance between any point in a room and an exit shall not exceed 150 ft (46 m).

(3) The travel distance between any point in a sleeping room and an exit access to that room shall not exceed 50 ft (15 m).

**17.6.2.6.3** The travel distance required by 17.6.2.6.2(1) and (2) shall be permitted to be increased by 50 ft (15 m) in buildings protected throughout by an approved, supervised automatic sprinkler system in accordance with Section 9.7.

**17.6.2.7 Discharge from Exits.** See 17.6.2.4.

**17.6.2.8 Illumination of Means of Egress.** Means of egress shall be illuminated in accordance with Section 7.8.

**17.6.2.9 Emergency Lighting.** (No requirements.)

**17.6.2.10 Marking of Means of Egress.** (No requirements.)

**17.6.3 Protection.**

**17.6.3.1 Protection of Vertical Openings.**

**17.6.3.1.1** For group day-care homes, the doorway between the level of exit discharge and any story below shall be equipped with a fire door assembly having a 20-minute fire protection rating.

**17.6.3.1.2** For group day-care homes where the story above the level of exit discharge is used for sleeping purposes, there shall be a fire door assembly having a 20-minute fire protection rating at the top or bottom of each stairway, unless otherwise permitted by 17.6.3.1.3.

**17.6.3.1.3** Approved, existing, self-closing, 1¾ in. (44 mm) thick, solid-bonded wood doors without rated frames shall be permitted to continue in use.

**17.6.3.2 Protection from Hazards.**

**17.6.3.2.1 Alcohol-Based Hand-Rub Dispensers.** Alcohol-based hand-rub dispensers shall be protected in accordance with 8.7.3.1, unless all of the following conditions are met:

(1) Dispensers shall be installed in rooms or spaces separated from corridors and exits.
(2) The maximum individual dispenser fluid capacity shall be as follows:
  (a) 0.32 gal (1.2 L) for dispensers in rooms
  (b) 0.53 gal (2.0 L) for dispensers in suites of rooms
(3) Dispensers shall be separated from each other by horizontal spacing of not less than 48 in. (1220 mm).
(4) Storage of quantities greater than 5 gal (18.9 L) in a single fire compartment shall meet the requirements of NFPA 30, *Flammable and Combustible Liquids Code.*
(5) Dispensers shall not be installed over or directly adjacent to an ignition source.
(6) Dispensers installed directly over carpeted floors shall be permitted only in sprinklered rooms or spaces.

**17.6.3.2.2 Reserved.**

**17.6.3.3 Interior Finish.**

**17.6.3.3.1 General.** Interior finish shall be in accordance with Section 10.2.

**17.6.3.3.2 Interior Wall and Ceiling Finish.**

**17.6.3.3.2.1** Interior wall and ceiling finish materials complying with Section 10.2 shall be Class A or Class B in exits.

**17.6.3.3.2.2** Interior wall and ceiling finish materials complying with Section 10.2 shall be Class A, Class B, or Class C in occupied spaces.

**17.6.3.3.3 Interior Floor Finish.** (No requirements.)

**17.6.3.4 Detection, Alarm, and Communications Systems.**

**17.6.3.4.1** Smoke alarms shall be installed within day-care homes in accordance with 9.6.2.10.

**17.6.3.4.2** Where a day-care home is located within a building of another occupancy, such as in an apartment building or office building, any corridors serving the day-care home shall be provided with a smoke detection system in accordance with Section 9.6.

**17.6.3.4.3** Single-station or multiple-station smoke alarms or smoke detectors shall be provided in all rooms used for sleeping in accordance with 9.6.2.10, other than as permitted by 17.6.3.4.4.

**17.6.3.4.4** Approved existing battery-powered smoke alarms, rather than house electrical service–powered smoke alarms required by 17.6.3.4.3, shall be permitted where the facility has testing, maintenance, and battery replacement programs that ensure reliability of power to the smoke alarms.

**17.6.3.5 Extinguishment Requirements.** Any required sprinkler systems shall be in accordance with Section 9.7 and, other than approved existing systems, shall be installed in accordance with 9.7.1.1(1), (2), or (3), as appropriate with respect to the scope of the installation standard.

**17.7 Operating Features.**

**17.7.1\* Emergency Plans.** Emergency plans shall be provided in accordance with Section 4.8.

**17.7.2 Emergency Egress and Relocation Drills.**

**17.7.2.1\*** Emergency egress and relocation drills shall be conducted in accordance with Section 4.7 and the applicable provisions of 17.7.2.2.

**17.7.2.2** Emergency egress and relocation drills shall be conducted as follows:

(1) Not less than one emergency egress and relocation drill shall be conducted every month the facility is in session, unless both of the following criteria are met:
   (a) In climates where the weather is severe, the monthly emergency egress and relocation drills shall be permitted to be deferred.
   (b) The required number of emergency egress and relocation drills shall be conducted, and not less than four shall be conducted before the drills are deferred.
(2) All occupants of the building shall participate in the drill.
(3) One additional emergency egress and relocation drill, other than for day-care occupancies that are open on a year-round basis, shall be required within the first 30 days of operation.

**17.7.3 Inspections.**

**17.7.3.1** Fire prevention inspections shall be conducted monthly by a trained senior member of the staff, after which a copy of the latest inspection report shall be posted in a conspicuous place in the day-care facility.

**17.7.3.2\*** It shall be the duty of site administrators and staff members to inspect all exit facilities daily to ensure that all stairways, doors, and other exits are in proper condition.

**17.7.3.3** Open plan buildings shall require extra surveillance to ensure that exit paths are maintained clear of obstruction and are obvious.

**17.7.3.4 Inspection of Door Openings.** Door openings shall be inspected in accordance with 7.2.1.15.

**17.7.4 Furnishings and Decorations.**

**17.7.4.1** Draperies, curtains, and other similar furnishings and decorations in day-care occupancies, other than in day-care homes, shall be in accordance with the provisions of 10.3.1.

**17.7.4.2** Clothing and personal effects shall not be stored in corridors, unless otherwise permitted by one of the following:

(1) This requirement shall not apply to corridors protected by an automatic sprinkler system in accordance with Section 9.7.
(2) This requirement shall not apply to corridor areas protected by a smoke detection system in accordance with Section 9.6.
(3) This requirement shall not apply to storage in metal lockers, provided that the required egress width is maintained.

**17.7.4.3** Artwork and teaching materials shall be permitted to be attached directly to the walls in accordance with the following:

(1) The artwork and teaching materials shall not exceed 20 percent of the wall area in a building that is not protected throughout by an approved automatic sprinkler system in accordance with Section 9.7.
(2) The artwork and teaching materials shall not exceed 50 percent of the wall area in a building that is protected throughout by an approved automatic sprinkler system in accordance with Section 9.7.

**17.7.4.4** The provision of 10.3.2 for cigarette ignition resistance of newly introduced upholstered furniture and mattresses shall not apply to day-care homes.

**17.7.5\* Day-Care Staff.** Adequate adult staff shall be on duty in the facility and alert at all times where clients are present.

## Chapter 18 New Health Care Occupancies

**18.1 General Requirements.**

**18.1.1 Application.**

**18.1.1.1 General.**

**18.1.1.1.1\*** The requirements of this chapter shall apply to new buildings or portions thereof used as health care occupancies. *(See 1.3.1.)*

**18.1.1.1.2 Administration.** The provisions of Chapter 1, Administration, shall apply.

**18.1.1.1.3 General.** The provisions of Chapter 4, General, shall apply.

**18.1.1.1.4** The requirements established by this chapter shall apply to the design of all new hospitals, nursing homes, and limited care facilities. The term *hospital*, wherever used in this *Code*, shall include general hospitals, psychiatric hospitals, and specialty hospitals. The term *nursing home*, wherever used in this *Code*, shall include nursing and convalescent homes, skilled nursing facilities, intermediate care facilities, and infirmaries in homes for the aged. Where requirements vary, the specific subclass of health care occupancy that shall apply is named in the paragraph pertaining thereto. The requirements established by Chapter 20 shall apply to all new ambulatory health care facilities. The operating feature requirements established by Section 18.7 shall apply to all health care occupancies.

**18.1.1.1.5** The health care facilities regulated by this chapter shall be those that provide sleeping accommodations for their occupants and are occupied by persons who are mostly incapable of self-preservation because of age, because of physical or mental disability, or because of security measures not under the occupants' control.

**18.1.1.1.6** Buildings, or sections of buildings, that primarily house patients who, in the opinion of the governing body of the facility and the governmental agency having jurisdiction, are capable of exercising judgment and appropriate physical action for self-preservation under emergency conditions shall be permitted to comply with chapters of this *Code* other than Chapter 18.

**18.1.1.1.7\*** It shall be recognized that, in buildings housing certain patients, it might be necessary to lock doors and bar windows to confine and protect building inhabitants.

**18.1.1.1.8** Buildings, or sections of buildings, that house older persons and that provide activities that foster continued independence but that do not include services distinctive to health care occupancies *(see 18.1.4.2)*, as defined in 3.3.188.7, shall be permitted to comply with the requirements of other chapters of this *Code*, such as Chapters 30 or 32.

**18.1.1.1.9** Facilities that do not provide housing on a 24-hour basis for their occupants shall be classified as other occupancies and shall be covered by other chapters of this *Code*.

**18.1.1.1.10\*** The requirements of this chapter shall apply based on the assumption that staff is available in all patient-occupied areas to perform certain fire safety functions as required in other paragraphs of this chapter.

**18.1.1.2\* Goals and Objectives.** The goals and objectives of Sections 4.1 and 4.2 shall be met with due consideration for functional requirements, which are accomplished by limiting the development and spread of a fire emergency to the room of fire origin and reducing the need for occupant evacuation, except from the room of fire origin.

**18.1.1.3 Total Concept.**

**18.1.1.3.1** All health care facilities shall be designed, constructed, maintained, and operated to minimize the possibility of a fire emergency requiring the evacuation of occupants.

**18.1.1.3.2** Because the safety of health care occupants cannot be ensured adequately by dependence on evacuation of the building, their protection from fire shall be provided by appropriate arrangement of facilities; adequate, trained staff; and development of operating and maintenance procedures composed of the following:

(1) Design, construction, and compartmentation
(2) Provision for detection, alarm, and extinguishment
(3) Fire prevention procedures and planning, training, and drilling programs for the isolation of fire, transfer of occupants to areas of refuge, or evacuation of the building

**18.1.1.4 Additions, Conversions, Modernization, Renovation, and Construction Operations.**

**18.1.1.4.1 Additions.** Additions shall be separated from any existing structure not conforming to the provisions within Chapter 19 by a fire barrier having not less than a 2-hour fire resistance rating and constructed of materials as required for the addition. *(See 4.6.7 and 4.6.11.)*

**18.1.1.4.1.1** Communicating openings in dividing fire barriers required by 18.1.1.4.1 shall be permitted only in corridors and shall be protected by approved self-closing fire door assemblies. *(See also Section 8.3.)*

**18.1.1.4.1.2** Doors in barriers required by 18.1.1.4.1 shall normally be kept closed, unless otherwise permitted by 18.1.1.4.1.3.

**18.1.1.4.1.3** Doors shall be permitted to be held open if they meet the requirements of 18.2.2.2.7.

**18.1.1.4.2 Changes of Use or Occupancy Classification.** Changes of use or occupancy classification shall comply with 4.6.11, unless otherwise permitted by one of the following:

(1) A change from a hospital to a nursing home or from a nursing home to a hospital shall not be considered a change in occupancy classification or a change in use.
(2) A change from a hospital or nursing home to a limited care facility shall not be considered a change in occupancy classification or a change in use.
(3) A change from a hospital or nursing home to an ambulatory health care facility shall not be considered a change in occupancy classification or a change in use.

**18.1.1.4.3 Rehabilitation.**

**18.1.1.4.3.1** For purposes of the provisions of this chapter, the following shall apply:

(1) A major rehabilitation shall involve the modification of more than 50 percent, or more than 4500 ft$^2$ (420 m$^2$), of the area of the smoke compartment.
(2) A minor rehabilitation shall involve the modification of not more than 50 percent, and not more than 4500 ft$^2$ (420 m$^2$), of the area of the smoke compartment.

**18.1.1.4.3.2** Work that is exclusively plumbing, mechanical, fire protection system, electrical, medical gas, or medical equipment work shall not be included in the computation of the modification area within the smoke compartment.

**18.1.1.4.3.3\*** Where major rehabilitation is done in a non-sprinklered smoke compartment, the automatic sprinkler requirements of 18.3.5 shall apply to the smoke compartment undergoing the rehabilitation, and, in cases where the building is not protected throughout by an approved automatic sprinkler system, the requirements of 18.4.3.2, 18.4.3.3, and 18.4.3.8 shall also apply.

**18.1.1.4.3.4\*** Where minor rehabilitation is done in a non-sprinklered smoke compartment, the requirements of 18.3.5.1 shall not apply, but, in such cases, the rehabilitation shall not reduce life safety below the level required for new buildings or below the level of the requirements of 18.4.3 for nonsprinklered smoke compartment rehabilitation. *(See 4.6.7.)*

**18.1.1.4.4 Construction, Repair, and Improvement Operations.** See 4.6.10.

**18.1.2 Classification of Occupancy.** See 6.1.5 and 18.1.4.2.

**18.1.3 Multiple Occupancies.**

**18.1.3.1** Multiple occupancies shall be in accordance with 6.1.14.

**18.1.3.2** Sections of health care facilities shall be permitted to be classified as other occupancies in accordance with the separated occupancies provisions of 6.1.14.4 and either 18.1.3.3 or 18.1.3.4.

**18.1.3.3*** Sections of health care facilities shall be permitted to be classified as other occupancies, provided that they meet both of the following conditions:

(1) They are not intended to provide services simultaneously for four or more inpatients for purposes of housing, treatment, or customary access by inpatients incapable of self-preservation.

(2) They are separated from areas of health care occupancies by construction having a minimum 2-hour fire resistance rating in accordance with Chapter 8.

**18.1.3.4 Contiguous Non-Health Care Occupancies.**

**18.1.3.4.1*** Ambulatory care facilities, medical clinics, and similar facilities that are contiguous to health care occupancies, but are primarily intended to provide outpatient services, shall be permitted to be classified as business occupancies or ambulatory health care facilities, provided that the facilities are separated from the health care occupancy by construction having a minimum 2-hour fire resistance rating, and the facility is not intended to provide services simultaneously for four or more inpatients who are incapable of self preservation.

**18.1.3.4.2** Ambulatory care facilities, medical clinics, and similar facilities that are contiguous to health care occupancies shall be permitted to be used for diagnostic and treatment services of inpatients who are capable of self-preservation.

**18.1.3.5** Where separated occupancies provisions are used in accordance with either 18.1.3.3 or 18.1.3.4, the most stringent construction type shall be provided throughout the building, unless a 2-hour separation is provided in accordance with 8.2.1.3, in which case the construction type shall be determined as follows:

(1) The construction type and supporting construction of the health care occupancy shall be based on the story on which it is located in the building in accordance with the provisions of 18.1.6 and Table 18.1.6.1.

(2) The construction type of the areas of the building enclosing the other occupancies shall be based on the applicable occupancy chapters of this *Code*.

**18.1.3.6** All means of egress from health care occupancies that traverse non-health care spaces shall conform to the requirements of this *Code* for health care occupancies, unless otherwise permitted by 18.1.3.7.

**18.1.3.7** Exit through a horizontal exit into other contiguous occupancies that do not conform to health care egress provisions, but that do comply with requirements set forth in the appropriate occupancy chapter of this *Code*, shall be permitted, provided that both of the following criteria apply:

(1) The occupancy does not contain high hazard contents.

(2) The horizontal exit complies with the requirements of 18.2.2.5.

**18.1.3.8** Egress provisions for areas of health care facilities that correspond to other occupancies shall meet the corresponding requirements of this *Code* for such occupancies, and, where the clinical needs of the occupant necessitate the locking of means of egress, staff shall be present for the supervised release of occupants during all times of use.

**18.1.3.9** Auditoriums, chapels, staff residential areas, or other occupancies provided in connection with health care facilities shall have means of egress provided in accordance with other applicable sections of this *Code*.

**18.1.3.10** Any area with a hazard of contents classified higher than that of the health care occupancy and located in the same building shall be protected as required by 18.3.2.

**18.1.3.11** Non-health care–related occupancies classified as containing high hazard contents shall not be permitted in buildings housing health care occupancies.

**18.1.4 Definitions.**

**18.1.4.1 General.** For definitions, see Chapter 3, Definitions.

**18.1.4.2 Special Definitions.** A list of special terms used in this chapter follows:

(1) **Ambulatory Health Care Occupancy.** See 3.3.188.1.

(2) **Hospital.** See 3.3.142.

(3) **Limited Care Facility.** See 3.3.88.2.

(4) **Nursing Home.** See 3.3.140.2.

**18.1.5 Classification of Hazard of Contents.** The classification of hazard of contents shall be as defined in Section 6.2.

**18.1.6 Minimum Construction Requirements.**

**18.1.6.1** Health care occupancies shall be limited to the building construction types specified in Table 18.1.6.1, unless otherwise permitted by 18.1.6.2 through 18.1.6.7. *(See 8.2.1.)*

**18.1.6.2** Any building of Type I(442), Type I(332), Type II(222), or Type II(111) construction shall be permitted to include roofing systems involving combustible supports, decking, or roofing, provided that all of the following criteria are met:

(1) The roof covering shall meet Class A requirements in accordance with ASTM E 108, *Standard Test Methods for Fire Tests of Roof Coverings*, or ANSI/UL 790, *Test Methods for Fire Tests of Roof Coverings*.

(2) The roof shall be separated from all occupied portions of the building by a noncombustible floor assembly having not less than a 2-hour fire resistance rating that includes not less than 2½ in. (63 mm) of concrete or gypsum fill.

(3) The structural elements supporting the 2-hour fire resistance–rated floor assembly specified in 18.1.6.2(2) shall be required to have only the fire resistance rating required of the building.

**18.1.6.3** Any building of Type I(442), Type I(332), Type II(222), or Type II(111) construction shall be permitted to include roofing systems involving combustible supports, decking, or roofing, provided that all of the following criteria are met:

(1) The roof covering shall meet Class A requirements in accordance with ASTM E 108, *Standard Test Methods for Fire Tests of Roof Coverings*, or ANSI/UL 790, *Test Methods for Fire Tests of Roof Coverings*.

(2) The roof/ceiling assembly shall be constructed with fire-retardant-treated wood meeting the requirements of NFPA 220, *Standard on Types of Building Construction*.

(3) The roof/ceiling assembly shall have the required fire resistance rating for the type of construction.

**18.1.6.4** Interior nonbearing walls in buildings of Type I or Type II construction shall be constructed of noncombustible or limited-combustible materials, unless otherwise permitted by 18.1.6.5.

**18.1.6.5** Interior nonbearing walls required to have a minimum 2-hour fire resistance rating shall be permitted to be of

**Table 18.1.6.1 Construction Type Limitations**

| Construction Type | Sprinklered[†] | Total Number of Stories of Building[‡] | | | |
|---|---|---|---|---|---|
| | | 1 | 2 | 3 | ≥4 |
| I (442) | Yes | X | X | X | X |
| | No | NP | NP | NP | NP |
| I (332) | Yes | X | X | X | X |
| | No | NP | NP | NP | NP |
| II (222) | Yes | X | X | X | X |
| | No | NP | NP | NP | NP |
| II (111) | Yes | X | X | X | NP |
| | No | NP | NP | NP | NP |
| II (000) | Yes | X | NP | NP | NP |
| | No | NP | NP | NP | NP |
| III (211) | Yes | X | NP | NP | NP |
| | No | NP | NP | NP | NP |
| III (200) | Yes | NP | NP | NP | NP |
| | No | NP | NP | NP | NP |
| IV (2HH) | Yes | X | NP | NP | NP |
| | No | NP | NP | NP | NP |
| V (111) | Yes | X | NP | NP | NP |
| | No | NP | NP | NP | NP |
| V (000) | Yes | NP | NP | NP | NP |
| | No | NP | NP | NP | NP |

X: Permitted. NP: Not permitted.

The total number of stories of the building is required to be determined as follows:

(1) The total number of stories is to be counted starting with the level of exit discharge and ending with the highest occupiable story of the building.

(2) Stories below the level of exit discharge are not counted as stories.

(3) Interstitial spaces used solely for building or process systems directly related to the level above or below are not considered a separate story.

(4) A mezzanine in accordance with 8.6.9 is not counted as a story.

[†]Sprinklered throughout by an approved, supervised automatic sprinkler system in accordance with Section 9.7. *(See 18.3.5.)*

[‡]Basements are not counted as stories.

fire-retardant-treated wood enclosed within noncombustible or limited-combustible materials, provided that such walls are not used as shaft enclosures.

**18.1.6.6** Fire-retardant-treated wood that serves as supports for the installation of fixtures and equipment shall be permitted to be installed behind noncombustible or limited-combustible sheathing.

**18.1.6.7** All buildings with more than one level below the level of exit discharge shall have all such lower levels separated from the level of exit discharge by not less than Type II(111) construction.

**18.1.7 Occupant Load.** The occupant load, in number of persons for whom means of egress and other provisions are required, either shall be determined on the basis of the occupant load factors of Table 7.3.1.2 that are characteristic of the use of the space or shall be determined as the maximum prob-

able population of the space under consideration, whichever is greater.

**18.2 Means of Egress Requirements.**

**18.2.1 General.** Every aisle, passageway, corridor, exit discharge, exit location, and access shall be in accordance with Chapter 7, unless otherwise modified by 18.2.2 through 18.2.11.

**18.2.2\* Means of Egress Components.**

**18.2.2.1 Components Permitted.** Components of means of egress shall be limited to the types described in 18.2.2.2 through 18.2.2.10.

**18.2.2.2 Doors.**

**18.2.2.2.1** Doors complying with 7.2.1 shall be permitted.

**18.2.2.2.2** Locks shall not be permitted on patient sleeping room doors, unless otherwise permitted by one of the following:

(1) Key-locking devices that restrict access to the room from the corridor and that are operable only by staff from the corridor side shall be permitted, provided that such devices do not restrict egress from the room.

(2) Locks complying with 18.2.2.2.5 shall be permitted.

**18.2.2.2.3** Doors not located in a required means of egress shall be permitted to be subject to locking.

**18.2.2.2.4** Doors within a required means of egress shall not be equipped with a latch or lock that requires the use of a tool or key from the egress side, unless otherwise permitted by one of the following:

(1) Locks complying with 18.2.2.2.5 shall be permitted.
(2)*Delayed-egress locks complying with 7.2.1.6.1 shall be permitted.
(3)*Access-controlled egress doors complying with 7.2.1.6.2 shall be permitted.
(4) Elevator lobby exit access door locking in accordance with 7.2.1.6.3 shall be permitted.

**18.2.2.2.5** Door-locking arrangements shall be permitted in accordance with either 18.2.2.2.5.1 or 18.2.2.2.5.2.

**18.2.2.2.5.1*** Door-locking arrangements shall be permitted where the clinical needs of patients require specialized security measures or where patients pose a security threat, provided that staff can readily unlock doors at all times in accordance with 18.2.2.2.6.

**18.2.2.2.5.2*** Door-locking arrangements shall be permitted where patient special needs require specialized protective measures for their safety, provided that all of the following criteria are met:

(1) Staff can readily unlock doors at all times in accordance with 18.2.2.2.6.
(2) A total (complete) smoke detection system is provided throughout the locked space in accordance with 9.6.2.9, or locked doors can be remotely unlocked at an approved, constantly attended location within the locked space.
(3)*The building is protected throughout by an approved, supervised automatic sprinkler system in accordance with 18.3.5.1.
(4) The locks are electrical locks that fail safely so as to release upon loss of power to the device.
(5) The locks release by independent activation of each of the following:
  (a) Activation of the smoke detection system required by 18.2.2.2.5.2(2)
  (b) Waterflow in the automatic sprinkler system required by 18.2.2.2.5.2(3)

**18.2.2.2.6** Doors that are located in the means of egress and are permitted to be locked under other provisions of 18.2.2.2.5 shall comply with both of the following:

(1) Provisions shall be made for the rapid removal of occupants by means of one of the following:
  (a) Remote control of locks from within the locked smoke compartment
  (b) Keying of all locks to keys carried by staff at all times
  (c) Other such reliable means available to the staff at all times
(2) Only one locking device shall be permitted on each door.

**18.2.2.2.7*** Any door in an exit passageway, stairway enclosure, horizontal exit, smoke barrier, or hazardous area enclosure (except boiler rooms, heater rooms, and mechanical equipment rooms) shall be permitted to be held open only by an automatic release device that complies with 7.2.1.8.2. The automatic sprinkler system and the fire alarm system, and the systems required by 7.2.1.8.2, shall be arranged to initiate the closing action of all such doors throughout the smoke compartment or throughout the entire facility.

**18.2.2.2.8** Where doors in a stair enclosure are held open by an automatic release device as permitted in 18.2.2.2.7, initiation of a door-closing action on any level shall cause all doors at all levels in the stair enclosure to close.

**18.2.2.2.9** High-rise health care occupancies shall comply with the re-entry provisions of 7.2.1.5.8.

**18.2.2.2.10** Horizontal-sliding doors shall be permitted in accordance with 18.2.2.2.10.1 or 18.2.2.2.10.2.

**18.2.2.2.10.1** Horizontal-sliding doors, as permitted by 7.2.1.14, that are not automatic-closing shall be limited to a single leaf and shall have a latch or other mechanism that ensures that the doors will not rebound into a partially open position if forcefully closed.

**18.2.2.2.10.2** Horizontal-sliding doors serving an occupant load of fewer than 10 shall be permitted, provided that all of the following criteria are met:

(1) The area served by the door has no high hazard contents.
(2) The door is readily operable from either side without special knowledge or effort.
(3) The force required to operate the door in the direction of door travel is not more than 30 lbf (133 N) to set the door in motion and is not more than 15 lbf (67 N) to close the door or open it to the minimum required width.
(4) The door assembly complies with any required fire protection rating and, where rated, is self-closing or automatic-closing by means of smoke detection in accordance with 7.2.1.8 and is installed in accordance with NFPA 80, *Standard for Fire Doors and Other Opening Protectives*.
(5) Where corridor doors are required to latch, the doors are equipped with a latch or other mechanism that ensures that the doors will not rebound into a partially open position if forcefully closed.

**18.2.2.3 Stairs.** Stairs complying with 7.2.2 shall be permitted.

**18.2.2.4 Smokeproof Enclosures.** Smokeproof enclosures complying with 7.2.3 shall be permitted.

**18.2.2.5 Horizontal Exits.** Horizontal exits complying with 7.2.4 and the modifications of 18.2.2.5.1 through 18.2.2.5.7 shall be permitted.

**18.2.2.5.1** Accumulation space shall be provided in accordance with 18.2.2.5.1.1 and 18.2.2.5.1.2.

**18.2.2.5.1.1** Not less than 30 net ft² (2.8 net m²) per patient in a hospital or nursing home, or not less than 15 net ft² (1.4 net m²) per resident in a limited care facility, shall be provided within the aggregated area of corridors, patient rooms, treatment rooms, lounge or dining areas, and other similar areas on each side of the horizontal exit.

**18.2.2.5.1.2** On stories not housing bedridden or litterborne patients, not less than 6 net ft² (0.56 net m²) per occupant shall be provided on each side of the horizontal exit for the total number of occupants in adjoining compartments.

**18.2.2.5.2**  The total egress capacity of the other exits (stairs, ramps, doors leading outside the building) shall not be reduced below one-third of that required for the entire area of the building.

**18.2.2.5.3**  A single door shall be permitted in a horizontal exit if all of the following conditions apply:

(1)  The exit serves one direction only.
(2)  Such door is a swinging door or a horizontal-sliding door complying with 7.2.1.14.
(3)  The door is not less than 41½ in. (1055 mm) in clear width.

**18.2.2.5.4**  A horizontal exit involving a corridor 8 ft (2440 mm) or more in width and serving as a means of egress from both sides of the doorway shall have the opening protected by a pair of swinging doors arranged to swing in opposite directions from each other, with each door having a clear width of not less than 41½ in. (1055 mm), or by a horizontal-sliding door that complies with 7.2.1.14 and provides a clear width of not less than 6 ft 11 in. (2110 mm).

**18.2.2.5.5**  A horizontal exit involving a corridor 6 ft (1830 mm) or more in width and serving as a means of egress from both sides of the doorway shall have the opening protected by a pair of swinging doors, arranged to swing in opposite directions from each other, with each door having a clear width of not less than 32 in. (810 mm), or by a horizontal-sliding door that complies with 7.2.1.14 and provides a clear width of not less than 64 in. (1625 mm).

**18.2.2.5.6**  An approved vision panel shall be required in each horizontal exit door.

**18.2.2.5.7**  Center mullions shall be prohibited in horizontal exit door openings.

**18.2.2.6  Ramps.**

**18.2.2.6.1**  Ramps complying with 7.2.5 shall be permitted.

**18.2.2.6.2**  Ramps enclosed as exits shall be of sufficient width to provide egress capacity in accordance with 18.2.3.

**18.2.2.7  Exit Passageways.**  Exit passageways complying with 7.2.6 shall be permitted.

**18.2.2.8  Fire Escape Ladders.**  Fire escape ladders complying with 7.2.9 shall be permitted.

**18.2.2.9  Alternating Tread Devices.**  Alternating tread devices complying with 7.2.11 shall be permitted.

**18.2.2.10  Areas of Refuge.**  Areas of refuge used as part of a required accessible means of egress shall comply with 7.2.12.

**18.2.3  Capacity of Means of Egress.**

**18.2.3.1**  The capacity of means of egress shall be in accordance with Section 7.3.

**18.2.3.2  Reserved.**

**18.2.3.3  Reserved.**

**18.2.3.4\***  Aisles, corridors, and ramps required for exit access in a hospital or nursing home shall be not less than 8 ft (2440 mm) in clear and unobstructed width, unless otherwise permitted by one of the following:

(1)\*Aisles, corridors, and ramps in adjunct areas not intended for the housing, treatment, or use of inpatients shall be not less than 44 in. (1120 mm) in clear and unobstructed width.

(2)\*Noncontinuous projections not more than 6 in. (150 mm) from the corridor wall, positioned not less than 38 in. (965 mm) above the floor, shall be permitted.
(3)\*Exit access within a room or suite of rooms complying with the requirements of 18.2.5 shall be permitted.
(4)  Projections into the required width shall be permitted for wheeled equipment, provided that all of the following conditions are met:
   (a)  The wheeled equipment does not reduce the clear unobstructed corridor width to less than 60 in. (1525 mm).
   (b)  The health care occupancy fire safety plan and training program address the relocation of the wheeled equipment during a fire or similar emergency.
   (c)\*The wheeled equipment is limited to the following:
      i.  Equipment in use and carts in use
      ii.  Medical emergency equipment not in use
      iii.  Patient lift and transport equipment
(5)\*Where the corridor width is at least 8 ft (2440 mm), projections into the required width shall be permitted for fixed furniture, provided that all of the following conditions are met:
   (a)  The fixed furniture is securely attached to the floor or to the wall.
   (b)  The fixed furniture does not reduce the clear unobstructed corridor width to less than 6 ft (1830 mm), except as permitted by 18.2.3.4(2).
   (c)  The fixed furniture is located only on one side of the corridor.
   (d)  The fixed furniture is grouped such that each grouping does not exceed an area of 50 ft² (4.6 m²).
   (e)  The fixed furniture groupings addressed in 18.2.3.4(5)(d) are separated from each other by a distance of at least 10 ft (3050 mm).
   (f)\*The fixed furniture is located so as to not obstruct access to building service and fire protection equipment.
   (g)  Corridors throughout the smoke compartment are protected by an electrically supervised automatic smoke detection system in accordance with 18.3.4, or the fixed furniture spaces are arranged and located to allow direct supervision by the facility staff from a nurses' station or similar space.
(6)\*Cross-corridor door openings in corridors with a required minimum width of 8 ft (2440 mm) shall have a clear width of not less than 6 ft 11 in. (2110 mm) for pairs of doors or a clear width of not less than 41½ in. (1055 mm) for a single door.

**18.2.3.5**  Aisles, corridors, and ramps required for exit access in a limited care facility or hospital for psychiatric care shall be not less than 6 ft (1830 mm) in clear and unobstructed width, unless otherwise permitted by one of the following:

(1)\*Aisles, corridors, and ramps in adjunct areas not intended for the housing, treatment, or use of inpatients shall be not less than 44 in. (1120 mm) in clear and unobstructed width.
(2)\*Noncontinuous projections not more than 6 in. (150 mm) from the corridor wall, positioned not less than 38 in. (965 mm) above the floor, shall be permitted.
(3)\*Exit access within a room or suite of rooms complying with the requirements of 18.2.5 shall be permitted.
(4)  Projections into the required width shall be permitted for wheeled equipment, provided that all of the following conditions are met:

(a) The wheeled equipment does not reduce the clear unobstructed corridor width to less than 60 in.(1525 mm).

(b) The health care occupancy fire safety plan and training program address the relocation of the wheeled equipment during a fire or similar emergency.

(c)*The wheeled equipment is limited to the following:

    i. Equipment in use and carts in use

    ii. Medical emergency equipment not in use

    iii. Patient lift and transport equipment

(5)*Cross-corridor door openings in corridors with a required minimum width of 6 ft (1830 mm) shall have a clear width of not less than 64 in. (1625 mm) for pairs of doors or a clear width of not less than 32 in. (810 mm) for a single door.

**18.2.3.6** The minimum clear width for doors in the means of egress from sleeping rooms; diagnostic and treatment areas, such as x-ray, surgery, or physical therapy; and nursery rooms shall be as follows:

(1) Hospitals and nursing homes — 41½ in. (1055 mm)

(2) Psychiatric hospitals and limited care facilities— 32 in. (810 mm)

**18.2.3.7** The requirements of 18.2.3.6 shall not apply where otherwise permitted by one of the following:

(1) Doors that are located so as not to be subject to use by any health care occupant shall be not less than 32 in. (810 mm) in clear width.

(2) Doors in exit stair enclosures shall be not less than 32 in. (810 mm) in clear width.

(3) Doors serving newborn nurseries shall be not less than 32 in. (810 mm) in clear width.

(4) Where a pair of doors is provided, all of the following criteria shall be met:

    (a) Not less than one of the doors shall provide not less than a 32 in. (810 mm) clear width opening.

    (b) A rabbet, bevel, or astragal shall be provided at the meeting edge.

    (c) The inactive door leaf shall have an automatic flush bolt to provide positive latching.

### 18.2.4 Number of Means of Egress.

**18.2.4.1** The number of means of egress shall be in accordance with Section 7.4.

**18.2.4.2** Not less than two exits shall be provided on every story.

**18.2.4.3** Not less than two separate exits shall be accessible from every part of every story.

**18.2.4.4*** Not less than two exits shall be accessible from each smoke compartment, and egress shall be permitted through an adjacent compartment(s), provided that the two required egress paths are arranged so that both do not pass through the same adjacent smoke compartment.

### 18.2.5 Arrangement of Means of Egress.

**18.2.5.1 General.** Arrangement of means of egress shall comply with Section 7.5.

**18.2.5.2 Dead-End Corridors.** Dead-end corridors shall not exceed 30 ft (9.1 m).

**18.2.5.3 Common Path of Travel.** Common path of travel shall not exceed 100 ft (30 m).

**18.2.5.4* Intervening Rooms or Spaces.** Every corridor shall provide access to not less than two approved exits in accordance with Sections 7.4 and 7.5 without passing through any intervening rooms or spaces other than corridors or lobbies.

### 18.2.5.5 Two Means of Egress.

**18.2.5.5.1** Sleeping rooms of more than 1000 ft$^2$ (93 m$^2$) shall have not less than two exit access doors remotely located from each other.

**18.2.5.5.2** Non-sleeping rooms of more than 2500 ft$^2$ (230 m$^2$) shall have not less than two exit access doors remotely located from each other.

### 18.2.5.6 Corridor Access.

**18.2.5.6.1*** Every habitable room shall have an exit access door leading directly to an exit access corridor, unless otherwise provided in 18.2.5.6.2, 18.2.5.6.3, and 18.2.5.6.4.

**18.2.5.6.2** Exit access from a patient sleeping room with not more than eight patient beds shall be permitted to pass through one intervening room to reach an exit access corridor, provided that the intervening room is equipped with an approved automatic smoke detection system in accordance with Section 9.6.

**18.2.5.6.3** Rooms having an exit door opening directly to the outside from the room at the finished ground level shall not be required to have an exit access door leading directly to an exit access corridor.

**18.2.5.6.4** Rooms within suites complying with 18.2.5.7 shall not be required to have an exit access door leading directly to an exit access corridor.

### 18.2.5.7 Suites.

### 18.2.5.7.1 General.

**18.2.5.7.1.1 Suite Permission.** Suites complying with 18.2.5.7 shall be permitted to be used to meet the corridor access requirements of 18.2.5.6.

**18.2.5.7.1.2* Suite Separation.** Suites shall be separated from the remainder of the building, and from other suites, by walls and doors meeting the requirements of 18.3.6.2 through 18.3.6.5.

**18.2.5.7.1.3 Suite Hazardous Contents Areas.**

**(A)*** Intervening rooms shall not be hazardous areas as defined by 18.3.2.

**(B)** Hazardous areas within a suite shall be separated from the remainder of the suite in accordance with 18.3.2.1, unless otherwise provided in 18.2.5.7.1.3(C).

**(C)*** Hazardous areas within a suite shall not be required to be separated from the remainder of the suite where complying with all of the following:

(1) The suite is primarily a hazardous area.

(2) The suite is protected by an approved automatic smoke detection system in accordance with Section 9.6.

(3) The suite is separated from the rest of the health care facility as required for a hazardous area by 18.3.2.1.

**18.2.5.7.1.4 Suite Subdivision.** The subdivision of suites shall be by means of noncombustible or limited-combustible partitions or partitions constructed with fire-retardant-treated wood enclosed with noncombustible or limited-combustible materials, and such partitions shall not be required to be fire rated.

**18.2.5.7.2 Sleeping Suites.** Sleeping suites shall be in accordance with the following:

(1) Sleeping suites for patient care shall comply with the provisions of 18.2.5.7.2.1 through 18.2.5.7.2.4.
(2) Sleeping suites not for patient care shall comply with the provisions of 18.2.5.7.4.

**18.2.5.7.2.1 Sleeping Suite Arrangement.**

**(A)*** Occupants of habitable rooms within sleeping suites shall have exit access to a corridor complying with 18.3.6, or to a horizontal exit, directly from the suite.

**(B)** Where two or more exit access doors are required from the suite by 18.2.5.5.1, one of the exit access doors shall be permitted to be directly to an exit stair, exit passageway, or exit door to the exterior.

**(C)** Sleeping suites shall be provided with constant staff supervision within the suite.

**(D)** Sleeping suites shall be arranged in accordance with one of the following:

(1)*Patient sleeping rooms within sleeping suites shall provide one of the following:
   (a) The patient sleeping rooms shall be arranged to allow for direct supervision from a normally attended location within the suite, such as is provided by glass walls, and cubicle curtains shall be permitted.
   (b) Any patient sleeping rooms without the direct supervision required by 18.2.5.7.2.1(D)(1)(a) shall be provided with smoke detection in accordance with Section 9.6 and 18.3.4.
(2) Sleeping suites shall be provided with a total coverage (complete) automatic smoke detection system in accordance with 9.6.2.9 and 18.3.4.

**18.2.5.7.2.2 Sleeping Suite Number of Means of Egress.**

**(A)** Sleeping suites of more than 1000 ft$^2$ (93 m$^2$) shall have not less than two exit access doors remotely located from each other.

**(B)*** One means of egress from the suite shall be directly to a corridor complying with 18.3.6.

**(C)*** For suites requiring two means of egress, one means of egress from the suite shall be permitted to be into another suite, provided that the separation between the suites complies with the corridor requirements of 18.3.6.2 through 18.3.6.5.

**18.2.5.7.2.3 Sleeping Suite Maximum Size.**

**(A) Reserved.**

**(B)** Sleeping suites shall not exceed 7500 ft$^2$ (700 m$^2$), unless otherwise provided in 18.2.5.7.2.3(C).

**(C)** Sleeping suites greater than 7500 ft$^2$ (700 m$^2$) and not exceeding 10,000 ft$^2$ (930 m$^2$) shall be permitted where both of the following are provided in the suite:

(1)*Direct visual supervision in accordance with 18.2.5.7.2.1(D)(1)(a)
(2) Total coverage (complete) automatic smoke detection in accordance with 9.6.2.9 and 18.3.4

**18.2.5.7.2.4 Sleeping Suite Travel Distance.**

**(A)** Travel distance between any point in a sleeping suite and an exit access door from that suite shall not exceed 100 ft (30 m).

**(B)** Travel distance between any point in a sleeping suite and an exit shall not exceed 200 ft (61 m).

**18.2.5.7.3 Patient Care Non-Sleeping Suites.** Non-sleeping suites shall be in accordance with the following:

(1) Non-sleeping suites for patient care shall comply with the provisions of 18.2.5.7.3.1 through 18.2.5.7.3.4.
(2) Non-sleeping suites not for patient care shall comply with the provisions of 18.2.5.7.4.

**18.2.5.7.3.1 Patient Care Non-Sleeping Suite Arrangement.**

**(A)** Occupants of habitable rooms within non-sleeping suites shall have exit access to a corridor complying with 18.3.6, or to a horizontal exit, directly from the suite.

**(B)** Where two or more exit access doors are required from the suite by 18.2.5.5.2, one of the exit access doors shall be permitted to be directly to an exit stair, exit passageway, or exit door to the exterior.

**18.2.5.7.3.2 Patient Care Non-Sleeping Suite Number of Means of Egress.**

**(A)** Non-sleeping suites of more than 2500 ft$^2$ (230 m$^2$) shall have not less than two exit access doors remotely located from each other.

**(B)*** One means of egress from the suite shall be directly to a corridor complying with 18.3.6.

**(C)*** For suites requiring two means of egress, one means of egress from the suite shall be permitted to be into another suite, provided that the separation between the suites complies with the corridor requirements of 18.3.6.2 through 18.3.6.5.

**18.2.5.7.3.3 Patient Care Non-Sleeping Suite Maximum Size.** Non-sleeping suites shall not exceed 10,000 ft$^2$ (930 m$^2$).

**18.2.5.7.3.4 Patient Care Non-Sleeping Suite Travel Distance.**

**(A)** Travel distance within a non-sleeping suite to an exit access door from the suite shall not exceed 100 ft (30 m).

**(B)** Travel distance between any point in a non-sleeping suite and an exit shall not exceed 200 ft (61 m).

**18.2.5.7.4 Non-Patient-Care Suites.** The egress provisions for non-patient-care suites shall be in accordance with the primary use and occupancy of the space.

**18.2.6 Travel Distance to Exits.**

**18.2.6.1** Travel distance shall be measured in accordance with Section 7.6.

**18.2.6.2** Travel distance shall comply with 18.2.6.2.1 through 18.2.6.2.4.

**18.2.6.2.1** The travel distance between any point in a room and an exit shall not exceed 200 ft (61 m).

**18.2.6.2.2 Reserved.**

**18.2.6.2.3** The travel distance between any point in a health care sleeping room and an exit access door in that room shall not exceed 50 ft (15 m).

**18.2.6.2.4** The travel distance within suites shall be in accordance with 18.2.5.7.

**18.2.7 Discharge from Exits.** Discharge from exits shall be arranged in accordance with Section 7.7.

**18.2.8 Illumination of Means of Egress.** Means of egress shall be illuminated in accordance with Section 7.8.

**18.2.9 Emergency Lighting.**

**18.2.9.1** Emergency lighting shall be provided in accordance with Section 7.9.

**18.2.9.2** Buildings equipped with, or in which patients require the use of, life-support systems *(see 18.5.1.3)* shall have emergency lighting equipment supplied by the life safety branch of the electrical system as described in NFPA 99, *Health Care Facilities Code.*

**18.2.10 Marking of Means of Egress.**

**18.2.10.1** Means of egress shall have signs in accordance with Section 7.10, unless otherwise permitted by 18.2.10.3 or 18.2.10.4.

**18.2.10.2 Reserved.**

**18.2.10.3** Where the path of egress travel is obvious, signs shall not be required at gates in outside secured areas.

**18.2.10.4** Access to exits within rooms or sleeping suites shall not be required to be marked where staff is responsible for relocating or evacuating occupants.

**18.2.10.5** Illumination of required exit and directional signs in buildings equipped with, or in which patients use, life-support systems *(see 18.5.1.3)* shall be provided as follows:

(1) Illumination shall be supplied by the life safety branch of the electrical system as described in NFPA 99, *Health Care Facilities Code.*
(2) Self-luminous exit signs complying with 7.10.4 shall be permitted.

**18.2.11 Special Means of Egress Features. (Reserved)**

**18.3 Protection.**

**18.3.1 Protection of Vertical Openings.** Any vertical opening shall be enclosed or protected in accordance with Section 8.6, unless otherwise modified by 18.3.1.1 through 18.3.1.8.

**18.3.1.1 Reserved.**

**18.3.1.2** Unprotected vertical openings in accordance with 8.6.9.1 shall be permitted.

**18.3.1.3** Subparagraph 8.6.7(1)(b) shall not apply to patient sleeping and treatment rooms.

**18.3.1.4** Multilevel patient sleeping areas in psychiatric facilities shall be permitted without enclosure protection between levels, provided that all of the following conditions are met:

(1) The entire normally occupied area, including all communicating floor levels, is sufficiently open and unobstructed so that a fire or other dangerous condition in any part is obvious to the occupants or supervisory personnel in the area.
(2) The egress capacity provides simultaneously for all the occupants of all communicating levels and areas, with all communicating levels in the same fire area being considered as a single floor area for purposes of determination of required egress capacity.
(3) The height between the highest and lowest finished floor levels does not exceed 13 ft (3960 mm), and the number of levels is permitted to be unrestricted.

**18.3.1.5** Unprotected openings in accordance with 8.6.6 shall not be permitted.

**18.3.1.6 Reserved.**

**18.3.1.7** A door in a stair enclosure shall be self-closing and shall normally be kept in the closed position, unless otherwise permitted by 18.3.1.8.

**18.3.1.8** Doors in stair enclosures shall be permitted to be held open under the conditions specified by 18.2.2.2.7 and 18.2.2.2.8.

**18.3.2 Protection from Hazards.**

**18.3.2.1\* Hazardous Areas.** Any hazardous areas shall be protected in accordance with Section 8.7, and the areas described in Table 18.3.2.1 shall be protected as indicated.

**Table 18.3.2.1 Hazardous Area Protection**

| Hazardous Area Description | Protection/ Separation† |
|---|---|
| Boiler and fuel-fired heater rooms | 1 hour |
| Central/bulk laundries larger than 100 ft² (9.3 m²) | 1 hour |
| Laboratories employing flammable or combustible materials in quantities less than those that would be considered a severe hazard | See 18.3.6.3.11. |
| Laboratories that use hazardous materials that would be classified as a severe hazard in accordance with NFPA 99, *Standard for Health Care Facilities* | 1 hour |
| Paint shops employing hazardous substances and materials in quantities less than those that would be classified as a severe hazard | 1 hour |
| Physical plant maintenance shops | 1 hour |
| Rooms with soiled linen in volume exceeding 64 gal (242 L) | 1 hour |
| Storage rooms larger than 50 ft² (4.6 m²) but not exceeding 100 ft² (9.3 m²) and storing combustible material | See 18.3.6.3.11. |
| Storage rooms larger than 100 ft² (9.3 m²) and storing combustible material | 1 hour |
| Rooms with collected trash in volume exceeding 64 gal (242 L) | 1 hour |

†Minimum fire resistance rating.

**18.3.2.2\* Laboratories.** Laboratories employing quantities of flammable, combustible, or hazardous materials that are considered as a severe hazard shall be protected in accordance with NFPA 99, *Health Care Facilities Code.*

**18.3.2.3 Anesthetizing Locations.** Anesthetizing locations shall be protected in accordance with NFPA 99, *Health Care Facilities Code.*

**18.3.2.4 Medical Gas.** Medical gas storage and administration areas shall be protected in accordance with NFPA 99, *Health Care Facilities Code.*

**18.3.2.5 Cooking Facilities.**

**18.3.2.5.1** Cooking facilities shall be protected in accordance with 9.2.3, unless otherwise permitted by 18.3.2.5.2, 18.3.2.5.3, or 18.3.2.5.4.

**18.3.2.5.2\*** Where residential cooking equipment is used for food warming or limited cooking, the equipment shall not be required to be protected in accordance with 9.2.3, and the presence of the equipment shall not require the area to be protected as a hazardous area.

**18.3.2.5.3\*** Within a smoke compartment, where residential or commercial cooking equipment is used to prepare meals for 30 or fewer persons, one cooking facility shall be permitted to be open to the corridor, provided that all of the following conditions are met:

(1) The portion of the health care facility served by the cooking facility is limited to 30 beds and is separated from other portions of the health care facility by a smoke barrier constructed in accordance with 18.3.7.3, 18.3.7.6, and 18.3.7.8.

(2) The cooktop or range is equipped with a range hood of a width at least equal to the width of the cooking surface, with grease baffles or other grease-collecting and clean-out capability.

(3)\*The hood systems have a minimum airflow of 500 cfm (14,000 L/min).

(4) The hood systems that are not ducted to the exterior additionally have a charcoal filter to remove smoke and odor.

(5) The cooktop or range complies with all of the following:

(a) The cooktop or range is protected with a fire suppression system listed in accordance with UL 300, *Standard for Fire Testing of Fire Extinguishing Systems for Protection of Commercial Cooking Equipment,* or is tested and meets all requirements of UL 300A, *Extinguishing System Units for Residential Range Top Cooking Surfaces,* in accordance with the applicable testing document's scope.

(b) A manual release of the extinguishing system is provided in accordance with NFPA 96, *Standard for Ventilation Control and Fire Protection of Commercial Cooking Operations,* Section 10.5.

(c) An interlock is provided to turn off all sources of fuel and electrical power to the cooktop or range when the suppression system is activated.

(6)\*The use of solid fuel for cooking is prohibited.

(7)\*Deep-fat frying is prohibited

(8) Portable fire extinguishers in accordance with NFPA 96 are located in all kitchen areas.

(9)\*A switch meeting all of the following is provided:

(a) A locked switch, or a switch located in a restricted location, is provided within the cooking facility that deactivates the cooktop or range.

(b) The switch is used to deactivate the cooktop or range whenever the kitchen is not under staff supervision.

(c) The switch is on a timer, not exceeding a 120-minute capacity, that automatically deactivates the cooktop or range, independent of staff action.

(10) Procedures for the use, inspection, testing, and maintenance of the cooking equipment are in accordance with Chapter 11 of NFPA 96 and the manufacturer's instructions and are followed.

(11)\*Not less than two AC-powered photoelectric smoke alarms, interconnected in accordance with 9.6.2.10.3, equipped with a silence feature, and in accordance with *NFPA 72, National Fire Alarm and Signaling Code,* are located not closer than 20 ft (6.1 m) from the cooktop or range.

(12) No smoke detector is located less than 20 ft (6.1 m) from the cooktop or range.

**18.3.2.5.4\*** Within a smoke compartment, residential or commercial cooking equipment that is used to prepare meals for 30 or fewer persons shall be permitted, provided that the cooking facility complies with all of the following conditions:

(1) The space containing the cooking equipment is not a sleeping room.

(2) The space containing the cooking equipment is separated from the corridor by partitions complying with 18.3.6.2 through 18.3.6.5.

(3) The requirements of 18.3.2.5.3(1) through (10) are met.

**18.3.2.5.5\*** Where cooking facilities are protected in accordance with 9.2.3, the presence of the cooking equipment shall not cause the room or space housing the equipment to be classified as a hazardous area with respect to the requirements of 18.3.2.1, and the room or space shall not be permitted to be open to the corridor.

**18.3.2.6\* Alcohol-Based Hand-Rub Dispensers.** Alcohol-based hand-rub dispensers shall be protected in accordance with 8.7.3.1, unless all of the following conditions are met:

(1) Where dispensers are installed in a corridor, the corridor shall have a minimum width of 6 ft (1830 mm).

(2) The maximum individual dispenser fluid capacity shall be as follows:

(a) 0.32 gal (1.2 L) for dispensers in rooms, corridors, and areas open to corridors

(b) 0.53 gal (2.0 L) for dispensers in suites of rooms

(3) Where aerosol containers are used, the maximum capacity of the aerosol dispenser shall be 18 oz. (0.51 kg) and shall be limited to Level 1 aerosols as defined in NFPA 30B, *Code for the Manufacture and Storage of Aerosol Products.*

(4) Dispensers shall be separated from each other by horizontal spacing of not less than 48 in. (1220 mm).

(5) Not more than an aggregate 10 gal (37.8 L) of alcohol-based hand-rub solution or 1135 oz (32.2 kg) of Level 1 aerosols, or a combination of liquids and Level 1 aerosols not to exceed, in total, the equivalent of 10 gal (37.8 L) or 1135 oz (32.2 kg), shall be in use outside of a storage cabinet in a single smoke compartment, except as otherwise provided in 18.3.2.6(6).

(6) One dispenser complying with 18.3.2.6(2) or (3) per room and located in that room shall not be included in the aggregated quantity addressed in 18.3.2.6(5).

(7) Storage of quantities greater than 5 gal (18.9 L) in a single smoke compartment shall meet the requirements of NFPA 30, *Flammable and Combustible Liquids Code.*

(8) Dispensers shall not be installed in the following locations:

(a) Above an ignition source within a 1 in. (25 mm) horizontal distance from each side of the ignition source

(b) To the side of an ignition source within a 1 in. (25 mm) horizontal distance from the ignition source

(c) Beneath an ignition source within a 1 in. (25 mm) vertical distance from the ignition source

(9) Dispensers installed directly over carpeted floors shall be permitted only in sprinklered smoke compartments.

(10) The alcohol-based hand-rub solution shall not exceed 95 percent alcohol content by volume.

(11) Operation of the dispenser shall comply with the following criteria:

(a) The dispenser shall not release its contents except when the dispenser is activated, either manually or automatically by touch-free activation.

(b) Any activation of the dispenser shall occur only when an object is placed within 4 in. (100 mm) of the sensing device.

(c) An object placed within the activation zone and left in place shall not cause more than one activation.

(d) The dispenser shall not dispense more solution than the amount required for hand hygiene consistent with label instructions.

(e) The dispenser shall be designed, constructed, and operated in a manner that ensures that accidental or malicious activation of the dispensing device is minimized.

(f) The dispenser shall be tested in accordance with the manufacturer's care and use instructions each time a new refill is installed.

**18.3.2.7 Heliports.** Buildings that house health care occupancies, as indicated in 18.1.1.1.4, and have rooftop heliports shall be protected in accordance with NFPA 418, *Standard for Heliports.*

**18.3.3 Interior Finish.**

**18.3.3.1 General.** Interior finish shall be in accordance with Section 10.2.

**18.3.3.2\* Interior Wall and Ceiling Finish.** Interior wall and ceiling finish materials complying with Section 10.2 shall be permitted throughout if Class A, except as indicated in 18.3.3.2.1 or 18.3.3.2.2.

**18.3.3.2.1** Walls and ceilings shall be permitted to have Class A or Class B interior finish in individual rooms having a capacity not exceeding four persons.

**18.3.3.2.2** Corridor wall finish not exceeding 48 in. (1220 mm) in height that is restricted to the lower half of the wall shall be permitted to be Class A or Class B.

**18.3.3.3 Interior Floor Finish.**

**18.3.3.3.1** Interior floor finish shall comply with Section 10.2.

**18.3.3.3.2** Interior floor finish in exit enclosures and exit access corridors and spaces not separated from them by walls complying with 18.3.6 shall be Class I or Class II.

**18.3.3.3.3** Interior floor finish shall comply with 10.2.7.1 or 10.2.7.2, as applicable.

**18.3.4 Detection, Alarm, and Communications Systems.**

**18.3.4.1 General.** Health care occupancies shall be provided with a fire alarm system in accordance with Section 9.6.

**18.3.4.2\* Initiation.**

**18.3.4.2.1** Initiation of the required fire alarm systems shall be by manual means in accordance with 9.6.2 and by means of any required sprinkler system waterflow alarms, detection devices, or detection systems, unless otherwise permitted by 18.3.4.2.2.

**18.3.4.2.2** Manual fire alarm boxes in patient sleeping areas shall not be required at exits if located at all nurses' control stations or other continuously attended staff location, provided that both of the following criteria are met:

(1) Such manual fire alarm boxes are visible and continuously accessible.

(2) Travel distances required by 9.6.2.5 are not exceeded.

**18.3.4.3 Notification.** Positive alarm sequence in accordance with 9.6.3.4 shall be permitted.

**18.3.4.3.1 Occupant Notification.** Occupant notification shall be accomplished automatically in accordance with 9.6.3, unless otherwise modified by the following:

(1) Paragraph 9.6.3.2.3 shall not be permitted to be used.

(2)\*In lieu of audible alarm signals, visible alarm-indicating appliances shall be permitted to be used in critical care areas.

**18.3.4.3.2 Emergency Forces Notification.**

**18.3.4.3.2.1** Fire department notification shall be accomplished in accordance with 9.6.4.

**18.3.4.3.2.2 Reserved.**

**18.3.4.3.3 Annunciation and Annunciation Zoning.**

**18.3.4.3.3.1** Annunciation and annunciation zoning shall be provided in accordance with 9.6.7, unless otherwise permitted by 18.3.4.3.3.2 or 18.3.4.3.3.3.

**18.3.4.3.3.2** The alarm zone shall be permitted to coincide with the permitted area for smoke compartments.

**18.3.4.3.3.3** The provision of 9.6.7.4.3, which permits sprinkler system waterflow to be annunciated as a single building zone, shall be prohibited.

**18.3.4.4 Fire Safety Functions.** Operation of any activating device in the required fire alarm system shall be arranged to accomplish automatically any control functions to be performed by that device. *(See 9.6.5.)*

**18.3.4.5 Detection.**

**18.3.4.5.1 General.** Detection systems, where required, shall be in accordance with Section 9.6.

**18.3.4.5.2 Detection in Spaces Open to Corridors.** See 18.3.6.1.

**18.3.4.5.3\* Nursing Homes.** An approved automatic smoke detection system shall be installed in corridors throughout smoke compartments containing patient sleeping rooms and in spaces open to corridors as permitted in nursing homes by 18.3.6.1, unless otherwise permitted by one of the following:

(1) Corridor systems shall not be required where each patient sleeping room is protected by an approved smoke detection system.

(2) Corridor systems shall not be required where patient room doors are equipped with automatic door-closing devices with integral smoke detectors on the room side installed in accordance with their listing, provided that the integral detectors provide occupant notification.

**18.3.5 Extinguishment Requirements.**

**18.3.5.1\*** Buildings containing health care occupancies shall be protected throughout by an approved, supervised automatic sprinkler system in accordance with Section 9.7, unless otherwise permitted by 18.3.5.5.

**18.3.5.2 Reserved.**

**18.3.5.3 Reserved.**

**18.3.5.4** The sprinkler system required by 18.3.5.1 shall be installed in accordance with 9.7.1.1(1).

**18.3.5.5** In Type I and Type II construction, alternative protection measures shall be permitted to be substituted for sprinkler protection without causing a building to be classified as nonsprinklered in specified areas where the authority having jurisdiction has prohibited sprinklers.

**18.3.5.6\*** Listed quick-response or listed residential sprinklers shall be used throughout smoke compartments containing patient sleeping rooms.

**18.3.5.7 Reserved.**

**18.3.5.8 Reserved.**

**18.3.5.9 Reserved.**

**18.3.5.10\*** Sprinklers shall not be required in clothes closets of patient sleeping rooms in hospitals where the area of the closet does not exceed 6 ft$^2$ (0.55 m$^2$), provided that the distance from the sprinkler in the patient sleeping room to the back wall of the closet does not exceed the maximum distance permitted by NFPA 13, *Standard for the Installation of Sprinkler Systems*.

**18.3.5.11\*** Sprinklers in areas where cubicle curtains are installed shall be in accordance with NFPA 13, *Standard for the Installation of Sprinkler Systems*.

**18.3.5.12** Portable fire extinguishers shall be provided in all health care occupancies in accordance with 9.7.4.1.

**18.3.6 Corridors.**

**18.3.6.1 Corridor Separation.** Corridors shall be separated from all other areas by partitions complying with 18.3.6.2 through 18.3.6.5 *(see also 18.2.5.4)*, unless otherwise permitted by one of the following:

(1) Spaces shall be permitted to be unlimited in area and open to the corridor, provided that all of the following criteria are met:
   (a)\*The spaces are not used for patient sleeping rooms, treatment rooms, or hazardous areas.
   (b) The corridors onto which the spaces open in the same smoke compartment are protected by an electrically supervised automatic smoke detection system in accordance with 18.3.4, or the smoke compartment in which the space is located is protected throughout by quick-response sprinklers.
   (c) The open space is protected by an electrically supervised automatic smoke detection system in accordance with 18.3.4, or the entire space is arranged and located to allow direct supervision by the facility staff from a nurses' station or similar space.
   (d) The space does not obstruct access to required exits.

(2) Waiting areas shall be permitted to be open to the corridor, provided that all of the following criteria are met:
   (a) The aggregate waiting area in each smoke compartment does not exceed 600 ft$^2$ (55.7 m$^2$).
   (b) Each area is protected by an electrically supervised automatic smoke detection system in accordance with 18.3.4, or each area is arranged and located to allow direct supervision by the facility staff from a nursing station or similar space.
   (c) The area does not obstruct access to required exits.

(3)\*This requirement shall not apply to spaces for nurses' stations.

(4) Gift shops not exceeding 500 ft$^2$ (46.4 m$^2$) shall be permitted to be open to the corridor or lobby.

(5) In a limited care facility, group meeting or multipurpose therapeutic spaces shall be permitted to open to the corridor, provided that all of the following criteria are met:
   (a) The space is not a hazardous area.
   (b) The space is protected by an electrically supervised automatic smoke detection system in accordance with 18.3.4, or the space is arranged and located to allow direct supervision by the facility staff from the nurses' station or similar location.
   (c) The space does not obstruct access to required exits.

(6) Cooking facilities in accordance with 18.3.2.5.3 shall be permitted to be open to the corridor.

**18.3.6.2\* Construction of Corridor Walls.**

**18.3.6.2.1** Corridor walls shall be permitted to terminate at the ceiling where the ceiling is constructed to limit the transfer of smoke.

**18.3.6.2.2** No fire resistance rating shall be required for corridor walls.

**18.3.6.2.3\*** Corridor walls shall form a barrier to limit the transfer of smoke.

**18.3.6.3\* Corridor Doors.**

**18.3.6.3.1\*** Doors protecting corridor openings shall be constructed to resist the passage of smoke, and the following also shall apply:

(1) Compliance with NFPA 80, *Standard for Fire Doors and Other Opening Protectives*, shall not be required.
(2) A clearance between the bottom of the door and the floor covering not exceeding 1 in. (25 mm) shall be permitted for corridor doors.
(3) Doors to toilet rooms, bathrooms, shower rooms, sink closets, and similar auxiliary spaces that do not contain flammable or combustible material shall not be required to be constructed to resist the passage of smoke.

**18.3.6.3.2 Reserved.**

**18.3.6.3.3 Reserved.**

**18.3.6.3.4 Reserved.**

**18.3.6.3.5** Doors shall be self-latching and provided with positive latching hardware.

**18.3.6.3.6** Doors to toilet rooms, bathrooms, shower rooms, sink closets, and similar auxiliary spaces that do not contain flammable or combustible materials shall not be required to meet the latching requirements of 18.3.6.3.5.

**18.3.6.3.7** Powered doors that comply with the requirements of 7.2.1.9 shall not be required to meet the latching requirements of 18.3.6.3.5, provided that both of the following criteria are met:

(1) The door is equipped with a means for keeping the door closed that is acceptable to the authority having jurisdiction.
(2) The device used is capable of keeping the door fully closed if a force of 5 lbf (22 N) is applied at the latch edge of a swinging door and applied in any direction to a sliding or folding door, whether or not power is applied.

**18.3.6.3.8** Corridor doors utilizing an inactive leaf shall have automatic flush bolts on the inactive leaf to provide positive latching.

**18.3.6.3.9 Roller Latches.**

**18.3.6.3.9.1** Roller latches shall be prohibited, except as permitted by 18.3.6.3.9.2

**18.3.6.3.9.2** Roller latches shall be permitted for acute psychiatric settings where patient special clinical needs require specialized protective measures for their safety, provided that the roller latches are capable of keeping the door fully closed if a force of 5 lbf (22 N) is applied at the latch edge of the door.

**18.3.6.3.10\*** Doors shall not be held open by devices other than those that release when the door is pushed or pulled.

**18.3.6.3.11** Door-closing devices shall not be required on doors in corridor wall openings other than those serving required exits, smoke barriers, or enclosures of vertical openings and hazardous areas.

**18.3.6.3.12\*** Nonrated, factory- or field applied protective plates, unlimited in height, shall be permitted.

**18.3.6.3.13** Dutch doors shall be permitted where they conform to 18.3.6.3 and meet all of the following criteria:

(1) Both the upper leaf and lower leaf are equipped with a latching device.
(2) The meeting edges of the upper and lower leaves are quipped with an astragal, a rabbet, or a bevel.
(3) Where protecting openings in enclosures around hazardous areas, the doors comply with NFPA 80, *Standard for Fire Doors and Other Opening Protectives.*

**18.3.6.4 Transfer Grilles.**

**18.3.6.4.1** Transfer grilles, regardless of whether they are protected by fusible link–operated dampers, shall not be used in corridor walls or doors, unless otherwise permitted by 18.3.6.4.2.

**18.3.6.4.2** Doors to toilet rooms, bathrooms, shower rooms, sink closets, and similar auxiliary spaces that do not contain flammable or combustible materials shall be permitted to have ventilating louvers or to be undercut.

**18.3.6.5 Openings.**

**18.3.6.5.1\*** In other than smoke compartments containing patient bedrooms, miscellaneous openings, such as mail slots, pharmacy pass-through windows, laboratory pass-through windows, and cashier pass-through windows, shall be permitted to be installed in vision panels or doors without special protection, provided that both of the following criteria are met:

(1) The aggregate area of openings per room does not exceed 80 in.$^2$ (0.05 m$^2$).
(2) The openings are installed at or below half the distance from the floor to the room ceiling.

**18.3.6.5.2 Reserved.**

**18.3.7\* Subdivision of Building Spaces.**

**18.3.7.1** Buildings containing health care facilities shall be subdivided by smoke barriers *(see 18.2.4.3)*, unless otherwise permitted by 18.3.7.2, as follows:

(1) To divide every story used by inpatients for sleeping or treatment into not less than two smoke compartments

(2) To divide every story having an occupant load of 50 or more persons, regardless of use, into not less than two smoke compartments
(3) To limit the size of each smoke compartment required by 18.3.7.1(1) and (2) to an area not exceeding 22,500 ft$^2$ (2100 m$^2$), unless the area is an atrium separated in accordance with 8.6.7, in which case no limitation in size is required
(4) To limit the travel distance from any point to reach a door in the required smoke barrier to a distance not exceeding 200 ft (61 m)

**18.3.7.2** The smoke barrier subdivision requirement of 18.3.7.1 shall not apply to any of the following:

(1) Stories that do not contain a health care occupancy located directly above the health care occupancy
(2) Areas that do not contain a health care occupancy and that are separated from the health care occupancy by a fire barrier complying with 7.2.4.3
(3) Stories that do not contain a health care occupancy and that are more than one story below the health care occupancy
(4) Stories located directly below a health care occupancy where such stories house mechanical equipment only and are separated from the story above by 2-hour fire resistance–rated construction
(5) Open-air parking structures protected throughout by an approved, supervised automatic sprinkler system in accordance with Section 9.7

**18.3.7.3** Any required smoke barrier shall be constructed in accordance with Section 8.5 and shall have a minimum 1-hour fire resistance rating, unless otherwise permitted by one of the following:

(1) This requirement shall not apply where an atrium is used, and both of the following criteria also shall apply:
    (a) Smoke barriers shall be permitted to terminate at an atrium wall constructed in accordance with 8.6.7(1)(c).
    (b) Not less than two separate smoke compartments shall be provided on each floor.
(2)\*Smoke dampers shall not be required in duct penetrations of smoke barriers in fully ducted heating, ventilating, and air-conditioning systems.

**18.3.7.4** Materials and methods of construction used for required smoke barriers shall not reduce the required fire resistance rating.

**18.3.7.5** Accumulation space shall be provided in accordance with 18.3.7.5.1 and 18.3.7.5.2.

**18.3.7.5.1** Not less than 30 net ft$^2$ (2.8 net m$^2$) per patient in a hospital or nursing home, or not less than 15 net ft$^2$ (1.4 net m$^2$) per resident in a limited care facility, shall be provided within the aggregate area of corridors, patient rooms, treatment rooms, lounge or dining areas, and other low hazard areas on each side of the smoke barrier.

**18.3.7.5.2** On stories not housing bedridden or litterborne patients, not less than 6 net ft$^2$ (0.56 net m$^2$) per occupant shall be provided on each side of the smoke barrier for the total number of occupants in adjoining compartments.

**18.3.7.6\*** Doors in smoke barriers shall be substantial doors, such as 1¾ in. (44 mm) thick, solid-bonded wood-core doors, or shall be of construction that resists fire for a minimum of 20 minutes, and shall meet the following requirements:

(1) Nonrated factory- or field-applied protective plates, unlimited in height, shall be permitted.
(2) Cross-corridor openings in smoke barriers shall be protected by a pair of swinging doors or a horizontal-sliding door complying with 7.2.1.14, unless otherwise permitted by 18.3.7.7.
(3) The swinging doors addressed by 18.3.7.6(2) shall be arranged so that each door swings in a direction opposite from the other.
(4) The minimum clear width for swinging doors shall be as follows:
  (a) Hospitals and nursing homes — 41½ in. (1055 mm)
  (b) Psychiatric hospitals and limited care facilities — 32 in. (810 mm)
(5) The minimum clear width opening for horizontal-sliding doors shall be as follows:
  (a) Hospitals and nursing homes — 6 ft 11 in. (2110 mm)
  (b) Psychiatric hospitals and limited care facilities — 64 in. (1625 mm)
(6) The clearance under the bottom of smoke barrier doors shall not exceed ¾ in. (19 mm).

**18.3.7.7** Cross-corridor openings in smoke barriers that are not in required means of egress from a health care space shall be permitted to be protected by a single-leaf door.

**18.3.7.8\*** Doors in smoke barriers shall comply with 8.5.4 and all of the following:

(1) The doors shall be self-closing or automatic-closing in accordance with 18.2.2.2.7.
(2) Latching hardware shall not be required.
(3) Stops shall be required at the head and sides of door frames.
(4) Rabbets, bevels, or astragals shall be required at the meeting edges of pairs of doors.
(5) Center mullions shall be prohibited.

**18.3.7.9\*** Vision panels consisting of fire-rated glazing in approved frames shall be provided in each cross-corridor swinging door and at each cross-corridor horizontal-sliding door in a smoke barrier.

**18.3.7.10** Vision panels in doors in smoke barriers, if provided, shall be of fire-rated glazing in approved frames.

**18.3.8 Special Protection Features. (Reserved)**

**18.4 Special Provisions.**

**18.4.1 Limited Access Buildings.** Limited access buildings or limited access portions of buildings shall not be used for patient sleeping rooms and shall comply with Section 11.7.

**18.4.2 High-Rise Buildings.** High-rise buildings shall comply with Section 11.8.

**18.4.3 Nonsprinklered Existing Smoke Compartment Rehabilitation.**

**18.4.3.1\* General.** Where a modification in a nonsprinklered smoke compartment is exempted by the provisions of 18.1.1.4.3.4 from the sprinkler requirement of 18.3.5.1, the requirements of 18.4.3.2 through 18.4.3.8 shall apply.

**18.4.3.2 Minimum Construction Requirements (Nonsprinklered Smoke Compartment Rehabilitation).** Health care occupancies in buildings not protected throughout by an approved, supervised automatic sprinkler system in accordance with 19.3.5.7 shall be limited to the building construction types specified in Table 18.4.3.2.

**18.4.3.3 Capacity of Means of Egress (Nonsprinklered Smoke Compartment Rehabilitation).** The capacity of the means of egress serving the modification area shall be as follows:

(1) ½ in. (13 mm) per person for horizontal travel, without stairs, by means such as doors, ramps, or level floor surfaces
(2) 0.6 in. (15 mm) per person for travel by means of stairs

**18.4.3.4 Travel Distance (Nonsprinklered Smoke Compartment Rehabilitation).**

**18.4.3.4.1** The travel distance between any room door required as an exit access and an exit shall not exceed the following:

(1) 150 ft (46 m) where the travel is wholly within smoke compartments protected throughout by an approved, supervised automatic sprinkler system in accordance with 19.3.5.7
(2) 100 ft (30 m) where the travel is not wholly within smoke compartments protected throughout by an approved, supervised automatic sprinkler system in accordance with 19.3.5.7

**18.4.3.4.2** The travel distance between any point in a room and an exit shall not exceed the following:

(1) 200 ft (61 m) where the travel is wholly within smoke compartments protected throughout by an approved supervised sprinkler system in accordance with 19.3.5.7
(2) 150 ft (46 m) where the travel is not wholly within smoke compartments protected throughout by an approved supervised sprinkler system in accordance with 19.3.5.7

**18.4.3.5 Hazardous Area Protection (Nonsprinklered Smoke Compartment Rehabilitation).** Where a new hazardous area is formed in an existing nonsprinklered smoke compartment, the hazardous area itself shall be protected as indicated in Table 18.4.3.5.

**18.4.3.6 Interior Finish (Nonsprinklered Smoke Compartment Rehabilitation).**

**18.4.3.6.1 General.** Interior finish within the modification area shall be in accordance with Section 10.2.

**18.4.3.6.2 Interior Wall and Ceiling Finish.** Newly installed interior wall and ceiling finish materials complying with Section 10.2 shall be permitted throughout nonsprinklered smoke compartments if the materials are Class A, except as otherwise permitted in 18.4.3.6.2.1 or 18.4.3.6.2.2.

**18.4.3.6.2.1** Walls and ceilings shall be permitted to have Class A or Class B interior finish in individual rooms having a capacity not exceeding four persons.

**18.4.3.6.2.2** Corridor wall finish not exceeding 48 in. (1220 mm) in height and restricted to the lower half of the wall shall be permitted to be Class A or Class B.

**18.4.3.6.3 Interior Floor Finish.**

**18.4.3.6.3.1** Newly installed interior floor finish shall comply with Section 10.2.

**18.4.3.6.3.2** The requirements for newly installed interior floor finish in exit enclosures and corridors not separated from them by walls complying with 19.3.5.7 shall be as follows:

**Table 18.4.3.2 Construction Type Limitations (Nonsprinklered Buildings)**

| Construction Type | Sprinklered | Total Number of Stories of Building[†] | | | |
|---|---|---|---|---|---|
| | | 1 | 2 | 3 | ≥4 |
| I (442) | Yes | NA | NA | NA | NA |
| | No | X | X | X | X |
| I (332) | Yes | NA | NA | NA | NA |
| | No | X | X | X | X |
| II (222) | Yes | NA | NA | NA | NA |
| | No | X | X | X | X |
| II (111) | Yes | NA | NA | NA | NA |
| | No | X | NP | NP | NP |
| II (000) | Yes | NA | NA | NA | NA |
| | No | NP | NP | NP | NP |
| III (211) | Yes | NA | NA | NA | NA |
| | No | NP | NP | NP | NP |
| III (200) | Yes | NA | NA | NA | NA |
| | No | NP | NP | NP | NP |
| IV (2HH) | Yes | NA | NA | NA | NA |
| | No | NP | NP | NP | NP |
| V (111) | Yes | NA | NA | NA | NA |
| | No | NP | NP | NP | NP |
| V (000) | Yes | NA | NA | NA | NA |
| | No | NP | NP | NP | NP |

NA: Not applicable. X: Permitted. NP: Not permitted.

The total number of stories of the building is required to be determined as follows:

(1) The total number of stories is to be counted starting with the level of exit discharge and ending with the highest occupiable story of the building.

(2) Stories below the level of exit discharge are not counted as stories.

(3) Interstitial spaces used solely for building or process systems directly related to the level above or below are not considered a separate story.

(4) A mezzanine in accordance with 8.6.9 is not counted as a story.

[†]Basements are not counted as stories.

(1) Unrestricted in smoke compartments protected throughout by an approved, supervised automatic sprinkler system in accordance with 19.3.5.7

(2) Not less than Class I in smoke compartments not protected throughout by an approved, supervised automatic sprinkler system in accordance with 19.3.5.7

**18.4.3.7 Corridors (Nonsprinklered Smoke Compartment Rehabilitation).**

**18.4.3.7.1 Construction of Corridor Walls.**

**18.4.3.7.1.1** Where the smoke compartment being modified is not protected throughout by an approved, supervised automatic sprinkler system in accordance with 19.3.5.7, corridor walls shall comply with all of the following, as modified by 18.4.3.7.1.2:

(1) They shall have a minimum ½-hour fire resistance rating.

(2) They shall be continuous from the floor to the underside of the floor or roof deck above.

(3) They shall resist the passage of smoke.

**18.4.3.7.1.2** The requirements of 18.4.3.7.1.1 shall be permitted to be modified for conditions permitted by 19.3.6.1(3) and (4) and 19.3.6.1(6) through (8).

**18.4.3.7.2 Corridor Doors.**

**18.4.3.7.2.1** Where the smoke compartment being modified is not protected throughout by an approved, supervised automatic sprinkler system in accordance with 19.3.5.7, all of the following shall apply:

(1) Doors protecting corridor openings shall be constructed of 1¾ in. (44 mm) thick, solid-bonded core wood or of construction that resists the passage of fire for a minimum of 20 minutes.

(2) Door frames shall be labeled or of steel construction.

(3) Existing roller latches demonstrated to keep the door closed against a force of 5 lbf (22 N) shall be permitted.

**18.4.3.7.2.2** Door-closing devices shall be required on doors in corridor wall openings serving smoke barriers or enclosures of exits, hazardous contents areas, or vertical openings.

**Table 18.4.3.5 Hazardous Area Protection (Nonsprinklered Buildings)**

| Hazardous Area Description | Protection[†]/Separation |
|---|---|
| Boiler and fuel-fired heater rooms | 1 hour and sprinklers |
| Central/bulk laundries larger than 100 ft² (9.3 m²) | 1 hour and sprinklers |
| Laboratories employing flammable or combustible materials in quantities less than those that would be considered a severe hazard | 1 hour or sprinklers *(Also see 18.4.3.7.2.2.)* |
| Laboratories that use hazardous materials that would be classified as a severe hazard in accordance with NFPA 99, *Standard for Health Care Facilities* | 1 hour and sprinklers |
| Paint shops employing hazardous substances and materials in quantities less than those that would be classified as a severe hazard | 1 hour and sprinklers |
| Physical plant maintenance shops | 1 hour and sprinklers |
| Soiled linen rooms | 1 hour and sprinklers |
| Storage rooms larger than 50 ft² (4.6 m²) but not exceeding 100 ft² (9.3 m²) and storing combustible material | 1 hour or sprinklers *(Also see 18.4.3.7.2.2.)* |
| Storage rooms larger than 100 ft² (9.3 m²) and storing combustible material | 1 hour and sprinklers |
| Trash collection rooms | 1 hour and sprinklers |

[†]Minimum fire resistance rating.

**18.4.3.8 Subdivision of Building Space (Nonsprinklered Smoke Compartment Rehabilitation).** Subparagraph 18.3.7.3(2) shall be permitted only where adjacent smoke compartments are protected throughout by an approved, supervised automatic sprinkler system in accordance with 18.3.5.4 and 18.3.5.6.

**18.5 Building Services.**

**18.5.1 Utilities.**

**18.5.1.1** Utilities shall comply with the provisions of Section 9.1.

**18.5.1.2** Power for alarms, emergency communications systems, and illumination of generator set locations shall be in accordance with the essential electrical system requirements of NFPA 99, *Health Care Facilities Code.*

**18.5.1.3** Any health care occupancy, as indicated in 18.1.1.1.4, that normally uses life-support devices shall have electrical systems designed and installed in accordance with NFPA 99, *Health Care Facilities Code,* unless the facility uses life-support equipment for emergency purposes only.

**18.5.2 Heating, Ventilating, and Air-Conditioning.**

**18.5.2.1** Heating, ventilating, and air-conditioning shall comply with the provisions of Section 9.2 and shall be installed in accordance with the manufacturer's specifications, unless otherwise modified by 18.5.2.2.

**18.5.2.2\*** Any heating device, other than a central heating plant, shall be designed and installed so that combustible ma-

terial cannot be ignited by the device or its appurtenances, and the following requirements shall also apply:

(1) If fuel-fired, such heating devices shall comply with the following:
  (a) They shall be chimney connected or vent connected.
  (b) They shall take air for combustion directly from outside.
  (c) They shall be designed and installed to provide for complete separation of the combustion system from the atmosphere of the occupied area.
(2) Any heating device shall have safety features to immediately stop the flow of fuel and shut down the equipment in case of either excessive temperatures or ignition failure.

**18.5.2.3** The requirements of 18.5.2.2 shall not apply where otherwise permitted by the following:

(1) Approved, suspended unit heaters shall be permitted in locations other than means of egress and patient sleeping areas, provided that both of the following criteria are met:
  (a) Such heaters are located high enough to be out of the reach of persons using the area.
  (b) Such heaters are equipped with the safety features required by 18.5.2.2.
(2) Direct-vent gas fireplaces, as defined in NFPA 54, *National Fuel Gas Code,* shall be permitted inside of smoke compartments containing patient sleeping areas, provided that all of the following criteria are met:
  (a) All such devices shall be installed, maintained, and used in accordance with 9.2.2.
  (b) No such device shall be located inside of a patient sleeping room.
  (c) The smoke compartment in which the direct-vent gas fireplace is located shall be protected throughout by an approved, supervised automatic sprinkler system in accordance with 9.7.1.1(1) with listed quick-response or listed residential sprinklers.
  (d)\*The direct-vent fireplace shall include a sealed glass front with a wire mesh panel or screen.
  (e)\*The controls for the direct-vent gas fireplace shall be locked or located in a restricted location.
  (f) Electrically supervised carbon monoxide detection in accordance with Section 9.8 shall be provided in the room where the fireplace is located.
(3) Solid fuel–burning fireplaces shall be permitted and used only in areas other than patient sleeping areas, provided that all of the following criteria are met:
  (a) Such areas are separated from patient sleeping spaces by construction having not less than a 1-hour fire resistance rating.
  (b) The fireplace complies with the provisions of 9.2.2.
  (c) The fireplace is equipped with both of the following:
    i. Hearth raised not less than 4 in. (100 mm)
    ii. Fireplace enclosure guaranteed against breakage up to a temperature of 650°F (343°C) and constructed of heat-tempered glass or other approved material
  (d) Electrically supervised carbon monoxide detection in accordance with Section 9.8 is provided in the room where the fireplace is located
(4) If, in the opinion of the authority having jurisdiction, special hazards are present, a lock on the enclosure specified in 18.5.2.3(3)(c)(ii) and other safety precautions shall be permitted to be required.

**18.5.3 Elevators, Escalators, and Conveyors.** Elevators, escalators, and conveyors shall comply with the provisions of Section 9.4.

**18.5.4 Rubbish Chutes, Incinerators, and Laundry Chutes.**

**18.5.4.1** Rubbish chutes, incinerators, and laundry chutes shall comply with the provisions of Section 9.5, unless otherwise specified in 18.5.4.2.

**18.5.4.2** The fire resistance rating of chute charging rooms shall not be required to exceed 1 hour.

**18.5.4.3** Any rubbish chute or linen chute, including pneumatic rubbish and linen systems, shall be provided with automatic extinguishing protection in accordance with Section 9.7. *(See Section 9.5.)*

**18.5.4.4** Any rubbish chute shall discharge into a trash collection room used for no other purpose and shall be protected in accordance with Section 8.7.

**18.5.4.5 Reserved.**

**18.5.4.6** Incinerators shall not be directly flue-fed, nor shall any floor-charging chute directly connect with the combustion chamber.

**18.6 Reserved.**

**18.7\* Operating Features.**

**18.7.1 Evacuation and Relocation Plan and Fire Drills.**

**18.7.1.1** The administration of every health care occupancy shall have, in effect and available to all supervisory personnel, written copies of a plan for the protection of all persons in the event of fire, for their evacuation to areas of refuge, and for their evacuation from the building when necessary.

**18.7.1.2** All employees shall be periodically instructed and kept informed with respect to their duties under the plan required by 18.7.1.1.

**18.7.1.3** A copy of the plan required by 18.7.1.1 shall be readily available at all times in the telephone operator's location or at the security center.

**18.7.1.4\*** Fire drills in health care occupancies shall include the transmission of a fire alarm signal and simulation of emergency fire conditions.

**18.7.1.5** Infirm or bedridden patients shall not be required to be moved during drills to safe areas or to the exterior of the building.

**18.7.1.6** Drills shall be conducted quarterly on each shift to familiarize facility personnel (nurses, interns, maintenance engineers, and administrative staff) with the signals and emergency action required under varied conditions.

**18.7.1.7** When drills are conducted between 9:00 p.m. and 6:00 a.m. (2100 hours and 0600 hours), a coded announcement shall be permitted to be used instead of audible alarms.

**18.7.1.8** Employees of health care occupancies shall be instructed in life safety procedures and devices.

**18.7.2 Procedure in Case of Fire.**

**18.7.2.1\* Protection of Patients.**

**18.7.2.1.1** For health care occupancies, the proper protection of patients shall require the prompt and effective response of health care personnel.

**18.7.2.1.2** The basic response required of staff shall include the following:

(1) Removal of all occupants directly involved with the fire emergency
(2) Transmission of an appropriate fire alarm signal to warn other building occupants and summon staff
(3) Confinement of the effects of the fire by closing doors to isolate the fire area
(4) Relocation of patients as detailed in the health care occupancy's fire safety plan

**18.7.2.2 Fire Safety Plan.** A written health care occupancy fire safety plan shall provide for all of the following:

(1) Use of alarms
(2) Transmission of alarms to fire department
(3) Emergency phone call to fire department
(4) Response to alarms
(5) Isolation of fire
(6) Evacuation of immediate area
(7) Evacuation of smoke compartment
(8) Preparation of floors and building for evacuation
(9) Extinguishment of fire

**18.7.2.3 Staff Response.**

**18.7.2.3.1** All health care occupancy personnel shall be instructed in the use of and response to fire alarms.

**18.7.2.3.2** All health care occupancy personnel shall be instructed in the use of the code phrase to ensure transmission of an alarm under any of the following conditions:

(1) When the individual who discovers a fire must immediately go to the aid of an endangered person
(2) During a malfunction of the building fire alarm system

**18.7.2.3.3** Personnel hearing the code announced shall first activate the building fire alarm using the nearest manual fire alarm box and then shall execute immediately their duties as outlined in the fire safety plan.

**18.7.3 Maintenance of Means of Egress.**

**18.7.3.1** Proper maintenance shall be provided to ensure the dependability of the method of evacuation selected.

**18.7.3.2** Health care occupancies that find it necessary to lock means of egress doors shall, at all times, maintain an adequate staff qualified to release locks and direct occupants from the immediate danger area to a place of safety in case of fire or other emergency.

**18.7.4\* Smoking.** Smoking regulations shall be adopted and shall include not less than the following provisions:

(1) Smoking shall be prohibited in any room, ward, or individual enclosed space where flammable liquids, combustible gases, or oxygen is used or stored and in any other hazardous location, and such areas shall be posted with signs that read NO SMOKING or shall be posted with the international symbol for no smoking.
(2) In health care occupancies where smoking is prohibited and signs are prominently placed at all major entrances, secondary signs with language that prohibits smoking shall not be required.
(3) Smoking by patients classified as not responsible shall be prohibited.
(4) The requirement of 18.7.4(3) shall not apply where the patient is under direct supervision.

(5) Ashtrays of noncombustible material and safe design shall be provided in all areas where smoking is permitted.

(6) Metal containers with self-closing cover devices into which ashtrays can be emptied shall be readily available to all areas where smoking is permitted.

**18.7.5 Furnishings, Mattresses, and Decorations.**

**18.7.5.1\*** Draperies, curtains, and other loosely hanging fabrics and films serving as furnishings or decorations in health care occupancies shall be in accordance with the provisions of 10.3.1 (*see 18.3.5.11*), and the following also shall apply:

(1) Such curtains shall include cubicle curtains.
(2) Such curtains shall not include curtains at showers and baths.
(3) Such draperies and curtains shall not include draperies and curtains at windows in patient sleeping rooms.
(4) Such draperies and curtains shall not include draperies and curtains in other rooms or areas where the draperies and curtains comply with both of the following:
 (a) Individual drapery or curtain panel area does not exceed 48 ft$^2$ (4.5 m$^2$)
 (b) Total area of drapery and curtain panels per room or area does not exceed 20 percent of the aggregate area of the wall on which they are located

**18.7.5.2** Newly introduced upholstered furniture within health care occupancies shall comply with one of the following provisions:

(1) The furniture shall meet the criteria specified in 10.3.2.1 and 10.3.3.
(2) The furniture shall be in a building protected throughout by an approved, supervised automatic sprinkler system in accordance with 9.7.1.1(1).

**18.7.5.3 Reserved.**

**18.7.5.4** Newly introduced mattresses within health care occupancies shall comply with one of the following provisions:

(1) The mattresses shall meet the criteria specified in 10.3.2.2 and 10.3.4.
(2) The mattresses shall be in a building protected throughout by an approved, supervised automatic sprinkler system in accordance with 9.7.1.1(1).

**18.7.5.5 Reserved.**

**18.7.5.6** Combustible decorations shall be prohibited in any health care occupancy, unless one of the following criteria is met:

(1) They are flame-retardant or are treated with approved fire-retardant coating that is listed and labeled for application to the material to which it is applied.
(2) The decorations meet the requirements of NFPA 701, *Standard Methods of Fire Tests for Flame Propagation of Textiles and Films.*
(3) The decorations exhibit a heat release rate not exceeding 100 kW when tested in accordance with NFPA 289, *Standard Method of Fire Test for Individual Fuel Packages*, using the 20 kW ignition source.
(4)\*The decorations, such as photographs, paintings, and other art, are attached directly to the walls, ceiling, and non-fire-rated doors in accordance with the following:
 (a) Decorations on non-fire-rated doors do not interfere with the operation or any required latching of the door and do not exceed the area limitations of 18.7.5.6(b), (c), or (d).

(b) Decorations do not exceed 20 percent of the wall, ceiling, and door areas inside any room or space of a smoke compartment that is not protected throughout by an approved automatic sprinkler system in accordance with Section 9.7.
(c) Decorations do not exceed 30 percent of the wall, ceiling, and door areas inside any room or space of a smoke compartment that is protected throughout by an approved supervised automatic sprinkler system in accordance with Section 9.7.
(d) Decorations do not exceed 50 percent of the wall, ceiling, and door areas inside patient sleeping rooms having a capacity not exceeding four persons, in a smoke compartment that is protected throughout by an approved, supervised automatic sprinkler system in accordance with Section 9.7.

**18.7.5.7 Soiled Linen and Trash Receptacles.**

**18.7.5.7.1** Soiled linen or trash collection receptacles shall not exceed 32 gal (121 L) in capacity and shall meet all of the following requirements:

(1) The average density of container capacity in a room or space shall not exceed 0.5 gal/ft$^2$ (20.4 L/m$^2$).
(2) A capacity of 32 gal (121 L) shall not be exceeded within any 64 ft$^2$ (6 m$^2$) area.
(3)\*Mobile soiled linen or trash collection receptacles with capacities greater than 32 gal (121 L) shall be located in a room protected as a hazardous area when not attended.
(4) Container size and density shall not be limited in hazardous areas.

**18.7.5.7.2\*** Containers used solely for recycling clean waste or for patient records awaiting destruction shall be permitted to be excluded from the requirements of 18.7.5.7.1 where all the following conditions are met:

(1) Each container shall be limited to a maximum capacity of 96 gal (363 L), except as permitted by 18.7.5.7.2(2) or (3).
(2)\*Containers with capacities greater than 96 gal (363 L) shall be located in a room protected as a hazardous area when not attended.
(3) Container size shall not be limited in hazardous areas.
(4) Containers for combustibles shall be labeled and listed as meeting the requirements of FM Approval Standard 6921, *Containers for Combustible Waste*; however, such testing, listing, and labeling shall not be limited to FM Approvals.

**18.7.5.7.3** The provisions of 10.3.9, applicable to containers for rubbish, waste, or linen, shall not apply.

**18.7.6 Maintenance and Testing.** See 4.6.12.

**18.7.7 Engineered Smoke Control Systems.**

**18.7.7.1** New engineered smoke control systems shall be designed, installed, tested, and maintained in accordance with NFPA 92, *Standard for Smoke Control Systems.*

**18.7.7.2** Test documentation shall be maintained on the premises at all times.

**18.7.8 Portable Space-Heating Devices.** Portable space-heating devices shall be prohibited in all health care occupancies, unless both of the following criteria are met:

(1) Such devices are permitted to be used only in nonsleeping staff and employee areas.
(2) The heating elements of such devices do not exceed 212°F (100°C).

**18.7.9 Construction, Repair, and Improvement Operations.**

**18.7.9.1** Construction, repair, and improvement operations shall comply with 4.6.10.

**18.7.9.2** The means of egress in any area undergoing construction, repair, or improvements shall be inspected daily for compliance with 7.1.10.1 and shall also comply with NFPA 241, *Standard for Safeguarding Construction, Alteration, and Demolition Operations.*

## Chapter 19   Existing Health Care Occupancies

### 19.1 General Requirements.

#### 19.1.1 Application.

##### 19.1.1.1 General.

**19.1.1.1.1\*** The requirements of this chapter shall apply to existing buildings or portions thereof currently occupied as health care occupancies, unless the authority having jurisdiction has determined equivalent safety has been provided in accordance with Section 1.4.

**19.1.1.1.2 Administration.** The provisions of Chapter 1, Administration, shall apply.

**19.1.1.1.3 General.** The provisions of Chapter 4, General, shall apply.

**19.1.1.1.4** The requirements established by this chapter shall apply to all existing hospitals, nursing homes, and limited care facilities. The term *hospital*, wherever used in this *Code*, shall include general hospitals, psychiatric hospitals, and specialty hospitals. The term *nursing home*, wherever used in this *Code*, shall include nursing and convalescent homes, skilled nursing facilities, intermediate care facilities, and infirmaries in homes for the aged. Where requirements vary, the specific subclass of health care occupancy that shall apply is named in the paragraph pertaining thereto. The requirements established by Chapter 21 shall apply to all existing ambulatory health care facilities. The operating features requirements established by Section 19.7 shall apply to all health care occupancies.

**19.1.1.1.5** The health care facilities regulated by this chapter shall be those that provide sleeping accommodations for their occupants and are occupied by persons who are mostly incapable of self-preservation because of age, because of physical or mental disability, or because of security measures not under the occupants' control.

**19.1.1.1.6** Buildings, or sections of buildings, that primarily house patients who, in the opinion of the governing body of the facility and the governmental agency having jurisdiction, are capable of exercising judgment and appropriate physical action for self-preservation under emergency conditions shall be permitted to comply with chapters of the *Code* other than Chapter 19.

**19.1.1.1.7\*** It shall be recognized that, in buildings housing certain patients, it might be necessary to lock doors and bar windows to confine and protect building inhabitants.

**19.1.1.1.8** Buildings, or sections of buildings, that house older persons and that provide activities that foster continued independence but do not include services distinctive to health care occupancies *(see 19.1.4.2)*, as defined in 3.3.188.7, shall be

permitted to comply with the requirements of other chapters of this *Code*, such as Chapters 31 or 33.

**19.1.1.1.9** Facilities that do not provide housing on a 24-hour basis for their occupants shall be classified as other occupancies and shall be covered by other chapters of this *Code*.

**19.1.1.1.10\*** The requirements of this chapter shall apply based on the assumption that staff is available in all patient-occupied areas to perform certain fire safety functions as required in other paragraphs of this chapter.

**19.1.1.2\* Goals and Objectives.** The goals and objectives of Sections 4.1 and 4.2 shall be met with due consideration for functional requirements, which are accomplished by limiting the development and spread of a fire emergency to the room of fire origin and reducing the need for occupant evacuation, except from the room of fire origin.

#### 19.1.1.3 Total Concept.

**19.1.1.3.1** All health care facilities shall be designed, constructed, maintained, and operated to minimize the possibility of a fire emergency requiring the evacuation of occupants.

**19.1.1.3.2** Because the safety of health care occupants cannot be ensured adequately by dependence on evacuation of the building, their protection from fire shall be provided by appropriate arrangement of facilities; adequate, trained staff; and development of operating and maintenance procedures composed of the following:

(1) Design, construction, and compartmentation
(2) Provision for detection, alarm, and extinguishment
(3) Fire prevention procedures and planning, training, and drilling programs for the isolation of fire, transfer of occupants to areas of refuge, or evacuation of the building

#### 19.1.1.4 Additions, Conversions, Modernization, Renovation, and Construction Operations.

**19.1.1.4.1 Additions.** Additions shall be separated from any existing structure not conforming to the provisions within Chapter 19 by a fire barrier having not less than a 2-hour fire resistance rating and constructed of materials as required for the addition. *(See 4.6.7 and 4.6.11.)*

**19.1.1.4.1.1** Communicating openings in dividing fire barriers required by 19.1.1.4.1 shall be permitted only in corridors and shall be protected by approved self-closing fire door assemblies. *(See also Section 8.3.)*

**19.1.1.4.1.2** Doors in barriers required by 19.1.1.4.1 shall normally be kept closed, unless otherwise permitted by 19.1.1.4.1.3.

**19.1.1.4.1.3** Doors shall be permitted to be held open if they meet the requirements of 19.2.2.2.7.

**19.1.1.4.2 Changes of Use or Occupancy Classification.** Changes of use or occupancy classification shall comply with 4.6.11, unless otherwise permitted by one of the following:

(1) A change from a hospital to a nursing home or from a nursing home to a hospital shall not be considered a change in occupancy classification or a change in use.
(2) A change from a hospital or nursing home to a limited care facility shall not be considered a change in occupancy classification or a change in use.
(3) A change from a hospital or nursing home to an ambulatory health care facility shall not be considered a change in occupancy classification or a change in use.

**19.1.1.4.3 Rehabilitation.**

**19.1.1.4.3.1** For purposes of the provisions of this chapter, the following shall apply:

(1) A major rehabilitation shall involve the modification of more than 50 percent, or more than 4500 ft² (420 m²), of the area of the smoke compartment.
(2) A minor rehabilitation shall involve the modification of not more than 50 percent, and not more than 4500 ft² (420 m²), of the area of the smoke compartment.

**19.1.1.4.3.2** Work that is exclusively plumbing, mechanical, fire protection system, electrical, medical gas, or medical equipment work shall not be included in the computation of the modification area within the smoke compartment.

**19.1.1.4.3.3\*** Where major rehabilitation is done in a non-sprinklered smoke compartment, the automatic sprinkler requirements of 18.3.5 shall apply to the smoke compartment undergoing the rehabilitation, and, in cases where the building is not protected throughout by an approved automatic sprinkler system, the requirements of 18.4.3.2, 18.4.3.3, and 18.4.3.8 shall also apply.

**19.1.1.4.3.4\*** Where minor rehabilitation is done in a nonsprinklered smoke compartment, the requirements of 18.3.5.1 shall not apply, but, in such cases, the rehabilitation shall not reduce life safety below the level required for new buildings or below the level of the requirements of 18.4.3 for nonsprinklered smoke compartment rehabilitation. *(See 4.6.7.)*

**19.1.1.4.4 Construction, Repair, and Improvement Operations.** See 4.6.10.

**19.1.2 Classification of Occupancy.** See 6.1.5 and 19.1.4.2.

**19.1.3 Multiple Occupancies.**

**19.1.3.1** Multiple occupancies shall be in accordance with 6.1.14.

**19.1.3.2** Sections of health care facilities shall be permitted to be classified as other occupancies in accordance with the separated occupancies provisions of 6.1.14.4 and either 19.1.3.3 or 19.1.3.4.

**19.1.3.3\*** Sections of health care facilities shall be permitted to be classified as other occupancies, provided that they meet all of the following conditions:

(1) They are not intended to provide services simultaneously for four or more inpatients for purposes of housing, treatment, or customary access by inpatients incapable of self-preservation.
(2) They are separated from areas of health care occupancies by construction having a minimum 2-hour fire resistance rating in accordance with Chapter 8.
(3) For other than previously approved occupancy separation arrangements, the entire building is protected throughout by an approved, supervised automatic sprinkler system in accordance with Section 9.7.

**19.1.3.4 Contiguous Non-Health Care Occupancies.**

**19.1.3.4.1\*** Ambulatory care facilities, medical clinics, and similar facilities that are contiguous to health care occupancies, but are primarily intended to provide outpatient services, shall be permitted to be classified as business occupancies or ambulatory health care facilities, provided that the facilities are separated from the health care occupancy by not less than 2-hour fire resistance–rated construction, and the facility is not intended to provide services simultaneously for four or more inpatients who are litterborne.

**19.1.3.4.2** Ambulatory care facilities, medical clinics, and similar facilities that are contiguous to health care occupancies shall be permitted to be used for diagnostic and treatment services of inpatients who are capable of self-preservation.

**19.1.3.5** Where separated occupancies provisions are used in accordance with either 19.1.3.3 or 19.1.3.4, the most stringent construction type shall be provided throughout the building, unless a 2-hour separation is provided in accordance with 8.2.1.3, in which case the construction type shall be determined as follows:

(1) The construction type and supporting construction of the health care occupancy shall be based on the story on which it is located in the building in accordance with the provisions of 19.1.6 and Table 19.1.6.1.
(2) The construction type of the areas of the building enclosing the other occupancies shall be based on the applicable occupancy chapters of this *Code*.

**19.1.3.6** All means of egress from health care occupancies that traverse non-health care spaces shall conform to the requirements of this *Code* for health care occupancies, unless otherwise permitted by 19.1.3.7.

**19.1.3.7** Exit through a horizontal exit into other contiguous occupancies that do not conform to health care egress provisions, but that do comply with requirements set forth in the appropriate occupancy chapter of this *Code*, shall be permitted, provided that both of the following criteria apply:

(1) The occupancy does not contain high hazard contents.
(2) The horizontal exit complies with the requirements of 19.2.2.5.

**19.1.3.8** Egress provisions for areas of health care facilities that correspond to other occupancies shall meet the corresponding requirements of this *Code* for such occupancies, and, where the clinical needs of the occupant necessitate the locking of means of egress, staff shall be present for the supervised release of occupants during all times of use.

**19.1.3.9** Auditoriums, chapels, staff residential areas, or other occupancies provided in connection with health care facilities shall have means of egress provided in accordance with other applicable sections of this *Code*.

**19.1.3.10** Any area with a hazard of contents classified higher than that of the health care occupancy and located in the same building shall be protected as required by 19.3.2.

**19.1.3.11** Non-health care–related occupancies classified as containing high hazard contents shall not be permitted in buildings housing health care occupancies.

**19.1.4 Definitions.**

**19.1.4.1 General.** For definitions, see Chapter 3, Definitions.

**19.1.4.2 Special Definitions.** A list of special terms used in this chapter follows:

(1) **Ambulatory Health Care Occupancy.** See 3.3.188.1.
(2) **Hospital.** See 3.3.142.
(3) **Limited Care Facility.** See 3.3.88.2.
(4) **Nursing Home.** See 3.3.140.2.

**19.1.5 Classification of Hazard of Contents.** The classification of hazard of contents shall be as defined in Section 6.2.

**19.1.6 Minimum Construction Requirements.**

**19.1.6.1** Health care occupancies shall be limited to the building construction types specified in Table 19.1.6.1, unless otherwise permitted by 19.1.6.2 through 19.1.6.7. *(See 8.2.1.)*

**19.1.6.2\*** Any building of Type I(442), Type I(332), Type II(222), or Type II(111) construction shall be permitted to include roofing systems involving combustible supports, decking, or roofing, provided that all of the following criteria are met:

(1) The roof covering shall meet Class C requirements in accordance with ASTM E 108, *Standard Test Methods for Fire Tests of Roof Coverings,* or ANSI/UL 790, *Test Methods for Fire Tests of Roof Coverings.*

(2) The roof shall be separated from all occupied portions of the building by a noncombustible floor assembly that includes not less than 2½ in. (63 mm) of concrete or gypsum fill.

(3) The attic or other space shall be either unoccupied or protected throughout by an approved automatic sprinkler system.

**19.1.6.3** Any building of Type I(442), Type I(332), Type II(222), or Type II(111) construction shall be permitted to include roofing systems involving combustible supports, decking, or roofing, provided that all of the following criteria are met:

(1) The roof covering shall meet Class A requirements in accordance with ASTM E 108, *Standard Test Methods for Fire Tests of Roof Coverings,* or ANSI/UL 790, *Test Methods for Fire Tests of Roof Coverings.*

**Table 19.1.6.1 Construction Type Limitations**

| Construction Type | Sprinklered[†] | Total Number of Stories of Building[‡] | | | |
|---|---|---|---|---|---|
| | | 1 | 2 | 3 | ≥4 |
| I (442) | Yes | X | X | X | X |
| | No | X | X | X | X |
| I (332) | Yes | X | X | X | X |
| | No | X | X | X | X |
| II (222) | Yes | X | X | X | X |
| | No | X | X | X | X |
| II (111) | Yes | X | X | X | NP |
| | No | X | NP | NP | NP |
| II (000) | Yes | X | X | NP | NP |
| | No | NP | NP | NP | NP |
| III (211) | Yes | X | X | NP | NP |
| | No | NP | NP | NP | NP |
| III (200) | Yes | X | NP | NP | NP |
| | No | NP | NP | NP | NP |
| IV (2HH) | Yes | X | X | NP | NP |
| | No | NP | NP | NP | NP |
| V (111) | Yes | X | X | NP | NP |
| | No | NP | NP | NP | NP |
| V (000) | Yes | X | NP | NP | NP |
| | No | NP | NP | NP | NP |

X: Permitted. NP: Not permitted.

The total number of stories of the building is to be determined as follows:

(1) The total number of stories is to be counted starting with the level of exit discharge and ending with the highest occupiable story of the building.

(2) Stories below the level of exit discharge are not counted as stories.

(3) Interstitial spaces used solely for building or process systems directly related to the level above or below are not considered a separate story.

(4) A mezzanine in accordance with 8.6.9 is not counted as a story.

†Sprinklered throughout by an approved, supervised automatic sprinkler system in accordance with Section 9.7. *(See 19.3.5.)*

‡Basements are not counted as stories.

(2) The roof/ceiling assembly shall be constructed with fire-retardant-treated wood meeting the requirements of NFPA 220, *Standard on Types of Building Construction.*

(3) The roof/ceiling assembly shall have the required fire resistance rating for the type of construction.

**19.1.6.4** Interior nonbearing walls in buildings of Type I or Type II construction shall be constructed of noncombustible or limited-combustible materials, unless otherwise permitted by 19.1.6.5.

**19.1.6.5** Interior nonbearing walls required to have a minimum 2-hour fire resistance rating shall be permitted to be fire-retardant-treated wood enclosed within noncombustible or limited-combustible materials, provided that such walls are not used as shaft enclosures.

**19.1.6.6** Fire-retardant-treated wood that serves as supports for the installation of fixtures and equipment shall be permitted to be installed behind noncombustible or limited-combustible sheathing.

**19.1.6.7** Each exterior wall of frame construction and all interior stud partitions shall be firestopped to cut off all concealed draft openings, both horizontal and vertical, between any cellar or basement and the first floor, and such firestopping shall consist of wood not less than 2 in. (51 mm) (nominal) thick or shall be of noncombustible material.

**19.1.7 Occupant Load.** The occupant load, in number of persons for whom means of egress and other provisions are required, either shall be determined on the basis of the occupant load factors of Table 7.3.1.2 that are characteristic of the use of the space or shall be determined as the maximum probable population of the space under consideration, whichever is greater.

**19.2 Means of Egress Requirements.**

**19.2.1 General.** Every aisle, passageway, corridor, exit discharge, exit location, and access shall be in accordance with Chapter 7, unless otherwise modified by 19.2.2 through 19.2.11.

**19.2.2 Means of Egress Components.**

**19.2.2.1 Components Permitted.** Components of means of egress shall be limited to the types described in 19.2.2.2 through 19.2.2.10.

**19.2.2.2 Doors.**

**19.2.2.2.1** Doors complying with 7.2.1 shall be permitted.

**19.2.2.2.2** Locks shall not be permitted on patient sleeping room doors, unless otherwise permitted by one of the following:

(1) Key-locking devices that restrict access to the room from the corridor and that are operable only by staff from the corridor side shall be permitted, provided that such devices do not restrict egress from the room.
(2) Locks complying with 19.2.2.2.5 shall be permitted.

**19.2.2.2.3** Doors not located in a required means of egress shall be permitted to be subject to locking.

**19.2.2.2.4** Doors within a required means of egress shall not be equipped with a latch or lock that requires the use of a tool or key from the egress side, unless otherwise permitted by one of the following:

(1) Locks complying with 19.2.2.2.5 shall be permitted.
(2)*Delayed-egress locks complying with 7.2.1.6.1 shall be permitted.

(3)*Access-controlled egress doors complying with 7.2.1.6.2 shall be permitted.
(4) Elevator lobby exit access door locking in accordance with 7.2.1.6.3 shall be permitted.
(5) Approved existing door-locking installations shall be permitted.

**19.2.2.2.5** Door-locking arrangements shall be permitted in accordance with either 19.2.2.2.5.1 or 19.2.2.2.5.2.

**19.2.2.2.5.1*** Door-locking arrangements shall be permitted where the clinical needs of patients require specialized security measures or where patients pose a security threat, provided that staff can readily unlock doors at all times in accordance with 19.2.2.2.6.

**19.2.2.2.5.2*** Door-locking arrangements shall be permitted where patient special needs require specialized protective measures for their safety, provided that all of the following are met:

(1) Staff can readily unlock doors at all times in accordance with 19.2.2.2.6.
(2) A total (complete) smoke detection system is provided throughout the locked space in accordance with 9.6.2.9, or locked doors can be remotely unlocked at an approved, constantly attended location within the locked space.
(3)*The building is protected throughout by an approved, supervised automatic sprinkler system in accordance with 19.3.5.1.
(4) The locks are electrical locks that fail safely so as to release upon loss of power to the device.
(5) The locks release by independent activation of each of the following:
    (a) Activation of the smoke detection system required by 19.2.2.2.5.2(2)
    (b) Waterflow in the automatic sprinkler system required by 19.2.2.2.5.2(3)

**19.2.2.2.6** Doors that are located in the means of egress and are permitted to be locked under other provisions of 19.2.2.2.5 shall comply with all of the following:

(1) Provisions shall be made for the rapid removal of occupants by means of one of the following:
    (a) Remote control of locks
    (b) Keying of all locks to keys carried by staff at all times
    (c) Other such reliable means available to the staff at all times
(2) Only one locking device shall be permitted on each door.
(3) More than one lock shall be permitted on each door, subject to approval of the authority having jurisdiction.

**19.2.2.2.7*** Any door in an exit passageway, stairway enclosure, horizontal exit, smoke barrier, or hazardous area enclosure shall be permitted to be held open only by an automatic release device that complies with 7.2.1.8.2. The automatic sprinkler system, if provided, and the fire alarm system, and the systems required by 7.2.1.8.2, shall be arranged to initiate the closing action of all such doors throughout the smoke compartment or throughout the entire facility.

**19.2.2.2.8** Where doors in a stair enclosure are held open by an automatic release device as permitted in 19.2.2.2.7, initiation of a door-closing action on any level shall cause all doors at all levels in the stair enclosure to close.

**19.2.2.2.9*** Existing health care occupancies shall be exempt from the re-entry provisions of 7.2.1.5.8.

**19.2.2.2.10** Horizontal-sliding doors shall be permitted in accordance with 19.2.2.2.10.1 or 19.2.2.2.10.2.

**19.2.2.2.10.1** Horizontal-sliding doors, as permitted by 7.2.1.14, that are not automatic-closing shall be limited to a single leaf and shall have a latch or other mechanism that ensures that the doors will not rebound into a partially open position if forcefully closed.

**19.2.2.2.10.2** Horizontal-sliding doors serving an occupant load of fewer than 10 shall be permitted, provided that all of the following criteria are met:

(1) The area served by the door has no high hazard contents.
(2) The door is readily operable from either side without special knowledge or effort.
(3) The force required to operate the door in the direction of door travel is not more than 30 lbf (133 N) to set the door in motion and is not more than 15 lbf (67 N) to close the door or open it to the minimum required width.
(4) The door assembly complies with any required fire protection rating and, where rated, is self-closing or automatic-closing by means of smoke detection in accordance with 7.2.1.8 and is installed in accordance with NFPA 80, *Standard for Fire Doors and Other Opening Protectives.*
(5) Where corridor doors are required to latch, the doors are equipped with a latch or other mechanism that ensures that the doors will not rebound into a partially open position if forcefully closed.

**19.2.2.3 Stairs.** Stairs complying with 7.2.2 shall be permitted.

**19.2.2.4 Smokeproof Enclosures.** Smokeproof enclosures complying with 7.2.3 shall be permitted.

**19.2.2.5 Horizontal Exits.** Horizontal exits complying with 7.2.4 and the modifications of 19.2.2.5.1 through 19.2.2.5.4 shall be permitted.

**19.2.2.5.1** Accumulation space shall be provided in accordance with 19.2.2.5.1.1 and 19.2.2.5.1.2.

**19.2.2.5.1.1** Not less than 30 net ft$^2$ (2.8 net m$^2$) per patient in a hospital or nursing home, or not less than 15 net ft$^2$ (1.4 net m$^2$) per resident in a limited care facility, shall be provided within the aggregated area of corridors, patient rooms, treatment rooms, lounge or dining areas, and other similar areas on each side of the horizontal exit.

**19.2.2.5.1.2** On stories not housing bedridden or litterborne patients, not less than 6 net ft$^2$ (0.56 net m$^2$) per occupant shall be provided on each side of the horizontal exit for the total number of occupants in adjoining compartments.

**19.2.2.5.2** The total egress capacity of the other exits (stairs, ramps, doors leading outside the building) shall not be reduced below one-third of that required for the entire area of the building.

**19.2.2.5.3\*** A door in a horizontal exit shall not be required to swing with egress travel as specified in 7.2.4.3.8(1).

**19.2.2.5.4** Door openings in horizontal exits shall be protected by one of the following methods:

(1) Such door openings shall be protected by a swinging door providing a clear width of not less than 32 in. (810 mm).

(2) Such door openings shall be protected by a horizontal-sliding door that complies with 7.2.1.14 and provides a clear width of not less than 32 in. (810 mm).
(3) Such door openings shall be protected by an existing 34 in. (865 mm) swinging door.

**19.2.2.6 Ramps.**

**19.2.2.6.1** Ramps complying with 7.2.5 shall be permitted.

**19.2.2.6.2** Ramps enclosed as exits shall be of sufficient width to provide egress capacity in accordance with 19.2.3.

**19.2.2.7 Exit Passageways.** Exit passageways complying with 7.2.6 shall be permitted.

**19.2.2.8 Fire Escape Ladders.** Fire escape ladders complying with 7.2.9 shall be permitted.

**19.2.2.9 Alternating Tread Devices.** Alternating tread devices complying with 7.2.11 shall be permitted.

**19.2.2.10 Areas of Refuge.** Areas of refuge used as part of a required accessible means of egress shall comply with 7.2.12.

**19.2.3 Capacity of Means of Egress.**

**19.2.3.1** The capacity of means of egress shall be in accordance with Section 7.3.

**19.2.3.2** The capacity of means of egress providing travel by means of stairs shall be 0.6 in. (15 mm) per person, and the capacity of means of egress providing horizontal travel (without stairs) by means such as doors, ramps, or horizontal exits shall be ½ in. (13 mm) per person, unless otherwise permitted by 19.2.3.3.

**19.2.3.3** The capacity of means of egress in health care occupancies protected throughout by an approved, supervised automatic sprinkler system in accordance with 19.3.5.7 shall be 0.3 in. (7.6 mm) per person for travel by means of stairs and 0.2 in. (5 mm) per person for horizontal travel without stairs.

**19.2.3.4\*** Any required aisle, corridor, or ramp shall be not less than 48 in. (1220 mm) in clear width where serving as means of egress from patient sleeping rooms, unless otherwise permitted by one of the following:

(1) Aisles, corridors, and ramps in adjunct areas not intended for the housing, treatment, or use of inpatients shall be not less than 44 in. (1120 mm) in clear and unobstructed width.
(2)\*Where corridor width is at least 6 ft (1830 mm), noncontinuous projections not more than 6 in. (150 mm) from the corridor wall, above the handrail height, shall be permitted.
(3) Exit access within a room or suite of rooms complying with the requirements of 19.2.5 shall be permitted.
(4) Projections into the required width shall be permitted for wheeled equipment, provided that all of the following conditions are met:
  (a) The wheeled equipment does not reduce the clear unobstructed corridor width to less than 60 in. (1525 mm).
  (b) The health care occupancy fire safety plan and training program address the relocation of the wheeled equipment during a fire or similar emergency.
  (c)\*The wheeled equipment is limited to the following:
    i. Equipment in use and carts in use
    ii. Medical emergency equipment not in use
    iii. Patient lift and transport equipment

(5)*Where the corridor width is at least 8 ft (2440 mm), projections into the required width shall be permitted for fixed furniture, provided that all of the following conditions are met:

(a) The fixed furniture is securely attached to the floor or to the wall.

(b) The fixed furniture does not reduce the clear unobstructed corridor width to less than 6 ft (1830 mm), except as permitted by 19.2.3.4(2).

(c) The fixed furniture is located only on one side of the corridor.

(d) The fixed furniture is grouped such that each grouping does not exceed an area of 50 ft$^2$ (4.6 m$^2$).

(e) The fixed furniture groupings addressed in 19.2.3.4(5)(d) are separated from each other by a distance of at least 10 ft (3050 mm).

(f)*The fixed furniture is located so as to not obstruct access to building service and fire protection equipment.

(g) Corridors throughout the smoke compartment are protected by an electrically supervised automatic smoke detection system in accordance with 19.3.4, or the fixed furniture spaces are arranged and located to allow direct supervision by the facility staff from a nurses' station or similar space.

(h) The smoke compartment is protected throughout by an approved, supervised automatic sprinkler system in accordance with 19.3.5.8.

**19.2.3.5** The aisle, corridor, or ramp shall be arranged to avoid any obstructions to the convenient removal of nonambulatory persons carried on stretchers or on mattresses serving as stretchers.

**19.2.3.6** The minimum clear width for doors in the means of egress from hospitals, nursing homes, limited care facilities, psychiatric hospital sleeping rooms, and diagnostic and treatment areas, such as x-ray, surgery, or physical therapy, shall be not less than 32 in. (810 mm) wide.

**19.2.3.7** The requirement of 19.2.3.6 shall not apply where otherwise permitted by the following:

(1) Existing 34 in. (865 mm) doors shall be permitted.
(2) Existing 28 in. (710 mm) corridor doors in facilities where the fire plans do not require evacuation by bed, gurney, or wheelchair shall be permitted.

**19.2.4 Number of Means of Egress.**

**19.2.4.1** The number of means of egress shall be in accordance with 7.4.1.1 and 7.4.1.3 through 7.4.1.6.

**19.2.4.2** Not less than two exits shall be provided on every story.

**19.2.4.3** Not less than two separate exits shall be accessible from every part of every story.

**19.2.4.4*** Not less than two exits shall be accessible from each smoke compartment, and egress shall be permitted through an adjacent compartment(s), provided that the two required egress paths are arranged so that both do not pass through the same adjacent smoke compartment.

**19.2.5 Arrangement of Means of Egress.**

**19.2.5.1 General.** Arrangement of means of egress shall comply with Section 7.5.

**19.2.5.2* Dead-End Corridors.** Existing dead-end corridors not exceeding 30 ft (9.1 m) shall be permitted. Existing dead-end corridors exceeding 30 ft (9.1 m) shall be permitted to continue in use if it is impractical and unfeasible to alter them.

**19.2.5.3 Reserved.**

**19.2.5.4* Intervening Rooms or Spaces.** Every corridor shall provide access to not less than two approved exits in accordance with Sections 7.4 and 7.5 without passing through any intervening rooms or spaces other than corridors or lobbies.

**19.2.5.5 Two Means of Egress.**

**19.2.5.5.1** Sleeping rooms of more than 1000 ft$^2$ (93 m$^2$) shall have not less than two exit access doors remotely located from each other.

**19.2.5.5.2** Non-sleeping rooms of more than 2500 ft$^2$ (230 m$^2$) shall have not less than two exit access doors remotely located from each other.

**19.2.5.6 Corridor Access.**

**19.2.5.6.1*** Every habitable room shall have an exit access door leading directly to an exit access corridor, unless otherwise provided in 19.2.5.6.2, 19.2.5.6.3, and 19.2.5.6.4.

**19.2.5.6.2** Exit access from a patient sleeping room with not more than eight patient beds shall be permitted to pass through one intervening room to reach an exit access corridor, provided that the intervening room is equipped with an approved automatic smoke detection system in accordance with Section 9.6, or the furnishings and furniture, in combination with all other combustibles within the area, are of such minimum quantity and arrangements that a fully developed fire is unlikely to occur.

**19.2.5.6.3** Rooms having an exit door opening directly to the outside from the room at the finished ground level shall not be required to have an exit access door leading directly to an exit access corridor.

**19.2.5.6.4** Rooms within suites complying with 19.2.5.7 shall not be required to have an exit access door leading directly to an exit access corridor.

**19.2.5.7 Suites.**

**19.2.5.7.1 General.**

**19.2.5.7.1.1 Suite Permission.** Suites complying with 19.2.5.7 shall be permitted to be used to meet the corridor access requirements of 19.2.5.6.

**19.2.5.7.1.2* Suite Separation.** Suites shall be separated from the remainder of the building, and from other suites, by one of the following:

(1) Walls and doors meeting the requirements of 19.3.6.2 through 19.3.6.5
(2) Existing approved barriers and doors that limit the transfer of smoke

**19.2.5.7.1.3 Suite Hazardous Contents Areas.**

**(A)*** Intervening rooms shall not be hazardous areas as defined by 19.3.2.

**(B)** Hazardous areas within a suite shall be separated from the remainder of the suite in accordance with 19.3.2.1, unless otherwise provided in 19.2.5.7.1.3(C) or 19.2.5.7.1.3(D).

**(C)*** Hazardous areas within a suite shall not be required to be separated from the remainder of the suite where complying with both of the following:

(1) The suite is primarily a hazardous area.

(2) The suite is separated from the rest of the health care facility as required for a hazardous area by 19.3.2.1.

**(D)\*** Spaces containing sterile surgical materials limited to a one-day supply in operating suites or similar spaces that are sprinklered in accordance with 19.3.5.7 shall be permitted to be open to the remainder of the suite without separation.

**19.2.5.7.1.4 Suite Subdivision.** The subdivision of suites shall be by means of noncombustible or limited-combustible partitions or partitions constructed with fire-retardant-treated wood enclosed with noncombustible or limited-combustible materials, and such partitions shall not be required to be fire rated.

**19.2.5.7.2 Sleeping Suites.** Sleeping suites shall be in accordance with the following:

(1) Sleeping suites for patient care shall comply with the provisions of 19.2.5.7.2.1 through 19.2.5.7.2.4.

(2) Sleeping suites not for patient care shall comply with the provisions of 19.2.5.7.4.

**19.2.5.7.2.1 Sleeping Suite Arrangement.**

**(A)\*** Occupants of habitable rooms within sleeping suites shall have exit access to a corridor complying with 19.3.6, or to a horizontal exit, directly from the suite.

**(B)** Where two or more exit access doors are required from the suite by 19.2.5.5.1, one of the exit access doors shall be permitted to be directly to an exit stair, exit passageway, or exit door to the exterior.

**(C)** Sleeping suites shall be provided with constant staff supervision within the suite.

**(D)** Sleeping suites shall be arranged in accordance with one of the following:

(1)\*Patient sleeping rooms within sleeping suites shall provide one of the following:

    (a) The patient sleeping rooms shall be arranged to allow for direct supervision from a normally attended location within the suite, such as is provided by glass walls, and cubicle curtains shall be permitted.

    (b) Any patient sleeping rooms without the direct supervision required by 19.2.5.7.2.1(D)(1)(a) shall be provided with smoke detection in accordance with Section 9.6 and 19.3.4.

(2) Sleeping suites shall be provided with a total coverage (complete) automatic smoke detection system in accordance with 9.6.2.9 and 19.3.4.

**19.2.5.7.2.2 Sleeping Suite Number of Means of Egress.**

**(A)** Sleeping suites of more than 1000 ft² (93 m²) shall have not less than two exit access doors remotely located from each other.

**(B)\*** One means of egress from the suite shall be directly to a corridor complying with 19.3.6.

**(C)\*** For suites requiring two means of egress, one means of egress from the suite shall be permitted to be into another suite, provided that the separation between the suites complies with the corridor requirements of 19.3.6.2 through 19.3.6.5

**19.2.5.7.2.3 Sleeping Suite Maximum Size.**

**(A)** Sleeping suites shall not exceed 5000 ft² (460 m²), unless otherwise provided in 19.2.5.7.2.3(B) or 19.2.5.7.2.3(C).

**(B)** Sleeping suites shall not exceed 7500 ft² (700 m²) where the smoke compartment is protected throughout by one of the following:

(1) Approved electrically supervised sprinkler system in accordance with 19.3.5.7 and total coverage (complete) automatic smoke detection in accordance with 9.6.2.9 and 19.3.4

(2) Approved electrically supervised sprinkler system protection complying with 19.3.5.8

**(C)** Sleeping suites greater than 7500 ft² (700 m²), and not exceeding 10,000 ft² (930 m²), shall be permitted where all of the following are provided in the suite:

(1)\*Direct visual supervision in accordance with 19.2.5.7.2.1(D)(1)(a)

(2) Total coverage (complete) automatic smoke detection in accordance with 9.6.2.9 and 19.3.4

(3) Approved electrically supervised sprinkler system protection complying with 19.3.5.8

**19.2.5.7.2.4 Sleeping Suite Travel Distance.**

**(A)** Travel distance between any point in a sleeping suite and an exit access door from that suite shall not exceed 100 ft (30 m).

**(B)** Travel distance between any point in a sleeping suite and an exit shall not exceed the following:

(1) 150 ft (46 m) if the building is not protected throughout by an approved electrically supervised sprinkler system complying with 19.3.5.7

(2) 200 ft (61 m) if the building is protected throughout by an approved electrically supervised sprinkler system complying with 19.3.5.7

**19.2.5.7.3 Patient Care Non-Sleeping Suites.** Non-sleeping suites shall be in accordance with the following:

(1) Non-sleeping suites for patient care shall comply with the provisions of 19.2.5.7.3.1 through 19.2.5.7.3.4.

(2) Non-sleeping suites not for patient care shall comply with the provisions of 19.2.5.7.4

**19.2.5.7.3.1 Patient Care Non-Sleeping Suite Arrangement.**

**(A)** Occupants of habitable rooms within non-sleeping suites shall have exit access to a corridor complying with 19.3.6, or to a horizontal exit, directly from the suite

**(B)** Where two or more exit access doors are required from the suite by 19.2.5.5.2, one of the exit access doors shall be permitted to be directly to an exit stair, exit passageway, or exit door to the exterior.

**19.2.5.7.3.2 Patient Care Non-Sleeping Suite Number of Means of Egress.**

**(A)** Non-sleeping suites of more than 2500 ft² (230 m²) shall have not less than two exit access doors remotely located from each other.

**(B)\*** One means of egress from the suite shall be directly to a corridor complying with 19.3.6.

**(C)\*** For suites requiring two means of egress, one means of egress from the suite shall be permitted to be into another suite, provided that the separation between the suites complies with the corridor requirements of 19.3.6.2 through 19.3.6.5.

**19.2.5.7.3.3 Patient Care Non-Sleeping Suite Maximum Size.** Non-sleeping suites shall not exceed 10,000 ft² (930 m²).

**19.2.5.7.3.4　Patient Care Non-Sleeping Suite Travel Distance.**

**(A)**　Travel distance within a non-sleeping suite to an exit access door from the suite shall not exceed 100 ft (30 m).

**(B)**　Travel distance between any point in a non-sleeping suite and an exit shall not exceed the following:

(1)　150 ft (46 m) if the building is not protected throughout by an approved electrically supervised sprinkler system complying with 19.3.5.7
(2)　200 ft (61 m) if the building is protected throughout by an approved electrically supervised sprinkler system complying with 19.3.5.7

**19.2.5.7.4　Non-Patient-Care Suites.**　The egress provisions for non-patient-care suites shall be in accordance with the primary use and occupancy of the space.

**19.2.6　Travel Distance to Exits.**

**19.2.6.1**　Travel distance shall be measured in accordance with Section 7.6.

**19.2.6.2**　Travel distance shall comply with 19.2.6.2.1 through 19.2.6.2.4.

**19.2.6.2.1**　The travel distance between any point in a room and an exit shall not exceed 150 ft (46 m), unless otherwise permitted by 19.2.6.2.2.

**19.2.6.2.2**　The maximum travel distance specified in 19.2.6.2.1 shall be permitted to be increased by 50 ft (15 m) in buildings protected throughout by an approved, supervised automatic sprinkler system in accordance with 19.3.5.7.

**19.2.6.2.3**　The travel distance between any point in a health care sleeping room and an exit access door in that room shall not exceed 50 ft (15 m).

**19.2.6.2.4**　The travel distance within suites shall be in accordance with 19.2.5.7.

**19.2.7　Discharge from Exits.**　Discharge from exits shall be arranged in accordance with Section 7.7.

**19.2.8　Illumination of Means of Egress.**　Means of egress shall be illuminated in accordance with Section 7.8.

**19.2.9　Emergency Lighting.**

**19.2.9.1**　Emergency lighting shall be provided in accordance with Section 7.9.

**19.2.9.2　Reserved.**

**19.2.10　Marking of Means of Egress.**

**19.2.10.1**　Means of egress shall have signs in accordance with Section 7.10, unless otherwise permitted by 19.2.10.2, 19.2.10.3, or 19.2.10.4.

**19.2.10.2**　Where the path of egress travel is obvious, signs shall not be required in one-story buildings with an occupant load of fewer than 30 persons.

**19.2.10.3**　Where the path of egress travel is obvious, signs shall not be required at gates in outside secured areas.

**19.2.10.4**　Access to exits within rooms or sleeping suites shall not be required to be marked where staff is responsible for relocating or evacuating occupants.

**19.2.11　Special Means of Egress Features. (Reserved)**

**19.3　Protection.**

**19.3.1　Protection of Vertical Openings.**　Any vertical opening shall be enclosed or protected in accordance with Section 8.6, unless otherwise modified by 19.3.1.1 through 19.3.1.8.

**19.3.1.1**　Where enclosure is provided, the construction shall have not less than a 1-hour fire resistance rating.

**19.3.1.2**　Unprotected vertical openings in accordance with 8.6.9.1 shall be permitted.

**19.3.1.3**　Subparagraph 8.6.7(1)(b) shall not apply to patient sleeping and treatment rooms.

**19.3.1.4**　Multilevel patient sleeping areas in psychiatric facilities shall be permitted without enclosure protection between levels, provided that all of the following conditions are met:

(1)　The entire normally occupied area, including all communicating floor levels, is sufficiently open and unobstructed so that a fire or other dangerous condition in any part is obvious to the occupants or supervisory personnel in the area.
(2)　The egress capacity provides simultaneously for all the occupants of all communicating levels and areas, with all communicating levels in the same fire area being considered as a single floor area for purposes of determination of required egress capacity.
(3)　The height between the highest and lowest finished floor levels does not exceed 13 ft (3960 mm), and the number of levels is permitted to be unrestricted.

**19.3.1.5**　Unprotected openings in accordance with 8.6.6 shall not be permitted.

**19.3.1.6**　Where a full enclosure of a stairway that is not a required exit is impracticable, the required enclosure shall be permitted to be limited to that necessary to prevent a fire originating in any story from spreading to any other story.

**19.3.1.7**　A door in a stair enclosure shall be self-closing and shall normally be kept in the closed position, unless otherwise permitted by 19.3.1.8.

**19.3.1.8**　Doors in stair enclosures shall be permitted to be held open under the conditions specified by 19.2.2.2.7 and 19.2.2.2.8.

**19.3.2　Protection from Hazards.**

**19.3.2.1　Hazardous Areas.**　Any hazardous areas shall be safeguarded by a fire barrier having a 1-hour fire resistance rating or shall be provided with an automatic extinguishing system in accordance with 8.7.1.

**19.3.2.1.1**　An automatic extinguishing system, where used in hazardous areas, shall be permitted to be in accordance with 19.3.5.9.

**19.3.2.1.2***　Where the sprinkler option of 19.3.2.1 is used, the areas shall be separated from other spaces by smoke partitions in accordance with Section 8.4.

**19.3.2.1.3**　The doors shall be self-closing or automatic-closing.

**19.3.2.1.4**　Doors in rated enclosures shall be permitted to have nonrated, factory- or field-applied protective plates extending not more than 48 in. (1220 mm) above the bottom of the door.

**19.3.2.1.5**　Hazardous areas shall include, but shall not be restricted to, the following:

(1)　Boiler and fuel-fired heater rooms

(2) Central/bulk laundries larger than 100 ft² (9.3 m²)

(3) Paint shops

(4) Repair shops

(5) Rooms with soiled linen in volume exceeding 64 gal (242 L)

(6) Rooms with collected trash in volume exceeding 64 gal (242 L)

(7) Rooms or spaces larger than 50 ft² (4.6 m²), including repair shops, used for storage of combustible supplies and equipment in quantities deemed hazardous by the authority having jurisdiction

(8) Laboratories employing flammable or combustible materials in quantities less than those that would be considered a severe hazard

**19.3.2.2\* Laboratories.** Laboratories employing quantities of flammable, combustible, or hazardous materials that are considered as a severe hazard shall be in accordance with Section 8.7 and the provisions of NFPA 99, *Health Care Facilities Code,* applicable to administration, maintenance, and testing.

**19.3.2.3 Anesthetizing Locations.** Anesthetizing locations shall be in accordance with Section 8.7 and the provisions of NFPA 99, *Health Care Facilities Code,* applicable to administration, maintenance, and testing.

**19.3.2.4 Medical Gas.** Medical gas storage and administration areas shall be in accordance with Section 8.7 and the provisions of NFPA 99, *Health Care Facilities Code,* applicable to administration, maintenance, and testing.

**19.3.2.5 Cooking Facilities.**

**19.3.2.5.1** Cooking facilities shall be protected in accordance with 9.2.3, unless otherwise permitted by 19.3.2.5.2, 19.3.2.5.3, or 19.3.2.5.4.

**19.3.2.5.2\*** Where residential cooking equipment is used for food warming or limited cooking, the equipment shall not be required to be protected in accordance with 9.2.3, and the presence of the equipment shall not require the area to be protected as a hazardous area.

**19.3.2.5.3\*** Within a smoke compartment, where residential or commercial cooking equipment is used to prepare meals for 30 or fewer persons, one cooking facility shall be permitted to be open to the corridor, provided that all of the following conditions are met:

(1) The portion of the health care facility served by the cooking facility is limited to 30 beds and is separated from other portions of the health care facility by a smoke barrier constructed in accordance with 19.3.7.3, 19.3.7.6, and 19.3.7.8.

(2) The cooktop or range is equipped with a range hood of a width at least equal to the width of the cooking surface, with grease baffles or other grease-collecting and clean-out capability.

(3)\*The hood systems have a minimum airflow of 500 cfm (14,000 L/min).

(4) The hood systems that are not ducted to the exterior additionally have a charcoal filter to remove smoke and odor.

(5) The cooktop or range complies with all of the following:

(a) The cooktop or range is protected with a fire suppression system listed in accordance with UL 300, *Standard for Fire Testing of Fire Extinguishing Systems for Protection of Commercial Cooking Equipment,* or is tested and meets all requirements of UL 300A, *Extinguishing System Units for Residential Range Top Cooking Surfaces,* in accordance with the applicable testing document's scope.

(b) A manual release of the extinguishing system is provided in accordance with NFPA 96, *Standard for Ventilation Control and Fire Protection of Commercial Cooking Operations,* Section 10.5.

(c) An interlock is provided to turn off all sources of fuel and electrical power to the cooktop or range when the suppression system is activated.

(6)\*The use of solid fuel for cooking is prohibited.

(7)\*Deep-fat frying is prohibited.

(8) Portable fire extinguishers in accordance with NFPA 96 are located in all kitchen areas.

(9)\*A switch meeting all of the following is provided:

(a) A locked switch, or a switch located in a restricted location, is provided within the cooking facility that deactivates the cooktop or range.

(b) The switch is used to deactivate the cooktop or range whenever the kitchen is not under staff supervision.

(c) The switch is on a timer, not exceeding a 120-minute capacity, that automatically deactivates the cooktop or range, independent of staff action.

(10) Procedures for the use, inspection, testing, and maintenance of the cooking equipment are in accordance with Chapter 11 of NFPA 96 and the manufacturer's instructions and are followed.

(11)\*Not less than two AC-powered photoelectric smoke alarms, interconnected in accordance with 9.6.2.10.3, equipped with a silence feature, and in accordance with *NFPA 72, National Fire Alarm and Signaling Code,* are located not closer than 20 ft (6.1 m) from the cooktop or range.

(12) No smoke detector is located less than 20 ft (6.1 m) from the cooktop or range.

(13) The smoke compartment is protected throughout by an approved, supervised automatic sprinkler system in accordance with Section 9.7.

**19.3.2.5.4\*** Within a smoke compartment, residential or commercial cooking equipment that is used to prepare meals for 30 or fewer persons shall be permitted, provided that the cooking facility complies with all of the following conditions:

(1) The space containing the cooking equipment is not a sleeping room.

(2) The space containing the cooking equipment shall be separated from the corridor by partitions complying with 19.3.6.2 through 19.3.6.5.

(3) The requirements of 19.3.2.5.3(1) through (10) and (13) are met.

**19.3.2.5.5\*** Where cooking facilities are protected in accordance with 9.2.3, the presence of the cooking equipment shall not cause the room or space housing the equipment to be classified as a hazardous area with respect to the requirements of 19.3.2.1, and the room or space shall not be permitted to be open to the corridor.

**19.3.2.6\* Alcohol-Based Hand-Rub Dispensers.** Alcohol-based hand-rub dispensers shall be protected in accordance with 8.7.3.1, unless all of the following conditions are met:

(1) Where dispensers are installed in a corridor, the corridor shall have a minimum width of 6 ft (1830 mm).

(2) The maximum individual dispenser fluid capacity shall be as follows:

   (a) 0.32 gal (1.2 L) for dispensers in rooms, corridors, and areas open to corridors

   (b) 0.53 gal (2.0 L) for dispensers in suites of rooms

(3) Where aerosol containers are used, the maximum capacity of the aerosol dispenser shall be 18 oz. (0.51 kg) and shall be limited to Level 1 aerosols as defined in NFPA 30B, *Code for the Manufacture and Storage of Aerosol Products.*

(4) Dispensers shall be separated from each other by horizontal spacing of not less than 48 in. (1220 mm).

(5) Not more than an aggregate 10 gal (37.8 L) of alcohol-based hand-rub solution or 1135 oz (32.2 kg) of Level 1 aerosols, or a combination of liquids and Level 1 aerosols not to exceed, in total, the equivalent of 10 gal (37.8 L) or 1135 oz (32.2 kg), shall be in use outside of a storage cabinet in a single smoke compartment, except as otherwise provided in 19.3.2.6(6).

(6) One dispenser complying with 19.3.2.6 (2) or (3) per room and located in that room shall not be included in the aggregated quantity addressed in 19.3.2.6(5).

(7) Storage of quantities greater than 5 gal (18.9 L) in a single smoke compartment shall meet the requirements of NFPA 30, *Flammable and Combustible Liquids Code.*

(8) Dispensers shall not be installed in the following locations:

   (a) Above an ignition source within a 1 in. (25 mm) horizontal distance from each side of the ignition source

   (b) To the side of an ignition source within a 1 in. (25 mm) horizontal distance from the ignition source

   (c) Beneath an ignition source within a 1 in. (25 mm) vertical distance from the ignition source

(9) Dispensers installed directly over carpeted floors shall be permitted only in sprinklered smoke compartments.

(10) The alcohol-based hand-rub solution shall not exceed 95 percent alcohol content by volume.

(11) Operation of the dispenser shall comply with the following criteria:

   (a) The dispenser shall not release its contents except when the dispenser is activated, either manually or automatically by touch-free activation.

   (b) Any activation of the dispenser shall occur only when an object is placed within 4 in. (100 mm) of the sensing device.

   (c) An object placed within the activation zone and left in place shall not cause more than one activation.

   (d) The dispenser shall not dispense more solution than the amount required for hand hygiene consistent with label instructions.

   (e) The dispenser shall be designed, constructed, and operated in a manner that ensures that accidental or malicious activation of the dispensing device is minimized.

   (f) The dispenser shall be tested in accordance with the manufacturer's care and use instructions each time a new refill is installed.

### 19.3.3 Interior Finish.

**19.3.3.1 General.** Interior finish shall be in accordance with Section 10.2.

**19.3.3.2\* Interior Wall and Ceiling Finish.** Existing interior wall and ceiling finish materials complying with Section 10.2 shall be permitted to be Class A or Class B.

**19.3.3.3 Interior Floor Finish.** No restrictions shall apply to existing interior floor finish.

### 19.3.4 Detection, Alarm, and Communications Systems.

**19.3.4.1 General.** Health care occupancies shall be provided with a fire alarm system in accordance with Section 9.6.

**19.3.4.2\* Initiation.**

**19.3.4.2.1** Initiation of the required fire alarm systems shall be by manual means in accordance with 9.6.2 and by means of any required sprinkler system waterflow alarms, detection devices, or detection systems, unless otherwise permitted by 19.3.4.2.2 through 19.3.4.2.4.

**19.3.4.2.2** Manual fire alarm boxes in patient sleeping areas shall not be required at exits if located at all nurses' control stations or other continuously attended staff location, provided that both of the following criteria are met:

(1) Such manual fire alarm boxes are visible and continuously accessible.

(2) Travel distances required by 9.6.2.5 are not exceeded.

**19.3.4.2.3** Fixed extinguishing systems protecting commercial cooking equipment in kitchens that are protected by a complete automatic sprinkler system shall not be required to initiate the fire alarm system.

**19.3.4.2.4** Detectors required by 19.7.5.3 and 19.7.5.5 shall not be required to initiate the fire alarm system.

**19.3.4.3 Notification.** Positive alarm sequence in accordance with 9.6.3.4 shall be permitted in health care occupancies protected throughout by an approved, supervised automatic sprinkler system in accordance with 9.7.1.1(1).

**19.3.4.3.1 Occupant Notification.** Occupant notification shall be accomplished automatically in accordance with 9.6.3, unless otherwise modified by the following:

(1)\*In lieu of audible alarm signals, visible alarm-indicating appliances shall be permitted to be used in critical care areas.

(2) Where visual devices have been installed in patient sleeping areas in place of an audible alarm, they shall be permitted where approved by the authority having jurisdiction.

**19.3.4.3.2 Emergency Forces Notification.**

**19.3.4.3.2.1** Fire department notification shall be accomplished in accordance with 9.6.4.

**19.3.4.3.2.2** Smoke detection devices or smoke detection systems equipped with reconfirmation features shall not be required to automatically notify the fire department, unless the alarm condition is reconfirmed after a period not exceeding 120 seconds.

**19.3.4.3.3 Reserved.**

**19.3.4.4 Fire Safety Functions.** Operation of any activating device in the required fire alarm system shall be arranged to accomplish automatically any control functions to be performed by that device. *(See 9.6.5.)*

**19.3.4.5 Detection.**

**19.3.4.5.1 Corridors.** An approved automatic smoke detection system in accordance with Section 9.6 shall be installed in all corridors of limited care facilities, unless otherwise permitted by one of the following:

(1) Where each patient sleeping room is protected by an approved smoke detection system, and a smoke detector is provided at smoke barriers and horizontal exits in accordance with Section 9.6, the corridor smoke detection system shall not be required on the patient sleeping room floors.

(2) Smoke compartments protected throughout by an approved, supervised automatic sprinkler system in accordance with 19.3.5.7 shall be permitted.

**19.3.4.5.2 Detection in Spaces Open to Corridors.** See 19.3.6.1.

**19.3.5 Extinguishment Requirements.**

**19.3.5.1** Buildings containing nursing homes shall be protected throughout by an approved, supervised automatic sprinkler system in accordance with Section 9.7, unless otherwise permitted by 19.3.5.5.

**19.3.5.2** High-rise buildings shall comply with 19.4.2.

**19.3.5.3** Where required by 19.1.6, buildings containing hospitals or limited care facilities shall be protected throughout by an approved, supervised automatic sprinkler system in accordance with Section 9.7, unless otherwise permitted by 19.3.5.5.

**19.3.5.4\*** The sprinkler system required by 19.3.5.1 or 19.3.5.3 shall be installed in accordance with 9.7.1.1(1).

**19.3.5.5** In Type I and Type II construction, alternative protection measures shall be permitted to be substituted for sprinkler protection in specified areas where the authority having jurisdiction has prohibited sprinklers, without causing a building to be classified as nonsprinklered.

**19.3.5.6 Reserved.**

**19.3.5.7\*** Where this *Code* permits exceptions for fully sprinklered buildings or smoke compartments, the sprinkler system shall meet all of the following criteria:

(1) It shall be in accordance with Section 9.7.
(2) It shall be installed in accordance with 9.7.1.1(1), unless it is an approved existing system.
(3) It shall be electrically connected to the fire alarm system.
(4) It shall be fully supervised.
(5) In Type I and Type II construction, where the authority having jurisdiction has prohibited sprinklers, approved alternative protection measures shall be permitted to be substituted for sprinkler protection in specified areas without causing a building to be classified as nonsprinklered.

**19.3.5.8\*** Where this *Code* permits exceptions for fully sprinklered buildings or smoke compartments and specifically references this paragraph, the sprinkler system shall meet all of the following criteria:

(1) It shall be installed throughout the building or smoke compartment in accordance with Section 9.7.
(2) It shall be installed in accordance with 9.7.1.1(1), unless it is an approved existing system.
(3) It shall be electrically connected to the fire alarm system.
(4) It shall be fully supervised.
(5) It shall be equipped with listed quick-response or listed residential sprinklers throughout all smoke compartments containing patient sleeping rooms.
(6)\*Standard-response sprinklers shall be permitted to be continued to be used in approved existing sprinkler systems where quick-response and residential sprinklers were not listed for use in such locations at the time of installation.

(7) Standard-response sprinklers shall be permitted for use in hazardous areas protected in accordance with 19.3.2.1.

**19.3.5.9** Isolated hazardous areas shall be permitted to be protected in accordance with 9.7.1.2. For new installations in existing health care occupancies, where more than two sprinklers are installed in a single area, waterflow detection shall be provided to sound the building fire alarm or to notify, by a signal, any constantly attended location, such as PBX, security, or emergency room, at which the necessary corrective action shall be taken.

**19.3.5.10\*** Sprinklers shall not be required in clothes closets of patient sleeping rooms in hospitals where the area of the closet does not exceed 6 ft$^2$ (0.55 m$^2$), provided that the distance from the sprinkler in the patient sleeping room to the back wall of the closet does not exceed the maximum distance permitted by NFPA 13, *Standard for the Installation of Sprinkler Systems.*

**19.3.5.11\*** Newly introduced cubicle curtains in sprinklered areas shall be installed in accordance with NFPA 13, *Standard for the Installation of Sprinkler Systems.*

**19.3.5.12** Portable fire extinguishers shall be provided in all health care occupancies in accordance with 9.7.4.1.

**19.3.6 Corridors.**

**19.3.6.1 Corridor Separation.** Corridors shall be separated from all other areas by partitions complying with 19.3.6.2 through 19.3.6.5 *(see also 19.2.5.4)*, unless otherwise permitted by one of the following:

(1) Smoke compartments protected throughout by an approved supervised automatic sprinkler system in accordance with 19.3.5.8 shall be permitted to have spaces that are unlimited in size and open to the corridor, provided that all of the following criteria are met:

(a)\*The spaces are not used for patient sleeping rooms, treatment rooms, or hazardous areas.
(b) The corridors onto which the spaces open in the same smoke compartment are protected by an electrically supervised automatic smoke detection system in accordance with 19.3.4, or the smoke compartment in which the space is located is protected throughout by quick-response sprinklers.
(c) The open space is protected by an electrically supervised automatic smoke detection system in accordance with 19.3.4, or the entire space is arranged and located to allow direct supervision by the facility staff from a nurses' station or similar space.
(d) The space does not obstruct access to required exits.

(2) In smoke compartments protected throughout by an approved, supervised automatic sprinkler system in accordance with 19.3.5.8, waiting areas shall be permitted to be open to the corridor, provided that all of the following criteria are met:

(a) The aggregate waiting area in each smoke compartment does not exceed 600 ft$^2$ (55.7 m$^2$).
(b) Each area is protected by an electrically supervised automatic smoke detection system in accordance with 19.3.4, or each area is arranged and located to allow direct supervision by the facility staff from a nursing station or similar space.
(c) The area does not obstruct access to required exits.

(3)\*This requirement shall not apply to spaces for nurses' stations.

(4) Gift shops not exceeding 500 ft$^2$ (46.4 m$^2$) shall be permitted to be open to the corridor or lobby, provided that one of the following criteria is met:
   (a) The building is protected throughout by an approved automatic sprinkler system in accordance with Section 9.7.
   (b) The gift shop is protected throughout by an approved automatic sprinkler system in accordance with Section 9.7, and storage is separately protected.

(5) Limited care facilities in smoke compartments protected throughout by an approved, supervised automatic sprinkler system in accordance with 19.3.5.8 shall be permitted to have group meeting or multipurpose therapeutic spaces open to the corridor, provided that all of the following criteria are met:
   (a) The space is not a hazardous area.
   (b) The space is protected by an electrically supervised automatic smoke detection system in accordance with 19.3.4, or the space is arranged and located to allow direct supervision by the facility staff from the nurses' station or similar location.
   (c) The space does not obstruct access to required exits.

(6) Cooking facilities in accordance with 19.3.2.5.3 shall be permitted to be open to the corridor.

(7) Spaces, other than patient sleeping rooms, treatment rooms, and hazardous areas, shall be permitted to be open to the corridor and unlimited in area, provided that all of the following criteria are met:
   (a) The space and the corridors onto which it opens, where located in the same smoke compartment, are protected by an electrically supervised automatic smoke detection system in accordance with 19.3.4.
   (b)*Each space is protected by automatic sprinklers, or the furnishings and furniture, in combination with all other combustibles within the area, are of such minimum quantity and arrangement that a fully developed fire is unlikely to occur.
   (c) The space does not obstruct access to required exits.

(8)*Waiting areas shall be permitted to be open to the corridor, provided that all of the following criteria are met:
   (a) Each area does not exceed 600 ft$^2$ (55.7 m$^2$).
   (b) The area is equipped with an electrically supervised automatic smoke detection system in accordance with 19.3.4.
   (c) The area does not obstruct any access to required exits.

(9) Group meeting or multipurpose therapeutic spaces, other than hazardous areas, that are under continuous supervision by facility staff shall be permitted to be open to the corridor, provided that all of the following criteria are met:
   (a) Each area does not exceed 1500 ft$^2$ (139 m$^2$).
   (b) Not more than one such space is permitted per smoke compartment.
   (c) The area is equipped with an electrically supervised automatic smoke detection system in accordance with 19.3.4.
   (d) The area does not obstruct access to required exits.

**19.3.6.2 Construction of Corridor Walls.**

**19.3.6.2.1** Corridor walls shall be continuous from the floor to the underside of the floor or roof deck above; through any concealed spaces, such as those above suspended ceilings; and through interstitial structural and mechanical spaces, unless otherwise permitted by 19.3.6.2.4 through 19.3.6.2.8.

**19.3.6.2.2\*** Corridor walls shall have a minimum ½-hour fire resistance rating.

**19.3.6.2.3\*** Corridor walls shall form a barrier to limit the transfer of smoke.

**19.3.6.2.4\*** In smoke compartments protected throughout by an approved, supervised automatic sprinkler system in accordance with 19.3.5.7, a corridor shall be permitted to be separated from all other areas by non-fire-rated partitions and shall be permitted to terminate at the ceiling where the ceiling is constructed to limit the transfer of smoke.

**19.3.6.2.5** Existing corridor partitions shall be permitted to terminate at ceilings that are not an integral part of a floor construction if 60 in. (1525 mm) or more of space exists between the top of the ceiling subsystem and the bottom of the floor or roof above, provided that all the following criteria are met:

(1) The ceiling is part of a fire-rated assembly tested to have a minimum 1-hour fire resistance rating in compliance with the provisions of Section 8.3.
(2) The corridor partitions form smoke-tight joints with the ceilings, and joint filler, if used, is noncombustible.
(3) Each compartment of interstitial space that constitutes a separate smoke area is vented, in a smoke emergency, to the outside by mechanical means having the capacity to provide not less than two air changes per hour but, in no case, a capacity less than 5000 ft$^3$/min (2.35 m$^3$/s).
(4) The interstitial space is not used for storage.
(5) The space is not used as a plenum for supply, exhaust, or return air, except as noted in 19.3.6.2.5(3).

**19.3.6.2.6\*** Existing corridor partitions shall be permitted to terminate at monolithic ceilings that resist the passage of smoke where there is a smoke-tight joint between the top of the partition and the bottom of the ceiling.

**19.3.6.2.7** Fixed fire window assemblies in accordance with Section 8.3 shall be permitted in corridor walls, unless otherwise permitted in 19.3.6.2.8.

**19.3.6.2.8** There shall be no restrictions in area and fire resistance of glass and frames in smoke compartments protected throughout by an approved, supervised automatic sprinkler system in accordance with 19.3.5.7.

**19.3.6.3\* Corridor Doors.**

**19.3.6.3.1\*** Doors protecting corridor openings in other than required enclosures of vertical openings, exits, or hazardous areas shall be doors constructed to resist the passage of smoke and shall be constructed of materials such as the following:

(1) 1¾ in. (44 mm) thick, solid-bonded core wood
(2) Material that resists fire for a minimum of 20 minutes

**19.3.6.3.2** The requirements of 19.3.6.3.1 shall not apply where otherwise permitted by either of the following:

(1) Doors to toilet rooms, bathrooms, shower rooms, sink closets, and similar auxiliary spaces that do not contain flammable or combustible materials shall not be required to comply with 19.3.6.3.1.
(2) In smoke compartments protected throughout by an approved, supervised automatic sprinkler system in accordance with 19.3.5.7, the door construction materials requirements of 19.3.6.3.1 shall not be mandatory, but the doors shall be constructed to resist the passage of smoke.

**19.3.6.3.3** Compliance with NFPA 80, *Standard for Fire Doors and Other Opening Protectives*, shall not be required.

**19.3.6.3.4** A clearance between the bottom of the door and the floor covering not exceeding 1 in. (25 mm) shall be permitted for corridor doors.

**19.3.6.3.5\*** Doors shall be provided with a means for keeping the door closed that is acceptable to the authority having jurisdiction, and the following requirements also shall apply:

(1) The device used shall be capable of keeping the door fully closed if a force of 5 lbf (22 N) is applied at the latch edge of the door.
(2) Roller latches shall be prohibited on corridor doors in buildings not fully protected by an approved automatic sprinkler system in accordance with 19.3.5.7.

**19.3.6.3.6** The requirements of 19.3.6.3.5 shall not apply where otherwise permitted by either of the following:

(1) Doors to toilet rooms, bathrooms, shower rooms, sink closets, and similar auxiliary spaces that do not contain flammable or combustible materials shall not be required to comply with 19.3.6.3.5.
(2) Existing roller latches demonstrated to keep the door closed against a force of 5 lbf (22 N) shall be permitted to be kept in service.

**19.3.6.3.7** Powered doors that comply with the requirements of 7.2.1.9 shall be considered as complying with the requirements of 19.3.6.3.5, provided that both of the following criteria are met:

(1) The door is equipped with a means for keeping the door closed that is acceptable to the authority having jurisdiction.
(2) The device used is capable of keeping the door fully closed if a force of 5 lbf (22 N) is applied at the latch edge of a swinging door and applied in any direction to a sliding or folding door, whether or not power is applied.

**19.3.6.3.8 Reserved.**

**19.3.6.3.9 Reserved.**

**19.3.6.3.10\*** Doors shall not be held open by devices other than those that release when the door is pushed or pulled.

**19.3.6.3.11** Door-closing devices shall not be required on doors in corridor wall openings other than those serving required exits, smoke barriers, or enclosures of vertical openings and hazardous areas.

**19.3.6.3.12\*** Nonrated, factory- or field-applied protective plates, unlimited in height, shall be permitted.

**19.3.6.3.13** Dutch doors shall be permitted where they conform to 19.3.6.3 and meet all of the following criteria:

(1) Both the upper leaf and lower leaf are equipped with a latching device.
(2) The meeting edges of the upper and lower leaves are equipped with an astragal, a rabbet, or a bevel.
(3) Where protecting openings in enclosures around hazardous areas, the doors comply with NFPA 80, *Standard for Fire Doors and Other Opening Protectives*.

**19.3.6.3.14** Door frames shall be labeled, shall be of steel construction, or shall be of other materials in compliance with the provisions of Section 8.3, unless otherwise permitted by 19.3.6.3.15.

**19.3.6.3.15** Door frames in smoke compartments protected throughout by an approved, supervised automatic sprinkler system in accordance with 19.3.5.7 shall not be required to comply with 19.3.6.3.14.

**19.3.6.3.16** Fixed fire window assemblies in accordance with Section 8.3 shall be permitted in corridor doors.

**19.3.6.3.17** Restrictions in area and fire resistance of glass and frames required by Section 8.3 shall not apply in smoke compartments protected throughout by an approved, supervised automatic sprinkler system in accordance with 19.3.5.7.

**19.3.6.4 Transfer Grilles.**

**19.3.6.4.1** Transfer grilles, regardless of whether they are protected by fusible link–operated dampers, shall not be used in corridor walls or doors.

**19.3.6.4.2** Doors to toilet rooms, bathrooms, shower rooms, sink closets, and similar auxiliary spaces that do not contain flammable or combustible materials shall be permitted to have ventilating louvers or to be undercut.

**19.3.6.5 Openings.**

**19.3.6.5.1\*** Miscellaneous openings, such as mail slots, pharmacy pass-through windows, laboratory pass-through windows, and cashier pass-through windows, shall be permitted to be installed in vision panels or doors without special protection, provided that both of the following criteria are met:

(1) The aggregate area of openings per room does not exceed 20 in.$^2$ (0.015 m$^2$).
(2) The openings are installed at or below half the distance from the floor to the room ceiling.

**19.3.6.5.2** The alternative requirements of 19.3.6.5.1 shall not apply where otherwise modified by the following:

(1) Openings in smoke compartments containing patient bedrooms shall not be permitted to be installed in vision panels or doors without special protection.
(2) For rooms protected throughout by an approved, supervised automatic sprinkler system in accordance with 19.3.5.7, the aggregate area of openings per room shall not exceed 80 in.$^2$ (0.05 m$^2$).

**19.3.7 Subdivision of Building Spaces.**

**19.3.7.1** Smoke barriers shall be provided to divide every story used for sleeping rooms for more than 30 patients into not less than two smoke compartments *(see 19.2.4.4)*, and the following also shall apply:

(1) The size of any such smoke compartment shall not exceed 22,500 ft$^2$ (2100 m$^2$), and the travel distance from any point to reach a door in the required smoke barrier shall not exceed 200 ft (61 m).
(2) Where neither the length nor width of the smoke compartment exceeds 150 ft (46 m), the travel distance to reach the smoke barrier door shall not be limited.
(3) The area of an atrium separated in accordance with 8.6.7 shall not be limited in size.

**19.3.7.2** For purposes of the requirements of 19.3.7, the number of health care occupants shall be determined by actual count of patient bed capacity.

**19.3.7.3** Any required smoke barrier shall be constructed in accordance with Section 8.5 and shall have a minimum ½-hour fire resistance rating, unless otherwise permitted by one of the following:

(1) This requirement shall not apply where an atrium is used, and both of the following criteria also shall apply:

    (a) Smoke barriers shall be permitted to terminate at an atrium wall constructed in accordance with 8.6.7(1)(c).

    (b) Not less than two separate smoke compartments shall be provided on each floor.

(2)*Smoke dampers shall not be required in duct penetrations of smoke barriers in fully ducted heating, ventilating, and air-conditioning systems where an approved, supervised automatic sprinkler system in accordance with 19.3.5.8 has been provided for smoke compartments adjacent to the smoke barrier.

**19.3.7.4 Reserved.**

**19.3.7.5** Accumulation space shall be provided in accordance with 19.3.7.5.1 and 19.3.7.5.2.

**19.3.7.5.1** Not less than 30 net ft$^2$ (2.8 net m$^2$) per patient in a hospital or nursing home, or not less than 15 net ft$^2$ (1.4 net m$^2$) per resident in a limited care facility, shall be provided within the aggregate area of corridors, patient rooms, treatment rooms, lounge or dining areas, and other low hazard areas on each side of the smoke barrier.

**19.3.7.5.2** On stories not housing bedridden or litterborne patients, not less than 6 net ft$^2$ (0.56 net m$^2$) per occupant shall be provided on each side of the smoke barrier for the total number of occupants in adjoining compartments.

**19.3.7.6** Openings in smoke barriers shall be protected using one of the following methods:

(1) Fire-rated glazing
(2) Wired glass panels in steel frames
(3) Doors, such as 1¾ in. (44 mm) thick, solid-bonded wood-core doors
(4) Construction that resists fire for a minimum of 20 minutes.

**19.3.7.6.1*** Nonrated factory- or field-applied protective plates, unlimited in height, shall be permitted.

**19.3.7.6.2** Doors shall be permitted to have fixed fire window assemblies in accordance with Section 8.5.

**19.3.7.7 Reserved.**

**19.3.7.8*** Doors in smoke barriers shall comply with 8.5.4 and all of the following:

(1) The doors shall be self-closing or automatic-closing in accordance with 19.2.2.2.7.
(2) Latching hardware shall not be required.
(3) The doors shall not be required to swing in the direction of egress travel.

**19.3.7.9** Door openings in smoke barriers shall be protected using one of the following methods:

(1) Swinging door providing a clear width of not less than 32 in. (810 mm)
(2) Horizontal-sliding door complying with 7.2.1.14 and providing a clear width of not less than 32 in. (810 mm)

**19.3.7.10** The requirement of 19.3.7.9 shall not apply to existing 34 in. (865 mm) doors.

**19.3.8 Special Protection Features. (Reserved)**

**19.4 Special Provisions.**

**19.4.1 Limited Access Buildings.** See Section 11.7 for requirements for limited access buildings.

**19.4.2 High-Rise Buildings.**

**19.4.2.1** All high-rise buildings containing health care occupancies shall be protected throughout by an approved, supervised automatic sprinkler system installed in accordance with Section 9.7 within 12 years of the adoption of this *Code*, except as otherwise provided in 19.4.2.2.

**19.4.2.2*** Where a jurisdiction adopts this edition of the *Code* and previously adopted the 2009 edition, the sprinklering required by 19.4.2.1 shall be installed within 9 years of the adoption of this *Code*.

**19.4.3 Nonsprinklered Existing Smoke Compartment Rehabilitation.** *(See 18.4.3.)*

**19.5 Building Services.**

**19.5.1 Utilities.**

**19.5.1.1** Utilities shall comply with the provisions of Section 9.1.

**19.5.1.2** Existing installations shall be permitted to be continued in service, provided that the systems do not present a serious hazard to life.

**19.5.2 Heating, Ventilating, and Air-Conditioning.**

**19.5.2.1** Heating, ventilating, and air-conditioning shall comply with the provisions of Section 9.2 and shall be installed in accordance with the manufacturer's specifications, unless otherwise modified by 19.5.2.2.

**19.5.2.2*** Any heating device, other than a central heating plant, shall be designed and installed so that combustible material cannot be ignited by the device or its appurtenances, and the following requirements also shall apply:

(1) If fuel-fired, such heating devices shall comply with the following:

    (a) They shall be chimney connected or vent connected.

    (b) They shall take air for combustion directly from the outside.

    (c) They shall be designed and installed to provide for complete separation of the combustion system from the atmosphere of the occupied area.

(2) Any heating device shall have safety features to immediately stop the flow of fuel and shut down the equipment in case of either excessive temperature or ignition failure.

**19.5.2.3** The requirements of 19.5.2.2 shall not apply where otherwise permitted by the following:

(1) Approved, suspended unit heaters shall be permitted in locations other than means of egress and patient sleeping areas, provided that both of the following criteria are met:

    (a) Such heaters are located high enough to be out of the reach of persons using the area.

    (b) Such heaters are equipped with the safety features required by 19.5.2.2(2).

(2) Direct-vent gas fireplaces, as defined in NFPA 54, *National Fuel Gas Code*, shall be permitted inside of smoke compartments containing patient sleeping areas, provided that all of the following criteria are met:

    (a) All such devices shall be installed, maintained, and used in accordance with 9.2.2.

(b) No such device shall be located inside of a patient sleeping room.

(c) The smoke compartment in which the direct-vent gas fireplace is located shall be protected throughout by an approved, supervised automatic sprinkler system in accordance with 9.7.1.1(1) with listed quick-response or listed residential sprinklers.

(d)*The direct-vent fireplace shall include a sealed glass front with a wire mesh panel or screen.

(e)*The controls for the direct-vent gas fireplace shall be locked or located in a restricted location.

(f) Electrically supervised carbon monoxide detection in accordance with Section 9.8 shall be provided in the room where the fireplace is located.

(3) Solid fuel–burning fireplaces shall be permitted and used only in areas other than patient sleeping areas, provided that all of the following criteria are met:

(a) Such areas are separated from patient sleeping spaces by construction having not less than a 1-hour fire resistance rating.

(b) The fireplace complies with the provisions of 9.2.2.

(c) The fireplace is equipped with a fireplace enclosure guaranteed against breakage up to a temperature of 650°F (343°C) and constructed of heat-tempered glass or other approved material.

(d) Electrically supervised carbon monoxide detection in accordance with Section 9.8 is provided in the room where the fireplace is located.

(4) If, in the opinion of the authority having jurisdiction, special hazards are present, a lock on the enclosure specified in 19.5.2.3(3)(c) and other safety precautions shall be permitted to be required.

**19.5.3 Elevators, Escalators, and Conveyors.** Elevators, escalators, and conveyors shall comply with the provisions of Section 9.4.

**19.5.4 Rubbish Chutes, Incinerators, and Laundry Chutes.**

**19.5.4.1** Existing rubbish chutes or linen chutes, including pneumatic rubbish and linen systems, that open directly onto any corridor shall be sealed by fire-resistive construction to prevent further use or shall be provided with a fire door assembly having a minimum 1-hour fire protection rating. All new chutes shall comply with Section 9.5.

**19.5.4.2 Reserved.**

**19.5.4.3** Any rubbish chute or linen chute, including pneumatic rubbish and linen systems, shall be provided with automatic extinguishing protection in accordance with Section 9.7. *(See Section 9.5.)*

**19.5.4.4** Any rubbish chute shall discharge into a trash collection room used for no other purpose and shall be protected in accordance with Section 8.7, unless otherwise provided in 19.5.4.5.

**19.5.4.5** Existing laundry chutes shall be permitted to discharge into the same room as rubish discharge chutes, provided that the room is protected by automatic sprinklers in accordance with 19.3.5.9 or 19.3.5.7.

**19.5.4.6** Existing flue-fed incinerators shall be sealed by fire-resistive construction to prevent further use.

**19.6 Reserved.**

**19.7* Operating Features.**

**19.7.1 Evacuation and Relocation Plan and Fire Drills.**

**19.7.1.1** The administration of every health care occupancy shall have, in effect and available to all supervisory personnel, written copies of a plan for the protection of all persons in the event of fire, for their evacuation to areas of refuge, and for their evacuation from the building when necessary.

**19.7.1.2** All employees shall be periodically instructed and kept informed with respect to their duties under the plan required by 19.7.1.1.

**19.7.1.3** A copy of the plan required by 19.7.1.1 shall be readily available at all times in the telephone operator's location or at the security center.

**19.7.1.4*** Fire drills in health care occupancies shall include the transmission of a fire alarm signal and simulation of emergency fire conditions.

**19.7.1.5** Infirm or bedridden patients shall not be required to be moved during drills to safe areas or to the exterior of the building.

**19.7.1.6** Drills shall be conducted quarterly on each shift to familiarize facility personnel (nurses, interns, maintenance engineers, and administrative staff) with the signals and emergency action required under varied conditions.

**19.7.1.7** When drills are conducted between 9:00 p.m. and 6:00 a.m. (2100 hours and 0600 hours), a coded announcement shall be permitted to be used instead of audible alarms.

**19.7.1.8** Employees of health care occupancies shall be instructed in life safety procedures and devices.

**19.7.2 Procedure in Case of Fire.**

**19.7.2.1* Protection of Patients.**

**19.7.2.1.1** For health care occupancies, the proper protection of patients shall require the prompt and effective response of health care personnel.

**19.7.2.1.2** The basic response required of staff shall include the following:

(1) Removal of all occupants directly involved with the fire emergency
(2) Transmission of an appropriate fire alarm signal to warn other building occupants and summon staff
(3) Confinement of the effects of the fire by closing doors to isolate the fire area
(4) Relocation of patients as detailed in the health care occupancy's fire safety plan

**19.7.2.2 Fire Safety Plan.** A written health care occupancy fire safety plan shall provide for all of the following:

(1) Use of alarms
(2) Transmission of alarms to fire department
(3) Emergency phone call to fire department
(4) Response to alarms
(5) Isolation of fire
(6) Evacuation of immediate area
(7) Evacuation of smoke compartment
(8) Preparation of floors and building for evacuation
(9) Extinguishment of fire

### 19.7.2.3 Staff Response.

**19.7.2.3.1** All health care occupancy personnel shall be instructed in the use of and response to fire alarms.

**19.7.2.3.2** All health care occupancy personnel shall be instructed in the use of the code phrase to ensure transmission of an alarm under any of the following conditions:

(1) When the individual who discovers a fire must immediately go to the aid of an endangered person
(2) During a malfunction of the building fire alarm system

**19.7.2.3.3** Personnel hearing the code announced shall first activate the building fire alarm using the nearest manual fire alarm box and then shall execute immediately their duties as outlined in the fire safety plan.

### 19.7.3 Maintenance of Means of Egress.

**19.7.3.1** Proper maintenance shall be provided to ensure the dependability of the method of evacuation selected.

**19.7.3.2** Health care occupancies that find it necessary to lock means of egress doors shall, at all times, maintain an adequate staff qualified to release locks and direct occupants from the immediate danger area to a place of safety in case of fire or other emergency.

**19.7.4\* Smoking.** Smoking regulations shall be adopted and shall include not less than the following provisions:

(1) Smoking shall be prohibited in any room, ward, or individual enclosed space where flammable liquids, combustible gases, or oxygen is used or stored and in any other hazardous location, and such areas shall be posted with signs that read NO SMOKING or shall be posted with the international symbol for no smoking.
(2) In health care occupancies where smoking is prohibited and signs are prominently placed at all major entrances, secondary signs with language that prohibits smoking shall not be required.
(3) Smoking by patients classified as not responsible shall be prohibited.
(4) The requirement of 19.7.4(3) shall not apply where the patient is under direct supervision.
(5) Ashtrays of noncombustible material and safe design shall be provided in all areas where smoking is permitted.
(6) Metal containers with self-closing cover devices into which ashtrays can be emptied shall be readily available to all areas where smoking is permitted.

### 19.7.5 Furnishings, Mattresses, and Decorations.

**19.7.5.1\*** Draperies, curtains, and other loosely hanging fabrics and films serving as furnishings or decorations in health care occupancies shall be in accordance with the provisions of 10.3.1 *(see 19.3.5.11)*, and the following also shall apply:

(1) Such curtains shall include cubicle curtains.
(2) Such curtains shall not include curtains at showers and baths.
(3) Such draperies and curtains shall not include draperies and curtains at windows in patient sleeping rooms in smoke compartments sprinklered in accordance with 19.3.5.
(4) Such draperies and curtains shall not include draperies and curtains in other rooms or areas where the draperies and curtains comply with all of the following:
　(a) Individual drapery or curtain panel area does not exceed 48 ft$^2$ (4.5 m$^2$).

　(b) Total area of drapery and curtain panels per room or area does not exceed 20 percent of the aggregate area of the wall on which they are located.
　(c) Smoke compartment in which draperies or curtains are located is sprinklered in accordance with 19.3.5.

**19.7.5.2** Newly introduced upholstered furniture within health care occupancies shall comply with one of the following provisions, unless otherwise provided in 19.7.5.3:

(1) The furniture shall meet the criteria specified in 10.3.2.1 and 10.3.3.
(2) The furniture shall be in a building protected throughout by an approved, supervised automatic sprinkler system in accordance with 9.7.1.1(1).

**19.7.5.3** The requirements of 19.7.5.2, 10.3.2.1, and 10.3.3 shall not apply to upholstered furniture belonging to the patient in sleeping rooms of nursing homes where the following criteria are met:

(1) A smoke detector shall be installed where the patient sleeping room is not protected by automatic sprinklers.
(2) Battery-powered single-station smoke detectors shall be permitted.

**19.7.5.4** Newly introduced mattresses within health care occupancies shall comply with one of the following provisions, unless otherwise provided in 19.7.5.5:

(1) The mattresses shall meet the criteria specified in 10.3.2.2 and 10.3.4.
(2) The mattresses shall be in a building protected throughout by an approved, supervised automatic sprinkler system in accordance with 9.7.1.1(1).

**19.7.5.5** The requirements of 19.7.5.4, 10.3.2.2, and 10.3.4 shall not apply to mattresses belonging to the patient in sleeping rooms of nursing homes where the following criteria are met:

(1) A smoke detector shall be installed where the patient sleeping room is not protected by automatic sprinklers.
(2) Battery-powered single-station smoke detectors shall be permitted.

**19.7.5.6** Combustible decorations shall be prohibited in any health care occupancy, unless one of the following criteria is met:

(1) They are flame-retardant or are treated with approved fire-retardant coating that is listed and labeled for application to the material to which it is applied.
(2) The decorations meet the requirements of NFPA 701, *Standard Methods of Fire Tests for Flame Propagation of Textiles and Films*.
(3) The decorations exhibit a heat release rate not exceeding 100 kW when tested in accordance with NFPA 289, *Standard Method of Fire Test for Individual Fuel Packages*, using the 20 kW ignition source.
(4)\*The decorations, such as photographs, paintings, and other art, are attached directly to the walls, ceiling, and non-fire-rated doors in accordance with the following:
　(a) Decorations on non-fire-rated doors do not interfere with the operation or any required latching of the door and do not exceed the area limitations of 19.7.5.6(b), (c), or (d).

(b) Decorations do not exceed 20 percent of the wall, ceiling, and door areas inside any room or space of a smoke compartment that is not protected throughout by an approved automatic sprinkler system in accordance with Section 9.7.

(c) Decorations do not exceed 30 percent of the wall, ceiling, and door areas inside any room or space of a smoke compartment that is protected throughout by an approved supervised automatic sprinkler system in accordance with Section 9.7.

(d) Decorations do not exceed 50 percent of the wall, ceiling, and door areas inside patient sleeping rooms, having a capacity not exceeding four persons, in a smoke compartment that is protected throughout by an approved, supervised automatic sprinkler system in accordance with Section 9.7.

(5)*They are decorations, such as photographs and paintings, in such limited quantities that a hazard of fire development or spread is not present.

**19.7.5.7 Soiled Linen and Trash Receptacles.**

**19.7.5.7.1** Soiled linen or trash collection receptacles shall not exceed 32 gal (121 L) in capacity and shall meet all of the following requirements:

(1) The average density of container capacity in a room or space shall not exceed 0.5 gal/ft$^2$ (20.4 L/m$^2$).
(2) A capacity of 32 gal (121 L) shall not be exceeded within any 64 ft$^2$ (6 m$^2$) area.
(3)*Mobile soiled linen or trash collection receptacles with capacities greater than 32 gal (121 L) shall be located in a room protected as a hazardous area when not attended.
(4) Container size and density shall not be limited in hazardous areas.

**19.7.5.7.2\*** Containers used solely for recycling clean waste or for patient records awaiting destruction shall be permitted to be excluded from the requirements of 19.7.5.7.1 where all the following conditions are met:

(1) Each container shall be limited to a maximum capacity of 96 gal (363 L), except as permitted by 19.7.5.7.2(2) or (3).
(2)*Containers with capacities greater than 96 gal (363 L) shall be located in a room protected as a hazardous area when not attended.
(3) Container size shall not be limited in hazardous areas.
(4) Containers for combustibles shall be labeled and listed as meeting the requirements of FM Approval Standard 6921, *Containers for Combustible Waste*; however, such testing, listing, and labeling shall not be limited to FM Approvals.

**19.7.5.7.3** The provisions of 10.3.9, applicable to containers for rubbish, waste, or linen, shall not apply.

**19.7.6 Maintenance and Testing.** See 4.6.12.

**19.7.7\* Engineered Smoke Control Systems.**

**19.7.7.1** Existing engineered smoke control systems, unless specifically exempted by the authority having jurisdiction, shall be tested in accordance with established engineering principles.

**19.7.7.2** Systems not meeting the performance requirements of the testing specified in 19.7.7.1 shall be continued in operation only with the specific approval of the authority having jurisdiction.

**19.7.8 Portable Space-Heating Devices.** Portable space-heating devices shall be prohibited in all health care occupancies, unless both of the following criteria are met:

(1) Such devices are used only in nonsleeping staff and employee areas.
(2) The heating elements of such devices do not exceed 212°F (100°C).

**19.7.9 Construction, Repair, and Improvement Operations.**

**19.7.9.1** Construction, repair, and improvement operations shall comply with 4.6.10.

**19.7.9.2** The means of egress in any area undergoing construction, repair, or improvements shall be inspected daily for compliance with 7.1.10.1 and shall also comply with NFPA 241, *Standard for Safeguarding Construction, Alteration, and Demolition Operations.*

## Chapter 20 New Ambulatory Health Care Occupancies

**20.1 General Requirements.**

**20.1.1 Application.**

**20.1.1.1 General.**

**20.1.1.1.1** The requirements of this chapter shall apply to new buildings or portions thereof used as ambulatory health care occupancies. (See 1.3.1.)

**20.1.1.1.2 Administration.** The provisions of Chapter 1, Administration, shall apply.

**20.1.1.1.3 General.** The provisions of Chapter 4, General, shall apply.

**20.1.1.1.4** Ambulatory health care facilities shall comply with the provisions of Chapter 38 and this chapter, whichever are more stringent.

**20.1.1.1.5** This chapter establishes life safety requirements, in addition to those required in Chapter 38, that shall apply to the design of all ambulatory health care occupancies as defined in 3.3.188.1.

**20.1.1.1.6** Buildings, or sections of buildings, that primarily house patients who, in the opinion of the governing body of the facility and the governmental agency having jurisdiction, are capable of exercising judgment and appropriate physical action for self-preservation under emergency conditions shall be permitted to comply with chapters of this *Code* other than Chapter 20.

**20.1.1.1.7** It shall be recognized that, in buildings providing treatment for certain types of patients or having detention rooms or a security section, it might be necessary to lock doors and bar windows to confine and protect building inhabitants. In such instances, the authority having jurisdiction shall make appropriate modifications to those sections of this *Code* that would otherwise require means of egress to be kept unlocked.

**20.1.1.1.8\*** The requirements of this chapter shall apply based on the assumption that staff is available in all patient-occupied areas to perform certain fire safety functions as required in other paragraphs of this chapter.

**20.1.1.2\* Goals and Objectives.** The goals and objectives of Sections 4.1 and 4.2 shall be met with due consideration for functional requirements, which are accomplished by limiting the development and spread of a fire emergency to the room of fire origin and reducing the need for occupant evacuation, except from the room of fire origin.

### 20.1.1.3 Total Concept.

**20.1.1.3.1** All ambulatory health care facilities shall be designed, constructed, maintained, and operated to minimize the possibility of a fire emergency requiring the evacuation of occupants.

**20.1.1.3.2** Because the safety of ambulatory health care occupants cannot be ensured adequately by dependence on evacuation of the building, their protection from fire shall be provided by appropriate arrangement of facilities; adequate, trained staff; and development of operating and maintenance procedures composed of the following:

(1) Design, construction, and compartmentation
(2) Provision for detection, alarm, and extinguishment
(3) Fire prevention and planning, training, and drilling programs for the isolation of fire, transfer of occupants to areas of refuge, or evacuation of the building

### 20.1.1.4 Additions, Conversions, Modernization, Renovation, and Construction Operations.

#### 20.1.1.4.1 Additions.

**20.1.1.4.1.1** Additions shall be separated from any existing structure not conforming to the provisions within Chapter 21 by a fire barrier having not less than a 2-hour fire resistance rating and constructed of materials as required for the addition. *(See 4.6.5 and 4.6.7.)*

**20.1.1.4.1.2** Doors in barriers required by 20.1.1.4.1.1 shall normally be kept closed, unless otherwise permitted by 20.1.1.4.1.3.

**20.1.1.4.1.3** Doors shall be permitted to be held open if they meet the requirements of 20.2.2.4.

**20.1.1.4.2 Changes of Occupancy.** A change from a hospital or nursing home to an ambulatory health care occupancy shall not be considered a change in occupancy or occupancy subclassification.

**20.1.1.4.3 Renovations, Alterations, and Modernizations.** See 4.6.7.

**20.1.1.4.4 Construction, Repair, and Improvement Operations.** See 4.6.10.

**20.1.2 Classification of Occupancy.** See 6.1.6 and 20.1.4.2.

### 20.1.3 Multiple Occupancies.

**20.1.3.1** Multiple occupancies shall be in accordance with 6.1.14.

**20.1.3.2\*** Sections of ambulatory health care facilities shall be permitted to be classified as other occupancies, provided that they meet both of the following conditions:

(1) They are not intended to serve ambulatory health care occupants for purposes of treatment or customary access by patients incapable of self-preservation.
(2) They are separated from areas of ambulatory health care occupancies by construction having a minimum 1-hour fire resistance rating.

**20.1.3.3** All means of egress from ambulatory health care occupancies that traverse nonambulatory health care spaces shall conform to the requirements of this *Code* for ambulatory health care occupancies, unless otherwise permitted by 20.1.3.4.

**20.1.3.4** Exit through a horizontal exit into other contiguous occupancies that do not conform to ambulatory health care egress provisions but that do comply with requirements set forth in the appropriate occupancy chapter of this *Code* shall be permitted, provided that the occupancy does not contain high hazard contents.

**20.1.3.5** Egress provisions for areas of ambulatory health care facilities that correspond to other occupancies shall meet the corresponding requirements of this *Code* for such occupancies, and, where the clinical needs of the occupant necessitate the locking of means of egress, staff shall be present for the supervised release of occupants during all times of use.

**20.1.3.6** Any area with a hazard of contents classified higher than that of the ambulatory health care occupancy and located in the same building shall be protected as required in 20.3.2.

**20.1.3.7** Non-health care–related occupancies classified as containing high hazard contents shall not be permitted in buildings housing ambulatory health care occupancies.

### 20.1.4 Definitions.

**20.1.4.1 General.** For definitions, see Chapter 3, Definitions.

**20.1.4.2 Definition — Ambulatory Health Care Occupancy.** See 3.3.188.1.

**20.1.5 Classification of Hazard of Contents.** The classification of hazard of contents shall be as defined in Section 6.2.

### 20.1.6 Minimum Construction Requirements.

**20.1.6.1** Ambulatory health care occupancies shall be limited to the building construction types specified in Table 20.1.6.1, unless otherwise permitted by 20.1.6.6. *(See 8.2.1.)*

**20.1.6.2** Any level below the level of exit discharge shall be separated from the level of exit discharge by not less than Type II(111), Type III(211), or Type V(111) construction *(see 8.2.1)*, unless both of the following criteria are met:

(1) Such levels are under the control of the ambulatory health care facility.
(2) Any hazardous spaces are protected in accordance with Section 8.7.

**20.1.6.3** Interior nonbearing walls in buildings of Type I or Type II construction shall be constructed of noncombustible or limited-combustible materials, unless otherwise permitted by 20.1.6.4.

**20.1.6.4** Interior nonbearing walls required to have a minimum 2-hour fire resistance rating shall be permitted to be fire-retardant-treated wood enclosed within noncombustible or limited-combustible materials, provided that such walls are not used as shaft enclosures.

**20.1.6.5** All buildings with more than one level below the level of exit discharge shall have all such lower levels separated from the level of exit discharge by not less than Type II(111) construction.

**20.1.6.6** Where new ambulatory health care occupancies are located in existing buildings, the authority having jurisdiction shall be permitted to accept construction systems of lesser fire resistance than those required by 20.1.6.1 through 20.1.6.5, provided that it can be demonstrated to the authority's satisfaction that prompt evacuation of the facility can be achieved in case of fire or that the exposing occupancies and materials of construction present no threat of fire penetration from

**Table 20.1.6.1 Construction Type Limitations**

| Construction Type | Sprinklered[†] | Stories in Height[‡] | |
|---|---|---|---|
| | | 1 | ≥2 |
| I (442) | Yes | X | X |
| | No | X | X |
| I (332) | Yes | X | X |
| | No | X | X |
| II (222) | Yes | X | X |
| | No | X | X |
| II (111) | Yes | X | X |
| | No | X | X |
| II (000) | Yes | X | X |
| | No | X | NP |
| III (211) | Yes | X | X |
| | No | X | X |
| III (200) | Yes | X | X |
| | No | X | NP |
| IV (2HH) | Yes | X | X |
| | No | X | X |
| V (111) | Yes | X | X |
| | No | X | X |
| V (000) | Yes | X | X |
| | No | X | NP |

X: Permitted. NP: Not permitted.
[†]Sprinklered throughout by an approved, supervised automatic sprinkler system in accordance with Section 9.7. *(See 20.3.5.)*
[‡]See 4.6.3.

such occupancy to the ambulatory health care facility or to the collapse of the structure.

**20.1.7 Occupant Load.** See 38.1.7.

**20.2 Means of Egress Requirements.**

**20.2.1 General.** Every aisle, passageway, corridor, exit discharge, exit location, and access shall be in accordance with Chapter 7, unless otherwise modified by 20.2.2 through 20.2.11.

**20.2.2 Means of Egress Components.**

**20.2.2.1** Components of means of egress shall be limited to the types described in 38.2.2.

**20.2.2.2** Special locking arrangements complying with 7.2.1.6 shall be permitted.

**20.2.2.3** Elevator lobby exit access door locking in accordance with 7.2.1.6.3 shall be permitted.

**20.2.2.4** Any door required to be self-closing shall be permitted to be held open only by an automatic release device that complies with 7.2.1.8.2. The required manual fire alarm system and the systems required by 7.2.1.8.2 shall be arranged to initiate the closing action of all such doors throughout the smoke compartment or throughout the entire facility.

**20.2.2.5** Where doors in a stair enclosure are held open by an automatic release device as permitted in 20.2.2.4, initiation of a door-closing action on any level shall cause all doors at all levels in the stair enclosure to close.

**20.2.3 Capacity of Means of Egress.**

**20.2.3.1** The capacity of any required means of egress shall be determined in accordance with the provisions of 38.2.3.

**20.2.3.2** The clear width of any corridor or passageway required for exit access shall be not less than 44 in. (1120 mm).

**20.2.3.3** Where minimum corridor width is 6 ft (1830 mm), projections not more than 6 in. (150 mm) from the corridor wall, above the handrail height, shall be permitted for the installation of hand-rub dispensing units in accordance with 20.3.2.6.

**20.2.3.4** Doors in the means of egress from diagnostic or treatment areas, such as x-ray, surgical, or physical therapy, shall provide a clear width of not less than 32 in. (810 mm).

**20.2.4 Number of Means of Egress.**

**20.2.4.1** The number of means of egress shall be in accordance with Section 7.4.

**20.2.4.2** Not less than two exits of the types described in 38.2.2 that are remotely located from each other shall be provided for each floor or fire section of the building.

**20.2.4.3** Any patient care room and any patient care suite of rooms of more than 2500 ft$^2$ (232 m$^2$) shall have not less than two exit access doors remotely located from each other.

**20.2.4.4** Not less than two exits of the types described in 38.2.2 shall be accessible from each smoke compartment.

**20.2.4.5** Egress from smoke compartments addressed in 20.2.4.4 shall be permitted through adjacent compartments but shall not require return through the compartment of fire origin.

**20.2.5 Arrangement of Means of Egress.** See 38.2.5.

**20.2.6 Travel Distance to Exits.**

**20.2.6.1** Travel distance shall be measured in accordance with Section 7.6.

**20.2.6.2** Travel distance shall comply with 20.2.6.2.1 and 20.2.6.2.2.

**20.2.6.2.1** The travel distance between any point in a room and an exit shall not exceed 150 ft (46 m).

**20.2.6.2.2** The maximum travel distance in 20.2.6.2.1 shall be permitted to be increased by 50 ft (15 m) in buildings protected throughout by an approved automatic sprinkler system in accordance with Section 9.7

**20.2.7 Discharge from Exits.** See 38.2.7.

**20.2.8 Illumination of Means of Egress.** Means of egress shall be illuminated in accordance with Section 7.8.

**20.2.9 Emergency Lighting and Essential Electrical Systems.**

**20.2.9.1** Emergency lighting shall be provided in accordance with Section 7.9.

**20.2.9.2** Where general anesthesia or life-support equipment is used, each ambulatory health care facility shall be provided with an essential electrical system in accordance with NFPA 99, *Health Care Facilities Code*, unless otherwise permitted by any of the following:

(1) Where battery-operated equipment is provided and acceptable to the authority having jurisdiction
(2) Where a facility uses life-support equipment for emergency purposes only

**20.2.10 Marking of Means of Egress.** Means of egress shall have signs in accordance with Section 7.10.

**20.2.11 Special Means of Egress Features.**

**20.2.11.1 Reserved.**

**20.2.11.2 Lockups.** Lockups in ambulatory health care occupancies shall comply with the requirements of 22.4.5.

**20.3 Protection.**

**20.3.1 Protection of Vertical Openings.** See 38.3.1.

**20.3.2 Protection from Hazards.** See 38.3.2.

**20.3.2.1 Doors.** Doors to hazardous areas shall be self-closing or automatic-closing in accordance with 20.2.2.4.

**20.3.2.2 Laboratories.** Laboratories employing quantities of flammable, combustible, or hazardous materials that are considered as a severe hazard shall be protected in accordance with NFPA 99, *Health Care Facilities Code*.

**20.3.2.3 Anesthetizing Locations.** Anesthetizing locations shall be protected in accordance with NFPA 99, *Health Care Facilities Code*.

**20.3.2.4 Cooking Facilities.** Cooking facilities shall be protected in accordance with 9.2.3, unless otherwise permitted by 20.3.2.5.

**20.3.2.5 Domestic Cooking Equipment.** Where domestic cooking equipment is used for food warming or limited cooking, protection or separation of food preparation facilities shall not be required.

**20.3.2.6\* Alcohol-Based Hand-Rub Dispensers.** Alcohol-based hand-rub dispensers shall be protected in accordance with 8.7.3.1, unless all of the following conditions are met:

(1) Where dispensers are installed in a corridor, the corridor shall have a minimum width of 6 ft (1830 mm).
(2) The maximum individual dispenser fluid capacity shall be as follows:
    (a) 0.32 gal (1.2 L) for dispensers in rooms, corridors, and areas open to corridors
    (b) 0.53 gal (2.0 L) for dispensers in suites of rooms
(3) Where aerosol containers are used, the maximum capacity of the aerosol dispenser shall be 18 oz (0.51 kg) and shall be limited to Level 1 aerosols as defined in NFPA 30B, *Code for the Manufacture and Storage of Aerosol Products*.
(4) Dispensers shall be separated from each other by horizontal spacing of not less than 48 in. (1220 mm).
(5) Not more than an aggregate 10 gal (37.8 L) of alcohol-based hand-rub solution or 1135 oz (32.2 kg) of Level 1 aerosols, or a combination of liquids and Level 1 aerosols not to exceed, in total, the equivalent of 10 gal (37.8 L) or 1135 oz (32.2 kg), shall be in use outside of a storage cabinet in a single smoke compartment, except as otherwise provided in 20.3.2.6(6).

(6) One dispenser per room complying with 20.3.2.6 (2) or (3), and located in the room, shall not be required to be included in the aggregated quantity specified in 20.3.2.6(5).
(7) Storage of quantities greater than 5 gal (18.9 L) in a single smoke compartment shall meet the requirements of NFPA 30, *Flammable and Combustible Liquids Code*.
(8) Dispensers shall not be installed in the following locations:
    (a) Above an ignition source within a 1 in. (25 mm) horizontal distance from each side of the ignition source
    (b) To the side of an ignition source within a 1 in. (25 mm) horizontal distance from the ignition source
    (c) Beneath an ignition source within a 1 in. (25 mm) vertical distance from the ignition source
(9) Dispensers installed directly over carpeted floors shall be permitted only in sprinklered smoke compartments.
(10) The alcohol-based hand-rub solution shall not exceed 95 percent alcohol content by volume.
(11) Operation of the dispenser shall comply with the following criteria:
    (a) The dispenser shall not release its contents except when the dispenser is activated, either manually or automatically by touch-free activation.
    (b) Any activation of the dispenser shall occur only when an object is placed within 4 in. (100 mm) of the sensing device.
    (c) An object placed within the activation zone and left in place shall not cause more than one activation.
    (d) The dispenser shall not dispense more solution than the amount required for hand hygiene consistent with label instructions.
    (e) The dispenser shall be designed, constructed, and operated in a manner that ensures that accidental or malicious activation of the dispensing device is minimized.
    (f) The dispenser shall be tested in accordance with the manufacturer's care and use instructions each time a new refill is installed.

**20.3.3 Interior Finish.** See 38.3.3.

**20.3.4 Detection, Alarm, and Communications Systems.**

**20.3.4.1 General.** Ambulatory health care facilities shall be provided with fire alarm systems in accordance with Section 9.6, except as modified by 20.3.4.2 through 20.3.4.4.

**20.3.4.2 Initiation.** Initiation of the required fire alarm systems shall be by manual means in accordance with 9.6.2 and by means of any detection devices or detection systems required.

**20.3.4.3 Notification.** Positive alarm sequence in accordance with 9.6.3.4 shall be permitted.

**20.3.4.3.1 Occupant Notification.** Occupant notification shall be accomplished automatically, without delay, in accordance with 9.6.3 upon operation of any fire alarm activating device.

**20.3.4.3.2 Emergency Forces Notification.**

**20.3.4.3.2.1** Fire department notification shall be accomplished in accordance with 9.6.4.

**20.3.4.3.2.2 Reserved.**

**20.3.4.4 Fire Safety Functions.** Operation of any activating device in the required fire alarm system shall be arranged to accomplish automatically, without delay, any control functions required to be performed by that device. *(See 9.6.5.)*

**20.3.5 Extinguishment Requirements.** See 38.3.5.

**20.3.5.1** Isolated hazardous areas shall be permitted to be protected in accordance with 9.7.1.2.

**20.3.5.2** Where more than two sprinklers are installed in a single area for protection in accordance with 9.7.1.2, water-flow detection shall be provided to sound the building fire alarm or to notify, by a signal, any constantly attended location, such as PBX, security, or emergency room, at which the necessary corrective action shall be taken.

**20.3.5.3** Portable fire extinguishers shall be provided in ambulatory health care facilities in accordance with 9.7.4.1.

**20.3.6 Corridors.**

**20.3.6.1 General.** See 38.3.6.

**20.3.6.2 Openings.**

**20.3.6.2.1** Miscellaneous openings, such as mail slots, pharmacy pass-through windows, laboratory pass-through windows, and cashier pass-through windows, shall be permitted to be installed in vision panels or doors without special protection, provided that both of the following criteria are met:

(1) The aggregate area of openings per room does not exceed 20 in.$^2$ (0.015 m$^2$).
(2) The openings are installed at or below half the distance from the floor to the room ceiling.

**20.3.6.2.2** For rooms protected throughout by an approved, supervised automatic sprinkler system in accordance with Section 9.7, the aggregate area of openings per room shall not exceed 80 in.$^2$ (0.05 m$^2$).

**20.3.7 Subdivision of Building Space.**

**20.3.7.1** Ambulatory health care facilities shall be separated from other tenants and occupancies and shall meet all of the following requirements:

(1) Walls shall have not less than a 1-hour fire resistance rating and shall extend from the floor slab below to the floor or roof slab above.
(2) Doors shall be constructed of not less than 1¾ in. (44 mm) thick, solid-bonded wood core or the equivalent and shall be equipped with positive latches.
(3) Doors shall be self-closing and shall be kept in the closed position, except when in use.
(4) Any windows in the barriers shall be of fixed fire window assemblies in accordance with Section 8.3.

**20.3.7.2** Every story of an ambulatory health care facility shall be divided into not less than two smoke compartments, unless otherwise permitted by one of the following:

(1) This requirement shall not apply to facilities of less than 5000 ft$^2$ (465 m$^2$) that are protected by an approved automatic smoke detection system.
(2) This requirement shall not apply to facilities of less than 10,000 ft$^2$ (929 m$^2$) that are protected throughout by an approved, supervised automatic sprinkler system installed in accordance with Section 9.7.
(3) An area in an adjoining occupancy shall be permitted to serve as a smoke compartment for an ambulatory health care facility if all of the following criteria are met:
    (a) The separating wall and both compartments meet the requirements of 20.3.7.

(b) The ambulatory health care facility is less than 22,500 ft$^2$ (2100 m$^2$).
(c) Access from the ambulatory health care facility to the other occupancy is unrestricted.

**20.3.7.3** Smoke compartments shall not exceed an area of 22,500 ft$^2$ (2100 m$^2$), and the travel distance from any point to reach a door in a smoke barrier shall not exceed 200 ft (61 m).

**20.3.7.4** The area of an atrium separated in accordance with 8.6.7 shall not be limited in size.

**20.3.7.5** Required smoke barriers shall be constructed in accordance with Section 8.5 and shall have a minimum 1-hour fire resistance rating, unless otherwise permitted by 20.3.7.6.

**20.3.7.6** Smoke dampers shall not be required in duct penetrations of smoke barriers in fully ducted heating, ventilating, and air-conditioning systems for buildings protected throughout by an approved, supervised automatic sprinkler system in accordance with Section 9.7.

**20.3.7.7** Windows in the smoke barrier shall be of fixed fire window assemblies in accordance with Section 8.3.

**20.3.7.8** Not less than 15 net ft$^2$ (1.4 net m$^2$) per ambulatory health care facility occupant shall be provided within the aggregate area of corridors, patient rooms, treatment rooms, lounges, and other low hazard areas on each side of the smoke compartment for the total number of occupants in adjoining compartments.

**20.3.7.9\*** Doors in smoke barriers shall be not less than 1¾ in. (44 mm) thick, solid-bonded wood core or the equivalent and shall be self-closing or automatic-closing in accordance with 20.2.2.4.

**20.3.7.10** Latching hardware shall not be required on smoke barrier cross-corridor doors.

**20.3.7.11** A vision panel consisting of fire-rated glazing in approved frames shall be provided in each cross-corridor swinging door and at each cross-corridor horizontal-sliding door in a smoke barrier.

**20.3.7.12** Vision panels in doors in smoke barriers, if provided, shall be of fire-rated glazing in approved frames.

**20.3.7.13\*** Rabbets, bevels, or astragals shall be required at the meeting edges, and stops shall be required at the head and sides of door frames in smoke barriers.

**20.3.7.14** Center mullions shall be prohibited in smoke barrier door openings.

**20.4 Special Provisions.** See Section 38.4.

**20.5 Building Services.**

**20.5.1 Utilities.** Utilities shall comply with the provisions of Section 9.1.

**20.5.2 Heating, Ventilating, and Air-Conditioning.**

**20.5.2.1** Heating, ventilating, and air-conditioning shall comply with the provisions of Section 9.2 and shall be installed in accordance with the manufacturer's specifications, unless otherwise modified by 20.5.2.2.

**20.5.2.2** If fuel-fired, heating devices shall comply with all of the following:

(1) They shall be chimney connected or vent connected.
(2) They shall take air for combustion directly from the outside.

(3) They shall be designed and installed to provide for complete separation of the combustion system from the atmosphere of the occupied area.

**20.5.2.2.1** Any heating device shall have safety features to immediately stop the flow of fuel and shut down the equipment in case of either excessive temperature or ignition failure.

**20.5.2.2.2** Approved, suspended unit heaters shall be permitted in locations other than means of egress and patient treatment areas, provided that both of the following criteria are met:

(1) Such heaters are located high enough to be out of the reach of persons using the area.
(2) Such heaters are equipped with the safety features required by 20.5.2.2.1.

**20.5.3 Elevators, Escalators, and Conveyors.** Elevators, escalators, and conveyors shall comply with the provisions of Section 9.4.

**20.5.4 Rubbish Chutes, Incinerators, and Laundry Chutes.** Rubbish chutes, incinerators, and laundry chutes shall comply with the provisions of Section 9.5.

**20.6 Reserved.**

**20.7* Operating Features.**

**20.7.1 Evacuation and Relocation Plan and Fire Drills.**

**20.7.1.1** The administration of every ambulatory health care facility shall have, in effect and available to all supervisory personnel, written copies of a plan for the protection of all persons in the event of fire, for their evacuation to areas of refuge, and for their evacuation from the building when necessary.

**20.7.1.2** All employees shall be periodically instructed and kept informed with respect to their duties under the plan required by 20.7.1.1.

**20.7.1.3** A copy of the plan required by 20.7.1.1 shall be readily available at all times in the telephone operator's location or at the security center.

**20.7.1.4*** Fire drills in ambulatory health care facilities shall include the transmission of a fire alarm signal and simulation of emergency fire conditions.

**20.7.1.5** Patients shall not be required to be moved during drills to safe areas or to the exterior of the building.

**20.7.1.6** Drills shall be conducted quarterly on each shift to familiarize facility personnel (nurses, interns, maintenance engineers, and administrative staff) with the signals and emergency action required under varied conditions.

**20.7.1.7** When drills are conducted between 9:00 p.m. and 6:00 a.m. (2100 hours and 0600 hours), a coded announcement shall be permitted to be used instead of audible alarms.

**20.7.1.8** Employees of ambulatory health care facilities shall be instructed in life safety procedures and devices.

**20.7.2 Procedure in Case of Fire.**

**20.7.2.1* Protection of Patients.**

**20.7.2.1.1** For ambulatory health care facilities, the proper protection of patients shall require the prompt and effective response of ambulatory health care personnel.

**20.7.2.1.2** The basic response required of staff shall include the following:

(1) Removal of all occupants directly involved with the fire emergency
(2) Transmission of an appropriate fire alarm signal to warn other building occupants and summon staff
(3) Confinement of the effects of the fire by closing doors to isolate the fire area
(4) Relocation of patients as detailed in the facility's fire safety plan

**20.7.2.2 Fire Safety Plan.** A written fire safety plan shall provide for all of the following:

(1) Use of alarms
(2) Transmission of alarms to fire department
(3) Response to alarms
(4) Isolation of fire
(5) Evacuation of immediate area
(6) Evacuation of smoke compartment
(7) Preparation of floors and building for evacuation
(8) Extinguishment of fire

**20.7.2.3 Staff Response.**

**20.7.2.3.1** All personnel shall be instructed in the use of and response to fire alarms.

**20.7.2.3.2** All health care personnel shall be instructed in the use of the code phrase to ensure transmission of an alarm under any of the following conditions:

(1) When the individual who discovers a fire must immediately go to the aid of an endangered person
(2) During a malfunction of the building fire alarm system

**20.7.2.3.3** Personnel hearing the code announced shall first activate the building fire alarm using the nearest fire alarm box and then shall execute immediately their duties as outlined in the fire safety plan.

**20.7.3 Maintenance of Exits.**

**20.7.3.1** Proper maintenance shall be provided to ensure the dependability of the method of evacuation selected.

**20.7.3.2** Ambulatory health care occupancies that find it necessary to lock exits shall, at all times, maintain an adequate staff qualified to release locks and direct occupants from the immediate danger area to a place of safety in case of fire or other emergency.

**20.7.4* Smoking.** Smoking regulations shall be adopted and shall include not less than the following provisions:

(1) Smoking shall be prohibited in any room, ward, or compartment where flammable liquids, combustible gases, or oxygen is used or stored and in any other hazardous location, and such areas shall be posted with signs that read NO SMOKING or shall be posted with the international symbol for no smoking.
(2) In ambulatory health care facilities where smoking is prohibited and signs are placed at all major entrances, secondary signs with language that prohibits smoking shall not be required.
(3) Smoking by patients classified as not responsible shall be prohibited.
(4) The requirement of 20.7.4(3) shall not apply where the patient is under direct supervision.
(5) Ashtrays of noncombustible material and safe design shall be provided in all areas where smoking is permitted.

(6) Metal containers with self-closing cover devices into which ashtrays can be emptied shall be readily available to all areas where smoking is permitted.

### 20.7.5 Furnishings, Mattresses, and Decorations.

**20.7.5.1\*** Draperies, curtains, and other loosely hanging fabrics and films serving as furnishings or decorations in ambulatory health care occupancies shall be in accordance with the provisions of 10.3.1, and the following also shall apply:

(1) Such curtains shall include cubicle curtains.
(2) Such curtains shall not include curtains at showers.

**20.7.5.2** Newly introduced upholstered furniture shall comply with 10.3.2.1 and one of the following provisions:

(1) The furniture shall meet the criteria specified in 10.3.3.
(2) The furniture shall be in a building protected throughout by an approved, supervised automatic sprinkler system in accordance with 9.7.1.1(1).

**20.7.5.3** Newly introduced mattresses shall comply with 10.3.2.2 and one of the following provisions:

(1) The mattresses shall meet the criteria specified in 10.3.4.
(2) The mattresses shall be in a building protected throughout by an approved, supervised automatic sprinkler system in accordance with 9.7.1.1(1).

**20.7.5.4** Combustible decorations shall be prohibited, unless one of the following criteria is met:

(1) They are flame-retardant.
(2) The decorations meet the requirements of NFPA 701, *Standard Methods of Fire Tests for Flame Propagation of Textiles and Films.*
(3) The decorations exhibit a heat release rate not exceeding 100 kW when tested in accordance with NFPA 289, *Standard Method of Fire Test for Individual Fuel Packages,* using the 20 kW ignition source.
(4) They are decorations, such as photographs and paintings, in such limited quantities that a hazard of fire development or spread is not present.

### 20.7.5.5 Soiled Linen and Trash Receptacles.

**20.7.5.5.1** Soiled linen or trash collection receptacles shall not exceed 32 gal (121 L) in capacity, and all of the following also shall apply:

(1) The average density of container capacity in a room or space shall not exceed 0.5 gal/ft$^2$ (20.4 L/m$^2$).
(2) A capacity of 32 gal (121 L) shall not be exceeded within any 64 ft$^2$ (6 m$^2$) area.
(3) Mobile soiled linen or trash collection receptacles with capacities greater than 32 gal (121 L) shall be located in a room protected as a hazardous area when not attended.
(4) Container size and density shall not be limited in hazardous areas.

**20.7.5.5.2** The provisions of 10.3.9, applicable to containers for rubbish, waste, or linen, shall not apply.

### 20.7.6 Maintenance and Testing. See 4.6.12.

### 20.7.7\* Engineered Smoke Control Systems.

**20.7.7.1** New engineered smoke control systems shall be tested in accordance with established engineering principles and shall meet the performance requirements of such testing prior to acceptance.

**20.7.7.2** Following acceptance, all engineered smoke control systems shall be tested periodically in accordance with recognized engineering principles.

**20.7.7.3** Test documentation shall be maintained on the premises at all times.

### 20.7.8 Portable Space-Heating Devices. Portable space-heating devices shall be prohibited in all ambulatory health care occupancies, unless both of the following criteria are met:

(1) Such devices are used only in nonsleeping staff and employee areas.
(2) The heating elements of such devices do not exceed 212°F (100°C).

### 20.7.9 Construction, Repair, and Improvement Operations.

**20.7.9.1** Construction, repair, and improvement operations shall comply with 4.6.10.

**20.7.9.2** The means of egress in any area undergoing construction, repair, or improvements shall be inspected daily for compliance with 7.1.10.1 and shall also comply with NFPA 241, *Standard for Safeguarding Construction, Alteration, and Demolition Operations.*

## Chapter 21    Existing Ambulatory Health Care Occupancies

### 21.1 General Requirements.

### 21.1.1 Application.

### 21.1.1.1 General.

**21.1.1.1.1** The requirements of this chapter shall apply to existing buildings or portions thereof currently occupied as an ambulatory health care occupancy.

**21.1.1.1.2 Administration.** The provisions of Chapter 1, Administration, shall apply.

**21.1.1.1.3 General.** The provisions of Chapter 4, General, shall apply.

**21.1.1.1.4** Ambulatory health care facilities shall comply with the provisions of Chapter 39 and this chapter, whichever are more stringent.

**21.1.1.1.5** This chapter establishes life safety requirements, in addition to those required in Chapter 39, that shall apply to the design of all ambulatory health care occupancies as defined in 3.3.188.1.

**21.1.1.1.6** Buildings, or sections of buildings, that primarily house patients who, in the opinion of the governing body of the facility and the governmental agency having jurisdiction, are capable of exercising judgment and appropriate physical action for self-preservation under emergency conditions shall be permitted to comply with chapters of this *Code* other than Chapter 21.

**21.1.1.1.7** It shall be recognized that, in buildings providing treatment for certain types of patients or having detention rooms or a security section, it might be necessary to lock doors and bar windows to confine and protect building inhabitants. In such instances, the authority having jurisdiction shall make appropriate modifications to those sections of this *Code* that would otherwise require means of egress to be kept unlocked.

**21.1.1.1.8\*** The requirements of this chapter shall apply based on the assumption that staff is available in all patient-occupied areas to perform certain fire safety functions as required in other paragraphs of this chapter.

**21.1.1.2\* Goals and Objectives.** The goals and objectives of Sections 4.1 and 4.2 shall be met with due consideration for functional requirements, which are accomplished by limiting the development and spread of a fire emergency to the room of fire origin and reducing the need for occupant evacuation, except from the room of fire origin.

**21.1.1.3 Total Concept.**

**21.1.1.3.1** All ambulatory health care facilities shall be designed, constructed, maintained, and operated to minimize the possibility of a fire emergency requiring the evacuation of occupants.

**21.1.1.3.2** Because the safety of ambulatory health care occupants cannot be ensured adequately by dependence on evacuation of the building, their protection from fire shall be provided by appropriate arrangement of facilities; adequate, trained staff; and development of operating and maintenance procedures composed of the following:

(1) Design, construction, and compartmentation
(2) Provision for detection, alarm, and extinguishment
(3) Fire prevention and planning, training, and drilling programs for the isolation of fire, transfer of occupants to areas of refuge, or evacuation of the building

**21.1.1.4 Additions, Conversions, Modernization, Renovation, and Construction Operations.**

**21.1.1.4.1 Additions.**

**21.1.1.4.1.1** Additions shall be separated from any existing structure not conforming to the provisions within Chapter 21 by a fire barrier having not less than a 2-hour fire resistance rating and constructed of materials as required for the addition. *(See 4.6.5 and 4.6.7.)*

**21.1.1.4.1.2** Doors in barriers required by 21.1.1.4.1.1 shall normally be kept closed, unless otherwise permitted by 21.1.1.4.1.3.

**21.1.1.4.1.3** Doors shall be permitted to be held open if they meet the requirements of 21.2.2.4.

**21.1.1.4.2 Changes of Occupancy.** A change from a hospital or nursing home to an ambulatory health care occupancy shall not be considered a change in occupancy or occupancy subclassification.

**21.1.1.4.3 Renovations, Alterations, and Modernizations.** See 4.6.7.

**21.1.1.4.4 Construction, Repair, and Improvement Operations.** See 4.6.10.

**21.1.2 Classification of Occupancy.** See 6.1.6 and 21.1.4.2.

**21.1.3 Multiple Occupancies.**

**21.1.3.1** Multiple occupancies shall be in accordance with 6.1.14.

**21.1.3.2\*** Sections of ambulatory health care facilities shall be permitted to be classified as other occupancies, provided that they meet both of the following conditions:

(1) They are not intended to serve ambulatory health care occupants for purposes of treatment or customary access by patients incapable of self-preservation.

(2) They are separated from areas of ambulatory health care occupancies by construction having a minimum 1-hour fire resistance rating.

**21.1.3.3** All means of egress from ambulatory health care occupancies that traverse nonambulatory health care spaces shall conform to the requirements of this *Code* for ambulatory health care occupancies, unless otherwise permitted by 21.1.3.4.

**21.1.3.4** Exit through a horizontal exit into other contiguous occupancies that do not conform with ambulatory health care egress provisions but that do comply with requirements set forth in the appropriate occupancy chapter of this *Code* shall be permitted, provided that the occupancy does not contain high hazard contents.

**21.1.3.5** Egress provisions for areas of ambulatory health care facilities that correspond to other occupancies shall meet the corresponding requirements of this *Code* for such occupancies, and, where the clinical needs of the occupant necessitate the locking of means of egress, staff shall be present for the supervised release of occupants during all times of use.

**21.1.3.6** Any area with a hazard of contents classified higher than that of the ambulatory health care occupancy and located in the same building shall be protected as required in 21.3.2.

**21.1.3.7** Non-health care–related occupancies classified as containing high hazard contents shall not be permitted in buildings housing ambulatory health care occupancies.

**21.1.4 Definitions.**

**21.1.4.1 General.** For definitions, see Chapter 3, Definitions.

**21.1.4.2 Definition — Ambulatory Health Care Occupancy.** See 3.3.188.1.

**21.1.5 Classification of Hazard of Contents.** The classification of hazard of contents shall be as defined in Section 6.2.

**21.1.6 Minimum Construction Requirements.**

**21.1.6.1** Ambulatory health care occupancies shall be limited to the building construction types specified in Table 21.1.6.1, unless otherwise permitted by 21.1.6.6. *(See 8.2.1.)*

**21.1.6.2** Any level below the level of exit discharge shall be separated from the level of exit discharge by not less than Type II(111), Type III(211), or Type V(111) construction *(see 8.2.1)*, unless both of the following criteria are met:

(1) Such levels are under the control of the ambulatory health care facility.
(2) Any hazardous spaces are protected in accordance with Section 8.7.

**21.1.6.3** Interior nonbearing walls in buildings of Type I or Type II construction shall be constructed of noncombustible or limited-combustible materials, unless otherwise permitted by 21.1.6.4.

**21.1.6.4** Interior nonbearing walls required to have a minimum 2-hour fire resistance rating shall be permitted to be fire-retardant-treated wood enclosed within noncombustible or limited-combustible materials, provided that such walls are not used as shaft enclosures.

**21.1.6.5** All buildings with more than one level below the level of exit discharge shall have all such lower levels separated from the level of exit discharge by not less than Type II(111) construction.

**Table 21.1.6.1 Construction Type Limitations**

| Construction Type | Sprinklered[†] | Stories in Height[‡] | |
|---|---|---|---|
| | | 1 | ≥2 |
| I (442) | Yes | X | X |
| | No | X | X |
| I (332) | Yes | X | X |
| | No | X | X |
| II (222) | Yes | X | X |
| | No | X | X |
| II (111) | Yes | X | X |
| | No | X | X |
| II (000) | Yes | X | X |
| | No | X | NP |
| III (211) | Yes | X | X |
| | No | X | X |
| III (200) | Yes | X | X |
| | No | X | NP |
| IV (2HH) | Yes | X | X |
| | No | X | X |
| V (111) | Yes | X | X |
| | No | X | X |
| V (000) | Yes | X | X |
| | No | X | NP |

X: Permitted. NP: Not permitted.

[†]Sprinklered throughout by an approved, supervised automatic sprinkler system in accordance with Section 9.7. *(See 21.3.5.)*

[‡]See 4.6.3.

**21.1.6.6** In existing buildings, the authority having jurisdiction shall be permitted to accept construction systems of lesser fire resistance than those required by 21.1.6.1 through 21.1.6.5, provided that it can be demonstrated to the authority's satisfaction that prompt evacuation of the facility can be achieved in case of fire or that the exposing occupancies and materials of construction present no threat of fire penetration from such occupancy to the ambulatory health care facility or to the collapse of the structure.

**21.1.7 Occupant Load.** See 39.1.7.

**21.2 Means of Egress Requirements.**

**21.2.1 General.** Every aisle, passageway, corridor, exit discharge, exit location, and access shall be in accordance with Chapter 7, unless otherwise modified by 21.2.2 through 21.2.11.

**21.2.2 Means of Egress Components.**

**21.2.2.1** Components of means of egress shall be limited to the types described in 39.2.2.

**21.2.2.2** Special locking arrangements complying with 7.2.1.6 shall be permitted.

**21.2.2.3** Elevator lobby exit access door locking in accordance with 7.2.1.6.3 shall be permitted.

**21.2.2.4** Any door required to be self-closing shall be permitted to be held open only by an automatic release device that complies with 7.2.1.8.2. The required manual fire alarm system and the systems required by 7.2.1.8.2 shall be arranged to initiate the closing action of all such doors throughout the smoke compartment or throughout the entire facility.

**21.2.2.5** Where doors in a stair enclosure are held open by an automatic release device as permitted in 21.2.2.4, initiation of a door-closing action on any level shall cause all doors at all levels in the stair enclosure to close.

**21.2.2.6*** A door in a horizontal exit shall not be required to swing in the direction of egress travel as specified in 7.2.4.3.8.1.

**21.2.3 Capacity of Means of Egress.**

**21.2.3.1** The capacity of any required means of egress shall be determined in accordance with the provisions of 39.2.3.

**21.2.3.2** The clear width of any corridor or passageway required for exit access shall be not less than 44 in. (1120 mm).

**21.2.3.3** Where minimum corridor width is 6 ft (1830 mm), projections not more than 6 in. (150 mm) from the corridor wall, above the handrail height, shall be permitted for the installation of hand-rub dispensing units in accordance with 21.3.2.6.

**21.2.3.4** Doors in the means of egress from diagnostic or treatment areas, such as x-ray, surgical, or physical therapy, shall provide a clear width of not less than 32 in. (810 mm), unless such doors are existing 34 in. (865 mm) doors.

**21.2.4 Number of Means of Egress.**

**21.2.4.1** The number of means of egress shall be in accordance with 7.4.1.1 and 7.4.1.3 through 7.4.1.6.

**21.2.4.2** Not less than two exits of the types described in 39.2.2 that are remotely located from each other shall be provided for each floor or fire section of the building.

**21.2.4.3** Any patient care room and any patient care suite of rooms of more than 2500 ft² (232 m²) shall have not less than two exit access doors remotely located from each other.

**21.2.4.4** Not less than two exits of the types described in 39.2.2 shall be accessible from each smoke compartment.

**21.2.4.5** Egress from smoke compartments addressed in 21.2.4.4 shall be permitted through adjacent compartments but shall not require return through the compartment of fire origin.

**21.2.5 Arrangement of Means of Egress.** See 39.2.5.

**21.2.6 Travel Distance to Exits.**

**21.2.6.1** Travel distance shall be measured in accordance with Section 7.6.

**21.2.6.2** Travel distance shall comply with 21.2.6.2.1 and 21.2.6.2.2.

**21.2.6.2.1** The travel distance between any point in a room and an exit shall not exceed 150 ft (46 m).

**21.2.6.2.2** The maximum travel distance in 21.2.6.2.1 shall be permitted to be increased by 50 ft (15 m) in buildings protected throughout by an approved automatic sprinkler system in accordance with Section 9.7.

**21.2.7 Discharge from Exits.** See 39.2.7.

**21.2.8 Illumination of Means of Egress.** Means of egress shall be illuminated in accordance with Section 7.8.

**21.2.9 Emergency Lighting and Essential Electrical Systems.**

**21.2.9.1** Emergency lighting shall be provided in accordance with Section 7.9.

**21.2.9.2** Where general anesthesia or life-support equipment is used, each ambulatory health care facility shall be provided with an essential electrical system in accordance with NFPA 99, *Health Care Facilities Code,* unless otherwise permitted by one of the following:

(1) Where battery-operated equipment is provided and acceptable to the authority having jurisdiction
(2) Where a facility uses life-support equipment for emergency purposes only

**21.2.10 Marking of Means of Egress.** Means of egress shall have signs in accordance with Section 7.10.

**21.2.11 Special Means of Egress Features.**

**21.2.11.1 Reserved.**

**21.2.11.2 Lockups.** Lockups in ambulatory health care occupancies, other than approved existing lockups, shall comply with the requirements of 23.4.5.

**21.3 Protection.**

**21.3.1 Protection of Vertical Openings.** See 39.3.1.

**21.3.2 Protection from Hazards.** See 39.3.2.

**21.3.2.1 Doors.** Doors to hazardous areas shall be self-closing or automatic-closing in accordance with 21.2.2.4.

**21.3.2.2 Laboratories.** Laboratories employing quantities of flammable, combustible, or hazardous materials that are considered as a severe hazard shall be protected in accordance with NFPA 99, *Health Care Facilities Code.*

**21.3.2.3 Anesthetizing Locations.** Anesthetizing locations shall be protected in accordance with NFPA 99, *Health Care Facilities Code.*

**21.3.2.4 Cooking Facilities.** Cooking facilities shall be protected in accordance with 9.2.3, unless otherwise permitted by 21.3.2.5.

**21.3.2.5 Domestic Cooking Equipment.** Where domestic cooking equipment is used for food warming or limited cooking, protection or separation of food preparation facilities shall not be required.

**21.3.2.6\* Alcohol-Based Hand-Rub Dispensers.** Alcohol-based hand-rub dispensers shall be protected in accordance with 8.7.3.1, unless all of the following conditions are met:

(1) Where dispensers are installed in a corridor, the corridor shall have a minimum width of 6 ft (1830 mm).
(2) The maximum individual dispenser fluid capacity shall be as follows:
   (a) 0.32 gal (1.2 L) for dispensers in rooms, corridors, and areas open to corridors
   (b) 0.53 gal (2.0 L) for dispensers in suites of rooms
(3) Where aerosol containers are used, the maximum capacity of the aerosol dispenser shall be 18 oz (0.51 kg) and shall be limited to Level 1 aerosols as defined in NFPA 30B, *Code for the Manufacture and Storage of Aerosol Products.*

(4) Dispensers shall be separated from each other by horizontal spacing of not less than 48 in. (1220 mm).
(5) Not more than an aggregate 10 gal (37.8 L) of alcohol-based hand-rub solution or 1135 oz (32.2 kg) of Level 1 aerosols, or a combination of liquids and Level 1 aerosols not to exceed, in total, the equivalent of 10 gal (37.8 L) or 1135 oz (32.2 kg), shall be in use outside of a storage cabinet in a single smoke compartment, except as otherwise provided in 21.3.2.6(6).
(6) One dispenser per room complying with 21.3.2.6(2) or (3), and located in the room, shall not be required to be included in the aggregated quantity specified in 21.3.2.6(5).
(7) Storage of quantities greater then 5 gal (18.9 L) in a single smoke compartment shall meet the requirements of NFPA 30, *Flammable and Combustible Liquids Code.*
(8) Dispensers shall not be installed in the following locations:
   (a) Above an ignition source within a 1 in. (25 mm) horizontal distance from each side of the ignition source
   (b) To the side of an ignition source within a 1 in. (25 mm) horizontal distance from the ignition source
   (c) Beneath an ignition source within a 1 in. (25 mm) vertical distance from the ignition source
(9) Dispensers installed directly over carpeted floors shall be permitted only in sprinklered smoke compartments.
(10) The alcohol-based hand-rub solution shall not exceed 95 percent alcohol content by volume.
(11) Operation of the dispenser shall comply with the following criteria:
   (a) The dispenser shall not release its contents except when the dispenser is activated, either manually or automatically by touch-free activation.
   (b) Any activation of the dispenser shall occur only when an object is placed within 4 in. (100 mm) of the sensing device.
   (c) An object placed within the activation zone and left in place shall not cause more than one activation.
   (d) The dispenser shall not dispense more solution than the amount required for hand hygiene consistent with label instructions.
   (e) The dispenser shall be designed, constructed, and operated in a manner that ensures that accidental or malicious activation of the dispensing device is minimized.
   (f) The dispenser shall be tested in accordance with the manufacturer's care and use instructions each time a new refill is installed.

**21.3.3 Interior Finish.** See 39.3.3.

**21.3.4 Detection, Alarm, and Communications Systems.**

**21.3.4.1 General.** Ambulatory health care facilities shall be provided with fire alarm systems in accordance with Section 9.6, except as modified by 21.3.4.2 through 21.3.4.4.

**21.3.4.2 Initiation.** Initiation of the required fire alarm systems shall be by manual means in accordance with 9.6.2 and by means of any detection devices or detection systems required.

**21.3.4.3 Notification.** Positive alarm sequence in accordance with 9.6.3.4 shall be permitted.

**21.3.4.3.1 Occupant Notification.** Occupant notification shall be accomplished automatically, without delay, in accordance with 9.6.3 upon operation of any fire alarm activating device.

**21.3.4.3.2 Emergency Forces Notification.**

**21.3.4.3.2.1** Fire department notification shall be accomplished in accordance with 9.6.4.

**21.3.4.3.2.2** Smoke detection devices or smoke detection systems equipped with reconfirmation features shall not be required to automatically notify the fire department, unless the alarm condition is reconfirmed after a period not exceeding 120 seconds.

**21.3.4.4 Fire Safety Functions.** Operation of any activating device in the required fire alarm system shall be arranged to accomplish automatically, without delay, any control functions required to be performed by that device. *(See 9.6.5.)*

**21.3.5 Extinguishment Requirements.** See 39.3.5.

**21.3.5.1** Isolated hazardous areas shall be permitted to be protected in accordance with 9.7.1.2.

**21.3.5.2** For new installations in existing ambulatory health care facilities, where more than two sprinklers are installed in a single area for protection in accordance with 9.7.1.2, waterflow detection shall be provided to sound the building fire alarm or to notify, by a signal, any constantly attended location, such as PBX, security, or emergency room, at which the necessary corrective action shall be taken.

**21.3.5.3** Portable fire extinguishers shall be provided in ambulatory health care facilities in accordance with 9.7.4.1.

**21.3.6 Corridors.** (No requirements.)

**21.3.7 Subdivision of Building Space.**

**21.3.7.1** Ambulatory health care facilities shall be separated from other tenants and occupancies and shall meet all of the following requirements:

(1) Walls shall have not less than a 1-hour fire resistance rating and shall extend from the floor slab below to the floor or roof slab above.
(2) Doors shall be constructed of not less than 1¾ in. (44 mm) thick, solid-bonded wood core or the equivalent and shall be equipped with positive latches.
(3) Doors shall be self-closing and shall be kept in the closed position, except when in use.
(4) Any windows in the barriers shall be of fixed fire window assemblies in accordance with Section 8.3.

**21.3.7.2** Every story of an ambulatory health care facility shall be divided into not less than two smoke compartments, unless otherwise permitted by one of the following:

(1) This requirement shall not apply to facilities of less than 5000 ft² (465 m²) that are protected by an approved automatic smoke detection system.
(2) This requirement shall not apply to facilities of less than 10,000 ft² (929 m²) that are protected throughout by an approved, supervised automatic sprinkler system installed in accordance with Section 9.7.
(3) An area in an adjoining occupancy shall be permitted to serve as a smoke compartment for an ambulatory health care facility if all of the following criteria are met:
    (a) The separating wall and both compartments meet the requirements of 21.3.7.
    (b) The ambulatory health care facility is less than 22,500 ft² (2100 m²).
    (c) Access from the ambulatory health care facility to the other occupancy is unrestricted.

**21.3.7.3 Reserved.**

**21.3.7.4 Reserved.**

**21.3.7.5** Required smoke barriers shall be constructed in accordance with Section 8.5 and shall have a minimum ½-hour fire resistance rating, unless otherwise permitted by 21.3.7.6.

**21.3.7.6** Smoke dampers shall not be required in duct penetrations of smoke barriers in fully ducted heating, ventilating, and air-conditioning systems where adjacent smoke compartments are protected throughout by an approved, supervised automatic sprinkler system in accordance with Section 9.7.

**21.3.7.7** Windows in the smoke barrier shall be of fixed fire window assemblies in accordance with Section 8.3.

**21.3.7.8 Reserved.**

**21.3.7.9\*** Doors in smoke barriers shall be not less than 1¾ in. (44 mm) thick, solid-bonded wood core or the equivalent and shall be self-closing or automatic-closing in accordance with 21.2.2.4.

**21.3.7.10** Latching hardware shall not be required on smoke barrier cross-corridor doors, and doors shall not be required to swing in the direction of egress travel.

**21.4 Special Provisions.** See Section 39.4.

**21.5 Building Services.**

**21.5.1 Utilities.**

**21.5.1.1** Utilities shall comply with the provisions of Section 9.1.

**21.5.1.2** Existing installations shall be permitted to be continued in service, provided that the systems do not present a serious hazard to life.

**21.5.2 Heating, Ventilating, and Air-Conditioning.**

**21.5.2.1** Heating, ventilating, and air-conditioning shall comply with the provisions of Section 9.2 and shall be in accordance with the manufacturer's specifications, unless otherwise modified by 21.5.2.2.

**21.5.2.2** If fuel-fired, heating devices shall comply with all of the following:

(1) They shall be chimney connected or vent connected.
(2) They shall take air for combustion directly from the outside.
(3) They shall be designed and installed to provide for complete separation of the combustion system from the atmosphere of the occupied area.

**21.5.2.2.1** Any heating device shall have safety features to immediately stop the flow of fuel and shut down the equipment in case of either excessive temperature or ignition failure.

**21.5.2.2.2** Approved, suspended unit heaters shall be permitted in locations other than means of egress and patient treatment areas, provided that both of the following criteria are met:

(1) Such heaters are located high enough to be out of the reach of persons using the area.
(2) Such heaters are equipped with the safety features required by 21.5.2.2.1.

**21.5.3 Elevators, Escalators, and Conveyors.** Elevators, escalators, and conveyors shall comply with the provisions of Section 9.4.

**21.5.4 Rubbish Chutes, Incinerators, and Laundry Chutes.** Rubbish chutes, incinerators, and laundry chutes shall comply with the provisions of Section 9.5.

**21.6 Reserved.**

**21.7\* Operating Features.**

**21.7.1 Evacuation and Relocation Plan and Fire Drills.**

**21.7.1.1** The administration of every ambulatory health care facility shall have, in effect and available to all supervisory personnel, written copies of a plan for the protection of all persons in the event of fire, for their evacuation to areas of refuge, and for their evacuation from the building when necessary.

**21.7.1.2** All employees shall be periodically instructed and kept informed with respect to their duties under the plan required by 21.7.1.1.

**21.7.1.3** A copy of the plan required by 21.7.1.1 shall be readily available at all times in the telephone operator's location or at the security center.

**21.7.1.4\*** Fire drills in ambulatory health care facilities shall include the transmission of a fire alarm signal and simulation of emergency fire conditions.

**21.7.1.5** Patients shall not be required to be moved during drills to safe areas or to the exterior of the building.

**21.7.1.6** Drills shall be conducted quarterly on each shift to familiarize facility personnel (nurses, interns, maintenance engineers, and administrative staff) with the signals and emergency action required under varied conditions.

**21.7.1.7** When drills are conducted between 9:00 p.m. and 6:00 a.m. (2100 hours and 0600 hours), a coded announcement shall be permitted to be used instead of audible alarms.

**21.7.1.8** Employees of ambulatory health care facilities shall be instructed in life safety procedures and devices.

**21.7.2 Procedure in Case of Fire.**

**21.7.2.1\* Protection of Patients.**

**21.7.2.1.1** For ambulatory health care facilities, the proper protection of patients shall require the prompt and effective response of ambulatory health care personnel.

**21.7.2.1.2** The basic response required of staff shall include the following:

(1) Removal of all occupants directly involved with the fire emergency
(2) Transmission of an appropriate fire alarm signal to warn other building occupants and summon staff
(3) Confinement of the effects of the fire by closing doors to isolate the fire area
(4) Relocation of patients as detailed in the facility's fire safety plan

**21.7.2.2 Fire Safety Plan.** A written fire safety plan shall provide for all of the following:

(1) Use of alarms
(2) Transmission of alarms to fire department
(3) Response to alarms
(4) Isolation of fire
(5) Evacuation of immediate area
(6) Evacuation of smoke compartment
(7) Preparation of floors and building for evacuation
(8) Extinguishment of fire

**21.7.2.3 Staff Response.**

**21.7.2.3.1** All personnel shall be instructed in the use of and response to fire alarms.

**21.7.2.3.2** All health care personnel shall be instructed in the use of the code phrase to ensure transmission of an alarm under any of the following conditions:

(1) When the individual who discovers a fire must immediately go to the aid of an endangered person
(2) During a malfunction of the building fire alarm system

**21.7.2.3.3** Personnel hearing the code announced shall first activate the building fire alarm using the nearest fire alarm box and then shall execute immediately their duties as outlined in the fire safety plan.

**21.7.3 Maintenance of Exits.**

**21.7.3.1** Proper maintenance shall be provided to ensure the dependability of the method of evacuation selected.

**21.7.3.2** Ambulatory health care occupancies that find it necessary to lock exits shall, at all times, maintain an adequate staff qualified to release locks and direct occupants from the immediate danger area to a place of safety in case of fire or other emergency.

**21.7.4\* Smoking.** Smoking regulations shall be adopted and shall include not less than the following provisions:

(1) Smoking shall be prohibited in any room, ward, or compartment where flammable liquids, combustible gases, or oxygen is used or stored and in any other hazardous location, and such areas shall be posted with signs that read NO SMOKING or shall be posted with the international symbol for no smoking.
(2) In ambulatory health care facilities where smoking is prohibited and signs are placed at all major entrances, secondary signs with language that prohibits smoking shall not be required.
(3) Smoking by patients classified as not responsible shall be prohibited.
(4) The requirement of 21.7.4(3) shall not apply where the patient is under direct supervision.
(5) Ashtrays of noncombustible material and safe design shall be provided in all areas where smoking is permitted.
(6) Metal containers with self-closing cover devices into which ashtrays can be emptied shall be readily available to all areas where smoking is permitted.

**21.7.5 Furnishings, Mattresses, and Decorations.**

**21.7.5.1\*** Draperies, curtains and other loosely hanging fabrics and films serving as furnishings or decorations in ambulatory health care occupancies shall be in accordance with the provisions of 10.3.1, and the following also shall apply:

(1) Such curtains shall include cubicle curtains.
(2) Such curtains shall not include curtains at showers.

**21.7.5.2** Newly introduced upholstered furniture shall comply with 10.3.2.1 and one of the following provisions:

(1) The furniture shall meet the criteria specified in 10.3.3.
(2) The furniture shall be in a building protected throughout by an approved, supervised automatic sprinkler system in accordance with 9.7.1.1(1).

**21.7.5.3** Newly introduced mattresses shall comply with 10.3.2.2 and one of the following provisions:

(1) The mattresses shall meet the criteria specified in 10.3.4.
(2) The mattresses shall be in a building protected throughout by an approved, supervised automatic sprinkler system in accordance with 9.7.1.1(1).

**21.7.5.4** Combustible decorations shall be prohibited, unless one of the following criteria is met:

(1) They are flame-retardant.
(2) The decorations meet the requirements of NFPA 701, *Standard Methods of Fire Tests for Flame Propagation of Textiles and Films.*
(3) The decorations exhibit a heat release rate not exceeding 100 kW when tested in accordance with NFPA 289, *Standard Method of Fire Test for Individual Fuel Packages,* using the 20 kW ignition source.
(4) They are decorations, such as photographs and paintings, in such limited quantities that a hazard of fire development or spread is not present.

**21.7.5.5 Soiled Linen and Trash Receptacles.**

**21.7.5.5.1** Soiled linen or trash collection receptacles shall not exceed 32 gal (121 L) in capacity, and all of the following also shall apply:

(1) The average density of container capacity in a room or space shall not exceed 0.5 gal/ft$^2$ (20.4 L/m$^2$).
(2) A capacity of 32 gal (121 L) shall not be exceeded within any 64 ft$^2$ (6 m$^2$) area.
(3) Mobile soiled linen or trash collection receptacles with capacities greater than 32 gal (121 L) shall be located in a room protected as a hazardous area when not attended.
(4) Container size and density shall not be limited in hazardous areas.

**21.7.5.5.2** The provisions of 10.3.9, applicable to containers for rubbish, waste, or linen, shall not apply.

**21.7.6 Maintenance and Testing.** See 4.6.12.

**21.7.7* Engineered Smoke Control Systems.**

**21.7.7.1** New engineered smoke control systems shall be tested in accordance with established engineering principles and shall meet the performance requirements of such testing prior to acceptance.

**21.7.7.2** Following acceptance, all engineered smoke control systems shall be tested periodically in accordance with recognized engineering principles.

**21.7.7.3** Test documentation shall be maintained on the premises at all times.

**21.7.8 Portable Space-Heating Devices.** Portable space-heating devices shall be prohibited in all ambulatory health care occupancies, unless both of the following criteria are met:

(1) Such devices are used only in nonsleeping staff and employee areas.
(2) The heating elements of such devices do not exceed 212°F (100°C).

**21.7.9 Construction, Repair, and Improvement Operations.**

**21.7.9.1** Construction, repair, and improvement operations shall comply with 4.6.10.

**21.7.9.2** The means of egress in any area undergoing construction, repair, or improvements shall be inspected daily for compliance with 7.1.10.1 and shall also comply with NFPA 241, *Standard for Safeguarding Construction, Alteration, and Demolition Operations.*

## Chapter 22 New Detention and Correctional Occupancies

**22.1 General Requirements.**

**22.1.1 Application.**

**22.1.1.1 General.**

**22.1.1.1.1** The requirements of this chapter shall apply to new buildings or portions thereof used as detention or correctional occupancies. *(See 1.3.1.)*

**22.1.1.1.2 Administration.** The provisions of Chapter 1, Administration, shall apply.

**22.1.1.1.3 General.** The provisions of Chapter 4, General, shall apply.

**22.1.1.1.4** This chapter establishes life safety requirements that shall apply to the design of all new detention and correctional facilities, other than the following:

(1) Use Condition I facilities protected as residential occupancies in accordance with 22.1.2.3
(2)*Facilities determined to have equivalent safety provided in accordance with Section 1.4

**22.1.1.1.5** Detention and correctional occupancies shall include those used for purposes such as correctional institutions, detention facilities, community residential centers, training schools, work camps, and substance abuse centers where occupants are confined or housed under some degree of restraint or security.

**22.1.1.1.6*** Detention and correctional occupancies shall include those that provide sleeping facilities for one or more residents and are occupied by persons who are generally prevented from taking self-preservation action because of security measures not under the occupants' control.

**22.1.1.1.7*** Lockups in other than detention and correctional occupancies and health care occupancies shall comply with the requirements of 22.4.5.

**22.1.1.2 Total Concept.**

**22.1.1.2.1** All detention and correctional facilities shall be designed, constructed, maintained, and operated to minimize the possibility of a fire emergency.

**22.1.1.2.2** Because the safety of all occupants in detention and correctional facilities cannot be adequately ensured solely by dependence on evacuation of the building, their protection from fire shall be provided by appropriate arrangement of facilities; adequate, trained staff; and development of operating, security, and maintenance procedures composed of the following:

(1) Design, construction, and compartmentation
(2) Provision for detection, alarm, and extinguishment
(3) Fire prevention and planning, training, and drilling programs for the isolation of fire and the transfer of occupants to areas of refuge, for evacuation of the building, or for protection of the occupants in place
(4) Provision of security to the degree necessary for the safety of the public and the occupants of the facility

**22.1.1.3 Additions.** Additions shall be separated from any existing structure not conforming with the provisions of Chapter 23 by a fire barrier having not less than a 2-hour fire resistance rating constructed to the requirements of the addition, and the following also shall apply:

(1) Doors in such partitions shall normally be kept closed.
(2) Doors in such partitions shall be permitted to be held open if they meet the requirements of 7.2.1.8.2.

**22.1.1.4 Modernizations or Renovations.**

**22.1.1.4.1** Modernizations and renovations shall be in accordance with 4.6.7, unless otherwise permitted by 22.1.1.4.2.

**22.1.1.4.2** In nonsprinklered existing buildings, modernizations or renovations shall be permitted to comply with the nonsprinklered options contained in 22.4.4 in lieu of the sprinkler requirement of 22.3.5.2.

**22.1.2 Classification of Occupancy.** See 6.1.7.

**22.1.2.1\*** For application of the life safety requirements of this chapter, the resident user category shall be divided into the groups specified in 22.1.2.1.1 through 22.1.2.1.5.

**22.1.2.1.1 Use Condition I — Free Egress.** Use Condition I shall be defined as a condition under which free movement is allowed from sleeping areas and other spaces where access or occupancy is permitted to the exterior via means of egress that meet the requirements of the *Code*.

**22.1.2.1.2 Use Condition II — Zoned Egress.** Use Condition II shall be defined as a condition under which free movement is allowed from sleeping areas and any other occupied smoke compartment to one or more other smoke compartments.

**22.1.2.1.3 Use Condition III — Zoned Impeded Egress.** Use Condition III shall be defined as a condition under which free movement is allowed within individual smoke compartments, such as within a residential unit comprised of individual sleeping rooms and a group activity space, with egress impeded by remote-controlled release of means of egress from such a smoke compartment to another smoke compartment.

**22.1.2.1.4 Use Condition IV — Impeded Egress.** Use Condition IV shall be defined as a condition under which free movement is restricted from an occupied space, and remote-controlled release is provided to allow movement from all sleeping rooms, activity spaces, and other occupied areas within the smoke compartment to another smoke compartment.

**22.1.2.1.5 Use Condition V — Contained.** Use Condition V shall be defined as a condition under which free movement is restricted from an occupied space, and staff-controlled manual release at each door is provided to allow movement from all sleeping rooms, activity spaces, and other occupied areas within the smoke compartment to another smoke compartment.

**22.1.2.2\*** To be classified as Use Condition III or Use Condition IV, the arrangement, accessibility, and security of the release mechanism(s) used for emergency egress shall be such that the minimum available staff, at any time, can promptly release the locks.

**22.1.2.3** Areas housing occupancies corresponding to Use Condition I shall conform to one of the following:

(1) Requirements of residential occupancies under this *Code*
(2)\*Requirements of this chapter for Use Condition II facilities, provided that the staffing requirements of Section 22.7 are met

**22.1.3\* Multiple Occupancies.**

**22.1.3.1** Multiple occupancies shall be in accordance with 6.1.14.

**22.1.3.2** Egress provisions for areas of detention and correctional facilities that correspond to other occupancies shall meet the corresponding requirements of this *Code* for such occupancies as modified by 22.1.3.2.1 and 22.1.3.2.2.

**22.1.3.2.1** Where security operations necessitate the locking of required means of egress, staff in the building shall be provided with a means for the supervised release of occupants during all times of use.

**22.1.3.2.2\*** Where security operations necessitate the locking of required means of egress, the following shall apply:

(1) Detention-grade hardware meeting the requirements of ASTM F 1577, *Standard Test Methods for Detention Locks for Swinging Doors*, shall be provided on swinging doors within the required means of egress.
(2) Sliding doors within the required means of egress shall be designed and engineered for detention and correctional use, and lock cylinders shall meet the cylinder test requirements of ASTM F 1577.

**22.1.3.3** Sections of detention and correctional facilities shall be permitted to be classified as other occupancies, provided that they meet both of the following conditions:

(1) They are not intended to serve residents for sleeping purposes.
(2) They are separated from areas of detention or correctional occupancies by construction having not less than a 2-hour fire resistance rating.

**22.1.3.4** All means of egress from detention and correctional occupancies that traverse other use areas shall, as a minimum, conform to the requirements of this *Code* for detention and correctional occupancies, unless otherwise permitted by 22.1.3.5.

**22.1.3.5** Egress through a horizontal exit into other contiguous occupancies that do not conform with detention and correctional occupancy egress provisions but that do comply with requirements set forth in the appropriate occupancy chapter of this *Code* shall be permitted, provided that both of the following criteria apply:

(1) The occupancy shall not contain high hazard contents.
(2) The horizontal exit shall comply with the requirements of 22.2.2.5.

**22.1.3.6** Any area with a hazard of contents classified higher than that of the detention or correctional occupancy and located in the same building shall be protected as required in 22.3.2.

**22.1.3.7** Nondetention- or noncorrectional-related occupancies classified as containing high hazard contents shall not be permitted in buildings housing detention or correctional occupancies.

**22.1.4 Definitions.**

**22.1.4.1 General.** For definitions, see Chapter 3, Definitions.

**22.1.4.2 Special Definitions.** A list of special terms used in this chapter follows:

(1) **Detention and Correctional Residential Housing Area.** See 3.3.21.1.
(2) **Sally Port (Security Vestibule).** See 3.3.236.

**22.1.5 Classification of Hazard of Contents.** The classification of hazard of contents shall be as defined in Section 6.2.

**22.1.6 Minimum Construction Requirements.**

**22.1.6.1** Detention and correctional occupancies shall be limited to the building construction types specified in Table 22.1.6.1. *(See 8.2.1.)*

**22.1.6.2** All interior walls and partitions in Type I or Type II construction shall be of noncombustible or limited-combustible materials.

**22.1.7 Occupant Load.** The occupant load, in number of persons for whom means of egress and other provisions are required, either shall be determined on the basis of the occupant load factors of Table 7.3.1.2 that are characteristic of the use of the space or shall be determined as the maximum probable population of the space under consideration, whichever is greater.

**22.2 Means of Egress Requirements.**

**22.2.1 General.** Means of egress shall comply with Chapter 7, unless otherwise provided or modified by Section 22.2.

**22.2.2 Means of Egress Components.**

**22.2.2.1 Components Permitted.** Components of means of egress shall be limited to the types described in 22.2.2.2 through 22.2.2.11.

**22.2.2.2 Doors.** Doors complying with 7.2.1 shall be permitted, unless otherwise provided by 22.2.11.

**22.2.2.3 Stairs.**

**22.2.2.3.1** Stairs shall be permitted as follows:

(1) Stairs complying with 7.2.2 shall be permitted.
(2) Noncombustible grated stair treads and landing floors shall be permitted.

**22.2.2.3.2** Spiral stairs complying with 7.2.2.2.3 shall be permitted for access to and between staff locations.

**22.2.2.4 Smokeproof Enclosures.** Smokeproof enclosures complying with 7.2.3 shall be permitted.

**22.2.2.5 Horizontal Exits.** Horizontal exits complying with 7.2.4 and the modifications of 22.2.2.5.1 and 22.2.2.5.2 shall be permitted.

**Table 22.1.6.1 Construction Type Limitations**

| Construction Type | Sprinklered[†] | Stories in Height [‡] | | | | | |
| | | 1 With Basement | 1 Without Basement | 2 | 3 | >3 But Not High-Rise | High-Rise |
|---|---|---|---|---|---|---|---|
| I (442) | Yes | X | X | X | X | X | X |
| | No | NP | NP | NP | NP | NP | NP |
| I (332) | Yes | X | X | X | X | X | X |
| | No | NP | NP | NP | NP | NP | NP |
| II (222) | Yes | X | X | X | X | X | X |
| | No | NP | NP | NP | NP | NP | NP |
| II (111) | Yes | X | X | X | NP | NP | NP |
| | No | NP | NP | NP | NP | NP | NP |
| II (000) | Yes | X | X | X | NP | NP | NP |
| | No | NP | NP | NP | NP | NP | NP |
| III (211) | Yes | X | X | X | NP | NP | NP |
| | No | NP | NP | NP | NP | NP | NP |
| III (200) | Yes | X | X | X | NP | NP | NP |
| | No | NP | NP | NP | NP | NP | NP |
| IV (2HH) | Yes | X | X | X | NP | NP | NP |
| | No | NP | NP | NP | NP | NP | NP |
| V (111) | Yes | X | X | X | NP | NP | NP |
| | No | NP | NP | NP | NP | NP | NP |
| V (000) | Yes | X | X | X | NP | NP | NP |
| | No | NP | NP | NP | NP | NP | NP |

X: Permitted for Use Conditions II, III, IV, and V. *(See 22.1.4.3 for Use Condition I.)*
NP: Not permitted.
[†]Sprinklered throughout by an approved, supervised automatic sprinkler system in accordance with 9.7.1.1(1). *(See 22.3.5.)*
[‡]See 4.6.3.

**22.2.2.5.1** Not less than 6 ft² (0.55 m²) of accessible space per occupant shall be provided on each side of the horizontal exit for the total number of people in adjoining compartments.

**22.2.2.5.2\*** Horizontal exits shall be permitted to comprise 100 percent of the exits required, provided that an exit, other than a horizontal exit, located in another (not necessarily adjacent) fire compartment is accessible without returning through the compartment of fire origin.

**22.2.2.6 Ramps.** Ramps complying with 7.2.5 shall be permitted.

**22.2.2.7 Exit Passageways.** Exit passageways complying with 7.2.6 shall be permitted.

**22.2.2.8 Reserved.**

**22.2.2.9 Fire Escape Ladders.** Fire escape ladders complying with 7.2.9 shall be permitted.

**22.2.2.10 Alternating Tread Devices.** Alternating tread devices complying with 7.2.11 shall be permitted.

**22.2.2.11 Areas of Refuge.** Areas of refuge complying with 7.2.12 shall be permitted.

**22.2.3 Capacity of Means of Egress.**

**22.2.3.1** The capacity of any required means of egress shall be in accordance with Section 7.3.

**22.2.3.2** Aisles, corridors, and ramps required for egress shall be not less than 48 in. (1220 mm) in width.

**22.2.3.3** Residents' sleeping room door widths shall be permitted to comply with 22.2.11.4.

**22.2.4 Number of Means of Egress.**

**22.2.4.1** The number of means of egress shall be in accordance with Section 7.4.

**22.2.4.2** Not less than two separate exits shall meet both of the following criteria:

(1) They shall be provided on every story.
(2) They shall be accessible from every part of every story, fire compartment, or smoke compartment; however, exit access travel shall be permitted to be common for the distances permitted as common path of travel by 22.2.5.3.

**22.2.4.3** Not less than one approved exit shall be accessible from each fire compartment and each required smoke compartment into which residents are potentially moved in a fire emergency, with the exits arranged so that egress is possible without returning through the zone of fire origin.

**22.2.5 Arrangement of Means of Egress.** See also Section 7.5.

**22.2.5.1** Every sleeping room shall have a door leading directly to an exit access corridor, unless otherwise permitted by one of the following:

(1) The requirement of 22.2.5.1 shall not apply if there is an exit door opening directly to the outside from a room at the finished ground level.
(2) One adjacent room, such as a day room, a group activity space, or other common space, shall be permitted to intervene, and the following also shall apply:
  (a) Where sleeping rooms directly adjoin a day room or group activity space that is used for access to an exit, such sleeping rooms shall be permitted to open directly to the day room or space.

(b) Sleeping rooms permitted to open directly to the day room or space shall be permitted to be separated in elevation by a one-half story or full story height.

**22.2.5.2** No exit or exit access shall contain a corridor, a hallway, or an aisle having a pocket or dead end exceeding 50 ft (15 m) for Use Condition II, Use Condition III, or Use Condition IV and 20 ft (6100 mm) for Use Condition V.

**22.2.5.3** A common path of travel shall not exceed 100 ft (30 m).

**22.2.5.4** A sally port shall be permitted in a means of egress where there are provisions for continuous and unobstructed travel through the sally port during an emergency egress condition.

**22.2.6 Travel Distance to Exits.** Travel distance shall comply with 22.2.6.1 through 22.2.6.7.

**22.2.6.1** Travel distance shall be measured in accordance with Section 7.6.

**22.2.6.2** The travel distance between any room door required as an exit access and an exit shall not exceed 150 ft (46 m).

**22.2.6.3 Reserved.**

**22.2.6.4** The travel distance between any point in a room and an exit shall not exceed 200 ft (61 m).

**22.2.6.5 Reserved.**

**22.2.6.6** The travel distance between any point in a sleeping room to the door in that room shall not exceed 50 ft (15 m), unless otherwise permitted by 22.2.6.7.

**22.2.6.7** The maximum travel distance limitation of 22.2.6.6 shall be permitted to be increased to 100 ft (30 m) in open dormitories, provided that both of the following criteria are met:

(1) The enclosing walls of the dormitory space shall be of smoke-tight construction.
(2) Not less than two exit access doors remotely located from each other shall be provided where travel distance to the exit access door from any point within the dormitory exceeds 50 ft (15 m).

**22.2.7 Discharge from Exits.**

**22.2.7.1** Exits shall be permitted to discharge into a fenced or walled courtyard, provided that not more than two walls of the courtyard are the building walls from which egress is being made.

**22.2.7.2** Enclosed yards or courts used for exit discharge in accordance with 22.2.7.1 shall be of sufficient size to accommodate all occupants at a distance of not less than 50 ft (15 m) from the building while providing a net area of 15 ft² (1.4 m²) per person.

**22.2.7.3** All exits shall be permitted to discharge through the level of exit discharge.

**22.2.7.4** The requirements of 7.7.2 shall be waived, provided that not more than 50 percent of the exits discharge into a single fire compartment separated from other compartments by construction having not less than a 1-hour fire resistance rating.

**22.2.8 Illumination of Means of Egress.** Means of egress shall be illuminated in accordance with Section 7.8.

**22.2.9 Emergency Lighting.** Emergency lighting shall be provided in accordance with Section 7.9.

**22.2.10 Marking of Means of Egress.** Exit marking shall be provided as follows:

(1) Exit signs shall be provided in areas accessible to the public in accordance with Section 7.10.
(2) Exit signs shall not be required in detention and correctional residential housing areas. *(See 3.3.21.1.)*

**22.2.11 Special Features.**

**22.2.11.1** Doors within means of egress shall be in accordance with Chapter 7, unless otherwise provided in 22.2.11.2 through 22.2.11.12.

**22.2.11.2** Doors shall be permitted to be locked in accordance with the applicable use condition.

**22.2.11.3** Where egress doors are locked with key-operated locks, the provisions of 22.7.7 shall apply.

**22.2.11.4\*** Doors to resident sleeping rooms shall be not less than 28 in. (710 mm) in clear width.

**22.2.11.5 Reserved.**

**22.2.11.6** Doors in a means of egress shall be permitted to be of the horizontal-sliding type, provided that the force necessary to slide the door to its fully open position does not exceed 50 lbf (222 N) where a force of 50 lbf (222 N) is simultaneously applied perpendicular to the door.

**22.2.11.7** Doors from areas of refuge to the exterior shall be permitted to be locked with key locks in lieu of locking methods described in 22.2.11.8, the keys to unlock such doors shall be maintained and available at the facility at all times, and the locks shall be operable from the outside.

**22.2.11.8\*** Any remote-control release used in a means of egress shall be provided with a reliable means of operation to release locks on all doors and shall be remotely located from the resident living areas, unless otherwise permitted by 22.2.11.8.2.

**22.2.11.8.1** The remote location of a remote-control release used in a means of egress shall provide sight and sound supervision of the resident living areas.

**22.2.11.8.2** Remote-control locking and unlocking of occupied rooms in Use Condition IV shall not be required, provided that both of the following criteria are met:

(1) Not more than 10 locks need to be unlocked to relocate all occupants from one smoke compartment to an area of refuge as promptly as is required where remote-control unlocking is used. *(See 22.3.7.9 for requirements for smoke barrier doors.)*
(2) Unlocking of all necessary locks is accomplished with not more than two separate keys.

**22.2.11.9 Remote-Control Release–Operated Doors.**

**22.2.11.9.1** All remote-control release–operated doors shall be provided with a redundant means of operation as follows:

(1) Power-operated sliding doors or power-operated locks shall be constructed so that, in the event of power failure, a manual mechanical means to release and open the doors is provided at each door, and either emergency power arranged in accordance with 22.2.11.9.2 is provided for the power operation or a remote-control manual mechanical release is provided.

(2) Mechanically operated sliding doors or mechanically operated locks shall be provided with a manual mechanical means at each door to release and open the door.

**22.2.11.9.2** The emergency power required by 23.2.11.9.1(1) shall be arranged to provide the required power automatically in the event of any interruption of normal power due to any of the following:

(1) Failure of a public utility or other outside electrical power supply
(2) Opening of a circuit breaker or fuse
(3) Manual act(s), including accidental opening of a switch controlling normal lighting facilities

**22.2.11.10** The provisions of 7.2.1.5.8 for stairway re-entry shall not apply.

**22.2.11.11** Doors unlocked by means of remote control under emergency conditions shall not automatically relock when closed, unless specific action is taken at the remote-control location to enable doors to relock.

**22.2.11.12** Emergency power shall be provided for all electric power–operated sliding doors and electric power–operated locks, unless otherwise permitted by 22.2.11.12.2.

**22.2.11.12.1** The emergency power shall be arranged to automatically operate within 10 seconds upon failure of normal power and to maintain the necessary power source for a minimum of 1½ hours.

**22.2.11.12.2** The emergency power specified in 22.2.11.12 shall not be required in facilities with 10 or fewer locks complying with 22.2.11.8.2.

**22.3 Protection.**

**22.3.1 Protection of Vertical Openings.** Any vertical opening shall be enclosed or protected in accordance with Section 8.6, unless otherwise permitted by one of the following:

(1) Unprotected vertical openings in accordance with 8.6.9.1 shall be permitted.
(2)\*In residential housing area smoke compartments, unprotected vertical openings shall be permitted in accordance with the conditions of 8.6.6, provided that the height between the lowest and highest finished floor levels does not exceed 23 ft (7010 mm), and the following also shall be permitted:
　(a) The number of levels shall not be restricted.
　(b) Residential housing areas subdivided in accordance with 22.3.8 shall be permitted to be considered as part of the communicating space.
　(c) The separation shall not be required to have a fire resistance rating. *[See 8.6.6(4)(b).]*

**22.3.2 Protection from Hazards.**

**22.3.2.1\*** Any hazardous area shall be protected in accordance with Section 8.7. The areas described in Table 22.3.2.1 shall be protected as indicated.

**22.3.2.2** Where Table 22.3.2.1 requires separations to be smoke resistant, the provision of 8.7.1.2 shall not apply.

**22.3.2.3** Hazardous areas determined by the authority having jurisdiction as not incidental to residents' housing shall be separated by 2-hour fire resistance–rated barriers in conjunction with automatic sprinkler protection.

**Table 22.3.2.1 Hazardous Area Protection**

| Hazardous Area Description | Separation/Protection[†] |
|---|---|
| Areas not incidental to resident housing | 2 hours |
| Boiler and fuel-fired heater rooms | 1 hour |
| Commercial cooking equipment | In accordance with 9.2.3 |
| Commissaries | Smoke resistant |
| Employee locker rooms | Smoke resistant |
| Hobby/handicraft shops | Smoke resistant |
| Laundries >100 ft² (>9.3 m²) | 1 hour |
| Maintenance shops | Smoke resistant |
| Padded cells | 1 hour |
| Soiled linen rooms | 1 hour |
| Storage rooms >50 ft² (>4.6 m²) but ≤100 ft² (≤9.3 m²) storing combustible material | Smoke resistant |
| Storage rooms >100 ft² (>9.3 m²) storing combustible materials | 1 hour |
| Trash collection rooms | 1 hour |

[†]Minimum fire resistance rating.

**22.3.2.4** Where cooking facilities are protected in accordance with 9.2.3, kitchens shall not be required to be provided with roomwide protection.

**22.3.3 Interior Finish.**

**22.3.3.1 General.** Interior finish shall be in accordance with Section 10.2.

**22.3.3.2 Interior Wall and Ceiling Finish.** Interior wall and ceiling finish materials complying with Section 10.2 shall be Class A or Class B in corridors, in exits, and in any space not separated from corridors and exits by partitions capable of retarding the passage of smoke; and Class A, Class B, or Class C in all other areas. The provisions of 10.2.8.1 shall not apply.

**22.3.3.3 Interior Floor Finish.**

**22.3.3.3.1** Interior floor finish shall comply with Section 10.2.

**22.3.3.3.2** Interior floor finish in exit enclosures and exit access corridors shall be not less than Class II. The provisions of 10.2.8.2 shall not apply.

**22.3.3.3.3** Interior floor finish shall comply with 10.2.7.1 or 10.2.7.2, as applicable.

**22.3.4 Detection, Alarm, and Communications Systems.**

**22.3.4.1 General.** Detention and correctional occupancies shall be provided with a fire alarm system in accordance with Section 9.6, except as modified by 22.3.4.2 through 22.3.4.4.3.

**22.3.4.2 Initiation.**

**22.3.4.2.1** Initiation of the required fire alarm system shall be by manual means in accordance with 9.6.2, by means of any required detection devices or detection systems, and by means of waterflow alarm in the sprinkler system required by 22.3.5.2, unless otherwise permitted by the following:

(1) Manual fire alarm boxes shall be permitted to be locked, provided that staff is present within the area when it is occupied and staff has keys readily available to unlock the boxes.
(2) Manual fire alarm boxes shall be permitted to be located in a staff location, provided that both of the following criteria are met:
  (a) The staff location is attended when the building is occupied.
  (b) The staff attendant has direct supervision of the sleeping area.

**22.3.4.2.2\*** Use of the provision of shall be permitted only as an exemption to 9.6.1.8.1(2) and (3).

**22.3.4.3 Notification.**

**22.3.4.3.1 Occupant Notification.** Occupant notification shall be accomplished automatically in accordance with 9.6.3, and the following also shall apply:

(1) A positive alarm sequence shall be permitted in accordance with 9.6.3.4.
(2)\*Any smoke detectors required by this chapter shall be permitted to be arranged to alarm at a constantly attended location only and shall not be required to accomplish general occupant notification.

**22.3.4.3.2 Emergency Forces Notification.**

**22.3.4.3.2.1** Fire department notification shall be accomplished in accordance with 9.6.4, unless otherwise permitted by one of the following:

(1) A positive alarm sequence shall be permitted in accordance with 9.6.3.4.
(2) Any smoke detectors required by this chapter shall not be required to transmit an alarm to the fire department.
(3) This requirement shall not apply where staff is provided at a constantly attended location that meets one of the following criteria:
  (a) It has the capability to promptly notify the fire department.
  (b) It has direct communication with a control room having direct access to the fire department.

**22.3.4.3.2.2** Where the provision of 22.3.4.3.2.1(3) is utilized, the fire plan, as required by 22.7.1.3, shall include procedures for logging of alarms and immediate notification of the fire department.

**22.3.4.4\* Detection.** An approved automatic smoke detection system shall be in accordance with Section 9.6, as modified by 22.3.4.4.1 through 22.3.4.4.3, throughout all resident sleeping areas and adjacent day rooms, activity rooms, or contiguous common spaces.

**22.3.4.4.1** Smoke detectors shall not be required in sleeping rooms with four or fewer occupants.

**22.3.4.4.2** Other arrangements and positioning of smoke detectors shall be permitted to prevent damage or tampering, or for other purposes.

**22.3.4.4.2.1** Other arrangements, as specified in 22.3.4.4.2, shall be capable of detecting any fire, and the placement of detectors shall be such that the speed of detection is equivalent to that provided by the spacing and arrangements required by the installation standards referenced in Section 9.6.

**22.3.4.4.2.2** Detectors shall be permitted to be located in exhaust ducts from cells, behind grilles, or in other locations.

**22.3.4.4.2.3** The equivalent performance of the design permitted by 22.3.4.4.2.2 shall be acceptable to the authority having jurisdiction in accordance with the equivalency concepts specified in Section 1.4.

**22.3.4.4.3\*** Smoke detectors shall not be required in Use Condition II open dormitories where staff is present within the dormitory whenever the dormitory is occupied.

**22.3.5 Extinguishment Requirements.**

**22.3.5.1** High-rise buildings shall comply with 22.4.3.

**22.3.5.2** All buildings classified as Use Condition II, Use Condition III, Use Condition IV, or Use Condition V shall be protected throughout by an approved, supervised automatic sprinkler system in accordance with 22.3.5.3.

**22.3.5.3** The automatic sprinkler system required by 22.3.5.2 shall meet all of the following criteria:

(1) It shall be in accordance with Section 9.7.
(2) It shall be installed in accordance with 9.7.1.1(1).
(3) It shall be electrically connected to the fire alarm system.
(4) It shall be fully supervised.

**22.3.5.4** Portable fire extinguishers shall be provided in accordance with 9.7.4.1, unless otherwise permitted by the following:

(1)\*Access to portable fire extinguishers shall be permitted to be locked.
(2)\*Portable fire extinguishers shall be permitted to be located at staff locations only.

**22.3.5.5** Standpipe and hose systems shall be provided in accordance with 9.7.4.2 as follows, unless otherwise permitted by 22.3.5.6:

(1) Class I standpipe systems shall be provided for any building three or more stories in height.
(2) Class III standpipe and hose systems shall be provided for all nonsprinklered buildings three or more stories in height.

**22.3.5.6** The requirements of 22.3.5.5 shall not apply where otherwise permitted by the following:

(1) Formed hose, 1 in. (25 mm) in diameter, on hose reels shall be permitted to provide Class II service.
(2) Separate Class I and Class II systems shall be permitted in lieu of a Class III system.

**22.3.6 Corridors.** See 22.3.8.

**22.3.7 Subdivision of Building Spaces.**

**22.3.7.1** Smoke barriers shall be provided to divide every story used for sleeping by residents, or any other story having an occupant load of 50 or more persons, into not less than two compartments, unless otherwise permitted by one of the following:

(1) Protection shall be permitted to be accomplished using horizontal exits. *(See 7.2.4.)*
(2)\*The requirement for subdivision of building space shall be permitted to be fulfilled by one of the following:
   (a) Smoke compartments having exit to a public way, where such exit serves only one area and has no openings to other areas
   (b) Building separated from the resident housing area by a 2-hour fire resistance rating or 50 ft (15 m) of open space

(c) Secured, open area having a holding space located 50 ft (15 m) from the housing area that provides 15 ft$^2$ (1.4 m$^2$) or more of refuge area for each person (resident, staff, visitors) potentially present at the time of a fire

**22.3.7.2** Doors used to access the areas specified in 22.3.7.1(2)(a), (b), and (c) shall meet the requirements for doors at smoke barriers for the applicable use condition.

**22.3.7.3** Where smoke barriers are required by 22.3.7.1, they shall be provided in accordance with both of the following criteria:

(1) They shall limit the occupant load to not more than 200 residents in any smoke compartment.
(2) They shall limit the travel distance to a door in a smoke barrier in accordance with both of the following criteria:
   (a) The distance from any room door required as exit access shall not exceed 150 ft (46 m).
   (b) The distance from any point in a room shall not exceed 200 ft (61 mm).

**22.3.7.4 Reserved.**

**22.3.7.5\*** Any required smoke barrier shall be constructed in accordance with Section 8.5, shall be of substantial construction, and shall have structural fire resistance.

**22.3.7.6** Openings in smoke barriers shall be protected in accordance with Section 8.5, unless otherwise permitted by the following:

(1)\*The total number of vision panels in any barrier shall not be restricted.
(2) Sliding doors in smoke barriers that are designed to normally be kept closed and are remotely operated from a continuously attended location shall not be required to be self-closing.

**22.3.7.7** Not less than 6 net ft$^2$ (0.55 net m$^2$) per occupant shall be provided on each side of the smoke barrier for the total number of occupants in adjoining compartments, and this space shall be readily available wherever occupants are moved across the smoke barrier in a fire emergency.

**22.3.7.8** Doors in smoke barriers shall meet all of the following criteria:

(1) The doors shall provide resistance to the passage of smoke.
(2) Swinging doors shall be self-latching, or the opening resistance of the door shall be not less than 5 lbf (22 N).
(3) Sliding doors shall be exempt from the latching requirement of 8.5.4.3.

**22.3.7.9** Doors in smoke barriers shall conform with the requirements for doors in means of egress as specified in Section 22.2 and shall have locking and release arrangements according to the applicable use condition. The provisions of 22.2.11.8.2 shall not be used for smoke barrier doors serving a smoke compartment containing more than 20 persons.

**22.3.7.10** Vision panels shall be provided in smoke barriers at points where the barrier crosses an exit access corridor.

**22.3.7.11** Smoke dampers shall be provided in accordance with 8.5.5, unless otherwise permitted by 22.3.7.12.

**22.3.7.12** Arrangements and positioning of smoke detectors required by 22.3.7.11 shall be permitted to prevent damage or tampering, or for other purposes, provided that both of the following criteria are met:

(1) Such arrangements shall be capable of detecting any fire.
(2) The placement of detectors shall be such that the speed of detection is equivalent to that provided by the spacing and arrangement required by *NFPA 72, National Fire Alarm and Signaling Code*, as referenced in 8.5.5.7.1.

**22.3.8\* Special Protection Features — Subdivision of Resident Housing Spaces.** Subdivision of facility spaces shall comply with Table 22.3.8.

**22.4 Special Provisions.**

**22.4.1 Limited Access Structures.** The provisions of Section 11.7 for limited access structures shall not apply.

**22.4.2 Underground Buildings.** See Section 11.7 for requirements for underground buildings.

**22.4.3 High-Rise Buildings.** High-rise buildings shall comply with 11.8.3.

**22.4.4 Nonsprinklered Existing Building Renovations.**

**22.4.4.1 General.** Modernizations or renovations of nonsprinklered existing buildings shall be permitted to meet the requirements of this chapter, as modified by 22.4.4.2 through 22.4.4.13, in lieu of the sprinkler requirement of 22.3.5.2.

**22.4.4.2 Minimum Construction Requirements (Nonsprinklered Buildings).**

**22.4.4.2.1** Detention and correctional occupancies in nonsprinklered buildings shall be limited to the building construction types specified in Table 22.4.4.2.1. *(See 8.2.1.)*

**22.4.4.2.2** A residential housing area complying with 22.4.4.6 shall be considered as one story in height for purposes of applying Table 22.4.4.2.1.

**22.4.4.3\* Horizontal Exit Duct Penetrations (Nonsprinklered Buildings).** Ducts shall be permitted to penetrate horizontal exits in accordance with 7.2.4.3.5(3) if protected by combination fire dampers/smoke leakage–rated dampers that meet the smoke damper actuation requirements of 8.5.5.

**Table 22.3.8 Subdivision of Resident Housing Spaces**

| Feature | Use Condition | | | |
|---|---|---|---|---|
| | **II** | **III** | **IV** | **V** |
| Room to room separation | NR | NR | NR | SR |
| Room face to corridor separation | NR | NR | NR | SR |
| Room face to common space separation | NR | NR    SR<br>≤50 ft    >50 ft<br>(≤15 m)†   (>15 m)† | NR    SR<br>≤50 ft    >50 ft<br>(≤15 m)†   (>15 m)† | SR |
| Common space to corridor separation | NR | NR | NR | SR |
| Total openings in solid room face where room face is required to be smoke resistant or fire rated‡ | 0.85 ft² (0.08 m²) | 0.85 ft² (0.08 m²) | 0.85 ft² (0.08 m²) | 0.85 ft² (0.08 m²) where meeting one of the following:<br>(1) Kept in closed position, except when in use by staff<br><br>(2) Closable from the inside<br><br>(3) Provided with smoke control |

NR: No requirement. SR: Smoke resistant.
Notes:
(1) Doors in openings in partitions required to be smoke resistant (SR) in accordance with Table 22.3.8 are required to be substantial doors of construction that resists the passage of smoke. Latches and door closers are not required on cell doors.
(2) Under Use Condition II, Use Condition III, or Use Condition IV, a space subdivided by open construction (any combination of grating doors and grating walls or solid walls) is permitted to be considered one room if housing not more than 16 persons. The perimeter walls of such space are required to be of smoke-resistant construction. Smoke detection is required to be provided in such space. Under Use Condition IV, common walls between sleeping areas within the space are required to be smoke resistant, and grating doors and fronts are permitted to be used. Under Use Condition II and Use Condition III, open dormitories are permitted to house more than 16 persons, as permitted by other sections of this chapter.
(3) Where barriers are required to be smoke resistant (SR), the provisions of Sections 8.4 and 8.5 do not apply.
†Travel distance through the common space to the exit access corridor.
‡"Total openings in solid room face" include all openings (e.g, undercuts, food passes, grilles), the total of which is not to exceed 0.85 ft² (0.08 m²). All openings are required to be 36 in. (915 mm) or less above the floor.

**Table 22.4.4.2.1 Construction Type Limitations — Nonsprinklered Buildings**

| Construction Type | Sprinklered | Stories in Height [†] | | | | | |
| | | 1 With Basement | 1 Without Basement | 2 | 3 | >3 But Not High-Rise | High-Rise |
|---|---|---|---|---|---|---|---|
| I (442) | Yes | NA | NA | NA | NA | NA | NA |
| | No | X | X | X | X | X | NP |
| I (332) | Yes | NA | NA | NA | NA | NA | NA |
| | No | X | X | X | X | X | NP |
| II (222) | Yes | NA | NA | NA | NA | NA | NA |
| | No | X | X | X | X | X | NP |
| II (111) | Yes | NA | NA | NA | NA | NA | NA |
| | No | X1 | X | X1 | NP | NP | NP |
| II (000) | Yes | NA | NA | NA | NA | NA | NA |
| | No | NP | NP | NP | NP | NP | NP |
| III (211) | Yes | NA | NA | NA | NA | NA | NA |
| | No | X1 | X1 | X1 | NP | NP | NP |
| III (200) | Yes | NA | NA | NA | NA | NA | NA |
| | No | NP | NP | NP | NP | NP | NP |
| IV (2HH) | Yes | NA | NA | NA | NA | NA | NA |
| | No | X1 | X1 | X1 | NP | NP | NP |
| V (111) | Yes | NA | NA | NA | NA | NA | NA |
| | No | X1 | X1 | X1 | NP | NP | NP |
| V (000) | Yes | NA | NA | NA | NA | NA | NA |
| | No | NP | NP | NP | NP | NP | NP |

NA: Not applicable. NP: Not permitted.

X: Permitted for Use Conditions II, III, IV, and V. *(See 22.1.4.3 for Use Condition I.)*

X1: Permitted for Use Conditions II, III, and IV. Use Condition V not permitted. *(See 22.1.4.3 for Use Condition I.)*

[†]See 4.6.3.

**22.4.4.4 Common Path of Travel (Nonsprinklered Buildings).** A common path of travel shall not exceed 50 ft (15 m).

**22.4.4.5 Travel Distance to Exits (Nonsprinklered Buildings).**

**22.4.4.5.1** The travel distance between any room door required as an exit access and an exit shall not exceed 100 ft (30 m).

**22.4.4.5.2** The travel distance between any point in a room and an exit shall not exceed 150 ft (46 m).

**22.4.4.6 Protection of Vertical Openings (Nonsprinklered Buildings).**

**22.4.4.6.1** Multilevel residential housing areas without enclosure protection between levels shall be permitted, provided that the conditions of 22.4.4.6.2 through 22.4.4.6.4 are met.

**22.4.4.6.2\*** The entire normally occupied area, including all communicating floor levels, shall be sufficiently open and unobstructed so that a fire or other dangerous condition in any part is obvious to the occupants or supervisory personnel in the area.

**22.4.4.6.3** Egress capacity shall simultaneously accommodate all occupants of all communicating levels and areas, with all communicating levels in the same fire area considered as a single floor area for purposes of determining required egress capacity.

**22.4.4.6.4\*** The height between the highest and lowest finished floor levels shall not exceed 13 ft (3960 mm). The number of levels shall not be restricted.

**22.4.4.7 Hazardous Areas (Nonsprinklered Buildings).** Any hazardous area shall be protected in accordance with Section 8.7. The areas described in Table 22.4.4.7 shall be protected as indicated.

**22.4.4.8 Interior Finish (Nonsprinklered Buildings).**

**22.4.4.8.1 Interior Wall and Ceiling Finish.** Interior wall and ceiling finish materials complying with Section 10.2 shall be Class A in corridors, in exits, and in any space not separated from corridors and exits by partitions capable of retarding the passage of smoke; and Class A, Class B, or Class C in all other areas.

**Table 22.4.4.7 Hazardous Area Protection — Nonsprinklered Buildings**

| Hazardous Area Description | Separation/Protection[†] |
|---|---|
| Areas not incidental to resident housing | 2 hours |
| Boiler and fuel-fired heater rooms | 2 hours or 1 hour and sprinklers |
| Central or bulk laundries >100 ft² (>9.3 m²) | 2 hours or 1 hour and sprinklers |
| Commercial cooking equipment | In accordance with 9.2.3 |
| Commissaries | 1 hour or sprinklers |
| Employee locker rooms | 1 hour or sprinklers |
| Hobby/handicraft shops | 1 hour or sprinklers |
| Maintenance shops | 1 hour or sprinklers |
| Padded cells | 2 hours or 1 hour and sprinklers |
| Soiled linen rooms | 2 hours or 1 hour and sprinklers |
| Storage rooms >50 ft² (>4.6 m²) but ≤100 ft² (≤9.3 m²) storing combustible material | 1 hour or sprinklers |
| Storage rooms >100 ft² (>9.3 m²) storing combustible materials | 2 hours or 1 hour and sprinklers |
| Trash collection rooms | 2 hours or 1 hour and sprinklers |

[†]Minimum fire resistance rating.

**22.4.4.8.2 Interior Floor Finish.**

**22.4.4.8.2.1** Interior floor finish shall comply with Section 10.2.

**22.4.4.8.2.2** Interior floor finish in exit enclosures and exit access corridors shall be not less than Class I.

**22.4.4.8.2.3** Interior floor finish shall comply with 10.2.7.1 or 10.2.7.2, as applicable.

**22.4.4.9 Detection, Alarm, and Communications Systems (Nonsprinklered Buildings).**

**22.4.4.9.1 Initiation.** Initiation of the fire alarm system required by 22.3.4.1 shall be by manual means in accordance with 9.6.2 and by means of any required detection devices or detection systems, unless otherwise permitted by the following:

(1) Manual fire alarm boxes shall be permitted to be locked, provided that staff is present within the area when it is occupied and staff has keys readily available to unlock the boxes.
(2) Manual fire alarm boxes shall be permitted to be located in a staff location, provided that both of the following criteria are met:
  (a) The staff location is attended when the building is occupied.
  (b) The staff attendant has direct supervision of the sleeping area.

**22.4.4.9.2 Detection.** An approved automatic smoke detection system shall be in accordance with Section 9.6, as modified by 22.4.4.9.2.1 and 22.4.4.9.2.2, throughout all resident sleeping areas and adjacent day rooms, activity rooms, or contiguous common spaces.

**22.4.4.9.2.1** Smoke detectors shall not be required in sleeping rooms with four or fewer occupants in Use Condition II or Use Condition III.

**22.4.4.9.2.2** Other arrangements and positioning of smoke detectors shall be permitted to prevent damage or tampering, or for other purposes. Such arrangements shall be capable of detecting any fire, and the placement of detectors shall be such that the speed of detection is equivalent to that provided by the spacing and arrangements required by the installation standards referenced in Section 9.6. Detectors shall be permitted to be located in exhaust ducts from cells, behind grilles, or in other locations. The equivalent performance of the design, however, shall be acceptable to the authority having jurisdiction in accordance with the equivalency concepts specified in Section 1.4.

**22.4.4.10 Subdivision of Building Spaces (Nonsprinklered Buildings).** Where smoke barriers are required by 22.3.7.1, they shall be provided in accordance with both of the following criteria:

(1) They shall limit the occupant load to not more than 200 residents in any smoke compartment.
(2) They shall limit the travel distance to a door in a smoke barrier in accordance with both of the following criteria:
  (a) The distance from any room door required as exit access shall not exceed 100 ft (30 m).
  (b) The distance from any point in a room shall not exceed 150 ft (46 m).

**22.4.4.11\* Subdivision of Resident Housing Spaces (Nonsprinklered Buildings).** Subdivision of facility spaces shall comply with Table 22.4.4.11.

**22.4.4.12 Limited Access Structures (Nonsprinklered Buildings).**

**22.4.4.12.1** Limited access structures used as detention and correctional occupancies shall comply with 22.4.4.12.2. The provisions of Section 11.7 for limited access structures shall not apply.

**22.4.4.12.2** Any one of the following means shall be provided to evacuate smoke from the smoke compartment of fire origin:

(1) Operable windows on not less than two sides of the building, spaced not more than 30 ft (9.1 m) apart, that provide openings with dimensions of not less than 22 in. (560 mm) in width and 24 in. (610 mm) in height
(2)\*Manual or automatic smoke vents
(3) Engineered smoke control system
(4) Mechanical exhaust system providing not less than six air changes per hour
(5) Other method acceptable to the authority having jurisdiction

**22.4.4.13\* Furnishings, Mattresses, and Decorations (Nonsprinklered Buildings).**

**22.4.4.13.1** Newly introduced upholstered furniture within detention and correctional occupancies shall meet the criteria specified in 10.3.2.1(2) and 10.3.3.

**22.4.4.13.2\*** Newly introduced mattresses within detention and correctional occupancies shall meet the criteria specified in 10.3.2.2 and 10.3.4.

**22.4.5 Lockups.**

**22.4.5.1 General.**

**22.4.5.1.1** Lockups in occupancies, other than detention and correctional occupancies and health care occupancies, where

**Table 22.4.4.11 Subdivision of Resident Housing Spaces — Nonsprinklered Buildings**

| Feature | Use Condition | | | | |
| --- | --- | --- | --- | --- | --- |
| | II | III | | IV | V |
| Room to room separation | NR | NR | | SR | FR(½) |
| Room face to corridor separation | SR | SR | | SR | FR |
| Room face to common space separation | NR | NR ≤50 ft (≤15 m)† | SR >50 ft (>15 m)† | SR | FR |
| Common space to corridor separation | FR | FR | | FR | FR |
| Total openings in solid room face where room face is required to be smoke resistant or fire rated‡ | 0.85 ft² (0.08 m²) | 0.85 ft² (0.08 m²) | | 0.85 ft² (0.08 m²) | 0.85 ft² (0.08 m²) where meeting one of the following: (1) Kept in closed position, except when in use by staff (2) Closable from the inside (3) Provided with smoke control |

NR: No requirement. SR: Smoke resistant. FR(½): Minimum ½-hour fire resistance rating. FR: Minimum 1-hour fire resistance rating.

Notes:

(1) Doors in openings in partitions required to be fire rated [FR(½), FR] in accordance with Table 22.4.4.11, in other than required enclosures of exits or hazardous areas, are required to be substantial doors of construction that resists fire for a minimum of 20 minutes. Vision panels with wired glass or glass with not less than 45-minute fire-rated glazing are permitted. Latches and door closers are not required on cell doors.

(2) Doors in openings in partitions required to be smoke resistant (SR) in accordance with Table 22.4.4.11 are required to be substantial doors of construction that resists the passage of smoke. Latches and door closers are not required on cell doors.

(3) Under Use Condition II, Use Condition III, or Use Condition IV, a space subdivided by open construction (any combination of grating doors and grating walls or solid walls) is permitted to be considered one room if housing not more than 16 persons. The perimeter walls of such space are required to be of smoke-resistant construction. Smoke detection is required to be provided in such space. Under Use Condition IV, common walls between sleeping areas within the space are required to be smoke resistant, and grating doors and fronts are permitted to be used. In Use Condition II and Use Condition III, open dormitories are permitted to house more than 16 persons, as permitted by other sections of this chapter.

(4) Where barriers are required to be smoke resistant (SR), the provisions of Sections 8.4 and 8.5 do not apply.

†Travel distance through the common space to the exit access corridor.

‡"Total openings in solid room face" include all openings (e.g, undercuts, food passes, grilles), the total of which is not to exceed 0.85 ft² (0.08 m²). All openings are required to be 36 in. (915 mm) or less above the floor.

the holding area has capacity for more than 50 detainees shall be classified as detention and correctional occupancies and shall comply with the requirements of Chapter 22.

**22.4.5.1.2** Lockups in occupancies, other than detention and correctional occupancies and health care occupancies, where any individual is detained for 24 or more hours shall be classified as detention and correctional occupancies and shall comply with the requirements of Chapter 22.

**22.4.5.1.3** Lockups in occupancies, other than detention and correctional occupancies and health care occupancies, where the holding area has capacity for not more than 50 detainees, and where no individual is detained for 24 hours or more, shall comply with 22.4.5.1.4 or 22.4.5.1.5.

**22.4.5.1.4** The lockup shall be permitted to comply with the requirements for the predominant occupancy in which the lockup is placed, provided that all of the following criteria are met:

(1)*Doors and other physical restraints to free egress by detainees can be readily released by staff within 2 minutes of the onset of a fire or similar emergency.

(2) Staff is in sufficient proximity to the lockup so as to be able to effect the 2-minute release required by 22.4.5.1.4(1) whenever detainees occupy the lockup.

(3) Staff is authorized to effect the release required by 22.4.5.1.4(1).

(4) Staff is trained and practiced in effecting the release required by 22.4.5.1.4(1).

(5) Where the release required by 22.4.5.1.4(1) is effected by means of remote release, detainees are not to be restrained from evacuating without the assistance of others.

**22.4.5.1.5** Where the lockup does not comply with all the criteria of 22.4.5.1.4, the requirements of 22.4.5.2 shall be met.

**22.4.5.1.6** The fire department with responsibility for responding to a building that contains a lockup shall be notified of the presence of the lockup.

**22.4.5.2 Alternate Provisions.**

**22.4.5.2.1** The requirements applicable to the predominant occupancy in which the lockup is placed shall be met.

**22.4.5.2.2** Where security operations necessitate the locking of required means of egress, the following shall apply:

(1) Detention-grade hardware meeting the requirements of ASTM F 1577, *Standard Test Methods for Detention Locks for Swinging Doors*, shall be provided on swinging doors within the required means of egress.
(2) Sliding doors within the required means of egress shall be designed and engineered for detention and correctional use, and lock cylinders shall meet the cylinder test requirements of ASTM F 1577.

**22.4.5.2.3** The lockup shall be provided with a complete smoke detection system in accordance with 9.6.2.9.

**22.4.5.2.4** Where the requirements applicable to the predominant occupancy do not mandate a fire alarm system, the lockup shall be provided with a fire alarm system meeting all of the following criteria:

(1) The alarm system shall be in accordance with Section 9.6.
(2) Initiation of the alarm system shall be accomplished by all of the following:
    (a) Manual fire alarm boxes in accordance with 9.6.2
    (b) Smoke detection system required by 22.4.5.2.3
    (c) Automatic sprinkler system required by the provisions applicable to the predominant occupancy
(3) Staff and occupant notification shall be provided automatically in accordance with 9.6.3.
(4) Emergency force notification shall be provided in accordance with 9.6.4.

**22.5 Building Services.**

**22.5.1 Utilities.**

**22.5.1.1** Utilities shall comply with the provisions of Section 9.1.

**22.5.1.2** Alarms, emergency communications systems, and the illumination of generator set locations shall be provided with emergency power in accordance with *NFPA 70, National Electrical Code*.

**22.5.2 Heating, Ventilating, and Air-Conditioning.**

**22.5.2.1** Heating, ventilating, and air-conditioning equipment shall comply with the provisions of Section 9.2 and shall be installed in accordance with the manufacturer's specifications, unless otherwise modified by 22.5.2.2.

**22.5.2.2** Portable space-heating devices shall be prohibited, unless otherwise permitted by 22.5.2.4.

**22.5.2.3** Any heating device, other than a central heating plant, shall be designed and installed so that combustible material cannot be ignited by the device or its appurtenances, and both of the following requirements also shall apply:

(1) If fuel-fired, such heating devices shall comply with all of the following:

(a) They shall be chimney connected or vent connected.
(b) They shall take air for combustion directly from outside.
(c) They shall be designed and installed to provide for complete separation of the combustion system from the atmosphere of the occupied area.

(2) The heating system shall have safety devices to immediately stop the flow of fuel and shut down the equipment in case of either excessive temperatures or ignition failure.

**22.5.2.4** Approved, suspended unit heaters shall be permitted in locations other than means of egress and sleeping areas, provided that both of the following criteria are met:

(1) Such heaters are located high enough to be out of the reach of persons using the area.
(2) Such heaters are vent connected and equipped with the safety devices required by 22.5.2.3(2).

**22.5.2.5** Combustion and ventilation air for boiler, incinerator, or heater rooms shall be taken directly from, and discharged directly to, the outside.

**22.5.3 Elevators, Escalators, and Conveyors.** Elevators, escalators, and conveyors shall comply with the provisions of Section 9.4.

**22.5.4 Rubbish Chutes, Incinerators, and Laundry Chutes.**

**22.5.4.1** Rubbish chutes, incinerators, and laundry chutes shall comply with the provisions of Section 9.5.

**22.5.4.2** Rubbish chutes and linen chutes, including pneumatic rubbish and linen systems, shall be provided with automatic extinguishing protection in accordance with Section 9.7.

**22.5.4.3** Trash chutes shall discharge into a trash collection room used for no other purpose and protected in accordance with Section 8.7.

**22.5.4.4** Incinerators shall not be directly flue-fed, and floor chutes shall not directly connect with the combustion chamber.

**22.6 Reserved.**

**22.7 Operating Features.**

**22.7.1 Attendants, Evacuation Plan, and Fire Drills.**

**22.7.1.1** Detention and correctional facilities, or those portions of facilities having such occupancy, shall be provided with 24-hour staffing, and the following requirements also shall apply:

(1) Staff shall be within three floors or a 300 ft (91 m) horizontal distance of the access door of each resident housing area.
(2) For Use Condition III, Use Condition IV, and Use Condition V, the arrangement shall be such that the staff involved starts the release of locks necessary for emergency evacuation or rescue and initiates other necessary emergency actions within 2 minutes of alarm.
(3) The following shall apply to areas in which all locks are unlocked remotely in compliance with 22.2.11.8:
    (a) Staff shall not be required to be within three floors or 300 ft (91 m) of the access door.
    (b) The 10-lock, manual key exemption of 22.2.11.8.2 shall not be permitted to be used in conjunction with the alternative requirement of 22.7.1.1(3)(a).

**22.7.1.2\*** Provisions shall be made so that residents in Use Condition III, Use Condition IV, and Use Condition V shall be able to notify staff of an emergency.

**22.7.1.3\*** The administration of every detention or correctional facility shall have, in effect and available to all supervisory personnel, written copies of a plan for the protection of all persons in the event of fire, for their evacuation to areas of refuge, and for evacuation from the building when necessary.

**22.7.1.3.1** All employees shall be instructed and drilled with respect to their duties under the plan.

**22.7.1.3.2** The plan shall be coordinated with, and reviewed by, the fire department legally committed to serve the facility.

**22.7.1.4** Employees of detention and correctional occupancies shall be instructed in the proper use of portable fire extinguishers and other manual fire suppression equipment.

**22.7.1.4.1** The training specified in 22.7.1.4 shall be provided to new staff promptly upon commencement of duty.

**22.7.1.4.2** Refresher training shall be provided to existing staff at not less than annual intervals.

**22.7.2 Combustible Personal Property.** Books, clothing, and other combustible personal property allowed in sleeping rooms shall be stored in closable metal lockers or an approved fire-resistant container.

**22.7.3 Heat-Producing Appliances.** The number of heat-producing appliances, such as toasters and hot plates, and the overall use of electrical power within a sleeping room shall be controlled by facility administration.

**22.7.4\* Furnishings, Mattresses, and Decorations.**

**22.7.4.1** Draperies and curtains, including privacy curtains, in detention and correctional occupancies shall be in accordance with the provisions of 10.3.1.

**22.7.4.2** Newly introduced upholstered furniture within detention and correctional occupancies shall be tested in accordance with the provisions of 10.3.2.1(2).

**22.7.4.3** Newly introduced mattresses within detention and correctional occupancies shall be tested in accordance with the provisions of 10.3.2.2.

**22.7.4.4** Combustible decorations shall be prohibited in any detention or correctional occupancy unless flame-retardant.

**22.7.4.5** Wastebaskets and other waste containers shall be of noncombustible or other approved materials. Waste containers with a capacity exceeding 20 gal (76 L) shall be provided with a noncombustible lid or lid of other approved material.

**22.7.5 Keys.** All keys necessary for unlocking doors installed in a means of egress shall be individually identified by both touch and sight.

**22.7.6 Portable Space-Heating Devices.** Portable space-heating devices shall be prohibited in all detention and correctional occupancies.

**22.7.7 Door Inspection.** Doors and door hardware in means of egress shall be inspected monthly by an appropriately trained person. The inspection shall be documented.

## Chapter 23   Existing Detention and Correctional Occupancies

**23.1 General Requirements.**

**23.1.1 Application.**

**23.1.1.1 General.**

**23.1.1.1.1** The requirements of this chapter shall apply to existing buildings or portions thereof currently occupied as detention or correctional occupancies.

**23.1.1.1.2 Administration.** The provisions of Chapter 1, Administration, shall apply.

**23.1.1.1.3 General.** The provisions of Chapter 4, General, shall apply.

**23.1.1.1.4** This chapter establishes life safety requirements that shall apply to all existing detention and correctional facilities, other than the following:

(1) Use Condition I facilities protected as residential occupancies in accordance with 23.1.2.3
(2)\*Facilities determined to have equivalent safety provided in accordance with Section 1.4

**23.1.1.1.5** Detention and correctional occupancies shall include those used for purposes such as correctional institutions, detention facilities, community residential centers, training schools, work camps, and substance abuse centers where occupants are confined or housed under some degree of restraint or security.

**23.1.1.1.6\*** Detention and correctional occupancies shall include those that provide sleeping facilities for one or more residents and are occupied by persons who are generally prevented from taking self-preservation action because of security measures not under the occupants' control.

**23.1.1.1.7\*** Lockups, other than approved existing lockups, in other than detention and correctional occupancies and health care occupancies shall comply with the requirements of 23.4.5.

**23.1.1.2 Total Concept.**

**23.1.1.2.1** All detention and correctional facilities shall be designed, constructed, maintained, and operated to minimize the possibility of a fire emergency.

**23.1.1.2.2** Because the safety of all occupants in detention and correctional facilities cannot be adequately ensured solely by dependence on evacuation of the building, their protection from fire shall be provided by appropriate arrangement of facilities; adequate, trained staff; and development of operating, security, and maintenance procedures composed of the following:

(1) Design, construction, and compartmentation
(2) Provision for detection, alarm, and extinguishment
(3) Fire prevention and planning, training, and drilling programs for the isolation of fire and the transfer of occupants to areas of refuge, for evacuation of the building, or for protection of the occupants in place
(4) Provision of security to the degree necessary for the safety of the public and the occupants of the facility

**23.1.1.3 Additions.** Additions shall be separated from any existing structure not conforming with the provisions of this chapter by a fire barrier having not less than a 2-hour fire resistance rating constructed to the requirements of the addition, and the following also shall apply:

(1) Doors in such partitions shall normally be kept closed.
(2) Doors shall be permitted to be held open if they meet the requirements of 7.2.1.8.2.

**23.1.1.4 Modernizations or Renovations.**

**23.1.1.4.1** Modernizations and renovations shall be in accordance with 4.6.7, unless otherwise permitted by 23.1.1.4.2.

**23.1.1.4.2** In nonsprinklered existing buildings, modernizations or renovations shall be permitted to comply with the nonsprinklered options contained in 22.4.4 in lieu of the sprinkler requirement of 22.3.5.2.

**23.1.2 Classification of Occupancy.** See 6.1.7.

**23.1.2.1\*** For application of the life safety requirements that follow, the resident user category shall be divided into the groups specified in 23.1.2.1.1 through 23.1.2.1.5.

**23.1.2.1.1 Use Condition I — Free Egress.** Use Condition I shall be defined as a condition under which free movement is allowed from sleeping areas and other spaces where access or occupancy is permitted to the exterior via means of egress meeting the requirements of this *Code*.

**23.1.2.1.2 Use Condition II — Zoned Egress.** Use Condition II shall be defined as a condition under which free movement is allowed from sleeping areas and any other occupied smoke compartment to one or more other smoke compartments.

**23.1.2.1.3 Use Condition III — Zoned Impeded Egress.** Use Condition III shall be defined as a condition under which free movement is allowed within individual smoke compartments, such as within a residential unit comprised of individual sleeping rooms and a group activity space, with egress impeded by remote-controlled release of means of egress from such a smoke compartment to another smoke compartment.

**23.1.2.1.4 Use Condition IV — Impeded Egress.** Use Condition IV shall be defined as a condition under which free movement is restricted from an occupied space, and remote-controlled release is provided to allow movement from all sleeping rooms, activity spaces, and other occupied areas within the smoke compartment to another smoke compartment.

**23.1.2.1.5 Use Condition V — Contained.** Use Condition V shall be defined as a condition under which free movement is restricted from an occupied space, and staff-controlled manual release at each door is provided to allow movement from all sleeping rooms, activity spaces, and other occupied areas within the smoke compartment to another smoke compartment.

**23.1.2.2\*** To be classified as Use Condition III or Use Condition IV, the arrangement, accessibility, and security of the release mechanism(s) used for emergency egress shall be such that the minimum available staff, at any time, can promptly release the locks.

**23.1.2.3** Areas housing occupancies corresponding to Use Condition I shall conform to one of the following:

(1) Requirements of residential occupancies under this *Code*
(2)\*Requirements of this chapter for Use Condition II facilities, provided that the staffing requirements of Section 23.7 are met

**23.1.3\* Multiple Occupancies.**

**23.1.3.1** Multiple occupancies shall be in accordance with 6.1.14.

**23.1.3.2** Egress provisions for areas of detention and correctional facilities that correspond to other occupancies shall meet the corresponding requirements of this *Code* for such occupancies as modified by 23.1.3.2.1.

**23.1.3.2.1\*** Where security operations necessitate the locking of required means of egress, staff in the building shall be provided with the means for the supervised release of occupants during all times of use.

**23.1.3.2.2 Reserved.**

**23.1.3.3** Sections of detention and correctional facilities shall be permitted to be classified as other occupancies, provided that they meet both of the following conditions:

(1) They are not intended to serve residents for sleeping purposes.
(2) They are separated from areas of detention or correctional occupancies by construction having not less than a 2-hour fire resistance rating.

**23.1.3.4** All means of egress from detention and correctional occupancies that traverse other use areas shall, as a minimum, conform to the requirements of this *Code* for detention and correctional occupancies, unless otherwise permitted by 23.1.3.5.

**23.1.3.5** Egress through a horizontal exit into other contiguous occupancies that do not conform to detention and correctional occupancy egress provisions but that do comply with requirements set forth in the appropriate occupancy chapter of this *Code* shall be permitted, provided that both of the following criteria apply:

(1) The occupancy shall not contain high hazard contents.
(2) The horizontal exit shall comply with the requirements of 23.2.2.5.

**23.1.3.6** Any area with a hazard of contents classified higher than that of the detention or correctional occupancy and located in the same building shall be protected as required in 23.3.2.

**23.1.3.7** Nondetention- or noncorrectional-related occupancies classified as containing high hazard contents shall not be permitted in buildings housing detention or correctional occupancies.

**23.1.4 Definitions.**

**23.1.4.1 General.** For definitions, see Chapter 3, Definitions.

**23.1.4.2 Special Definitions.** A list of special terms used in this chapter follows:

(1) **Detention and Correctional Residential Housing Area.** See 3.3.21.1.
(2) **Sally Port (Security Vestibule).** See 3.3.236.

**23.1.5 Classification of Hazard of Contents.** The classification of hazard of contents shall be as defined in Section 6.2.

**23.1.6 Minimum Construction Requirements.**

**23.1.6.1** Detention and correctional occupancies shall be limited to the building construction types specified in Table 23.1.6.1. *(See 8.2.1.)*

**23.1.6.2** A residential housing area complying with 23.3.1.2 shall be considered as one story in height for purposes of applying 23.1.6.1.

**23.1.7 Occupant Load.** The occupant load, in number of persons for whom means of egress and other provisions are required, either shall be determined on the basis of the occupant load factors of Table 7.3.1.2 that are characteristic of the use of the space or shall be determined as the maximum probable population of the space under consideration, whichever is greater.

**23.2 Means of Egress Requirements.**

**23.2.1 General.** Means of egress shall comply with Chapter 7, unless otherwise provided or modified by Section 23.2.

**Table 23.1.6.1 Construction Type Limitations**

| Construction Type | Sprinklered[a] | Stories in Height[b] | | | | | |
|---|---|---|---|---|---|---|---|
| | | 1 With Basement | 1 Without Basement | 2 | 3 | >3 But Not High-Rise | High-Rise |
| I (442)[c, d] | Yes | X | X | X | X | X | X |
| | No | X | X | X | X | X | NP |
| I (332)[c, d] | Yes | X | X | X | X | X | X |
| | No | X | X | X | X | X | NP |
| II (222)[c, d] | Yes | X | X | X | X | X | X |
| | No | X | X | X | X | X | NP |
| II (111)[c, d] | Yes | X | X | X | X | X | X |
| | No | X1 | X | X1 | NP | NP | NP |
| II (000)[d] | Yes | X | X | X | X | X | X |
| | No | X1 | X1 | NP | NP | NP | NP |
| III (211)[d] | Yes | X | X | X | X | X | X |
| | No | X1 | X | X1 | NP | NP | NP |
| III (200)[d] | Yes | X | X | X | X | X | X |
| | No | X1 | X1 | NP | NP | NP | NP |
| IV (2HH)[d] | Yes | X | X | X | X | X | X |
| | No | X1 | X | X1 | NP | NP | NP |
| V (111)[d] | Yes | X | X | X | X | X | X |
| | No | X1 | X | X1 | NP | NP | NP |
| V (000)[d] | Yes | X | X | X | X | X | X |
| | No | X1 | X1 | NP | NP | NP | NP |

NP: Not permitted.

X: Permitted for Use Conditions II, III, IV, and V. *(See 23.1.4.3 for Use Condition I.)*

X1: Permitted for Use Conditions II, III, and IV. Use Condition V not permitted. *(See 23.1.4.3 for Use Condition I.)*

[a]Entire building is protected throughout by an approved, supervised automatic sprinkler system in accordance with 9.7.1.1(1). *(See 23.3.5.)*

[b]See 4.6.3.

[c]Any building of Type I, Type II(222), or Type II(111) construction is permitted to include roofing systems involving combustible or steel supports, decking, or roofing, provided that all of the following are met:

  (1) The roof covering meets not less than Class C requirements in accordance with ASTM E 108, *Standard Test Methods for Fire Tests of Roof Coverings*, or ANSI/UL 790, *Test Methods for Fire Tests of Roof Coverings*.

  (2) The roof is separated from all occupied portions of the building by a noncombustible floor assembly that includes not less than 2½ in. (64 mm) of concrete or gypsum fill, and the attic or other space so developed meets one of the following requirements:

    (a) It is unoccupied.

    (b) It is protected throughout by an approved automatic sprinkler system.

[d]In determining building construction type, exposed steel roof members located 16 ft (4875 mm) or more above the floor of the highest cell are permitted to be disregarded.

**23.2.2 Means of Egress Components.**

**23.2.2.1 Components Permitted.** Components of means of egress shall be limited to the types described in 23.2.2.2 through 23.2.2.11.

**23.2.2.2 Doors.** Doors complying with 7.2.1 shall be permitted, unless otherwise provided in 23.2.11.

**23.2.2.3 Stairs.**

**23.2.2.3.1** Stairs shall be permitted as follows:

(1) Stairs complying with 7.2.2 shall be permitted.

(2) Noncombustible grated stair treads and landing floors shall be permitted.

**23.2.2.3.2** Spiral stairs complying with 7.2.2.2.3 shall be permitted for access to and between staff locations.

**23.2.2.4 Smokeproof Enclosures.** Smokeproof enclosures complying with 7.2.3 shall be permitted.

**23.2.2.5 Horizontal Exits.** Horizontal exits complying with 7.2.4 and the modifications of 23.2.2.5.1 through 23.2.2.5.4 shall be permitted.

**23.2.2.5.1** Not less than 6 ft² (0.55 m²) of accessible space per occupant shall be provided on each side of the horizontal exit for the total number of people in adjoining compartments.

**23.2.2.5.2\*** Horizontal exits shall be permitted to comprise 100 percent of the exits required, provided that an exit, other than a horizontal exit, located in another (not necessarily adjacent) fire compartment is accessible without returning through the compartment of fire origin.

**23.2.2.5.3\*** Ducts shall be permitted to penetrate horizontal exits in accordance with 7.2.4.3.5(3) if protected by combination fire dampers/smoke leakage–rated dampers that meet the smoke damper actuation requirements of 8.5.5.

**23.2.2.5.4** A door in a horizontal exit shall not be required to swing with egress travel as specified in 7.2.4.3.8(1).

**23.2.2.6 Ramps.** Ramps complying with 7.2.5 shall be permitted.

**23.2.2.7 Exit Passageways.** Exit passageways complying with 7.2.6 shall be permitted.

**23.2.2.8 Fire Escape Stairs.** Fire escape stairs complying with 7.2.8 shall be permitted.

**23.2.2.9 Fire Escape Ladders.** Fire escape ladders complying with 7.2.9 shall be permitted.

**23.2.2.10 Alternating Tread Devices.** Alternating tread devices complying with 7.2.11 shall be permitted.

**23.2.2.11 Areas of Refuge.** Areas of refuge complying with 7.2.12 shall be permitted.

**23.2.3 Capacity of Means of Egress.**

**23.2.3.1** The capacity of any required means of egress shall be in accordance with Section 7.3.

**23.2.3.2** Aisles, corridors, and ramps required for egress shall be not less than 36 in. (915 mm) in width.

**23.2.3.3** Residents' sleeping room door widths shall be permitted to comply with 23.2.11.4.

**23.2.4 Number of Means of Egress.**

**23.2.4.1** The number of means of egress shall be in accordance with 7.4.1.1 and 7.4.1.3 through 7.4.1.6.

**23.2.4.2\*** Not less than two separate exits shall meet both of the following criteria:

(1) They shall be provided on every story.
(2) They shall be accessible from every part of every story, fire compartment, or smoke compartment; however, exit access travel shall be permitted to be common for the distances permitted as common path of travel by 23.2.5.3.

**23.2.4.3\*** Not less than one approved exit shall be accessible from each fire compartment and each required smoke compartment into which residents are potentially moved in a fire emergency, with the exits arranged so that egress is possible without returning through the zone of fire origin.

**23.2.5 Arrangement of Means of Egress.** See also Section 7.5.

**23.2.5.1** Every sleeping room shall have a door leading directly to an exit access corridor, unless otherwise permitted by one of the following:

(1) The requirement of 23.2.5.1 shall not apply if there is an exit door opening directly to the outside from a room at the finished ground level.
(2) One adjacent room, such as a day room, a group activity space, or other common space, shall be permitted to intervene, and the following also shall apply:
 (a) Where sleeping rooms directly adjoin a day room or group activity space that is used for access to an exit, such sleeping rooms shall be permitted to open directly to the day room or space.
 (b) Sleeping rooms permitted to open directly to the day room or space shall be permitted to be separated in elevation by a one-half story or full story height.

**23.2.5.2\*** Existing dead-end corridors are undesirable and shall be altered wherever possible so that exits are accessible in not less than two different directions from all points in aisles, passageways, and corridors.

**23.2.5.3** A common path of travel shall not exceed 50 ft (15 m), unless otherwise permitted by one of the following:

(1) A common path of travel shall be permitted for the first 100 ft (30 m) in smoke compartments protected throughout by an approved automatic sprinkler system in accordance with 23.3.5.3.
(2) A common path of travel shall be permitted to exceed 50 ft (15 m) in multilevel residential housing units in which each floor level, considered separately, has not less than one-half of its individual required egress capacity accessible by exit access leading directly out of that level without traversing another communicating floor level.
(3)\*Approved existing common paths of travel that exceed 50 ft (15 m) shall be permitted to continue to be used.

**23.2.5.4** A sally port shall be permitted in a means of egress where there are provisions for continuous and unobstructed travel through the sally port during an emergency egress condition.

**23.2.6 Travel Distance to Exits.** Travel distance shall comply with 23.2.6.1 through 23.2.6.7.

**23.2.6.1** Travel distance shall be measured in accordance with Section 7.6.

**23.2.6.2** The travel distance between any room door required as an exit access and an exit or smoke barrier shall not exceed 100 ft (30 m), unless otherwise permitted by 23.2.6.3.

**23.2.6.3** The maximum travel distance limitations of 23.2.6.2 shall be permitted to be increased by 50 ft (15 m) in buildings protected throughout by an approved automatic sprinkler system in accordance with 23.3.5.3 or a smoke control system.

**23.2.6.4** The travel distance between any point in a room and an exit or smoke barrier shall not exceed 150 ft (46 m), unless otherwise permitted by 23.2.6.5.

**23.2.6.5** The maximum travel distance limitations of 23.2.6.4 shall be permitted to be increased by 50 ft (15 m) in buildings protected throughout by an approved automatic sprinkler system in accordance with 23.3.5.3 or a smoke control system.

**23.2.6.6** The travel distance between any point in a sleeping room to the door of that room shall not exceed 50 ft (15 m), unless otherwise permitted by 23.2.6.7.

**23.2.6.7** The maximum travel distance limitations of 23.2.6.6 shall be permitted to be increased to 100 ft (30 m) in open dormitories, provided that both of the following criteria are met:

(1) The enclosing walls of the dormitory space shall be of smoke-tight construction.
(2) Not less than two exit access doors remotely located from each other shall be provided where travel distance to the exit access door from any point within the dormitory exceeds 50 ft (15 m).

### 23.2.7 Discharge from Exits.

**23.2.7.1** Exits shall be permitted to discharge into a fenced or walled courtyard, provided that not more than two walls of the courtyard are the building walls from which egress is being made.

**23.2.7.2** Enclosed yards or courts used for exit discharge in accordance with 23.2.7.1 shall be of sufficient size to accommodate all occupants at a distance of not less than 50 ft (15 m) from the building while providing a net area of 15 ft$^2$ (1.4 m$^2$) per person.

**23.2.7.3** All exits shall be permitted to discharge through the level of exit discharge.

**23.2.7.4** The requirements of 7.7.2 shall be waived, provided that not more than 50 percent of the exits discharge into a single fire compartment separated from other compartments by construction having not less than a 1-hour fire resistance rating.

**23.2.7.5** Where all exits are permitted to discharge through areas on the level of discharge, all of the following criteria shall be met:

(1) A smoke barrier shall be provided to divide that level into not less than two compartments, with not less than one exit discharging into each compartment.
(2) Each smoke compartment shall have an exit discharge to the building exterior.
(3) The level of discharge shall be provided with automatic sprinkler protection.
(4) Any other portion of the level of discharge with access to the discharge area shall be provided with automatic sprinkler protection or shall be separated from the discharge area in accordance with the requirements for the enclosure of exits. *(See 7.1.3.2.1.)*

**23.2.8 Illumination of Means of Egress.** Means of egress shall be illuminated in accordance with Section 7.8.

### 23.2.9 Emergency Lighting.

**23.2.9.1** Emergency lighting shall be provided in accordance with Section 7.9, unless otherwise permitted by 23.2.9.2.

**23.2.9.2** Emergency lighting of not less than a 1-hour duration shall be permitted to be provided.

**23.2.10 Marking of Means of Egress.** Exit marking shall be provided as follows:

(1) Exit signs shall be provided in areas accessible to the public in accordance with Section 7.10.
(2) Exit signs shall not be required in detention and correctional residential housing areas. *(See 3.3.21.1.)*

### 23.2.11 Special Features.

**23.2.11.1** Doors within means of egress shall be in accordance with Chapter 7, unless otherwise provided in 23.2.11.2 through 23.2.11.10.

**23.2.11.2** Doors shall be permitted to be locked in accordance with the applicable use condition.

**23.2.11.3** Where egress doors are locked with key-operated locks, the provisions of 23.7.7 shall apply.

**23.2.11.4\*** Doors to resident sleeping rooms shall be not less than 28 in. (710 mm) in clear width.

**23.2.11.5** Existing doors to resident sleeping rooms housing four or fewer residents shall be permitted to be not less than 19 in. (485 mm) in clear width.

**23.2.11.6** Doors in a means of egress shall be permitted to be of the horizontal-sliding type, provided that the force necessary to slide the door to its fully open position does not exceed 50 lbf (222 N) where a force of 50 lbf (222 N) is simultaneously applied perpendicular to the door.

**23.2.11.7** Doors from areas of refuge to the exterior shall be permitted to be locked with key locks in lieu of locking methods described in 23.2.11.8, the keys to unlock such doors shall be maintained and available at the facility at all times, and the locks shall be operable from the outside.

**23.2.11.8\*** Any remote-control release used in a means of egress shall be provided with a reliable means of operation to release locks on all doors and shall be remotely located from the resident living area, unless otherwise permitted by 23.2.11.8.2.

**23.2.11.8.1** The remote location of a remote-control release used in a means of egress shall provide sight and sound supervision of the resident living areas.

**23.2.11.8.2** Remote-control locking and unlocking of occupied rooms in Use Condition IV shall not be required, provided that both of the following criteria are met:

(1) Not more than 10 locks need to be unlocked to relocate all occupants from one smoke compartment to an area of refuge as promptly as is required where remote-control unlocking is used. *(See 23.3.7.9 for requirements for smoke barrier doors.)*
(2) Unlocking of all necessary locks is accomplished with not more than two separate keys.

**23.2.11.9 Remote-Control Release–Operated Doors.**

**23.2.11.9.1** All remote-control release–operated doors shall be provided with a redundant means of operation as follows:

(1) Power-operated sliding doors or power-operated locks shall be constructed so that, in the event of power failure, a manual mechanical means to release and open the doors is provided at each door, and either emergency power arranged in accordance with 23.2.11.9.2 is provided for the power operation or a remote-control manual mechanical release is provided.
(2) A combination of the emergency power–operated release of selected individual doors and remote-control manual mechanical ganged release specified in 23.2.11.9(1) shall be permitted without mechanical release means at each door.
(3) Mechanically operated sliding doors or mechanically operated locks shall be provided with a manual mechanical means at each door to release and open the door.

**23.2.11.9.2** The emergency power required by 23.2.11.9.1(1) shall be arranged to provide the required power automatically in the event of any interruption of normal power due to any of the following:

(1) Failure of a public utility or other outside electrical power supply
(2) Opening of a circuit breaker or fuse
(3) Manual act(s), including accidental opening of a switch controlling normal lighting facilities

**23.2.11.10** The provisions of 7.2.1.5.8 for stairway re-entry shall not apply.

### 23.3 Protection.

#### 23.3.1 Protection of Vertical Openings.

**23.3.1.1** Any vertical opening shall be enclosed or protected in accordance with Section 8.6, unless otherwise permitted by one of the following:

(1) Unprotected vertical openings in accordance with 8.6.9.1 shall be permitted.
(2) In residential housing area smoke compartments protected throughout by an approved automatic sprinkler system in accordance with 23.3.5.3, unprotected vertical openings shall be permitted in accordance with the conditions of 8.6.6, provided that the height between the lowest and highest finished floor levels does not exceed 23 ft (7010 mm), and the following also shall be permitted:
    (a) The number of levels shall not be restricted.
    (b) Residential housing areas subdivided in accordance with 23.3.8 shall be permitted to be considered as part of the communicating space.
    (c) The separation shall not be required to have a fire resistance rating. *[See 8.6.6(4)(b).]*
(3) The requirement of 23.3.1.1 shall not apply to multilevel residential housing areas in accordance with 23.3.1.2.
(4) Where full enclosure is impractical, the required enclosure shall be permitted to be limited to that necessary to prevent a fire originating in any story from spreading to any other story.
(5) Enclosures in detention and correctional occupancies shall have a minimum 1-hour fire resistance rating and shall be protected throughout by an approved automatic sprinkler system in accordance with 23.3.5.3.

**23.3.1.2** Multilevel residential housing areas without enclosure protection between levels shall be permitted, provided that the conditions of 23.3.1.2.1 through 23.3.1.2.3 are met.

**23.3.1.2.1\*** The entire normally occupied area, including all communicating floor levels, shall be sufficiently open and unobstructed so that a fire or other dangerous condition in any part is obvious to the occupants or supervisory personnel in the area.

**23.3.1.2.2** Egress capacity shall simultaneously accommodate all occupants of all communicating levels and areas, with all communicating levels in the same fire area considered as a single floor area for purposes of determining required egress capacity.

**23.3.1.2.3\*** The height between the highest and lowest finished floor levels shall not exceed 13 ft (3960 mm). The number of levels shall not be restricted.

**23.3.1.3\*** A multitiered, open cell block shall be considered as a one-story building where one of the following criteria is met:

(1) A smoke control system is provided to maintain the level of smoke from potential cell fires at not less than 60 in. (1525 mm) above the floor level of any occupied tier involving space that is classified as follows:
    (a) Use Condition IV or Use Condition V
    (b) Use Condition III, unless all persons housed in such space can pass through a free access smoke barrier or freely pass below the calculated smoke level with not more than 50 ft (15 m) of travel from their cells
(2) The entire building, including cells, is provided with complete automatic sprinkler protection in accordance with 23.3.5.3.

#### 23.3.2 Protection from Hazards.

**23.3.2.1\*** Any hazardous area shall be protected in accordance with Section 8.7. The areas described in Table 23.3.2.1 shall be protected as indicated.

**Table 23.3.2.1 Hazardous Area Protection**

| Hazardous Area Description | Separation/Protection[†] |
|---|---|
| Areas not incidental to resident housing | 2 hours |
| Boiler and fuel-fired heater rooms | 1 hour or sprinklers |
| Central or bulk laundries >100 ft² (>9.3 m²) | 1 hour or sprinklers |
| Commercial cooking equipment | In accordance with 9.2.3 |
| Commissaries | 1 hour or sprinklers |
| Employee locker rooms | 1 hour or sprinklers |
| Hobby/handicraft shops | 1 hour or sprinklers |
| Maintenance shops | 1 hour or sprinklers |
| Padded cells | 1 hour and sprinklers |
| Soiled linen rooms | 1 hour or sprinklers |
| Storage rooms >50 ft² (>4.6 m²) storing combustible material | 1 hour or sprinklers |
| Trash collection rooms | 1 hour or sprinklers |

[†]Minimum fire resistance rating.

#### 23.3.2.2 Reserved.

**23.3.2.3** Hazardous areas determined by the authority having jurisdiction as not incidental to residents' housing shall be separated by 2-hour fire resistance–rated barriers in conjunction with automatic sprinkler protection.

**23.3.2.4** Where cooking facilities are protected in accordance with 9.2.3, kitchens shall not be required to be provided with roomwide protection.

#### 23.3.3 Interior Finish.

**23.3.3.1 General.** Interior finish shall be in accordance with Section 10.2.

**23.3.3.2 Interior Wall and Ceiling Finish.** Interior wall and ceiling finish materials complying with Section 10.2 shall be Class A or Class B in corridors, in exits, and in any space not separated from corridors and exits by partitions capable of retarding the passage of smoke; and Class A, Class B, or Class C in all other areas.

**23.3.3.3 Interior Floor Finish.**

**23.3.3.3.1** Interior floor finish complying with Section 10.2 shall be Class I or Class II in corridors and exits.

**23.3.3.3.2** Existing floor finish material of Class A or Class B in nonsprinklered smoke compartments and Class A, Class B, or Class C in sprinklered smoke compartments shall be permitted to be continued to be used, provided that it has been evaluated based on tests performed in accordance with 10.2.3.

**23.3.4 Detection, Alarm, and Communications Systems.**

**23.3.4.1 General.** Detention and correctional occupancies shall be provided with a fire alarm system in accordance with Section 9.6, except as modified by 23.3.4.2 through 23.3.4.4.4.

**23.3.4.2 Initiation.**

**23.3.4.2.1** Initiation of the required fire alarm system shall be by manual means in accordance with 9.6.2 and by means of any required detection devices or detection systems, unless otherwise permitted by the following:

(1) Manual fire alarm boxes shall be permitted to be locked, provided that staff is present within the area when it is occupied and staff has keys readily available to unlock the boxes.
(2) Manual fire alarm boxes shall be permitted to be located in a staff location, provided that both of the following criteria are met:
 (a) The staff location is attended when the building is occupied.
 (b) The staff attendant has direct supervision of the sleeping area.

**23.3.4.2.2\*** Use of the provision of 9.6.1.8.1.3 shall be permitted only as an exemption to 9.6.1.8.1(2) and (3).

**23.3.4.3 Notification.**

**23.3.4.3.1 Occupant Notification.** Occupant notification shall be accomplished automatically in accordance with 9.6.3, and the following also shall apply:

(1) A positive alarm sequence shall be permitted in accordance with 9.6.3.4.
(2)\*Any smoke detectors required by this chapter shall be permitted to be arranged to alarm at a constantly attended location only and shall not be required to accomplish general occupant notification.

**23.3.4.3.2 Emergency Forces Notification.**

**23.3.4.3.2.1** Fire department notification shall be accomplished in accordance with 9.6.4, unless otherwise permitted by one of the following:

(1) A positive alarm sequence shall be permitted in accordance with 9.6.3.4.
(2) Any smoke detectors required by this chapter shall not be required to transmit an alarm to the fire department.
(3) This requirement shall not apply where staff is provided at a constantly attended location that meets one of the following criteria:
 (a) It has the capability to promptly notify the fire department.
 (b) It has direct communication with a control room having direct access to the fire department.

**23.3.4.3.2.2** Where the provision of 23.3.4.3.2.1(3) is utilized, the fire plan, as required by 23.7.1.3, shall include procedures for logging of alarms and immediate notification of the fire department.

**23.3.4.4 Detection.** An approved automatic smoke detection system shall be in accordance with Section 9.6, as modified by 23.3.4.4.1 through 23.3.4.4.4, throughout all resident housing areas.

**23.3.4.4.1** Smoke detectors shall not be required in sleeping rooms with four or fewer occupants in Use Condition II or Use Condition III.

**23.3.4.4.2** Other arrangements and positioning of smoke detectors shall be permitted to prevent damage or tampering, or for other purposes.

**23.3.4.4.2.1** Other arrangements, as specified in 23.3.4.4.2, shall be capable of detecting any fire, and the placement of detectors shall be such that the speed of detection is equivalent to that provided by the spacing and arrangements required by the installation standards referenced in Section 9.6.

**23.3.4.4.2.2** Detectors shall be permitted to be located in exhaust ducts from cells, behind grilles, or in other locations.

**23.3.4.4.2.3** The equivalent performance of the design permitted by 23.3.4.4.2.2 shall be acceptable to the authority having jurisdiction in accordance with the equivalency concepts specified in Section 1.4.

**23.3.4.4.3\*** Smoke detectors shall not be required in Use Condition II open dormitories where staff is present within the dormitory whenever the dormitory is occupied and the building is protected throughout by an approved, supervised automatic sprinkler system in accordance with 23.3.5.3.

**23.3.4.4.4** In smoke compartments protected throughout by an approved automatic sprinkler system in accordance with 23.3.5.3, smoke detectors shall not be required, except in corridors, common spaces, and sleeping rooms with more than four occupants.

**23.3.5 Extinguishment Requirements.**

**23.3.5.1** High-rise buildings shall comply with 23.4.3.

**23.3.5.2\*** Where required by Table 23.1.6.1, facilities shall be protected throughout by an approved, supervised automatic sprinkler system in accordance with 23.3.5.3.

**23.3.5.3** Where this *Code* permits exceptions for fully sprinklered detention and correctional occupancies or sprinklered smoke compartments, the sprinkler system shall meet all of the following criteria:

(1) It shall be in accordance with Section 9.7.
(2) It shall be installed in accordance with 9.7.1.1(1).
(3) It shall be electrically connected to the fire alarm system.
(4) It shall be fully supervised.

**23.3.5.4** Portable fire extinguishers shall be provided in accordance with 9.7.4.1, unless otherwise permitted by the following:

(1)\*Access to portable fire extinguishers shall be permitted to be locked.
(2)\*Portable fire extinguishers shall be permitted to be located at staff locations only.

**23.3.5.5** Standpipe and hose systems shall be provided in accordance with 9.7.4.2 as follows, unless otherwise permitted by 23.3.5.6:

(1) Class I standpipe systems shall be provided for any building three or more stories in height.
(2) Class III standpipe and hose systems shall be provided for all nonsprinklered buildings three or more stories in height.

**23.3.5.6** The requirements of 23.3.5.5 shall not apply where otherwise permitted by the following:

(1) Formed hose, 1 in. (25 mm) in diameter, on hose reels shall be permitted to provide Class II service.
(2) Separate Class I and Class II systems shall be permitted in lieu of a Class III system.

**23.3.6 Corridors.** See 23.3.8.

**23.3.7 Subdivision of Building Spaces.**

**23.3.7.1\*** Smoke barriers shall be provided to divide every story used for sleeping by 10 or more residents, or any other story having an occupant load of 50 or more persons, into not less than two compartments, unless otherwise permitted by one of the following:

(1) Protection shall be permitted to be accomplished using horizontal exits. *(See 7.2.4.)*
(2)\*The requirement for subdivision of building space shall be permitted to be fulfilled by one of the following:
  (a) Smoke compartments having exit to a public way, where such exit serves only one area and has no openings to other areas
  (b) Building separated from the resident housing area by a 2-hour fire resistance rating or 50 ft (15 m) of open space
  (c) Secured, open area having a holding space located 50 ft (15 m) from the housing area that provides 15 ft$^2$ (1.4 m$^2$) or more of refuge area for each person (resident, staff, visitors) potentially present at the time of a fire

**23.3.7.2** Doors used to access the areas specified in 23.3.7.1(2)(a), (b), and (c) shall meet the requirements for doors at smoke barriers for the applicable use condition.

**23.3.7.3** Where smoke barriers are required by 23.3.7.1, they shall be provided in accordance with both of the following criteria:

(1) They shall limit the occupant load to not more than 200 residents in any smoke compartment.
(2)\*They shall limit the travel distance to a door in a smoke barrier, unless otherwise permitted by 23.3.7.4, in accordance with both of the following criteria:
  (a) The distance from any room door required as exit access shall not exceed 100 ft (30 m).
  (b) The distance from any point in a room shall not exceed 150 ft (46 m).

**23.3.7.4** The maximum travel distance to a door in a smoke barrier shall be permitted to be increased by 50 ft (15 m) in smoke compartments protected throughout by an approved automatic sprinkler system in accordance with 23.3.5.3 or an automatic smoke control system.

**23.3.7.5\*** Any required smoke barrier shall be constructed in accordance with Section 8.5, shall be of substantial construction, and shall have structural fire resistance.

**23.3.7.6** Openings in smoke barriers shall be protected in accordance with Section 8.5, unless otherwise permitted by the following:

(1)\*The total number of vision panels in any barrier shall not be restricted.
(2) Sliding doors in smoke barriers that are designed to normally be kept closed and are remotely operated from a continuously attended location shall not be required to be self-closing.

**23.3.7.7** Not less than 6 net ft$^2$ (0.55 net m$^2$) per occupant shall be provided on each side of the smoke barrier for the total number of occupants in adjoining compartments, and this space shall be readily available wherever occupants are moved across the smoke barrier in a fire emergency.

**23.3.7.8** Doors in smoke barriers shall meet all of the following criteria:

(1) The doors shall provide resistance to the passage of smoke.
(2) Swinging doors shall be self-latching, or the opening resistance of the door shall be not less than 5 lbf (22 N).
(3) Sliding doors shall be exempt from the latching requirement of 8.5.4.3.
(4) The doors shall not be required to swing in the direction of egress travel.

**23.3.7.9** Doors in smoke barriers shall conform with the requirements for doors in means of egress as specified in Section 23.2 and shall have locking and release arrangements according to the applicable use condition. The provisions of 23.2.11.8.2 shall not be used for smoke barrier doors serving a smoke compartment containing more than 20 persons.

**23.3.7.10** Vision panels shall be provided in smoke barriers at points where the barrier crosses an exit access corridor.

**23.3.7.11** Smoke dampers shall be provided in accordance with 8.5.5, unless otherwise permitted by 23.3.7.12.

**23.3.7.12** Arrangements and positioning of smoke detectors required by 23.3.7.11 shall be permitted to prevent damage or tampering, or for other purposes, provided that both of the following criteria are met:

(1) Such arrangements shall be capable of detecting any fire.
(2) The placement of detectors shall be such that the speed of detection is equivalent to that provided by the spacing and arrangement required by *NFPA 72, National Fire Alarm and Signaling Code*, as referenced in 8.5.5.7.1.

**23.3.8\* Special Protection Features— Subdivision of Resident Housing Spaces.** Subdivision of facility spaces shall comply with Table 23.3.8.

**23.4 Special Provisions.**

**23.4.1 Limited Access Structures.**

**23.4.1.1** Limited access structures used as detention and correctional occupancies shall comply with 23.4.1.2, unless otherwise permitted by one of the following:

(1) The provisions of Section 11.7 for limited access structures shall not apply.
(2) The requirement of 23.4.1.1 shall not apply to buildings protected throughout by an approved automatic sprinkler system in accordance with 23.3.5.3.

**Table 23.3.8 Subdivision of Resident Housing Spaces**

| | Use Condition | | | | | | | |
|---|---|---|---|---|---|---|---|---|
| | **II** | | **III** | | **IV** | | **V** | |
| **Feature** | **NS** | **AS** | **NS** | **AS** | **NS** | **AS** | **NS** | **AS** |
| Room to room separation | NR | NR | NR | NR | SR | NR | SR | SR[a] |
| Room face to corridor separation | NR | NR | SR[b] | NR | SR[b] | NR | FR[b] | SR[a] |
| Room face to common space separation | NR | NR | NR (≤50 ft (≤15 m)[c]) / SR[b] (>50 ft (>15 m)[c]) | NR (≤50 ft (≤15 m)[c]) / SR[b] (>50 ft (>15 m)[c]) | SR[b] | NR (≤50 ft (≤15 m)[c]) / SR[a] (>50 ft (>15 m)[c]) | SR[b] | SR[a] |
| Common space to corridor separation | SR | NR | SR | NR | SR | NR | FR | SR[a] |
| Total openings in solid room face where room face is required to be smoke resistant or fire rated[d] | 0.85 ft² (0.08 m²) | | 0.85 ft² (0.08 m²) | | 0.85 ft² (0.08 m²) | | 0.85 ft² (0.08 m²) where meeting one of the following: (1) Kept in closed position, except when in use by staff (2) Closable from the inside (3) Provided with smoke control | |

NS: Not protected by automatic sprinklers. AS: Protected by automatic sprinklers. NR: No requirement. SR: Smoke resistant. FR: Minimum 1-hour fire resistance rating.

Notes:

(1) Doors in openings in partitions required to be fire rated (FR) in accordance with Table 23.3.8, in other than required enclosures of exits or hazardous areas, are required to be substantial doors of construction that resists fire for a minimum of 20 minutes. Vision panels with wired glass or glass with not less than 45-minute fire-rated glazing are permitted. Latches and door closers are not required on cell doors.

(2) Doors in openings in partitions required to be smoke resistant (SR) in accordance with Table 23.3.8 are required to be substantial doors of construction that resists the passage of smoke. Latches and door closers are not required on cell doors.

(3) Under Use Condition II, Use Condition III, or Use Condition IV, a space subdivided by open construction (any combination of grating doors and grating walls or solid walls) is permitted to be considered one room if housing not more than 16 persons. The perimeter walls of such space are required to be of smoke-resistant construction. Smoke detection is required to be provided in such space. Under Use Condition IV, common walls between sleeping areas within the space are required to be smoke resistant, and grating doors and fronts are permitted to be used. Under Use Condition II and Use Condition III, open dormitories are permitted to house more than 16 persons, as permitted by other sections of this chapter.

(4) Where barriers are required to be smoke resistant (SR), the provisions of Sections 8.4 and 8.5 do not apply.

[a]Might be no requirement (NR) where one of the following is provided:
  (1) Approved automatic smoke detection system installed in all corridors and common spaces
  (2) Multitiered cell blocks meeting the requirements of 23.3.1.3
[b]Might be no requirement (NR) in multitiered, open cell blocks meeting the requirements of 23.3.1.3.
[c]Travel distance through the common space to the exit access corridor.
[d]"Total openings in solid room face" include all openings (e.g., undercuts, food passes, grilles), the total of which is not to exceed 0.85 ft² (0.08 m²). All openings are required to be 36 in. (915 mm) or less above the floor.

**23.4.1.2** Any one of the following means shall be provided to evacuate smoke from the smoke compartment of fire origin:

(1) Operable windows on not less than two sides of the building, spaced not more than 30 ft (9.1 m) apart, that provide openings with dimensions of not less than 22 in. (560 mm) in width and 24 in. (610 mm) in height
(2)*Manual or automatic smoke vents
(3) Engineered smoke control system
(4) Mechanical exhaust system providing not less than six air changes per hour
(5) Other method acceptable to the authority having jurisdiction

**23.4.2 Underground Buildings.** See Section 11.7 for requirements for underground buildings.

**23.4.3 High-Rise Buildings.** Existing high-rise buildings shall be protected throughout by an approved, supervised automatic sprinkler system in accordance with 23.3.5.3. A sprinkler control valve and a waterflow device shall be provided for each floor.

**23.4.4 Reserved.**

**23.4.5 Lockups.**

**23.4.5.1 General.**

**23.4.5.1.1** Lockups in occupancies, other than detention and correctional occupancies and health care occupancies, where the holding area has capacity for more than 50 detainees shall be classified as detention and correctional occupancies and shall comply with the requirements of Chapter 23.

**23.4.5.1.2** Lockups in occupancies, other than detention and correctional occupancies and health care occupancies, where any individual is detained for 24 or more hours shall be classified as detention and correctional occupancies and shall comply with the requirements of Chapter 23.

**23.4.5.1.3** Lockups in occupancies, other than detention and correctional occupancies and health care occupancies, where the holding area has capacity for not more than 50 detainees, and where no individual is detained for 24 hours or more, shall comply with 23.4.5.1.4 or 23.4.5.1.5.

**23.4.5.1.4** The lockup shall be permitted to comply with the requirements for the predominant occupancy in which the lockup is placed, provided that all of the following criteria are met:

(1)*Doors and other physical restraints to free egress by detainees can be readily released by staff within 2 minutes of the onset of a fire or similar emergency.
(2) Staff is in sufficient proximity to the lockup so as to be able to effect the 2-minute release required by 23.4.5.1.4(1) whenever detainees occupy the lockup.
(3) Staff is authorized to effect the release required by 23.4.5.1.4(1).
(4) Staff is trained and practiced in effecting the release required by 23.4.5.1.4(1).
(5) Where the release required by 23.4.5.1.4(1) is effected by means of remote release, detainees are not to be restrained from evacuating without the assistance of others.

**23.4.5.1.5** Where the lockup does not comply with all the criteria of 23.4.5.1.4, the requirements of 23.4.5.2 shall be met.

**23.4.5.1.6** The fire department with responsibility for responding to a building that contains a lockup shall be notified of the presence of the lockup.

**23.4.5.2 Alternate Provisions.**

**23.4.5.2.1** The requirements applicable to the predominant occupancy in which the lockup is placed shall be met.

**23.4.5.2.2** Where security operations necessitate the locking of required means of egress, the following shall apply:

(1) Detention-grade hardware meeting the requirements of ASTM F 1577, *Standard Test Methods for Detention Locks for Swinging Doors*, shall be provided on swinging doors within the required means of egress.
(2) Sliding doors within the required means of egress shall be designed and engineered for detention and correctional use, and lock cylinders shall meet the cylinder test requirements of ASTM F 1577.

**23.4.5.2.3** The lockup shall be provided with a complete smoke detection system in accordance with 9.6.2.9.

**23.4.5.2.4** Where the requirements applicable to the predominant occupancy do not require a fire alarm system, the

lockup shall be provided with a fire alarm system meeting all of the following criteria:

(1) The alarm system shall be in accordance with Section 9.6.
(2) Initiation of the alarm system shall be accomplished by all of the following:
    (a) Manual fire alarm boxes in accordance with 9.6.2
    (b) Smoke detection system required by 23.4.5.2.3
    (c) Automatic sprinkler system required by the provisions applicable to the predominant occupancy
(3) Staff and occupant notification shall be provided automatically in accordance with 9.6.3.
(4) Emergency force notification shall be provided in accordance with 9.6.4.

**23.5 Building Services.**

**23.5.1 Utilities.**

**23.5.1.1** Utilities shall comply with the provisions of Section 9.1.

**23.5.1.2** Alarms, emergency communications systems, and the illumination of generator set installations shall be provided with emergency power in accordance with *NFPA 70, National Electrical Code*, unless otherwise permitted by 23.5.1.3.

**23.5.1.3** Systems complying with earlier editions of *NFPA 70, National Electrical Code*, and not presenting a life safety hazard shall be permitted to continue to be used.

**23.5.2 Heating, Ventilating, and Air-Conditioning.**

**23.5.2.1** Heating, ventilating, and air-conditioning equipment shall comply with the provisions of Section 9.2 and shall be installed in accordance with the manufacturer's specifications, unless otherwise permitted by one of the following:

(1) The requirement of 23.5.2.1 shall not apply where otherwise modified by 23.5.2.2.
(2) Systems complying with earlier editions of the applicable codes and not presenting a life safety hazard shall be permitted to continue to be used.

**23.5.2.2** Portable space-heating devices shall be prohibited, unless otherwise permitted by 23.5.2.4.

**23.5.2.3** Any heating device, other than a central heating plant, shall be designed and installed so that combustible material cannot be ignited by the device or its appurtenances, and both of the following requirements also shall apply:

(1) If fuel-fired, such heating devices shall comply with all of the following:
    (a) They shall be chimney connected or vent connected.
    (b) They shall take air for combustion directly from outside.
    (c) They shall be designed and installed to provide for complete separation of the combustion system from the atmosphere of the occupied area.
(2) The heating system shall have safety devices to immediately stop the flow of fuel and shut down the equipment in case of either excessive temperatures or ignition failure.

**23.5.2.4** Approved, suspended unit heaters shall be permitted in locations other than means of egress and sleeping areas, provided that both of the following criteria are met:

(1) Such heaters are located high enough to be out of the reach of persons using the area.
(2) Such heaters are vent connected and equipped with the safety devices required by 23.5.2.3(2).

**23.5.2.5** Combustion and ventilation air for boiler, incinerator, or heater rooms shall be taken directly from, and discharged directly to, the outside.

**23.5.3 Elevators, Escalators, and Conveyors.** Elevators, escalators, and conveyors shall comply with the provisions of Section 9.4.

**23.5.4 Rubbish Chutes, Incinerators, and Laundry Chutes.**

**23.5.4.1** Rubbish chutes, incinerators, and laundry chutes shall comply with the provisions of Section 9.5.

**23.5.4.2** Rubbish chutes and linen chutes, including pneumatic rubbish and linen systems, shall be provided with automatic extinguishing protection in accordance with Section 9.7.

**23.5.4.3** Trash chutes shall discharge into a trash collection room used for no other purpose and protected in accordance with Section 8.7.

**23.5.4.4** Incinerators shall not be directly flue-fed, and floor chutes shall not directly connect with the combustion chamber.

**23.6 Reserved.**

**23.7 Operating Features.**

**23.7.1 Attendants, Evacuation Plan, and Fire Drills.**

**23.7.1.1** Detention and correctional facilities, or those portions of facilities having such occupancy, shall be provided with 24-hour staffing, and the following requirements also shall apply:

(1) Staff shall be within three floors or a 300 ft (91 m) horizontal distance of the access door of each resident housing area.

(2) For Use Condition III, Use Condition IV, and Use Condition V, the arrangement shall be such that the staff involved starts the release of locks necessary for emergency evacuation or rescue and initiates other necessary emergency actions within 2 minutes of alarm.

(3) The following shall apply to areas in which all locks are unlocked remotely in compliance with 23.2.11.8:

  (a) Staff shall not be required to be within three floors or 300 ft (91 m) of the access door.

  (b) The 10-lock, manual key exemption of 23.2.11.8.2 shall not be permitted to be used in conjunction with the alternative requirement of 23.7.1.1(3)(a).

**23.7.1.2\*** Provisions shall be made so that residents in Use Condition III, Use Condition IV, and Use Condition V shall be able to notify staff of an emergency.

**23.7.1.3\*** The administration of every detention or correctional facility shall have, in effect and available to all supervisory personnel, written copies of a plan for the protection of all persons in the event of fire, for their evacuation to areas of refuge, and for evacuation from the building when necessary.

**23.7.1.3.1** All employees shall be instructed and drilled with respect to their duties under the plan.

**23.7.1.3.2** The plan shall be coordinated with, and reviewed by, the fire department legally committed to serve the facility.

**23.7.1.4** Employees of detention and correctional occupancies shall be instructed in the proper use of portable fire extinguishers and other manual fire suppression equipment.

**23.7.1.4.1** The training specified in 23.7.1.4 shall be provided to new staff promptly upon commencement of duty.

**23.7.1.4.2** Refresher training shall be provided to existing staff at not less than annual intervals.

**23.7.2 Combustible Personal Property.** Books, clothing, and other combustible personal property allowed in sleeping rooms shall be stored in closable metal lockers or an approved fire-resistant container.

**23.7.3 Heat-Producing Appliances.** The number of heat-producing appliances, such as toasters and hot plates, and the overall use of electrical power within a sleeping room shall be controlled by facility administration.

**23.7.4\* Furnishings, Mattresses, and Decorations.**

**23.7.4.1** Draperies and curtains, including privacy curtains, in detention and correctional occupancies shall be in accordance with the provisions of 10.3.1.

**23.7.4.2** Newly introduced upholstered furniture within detention and correctional occupancies shall meet the criteria specified in 10.3.2.1(2) and 10.3.3.

**23.7.4.3\*** Newly introduced mattresses within detention and correctional occupancies shall meet the criteria specified in 10.3.2.2 and 10.3.4.

**23.7.4.4** Combustible decorations shall be prohibited in any detention or correctional occupancy unless flame-retardant.

**23.7.4.5** Wastebaskets and other waste containers shall be of noncombustible or other approved materials. Waste containers with a capacity exceeding 20 gal (76 L) shall be provided with a noncombustible lid or lid of other approved material.

**23.7.5 Keys.** All keys necessary for unlocking doors installed in a means of egress shall be individually identified by both touch and sight.

**23.7.6 Portable Space-Heating Devices.** Portable space-heating devices shall be prohibited in all detention and correctional occupancies.

**23.7.7 Door Inspection.** Doors and door hardware in means of egress shall be inspected monthly by an appropriately trained person. The inspection shall be documented.

## Chapter 24 One- and Two-Family Dwellings

**24.1 General Requirements.**

**24.1.1 Application.**

**24.1.1.1\*** The requirements of this chapter shall apply to one- and two-family dwellings, which shall include those buildings containing not more than two dwelling units in which each dwelling unit is occupied by members of a single family with not more than three outsiders, if any, accommodated in rented rooms.

**24.1.1.2 Administration.** The provisions of Chapter 1, Administration, shall apply.

**24.1.1.3 General.** The provisions of Chapter 4, General, shall apply.

**24.1.1.4** The requirements of this chapter shall apply to new buildings and to existing or modified buildings according to the provisions of 1.3.1 of this *Code.*

**24.1.2 Classification of Occupancy.** See 6.1.8 and 24.1.1.1.

### 24.1.3 Multiple Occupancies.

**24.1.3.1** Multiple occupancies shall be in accordance with 6.1.14.

**24.1.3.2** No dwelling unit of a residential occupancy shall have its sole means of egress pass through any nonresidential occupancy in the same building, unless otherwise permitted by 24.1.3.2.1 or 24.1.3.2.2.

**24.1.3.2.1** In buildings that are protected by an automatic sprinkler system in accordance with Section 9.7, dwelling units of a residential occupancy shall be permitted to have their sole means of egress pass through a nonresidential occupancy in the same building, provided that all of the following criteria are met:

(1) The dwelling unit of the residential occupancy shall comply with Chapter 24.
(2) The sole means of egress from the dwelling unit of the residential occupancy shall not pass through a high hazard contents area, as defined in 6.2.2.4.

**24.1.3.2.2** In buildings that are not protected by an automatic sprinkler system in accordance with Section 9.7, dwelling units of a residential occupancy shall be permitted to have their sole means of egress pass through a nonresidential occupancy in the same building, provided that all of the following criteria are met:

(1) The sole means of egress from the dwelling unit of the residential occupancy to the exterior shall be separated from the remainder of the building by fire barriers having a minimum 1-hour fire resistance rating.
(2) The dwelling unit of the residential occupancy shall comply with Chapter 24.
(3) The sole means of egress from the dwelling unit of the residential occupancy shall not pass through a high hazard contents area, as defined in 6.2.2.4.

**24.1.3.3** Multiple dwelling units of a residential occupancy shall be permitted to be located above a nonresidential occupancy only where one of the following conditions exists:

(1) Where the dwelling unit of the residential occupancy and exits therefrom are separated from the nonresidential occupancy by construction having a minimum 1-hour fire resistance rating
(2) Where the nonresidential occupancy is protected throughout by an approved, supervised automatic sprinkler system in accordance with Section 9.7
(3) Where the nonresidential occupancy is protected by an automatic fire detection system in accordance with Section 9.6

### 24.1.4 Definitions.

**24.1.4.1 General.** For definitions, see Chapter 3, Definitions.

**24.1.4.2 Special Definitions.** Special terms applicable to this chapter are defined in Chapter 3 of this *Code*. Where necessary, other terms are defined in the text.

**24.1.5 Classification of Hazard of Contents.** The contents of residential occupancies shall be classified as ordinary hazard in accordance with 6.2.2.

**24.1.6 Minimum Construction Requirements.** (No special requirements.)

**24.1.7 Occupant Load.** (No requirements.)

### 24.2* Means of Escape Requirements.

**24.2.1 General.** The provisions of Chapter 7 shall not apply to means of escape, unless specifically referenced in this chapter.

**24.2.2 Number and Types of Means of Escape.**

**24.2.2.1 Number of Means of Escape.**

**24.2.2.1.1** In dwellings or dwelling units of two rooms or more, every sleeping room and every living area shall have not less than one primary means of escape and one secondary means of escape.

**24.2.2.1.2** A secondary means of escape shall not be required where one of the following conditions is met:

(1) The bedroom or living area has a door leading directly to the outside of the building at or to the finished ground level.
(2) The dwelling unit is protected throughout by an approved automatic sprinkler system in accordance with 24.3.5.

**24.2.2.2 Primary Means of Escape.** The primary means of escape shall be a door, stairway, or ramp providing a means of unobstructed travel to the outside of the dwelling unit at street or the finished ground level.

**24.2.2.3 Secondary Means of Escape.** The secondary means of escape, other than an existing approved means of escape, shall be one of the means specified in 24.2.2.3.1 through 24.2.2.3.4.

**24.2.2.3.1** It shall be a door, stairway, passage, or hall providing a way of unobstructed travel to the outside of the dwelling at street or the finished ground level that is independent of and remote from the primary means of escape.

**24.2.2.3.2** It shall be a passage through an adjacent nonlockable space, independent of and remote from the primary means of escape, to any approved means of escape.

**24.2.2.3.3*** It shall be an outside window or door operable from the inside without the use of tools, keys, or special effort and shall provide a clear opening of not less than 5.7 ft² (0.53 m²). The width shall be not less than 20 in. (510 mm), and the height shall be not less than 24 in. (610 mm). The bottom of the opening shall be not more than 44 in. (1120 mm) above the floor. Such means of escape shall be acceptable where one of the following criteria is met:

(1) The window shall be within 20 ft (6100 mm) of the finished ground level.
(2) The window shall be directly accessible to fire department rescue apparatus as approved by the authority having jurisdiction.
(3) The window or door shall open onto an exterior balcony.
(4) Windows having a sill height below the adjacent finished ground level shall be provided with a window well meeting all of the following criteria:
  (a) The window well shall have horizontal dimensions that allow the window to be fully opened.
  (b) The window well shall have an accessible net clear opening of not less than 9 ft² (0.82 m²) with a length and width of not less than 36 in. (915 mm).
  (c) A window well with a vertical depth of more than 44 in. (1120 mm) shall be equipped with an approved permanently affixed ladder or with steps meeting both of the following criteria:

i. The ladder or steps shall not encroach more than 6 in. (150 mm) into the required dimensions of the window well.

ii. The ladder or steps shall not be obstructed by the window.

**24.2.2.3.4** It shall be a bulkhead complying with 24.2.7 and meeting the minimum area requirements of 24.2.2.3.

**24.2.2.3.5** Ladders or steps that comply with the requirements of 24.2.2.3.3(4)(c) shall be exempt from the requirements of 7.2.2.

**24.2.2.4 Two Primary Means of Escape.** In buildings, other than existing buildings and other than those protected throughout by an approved, supervised automatic sprinkler system in accordance with 24.3.5, every story more than 2000 ft$^2$ (185 m$^2$) in area within the dwelling unit shall be provided with two primary means of escape remotely located from each other.

**24.2.3 Arrangement of Means of Escape.** Any required path of travel in a means of escape from any room to the outside shall not pass through another room or apartment not under the immediate control of the occupant of the first room or through a bathroom or other space subject to locking.

**24.2.4 Doors.**

**24.2.4.1** Doors in the path of travel of a means of escape, other than bathroom doors in accordance with 24.2.4.2 and doors serving a room not exceeding 70 ft$^2$ (6.5 m$^2$), shall be not less than 28 in. (710 mm) wide.

**24.2.4.2** Bathroom doors and doors serving a room not exceeding 70 ft$^2$ (6.5 m$^2$) shall be not less than 24 in. (610 mm) wide.

**24.2.4.3** Doors shall be not less than 6 ft 6 in. (1980 mm) in nominal height.

**24.2.4.4** Every closet door latch shall be such that children can open the door from inside the closet.

**24.2.4.5** Every bathroom door shall be designed to allow opening from the outside during an emergency when locked.

**24.2.4.6** Doors shall be swinging or sliding.

**24.2.4.7\*** No door in any means of escape shall be locked against egress when the building is occupied. All locking devices that impede or prohibit egress or that cannot be easily disengaged shall be prohibited.

**24.2.4.8** Floor levels at doors in the primary means of escape shall comply with 7.2.1.3, unless otherwise permitted by any of the following:

(1) In existing buildings, where the door discharges to the outside or to an exterior balcony or exterior exit access, the floor level outside the door shall be permitted to be one step lower than the inside, but shall not be in excess of 8 in. (205 mm).

(2) In new buildings, where the door discharges to the outside or to an exterior exit access, an exterior landing with not more than a 7 in. (180 mm) drop below the door threshold and a minimum dimension of 36 in. (915 mm) or the width of the door leaf, whichever is smaller, shall be permitted.

(3) A door at the top of an interior stair shall be permitted to open directly onto a stair, provided that the door does not swing over the stair and the door serves an area with an occupant load of fewer than 50 persons.

**24.2.4.9** Forces to open doors shall comply with 7.2.1.4.5.

**24.2.4.10** Latching devices for doors shall comply with 7.2.1.5.10.

**24.2.5 Stairs, Guards, and Ramps.**

**24.2.5.1** Stairs, guards, ramps, and handrails shall be in accordance with 7.2.2 for stairs, 7.2.2.4 for guards, and 7.2.5 for ramps, as modified by 24.2.5.1.1 through 24.2.5.1.3.

**24.2.5.1.1** The provisions of 7.2.2.5, 7.2.5.5, and 7.7.3 shall not apply.

**24.2.5.1.2** If serving as a secondary means of escape, stairs complying with the fire escape requirements of Table 7.2.8.4.1(a) or Table 7.2.8.4.1(b) shall be permitted.

**24.2.5.1.3** If serving as a secondary means of escape, ramps complying with the existing ramp requirements of Table 7.2.5.2(b) shall be permitted.

**24.2.5.2** Interior stairways shall be provided with means capable of providing artificial light at the minimum level specified by 7.8.1.3 for exit stairs, measured at the center of treads and on landing surfaces within 24 in. (610 mm) of step nosings.

**24.2.5.3** For interior stairways, manual lighting controls shall be reachable and operable without traversing any step of the stair.

**24.2.5.4** The clear width of stairs, landings, ramps, balconies, and porches shall be not less than 36 in. (915 mm), measured in accordance with 7.3.2.

**24.2.5.5** Spiral stairs and winders in accordance with 7.2.2.2.3 and 7.2.2.2.4 shall be permitted within a single dwelling unit.

**24.2.5.6** No sleeping rooms or living areas shall be accessible only by a ladder, a stair ladder, an alternating tread device or folding stairs, or through a trap door.

**24.2.6 Hallways.**

**24.2.6.1** The width of hallways, other than existing approved hallways, which shall be permitted to continue to be used, shall be not less than 36 in. (915 mm).

**24.2.6.2** The height of hallways, other than existing approved hallways, which shall be permitted to continue to be used, shall be not less than 7 ft (2135 mm) nominal, with clearance below projections from the ceiling of not less than 6 ft 8 in. (2030 mm) nominal.

**24.2.7 Bulkheads.**

**24.2.7.1 Bulkhead Enclosures.** Where provided, bulkhead enclosures shall provide direct access to the basement from the exterior.

**24.2.7.2 Bulkhead Enclosure Stairways.** Stairways serving bulkhead enclosures that are not part of the required primary means of escape, and that provide access from the outside finished ground level to the basement, shall be exempt from the provisions of 24.2.5.1 when the maximum height from the basement finished floor level to the finished ground level adjacent to the stairway does not exceed 8 ft (2440 mm), and the finished ground level opening to the stairway is covered by a bulkhead enclosure with hinged doors or other approved means.

**24.3 Protection.**

**24.3.1 Protection of Vertical Openings.** (No requirements.)

**24.3.2 Reserved.**

**24.3.3 Interior Finish.**

**24.3.3.1 General.** Interior finish shall be in accordance with Section 10.2.

**24.3.3.2 Interior Wall and Ceiling Finish.** Interior wall and ceiling finish materials complying with Section 10.2 shall be Class A, Class B, or Class C.

**24.3.3.3 Interior Floor Finish.** (No requirements.)

**24.3.3.4 Contents and Furnishings.** Contents and furnishings shall not be required to comply with Section 10.3.

**24.3.4 Detection, Alarm, and Communications Systems.**

**24.3.4.1** Smoke alarms or a smoke detection system shall be provided in accordance with either 24.3.4.1.1 or 24.3.4.1.2, as modified by 24.3.4.1.3.

**24.3.4.1.1\*** Smoke alarms shall be installed in accordance with 9.6.2.10 in all of the following locations:

(1) All sleeping rooms
(2)\*Outside of each separate sleeping area, in the immediate vicinity of the sleeping rooms
(3) On each level of the dwelling unit, including basements

**24.3.4.1.2** Dwelling units shall be protected by an approved smoke detection system in accordance with Section 9.6 and equipped with an approved means of occupant notification.

**24.3.4.1.3** In existing one- and two-family dwellings, approved smoke alarms powered by batteries shall be permitted.

**24.3.4.2 Carbon Monoxide and Carbon Monoxide Detection Systems.**

**24.3.4.2.1** Carbon monoxide alarms or carbon monoxide detectors in accordance with Section 9.8 and 24.3.4.2 shall be provided in new one- and two-family dwellings where either of the following conditions exists:

(1) Dwelling units with communicating attached garages, unless otherwise exempted by 24.3.4.2.3
(2) Dwelling units containing fuel-burning appliances

**24.3.4.2.2\*** Where required by 24.3.4.2.1, carbon monoxide alarms or carbon monoxide detectors shall be installed in the following locations:

(1) Outside of each separate dwelling unit sleeping area in the immediate vicinity of the sleeping rooms
(2) On every occupiable level of a dwelling unit, including basements, and excluding attics and crawl spaces

**24.3.4.2.3** Carbon monoxide alarms and carbon monoxide detectors as specified in 24.3.4.2.1(1) shall not be required in the following locations:

(1) In garages
(2) Within dwelling units with communicating attached garages that are open parking structures as defined by the building code
(3) Within dwelling units with communicating attached garages that are mechanically ventilated in accordance with the mechanical code

**24.3.5\* Extinguishment Requirements.**

**24.3.5.1** All new one- and two-family dwellings shall be protected throughout by an approved automatic sprinkler system in accordance with 24.3.5.2.

**24.3.5.2** Where an automatic sprinkler system is installed, either for total or partial building coverage, the system shall be in accordance with Section 9.7; in buildings of four or fewer stories in height above grade plane, systems in accordance with NFPA 13R, *Standard for the Installation of Sprinkler Systems in Residential Occupancies up to and Including Four Stories in Height*, and with NFPA 13D, *Standard for the Installation of Sprinkler Systems in One- and Two-Family Dwellings and Manufactured Homes*, shall also be permitted.

**24.4 Reserved.**

**24.5 Building Services.**

**24.5.1 Heating, Ventilating, and Air-Conditioning.**

**24.5.1.1** Heating, ventilating, and air-conditioning equipment shall comply with the provisions of Section 9.2.

**24.5.1.2** Unvented fuel-fired heaters shall not be used unless they are listed and approved.

**24.5.2 Reserved.**

# Chapter 25    Reserved

# Chapter 26    Lodging or Rooming Houses

**26.1 General Requirements.**

**26.1.1 Application.**

**26.1.1.1\*** The requirements of this chapter shall apply to buildings that provide sleeping accommodations for 16 or fewer persons on either a transient or permanent basis, with or without meals, but without separate cooking facilities for individual occupants, except as provided in Chapter 24.

**26.1.1.2 Administration.** The provisions of Chapter 1, Administration, shall apply.

**26.1.1.3 General.** The provisions of Chapter 4, General, shall apply.

**26.1.1.4** The requirements of this chapter shall apply to new buildings and to existing or modified buildings according to the provisions of 1.3.1 of this *Code*.

**26.1.2 Classification of Occupancy.** See 6.1.8 and 26.1.1.1.

**26.1.3 Multiple Occupancies.**

**26.1.3.1** Multiple occupancies shall be in accordance with 6.1.14.

**26.1.3.2** No lodging or rooming house shall have its sole means of egress pass through any nonresidential occupancy in the same building, unless otherwise permitted by 26.1.3.2.1 or 26.1.3.2.2.

**26.1.3.2.1** In buildings that are protected by an automatic sprinkler system in accordance with Section 9.7, lodging or rooming houses shall be permitted to have their sole means of egress pass through a nonresidential occupancy in the same building, provided that both of the following criteria are met:

(1) The lodging or rooming house shall comply with Chapter 26.

(2) The sole means of egress from the lodging or rooming house shall not pass through a high hazard contents area, as defined in 6.2.2.4.

**26.1.3.2.2** In buildings that are not protected by an automatic sprinkler system in accordance with Section 9.7, lodging or rooming houses shall be permitted to have their sole means of egress pass through a nonresidential occupancy in the same building, provided that all of the following criteria are met:

(1) The sole means of egress from the lodging or rooming house to the exterior shall be separated from the remainder of the building by fire barriers having a minimum 1-hour fire resistance rating.

(2) The lodging or rooming house shall comply with Chapter 26.

(3) The sole means of egress from the lodging or rooming house shall not pass through a high hazard contents area, as defined in 6.2.2.4.

**26.1.3.3** Lodging or rooming houses shall be permitted to be located above a nonresidential occupancy only where one of the following conditions exists:

(1) Where the lodging or rooming house and exits therefrom are separated from the nonresidential occupancy by construction having a minimum 1-hour fire resistance rating

(2) Where the nonresidential occupancy is protected throughout by an approved, supervised automatic sprinkler system in accordance with Section 9.7

(3) Where the lodging or rooming house is located above a nonresidential occupancy, and the nonresidential occupancy is protected by an automatic fire detection system in accordance with Section 9.6

**26.1.4 Definitions.**

**26.1.4.1 General.** For definitions, see Chapter 3, Definitions.

**26.1.4.2 Special Definitions.** Special terms applicable to this chapter are defined in Chapter 3. Where necessary, other terms are defined in the text.

**26.1.5 Classification of Hazard of Contents.** The contents of residential occupancies shall be classified as ordinary hazard in accordance with 6.2.2.

**26.1.6 Minimum Construction Requirements.** (No special requirements.)

**26.1.7 Occupant Load.** See 26.1.1.1.

**26.2 Means of Escape Requirements.**

**26.2.1 Number and Types of Means of Escape.**

**26.2.1.1 Primary Means of Escape.**

**26.2.1.1.1** Every sleeping room and living area shall have access to a primary means of escape complying with Chapter 24 and located to provide a safe path of travel to the outside.

**26.2.1.1.2** Where the sleeping room is above or below the level of exit discharge, the primary means of escape shall be an interior stair in accordance with 26.2.2, an exterior stair, a horizontal exit in accordance with 7.2.4, or an existing fire escape stair in accordance with 7.2.8.

**26.2.1.2 Secondary Means of Escape.** In addition to the primary route, each sleeping room and living area shall have a second means of escape in accordance with 24.2.2, unless the sleeping room or living area has a door leading directly outside the building with access to the finished ground level or to

a stairway that meets the requirements for exterior stairs in 26.2.1.1.2.

**26.2.1.3 Two Primary Means of Escape.** In other than existing buildings and those protected throughout by an approved, supervised automatic sprinkler system in accordance with 26.3.6, every story more than 2000 ft$^2$ (185 m$^2$) in area, or with travel distance to the primary means of escape more than 75 ft (23 m), shall be provided with two primary means of escape remotely located from each other.

**26.2.2 Stairways.**

**26.2.2.1** Interior stairways, other than those in accordance with 26.2.2.2 or 26.2.2.3, shall comply with 7.2.2.5.3 and shall be enclosed by fire barriers having a minimum ½-hour fire resistance rating, with all openings protected with smoke-actuated automatic-closing or self-closing doors having a fire resistance comparable to that required for the enclosure.

**26.2.2.2** Where an interior stair connects the street floor with the story next above or below only, but not with both, the interior stair shall be required to be enclosed only on the street floor.

**26.2.2.3** Stairways shall be permitted to be unenclosed in accordance with 26.3.1.1.2 and 26.3.1.1.3.

**26.2.2.4** Winders in accordance with 7.2.2.2.4 shall be permitted.

**26.2.3 Doors.**

**26.2.3.1** Doors in a means of escape, other than bathroom doors in accordance with 26.2.3.2, and paths of travel in a means of escape shall be not less than 28 in. (710 mm) wide.

**26.2.3.2** Bathroom doors shall be not less than 24 in. (610 mm) wide.

**26.2.3.3** Every closet door latch shall be such that it can be readily opened from the inside in case of emergency.

**26.2.3.4** Every bathroom door shall be designed to allow opening from the outside during an emergency when locked.

**26.2.3.5** Door-locking arrangements shall comply with either 26.2.3.5.1 or 26.2.3.5.2.

**26.2.3.5.1\*** No door in any means of escape shall be locked against egress when the building is occupied.

**26.2.3.5.2** Delayed-egress locks complying with 7.2.1.6.1 shall be permitted, provided that not more than one such device is located in any one escape path.

**26.2.3.6** Doors serving a single dwelling unit shall be permitted to be provided with a lock in accordance with 7.2.1.5.7.

**26.3 Protection.**

**26.3.1 Protection of Vertical Openings.**

**26.3.1.1** Vertical openings shall comply with 26.3.1.1.1, 26.3.1.1.2, or 26.3.1.1.3.

**26.3.1.1.1** Vertical openings shall be protected so that no primary escape route is exposed to an unprotected vertical opening.

**26.3.1.1.1.1** The vertical opening shall be considered protected if the opening is cut off and enclosed in a manner that provides a smoke- and fire-resisting capability of not less than ½ hour.

**26.3.1.1.1.2** Any doors or openings shall have a smoke- and fire-resisting capability equivalent to that of the enclosure and shall be automatic-closing on detection of smoke or shall be self-closing.

**26.3.1.1.2** In buildings three or fewer stories in height that are protected throughout by an approved automatic sprinkler system in accordance with 26.3.6, unprotected vertical openings shall be permitted, provided that a primary means of escape from each sleeping area is provided that does not pass through a portion of a lower floor, unless such portion is separated from all spaces on that floor by construction having a minimum ½-hour fire resistance rating.

**26.3.1.1.3** Stair enclosures shall not be required in buildings two or fewer stories in height where both of the following conditions exist:

(1) The building is protected throughout by an approved, supervised automatic sprinkler system in accordance with 26.3.6.1.
(2) The allowance of 24.2.2.1.2 to omit a secondary means of escape is not used.

**26.3.1.2*** Exterior stairs shall be protected against blockage caused by fire within the building.

**26.3.2 Reserved.**

**26.3.3 Interior Finish.**

**26.3.3.1 General.** Interior finish shall be in accordance with Section 10.2.

**26.3.3.2 Interior Wall and Ceiling Finish.** Interior wall and ceiling finish materials complying with Section 10.2 shall be Class A, Class B, or Class C.

**26.3.3.3 Interior Floor Finish.**

**26.3.3.3.1** Newly installed interior floor finish shall comply with Section 10.2.

**26.3.3.3.2** Newly installed interior floor finish shall comply with 10.2.7.1 or 10.2.7.2, as applicable.

**26.3.4 Detection, Alarm, and Communications Systems.**

**26.3.4.1 General.**

**26.3.4.1.1** Lodging and rooming houses, other than those meeting 26.3.4.1.2, shall be provided with a fire alarm system in accordance with Section 9.6.

**26.3.4.1.2** A fire alarm system in accordance with Section 9.6 shall not be required in existing lodging and rooming houses that have an existing smoke detection system meeting or exceeding the requirements of 26.3.4.5.1 where that detection system includes not less than one manual fire alarm box per floor arranged to initiate the smoke detection alarm.

**26.3.4.2 Initiation.** Initiation of the required fire alarm system shall be by manual means in accordance with 9.6.2, or by alarm initiation in accordance with 9.6.2.1(3) in buildings protected throughout by an approved automatic sprinkler system in accordance with 26.3.6.

**26.3.4.3 Notification.** Occupant notification shall be provided automatically in accordance with 9.6.3, as modified by 26.3.4.3.1 and 26.3.4.3.2.

**26.3.4.3.1*** Visible signals for the hearing impaired shall not be required where the proprietor resides in the building and there are five or fewer rooms for rent.

**26.3.4.3.2** Positive alarm sequence in accordance with 9.6.3.4 shall be permitted.

**26.3.4.4 Detection. (Reserved)**

**26.3.4.5 Smoke Alarms.**

**26.3.4.5.1** Approved single-station smoke alarms, other than existing smoke alarms meeting the requirements of 26.3.4.5.3, shall be installed in accordance with 9.6.2.10 in every sleeping room.

**26.3.4.5.2** In other than existing buildings, the smoke alarms required by 26.3.4.5.1 shall be interconnected in accordance with 9.6.2.10.3.

**26.3.4.5.3** Existing battery-powered smoke alarms, rather than house electric-powered smoke alarms, shall be permitted where the facility has demonstrated to the authority having jurisdiction that the testing, maintenance, and battery replacement programs will ensure reliability of power to the smoke alarms.

**26.3.4.6 Carbon Monoxide Alarms and Carbon Monoxide Detection Systems.**

**26.3.4.6.1** Carbon monoxide alarms or carbon monoxide detectors in accordance with Section 9.8 and 26.3.4.6 shall be provided in new lodging or rooming houses where either of the following conditions exists:

(1) Lodging or rooming houses with communicating attached garages, unless otherwise exempted by 26.3.4.6.3
(2) Lodging or rooming houses containing fuel-burning appliances

**26.3.4.6.2*** Where required by 26.3.4.6.1, carbon monoxide alarms or carbon monoxide detectors shall be installed in the following locations:

(1) Outside of each separate sleeping area in the immediate vicinity of the sleeping rooms
(2) On every occupiable level, including basements, and excluding attics and crawl spaces

**26.3.4.6.3** Carbon monoxide alarms and carbon monoxide detectors as specified in 26.3.4.6.1(1) shall not be required in the following locations:

(1) In garages
(2) Within lodging or rooming houses with communicating attached garages that are open parking structures as defined by the building code
(3) Within lodging or rooming houses with communicating attached garages that are mechanically ventilated in accordance with the mechanical code

**26.3.4.7*** **Protection of Fire Alarm System.** The provision of 9.6.1.8.1.3 shall not apply to the smoke detection required at each fire alarm control unit by 9.6.1.8.1(1).

**26.3.5 Separation of Sleeping Rooms.**

**26.3.5.1** All sleeping rooms shall be separated from escape route corridors by smoke partitions in accordance with Section 8.4.

**26.3.5.2** There shall be no louvers or operable transoms in corridor walls.

**26.3.5.3** Air passages shall not penetrate corridor walls, unless they are properly installed heating and utility installations other than transfer grilles.

**26.3.5.4** Transfer grilles shall be prohibited in corridor walls.

**26.3.5.5** Doors shall be provided with latches or other mechanisms suitable for keeping the doors closed.

**26.3.5.6** Doors shall not be arranged to prevent the occupant from closing the door.

**26.3.5.7** In buildings other than those protected throughout by an approved automatic sprinkler system in accordance with 26.3.6, doors shall be self-closing or automatic-closing upon detection of smoke.

**26.3.6 Extinguishment Requirements.**

**26.3.6.1** All new lodging or rooming houses, other than those meeting the requirements of 26.3.6.2, shall be protected throughout by an approved automatic sprinkler system in accordance with 26.3.6.3.

**26.3.6.2** An automatic sprinkler system shall not be required where every sleeping room has a door opening directly to the outside of the building at street or the finished ground level, or has a door opening directly to the outside leading to an exterior stairway that meets the requirements of 26.2.1.1.2.

**26.3.6.3** Where an automatic sprinkler system is required or is used as an alternative method of protection, either for total or partial building coverage, the system shall be in accordance with Section 9.7 and 26.3.6.3.1 through 26.3.6.3.6.

**26.3.6.3.1** Activation of the automatic sprinkler system shall actuate the fire alarm system in accordance with Section 9.6.

**26.3.6.3.2** In buildings four or fewer stories above grade plane, systems in accordance with NFPA 13R, *Standard for the Installation of Sprinkler Systems in Residential Occupancies up to and Including Four Stories in Height*, shall be permitted.

**26.3.6.3.3\*** Systems in accordance with NFPA 13D, *Standard for the Installation of Sprinkler Systems in One- and Two-Family Dwellings and Manufactured Homes*, shall be permitted where all of the following requirements are met:

(1) The lodging or rooming house shall not be part of a mixed occupancy.
(2) Entrance foyers shall be sprinklered.
(3) Lodging or rooming houses with sleeping accommodations for more than eight occupants shall be treated as two-family dwellings with regard to the water supply.

**26.3.6.3.4** In buildings sprinklered in accordance with NFPA 13, *Standard for the Installation of Sprinkler Systems*, closets less than 12 ft² (1.1 m²) in area in individual dwelling units shall not be required to be sprinklered.

**26.3.6.3.5** In buildings sprinklered in accordance with NFPA 13, *Standard for the Installation of Sprinkler Systems*, closets that contain equipment such as washers, dryers, furnaces, or water heaters shall be sprinklered, regardless of size.

**26.3.6.3.6** In existing lodging or rooming houses, sprinkler installations shall not be required in closets not exceeding 24 ft² (2.2 m²) and in bathrooms not exceeding 55 ft² (5.1 m²).

**26.4 Reserved.**

**26.5 Building Services.**

**26.5.1 Utilities.** Utilities shall comply with the provisions of Section 9.1.

**26.5.2 Heating, Ventilating, and Air-Conditioning.**

**26.5.2.1** Heating, ventilating, and air-conditioning equipment shall comply with the provisions of Section 9.2.

**26.5.2.2** Unvented fuel-fired heaters, other than gas space heaters in compliance with NFPA 54, *National Fuel Gas Code*, shall not be used.

**26.5.3 Elevators, Escalators, and Conveyors.** Elevators, escalators, and conveyors shall comply with the provisions of Section 9.4.

**26.6 Reserved.**

**26.7 Operating Features.**

**26.7.1 Contents and Furnishings.**

**26.7.1.1** Contents and furnishings shall not be required to comply with Section 10.3.

**26.7.1.2** Furnishings or decorations of an explosive or highly flammable character shall not be used.

**26.7.1.3** Fire-retardant coatings shall be maintained to retain the effectiveness of the treatment under service conditions encountered in actual use.

# Chapter 27   Reserved

# Chapter 28   New Hotels and Dormitories

**28.1 General Requirements.**

**28.1.1 Application.**

**28.1.1.1** The requirements of this chapter shall apply to new buildings or portions thereof used as hotel or dormitory occupancies. *(See 1.3.1.)*

**28.1.1.2 Administration.** The provisions of Chapter 1, Administration, shall apply.

**28.1.1.3 General.** The provisions of Chapter 4, General, shall apply.

**28.1.1.4** Any dormitory divided into suites of rooms, with one or more bedrooms opening into a living room or study that has a door opening into a common corridor serving a number of suites, shall be classified as an apartment building.

**28.1.1.5** The term *hotel*, wherever used in this *Code*, shall include a hotel, an inn, a club, a motel, a bed and breakfast, or any other structure meeting the definition of hotel.

**28.1.2 Classification of Occupancy.** See 6.1.8 and 28.1.4.2.

**28.1.3 Multiple Occupancies.**

**28.1.3.1** Multiple occupancies shall be in accordance with 6.1.14.

**28.1.3.2** No hotel or dormitory shall have its sole means of egress pass through any nonresidential occupancy in the same building, unless otherwise permitted by 28.1.3.2.1 or 28.1.3.2.2.

**28.1.3.2.1** In buildings that are protected by an automatic sprinkler system in accordance with Section 9.7, hotels and dormitories shall be permitted to have their sole means of egress pass through a nonresidential occupancy in the same building, provided that both of the following criteria are met:

(1) The hotel or dormitory shall comply with Chapter 28.
(2) The sole means of egress from the hotel or dormitory shall not pass through a high hazard contents area, as defined in 6.2.2.4.

**28.1.3.2.2**  In buildings that are not protected by an automatic sprinkler system in accordance with Section 9.7, hotels and dormitories shall be permitted to have their sole means of egress pass through a nonresidential occupancy in the same building, provided that all of the following criteria are met:

(1) The sole means of egress from the hotel or dormitory to the exterior shall be separated from the remainder of the building by fire barriers having a minimum 1-hour fire resistance rating.
(2) The hotel or dormitory shall comply with Chapter 28.
(3) The sole means of egress from the hotel or dormitory shall not pass through a high hazard contents area, as defined in 6.2.2.4.

**28.1.4 Definitions.**

**28.1.4.1 General.** For definitions, see Chapter 3, Definitions.

**28.1.4.2 Special Definitions.** A list of special terms used in this chapter follows:

(1) **Dormitory.** See 3.3.64.
(2) **Guest Room.** See 3.3.130.
(3) **Guest Suite.** See 3.3.272.1.
(4) **Hotel.** See 3.3.143.

**28.1.5 Classification of Hazard of Contents.**

**28.1.5.1**  The contents of residential occupancies shall be classified as ordinary hazard in accordance with 6.2.2.

**28.1.5.2**  For the design of automatic sprinkler systems, the classification of contents in NFPA 13, *Standard for the Installation of Sprinkler Systems*, shall apply.

**28.1.6 Minimum Construction Requirements.** (No special requirements.)

**28.1.7 Occupant Load.** The occupant load, in number of persons for whom means of egress and other provisions are required, shall be determined on the basis of the occupant load factors of Table 7.3.1.2 that are characteristic of the use of the space or shall be determined as the maximum probable population of the space under consideration, whichever is greater.

**28.2 Means of Egress Requirements.**

**28.2.1 General.**

**28.2.1.1**  Means of egress from guest rooms or guest suites to the outside of the building shall be in accordance with Chapter 7 and this chapter.

**28.2.1.2**  Means of escape within the guest room or guest suite shall comply with the provisions of Section 24.2 for one- and two-family dwellings.

**28.2.1.3**  For the purpose of application of the requirements of Chapter 24, the terms *guest room* and *guest suite* shall be synonymous with the terms *dwelling unit* or *living unit*.

**28.2.2 Means of Egress Components.**

**28.2.2.1 General.**

**28.2.2.1.1**  Components of means of egress shall be limited to the types described in 28.2.2.2 through 28.2.2.12.

**28.2.2.1.2**  In buildings, other than high-rise buildings, that are protected throughout by an approved, supervised automatic sprinkler system in accordance with 28.3.5, exit enclosures shall have a minimum 1-hour fire resistance rating, and doors shall have a minimum 1-hour fire protection rating.

**28.2.2.2 Doors.**

**28.2.2.2.1**  Doors complying with 7.2.1 shall be permitted.

**28.2.2.2.2**  Door-locking arrangements shall comply with 28.2.2.2.2.1, 28.2.2.2.2.2, 28.2.2.2.2.3, or 28.2.2.2.2.4.

**28.2.2.2.2.1**  No door in any means of egress shall be locked against egress when the building is occupied.

**28.2.2.2.2.2**  Delayed-egress locks complying with 7.2.1.6.1 shall be permitted, provided that not more than one such device is located in any one egress path.

**28.2.2.2.2.3**  Access-controlled egress doors complying with 7.2.1.6.2 shall be permitted.

**28.2.2.2.2.4**  Elevator lobby exit access door locking in accordance with 7.2.1.6.3 shall be permitted.

**28.2.2.2.3**  Revolving doors complying with 7.2.1.10 shall be permitted.

**28.2.2.3 Stairs.** Stairs complying with 7.2.2 shall be permitted.

**28.2.2.4 Smokeproof Enclosures.** Smokeproof enclosures complying with 7.2.3 shall be permitted.

**28.2.2.5 Horizontal Exits.** Horizontal exits complying with 7.2.4 shall be permitted.

**28.2.2.6 Ramps.** Ramps complying with 7.2.5 shall be permitted.

**28.2.2.7 Exit Passageways.** Exit passageways complying with 7.2.6 shall be permitted.

**28.2.2.8 Reserved.**

**28.2.2.9 Reserved.**

**28.2.2.10 Fire Escape Ladders.** Fire escape ladders complying with 7.2.9 shall be permitted.

**28.2.2.11 Alternating Tread Devices.** Alternating tread devices complying with 7.2.11 shall be permitted.

**28.2.2.12 Areas of Refuge.**

**28.2.2.12.1**  Areas of refuge complying with 7.2.12 shall be permitted, as modified by 28.2.2.12.2.

**28.2.2.12.2\***  In buildings protected throughout by an approved, supervised automatic sprinkler system in accordance with 28.3.5, the two accessible rooms or spaces separated from each other by smoke-resistant partitions in accordance with the definition of area of refuge in 3.3.22 shall not be required.

**28.2.3 Capacity of Means of Egress.**

**28.2.3.1**  The capacity of means of egress shall be in accordance with Section 7.3.

**28.2.3.2**  Street floor exits shall be sufficient for the occupant load of the street floor plus the required capacity of stairs and ramps discharging onto the street floor.

**28.2.3.3\***  Corridors, other than those within individual guest rooms or individual guest suites, shall be of sufficient width to accommodate the required occupant load and shall be not less than 44 in. (1120 mm).

## 28.2.4 Number of Means of Egress.

**28.2.4.1** Means of egress shall comply with all of the following, except as otherwise permitted by 28.2.4.2 and 28.2.4.3:

(1) The number of means of egress shall be in accordance with Section 7.4.
(2) Not less than two separate exits shall be provided on every story.
(3) Not less than two separate exits shall be accessible from every part of every story.

**28.2.4.2** Exit access, as required by 28.2.4.1(3), shall be permitted to include a single exit access path for the distances permitted as common paths of travel by 28.2.5.

**28.2.4.3** A single exit shall be permitted in buildings where the total number of stories does not exceed four, provided that all of the following conditions are met:

(1) There are four or fewer guest rooms or guest suites per story.
(2) The building is protected throughout by an approved, supervised automatic sprinkler system in accordance with 28.3.5.
(3) The exit stairway does not serve more than one-half of a story below the level of exit discharge.
(4) The travel distance from the entrance door of any guest room or guest suite to an exit does not exceed 35 ft (10.7 m).
(5) The exit stairway is completely enclosed or separated from the rest of the building by barriers having a minimum 1-hour fire resistance rating.
(6) All openings between the exit stairway enclosure and the building are protected with self-closing door assemblies having a minimum 1-hour fire protection rating.
(7) All corridors serving as access to exits have a minimum 1-hour fire resistance rating.
(8) Horizontal and vertical separation having a minimum ½-hour fire resistance rating is provided between guest rooms or guest suites.

## 28.2.5 Arrangement of Means of Egress.

**28.2.5.1** Access to all required exits shall be in accordance with Section 7.5, as modified by 28.2.5.2.

**28.2.5.2** The distance between exits addressed by 7.5.1.3 shall not apply to common nonlooped exit access corridors in buildings that have corridor doors from the guest room or guest suite that are arranged such that the exits are located in opposite directions from such doors.

**28.2.5.3** In buildings not protected throughout by an approved, supervised automatic sprinkler system in accordance with 28.3.5, common paths of travel shall not exceed 35 ft (10.7 m); travel within a guest room or guest suite shall not be included when calculating common path of travel.

**28.2.5.4** In buildings protected throughout by an approved, supervised automatic sprinkler system in accordance with 28.3.5, common path of travel shall not exceed 50 ft (15 m); travel within a guest room or guest suite shall not be included when determining common path of travel.

**28.2.5.5** In buildings not protected throughout by an approved, automatic sprinkler system in accordance with 28.3.5, dead-end corridors shall not exceed 35 ft (10.7 m).

**28.2.5.6** In buildings protected throughout by an approved, supervised automatic sprinkler system in accordance with 28.3.5, dead-end corridors shall not exceed 50 ft (15 m).

**28.2.5.7** Any guest room or any guest suite of rooms in excess of 2000 ft² (185 m²) shall be provided with not less than two exit access doors remotely located from each other.

## 28.2.6 Travel Distance to Exits.

**28.2.6.1** Travel distance within a guest room or guest suite to a corridor door shall not exceed 75 ft (23 m) in buildings not protected by an approved, supervised automatic sprinkler system in accordance with 28.3.5.

**28.2.6.2** Travel distance within a guest room or guest suite to a corridor door shall not exceed 125 ft (38 m) in buildings protected by an approved, supervised automatic sprinkler system in accordance with 28.3.5.

**28.2.6.3** Travel distance from the corridor door of any guest room or guest suite to the nearest exit shall comply with 28.2.6.3.1, 28.2.6.3.2, or 28.2.6.3.3.

**28.2.6.3.1** Travel distance from the corridor door of any guest room or guest suite to the nearest exit, measured in accordance with Section 7.6, shall not exceed 100 ft (30 m).

**28.2.6.3.2** Travel distance from the corridor door of any guest room or guest suite to the nearest exit, measured in accordance with Section 7.6, shall not exceed 200 ft (61 m) for exterior ways of exit access arranged in accordance with 7.5.3.

**28.2.6.3.3** Travel distance from the corridor door of any guest room or guest suite to the nearest exit shall comply with 28.2.6.3.3.1 and 28.2.6.3.3.2.

**28.2.6.3.3.1** Travel distance from the corridor door of any guest room or guest suite to the nearest exit shall be measured in accordance with Section 7.6 and shall not exceed 200 ft (61 m) where the exit access and any portion of the building that is tributary to the exit access are protected throughout by an approved, supervised automatic sprinkler system in accordance with 28.3.5.

**28.2.6.3.3.2** Where the building is not protected throughout by an approved, supervised automatic sprinkler system, the 200 ft (61 m) travel distance shall be permitted within any portion of the building that is protected by an approved, supervised automatic sprinkler system, provided that the sprinklered portion of the building is separated from any nonsprinklered portion by fire barriers having a fire resistance rating as follows:

(1) Minimum 1-hour fire resistance rating for buildings three or fewer stories in height
(2) Minimum 2-hour fire resistance rating for buildings four or more stories in height

## 28.2.7 Discharge from Exits.

**28.2.7.1** Exit discharge shall comply with Section 7.7.

**28.2.7.2\*** Any required exit stair that is located so that it is necessary to pass through the lobby or other open space to reach the outside of the building shall be continuously enclosed down to a level of exit discharge or to a mezzanine within a lobby at a level of exit discharge.

**28.2.7.3** The distance of travel from the termination of the exit enclosure to an exterior door leading to a public way shall not exceed 100 ft (30 m).

## 28.2.8 Illumination of Means of Egress.
Means of egress shall be illuminated in accordance with Section 7.8.

## 28.2.9 Emergency Lighting.

**28.2.9.1** Emergency lighting in accordance with Section 7.9 shall be provided.

**28.2.9.2** The requirement of 28.2.9.1 shall not apply where each guest room or guest suite has an exit direct to the outside of the building at street or the finished ground level.

**28.2.10 Marking of Means of Egress.** Means of egress shall have signs in accordance with Section 7.10.

**28.2.11 Special Means of Egress Features.**

**28.2.11.1 Reserved.**

**28.2.11.2 Lockups.** Lockups in hotel and dormitory occupancies shall comply with the requirements of 22.4.5.

**28.2.11.3 Normally Unoccupied Building Service Equipment Support Areas.** The use of Section 7.13 shall be prohibited.

**28.3 Protection.**

**28.3.1 Protection of Vertical Openings.**

**28.3.1.1** Vertical openings shall comply with 28.3.1.1.1 through 28.3.1.2.

**28.3.1.1.1** Vertical openings shall be enclosed or protected in accordance with Section 8.6.

**28.3.1.1.2** Vertical openings in accordance with 8.6.9.1 shall be permitted.

**28.3.1.1.3** In buildings, other than high-rise buildings, that are protected throughout by an approved, supervised automatic sprinkler system in accordance with 28.3.5, the walls enclosing vertical openings shall have a minimum 1-hour fire resistance rating, and doors shall have a minimum 1-hour fire protection rating.

**28.3.1.2** No floor below the level of exit discharge used only for storage, heating equipment, or purposes other than residential occupancy shall have unprotected openings to floors used for residential purposes.

**28.3.2 Protection from Hazards.**

**28.3.2.1 General.** All rooms containing high-pressure boilers, refrigerating machinery, transformers, or other service equipment subject to possible explosion shall not be located directly under or directly adjacent to exits and shall be effectively cut off from other parts of the building as specified in Section 8.7.

**28.3.2.2 Hazardous Areas.**

**28.3.2.2.1** Any hazardous area shall be protected in accordance with Section 8.7.

**28.3.2.2.2** The areas described in Table 28.3.2.2.2 shall be protected as indicated.

**28.3.2.2.3** Where sprinkler protection without fire-rated separation is used, areas shall be separated from other spaces by smoke partitions complying with Section 8.4.

**28.3.3 Interior Finish.**

**28.3.3.1 General.** Interior finish shall be in accordance with Section 10.2.

**28.3.3.2 Interior Wall and Ceiling Finish.** Interior wall and ceiling finish materials complying with Section 10.2 shall be permitted as follows:

(1) Exit enclosures — Class A
(2) Lobbies and corridors — Class A or Class B
(3) Other spaces — Class A, Class B, or Class C

**Table 28.3.2.2.2 Hazardous Area Protection**

| Hazardous Area Description | Separation/Protection[a] |
|---|---|
| Boiler and fuel-fired heater rooms serving more than a single guest room or guest suite | 1 hour and sprinklers |
| Employee locker rooms | 1 hour or sprinklers |
| Gift or retail shops | 1 hour or sprinklers |
| Bulk laundries | 1 hour and sprinklers |
| Guest laundries ≤100 ft² (≤9.3 m²) outside of guest rooms or guest suites | 1 hour or sprinklers[b] |
| Guest laundries >100 ft² (>9.3 m²) outside of guest rooms or guest suites | 1 hour and sprinklers |
| Maintenance shops | 1 hour and sprinklers |
| Storage rooms[c] | 1 hour or sprinklers |
| Trash collection rooms | 1 hour and sprinklers |

[a]Minimum fire resistance rating.
[b]Where sprinklers are provided, the separation specified in 8.7.1.2 and 28.3.2.2.3 is not required.
[c]Where storage areas not exceeding 24 ft² (2.2 m²) are directly accessible from the guest room or guest suite, no separation or protection is required.

**28.3.3.3 Interior Floor Finish.**

**28.3.3.3.1** Interior floor finish shall comply with Section 10.2.

**28.3.3.3.2** Interior floor finish in exit enclosures and exit access corridors and spaces not separated from them by walls complying with 28.3.6.1 shall be not less than Class II.

**28.3.3.3.3** Interior floor finish shall comply with 10.2.7.1 or 10.2.7.2, as applicable.

**28.3.4 Detection, Alarm, and Communications Systems.**

**28.3.4.1 General.** A fire alarm system in accordance with Section 9.6, except as modified by 28.3.4.2 through 28.3.4.6, shall be provided.

**28.3.4.2 Initiation.** The required fire alarm system shall be initiated by each of the following:

(1) Manual means in accordance with 9.6.2
(2) Manual fire alarm box located at the hotel desk or other convenient central control point under continuous supervision by responsible employees
(3) Required automatic sprinkler system
(4) Required automatic detection system other than sleeping room smoke detectors

**28.3.4.3 Notification.**

**28.3.4.3.1\*** Occupant notification shall be provided automatically in accordance with 9.6.3.

**28.3.4.3.2** Positive alarm sequence in accordance with 9.6.3.4 shall be permitted.

**28.3.4.3.3\*** Guest rooms and guest suites specifically required and equipped to accommodate hearing-impaired individuals shall be provided with a visible notification appliance.

**28.3.4.3.4** In occupiable areas, other than guest rooms and guest suites, visible notification appliances shall be provided.

**28.3.4.3.5** Annunciation and annunciation zoning in accordance with 9.6.7 shall be provided in buildings three or more stories in height or having more than 50 guest rooms or guest suites. Annunciation shall be provided at a location readily accessible from the primary point of entry for emergency response personnel.

**28.3.4.3.6** Emergency forces notification shall be provided in accordance with 9.6.4.

**28.3.4.4 Detection.** A corridor smoke detection system in accordance with Section 9.6 shall be provided in buildings other than those protected throughout by an approved, supervised automatic sprinkler system in accordance with 28.3.5.3.

**28.3.4.5* Smoke Alarms.** An approved single-station smoke alarm shall be installed in accordance with 9.6.2.10 in every guest room and every living area and sleeping room within a guest suite.

**28.3.4.6 Carbon Monoxide Alarms and Carbon Monoxide Detection Systems.**

**28.3.4.6.1** Carbon monoxide alarms or carbon monoxide detectors in accordance with Section 9.8 and 28.3.4.6 shall be provided in new hotels and dormitories where either of the following conditions exists:

(1) Guest rooms or guest suites with communicating attached garages, unless otherwise exempted by 28.3.4.6.3
(2) Guest rooms or guest suites containing a permanently installed fuel-burning appliance

**28.3.4.6.2** Where required by 28.3.4.6.1, carbon monoxide alarms or carbon monoxide detectors shall be installed in the following locations:

(1) Outside of each separate guest room or guest suite sleeping area in the immediate vicinity of the sleeping rooms
(2) On every occupiable level of a guest room and guest suite

**28.3.4.6.3** Carbon monoxide alarms and carbon monoxide detectors as specified in 28.3.4.6.1(1) shall not be required in the following locations:

(1) In garages
(2) Within guest rooms or guest suites with communicating attached garages that are open parking structures as defined by the building code
(3) Within guest rooms or guest suites with communicating attached garages that are mechanically ventilated in accordance with the mechanical code

**28.3.4.6.4** Carbon monoxide alarms or carbon monoxide detectors shall be provided in areas other than guest rooms and guest suites in accordance with Section 9.8, as modified by 28.3.4.6.5.

**28.3.4.6.5** Carbon monoxide alarms or carbon monoxide detectors shall be installed in accordance with the manufacturer's published instructions in the locations specified as follows:

(1) On the ceilings of rooms containing permanently installed fuel-burning appliances
(2) Centrally located within occupiable spaces served by the first supply air register from a permanently installed, fuel-burning HVAC system
(3) Centrally located within occupiable spaces adjacent to a communicating attached garage

**28.3.5 Extinguishment Requirements.**

**28.3.5.1** All buildings, other than those complying with 28.3.5.2, shall be protected throughout by an approved, supervised automatic sprinkler system in accordance with 28.3.5.3.

**28.3.5.2** Automatic sprinkler protection shall not be required in buildings where all guest sleeping rooms or guest suites have a door opening directly to either of the following:

(1) Outside at the street or the finished ground level
(2) Exterior exit access arranged in accordance with 7.5.3 in buildings three or fewer stories in height

**28.3.5.3** Where an automatic sprinkler system is installed, either for total or partial building coverage, the system shall be in accordance with Section 9.7, as modified by 28.3.5.4. In buildings four or fewer stories above grade plane, systems in accordance with NFPA 13R, *Standard for the Installation of Sprinkler Systems in Residential Occupancies up to and Including Four Stories in Height*, shall be permitted.

**28.3.5.4** The provisions for draft stops and closely spaced sprinklers in NFPA 13, *Standard for the Installation of Sprinkler Systems*, shall not be required for openings complying with 8.6.9.1 where the opening is within the guest room or guest suite.

**28.3.5.5 Reserved.**

**28.3.5.6** Listed quick-response or listed residential sprinklers shall be used throughout guest rooms and guest room suites.

**28.3.5.7** Open parking structures that comply with NFPA 88A, *Standard for Parking Structures*, and are contiguous with hotels or dormitories shall be exempt from the sprinkler requirements of 28.3.5.1.

**28.3.5.8** In buildings other than those protected throughout with an approved, supervised automatic sprinkler system in accordance with 28.3.5.3, portable fire extinguishers shall be provided as specified in 9.7.4.1 in hazardous areas addressed by 28.3.2.2.

**28.3.6 Corridors.**

**28.3.6.1 Walls.**

**28.3.6.1.1** Exit access corridor walls shall comply with 28.3.6.1.2 or 28.3.6.1.3.

**28.3.6.1.2** In buildings not complying with 28.3.6.1.3, exit access corridor walls shall consist of fire barriers in accordance with Section 8.3 that have not less than a 1-hour fire resistance rating.

**28.3.6.1.3** In buildings protected throughout by an approved, supervised automatic sprinkler system in accordance with 28.3.5, corridor walls shall have a minimum ½-hour fire resistance rating.

**28.3.6.2 Doors.**

**28.3.6.2.1** Doors that open onto exit access corridors shall have not less than a 20-minute fire protection rating in accordance with Section 8.3.

**28.3.6.2.2 Reserved.**

**28.3.6.2.3** Doors that open onto exit access corridors shall be self-closing and self-latching.

**28.3.6.3 Unprotected Openings.**

**28.3.6.3.1** Unprotected openings, other than those from spaces complying with 28.3.6.3.2, shall be prohibited in exit access corridor walls and doors.

**28.3.6.3.2** Spaces shall be permitted to be unlimited in area and open to the corridor, provided that all of the following criteria are met:

(1) The space is not used for guest rooms or guest suites or hazardous areas.
(2) The building is protected throughout by an approved, supervised automatic sprinkler system in accordance with 28.3.5.
(3) The space does not obstruct access to required exits.

**28.3.6.4 Transoms, Louvers, or Transfer Grilles.** Transoms, louvers, or transfer grilles shall be prohibited in walls or doors of exit access corridors.

**28.3.7 Subdivision of Building Spaces.** Buildings shall be subdivided in accordance with 28.3.7.1 or 28.3.7.2.

**28.3.7.1** In buildings not protected throughout by an approved, supervised automatic sprinkler system, each hotel guest room, including guest suites, and dormitory room shall be separated from other guest rooms or dormitory rooms by walls and floors constructed as fire barriers having a minimum 1-hour fire resistance rating.

**28.3.7.2** In buildings protected throughout by an approved, supervised automatic sprinkler system, each hotel guest room, including guest suites, and dormitory room shall be separated from other guest rooms or dormitory rooms by walls and floors constructed as fire barriers having a minimum ½-hour fire resistance rating.

**28.3.7.3** Doors in the barriers required by 28.3.7.1 and 28.3.7.2 shall have a fire protection rating of not less than 20 minutes and shall not be required to be self-closing.

**28.3.8 Special Protection Features. (Reserved)**

**28.4 Special Provisions.**

**28.4.1 High-Rise Buildings.**

**28.4.1.1** High-rise buildings shall comply with Section 11.8.

**28.4.1.2\*** Emergency plans in accordance with Section 4.8 shall be provided and shall include all of the following:

(1) Egress procedures
(2) Methods
(3) Preferred evacuation routes for each event, including appropriate use of elevators

**28.4.2 Reserved.**

**28.5 Building Services.**

**28.5.1 Utilities.** Utilities shall comply with the provisions of Section 9.1.

**28.5.2 Heating, Ventilating, and Air-Conditioning.**

**28.5.2.1** Heating, ventilating, and air-conditioning equipment shall comply with the provisions of Section 9.2, except as otherwise required in this chapter.

**28.5.2.2** Unvented fuel-fired heaters, other than gas space heaters in compliance with NFPA 54, *National Fuel Gas Code,* shall not be used.

**28.5.3 Elevators, Escalators, and Conveyors.**

**28.5.3.1** Elevators, escalators, and conveyors shall comply with the provisions of Section 9.4.

**28.5.3.2\*** In high-rise buildings, one elevator shall be provided with a protected power supply and shall be available for use by the fire department in case of emergency.

**28.5.4 Rubbish Chutes, Incinerators, and Laundry Chutes.** Rubbish chutes, incinerators, and laundry chutes shall comply with the provisions of Section 9.5.

**28.6 Reserved.**

**28.7 Operating Features.**

**28.7.1 Hotel Emergency Organization.**

**28.7.1.1\*** Employees of hotels shall be instructed and drilled in the duties they are to perform in the event of fire, panic, or other emergency.

**28.7.1.2\*** Drills of the emergency organization shall be held at quarterly intervals and shall cover such points as the operation and maintenance of the available first aid fire appliances, the testing of devices to alert guests, and a study of instructions for emergency duties.

**28.7.2 Emergency Duties.** Upon discovery of a fire, employees shall carry out all of the following duties:

(1) Activation of the facility fire protection signaling system, if provided
(2) Notification of the public fire department
(3) Other action as previously instructed

**28.7.3 Drills in Dormitories.** Emergency egress and relocation drills in accordance with Section 4.7 shall be held with sufficient frequency to familiarize occupants with all types of hazards and to establish conduct of the drill as a matter of routine. Drills shall be conducted during peak occupancy periods and shall include suitable procedures to ensure that all persons subject to the drill participate.

**28.7.4 Emergency Instructions for Residents or Guests.**

**28.7.4.1\*** A floor diagram reflecting the actual floor arrangement, exit locations, and room identification shall be posted in a location and manner acceptable to the authority having jurisdiction on, or immediately adjacent to, every guest room door in hotels and in every resident room in dormitories.

**28.7.4.2\*** Fire safety information shall be provided to allow guests to make the decision to evacuate to the outside, to evacuate to an area of refuge, to remain in place, or to employ any combination of the three options.

**28.7.5 Emergency Plans.** Emergency plans in accordance with Section 4.8 shall be provided.

**28.7.6 Contents and Furnishings.**

**28.7.6.1** New draperies, curtains, and other similar loosely hanging furnishings and decorations shall be flame resistant as demonstrated by testing in accordance with NFPA 701, *Standard Methods of Fire Tests for Flame Propagation of Textiles and Films.*

**28.7.6.2 Upholstered Furniture and Mattresses.**

**28.7.6.2.1** Newly introduced upholstered furniture shall meet the criteria specified in 10.3.2.1 and 10.3.3.

**28.7.6.2.2** Newly introduced mattresses shall meet the criteria specified in 10.3.2.2 and 10.3.4.

**28.7.6.3** Furnishings or decorations of an explosive or highly flammable character shall not be used.

**28.7.6.4** Fire-retardant coatings shall be maintained to retain the effectiveness of the treatment under service conditions encountered in actual use.

## Chapter 29   Existing Hotels and Dormitories

**29.1 General Requirements.**

**29.1.1 Application.**

**29.1.1.1** The requirements of this chapter shall apply to existing buildings or portions thereof currently occupied as hotel or dormitory occupancies, unless meeting the requirement of 29.1.1.4.

**29.1.1.2 Administration.** The provisions of Chapter 1, Administration, shall apply.

**29.1.1.3 General.** The provisions of Chapter 4, General, shall apply.

**29.1.1.4** Any dormitory divided into suites of rooms, with one or more bedrooms opening into a living room or study that has a door opening into a common corridor serving a number of suites, shall be classified as an apartment building.

**29.1.1.5** The term *hotel*, wherever used in this *Code*, shall include a hotel, an inn, a club, a motel, a bed and breakfast, or any other structure meeting the definition of hotel.

**29.1.2 Classification of Occupancy.** See 6.1.8 and 29.1.4.2.

**29.1.3 Multiple Occupancies.**

**29.1.3.1** Multiple occupancies shall be in accordance with 6.1.14.

**29.1.3.2** No hotel or dormitory shall have its sole means of egress pass through any nonresidential occupancy in the same building, unless otherwise permitted by 29.1.3.2.1 or 29.1.3.2.2.

**29.1.3.2.1** In buildings that are protected by an automatic sprinkler system in accordance with Section 9.7, hotels and dormitories shall be permitted to have their sole means of egress pass through a nonresidential occupancy in the same building, provided that both of the following criteria are met:

(1) The hotel or dormitory shall comply with Chapter 29.
(2) The sole means of egress from the hotel or dormitory shall not pass through a high hazard contents area, as defined in 6.2.2.4.

**29.1.3.2.2** In buildings that are not protected by an automatic sprinkler system in accordance with Section 9.7, hotels and dormitories shall be permitted to have their sole means of egress pass through a nonresidential occupancy in the same building, provided that all of the following criteria are met:

(1) The sole means of egress from the hotel or dormitory to the exterior shall be separated from the remainder of the building by fire barriers having a minimum 1-hour fire resistance rating.
(2) The hotel or dormitory shall comply with Chapter 29.
(3) The sole means of egress from the hotel or dormitory shall not pass through a high hazard contents area, as defined in 6.2.2.4.

**29.1.4 Definitions.**

**29.1.4.1 General.** For definitions, see Chapter 3, Definitions.

**29.1.4.2 Special Definitions.** A list of special terms used in this chapter follows:

(1) **Dormitory.** See 3.3.64.
(2) **Guest Room.** See 3.3.130.
(3) **Guest Suite.** See 3.3.272.1.
(4) **Hotel.** See 3.3.143.

**29.1.5 Classification of Hazard of Contents.**

**29.1.5.1** The contents of residential occupancies shall be classified as ordinary hazard in accordance with 6.2.2.

**29.1.5.2** For the design of automatic sprinkler systems, the classification of contents in NFPA 13, *Standard for the Installation of Sprinkler Systems*, shall apply.

**29.1.6 Minimum Construction Requirements.** (No special requirements.)

**29.1.7 Occupant Load.** The occupant load, in number of persons for whom means of egress and other provisions are required, shall be determined on the basis of the occupant load factors of Table 7.3.1.2 that are characteristic of the use of the space or shall be determined as the maximum probable population of the space under consideration, whichever is greater.

**29.2 Means of Egress Requirements.**

**29.2.1 General.**

**29.2.1.1** Means of egress from guest rooms or guest suites to the outside of the building shall be in accordance with Chapter 7 and this chapter.

**29.2.1.2** Means of escape within the guest room or guest suite shall comply with the provisions of Section 24.2 for one- and two-family dwellings.

**29.2.1.3** For the purpose of application of the requirements of Chapter 24, the terms *guest room* and *guest suite* shall be synonymous with the terms *dwelling unit* or *living unit*.

**29.2.2 Means of Egress Components.**

**29.2.2.1 General.**

**29.2.2.1.1** Components of means of egress shall be limited to the types described in 29.2.2.2 through 29.2.2.12.

**29.2.2.1.2** In buildings, other than high-rise buildings, that are protected throughout by an approved automatic sprinkler system in accordance with 29.3.5, exit enclosures shall have a minimum 1-hour fire resistance rating, and doors shall have a minimum 1-hour fire protection rating.

**29.2.2.2 Doors.**

**29.2.2.2.1** Doors complying with 7.2.1 shall be permitted.

**29.2.2.2.2** Door-locking arrangements shall comply with 29.2.2.2.2.1, 29.2.2.2.2.2, 29.2.2.2.2.3, or 29.2.2.2.2.4.

**29.2.2.2.2.1** No door in any means of egress shall be locked against egress when the building is occupied.

**29.2.2.2.2.2** Delayed-egress locks complying with 7.2.1.6.1 shall be permitted, provided that not more than one such device is located in any one egress path.

**29.2.2.2.2.3** Access-controlled egress doors complying with 7.2.1.6.2 shall be permitted.

**29.2.2.2.2.4** Elevator lobby exit access door locking in accordance with 7.2.1.6.3 shall be permitted.

**29.2.2.2.3** Revolving doors complying with 7.2.1.10 shall be permitted.

**29.2.2.3 Stairs.** Stairs complying with 7.2.2 shall be permitted.

**29.2.2.4 Smokeproof Enclosures.** Smokeproof enclosures complying with 7.2.3 shall be permitted.

**29.2.2.5 Horizontal Exits.** Horizontal exits complying with 7.2.4 shall be permitted.

**29.2.2.6 Ramps.** Ramps complying with 7.2.5 shall be permitted.

**29.2.2.7 Exit Passageways.** Exit passageways complying with 7.2.6 shall be permitted.

**29.2.2.8\* Escalators.** Escalators previously approved as a component in a means of egress shall be permitted to continue to be considered in compliance.

**29.2.2.9 Fire Escape Stairs.** Fire escape stairs complying with 7.2.8 shall be permitted.

**29.2.2.10 Fire Escape Ladders.** Fire escape ladders complying with 7.2.9 shall be permitted.

**29.2.2.11 Alternating Tread Devices.** Alternating tread devices complying with 7.2.11 shall be permitted.

**29.2.2.12 Areas of Refuge.**

**29.2.2.12.1** Areas of refuge complying with 7.2.12 shall be permitted, as modified by 29.2.2.12.2.

**29.2.2.12.2\*** In buildings protected throughout by an approved, supervised automatic sprinkler system in accordance with 29.3.5, the two accessible rooms or spaces separated from each other by smoke-resistive partitions in accordance with the definition of area of refuge in 3.3.22 shall not be required.

**29.2.3 Capacity of Means of Egress.**

**29.2.3.1** The capacity of means of egress shall be in accordance with Section 7.3.

**29.2.3.2** Street floor exits shall be sufficient for the occupant load of the street floor plus the required capacity of stairs and ramps discharging onto the street floor.

**29.2.4 Number of Means of Egress.**

**29.2.4.1** Means of egress shall comply with all of the following, except as otherwise permitted by 29.2.4.2 and 29.2.4.3:

(1) The number of means of egress shall be in accordance with 7.4.1.1 and 7.4.1.3 through 7.4.1.6.
(2) Not less than two separate exits shall be accessible from every part of every story, including stories below the level of exit discharge and stories occupied for public purposes.

**29.2.4.2** Exit access, as required by 29.2.4.1(2), shall be permitted to include a single exit access path for the distances permitted as common paths of travel by 29.2.5.

**29.2.4.3** A single exit shall be permitted in buildings where the total number of stories does not exceed four, provided that all of the following conditions are met:

(1) There are four or fewer guest rooms or guest suites per story.
(2) The building is protected throughout by an approved, supervised automatic sprinkler system in accordance with 29.3.5.

(3) The exit stairway does not serve more than one-half of a story below the level of exit discharge.
(4) The travel distance from the entrance door of any guest room or guest suite to an exit does not exceed 35 ft (10.7 m).
(5) The exit stairway is completely enclosed or separated from the rest of the building by barriers having a minimum 1-hour fire resistance rating.
(6) All openings between the exit stairway enclosure and the building are protected with self-closing door assemblies having a minimum 1-hour fire protection rating.
(7) All corridors serving as access to exits have a minimum 1-hour fire resistance rating.
(8) Horizontal and vertical separation having a minimum ½-hour fire resistance rating is provided between guest rooms or guest suites.

**29.2.5 Arrangement of Means of Egress.**

**29.2.5.1** Access to all required exits shall be in accordance with Section 7.5.

**29.2.5.2 Reserved.**

**29.2.5.3** In buildings not protected throughout by an approved, supervised automatic sprinkler system in accordance with 29.3.5, common paths of travel shall not exceed 35 ft (10.7 m); travel within a guest room or guest suite shall not be included when calculating common path of travel.

**29.2.5.4** In buildings protected throughout by an approved, supervised automatic sprinkler system in accordance with 29.3.5, common path of travel shall not exceed 50 ft (15 m); travel within a guest room or guest suite shall not be included when determining common path of travel.

**29.2.5.5** Dead-end corridors shall not exceed 50 ft (15 m).

**29.2.6 Travel Distance to Exits.**

**29.2.6.1** Travel distance within a guest room or guest suite to a corridor door shall not exceed 75 ft (23 m) in buildings not protected by an approved, supervised automatic sprinkler system in accordance with 29.3.5.

**29.2.6.2** Travel distance within a guest room or guest suite to a corridor door shall not exceed 125 ft (38 m) in buildings protected by an approved, supervised automatic sprinkler system in accordance with 29.3.5.

**29.2.6.3** Travel distance from the corridor door of any guest room or guest suite to the nearest exit shall comply with 29.2.6.3.1, 29.2.6.3.2, or 29.2.6.3.3.

**29.2.6.3.1** Travel distance from the corridor door of any guest room or guest suite to the nearest exit, measured in accordance with Section 7.6, shall not exceed 100 ft (30 m).

**29.2.6.3.2** Travel distance from the corridor door of any guest room or guest suite to the nearest exit, measured in accordance with Section 7.6, shall not exceed 200 ft (61 m) for exterior ways of exit access arranged in accordance with 7.5.3.

**29.2.6.3.3** Travel distance from the corridor door of any guest room or guest suite to the nearest exit shall comply with 29.2.6.3.3.1 and 29.2.6.3.3.2.

**29.2.6.3.3.1** Travel distance from the corridor door of any guest room or guest suite to the nearest exit shall be measured in accordance with Section 7.6 and shall not exceed 200 ft (61 m) where the exit access and any portion of the building that is tributary to the exit access are protected throughout by an approved, supervised automatic sprinkler system in accordance with 29.3.5.

**29.2.6.3.3.2** Where the building is not protected throughout by an approved, supervised automatic sprinkler system, the 200 ft (61 m) travel distance shall be permitted within any portion of the building that is protected by an approved, supervised automatic sprinkler system, provided that the sprinklered portion of the building is separated from any nonsprinklered portion by fire barriers having a fire resistance rating as follows:

(1) Minimum 1-hour fire resistance rating for buildings three or fewer stories in height
(2) Minimum 2-hour fire resistance rating for buildings four or more stories in height

**29.2.7 Discharge from Exits.**

**29.2.7.1** Exit discharge shall comply with Section 7.7.

**29.2.7.2\*** Any required exit stair that is located so that it is necessary to pass through the lobby or other open space to reach the outside of the building shall be continuously enclosed down to a level of exit discharge or to a mezzanine within a lobby at a level of exit discharge.

**29.2.7.3** The distance of travel from the termination of the exit enclosure to an exterior door leading to a public way shall not exceed 150 ft (46 m) in buildings protected throughout by an approved automatic sprinkler system in accordance with 29.3.5 and shall not exceed 100 ft (30 m) in all other buildings.

**29.2.8 Illumination of Means of Egress.** Means of egress shall be illuminated in accordance with Section 7.8.

**29.2.9 Emergency Lighting.**

**29.2.9.1** Emergency lighting in accordance with Section 7.9 shall be provided in all buildings with more than 25 rooms.

**29.2.9.2** The requirement of 29.2.9.1 shall not apply where each guest room or guest suite has an exit direct to the outside of the building at street or the finished ground level.

**29.2.10 Marking of Means of Egress.** Means of egress shall have signs in accordance with Section 7.10.

**29.2.11 Special Means of Egress Features.**

**29.2.11.1 Reserved.**

**29.2.11.2 Lockups.** Lockups in hotel and dormitory occupancies, other than approved existing lockups, shall comply with the requirements of 23.4.5.

**29.2.11.3 Normally Unoccupied Building Service Equipment Support Areas.** The use of Section 7.13 shall be prohibited.

**29.3 Protection.**

**29.3.1 Protection of Vertical Openings.**

**29.3.1.1** Vertical openings shall comply with 29.3.1.1.1 through 29.3.1.2.

**29.3.1.1.1** Vertical openings shall be enclosed or protected in accordance with Section 8.6.

**29.3.1.1.2** Vertical openings in accordance with 8.6.9.1 shall be permitted.

**29.3.1.1.3** In buildings, other than high-rise buildings, that are protected throughout by an approved automatic sprinkler system in accordance with 29.3.5, and in which exits and required ways of travel thereto are adequately safeguarded against fire and smoke within the building, or where every individual room has direct access to an exterior exit without passing through any public corridor, the protection of vertical openings that are not part of required exits shall not be required where approved by the authority having jurisdiction and where such openings do not endanger required means of egress.

**29.3.1.1.4** In buildings two or fewer stories in height, unprotected openings shall be permitted by the authority having jurisdiction to continue to be used where the building is protected throughout by an approved automatic sprinkler system in accordance with 29.3.5.

**29.3.1.2** No floor below the level of exit discharge used only for storage, heating equipment, or purposes other than residential occupancy shall have unprotected openings to floors used for residential purposes.

**29.3.2 Protection from Hazards.**

**29.3.2.1 General.** All rooms containing high-pressure boilers, refrigerating machinery, transformers, or other service equipment subject to possible explosion shall not be located directly under or directly adjacent to exits and shall be effectively cut off from other parts of the building as specified in Section 8.7.

**29.3.2.2 Hazardous Areas.**

**29.3.2.2.1** Any hazardous area shall be protected in accordance with Section 8.7.

**29.3.2.2.2** The areas described in Table 29.3.2.2.2 shall be protected as indicated.

**29.3.2.2.3** Where sprinkler protection without fire-rated separation is used, areas shall be separated from other spaces by smoke partitions complying with Section 8.4.

**Table 29.3.2.2.2 Hazardous Area Protection**

| Hazardous Area Description | Separation/Protection[a] |
|---|---|
| Boiler and fuel-fired heater rooms serving more than a single guest room or guest suite | 1 hour or sprinklers |
| Employee locker rooms | 1 hour or sprinklers |
| Gift or retail shops >100 ft² (>9.3 m²) | 1 hour or sprinklers[b] |
| Bulk laundries | 1 hour or sprinklers |
| Guest laundries >100 ft² (>9.3 m²) outside of guest rooms or guest suites | 1 hour or sprinklers[b] |
| Maintenance shops | 1 hour and sprinklers |
| Rooms or spaces used for storage of combustible supplies and equipment in quantities deemed hazardous by the authority having jurisdiction[c] | 1 hour or sprinklers |
| Trash collection rooms | 1 hour and sprinklers |

[a]Minimum fire resistance rating.
[b]Where sprinklers are provided, the separation specified in 8.7.1.2 and 29.3.2.2.3 shall not be required.
[c]Where storage areas not exceeding 24 ft² (2.2 m²) are directly accessible from the guest room or guest suite, no separation or protection is required.

### 29.3.3 Interior Finish.

**29.3.3.1 General.** Interior finish shall be in accordance with Section 10.2.

**29.3.3.2 Interior Wall and Ceiling Finish.** Interior wall and ceiling finish materials complying with Section 10.2 shall be permitted as follows:

(1) Exit enclosures — Class A or Class B
(2) Lobbies and corridors — Class A or Class B
(3) Other spaces — Class A, Class B, or Class C

**29.3.3.3 Interior Floor Finish.** In nonsprinklered buildings, newly installed interior floor finish in exits and exit access corridors shall be not less than Class II in accordance with 10.2.7.

### 29.3.4 Detection, Alarm, and Communications Systems.

**29.3.4.1 General.** A fire alarm system in accordance with Section 9.6, except as modified by 29.3.4.2 through 29.3.4.5, shall be provided in buildings, other than those where each guest room has exterior exit access in accordance with 7.5.3 and the building is three or fewer stories in height.

**29.3.4.2 Initiation.** The required fire alarm system shall be initiated by each of the following:

(1) Manual means in accordance with 9.6.2, unless there are other effective means to activate the fire alarm system, such as complete automatic sprinkler or automatic detection systems, with manual fire alarm box in accordance with 29.3.4.2(2) required
(2) Manual fire alarm box located at the hotel desk or other convenient central control point under continuous supervision by responsible employees
(3) Required automatic sprinkler system
(4) Required automatic detection system other than sleeping room smoke detectors

**29.3.4.3 Notification.**

**29.3.4.3.1** Occupant notification shall be provided automatically in accordance with 9.6.3.

**29.3.4.3.2** Positive alarm sequence in accordance with 9.6.3.4, and a presignal system in accordance with 9.6.3.3, shall be permitted.

**29.3.4.3.3 Reserved.**

**29.3.4.3.4 Reserved.**

**29.3.4.3.5 Reserved.**

**29.3.4.3.6\*** Where the existing fire alarm system does not provide for automatic emergency forces notification in accordance with 9.6.4, provisions shall be made for the immediate notification of the public fire department by telephone or other means in case of fire, and, where there is no public fire department, notification shall be made to the private fire brigade.

**29.3.4.3.7** Where a new fire alarm system is installed or the existing fire alarm system is replaced, emergency forces notification shall be provided in accordance with 9.6.4.

**29.3.4.4 Detection. (Reserved)**

**29.3.4.5\* Smoke Alarms.** An approved single-station smoke alarm shall be installed in accordance with 9.6.2.10 in every guest room and every living area and sleeping room within a guest suite.

**29.3.4.5.1** The smoke alarms shall not be required to be interconnected.

**29.3.4.5.2** Single-station smoke alarms without a secondary (standby) power source shall be permitted.

### 29.3.5 Extinguishment Requirements.

**29.3.5.1** All high-rise buildings, other than those where each guest room or guest suite has exterior exit access in accordance with 7.5.3, shall be protected throughout by an approved, supervised automatic sprinkler system in accordance with 29.3.5.3.

**29.3.5.2 Reserved.**

**29.3.5.3\*** Where an automatic sprinkler system is installed, either for total or partial building coverage, the system shall be in accordance with Section 9.7, as modified by 29.3.5.4 and 29.3.5.5. In buildings four or fewer stories above grade plane, systems in accordance with NFPA 13R, *Standard for the Installation of Sprinkler Systems in Residential Occupancies up to and Including Four Stories in Height*, shall be permitted.

**29.3.5.4** The provisions for draft stops and closely spaced sprinklers in NFPA 13, *Standard for the Installation of Sprinkler Systems*, shall not be required for openings complying with 8.6.9.1 where the opening is within the guest room or guest suite.

**29.3.5.5** In guest rooms and in guest room suites, sprinkler installations shall not be required in closets not exceeding 24 ft$^2$ (2.2 m$^2$) and in bathrooms not exceeding 55 ft$^2$ (5.1 m$^2$).

**29.3.5.6 Reserved.**

**29.3.5.7 Reserved.**

**29.3.5.8** In buildings other than those protected throughout with an approved, supervised automatic sprinkler system in accordance with 29.3.5.3, portable fire extinguishers shall be provided as specified in 9.7.4.1 in hazardous areas addressed by 29.3.2.2.

### 29.3.6 Corridors.

**29.3.6.1 Walls.**

**29.3.6.1.1** Exit access corridor walls shall comply with either 29.3.6.1.2 or 29.3.6.1.3.

**29.3.6.1.2** In buildings not complying with 29.3.6.1.3, exit access corridor walls shall consist of fire barriers in accordance with 8.2.3 having a minimum ½-hour fire resistance rating.

**29.3.6.1.3** In buildings protected throughout by an approved automatic sprinkler system in accordance with 29.3.5, no fire resistance rating shall be required, but the walls and all openings therein shall resist the passage of smoke.

**29.3.6.2 Doors.**

**29.3.6.2.1** Doors that open onto exit access corridors, other than those complying with 8.3.4 or in buildings meeting the requirements of 29.3.6.2.2, shall have a minimum 20-minute fire protection rating in accordance with Section 8.3.

**29.3.6.2.2** Where automatic sprinkler protection is provided in the corridor in accordance with 31.3.5.8 through 31.3.5.9, doors shall not be required to have a fire protection rating but shall resist the passage of smoke and be equipped with latches to keep doors tightly closed.

**29.3.6.2.3** Doors that open onto exit access corridors shall be self-closing and self-latching.

**29.3.6.3 Unprotected Openings.**

**29.3.6.3.1** Unprotected openings, other than those from spaces complying with 29.3.6.3.2, shall be prohibited in exit access corridor walls and doors.

**29.3.6.3.2** Spaces shall be permitted to be unlimited in area and open to the corridor, provided that all of the following criteria are met:

(1) The space is not used for guest rooms or guest suites or hazardous areas.
(2) The space is protected throughout by an approved automatic sprinkler system in accordance with 29.3.5.
(3) The space does not obstruct access to required exits.

**29.3.6.4 Transoms, Louvers, or Transfer Grilles.**

**29.3.6.4.1** Transoms, louvers, or transfer grilles shall be prohibited in walls or doors of exit access corridors, unless meeting the requirements of 29.3.6.4.2, 29.3.6.4.3, or 29.3.6.4.4.

**29.3.6.4.2** Existing transoms shall be permitted but shall be fixed in the closed position and shall be covered or otherwise protected to provide a fire resistance rating not less than that of the wall in which they are installed.

**29.3.6.4.3** The requirement of 29.3.6.4.1 shall not apply where a corridor smoke detection system is provided that, when sensing smoke, sounds the building alarm and shuts down return or exhaust fans that draw air into the corridor from the guest rooms. The transfer grille or louver shall be located in the lower one-third of the wall or door height.

**29.3.6.4.4** The requirement of 29.3.6.4.1 shall not apply to buildings protected throughout by an approved automatic sprinkler system complying with 29.3.5 or buildings with corridor sprinkler protection in accordance with 31.3.5.8 through 31.3.5.9. The transfer grille or louver shall be located in the lower one-third of the wall or door height.

**29.3.7 Subdivision of Building Spaces.** In buildings other than those meeting the requirements of 29.3.7.1, 29.3.7.2, or 29.3.7.3, every guest room floor shall be divided into not less than two smoke compartments of approximately the same size by smoke partitions in accordance with Section 8.4.

**29.3.7.1** Smoke partitions shall not be required in buildings protected throughout by an approved automatic sprinkler system in accordance with 29.3.5 or a corridor sprinkler system conforming to 31.3.5.8 through 31.3.5.9.

**29.3.7.2** Smoke partitions shall not be required where each guest room is provided with exterior ways of exit access arranged in accordance with 7.5.3.

**29.3.7.3** Smoke partitions shall not be required where the aggregate corridor length on each floor is not more than 150 ft (46 m).

**29.3.7.4** Additional smoke partitions shall be provided so that the travel distance from a guest room corridor door to a smoke partition shall not exceed 150 ft (46 m).

**29.3.8 Special Protection Features. (Reserved)**

**29.4 Special Provisions.**

**29.4.1 High-Rise Buildings.**

**29.4.1.1** High-rise buildings shall comply with 29.3.5.1.

**29.4.1.2\*** Emergency plans in accordance with Section 4.8 shall be provided and shall include all of the following:

(1) Egress procedures
(2) Methods
(3) Preferred evacuation routes for each event, including appropriate use of elevators

**29.4.2 Reserved.**

**29.5 Building Services.**

**29.5.1 Utilities.** Utilities shall comply with the provisions of Section 9.1.

**29.5.2 Heating, Ventilating, and Air-Conditioning.**

**29.5.2.1** Heating, ventilating, and air-conditioning equipment shall comply with the provisions of Section 9.2, except as otherwise required in this chapter.

**29.5.2.2** Unvented fuel-fired heaters, other than gas space heaters in compliance with NFPA 54, *National Fuel Gas Code*, shall not be used.

**29.5.3 Elevators, Escalators, and Conveyors.** Elevators, escalators, and conveyors shall comply with the provisions of Section 9.4.

**29.5.4 Rubbish Chutes, Incinerators, and Laundry Chutes.** Rubbish chutes, incinerators, and laundry chutes shall comply with the provisions of Section 9.5.

**29.6 Reserved.**

**29.7 Operating Features.**

**29.7.1 Hotel Emergency Organization.**

**29.7.1.1\*** Employees of hotels shall be instructed and drilled in the duties they are to perform in the event of fire, panic, or other emergency.

**29.7.1.2\*** Drills of the emergency organization shall be held at quarterly intervals and shall cover such points as the operation and maintenance of the available first aid fire appliances, the testing of devices to alert guests, and a study of instructions for emergency duties.

**29.7.2 Emergency Duties.** Upon discovery of a fire, employees shall carry out all of the following duties:

(1) Activation of the facility fire protection signaling system, if provided
(2) Notification of the public fire department
(3) Other action as previously instructed

**29.7.3 Drills in Dormitories.** Emergency egress and relocation drills in accordance with Section 4.7 shall be held with sufficient frequency to familiarize occupants with all types of hazards and to establish conduct of the drill as a matter of routine. Drills shall be conducted during peak occupancy periods and shall include suitable procedures to ensure that all persons subject to the drill participate.

**29.7.4 Emergency Instructions for Residents or Guests.**

**29.7.4.1\*** A floor diagram reflecting the actual floor arrangement, exit locations, and room identification shall be posted in a location and manner acceptable to the authority having jurisdiction on, or immediately adjacent to, every guest room door in hotels and in every resident room in dormitories.

**29.7.4.2\*** Fire safety information shall be provided to allow guests to make the decision to evacuate to the outside, to evacuate to an area of refuge, to remain in place, or to employ any combination of the three options.

**29.7.5 Emergency Plans.** Emergency plans in accordance with Section 4.8 shall be provided.

**29.7.6 Contents and Furnishings.**

**29.7.6.1** New draperies, curtains, and other similar loosely hanging furnishings and decorations shall be flame resistant as demonstrated by testing in accordance with NFPA 701, *Standard Methods of Fire Tests for Flame Propagation of Textiles and Films.*

**29.7.6.2 Upholstered Furniture and Mattresses.**

**29.7.6.2.1** Newly introduced upholstered furniture shall meet the criteria specified in 10.3.2.1 and 10.3.3.

**29.7.6.2.2** Newly introduced mattresses shall meet the criteria specified in 10.3.2.2 and 10.3.4.

**29.7.6.3** Furnishings or decorations of an explosive or highly flammable character shall not be used.

**29.7.6.4** Fire-retardant coatings shall be maintained to retain the effectiveness of the treatment under service conditions encountered in actual use.


# Chapter 30    New Apartment Buildings

**30.1 General Requirements.**

**30.1.1 Application.**

**30.1.1.1** The requirements of this chapter shall apply to new buildings or portions thereof used as apartment occupancies. *(See 1.3.1.)*

**30.1.1.2 Administration.** The provisions of Chapter 1, Administration, shall apply.

**30.1.1.3 General.** The provisions of Chapter 4, General, shall apply.

**30.1.1.4** The term *apartment building*, wherever used in this *Code*, shall include an apartment house, a tenement, a garden apartment, or any other structure meeting the definition of apartment building.

**30.1.2 Classification of Occupancy.** See 6.1.8 and 30.1.4.2.

**30.1.3 Multiple Occupancies.**

**30.1.3.1** Multiple occupancies shall be in accordance with 6.1.14.

**30.1.3.2** No dwelling unit of an apartment building shall have its sole means of egress pass through any nonresidential occupancy in the same building, unless otherwise permitted by 30.1.3.2.1 or 30.1.3.2.2.

**30.1.3.2.1** In buildings that are protected by an automatic sprinkler system in accordance with Section 9.7, dwelling units of an apartment building shall be permitted to have their sole means of egress pass through a nonresidential occupancy in the same building, provided that both of the following criteria are met:

(1) The dwelling unit of the apartment building shall comply with Chapter 30.

(2) The sole means of egress from the dwelling unit of the apartment building shall not pass through a high hazard contents area, as defined in 6.2.2.4.

**30.1.3.2.2** In buildings that are not protected by an automatic sprinkler system in accordance with Section 9.7, dwelling units of an apartment building shall be permitted to have their sole means of egress pass through a nonresidential occupancy in the same building, provided that all of the following criteria are met:

(1) The sole means of egress from the dwelling unit of the apartment building to the exterior shall be separated from the remainder of the building by fire barriers having a minimum 1-hour fire resistance rating.

(2) The dwelling unit of the apartment building shall comply with Chapter 30.

(3) The sole means of egress from the dwelling unit of the apartment building shall not pass through a high hazard contents area, as defined in 6.2.2.4.

**30.1.3.3** Multiple dwelling units shall be permitted to be located above a nonresidential occupancy only where one of the following conditions exists:

(1) Where the dwelling units of the residential occupancy and exits therefrom are separated from the nonresidential occupancy by construction having a minimun 1-hour fire resistance rating

(2) Where the nonresidential occupancy is protected throughout by an approved, supervised automatic sprinkler system in accordance with Section 9.7

**30.1.4 Definitions.**

**30.1.4.1 General.** For definitions, see Chapter 3, Definitions.

**30.1.4.2 Special Definitions.**

**30.1.4.2.1 General.** Special terms applicable to this chapter are defined in Chapter 3. Where necessary, other terms are defined in the text.

**30.1.4.2.2 Apartment Building.** See 3.3.36.3.

**30.1.5 Classification of Hazard of Contents.** The contents of residential occupancies shall be classified as ordinary hazard in accordance with 6.2.2.

**30.1.6 Minimum Construction Requirements.** (No special requirements.)

**30.1.7 Occupant Load.** The occupant load, in number of persons for whom means of egress and other provisions are required, shall be determined on the basis of the occupant load factors of Table 7.3.1.2 that are characteristic of the use of the space or shall be determined as the maximum probable population of the space under consideration, whichever is greater.

**30.2 Means of Egress Requirements.**

**30.2.1 General.**

**30.2.1.1** Means of egress from dwelling units to the outside of the building shall be in accordance with Chapter 7 and this chapter.

**30.2.1.2** Means of escape within the dwelling unit shall comply with the provisions of Section 24.2 for one- and two-family dwellings.

**30.2.2 Means of Egress Components.**

**30.2.2.1 General.**

**30.2.2.1.1** Components of means of egress shall be limited to the types described in 30.2.2.2 through 30.2.2.12.

**30.2.2.1.2** In buildings protected throughout by an approved, supervised automatic sprinkler system in accordance with 30.3.5, exit enclosures shall have a minimum 1-hour fire resistance rating, and doors shall have a minimum 1-hour fire protection rating.

**30.2.2.2 Doors.**

**30.2.2.2.1** Doors complying with 7.2.1 shall be permitted.

**30.2.2.2.2** Door-locking arrangements shall comply with 30.2.2.2.2.1, 30.2.2.2.2.2, 30.2.2.2.2.3, or 30.2.2.2.2.4.

**30.2.2.2.2.1*** No door in any means of egress shall be locked against egress when the building is occupied.

**30.2.2.2.2.2** Delayed-egress locks complying with 7.2.1.6.1 shall be permitted, provided that not more than one such device is located in any one egress path.

**30.2.2.2.2.3** Access-controlled egress doors complying with 7.2.1.6.2 shall be permitted.

**30.2.2.2.2.4** Elevator lobby exit access door locking in accordance with 7.2.1.6.3 shall be permitted.

**30.2.2.2.3** Revolving doors complying with 7.2.1.10 shall be permitted.

**30.2.2.2.4** Apartment occupancies shall be exempt from the re-entry provisions of 7.2.1.5.8 where the exit enclosure serves directly only one dwelling unit per floor, and such exit is a smokeproof enclosure in accordance with 7.2.3.

**30.2.2.3 Stairs.**

**30.2.2.3.1** Stairs complying with 7.2.2 shall be permitted.

**30.2.2.3.2 Reserved.**

**30.2.2.3.3** Spiral stairs complying with 7.2.2.2.3 shall be permitted within a single dwelling unit.

**30.2.2.3.4** Winders complying with 7.2.2.2.4 shall be permitted within a single dwelling unit.

**30.2.2.4 Smokeproof Enclosures.** Smokeproof enclosures complying with 7.2.3 shall be permitted.

**30.2.2.5 Horizontal Exits.** Horizontal exits complying with 7.2.4 shall be permitted.

**30.2.2.6 Ramps.** Ramps complying with 7.2.5 shall be permitted.

**30.2.2.7 Exit Passageways.** Exit passageways complying with 7.2.6 shall be permitted.

**30.2.2.8 Reserved.**

**30.2.2.9 Reserved.**

**30.2.2.10 Fire Escape Ladders.** Fire escape ladders complying with 7.2.9 shall be permitted.

**30.2.2.11 Alternating Tread Devices.** Alternating tread devices complying with 7.2.11 shall be permitted.

**30.2.2.12 Areas of Refuge.**

**30.2.2.12.1** Areas of refuge complying with 7.2.12 shall be permitted, as modified by 30.2.2.12.2.

**30.2.2.12.2*** In buildings protected throughout by an approved, supervised automatic sprinkler system in accordance with 30.3.5, the two accessible rooms or spaces separated from each other by smoke-resistive partitions in accordance with the definition of area of refuge in 3.3.22 shall not be required.

**30.2.3 Capacity of Means of Egress.**

**30.2.3.1** The capacity of means of egress shall be in accordance with Section 7.3.

**30.2.3.2** Street floor exits shall be sufficient for the occupant load of the street floor plus the required capacity of stairs and ramps discharging onto the street floor.

**30.2.3.3** Corridors with a required capacity of more than 50 persons, as defined in Section 7.3, shall be of sufficient width to accommodate the required occupant load but have a width of not less than 44 in. (1120 mm).

**30.2.3.4** Corridors with a required capacity of not more than 50 persons, as defined in Section 7.3, shall be not less than 36 in. (915 mm) in width.

**30.2.4 Number of Means of Egress.**

**30.2.4.1** The number of means of egress shall comply with Section 7.4.

**30.2.4.2** The minimum number of exits shall comply with 30.2.4.3, 30.2.4.4, or 30.2.4.6.

**30.2.4.3** Every dwelling unit shall have access to at least two separate exits remotely located from each other as required by 7.5.1.

**30.2.4.4** Dwelling units shall be permitted to have access to a single exit, provided that one of the following conditions is met:

(1) The dwelling unit has an exit door opening directly to the street or yard at the finished ground level.
(2) The dwelling unit has direct access to an outside stair that complies with 7.2.2 and serves a maximum of two units, both of which are located on the same story.
(3) The dwelling unit has direct access to an interior stair that serves only that unit and is separated from all other portions of the building by fire barriers having a minimum 1-hour fire resistance rating, with no opening therein.

**30.2.4.5 Reserved.**

**30.2.4.6** A single exit shall be permitted in buildings where the total number of stories does not exceed four, provided that all of the following conditions are met:

(1) There are four or fewer dwelling units per story.
(2) The building is protected throughout by an approved, supervised automatic sprinkler system in accordance with 30.3.5.
(3) The exit stairway does not serve more than one-half story below the level of exit discharge.
(4) The travel distance from the entrance door of any dwelling unit to an exit does not exceed 35 ft (10.7 m).
(5) The exit stairway is completely enclosed or separated from the rest of the building by barriers having a minimum 1-hour fire resistance rating.

(6) All openings between the exit stairway enclosure and the building are protected with self-closing door assemblies having a minimum 1-hour fire protection rating.

(7) All corridors serving as access to exits have a minimum 1-hour fire resistance rating.

(8) Horizontal and vertical separation having a minimum ½-hour fire resistance rating is provided between dwelling units.

**30.2.5 Arrangement of Means of Egress.**

**30.2.5.1** Access to all required exits shall be in accordance with Section 7.5, as modified by 30.2.5.2.

**30.2.5.2** The distance between exits addressed by 7.5.1.3 shall not apply to nonlooped exit access corridors in buildings that have corridor doors from the dwelling units that are arranged such that the exits are located in opposite directions from such doors.

**30.2.5.3** Common path of travel shall comply with 30.2.5.3.1 or 30.2.5.3.2.

**30.2.5.3.1** No common path of travel shall exceed 35 ft (10.7 m) in buildings not protected throughout by an approved, supervised automatic sprinkler system installed in accordance with 30.3.5. Travel within a dwelling unit shall not be included when calculating common path of travel.

**30.2.5.3.2** No common path of travel shall exceed 50 ft (15 m) in buildings protected throughout by an approved, supervised automatic sprinkler system installed in accordance with 30.3.5. Travel within a dwelling unit shall not be included when determining common path of travel.

**30.2.5.4** Dead-end corridors shall be limited in accordance with either 30.2.5.4.1 or 30.2.5.4.2.

**30.2.5.4.1** Dead-end corridors shall not exceed 35 ft (10.7 m) in buildings not protected throughout by an approved automatic sprinkler system in accordance with 30.3.5.

**30.2.5.4.2** Dead-end corridors shall not exceed 50 ft (15 m) in buildings protected throughout by an approved, supervised automatic sprinkler system in accordance with 30.3.5.

**30.2.6 Travel Distance to Exits.** Travel distance shall be measured in accordance with Section 7.6.

**30.2.6.1** Travel distance within a dwelling unit (apartment) to a corridor door shall not exceed 75 ft (23 m) in buildings not protected throughout by an approved, supervised automatic sprinkler system installed in accordance with 30.3.5.

**30.2.6.2** Travel distance within a dwelling unit (apartment) to a corridor door shall not exceed 125 ft (38 m) in buildings protected throughout by an approved, supervised automatic sprinkler system installed in accordance with 30.3.5.

**30.2.6.3** The travel distance from a dwelling unit (apartment) entrance door to the nearest exit shall be limited in accordance with 30.2.6.3.1, 30.2.6.3.2, or 30.2.6.3.3.

**30.2.6.3.1** The travel distance from a dwelling unit (apartment) entrance door to the nearest exit shall not exceed 100 ft (30 m).

**30.2.6.3.2** In buildings protected throughout by an approved, supervised automatic sprinkler system installed in accordance with 30.3.5, the travel distance from a dwelling unit (apartment) entrance door to the nearest exit shall not exceed 200 ft (61 m).

**30.2.6.3.3** The travel distance from a dwelling unit (apartment) entrance door to the nearest exit shall not exceed 200 ft (61 m) for exterior ways of exit access arranged in accordance with 7.5.3.

**30.2.6.4** The travel distance, from areas other than those within living units, to an exit, shall not exceed 200 ft (61 m), or 250 ft (76 m) in buildings protected throughout by an approved, supervised automatic sprinkler system installed in accordance with 30.3.5.5.

**30.2.7 Discharge from Exits.** Exit discharge shall comply with Section 7.7.

**30.2.8 Illumination of Means of Egress.** Means of egress shall be illuminated in accordance with Section 7.8.

**30.2.9 Emergency Lighting.** Emergency lighting in accordance with Section 7.9 shall be provided in all buildings four or more stories in height, or with more than 12 dwelling units, unless every dwelling unit has a direct exit to the outside of the building at the finished ground level.

**30.2.10 Marking of Means of Egress.** Means of egress shall have signs in accordance with Section 7.10 in all buildings requiring more than one exit.

**30.2.11 Special Means of Egress Features.**

**30.2.11.1 Reserved.**

**30.2.11.2 Lockups.** Lockups in apartment buildings shall comply with the requirements of 22.4.5.

**30.2.11.3 Normally Unoccupied Building Service Equipment Support Areas.** The use of Section 7.13 shall be prohibited.

**30.3 Protection.**

**30.3.1 Protection of Vertical Openings.**

**30.3.1.1** Vertical openings shall comply with 30.3.1.1.1 through 30.3.1.3.

**30.3.1.1.1** Vertical openings shall be enclosed or protected in accordance with Section 8.6.

**30.3.1.1.2** Where the provisions of 8.6.6 are used, the requirements of 30.3.5.7 shall be met.

**30.3.1.1.3** Vertical openings in accordance with 8.6.9.1 shall be permitted.

**30.3.1.1.4** In buildings protected throughout by an approved, supervised automatic sprinkler system in accordance with 30.3.5, walls enclosing vertical openings shall have a minimum 1-hour fire resistance rating, and the doors shall have a minimum 1-hour fire protection rating.

**30.3.1.2** No floor below the level of exit discharge used only for storage, heating equipment, or purposes other than residential occupancy and open to the public shall have unprotected openings to floors used for residential purposes.

**30.3.1.3** Within any individual dwelling unit, unless protected by an approved automatic sprinkler system in accordance with 30.3.5, vertical openings more than one story above or below the entrance floor level of the dwelling unit shall not be permitted.

**30.3.2 Protection from Hazards.**

**30.3.2.1 Hazardous Areas.** Any hazardous area shall be protected in accordance with Section 8.7.

**30.3.2.1.1** The areas described in Table 30.3.2.1.1 shall be protected as indicated.

**Table 30.3.2.1.1 Hazardous Area Protection**

| Hazardous Area Description | Separation/Protection[†] |
|---|---|
| Boiler and fuel-fired heater rooms serving more than a single dwelling unit | 1 hour and sprinklers |
| Employee locker rooms | 1 hour or sprinklers |
| Gift or retail shops | 1 hour or sprinklers |
| Bulk laundries | 1 hour and sprinklers |
| Laundries ≤100 ft$^2$ (≤9.3 m$^2$) outside of dwelling units | 1 hour or sprinklers[‡] |
| Laundries >100 ft$^2$ (>9.3 m$^2$) outside of dwelling units | 1 hour and sprinklers |
| Maintenance shops | 1 hour and sprinklers |
| Storage rooms outside of dwelling units | 1 hour or sprinklers |
| Trash collection rooms | 1 hour and sprinklers |

[†]Minimum fire resistance rating.
[‡]Where sprinklers are provided, the separation specified in 8.7.1.2 and 30.3.2.1.2 is not required.

**30.3.2.1.2** Where sprinkler protection without fire-rated separation is used, areas shall be separated from other spaces by smoke partitions complying with Section 8.4.

**30.3.2.2 Reserved.**

**30.3.3 Interior Finish.**

**30.3.3.1 General.** Interior finish shall be in accordance with Section 10.2.

**30.3.3.2 Interior Wall and Ceiling Finish.** Interior wall and ceiling finish materials complying with Section 10.2 shall be permitted as follows:

(1) Exit enclosures — Class A
(2) Lobbies and corridors — Class A or Class B
(3) Other spaces — Class A, Class B, or Class C

**30.3.3.3 Interior Floor Finish.**

**30.3.3.3.1** Interior floor finish shall comply with Section 10.2.

**30.3.3.3.2** Interior floor finish in exit enclosures and exit access corridors and spaces not separated from them by walls complying with 30.3.6 shall be not less than Class II.

**30.3.3.3.3** Interior floor finish shall comply with 10.2.7.1 or 10.2.7.2, as applicable.

**30.3.4 Detection, Alarm, and Communications Systems.**

**30.3.4.1 General.**

**30.3.4.1.1** Apartment buildings four or more stories in height or with more than 11 dwelling units, other than those meeting the requirements of 30.3.4.1.2, shall be provided with a fire alarm system in accordance with Section 9.6, except as modified by 30.3.4.2 through 30.3.4.5.

**30.3.4.1.2** A fire alarm system shall not be required in buildings where each dwelling unit is separated from other contiguous dwelling units by fire barriers *(see Section 8.3)* having a minimum 1-hour fire resistance rating, and where each dwelling unit has either its own independent exit or its own independent stairway or ramp discharging at the finished ground level.

**30.3.4.2 Initiation.**

**30.3.4.2.1** Initiation of the required fire alarm system shall be by manual means in accordance with 9.6.2, unless the building complies with 30.3.4.2.2.

**30.3.4.2.2** Initiation of the required fire alarm system by manual means shall not be required in buildings four or fewer stories in height, containing not more than 16 dwelling units, and protected throughout by an approved, supervised automatic sprinkler system installed in accordance with 30.3.5.1.

**30.3.4.2.3** In buildings protected throughout by an approved, supervised automatic sprinkler system in accordance with 30.3.5, required fire alarm systems shall be initiated upon operation of the automatic sprinkler system.

**30.3.4.3 Notification.**

**30.3.4.3.1** Occupant notification shall be provided automatically in accordance with Section 9.6, and both of the following shall also apply:

(1) Visible signals shall be installed in units designed for the hearing impaired.
(2) Positive alarm sequence in accordance with 9.6.3.4 shall be permitted.

**30.3.4.3.2** Annunciation, and annunciation zoning, in accordance with 9.6.7 shall be provided, unless the building complies with either 30.3.4.3.3 or 30.3.4.3.4. Annunciation shall be provided at a location readily accessible from the primary point of entry for emergency response personnel.

**30.3.4.3.3** Annunciation, and annunciation zoning, shall not be required in buildings two or fewer stories in height and having not more than 50 dwelling units.

**30.3.4.3.4** Annunciation, and annunciation zoning, shall not be required in buildings four or fewer stories in height containing not more than 16 dwelling units and protected throughout by an approved, supervised automatic sprinkler system installed in accordance with 30.3.5.1.

**30.3.4.3.5** Fire department notification shall be accomplished in accordance with 9.6.4.

**30.3.4.4 Detection. (Reserved)**

**30.3.4.5\* Smoke Alarms.** Smoke alarms shall be installed in accordance with 9.6.2.10 in every sleeping area, outside every sleeping area in the immediate vicinity of the bedrooms, and on all levels of the dwelling unit, including basements.

**30.3.4.6 Carbon Monoxide Alarms and Carbon Monoxide Detection Systems.**

**30.3.4.6.1** Carbon monoxide alarms or carbon monoxide detectors in accordance with Section 9.8 and 30.3.4.6 shall be provided in new apartment buildings where either of the following conditions exists:

(1) Dwelling units with communicating attached garages, unless otherwise exempted by 30.3.4.6.3
(2) Dwelling units containing a permanently installed fuel-burning appliance

**30.3.4.6.2** Where required by 30.3.4.6.1, carbon monoxide alarms or carbon monoxide detectors shall be installed in the following locations:

(1) Outside of each separate dwelling unit sleeping area in the immediate vicinity of the sleeping rooms
(2) On every occupiable level of a dwelling unit

**30.3.4.6.3** Carbon monoxide alarms and carbon monoxide detectors as specified in 30.3.4.6.1(1) shall not be required in the following locations:

(1) In garages
(2) Within dwelling units with communicating attached garages that are open parking structures as defined by the building code
(3) Within dwelling units with communicating attached garages that are mechanically ventilated in accordance with the mechanical code

**30.3.4.6.4** Carbon monoxide alarms or carbon monoxide detectors shall be provided in areas other than dwelling units in accordance with Section 9.8, as modified by 30.3.4.7.5.

**30.3.4.6.5** Carbon monoxide alarms or carbon monoxide detectors shall be installed in accordance with the manufacturer's published instructions in the locations specified as follows:

(1) On the ceilings of rooms containing permanently installed fuel-burning appliances
(2) Centrally located within occupiable spaces served by the first supply air register from a permanently installed, fuel-burning HVAC system
(3) Centrally located within occupiable spaces adjacent to a communicating attached garage

**30.3.5 Extinguishment Requirements.**

**30.3.5.1** All buildings shall be protected throughout by an approved, supervised automatic sprinkler system installed in accordance with 30.3.5.2.

**30.3.5.2** Where an automatic sprinkler system is installed, either for total or partial building coverage, the system shall be installed in accordance with Section 9.7, as modified by 30.3.5.3 and 30.3.5.4. In buildings four or fewer stories above grade plane, systems in accordance with NFPA 13R, *Standard for the Installation of Sprinkler Systems in Residential Occupancies up to and Including Four Stories in Height*, shall be permitted.

**30.3.5.3\*** In buildings sprinklered in accordance with NFPA 13, *Standard for the Installation of Sprinkler Systems*, closets less than 12 ft$^2$ (1.1 m$^2$) in area in individual dwelling units shall not be required to be sprinklered. Closets that contain equipment such as washers, dryers, furnaces, or water heaters shall be sprinklered, regardless of size.

**30.3.5.4** The draft stop and closely spaced sprinkler requirements of NFPA 13, *Standard for the Installation of Sprinkler Systems*, shall not be required for convenience openings complying with 8.6.9.1 where the convenience opening is within the dwelling unit.

**30.3.5.5** Listed quick-response or listed residential sprinklers shall be used throughout all dwelling units.

**30.3.5.6** Open parking structures complying with NFPA 88A, *Standard for Parking Structures*, that are contiguous with apartment buildings shall be exempt from the sprinkler requirements of 30.3.5.1.

**30.3.5.7** Buildings with unprotected openings in accordance with 8.6.6 shall be protected throughout by an approved, supervised automatic sprinkler system in accordance with 30.3.5.

**30.3.5.8 Reserved.**

**30.3.5.9 Reserved.**

**30.3.5.10 Reserved.**

**30.3.5.11 Reserved.**

**30.3.5.12** Portable fire extinguishers in accordance with 9.7.4.1 shall be provided in hazardous areas addressed by 30.3.2.1, unless the building is protected throughout with an approved, supervised automatic sprinkler system in accordance with 30.3.5.2.

**30.3.6 Corridors.**

**30.3.6.1 Walls.** Exit access corridor walls shall comply with 30.3.6.1.1 or 30.3.6.1.2.

**30.3.6.1.1** In buildings not complying with 30.3.6.1.2, exit access corridor walls shall consist of fire barriers in accordance with Section 8.3 that have not less than a 1-hour fire resistance rating.

**30.3.6.1.2** In buildings protected throughout by an approved, supervised automatic sprinkler system in accordance with 30.3.5.2, corridor walls shall have a minimum ½-hour fire resistance rating.

**30.3.6.2 Doors.**

**30.3.6.2.1** Doors that open onto exit access corridors shall have not less than a 20-minute fire protection rating in accordance with Section 8.3.

**30.3.6.2.2 Reserved.**

**30.3.6.2.3** Doors that open onto exit access corridors shall be self-closing and self-latching.

**30.3.6.3 Unprotected Openings.**

**30.3.6.3.1** Unprotected openings, other than those from spaces complying with 30.3.6.3.2, shall be prohibited in exit access corridor walls and doors.

**30.3.6.3.2** Spaces shall be permitted to be unlimited in area and open to the corridor, provided that the following criteria are met:

(1) The space is not used for guest rooms or guest suites or hazardous areas.
(2) The building is protected throughout by an approved, supervised automatic sprinkler system in accordance with 30.3.5.
(3) The space does not obstruct access to required exits.

**30.3.6.4 Transoms, Louvers, or Transfer Grilles.** Transoms, louvers, or transfer grilles shall be prohibited in walls or doors of exit access corridors.

**30.3.7 Subdivisions of Building Spaces.** Buildings shall be subdivided in accordance with 30.3.7.1 or 30.3.7.2.

**30.3.7.1** In buildings not meeting the requirement of 30.3.7.2, dwelling units shall be separated from each other by walls and floors constructed as fire barriers having a minimum 1-hour fire resistance rating.

**30.3.7.2** In buildings protected throughout by an approved, supervised automatic sprinkler system, dwelling units shall be separated from each other by walls and floors constructed as fire barriers having a minimum ½-hour fire resistance rating.

### 30.3.8 Special Protection Features. (Reserved)

### 30.4 Special Provisions.

#### 30.4.1 High-Rise Buildings.

**30.4.1.1** High-rise buildings shall comply with Section 11.8. The provisions of 30.3.5.3 and 30.3.4.5 shall be permitted.

**30.4.1.2\*** Emergency plans in accordance with Section 4.8 shall be provided and shall include all of the following:

(1) Egress procedures
(2) Methods
(3) Preferred evacuation routes for each event, including appropriate use of elevators

#### 30.4.2 Reserved.

### 30.5 Building Services.

**30.5.1 Utilities.** Utilities shall comply with the provisions of Section 9.1.

#### 30.5.2 Heating, Ventilating, and Air-Conditioning.

**30.5.2.1** Heating, ventilating, and air-conditioning equipment shall comply with the provisions of Section 9.2.

**30.5.2.2** Unvented fuel-fired heaters, other than gas space heaters in compliance with NFPA 54, *National Fuel Gas Code*, shall not be used.

**30.5.3 Elevators, Escalators, and Conveyors.** Elevators, escalators, and conveyors shall comply with the provisions of Section 9.4.

**30.5.4 Rubbish Chutes, Incinerators, and Laundry Chutes.** Rubbish chutes, incinerators, and laundry chutes shall comply with the provisions of Section 9.5.

### 30.6 Reserved.

### 30.7 Operating Features.

**30.7.1 Emergency Instructions for Residents of Apartment Buildings.** Emergency instructions shall be provided annually to each dwelling unit to indicate the location of alarms, egress paths, and actions to be taken, both in response to a fire in the dwelling unit and in response to the sounding of the alarm system.

#### 30.7.2 Contents and Furnishings.

**30.7.2.1** Contents and furnishings shall not be required to comply with Section 10.3.

**30.7.2.2** Furnishings or decorations of an explosive or highly flammable character shall not be used outside of dwelling units.

**30.7.2.3** Fire-retardant coatings shall be maintained to retain the effectiveness of the treatment under service conditions encountered in actual use.

## Chapter 31   Existing Apartment Buildings

### 31.1\* General Requirements.

#### 31.1.1 Application.

**31.1.1.1** The requirements of this chapter shall apply to existing buildings or portions thereof currently occupied as apartment occupancies. In addition, the building shall meet the requirements of one of the following options:

(1) Option 1, buildings without fire suppression or detection systems
(2) Option 2, buildings provided with a complete approved automatic fire detection and notification system in accordance with 31.3.4.4
(3) Option 3, buildings provided with approved automatic sprinkler protection in selected areas, as described in 31.3.5.8
(4) Option 4, buildings protected throughout by an approved automatic sprinkler system

**31.1.1.2 Administration.** The provisions of Chapter 1, Administration, shall apply.

**31.1.1.3 General.** The provisions of Chapter 4, General, shall apply.

**31.1.1.4** The term *apartment building*, wherever used in this *Code*, shall include an apartment house, a tenement, a garden apartment, or any other structure meeting the definition of apartment building.

**31.1.2 Classification of Occupancy.** See 6.1.8 and 31.1.4.2.

**31.1.3 Multiple Occupancies.**

**31.1.3.1** Multiple occupancies shall be in accordance with 6.1.14.

**31.1.3.2** No dwelling unit of an apartment building shall have its sole means of egress pass through any nonresidential occupancy in the same building, unless otherwise permitted by 31.1.3.2.1 or 31.1.3.2.2.

**31.1.3.2.1** In buildings that are protected by an automatic sprinkler system in accordance with Section 9.7, dwelling units of an apartment building shall be permitted to have their sole means of egress pass through a nonresidential occupancy in the same building, provided that all of the following criteria are met:

(1) The dwelling unit of the apartment building shall comply with Chapter 31.
(2) The sole means of egress from the dwelling unit of the apartment building shall not pass through a high hazard contents area, as defined in 6.2.2.4.

**31.1.3.2.2** In buildings that are not protected by an automatic sprinkler system in accordance with Section 9.7, dwelling units of an apartment building shall be permitted to have their sole means of egress pass through a nonresidential occupancy in the same building, provided that all of the following criteria are met:

(1) The sole means of egress from the dwelling unit of the apartment building to the exterior shall be separated from the remainder of the building by fire barriers having a minimum 1-hour fire resistance rating.
(2) The dwelling unit of the apartment building shall comply with Chapter 31.
(3) The sole means of egress from the dwelling unit of the apartment building shall not pass through a high hazard contents area, as defined in 6.2.2.4.

**31.1.3.3** Multiple dwelling units shall be permitted to be located above a nonresidential occupancy only where one of the following conditions exists:

(1) Where the dwelling units of the residential occupancy and exits therefrom are separated from the nonresidential occupancy by construction having a minimum 1-hour fire resistance rating

(2) Where the nonresidential occupancy is protected throughout by an approved, supervised automatic sprinkler system in accordance with Section 9.7

(3) Where not more than two dwelling units are located above a nonresidential occupancy that is protected by an automatic fire detection system in accordance with Section 9.6

### 31.1.4 Definitions.

**31.1.4.1 General.** For definitions, see Chapter 3, Definitions.

**31.1.4.2 Special Definitions.**

**31.1.4.2.1 General.** Special terms applicable to this chapter are defined in Chapter 3. Where necessary, other terms are defined in the text.

**31.1.4.2.2 Apartment Building.** See 3.3.36.3.

**31.1.5 Classification of Hazard of Contents.** The contents of residential occupancies shall be classified as ordinary hazard in accordance with 6.2.2.

**31.1.6 Minimum Construction Requirements.** (No special requirements.)

**31.1.7 Occupant Load.** The occupant load, in number of persons for whom means of egress and other provisions are required, shall be determined on the basis of the occupant load factors of Table 7.3.1.2 that are characteristic of the use of the space or shall be determined as the maximum probable population of the space under consideration, whichever is greater.

### 31.2 Means of Egress Requirements.

### 31.2.1 General.

**31.2.1.1** Means of egress from dwelling units to the outside of the building shall be in accordance with Chapter 7 and this chapter.

**31.2.1.2** Means of escape within the dwelling unit shall comply with the provisions of Section 24.2 for one- and two-family dwellings.

### 31.2.2 Means of Egress Components.

**31.2.2.1 General.**

**31.2.2.1.1** Components of means of egress shall be limited to the types described in 31.2.2.2 through 31.2.2.12.

**31.2.2.1.2** In buildings using Option 4, exit enclosures shall have a minimum 1-hour fire resistance rating, and doors shall have a minimum 1-hour fire protection rating.

**31.2.2.1.3** In non-high-rise buildings using Option 2, Option 3, or Option 4, exit stair doors shall be permitted to be 1¾ in. (44 mm) thick, solid-bonded wood-core doors that are self-closing and self-latching and in wood frames not less than ¾ in. (19 mm) thick.

**31.2.2.2 Doors.**

**31.2.2.2.1** Doors complying with 7.2.1 shall be permitted.

**31.2.2.2.2** Door-locking arrangements shall comply with 30.2.2.2.2.1, 30.2.2.2.2.2, 30.2.2.2.2.3, or 31.2.2.2.2.4.

**31.2.2.2.2.1** No door in any means of egress shall be locked against egress when the building is occupied.

**31.2.2.2.2.2** Delayed-egress locks complying with 7.2.1.6.1 shall be permitted, provided that not more than one such device is located in any one egress path.

**31.2.2.2.2.3** Access-controlled egress doors complying with 7.2.1.6.2 shall be permitted.

**31.2.2.2.2.4** Elevator lobby exit access door locking in accordance with 7.2.1.6.3 shall be permitted.

**31.2.2.2.3** Revolving doors complying with 7.2.1.10 shall be permitted.

**31.2.2.2.4** Apartment occupancies protected throughout by an approved, supervised automatic sprinkler system shall be exempt from the re-entry provisions of 7.2.1.5.8 where the exit enclosure serves directly only one dwelling unit per floor, and such exit is a smokeproof enclosure in accordance with 7.2.3.

**31.2.2.3 Stairs.**

**31.2.2.3.1** Stairs complying with 7.2.2 shall be permitted.

**31.2.2.3.2** Within any individual dwelling unit, unless protected by an approved automatic sprinkler system in accordance with 31.3.5, stairs more than one story above or below the entrance floor level of the dwelling unit shall not be permitted.

**31.2.2.3.3** Spiral stairs complying with 7.2.2.2.3 shall be permitted within a single dwelling unit.

**31.2.2.3.4** Winders complying with 7.2.2.2.4 shall be permitted.

**31.2.2.4 Smokeproof Enclosures.** Smokeproof enclosures complying with 7.2.3 shall be permitted. *(See also 31.2.11.1.)*

**31.2.2.5 Horizontal Exits.** Horizontal exits complying with 7.2.4 shall be permitted.

**31.2.2.6 Ramps.** Ramps complying with 7.2.5 shall be permitted.

**31.2.2.7 Exit Passageways.** Exit passageways complying with 7.2.6 shall be permitted.

**31.2.2.8\* Escalators.** Escalators previously approved as a component in the means of egress shall be permitted to continue to be considered as in compliance.

**31.2.2.9 Fire Escape Stairs.** Fire escape stairs complying with 7.2.8 shall be permitted.

**31.2.2.10 Fire Escape Ladders.** Fire escape ladders complying with 7.2.9 shall be permitted.

**31.2.2.11 Alternating Tread Devices.** Alternating tread devices complying with 7.2.11 shall be permitted.

**31.2.2.12 Areas of Refuge.**

**31.2.2.12.1** Areas of refuge complying with 7.2.12 shall be permitted, as modified by 31.2.2.12.2.

**31.2.2.12.2\*** In buildings protected throughout by an approved, supervised automatic sprinkler system in accordance with 31.3.5, the two accessible rooms or spaces separated from each other by smoke-resistive partitions in accordance with the definition of area of refuge in 3.3.22 shall not be required.

### 31.2.3 Capacity of Means of Egress.

**31.2.3.1** The capacity of means of egress shall be in accordance with Section 7.3.

**31.2.3.2** Street floor exits shall be sufficient for the occupant load of the street floor plus the required capacity of stairs and ramps discharging onto the street floor.

**31.2.4 Number of Means of Egress.**

**31.2.4.1** The number of means of egress shall comply with 7.4.1.1 and 7.4.1.3 through 7.4.1.6.

**31.2.4.2** The minimum number of exits shall comply with 31.2.4.3, 31.2.4.4, 31.2.4.5, 31.2.4.6, or 31.2.4.7.

**31.2.4.3** Every dwelling unit shall have access to not less than two separate exits remotely located from each other as required by 7.5.1.

**31.2.4.4** Dwelling units shall be permitted to have access to a single exit, provided that one of the following conditions is met:

(1) The dwelling unit has an exit door opening directly to the street or yard at the finished ground level.
(2) The dwelling unit has direct access to an outside stair that complies with 7.2.2 and serves not more than two units, both located on the same story.
(3) The dwelling unit has direct access to an interior stair that serves only that unit and is separated from all other portions of the building by fire barriers having a minimum 1-hour fire resistance rating, with no opening therein.

**31.2.4.5** A single exit shall be permitted in buildings where the total number of stories does not exceed four, provided that all of the following conditions are met:

(1) The building is protected throughout by an approved, supervised automatic sprinkler system in accordance with 31.3.5.
(2) The exit stairway does not serve more than one-half of a story below the level of exit discharge.
(3) The travel distance from the entrance door of any dwelling unit to an exit does not exceed 35 ft (10.7 m).
(4) The exit stairway is completely enclosed or separated from the rest of the building by barriers having a minimum 1-hour fire resistance rating.
(5) All openings between the exit stairway enclosure and the building are protected with self-closing doors having a minimum 1-hour fire protection rating.
(6) All corridors serving as access to exits have a minimum ½-hour fire resistance rating.
(7) Horizontal and vertical separation having a minimum ½-hour fire resistance rating is provided between dwelling units.

**31.2.4.6\*** A single exit shall be permitted in buildings not exceeding three stories in height, provided that all of the following conditions are met:

(1) The exit stairway does not serve more than one-half of a story below the level of exit discharge.
(2) The travel distance from the entrance door of any dwelling unit to an exit does not exceed 35 ft (10.7 m).
(3) The exit stairway is completely enclosed or separated from the rest of the building by barriers having a minimum 1-hour fire resistance rating.
(4) All openings between the exit stairway enclosure and the building are protected with self-closing doors having a minimum 1-hour fire protection rating.
(5) All corridors serving as access to exits have a minimum ½-hour fire resistance rating.
(6) Horizontal and vertical separation having a minimum ½-hour fire resistance rating is provided between dwelling units.

**31.2.4.7** A building of any height with not more than four dwelling units per floor, with a smokeproof enclosure in accor-

dance with the requirements of 7.2.3 or outside stair as the exit, where such exit is immediately accessible to all dwelling units served thereby, shall be permitted to have a single exit. The term *immediately accessible* means that the travel distance from the entrance door of any dwelling unit to an exit shall not exceed 20 ft (6100 mm).

**31.2.5 Arrangement of Means of Egress.**

**31.2.5.1** Access to all required exits shall be in accordance with Section 7.5.

**31.2.5.2 Reserved.**

**31.2.5.3** Common path of travel shall comply with 31.2.5.3.1 or 31.2.5.3.2.

**31.2.5.3.1** No common path of travel shall exceed 35 ft (10.7 m) in buildings not protected throughout by an approved, supervised automatic sprinkler system installed in accordance with 31.3.5. Travel within a dwelling unit shall not be included when calculating common path of travel.

**31.2.5.3.2** No common path of travel shall exceed 50 ft (15 m) in buildings protected throughout by an approved, supervised automatic sprinkler system installed in accordance with 31.3.5. Travel within a dwelling unit shall not be included when calculating common path of travel.

**31.2.5.4** Dead-end corridors shall not exceed 50 ft (15 m).

**31.2.6 Travel Distance to Exits.** Travel distance shall be measured in accordance with Section 7.6.

**31.2.6.1** Travel distance within a dwelling unit (apartment) to a corridor door shall not exceed the following limits:

(1) For buildings using Option 1 or Option 3, 75 ft (23 m)
(2) For buildings using Option 2 or Option 4, 125 ft (38 m)

**31.2.6.2** The travel distance from a dwelling unit (apartment) entrance door to the nearest exit shall not exceed the following limits, as modified by 31.2.6.3:

(1) For buildings using Option 1, 100 ft (30 m)
(2) For buildings using Option 2 or Option 3, 150 ft (46 m)
(3) For buildings using Option 4, 200 ft (61 m)

**31.2.6.3** Travel distance to exits shall not exceed 200 ft (61 m) for exterior ways of exit access arranged in accordance with 7.5.3.

**31.2.6.4** The travel distance, from areas other than those within living units, to an exit shall not exceed 200 ft (61 m), or 250 ft (76 m) in buildings protected throughout by an approved, supervised automatic sprinkler system installed in accordance with 31.3.5.

**31.2.7 Discharge from Exits.** Exit discharge shall comply with Section 7.7.

**31.2.8 Illumination of Means of Egress.** Means of egress shall be illuminated in accordance with Section 7.8.

**31.2.9 Emergency Lighting.** Emergency lighting in accordance with Section 7.9 shall be provided in all buildings four or more stories in height or with more than 12 dwelling units, unless every dwelling unit has a direct exit to the outside of the building at grade level.

**31.2.10 Marking of Means of Egress.** Means of egress shall have signs in accordance with Section 7.10 in all buildings requiring more than one exit.

**31.2.11 Special Means of Egress Features.**

**31.2.11.1\* High-Rise Buildings.** In high-rise buildings using Option 1, Option 2, or Option 3, smokeproof enclosures shall be provided in accordance with 7.2.3.

**31.2.11.2 Lockups.** Lockups in apartment buildings, other than approved existing lockups, shall comply with the requirements of 23.4.5.

**31.2.11.3 Normally Unoccupied Building Service Equipment Support Areas.** The use of Section 7.13 shall be prohibited.

**31.3 Protection.**

**31.3.1 Protection of Vertical Openings.**

**31.3.1.1** Vertical openings shall comply with 31.3.1.1.1 through 31.3.1.2.

**31.3.1.1.1** Vertical openings shall be enclosed or protected in accordance with Section 8.6.

**31.3.1.1.2 Reserved.**

**31.3.1.1.3** Vertical openings in accordance with 8.6.9.1 shall be permitted.

**31.3.1.1.4** In buildings protected throughout by an approved automatic sprinkler system in accordance with 31.3.5, and in which exits and required ways of travel thereto are adequately safeguarded against fire and smoke within the building, or where every individual room has direct access to an exterior exit without passing through any public corridor, the protection of vertical openings that are not part of required exits shall not be required.

**31.3.1.2** No floor below the level of exit discharge used only for storage, heating equipment, or purposes other than residential occupancy and open to the public shall have unprotected openings to floors used for residential purposes.

**31.3.2 Protection from Hazards.**

**31.3.2.1 Hazardous Areas.** Any hazardous area shall be protected in accordance with Section 8.7.

**31.3.2.1.1** The areas described in Table 31.3.2.1.1 shall be protected as indicated.

**31.3.2.1.2** Where sprinkler protection without fire-rated separation is used, areas shall be separated from other spaces by smoke partitions complying with Section 8.4.

**31.3.2.2 Reserved.**

**31.3.3 Interior Finish.**

**31.3.3.1 General.** Interior finish shall be in accordance with Section 10.2.

**31.3.3.2 Interior Wall and Ceiling Finish.** Interior wall and ceiling finish materials complying with Section 10.2 shall be permitted as follows:

(1) Exit enclosures — Class A or Class B
(2) Lobbies and corridors — Class A or Class B
(3) Other spaces — Class A, Class B, or Class C

**31.3.3.3 Interior Floor Finish.** In buildings utilizing Option 1 or Option 2, newly installed interior floor finish in exits and exit access corridors shall be not less than Class II in accordance with 10.2.7.

**Table 31.3.2.1.1 Hazardous Area Protection**

| Hazardous Area Description | Separation/Protection[†] |
|---|---|
| Boiler and fuel-fired heater rooms serving more than a single dwelling unit | 1 hour or sprinklers |
| Employee locker rooms | 1 hour or sprinklers |
| Gift or retail shops >100 ft² (>9.3 m²) | 1 hour or sprinklers[‡] |
| Bulk laundries | 1 hour or sprinklers |
| Laundries >100 ft² (>9.3 m²) outside of dwelling units | 1 hour or sprinklers[‡] |
| Maintenance shops | 1 hour or sprinklers |
| Rooms or spaces used for storage of combustible supplies and equipment in quantities deemed hazardous by the authority having jurisdiction | 1 hour or sprinklers |
| Trash collection rooms | 1 hour or sprinklers |

[†]Minimum fire resistance rating.
[‡]Where sprinklers are provided, the separation specified in 8.7.1.2 and 31.3.2.1.2 is not required.

**31.3.4 Detection, Alarm, and Communications Systems.**

**31.3.4.1 General.**

**31.3.4.1.1** Apartment buildings four or more stories in height or with more than 11 dwelling units, other than those meeting the requirements of 31.3.4.1.2, shall be provided with a fire alarm system in accordance with Section 9.6, except as modified by 31.3.4.2 through 31.3.4.5.

**31.3.4.1.2** A fire alarm system shall not be required where each dwelling unit is separated from other contiguous dwelling units by fire barriers *(see Section 8.3)* having a minimum ½-hour fire resistance rating, and where each dwelling unit has either its own independent exit or its own independent stairway or ramp discharging at the finished ground level.

**31.3.4.2 Initiation.**

**31.3.4.2.1** Initiation of the required fire alarm system shall be by manual means in accordance with 9.6.2, unless the building complies with 31.3.4.2.2.

**31.3.4.2.2** Initiation of the required fire alarm system by manual means shall not be required in buildings four or fewer stories in height, containing not more than 16 dwelling units, and protected throughout by an approved, supervised automatic sprinkler system installed in accordance with 31.3.5.2.

**31.3.4.2.3** In buildings using Option 2, the required fire alarm system shall be initiated by the automatic fire detection system in addition to the manual initiation means of 31.3.4.2.1.

**31.3.4.2.4** In buildings using Option 3, the required fire alarm system shall be initiated upon operation of the automatic sprinkler system in addition to the manual initiation means of 31.3.4.2.1.

**31.3.4.2.5** In buildings using Option 4, the required fire alarm system shall be initiated upon operation of the automatic sprinkler system in addition to the manual initiation means of 31.3.4.2.1.

### 31.3.4.3 Notification.

**31.3.4.3.1** Occupant notification shall be provided automatically in accordance with Section 9.6, and all of the following shall also apply:

(1) Visible signals shall be installed in units designed for the hearing impaired.
(2) Positive alarm sequence in accordance with 9.6.3.4 shall be permitted.
(3) Existing approved presignal systems shall be permitted in accordance with 9.6.3.3.

**31.3.4.3.2** An annunciator panel, whose location shall be approved by the authority having jurisdiction, connected with the required fire alarm system shall be provided, unless the building meets the requirements of 31.3.4.3.3 or 31.3.4.3.4.

**31.3.4.3.3** Annunciation shall not be required in buildings two or fewer stories in height and having not more than 50 rooms.

**31.3.4.3.4** Annunciation shall not be required in buildings four or fewer stories in height containing not more than 16 dwelling units and protected throughout by an approved, supervised automatic sprinkler system installed in accordance with 31.3.5.2.

**31.3.4.3.5** Fire department notification shall be accomplished in accordance with 9.6.4.

### 31.3.4.4 Detection.

**31.3.4.4.1\*** In buildings using Option 2, a complete automatic fire detection system in accordance with 9.6.1.3 and 31.3.4.4.2 shall be required.

**31.3.4.4.2** Automatic fire detection devices shall be installed as follows:

(1) Smoke detectors shall be installed in all common areas and work spaces outside the living unit, such as exit stairs, egress corridors, lobbies, storage rooms, equipment rooms, and other tenantless spaces in environments that are suitable for proper smoke detector operation.
(2) Heat detectors shall be located within each room of the living unit.

### 31.3.4.5 Smoke Alarms.

**31.3.4.5.1\*** In buildings other than those equipped throughout with an existing, complete automatic smoke detection system, smoke alarms shall be installed in accordance with 9.6.2.10, as modified by 31.3.4.5.2, outside every sleeping area in the immediate vicinity of the bedrooms and on all levels of the dwelling unit, including basements.

**31.3.4.5.2** Smoke alarms required by 31.3.4.5.1 shall not be required to be provided with a secondary (standby) power source.

**31.3.4.5.3** In buildings other than those equipped throughout with an existing, complete automatic smoke detection system or a complete, supervised automatic sprinkler system in accordance with 31.3.5, smoke alarms shall be installed in every sleeping area in accordance with 9.6.2.10, as modified by 31.3.4.5.4.

**31.3.4.5.4** Smoke alarms required by 31.3.4.5.3 shall be permitted to be battery powered.

### 31.3.5 Extinguishment Requirements.

**31.3.5.1 Reserved.**

**31.3.5.2\*** Where an automatic sprinkler system is installed, either for total or partial building coverage, the system shall be installed in accordance with Section 9.7, as modified by 31.3.5.3 and 31.3.5.4. In buildings four or fewer stories above grade plane, systems in accordance with NFPA 13R, *Standard for the Installation of Sprinkler Systems in Residential Occupancies up to and Including Four Stories in Height*, shall be permitted.

**31.3.5.3** In individual dwelling units, sprinkler installation shall not be required in closets not exceeding 24 ft² (2.2 m²) and in bathrooms not exceeding 55 ft² (5.1 m²). Closets that contain equipment such as washers, dryers, furnaces, or water heaters shall be sprinklered, regardless of size.

**31.3.5.4** The draft stop and closely spaced sprinkler requirements of NFPA 13, *Standard for the Installation of Sprinkler Systems*, shall not be required for convenience openings complying with 8.6.9.1 where the convenience opening is within the dwelling unit.

**31.3.5.5 Reserved.**

**31.3.5.6 Reserved.**

**31.3.5.7 Reserved.**

**31.3.5.8** Buildings using Option 3 shall be provided with automatic sprinkler protection installed in accordance with 31.3.5.8.1 through 31.3.5.8.4.

**31.3.5.8.1** Automatic sprinklers shall be installed in the corridor, along the corridor ceiling, utilizing the maximum spacing requirements of the standards referenced by Section 9.7.

**31.3.5.8.2** An automatic sprinkler shall be installed within every dwelling unit that has a door opening to the corridor, with such sprinkler positioned over the center of the door, unless the door to the dwelling unit has not less than a 20-minute fire protection rating and is self-closing.

**31.3.5.8.3** The workmanship and materials of the sprinkler installation specified in 31.3.5.8 shall meet the requirements of Section 9.7.

**31.3.5.8.4** Where Option 3 is being used to permit the use of 1¾ in. (44 mm) thick, solid-bonded wood-core doors in accordance with 31.2.2.1.3, sprinklers shall be provided within the exit enclosures in accordance with NFPA 13, *Standard for the Installation of Sprinkler Systems*.

**31.3.5.9** Buildings using Option 4 shall be protected throughout by an approved automatic sprinkler system in accordance with 31.3.5.2 and meeting the requirements of Section 9.7 for supervision for buildings seven or more stories in height.

**31.3.5.10\*** Where sprinklers are being used as an option to any requirement in this *Code*, the sprinklers shall be installed throughout the space in accordance with the requirements of that option.

### 31.3.5.11 High-Rise Building Sprinklers.

**31.3.5.11.1** All high-rise buildings, other than those meeting 31.3.5.11.2 or 31.3.5.11.3, shall be protected throughout by an approved, supervised automatic sprinkler system in accordance with 31.3.5.2.

**31.3.5.11.2** An automatic sprinkler system shall not be required where every dwelling unit has exterior exit access in accordance with 7.5.3.

**31.3.5.11.3\*** An automatic sprinkler system shall not be required in buildings having an approved, engineered life safety system in accordance with 31.3.5.11.4.

**31.3.5.11.4** Where required by 31.3.5.11.3, an engineered life safety system shall be developed by a registered professional engineer experienced in fire and life safety system design, shall be approved by the authority having jurisdiction, and shall include any or all of the following:

(1) Partial automatic sprinkler protection
(2) Smoke detection systems
(3) Smoke control systems
(4) Compartmentation
(5) Other approved systems

**31.3.5.12** Portable fire extinguishers in accordance with 9.7.4.1 shall be provided in hazardous areas addressed by 31.3.2.1, unless the building is protected throughout with an approved, supervised automatic sprinkler system in accordance with 31.3.5.2.

### 31.3.6 Corridors.

**31.3.6.1\* Walls.** Exit access corridor walls shall consist of fire barriers in accordance with Section 8.3 having a minimum ½-hour fire resistance rating.

### 31.3.6.2 Doors.

**31.3.6.2.1** Doors that open onto exit access corridors, other than those complying with 8.3.4 or in buildings meeting the requirement of 31.3.6.2.2, shall have not less than a 20-minute fire protection rating in accordance with Section 8.3.

**31.3.6.2.2** In buildings using Option 3 or Option 4, doors shall be constructed to resist the passage of smoke.

**31.3.6.2.3** Doors that open onto exit access corridors shall be self-closing and self-latching.

### 31.3.6.3 Unprotected Openings.

**31.3.6.3.1** Unprotected openings, other than those from spaces complying with 31.3.6.3.2, shall be prohibited in exit access corridor walls and doors.

**31.3.6.3.2** Spaces shall be permitted to be unlimited in area and open to the corridor, provided that all of the following criteria are met:

(1) The space is not used for guest rooms or guest suites or hazardous areas.
(2) The building is protected throughout by an approved, supervised automatic sprinkler system in accordance with 31.3.5.2.
(3) The space does not obstruct access to required exits.

**31.3.6.4 Transoms, Louvers, or Transfer Grilles.** Transoms, louvers, or transfer grilles shall be prohibited in walls or doors of exit access corridors.

**31.3.7 Subdivision of Building Spaces — Smoke Partitions.** In buildings other than those meeting the requirements of 31.3.7.1, 31.3.7.2, 31.3.7.3, 31.3.7.4, or 31.3.7.5, both of the following criteria shall be met:

(1) Smoke partitions in accordance with Section 8.4 shall be provided in exit access corridors to establish not less than two compartments of approximately equal size.
(2) The length of each smoke compartment, measured along the corridor, shall not exceed 200 ft (61 m).

**31.3.7.1** Smoke partitions shall not be required in buildings using Option 4.

**31.3.7.2** Smoke partitions shall not be required in buildings having exterior exit access in accordance with 7.5.3 that provides access to two exits.

**31.3.7.3** Smoke partitions shall not be required in buildings complying with 31.2.4.4, 31.2.4.5, 31.2.4.6, or 31.2.4.7.

**31.3.7.4** Smoke partitions shall not be required in buildings with exits not more than 50 ft (15 m) apart.

**31.3.7.5** Smoke partitions shall not be required where each dwelling unit has direct access to the exterior at the finished ground level.

**31.3.8 Special Protection Features. (Reserved)**

### 31.4 Special Provisions.

### 31.4.1 High-Rise Buildings.

**31.4.1.1** High-rise buildings shall comply with 31.2.11.1 and 31.3.5.11.

**31.4.1.2\*** Emergency plans in accordance with Section 4.8 shall be provided and shall include all of the following:

(1) Egress procedures
(2) Methods
(3) Preferred evacuation routes for each event, including appropriate use of elevators

### 31.4.2 Reserved.

### 31.5 Building Services.

**31.5.1 Utilities.** Utilities shall comply with the provisions of Section 9.1.

### 31.5.2 Heating, Ventilating, and Air-Conditioning.

**31.5.2.1** Heating, ventilating, and air-conditioning equipment shall comply with the provisions of Section 9.2.

**31.5.2.2** Unvented fuel-fired heaters, other than gas space heaters in compliance with NFPA 54, *National Fuel Gas Code*, shall not be used.

**31.5.3 Elevators, Escalators, and Conveyors.** Elevators, escalators, and conveyors shall comply with the provisions of Section 9.4.

**31.5.4 Rubbish Chutes, Incinerators, and Laundry Chutes.** Rubbish chutes, incinerators, and laundry chutes shall comply with the provisions of Section 9.5.

### 31.6 Reserved.

### 31.7 Operating Features.

**31.7.1 Emergency Instructions for Residents of Apartment Buildings.** Emergency instructions shall be provided annually to each dwelling unit to indicate the location of alarms, egress paths, and actions to be taken, both in response to a fire in the dwelling unit and in response to the sounding of the alarm system.

**31.7.2 Contents and Furnishings.**

**31.7.2.1** Contents and furnishings shall not be required to comply with Section 10.3.

**31.7.2.2** Furnishings or decorations of an explosive or highly flammable character shall not be used outside of dwelling units.

**31.7.2.3** Fire-retardant coatings shall be maintained to retain the effectiveness of the treatment under service conditions encountered in actual use.

## Chapter 32   New Residential Board and Care Occupancies

### 32.1 General Requirements.

**32.1.1 Application.**

**32.1.1.1 General.** The requirements of this chapter shall apply to new buildings or portions thereof used as residential board and care occupancies. *(See 1.3.1.)*

**32.1.1.2 Administration.** The provisions of Chapter 1, Administration, shall apply.

**32.1.1.3 General.** The provisions of Chapter 4, General, shall apply.

**32.1.1.4 Reserved.**

**32.1.1.5 Chapter Sections.** This chapter is divided into five sections as follows:

(1) Section 32.1 — General Requirements
(2) Section 32.2 — Small Facilities (that is, sleeping accommodations for not more than 16 residents)
(3) Section 32.3 — Large Facilities (that is, sleeping accommodations for more than 16 residents)
(4) Section 32.4 — Suitability of an Apartment Building to House a Board and Care Occupancy *(Sections 32.5 and 32.6 are reserved.)*
(5) Section 32.7 — Operating Features

**32.1.1.6 Conversion.** For the purposes of this chapter, exceptions for conversions shall apply only for a change of occupancy from an existing residential or health care occupancy to a residential board and care occupancy.

**32.1.2 Classification of Occupancy.** See 6.1.9 and 32.1.3.

**32.1.3 Multiple Occupancies.**

**32.1.3.1** Multiple occupancies shall comply with 6.1.14 and 32.1.3 in buildings other than those meeting the requirement of 32.1.3.2.

**32.1.3.2** The requirement of 32.1.3.1 shall not apply to apartment buildings housing residential board and care occupancies in conformance with Section 32.4. In such facilities, any safeguards required by Section 32.4 that are more restrictive than those for other housed occupancies shall apply only to the extent prescribed by Section 32.4.

**32.1.3.3** No board and care occupancy shall have its sole means of egress or means of escape pass through any nonresidential or non-health care occupancy in the same building.

**32.1.3.4** No board and care occupancy shall be located above a nonresidential or non-health care occupancy, unless the board and care occupancy and exits therefrom are separated from the nonresidential or non-health care occupancy by construction having a minimum 2-hour fire resistance rating.

**32.1.4 Definitions.**

**32.1.4.1 General.** For definitions, see Chapter 3, Definitions.

**32.1.4.2 Special Definitions.** A list of special terms used in this chapter follows:

(1) **Personal Care.** See 3.3.206.
(2) **Point of Safety.** See 3.3.211.
(3) **Residential Board and Care Occupancy.** See 3.3.188.12.
(4) **Residential Board and Care Resident.** See 3.3.231.
(5) **Staff (Residential Board and Care).** See 3.3.261.
(6) **Thermal Barrier.** See 3.3.31.3.

**32.1.5 Acceptability of Means of Egress or Escape.** No means of escape or means of egress shall be considered as complying with the minimum criteria for acceptance, unless emergency evacuation drills are regularly conducted using that route in accordance with the requirements of 32.7.3.

**32.1.6\* Fire Resistance–Rated Assemblies.** Fire resistance–rated assemblies shall comply with Section 8.3.

**32.1.7 Reserved.**

**32.1.8 Reserved.**

**32.2 Small Facilities.**

**32.2.1 General.**

**32.2.1.1 Scope.**

**32.2.1.1.1** Section 32.2 shall apply to residential board and care occupancies providing sleeping accommodations for not more than 16 residents.

**32.2.1.1.2** Where there are sleeping accommodations for more than 16 residents, the occupancy shall be classified as a large facility in accordance with Section 32.3.

**32.2.1.2 Reserved.**

**32.2.1.3 Minimum Construction Requirements.** (No requirements.)

**32.2.2 Means of Escape.** Designated means of escape shall be continuously maintained free of all obstructions or impediments to full instant use in the case of fire or emergency.

**32.2.2.1 Reserved.**

**32.2.2.2 Primary Means of Escape.**

**32.2.2.2.1** Every sleeping room and living area shall have access to a primary means of escape located to provide a safe path of travel to the outside.

**32.2.2.2.2** Where sleeping rooms or living areas are above or below the level of exit discharge, the primary means of escape shall be an interior stair in accordance with 32.2.2.4, an exterior stair, a horizontal exit, or a fire escape stair.

**32.2.2.3 Secondary Means of Escape.**

**32.2.2.3.1** Sleeping rooms, other than those complying with 32.2.2.3.2, and living areas in facilities without a sprinkler system installed in accordance with 32.2.3.5 shall have a second means of escape consisting of one of the following:

(1) Door, stairway, passage, or hall providing a way of unobstructed travel to the outside of the dwelling at street or the finished ground level that is independent of, and remotely located from, the primary means of escape

(2) Passage through an adjacent nonlockable space independent of, and remotely located from, the primary means of escape to any approved means of escape

(3)*Outside window or door operable from the inside, without the use of tools, keys, or special effort, that provides a clear opening of not less than 5.7 ft$^2$ (0.53 m$^2$), with the width not less than 20 in. (510 mm), the height not less than 24 in. (610 mm), and the bottom of the opening not more than 44 in. (1120 mm) above the floor, with such means of escape acceptable, provided that one of the following criteria is met:

(a) The window is within 20 ft (6100 mm) of the finished ground level.

(b) The window is directly accessible to fire department rescue apparatus, as approved by the authority having jurisdiction.

(c) The window or door opens onto an exterior balcony.

(4) Windows having a sill height below the adjacent finished ground level that are provided with a window well meeting the following criteria:

(a) The window well has horizontal dimensions that allow the window to be fully opened.

(b) The window well has an accessible net clear opening of not less than 9 ft$^2$ (0.84 m$^2$), with a length and width of not less than 36 in. (915 mm).

(c) A window well with a vertical depth of more than 44 in. (1120 mm) is equipped with an approved permanently affixed ladder or with steps meeting the following criteria:

i. The ladder or steps do not encroach more than 6 in. (150 mm) into the required dimensions of the window well.

ii. The ladder or steps are not obstructed by the window.

**32.2.2.3.2** Sleeping rooms that have a door leading directly to the outside of the building with access to the finished ground level or to an exterior stairway meeting the requirements of 32.2.2.6.3 shall be considered as meeting all the requirements for a second means of escape.

**32.2.2.4 Interior Stairs Used for Primary Means of Escape.** Interior stairs shall be protected in accordance with 32.2.2.4.1 through 32.2.2.4.4, unless they meet the requirement of 32.2.2.4.5, 32.2.2.4.6, or 32.2.2.4.7.

**32.2.2.4.1** Interior stairs shall be enclosed with fire barriers in accordance with Section 8.3 having a minimum ½-hour resistance rating.

**32.2.2.4.2** Stairs shall comply with 7.2.2.5.3.

**32.2.2.4.3** The entire primary means of escape shall be arranged so that occupants are not required to pass through a portion of a lower story, unless that route is separated from all spaces on that story by construction having a minimum ½-hour fire resistance rating.

**32.2.2.4.4** In buildings of construction other than Type II(000), Type III(200), or Type V(000), the supporting construction shall be protected to afford the required fire resistance rating of the supported wall.

**32.2.2.4.5** Stairs that connect a story at street level to only one other story shall be permitted to be open to the story that is not at street level.

**32.2.2.4.6** In buildings three or fewer stories in height and protected by an approved automatic sprinkler system in accor-

dance with 32.2.3.5, stair enclosures shall not be required, provided that there still remains a primary means of escape from each sleeping area that does not require occupants to pass through a portion of a lower floor, unless that route is separated from all spaces on that floor by construction having a minimum ½-hour fire resistance rating.

**32.2.2.4.7** Stairs serving a maximum of two stories in buildings protected with an approved automatic sprinkler system in accordance with 32.2.3.5 shall be permitted to be unenclosed.

**32.2.2.5 Doors.**

**32.2.2.5.1** Doors, other than those meeting the requirements of 32.2.2.5.1.1 and 32.2.2.5.1.2, and paths of travel to a means of escape shall be not less than 32 in. (810 mm) wide.

**32.2.2.5.1.1** Bathroom doors shall be not less than 24 in. (610 mm) wide.

**32.2.2.5.1.2** In conversions (*see 32.1.1.6*), 28 in. (710 mm) doors shall be permitted.

**32.2.2.5.2** Doors shall be swinging or sliding.

**32.2.2.5.3** Every closet door latch shall be readily opened from the inside.

**32.2.2.5.4** Every bathroom door shall be designed to allow opening from the outside during an emergency when locked.

**32.2.2.5.5** No door in any means of escape, other than those meeting the requirement of 32.2.2.5.5.1 or 32.2.2.5.5.2, shall be locked against egress when the building is occupied.

**32.2.2.5.5.1** Delayed-egress locks complying with 7.2.1.6.1 shall be permitted on exterior doors only.

**32.2.2.5.5.2** Access-controlled egress locks complying with 7.2.1.6.2 shall be permitted.

**32.2.2.5.6** Forces to open doors shall comply with 7.2.1.4.5.

**32.2.2.5.7** Door-latching devices shall comply with 7.2.1.5.10.

**32.2.2.5.8** Floor levels at doors shall comply with 7.2.1.3.

**32.2.2.6 Stairs.**

**32.2.2.6.1** Stairs shall comply with 7.2.2, unless otherwise specified in this chapter.

**32.2.2.6.2** Winders complying with 7.2.2.2.4 shall be permitted only in conversions.

**32.2.2.6.3*** Exterior stairs shall be protected against blockage caused by fire within the building.

**32.2.3 Protection.**

**32.2.3.1 Protection of Vertical Openings.**

**32.2.3.1.1 Reserved.**

**32.2.3.1.2** Vertical openings, other than those meeting the requirement of 32.2.3.1.4, shall be separated by smoke partitions in accordance with Section 8.4 having a minimum ½-hour fire resistance rating.

**32.2.3.1.3 Reserved.**

**32.2.3.1.4** Stairs shall be permitted to be open where complying with 32.2.2.4.6 or 32.2.2.4.7.

**32.2.3.2 Hazardous Areas.**

**32.2.3.2.1*** Any space where there is storage or activity having fuel conditions exceeding those of a one- or two-family dwelling and that possesses the potential for a fully involved fire shall be protected in accordance with 32.2.3.2.4 and 32.2.3.2.5.

**32.2.3.2.2** Spaces requiring protection in accordance with 32.2.3.2.1 shall include, but shall not be limited to, areas for cartoned storage, food or household maintenance items in wholesale or institutional-type quantities and concentrations, or mass storage of residents' belongings.

**32.2.3.2.3 Reserved.**

**32.2.3.2.4** Any hazardous area that is on the same floor as, and is in or abuts, a primary means of escape or a sleeping room shall be protected by one of the following means:

(1) Protection shall be an enclosure having a minimum 1-hour fire resistance rating, in accordance with 8.2.3, and an automatic fire detection system connected to the fire alarm system provided in 32.2.3.4.1.
(2) Protection shall be automatic sprinkler protection, in accordance with 32.2.3.5, and a smoke partition, in accordance with Section 8.4, located between the hazardous area and the sleeping area or primary escape route, with any doors in such separation self-closing or automatic-closing in accordance with 7.2.1.8.

**32.2.3.2.5** Other hazardous areas shall be protected by one of the following:

(1) Enclosure having a minimum ½-hour fire resistance rating, with a self-closing or automatic-closing door in accordance with 7.2.1.8 that is equivalent to minimum 1¾ in. (44 mm) thick, solid-bonded wood-core construction, and protected by an automatic fire detection system connected to the fire alarm system provided in 32.2.3.4.1
(2) Automatic sprinkler protection in accordance with 32.2.3.5, regardless of enclosure

**32.2.3.3 Interior Finish.**

**32.2.3.3.1 General.** Interior finish shall be in accordance with Section 10.2.

**32.2.3.3.2 Interior Wall and Ceiling Finish.** Interior wall and ceiling finish materials complying with Section 10.2 shall be Class A, Class B, or Class C.

**32.2.3.3.3 Interior Floor Finish.**

**32.2.3.3.3.1** Interior floor finish shall comply with Section 10.2.

**32.2.3.3.3.2** Interior floor finish shall comply with 10.2.7.1 or 10.2.7.2, as applicable.

**32.2.3.4 Detection, Alarm, and Communications Systems.**

**32.2.3.4.1 Fire Alarm Systems.** A manual fire alarm system shall be provided in accordance with Section 9.6.

**32.2.3.4.2 Occupant Notification.** Occupant notification shall be provided automatically, without delay, in accordance with 9.6.3.

**32.2.3.4.3 Smoke Alarms.**

**32.2.3.4.3.1** Approved smoke alarms shall be provided in accordance with 9.6.2.10.

**32.2.3.4.3.2** Smoke alarms shall be installed on all levels, including basements but excluding crawl spaces and unfinished attics.

**32.2.3.4.3.3** Additional smoke alarms shall be installed in all living areas, as defined in 3.3.21.5.

**32.2.3.4.3.4** Each sleeping room shall be provided with an approved smoke alarm in accordance with 9.6.2.10.

**32.2.3.5\* Extinguishment Requirements.**

**32.2.3.5.1\*** All facilities, other than those meeting the requirement of 32.2.3.5.2, shall be protected throughout by an approved automatic sprinkler system, installed in accordance with 32.2.3.5.3, using quick-response or residential sprinklers.

**32.2.3.5.2\*** In conversions, sprinklers shall not be required in small board and care homes serving eight or fewer residents when all occupants have the ability as a group to move reliably to a point of safety within 3 minutes.

**32.2.3.5.3** Where an automatic sprinkler system is installed, for either total or partial building coverage, all of the following requirements shall be met:

(1) The system shall be in accordance with NFPA 13, *Standard for the Installation of Sprinkler Systems*, and shall initiate the fire alarm system in accordance with 32.2.3.4.1.
(2) The adequacy of the water supply shall be documented to the authority having jurisdiction.

**32.2.3.5.3.1** In buildings four or fewer stories above grade plane, systems in accordance with NFPA 13R, *Standard for the Installation of Sprinkler Systems in Residential Occupancies up to and Including Four Stories in Height*, shall be permitted. All habitable areas, closets, roofed porches, roofed decks, and roofed balconies shall be sprinklered.

**32.2.3.5.3.2\*** An automatic sprinkler system with a 30-minute water supply, and complying with all of the following requirements and with NFPA 13D, *Standard for the Installation of Sprinkler Systems in One- and Two-Family Dwellings and Manufactured Homes*, shall be permitted:

(1) All habitable areas, closets, roofed porches, roofed decks, and roofed balconies shall be sprinklered.
(2) Facilities with more than eight residents shall be treated as two-family dwellings with regard to water supply.

**32.2.3.5.4** Automatic sprinkler systems installed in accordance with NFPA 13, *Standard for the Installation of Sprinkler Systems*, and NFPA 13R, *Standard for the Installation of Sprinkler Systems in Residential Occupancies up to and Including Four Stories in Height*, shall be provided with electrical supervision in accordance with 9.7.2.

**32.2.3.5.5** Automatic sprinkler systems installed in accordance with NFPA 13D, *Standard for the Installation of Sprinkler Systems in One- and Two-Family Dwellings and Manufactured Homes*, shall be provided with valve supervision by one of the following methods:

(1) Single listed control valve that shuts off both domestic and sprinkler systems and separate shutoff for the domestic system only
(2) Electrical supervision in accordance with 9.7.2
(3) Valve closure that causes the sounding of an audible signal in the facility

**32.2.3.5.6** Sprinkler piping serving not more than six sprinklers for any isolated hazardous area shall be permitted to be installed in accordance with 9.7.1.2 and shall meet all of the following requirements:

(1) In new installations, where more than two sprinklers are installed in a single area, waterflow detection shall be provided to initiate the fire alarm system required by 32.2.3.4.1.
(2) The duration of water supplies shall be as required by 32.2.3.5.3.2.

**32.2.3.5.7** Attics shall be protected in accordance with 32.2.3.5.7.1 or 32.2.3.5.7.2.

**32.2.3.5.7.1** Where an automatic sprinkler system is required by 32.2.3.5, attics used for living purposes, storage, or fuel-fired equipment shall be protected with automatic sprinklers that are part of the required, approved automatic sprinkler system in accordance with 9.7.1.1.

**32.2.3.5.7.2** Where an automatic sprinkler system is required by 32.2.3.5, attics not used for living purposes, storage, or fuel-fired equipment shall meet one of the following criteria:

(1) Attics shall be protected throughout by a heat detection system arranged to activate the building fire alarm system in accordance with Section 9.6.
(2) Attics shall be protected with automatic sprinklers that are part of the required, approved automatic sprinkler system in accordance with 9.7.1.1.
(3) Attics shall be of noncombustible or limited-combustible construction.
(4) Attics shall be constructed of fire-retardant-treated wood in accordance with NFPA 703, *Standard for Fire Retardant–Treated Wood and Fire-Retardant Coatings for Building Materials.*

**32.2.3.5.8** Systems installed in accordance with NFPA 13D, *Standard for the Installation of Sprinkler Systems in One- and Two-Family Dwellings and Manufactured Homes,* shall be inspected, tested, and maintained in accordance with 32.2.3.5.8.1 through 32.2.3.5.8.15, which reference specific sections of NFPA 25, *Standard for the Inspection, Testing, and Maintenance of Water-Based Fire Protection Systems.* The frequency of the inspection, test, or maintenance shall be in accordance with this *Code,* whereas the purpose and procedure shall be from NFPA 25.

**32.2.3.5.8.1** Control valves shall be inspected monthly in accordance with 13.3.2 of NFPA 25, *Standard for the Inspection, Testing, and Maintenance of Water-Based Fire Protection Systems.*

**32.2.3.5.8.2** Gages shall be inspected monthly in accordance with 13.2.7.1 of NFPA 25, *Standard for the Inspection, Testing, and Maintenance of Water-Based Fire Protection Systems.*

**32.2.3.5.8.3** Alarm devices shall be inspected quarterly in accordance with 5.2.6 of NFPA 25, *Standard for the Inspection, Testing, and Maintenance of Water-Based Fire Protection Systems.*

**32.2.3.5.8.4** Alarm devices shall be tested semiannually in accordance with 5.3.3 of NFPA 25, *Standard for the Inspection, Testing, and Maintenance of Water-Based Fire Protection Systems.*

**32.2.3.5.8.5** Valve supervisory switches shall be tested semiannually in accordance with 13.3.3.5 of NFPA 25, *Standard for the Inspection, Testing, and Maintenance of Water-Based Fire Protection Systems.*

**32.2.3.5.8.6** Visible sprinklers shall be inspected annually in accordance with 5.2.1 of NFPA 25, *Standard for the Inspection, Testing, and Maintenance of Water-Based Fire Protection Systems.*

**32.2.3.5.8.7** Visible pipe shall be inspected annually in accordance with 5.2.2 of NFPA 25, *Standard for the Inspection, Testing, and Maintenance of Water-Based Fire Protection Systems.*

**32.2.3.5.8.8** Visible pipe hangers shall be inspected annually in accordance with 5.2.3 of NFPA 25, *Standard for the Inspection, Testing, and Maintenance of Water-Based Fire Protection Systems.*

**32.2.3.5.8.9** Buildings shall be inspected annually prior to the onset of freezing weather to ensure that there is adequate heat wherever water-filled piping is run in accordance with 5.2.5 of NFPA 25, *Standard for the Inspection, Testing, and Maintenance of Water-Based Fire Protection Systems.*

**32.2.3.5.8.10** A representative sample of fast-response sprinklers shall be tested once the sprinklers in the system are 20 years old in accordance with 5.3.1.1.1.2 of NFPA 25, *Standard for the Inspection, Testing, and Maintenance of Water-Based Fire Protection Systems.* If the sample fails the test, all of the sprinklers represented by that sample shall be replaced. If the sprinklers pass the test, the test shall be repeated every 10 years thereafter.

**32.2.3.5.8.11** A representative sample of dry-pendent sprinklers shall be tested once the sprinklers in the system are 10 years old in accordance with 5.3.1.1.1.5 of NFPA 25, *Standard for the Inspection, Testing, and Maintenance of Water-Based Fire Protection Systems.* If the sample fails the test, all of the sprinklers represented by that sample shall be replaced. If the sprinklers pass the test, the test shall be repeated every 10 years thereafter.

**32.2.3.5.8.12** Antifreeze solutions shall be tested annually in accordance with 5.3.4 of NFPA 25, *Standard for the Inspection, Testing, and Maintenance of Water-Based Fire Protection Systems.*

**32.2.3.5.8.13** Control valves shall be operated through their full range and returned to normal annually in accordance with 13.3.3.1 of NFPA 25, *Standard for the Inspection, Testing, and Maintenance of Water-Based Fire Protection Systems.*

**32.2.3.5.8.14** Operating stems of OS&Y valves shall be lubricated annually in accordance with 13.3.4 of NFPA 25, *Standard for the Inspection, Testing, and Maintenance of Water-Based Fire Protection Systems.*

**32.2.3.5.8.15** Dry-pipe systems that extend into the unheated portions of the building shall be inspected, tested, and maintained in accordance with 13.4.4 of NFPA 25, *Standard for the Inspection, Testing, and Maintenance of Water-Based Fire Protection Systems.*

**32.2.3.6 Construction of Corridor Walls.**

**32.2.3.6.1** Corridor walls, other than those meeting the provisions of 32.2.3.6.2, shall meet all of the following requirements:

(1) Walls separating sleeping rooms shall have a minimum ½-hour fire resistance rating. The minimum ½-hour fire resistance rating shall be considered to be achieved if the partitioning is finished on both sides with lath and plaster or materials providing a 15-minute thermal barrier.
(2) Sleeping room doors shall be substantial doors, such as those of 1¾ in. (44 mm) thick, solid-bonded wood-core construction or of other construction of equal or greater stability and fire integrity.
(3) Any vision panels shall be fixed fire window assemblies in accordance with 8.3.4 or shall be wired glass not exceeding 9 ft² (0.84 m²) each in area and installed in approved frames.

**32.2.3.6.2** The requirements of 32.2.3.6.1 shall not apply to corridor walls that are smoke partitions in accordance with Section 8.4 where the facility is protected in accordance with 32.2.3.5, and all of the following shall also apply:

(1) In such instances, there shall be no limitation on the type or size of glass panels.
(2) Door closing shall comply with 32.2.3.6.4.

**32.2.3.6.3** No louvers, operable transoms, or other air passages shall penetrate the wall, except properly installed heating and utility installations other than transfer grilles, which shall be prohibited.

**32.2.3.6.4** Doors shall meet all of the following requirements:

(1) Doors shall be provided with latches or other mechanisms suitable for keeping the doors closed.
(2) No doors shall be arranged to prevent the occupant from closing the door.
(3) Doors shall be self-closing or automatic-closing in accordance with 7.2.1.8 in buildings other than those protected throughout by an approved automatic sprinkler system in accordance with 32.2.3.5.

**32.2.4 Reserved.**

**32.2.5 Building Services.**

**32.2.5.1 Utilities.** Utilities shall comply with Section 9.1.

**32.2.5.2 Heating, Ventilating, and Air-Conditioning.**

**32.2.5.2.1** Heating, ventilating, and air-conditioning equipment shall comply with 9.2.1 and 9.2.2, unless otherwise required in this chapter.

**32.2.5.2.2** No stove or combustion heater shall be located to block escape in case of fire caused by the malfunction of the stove or heater.

**32.2.5.2.3** Unvented fuel-fired heaters shall not be used in any residential board and care facility.

**32.2.5.3 Elevators, Escalators, and Conveyors.** Elevators, escalators, and conveyors shall comply with Section 9.4.

**32.3 Large Facilities.**

**32.3.1 General.**

**32.3.1.1 Scope.**

**32.3.1.1.1** Section 32.3 shall apply to residential board and care occupancies providing sleeping accommodations for more than 16 residents.

**32.3.1.1.2** Facilities having sleeping accommodations for not more than 16 residents shall comply with Section 32.2.

**32.3.1.2 Reserved.**

**32.3.1.3 Minimum Construction Requirements.** Large board and care facilities shall be limited to the building construction types specified in Table 32.3.1.3 *(see 8.2.1)*, based on the number of stories in height as defined in 4.6.3.

**32.3.1.4 Occupant Load.** The occupant load, in number of persons for whom means of egress and other provisions are required, shall be determined on the basis of the occupant load factors of Table 7.3.1.2 that are characteristic of the use of the space, or shall be determined as the maximum probable population of the space under consideration, whichever is greater.

**32.3.2 Means of Egress.**

**32.3.2.1 General.**

**32.3.2.1.1** Means of egress from resident rooms and resident dwelling units to the outside of the building shall be in accordance with Chapter 7 and this chapter.

**32.3.2.1.2** Means of escape within the resident room or resident dwelling unit shall comply with Section 24.2 for one- and two-family dwellings.

**32.3.2.2 Means of Egress Components.**

**32.3.2.2.1 Components Permitted.** Components of means of egress shall be limited to the types described in 32.3.2.2.2 through 32.3.2.2.10.

**32.3.2.2.2 Doors.** Doors in means of egress shall meet all of the following criteria:

(1) Doors complying with 7.2.1 shall be permitted.
(2) Doors within individual rooms and suites of rooms shall be permitted to be swinging or sliding.
(3) No door, other than those meeting the requirement of 32.3.2.2.2(4) or (5), shall be equipped with a lock or latch that requires the use of a tool or key from the egress side.
(4) Delayed-egress locks in accordance with 7.2.1.6.1 shall be permitted.
(5) Access-controlled egress doors in accordance with 7.2.1.6.2 shall be permitted.
(6) Doors located in the means of egress that are permitted to be locked under other provisions of Chapter 32, other than those meeting the requirement of 32.3.2.2.2(4) or (5), shall have adequate provisions made for the rapid removal of occupants by means such as remote control of locks, keying of all locks to keys carried by staff at all times, or other such reliable means available to staff at all times.
(7) Only one such locking device, as described in 32.3.2.2.2(6), shall be permitted on each door.

**32.3.2.2.3 Stairs.** Stairs complying with 7.2.2 shall be permitted.

**32.3.2.2.4 Smokeproof Enclosures.** Smokeproof enclosures complying with 7.2.3 shall be permitted.

**32.3.2.2.5 Horizontal Exits.** Horizontal exits complying with 7.2.4 shall be permitted.

**32.3.2.2.6 Ramps.** Ramps complying with 7.2.5 shall be permitted.

**32.3.2.2.7 Exit Passageways.** Exit passageways complying with 7.2.6 shall be permitted.

**32.3.2.2.8 Fire Escape Ladders.** Fire escape ladders complying with 7.2.9 shall be permitted.

**32.3.2.2.9 Alternating Tread Devices.** Alternating tread devices complying with 7.2.11 shall be permitted.

**32.3.2.2.10 Areas of Refuge.** Areas of refuge complying with 7.2.12 shall be permitted.

**32.3.2.3 Capacity of Means of Egress.**

**32.3.2.3.1** The capacity of means of egress shall be in accordance with Section 7.3.

**32.3.2.3.2** Street floor exits shall be sufficient for the occupant load of the street floor plus the required capacity of stairs and ramps discharging onto the street floor.

**32.3.2.3.3** The width of corridors shall be sufficient for the occupant load served but shall be not less than 60 in. (1525 mm).

**32.3.2.4 Number of Means of Egress.**

**32.3.2.4.1** Means of egress shall comply with the following, except as otherwise permitted by 32.3.2.4.2:

(1) The number of means of egress shall be in accordance with Section 7.4.
(2) Not less than two separate exits shall be provided on every story.
(3) Not less than two separate exits shall be accessible from every part of every story.

**Table 32.3.1.3 Construction Type Limitations**

| Construction Type | Sprinklered[a] | Stories in Height[b] | | | | |
|---|---|---|---|---|---|---|
| | | 1 | 2 | 3 | 4–12 | >12 |
| I (442)[c, d] | Yes | X | X | X | X | X |
| | No | NP | NP | NP | NP | NP |
| I (332)[c, d] | Yes | X | X | X | X | X |
| | No | NP | NP | NP | NP | NP |
| II (222)[c, d] | Yes | X | X | X | X | NP |
| | No | NP | NP | NP | NP | NP |
| II (111)[c, d] | Yes | X | X | X | NP | NP |
| | No | NP | NP | NP | NP | NP |
| II (000) | Yes | X | X | NP | NP | NP |
| | No | NP | NP | NP | NP | NP |
| III (211) | Yes | X | X | NP | NP | NP |
| | No | NP | NP | NP | NP | NP |
| III (200) | Yes | X | NP | NP | NP | NP |
| | No | NP | NP | NP | NP | NP |
| IV (2HH) | Yes | X | X | NP | NP | NP |
| | No | NP | NP | NP | NP | NP |
| V (111) | Yes | X | X | NP | NP | NP |
| | No | NP | NP | NP | NP | NP |
| V (000) | Yes | X | NP | NP | NP | NP |
| | No | NP | NP | NP | NP | NP |

X: Permitted. NP: Not permitted.

[a]Building protected throughout by an approved automatic sprinkler system installed in accordance with 9.7.1.1(1), and provided with quick-response or residential sprinklers throughout. *(See 32.3.3.5.)*

[b]See 4.6.3.

[c]Any building of Type I, Type II(222), or Type II(111) construction is permitted to include roofing systems involving combustible supports, decking, or roofing, provided that all of the following criteria are met:

(1) The roof covering meets Class A requirements in accordance with ASTM E 108, *Standard Test Methods for Fire Tests of Roof Coverings,* or ANSI/UL 790, *Test Methods for Fire Tests of Roof Coverings.*

(2) The roof is separated from all occupied portions of the building by a noncombustible floor assembly having not less than a 2-hour fire resistance rating that includes not less than 2½ in. (63 mm) of concrete or gypsum fill.

(3) The structural elements supporting the 2-hour fire resistance–rated floor assembly specified in item (2) are required to have only the fire resistance rating required of the building.

[d]Any building of Type I, Type II(222), or Type II(111) construction is permitted to include roofing systems involving combustible supports, decking, or roofing, provided that all of the following criteria are met:

(1) The roof covering meets Class A requirements in accordance with ASTM E 108, *Standard Test Methods for Fire Tests of Roof Coverings,* or ANSI/UL 790, *Test Methods for Fire Tests of Roof Coverings.*

(2) The roof/ceiling assembly is constructed with fire-retardant-treated wood meeting the requirements of NFPA 220, *Standard on Types of Building Construction.*

(3) The roof/ceiling assembly has the required fire resistance rating for the type of construction.

**32.3.2.4.2** Exit access, as required by 32.3.2.4.1(3), shall be permitted to include a single exit access path for the distances permitted as common paths of travel by 32.3.2.5.2.

**32.3.2.5 Arrangement of Means of Egress.**

**32.3.2.5.1** Access to all required exits shall be in accordance with Section 7.5.

**32.3.2.5.2** Common paths of travel shall not exceed 75 ft (23 m).

**32.3.2.5.3 Reserved.**

**32.3.2.5.4** Dead-end corridors shall not exceed 30 ft (9.1 mm).

**32.3.2.5.5** Any room, or any suite of rooms, exceeding 2000 ft² (185 m²) shall be provided with not less than two exit access doors located remotely from each other.

**32.3.2.6 Travel Distance to Exits.** Travel distance from any point in a room to the nearest exit, measured in accordance with Section 7.6, shall not exceed 250 ft (76 m).

**32.3.2.7 Discharge from Exits.** Exit discharge shall comply with Section 7.7.

**32.3.2.8 Illumination of Means of Egress.** Means of egress shall be illuminated in accordance with Section 7.8.

**32.3.2.9 Emergency Lighting.** Emergency lighting in accordance with Section 7.9 shall be provided, unless each sleeping room has a direct exit to the outside at the finished ground level.

**32.3.2.10 Marking of Means of Egress.** Means of egress shall be marked in accordance with Section 7.10.

**32.3.2.11 Special Means of Egress Features.**

**32.3.2.11.1 Reserved.**

**32.3.2.11.2 Lockups.** Lockups in residential board and care occupancies shall comply with the requirements of 22.4.5.

**32.3.3 Protection.**

**32.3.3.1 Protection of Vertical Openings.**

**32.3.3.1.1** Vertical openings shall be enclosed or protected in accordance with Section 8.6.

**32.3.3.1.2** Unenclosed vertical openings in accordance with 8.6.9.1 shall be permitted.

**32.3.3.1.3** No floor below the level of exit discharge used only for storage, heating equipment, or purposes other than residential occupancy shall have unprotected openings to floors used for residential occupancy.

**32.3.3.2 Protection from Hazards.**

**32.3.3.2.1** Hazardous areas shall be protected in accordance with Section 8.7.

**32.3.3.2.2** The areas described in Table 32.3.3.2.2 shall be protected as indicated.

**32.3.3.3\* Interior Finish.**

**32.3.3.3.1 General.** Interior finish shall be in accordance with Section 10.2.

**Table 32.3.3.2.2 Hazardous Area Protection**

| Hazardous Area Description | Separation/Protection[†] |
|---|---|
| Boiler and fuel-fired heater rooms | 1 hour |
| Central/bulk laundries larger than 100 ft² (9.3 m²) | 1 hour |
| Paint shops employing hazardous substances and materials in quantities less than those that would be classified as a severe hazard | 1 hour |
| Physical plant maintenance shops | 1 hour |
| Soiled linen rooms | 1 hour |
| Storage rooms larger than 50 ft² (4.6 m²), but not exceeding 100 ft² (9.3 m²), storing combustible material | Smoke partition |
| Storage rooms larger than 100 ft² (9.3 m²) storing combustible material | 1 hour |
| Trash collection rooms | 1 hour |

[†]Minimum fire resistance rating.

**32.3.3.3.2 Interior Wall and Ceiling Finish.** Interior wall and ceiling finish materials complying with Section 10.2 shall be in accordance with the following:

(1) Exit enclosures — Class A
(2) Lobbies and corridors — Class B
(3) Rooms and enclosed spaces — Class B

**32.3.3.3.3 Interior Floor Finish.**

**32.3.3.3.3.1** Interior floor finish shall comply with Section 10.2.

**32.3.3.3.3.2** Interior floor finish in exit enclosures and exit access corridors and spaces not separated from them by walls complying with 32.3.3.6 shall be not less than Class II.

**32.3.3.3.3.3** Interior floor finish shall comply with 10.2.7.1 or 10.2.7.2, as applicable.

**32.3.3.4 Detection, Alarm, and Communications Systems.**

**32.3.3.4.1 General.** A fire alarm system shall be provided in accordance with Section 9.6.

**32.3.3.4.2 Initiation.** The required fire alarm system shall be initiated by each of the following:

(1) Manual means in accordance with 9.6.2
(2) Manual fire alarm box located at a convenient central control point under continuous supervision of responsible employees
(3) Required automatic sprinkler system
(4) Required detection system

**32.3.3.4.3 Annunciator Panel.** An annunciator panel, connected to the fire alarm system, shall be provided at a location readily accessible from the primary point of entry for emergency response personnel.

**32.3.3.4.4 Occupant Notification.** Occupant notification shall be provided automatically, without delay, in accordance with 9.6.3.

**32.3.3.4.5 High-Rise Buildings.** High-rise buildings shall be provided with an approved emergency voice communication/alarm system in accordance with 11.8.4.

**32.3.3.4.6\* Emergency Forces Notification.** Emergency forces notification shall meet the following requirements:

(1) Fire department notification shall be accomplished in accordance with 9.6.4.
(2) Smoke detection devices or smoke detection systems shall be permitted to initiate a positive alarm sequence in accordance with 9.6.3.4 for not more than 120 seconds.

**32.3.3.4.7 Smoke Alarms.** Approved smoke alarms shall be installed in accordance with 9.6.2.10 inside every sleeping room, outside every sleeping area in the immediate vicinity of the bedrooms, and on all levels within a resident unit.

**32.3.3.4.8 Smoke Detection Systems.**

**32.3.3.4.8.1** Corridors and spaces open to the corridors, other than those meeting the requirement of 32.3.3.4.8.3, shall be provided with smoke detectors that comply with *NFPA 72, National Fire Alarm and Signaling Code*, and are arranged to initiate an alarm that is audible in all sleeping areas.

**32.3.3.4.8.2 Reserved.**

**32.3.3.4.8.3** Smoke detection systems shall not be required in unenclosed corridors, passageways, balconies, colonnades, or other arrangements with one or more sides along the long dimension fully or extensively open to the exterior at all times.

**32.3.3.5 Extinguishment Requirements.**

**32.3.3.5.1 General.** All buildings shall be protected throughout by an approved automatic sprinkler system installed in accordance with 9.7.1.1(1) and provided with quick-response or residential sprinklers throughout.

**32.3.3.5.2 Reserved.**

**32.3.3.5.3 Reserved.**

**32.3.3.5.4 Reserved.**

**32.3.3.5.5 Supervision.** Automatic sprinkler systems shall be provided with electrical supervision in accordance with 9.7.2.

**32.3.3.5.6 Reserved.**

**32.3.3.5.7 Portable Fire Extinguishers.** Portable fire extinguishers shall be provided in accordance with 9.7.4.1.

**32.3.3.6\* Corridors and Separation of Sleeping Rooms.**

**32.3.3.6.1** Access shall be provided from every resident use area to at least one means of egress that is separated from all sleeping rooms by walls complying with 32.3.3.6.3 through 32.3.3.6.6.

**32.3.3.6.2** Sleeping rooms shall be separated from corridors, living areas, and kitchens by walls complying with 32.3.3.6.3 through 32.3.3.6.6.

**32.3.3.6.3** Walls required by 32.3.3.6.1 or 32.3.3.6.2 shall be smoke partitions in accordance with Section 8.4 having a minimum ½-hour fire resistance rating.

**32.3.3.6.4** Doors protecting corridor openings shall not be required to have a fire protection rating, but shall be constructed to resist the passage of smoke.

**32.3.3.6.5** Door-closing devices shall not be required on doors in corridor wall openings, other than those serving exit enclosures, smoke barriers, enclosures of vertical openings, and hazardous areas.

**32.3.3.6.6** No louvers, transfer grilles, operable transoms, or other air passages, other than properly installed heating and utility installations, shall penetrate the walls or doors specified in 32.3.3.6.

**32.3.3.7 Subdivision of Building Spaces.** Buildings shall be subdivided by smoke barriers in accordance with 32.3.3.7.1 through 32.3.3.7.21.

**32.3.3.7.1** Every story shall be divided into not less than two smoke compartments, unless it meets the requirement of 32.3.3.7.4, 32.3.3.7.5, 32.3.3.7.6, or 32.3.3.7.7.

**32.3.3.7.2** Each smoke compartment shall have an area not exceeding 22,500 ft$^2$ (2100 m$^2$).

**32.3.3.7.3** The travel distance from any point to reach a door in the required smoke barrier shall be limited to a distance of 200 ft (61 m).

**32.3.3.7.4** Smoke barriers shall not be required on stories that do not contain a board and care occupancy located above the board and care occupancy.

**32.3.3.7.5** Smoke barriers shall not be required in areas that do not contain a board and care occupancy and that are sepa-

rated from the board and care occupancy by a fire barrier complying with Section 8.3.

**32.3.3.7.6** Smoke barriers shall not be required on stories that do not contain a board and care occupancy and that are more than one story below the board and care occupancy.

**32.3.3.7.7** Smoke barriers shall not be required in open parking structures protected throughout by an approved, supervised automatic sprinkler system in accordance with 32.3.3.5.

**32.3.3.7.8** Smoke barriers shall be constructed in accordance with Section 8.5 and shall have a minimum 1-hour fire resistance rating, unless they meet the requirement of 32.3.3.7.9 or 32.3.3.7.10.

**32.3.3.7.9** Where an atrium is used, smoke barriers shall be permitted to terminate at an atrium wall constructed in accordance with 8.6.7(1)(c), in which case not less than two separate smoke compartments shall be provided on each floor.

**32.3.3.7.10\*** Dampers shall not be required in duct penetrations of smoke barriers in fully ducted heating, ventilating, and air-conditioning systems.

**32.3.3.7.11** Not less than 15 net ft$^2$ (1.4 net m$^2$) per resident shall be provided within the aggregate area of corridors, lounge or dining areas, and other low hazard areas on each side of the smoke barrier.

**32.3.3.7.12** On stories not housing residents, not less than 6 net ft$^2$ (0.56 net m$^2$) per occupant shall be provided on each side of the smoke barrier for the total number of occupants in adjoining compartments.

**32.3.3.7.13\*** Doors in smoke barriers shall be substantial doors, such as 1¾ in. (44 mm) thick, solid-bonded wood-core doors, or shall be of construction that resists fire for a minimum of 20 minutes.

**32.3.3.7.14** Nonrated factory- or field-applied protective plates extending not more than 48 in. (1220 mm) above the bottom of the door shall be permitted.

**32.3.3.7.15** Cross-corridor openings in smoke barriers shall be protected by a pair of swinging doors or a horizontal-sliding door complying with 7.2.1.14.

**32.3.3.7.16** Swinging doors shall be arranged so that each door swings in a direction opposite from the other.

**32.3.3.7.17\*** Doors in smoke barriers shall comply with 8.5.4 and shall be self-closing or automatic-closing in accordance with 7.2.1.8.

**32.3.3.7.18\*** Vision panels consisting of fire-rated glazing or wired glass panels in approved frames shall be provided in each cross-corridor swinging door and in each cross-corridor horizontal-sliding door in a smoke barrier.

**32.3.3.7.19** Rabbets, bevels, or astragals shall be required at the meeting edges, and stops shall be required at the head and sides of door frames in smoke barriers.

**32.3.3.7.20** Positive latching hardware shall not be required.

**32.3.3.7.21** Center mullions shall be prohibited.

**32.3.3.8\* Cooking Facilities.** Cooking facilities, other than those within individual residential units, shall be protected in accordance with 9.2.3.

**32.3.3.9 Standpipes.**

**32.3.3.9.1 General.** Where required, standpipe and hose systems shall be installed and maintained in accordance with 9.7.4.2.

**32.3.3.9.2 In High-Rise Buildings.** Class I standpipe systems shall be installed throughout all high-rise buildings.

**32.3.3.9.3 Roof Outlets.** Roof outlets shall not be required on roofs having a slope of 3 in 12 or greater.

**32.3.4 Special Provisions.**

**32.3.4.1 High-Rise Buildings.** High-rise buildings shall comply with Section 11.8.

**32.3.4.2 Reserved.**

**32.3.5 Reserved.**

**32.3.6 Building Services.**

**32.3.6.1 Utilities.** Utilities shall comply with Section 9.1.

**32.3.6.2 Heating, Ventilating, and Air-Conditioning.**

**32.3.6.2.1** Heating, ventilating, and air-conditioning equipment shall comply with Section 9.2.

**32.3.6.2.2** No stove or combustion heater shall be located such that it blocks escape in case of fire caused by the malfunction of the stove or heater.

**32.3.6.2.3** Unvented fuel-fired heaters shall not be used in any board and care occupancy.

**32.3.6.3 Elevators, Dumbwaiters, and Vertical Conveyors.**

**32.3.6.3.1** Elevators, dumbwaiters, and vertical conveyors shall comply with Section 9.4.

**32.3.6.3.2\*** In high-rise buildings, one elevator shall be provided with a protected power supply and shall be available for use by the fire department in case of emergency.

**32.3.6.4 Rubbish Chutes, Incinerators, and Laundry Chutes.** Rubbish chutes, incinerators, and laundry chutes shall comply with Section 9.5.

**32.4\* Suitability of an Apartment Building to House a Board and Care Occupancy.**

**32.4.1 General.**

**32.4.1.1 Scope.**

**32.4.1.1.1** Section 32.4 shall apply to apartment buildings that have one or more individual apartments used as a board and care occupancy. *(See 32.1.3.2.)*

**32.4.1.1.2** The provisions of Section 32.4 shall be used to determine the suitability of apartment buildings, other than those complying with 32.4.1.1.4, to house a residential board and care facility.

**32.4.1.1.3** The suitability of apartment buildings not used for board and care occupancies shall be determined in accordance with Chapter 30.

**32.4.1.1.4** If a new board and care occupancy is created in an existing apartment building, the suitability of such a building for apartments not used for board and care occupancies shall be determined in accordance with Chapter 31.

**32.4.1.2 Requirements for Individual Apartments.** Requirements for individual apartments used as residential board and care occupancies shall be as specified in Section 32.2. Egress from the apartment into the common building corridor shall be considered acceptable egress from the board and care facility.

**32.4.1.3\* Additional Requirements.** Apartment buildings housing board and care facilities shall comply with the requirements of Chapter 30 and the additional requirements of Section 32.4, unless the authority having jurisdiction has determined that equivalent safety for housing a residential board and care facility is provided in accordance with Section 1.4.

**32.4.1.4 Minimum Construction Requirements.**

**32.4.1.4.1** In addition to the requirements of Chapter 30, apartment buildings, other than those complying with 32.4.1.4.2, housing residential board and care facilities shall meet the construction requirements of 32.3.1.3.

**32.4.1.4.2** If a new board and care occupancy is created in an existing apartment building, the construction requirements of 19.1.6 shall apply.

**32.4.2 Means of Egress.**

**32.4.2.1** The requirements of Section 30.2 shall apply only to the parts of means of egress serving the apartment(s) used as a residential board and care occupancy, as modified by 32.4.2.2.

**32.4.2.2** If a new board and care occupancy is created in an existing apartment building, the requirements of Section 31.2 shall apply to the parts of the means of egress serving the apartment(s) used as a residential board and care occupancy.

**32.4.3 Protection.**

**32.4.3.1 Interior Finish.**

**32.4.3.1.1** The requirements of 30.3.3 shall apply only to the parts of means of egress serving the apartment(s) used as a residential board and care occupancy, as modified by 32.4.3.1.2.

**32.4.3.1.2** If a new board and care occupancy is created in an existing apartment building, the requirements of 31.3.3 shall apply to the parts of the means of egress serving the apartment(s) used as a residential board and care occupancy.

**32.4.3.2 Construction of Corridor Walls.**

**32.4.3.2.1** The requirements of 30.3.6 shall apply only to corridors serving the residential board and care facility, including that portion of the corridor wall separating the residential board and care facility from the common corridor, as modified by 32.4.3.2.2.

**32.4.3.2.2** If a new board and care occupancy is created in an existing apartment building, the requirements of 31.3.6 shall apply to the corridor serving the residential board and care facility.

**32.4.3.3 Subdivision of Building Spaces. (Reserved)**

**32.5 Reserved.**

**32.6 Reserved.**

**32.7 Operating Features.**

**32.7.1 Emergency Plan.**

**32.7.1.1** The administration of every residential board and care facility shall have, in effect and available to all supervisory personnel, written copies of a plan for protecting all persons in the event of fire, for keeping persons in place, for evacuating persons to areas of refuge, and for evacuating persons from the building when necessary.

**32.7.1.2** The emergency plan shall include special staff response, including the fire protection procedures needed to ensure the safety of any resident, and shall be amended or revised whenever any resident with unusual needs is admitted to the home.

**32.7.1.3** All employees shall be periodically instructed and kept informed with respect to their duties and responsibilities under the plan, and such instruction shall be reviewed by the staff not less than every 2 months.

**32.7.1.4** A copy of the plan shall be readily available at all times within the facility.

**32.7.2 Resident Training.**

**32.7.2.1** All residents participating in the emergency plan shall be trained in the proper actions to be taken in the event of fire.

**32.7.2.2** The training required by 32.7.2.1 shall include actions to be taken if the primary escape route is blocked.

**32.7.2.3** If a resident is given rehabilitation or habilitation training, training in fire prevention and the actions to be taken in the event of a fire shall be a part of the training program.

**32.7.2.4** Residents shall be trained to assist each other in case of fire to the extent that their physical and mental abilities permit them to do so without additional personal risk.

**32.7.3 Emergency Egress and Relocation Drills.** Emergency egress and relocation drills shall be conducted in accordance with 32.7.3.1 through 32.7.3.6.

**32.7.3.1** Emergency egress and relocation drills shall be conducted not less than six times per year on a bimonthly basis, with not less than two drills conducted during the night when residents are sleeping, as modified by 32.7.3.5 and 32.7.3.6.

**32.7.3.2** The emergency drills shall be permitted to be announced to the residents in advance.

**32.7.3.3** The drills shall involve the actual evacuation of all residents to an assembly point, as specified in the emergency plan, and shall provide residents with experience in egressing through all exits and means of escape required by the *Code*.

**32.7.3.4** Exits and means of escape not used in any drill shall not be credited in meeting the requirements of this *Code* for board and care facilities.

**32.7.3.5** Actual exiting from windows shall not be required to comply with 32.7.3; opening the window and signaling for help shall be an acceptable alternative.

**32.7.3.6** Residents who cannot meaningfully assist in their own evacuation or who have special health problems shall not be required to actively participate in the drill. Section 18.7 shall apply in such instances.

**32.7.4 Smoking.**

**32.7.4.1\*** Smoking regulations shall be adopted by the administration of board and care occupancies.

**32.7.4.2** Where smoking is permitted, noncombustible safety-type ashtrays or receptacles shall be provided in convenient locations.

**32.7.5\* Furnishings, Mattresses, and Decorations.**

**32.7.5.1** New draperies, curtains, and other similar loosely hanging furnishings and decorations shall comply with 32.7.5.1.1 and 32.7.5.1.2.

**32.7.5.1.1** New draperies, curtains, and other similar loosely hanging furnishings and decorations in board and care facilities shall be in accordance with the provisions of 10.3.1, unless otherwise permitted by 32.7.5.1.2.

**32.7.5.1.2** In other than common areas, new draperies, curtains, and other similar loosely hanging furnishings and decorations shall not be required to comply with 32.7.5.1.1 where the building is protected throughout by an approved automatic sprinkler system installed in accordance with 32.2.3.5 for small facilities or 32.3.3.5 for large facilities.

**32.7.5.2\*** New upholstered furniture within board and care facilities shall comply with 32.7.5.2.1 or 32.7.5.2.2.

**32.7.5.2.1** New upholstered furniture shall be tested in accordance with the provisions of 10.3.2.1(1) and 10.3.3.

**32.7.5.2.2** Upholstered furniture belonging to residents in sleeping rooms shall not be required to be tested, provided that a smoke alarm is installed in such rooms; battery-powered single-station smoke alarms shall be permitted in such rooms.

**32.7.5.3\*** Newly introduced mattresses within board and care facilities shall comply with 32.7.5.3.1 or 32.7.5.3.2.

**32.7.5.3.1** Newly introduced mattresses shall be tested in accordance with the provisions of 10.3.2.2 and 10.3.4.

**32.7.5.3.2** Mattresses belonging to residents in sleeping rooms shall not be required to be tested, provided that a smoke alarm is installed in such rooms; battery-powered single-station smoke alarms shall be permitted in such rooms.

**32.7.6 Staff.** Staff shall be on duty and in the facility at all times when residents requiring evacuation assistance are present.

**32.7.7 Inspection of Door Openings.** Door assemblies for which the door leaf is required to swing in the direction of egress travel shall be inspected and tested not less than annually in accordance with 7.2.1.15.

## Chapter 33    Existing Residential Board and Care Occupancies

**33.1 General Requirements.**

**33.1.1\* Application.**

**33.1.1.1 General.** The requirements of this chapter shall apply to existing buildings or portions thereof currently occupied as residential board and care occupancies.

**33.1.1.2 Administration.** The provisions of Chapter 1, Administration, shall apply.

**33.1.1.3 General.** The provisions of Chapter 4, General, shall apply.

**33.1.1.4\* Chapter 32 Compliance.** Any facility meeting the requirements of Chapter 32 shall not be required to meet those of Chapter 33.

**33.1.1.5 Chapter Sections.** This chapter is divided into five sections as follows:

(1) Section 33.1 — General Requirements
(2) Section 33.2 — Small Facilities (that is, sleeping accommodations for not more than 16 residents)

(3) Section 33.3 — Large Facilities (that is, sleeping accommodations for more than 16 residents)

(4) Section 33.4 — Suitability of an Apartment Building to House a Board and Care Occupancy (Sections 33.5 and 33.6 are reserved.)

(5) Section 33.7 — Operating Features

**33.1.1.6 Conversion.** For the purposes of this chapter, exceptions for conversions shall apply only for a change of occupancy from an existing residential or health care occupancy to a residential board and care occupancy.

**33.1.2 Classification of Occupancy.** See 6.1.9 and 33.1.3.

**33.1.3 Multiple Occupancies.**

**33.1.3.1** Multiple occupancies shall comply with 6.1.14 in buildings other than those meeting the requirement of 33.1.3.2.

**33.1.3.2** The requirement of 33.1.3.1 shall not apply to apartment buildings housing residential board and care occupancies in conformance with Section 33.4. In such facilities, any safeguards required by Section 33.4 that are more restrictive than those for other housed occupancies shall apply only to the extent prescribed by Section 33.4.

**33.1.3.3** No board and care occupancy shall have its sole means of egress or means of escape pass through any nonresidential or non-health care occupancy in the same building.

**33.1.3.4** No board and care occupancy shall be located above a nonresidential or non-health care occupancy, unless one of the following conditions is met:

(1) The board and care occupancy and exits therefrom are separated from the nonresidential or non-health care occupancy by construction having a minimum 2-hour fire resistance rating.

(2) The nonresidential or non-health care occupancy is protected throughout by an approved, supervised automatic sprinkler system in accordance with Section 9.7 and is separated therefrom by construction having a minimum 1-hour fire resistance rating.

**33.1.4 Definitions.**

**33.1.4.1 General.** For definitions, see Chapter 3, Definitions.

**33.1.4.2 Special Definitions.** A list of special terms used in this chapter follows:

(1) **Evacuation Capability.** See 3.3.76.
(2) **Impractical Evacuation Capability.** See 3.3.76.1.
(3) **Personal Care.** See 3.3.206.
(4) **Point of Safety.** See 3.3.211.
(5) **Prompt Evacuation Capability.** See 3.3.76.2.
(6) **Residential Board and Care Occupancy.** See 3.3.188.12.
(7) **Residential Board and Care Resident.** See 3.3.231.
(8) **Slow Evacuation Capability.** See 3.3.76.3.
(9) **Staff (Residential Board and Care).** See 3.3.261.
(10) **Thermal Barrier.** See 3.3.31.3.

**33.1.5 Acceptability of Means of Egress or Escape.** No means of escape or means of egress shall be considered as complying with the minimum criteria for acceptance, unless emergency evacuation drills are regularly conducted using that route in accordance with the requirements of 33.7.3.

**33.1.6\* Fire Resistance–Rated Assemblies.** Fire resistance–rated assemblies shall comply with Section 8.3.

**33.1.7 Changes in Facility Size.** A change in facility size from small to large shall be considered a change in occupancy sub-classification and shall require compliance with the provisions applicable to new construction.

**33.1.8\* Changes in Group Evacuation Capability.** A change in evacuation capability to a slower level shall be permitted where the facility conforms to one of the following requirements:

(1) The requirements of Chapter 32 applicable to new board and care facilities.

(2) The requirements of Chapter 33 applicable to existing board and care facilities for the new evacuation capability, provided that the building is protected throughout by an approved, supervised automatic sprinkler system complying with 32.3.3.5.

**33.2 Small Facilities.**

**33.2.1 General.**

**33.2.1.1 Scope.**

**33.2.1.1.1** Section 33.2 shall apply to residential board and care occupancies providing sleeping accommodations for not more than 16 residents.

**33.2.1.1.2** Where there are sleeping accommodations for more than 16 residents, the occupancy shall be classified as a large facility in accordance with Section 33.3.

**33.2.1.2 Requirements Based on Evacuation Capability.**

**33.2.1.2.1** Small facilities, other than those meeting the requirement of 33.2.1.2.1.1 or 33.2.1.2.1.2, shall comply with the requirements of Section 33.2, as indicated for the appropriate evacuation capability; the ability of all occupants, residents, staff, and family members shall be considered in determining evacuation capability.

**33.2.1.2.1.1\*** Facilities where the authority having jurisdiction has determined equivalent safety is provided in accordance with Section 1.4 shall not be required to comply with Section 33.2.

**33.2.1.2.1.2** Facilities that were previously approved as complying with the requirements for a large facility having the same evacuation capability shall not be required to comply with Section 33.2.

**33.2.1.2.2** Facility management shall furnish to the authority having jurisdiction, upon request, an evacuation capability determination using a procedure acceptable to the authority having jurisdiction; where such documentation is not furnished, the evacuation capability shall be classified as impractical.

**33.2.1.3 Minimum Construction Requirements.**

**33.2.1.3.1 Prompt Evacuation Capability.** (No special requirements.)

**33.2.1.3.2 Slow Evacuation Capability.**

**33.2.1.3.2.1** The facility shall be housed in a building where the interior is fully sheathed with lath and plaster or other material providing a minimum 15-minute thermal barrier, as modified by 33.2.1.3.2.3 through 33.2.1.3.2.7, including all portions of bearing walls, bearing partitions, floor construction, and roofs.

**33.2.1.3.2.2** All columns, beams, girders, and trusses shall be encased or otherwise protected with construction having a minimum ½-hour fire resistance rating.

**33.2.1.3.2.3** Exposed steel or wood columns, girders, and beams (but not joists) located in the basement shall be permitted.

**33.2.1.3.2.4** Buildings of Type I, Type II(222), Type II(111), Type III(211), Type IV, or Type V(111) construction shall not be required to meet the requirements of 33.2.1.3.2. *(See 8.2.1.)*

**33.2.1.3.2.5** Areas protected by approved automatic sprinkler systems in accordance with 33.2.3.5 shall not be required to meet the requirements of 33.2.1.3.2.

**33.2.1.3.2.6** Unfinished, unused, and essentially inaccessible loft, attic, or crawl spaces shall not be required to meet the requirements of 33.2.1.3.2.

**33.2.1.3.2.7** Where the facility has demonstrated to the authority having jurisdiction that the group is capable of evacuating the building in 8 minutes or less, or where the group achieves an E-score of 3 or less using the board and care occupancies evacuation capability determination methodology of NFPA 101A, *Guide on Alternative Approaches to Life Safety*, the requirements of 33.2.1.3.2 shall not apply.

**33.2.1.3.3 Impractical Evacuation Capability.** Nonsprinklered buildings shall be of any construction type in accordance with 8.2.1, other than Type II(000), Type III(200), or Type V(000) construction. Buildings protected throughout by an approved, supervised automatic sprinkler system in accordance with 33.2.3.5 shall be permitted to be of any type of construction.

**33.2.2 Means of Escape.** Designated means of escape shall be continuously maintained free of all obstructions or impediments to full instant use in the case of fire or emergency.

**33.2.2.1 Number of Means of Escape.**

**33.2.2.1.1** Each normally occupied story of the facility shall have not less than two remotely located means of escape that do not involve using windows, unless the facility meets the requirement of 33.2.2.1.4 or 33.2.2.1.5.

**33.2.2.1.2** Not less than one of the means of escape required by 33.2.2.1.1 shall be in accordance with 33.2.2.2.

**33.2.2.1.3** The provisions of Chapter 7 shall not apply to means of escape, unless specifically referenced in this chapter.

**33.2.2.1.4** In prompt evacuation capability facilities, one means of escape shall be permitted to involve windows complying with 33.2.2.3.1(3).

**33.2.2.1.5** A second means of escape from each story shall not be required where the entire building is protected throughout by an approved automatic sprinkler system complying with 33.2.3.5 and the facility has two means of escape; this provision shall not be permitted to be used in conjunction with 33.2.2.3.3.

**33.2.2.2 Primary Means of Escape.**

**33.2.2.2.1** Every sleeping room and living area shall have access to a primary means of escape located to provide a safe path of travel to the outside.

**33.2.2.2.2** Where sleeping rooms or living areas are above or below the level of exit discharge, the primary means of escape shall be an interior stair in accordance with 33.2.2.4, an exterior stair, a horizontal exit, or a fire escape stair.

**33.2.2.2.3** In slow and impractical evacuation capability facilities, the primary means of escape for each sleeping room shall not be exposed to living areas and kitchens, unless the building is protected by an approved automatic sprinkler system in

accordance with 33.2.3.5 utilizing quick-response or residential sprinklers throughout.

**33.2.2.2.4** Standard-response sprinklers shall be permitted for use in hazardous areas in accordance with 33.2.3.2.

**33.2.2.3 Secondary Means of Escape.**

**33.2.2.3.1** In addition to the primary route, each sleeping room shall have a second means of escape consisting of one of the following, unless the provisions of 33.2.2.3.2, 33.2.2.3.3, or 33.2.2.3.4 are met:

(1) Door, stairway, passage, or hall providing a way of unobstructed travel to the outside of the dwelling at street or the finished ground level that is independent of, and remotely located from, the primary means of escape

(2) Passage through an adjacent nonlockable space independent of, and remotely located from, the primary means of escape to any approved means of escape

(3)\*Outside window or door operable from the inside, without the use of tools, keys, or special effort, that provides a clear opening of not less than 5.7 ft$^2$ (0.53 m$^2$), with the width not less than 20 in. (510 mm), the height not less than 24 in. (610 mm), and the bottom of the opening not more than 44 in. (1120 mm) above the floor, with such means of escape acceptable, provided that one of the following criteria is met:

  (a) The window is within 20 ft (6100 mm) of the finished ground level.

  (b) The window is directly accessible to fire department rescue apparatus, as approved by the authority having jurisdiction.

  (c) The window or door opens onto an exterior balcony.

(4) Windows having a sill height below the adjacent finished ground level that are provided with a window well meeting the following criteria:

  (a) The window well has horizontal dimensions that allow the window to be fully opened.

  (b) The window well has an accessible net clear opening of not less than 9 ft$^2$ (0.84 m$^2$), with a length and width of not less than 36 in. (915 mm).

  (c) A window well with a vertical depth of more than 44 in. (1120 mm) is equipped with an approved permanently affixed ladder or with steps meeting the following criteria:

    i. The ladder or steps do not encroach more than 6 in. (150 mm) into the required dimensions of the window well.

    ii. The ladder or steps are not obstructed by the window.

**33.2.2.3.2** Sleeping rooms that have a door leading directly to the outside of the building with access to the finished ground level or to a stairway that meets the requirements of exterior stairs in 33.2.2.2.2 shall be considered as meeting all the requirements for a second means of escape.

**33.2.2.3.3** A second means of escape from each sleeping room shall not be required where the facility is protected throughout by an approved automatic sprinkler system in accordance with 33.2.3.5.

**33.2.2.3.4** Existing approved means of escape shall be permitted to continue to be used.

**33.2.2.4 Interior Stairs Used for Primary Means of Escape.** Interior stairs used for primary means of escape shall comply with 33.2.2.4.1 through 33.2.2.4.9.

**33.2.2.4.1** Interior stairs shall be enclosed with fire barriers in accordance with Section 8.3 having a minimum ½-hour fire resistance rating and shall comply with 7.2.2.5.3.

**33.2.2.4.2 Reserved.**

**33.2.2.4.3** The entire primary means of escape shall be arranged so that it is not necessary for occupants to pass through a portion of a lower story, unless that route is separated from all spaces on that story by construction having a minimum ½-hour fire resistance rating.

**33.2.2.4.4** In buildings of construction other than Type II(000), Type III(200), or Type V(000), the supporting construction shall be protected to afford the required fire resistance rating of the supported wall.

**33.2.2.4.5** Stairs that connect a story at street level to only one other story shall be permitted to be open to the story that is not at street level.

**33.2.2.4.6** Stair enclosures shall not be required in buildings three or fewer stories in height that house prompt or slow evacuation capability facilities, provided that both of the following criteria are met:

(1) The building is protected by an approved automatic sprinkler system in accordance with 33.2.3.5 that uses quick-response or residential sprinklers.
(2) A primary means of escape from each sleeping area exists that does not pass through a portion of a lower floor, unless that route is separated from all spaces on that floor by construction having a minimum ½-hour fire resistance rating.

**33.2.2.4.7** Stair enclosures shall not be required in buildings that are two or fewer stories in height, that house prompt evacuation capability facilities with not more than eight residents, and that are protected by an approved automatic sprinkler system in accordance with 33.2.3.5 that uses quick-response or residential sprinklers.

**33.2.2.4.8** The provisions of 33.2.2.3.3, 33.2.3.4.3.6, or 33.2.3.4.3.7 shall not be used in conjunction with 33.2.2.4.7.

**33.2.2.4.9** Stairs shall be permitted to be open at the topmost story only where all of the following criteria are met:

(1) The building is three or fewer stories in height.
(2) The building houses prompt or slow evacuation capability facilities.
(3) The building is protected by an approved automatic sprinkler system in accordance with 33.2.3.5.
(4) The entire primary means of escape of which the stairs are a part is separated from all portions of lower stories.

**33.2.2.5 Doors.**

**33.2.2.5.1** Doors, other than bathroom doors addressed in 33.2.2.5.1.1, and paths of travel to a means of escape shall be not less than 28 in. (710 mm) wide.

**33.2.2.5.1.1** Bathroom doors shall be not less than 24 in. (610 mm) wide.

**33.2.2.5.1.2 Reserved.**

**33.2.2.5.2** Doors shall be swinging or sliding.

**33.2.2.5.3** Every closet door latch shall be readily opened from the inside.

**33.2.2.5.4** Every bathroom door shall be designed to allow opening from the outside during an emergency when locked.

**33.2.2.5.5** No door in any means of escape, other than those meeting the requirement of 33.2.2.5.5.1 or 33.2.2.5.5.2, shall be locked against egress when the building is occupied.

**33.2.2.5.5.1** Delayed-egress locks complying with 7.2.1.6.1 shall be permitted on exterior doors only.

**33.2.2.5.5.2** Access-controlled egress locks complying with 7.2.1.6.2 shall be permitted.

**33.2.2.5.6** Forces to open doors shall comply with 7.2.1.4.5.

**33.2.2.5.7** Door-latching devices shall comply with 7.2.1.5.10.

**33.2.2.6 Stairs.**

**33.2.2.6.1** Stairs shall comply with 7.2.2, unless otherwise specified in this chapter.

**33.2.2.6.2** Winders complying with 7.2.2.2.4 shall be permitted.

**33.2.2.6.3\*** Exterior stairs shall be protected against blockage caused by fire within the building.

**33.2.3 Protection.**

**33.2.3.1 Protection of Vertical Openings.**

**33.2.3.1.1** Vertical openings, other than stairs complying with 33.2.2.4.5, 33.2.2.4.6, or 33.2.2.4.7, shall be protected so as not to expose a primary means of escape.

**33.2.3.1.2** Vertical openings required to be protected by 33.2.3.1.1 shall be considered protected where separated by smoke partitions in accordance with Section 8.4 that resist the passage of smoke from one story to any primary means of escape on another story.

**33.2.3.1.3** Smoke partitions used to protect vertical openings shall have a minimum ½-hour fire resistance rating.

**33.2.3.1.4** Any doors or openings to the protected vertical opening shall be capable of resisting fire for a minimum of 20 minutes.

**33.2.3.2 Hazardous Areas.**

**33.2.3.2.1** Any space where there is storage or activity having fuel conditions exceeding those of a one- or two-family dwelling and that possesses the potential for a fully involved fire shall be protected in accordance with 33.2.3.2.4 and 33.2.3.2.5.

**33.2.3.2.2** Spaces requiring protection in accordance with 33.2.3.2.1 shall include, but shall not be limited to, areas for cartoned storage, food or household maintenance items in wholesale or institutional-type quantities and concentrations, or mass storage of residents' belongings.

**33.2.3.2.3** Areas containing approved, properly installed and maintained furnaces and heating equipment; furnace rooms; and cooking and laundry facilities shall not be classified as hazardous areas solely on the basis of such equipment.

**33.2.3.2.4** Any hazardous area that is on the same floor as, and is in or abuts, a primary means of escape or a sleeping room shall be protected by one of the following means:

(1) Protection shall be an enclosure having a minimum 1-hour fire resistance rating, with self-closing or automatic-closing fire doors in accordance with 7.2.1.8 having a minimum ¾-hour fire protection rating.

(2) Protection shall be automatic sprinkler protection, in accordance with 33.2.3.5, and a smoke partition, in accordance with Section 8.4, located between the hazardous area and the sleeping area or primary escape route, with any doors in such separation self-closing or automatic-closing in accordance with 7.2.1.8.

**33.2.3.2.5** Other hazardous areas shall be protected by one of the following:

(1) Enclosure having a minimum ½-hour fire resistance rating, with self-closing or automatic-closing doors in accordance with 7.2.1.8 equivalent to minimum 1¾ in. (44 mm) thick, solid-bonded wood-core construction
(2) Automatic sprinkler protection in accordance with 33.2.3.5, regardless of enclosure

### 33.2.3.3 Interior Finish.

**33.2.3.3.1 General.** Interior finish shall be in accordance with Section 10.2.

**33.2.3.3.2 Interior Wall and Ceiling Finish.** Interior wall and ceiling finish materials complying with Section 10.2 shall be as follows:

(1) Class A or Class B in facilities other than those having prompt evacuation capability
(2) Class A, Class B, or Class C in facilities having prompt evacuation capability

**33.2.3.3.3 Interior Floor Finish.** (No requirements.)

### 33.2.3.4 Detection, Alarm, and Communications Systems.

**33.2.3.4.1 Fire Alarm Systems.** A manual fire alarm system shall be provided in accordance with Section 9.6, unless the provisions of 33.2.3.4.1.1 or 33.2.3.4.1.2 are met.

**33.2.3.4.1.1** A fire alarm system shall not be required where interconnected smoke alarms complying with 33.2.3.4.3 and not less than one manual fire alarm box per floor arranged to continuously sound the smoke detector alarms are provided.

**33.2.3.4.1.2** Other manually activated continuously sounding alarms acceptable to the authority having jurisdiction shall be permitted in lieu of a fire alarm system.

**33.2.3.4.2 Occupant Notification.** Occupant notification shall be in accordance with 9.6.3.

**33.2.3.4.3\* Smoke Alarms.**

**33.2.3.4.3.1** Approved smoke alarms shall be provided in accordance with 9.6.2.10, unless otherwise indicated in 33.2.3.4.3.6 and 33.2.3.4.3.7.

**33.2.3.4.3.2** Smoke alarms shall be installed on all levels, including basements but excluding crawl spaces and unfinished attics.

**33.2.3.4.3.3** Additional smoke alarms shall be installed for living rooms, dens, day rooms, and similar spaces.

**33.2.3.4.3.4 Reserved.**

**33.2.3.4.3.5** Smoke alarms shall be powered from the building electrical system and, when activated, shall initiate an alarm that is audible in all sleeping areas.

**33.2.3.4.3.6** Smoke alarms in accordance with 33.2.3.4.3.1 shall not be required where buildings are protected throughout by an approved automatic sprinkler system, in accordance with 33.2.3.5, that uses quick-response or residential sprinklers, and

are protected with approved smoke alarms installed in each sleeping room, in accordance with 9.6.2.10, that are powered by the building electrical system.

**33.2.3.4.3.7** Smoke alarms in accordance with 33.2.3.4.3.1 shall not be required where buildings are protected throughout by an approved automatic sprinkler system, in accordance with 33.2.3.5, that uses quick-response or residential sprinklers, with existing battery-powered smoke alarms in each sleeping room, and where, in the opinion of the authority having jurisdiction, the facility has demonstrated that testing, maintenance, and a battery replacement program ensure the reliability of power to the smoke alarms.

### 33.2.3.5\* Extinguishment Requirements.

**33.2.3.5.1 Reserved.**

**33.2.3.5.2 Reserved.**

**33.2.3.5.3** Where an automatic sprinkler system is installed, for either total or partial building coverage, all of the following requirements shall be met:

(1) The system shall be in accordance with Section 9.7 and shall initiate the fire alarm system in accordance with 33.2.3.4.1, as modified by 33.2.3.5.3.1 through 33.2.3.5.3.6.
(2) The adequacy of the water supply shall be documented to the authority having jurisdiction.

**33.2.3.5.3.1\*** In prompt evacuation capability facilities, all of the following shall apply:

(1) An automatic sprinkler system in accordance with NFPA 13D, *Standard for the Installation of Sprinkler Systems in One- and Two-Family Dwellings and Manufactured Homes*, shall be permitted.
(2) Automatic sprinklers shall not be required in closets not exceeding 24 ft² (2.2 m²) and in bathrooms not exceeding 55 ft² (5.1 m²), provided that such spaces are finished with lath and plaster or materials providing a 15-minute thermal barrier.

**33.2.3.5.3.2** In slow and impractical evacuation capability facilities, all of the following shall apply:

(1) An automatic sprinkler system in accordance with NFPA 13D, *Standard for the Installation of Sprinkler Systems in One- and Two-Family Dwellings and Manufactured Homes*, with a 30-minute water supply, shall be permitted.
(2) All habitable areas and closets shall be sprinklered.
(3) Automatic sprinklers shall not be required in bathrooms not exceeding 55 ft² (5.1 m²), provided that such spaces are finished with lath and plaster or materials providing a 15-minute thermal barrier.

**33.2.3.5.3.3** In prompt and slow evacuation capability facilities, where an automatic sprinkler system is in accordance with NFPA 13, *Standard for the Installation of Sprinkler Systems*, sprinklers shall not be required in closets not exceeding 24 ft² (2.2 m²) and in bathrooms not exceeding 55 ft² (5.1 m²), provided that such spaces are finished with lath and plaster or materials providing a 15-minute thermal barrier.

**33.2.3.5.3.4** In prompt and slow evacuation capability facilities in buildings four or fewer stories above grade plane, systems in accordance with NFPA 13R, *Standard for the Installation of Sprinkler Systems in Residential Occupancies up to and Including Four Stories in Height*, shall be permitted.

**33.2.3.5.3.5** In impractical evacuation capability facilities in buildings four or fewer stories above grade plane, systems in accordance with NFPA 13R, *Standard for the Installation of Sprinkler Systems in Residential Occupancies up to and Including Four Stories in Height,* shall be permitted. All habitable areas and closets shall be sprinklered. Automatic sprinklers shall not be required in bathrooms not exceeding 55 ft² (5.1 m²), provided that such spaces are finished with lath and plaster or materials providing a 15-minute thermal barrier.

**33.2.3.5.3.6** Initiation of the fire alarm system shall not be required for existing installations in accordance with 33.2.3.5.6.

**33.2.3.5.3.7** All impractical evacuation capability facilities shall be protected throughout by an approved, supervised automatic sprinkler system in accordance with 33.2.3.5.3.

**33.2.3.5.4 Reserved.**

**33.2.3.5.5 Reserved.**

**33.2.3.5.6** Sprinkler piping serving not more than six sprinklers for any isolated hazardous area shall be permitted to be installed in accordance with 9.7.1.2 and shall meet all of the following requirements:

(1) In new installations, where more than two sprinklers are installed in a single area, waterflow detection shall be provided to initiate the fire alarm system required by 33.2.3.4.1.

(2) The duration of water supplies shall be as required for the sprinkler systems addressed in 33.2.3.5.3.

**33.2.3.5.7** Attics shall be protected in accordance with 33.2.3.5.7.1 or 33.2.3.5.7.2.

**33.2.3.5.7.1** Where an automatic sprinkler system is installed, attics used for living purposes, storage, or fuel-fired equipment shall be protected with automatic sprinklers that are part of the required, approved automatic sprinkler system in accordance with 9.7.1.1.

**33.2.3.5.7.2** Where an automatic sprinkler system is installed, attics not used for living purposes, storage, or fuel-fired equipment shall meet one of the following criteria:

(1) Attics shall be protected throughout by a heat detection system arranged to activate the building fire alarm system in accordance with Section 9.6.

(2) Attics shall be protected with automatic sprinklers that are part of the required, approved automatic sprinkler system in accordance with 9.7.1.1.

(3) Attics shall be of noncombustible or limited-combustible construction.

(4) Attics shall be constructed of fire-retardant-treated wood in accordance with NFPA 703, *Standard for Fire Retardant–Treated Wood and Fire-Retardant Coatings for Building Materials.*

**33.2.3.5.8** Systems installed in accordance with NFPA 13D, *Standard for the Installation of Sprinkler Systems in One- and Two-Family Dwellings and Manufactured Homes,* shall be inspected, tested, and maintained in accordance with 33.2.3.5.8.1 through 33.2.3.5.8.15, which reference specific sections of NFPA 25, *Standard for the Inspection, Testing, and Maintenance of Water-Based Fire Protection Systems.* The frequency of the inspection, test, or maintenance shall be in accordance with this *Code,* whereas the purpose and procedure shall be from NFPA 25.

**33.2.3.5.8.1** Control valves shall be inspected monthly in accordance with 13.3.2 of NFPA 25, *Standard for the Inspection, Testing, and Maintenance of Water-Based Fire Protection Systems.*

**33.2.3.5.8.2** Gages shall be inspected monthly in accordance with 13.2.7.1 of NFPA 25, *Standard for the Inspection, Testing, and Maintenance of Water-Based Fire Protection Systems.*

**33.2.3.5.8.3** Alarm devices shall be inspected quarterly in accordance with 5.2.6 of NFPA 25, *Standard for the Inspection, Testing, and Maintenance of Water-Based Fire Protection Systems.*

**33.2.3.5.8.4** Alarm devices shall be tested semiannually in accordance with 5.3.3 of NFPA 25, *Standard for the Inspection, Testing, and Maintenance of Water-Based Fire Protection Systems.*

**33.2.3.5.8.5** Valve supervisory switches shall be tested semiannually in accordance with 13.3.3.5 of NFPA 25, *Standard for the Inspection, Testing, and Maintenance of Water-Based Fire Protection Systems.*

**33.2.3.5.8.6** Visible sprinklers shall be inspected annually in accordance with 5.2.1 of NFPA 25, *Standard for the Inspection, Testing, and Maintenance of Water-Based Fire Protection Systems.*

**33.2.3.5.8.7** Visible pipe shall be inspected annually in accordance with 5.2.2 of NFPA 25, *Standard for the Inspection, Testing, and Maintenance of Water-Based Fire Protection Systems.*

**33.2.3.5.8.8** Visible pipe hangers shall be inspected annually in accordance with 5.2.3 of NFPA 25, *Standard for the Inspection, Testing, and Maintenance of Water-Based Fire Protection Systems.*

**33.2.3.5.8.9** Buildings shall be inspected annually prior to the onset of freezing weather to ensure that there is adequate heat wherever water-filled piping is run in accordance with 5.2.5 of NFPA 25, *Standard for the Inspection, Testing, and Maintenance of Water-Based Fire Protection Systems.*

**33.2.3.5.8.10** A representative sample of fast-response sprinklers shall be tested once the sprinklers in the system are 20 years old in accordance with 5.3.1.1.1.2 of NFPA 25, *Standard for the Inspection, Testing, and Maintenance of Water-Based Fire Protection Systems.* If the sample fails the test, all of the sprinklers represented by that sample shall be replaced. If the sprinklers pass the test, the test shall be repeated every 10 years thereafter.

**33.2.3.5.8.11** A representative sample of dry-pendent sprinklers shall be tested once the sprinklers in the system are 10 years old in accordance with 5.3.1.1.1.5 of NFPA 25, *Standard for the Inspection, Testing, and Maintenance of Water-Based Fire Protection Systems.* If the sample fails the test, all of the sprinklers represented by that sample shall be replaced. If the sprinklers pass the test, the test shall be repeated every 10 years thereafter.

**33.2.3.5.8.12** Antifreeze solutions shall be tested annually in accordance with 5.3.4 of NFPA 25, *Standard for the Inspection, Testing, and Maintenance of Water-Based Fire Protection Systems.*

**33.2.3.5.8.13** Control valves shall be operated through their full range and returned to normal annually in accordance with 13.3.3.1 of NFPA 25, *Standard for the Inspection, Testing, and Maintenance of Water-Based Fire Protection Systems.*

**33.2.3.5.8.14** Operating stems of OS&Y valves shall be lubricated annually in accordance with 13.3.4 of NFPA 25, *Standard for the Inspection, Testing, and Maintenance of Water-Based Fire Protection Systems.*

**33.2.3.5.8.15** Dry-pipe systems that extend into the unheated portions of the building shall be inspected, tested, and maintained in accordance with 13.4.4 of NFPA 25, *Standard for the Inspection, Testing, and Maintenance of Water-Based Fire Protection Systems.*

**33.2.3.6 Construction of Corridor Walls.**

**33.2.3.6.1** Unless otherwise indicated in 33.2.3.6.1.1 through 33.2.3.6.1.4, corridor walls shall meet all of the following requirements:

(1) Walls separating sleeping rooms shall have a minimum ½-hour fire resistance rating. The minimum ½-hour fire resistance rating shall be considered to be achieved if the partitioning is finished on both sides with lath and plaster or materials providing a 15-minute thermal barrier.
(2) Sleeping room doors shall be substantial doors, such as those of 1¾ in. (44 mm) thick, solid-bonded wood-core construction or of other construction of equal or greater stability and fire integrity.
(3) Any vision panels shall be fixed fire window assemblies in accordance with 8.3.4 or shall be wired glass not exceeding 9 ft² (0.84 m²) each in area and installed in approved frames.

**33.2.3.6.1.1** In prompt evacuation capability facilities, all sleeping rooms shall be separated from the escape route by smoke partitions in accordance with Section 8.4, and door closing shall be regulated by 33.2.3.6.4.

**33.2.3.6.1.2** The requirement of 33.2.3.6.1 shall not apply to corridor walls that are smoke partitions in accordance with Section 8.4 and that are protected by automatic sprinklers in accordance with 33.2.3.5 on both sides of the wall and door, and all of the following shall also apply:

(1) In such instances, there shall be no limitation on the type or size of glass panels.
(2) Door closing shall comply with 33.2.3.6.4.

**33.2.3.6.1.3** Sleeping arrangements that are not located in sleeping rooms shall be permitted for nonresident staff members, provided that the audibility of the alarm in the sleeping area is sufficient to awaken staff who might be sleeping.

**33.2.3.6.1.4** In previously approved facilities, where the facility has demonstrated to the authority having jurisdiction that the group is capable of evacuating the building in 8 minutes or less, or where the group achieves an E-score of 3 or less using the board and care occupancies evacuation capability determination methodology of NFPA 101A, *Guide on Alternative Approaches to Life Safety*, sleeping rooms shall be separated from escape routes by walls and doors that are smoke resistant.

**33.2.3.6.2 Reserved.**

**33.2.3.6.3** No louvers, operable transoms, or other air passages shall penetrate the wall, except properly installed heating and utility installations other than transfer grilles, which shall be prohibited.

**33.2.3.6.4** Doors shall meet all of the following requirements:

(1) Doors shall be provided with latches or other mechanisms suitable for keeping the doors closed.
(2) No doors shall be arranged to prevent the occupant from closing the door.
(3) Doors shall be self-closing or automatic-closing in accordance with 7.2.1.8 in buildings other than those protected throughout by an approved automatic sprinkler system in accordance with 33.2.3.5.3.

**33.2.4 Reserved.**

**33.2.5 Building Services.**

**33.2.5.1 Utilities.** Utilities shall comply with Section 9.1.

**33.2.5.2 Heating, Ventilating, and Air-Conditioning.**

**33.2.5.2.1** Heating, ventilating, and air-conditioning equipment shall comply with the provisions of 9.2.1 and 9.2.2, except as otherwise required in this chapter.

**33.2.5.2.2** No stove or combustion heater shall be located to block escape in case of fire caused by the malfunction of the stove or heater.

**33.2.5.2.3** Unvented fuel-fired heaters shall not be used in any residential board and care facility.

**33.3 Large Facilities.**

**33.3.1 General.**

**33.3.1.1 Scope.**

**33.3.1.1.1** Section 33.3 shall apply to residential board and care occupancies providing sleeping accommodations for more than 16 residents.

**33.3.1.1.2** Facilities having sleeping accommodations for not more than 16 residents shall be evaluated in accordance with Section 33.2.

**33.3.1.1.3** Facilities meeting the requirements of Section 33.3 shall be considered to have met the requirements of Section 33.2 for the appropriate evacuation capability classification, except as amended in Section 33.3.

**33.3.1.2 Requirements Based on Evacuation Capability.**

**33.3.1.2.1 Prompt and Slow.** Large facilities classified as prompt or slow evacuation capability, other than those meeting the requirement of 33.3.1.2.1.1 or 33.3.1.2.1.2, shall comply with the requirements of Section 33.3, as indicated for the appropriate evacuation capability.

**33.3.1.2.1.1\*** Facilities where the authority having jurisdiction has determined equivalent safety is provided in accordance with Section 1.4 shall not be required to comply with the requirements of Section 33.3, as indicated for the appropriate evacuation capability.

**33.3.1.2.1.2** Facilities that were previously approved as complying with 33.3.1.2.2 shall not be required to comply with the requirements of Section 33.3, as indicated for the appropriate evacuation capability.

**33.3.1.2.2\* Impractical.** Large facilities classified as impractical evacuation capability shall meet the requirements of Section 33.3 for impractical evacuation capability, or the requirements for limited care facilities in Chapter 19, unless the authority having jurisdiction has determined equivalent safety is provided in accordance with Section 1.4.

**33.3.1.2.3 Evacuation Capability Determination.**

**33.3.1.2.3.1** Facility management shall furnish to the authority having jurisdiction, upon request, an evacuation capability determination using a procedure acceptable to the authority having jurisdiction.

**33.3.1.2.3.2** Where the documentation required by 33.3.1.2.3.1 is not furnished, the evacuation capability shall be classified as impractical.

**33.3.1.3 Minimum Construction Requirements.** Large facilities shall be limited to the building construction types specified in Table 33.3.1.3. *(See 8.2.1.)*

**Table 33.3.1.3 Construction Type Limitations**

| Construction Type | Sprinklered[a] | Stories in Height[b] | | | | | | |
|---|---|---|---|---|---|---|---|---|
| | | 1[c] | 2 | 3 | 4 | 5 | 6 | >6 |
| I (442)[d, e] | Yes | X | X | X | X | X | X | X |
| | No | X | X | X | X | X | X | X |
| I (332)[d, e] | Yes | X | X | X | X | X | X | X |
| | No | X | X | X | X | X | X | X |
| II (222)[d, e] | Yes | X | X | X | X | X | X | X |
| | No | X | X | X | X | X | X | X |
| II (111)[d, e] | Yes | X | X | X | X | X | X | X |
| | No | X | X | X | X | X | X | NP |
| II (000) | Yes | X | X | X2 | X2 | X2 | X2 | NP |
| | No | X1 | X1 | NP | NP | NP | NP | NP |
| III (211) | Yes | X | X | X | X | X | X | X |
| | No | X | X | X | X | X | X | NP |
| III (200) | Yes | X | X | X2 | X2 | X2 | X2 | NP |
| | No | X1 | X1 | NP | NP | NP | NP | NP |
| IV (2HH) | Yes | X | X | X | X | X | X | X |
| | No | X | X | NP | NP | NP | NP | NP |
| V (111) | Yes | X | X | X2 | X2 | X2 | X2 | NP |
| | No | X | X | NP | NP | NP | NP | NP |
| V (000) | Yes | X | X | X2 | X2 | NP | NP | NP |
| | No | X1 | X1 | NP | NP | NP | NP | NP |

NP: Not permitted.

X: Permitted.

X1: Permitted if the interior walls are covered with lath and plaster or materials providing a 15-minute thermal barrier.

X2: Permitted if the interior walls are covered with lath and plaster or materials providing a 15-minute thermal barrier, and protected throughout by an approved automatic sprinkler system installed in accordance with 33.3.3.5.

[a]Building protected throughout by an approved, supervised automatic sprinkler system installed in accordance with Section 9.7. (*See 33.3.3.5.*)

[b]See 4.6.3.

[c]One-story prompt evacuation capability facilities having 30 or fewer residents, with egress directly to the exterior at the finished ground level, are permitted to be of any construction type.

[d]Any building of Type I, Type II(222), or Type II(111) construction is permitted to include roofing systems involving combustible supports, decking, or roofing, provided that all of the following criteria are met:

(1) The roof covering meets Class A requirements in accordance with ASTM E 108, *Standard Test Methods for Fire Tests of Roof Coverings*, or ANSI/UL 790, *Test Methods for Fire Tests of Roof Coverings*.

(2) The roof is separated from all occupied portions of the building by a noncombustible floor assembly having not less than a 2-hour fire resistance rating that includes not less than 2½ in. (63 mm) of concrete or gypsum fill, and the attic or other space so developed is either unused or protected throughout by an approved automatic sprinkler system in accordance with 33.3.3.5.1.

[e]Any building of Type I, Type II(222), or Type II(111) construction is permitted to include roofing systems involving combustible supports, decking, or roofing, provided that all of the following criteria are met:

(1) The roof covering meets Class A requirements in accordance with ASTM E 108, *Standard Test Methods for Fire Tests of Roof Coverings*, or ANSI/UL 790, *Test Methods for Fire Tests of Roof Coverings*.

(2) The roof/ceiling assembly is constructed with fire-retardant-treated wood meeting the requirements of NFPA 220, *Standard on Types of Building Construction*.

(3) The roof/ceiling assembly has the required fire resistance rating for the type of construction.

**33.3.1.4 Occupant Load.** The occupant load, in number of persons for whom means of egress and other provisions are required, shall be determined on the basis of the occupant load factors of Table 7.3.1.2 that are characteristic of the use of the space, or shall be determined as the maximum probable population of the space under consideration, whichever is greater.

### 33.3.2 Means of Egress.

#### 33.3.2.1 General.

**33.3.2.1.1** Means of egress from resident rooms and resident dwelling units to the outside of the building shall be in accordance with Chapter 7 and this chapter.

**33.3.2.1.2** Means of escape within the resident room or resident dwelling unit shall comply with Section 24.2 for one- and two-family dwellings.

#### 33.3.2.2 Means of Egress Components.

**33.3.2.2.1 Components Permitted.** Components of means of egress shall be limited to the types described in 33.3.2.2.2 through 33.3.2.2.10.

**33.3.2.2.2 Doors.** Doors in means of egress shall be as follows:

(1) Doors complying with 7.2.1 shall be permitted.
(2) Doors within individual rooms and suites of rooms shall be permitted to be swinging or sliding.
(3) No door in any means of egress, other than those meeting the requirement of 33.3.2.2.2(4) or (5), shall be locked against egress when the building is occupied.
(4) Delayed-egress locks in accordance with 7.2.1.6.1 shall be permitted.
(5) Access-controlled egress doors in accordance with 7.2.1.6.2 shall be permitted.
(6) Revolving doors complying with 7.2.1.10 shall be permitted.

**33.3.2.2.3 Stairs.** Stairs complying with 7.2.2 shall be permitted.

**33.3.2.2.4 Smokeproof Enclosures.** Smokeproof enclosures complying with 7.2.3 shall be permitted.

**33.3.2.2.5 Horizontal Exits.** Horizontal exits complying with 7.2.4 shall be permitted.

**33.3.2.2.6 Ramps.** Ramps complying with 7.2.5 shall be permitted.

**33.3.2.2.7 Exit Passageways.** Exit passageways complying with 7.2.6 shall be permitted.

**33.3.2.2.8 Fire Escape Ladders.** Fire escape ladders complying with 7.2.9 shall be permitted.

**33.3.2.2.9 Alternating Tread Devices.** Alternating tread devices complying with 7.2.11 shall be permitted.

**33.3.2.2.10 Areas of Refuge.** Areas of refuge complying with 7.2.12 shall be permitted.

#### 33.3.2.3 Capacity of Means of Egress.

**33.3.2.3.1** The capacity of means of egress shall be in accordance with Section 7.3.

**33.3.2.3.2** Street floor exits shall be sufficient for the occupant load of the street floor plus the required capacity of stairs and ramps discharging onto the street floor.

**33.3.2.3.3** The width of corridors serving an occupant load of 50 or more in facilities having prompt or slow evacuation capability, and all facilities having impractical evacuation capability, shall be sufficient for the occupant load served but shall be not less than 44 in. (1120 mm).

**33.3.2.3.4** The width of corridors serving an occupant load of less than 50 in facilities having prompt or slow evacuation capability shall be not less than 36 in. (915 mm).

#### 33.3.2.4 Number of Means of Egress.

**33.3.2.4.1** Means of egress shall comply with the following, except as otherwise permitted by 33.3.2.4.2:

(1) The number of means of egress shall be in accordance with 7.4.1.1 and 7.4.1.3 through 7.4.1.6.
(2) Not less than two separate exits shall be provided on every story.
(3) Not less than two separate exits shall be accessible from every part of every story.

**33.3.2.4.2** Exit access, as required by 33.3.2.4.1(3), shall be permitted to include a single exit access path for the distances permitted as common paths of travel by 33.3.2.5.2 and 33.3.2.5.3.

#### 33.3.2.5 Arrangement of Means of Egress.

**33.3.2.5.1** Access to all required exits shall be in accordance with Section 7.5.

**33.3.2.5.2** Common paths of travel shall not exceed 110 ft (33.5 m) in buildings not protected throughout by an automatic sprinkler system in accordance with 33.3.3.5.

**33.3.2.5.3** In buildings protected throughout by automatic sprinkler systems in accordance with 33.3.3.5, common paths of travel shall not exceed 160 ft (48.8 m).

**33.3.2.5.4** Dead-end corridors shall not exceed 50 ft (15 m).

#### 33.3.2.6 Travel Distance to Exits.

**33.3.2.6.1** Travel distance from the door within a room, suite, or living unit to a corridor door shall not exceed 75 ft (23 m) in buildings not protected throughout by an approved automatic sprinkler system in accordance with 33.3.3.5.

**33.3.2.6.2** Travel distance from the door within a room, suite, or living unit to a corridor door shall not exceed 125 ft (38 m) in buildings protected throughout by an approved automatic sprinkler system in accordance with 33.3.3.5.

**33.3.2.6.3** Travel distance from the corridor door of any room to the nearest exit shall be in accordance with 33.3.2.6.3.1, 33.3.2.6.3.2, or 33.3.2.6.3.3.

**33.3.2.6.3.1** Travel distance from the corridor door of any room to the nearest exit, measured in accordance with Section 7.6, shall not exceed 100 ft (30 m).

**33.3.2.6.3.2** Travel distance to exits shall not exceed 200 ft (61 m) for exterior ways of exit access arranged in accordance with 7.5.3.

**33.3.2.6.3.3** Travel distance to exits shall not exceed 200 ft (61 m) if the exit access and any portion of the building that is tributary to the exit access are protected throughout by approved automatic sprinkler systems in accordance with 33.3.3.5. In addition, the portion of the building in which 200 ft (61 m) travel distance is permitted shall be separated from the remainder of the building by construction having a minimum 1-hour fire resistance rating, for buildings three or fewer stories in height, and a minimum 2-hour fire resistance rating for buildings four or more stories in height.

**33.3.2.7 Discharge from Exits.** Exit discharge shall comply with Section 7.7.

**33.3.2.8 Illumination of Means of Egress.** Means of egress shall be illuminated in accordance with Section 7.8.

**33.3.2.9 Emergency Lighting.** Emergency lighting in accordance with Section 7.9 shall be provided in all facilities meeting any of the following criteria:

(1) Facilities having an impractical evacuation capability
(2) Facilities having a prompt or slow evacuation capability with more than 25 rooms, unless each room has a direct exit to the outside of the building at the finished ground level

**33.3.2.10 Marking of Means of Egress.** Means of egress shall be marked in accordance with Section 7.10.

**33.3.2.11 Special Means of Egress Features.**

**33.3.2.11.1 Reserved.**

**33.3.2.11.2 Lockups.** Lockups in residential board and care occupancies, other than approved existing lockups, shall comply with the requirements of 23.4.5.

**33.3.3 Protection.**

**33.3.3.1 Protection of Vertical Openings.**

**33.3.3.1.1** Vertical openings shall comply with 33.3.3.1.1.1, 33.3.3.1.1.2, or 33.3.3.1.1.3.

**33.3.3.1.1.1** Vertical openings shall be enclosed or protected in accordance with Section 8.6.

**33.3.3.1.1.2** Unprotected vertical openings not part of required egress shall be permitted by the authority having jurisdiction where such openings do not endanger required means of egress, provided that the building is protected throughout by an approved automatic sprinkler system in accordance with 33.3.3.5, and the exits and required ways of travel thereto are adequately safeguarded against fire and smoke within the building, or where every individual room has direct access to an exterior exit without passing through a public corridor.

**33.3.3.1.1.3** In buildings two or fewer stories in height, unprotected vertical openings shall be permitted by the authority having jurisdiction, provided that the building is protected throughout by an approved automatic sprinkler system in accordance with 33.3.3.5.

**33.3.3.1.2 Reserved.**

**33.3.3.1.3** No floor below the level of exit discharge and used only for storage, heating equipment, or purposes other than residential occupancy shall have unprotected openings to floors used for residential occupancy.

**33.3.3.2 Protection from Hazards.**

**33.3.3.2.1** Rooms containing high-pressure boilers, refrigerating machinery, transformers, or other service equipment subject to possible explosion shall not be located directly under or adjacent to exits, and such rooms shall be effectively separated from other parts of the building as specified in Section 8.7.

**33.3.3.2.2** Hazardous areas, which shall include, but shall not be limited to, the following, shall be separated from other parts of the building by construction having a minimum 1-hour fire resistance rating, with communicating openings protected by approved self-closing fire doors, or such areas shall be equipped with automatic fire-extinguishing systems:

(1) Boiler and heater rooms
(2) Laundries
(3) Repair shops
(4) Rooms or spaces used for storage of combustible supplies and equipment in quantities deemed hazardous by the authority having jurisdiction

**33.3.3.2.3** In facilities having impractical evacuation capability, hazardous areas shall be separated from other parts of the building by smoke partitions in accordance with Section 8.4.

**33.3.3.3 Interior Finish.**

**33.3.3.3.1 General.** Interior finish shall be in accordance with Section 10.2.

**33.3.3.3.2 Interior Wall and Ceiling Finish.** Interior wall and ceiling finish materials complying with Section 10.2 shall be Class A or Class B.

**33.3.3.3.3 Interior Floor Finish.** Interior floor finish, other than approved existing floor coverings, shall be Class I or Class II in corridors or exits.

**33.3.3.4 Detection, Alarm, and Communications Systems.**

**33.3.3.4.1 General.** A fire alarm system in accordance with Section 9.6 shall be provided, unless all of the following conditions are met:

(1) The facility has an evacuation capability of prompt or slow.
(2) Each sleeping room has exterior exit access in accordance with 7.5.3.
(3) The building does not exceed three stories in height.

**33.3.3.4.2 Initiation.** The required fire alarm system shall be initiated by each of the following means:

(1) Manual means in accordance with 9.6.2, unless there are other effective means (such as a complete automatic sprinkler or detection system) for notification of fire as required
(2) Manual fire alarm box located at a convenient central control point under continuous supervision of responsible employees
(3) Automatic sprinkler system, other than that not required by another section of this *Code*
(4) Required detection system, other than sleeping room smoke alarms

**33.3.3.4.3 Reserved.**

**33.3.3.4.4 Occupant Notification.** Occupant notification shall be provided automatically, without delay, by internal audible alarm in accordance with 9.6.3.

**33.3.3.4.5 Reserved.**

**33.3.3.4.6 Emergency Forces Notification.**

**33.3.3.4.6.1\*** Where the existing fire alarm system does not provide for automatic emergency forces notification in accordance with 9.6.4, provisions shall be made for the immediate notification of the public fire department by either telephone or other means, or, where there is no public fire department, notification shall be made to the private fire brigade.

**33.3.3.4.6.2** Where a new fire alarm system is installed, or the existing fire alarm system is replaced, emergency forces notification shall be provided in accordance with 9.6.4.

**33.3.3.4.7 Smoke Alarms.** Smoke alarms shall be provided in accordance with 33.3.3.4.7.1, 33.3.3.4.7.2, or 33.3.3.4.7.3.

**33.3.3.4.7.1** Each sleeping room shall be provided with an approved smoke alarm in accordance with 9.6.2.10 that is powered from the building electrical system.

**33.3.3.4.7.2** Existing battery-powered smoke alarms, rather than building electrical service–powered smoke alarms, shall be accepted where, in the opinion of the authority having jurisdiction, the facility has demonstrated that testing, maintenance, and battery replacement programs ensure the reliability of power to the smoke alarms.

**33.3.3.4.7.3** Sleeping room smoke alarms shall not be required in facilities having an existing corridor smoke detection system that complies with Section 9.6 and is connected to the building fire alarm system.

**33.3.3.4.8 Smoke Detection Systems.**

**33.3.3.4.8.1** All living areas, as defined in 3.3.21.5, and all corridors shall be provided with smoke detectors that comply with *NFPA 72, National Fire Alarm and Signaling Code*, and are arranged to initiate an alarm that is audible in all sleeping areas, as modified by 33.3.3.4.8.2 and 33.3.3.4.8.3.

**33.3.3.4.8.2** Smoke detection systems shall not be required in living areas of buildings having a prompt or slow evacuation capability protected throughout by an approved automatic sprinkler system installed in accordance with 33.3.3.5.

**33.3.3.4.8.3** Smoke detection systems shall not be required in unenclosed corridors, passageways, balconies, colonnades, or other arrangements with one or more sides along the long dimension fully or extensively open to the exterior at all times.

**33.3.3.5 Extinguishment Requirements.**

**33.3.3.5.1\* General.** Where an automatic sprinkler system is installed, for either total or partial building coverage, the system shall be installed in accordance with Section 9.7, as modified by 33.3.3.5.1.1, 33.3.3.5.1.2, and 33.3.3.5.1.3.

**33.3.3.5.1.1** In buildings four or fewer stories above grade plane, systems in accordance with NFPA 13R, *Standard for the Installation of Sprinkler Systems in Residential Occupancies up to and Including Four Stories in Height*, shall be permitted.

**33.3.3.5.1.2** In facilities having prompt or slow evacuation capability, automatic sprinklers shall not be required in closets not exceeding 24 ft$^2$ (2.2 m$^2$) and in bathrooms not exceeding 55 ft$^2$ (5.1 m$^2$), provided that such spaces are finished with noncombustible or limited-combustible materials.

**33.3.3.5.1.3** Initiation of the fire alarm system shall not be required for existing installations in accordance with 33.3.3.5.6.

**33.3.3.5.2 Impractical Evacuation Capability.** All facilities having impractical evacuation capability shall be protected throughout by an approved, supervised automatic sprinkler system in accordance with 9.7.1.1(1).

**33.3.3.5.3 High-Rise Buildings.** All high-rise buildings shall be protected throughout by an approved, supervised automatic sprinkler system in accordance with 33.3.3.5. Such systems shall initiate the fire alarm system in accordance with Section 9.6.

**33.3.3.5.4** Attics shall be protected in accordance with 33.3.3.5.4.1 or 33.3.3.5.4.2.

**33.3.3.5.4.1** Where an automatic sprinkler system is installed, attics used for living purposes, storage, or fuel-fired equipment shall be protected with automatic sprinklers that are part

of the required, approved automatic sprinkler system in accordance with 9.7.1.1.

**33.3.3.5.4.2** Where an automatic sprinkler system is installed, attics not used for living purposes, storage, or fuel-fired equipment shall meet one of the following criteria:

(1) Attics shall be protected throughout by a heat detection system arranged to activate the building fire alarm system in accordance with Section 9.6.
(2) Attics shall be protected with automatic sprinklers that are part of the required, approved automatic sprinkler system in accordance with 9.7.1.1.
(3) Attics shall be of noncombustible or limited-combustible construction.
(4) Attics shall be constructed of fire-retardant-treated wood in accordance with NFPA 703, *Standard for Fire Retardant–Treated Wood and Fire-Retardant Coatings for Building Materials.*

**33.3.3.5.5 Supervision.** Automatic sprinkler systems shall be supervised in accordance with Section 9.7; waterflow alarms shall not be required to be transmitted off-site.

**33.3.3.5.6 Domestic Water Supply Option.** Sprinkler piping serving not more than six sprinklers for any isolated hazardous area in accordance with 9.7.1.2 shall be permitted; in new installations where more than two sprinklers are installed in a single area, waterflow detection shall be provided to initiate the fire alarm system required by 33.3.3.4.1.

**33.3.3.5.7 Portable Fire Extinguishers.** Portable fire extinguishers in accordance with 9.7.4.1 shall be provided near hazardous areas.

**33.3.3.6 Corridors and Separation of Sleeping Rooms.**

**33.3.3.6.1** Access shall be provided from every resident use area to not less than one means of egress that is separated from all other rooms or spaces by walls complying with 33.3.3.6.3 through 33.3.3.6.6.3, unless otherwise indicated in 33.3.3.6.1.1 through 33.3.3.6.1.3.

**33.3.3.6.1.1** Rooms or spaces, other than sleeping rooms, protected throughout by an approved automatic sprinkler system in accordance with 33.3.3.5 shall not be required to comply with 33.3.3.6.1.

**33.3.3.6.1.2** Prompt evacuation capability facilities in buildings two or fewer stories in height, where not less than one required means of egress from each sleeping room provides a path of travel to the outside without traversing any corridor or other spaces exposed to unprotected vertical openings, living areas, and kitchens, shall not be required to comply with 33.3.3.6.1.

**33.3.3.6.1.3** Rooms or spaces, other than sleeping rooms, provided with a smoke detection and alarm system connected to activate the building evacuation alarm shall not be required to comply with 33.3.3.6.1. Furnishings, finishes, and furniture, in combination with all other combustibles within the spaces, shall be of minimum quantity and arranged so that a fully developed fire is unlikely to occur.

**33.3.3.6.2** Sleeping rooms shall be separated from corridors, living areas, and kitchens by walls complying with 33.3.3.6.3 through 33.3.3.6.6.3.

**33.3.3.6.3** Walls required by 33.3.3.6.1 or 33.3.3.6.2 shall comply with 33.3.3.6.3.1, 33.3.3.6.3.2, or 33.3.3.6.3.3.

**33.3.3.6.3.1** Walls shall have a minimum ½-hour fire resistance rating.

**33.3.3.6.3.2** In buildings protected throughout by an approved automatic sprinkler system in accordance with 33.3.3.5, walls shall be smoke partitions in accordance with Section 8.4, and the provisions of 8.4.3.5 shall not apply.

**33.3.3.6.3.3** In buildings two or fewer stories in height that are classified as prompt evacuation capability and that house not more than 30 residents, walls shall be smoke partitions in accordance with Section 8.4, and the provisions of 8.4.3.5 shall not apply.

**33.3.3.6.4** Doors in walls required by 33.3.3.6.1 or 33.3.3.6.2 shall comply with 33.3.3.6.4.1, 33.3.3.6.4.2, 33.3.3.6.4.3, or 33.3.3.6.4.4.

**33.3.3.6.4.1** Doors shall have a minimum 20-minute fire protection rating.

**33.3.3.6.4.2** Solid-bonded wood-core doors of not less than 1¾ in. (44 mm) thickness shall be permitted to continue in use.

**33.3.3.6.4.3** In buildings protected throughout by an approved automatic sprinkler system in accordance with 33.3.3.5, doors that are nonrated shall be permitted to continue in use.

**33.3.3.6.4.4** Where automatic sprinkler protection is provided in the corridor in accordance with 31.3.5.8, all of the following requirements shall be met:

(1) Doors shall not be required to have a fire protection rating, but shall be in accordance with 8.4.3.
(2) The provisions of 8.4.3.5 shall not apply.
(3) Doors shall be equipped with latches for keeping the doors tightly closed.

**33.3.3.6.5** Where walls and doors are required by 33.3.3.6.1 and 33.3.3.6.2, all of the following requirements shall be met:

(1) Such walls and doors shall be constructed as smoke partitions in accordance with Section 8.4.
(2) The provisions of 8.4.3.5 shall not apply.
(3) No louvers, transfer grilles, operable transoms, or other air passages shall penetrate such walls or doors, except properly installed heating and utility installations.

**33.3.3.6.6** Doors in walls required by 33.3.3.6.1 and 33.3.3.6.2 shall comply with 33.3.3.6.6.1, 33.3.3.6.6.2, or 33.3.3.6.6.3.

**33.3.3.6.6.1** Doors shall be self-closing or automatic-closing in accordance with 7.2.1.8, and doors in walls separating sleeping rooms from corridors shall be automatic-closing in accordance with 7.2.1.8.2.

**33.3.3.6.6.2** Doors to sleeping rooms that have occupant-control locks such that access is normally restricted to the occupants or staff personnel shall be permitted to be self-closing.

**33.3.3.6.6.3** In buildings protected throughout by an approved automatic sprinkler system installed in accordance with 33.3.3.5, doors, other than doors to hazardous areas, vertical openings, and exit enclosures, shall not be required to be self-closing or automatic-closing.

**33.3.3.7 Subdivision of Building Spaces.** The requirements of 33.3.3.7.1 through 33.3.3.7.7 shall be met for all sleeping floors, unless otherwise permitted by 33.3.3.7.8.

**33.3.3.7.1** Every sleeping room floor shall be divided into not less than two smoke compartments of approximately the same size, with smoke barriers in accordance with Section 8.5, unless otherwise indicated in 33.3.3.7.4, 33.3.3.7.5, and 33.3.3.7.6.

**33.3.3.7.2** Smoke dampers shall not be required.

**33.3.3.7.3** Additional smoke barriers shall be provided such that the travel distance from a sleeping room corridor door to a smoke barrier shall not exceed 150 ft (46 m).

**33.3.3.7.4** Smoke barriers shall not be required in buildings having prompt or slow evacuation capability where protected throughout by an approved automatic sprinkler system installed in accordance with 33.3.3.5.

**33.3.3.7.5** Smoke barriers shall not be required in buildings having prompt or slow evacuation capability where each sleeping room is provided with exterior ways of exit access arranged in accordance with 7.5.3.

**33.3.3.7.6** Smoke barriers shall not be required in buildings having prompt or slow evacuation capability where the aggregate corridor length on each floor is not more than 150 ft (46 m).

**33.3.3.7.7** Positive latching hardware shall not be required on smoke barrier doors.

**33.3.3.7.8** Smoke partitions in accordance with Section 8.4 shall be permitted in lieu of smoke barriers on stories used for sleeping by not more than 30 residents.

**33.3.4 Special Provisions. (Reserved)**

**33.3.5 Reserved.**

**33.3.6 Building Services.**

**33.3.6.1 Utilities.** Utilities shall comply with the provisions of Section 9.1.

**33.3.6.2 Heating, Ventilating, and Air-Conditioning.**

**33.3.6.2.1** Heating, ventilating, and air-conditioning equipment shall comply with the provisions of Section 9.2.

**33.3.6.2.2** No stove or combustion heater shall be located such that it blocks escape in case of fire caused by the malfunction of the stove or heater.

**33.3.6.2.3** Unvented fuel-fired heaters shall not be used in any board and care occupancy.

**33.3.6.3 Elevators, Dumbwaiters, and Vertical Conveyors.** Elevators, dumbwaiters, and vertical conveyors shall comply with Section 9.4.

**33.3.6.4 Rubbish Chutes, Incinerators, and Laundry Chutes.** Rubbish chutes, incinerators, and laundry chutes shall comply with the provisions of Section 9.5.

**33.4\* Suitability of an Apartment Building to House a Board and Care Occupancy.**

**33.4.1 General.**

**33.4.1.1 Scope.**

**33.4.1.1.1** Section 33.4 shall apply to apartment buildings that have one or more individual apartments used as a board and care occupancy. *(See 33.1.3.2.)*

**33.4.1.1.2** The provisions of Section 33.4 shall be used to determine the suitability of apartment buildings to house a residential board and care facility.

**33.4.1.1.3** The suitability of existing apartment buildings not used for board and care occupancies shall be determined in accordance with Chapter 31.

**33.4.1.2 Requirements for Individual Apartments.** Requirements for individual apartments used as residential board and care occupancies shall be as specified in Section 33.2. Egress from the apartment into the common building corridor shall be considered acceptable egress from the board and care facility.

**33.4.1.3 Additional Requirements.**

**33.4.1.3.1\*** Apartment buildings housing board and care facilities shall comply with the requirements of Section 33.4, unless the authority having jurisdiction has determined that equivalent safety for housing a residential board and care facility is provided in accordance with Section 1.4.

**33.4.1.3.2** All facilities shall meet the requirements of Chapter 31 and the additional requirements of Section 33.4.

**33.4.1.4 Minimum Construction Requirements.** In addition to the requirements of Chapter 31, apartment buildings housing residential board and care facilities for groups classified as prompt or slow evacuation capability shall meet the construction requirements of 33.3.1.3, and those for groups classified as impractical evacuation capability shall meet the construction requirements of 19.1.6.

**33.4.2 Means of Egress.** The requirements of Section 31.2 shall apply only to the parts of means of egress serving the apartment(s) used as a residential board and care occupancy.

**33.4.3 Protection.**

**33.4.3.1 Interior Finish.** The requirements of 31.3.3 shall apply only to the parts of means of egress serving the apartment(s) used as a residential board and care occupancy.

**33.4.3.2 Construction of Corridor Walls.** The requirements of 31.3.6 shall apply only to corridors serving the residential board and care facility, including that portion of the corridor wall separating the residential board and care facility from the common corridor.

**33.4.3.3 Subdivision of Building Spaces.** The requirements of 31.3.7 shall apply to those stories with an apartment(s) used as a residential board and care occupancy.

**33.5 Reserved.**

**33.6 Reserved.**

**33.7 Operating Features.**

**33.7.1 Emergency Plan.**

**33.7.1.1** The administration of every residential board and care facility shall have, in effect and available to all supervisory personnel, written copies of a plan for protecting all persons in the event of fire, for keeping persons in place, for evacuating persons to areas of refuge, and for evacuating persons from the building when necessary.

**33.7.1.2** The emergency plan shall include special staff response, including the fire protection procedures needed to ensure the safety of any resident, and shall be amended or revised whenever any resident with unusual needs is admitted to the home.

**33.7.1.3** All employees shall be periodically instructed and kept informed with respect to their duties and responsibilities under the plan, and such instruction shall be reviewed by the staff not less than every 2 months.

**33.7.1.4** A copy of the plan shall be readily available at all times within the facility.

**33.7.2 Resident Training.**

**33.7.2.1** All residents participating in the emergency plan shall be trained in the proper actions to be taken in the event of fire.

**33.7.2.2** The training required by 32.7.2.1 shall include actions to be taken if the primary escape route is blocked.

**33.7.2.3** If the resident is given rehabilitation or habilitation training, training in fire prevention and the actions to be taken in the event of a fire shall be a part of the training program.

**33.7.2.4** Residents shall be trained to assist each other in case of fire to the extent that their physical and mental abilities permit them to do so without additional personal risk.

**33.7.3 Emergency Egress and Relocation Drills.** Emergency egress and relocation drills shall be conducted in accordance with 33.7.3.1 through 33.7.3.6.

**33.7.3.1** Emergency egress and relocation drills shall be conducted not less than six times per year on a bimonthly basis, with not less than two drills conducted during the night when residents are sleeping, as modified by 33.7.3.5 and 33.7.3.6.

**33.7.3.2** The emergency drills shall be permitted to be announced to the residents in advance.

**33.7.3.3** The drills shall involve the actual evacuation of all residents to an assembly point, as specified in the emergency plan, and shall provide residents with experience in egressing through all exits and means of escape required by this *Code*.

**33.7.3.4** Exits and means of escape not used in any drill shall not be credited in meeting the requirements of this *Code* for board and care facilities.

**33.7.3.5** Actual exiting from windows shall not be required to comply with 33.7.3; opening the window and signaling for help shall be an acceptable alternative.

**33.7.3.6** If the board and care facility has an evacuation capability classification of impractical, those residents who cannot meaningfully assist in their own evacuation or who have special health problems shall not be required to actively participate in the drill.

**33.7.4 Smoking.**

**33.7.4.1\*** Smoking regulations shall be adopted by the administration of board and care occupancies.

**33.7.4.2** Where smoking is permitted, noncombustible safety-type ashtrays or receptacles shall be provided in convenient locations.

**33.7.5\* Furnishings, Mattresses, and Decorations.**

**33.7.5.1** New draperies, curtains, and other similar loosely hanging furnishings and decorations shall comply with 33.7.5.1.1 and 33.7.5.1.2.

**33.7.5.1.1** New draperies, curtains, and other similar loosely hanging furnishings and decorations in board and care facilities shall be in accordance with the provisions of 10.3.1, unless otherwise permitted by 33.7.5.1.2.

**33.7.5.1.2** In other than common areas, new draperies, curtains, and other similar loosely hanging furnishings and decorations shall not be required to comply with 33.7.5.1.1 where the building is protected throughout by an approved automatic sprinkler system installed in accordance with 33.2.3.5 for small facilities or 33.3.3.5 for large facilities.

**33.7.5.2\*** New upholstered furniture within board and care facilities shall comply with 33.7.5.2.1 or 33.7.5.2.2.

**33.7.5.2.1** New upholstered furniture shall be tested in accordance with the provisions of 10.3.2.1(1) and 10.3.3.

**33.7.5.2.2** Upholstered furniture belonging to residents in sleeping rooms shall not be required to be tested, provided that a smoke alarm is installed in such rooms; battery-powered single-station smoke alarms shall be permitted in such rooms.

**33.7.5.3\*** Newly introduced mattresses within board and care facilities shall comply with 33.7.5.3.1 or 33.7.5.3.2.

**33.7.5.3.1** Newly introduced mattresses shall be tested in accordance with the provisions of 10.3.2.2 and 10.3.4.

**33.7.5.3.2** Mattresses belonging to residents in sleeping rooms shall not be required to be tested, provided that a smoke alarm is installed in such rooms; battery-powered single-station smoke alarms shall be permitted in such rooms.

**33.7.6 Staff.** Staff shall be on duty and in the facility at all times when residents requiring evacuation assistance are present.

**33.7.7 Inspection of Door Openings.** Door assemblies for which the door leaf is required to swing in the direction of egress travel shall be inspected and tested not less than annually in accordance with 7.2.1.15.

## Chapter 34    Reserved

## Chapter 35    Reserved

## Chapter 36    New Mercantile Occupancies

### 36.1 General Requirements.

### 36.1.1 Application.

**36.1.1.1** The requirements of this chapter shall apply to new buildings or portions thereof used as mercantile occupancies. *(See 1.3.1.)*

**36.1.1.2 Administration.** The provisions of Chapter 1, Administration, shall apply.

**36.1.1.3 General.** The provisions of Chapter 4, General, shall apply.

**36.1.1.4** The provisions of this chapter shall apply to life safety requirements for all new mercantile buildings. Specific requirements shall apply to suboccupancy groups, such as Class A, Class B, and Class C mercantile occupancies; covered malls; and bulk merchandising retail buildings, and are contained in paragraphs pertaining thereto.

**36.1.1.5** Additions to existing buildings shall comply with 36.1.1.5.1, 36.1.1.5.2, and 36.1.1.5.3.

**36.1.1.5.1** Additions to existing buildings shall conform to the requirements of 4.6.7.

**36.1.1.5.2** Existing portions of the structure shall not be required to be modified, provided that the new construction has not diminished the fire safety features of the facility.

**36.1.1.5.3** Existing portions shall be upgraded if the addition results in a change of mercantile subclassification. *(See 36.1.2.2.)*

**36.1.1.6** When a mercantile occupancy changes from Class C to Class A or Class B, or from Class B to Class A, the provisions of this chapter shall apply.

### 36.1.2 Classification of Occupancy.

**36.1.2.1 General.** Mercantile occupancies shall include all buildings and structures or parts thereof with occupancy as defined in 6.1.10.

**36.1.2.2 Subclassification of Occupancy.**

**36.1.2.2.1** Mercantile occupancies shall be subclassified as follows:

(1) Class A, all mercantile occupancies having an aggregate gross area of more than 30,000 ft$^2$ (2800 m$^2$) or occupying more than three stories for sales purposes
(2) Class B, as follows:
   (a) All mercantile occupancies of more than 3000 ft$^2$ (280 m$^2$), but not more than 30,000 ft$^2$ (2800 m$^2$), aggregate gross area and occupying not more than three stories for sales purposes
   (b) All mercantile occupancies of not more than 3000 ft$^2$ (280 m$^3$) gross area and occupying two or three stories for sales purposes
(3) Class C, all mercantile occupancies of not more than 3000 ft$^2$ (280 m$^2$) gross area and used for sales purposes occupying one story only

**36.1.2.2.2** For the purpose of the classification required in 36.1.2.2.1, the requirements of 36.1.2.2.2.1, 36.1.2.2.2.2, and 36.1.2.2.2.3 shall be met.

**36.1.2.2.2.1** The aggregate gross area shall be the total gross area of all floors used for mercantile purposes.

**36.1.2.2.2.2** Where a mercantile occupancy is divided into sections, regardless of fire separation, the aggregate gross area shall include the area of all sections used for sales purposes.

**36.1.2.2.2.3** Areas of floors not used for sales purposes, such as an area used only for storage and not open to the public, shall not be counted for the purposes of the classifications in 36.1.2.2.1(1), (2), and (3), but means of egress shall be provided for such nonsales areas in accordance with their occupancy, as specified by other chapters of this *Code*.

**36.1.2.2.3** Mezzanines shall comply with 8.6.10.

**36.1.2.2.4** Where a number of tenant spaces under different management are located in the same building, the aggregate gross area for subclassification shall be one of the following:

(1) Where tenant spaces are not separated, the aggregate gross floor area of all such tenant spaces shall be used in determining classification per 36.1.2.2.1.
(2) Where individual tenant spaces are separated by fire barriers with a 2-hour fire resistance rating, each tenant space shall be individually classified.

(3) Where tenant spaces are separated by fire barriers with a 1-hour fire resistance rating, and the building is protected throughout by an approved, supervised automatic sprinkler system in accordance with 9.7.1.1(1), each tenant space shall be individually classified.

(4) The tenant spaces in a mall building in accordance with 36.4.4 shall be classified individually.

### 36.1.3 Multiple Occupancies.

#### 36.1.3.1 General.

**36.1.3.1.1** All multiple occupancies shall be in accordance with 6.1.14 and 36.1.3.

**36.1.3.1.2** Where there are differences in the specific requirements in this chapter and provisions for mixed occupancies or separated occupancies as specified in 6.1.14.3 and 6.1.14.4, the requirements of this chapter shall apply.

#### 36.1.3.2 Combined Mercantile Occupancies and Parking Structures.

**36.1.3.2.1** The fire barrier separating parking structures from a building classified as a mercantile occupancy shall be a fire barrier having a minimum 2-hour fire resistance rating.

**36.1.3.2.2** Openings in the fire barrier required by 36.1.3.2.1 shall not be required to be protected with fire protection–rated opening protectives in enclosed parking structures that are protected throughout by an approved, supervised automatic sprinkler system in accordance with 9.7.1.1(1), or in open parking structures, provided that all of the following conditions are met:

(1) The openings do not exceed 25 percent of the area of the fire barrier in which they are located.
(2) The openings are used as a public entrance and for associated sidelight functions.
(3) The building containing the mercantile occupancy is protected throughout by an approved, supervised automatic sprinkler system in accordance with 9.7.1.1(1).
(4)*Means are provided to prevent spilled fuel from accumulating adjacent to the openings and entering the building.
(5) Physical means are provided to prevent vehicles from being parked or driven within 10 ft (3050 mm) of the openings.
(6) The openings are protected as a smoke partition in accordance with Section 8.4, with no minimum fire protection rating required.

### 36.1.4 Definitions.

**36.1.4.1 General.** For definitions, see Chapter 3, Definitions.

**36.1.4.2 Special Definitions.** A list of special terms used in this chapter follows:

(1) **Anchor Building.** See 3.3.36.2.
(2) **Bulk Merchandising Retail Building.** See 3.3.36.4.
(3) **Gross Leasable Area.** See 3.3.21.3.
(4) **Major Tenant.** See 3.3.166.
(5) **Mall.** See 3.3.167.
(6) **Mall Building.** See 3.3.36.9.
(7) **Open-Air Mercantile Operation.** See 3.3.197.

### 36.1.5 Classification of Hazard of Contents.

**36.1.5.1** The contents of mercantile occupancies shall be classified in accordance with Section 6.2.

**36.1.5.2** Mercantile occupancies classified as high hazard in accordance with Section 6.2 shall meet all of the following additional requirements:

(1) Exits shall be located so that not more than 75 ft (23 m) of travel from any point is needed to reach the nearest exit.
(2) From every point, there shall be not less than two exits accessible by travel in different directions (no common path of travel).
(3) All vertical openings shall be enclosed.

### 36.1.6 Minimum Construction Requirements. (No special requirements.)

**36.1.7 Occupant Load.** The occupant load, in number of persons for whom means of egress and other provisions are required, shall be determined on the basis of the occupant load factors of Table 7.3.1.2 that are characteristic of the use of the space, or shall be determined as the maximum probable population of the space under consideration, whichever is greater.

### 36.2 Means of Egress Requirements.

#### 36.2.1 General.

**36.2.1.1** All means of egress shall be in accordance with Chapter 7 and this chapter.

**36.2.1.2** No inside open stairway or inside open ramp shall be permitted to serve as a component of the required means of egress system for more than one floor.

**36.2.1.3** Where there are two or more floors below the street floor, the same stairway or other exit shall be permitted to serve all floors, but all required exits from such areas shall be independent of any open stairways between the street floor and the floor below it.

**36.2.1.4** Where exits from the upper floor also serve as an entrance from a principal street, the upper floor shall be classified as a street floor in accordance with the definition of street floor in 3.3.270 and shall be subject to the requirements of this chapter for street floors.

**36.2.1.5** High hazard mercantile occupancies shall be arranged in accordance with 36.1.5.2.

#### 36.2.2 Means of Egress Components.

**36.2.2.1 Components Permitted.** Components of means of egress shall be limited to the types described in 36.2.2.2 through 36.2.2.12.

#### 36.2.2.2 Doors.

**36.2.2.2.1** Doors complying with 7.2.1 shall be permitted.

**36.2.2.2.2*** Locks complying with 7.2.1.5.5 shall be permitted only on principal entrance/exit doors.

**36.2.2.2.3** Elevator lobby exit access door-locking arrangements in accordance with 7.2.1.6.3 shall be permitted.

**36.2.2.2.4 Reserved.**

**36.2.2.2.5** Delayed-egress locks complying with 7.2.1.6.1 shall be permitted.

**36.2.2.2.6** Access-controlled egress doors complying with 7.2.1.6.2 shall be permitted in buildings protected throughout by an approved, supervised fire detection system in accordance with Section 9.6 or an approved automatic sprinkler system in accordance with 9.7.1.1(1).

**36.2.2.2.7** Horizontal or vertical security grilles or doors complying with 7.2.1.4.1(3) shall be permitted to be used as a part of the required means of egress from a tenant space.

**36.2.2.2.8** All doors at the foot of stairs from upper floors or at the head of stairs leading to floors below the street floor shall swing in the direction of egress travel.

**36.2.2.2.9** Revolving doors complying with 7.2.1.10 shall be permitted.

**36.2.2.3 Stairs.**

**36.2.2.3.1** Stairs complying with 7.2.2 shall be permitted.

**36.2.2.3.2** Spiral stairs complying with 7.2.2.2.3 shall be permitted.

**36.2.2.4 Smokeproof Enclosures.** Smokeproof enclosures complying with 7.2.3 shall be permitted.

**36.2.2.5 Horizontal Exits.** Horizontal exits complying with 7.2.4 shall be permitted.

**36.2.2.6 Ramps.** Ramps complying with 7.2.5 shall be permitted.

**36.2.2.7 Exit Passageways.**

**36.2.2.7.1** Exit passageways complying with 7.2.6 shall be permitted.

**36.2.2.7.2\*** Exit passageways in a mall building shall be permitted to accommodate the following occupant loads independently:

(1) Portion of the occupant load assigned to the exit passageway from only the mall/pedestrian way
(2) Largest occupant load assigned to the exit passageway from a single tenant space

**36.2.2.8 Reserved.**

**36.2.2.9 Reserved.**

**36.2.2.10 Fire Escape Ladders.** Fire escape ladders complying with 7.2.9 shall be permitted.

**36.2.2.11 Alternating Tread Devices.** Alternating tread devices complying with 7.2.11 shall be permitted.

**36.2.2.12 Areas of Refuge.**

**36.2.2.12.1** Areas of refuge complying with 7.2.12 shall be permitted.

**36.2.2.12.2** In buildings protected throughout by an approved, supervised automatic sprinkler system in accordance with 9.7.1.1(1), two rooms or spaces separated from each other by smoke-resistant partitions in accordance with the definition of area of refuge in 3.3.22 shall not be required.

**36.2.3 Capacity of Means of Egress.**

**36.2.3.1** The capacity of means of egress shall be in accordance with Section 7.3.

**36.2.3.2** In Class A and Class B mercantile occupancies, street floor exits shall be sufficient for the occupant load of the street floor plus the required capacity of stairs and ramps discharging through the street floor.

**36.2.4 Number of Means of Egress.**

**36.2.4.1** Means of egress shall comply with all of the following, except as otherwise permitted by 36.2.4.2 through 36.2.4.5:

(1) The number of means of egress shall be in accordance with Section 7.4.
(2) Not less than two separate exits shall be provided on every story.

(3) Not less than two separate exits shall be accessible from every part of every story.

**36.2.4.2** Exit access, as required by 36.2.4.1(3), shall be permitted to include a single exit access path for the distances permitted as common paths of travel by 36.2.5.3.

**36.2.4.3** A single means of egress shall be permitted in a Class C mercantile occupancy, provided that the travel distance to the exit or to a mall pedestrian way (*see 36.4.4.2*) does not exceed 75 ft (23 m).

**36.2.4.4** A single means of egress shall be permitted in a Class C mercantile occupancy, provided that the travel distance to the exit or to a mall does not exceed 100 ft (30 m), and the story on which the occupancy is located, and all communicating levels that are traversed to reach the exit or mall, are protected throughout by an approved, supervised automatic sprinkler system in accordance with 9.7.1.1(1).

**36.2.4.5** A single means of egress to an exit or to a mall shall be permitted from a mezzanine within any Class A, Class B, or Class C mercantile occupancy, provided that the common path of travel does not exceed 75 ft (23 m), or does not exceed 100 ft (30 m) if protected throughout by an approved, supervised automatic sprinkler system in accordance with 9.7.1.1(1).

**36.2.5 Arrangement of Means of Egress.**

**36.2.5.1** Means of egress shall be arranged in accordance with Section 7.5.

**36.2.5.2** Dead-end corridors shall comply with 36.2.5.2.1 or 36.2.5.2.2.

**36.2.5.2.1** In buildings protected throughout by an approved, supervised automatic sprinkler system in accordance with 9.7.1.1(1), dead-end corridors shall not exceed 50 ft (15 m).

**36.2.5.2.2** In all buildings not complying with 36.2.5.2.1, dead-end corridors shall not exceed 20 ft (6100 mm).

**36.2.5.3** Common paths of travel shall be limited by any of the following:

(1) Common paths of travel shall not exceed 75 ft (23 m) in mercantile occupancies classified as low or ordinary hazard.
(2) Common paths of travel shall not exceed 100 ft (30 m) in mercantile occupancies classified as low or ordinary hazard where the building is protected throughout by an approved, supervised automatic sprinkler system in accordance with 9.7.1.1(1).
(3) Common paths of travel shall not be permitted in mercantile occupancies classified as high hazard.

**36.2.5.4** Aisles leading to each exit shall be required, and the aggregate width of such aisles shall be not less than the required width of the exit.

**36.2.5.5** Required aisles shall be not less than 36 in. (915 mm) in clear width.

**36.2.5.6** In Class A mercantile occupancies, not less than one aisle of a 60 in. (1525 mm) minimum clear width shall lead directly to an exit.

**36.2.5.7** In mercantile occupancies other than bulk merchandising retail buildings, if the only means of customer entrance is through one exterior wall of the building, one-half of the required egress width from the street floor shall be located in such wall. Means of egress from floors above or below the street floor shall be arranged in accordance with Section 7.5.

**36.2.5.8** Not less than one-half of the required exits shall be located so as to be reached without passing through checkout stands.

**36.2.5.9** Checkout stands or associated railings or barriers shall not obstruct exits, required aisles, or approaches thereto.

**36.2.5.10\*** Where wheeled carts or buggies are used by customers, adequate provision shall be made for the transit and parking of such carts to minimize the possibility that they might obstruct means of egress.

**36.2.5.11** Exit access in Class A and Class B mercantile occupancies that are protected throughout by an approved, supervised automatic sprinkler system in accordance with 9.7.1.1(1), and exit access in all Class C mercantile occupancies, shall be permitted to pass through storerooms, provided that all of the following conditions are met:

(1) Not more than 50 percent of exit access shall be provided through the storeroom.
(2) The storeroom shall not be subject to locking.
(3) The main aisle through the storeroom shall be not less than 44 in. (1120 mm) wide.
(4) The path of travel through the storeroom shall be defined, direct, and continuously maintained in an unobstructed condition.

**36.2.6 Travel Distance to Exits.** Travel distance shall be as specified in 36.2.6.1, 36.2.6.2, and 36.2.6.3 and shall be measured in accordance with Section 7.6.

**36.2.6.1** In mercantile occupancies classified as ordinary hazard, travel distance shall not exceed 150 ft (46 m).

**36.2.6.2** In mercantile occupancies classified as ordinary hazard in buildings protected throughout by an approved, supervised automatic sprinkler system in accordance with 9.7.1.1(1), travel distance shall not exceed 250 ft (76 m).

**36.2.6.3** In mercantile occupancies classified as high hazard, travel distance shall not exceed 75 ft (23 m).

**36.2.7 Discharge from Exits.**

**36.2.7.1** Exit discharge shall comply with Section 7.7 and 36.2.7.2.

**36.2.7.2\*** Fifty percent of the exits shall be permitted to discharge through the level of exit discharge in accordance with 7.7.2 only where the building is protected throughout by an approved, supervised automatic sprinkler system in accordance with 9.7.1.1(1).

**36.2.8 Illumination of Means of Egress.** Means of egress shall be illuminated in accordance with Section 7.8.

**36.2.9 Emergency Lighting.** Class A and Class B mercantile occupancies and mall buildings shall have emergency lighting facilities in accordance with Section 7.9.

**36.2.10 Marking of Means of Egress.** Where an exit is not immediately apparent from all portions of the sales area, means of egress shall have signs in accordance with Section 7.10.

**36.2.11 Special Means of Egress Features.**

**36.2.11.1 Reserved.**

**36.2.11.2 Lockups.** Lockups in mercantile occupancies shall comply with the requirements of 22.4.5.

**36.3 Protection.**

**36.3.1 Protection of Vertical Openings.** Any vertical opening shall be protected in accordance with Section 8.6, except under any of the following conditions:

(1) In Class A or Class B mercantile occupancies protected throughout by an approved, supervised automatic sprinkler system in accordance with 9.7.1.1(1), unprotected vertical openings shall be permitted at one of the following locations:
   (a) Between any two floors
   (b) Among the street floor, the first adjacent floor below, and the adjacent floor (or mezzanine) above
(2) In Class C mercantile occupancies, unprotected openings shall be permitted between the street floor and the mezzanine.
(3) The draft stop and closely spaced sprinkler requirements of NFPA 13, *Standard for the Installation of Sprinkler Systems*, shall not be required for unenclosed vertical openings permitted in 36.3.1(1) and (2).

**36.3.2 Protection from Hazards.**

**36.3.2.1\* General.** Hazardous areas shall be protected in accordance with 36.3.2.1.1 or 36.3.2.1.2.

**36.3.2.1.1\*** Hazardous areas shall be protected in accordance with Section 8.7.

**36.3.2.1.2** In general storage and stock areas protected by an automatic extinguishing system in accordance with 9.7.1.1(1) or 9.7.1.2, an enclosure shall be exempt from the provisions of 8.7.1.2.

**36.3.2.2\* High Hazard Contents Areas.** High hazard contents areas, as classified in Section 6.2, shall meet all of the following criteria:

(1) The area shall be separated from other parts of the building by fire barriers having a minimum 1-hour fire resistance rating, with all openings therein protected by self-closing fire door assemblies having a minimum ¾-hour fire protection rating.
(2) The area shall be protected by an automatic extinguishing system in accordance with 9.7.1.1(1) or 9.7.1.2.
(3) In high hazard areas, all vertical openings shall be enclosed.

**36.3.2.3 Cooking Equipment.** Cooking equipment shall be protected in accordance with 9.2.3, unless the cooking equipment is one of the following types:

(1) Outdoor equipment
(2) Portable equipment not flue-connected
(3) Equipment used only for food warming

**36.3.3 Interior Finish.**

**36.3.3.1 General.** Interior finish shall be in accordance with Section 10.2.

**36.3.3.2 Interior Wall and Ceiling Finish.** Interior wall and ceiling finish materials complying with Section 10.2 shall be Class A, Class B, or Class C.

**36.3.3.3 Interior Floor Finish.**

**36.3.3.3.1** Interior floor finish shall comply with Section 10.2.

**36.3.3.3.2** Interior floor finish in exit enclosures shall be Class I or Class II.

**36.3.3.3.3** Interior floor finish shall comply with 10.2.7.1 or 10.2.7.2, as applicable.

**36.3.4 Detection, Alarm, and Communications Systems.**

**36.3.4.1 General.** Class A mercantile occupancies shall be provided with a fire alarm system in accordance with Section 9.6.

**36.3.4.2 Initiation.** Initiation of the required fire alarm system shall be by any one of the following means:

(1) Manual means in accordance with 9.6.2.1(1)
(2) Approved automatic fire detection system that complies with 9.6.2.1(2) and provides protection throughout the building, plus a minimum of one manual fire alarm box in accordance with 9.6.2.6
(3) Approved automatic sprinkler system that complies with 9.6.2.1(3) and provides protection throughout the building, plus a minimum of one manual fire alarm box in accordance with 9.6.2.6

**36.3.4.3 Notification.**

**36.3.4.3.1 Occupant Notification.** During all times that the mercantile occupancy is occupied, the required fire alarm system, once initiated, shall perform one of the following functions:

(1) It shall activate an alarm in accordance with 9.6.3 throughout the mercantile occupancy.
(2) Positive alarm sequence in accordance with 9.6.3.4 shall be permitted.

**36.3.4.3.2 Emergency Forces Notification.** Emergency forces notification shall be provided and shall include notifying both of the following:

(1) Fire department in accordance with 9.6.4
(2) Local emergency organization, if provided

**36.3.5 Extinguishment Requirements.**

**36.3.5.1** Mercantile occupancies shall be protected by an approved automatic sprinkler system in accordance with 9.7.1.1(1) in any of the following specified locations:

(1) Throughout all mercantile occupancies three or more stories in height
(2) Throughout all mercantile occupancies exceeding 12,000 ft² (1115 m²) in gross area
(3) Throughout stories below the level of exit discharge where such stories have an area exceeding 2500 ft² (232 m²) and are used for the sale, storage, or handling of combustible goods and merchandise
(4) Throughout multiple occupancies protected as mixed occupancies in accordance with 6.1.14 where the conditions of 36.3.5.1(1), (2), or (3) apply to the mercantile occupancy

**36.3.5.2** Automatic sprinkler systems in Class A mercantile occupancies shall be supervised in accordance with 9.7.2.

**36.3.5.3** Portable fire extinguishers shall be provided in all mercantile occupancies in accordance with 9.7.4.1.

**36.3.6 Corridors.**

**36.3.6.1\*** Where access to exits is provided by corridors, such corridors shall be separated from use areas by fire barriers in accordance with Section 8.3 having a minimum 1-hour fire resistance rating, except under any of the following conditions:

(1) Where exits are available from an open floor area
(2) Within a space occupied by a single tenant

(3) Within buildings protected throughout by an approved, supervised automatic sprinkler system in accordance with 9.7.1.1(1)

**36.3.6.2** Openings in corridor walls required by 36.3.6.1 to have a fire resistance rating shall be protected in accordance with Section 8.3.

**36.3.7 Subdivision of Building Spaces.** (No special requirements.)

**36.4 Special Provisions.**

**36.4.1 Limited Access or Underground Buildings.** See Section 11.7.

**36.4.2 High-Rise Buildings.** High-rise buildings shall comply with the requirements of Section 11.8.

**36.4.3 Open-Air Mercantile Operations.**

**36.4.3.1** Open-air mercantile operations, such as open-air markets, gasoline filling stations, roadside stands for the sale of farm produce, and other outdoor mercantile operations, shall be arranged and conducted to maintain free and unobstructed ways of travel at all times.

**36.4.3.2** Ways of travel shall allow prompt escape from any point of danger in case of fire or other emergency, with no dead ends in which persons might be trapped due to display stands, adjoining buildings, fences, vehicles, or other obstructions.

**36.4.3.3** Mercantile operations that are conducted in roofed-over areas shall be treated as mercantile buildings, provided that canopies over individual small stands to protect merchandise from the weather are not construed as constituting buildings for the purpose of this *Code*.

**36.4.4 Mall Buildings.** The provisions of 36.4.4 shall apply to mall buildings three or fewer stories in height and any number of anchor buildings. *(See 3.3.36.9.)*

**36.4.4.1 General.** The mall building shall be treated as a single building for the purpose of calculation of means of egress and shall be subject to the requirements for appropriate occupancies, except as modified by the provisions of 36.4.4; and the mall shall be of a clear width not less than that needed to accommodate egress requirements as set forth in other sections of this *Code*.

**36.4.4.2 Pedestrian Way.** The mall shall be permitted to be considered a pedestrian way, provided that the criteria of 36.4.4.2.1 and 36.4.4.2.2 are met.

**36.4.4.2.1** The travel distance within a tenant space to an exit or to the mall shall not exceed the maximum travel distance permitted by the occupancy chapter.

**36.4.4.2.2** An additional 200 ft (61 m) shall be permitted for travel through the mall space, provided that all the following requirements are met:

(1) The mall shall be of a clear width not less than that needed to accommodate egress requirements, as set forth in other sections of this chapter, but shall be not less than 20 ft (6100 mm) wide in its narrowest dimension.
(2) On each side of the mall floor area, the mall shall be provided with an unobstructed exit access of not less than 10 ft (3050 mm) in clear width parallel to, and adjacent to, the mall tenant front.
(3)\*The exit access specified in 36.4.4.2.2(2) shall lead to an exit having a width of not less than 66 in. (1675 mm).

(4) The mall, and all buildings connected thereto, except open parking structures, shall be protected throughout by an approved, supervised automatic sprinkler system in accordance with 9.7.1.1(1), which shall be installed in such a manner that any portion of the system serving tenant spaces can be taken out of service without affecting the operation of the portion of the system serving the mall.

(5)*Walls dividing tenant spaces from each other shall have a fire resistance rating of not less than 1 hour, and all of the following also shall apply:

   (a) The partition shall extend to the underside of the ceiling or to the roof or floor above.

   (b) No separation shall be required between a tenant space and the mall.

(6)*Malls with a floor opening connecting more than two levels shall be provided with a smoke control system.

### 36.4.4.3 Means of Egress Details.

**36.4.4.3.1** Dead ends not exceeding a length equal to twice the width of the mall, measured at the narrowest location within the dead-end portion of the mall, shall be permitted.

**36.4.4.3.2** Every story of a mall building shall be provided with the number of means of egress specified by Section 7.4 and as modified by 36.4.4.3.2.1 or 36.4.4.3.2.2.

**36.4.4.3.2.1** Exit access travel shall be permitted to be common for the distances permitted as common paths of travel by 36.2.5.3.

**36.4.4.3.2.2** A single means of egress shall be permitted in a Class C mercantile occupancy or a business occupancy, provided that the travel distance to the exit or to a mall pedestrian way *(see 36.4.4.2)* does not exceed 100 ft (30 m).

**36.4.4.3.3** Every floor of a mall shall be provided with the number of means of egress specified by Section 7.4, with not less than two means of egress remotely located from each other.

**36.4.4.3.4** Class A and Class B mercantile occupancies connected to a mall shall be provided with the number of means of egress required by Section 7.4, with not less than two means of egress remotely located from one another.

**36.4.4.3.5*** Each individual anchor building shall have means of egress independent of the mall.

**36.4.4.3.6** Each individual major tenant of a mall building shall have a minimum of one-half of its required means of egress independent of the mall.

**36.4.4.3.7** Every mall shall be provided with unobstructed exit access parallel to, and adjacent to, the mall tenant fronts and extending to each mall exit.

**36.4.4.3.8** Each assembly occupancy with an occupant load of 500 or more shall have not less than one-half of its required means of egress independent of the mall.

**36.4.4.3.9** Emergency lighting shall be provided in accordance with 36.2.9.

### 36.4.4.4 Detection, Alarm, and Communications Systems.

**36.4.4.4.1 General.** Malls shall be provided with a fire alarm system in accordance with Section 9.6.

**36.4.4.4.2 Initiation.** Initiation of the required fire alarm system shall be by means of the required automatic sprinkler system in accordance with 9.6.2.1(3).

### 36.4.4.4.3 Notification.

**36.4.4.4.3.1 Occupant Notification.** During all times that the mall is occupied, the required fire alarm system, once initiated, shall perform one of the following functions:

(1) It shall activate a general alarm in accordance with 9.6.3 throughout the mall, and positive alarm sequence in accordance with 9.6.3.4 shall be permitted.

(2) Occupant notification shall be made via a voice communication or public address system in accordance with 9.6.3.9.2

**36.4.4.4.3.2*** Visible signals shall not be required in malls. *(See 9.6.3.5.7 and 9.6.3.5.8.)*

**36.4.4.4.3.3 Emergency Forces Notification.** Emergency forces notification shall be provided and shall include notifying all of the following:

(1) Fire department in accordance with 9.6.4

(2) Local emergency organization, if provided

**36.4.4.4.4 Emergency Control.** The fire alarm system shall be arranged to automatically actuate smoke management or smoke control systems in accordance with 9.6.5.2(3).

**36.4.4.5 Tenant Spaces.** Each individual tenant space shall have means of egress to the outside or to the mall, based on occupant load calculated by using Table 7.3.1.2.

**36.4.4.6 Exit Passageways.** Exit passageways shall comply with 36.4.4.6.1 and 36.4.4.6.2.

**36.4.4.6.1** Exit passageways in a mall building shall be permitted to accommodate the following occupant loads independently:

(1) Portion of the occupant load assigned to the exit passageway from only the mall

(2) Largest occupant load assigned to the exit passageway from a single tenant space

**36.4.4.6.2*** Rooms housing building service equipment, janitor closets, and service elevators shall be permitted to open directly onto exit passageways, provided that all of the following criteria are met:

(1) The required fire resistance rating between such rooms or areas and the exit passageway shall be maintained in accordance with 7.1.3.2.

(2) Such rooms or areas shall be protected by an approved, supervised automatic sprinkler system in accordance with 9.7.1.1(1), but the exceptions in NFPA 13, *Standard for the Installation of Sprinkler Systems*, allowing the omission of sprinklers from such rooms shall not be permitted.

(3) Service elevators opening into the exit passageway shall not open into areas other than exit passageways.

(4) Where exit stair enclosures discharge into the exit passageway, the provisions of 7.2.1.5.8 shall apply, regardless of the number of stories served.

**36.4.4.7 Plastic Signs.** Within every store or level, and from sidewall to sidewall of each tenant space facing the mall, plastic signs shall comply with all of the following:

(1) Plastic signs shall not exceed 20 percent of the wall area facing the mall.

(2) Plastic signs shall not exceed a height of 36 in. (915 mm), except if the sign is vertical, in which case the height shall not exceed 8 ft (2440 mm) and the width shall not exceed 36 in. (915 mm).

(3) Plastic signs shall be located a minimum distance of 18 in. (455 mm) from adjacent tenants.

(4) Plastics, other than foamed plastics, shall meet one of the following criteria:

    (a) They shall be light-transmitting plastics.

    (b) They shall have a self-ignition temperature of 650°F (343°C) or greater when tested in accordance with ASTM D 1929, *Standard Test Method for Determining Ignition Temperatures of Plastic*, and a flame spread index not greater than 75 and a smoke developed index not greater than 450 when tested in the manner intended for use in accordance with ASTM E 84, *Standard Test Method for Surface Burning Characteristics of Building Materials*, or ANSI/UL 723, *Standard for Test for Surface Burning Characteristics of Building Materials*.

(5) The edges and backs of plastic signs in the mall shall be fully encased in metal.

(6) Foamed plastics shall have a maximum heat release rate of 150 kW when tested in accordance with ANSI/UL 1975, *Standard for Fire Tests for Foamed Plastics Used for Decorative Purposes*, or in accordance with NFPA 289, *Standard Method of Fire Test for Individual Fuel Packages*, using the 20 kW ignition source.

(7) Foamed plastics shall comply with all of the following:

    (a) The density of foamed plastic signs shall be not less than 20 lb/ft$^3$ (320 kg/m$^3$).

    (b) The thickness of foamed plastic signs shall be not greater than ½ in. (13 mm).

**36.4.4.8 Kiosks.** Kiosks and similar structures (temporary or permanent) shall not be considered as tenant spaces and shall meet all of the following requirements:

(1) Combustible kiosks and similar structures shall be constructed of any of the following materials:

    (a) Fire-retardant-treated wood complying with the requirements for fire-retardant-impregnated wood in NFPA 703, *Standard for Fire Retardant–Treated Wood and Fire-Retardant Coatings for Building Materials*

    (b) Light-transmitting plastics complying with the building code

    (c) Foamed plastics having a maximum heat release rate not greater than 100 kW when tested in accordance with ANSI/UL 1975, *Standard for Fire Tests for Foamed Plastics Used for Decorative Purposes*, or in accordance with NFPA 289, *Standard Method of Fire Test for Individual Fuel Packages*, using the 20 kW ignition source

    (d) Metal composite material (MCM) having a flame spread index not greater than 25 and a smoke developed index not greater than 450 in accordance with ASTM E 84, *Standard Test Method for Surface Burning Characteristics of Building Materials*, or ANSI/UL 723, *Standard for Test for Surface Burning Characteristics of Building Materials*, when tested as an assembly in the maximum thickness intended for use

    (e) Textiles and films meeting the flame propagation performance criteria contained in NFPA 701, *Standard Methods of Fire Tests for Flame Propagation of Textiles and Films*

(2) Kiosks or similar structures located within the mall shall be protected with approved fire suppression and detection devices.

(3) The minimum horizontal separation between kiosks, or groups of kiosks, and other structures within the mall shall be 20 ft (6100 mm).

(4) Each kiosk, or group of kiosks, or similar structure shall have a maximum area of 300 ft$^2$ (27.8 m$^2$).

**36.4.4.9\* Smoke Control.** Smoke control in accordance with Section 9.3 and complying with 8.6.7(5) shall be provided in a mall with floor openings connecting more than two levels.

**36.4.4.10 Automatic Extinguishing Systems.**

**36.4.4.10.1** The mall building and all anchor buildings shall be protected throughout by an approved, supervised automatic sprinkler system in accordance with 9.7.1.1(1) and 36.4.4.10.2.

**36.4.4.10.2** The system shall be installed in such a manner that any portion of the system serving tenant spaces can be taken out of service without affecting the operation of the portion of the system serving the mall.

**36.4.5 Bulk Merchandising Retail Buildings.** New bulk merchandising retail buildings exceeding 12,000 ft$^2$ (1115 m$^2$) in area shall comply with the requirements of this chapter, as modified by 36.4.5.1 through 36.4.5.6.2.

**36.4.5.1 Minimum Construction Requirements.** Bulk merchandising retail buildings shall have a distance of not less than 16 ft (4875 mm) from the floor to the ceiling, from the floor to the floor above, or from the floor to the roof of any story.

**36.4.5.2 Means of Egress Requirements.**

**36.4.5.2.1** All means of egress shall be in accordance with Chapter 7 and this chapter.

**36.4.5.2.2** Not less than 50 percent of the required egress capacity shall be located independent of the main entrance/exit doors.

**36.4.5.3 Storage, Arrangement, Protection, and Quantities of Hazardous Commodities.** The storage, arrangement, protection, and quantities of hazardous commodities shall be in accordance with the applicable provisions of the following:

(1) The fire code *(see 3.3.94)*

(2) NFPA 13, *Standard for the Installation of Sprinkler Systems*

(3) NFPA 30, *Flammable and Combustible Liquids Code*

(4) NFPA 30B, *Code for the Manufacture and Storage of Aerosol Products*

(5) NFPA 400, *Hazardous Materials Code*, Chapter 14, for organic peroxide formulations

(6) NFPA 400, *Hazardous Materials Code*, Chapter 15, for oxidizer solids and liquids

(7) NFPA 400, *Hazardous Materials Code*, various chapters, depending on characteristics of a particular pesticide

(8) NFPA 1124, *Code for the Manufacture, Transportation, Storage, and Retail Sales of Fireworks and Pyrotechnic Articles*

**36.4.5.4 Detection, Alarm, and Communications Systems.**

**36.4.5.4.1 General.** Bulk merchandising retail buildings shall be provided with a fire alarm system in accordance with Section 9.6.

**36.4.5.4.2 Initiation.** Initiation of the required fire alarm system shall be by means of the required approved automatic sprinkler system *(see 36.4.5.5)* in accordance with 9.6.2.1(3).

**36.4.5.4.3 Occupant Notification.** During all times that the mercantile occupancy is occupied, the required fire alarm system, once initiated, shall activate an alarm in accordance with 9.6.3 throughout the mercantile occupancy, and positive alarm sequence in accordance with 9.6.3.4 shall be permitted.

**36.4.5.4.4 Emergency Forces Notification.** Emergency forces notification shall be provided and shall include notifying both of the following:

(1) Fire department in accordance with 9.6.4
(2) Local emergency organization, if provided

**36.4.5.5 Extinguishing Requirements.** Bulk merchandising retail buildings shall be protected throughout by an approved, supervised automatic sprinkler system in accordance with 9.7.1.1(1) and the applicable provisions of the following:

(1) The fire code (*see 3.3.94*)
(2) NFPA 13, *Standard for the Installation of Sprinkler Systems*
(3) NFPA 30, *Flammable and Combustible Liquids Code*
(4) NFPA 30B, *Code for the Manufacture and Storage of Aerosol Products*

**36.4.5.6 Emergency Plan and Employee Training.**

**36.4.5.6.1** There shall be in effect an approved written plan for the emergency egress and relocation of occupants.

**36.4.5.6.2** All employees shall be instructed and periodically drilled with respect to their duties under the plan.

**36.4.6 Retail Sales of Consumer Fireworks, 1.4G.** Mercantile occupancies in which the retail sale of consumer fireworks, 1.4G, is conducted shall comply with NFPA 1124, *Code for the Manufacture, Transportation, Storage, and Retail Sales of Fireworks and Pyrotechnic Articles.*

**36.5 Building Services.**

**36.5.1 Utilities.** Utilities shall comply with the provisions of Section 9.1.

**36.5.2 Heating, Ventilating, and Air-Conditioning.** Heating, ventilating, and air-conditioning equipment shall comply with the provisions of Section 9.2.

**36.5.3 Elevators, Escalators, and Conveyors.** Elevators, escalators, and conveyors shall comply with the provisions of Section 9.4.

**36.5.4 Rubbish Chutes, Incinerators, and Laundry Chutes.** Rubbish chutes, incinerators, and laundry chutes shall comply with the provisions of Section 9.5.

**36.6 Reserved.**

**36.7 Operating Features.**

**36.7.1 Emergency Plans.** Emergency plans complying with Section 4.8 shall be provided in high-rise buildings.

**36.7.2 Drills.** In every Class A or Class B mercantile occupancy, employees shall be periodically trained in accordance with Section 4.7.

**36.7.3 Extinguisher Training.** Employees of mercantile occupancies shall be periodically instructed in the use of portable fire extinguishers.

**36.7.4 Food Service Operations.** Food service operations shall comply with 12.7.2.

**36.7.5 Upholstered Furniture and Mattresses.** The provisions of 10.3.2 shall not apply to upholstered furniture and mattresses.

**36.7.6 Soiled Linen and Trash Receptacles.** The requirements of 10.3.9 for containers for rubbish, waste, or linen with a capacity of 20 gal (75.7 L) or more shall not apply.

## Chapter 37   Existing Mercantile Occupancies

**37.1 General Requirements.**

**37.1.1 Application.**

**37.1.1.1** The requirements of this chapter shall apply to existing buildings or portions thereof currently occupied as mercantile occupancies.

**37.1.1.2 Administration.** The provisions of Chapter 1, Administration, shall apply.

**37.1.1.3 General.** The provisions of Chapter 4, General, shall apply.

**37.1.1.4** The provisions of this chapter shall apply to life safety requirements for all existing mercantile buildings. Specific requirements shall apply to suboccupancy groups, such as Class A, Class B, and Class C mercantile occupancies; covered malls; and bulk merchandising retail buildings, and are contained in paragraphs pertaining thereto.

**37.1.1.5** Additions to existing buildings shall comply with 37.1.1.5.1, 37.1.1.5.2, and 37.1.1.5.3.

**37.1.1.5.1** Additions to existing buildings shall conform to the requirements of 4.6.7.

**37.1.1.5.2** Existing portions of the structure shall not be required to be modified, provided that the new construction has not diminished the fire safety features of the facility.

**37.1.1.5.3** Existing portions shall be upgraded if the addition results in a change of mercantile subclassification. *(See 37.1.2.2.)*

**37.1.1.6** When a change in mercantile occupancy subclassification occurs, either of the following requirements shall be met:

(1) When a mercantile occupancy changes from Class A to Class B or Class C, or from Class B to Class C, the provisions of this chapter shall apply.
(2) When a mercantile occupancy changes from Class C to Class A or Class B, or from Class B to Class A, the provisions of Chapter 36 shall apply.

**37.1.2 Classification of Occupancy.**

**37.1.2.1 General.** Mercantile occupancies shall include all buildings and structures or parts thereof with occupancy as defined in 6.1.10.

**37.1.2.2 Subclassification of Occupancy.**

**37.1.2.2.1** Mercantile occupancies shall be subclassified as follows:

(1) Class A, all mercantile occupancies having an aggregate gross area of more than 30,000 ft$^2$ (2800 m$^2$) or occupying more than three stories for sales purposes
(2) Class B, as follows:
   (a) All mercantile occupancies of more than 3000 ft$^2$ (280 m$^2$), but not more than 30,000 ft$^2$ (2800 m$^2$), aggregate gross area and occupying not more than three stories for sales purposes
   (b) All mercantile occupancies of not more than 3000 ft$^2$ (280 m$^2$) gross area and occupying two or three stories for sales purposes
(3) Class C, all mercantile occupancies of not more than 3000 ft$^2$ (280 m$^2$) gross area used for sales purposes and occupying one story only, excluding mezzanines

**37.1.2.2.2** For the purpose of the classification required in 37.1.2.2.1, the requirements of 37.1.2.2.2.1, 37.1.2.2.2.2, and 37.1.2.2.2.3 shall be met.

**37.1.2.2.2.1** The aggregate gross area shall be the total gross area of all floors used for mercantile purposes.

**37.1.2.2.2.2** Where a mercantile occupancy is divided into sections, regardless of fire separation, the aggregate gross area shall include the area of all sections used for sales purposes.

**37.1.2.2.2.3** Areas of floors not used for sales purposes, such as an area used only for storage and not open to the public, shall not be counted for the purposes of the classifications in 37.1.2.2.1(1), (2), and (3), but means of egress shall be provided for such nonsales areas in accordance with their occupancy, as specified by other chapters of this *Code*.

**37.1.2.2.3** The floor area of a mezzanine, or the aggregate floor area of multiple mezzanines, shall not exceed one-half of the floor area of the room or story in which the mezzanines are located; otherwise, such mezzanine or aggregated mezzanines shall be treated as floors.

**37.1.2.2.4** Where a number of tenant spaces under different management are located in the same building, the aggregate gross area for subclassification shall be one of the following:

(1) Where tenant spaces are not separated, the aggregate gross floor area of all such tenant spaces shall be used in determining classification per 37.1.2.2.1.
(2) Where individual tenant spaces are separated by fire barriers with a 1-hour fire resistance rating, each tenant space shall be individually classified.
(3) The tenant spaces in a mall building in accordance with 37.4.4 shall be classified individually.

### 37.1.3 Multiple Occupancies.

#### 37.1.3.1 General.

**37.1.3.1.1** All multiple occupancies shall be in accordance with 6.1.14 and 37.1.3.

**37.1.3.1.2** Where there are differences in the specific requirements in this chapter and provisions for mixed occupancies or separated occupancies as specified in 6.1.14.3 and 6.1.14.4, the requirements of this chapter shall apply.

#### 37.1.3.2 Combined Mercantile Occupancies and Parking Structures.

**37.1.3.2.1** The fire barrier separating parking structures from a building classified as a mercantile occupancy shall be a fire barrier having a minimum 2-hour fire resistance rating.

**37.1.3.2.2** Openings in the fire barrier required by 37.1.3.2.1 shall not be required to be protected with fire protection–rated opening protectives in enclosed parking structures that are protected throughout by an approved, supervised automatic sprinkler system in accordance with 9.7.1.1(1), or in open parking structures, provided that all of the following conditions are met:

(1) The openings do not exceed 25 percent of the area of the fire barrier in which they are located.
(2) The openings are used as a public entrance and for associated sidelight functions.
(3) The building containing the mercantile occupancy is protected throughout by an approved, supervised automatic sprinkler system in accordance with 9.7.1.1(1).

(4)*Means are provided to prevent spilled fuel from accumulating adjacent to the openings and entering the building.
(5) Physical means are provided to prevent vehicles from being parked or driven within 10 ft (3050 mm) of the openings.
(6) The openings are protected as a smoke partition in accordance with Section 8.4, with no minimum fire protection rating required.

### 37.1.4 Definitions.

**37.1.4.1 General.** For definitions, see Chapter 3, Definitions.

**37.1.4.2 Special Definitions.** A list of special terms used in this chapter follows:

(1) **Anchor Building.** See 3.3.36.2.
(2) **Bulk Merchandising Retail Building.** See 3.3.36.4.
(3) **Gross Leasable Area.** See 3.3.21.3.
(4) **Major Tenant.** See 3.3.166.
(5) **Mall.** See 3.3.167.
(6) **Mall Building.** See 3.3.36.9.
(7) **Open-Air Mercantile Operation.** See 3.3.197.

### 37.1.5 Classification of Hazard of Contents.

**37.1.5.1** The contents of mercantile occupancies shall be classified in accordance with Section 6.2.

**37.1.5.2** Mercantile occupancies classified as high hazard in accordance with Section 6.2 shall meet all of the following additional requirements:

(1) Exits shall be located so that not more than 75 ft (23 m) of travel from any point is needed to reach the nearest exit.
(2) From every point, there shall be not less than two exits accessible by travel in different directions (no common path of travel).
(3) All vertical openings shall be enclosed.

**37.1.6 Minimum Construction Requirements.** (No special requirements.)

**37.1.7 Occupant Load.** The occupant load, in number of persons for whom means of egress and other provisions are required, shall be determined on the basis of the occupant load factors of Table 7.3.1.2 that are characteristic of the use of the space, or shall be determined as the maximum probable population of the space under consideration, whichever is greater.

### 37.2 Means of Egress Requirements.

#### 37.2.1 General.

**37.2.1.1** All means of egress shall be in accordance with Chapter 7 and this chapter.

**37.2.1.2** No inside open stairway, inside open escalator, or inside open ramp shall be permitted to serve as a component of the required means of egress system for more than one floor.

**37.2.1.3** Where there are two or more floors below the street floor, the same stairway or other exit shall be permitted to serve all floors, but all required exits from such areas shall be independent of any open stairways between the street floor and the floor below it.

**37.2.1.4** Where exits from the upper floor also serve as an entrance from a principal street, the upper floor shall be classified as a street floor in accordance with the definition of street floor in 3.3.270 and shall be subject to the requirements of this chapter for street floors.

**37.2.1.5** High hazard mercantile occupancies shall be arranged in accordance with 37.1.5.2.

**37.2.2 Means of Egress Components.**

**37.2.2.1 Components Permitted.** Components of means of egress shall be limited to the types described in 37.2.2.2 through 37.2.2.12.

**37.2.2.2 Doors.**

**37.2.2.2.1** Doors complying with 7.2.1 shall be permitted.

**37.2.2.2.2\*** Locks complying with 7.2.1.5.5 shall be permitted only on principal entrance/exit doors.

**37.2.2.2.3** Elevator lobby exit access door-locking arrangements in accordance with 7.2.1.6.3 shall be permitted.

**37.2.2.2.4** The re-entry provisions of 7.2.1.5.8 shall not apply. [See 7.2.1.5.8.2(1).]

**37.2.2.2.5** Delayed-egress locks complying with 7.2.1.6.1 shall be permitted.

**37.2.2.2.6** Access-controlled egress doors complying with 7.2.1.6.2 shall be permitted in buildings protected throughout by an approved, supervised fire detection system in accordance with Section 9.6 or an approved automatic sprinkler system in accordance with 9.7.1.1(1).

**37.2.2.2.7** Horizontal or vertical security grilles or doors complying with 7.2.1.4.1(3) shall be permitted to be used as part of the required means of egress from a tenant space.

**37.2.2.2.8** All doors at the foot of stairs from upper floors or at the head of stairs leading to floors below the street floor shall swing in the direction of egress travel.

**37.2.2.2.9** Revolving doors complying with 7.2.1.10 shall be permitted.

**37.2.2.2.10** In Class C mercantile occupancies, doors shall be permitted to swing inward against the direction of egress travel where such doors serve only the street floor area.

**37.2.2.3 Stairs.**

**37.2.2.3.1** Stairs complying with 7.2.2 shall be permitted.

**37.2.2.3.2** Spiral stairs complying with 7.2.2.2.3 shall be permitted.

**37.2.2.3.3** Winders complying with 7.2.2.2.4 shall be permitted.

**37.2.2.4 Smokeproof Enclosures.** Smokeproof enclosures complying with 7.2.3 shall be permitted.

**37.2.2.5 Horizontal Exits.** Horizontal exits complying with 7.2.4 shall be permitted.

**37.2.2.6 Ramps.** Ramps complying with 7.2.5 shall be permitted.

**37.2.2.7 Exit Passageways.**

**37.2.2.7.1** Exit passageways complying with 7.2.6 shall be permitted.

**37.2.2.7.2\*** Exit passageways in a mall building shall be permitted to accommodate the following occupant loads independently:

(1) Portion of the occupant load assigned to the exit passageway from only the mall/pedestrian way
(2) Largest occupant load assigned to the exit passageway from a single tenant space

**37.2.2.8 Escalators and Moving Walks.** Escalators and moving walks complying with 7.2.7 shall be permitted.

**37.2.2.9 Fire Escape Stairs.** Fire escape stairs complying with 7.2.8 shall be permitted.

**37.2.2.10 Fire Escape Ladders.** Fire escape ladders complying with 7.2.9 shall be permitted.

**37.2.2.11 Alternating Tread Devices.** Alternating tread devices complying with 7.2.11 shall be permitted.

**37.2.2.12 Areas of Refuge.**

**37.2.2.12.1** Areas of refuge complying with 7.2.12 shall be permitted.

**37.2.2.12.2** In buildings protected throughout by an approved, supervised automatic sprinkler system in accordance with 9.7.1.1(1), two rooms or spaces separated from each other by smoke-resistant partitions in accordance with the definition of area of refuge in 3.3.22 shall not be required.

**37.2.3 Capacity of Means of Egress.**

**37.2.3.1** The capacity of means of egress shall be in accordance with Section 7.3.

**37.2.3.2** In Class A and Class B mercantile occupancies, street floor exits shall be sufficient for the occupant load of the street floor plus the required capacity of stairs, ramps, escalators, and moving walks discharging through the street floor.

**37.2.4 Number of Means of Egress.**

**37.2.4.1** Means of egress shall comply with all of the following, except as otherwise permitted by 37.2.4.2 through 37.2.4.5:

(1) The number of means of egress shall be in accordance with Section 7.4.
(2) Not less than two separate exits shall be provided on every story.
(3) Not less than two separate exits shall be accessible from every part of every story.

**37.2.4.2** Exit access as required by 37.2.4.1(3) shall be permitted to include a single exit access path for the distances permitted as common paths of travel by 37.2.5.3.

**37.2.4.3** A single means of egress shall be permitted in a Class C mercantile occupancy, provided that the travel distance to the exit or to a mall pedestrian way (see 37.4.4.2) does not exceed 75 ft (23 m).

**37.2.4.4** A single means of egress shall be permitted in a Class C mercantile occupancy, provided that the travel distance to the exit or to a mall does not exceed 100 ft (30 m), and the story on which the occupancy is located, and all communicating levels that are traversed to reach the exit or mall, are protected throughout by an approved, supervised automatic sprinkler system in accordance with 9.7.1.1(1).

**37.2.4.5** A single means of egress to an exit or to a mall shall be permitted from a mezzanine within any Class A, Class B, or Class C mercantile occupancy, provided that the common path of travel does not exceed 75 ft (23 m), or does not exceed 100 ft (30 m) if protected throughout by an approved, supervised automatic sprinkler system in accordance with 9.7.1.1(1).

**37.2.5 Arrangement of Means of Egress.**

**37.2.5.1** Means of egress shall be arranged in accordance with Section 7.5.

**37.2.5.2\*** Dead-end corridors shall not exceed 50 ft (15 m).

**37.2.5.3\*** Common paths of travel shall be limited in accordance with 37.2.5.3.1 or 37.2.5.3.2.

**37.2.5.3.1** In buildings protected throughout by an approved, supervised automatic sprinkler system in accordance with 9.7.1.1(1), common paths of travel shall not exceed 100 ft (30 m).

**37.2.5.3.2** In buildings not complying with 37.2.5.3.1, common paths of travel shall not exceed 75 ft (23 m).

**37.2.5.4** Aisles leading to each exit shall be required, and the aggregate width of such aisles shall be not less than the required width of the exit.

**37.2.5.5** Required aisles shall be not less than 28 in. (710 mm) in clear width.

**37.2.5.6** In Class A mercantile occupancies, not less than one aisle of a 60 in. (1525 mm) minimum clear width shall lead directly to an exit.

**37.2.5.7** In mercantile occupancies other than bulk merchandising retail buildings, if the only means of customer entrance is through one exterior wall of the building, one-half of the required egress width from the street floor shall be located in such wall. Means of egress from floors above or below the street floor shall be arranged in accordance with Section 7.5.

**37.2.5.8** Not less than one-half of the required exits shall be located so as to be reached without passing through checkout stands.

**37.2.5.9** Checkout stands or associated railings or barriers shall not obstruct exits, required aisles, or approaches thereto.

**37.2.5.10\*** Where wheeled carts or buggies are used by customers, adequate provision shall be made for the transit and parking of such carts to minimize the possibility that they might obstruct means of egress.

**37.2.5.11** Exit access in Class A mercantile occupancies that are protected throughout by an approved, supervised automatic sprinkler system in accordance with 9.7.1.1(1), and exit access in all Class B and Class C mercantile occupancies, shall be permitted to pass through storerooms, provided that all of the following conditions are met:

(1) Not more than 50 percent of exit access shall be provided through the storeroom.
(2) The storeroom shall not be subject to locking.
(3) The main aisle through the storeroom shall be not less than 44 in. (1120 mm) wide.
(4) The path of travel through the storeroom shall be defined, direct, and continuously maintained in an unobstructed condition.

**37.2.6 Travel Distance to Exits.** Travel distance shall be as specified in 37.2.6.1 and 37.2.6.2 and shall be measured in accordance with Section 7.6.

**37.2.6.1** In buildings protected throughout by an approved, supervised automatic sprinkler system in accordance with 9.7.1.1(1), travel distance shall not exceed 250 ft (76 m).

**37.2.6.2** In buildings not complying with 37.2.6.1, the travel distance shall not exceed 150 ft (46 m).

**37.2.7 Discharge from Exits.**

**37.2.7.1** Exit discharge shall comply with Section 7.7 and 37.2.7.2.

**37.2.7.2\*** Fifty percent of the exits shall be permitted to discharge through the level of exit discharge in accordance with 7.7.2 only where the building is protected throughout by an approved automatic sprinkler system in accordance with 9.7.1.1(1).

**37.2.8 Illumination of Means of Egress.** Means of egress shall be illuminated in accordance with Section 7.8.

**37.2.9 Emergency Lighting.** Class A and Class B mercantile occupancies and mall buildings shall have emergency lighting facilities in accordance with Section 7.9.

**37.2.10 Marking of Means of Egress.** Where an exit is not immediately apparent from all portions of the sales area, means of egress shall have signs in accordance with Section 7.10.

**37.2.11 Special Means of Egress Features.**

**37.2.11.1 Reserved.**

**37.2.11.2 Lockups.** Lockups in mercantile occupancies, other than approved existing lockups, shall comply with the requirements of 23.4.5.

**37.3 Protection.**

**37.3.1 Protection of Vertical Openings.** Any vertical opening shall be protected in accordance with Section 8.6, except under any of the following conditions:

(1) In Class A or Class B mercantile occupancies protected throughout by an approved, supervised automatic sprinkler system in accordance with 9.7.1.1(1), unprotected vertical openings shall be permitted at one of the following locations:
  (a) Between any two floors
  (b) Among the street floor, the first adjacent floor below, and the adjacent floor (or mezzanine) above
(2) In Class C mercantile occupancies, unprotected openings shall be permitted between the street floor and the mezzanine.
(3) The draft stop and closely spaced sprinkler requirements of NFPA 13, *Standard for the Installation of Sprinkler Systems*, shall not be required for unenclosed vertical openings permitted in 37.3.1(1) and (2).

**37.3.2 Protection from Hazards.**

**37.3.2.1\* General.** Hazardous areas shall be protected in accordance with 37.3.2.1.1 or 37.3.2.1.2.

**37.3.2.1.1\*** Hazardous areas shall be protected in accordance with Section 8.7.

**37.3.2.1.2** In general storage and stock areas protected by an automatic extinguishing system in accordance with 9.7.1.1(1) or 9.7.1.2, an enclosure shall be exempt from the provisions of 8.7.1.2.

**37.3.2.2\* High Hazard Contents Areas.** High hazard contents areas, as classified in Section 6.2, shall meet all of the following criteria:

(1) The area shall be separated from other parts of the building by fire barriers having a minimum 1-hour fire resistance rating, with all openings therein protected by self-closing fire door assemblies having a minimum ¾-hour fire protection rating.
(2) The area shall be protected by an automatic extinguishing system in accordance with 9.7.1.1(1) or 9.7.1.2.

**37.3.2.3 Cooking Equipment.** Cooking equipment shall be protected in accordance with 9.2.3, unless the cooking equipment is one of the following types:

(1) Outdoor equipment
(2) Portable equipment not flue-connected
(3) Equipment used only for food warming

**37.3.3 Interior Finish.**

**37.3.3.1 General.** Interior finish shall be in accordance with Section 10.2.

**37.3.3.2 Interior Wall and Ceiling Finish.** Interior wall and ceiling finish materials complying with Section 10.2 shall be Class A, Class B, or Class C.

**37.3.3.3 Interior Floor Finish.** (No requirements.)

**37.3.4 Detection, Alarm, and Communications Systems.**

**37.3.4.1 General.** Class A mercantile occupancies shall be provided with a fire alarm system in accordance with Section 9.6.

**37.3.4.2 Initiation.** Initiation of the required fire alarm system shall be by one of the following means:

(1) Manual means per 9.6.2.1(1)
(2) Approved automatic fire detection system that complies with 9.6.2.1(2) and provides protection throughout the building, plus a minimum of one manual fire alarm box in accordance with 9.6.2.6
(3) Approved automatic sprinkler system that complies with 9.6.2.1(3) and provides protection throughout the building, plus a minimum of one manual fire alarm box in accordance with 9.6.2.6

**37.3.4.3 Notification.**

**37.3.4.3.1 Occupant Notification.** During all times that the mercantile occupancy is occupied, the required fire alarm system, once initiated, shall perform one of the following functions:

(1) It shall activate an alarm in accordance with 9.6.3 throughout the mercantile occupancy, and both of the following also shall apply:
   (a) Positive alarm sequence in accordance with 9.6.3.4 shall be permitted.
   (b) A presignal system in accordance with 9.6.3.3 shall be permitted.
(2) Occupant notification shall be made via a voice communication or public address system in accordance with 9.6.3.9.2

**37.3.4.3.2 Emergency Forces Notification.** Emergency forces notification shall be provided and shall include notifying both of the following:

(1) Fire department in accordance with 9.6.4
(2) Local emergency organization, if provided

**37.3.5 Extinguishment Requirements.**

**37.3.5.1** Mercantile occupancies, other than one-story buildings that meet the requirements of a street floor, as defined in 3.3.270, shall be protected by an approved automatic sprinkler system in accordance with 9.7.1.1(1) in any of the following specified locations:

(1) Throughout all mercantile occupancies with a story over 15,000 ft² (1400 m²) in area
(2) Throughout all mercantile occupancies exceeding 30,000 ft² (2800 m²) in gross area

(3) Throughout stories below the level of exit discharge where such stories have an area exceeding 2500 ft² (232 m²) and are used for the sale, storage, or handling of combustible goods and merchandise
(4) Throughout multiple occupancies protected as mixed occupancies in accordance with 6.1.14 where the conditions of 37.3.5.1(1), (2), or (3) apply to the mercantile occupancy

**37.3.5.2 Reserved.**

**37.3.5.3** Portable fire extinguishers shall be provided in all mercantile occupancies in accordance with 9.7.4.1.

**37.3.6 Corridors.** (No requirements.)

**37.3.7 Subdivision of Building Spaces.** (No special requirements.)

**37.4 Special Provisions.**

**37.4.1 Limited Access or Underground Buildings.** See Section 11.7.

**37.4.2 High-Rise Buildings.** (No additional requirements.)

**37.4.3 Open-Air Mercantile Operations.**

**37.4.3.1** Open-air mercantile operations, such as open-air markets, gasoline filling stations, roadside stands for the sale of farm produce, and other outdoor mercantile operations, shall be arranged and conducted to maintain free and unobstructed ways of travel at all times.

**37.4.3.2** Ways of travel shall allow prompt escape from any point of danger in case of fire or other emergency, with no dead ends in which persons might be trapped due to display stands, adjoining buildings, fences, vehicles, or other obstructions.

**37.4.3.3** Mercantile operations that are conducted in roofed-over areas shall be treated as mercantile buildings, provided that canopies over individual small stands to protect merchandise from the weather are not construed as constituting buildings for the purpose of this *Code*.

**37.4.4 Mall Buildings.** The provisions of 37.4.4 shall apply to mall buildings and any number of anchor buildings. *(See 3.3.36.9.)*

**37.4.4.1 General.** The mall building shall be treated as a single building for the purpose of calculation of means of egress and shall be subject to the requirements for appropriate occupancies, except as modified by the provisions of 37.4.4; and the mall shall be of a clear width not less than that needed to accommodate egress requirements as set forth in other sections of this *Code*.

**37.4.4.2 Pedestrian Way.** The mall shall be permitted to be considered a pedestrian way, provided that the criteria of 37.4.4.2.1 and 37.4.4.2.2 are met.

**37.4.4.2.1** The travel distance within a tenant space to an exit or to the mall shall not exceed the maximum travel distance permitted by the occupancy chapter.

**37.4.4.2.2** An additional 200 ft (61 m) shall be permitted for travel through the mall space, provided that all the following requirements are met:

(1) The mall shall be of a clear width not less than that needed to accommodate egress requirements, as set forth in other sections of this chapter, but shall be not less than 20 ft (6100 mm) wide in its narrowest dimension.

(2) On each side of the mall floor area, the mall shall be provided with an unobstructed exit access of not less than 10 ft (3050 mm) in clear width parallel to, and adjacent to, the mall tenant front.

(3)*The exit access specified in 37.4.4.2.2(2) shall lead to an exit having a width of not less than 66 in. (1675 mm).

(4) The mall, and all buildings connected thereto, except open parking structures, shall be protected throughout by an approved, supervised automatic sprinkler system in accordance with 9.7.1.1(1).

(5) Walls dividing tenant spaces from each other shall extend from the floor to the underside of the roof deck, to the floor deck above, or to the ceiling where the ceiling is constructed to limit the transfer of smoke, and all of the following also shall apply:

(a) Where the tenant areas are provided with an engineered smoke control system, walls shall not be required to divide tenant spaces from each other.

(b) No separation shall be required between a tenant space and the mall.

(6)*Malls with a floor opening connecting more than two levels shall be provided with a smoke control system.

### 37.4.4.3 Means of Egress Details.

**37.4.4.3.1** Dead ends not exceeding a length equal to twice the width of the mall, measured at the narrowest location within the dead-end portion of the mall, shall be permitted.

**37.4.4.3.2** Every story of a covered mall building shall be provided with the number of means of egress specified by Section 7.4 and as modified by 37.4.4.3.2.1 or 37.4.4.3.2.2.

**37.4.4.3.2.1** Exit access travel shall be permitted to be common for the distances permitted as common paths of travel by 37.2.5.3.

**37.4.4.3.2.2** A single means of egress shall be permitted in a Class C mercantile occupancy or a business occupancy, provided that the travel distance to the exit or to a mall pedestrian way *(see 37.4.2)* does not exceed 100 ft (30 m).

**37.4.4.3.3** Every floor of a mall shall be provided with the number of means of egress specified by Section 7.4, with not less than two means of egress remotely located from each other.

**37.4.4.3.4** Class A and Class B mercantile occupancies connected to a mall shall be provided with the number of means of egress required by Section 7.4, with not less than two means of egress remotely located from one another.

**37.4.4.3.5*** Each individual anchor building shall have means of egress independent of the mall.

**37.4.4.3.6** Each individual major tenant of a mall building shall have a minimum of one-half of its required means of egress independent of the mall.

**37.4.4.3.7** Every mall shall be provided with unobstructed exit access parallel to, and adjacent to, the mall tenant fronts and extending to each mall exit.

### 37.4.4.3.8 Reserved.

**37.4.4.3.9** Emergency lighting shall be provided in accordance with 37.2.9.

### 37.4.4.4 Detection, Alarm, and Communications Systems.

**37.4.4.4.1 General.** Malls shall be provided with a fire alarm system in accordance with Section 9.6.

**37.4.4.4.2 Initiation.** Initiation of the required fire alarm system shall be by means of the required automatic sprinkler system in accordance with 9.6.2.1(3).

### 37.4.4.4.3 Notification.

**37.4.4.4.3.1 Occupant Notification.** During all times that the mall is occupied, the required fire alarm system, once initiated, shall perform one of the following functions:

(1) It shall activate an alarm in accordance with 9.6.3 throughout the mall, and positive alarm sequence in accordance with 9.6.3.4 shall be permitted.

(2) Occupant notification shall be permitted to be made via a voice communication or public address system in accordance with 9.6.3.9.2.

**37.4.4.4.3.2** *(See 9.6.3.5.3.)*

**37.4.4.4.3.3 Emergency Forces Notification.** Emergency forces notification shall be provided and shall include notifying all of the following:

(1) Fire department in accordance with 9.6.4
(2) Local emergency organization, if provided

**37.4.4.4.4 Emergency Control.** The fire alarm system shall be arranged to automatically actuate smoke management or smoke control systems in accordance with 9.6.5.2(3).

**37.4.4.5 Tenant Spaces.** Each individual tenant space shall have means of egress to the outside or to the mall based on occupant load calculated by using Table 7.3.1.2.

**37.4.4.6 Exit Passageways.** Exit passageways shall comply with 37.4.4.6.1 and 37.4.4.6.2.

**37.4.4.6.1** Exit passageways in a mall building shall be permitted to accommodate the following occupant loads independently:

(1) Portion of the occupant load assigned to the exit passageway from only the mall
(2) Largest occupant load assigned to the exit passageway from a single tenant space

**37.4.4.6.2*** Rooms housing building service equipment, janitor closets, and service elevators shall be permitted to open directly onto exit passageways, provided that all of the following criteria are met:

(1) The required fire resistance rating between such rooms or areas and the exit passageway shall be maintained in accordance with 7.1.3.2.

(2) Such rooms or areas shall be protected by an approved automatic sprinkler system in accordance with 9.7.1.1(1), but the exceptions in NFPA 13, *Standard for the Installation of Sprinkler Systems*, allowing the omission of sprinklers from such rooms shall not be permitted.

(3) Service elevators opening into the exit passageway shall not open into areas other than exit passageways.

(4) Where exit stair enclosures discharge into the exit passageway, the provisions of 7.2.1.5.8 shall apply, regardless of the number of stories served.

**37.4.4.7 Plastic Signs.** Within every store or level, and from sidewall to sidewall of each tenant space facing the mall, plastic signs shall comply with all of the following:

(1) Plastic signs shall not exceed 20 percent of the wall area facing the mall.

(2) Plastic signs shall not exceed a height of 36 in. (915 mm), except if the sign is vertical, in which case the height shall not exceed 8 ft (2440 mm) and the width shall not exceed 36 in. (915 mm).

(3) Plastic signs shall be located a minimum distance of 18 in. (455 mm) from adjacent tenants.

(4) Plastics, other than foamed plastics, shall meet one of the following criteria:

(a) They shall be light-transmitting plastics.

(b) They shall have a self-ignition temperature of 650°F (343°C) or greater when tested in accordance with ASTM D 1929, *Standard Test Method for Determining Ignition Temperatures of Plastic*, and a flame spread index not greater than 75 and a smoke developed index not greater than 450 when tested in the manner intended for use in accordance with ASTM E 84, *Standard Test Method for Surface Burning Characteristics of Building Materials*, or ANSI/UL 723, *Standard for Test for Surface Burning Characteristics of Building Materials*.

(5) The edges and backs of plastic signs in the mall shall be fully encased in metal.

(6) Foamed plastics shall have a maximum heat release rate of 150 kW when tested in accordance with ANSI/UL 1975, *Standard for Fire Tests for Foamed Plastics Used for Decorative Purposes*, or in accordance with NFPA 289, *Standard Method of Fire Test for Individual Fuel Packages*, using the 20 kW ignition source.

(7) Foamed plastics shall comply with all of the following:

(a) The density of foamed plastic signs shall be not less than 20 lb/ft$^3$ (320 kg/m$^3$).

(b) The thickness of foamed plastic signs shall be not greater than ½ in. (13 mm).

**37.4.4.8 Kiosks.** Kiosks and similar structures (temporary or permanent) shall not be considered as tenant spaces and shall meet all of the following requirements:

(1) Combustible kiosks and similar structures shall be constructed of any of the following materials:

(a) Fire-retardant-treated wood complying with the requirements for fire-retardant-impregnated wood in NFPA 703, *Standard for Fire Retardant–Treated Wood and Fire-Retardant Coatings for Building Materials*

(b) Light-transmitting plastics complying with the building code

(c) Foamed plastics having a maximum heat release rate not greater than 100 kW when tested in accordance with ANSI/UL 1975, *Standard for Fire Tests for Foamed Plastics Used for Decorative Purposes*, or in accordance with NFPA 289, *Standard Method of Fire Test for Individual Fuel Packages*, using the 20 kW ignition source

(d) Metal composite material (MCM) having a flame spread index not greater than 25 and a smoke developed index not greater than 450 in accordance with ASTM E 84, *Standard Test Method for Surface Burning Characteristics of Building Materials*, or ANSI/UL 723, *Standard for Test for Surface Burning Characteristics of Building Materials*, when tested as an assembly in the maximum thickness intended for use

(e) Textiles and films meeting the flame propagation performance criteria contained in NFPA 701, *Standard Methods of Fire Tests for Flame Propagation of Textiles and Films*

(2) Kiosks or similar structures located within the mall shall be protected with approved fire suppression and detection devices.

(3) The minimum horizontal separation between kiosks, or groups of kiosks, and other structures within the mall shall be 20 ft (6100 mm).

(4) Each kiosk, or group of kiosks, or similar structure shall have a maximum area of 300 ft$^2$ (27.8 m$^2$).

**37.4.5 Bulk Merchandising Retail Buildings.** Existing bulk merchandising retail buildings exceeding 15,000 ft$^2$ (1400 m$^2$) in area shall comply with the requirements of this chapter, as modified by 37.4.5.1 through 37.4.5.6.2.

**37.4.5.1 Minimum Construction Requirements.** (No requirements.)

**37.4.5.2 Means of Egress Requirements.**

**37.4.5.2.1** All means of egress shall be in accordance with Chapter 7 and this chapter.

**37.4.5.2.2** Not less than 50 percent of the required egress capacity shall be located independent of the main entrance/exit doors.

**37.4.5.3 Storage, Arrangement, Protection, and Quantities of Hazardous Commodities.** The storage, arrangement, protection, and quantities of hazardous commodities shall be in accordance with the applicable provisions of the following:

(1) The fire code *(see 3.3.94)*
(2) NFPA 13, *Standard for the Installation of Sprinkler Systems*
(3) NFPA 30, *Flammable and Combustible Liquids Code*
(4) NFPA 30B, *Code for the Manufacture and Storage of Aerosol Products*
(5) NFPA 400, *Hazardous Materials Code*, Chapter 14, for organic peroxide formulations
(6) NFPA 400, *Hazardous Materials Code*, Chapter 15, for oxidizer solids and liquids
(7) NFPA 400, *Hazardous Materials Code*, various chapters, depending on characteristics of a particular pesticide
(8) NFPA 1124, *Code for the Manufacture, Transportation, Storage, and Retail Sales of Fireworks and Pyrotechnic Articles*

**37.4.5.4 Detection, Alarm, and Communications Systems.**

**37.4.5.4.1 General.** Bulk merchandising retail buildings shall be provided with a fire alarm system in accordance with Section 9.6.

**37.4.5.4.2 Initiation.** Initiation of the required fire alarm system shall be by means of the required approved automatic sprinkler system *(see 37.4.5.5)* in accordance with 9.6.2.1(3).

**37.4.5.4.3 Occupant Notification.** During all times that the mercantile occupancy is occupied, the required fire alarm system, once initiated, shall perform one of the following functions:

(1) It shall activate an alarm in accordance with 9.6.3 throughout the mercantile occupancy, and positive alarm sequence in accordance with 9.6.3.4 shall be permitted.
(2) Occupant notification shall be permitted to be made via a voice communication or public address system in accordance with 9.6.3.9.2.

**37.4.5.4.4 Emergency Forces Notification.** Emergency forces notification shall be provided and shall include notifying both of the following:

(1) Fire department in accordance with 9.6.4
(2) Local emergency organization, if provided

**37.4.5.5 Extinguishing Requirements.** Bulk merchandising retail buildings shall be protected throughout by an approved, supervised automatic sprinkler system in accordance with 9.7.1.1(1) and the applicable provisions of the following:

(1) The fire code (*see 3.3.94*)
(2) NFPA 13, *Standard for the Installation of Sprinkler Systems*
(3) NFPA 30, *Flammable and Combustible Liquids Code*
(4) NFPA 30B, *Code for the Manufacture and Storage of Aerosol Products*

**37.4.5.6 Emergency Plan and Employee Training.**

**37.4.5.6.1** There shall be in effect an approved written plan for the emergency egress and relocation of occupants.

**37.4.5.6.2** All employees shall be instructed and periodically drilled with respect to their duties under the plan.

**37.4.6 Retail Sales of Consumer Fireworks, 1.4G.** Mercantile occupancies in which the retail sale of consumer fireworks, 1.4G, is conducted, other than approved existing facilities, shall comply with NFPA 1124, *Code for the Manufacture, Transportation, Storage, and Retail Sales of Fireworks and Pyrotechnic Articles.*

**37.5 Building Services.**

**37.5.1 Utilities.** Utilities shall comply with the provisions of Section 9.1.

**37.5.2 Heating, Ventilating, and Air-Conditioning.** Heating, ventilating, and air-conditioning equipment shall comply with the provisions of Section 9.2.

**37.5.3 Elevators, Escalators, and Conveyors.** Elevators, escalators, and conveyors shall comply with the provisions of Section 9.4.

**37.5.4 Rubbish Chutes, Incinerators, and Laundry Chutes.** Rubbish chutes, incinerators, and laundry chutes shall comply with the provisions of Section 9.5.

**37.6 Reserved.**

**37.7 Operating Features.**

**37.7.1 Emergency Plans.** Emergency plans complying with Section 4.8 shall be provided in high-rise buildings.

**37.7.2 Drills.** In every Class A or Class B mercantile occupancy, employees shall be periodically trained in accordance with Section 4.7.

**37.7.3 Extinguisher Training.** Employees of mercantile occupancies shall be periodically instructed in the use of portable fire extinguishers.

**37.7.4 Food Service Operations.** Food service operations shall comply with 13.7.2.

**37.7.5 Upholstered Furniture and Mattresses.** The provisions of 10.3.2 shall not apply to upholstered furniture and mattresses.

**37.7.6 Soiled Linen and Trash Receptacles.** The requirements of 10.3.9 for containers for rubbish, waste, or linen with a capacity of 20 gal (75.7 L) or more shall not apply.

## Chapter 38  New Business Occupancies

**38.1 General Requirements.**

**38.1.1 Application.**

**38.1.1.1** The requirements of this chapter shall apply to new buildings or portions thereof used as business occupancies. (*See 1.3.1.*)

**38.1.1.2 Administration.** The provisions of Chapter 1, Administration, shall apply.

**38.1.1.3 General.** The provisions of Chapter 4, General, shall apply.

**38.1.1.4** The provisions of this chapter shall apply to life safety requirements for all new business buildings.

**38.1.1.5** Additions to existing buildings shall conform to the requirements of 4.6.7. Existing portions of the structure shall not be required to be modified, provided that the new construction has not diminished the fire safety features of the facility.

**38.1.2 Classification of Occupancy.** Business occupancies shall include all buildings and structures or parts thereof with occupancy as defined in 6.1.11.

**38.1.3 Multiple Occupancies.**

**38.1.3.1 General.**

**38.1.3.1.1** All multiple occupancies shall be in accordance with 6.1.14 and 38.1.3.

**38.1.3.1.2** Where there are differences in the specific requirements in this chapter and provisions for mixed occupancies or separated occupancies as specified in 6.1.14.3 and 6.1.14.4, the requirements of this chapter shall apply.

**38.1.3.2 Combined Business Occupancies and Parking Structures.**

**38.1.3.2.1** The fire barrier separating parking structures from a building classified as a business occupancy shall be a fire barrier having a minimum 2-hour fire resistance rating.

**38.1.3.2.2** Openings in the fire barrier required by 38.1.3.2.1 shall not be required to be protected with fire protection–rated opening protectives in enclosed parking structures that are protected throughout by an approved, supervised automatic sprinkler system in accordance with 9.7.1.1(1), or in open parking structures, provided that all of the following conditions are met:

(1) The openings do not exceed 25 percent of the area of the fire barrier in which they are located.
(2) The openings are used as a public entrance and for associated sidelight functions.
(3) The building containing the business occupancy is protected throughout by an approved, supervised automatic sprinkler system in accordance with 9.7.1.1(1).
(4)*Means are provided to prevent spilled fuel from accumulating adjacent to the openings and entering the building.
(5) Physical means are provided to prevent vehicles from being parked or driven within 10 ft (3050 mm) of the openings.
(6) The openings are protected as a smoke partition in accordance with Section 8.4, with no minimum fire protection rating required.

**38.1.4 Definitions.**

**38.1.4.1 General.** For definitions, see Chapter 3, Definitions.

**38.1.4.2 Special Definitions.** Special terms applicable to this chapter are defined in Chapter 3.

**38.1.5 Classification of Hazard of Contents.** The contents of business occupancies shall be classified as ordinary hazard in accordance with Section 6.2.

**38.1.6 Minimum Construction Requirements.** (No requirements.)

**38.1.7 Occupant Load.** The occupant load, in number of persons for whom means of egress and other provisions are required, shall be determined on the basis of the occupant load factors of Table 7.3.1.2 that are characteristic of the use of the space, or shall be determined as the maximum probable population of the space under consideration, whichever is greater.

**38.2 Means of Egress Requirements.**

**38.2.1 General.**

**38.2.1.1** All means of egress shall be in accordance with Chapter 7 and this chapter.

**38.2.1.2** If, owing to differences in grade, any street floor exits are at points above or below the street or the finished ground level, such exits shall comply with the provisions for exits from upper floors or floors below the street floor.

**38.2.1.3** Stairs and ramps serving two or more floors below a street floor occupied for business use shall be permitted in accordance with 38.2.1.3.1 and 38.2.1.3.2.

**38.2.1.3.1** Where two or more floors below the street floor are occupied for business use, the same stairs or ramps shall be permitted to serve each.

**38.2.1.3.2** An inside open stairway or inside open ramp shall be permitted to serve as a required egress facility from not more than one floor level below the street floor.

**38.2.1.4** Floor levels that are below the street floor; are used only for storage, heating, and other service equipment; and are not subject to business occupancy shall have means of egress in accordance with Chapter 42.

**38.2.2 Means of Egress Components.**

**38.2.2.1 Components Permitted.** Means of egress components shall be limited to the types described in 38.2.2.2 through 38.2.2.12.

**38.2.2.2 Doors.**

**38.2.2.2.1** Doors complying with 7.2.1 shall be permitted.

**38.2.2.2.2*** Locks complying with 7.2.1.5.5 shall be permitted only on principal entrance/exit doors.

**38.2.2.2.3** Elevator lobby exit access door-locking arrangements in accordance with 7.2.1.6.3 shall be permitted.

**38.2.2.2.4 Reserved.**

**38.2.2.2.5** Delayed-egress locks complying with 7.2.1.6.1 shall be permitted.

**38.2.2.2.6** Access-controlled egress doors complying with 7.2.1.6.2 shall be permitted.

**38.2.2.2.7** Horizontal or vertical security grilles or doors complying with 7.2.1.4.1(3) shall be permitted to be used as part of the required means of egress from a tenant space.

**38.2.2.2.8 Reserved.**

**38.2.2.2.9** Revolving doors complying with 7.2.1.10 shall be permitted.

**38.2.2.3 Stairs.**

**38.2.2.3.1** Stairs complying with 7.2.2 shall be permitted.

**38.2.2.3.2** Spiral stairs complying with 7.2.2.2.3 shall be permitted.

**38.2.2.4 Smokeproof Enclosures.** Smokeproof enclosures complying with 7.2.3 shall be permitted.

**38.2.2.5 Horizontal Exits.** Horizontal exits complying with 7.2.4 shall be permitted.

**38.2.2.6 Ramps.** Ramps complying with 7.2.5 shall be permitted.

**38.2.2.7 Exit Passageways.** Exit passageways complying with 7.2.6 shall be permitted.

**38.2.2.8 Reserved.**

**38.2.2.9 Reserved.**

**38.2.2.10 Fire Escape Ladders.** Fire escape ladders complying with 7.2.9 shall be permitted.

**38.2.2.11 Alternating Tread Devices.** Alternating tread devices complying with 7.2.11 shall be permitted.

**38.2.2.12 Areas of Refuge.**

**38.2.2.12.1** Areas of refuge complying with 7.2.12 shall be permitted.

**38.2.2.12.2** In buildings protected throughout by an approved, supervised automatic sprinkler system in accordance with 9.7.1.1(1), two rooms or spaces separated from each other by smoke-resistant partitions in accordance with the definition of area of refuge in 3.3.22 shall not be required.

**38.2.3 Capacity of Means of Egress.**

**38.2.3.1** The capacity of means of egress shall be in accordance with Section 7.3.

**38.2.3.2*** The clear width of any corridor or passageway serving an occupant load of 50 or more shall be not less than 44 in. (1120 mm).

**38.2.3.3** Street floor exits shall be sufficient for the occupant load of the street floor plus the required capacity of open stairs and ramps discharging through the street floor.

**38.2.4 Number of Means of Egress.**

**38.2.4.1** Means of egress shall comply with all of the following, except as otherwise permitted by 38.2.4.2 through 38.2.4.6:

(1) The number of means of egress shall be in accordance with Section 7.4.
(2) Not less than two separate exits shall be provided on every story.
(3) Not less than two separate exits shall be accessible from every part of every story.

**38.2.4.2** Exit access, as required by 38.2.4.1(3), shall be permitted to include a single exit access path for the distances permitted as common paths of travel by 38.2.5.3.

**38.2.4.3** A single exit shall be permitted for a room or area with a total occupant load of less than 100 persons, provided that all of the following criteria are met:

(1) The exit shall discharge directly to the outside at the level of exit discharge for the building.
(2) The total distance of travel from any point, including travel within the exit, shall not exceed 100 ft (30 m).

(3) The total distance of travel specified in 38.2.4.3(2) shall be on the same story, or, if traversing of stairs is necessary, such stairs shall not exceed 15 ft (4570 mm) in height, and both of the following also shall apply:

    (a) Interior stairs shall be provided with complete enclosures to separate them from any other part of the building, with no door openings therein.

    (b) A single outside stair in accordance with 7.2.2 shall be permitted to serve all stories permitted within the 15 ft (4570 mm) vertical travel limitation.

**38.2.4.4** Any business occupancy three or fewer stories in height, and not exceeding an occupant load of 30 people per story, shall be permitted a single separate exit to each story, provided that all of the following criteria are met:

(1) The exit shall discharge directly to the outside.
(2) The total travel distance to the outside of the building shall not exceed 100 ft (30 m).
(3) Interior exit stairs shall be enclosed in accordance with 7.1.3.2, and both of the following also shall apply:

    (a) The stair shall serve as an exit from no other stories.
    (b) A single outside stair in accordance with 7.2.2 shall be permitted to service all stories.

**38.2.4.5** A single means of egress shall be permitted from a mezzanine within a business occupancy, provided that the common path of travel does not exceed 75 ft (23 m), or 100 ft (30 m) if protected throughout by an approved, supervised automatic sprinkler system in accordance with 9.7.1.1(1).

**38.2.4.6** A single exit shall be permitted for a single-tenant space or building two or fewer stories in height, provided that both of the following criteria are met:

(1) The building is protected throughout by an approved, supervised automatic sprinkler system in accordance with 9.7.1.1(1).
(2) The total travel to the outside does not exceed 100 ft (30 m).

**38.2.5 Arrangement of Means of Egress.**

**38.2.5.1** Means of egress shall be arranged in accordance with Section 7.5.

**38.2.5.2** Dead-end corridors shall be permitted in accordance with 38.2.5.2.1 or 38.2.5.2.2.

**38.2.5.2.1** In buildings protected throughout by an approved, supervised automatic sprinkler system in accordance with 9.7.1.1(1), dead-end corridors shall not exceed 50 ft (15 m).

**38.2.5.2.2** In buildings other than those complying with 38.2.5.2.1, dead-end corridors shall not exceed 20 ft (6100 mm).

**38.2.5.3** Limitations on common path of travel shall be in accordance with 38.2.5.3.1, 38.2.5.3.2, and 38.2.5.3.3.

**38.2.5.3.1** Common path of travel shall not exceed 100 ft (30 m) in a building protected throughout by an approved, supervised automatic sprinkler system in accordance with 9.7.1.1(1).

**38.2.5.3.2** Common path of travel shall not exceed 100 ft (30 m) within a single tenant space having an occupant load not exceeding 30 persons.

**38.2.5.3.3** In buildings other than those complying with 38.2.5.3.1 or 38.2.5.3.2, common path of travel shall not exceed 75 ft (23 m).

**38.2.6 Travel Distance to Exits.** Travel distance shall comply with 38.2.6.1 through 38.2.6.3.

**38.2.6.1** Travel distance shall be measured in accordance with Section 7.6.

**38.2.6.2** Travel distance to an exit shall not exceed 200 ft (61 m) from any point in a building, unless otherwise permitted by 38.2.6.3.

**38.2.6.3** Travel distance shall not exceed 300 ft (91 m) in business occupancies protected throughout by an approved, supervised automatic sprinkler system in accordance with Section 9.7.

**38.2.7 Discharge from Exits.** Exit discharge shall comply with Section 7.7.

**38.2.8 Illumination of Means of Egress.** Means of egress shall be illuminated in accordance with Section 7.8.

**38.2.9 Emergency Lighting.**

**38.2.9.1** Emergency lighting shall be provided in accordance with Section 7.9 in any building where any one of the following conditions exists:

(1) The building is three or more stories in height.
(2) The occupancy is subject to 50 or more occupants above or below the level of exit discharge.
(3) The occupancy is subject to 300 or more total occupants.

**38.2.9.2** Emergency lighting in accordance with Section 7.9 shall be provided for all underground and limited access structures, as defined in 3.3.271.11 and 3.3.271.3, respectively.

**38.2.10 Marking of Means of Egress.** Means of egress shall have signs in accordance with Section 7.10.

**38.2.11 Special Means of Egress Features.**

**38.2.11.1 Reserved.**

**38.2.11.2 Lockups.** Lockups in business occupancies shall comply with the requirements of 22.4.5.

**38.3 Protection.**

**38.3.1 Protection of Vertical Openings.**

**38.3.1.1** Vertical openings shall be enclosed or protected in accordance with Section 8.6, unless otherwise permitted by any of the following:

(1) Unenclosed vertical openings in accordance with 8.6.9.1 shall be permitted.
(2) Exit access stairs in accordance with 38.2.4.6 shall be permitted to be unenclosed.

**38.3.1.2** Floors that are below the street floor and are used for storage or other than a business occupancy shall have no unprotected openings to business occupancy floors.

**38.3.2 Protection from Hazards.**

**38.3.2.1\* General.** Hazardous areas including, but not limited to, areas used for general storage, boiler or furnace rooms, and maintenance shops that include woodworking and painting areas shall be protected in accordance with Section 8.7.

**38.3.2.2\* High Hazard Contents Areas.** High hazard contents areas, as classified in Section 6.2, shall meet all of the following criteria:

(1) The area shall be separated from other parts of the building by fire barriers having a minimum 1-hour fire resistance rating, with all openings therein protected by self-closing fire door assemblies having a minimum ¾-hour fire protection rating.
(2) The area shall be protected by an automatic extinguishing system in accordance with 9.7.1.1(1) or 9.7.1.2.

**38.3.2.3 Cooking Equipment.** Cooking equipment shall be protected in accordance with 9.2.3, unless the cooking equipment is one of the following types:

(1) Outdoor equipment
(2) Portable equipment not flue-connected
(3) Equipment used only for food warming

**38.3.3 Interior Finish.**

**38.3.3.1 General.** Interior finish shall be in accordance with Section 10.2.

**38.3.3.2 Interior Wall and Ceiling Finish.**

**38.3.3.2.1** Interior wall and ceiling finish material complying with Section 10.2 shall be Class A or Class B in exits and in exit access corridors.

**38.3.3.2.2** Interior wall and ceiling finishes shall be Class A, Class B, or Class C in areas other than those specified in 38.3.3.2.1.

**38.3.3.3 Interior Floor Finish.**

**38.3.3.3.1** Interior floor finish shall comply with Section 10.2.

**38.3.3.3.2** Interior floor finish in exit enclosures shall be Class I or Class II.

**38.3.3.3.3** Interior floor finish shall comply with 10.2.7.1 or 10.2.7.2, as applicable.

**38.3.4 Detection, Alarm, and Communications Systems.**

**38.3.4.1 General.** A fire alarm system in accordance with Section 9.6 shall be provided in all business occupancies where any one of the following conditions exists:

(1) The building is three or more stories in height.
(2) The occupancy is subject to 50 or more occupants above or below the level of exit discharge.
(3) The occupancy is subject to 300 or more total occupants.

**38.3.4.2 Initiation.** Initiation of the required fire alarm system shall be by one of the following means:

(1) Manual means in accordance with 9.6.2.1(1)
(2) Means of an approved automatic fire detection system that complies with 9.6.2.1(2) and provides protection throughout the building
(3) Means of an approved automatic sprinkler system that complies with 9.6.2.1(3) and provides protection throughout the building

**38.3.4.3 Occupant Notification.** During all times that the building is occupied, the required fire alarm system, once initiated, shall activate a general alarm in accordance with 9.6.3 throughout the building, and positive alarm sequence in accordance with 9.6.3.4 shall be permitted.

**38.3.4.4 Emergency Forces Notification.** Emergency forces notification shall be provided and shall include notifying both of the following:

(1) Fire department in accordance with 9.6.4
(2) Local emergency organization, if provided

**38.3.5 Extinguishment Requirements.** Portable fire extinguishers shall be provided in every business occupancy in accordance with 9.7.4.1.

**38.3.6 Corridors.**

**38.3.6.1\*** Where access to exits is provided by corridors, such corridors shall be separated from use areas by fire barriers in accordance with Section 8.3 having a minimum 1-hour fire resistance rating, unless one of the following conditions exists:

(1)\*Where exits are available from an open floor area
(2)\*Within a space occupied by a single tenant
(3) Within buildings protected throughout by an approved, supervised automatic sprinkler system in accordance with 9.7.1.1(1)

**38.3.6.2** Openings in corridor walls required by 38.3.6.1 to have a fire resistance rating shall be protected in accordance with Section 8.3.

**38.3.7 Subdivision of Building Spaces.** (No special requirements.)

**38.4 Special Provisions.**

**38.4.1 Limited Access or Underground Buildings.** See Section 11.7.

**38.4.2 High-Rise Buildings.** High-rise buildings shall comply with Section 11.8.

**38.4.3 Air Traffic Control Towers.**

**38.4.3.1** Air traffic control towers shall comply with the requirements of this chapter and Section 11.3.

**38.4.3.2** The requirements of Section 11.8 shall not apply to air traffic control towers.

**38.5 Building Services.**

**38.5.1 Utilities.** Utilities shall comply with the provisions of Section 9.1.

**38.5.2 Heating, Ventilating, and Air-Conditioning.** Heating, ventilating, and air-conditioning equipment shall comply with the provisions of Section 9.2.

**38.5.3 Elevators, Escalators, and Conveyors.** Elevators, escalators, and conveyors shall comply with the provisions of Section 9.4.

**38.5.4 Rubbish Chutes, Incinerators, and Laundry Chutes.** Rubbish chutes, incinerators, and laundry chutes shall comply with the provisions of Section 9.5.

**38.6 Reserved.**

**38.7 Operating Features.**

**38.7.1 Emergency Plans.** Emergency plans complying with Section 4.8 shall be provided in high-rise buildings.

**38.7.2 Drills.** In all business occupancy buildings occupied by more than 500 persons, or by more than 100 persons above or below the street level, employees and supervisory personnel shall be periodically instructed in accordance with Section 4.7 and shall hold drills periodically where practicable.

**38.7.3 Extinguisher Training.** Designated employees of business occupancies shall be periodically instructed in the use of portable fire extinguishers.

**38.7.4 Food Service Operations.** Food service operations shall comply with 12.7.2.

**38.7.5 Upholstered Furniture and Mattresses.** The provisions of 10.3.2 shall not apply to upholstered furniture and mattresses.

**38.7.6 Soiled Linen and Trash Receptacles.** The requirements of 10.3.9 for containers for rubbish, waste, or linen with a capacity of 20 gal (75.7 L) or more shall not apply.

## Chapter 39 Existing Business Occupancies

**39.1 General Requirements.**

**39.1.1 Application.**

**39.1.1.1** The requirements of this chapter shall apply to existing buildings or portions thereof currently occupied as business occupancies.

**39.1.1.2 Administration.** The provisions of Chapter 1, Administration, shall apply.

**39.1.1.3 General.** The provisions of Chapter 4, General, shall apply.

**39.1.1.4** The provisions of this chapter shall apply to life safety requirements for existing business buildings. Specific requirements shall apply to high-rise buildings (*see definition in 3.3.36.7*) and are contained in paragraphs pertaining thereto.

**39.1.2 Classification of Occupancy.** Business occupancies shall include all buildings and structures or parts thereof with occupancy as defined in 6.1.11.

**39.1.3 Multiple Occupancies.**

**39.1.3.1 General.**

**39.1.3.1.1** All multiple occupancies shall be in accordance with 6.1.14 and 39.1.3.

**39.1.3.1.2** Where there are differences in the specific requirements in this chapter and provisions for mixed occupancies or separated occupancies as specified in 6.1.14.3 and 6.1.14.4, the requirements of this chapter shall apply.

**39.1.3.2 Combined Business Occupancies and Parking Structures.**

**39.1.3.2.1** The fire barrier separating parking structures from a building classified as a business occupancy shall be a fire barrier having a minimum 2-hour fire resistance rating.

**39.1.3.2.2** Openings in the fire barrier required by 39.1.3.2.1 shall not be required to be protected with fire protection–rated opening protectives in enclosed parking structures that are protected throughout by an approved, supervised automatic sprinkler system in accordance with 9.7.1.1(1), or in open parking structures, provided that all of the following conditions are met:

(1) The openings do not exceed 25 percent of the area of the fire barrier in which they are located.
(2) The openings are used as a public entrance and for associated sidelight functions.

(3) The building containing the business occupancy is protected throughout by an approved, supervised automatic sprinkler system in accordance with 9.7.1.1(1).
(4)*Means are provided to prevent spilled fuel from accumulating adjacent to the openings and entering the building.
(5) Physical means are provided to prevent vehicles from being parked or driven within 10 ft (3050 mm) of the openings.
(6) The openings are protected as a smoke partition in accordance with Section 8.4, with no minimum fire protection rating required.

**39.1.4 Definitions.**

**39.1.4.1 General.** For definitions, see Chapter 3, Definitions.

**39.1.4.2 Special Definitions.** Special terms applicable to this chapter are defined in Chapter 3.

**39.1.5 Classification of Hazard of Contents.** The contents of business occupancies shall be classified as ordinary hazard in accordance with Section 6.2.

**39.1.6 Minimum Construction Requirements.** (No requirements.)

**39.1.7 Occupant Load.** The occupant load, in number of persons for whom means of egress and other provisions are required, shall be determined on the basis of the occupant load factors of Table 7.3.1.2 that are characteristic of the use of the space, or shall be determined as the maximum probable population of the space under consideration, whichever is greater.

**39.2 Means of Egress Requirements.**

**39.2.1 General.**

**39.2.1.1** All means of egress shall be in accordance with Chapter 7 and this chapter.

**39.2.1.2** If, owing to differences in grade, any street floor exits are at points above or below the street or the finished ground level, such exits shall comply with the provisions for exits from upper floors or floors below the street floor.

**39.2.1.3** Stairs and ramps serving two or more floors below a street floor occupied for business use shall be permitted in accordance with 39.2.1.3.1 and 39.2.1.3.2.

**39.2.1.3.1** Where two or more floors below the street floor are occupied for business use, the same stairs, escalators, or ramps shall be permitted to serve each.

**39.2.1.3.2** An inside open stairway, inside open escalator, or inside open ramp shall be permitted to serve as a required egress facility from not more than one floor level below the street floor.

**39.2.1.4** Floor levels that are below the street floor; are used only for storage, heating, and other service equipment; and are not subject to business occupancy shall have means of egress in accordance with Chapter 42.

**39.2.2 Means of Egress Components.**

**39.2.2.1 Components Permitted.** Means of egress components shall be limited to the types described in 39.2.2.2 through 39.2.2.12.

**39.2.2.2 Doors.**

**39.2.2.2.1** Doors complying with 7.2.1 shall be permitted.

**39.2.2.2.2*** Locks complying with 7.2.1.5.5 shall be permitted only on principal entrance/exit doors.

**39.2.2.2.3** Elevator lobby exit access door-locking arrangements in accordance with 7.2.1.6.3 shall be permitted.

**39.2.2.2.4** The re-entry provisions of 7.2.1.5.8 shall not apply to any of the following:

(1) Existing business occupancies that are not high-rise buildings
(2) Existing high-rise business occupancy buildings that are protected throughout by an approved automatic sprinkler system in accordance with 9.7.1.1(1)
(3) Existing high-rise business occupancy buildings having approved existing means for providing stair re-entry

**39.2.2.2.5** Delayed-egress locks complying with 7.2.1.6.1 shall be permitted.

**39.2.2.2.6** Access-controlled egress doors complying with 7.2.1.6.2 shall be permitted.

**39.2.2.2.7** Horizontal or vertical security grilles or doors complying with 7.2.1.4(3) shall be permitted to be used as part of the required means of egress from a tenant space.

**39.2.2.2.8** Approved existing horizontal-sliding or vertical-rolling fire doors shall be permitted in the means of egress where they comply with all of the following conditions:

(1) They are held open by fusible links.
(2) The fusible links are rated at not less than 165°F (74°C).
(3) The fusible links are located not more than 10 ft (3050 mm) above the floor.
(4) The fusible links are in immediate proximity to the door opening.
(5) The fusible links are not located above a ceiling.
(6) The door is not credited with providing any protection under this *Code*.

**39.2.2.2.9** Revolving doors complying with 7.2.1.10 shall be permitted.

**39.2.2.3 Stairs.**

**39.2.2.3.1** Stairs complying with 7.2.2 shall be permitted.

**39.2.2.3.2** Spiral stairs complying with 7.2.2.2.3 shall be permitted.

**39.2.2.3.3** Winders complying with 7.2.2.2.4 shall be permitted.

**39.2.2.4 Smokeproof Enclosures.** Smokeproof enclosures complying with 7.2.3 shall be permitted.

**39.2.2.5 Horizontal Exits.** Horizontal exits complying with 7.2.4 shall be permitted.

**39.2.2.6 Ramps.** Ramps complying with 7.2.5 shall be permitted.

**39.2.2.7 Exit Passageways.** Exit passageways complying with 7.2.6 shall be permitted.

**39.2.2.8 Escalators and Moving Walks.** Escalators and moving walks complying with 7.2.7 shall be permitted.

**39.2.2.9 Fire Escape Stairs.** Fire escape stairs complying with 7.2.8 shall be permitted.

**39.2.2.10 Fire Escape Ladders.** Fire escape ladders complying with 7.2.9 shall be permitted.

**39.2.2.11 Alternating Tread Devices.** Alternating tread devices complying with 7.2.11 shall be permitted.

**39.2.2.12 Areas of Refuge.**

**39.2.2.12.1** Areas of refuge complying with 7.2.12 shall be permitted.

**39.2.2.12.2** In buildings protected throughout by an approved, supervised automatic sprinkler system in accordance with 9.7.1.1(1), two rooms or spaces separated from each other by smoke-resistant partitions in accordance with the definition of area of refuge in 3.3.22 shall not be required.

**39.2.3 Capacity of Means of Egress.**

**39.2.3.1** The capacity of means of egress shall be in accordance with Section 7.3.

**39.2.3.2** The clear width of any corridor or passageway serving an occupant load of 50 or more shall be not less than 44 in. (1120 mm).

**39.2.3.3** Street floor exits shall be sufficient for the occupant load of the street floor plus the required capacity of open stairs, ramps, escalators, and moving walks discharging through the street floor.

**39.2.4 Number of Means of Egress.**

**39.2.4.1** Means of egress shall comply with all of the following, except as otherwise permitted by 39.2.4.2 through 39.2.4.6:

(1) The number of means of egress shall be in accordance with 7.4.1.1 and 7.4.1.3 through 7.4.1.6.
(2) Not less than two separate exits shall be provided on every story.
(3) Not less than two separate exits shall be accessible from every part of every story.

**39.2.4.2** Exit access, as required by 39.2.4.1(3), shall be permitted to include a single exit access path for the distances permitted as common paths of travel by 39.2.5.3.

**39.2.4.3** A single exit shall be permitted for a room or area with a total occupant load of less than 100 persons, provided that all of the following criteria are met:

(1) The exit shall discharge directly to the outside at the level of exit discharge for the building.
(2) The total distance of travel from any point, including travel within the exit, shall not exceed 100 ft (30 m).
(3) The total distance of travel specified in 39.2.4.3(2) shall be on the same story, or, if traversing of stairs is necessary, such stairs shall not exceed 15 ft (4570 mm) in height, and both of the following also shall apply:
   (a) Interior stairs shall be provided with complete enclosures to separate them from any other part of the building, with no door openings therein.
   (b) A single outside stair in accordance with 7.2.2 shall be permitted to serve all stories permitted within the 15 ft (4570 mm) vertical travel limitation.

**39.2.4.4** Any business occupancy three or fewer stories in height, and not exceeding an occupant load of 30 people per story, shall be permitted a single separate exit to each story, provided that all of the following criteria are met:

(1) The exit shall discharge directly to the outside.
(2) The total travel distance to the outside of the building shall not exceed 100 ft (30 m).
(3) Interior exit stairs shall be enclosed in accordance with 7.1.3.2, and both of the following also shall apply:

(a) The stair shall serve as an exit from no other stories.

(b) A single outside stair in accordance with 7.2.2 shall be permitted to service all stories.

**39.2.4.5** A single means of egress shall be permitted from a mezzanine within a business occupancy, provided that the common path of travel does not exceed 75 ft (23 m), or 100 ft (30 m) if protected throughout by an approved automatic sprinkler system in accordance with 9.7.1.1(1).

**39.2.4.6** A single exit shall be permitted for a single-tenant space or building two or fewer stories in height, provided that both of the following criteria are met:

(1) The building is protected throughout by an approved, supervised automatic sprinkler system in accordance with 9.7.1.1(1).

(2) The total travel to the outside does not exceed 100 ft (30 m).

**39.2.4.7** A single exit shall be permitted for a single-tenant building three or fewer stories in height and not exceeding an occupant load of 15 people per story, provided that all of the following criteria are met:

(1) The building is protected throughout by an approved, supervised automatic sprinkler system in accordance with 9.7.1.1(1) and an automatic smoke detection system in accordance with Section 9.6.

(2) Activation of the building sprinkler and smoke detection system shall provide occupant notification throughout the building.

(3) The total travel to the outside does not exceed 100 ft (30 m).

### 39.2.5 Arrangement of Means of Egress.

**39.2.5.1** Means of egress shall be arranged in accordance with Section 7.5.

**39.2.5.2\*** Dead-end corridors shall not exceed 50 ft (15 m).

**39.2.5.3\*** Limitations on common path of travel shall be in accordance with 39.2.5.3.1, 39.2.5.3.2, and 39.2.5.3.3.

**39.2.5.3.1** Common path of travel shall not exceed 100 ft (30 m) on a story protected throughout by an approved automatic sprinkler system in accordance with 9.7.1.1(1).

**39.2.5.3.2** Common path of travel shall not be limited in a single-tenant space with an occupant load not exceeding 30 people.

**39.2.5.3.3** In buildings other than those complying with 39.2.5.3.1 or 39.2.5.3.2, common path of travel shall not exceed 75 ft (23 m).

**39.2.6 Travel Distance to Exits.** Travel distance shall comply with 39.2.6.1 through 39.2.6.3.

**39.2.6.1** Travel distance shall be measured in accordance with Section 7.6.

**39.2.6.2** Travel distance to an exit shall not exceed 200 ft (61 m) from any point in a building, unless otherwise permitted by 39.2.6.3.

**39.2.6.3** Travel distance shall not exceed 300 ft (91 m) in business occupancies protected throughout by an approved, supervised automatic sprinkler system in accordance with Section 9.7.

**39.2.7 Discharge from Exits.** Exit discharge shall comply with Section 7.7.

**39.2.8 Illumination of Means of Egress.** Means of egress shall be illuminated in accordance with Section 7.8.

### 39.2.9 Emergency Lighting.

**39.2.9.1** Emergency lighting shall be provided in accordance with Section 7.9 in any building where any one of the following conditions exists:

(1) The building is three or more stories in height.

(2) The occupancy is subject to 100 or more occupants above or below the level of exit discharge.

(3) The occupancy is subject to 1000 or more total occupants.

**39.2.9.2** Emergency lighting in accordance with Section 7.9 shall be provided for all underground and limited access structures, as defined in 3.3.271.11 and 3.3.271.3, respectively.

**39.2.10 Marking of Means of Egress.** Means of egress shall have signs in accordance with Section 7.10.

### 39.2.11 Special Means of Egress Features.

**39.2.11.1 Reserved.**

**39.2.11.2 Lockups.** Lockups in business occupancies, other than approved existing lockups, shall comply with the requirements of 23.4.5.

### 39.3 Protection.

### 39.3.1 Protection of Vertical Openings.

**39.3.1.1** Vertical openings shall be enclosed or protected in accordance with Section 8.6, unless otherwise permitted by any of the following:

(1) Unenclosed vertical openings in accordance with 8.6.9.1 or 39.2.4.7 shall be permitted.

(2) Exit access stairs in accordance with 39.2.4.6 or 39.2.4.7 shall be permitted to be unenclosed.

(3) Unprotected vertical openings shall be permitted in buildings complying with all of the following:

(a) Where protected throughout by an approved automatic sprinkler system in accordance with 9.7.1.1(1)

(b) Where no unprotected vertical opening serves as any part of any required means of egress

(c) Where required exits consist of exit doors that discharge directly to the finished ground level in accordance with 7.2.1, outside stairs in accordance with 7.2.2, smokeproof enclosures in accordance with 7.2.3, or horizontal exits in accordance with 7.2.4

**39.3.1.2** Floors that are below the street floor and are used for storage or other than a business occupancy shall have no unprotected openings to business occupancy floors.

### 39.3.2 Protection from Hazards.

**39.3.2.1\* General.** Hazardous areas including, but not limited to, areas used for general storage, boiler or furnace rooms, and maintenance shops that include woodworking and painting areas shall be protected in accordance with Section 8.7.

**39.3.2.2\* High Hazard Contents Areas.** High hazard contents areas, as classified in Section 6.2, shall meet all of the following criteria:

(1) The area shall be separated from other parts of the building by fire barriers having a minimum 1-hour fire resistance rating, with all openings therein protected by self-closing fire door assemblies having a minimum ¾-hour fire protection rating.

(2) The area shall be protected by an automatic extinguishing system in accordance with 9.7.1.1(1) or 9.7.1.2.

**39.3.2.3 Cooking Equipment.** Cooking equipment shall be protected in accordance with 9.2.3, unless the cooking equipment is one of the following types:

(1) Outdoor equipment
(2) Portable equipment not flue-connected
(3) Equipment used only for food warming

**39.3.3 Interior Finish.**

**39.3.3.1 General.** Interior finish shall be in accordance with Section 10.2.

**39.3.3.2 Interior Wall and Ceiling Finish.**

**39.3.3.2.1** Interior wall and ceiling finish materials complying with Section 10.2 shall be Class A or Class B in exits and in exit access corridors.

**39.3.3.2.2** Interior wall and ceiling finishes shall be Class A, Class B, or Class C in areas other than those specified in 39.3.3.2.1.

**39.3.3.3 Interior Floor Finish.** (No requirements.)

**39.3.4 Detection, Alarm, and Communications Systems.**

**39.3.4.1 General.** A fire alarm system in accordance with Section 9.6 shall be provided in all business occupancies where any one of the following conditions exists:

(1) The building is three or more stories in height.
(2) The occupancy is subject to 100 or more occupants above or below the level of exit discharge.
(3) The occupancy is subject to 1000 or more total occupants.

**39.3.4.2 Initiation.** Initiation of the required fire alarm system shall be by one of the following means:

(1) Manual means in accordance with 9.6.2.1(1)
(2) Means of an approved automatic fire detection system that complies with 9.6.2.1(2) and provides protection throughout the building
(3) Means of an approved automatic sprinkler system that complies with 9.6.2.1(3) and provides protection throughout the building

**39.3.4.3 Occupant Notification.** During all times that the building is occupied *(see 7.2.1.1.3)*, the required fire alarm system, once initiated, shall perform one of the following functions:

(1) It shall activate a general alarm in accordance with 9.6.3 throughout the building, and both of the following also shall apply:
   (a) Positive alarm sequence in accordance with 9.6.3.4 shall be permitted.
   (b) A presignal system in accordance with 9.6.3.3 shall be permitted.
(2) Occupant notification shall be permitted to be made via a voice communication or public address system in accordance with 9.6.3.9.2.

**39.3.4.4 Emergency Forces Notification.** Emergency forces notification shall be accomplished in accordance with 9.6.4 when the existing fire alarm system is replaced.

**39.3.5 Extinguishment Requirements.** Portable fire extinguishers shall be provided in every business occupancy in accordance with 9.7.4.1.

**39.3.6 Corridors.** (No requirements.)

**39.3.7 Subdivision of Building Spaces.** (No special requirements.)

**39.4 Special Provisions.**

**39.4.1 Limited Access or Underground Buildings.** See Section 11.7.

**39.4.2 High-Rise Buildings.**

**39.4.2.1** All high-rise business occupancy buildings shall be provided with a reasonable degree of safety from fire, and such degree of safety shall be accomplished by one of the following means:

(1) Installation of a complete, approved, supervised automatic sprinkler system in accordance with 9.7.1.1(1)
(2) Installation of an engineered life safety system complying with all of the following:
   (a) The engineered life safety system shall be developed by a registered professional engineer experienced in fire and life safety systems design.
   (b) The life safety system shall be approved by the authority having jurisdiction and shall be permitted to include any or all of the following systems:
      i. Partial automatic sprinkler protection
      ii. Smoke detection alarms
      iii. Smoke control
      iv. Compartmentation
      v. Other approved systems

**39.4.2.2\*** A limited, but reasonable, time shall be permitted for compliance with any part of 39.4.2.1, commensurate with the magnitude of expenditure and the disruption of services.

**39.4.2.3** In addition to the requirements of 39.4.2.1 and 39.4.2.2, all buildings, regardless of height, shall comply with all other applicable provisions of this chapter.

**39.4.3 Air Traffic Control Towers.**

**39.4.3.1** Air traffic control towers shall comply with the requirements of this chapter and Section 11.3.

**39.4.3.2** The requirements of Section 11.8 shall not apply to air traffic control towers.

**39.5 Building Services.**

**39.5.1 Utilities.** Utilities shall comply with the provisions of Section 9.1.

**39.5.2 Heating, Ventilating, and Air-Conditioning.** Heating, ventilating, and air-conditioning equipment shall comply with the provisions of Section 9.2.

**39.5.3 Elevators, Escalators, and Conveyors.** Elevators, escalators, and conveyors shall comply with the provisions of Section 9.4.

**39.5.4 Rubbish Chutes, Incinerators, and Laundry Chutes.** Rubbish chutes, incinerators, and laundry chutes shall comply with the provisions of Section 9.5.

**39.6 Reserved.**

**39.7 Operating Features.**

**39.7.1 Emergency Plans.** Emergency plans complying with Section 4.8 shall be provided in high-rise buildings.

**39.7.2 Drills.** In all business occupancy buildings occupied by more than 500 persons, or by more than 100 persons above or below the street level, employees and supervisory personnel shall be periodically instructed in accordance with Section 4.7 and shall hold drills periodically where practicable.

**39.7.3 Extinguisher Training.** Designated employees of business occupancies shall be periodically instructed in the use of portable fire extinguishers.

**39.7.4 Food Service Operations.** Food service operations shall comply with 13.7.2.

**39.7.5 Upholstered Furniture and Mattresses.** The provisions of 10.3.2 shall not apply to upholstered furniture and mattresses.

**39.7.6 Soiled Linen and Trash Receptacles.** The requirements of 10.3.9 for containers for rubbish, waste, or linen with a capacity of 20 gal (75.7 L) or more shall not apply.

## Chapter 40 Industrial Occupancies

**40.1 General Requirements.**

**40.1.1 Application.**

**40.1.1.1** The requirements of this chapter shall apply to both new and existing industrial occupancies.

**40.1.1.2 Administration.** The provisions of Chapter 1, Administration, shall apply.

**40.1.1.3 General.** The provisions of Chapter 4, General, shall apply.

**40.1.1.4** Industrial occupancies shall include factories making products of all kinds and properties used for operations such as processing, assembling, mixing, packaging, finishing or decorating, repairing, and similar operations.

**40.1.1.5** Incidental high hazard operations protected in accordance with Section 8.7 and 40.3.2 in occupancies containing low or ordinary hazard contents shall not be the basis for high hazard industrial occupancy classification.

**40.1.2 Classification of Occupancy.** Classification of occupancy shall be in accordance with 6.1.12.

**40.1.2.1 Subclassification of Occupancy.** Each industrial occupancy shall be subclassified according to its use as described in 40.1.2.1.1, 40.1.2.1.2, and 40.1.2.1.3.

**40.1.2.1.1 General Industrial Occupancy.** General industrial occupancies shall include all of the following:

(1) Industrial occupancies that conduct ordinary and low hazard industrial operations in buildings of conventional design that are usable for various types of industrial processes
(2) Industrial occupancies that include multistory buildings where floors are occupied by different tenants, or buildings that are usable for such occupancy and, therefore, are subject to possible use for types of industrial processes with a high density of employee population

**40.1.2.1.2 Special-Purpose Industrial Occupancy.** Special-purpose industrial occupancies shall include all of the following:

(1) Industrial occupancies that conduct ordinary and low hazard industrial operations in buildings designed for, and that are usable only for, particular types of operations
(2) Industrial occupancies that are characterized by a relatively low density of employee population, with much of the area occupied by machinery or equipment

**40.1.2.1.3\* High Hazard Industrial Occupancy.** High hazard industrial occupancies shall include all of the following:

(1) Industrial occupancies that conduct industrial operations that use high hazard materials or processes or house high hazard contents
(2) Industrial occupancies in which incidental high hazard operations in low or ordinary hazard occupancies that are protected in accordance with Section 8.7 and 40.3.2 are not required to be the basis for overall occupancy classification

**40.1.2.2 Change of Industrial Occupancy Subclassification.** A change from one subclassification of industrial occupancy to another shall comply with Chapter 43.

**40.1.3 Multiple Occupancies.** All multiple occupancies shall be in accordance with 6.1.14.

**40.1.4 Definitions.**

**40.1.4.1 General.** For definitions, see Chapter 3, Definitions.

**40.1.4.2 Special Definitions.** Special terms applicable to this chapter are defined in Chapter 3.

**40.1.5 Classification of Hazard of Contents.** Classification of hazard of contents shall be in accordance with Section 6.2.

**40.1.6 Minimum Construction Requirements.** (No requirements.)

**40.1.7\* Occupant Load.** The occupant load, in number of persons for whom means of egress and other provisions are required, shall be determined on the basis of the occupant load factors of Table 7.3.1.2 that are characteristic of the use of the space, or shall be determined as the maximum probable population of the space under consideration, whichever is greater.

**40.2 Means of Egress Requirements.**

**40.2.1 General.**

**40.2.1.1** Each required means of egress shall be in accordance with the applicable portions of Chapter 7.

**40.2.1.2\*** Normally unoccupied utility chases that are secured from unauthorized access and are used exclusively for routing of electrical, mechanical, or plumbing equipment shall not be required to comply with the provisions of Chapter 7

**40.2.2 Means of Egress Components.**

**40.2.2.1 Components Permitted.** Components of means of egress shall be limited to the types described in 40.2.2.2 through 40.2.2.13.

**40.2.2.2 Doors.**

**40.2.2.2.1** Doors complying with 7.2.1 shall be permitted.

**40.2.2.2.2** Delayed-egress locks complying with 7.2.1.6.1 shall be permitted.

**40.2.2.2.3** Access-controlled egress doors complying with 7.2.1.6.2 shall be permitted.

**40.2.2.2.4** Approved existing horizontal-sliding fire doors shall be permitted in the means of egress where they comply with all of the following conditions:

(1) They are held open by fusible links.
(2) The fusible links are rated at not less than 165°F (74°C).
(3) The fusible links are located not more than 10 ft (3050 mm) above the floor.

(4) The fusible links are in immediate proximity to the door opening.
(5) The fusible links are not located above a ceiling.
(6) The door is not credited with providing any protection under this *Code*.

#### 40.2.2.3 Stairs.

**40.2.2.3.1** Stairs shall comply with 7.2.2 and shall be permitted to be modified by any of the following:

(1) Noncombustible grated stair treads and noncombustible grated landing floors shall be permitted.
(2) Industrial equipment access stairs in accordance with 40.2.5.2 shall be permitted.

**40.2.2.3.2** Spiral stairs complying with 7.2.2.2.3 shall be permitted.

**40.2.2.3.3** Existing winders complying with 7.2.2.2.4 shall be permitted.

#### 40.2.2.4 Smokeproof Enclosures. Smokeproof enclosures complying with 7.2.3 shall be permitted.

#### 40.2.2.5 Horizontal Exits.

**40.2.2.5.1** Horizontal exits complying with 7.2.4 shall be permitted.

**40.2.2.5.2\*** In horizontal exits where the opening is protected by a fire door assembly on each side of the wall in which it is located, one fire door shall be of the swinging type, as provided in 7.2.4.3.7, and the other shall be permitted to be an automatic-sliding fire door that shall be kept open whenever the building is occupied.

#### 40.2.2.6 Ramps. Ramps shall comply with 7.2.5, except that industrial equipment access ramps shall be permitted to be in accordance with 40.2.5.2.

#### 40.2.2.7 Exit Passageways. Exit passageways complying with 7.2.6 shall be permitted.

#### 40.2.2.8 Escalators and Moving Walks. Existing previously approved escalators and moving walks complying with 7.2.7 and located within the required means of egress shall be permitted.

#### 40.2.2.9 Fire Escape Stairs. Existing fire escape stairs complying with 7.2.8 shall be permitted.

#### 40.2.2.10 Fire Escape Ladders.

**40.2.2.10.1** Fire escape ladders complying with 7.2.9 shall be permitted.

**40.2.2.10.2** Fixed industrial stairs in accordance with the minimum requirements for fixed stairs in ANSI A1264.1, *Safety Requirements for Workplace Floor and Wall Openings, Stairs and Railing Systems*, shall be permitted where fire escape ladders are permitted in accordance with 7.2.9.1.

#### 40.2.2.11 Slide Escapes.

**40.2.2.11.1** Approved slide escapes complying with 7.2.10 shall be permitted as components in 100 percent of the required means of egress for both new and existing high hazard industrial occupancies.

**40.2.2.11.2** Slide escapes permitted by 40.2.2.11.1 shall be counted as means of egress only where regularly used in emergency egress drills to ensure that occupants are familiar with their use through practice.

#### 40.2.2.12 Alternating Tread Devices. Alternating tread devices complying with 7.2.11 shall be permitted.

#### 40.2.2.13 Areas of Refuge. Areas of refuge complying with 7.2.12 shall be permitted.

#### 40.2.3 Capacity of Means of Egress. Capacity of means of egress shall comply with either 40.2.3.1 or 40.2.3.2.

**40.2.3.1** The capacity of means of egress shall be in accordance with Section 7.3.

**40.2.3.2** In industrial occupancies, means of egress shall be sized to accommodate the occupant load as determined in accordance with 40.1.7; spaces not subject to human occupancy because of the presence of machinery or equipment shall not be included in the computation.

#### 40.2.4 Number of Means of Egress. See also Section 7.4.

**40.2.4.1** The number of means of egress shall comply with either 40.2.4.1.1 or 40.2.4.1.2.

**40.2.4.1.1** Not less than two means of egress shall be provided from every story or section, and not less than one exit shall be reached without traversing another story.

**40.2.4.1.2** A single means of egress shall be permitted from any story or section in low and ordinary hazard industrial occupancies, provided that the exit can be reached within the distance permitted as a common path of travel.

**40.2.4.2** In new buildings, floors or portions thereof with an occupant load of more than 500 shall have the minimum number of separate and remote means of egress specified by 7.4.1.2.

**40.2.4.3** Areas with high hazard contents shall comply with Section 7.11.

#### 40.2.5 Arrangement of Means of Egress. Means of egress, arranged in accordance with Section 7.5, shall not exceed that provided by Table 40.2.5.

#### 40.2.5.1 Ancillary Facilities.

**40.2.5.1.1\*** New ancillary facilities shall be arranged to allow travel in independent directions after leaving the ancillary facility so that both means of egress paths do not become compromised by the same fire or similar emergency.

**40.2.5.1.2\*** New ancillary facilities in special-purpose industrial occupancies where delayed evacuation is anticipated shall have not less than a 2-hour fire resistance–rated separation from the predominant industrial occupancy, and shall have one means of egress that is separated from the predominant industrial occupancy by 2-hour fire resistance–rated construction.

#### 40.2.5.2 Industrial Equipment Access.

**40.2.5.2.1** Industrial equipment access doors, walkways, platforms, ramps, and stairs that serve as a component of the means of egress from the involved equipment shall be permitted in accordance with the applicable provisions of Chapter 7, as modified by Table 40.2.5.2.1.

**40.2.5.2.2** Any means of egress component permitted by 40.2.5.2.1 shall serve not more than 20 people.

#### 40.2.6 Travel Distance to Exits. Travel distance, measured in accordance with Section 7.6, shall not exceed that provided by Table 40.2.6.

#### 40.2.7 Discharge from Exits. Discharge from exits shall be in accordance with Section 7.7.

**Table 40.2.5 Arrangement of Means of Egress**

| Level of Protection | General Industrial Occupancy | | Special-Purpose Industrial Occupancy | | High Hazard Industrial Occupancy |
| --- | --- | --- | --- | --- | --- |
| | ft | m | ft | m | |
| **Dead-End Corridor** | | | | | |
| Protected throughout by an approved, supervised automatic sprinkler system in accordance with 9.7.1.1(1) | 50 | 15 | 50 | 15 | Prohibited, except as permitted by 7.11.4 |
| Not protected throughout by an approved, supervised automatic sprinkler system in accordance with 9.7.1.1(1) | 50 | 15 | 50 | 15 | Prohibited, except as permitted by 7.11.4 |
| **Common Path of Travel** | | | | | |
| Protected throughout by an approved, supervised automatic sprinkler system in accordance with 9.7.1.1(1) | 100 | 30 | 100 | 30 | Prohibited, except as permitted by 7.11.4 |
| Not protected throughout by an approved, supervised automatic sprinkler system in accordance with 9.7.1.1(1) | 50 | 15 | 50 | 15 | Prohibited, except as permitted by 7.11.4 |

**Table 40.2.5.2.1 Industrial Equipment Access Dimensional Criteria**

| Feature | Dimensional Criteria |
| --- | --- |
| Minimum horizontal dimension of any walkway, landing, or platform | 22 in. (560 mm) clear |
| Minimum stair or ramp width | 22 in. (560 mm) clear between rails |
| Minimum tread width | 22 in. (560 mm) clear |
| Minimum tread depth | 10 in. (255 mm) |
| Maximum riser height | 9 in. (230 mm) |
| Handrails are permitted to terminate, at the required height, at a point directly above the top and bottom risers. | |
| Maximum height between landings | 12 ft (3660 mm) |
| Minimum headroom | 6 ft 8 in. (2030 mm) |
| Minimum width of door openings | 22 in. (560 mm) clear |

**Table 40.2.6 Maximum Travel Distance to Exits**

| Level of Protection | General Industrial Occupancy | | Special-Purpose Industrial Occupancy | | High Hazard Industrial Occupancy | |
| --- | --- | --- | --- | --- | --- | --- |
| | ft | m | ft | m | ft | m |
| Protected throughout by an approved, supervised automatic sprinkler system in accordance with 9.7.1.1(1) | 250[†] | 76[†] | 400 | 122 | 75 | 23 |
| Not protected throughout by an approved, supervised automatic sprinkler system in accordance with 9.7.1.1(1) | 200 | 61 | 300 | 91 | NP | NP |

NP: Not permitted.

[†]In one-story buildings, a travel distance of 400 ft (122 m) is permitted, provided that a performance-based analysis demonstrates that safe egress can be accomplished.

**40.2.8 Illumination of Means of Egress.** Means of egress shall be illuminated in accordance with Section 7.8 or with natural lighting that provides the required level of illumination in structures occupied only during daylight hours.

**40.2.9\* Emergency Lighting.**

**40.2.9.1** Emergency lighting shall be provided in accordance with Section 7.9, except as otherwise exempted by 40.2.9.2.

**40.2.9.2** Emergency lighting shall not be required for any of the following:

(1) Special-purpose industrial occupancies without routine human habitation
(2) Structures occupied only during daylight hours, with skylights or windows arranged to provide the required level of illumination on all portions of the means of egress during such hours

**40.2.10 Marking of Means of Egress.** Means of egress shall have signs in accordance with Section 7.10.

**40.2.11 Special Means of Egress Features.**

**40.2.11.1 Reserved.**

**40.2.11.2 Lockups.**

**40.2.11.2.1** Lockups in new industrial occupancies shall comply with the requirements of 22.4.5.

**40.2.11.2.2** Lockups in existing industrial occupancies, other than approved existing lockups, shall comply with the requirements of 23.4.5.

**40.3 Protection.**

**40.3.1 Protection of Vertical Openings.** Any vertical opening shall be protected in accordance with Section 8.6, unless otherwise permitted by one of the following:

(1) In special-purpose industrial and high hazard industrial occupancies where unprotected vertical openings exist

and are necessary to manufacturing operations, such openings shall be permitted beyond the specified limits, provided that every floor level has direct access to one or more enclosed stairs or other exits protected against obstruction by any fire or smoke in the open areas connected by the unprotected vertical openings.

(2) Approved existing open stairs, existing open ramps, and existing escalators shall be permitted where connecting only two floor levels.

(3) Approved, existing, unprotected vertical openings in buildings with low or ordinary hazard contents that are protected throughout by an approved automatic sprinkler system in accordance with 9.7.1.1(1) shall be permitted, provided that the following conditions exist:

    (a) The vertical opening does not serve as a required exit.

    (b) All required exits consist of outside stairs in accordance with 7.2.2, smokeproof enclosures in accordance with 7.2.3, or horizontal exits in accordance with 7.2.4.

(4) Vertical openings in accordance with 8.6.9.1 shall be permitted.

**40.3.2\* Protection from Hazards.**

**40.3.2.1** All high hazard industrial occupancies, operations, or processes shall have approved, supervised automatic extinguishing systems in accordance with Section 9.7 or other protection appropriate to the particular hazard, such as explosion venting or suppression.

**40.3.2.2** Protection in accordance with 40.3.2.1 shall be provided for any area subject to an explosion hazard in order to minimize danger to occupants in case of fire or other emergency before they have time to use exits to escape.

**40.3.2.3** Activation of the fire-extinguishing or suppression system required by 40.3.2.1 shall initiate the required building fire alarm system in accordance with 40.3.4.3.4.

**40.3.2.4** Hazardous areas in industrial occupancies protected by approved automatic extinguishing systems in accordance with Section 9.7 shall be exempt from the smoke-resisting enclosure requirement of 8.7.1.2.

**40.3.3 Interior Finish.**

**40.3.3.1 General.** Interior finish shall be in accordance with Section 10.2.

**40.3.3.2 Interior Wall and Ceiling Finish.** Interior wall and ceiling finish materials complying with Section 10.2 shall be Class A, Class B, or Class C in operating areas and shall be as required by 7.1.4 in exit enclosures.

**40.3.3.3 Interior Floor Finish.**

**40.3.3.3.1** Interior floor finish in exit enclosures and in exit access corridors shall be not less than Class II.

**40.3.3.3.2** Interior floor finish in areas other than those specified in 40.3.3.3.1 shall not be required to comply with Section 10.2.

**40.3.4 Detection, Alarm, and Communications Systems.**

**40.3.4.1 General.** A fire alarm system shall be required in accordance with Section 9.6 for industrial occupancies, unless the total occupant load of the building is under 100 persons and unless, of these, fewer than 25 persons are above or below the level of exit discharge.

**40.3.4.2 Initiation.** Initiation of the required fire alarm system shall be by any of the following means:

(1) Manual means in accordance with 9.6.2.1(1)

(2) Approved automatic fire detection system in accordance with 9.6.2.1(2) throughout the building, plus a minimum of one manual fire alarm box in accordance with 9.6.2.6

(3) Approved, supervised automatic sprinkler system in accordance with 9.6.2.1(3) throughout the building, plus a minimum of one manual fire alarm box in accordance with 9.6.2.6

**40.3.4.3 Notification.**

**40.3.4.3.1** The required fire alarm system shall meet one of the following criteria:

(1) It shall provide occupant notification in accordance with 9.6.3.

(2) It shall sound an audible and visible signal in a constantly attended location for the purposes of initiating emergency action.

**40.3.4.3.2** Positive alarm sequence in accordance with 9.6.3.4 shall be permitted.

**40.3.4.3.3** Existing presignal systems in accordance with 9.6.3.3 shall be permitted.

**40.3.4.3.4** In high hazard industrial occupancies, as described in 40.1.2.1.3, the required fire alarm system shall automatically initiate an occupant evacuation alarm signal in accordance with 9.6.3.

**40.3.5 Extinguishment Requirements.** (No requirements.)

**40.3.6 Corridors.** The provisions of 7.1.3.1 shall not apply.

**40.4 Special Provisions — High-Rise Buildings.** New high-rise industrial occupancies shall comply with Section 11.8.

**40.4.1** The provisions of 11.8.5.2.4(2) for jockey pumps and 11.8.5.2.4(3) for air compressors serving dry-pipe and pre-action systems shall not apply to special-purpose industrial occupancies.

**40.5 Building Services.**

**40.5.1 Utilities.** Utilities shall comply with the provisions of Section 9.1.

**40.5.2 Heating, Ventilating, and Air-Conditioning.** Heating, ventilating, and air-conditioning equipment shall comply with the provisions of Section 9.2.

**40.5.3 Elevators, Escalators, and Conveyors.** Elevators, escalators, and conveyors shall comply with the provisions of Section 9.4.

**40.5.4 Rubbish Chutes, Incinerators, and Laundry Chutes.** Rubbish chutes, incinerators, and laundry chutes shall comply with the provisions of Section 9.5.

**40.6\* Special Provisions for Aircraft Servicing Hangars.**

**40.6.1** The requirements of Sections 40.1 through 40.5 shall be met, except as modified by 40.6.1.1 through 40.6.1.4.

**40.6.1.1** There shall be not less than two means of egress from each aircraft servicing area.

**40.6.1.2** Exits from aircraft servicing areas shall be provided at intervals not exceeding 150 ft (46 m) on all exterior walls.

**40.6.1.3** Where horizontal exits are provided, doors shall be provided in the horizontal exit fire barrier at intervals not exceeding 100 ft (30 m).

**40.6.1.4** Where dwarf, or "smash," doors are provided in doors that accommodate aircraft, such doors shall be permitted for compliance with 40.6.1.1 through 40.6.1.3.

**40.6.2** Means of egress from mezzanine floors in aircraft servicing areas shall be arranged so that the travel distance to the nearest exit from any point on the mezzanine does not exceed 75 ft (23 m), and such means of egress shall lead directly to a properly enclosed stair discharging directly to the exterior, to a suitable cutoff area, or to outside stairs.

**40.6.3** Dead ends shall not exceed 50 ft (15 m) for other than high hazard contents areas and shall not be permitted for high hazard contents areas.

**40.7 Operating Features.**

**40.7.1 Upholstered Furniture and Mattresses.** The provisions of 10.3.2 shall not apply to upholstered furniture and mattresses.

**40.7.2 Soiled Linen and Trash Receptacles.** The requirements of 10.3.9 for containers for rubbish, waste, or linen with a capacity of 20 gal (75.7 L) or more shall not apply.

## Chapter 41   Reserved

## Chapter 42   Storage Occupancies

**42.1 General Requirements.**

**42.1.1 Application.**

**42.1.1.1** The requirements of this chapter shall apply to both new and existing storage occupancies.

**42.1.1.2 Administration.** The provisions of Chapter 1, Administration, shall apply.

**42.1.1.3 General.** The provisions of Chapter 4, General, shall apply.

**42.1.1.4** Storage occupancies shall include all buildings or structures used primarily for the storage or sheltering of goods, merchandise, products, or vehicles.

**42.1.2 Classification of Occupancy.**

**42.1.2.1** Storage occupancies shall include all buildings and structures or parts thereof with occupancy as defined in 6.1.13.

**42.1.2.2** Incidental storage in another occupancy shall not be the basis for overall occupancy classification.

**42.1.2.3** Storage occupancies or areas of storage occupancies that are used for the purpose of packaging, labeling, sorting, special handling, or other operations requiring an occupant load greater than that normally contemplated for storage shall be classified as industrial occupancies. *(See Chapter 40.)*

**42.1.3 Multiple Occupancies.** All multiple occupancies shall be in accordance with 6.1.14.

**42.1.4 Definitions.**

**42.1.4.1 General.** For definitions, see Chapter 3, Definitions.

**42.1.4.2 Special Definitions.** Special terms applicable to this chapter are defined in Chapter 3.

**42.1.5 Classification of Hazard of Contents.** Contents of storage occupancies shall be classified as low hazard, ordinary hazard, or high hazard in accordance with Section 6.2, depending on the character of the materials stored, their packaging, and other factors.

**42.1.6 Minimum Construction Requirements.** (No requirements.)

**42.1.7* Occupant Load.** The occupant load, in number of persons for whom means of egress and other provisions are required, shall be determined on the basis of the maximum probable population of the space under consideration.

**42.2 Means of Egress Requirements.**

**42.2.1 General.**

**42.2.1.1** Each required means of egress shall be in accordance with the applicable portions of Chapter 7.

**42.2.1.2*** Normally unoccupied utility chases that are secured from unauthorized access and are used exclusively for routing of electrical, mechanical, or plumbing equipment shall not be required to comply with the provisions of Chapter 7.

**42.2.2 Means of Egress Components.**

**42.2.2.1 Components Permitted.** Components of means of egress shall be limited to the types described in 42.2.2.2 through 42.2.2.12.

**42.2.2.2 Doors.**

**42.2.2.2.1** Doors complying with 7.2.1 shall be permitted.

**42.2.2.2.2** Delayed-egress locks complying with 7.2.1.6.1 shall be permitted.

**42.2.2.2.3** Access-controlled egress doors complying with 7.2.1.6.2 shall be permitted.

**42.2.2.2.4** Approved existing horizontal-sliding fire doors shall be permitted in the means of egress where they comply with all of the following conditions:

(1) They are held open by fusible links.
(2) The fusible links are rated at not less than 165°F (74°C).
(3) The fusible links are located not more than 10 ft (3050 mm) above the floor.
(4) The fusible links are in immediate proximity to the door opening.
(5) The fusible links are not located above a ceiling.
(6) The door is not credited with providing any protection under this *Code.*

**42.2.2.3 Stairs.**

**42.2.2.3.1** Stairs shall comply with 7.2.2 and shall be permitted to be modified by any of the following:

(1) Noncombustible grated stair treads and noncombustible grated landing floors shall be permitted.
(2) Industrial equipment access stairs in accordance with 40.2.5.2 shall be permitted.

**42.2.2.3.2** Spiral stairs complying with 7.2.2.2.3 shall be permitted.

**42.2.2.3.3** Existing winders complying with 7.2.2.2.4 shall be permitted.

**42.2.2.4 Smokeproof Enclosures.** Smokeproof enclosures complying with 7.2.3 shall be permitted.

**42.2.2.5 Horizontal Exits.**

**42.2.2.5.1** Horizontal exits complying with 7.2.4 shall be permitted.

**42.2.2.5.2\*** In horizontal exits where the opening is protected by a fire door assembly on each side of the wall in which it is located, one fire door shall be of the swinging type, as provided in 7.2.4.3.7, and the other shall be permitted to be an automatic-sliding fire door that shall be kept open whenever the building is occupied.

**42.2.2.6 Ramps.**

**42.2.2.6.1** Ramps complying with 7.2.5 shall be permitted.

**42.2.2.6.2** Industrial equipment access ramps in accordance with 40.2.5.2 shall be permitted.

**42.2.2.7 Exit Passageways.** Exit passageways complying with 7.2.6 shall be permitted.

**42.2.2.8 Fire Escape Stairs.** Existing fire escape stairs complying with 7.2.8 shall be permitted.

**42.2.2.9 Fire Escape Ladders.**

**42.2.2.9.1** Fire escape ladders complying with 7.2.9 shall be permitted.

**42.2.2.9.2** Fixed industrial stairs in accordance with the minimum requirements for fixed stairs in ANSI A1264.1, *Safety Requirements for Workplace Floor and Wall Openings, Stairs and Railing Systems,* shall be permitted where fire escape ladders are permitted in accordance with 7.2.9.1.

**42.2.2.10 Slide Escapes.** Existing slide escapes complying with 7.2.10 shall be permitted.

**42.2.2.11 Alternating Tread Devices.** Alternating tread devices complying with 7.2.11 shall be permitted.

**42.2.2.12 Areas of Refuge.** Areas of refuge complying with 7.2.12 shall be permitted.

**42.2.3 Capacity of Means of Egress.** The capacity of means of egress shall be in accordance with Section 7.3.

**42.2.4 Number of Means of Egress.** See also Section 7.4.

**42.2.4.1** The number of means of egress shall comply with any of the following:

(1) In low hazard storage occupancies, a single means of egress shall be permitted from any story or section.

(2) In ordinary hazard storage occupancies, a single means of egress shall be permitted from any story or section, provided that the exit can be reached within the distance permitted as a common path of travel.

(3) All buildings or structures not complying with 42.2.4.1(1) or (2) and used for storage, and every section thereof considered separately, shall have not less than two separate means of egress as remotely located from each other as practicable.

**42.2.4.2** In new buildings, floors or portions thereof with an occupant load of more than 500 persons shall have the minimum number of separate and remote means of egress specified by 7.4.1.2.

**42.2.4.3** Areas with high hazard contents shall comply with Section 7.11.

**42.2.5 Arrangement of Means of Egress.** Means of egress, arranged in accordance with Section 7.5, shall not exceed that provided by Table 42.2.5.

**Table 42.2.5 Arrangements of Means of Egress**

| Level of Protection | Low Hazard Storage Occupancy | Ordinary Hazard Storage Occupancy | | High Hazard Storage Occupancy |
|---|---|---|---|---|
| | | ft | m | |
| **Dead-End Corridor** | | | | |
| Protected throughout by an approved, supervised automatic sprinkler system in accordance with 9.7.1.1(1) | NL | 100 | 30 | Prohibited, except as permitted by 7.11.4 |
| Not protected throughout by an approved, supervised automatic sprinkler system in accordance with 9.7.1.1(1) | NL | 50 | 15 | Prohibited, except as permitted by 7.11.4 |
| **Common Path of Travel** | | | | |
| Protected throughout by an approved, supervised automatic sprinkler system in accordance with 9.7.1.1(1) | NL | 100 | 30 | Prohibited, except as permitted by 7.11.4 |
| Not protected throughout by an approved, supervised automatic sprinkler system in accordance with 9.7.1.1(1) | NL | 50 | 15 | Prohibited, except as permitted by 7.11.4 |

NL: Not limited.

**42.2.6\* Travel Distance to Exits.** Travel distance, measured in accordance with Section 7.6, shall not exceed that provided by Table 42.2.6.

**42.2.7 Discharge from Exits.** Discharge from exits shall be in accordance with Section 7.7.

**42.2.8 Illumination of Means of Egress.**

**42.2.8.1** Means of egress shall be illuminated in accordance with Section 7.8.

**42.2.8.2** In structures occupied only during daylight hours, means of egress shall be permitted to be illuminated with windows arranged to provide the required level of illumination on all portions of the means of egress during such hours, when approved by the authority having jurisdiction.

**Table 42.2.6 Maximum Travel Distance to Exits**

| Level of Protection | Low Hazard Storage Occupancy | Ordinary Hazard Storage Occupancy | | High Hazard Storage Occupancy | |
|---|---|---|---|---|---|
| | | ft | m | ft | m |
| Protected throughout by an approved, supervised automatic sprinkler system in accordance with 9.7.1.1(1) | NL | 400 | 122 | 100 | 30 |
| Not protected throughout by an approved, supervised automatic sprinkler system in accordance with 9.7.1.1(1) | NL | 200 | 61 | 75 | 23 |
| Flammable and combustible liquid products stored and protected in accordance with NFPA 30, *Flammable and Combustible Liquids Code* | NA | NA | NA | 150 | 46 |

NL: Not limited. NA: Not applicable.

**42.2.9 Emergency Lighting.** Emergency lighting shall be provided in normally occupied storage occupancies in accordance with Section 7.9, except for spaces occupied only during daylight hours with natural illumination in accordance with 42.2.8.2.

**42.2.10 Marking of Means of Egress.** Means of egress shall have signs in accordance with Section 7.10.

**42.2.11 Special Means of Egress Features.**

**42.2.11.1 Reserved.**

**42.2.11.2 Lockups.**

**42.2.11.2.1** Lockups in new storage occupancies shall comply with the requirements of 22.4.5.

**42.2.11.2.2** Lockups in existing storage occupancies, other than approved existing lockups, shall comply with the requirements of 23.4.5.

**42.3 Protection.**

**42.3.1 Protection of Vertical Openings.** Any vertical opening shall be protected in accordance with Section 8.6, unless otherwise permitted by one of the following:

(1) Existing open stairs, existing open ramps, and existing open escalators shall be permitted where connecting only two floor levels.

(2) Existing unprotected vertical openings in buildings with low or ordinary hazard contents, and protected throughout by an approved automatic sprinkler system in accordance with 9.7.1.1(1), shall be permitted where they do not serve as required exits, and where all required exits consist of outside stairs in accordance with 7.2.2, smokeproof enclosures in accordance with 7.2.3, or horizontal exits in accordance with 7.2.4.

**42.3.2 Protection from Hazards.** (No requirements.) *(See also Section 8.7.)*

**42.3.3 Interior Finish.**

**42.3.3.1 General.** Interior finish shall be in accordance with Section 10.2.

**42.3.3.2 Interior Wall and Ceiling Finish.** Interior wall and ceiling finish materials complying with Section 10.2 shall be Class A, Class B, or Class C in storage areas and shall be as required by 7.1.4 in exit enclosures.

**42.3.3.3 Interior Floor Finish.**

**42.3.3.3.1** Interior floor finish in exit enclosures shall be not less than Class II.

**42.3.3.3.2** Interior floor finish in areas other than those specified in 42.3.3.3.1 shall not be required to comply with Section 10.2.

**42.3.4 Detection, Alarm, and Communications Systems.**

**42.3.4.1 General.** A fire alarm system shall be required in accordance with Section 9.6 for storage occupancies, except as modified by 42.3.4.1.1, 42.3.4.1.2, and 42.3.4.1.3.

**42.3.4.1.1** Storage occupancies limited to low hazard contents shall not be required to have a fire alarm system.

**42.3.4.1.2** Storage occupancies with ordinary or high hazard contents not exceeding an aggregate floor area of 100,000 ft$^2$ (9300 m$^2$) shall not be required to have a fire alarm system.

**42.3.4.1.3** Storage occupancies protected throughout by an approved automatic sprinkler system in accordance with Section 9.7 shall not be required to have a fire alarm system.

**42.3.4.2 Initiation.** Initiation of the required fire alarm system shall be by any of the following means:

(1) Manual means in accordance with 9.6.2.1(1)
(2) Approved automatic fire detection system in accordance with 9.6.2.1(2) throughout the building, plus a minimum of one manual fire alarm box in accordance with 9.6.2.6
(3) Approved, supervised automatic sprinkler system in accordance with 9.6.2.1(3) throughout the building, plus a minimum of one manual fire alarm box in accordance with 9.6.2.6

**42.3.4.3 Notification.**

**42.3.4.3.1** The required fire alarm system shall meet one of the following criteria:

(1) It shall provide occupant notification in accordance with 9.6.3.
(2) It shall sound an audible and visible signal in a constantly attended location for the purposes of initiating emergency action.

**42.3.4.3.2** Positive alarm sequence in accordance with 9.6.3.4 shall be permitted.

**42.3.4.3.3** Existing presignal systems in accordance with 9.6.3.3 shall be permitted.

**42.3.4.3.4** In high hazard storage occupancies, the required fire alarm system shall automatically initiate an occupant evacuation alarm signal in accordance with 9.6.3.

**42.3.5 Extinguishment Requirements.** (No requirements.)

**42.3.6 Corridors.** The provisions of 7.1.3.1 shall not apply.

**42.4 Special Provisions — High-Rise Buildings.** New high-rise storage occupancies shall comply with Section 11.8.

**42.5 Building Services.**

**42.5.1 Utilities.** Utilities shall comply with the provisions of Section 9.1.

**42.5.2 Heating, Ventilating, and Air-Conditioning.** Heating, ventilating, and air-conditioning equipment shall comply with the provisions of Section 9.2.

**42.5.3 Elevators, Escalators, and Conveyors.** Elevators, escalators, and conveyors shall comply with the provisions of Section 9.4.

**42.5.4 Rubbish Chutes, Incinerators, and Laundry Chutes.** Rubbish chutes, incinerators, and laundry chutes shall comply with the provisions of Section 9.5.

**42.6* Special Provisions for Aircraft Storage Hangars.**

**42.6.1** The requirements of Sections 42.1 through 42.5 shall be met, except as modified by 42.6.1.1 through 42.6.3.

**42.6.1.1** There shall be not less than two means of egress from each aircraft storage area.

**42.6.1.2** Exits from aircraft storage areas shall be provided at intervals not exceeding 150 ft (46 m) on all exterior walls.

**42.6.1.3** Where horizontal exits are provided, doors shall be provided in the horizontal exit fire barrier at intervals not exceeding 100 ft (30 m).

**42.6.1.4** Where dwarf, or "smash," doors are provided in doors that accommodate aircraft, such doors shall be permitted for compliance with 42.6.1.1, 42.6.1.2, and 42.6.1.3.

**42.6.2** Means of egress from mezzanine floors in aircraft storage areas shall be arranged so that the travel distance to the nearest exit from any point on the mezzanine does not exceed 75 ft (23 m), and such means of egress shall lead directly to a properly enclosed stair discharging directly to the exterior, to a suitable cutoff area, or to outside stairs.

**42.6.3** Dead ends shall not exceed 50 ft (15 m) for other than high hazard contents areas and shall not be permitted for high hazard contents areas.

**42.7* Special Provisions for Grain Handling, Processing, Milling, or Other Bulk Storage Facilities.**

**42.7.1 General.** The requirements of Sections 42.1 through 42.5 shall be met, except as modified by 42.7.2 through 42.7.4.2.

**42.7.2 Number of Means of Egress.** There shall be not less than two means of egress from all working levels of the head house, as modified by 42.7.2.1, 42.7.2.2, and 42.7.2.3.

**42.7.2.1** One of the two means of egress shall be a stair to the level of exit discharge, and, if this means of egress is interior to the structure, it shall be enclosed by a dust-resistant, 1-hour fire resistance–rated enclosure in accordance with 7.1.3.2. Exterior stair means of egress shall be protected from the structure by a 1-hour fire resistance–rated wall that extends at least 10 ft (3050 mm) beyond the stair.

**42.7.2.2** The second means of egress shall be one of the following:

(1) Exterior stair or basket ladder–type fire escape that is accessible from all working levels of the structure and provides a passage to the finished ground level
(2) Exterior stair or basket ladder–type fire escape that is accessible from all working levels of the structure, provides access to adjoining structures, and provides a continuous path to the means of egress described in 42.7.3

**42.7.2.3** Stair enclosures in existing structures shall be permitted to have non-fire-rated dust-resistant enclosures.

**42.7.3 Fire Escapes.** An exterior stair or basket ladder–type fire escape shall provide passage to the finished ground level from the top of the end of an adjoining structure, such as a silo, conveyor, gallery, or gantry.

**42.7.4 Underground Spaces.**

**42.7.4.1 Number of Means of Egress.**

**42.7.4.1.1** Underground spaces shall have not less than two means of egress, one of which shall be permitted to be a means of escape, except as permitted in 42.7.4.1.2.

**42.7.4.1.2** Where the horizontal travel distance to the means of egress is less than 50 ft (15 m) in normally unoccupied spaces, a single means of egress shall be permitted.

**42.7.4.2 Travel Distance to Exits.** Travel distance, measured in accordance with Section 7.6, shall not exceed that provided by Table 42.7.4.2.

**Table 42.7.4.2 Maximum Travel Distance to Means of Escape or Exits**

| Level of Protection | Travel Distance | |
| --- | --- | --- |
| | ft | m |
| Protected throughout by an approved, supervised automatic sprinkler system in accordance with 9.7.1.1(1) | 400 | 122 |
| Not protected throughout by an approved, supervised automatic sprinkler system in accordance with 9.7.1.1(1) | 200 | 61 |
| Existing structures | Unlimited | |

**42.8 Special Provisions for Parking Structures.**

**42.8.1 General Requirements.**

**42.8.1.1 Application.** The provisions of 42.8.1 through 42.8.5.4 shall apply to parking structures of the closed or open type, above or below grade plane, but shall not apply to assisted mechanical-type or automated-type parking facilities that are not occupied by customers. The requirements of Sections 42.1 through 42.7 shall not apply.

#### 42.8.1.2 Multiple Occupancies.

**42.8.1.2.1** Where both parking and repair operations are conducted in the same building, the entire building shall comply with Chapter 40, except as modified by 42.8.1.2.2.

**42.8.1.2.2** Where the parking and repair sections are separated by not less than 1-hour fire-rated construction, the parking and repair sections shall be permitted to be treated separately.

**42.8.1.2.3** In areas where repair operations are conducted, the means of egress shall comply with Chapter 40.

**42.8.1.3 Open Parking Structures.** Open parking structures shall comply with 42.8.1.3.1 through 42.8.1.3.3.

**42.8.1.3.1** Each parking level shall have wall openings open to the atmosphere for an area of not less than 1.4 ft$^2$ for each linear foot (0.4 m$^2$ for each linear meter) of its exterior perimeter. [**88A:** 5.5.1]

**42.8.1.3.2** The openings addressed in 42.8.1.3.1 shall be distributed over 40 percent of the building perimeter or uniformly over two opposing sides. [**88A:** 5.5.2]

**42.8.1.3.3** Interior wall lines and column lines shall be at least 20 percent open, with openings distributed to provide ventilation. [**88A:** 5.5.3]

**42.8.1.4 Classification of Occupancy.** Incidental vehicle parking in another occupancy shall not be the basis for overall occupancy classification.

**42.8.1.5 Classification of Hazard of Contents.** Parking structures used only for the storage of vehicles shall be classified as ordinary hazard in accordance with Section 6.2.

**42.8.1.6 Minimum Construction Requirements.** (No requirements.)

**42.8.1.7 Occupant Load.** (No requirements.)

#### 42.8.2 Means of Egress Requirements.

**42.8.2.1 General.** Means of egress shall be in accordance with Chapter 7 and 42.8.2.

**42.8.2.2 Means of Egress Components.**

**42.8.2.2.1 Components Permitted.** Components of means of egress shall be limited to the types described in 42.8.2.2.2 through 42.8.2.2.9.

**42.8.2.2.2 Doors.**

**42.8.2.2.2.1** Doors complying with 7.2.1 shall be permitted.

**42.8.2.2.2.2** Special locking arrangements complying with 7.2.1.6 shall be permitted.

**42.8.2.2.2.3** An opening for the passage of automobiles shall be permitted to serve as an exit from a street floor, provided that no door or shutter is installed therein.

**42.8.2.2.3 Stairs.**

**42.8.2.2.3.1** Stairs complying with 7.2.2 shall be permitted, unless otherwise permitted by 42.8.2.2.3.2.

**42.8.2.2.3.2** In open parking structures, stairs complying with 7.2.2.5.1 shall not be required.

**42.8.2.2.3.3** Existing winders complying with 7.2.2.2.4 shall be permitted.

**42.8.2.2.3.4** Paragraph 7.2.2.4.5.3(2) shall not apply to guards for parking garages that are accessible to the general public.

**42.8.2.2.4 Smokeproof Enclosures.** Smokeproof enclosures complying with 7.2.3 shall be permitted.

**42.8.2.2.5 Horizontal Exits.** Horizontal exits complying with 7.2.4 shall be permitted.

**42.8.2.2.6 Ramps.**

**42.8.2.2.6.1** Ramps shall be permitted in accordance with any of the following conditions:

(1) Ramps complying with 7.2.5 shall be permitted and shall not be subject to normal vehicular traffic where used as an exit.
(2) In a ramp-type open parking structure with open vehicle ramps not subject to closure, the ramp shall be permitted to serve in lieu of the second means of egress from floors above the level of exit discharge, provided that the ramp discharges directly outside at the street level.
(3) For parking structures extending only one floor level below the level of exit discharge, a vehicle ramp leading directly to the outside shall be permitted to serve in lieu of the second means of egress, provided that no door or shutter is installed therein.

**42.8.2.2.6.2** Paragraph 7.2.4.5.3(2) shall not apply to guards for parking structures that are accessible to the general public.

**42.8.2.2.7 Exit Passageways.** Exit passageways complying with 7.2.6 shall be permitted.

**42.8.2.2.8 Fire Escape Stairs.** Fire escape stairs complying with 7.2.8 shall be permitted for existing parking structures only.

**42.8.2.2.9 Areas of Refuge.**

**42.8.2.2.9.1** Areas of refuge complying with 7.2.12 shall be permitted, as modified by 42.8.2.2.9.2.

**42.8.2.2.9.2** In open-air parking structures, the area of refuge requirements of 7.2.12.1.2(2) shall not apply.

**42.8.2.3 Capacity of Means of Egress.** See also 42.8.2.4 and 42.8.2.5.

**42.8.2.4 Number of Means of Egress.** See also Section 7.4.

**42.8.2.4.1** Not less than two means of egress shall be provided from every floor or section of every parking structure.

**42.8.2.4.2** In new buildings, floors or portions thereof with an occupant load of more than 500 persons shall have the minimum number of separate and remote means of egress specified by 7.4.1.2.

**42.8.2.5 Arrangement of Means of Egress.** See also Section 7.5.

**42.8.2.5.1** A common path of travel shall be permitted for the first 50 ft (15 m) from any point in the parking structure.

**42.8.2.5.2** Dead ends shall not exceed 50 ft (15 m).

**42.8.2.5.3** Where fuel-dispensing devices are located within a parking structure, 42.8.2.5.3.1 and 42.8.2.5.3.2 shall apply.

**42.8.2.5.3.1** Travel away from the fuel-dispensing device in any direction shall lead to an exit with no dead end in which occupants might be trapped by fire.

**42.8.2.5.3.2** Within closed parking structures containing fuel-dispensing devices, exits shall be arranged and located to meet all of the following additional requirements:

(1) Exits shall lead to the outside of the building on the same level or to stairs, with no upward travel permitted, unless direct outside exits are available from that floor.

(2) Any story below the story at which fuel is being dispensed shall have exits leading directly to the outside via outside stairs or doors at the finished ground level.

**42.8.2.6 Travel Distance to Exits.**

**42.8.2.6.1** Travel distance, measured in accordance with Section 7.6, shall not exceed that provided by Table 42.8.2.6.1, except as otherwise permitted in 42.8.2.6.2.

**Table 42.8.2.6.1 Maximum Travel Distance to Exits**

| Level of Protection | Enclosed Parking Structure | | Open Parking Structure | | Parking Structure Open Not Less than 50% on All Sides | |
|---|---|---|---|---|---|---|
| | ft | m | ft | m | ft | m |
| Protected throughout by an approved, supervised automatic sprinkler system in accordance with 9.7.1.1(1) | 200 | 61 | 400 | 122 | 400 | 122 |
| Not protected throughout by an approved, supervised automatic sprinkler system in accordance with 9.7.1.1(1) | 150 | 46 | 300 | 91 | 400 | 122 |

**42.8.2.6.2** In open parking structures, travel distance shall comply with one of the following:

(1) The travel distance to an exit shall not exceed the travel distance specified in Table 42.8.2.6.1.
(2) The travel distance to a stair that does not meet the provisions for an exit enclosure shall not exceed the travel distance specified in Table 42.8.2.6.1, and travel along the stair shall not be limited.

**42.8.2.7 Discharge from Exits.** Exit discharge shall comply with Section 7.7.

**42.8.2.8 Illumination of Means of Egress.** Means of egress shall be illuminated in accordance with Section 7.8 or with natural lighting that provides the required level of illumination in structures occupied only during daylight hours.

**42.8.2.9 Emergency Lighting.** Parking structures shall be provided with emergency lighting in accordance with Section 7.9, except for structures occupied only during daylight hours and arranged to provide the required level of illumination of all portions of the means of egress by natural means.

**42.8.2.10 Marking of Means of Egress.** Means of egress shall have signs in accordance with Section 7.10.

**42.8.2.11 Special Means of Egress Features. (Reserved)**

**42.8.3 Protection.**

**42.8.3.1 Protection of Vertical Openings.**

**42.8.3.1.1 Vertical Openings in Enclosed Parking Structures.**

**42.8.3.1.1.1** Vertical openings through floors in buildings four or more stories in height shall be enclosed with walls or partitions having a minimum 2-hour fire resistance rating.

**42.8.3.1.1.2** For buildings three or fewer stories in height, the walls or partitions required by 42.8.3.1.1.1 shall have a minimum 1-hour fire resistance rating.

**42.8.3.1.1.3** Ramps in enclosed parking structures shall not be required to be enclosed when one of the following safeguards is provided:

(1) An approved automatic sprinkler system fully protecting the enclosed parking structure
(2) An approved, supervised automatic fire detection system installed throughout the enclosed parking structure, and a mechanical ventilation system capable of providing a minimum of 1 ft$^3$/min/ft$^2$ (300 L/min/m$^2$) of floor area during hours of normal operation
(3)*Where a parking structure consists of sprinklered enclosed parking levels, and sprinklered or non-sprinklered open parking levels

**42.8.3.1.1.4** Sprinkler systems provided in accordance with 42.8.3.1.1.3(1) or (3) shall be supervised in accordance with 9.7.2.

**42.8.3.1.2 Open Parking Structures.** Unprotected vertical openings through floors in open parking structures shall be permitted. [**88A:**5.4.8]

**42.8.3.2 Protection from Hazards.** (No requirements.)

**42.8.3.3 Interior Finish.**

**42.8.3.3.1 General.** Interior finish shall be in accordance with Section 10.2.

**42.8.3.3.2 Interior Wall and Ceiling Finish.** Interior wall and ceiling finish materials complying with Section 10.2 shall be Class A, Class B, or Class C in parking structures and shall be as required by 7.1.4 in exit enclosures.

**42.8.3.3.3 Interior Floor Finish.**

**42.8.3.3.3.1** Interior floor finish in exit enclosures shall be not less than Class II.

**42.8.3.3.3.2** Interior floor finish in areas other than those specified in 42.8.3.3.3.1 shall not be required to comply with Section 10.2.

**42.8.3.4 Detection, Alarm, and Communications Systems.**

**42.8.3.4.1 General.** A fire alarm system shall be required in accordance with Section 9.6 for parking structures, except as modified by 42.3.4.1.1, 42.3.4.1.2, and 42.3.4.1.3.

**42.8.3.4.1.1** Parking structures not exceeding an aggregate floor area of 100,000 ft$^2$ (9300 m$^2$) shall not be required to have a fire alarm system.

**42.8.3.4.1.2** Open parking structures shall not be required to have a fire alarm system.

**42.8.3.4.1.3** Parking structures protected throughout by an approved automatic sprinkler system in accordance with Section 9.7 shall not be required to have a fire alarm system.

**42.8.3.4.2 Initiation.** Initiation of the required fire alarm system shall be by one of the following means:

(1) Manual means in accordance with 9.6.2.1(1)
(2) Approved automatic fire detection system in accordance with 9.6.2.1(2) throughout the building, plus a minimum of one manual fire alarm box in accordance with 9.6.2.6
(3) Approved, supervised automatic sprinkler system in accordance with 9.6.2.1(3) throughout the building, plus a minimum of one manual fire alarm box in accordance with 9.6.2.6

**42.8.3.4.3 Notification.**

**42.8.3.4.3.1** The required fire alarm system shall sound an audible alarm in a continuously attended location for purposes of initiating emergency action.

**42.8.3.4.3.2** Positive alarm sequence in accordance with 9.6.3.4 shall be permitted.

**42.8.3.4.3.3** Existing presignal systems in accordance with 9.6.3.3 shall be permitted.

**42.8.3.5 Extinguishing Requirements.** (No requirements.)

**42.8.3.6 Corridors.** The provisions of 7.1.3.1 shall not apply.

**42.8.4 Special Provisions — High-Rise Buildings.** (No requirements.)

**42.8.5 Building Services.**

**42.8.5.1 Utilities.** Utilities shall comply with the provisions of Section 9.1.

**42.8.5.2 Heating, Ventilating, and Air-Conditioning.** Heating, ventilating, and air-conditioning equipment shall comply with the provisions of Section 9.2.

**42.8.5.3 Elevators, Escalators, and Conveyors.** Elevators, escalators, and conveyors shall comply with the provisions of Section 9.4.

**42.8.5.4 Rubbish Chutes, Incinerators, and Laundry Chutes.** Rubbish chutes, incinerators, and laundry chutes shall comply with the provisions of Section 9.5.

**42.9 Operating Features.**

**42.9.1 Upholstered Furniture and Mattresses.** The provisions of 10.3.2 shall not apply to upholstered furniture and mattresses.

**42.9.2 Soiled Linen and Trash Receptacles.** The requirements of 10.3.9 for containers for rubbish, waste, or linen with a capacity of 20 gal (75.7 L) or more shall not apply.

# Chapter 43   Building Rehabilitation

**43.1 General.**

**43.1.1 Classification of Rehabilitation Work Categories.** Rehabilitation work on existing buildings shall be classified as one of the following work categories:

(1) Repair
(2) Renovation
(3) Modification
(4) Reconstruction
(5) Change of use or occupancy classification
(6) Addition

**43.1.2 Applicable Requirements.**

**43.1.2.1** Any building undergoing repair, renovation, modification, or reconstruction *(see 43.2.2.1.1 through 43.2.2.1.4)* shall comply with both of the following:

(1) Requirements of the applicable existing occupancy chapters *(see Chapters 13, 15, 17, 19, 21, 23, 24, 26, 29, 31, 33, 37, 39, 40, and 42)*
(2) Requirements of the applicable section of this chapter *(see Sections 43.3, 43.4, 43.5, and 43.6)*

**43.1.2.2** Any building undergoing change of use or change of occupancy classification *(see 43.2.2.1.5 and 43.2.2.1.6)* shall comply with the requirements of Section 43.7.

**43.1.2.3** Any building undergoing addition *(see 43.2.2.1.7)* shall comply with the requirements of Section 43.8.

**43.1.2.4** Historic buildings undergoing rehabilitation shall comply with the requirements of Section 43.10.

**43.1.2.5** Nothing in this chapter shall be interpreted as excluding the use of the performance-based option of Chapter 5.

**43.1.3 Multiple Rehabilitation Work Categories.**

**43.1.3.1** Work of more than one rehabilitation work category shall be permitted to be part of a single work project.

**43.1.3.2** Where a project includes one category of rehabilitation work in one building area and another category of rehabilitation work in a separate area of the building, each project area shall comply with the requirements of the respective category of rehabilitation work.

**43.1.3.3** Where a project consisting of modification and reconstruction is performed in the same work area, or in contiguous work areas, the project shall comply with the requirements applicable to reconstruction, unless otherwise specified in 43.1.3.4.

**43.1.3.4** Where the reconstruction work area is less than 10 percent of the modification work area, the two shall be considered as independent work areas, and the respective requirements shall apply.

**43.1.4 Compliance.**

**43.1.4.1** Repairs, renovations, modifications, reconstruction, changes of use or occupancy classification, and additions shall conform to the specific requirements for each category in other sections of this chapter.

**43.1.4.2** This chapter shall not prevent the use of any alternative material, alternative design, or alternative method of construction not specifically prescribed herein, provided that the alternative has been deemed to be equivalent and its use authorized by the authority having jurisdiction in accordance with Section 1.4.

**43.1.4.3** Where compliance with this chapter, or with any provision required by this chapter, is technically infeasible or would impose undue hardship because of structural, construction, or dimensional difficulties, the authority having jurisdiction shall be authorized to accept alternative materials, design features, or operational features.

**43.1.4.4** Elements, components, and systems of existing buildings with features that exceed the requirements of this *Code* for new construction, and not otherwise required as part of previously documented, approved, alternative arrangements, shall not be prevented by this chapter from being modified, provided that such elements, components, and systems remain in compliance with the applicable *Code* provisions for new construction.

**43.1.4.5** Work mandated by any accessibility, property, housing, or fire code; mandated by the existing building requirements of this *Code*; or mandated by any licensing rule or ordinance, adopted pursuant to law, shall conform only to the requirements of that code, rule, or ordinance and shall not be required to conform to this chapter, unless the code requiring such work so provides.

**43.2 Special Definitions.**

**43.2.1 General.** The words and terms used in Chapter 43 shall be defined as detailed in 43.2.2, unless the context clearly indicates otherwise.

**43.2.2 Special Definitions.**

**43.2.2.1 Categories of Rehabilitation Work.** The nature and extent of rehabilitation work undertaken in an existing building.

**43.2.2.1.1 Repair.** The patching, restoration, or painting of materials, elements, equipment, or fixtures for the purpose of maintaining such materials, elements, equipment, or fixtures in good or sound condition.

**43.2.2.1.2 Renovation.** The replacement in kind, strengthening, or upgrading of building elements, materials, equipment, or fixtures, that does not result in a reconfiguration of the building spaces within.

**43.2.2.1.3 Modification.** The reconfiguration of any space; the addition, relocation, or elimination of any door or window; the addition or elimination of load-bearing elements; the reconfiguration or extension of any system; or the installation of any additional equipment.

**43.2.2.1.4\* Reconstruction.** The reconfiguration of a space that affects an exit or a corridor shared by more than one occupant space; or the reconfiguration of a space such that the rehabilitation work area is not permitted to be occupied because existing means of egress and fire protection systems, or their equivalent, are not in place or continuously maintained.

**43.2.2.1.5 Change of Use.** A change in the purpose or level of activity within a structure that involves a change in application of the requirements of the *Code*.

**43.2.2.1.6 Change of Occupancy Classification.** The change in the occupancy classification of a structure or portion of a structure.

**43.2.2.1.7 Addition.** An increase in the building area, aggregate floor area, building height, or number of stories of a structure.

**43.2.2.2\* Equipment or Fixture.** Any plumbing, heating, electrical, ventilating, air-conditioning, refrigerating, and fire protection equipment; and elevators, dumbwaiters, escalators, boilers, pressure vessels, or other mechanical facilities or installations related to building services.

**43.2.2.3 Load-Bearing Element.** Any column, girder, beam, joist, truss, rafter, wall, floor, or roof sheathing that supports any vertical load in addition to its own weight, or any lateral load.

**43.2.2.4 Rehabilitation Work Area.** That portion of a building affected by any renovation, modification, or reconstruction work as initially intended by the owner, and indicated as such in the permit, but excluding other portions of the building where inci-

dental work entailed by the intended work must be performed, and excluding portions of the building where work not initially intended by the owner is specifically required.

**43.2.2.5 Technically Infeasible.** A change to a building that has little likelihood of being accomplished because the existing structural conditions require the removal or alteration of a load-bearing member that is an essential part of the structural frame, or because other existing physical or site constraints prohibit modification or addition of elements, spaces, or features that are in full and strict compliance with applicable requirements.

**43.3 Repairs.**

**43.3.1 General Requirements.**

**43.3.1.1** A repair, as defined in 43.2.2.1.1, in other than historic buildings shall comply with the requirements of Section 43.3.

**43.3.1.2** Repairs in historic buildings shall comply with the requirements of one of the following:

(1) Section 43.3
(2) Section 43.3, as modified by Section 43.10

**43.3.1.3** The work shall be done using like materials or materials permitted by other sections of this *Code*.

**43.3.1.4** The work shall not make the building less conforming with the other sections of this *Code*, or with any previously approved alternative arrangements, than it was before the repair was undertaken.

**43.4 Renovations.**

**43.4.1 General Requirements.**

**43.4.1.1** A renovation, as defined in 43.2.2.1.2, in other than historic buildings shall comply with the requirements of Section 43.4.

**43.4.1.2** Renovations in historic buildings shall comply with the requirements of one of the following:

(1) Section 43.4
(2) Section 43.4, as modified by Section 43.10

**43.4.1.3** All new work shall comply with the requirements of this *Code* applicable to existing buildings.

**43.4.1.4** The work shall not make the building less conforming with other sections of this *Code*, or with any previous approved alternative arrangements, than it was before the renovation was undertaken, unless otherwise specified in 43.4.1.5.

**43.4.1.5** Minor reductions in the clear opening dimensions of replacement doors and windows that result from the use of different materials shall be permitted, unless such reductions are prohibited.

**43.4.2 Capacity of Means of Egress.** The capacity of means of egress, determined in accordance with Section 7.3, shall be sufficient for the occupant load thereof, unless one of the following conditions exists:

(1) The authority having jurisdiction shall be permitted to establish the occupant load as the number of persons for which existing means of egress is adequate, provided that measures are established to prevent occupancy by a greater number of persons.
(2)\*The egress capacity shall have been previously approved as being adequate.

**43.4.3 Interior Finish Requirements.** New interior finish materials shall meet the requirements for new construction.

**43.4.4 Other Requirements.** The reconfiguration or extension of any system, or the installation of any additional equipment, shall comply with Section 43.5.

**43.5 Modifications.**

**43.5.1 General Requirements.**

**43.5.1.1** A modification, as defined in 43.2.2.1.3, in other than historic buildings shall comply with both of the following:

(1) Section 43.5
(2) Section 43.4

**43.5.1.2** Modifications in historic buildings shall comply with the requirements of one of the following:

(1) 43.5.1.1(1) and (2)
(2) 43.5.1.1(1) and (2), as modified by Section 43.10

**43.5.1.3** Newly constructed elements, components, and systems shall comply with the requirements of other sections of this *Code* applicable to new construction.

**43.5.2 Extensive Modifications.**

**43.5.2.1** The modification of an entire building or an entire occupancy within a building shall be considered as a reconstruction and shall comply with the requirements of Section 43.6 for the applicable occupancy, unless otherwise specified in 43.5.2.2.

**43.5.2.2** Modification work that is exclusively electrical, plumbing, mechanical, fire protection system, or structural work shall not be considered a reconstruction, regardless of its extent.

**43.5.2.3** Where the total area of all the rehabilitation work areas included in a modification exceeds 50 percent of the area of the building, the work shall be considered as a reconstruction and shall comply with the requirements of Section 43.6 for the applicable occupancy, unless otherwise specified in 43.5.2.4.

**43.5.2.4** Rehabilitation work areas in which the modification work is exclusively plumbing, mechanical, fire protection system, or electrical work shall not be included in the computation of total area of all rehabilitation work areas.

**43.6 Reconstruction.**

**43.6.1 General Requirements.**

**43.6.1.1** A reconstruction, as defined in 43.2.2.1.4, in other than historic buildings shall comply with all of the following:

(1) Section 43.6
(2) Section 43.5, except that any stairway replacing an existing stairway shall be permitted to comply with 7.2.2.2.1.1(3)
(3) Section 43.4

**43.6.1.2** Reconstruction work in historic buildings shall comply with the requirements of one of the following:

(1) 43.6.1.1(1), (2), and (3)
(2) 43.6.1.1(1), (2), and (3), as modified by Section 43.10

**43.6.1.3** Wherever the term *rehabilitation work area* is used in Section 43.6, it shall include only the area affected by reconstruction work and areas covered by 43.5.2.

**43.6.1.4** Other rehabilitation work areas affected exclusively by renovation or modification work shall not be included in the rehabilitation work area required to comply with Section 43.6.

**43.6.2 Means of Egress.**

**43.6.2.1 General.** The means of egress shall comply with the requirements applicable to the existing occupancy *[see 43.1.2.1(1)]*, as modified by 43.6.2.

**43.6.2.2\* Illumination, Emergency Lighting, and Marking of Means of Egress.**

**43.6.2.2.1** Means of egress in rehabilitation work areas shall be provided with illumination, emergency lighting, and marking of means of egress in accordance with the requirements of other sections of this *Code* applicable to new construction for the occupancy.

**43.6.2.2.2** Where the reconstruction rehabilitation work area on any floor exceeds 50 percent of that floor area, means of egress throughout the floor shall be provided with illumination, emergency lighting, and marking of means of egress in accordance with the requirements of other sections of this *Code* applicable to new construction for the occupancy, unless otherwise specified in 43.6.2.2.4.

**43.6.2.2.3** In a building with rehabilitation work areas involving more than 50 percent of the aggregate floor area within the building, the means of egress within the rehabilitation work area and the means of egress, including the exit and exit discharge paths, serving the rehabilitation work area shall be provided with illumination, emergency lighting, and marking of means of egress in accordance with the requirements of other sections of this *Code* applicable to new construction for the occupancy, unless otherwise specified in 43.6.2.2.4.

**43.6.2.2.4** Means of egress within a tenant space that is entirely outside the rehabilitation work area shall be permitted to comply with the requirements for illumination, emergency lighting, and marking of means of egress applicable to the existing occupancy in lieu of the requirements for illumination and emergency lighting applicable to new construction required by 43.6.2.2.2 and 43.6.2.2.3.

**43.6.3 Fire Barriers and Smoke Barriers.**

**43.6.3.1** In small residential board and care occupancies and one- and two-family dwellings where the rehabilitation work area is in any attached dwelling unit, walls separating the dwelling units, where such walls are not continuous from the foundation to the underside of the roof sheathing, shall be constructed to provide a continuous fire separation using construction materials that are consistent with the existing wall or that comply with the requirements for new buildings of the occupancy involved.

**43.6.3.2** The following shall apply to work required by 43.6.3.1:

(1) It shall be performed on the side of the wall of the dwelling unit that is part of the rehabilitation work area.
(2) It shall not be required to be continuous through concealed floor spaces.

**43.6.4 Extinguishing Systems.**

**43.6.4.1** In a building with rehabilitation work areas involving over 50 percent of the aggregate building area, automatic sprinkler systems shall be provided on the highest floor containing a rehabilitation work area and on all floors below in accordance with the requirements of other sections of this *Code* applicable to new construction for the occupancy.

**43.6.4.2** On any story with rehabilitation work areas involving over 50 percent of the area of the story, a sprinkler system shall be provided throughout the story in accordance with the requirements of other sections of this *Code* applicable to new construction for the occupancy.

**43.6.4.3** Where sprinklers are installed in an elevator hoistway or elevator machine room as part of the rehabilitation work, the elevators shall comply with the fire fighters' emergency operations requirements of ASME A17.1/CSA B44, *Safety Code for Elevators and Escalators.*

**43.6.4.4** Any rehabilitation work areas in a building that is required to be provided with a standpipe system by other sections of this *Code* shall be provided with standpipes up to and including the highest rehabilitation work area floor.

**43.6.4.5** The standpipes required by 43.6.4.4 shall be located and installed in accordance with NFPA 14, *Standard for the Installation of Standpipe and Hose Systems,* unless otherwise provided in 43.6.4.6 and 43.6.4.7.

**43.6.4.6** No pump shall be required, provided that the following criteria are met:

(1) The standpipes are capable of accepting delivery by fire department apparatus of a minimum of 250 gpm at 65 psi (945 L/min at 4.5 bar) to the topmost floor in buildings equipped throughout with an automatic sprinkler system or a minimum of 500 gpm at 65 psi (1890 L/min at 4.5 bar) to the topmost floor in other buildings.
(2) Where the standpipe terminates below the topmost floor, the standpipe is designed to meet the flow/pressure requirements of 43.6.4.6(1) for possible future extension of the standpipe.

**43.6.4.7** In other than high-rise buildings, the required interconnection of the standpipes for a wet system shall be permitted at the lowest level of the rehabilitation work area.

**43.6.5 Fire Alarm Systems — Smoke Alarms.**

**43.6.5.1** In lodging or rooming houses, hotels and dormitories, and apartment buildings, individual sleeping rooms, guest rooms, and dwelling units within any rehabilitation work area shall be provided with smoke alarms complying with the requirements of other sections of this *Code* applicable to new construction for the occupancy.

**43.6.5.2** Where the rehabilitation work area is located in residential board and care occupancies or one- and two-family dwelling units, smoke alarms complying with the requirements of other sections of this *Code* applicable to new construction for the occupancy shall be provided.

**43.6.6 Elevators.** In high-rise buildings, where the rehabilitation work area is one entire floor, or where the rehabilitation work area is 20 percent or more of the occupied floor area of the building, all floors shall be accessible by at least one elevator.

**43.7 Change of Use or Occupancy Classification.**

**43.7.1 Change of Use.**

**43.7.1.1** A change of use that does not involve a change of occupancy classification shall comply with the requirements applicable to the new use in accordance with the applicable existing occupancy chapter, unless the change of use creates a hazardous contents area as addressed in 43.7.1.2.

**43.7.1.2** A change of use that does not involve a change of occupancy classification but that creates a hazardous area shall comply with one of the following:

(1) The change of use shall comply with the requirements applicable to the new use in accordance with the applicable occupancy chapter for new construction.
(2) For existing health care occupancies protected throughout by an approved, supervised automatic sprinkler system in accordance with 9.7.1.1(1), where a change in use of a room or space not exceeding 250 ft² (23.2 m²) results in a room or space that is described by 19.3.2.1.5(7), the requirements for new construction shall not apply, provided that the enclosure meets the requirements of 19.3.2.1.2 through 19.3.2.1.4.

**43.7.1.3** Any repair, renovation, modification, or reconstruction work undertaken in connection with a change of use that does not involve a change of occupancy classification shall comply with the requirements of Sections 43.3, 43.4, 43.5, and 43.6, respectively.

**43.7.2 Change of Occupancy Classification.** Where the occupancy classification of an existing building or portion of an existing building is changed, in other than historic buildings, the building shall meet the requirements of 43.7.2.1 or 43.7.2.3.

**43.7.2.1** Where a change of occupancy classification creates other than an assembly occupancy, and the change occurs within the same hazard classification category or to an occupancy classification of a lesser hazard classification category (i.e., a higher hazard category number), as addressed by Table 43.7.3, the building shall meet both of the following:

(1) Requirements of the applicable existing occupancy chapters for the occupancy created by the change (*see Chapters 15, 17, 19, 21, 23, 24, 26, 29, 31, 33, 37, 39, 40, and 42*)
(2)*Requirements for automatic sprinkler and detection, alarm, and communications systems and requirements for hazardous areas applicable to new construction for the occupancy created by the change (*see Chapters 14, 16, 18, 20, 22, 24, 26, 28, 30, 32, 36, 38, 40, and 42*)

**43.7.2.2** Where a change of occupancy classification creates an assembly occupancy, and the change occurs within the same hazard classification category or to an occupancy classification of a lesser hazard classification category (i.e., a higher number), as addressed by 43.7.3, the building shall meet both of the following:

(1) Requirements of Chapter 13 for existing assembly occupancies
(2) Requirements for automatic sprinkler and detection, alarm, and communications systems, requirements for hazardous areas, and requirements for main entrance/exit of Chapter 12 for new assembly occupancies

**43.7.2.3** Where a change of occupancy classification occurs to an occupancy classification of a higher hazard classification category (i.e., a lower hazard category number), as addressed by Table 43.7.3, the building shall comply with the requirements of the occupancy chapters applicable to new construction for the occupancy created by the change. *(See Chapters 12, 14, 16, 18, 20, 22, 24, 26, 28, 30, 32, 36, 38, 40, and 42.)*

**43.7.2.4** In historic buildings where a change of occupancy classification occurs within the same hazard classification category or to an occupancy classification in a lesser hazard classification category (i.e., a higher hazard category number), as addressed by Table 43.7.3, the building shall meet the requirements of one of the following:

(1) 43.7.2.1 or 43.7.2.2, as applicable
(2) 43.7.2.1 or 43.7.2.2, as applicable, as modified by Section 43.10

**43.7.2.5** In historic buildings where a change of occupancy classification occurs to an occupancy classification in a higher hazard classification category (i.e., a lower hazard category number), as addressed by Table 43.7.3, the building shall meet the requirements of one of the following:

(1) 43.7.2.3
(2) 43.7.2.3, as modified by Section 43.10

**43.7.3\* Hazard Category Classifications.** The relative degree of hazard between different occupancy classifications shall be as set forth in the hazard category classifications of Table 43.7.3.

**Table 43.7.3 Hazard Categories and Classifications**

| Hazard Category | Occupancy Classification |
|---|---|
| 1 (highest hazard) | Industrial or storage occupancies with high hazard contents |
| 2 | Health care, detention and correctional, residential board and care |
| 3 | Assembly, educational, day care, ambulatory health care, residential, mercantile, business, general and special-purpose industrial, ordinary hazard storage |
| 4 (lowest hazard) | Industrial or storage occupancies with low hazard contents |

**43.8 Additions.**

**43.8.1 General Requirements.**

**43.8.1.1** Where an addition, as defined in 43.2.2.1.7, is made to a building, both of the following criteria shall be met:

(1) The addition shall comply with other sections of this *Code* applicable to new construction for the occupancy.
(2) The existing portion of the building shall comply with the requirements of this *Code* applicable to existing buildings for the occupancy.

**43.8.1.2** An addition shall not create or extend any nonconformity with regard to fire safety or the means of egress in the existing building for which the addition is constructed.

**43.8.1.3** Any repair, renovation, alteration, or reconstruction work within an existing building to which an addition is being made shall comply with the requirements of Sections 43.3, 43.4, 43.5, and 43.6.

**43.8.2 Heights.** No addition shall increase the height of an existing building beyond that permitted under the applicable provisions for new building construction.

**43.8.3 Fire Protection Systems.** In other than one- and two-family dwellings, existing compartment areas without an approved separation from the addition shall be protected by an approved automatic sprinkler system where the combined areas would be required to be sprinklered by the provisions applicable to new construction for the occupancy.

**43.8.4 Smoke Alarms.** Where an addition is made to a one- or two-family dwelling or a small residential board and care occupancy, interconnected smoke alarms, powered by the electrical system, meeting the requirements of the other sections of this *Code* shall be installed and maintained in the addition.

**43.9 Reserved.**

**43.10 Historic Buildings.**

**43.10.1 General Requirements.** Historic buildings undergoing rehabilitation shall comply with the requirements of one of the following:

(1) Section 43.10
(2) Sections 43.3, 43.4, 43.5, 43.6, and 43.7, as they relate, respectively, to repair, renovation, modification, reconstruction, and change of use or occupancy classification
(3) NFPA 914, *Code for Fire Protection of Historic Structures*

**43.10.2 Evaluation.** A historic building undergoing modification, reconstruction, or change of occupancy classification in accordance with the requirements of Chapter 43 shall be investigated and evaluated as follows:

(1) A written report shall be prepared for such a building and filed with the authority having jurisdiction by a registered design professional.
(2) If the subject matter of the report does not require an evaluation by a registered design professional, the authority having jurisdiction shall be permitted to allow the report to be prepared by a licensed building contractor, electrician, plumber, or mechanical contractor responsible for the work.
(3) The licensed person preparing the report shall be knowledgeable in historic preservation, or the report shall be coauthored by a preservation professional.
(4) The report shall identify each required safety feature in compliance with Chapter 43 and where compliance with other chapters of this *Code* would be damaging to the contributing historic features.
(5) The report shall describe each feature not in compliance with this *Code* and demonstrate how the intent of this *Code* is met in providing an equivalent level of safety.
(6) The local preservation official shall be permitted to review and comment on the written report or shall be permitted to request review comments on the report from the historic preservation officer.
(7) Unless it is determined by the authority having jurisdiction that a report is required to protect the health and safety of the public, the submission of a report shall not be required for a building that is being rehabilitated for the personal use of the owner or a member of the owner's immediate family and is not intended for any use or occupancy by the public.

**43.10.3 Repairs.** Repairs to any portion of a historic building shall be permitted to be made with original or like materials and original methods of construction, except as otherwise provided in Section 43.10.

**43.10.4 Repair, Renovation, Modification, or Reconstruction.**

**43.10.4.1 General.** Historic buildings undergoing repair, renovation, modification, or reconstruction shall comply with the applicable requirements of Sections 43.3, 43.4, 43.5, and 43.6, except as specifically permitted in 43.10.4.

**43.10.4.2 Replacement.** Replacements shall meet the following criteria:

(1) Replacement of existing or missing features using original or like materials shall be permitted.
(2) Partial replacement for repairs that match the original in configuration, height, and size shall be permitted.
(3) Replacements shall not be required to meet the requirements of this *Code* that specify material standards, details of installation and connection, joints, or penetrations; or continuity of any element, component, or system in the building.

**43.10.4.3 Means of Egress.** Existing door openings, window openings intended for emergency egress, and corridor and stairway widths narrower than those required for nonhistoric buildings under this *Code* shall be permitted, provided that one of the following criteria is met:

(1) In the opinion of the authority having jurisdiction, sufficient width and height exists for a person to pass through the opening or traverse the exit, and the capacity of the egress system is adequate for the occupant load.
(2) Other operational controls to limit the number of occupants are approved by the authority having jurisdiction.

**43.10.4.4 Door Swing.** Where approved by the authority having jurisdiction, existing front doors shall not be required to swing in the direction of egress travel, provided that other approved exits have sufficient egress capacity to serve the total occupant load.

**43.10.4.5 Transoms.** In fully sprinklered buildings of hotel and dormitory occupancies, apartment occupancies, and residential board and care occupancies, existing transoms in corridors and other fire resistance–rated walls shall be permitted to remain in use, provided that the transoms are fixed in the closed position.

**43.10.4.6 Interior Finishes.**

**43.10.4.6.1** Existing interior wall and ceiling finishes, in other than exits, shall be permitted to remain in place where it is demonstrated that such finishes are the historic finish.

**43.10.4.6.2** Interior wall and ceiling finishes in exits, other than in one- and two-family dwellings, shall meet one of the following criteria:

(1) The material shall be Class A, Class B, or Class C in accordance with Section 10.2 of this *Code.*
(2) Existing materials not meeting the minimum Class C flame spread index shall be surfaced with an approved fire-retardant paint or finish.
(3) Existing materials not meeting the minimum Class C flame spread index shall be permitted to be continued in use, provided that the building is protected throughout by an approved automatic sprinkler system.

**43.10.4.7 Stairway Enclosure.**

**43.10.4.7.1** Stairways shall be permitted to be unenclosed in a historic building where such stairways serve only one adjacent floor.

**43.10.4.7.2** In buildings of three or fewer stories in height, exit enclosure construction shall limit the spread of smoke by the use of tight-fitting doors and solid elements; however, such elements shall not be required to have a fire rating.

**43.10.4.8 One-Hour Fire-Rated Assemblies.** Existing walls and ceilings shall be exempt from the minimum 1-hour fire resistance–rated construction requirements of other sections of this *Code* where the existing wall and ceiling are of wood lath and plaster construction in good condition.

**43.10.4.9 Stairway Handrails and Guards.**

**43.10.4.9.1** Existing grand stairways shall be exempt from the handrail and guard requirements of other sections of this *Code.*

**43.10.4.9.2** Existing handrails and guards on grand staircases shall be permitted to remain in use, provided that they are not structurally dangerous.

**43.10.4.10 Exit Signs.** The authority having jurisdiction shall be permitted to accept alternative exit sign or directional exit sign location, provided that signs installed in compliance with other sections of this *Code* would have an adverse effect on the historic character and such alternative signs identify the exits and egress path.

**43.10.4.11 Sprinkler Systems.**

**43.10.4.11.1** Historic buildings that do not conform to the construction requirements specified in other chapters of this *Code* for the applicable occupancy or use and that, in the opinion of the authority having jurisdiction, constitute a fire safety hazard shall be protected throughout by an approved automatic sprinkler system.

**43.10.4.11.2** The automatic sprinkler system required by 43.10.4.11.1 shall not be used as a substitute for, or serve as an alternative to, the required number of exits from the facility.

**43.10.5 Change of Occupancy.**

**43.10.5.1 General.** Historic buildings undergoing a change of occupancy shall comply with the applicable provisions of Section 43.7, except as otherwise permitted by 43.10.5.

**43.10.5.2 Means of Egress.** Existing door openings, window openings intended for emergency egress, and corridor and stairway widths narrower than those required for nonhistoric buildings under this *Code* shall be permitted, provided that one of the following criteria is met:

(1) In the opinion of the authority having jurisdiction, sufficient width and height exists for a person to pass through the opening or traverse the exit, and the capacity of the egress system is adequate for the occupant load.
(2) Other operational controls to limit the number of occupants are approved by the authority having jurisdiction.

**43.10.5.3 Door Swing.** Where approved by the authority having jurisdiction, existing front doors shall not be required to swing in the direction of egress travel, provided that other approved exits have sufficient capacity to serve the total occupant load.

**43.10.5.4 Transoms.** In corridor walls required to be fire rated by this *Code,* existing transoms shall be permitted to remain in use, provided that the transoms are fixed in the closed position and one of the following criteria is met:

(1) An automatic sprinkler shall be installed on each side of the transom.
(2) Fixed wired glass set in a steel frame or other approved glazing shall be installed on one side of the transom.

**43.10.5.5 Interior Finishes.** Existing interior wall and ceiling finishes shall meet one of the following criteria:

(1) The material shall comply with the requirements for flame spread index of other sections of this *Code* applicable to the occupancy.

(2) Materials not complying with 43.10.5.5(1) shall be permitted to be surfaced with an approved fire-retardant paint or finish.

(3) Materials not complying with 43.10.5.5(1) shall be permitted to be continued in use, provided that the building is protected throughout by an approved automatic sprinkler system, and the nonconforming materials are substantiated as being historic in character.

**43.10.5.6 One-Hour Fire-Rated Assemblies.** Existing walls and ceilings shall be exempt from the minimum 1-hour fire resistance–rated construction requirements of other sections of this *Code* where the existing wall and ceiling are of wood lath and plaster construction in good condition.

**43.10.5.7 Stairs and Handrails.**

**43.10.5.7.1** Existing stairs and handrails shall comply with the requirements of this *Code*, unless otherwise specified in 43.10.5.7.2.

**43.10.5.7.2** The authority having jurisdiction shall be permitted to accept alternatives for grand stairways and associated handrails where the alternatives are approved as meeting the intent of this *Code*.

**43.10.5.8 Exit Signs.** The authority having jurisdiction shall be permitted to accept alternative exit sign or directional exit sign location, provided that signs installed in compliance with other sections of this *Code* would have an adverse effect on the historic character and such alternative signs identify the exits and egress path.

**43.10.5.9 Exit Stair Live Load.** Existing historic stairways in buildings changed to hotel and dormitory occupancies and apartment occupancies shall be permitted to be continued in use, provided that the stairway can support a 75 lb/ft² (3600 N/m²) live load.

## Annex A   Explanatory Material

*Annex A is not a part of the requirements of this NFPA document but is included for informational purposes only. This annex contains explanatory material, numbered to correspond with the applicable text paragraphs.*

**A.1.1** The following is a suggested procedure for determining the *Code* requirements for a building or structure:

(1) Determine the occupancy classification by referring to the occupancy definitions in Chapter 6 and the occupancy Chapters 12 through 42. *(See 6.1.14 for buildings with more than one use.)*

(2) Determine if the building or structure is new or existing. *(See the definitions in Chapter 3.)*

(3) Determine the occupant load. *(See 7.3.1.)*

(4) Determine the hazard of contents. *(See Section 6.2.)*

(5) Refer to the applicable occupancy chapter of the *Code*, Chapters 12 through 42. *[See Chapters 1 through 4 and Chapters 6 through 11, as needed, for general information (such as definitions) or as directed by the occupancy chapter.]*

(6) Determine the occupancy subclassification or special use condition, if any, by referring to Chapters 16 and 17, day-care occupancies; Chapters 18 and 19, health care occupancies; Chapters 22 and 23, detention and correctional occupancies; Chapters 28 and 29, hotels and dormitories; Chapters 32 and 33, residential board and care occupancies; Chapters 36 and 37, mercantile occupancies; and

Chapter 40, industrial occupancies, which contain subclassifications or special use definitions.

(7) Proceed through the applicable occupancy chapter to verify compliance with each referenced section, subsection, paragraph, subparagraph, and referenced codes, standards, and other documents.

(8) Where two or more requirements apply, refer to the occupancy chapter, which generally takes precedence over the base Chapters 1 through 4 and Chapters 6 through 11.

(9) Where two or more occupancy chapters apply, such as in a mixed occupancy *(see 6.1.14)*, apply the most restrictive requirements.

**A.1.1.5** Life safety in buildings includes more than safety from fire. Although fire safety has been the long-standing focus of NFPA *101*, its widely known title, *Life Safety Code*, and its technical requirements respond to a wider range of concerns, including, for example, crowd safety.

**A.1.1.6(1)** This *Code* is intended to be adopted and used as part of a comprehensive program of building regulations that include building, mechanical, plumbing, electrical, fuel gas, fire prevention, and land use regulations.

**A.1.2** The *Code* endeavors to avoid requirements that might involve unreasonable hardships or unnecessary inconvenience or interference with the normal use and occupancy of a building but provides for fire safety consistent with the public interest.

Protection of occupants is achieved by the combination of prevention, protection, egress, and other features, with due regard to the capabilities and reliability of the features involved. The level of life safety from fire is defined through requirements directed at the following:

(1) Prevention of ignition
(2) Detection of fire
(3) Control of fire development
(4) Confinement of the effects of fire
(5) Extinguishment of fire
(6) Provision of refuge or evacuation facilities, or both
(7) Staff reaction
(8) Provision of fire safety information to occupants

**A.1.3.1** Various chapters contain specific provisions for existing buildings and structures that might differ from those for new construction.

**A.1.4** Before a particular mathematical fire model or evaluation system is used, its purpose and limitations need to be known. The technical documentation should clearly identify any assumptions included in the evaluation. Also, it is the intent of the Committee on Safety to Life to recognize that future editions of this *Code* are a further refinement of this edition and earlier editions. The changes in future editions will reflect the continuing input of the fire protection/life safety community in its attempt to meet the purpose stated in this *Code*.

**A.1.4.3** An equivalent method of protection provides an equal or greater level of safety. It is not a waiver or deletion of a *Code* requirement.

The prescriptive provisions of this *Code* provide specific requirements for broad classifications of buildings and structures. These requirements are stated in terms of fixed values, such as maximum travel distance, minimum fire resistance ratings, and minimum features of required systems, such as detection, alarm, suppression, and ventilation, and not in terms of overall building or system performance.

However, the equivalency clause in 1.4.3 permits the use of alternative systems, methods, or devices to meet the intent of the prescribed code provisions where approved as being equivalent. Equivalency provides an opportunity for a performance-based design approach. Through the rigor of a performance based design, it can be demonstrated whether a building design is satisfactory and complies with the implicit or explicit intent of the applicable code requirement.

When employing the equivalency clause, it is important to clearly identify the prescriptive-based code provision being addressed (scope), to provide an interpretation of the intent of the provision (goals and objectives), to provide an alternative approach (proposed design), and to provide appropriate support for the suggested alternative (evaluation of proposed designs).

Performance resulting from proposed designs can be compared to the performance of the design features required by this *Code*. Using prescribed features as a baseline for comparison, it can then be demonstrated in the evaluation whether a proposed design offers the intended level of performance. A comparison of safety provided can be used as the basis for establishing equivalency.

**A.2.1(1)** For example, NFPA 10, *Standard for Portable Fire Extinguishers*, is referenced in Chapter 2. This does not mean that all buildings must have portable fire extinguishers. Portable fire extinguishers are mandatory only to the extent called for elsewhere in the *Code*.

**A.2.1(3)** The Committee on Safety to Life recognizes that it is sometimes impractical to continually upgrade existing buildings or installations to comply with all the requirements of the referenced publications included in Chapter 2.

**A.2.2** It is possible that governing authorities have adopted a code or standard other than one that is listed in Chapter 2. Where such is the case, and where a provision of a code or standard is referenced by this *Code* but the text of the requirement is not extracted into this *Code*, the code or standard adopted by the governing authority is permitted to be utilized where it is deemed by the authority having jurisdiction to adequately address the issue or condition of concern. Where the adopted code or standard does not address the issue, the requirement from the referenced code or standard should be applied by the authority having jurisdiction, unless the governing authority has established other procedures, policies, or guidelines. Where the text of a requirement is extracted from another NFPA code or standard and appears in this *Code*, it is the intent that the requirement be met as if it had originated in this *Code*, regardless of whether the governing authority has adopted the code or standard from which the text is extracted.

**A.3.2.1 Approved.** The National Fire Protection Association does not approve, inspect, or certify any installations, procedures, equipment, or materials; nor does it approve or evaluate testing laboratories. In determining the acceptability of installations, procedures, equipment, or materials, the authority having jurisdiction may base acceptance on compliance with NFPA or other appropriate standards. In the absence of such standards, said authority may require evidence of proper installation, procedure, or use. The authority having jurisdiction may also refer to the listings or labeling practices of an organization that is concerned with product evaluations and is thus in a position to determine compliance with appropriate standards for the current production of listed items.

**A.3.2.2 Authority Having Jurisdiction (AHJ).** The phrase "authority having jurisdiction," or its acronym AHJ, is used in NFPA documents in a broad manner, since jurisdictions and approval agencies vary, as do their responsibilities. Where public safety is primary, the authority having jurisdiction may be a federal, state, local, or other regional department or individual such as a fire chief; fire marshal; chief of a fire prevention bureau, labor department, or health department; building official; electrical inspector; or others having statutory authority. For insurance purposes, an insurance inspection department, rating bureau, or other insurance company representative may be the authority having jurisdiction. In many circumstances, the property owner or his or her designated agent assumes the role of the authority having jurisdiction; at government installations, the commanding officer or departmental official may be the authority having jurisdiction.

**A.3.2.3 Code.** The decision to designate a standard as a "code" is based on such factors as the size and scope of the document, its intended use and form of adoption, and whether it contains substantial enforcement and administrative provisions.

**A.3.2.5 Listed.** The means for identifying listed equipment may vary for each organization concerned with product evaluation; some organizations do not recognize equipment as listed unless it is also labeled. The authority having jurisdiction should utilize the system employed by the listing organization to identify a listed product.

**A.3.3.4 Actuating Member or Bar.** The active surface of the actuating bar needs to be visually and physically distinct from the rest of the device. The actuating bar is also called a cross bar or push pad.

**A.3.3.11 Aisle Accessway.** *Aisle accessway* is the term used for the previously unnamed means of egress component leading to an aisle or other means of egress. For example, circulation space between parallel rows of seats having a width of 12 in. to 24 in. (305 mm to 610 mm) and a length not exceeding 100 ft (30 m) is an aisle accessway. Some of the circulation space between tables or seats in restaurants might be considered aisle accessway.

Depending on the width of aisle accessway, which is influenced by its length and expected utilization, the movement of a person through the aisle accessway might require others to change their individual speed of movement, alter their postures, move their chairs out of the way, or proceed ahead of the person.

**A.3.3.21.2.1 Gross Floor Area.** Where the term *floor area* is used, it should be understood to be gross floor area, unless otherwise specified.

**A.3.3.21.4 Hazardous Area.** Hazardous areas include areas for the storage or use of combustibles or flammables; toxic, noxious, or corrosive materials; or heat-producing appliances.

**A.3.3.21.6 Normally Unoccupied Building Service Equipment Support Area.** Normally unoccupied building service support areas are often found in attics, crawl spaces, chases, and interstitial areas where the space is vacant or intended exclusively for routing ductwork, cables, conduits, piping, and similar services and is rarely accessed. In such spaces, it is often difficult or impossible to fully comply with the egress requirements of Chapter 7. Where portions of such spaces are routinely visited for storage, maintenance, testing, or inspection, that portion is excluded from this definition, but the remainder of the space might be considered a normally unoccupied building service equipment support area. Storage and fuel-fired equipment would not be expected to be permitted in these locations. Roofs are not considered to be normally unoccupied building service equipment support areas.

**A.3.3.22 Area of Refuge.** An area of refuge has a temporary use during egress. It generally serves as a staging area that provides relative safety to its occupants while potential emergencies are assessed, decisions are made, and mitigating activities are begun. Taking refuge within such an area is, thus, a stage of the total egress process, a stage between egress from the immediately threatened area and egress to a public way.

An area of refuge might be another building connected by a bridge or balcony, a compartment of a subdivided story, an elevator lobby, or an enlarged story-level exit stair landing. An area of refuge is accessible by means of horizontal travel or, as a minimum, via an accessible route meeting the requirements of ICC/ANSI A117.1, *American National Standard for Accessible and Usable Buildings and Facilities.*

This *Code* recognizes any floor in a building protected throughout by an approved, supervised automatic sprinkler system as an area of refuge. This recognition acknowledges the ability of a properly designed and functioning automatic sprinkler system to control a fire at its point of origin and to limit the production of toxic products to a level that is not life threatening.

The requirement for separated rooms or spaces can be met on an otherwise undivided floor by enclosing the elevator lobby with ordinary glass or other simple enclosing partitions that are smoke resisting.

For some occupancies, one accessible room or space is permitted.

**A.3.3.27 Atrium.** As defined in NFPA 92, *Standard for Smoke Control Systems,* a large-volume space is an uncompartmented space, generally two or more stories high, within which smoke from a fire either in the space or in a communicating space can move and accumulate without restriction. Atria and covered malls are examples of large-volume spaces.

**A.3.3.28 Attic.** The attic space might be used for storage. The concealed rafter space between the ceiling membrane and the roof sheathing that are attached to the rafters is not considered an attic.

**A.3.3.31.1 Fire Barrier.** A fire barrier might be vertically or horizontally aligned, such as a wall or floor assembly.

**A.3.3.31.2 Smoke Barrier.** A smoke barrier might be vertically or horizontally aligned, such as a wall, floor, or ceiling assembly. A smoke barrier might or might not have a fire resistance rating. Application of smoke barrier criteria where required elsewhere in the *Code* should be in accordance with Section 8.3.

**A.3.3.31.3 Thermal Barrier.** Finish ratings, as published in the UL *Fire Resistance Directory,* are one way of determining thermal barrier.

**A.3.3.33 Birth Center.** A birth center is a low-volume service for healthy, childbearing women, and their families, who are capable of ambulation in the event of fire or fire-threatening events. Birth center mothers and babies have minimal analgesia, receive no general or regional anesthesia, and are capable of ambulation, even in second-stage labor.

**A.3.3.36 Building.** The term *building* is to be understood as if followed by the words *or portions thereof. (See also Structure, A.3.3.271.)*

**A.3.3.36.3 Apartment Building.** The *Code* specifies that, wherever there are three or more living units in a building, the building is considered an apartment building and is required to comply with either Chapter 30 or Chapter 31, as appropriate. Townhouse units are considered to be apartment buildings if there are three or more units in the building. The type of wall required between units in order to consider them to be separate buildings is normally established by the authority having jurisdiction. If the units are separated by a wall of sufficient fire resistance and structural integrity to be considered as separate buildings, then the provisions of Chapter 24 apply to each townhouse. Condominium status is a form of ownership, not occupancy; for example, there are condominium warehouses, condominium apartments, and condominium offices.

**A.3.3.36.5 Existing Building.** With respect to judging whether a building should be considered existing, the deciding factor is not when the building was designed or when construction started but, rather, the date plans were approved for construction by the appropriate authority having jurisdiction.

**A.3.3.36.6 Flexible Plan and Open Plan Educational or Day-Care Building.** Flexible plan buildings have movable corridor walls and movable partitions of full-height construction with doors leading from rooms to corridors. Open plan buildings have rooms and corridors delineated by tables, chairs, desks, bookcases, counters, low-height partitions, or similar furnishings. It is the intent that low-height partitions not exceed 60 in. (1525 mm).

**A.3.3.36.7 High-Rise Building.** It is the intent of this definition that, in determining the level from which the highest occupiable floor is to be measured, the enforcing agency should exercise reasonable judgment, including consideration of overall accessibility to the building by fire department personnel and vehicular equipment. Where a building is situated on a sloping terrain and there is building access on more than one level, the enforcing agency might select the level that provides the most logical and adequate fire department access.

**A.3.3.36.8 Historic Building.** Designation for a historic building might be in an official national, regional, or local historic register, listing, or inventory.

**A.3.3.36.9 Mall Building.** A mall building might enclose one or more uses, such as retail and wholesale stores, drinking and dining establishments, entertainment and amusement facilities, transportation facilities, offices, and other similar uses.

**A.3.3.36.10 Special Amusement Building.** Special amusement buildings include amusements such as a haunted house, a roller coaster–type ride within a building, a multilevel play structure within a building, a submarine ride, and similar amusements where the occupants are not in the open air.

**A.3.3.37 Building Code.** Where no building code has been adopted, *NFPA 5000, Building Construction and Safety Code,* should be used where the building code is referenced in this *Code.*

**A.3.3.41 Cellular or Foamed Plastic.** Cellular or foamed plastic might contain foamed and unfoamed polymeric or monomeric precursors (prepolymer, if used), plasticizers, fillers, extenders, catalysts, blowing agents, colorants, stabilizers, lubricants, surfactants, pigments, reaction control agents, processing aids, and flame retardants.

**A.3.3.47 Common Path of Travel.** Common path of travel is measured in the same manner as travel distance but terminates at that point where two separate and distinct routes become available. Paths that merge are common paths of travel.

**A.3.3.48.1 Fire Compartment.** Additional fire compartment information is contained in 8.2.2.

In the provisions for fire compartments utilizing the outside walls of a building, it is not intended that the outside wall be specifically fire resistance rated, unless required by other standards. Likewise, it is not intended that outside windows or doors be protected, unless specifically required for exposure protection by another section of this *Code* or by other standards.

**A.3.3.48.2 Smoke Compartment.** Where smoke compartments using the outside walls or the roof of a building are provided, it is not intended that outside walls or roofs, or any openings therein, be capable of resisting the passage of smoke. Application of smoke compartment criteria where required elsewhere in the *Code* should be in accordance with Section 8.5.

**A.3.3.49 Consumer Fireworks, 1.4G.** Consumer Fireworks, 1.4G contain limited quantities of pyrotechnic composition per unit and do not pose a mass explosion hazard where stored; therefore, they are not required to be stored in a magazine.

Consumer Fireworks, 1.4G are normally classed as Explosive, 1.4G and described as Fireworks UN0336 by the U.S. Department of Transportation (U.S. DOT). *(See Annex C of NFPA 1124, Code for the Manufacture, Transportation, Storage, and Retail Sales of Fireworks and Pyrotechnic Articles.)*

**A.3.3.52 Critical Radiant Flux.** Critical radiant flux is the property determined by the test procedure of NFPA 253, *Standard Method of Test for Critical Radiant Flux of Floor Covering Systems Using a Radiant Heat Energy Source.* The unit of measurement of critical radiant flux is watts per square centimeter (W/cm$^2$).

**A.3.3.61.2 Stair Descent Device.** A stair descent device typically requires the assistance of a trained operator.

**A.3.3.64 Dormitory.** Rooms within dormitories intended for the use of individuals for combined living and sleeping purposes are guest rooms or guest suites. Examples of dormitories are college dormitories, fraternity and sorority houses, and military barracks.

**A.3.3.66 Dwelling Unit.** It is not the intent of the *Code* that the list of spaces in the definition of the term *dwelling unit* in 3.3.66 is to be all inclusive. It is the intent of the *Code* that the list of spaces is a minimal set of criteria that must be provided to be considered a dwelling unit, and, therefore, the dwelling unit can contain other spaces that are typical to a single-family dwelling.

**A.3.3.66.1 One- and Two-Family Dwelling Unit.** The application statement of 24.1.1.1 limits each dwelling unit to being "occupied by members of a single family with not more than three outsiders." The *Code* does not define the term *family.* The definition of family is subject to federal, state, and local regulations and might not be restricted to a person or a couple (two people) and their children. The following examples aid in differentiating between a single-family dwelling and a lodging or rooming house:

(1) An individual or a couple (two people) who rent a house from a landlord and then sublease space for up to three individuals should be considered a family renting to a maximum of three outsiders, and the house should be regulated as a single-family dwelling in accordance with Chapter 24.

(2) A house rented from a landlord by an individual or a couple (two people) in which space is subleased to 4 or more individuals, but not more than 16, should be considered and regulated as a lodging or rooming house in accordance with Chapter 26.

(3) A residential building that is occupied by 4 or more individuals, but not more than 16, each renting from a landlord, without separate cooking facilities, should be considered and regulated as a lodging or rooming house in accordance with Chapter 26.

**A.3.3.68 Electroluminescent.** This light source is typically contained inside the device.

**A.3.3.76 Evacuation Capability.** The evacuation capability of the residents and staff is a function of both the ability of the residents to evacuate and the assistance provided by the staff. It is intended that the evacuation capability be determined by the procedure acceptable to the authority having jurisdiction. It is also intended that the timing of drills, the rating of residents, and similar actions related to determining the evacuation capability be performed by persons approved by or acceptable to the authority having jurisdiction. The evacuation capability can be determined by the use of the definitions in 3.3.76, the application of NFPA 101A, *Guide on Alternative Approaches to Life Safety,* Chapter 6, or a program of drills (timed).

Where drills are used in determining evacuation capability, it is suggested that the facility conduct and record fire drills six times per year on a bimonthly basis, with a minimum of two drills conducted during the night when residents are sleeping, and that the facility conduct the drills in consultation with the authority having jurisdiction. Records should indicate the time taken to reach a point of safety, date and time of day, location of simulated fire origin, escape paths used, and comments relating to residents who resisted or failed to participate in the drills.

Translation of drill times to evacuation capability is determined as follows:

(1) 3 minutes or less — prompt
(2) Over 3 minutes, but not in excess of 13 minutes— slow
(3) More than 13 minutes — impractical

Evacuation capability, in all cases, is based on the time of day or night when evacuation of the facility would be most difficult, such as when residents are sleeping or fewer staff are present.

Evacuation capability determination is considered slow if the following conditions are met:

(1) All residents are able to travel to centralized dining facilities without continuous staff assistance.
(2) There is continuous staffing whenever there are residents in the facility.

**A.3.3.79 Existing.** See *Existing Building,* A.3.3.36.5.

**A.3.3.81 Exit.** Exits include exterior exit doors, exit passageways, horizontal exits, exit stairs, and exit ramps. In the case of a stairway, the exit includes the stair enclosure, the door to the stair enclosure, the stairs and landings inside the enclosure, the door from the stair enclosure to the outside or to the level of exit discharge, and any exit passageway and its associated doors, if such are provided, so as to discharge the stair directly to the outside. In the case of a door leading directly from the street floor to the street or open air, the exit comprises only the door. *(See also 7.2.2.6.3.1 and A.7.2.2.6.3.1.)*

Doors of small individual rooms, as in hotels, while constituting exit access from the room, are not referred to as exits, except where they lead directly to the outside of the building from the street floor.

**A.3.3.81.1 Horizontal Exit.** Horizontal exits should not be confused with egress through doors in smoke barriers. Doors in smoke barriers are designed only for temporary protection against smoke, whereas horizontal exits provide protection against serious fire for a relatively long period of time in addition to providing immediate protection from smoke. *(See 7.2.4.)*

**A.3.3.83.1 Level of Exit Discharge.** Low occupancy, ancillary spaces with exit doors discharging directly to the outside, such as mechanical equipment rooms or storage areas, that are located on levels other than main occupiable floors should not be considered in the determination of level of exit discharge.

**A.3.3.86 Exposure Fire.** An exposure fire usually refers to a fire that starts outside a building, such as a wildlands fire or vehicle fire, and that, consequently, exposes the building to a fire.

**A.3.3.88.2 Limited Care Facility.** Limited care facilities and residential board and care occupancies both provide care to people with physical and mental limitations. However, the goals and programs of the two types of occupancies differ greatly. The requirements in this *Code* for limited care facilities are based on the assumption that these are medical facilities, that they provide medical care and treatment, and that the patients are not trained to respond to the fire alarm; that is, the patients do not participate in fire drills but, rather, await rescue. *(See Section 18.7.)*

The requirements for residential board and care occupancies are based on the assumption that the residents are provided with personal care and activities that foster continued independence, that the residents are encouraged and taught to overcome their limitations, and that most residents, including all residents in prompt and slow homes, are trained to respond to fire drills to the extent they are able. Residents are required to participate in fire drills. *(See Section 32.7.)*

**A.3.3.90.2 Interior Finish.** Interior finish is not intended to apply to surfaces within spaces such as those that are concealed or inaccessible. Furnishings that, in some cases, might be secured in place for functional reasons should not be considered as interior finish.

**A.3.3.90.3 Interior Floor Finish.** Interior floor finish includes coverings applied over a normal finished floor or stair treads and risers.

**A.3.3.94 Fire Code.** Where no fire code has been adopted, NFPA 1, *Fire Code*, should be used where the fire code is referenced in this *Code*.

**A.3.3.99 Fire Model.** Due to the complex nature of the principles involved, models are often packaged as computer software. Any relevant input data, assumptions, and limitations needed to properly implement the model will be attached to the fire models.

**A.3.3.103 Fire Scenario.** A fire scenario defines the conditions under which a proposed design is expected to meet the fire safety goals. Factors typically include fuel characteristics, ignition sources, ventilation, building characteristics, and occupant locations and characteristics. The term *fire scenario* includes more than the characteristics of the fire itself but excludes design specifications and any characteristics that do not vary from one fire to another; the latter are called assumptions. The term *fire scenario* is used here to mean only those specifications required to calculate the fire's development and effects, but, in other contexts, the term might be used to mean both the initial specifications and the subsequent development and effects (i.e., a complete description of fire from conditions prior to ignition to conditions following extinguishment).

**A.3.3.110 Flame Spread.** See Section 10.2.

**A.3.3.125 Grandstand.** Where the term *grandstand* is preceded by an adjective denoting a material, it means a grandstand the essential members of which, exclusive of seating, are of the material designated.

**A.3.3.136 Heat Release Rate (HRR).** The heat release rate of a fuel is related to its chemistry, physical form, and availability of oxidant and is ordinarily expressed as British thermal units per second (Btu/s) or kilowatts (kW).

Chapters 40 and 42 include detailed provisions on high hazard industrial and storage occupancies.

**A.3.3.140.1 Day-Care Home.** A day-care home is generally located within a dwelling unit.

**A.3.3.143 Hotel.** So-called apartment hotels should be classified as hotels, because they are potentially subject to the same transient occupancy as hotels. Transients are those who occupy accommodations for less than 30 days.

**A.3.3.144.1 Externally Illuminated.** The light source is typically a dedicated incandescent or fluorescent source.

**A.3.3.144.2 Internally Illuminated.** The light source is typically incandescent, fluorescent, electroluminescent, photoluminescent, or self-luminous or is a light-emitting diode(s).

**A.3.3.162.1 Fuel Load.** Fuel load includes interior finish and trim.

**A.3.3.170 Means of Egress.** A means of egress comprises the vertical and horizontal travel and includes intervening room spaces, doorways, hallways, corridors, passageways, balconies, ramps, stairs, elevators, enclosures, lobbies, escalators, horizontal exits, courts, and yards.

**A.3.3.172 Membrane.** For the purpose of fire protection features, a membrane can consist of materials such as gypsum board, plywood, glass, or fabric. For the purpose of membrane structures, a membrane consists of thin, flexible, water-impervious material capable of being supported by an air pressure of 1½ in. (38 mm) water column.

**A.3.3.178 Modification.** Modification does not include repair or replacement of interior finishes.

**A.3.3.187 Objective.** Objectives define a series of actions necessary to make the achievement of a goal more likely. Objectives are stated in more specific terms than goals and are measured on a more quantitative, rather than qualitative, basis.

**A.3.3.188.1 Ambulatory Health Care Occupancy.** It is not the intent that occupants be considered to be incapable of self-preservation just because they are in a wheelchair or use assistive walking devices, such as a cane, a walker, or crutches. Rather, it is the intent to address emergency care centers that receive patients who have been rendered incapable of self-preservation due to the emergency, such as being rendered unconscious as a result of an accident or being unable to move due to sudden illness.

**A.3.3.188.2 Assembly Occupancy.** Assembly occupancies might include the following:

(1) Armories
(2) Assembly halls

(3) Auditoriums
(4) Bowling lanes
(5) Club rooms
(6) College and university classrooms, 50 persons and over
(7) Conference rooms
(8) Courtrooms
(9) Dance halls
(10) Drinking establishments
(11) Exhibition halls
(12) Gymnasiums
(13) Libraries
(14) Mortuary chapels
(15) Motion picture theaters
(16) Museums
(17) Passenger stations and terminals of air, surface, underground, and marine public transportation facilities
(18) Places of religious worship
(19) Pool rooms
(20) Recreation piers
(21) Restaurants
(22) Skating rinks
(23) Special amusement buildings, regardless of occupant load
(24) Theaters

Assembly occupancies are characterized by the presence or potential presence of crowds with attendant panic hazard in case of fire or other emergency. They are generally open or occasionally open to the public, and the occupants, who are present voluntarily, are not ordinarily subject to discipline or control. Such buildings are ordinarily occupied by able-bodied persons and are not used for sleeping purposes. Special conference rooms, snack areas, and other areas incidental to, and under the control of, the management of other occupancies, such as offices, fall under the 50-person limitation.

Restaurants and drinking establishments with an occupant load of fewer than 50 persons should be classified as mercantile occupancies.

For special amusement buildings, see 12.4.7 and 13.4.7.

**A.3.3.188.3 Business Occupancy.** Business occupancies include the following:

(1) Air traffic control towers (ATCTs)
(2) City halls
(3) College and university instructional buildings, classrooms under 50 persons, and instructional laboratories
(4) Courthouses
(5) Dentists' offices
(6) Doctors' offices
(7) General offices
(8) Outpatient clinics (ambulatory)
(9) Town halls

Doctors' and dentists' offices are included, unless of such character as to be classified as ambulatory health care occupancies. (See 3.3.188.1.)

Birth centers should be classified as business occupancies if they are occupied by fewer than four patients, not including infants, at any one time; do not provide sleeping facilities for four or more occupants; and do not provide treatment procedures that render four or more patients, not including infants, incapable of self-preservation at any one time. For birth centers occupied by patients not meeting these parameters, see Chapter 18 or Chapter 19, as appropriate.

Service facilities common to city office buildings, such as newsstands, lunch counters serving fewer than 50 persons,

barber shops, and beauty parlors are included in the business occupancy group.

City halls, town halls, and courthouses are included in this occupancy group, insofar as their principal function is the transaction of public business and the keeping of books and records. Insofar as they are used for assembly purposes, they are classified as assembly occupancies.

**A.3.3.188.4 Day-Care Occupancy.** Day-care occupancies include the following:

(1) Adult day-care occupancies, except where part of a health care occupancy
(2) Child day-care occupancies
(3) Day-care homes
(4) Kindergarten classes that are incidental to a child day-care occupancy
(5) Nursery schools

In areas where public schools offer only half-day kindergarten programs, many child day-care occupancies offer state-approved kindergarten classes for children who need full-day care. Because these classes are normally incidental to the day-care occupancy, the requirements of the day-care occupancy should be followed.

**A.3.3.188.5 Detention and Correctional Occupancy.** Detention and correctional occupancies include the following:

(1) Adult and juvenile substance abuse centers
(2) Adult and juvenile work camps
(3) Adult community residential centers
(4) Adult correctional institutions
(5) Adult local detention facilities
(6) Juvenile community residential centers
(7) Juvenile detention facilities
(8) Juvenile training schools
    See A.22.1.1.1.6 and A.23.1.1.1.6.

**A.3.3.188.6 Educational Occupancy.** Educational occupancies include the following:

(1) Academies
(2) Kindergartens
(3) Schools

An educational occupancy is distinguished from an assembly occupancy in that the same occupants are regularly present.

**A.3.3.188.7 Health Care Occupancy.** Health care occupancies include the following:

(1) Hospitals
(2) Limited care facilities
(3) Nursing homes

Occupants of health care occupancies typically have physical or mental illness, disease, or infirmity. They also include infants, convalescents, or infirm aged persons. It is not the intent to consider occupants incapable of self-preservation because they are in a wheelchair or use assistive walking devices, such as a cane, a walker, or crutches.

**A.3.3.188.8 Industrial Occupancy.** Industrial occupancies include the following:

(1) Drycleaning plants
(2) Factories of all kinds
(3) Food processing plants
(4) Gas plants
(5) Hangars (for servicing/maintenance)

(6) Laundries
(7) Power plants
(8) Pumping stations
(9) Refineries
(10) Sawmills
(11) Telephone exchanges

In evaluating the appropriate classification of laboratories, the authority having jurisdiction should treat each case individually, based on the extent and nature of the associated hazards. Some laboratories are classified as occupancies other than industrial; for example, a physical therapy laboratory or a computer laboratory.

**A.3.3.188.8.1 General Industrial Occupancy.** General industrial occupancies include multistory buildings where floors are occupied by different tenants or buildings suitable for such occupancy and, therefore, are subject to possible use for types of industrial processes with a high density of employee population.

**A.3.3.188.8.2 High Hazard Industrial Occupancy.** A high hazard industrial occupancy includes occupancies where gasoline and other flammable liquids are handled, used, or stored under such conditions that involve possible release of flammable vapors; where grain dust, wood flour or plastic dust, aluminum or magnesium dust, or other explosive dusts are produced; where hazardous chemicals or explosives are manufactured, stored, or handled; where materials are processed or handled under conditions that might produce flammable flyings; and where other situations of similar hazard exist. Chapters 40 and 42 include detailed provisions on high hazard industrial and storage occupancies.

**A.3.3.188.9 Mercantile Occupancy.** Mercantile occupancies include the following:

(1) Auction rooms
(2) Department stores
(3) Drugstores
(4) Restaurants with fewer than 50 persons
(5) Shopping centers
(6) Supermarkets

Office, storage, and service facilities incidental to the sale of merchandise and located in the same building should be considered part of the mercantile occupancy classification.

**A.3.3.188.12 Residential Board and Care Occupancy.** The following are examples of facilities that are classified as residential board and care occupancies:

(1) Group housing arrangement for physically or mentally handicapped persons who normally attend school in the community, attend worship in the community, or otherwise use community facilities
(2) Group housing arrangement for physically or mentally handicapped persons who are undergoing training in preparation for independent living, for paid employment, or for other normal community activities
(3) Group housing arrangement for the elderly that provides personal care services but that does not provide nursing care
(4) Facilities for social rehabilitation, alcoholism, drug abuse, or mental health problems that contain a group housing arrangement and that provide personal care services but do not provide acute care
(5) Assisted living facilities
(6) Other group housing arrangements that provide personal care services but not nursing care

**A.3.3.188.13 Residential Occupancy.** Residential occupancies are treated as separate occupancies in this *Code* as follows:

(1) One- and two-family dwellings (Chapter 24)
(2) Lodging or rooming houses (Chapter 26)
(3) Hotels, motels, and dormitories (Chapters 28 and 29)
(4) Apartment buildings (Chapters 30 and 31)

**A.3.3.188.15 Storage Occupancy.** Storage occupancies include the following:

(1) Barns
(2) Bulk oil storage
(3) Cold storage
(4) Freight terminals
(5) Grain elevators
(6) Hangars (for storage only)
(7) Parking structures
(8) Truck and marine terminals
(9) Warehouses

Storage occupancies are characterized by the presence of relatively small numbers of persons in proportion to the area.

**A.3.3.204 Performance Criteria.** Performance criteria are stated in engineering terms. Engineering terms include temperatures, radiant heat flux, and levels of exposure to fire products. Performance criteria provide threshold values used to evaluate a proposed design.

**A.3.3.206 Personal Care.** Personal care involves responsibility for the safety of the resident while inside the building. Personal care might include daily awareness by management of the resident's functioning and whereabouts, making and reminding a resident of appointments, the ability and readiness for intervention in the event of a resident experiencing a crisis, supervision in the areas of nutrition and medication, and actual provision of transient medical care.

**A.3.3.207 Photoluminescent.** The released light is normally visible for a limited time if the ambient light sources are removed or partially obscured.

**A.3.3.209 Platform.** Platforms also include the head tables for special guests; the raised area for lecturers and speakers; boxing and wrestling rings; theater-in-the-round; and for similar purposes wherein there are no overhead drops, pieces of scenery, or stage effects other than lighting and a screening valance.

A platform is not intended to be prohibited from using a curtain as a valance to screen or hide the electric conduit, lighting track, or similar fixtures, nor is a platform prohibited from using curtains that are used to obscure the back wall of the stage; from using a curtain between the auditorium and the stage (grand or house curtain); from using a maximum of four leg drops; or from using a valance to screen light panels, plumbing, and similar equipment from view.

**A.3.3.216 Proposed Design.** The design team might develop a number of trial designs that will be evaluated to determine whether they meet the performance criteria. One of the trial designs will be selected from those that meet the performance criteria for submission to the authority having jurisdiction as the proposed design.

The proposed design is not necessarily limited to fire protection systems and building features. It also includes any component of the proposed design that is installed, established, or maintained for the purpose of life safety, without which the proposed design could fail to achieve specified performance criteria. Therefore, the proposed design often includes emergency procedures and organizational structures that are needed to meet the performance criteria specified for the proposed design.

**A.3.3.219 Ramp.** See 7.2.5.

**A.3.3.222 Reconstruction.** It is not the intent that a corridor, an aisle, or a circulation space within a suite be considered as a corridor that is shared by more than one occupant space. The suite should be considered as only one occupant space. The following situations should be considered to involve more than one occupant space:

(1) Work affecting a corridor that is common to multiple guest rooms on a floor of a hotel occupancy
(2) Work affecting a corridor that is common to multiple living units on a floor of an apartment building occupancy
(3) Work affecting a corridor that is common to multiple tenants on a floor of a business occupancy

**A.3.3.237.1 Festival Seating.** Festival seating describes situations in assembly occupancies where live entertainment events are held that are expected to result in overcrowding and high audience density that can compromise public safety. It is not the intent to apply the term *festival seating* to exhibitions; sports events; dances; conventions; and bona fide political, religious, and educational events. Assembly occupancies with 15 ft² (1.4 m²) or more per person should not be considered festival seating.

**A.3.3.239 Self-Luminous.** An example of a self-contained power source is tritium gas. Batteries do not qualify as a self-contained power source. The light source is typically contained inside the device.

**A.3.3.240 Self-Preservation (Day-Care Occupancy).** Examples of clients who are incapable of self-preservation include infants, clients who are unable to use stairs because of confinement to a wheelchair or other physical disability, and clients who cannot follow directions or a group to the outside of a facility due to mental or behavioral disorders. It is the intent of this *Code* to classify children under the age of 24 months as incapable of self-preservation. Examples of direct intervention by staff members include carrying a client, pushing a client outside in a wheelchair, and guiding a client by direct hand-holding or continued bodily contact. If clients cannot exit the building by themselves with minimal intervention from staff members, such as verbal orders, classification as incapable of self-preservation should be considered.

**A.3.3.247 Situation Awareness.** Situation awareness (also called situational awareness), described in a simpler fashion, is being aware of what is happening around you and understanding what that information means to you now and in the future. This definition, and the more formal definition, come from the extensive work of human factors (ergonomics) experts in situation awareness, most notably Mica R. Endsley (Endsley, Bolte and Jones, *Designing for Situation Awareness: An approach to user-centered design*, CRC Press, Taylor and Francis, Boca Raton, FL, 2003). Within the *Code*, and the standards it references, are long-standing requirements for systems and facilities that enhance situation awareness. Included are fire/smoke detection, alarm, and communication systems plus the system status panels in emergency command centers; supervisory systems for various especially critical components (e.g., certain valves) of fire protection systems; waterflow indicators; certain signs; and the availability of trained staff, notably in health care occupancies. Serious failures of situation aware-

ness have been identified as central to unfortunate outcomes in various emergencies; for example, typical responses of people to developing fires also exhibit situation awareness problems as incorrect assumptions are made about the rapidity of fire growth or the effect of opening a door. Good situation awareness is critical to decision making, which, in turn, is critical to performance during an emergency.

**A.3.3.254 Smoke Partition.** A smoke partition is not required to have a fire resistance rating.

**A.3.3.255 Smokeproof Enclosure.** For further guidance, see the following publications:

(1) ASHRAE *Handbook and Product Directory — Fundamentals*
(2) *Principles of Smoke Management*, by Klote and Milke
(3) NFPA 105, *Standard for Smoke Door Assemblies and Other Opening Protectives*

**A.3.3.260.1 Design Specification.** Design specifications include both hardware and human factors, such as the conditions produced by maintenance and training. For purposes of performance-based design, the design specifications of interest are those that affect the ability of the building to meet the stated goals and objectives.

**A.3.3.263.2 Outside Stair.** See 7.2.2.6.

**A.3.3.267 Stories in Height.** Stories below the level of exit discharge are not counted as stories for determining the stories in height of a building.

**A.3.3.268.1 Occupiable Story.** Stories used exclusively for mechanical equipment rooms, elevator penthouses, and similar spaces are not occupiable stories.

**A.3.3.270 Street Floor.** Where, due to differences in street levels, two or more stories are accessible from the street, each is a street floor. Where there is no floor level within the specified limits for a street floor above or below the finished ground level, the building has no street floor.

**A.3.3.271 Structure.** The term *structure* is to be understood as if followed by the words *or portion thereof.* (*See also Building, A.3.3.36.*)

**A.3.3.271.2 Air-Supported Structure.** A cable-restrained air-supported structure is one in which the uplift is resisted by cables or webbing that is anchored by various methods to the membrane or that might be an integral part of the membrane. An air-supported structure is not a tensioned-membrane structure.

**A.3.3.271.6 Open Structure.** Open structures are often found in oil refining, chemical processing, or power plants. Roofs or canopies without enclosing walls are not considered an enclosure.

**A.3.3.271.7 Parking Structure.** A parking structure is permitted to be enclosed or open, use ramps, and use mechanical control push-button-type elevators to transfer vehicles from one floor to another. Motor vehicles are permitted to be parked by the driver or an attendant or are permitted to be parked mechanically by automated facilities. Where automated-type parking is provided, the operator of those facilities is permitted either to remain at the entry level or to travel to another level. Motor fuel is permitted to be dispensed, and motor vehicles are permitted to be serviced in a parking structure in accordance with NFPA 30A. [**88A,** 2011]

**A.3.3.271.11 Underground Structure.** In determining openings in exterior walls, doors or access panels are permitted to be included. Windows are also permitted to be included, provided that they are openable or provide a breakable glazed area.

**A.3.3.278 Tent.** A tent might also include a temporary tensioned-membrane structure.

**A.3.3.285 Vertical Opening.** Vertical openings might include items such as stairways; hoistways for elevators, dumbwaiters, and inclined and vertical conveyors; shaftways used for light, ventilation, or building services; or expansion joints and seismic joints used to allow structural movements.

**A.3.3.288 Wall or Ceiling Covering.** Wall or ceiling coverings with ink or top coat layers added as part of the manufacturing process are included in this definition. The term "polymeric" is intended to include "vinyl."

**A.4.1** The goals in Section 4.1 reflect the scope of this *Code* (*see Section 1.1*). Other fire safety goals that are outside the scope of this *Code* might also need to be considered, such as property protection and continuity of operations. Compliance with this *Code* can assist in meeting goals outside the scope of the *Code*.

**A.4.1.1** Reasonable safety risk is further defined by subsequent language in this *Code*.

**A.4.1.1(1)** The phrase "intimate with the initial fire development" refers to the person(s) at the ignition source or first materials burning, not to all persons within the same room or area.

**A.4.1.2** "Comparable emergencies" refers to incidents where the hazard involves thermal attributes similar to fires or airborne contaminants similar to smoke, such that features mandated by this *Code* can be expected to mitigate the hazard. Examples of such incidents might be explosions and hazardous material releases. The *Code* recognizes that features mandated by this *Code* might be less effective against such hazards than against fires.

**A.4.1.3** An assembly occupancy is an example of an occupancy where the goal of providing for reasonably safe emergency and nonemergency crowd movement has applicability. A detention or correctional occupancy is an example of an occupancy where emergency and nonemergency crowd movement is better addressed by detention and correctional facilities specialists than by this *Code*.

**A.4.3** Additional assumptions that need to be identified for a performance-based design are addressed in Chapter 5.

**A.4.3.1** Protection against certain terrorist acts will generally require protection methods beyond those required by this *Code*.

**A.4.5.4** Fire alarms alert occupants to initiate emergency procedures, facilitate orderly conduct of fire drills, and initiate response by emergency services.

**A.4.5.5** Systems encompass facilities or equipment and people. Included are fire/smoke detection, alarm, and communication systems plus the system status panels in emergency command centers; supervisory systems for various especially critical components (e.g., certain valves) of fire protection systems; certain signs; and the availability of trained staff, notably in health care occupancies.

**A.4.6.4.2** See A.4.6.5.

**A.4.6.5** In existing buildings, it is not always practical to strictly apply the provisions of this *Code*. Physical limitations can cause the need for disproportionate effort or expense with little increase in life safety. In such cases, the authority

having jurisdiction needs to be satisfied that reasonable life safety is ensured.

In existing buildings, it is intended that any condition that represents a serious threat to life be mitigated by the application of appropriate safeguards. It is not intended to require modifications for conditions that do not represent a significant threat to life, even though such conditions are not literally in compliance with the *Code*.

An example of what is intended by 4.6.5 would be a historic ornamental guardrail baluster with spacing that does not comply with the 4 in. (100 mm) requirement. Because reducing the spacing would have minimal impact on life safety but could damage the historic character of the guardrail, the existing spacing might be approved by the authority having jurisdiction.

**A.4.6.7.4** In some cases, the requirements for new construction are less restrictive, and it might be justifiable to permit an existing building to use the less restrictive requirements. However, extreme care needs to be exercised when granting such permission, because the less restrictive provision might be the result of a new requirement elsewhere in the *Code*. For example, in editions of the *Code* prior to 1991, corridors in new health care occupancies were required to have a 1-hour fire resistance rating. Since 1991, such corridors have been required only to resist the passage of smoke. However, this provision is based on the new requirement that all new health care facilities be protected throughout by automatic sprinklers. (*See A.4.6.7.5.*)

**A.4.6.7.5** An example of what is intended by 4.6.7.4 and 4.6.7.5 follows. In a hospital that has 6 ft (1830 mm) wide corridors, such corridors cannot be reduced in width, even though the provisions for existing hospitals do not require 6 ft (1830 mm) wide corridors. However, if a hospital has 10 ft (3050 mm) wide corridors, they are permitted to be reduced to 8 ft (2440 mm) in width, which is the requirement for new construction. If the hospital corridor is 36 in. (915 mm) wide, it would have to be increased to 48 in. (1220 mm), which is the requirement for existing hospitals.

**A.4.6.10.1** Fatal fires have occurred when, for example, a required stair has been closed for repairs or removed for rebuilding, or when a required automatic sprinkler system has been shut off to change piping.

**A.4.6.10.2** See also NFPA 241, *Standard for Safeguarding Construction, Alteration, and Demolition Operations*.

**A.4.6.12.3** Examples of such features include automatic sprinklers, fire alarm systems, standpipes, and portable fire extinguishers. The presence of a life safety feature, such as sprinklers or fire alarm devices, creates a reasonable expectation by the public that these safety features are functional. When systems are inoperable or taken out of service but the devices remain, they present a false sense of safety. Also, before taking any life safety features out of service, extreme care needs to be exercised to ensure that the feature is not required, was not originally provided as an alternative or equivalent, or is no longer required due to other new requirements in the current *Code*. It is not intended that the entire system or protection feature be removed. Instead, components such as sprinklers, initiating devices, notification appliances, standpipe hose, and exit systems should be removed to reduce the likelihood of relying on inoperable systems or features. Conversely, equipment, such as fire or smoke dampers, that is not obvious to the public should be able to be taken out of service if no longer required by this *Code*.

**A.4.6.13** The provisions of 4.6.13 do not require inherently noncombustible materials to be tested in order to be classified as noncombustible materials.

**A.4.6.13.1(1)** Examples of such materials include steel, concrete, masonry, and glass.

**A.4.6.14** Materials subject to increase in combustibility or flame spread index beyond the limits herein established through the effects of age, moisture, or other atmospheric condition are considered combustible. (See NFPA 259, *Standard Test Method for Potential Heat of Building Materials*, and NFPA 220, *Standard on Types of Building Construction*.)

**A.4.7** The purpose of emergency egress and relocation drills is to educate the participants in the fire safety features of the building, the egress facilities available, and the procedures to be followed. Speed in emptying buildings or relocating occupants, while desirable, is not the only objective. Prior to an evaluation of the performance of an emergency egress and relocation drill, an opportunity for instruction and practice should be provided. This educational opportunity should be presented in a nonthreatening manner, with consideration given to the prior knowledge, age, and ability of audience.

The usefulness of an emergency egress and relocation drill, and the extent to which it can be performed, depends on the character of the occupancy.

In buildings where the occupant load is of a changing character, such as hotels or department stores, no regularly organized emergency egress and relocation drill is possible. In such cases, the emergency egress and relocation drills are to be limited to the regular employees, who can be thoroughly schooled in the proper procedure and can be trained to properly direct other occupants of the building in case of emergency evacuation or relocation. In occupancies such as hospitals, regular employees can be rehearsed in the proper procedure in case of fire; such training is always advisable in all occupancies, regardless of whether regular emergency egress and relocation drills can be held.

**A.4.7.2** If an emergency egress and relocation drill is considered merely as a routine exercise from which some persons are allowed to be excused, there is a grave danger that, in an actual emergency, the evacuation and relocation will not be successful. However, there might be circumstances under which all occupants do not participate in an emergency egress and relocation drill; for example, infirm or bedridden patients in a health care occupancy.

**A.4.7.4** Fire is always unexpected. If the drill is always held in the same way at the same time, it loses much of its value. When, for some reason during an actual fire, it is not possible to follow the usual routine of the emergency egress and relocation drill to which occupants have become accustomed, confusion and panic might ensue. Drills should be carefully planned to simulate actual fire conditions. Not only should drills be held at varying times, but different means of exit or relocation areas should be used, based on an assumption that fire or smoke might prevent the use of normal egress and relocation avenues.

**A.4.7.6** The written record required by this paragraph should include such details as the date, time, participants, location, and results of that drill.

**A.4.8.2.1** Items to be considered in preparing an emergency plan should include the following:

(1) Purpose of plan
(2) Building description, including certificate of occupancy
(3) Appointment, organization, and contact details of designated building staff to carry out the emergency duties
(4) Identification of events (man-made and natural) considered life safety hazards impacting the building
(5) Responsibilities matrix (role-driven assignments)
(6) Policies and procedures for those left behind to operate critical equipment
(7) Specific procedures to be used for each type of emergency
(8) Requirements and responsibilities for assisting people with disabilities
(9) Procedures for accounting for employees
(10) Training of building staff, building emergency response teams, and other occupants in their responsibilities
(11) Documents, including diagrams, showing the type, location, and operation of the building emergency features, components, and systems
(12) Practices for controlling life safety hazards in the building
(13) Inspection and maintenance of building facilities that provide for the safety of occupants
(14) Conducting fire and evacuation drills
(15) Interface between key building management and emergency responders
(16) Names or job titles of persons who can be contacted for further information or explanation of duties
(17) Post-event (including drill) critique/evaluation, as addressed in 5.14 of *NFPA 1600, Standard on Disaster/Emergency Management and Business Continuity Programs*
(18) Means to update the plan, as necessary

**A.4.8.2.1(3)** It is assumed that a majority of buildings will use a total evacuation strategy during a fire. It should be noted that evacuation from a building could occur for reasons other than a fire, but such other reasons are not the primary focus of the *Code*. As used herein, total evacuation is defined as the process in which all, or substantially all, occupants leave a building or facility in either an unmanaged or managed sequence or order. An alternative to total evacuation is partial evacuation, which can be defined as the process in which a select portion of a building or facility is cleared or emptied of its occupants while occupants in other portions mostly carry on normal activity. In either case, the evacuation process can be ordered or managed in accordance with an established priority in which some or all occupants of a building or facility clear their area and utilize means of egress routes. This is typically done so that the more-endangered occupants are removed before occupants in less-endangered areas. Alternative terms describing this sequencing or ordering of evacuation are *staged evacuation* and *phased evacuation*.

Table A.4.8.2.1(3) illustrates options for extent of management and extent of evacuation. Some of the options shown might not be appropriate. As noted in Table A.4.8.2.1(3), either total or partial evacuation can include staged (zoned) evacuation or phased evacuation, which is referred to as managed or controlled evacuation. It should also be noted that the evacuation process might not include relocation to the outside of the building but might instead include relocation to an area of refuge or might defend the occupants in place to minimize the need for evacuation.

The different methods of evacuation are also used in several contexts throughout the *Code*. Though most of the methods of evacuation are not specifically defined or do not have established criteria, various sections of the *Code* promulgate

**Table A.4.8.2.1(3) Occupant Evacuation Strategies**

| Extent of Evacuation | Extent of Management | |
| --- | --- | --- |
| | Managed Sequence | Unmanaged Sequence |
| No evacuation | No movement — remain in place upon direction | No movement — remain in place per prior instruction |
| Partial evacuation | Managed or controlled partial evacuation In-building relocation on same floor In-building relocation to different floors Occupants of some floors leave building | Unmanaged or uncontrolled partial evacuation |
| Total evacuation | Managed or controlled total evacuation | Unmanaged or uncontrolled total evacuation |

them as alternatives to total evacuation. The following sections discuss these alternatives in more detail:

(1) Section 4.7 — Provides requirements for fire and relocation drills
(2) 7.2.12 — Provides requirements for area of refuge
(3) 7.2.4 — Provides requirements for horizontal exits
(4) 9.6.3.6 — Provides the alarm signal requirements for different methods of evacuation
(5) 9.6.3.9 — Permits automatically transmitted or live voice evacuation or relocation instructions to occupants and requires them in accordance with *NFPA 72, National Fire Alarm and Signaling Code*
(6) 14.3.4.2.3 (also Chapter 15) — Describes alternative protection systems in educational occupancies
(7) 18.1.1.2/18.1.1.3/Section 18.7 (also Chapter 19) — Provide methods of evacuation for health care occupancies
(8) Chapters 22 and 23 — Provide methods of evacuation for detention and correctional occupancies, including the five groups of resident user categories
(9) Chapters 32 and 33 — Provide method of evacuation for residential board and care occupancies
(10) 32.1.5/33.1.5 — For residential board and care occupancies, state that "no means of escape or means of egress shall be considered as complying with the minimum criteria for acceptance, unless emergency evacuation drills are regularly conducted"
(11) 40.2.5.1.2 — For industrial occupancies, states that "ancillary facilities in special-purpose industrial occupancies where delayed evacuation is anticipated shall have not less than a 2-hour fire resistance–rated separation from the predominant industrial occupancy and shall have one means of egress that is separated from the predominant industrial occupancy by 2-hour fire resistance–rated construction"

The method of evacuation should be accomplished in the context of the physical facilities, the type of activities undertaken, and the provisions for the capabilities of occupants (and staff, if available). Therefore, in addition to meeting the requirements of the *Code*, or when establishing an equivalency or a performance-based design, the following recommendations and general guidance information should be taken into account when designing, selecting, executing, and maintaining a method of evacuation:

(1) When choosing a method of evacuation, the available safe egress time (ASET) must always be greater than the required safe egress time (RSET).
(2) The occupants' characteristics will drive the method of evacuation. For example, occupants might be incapable of evacuating themselves because of age, physical or mental disabilities, physical restraint, or a combination thereof. However, some buildings might be staffed with people who could assist in evacuating. Therefore, the method of evacuation is dependent on the ability of occupants to move as a group, with or without assistance. For more information, see the definitions under the term *Evacuation Capability* in Chapter 3.
(3) An alternative method of evacuation might or might not have a faster evacuation time than a total evacuation. However, the priority of evacuation should be such that the occupants in the most danger are given a higher priority. This prioritization will ensure that occupants more intimate with the fire will have a faster evacuation time.
(4) Design, construction, and compartmentation are also variables in choosing a method of evacuation. The design, construction, and compartmentation should limit the development and spread of a fire and smoke and reduce the need for occupant evacuation. The fire should be limited to the room or compartment of fire origin. Therefore, the following factors need to be considered:
    (a) Overall fire resistance rating of the building
    (b) Fire-rated compartmentation provided with the building
    (c) Number and arrangement of the means of egress
(5) Fire safety systems should be installed that compliment the method of evacuation, and should include consideration of the following:
    (a) Detection of fire
    (b) Control of fire development
    (c) Confinement of the effects of fire
    (d) Extinguishment of fire
    (e) Provision of refuge or evacuation facilities, or both
(6) One of the most important fire safety systems is the fire alarm and communication system, particularly the notification system. The fire alarm system should be in accordance with *NFPA 72, National Fire Alarm and Signaling Code*, and should take into account the following:
    (a) Initial notification of only the occupants in the affected zone(s) (e.g., zone of fire origin and adjacent zones)
    (b) Provisions to notify occupants in other unaffected zones to allow orderly evacuation of the entire building
    (c) Need for live voice communication
    (d) Reliability of the fire alarm and communication system
(7) The capabilities of the staff assisting in the evacuation process should be considered in determining the method of evacuation.

(8) The ability of the fire department to interact with the evacuation should be analyzed. It is important to determine if the fire department can assist in the evacuation or if fire department operations hinder the evacuation efforts.

(9) Evacuation scenarios for hazards that are normally outside of the scope of the *Code* should be considered to the extent practicable. (*See 4.3.1.*)

(10) Consideration should be given to the desire of the occupants to self-evacuate, especially if the nature of the building or the fire warrants evacuation in the minds of the occupants. Self-evacuation might also be initiated by communication between the occupants themselves through face-to-face contact, mobile phones, and so forth.

(11) An investigation period, a delay in the notification of occupants after the first activation of the fire alarm, could help to reduce the number of false alarms and unnecessary evacuations. However, a limit to such a delay should be established before a general alarm is sounded, such as positive alarm sequence, as defined in *NFPA 72, National Fire Alarm and Signaling Code.*

(12) Consideration should be given to the need for an evacuation that might be necessary for a scenario other than a fire (e.g., bomb threat, earthquake).

(13) Contingency plans should be established in the event the fire alarm and communication system fail, which might facilitate the need for total evacuation.

(14) The means of egress systems should be properly maintained to ensure the dependability of the method of evacuation.

(15) Fire prevention policies or procedures, or both, should be implemented that reduce the chance of a fire (e.g., limiting smoking or providing fire-safe trash cans).

(16) The method of evacuation should be properly documented, and written forms of communication should be provided to all of the occupants, which might include sign postings throughout the building. Consideration should be given to the development of documentation for an operation and maintenance manual or a fire emergency plan, or both.

(17) Emergency egress drills should be performed on a regular basis. For more information, see Section 4.7.

(18) The authority having jurisdiction should also be consulted when developing the method of evacuation.

Measures should be in place and be employed to sequence or control the order of a total evacuation, so that such evacuations proceed in a reasonably safe, efficient manner. Such measures include special attention to the evacuation capabilities and needs of occupants with disabilities, either permanent or temporary. For comprehensive guidance on facilitating life safety for such populations, go to www.nfpa.org. For specific guidance on stair descent devices, see A.7.2.12.2.3(2).

In larger buildings, especially high-rise buildings, it is recommended that all evacuations — whether partial or total — be managed to sequence or control the order in which certain occupants are evacuated from their origin areas and to make use of available means of egress. In high-rise buildings, the exit stairs, at any level, are designed to accommodate the egress flow of only a very small portion of the occupants — from only one or a few stories, and within a relatively short time period — on the order of a few minutes. In case of a fire, only the immediately affected floor(s) should be given priority use of the means of egress serving that floor(s). Other floors should then be given priority use of the means of egress, depending on the anticipated spread of the fire and its combus-

tion products and for the purpose of clearing certain floors to facilitate eventual fire service operations. Typically, this means that the one or two floors above and below a fire floor will have secondary priority immediately after the fire floor. Depending on where combustion products move, for example, upwards through a building with cool-weather stack effect, the next priority floors will be the uppermost occupied floors in the building.

Generally, in order to minimize evacuation time for most or all of a relatively tall building to be evacuated, occupants from upper floors should have priority use of exit stairs. For people descending many stories of stairs, this priority will maximize their opportunity to take rest stops without unduly extending their overall time to evacuate a building. Thus, the precedence behavior of evacuees should be that people already in an exit stair should normally not defer to people attempting to enter the exit stair from lower floors, except for those lower floors most directly impacted by a fire or other imminent danger. Notably, this is contrary to the often observed behavior of evacuees in high-rise building evacuations where lower floor precedence behavior occurs. (Similarly, in the most commonly observed behavior of people normally disembarking a passenger airliner, people within the aisle defer to people entering the aisle, so that the areas closest to the exit typically clear first.) Changing, and generally managing, the sequence or order in which egress occurs will require effectively informing building occupants and evaluating resulting performance in a program of education, training, and drills.

When designing the method of evacuation for a complex building, all forms of egress should be considered. For example, consideration could be given to an elevator evacuation system. An elevator evacuation system involves an elevator design that provides protection from fire effects so that elevators can be used safely for egress. See 7.2.13 and A.7.2.12.2.4 for more information.

For further guidance, see the following publications:

(1) *NFPA Fire Protection Handbook*, 19th edition, Section 2, Chapter 2, which provides good methodology for managing exposures and determining the method of evacuation

(2) *NFPA Fire Protection Handbook*, 19th edition, Section 13, which provides further commentary on methods of evacuation for different occupancies

(3) *SFPE Handbook of Fire Protection Engineering*, Section 3, Chapter 13, which provides an overview of some of the research on methods of evacuation

**A.5.1.1** Chapter 5 provides requirements for the evaluation of a performance-based life safety design. The evaluation process is summarized in Figure A.5.1.1.

*Code Criteria.* On the left side of Figure A.5.1.1 is input from the *Code.* The life safety goals have been stated in Section 4.1. The objectives necessary to achieve these goals are stated in Section 4.2. Section 5.2 specifies the performance criteria that are to be used to determine whether the objectives have been met.

*Input.* At the top of Figure A.5.1.1 is the input necessary to evaluate a life safety design.

The design specifications are to include certain retained prescriptive requirements, as specified in Section 5.3. All assumptions about the life safety design and the response of the building and its occupants to a fire are to be clearly stated as indicated in Section 5.4. Scenarios are used to assess the adequacy of the design. Eight sets of initiating events are specified for which the ensuing outcomes are to be satisfactory.

*Performance Assessment.* Appropriate methods for assessing performance are to be used per Section 5.6. Safety factors are to be applied to account for uncertainties in the assessment, as stated in Section 5.7. If the resulting predicted outcome of the scenarios is bounded by the performance criteria, the objectives have been met, and the life safety design is considered to be in compliance with this *Code.* Although not part of this *Code,* a design that fails to comply can be changed and reassessed, as indicated on the right side of Figure A.5.1.1.

*Documentation.* The approval and acceptance of a life safety design are dependent on the quality of the documentation of the process. Section 5.8 specifies a minimum set of documentation that is to accompany a submission.

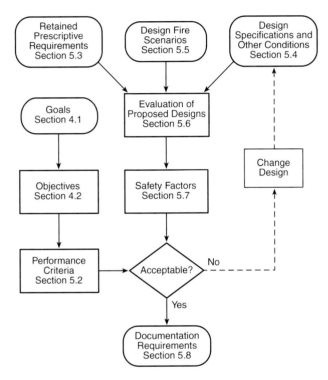

**FIGURE A.5.1.1  Performance-Based *Life Safety Code* Compliance Process.**

The performance option of this *Code* establishes acceptable levels of risk to occupants of buildings and structures as addressed in Section 1.1. While the performance option of this *Code* does contain goals, objectives, and performance criteria necessary to provide an acceptable level of risk to occupants, it does not describe how to meet the goals, objectives, and performance criteria. Design and engineering are needed to develop solutions that meet the provisions of Chapter 5. The *SFPE Engineering Guide to Performance-Based Fire Protection Analysis and Design of Buildings* provides a framework for these assessments. Other useful references include the *Australian Fire Engineering Guidelines* and the *British Standard Firesafety Engineering in Buildings.*

**A.5.1.4**  A third-party reviewer is a person or group of persons chosen by the authority having jurisdiction to review proposed performance-based designs. The *SFPE Guidelines for Peer Review in the Fire Protection Design Process* provides a method for the initiation, scope, conduct, and report of a peer review of a fire protection engineering design.

**A.5.1.6**  For guidance on reviewing performance-based designs, see the *SFPE Enforcer's Guide to Performance-Based Design Review.* Additional guidance on reviewing designs in which fire risk assessment is used can be found in NFPA 551, *Guide for the Evaluation of Fire Risk Assessments.*

**A.5.1.7**  Continued compliance with the goals and objectives of the *Code* involves many factors. The building construction — including openings, interior finish, and fire- and smoke-resistive construction — and the building and fire protection systems need to retain at least the same level of performance as is provided for the original design parameters. The use and occupancy should not change to the degree that assumptions made about the occupant characteristics, combustibility of furnishings, and existence of trained personnel are no longer valid. In addition, actions provided by other personnel, such as emergency responders, should not be diminished below the documented assumed levels. Also, actions needed to maintain reliability of systems at the anticipated level need to meet the initial design criteria.

**A.5.2.2**  One of the methods that follows can be used to avoid exposing occupants to untenable conditions.

**Method 1.**  The design team can set detailed performance criteria that ensure that occupants are not incapacitated by fire effects. The *SFPE Engineering Guide to Performance-Based Fire Protection Analysis and Design of Buildings* describes a process of establishing tenability limits.

The guide references D. A. Purser, "Toxicity Assessment of Combustion Products," Chapter 2/6, *SFPE Handbook of Fire Protection Engineering,* which describes a fractional effective dose (FED) calculation approach, which is also contained in NFPA 269, *Standard Test Method for Developing Toxic Potency Data for Use in Fire Hazard Modeling.* FED addresses the effects of carbon monoxide, hydrogen cyanide, carbon dioxide, hydrogen chloride, hydrogen bromide, and anoxia. It is possible to use the test data, combined with laboratory experience, to estimate the FED value that leads to the survival of virtually all people. This value is about 0.8.

There is a relationship between exposures leading to death and those leading to incapacitation. Kaplan [Kaplan and Hartzell, *Journal of Fire Sciences,* 2:286–305 (1984)] found that rodent susceptibility is similar to that of humans and that for the narcotic gases, CO and HCN, incapacitation is estimated to occur at one-third to one-half the lethal exposure. A set of very large statistical studies on human lethality associated with carbon monoxide involving almost 5000 fatalities (Hirschler et al., "Carbon monoxide and human lethality: Fire and non-fire studies," Elsevier, 1993) showed that the vast majority of fire deaths are attributable to carbon monoxide poisoning, which results in lethality at levels as low as 25 percent carboxyhemoglobin (much lower than previously believed) without requiring the effect of additional toxicants. This work was also confirmed by Gann [Gann et al., *Fire and Materials,* 18:193 (1994)], who also found that carbon monoxide dominates the lethality of fire smoke, since most fire deaths occur remote from the fire room in fires that have proceeded past flashover. Thus, if an FED value of 0.8 were used for a nonlethal exposure, an FED of 0.3 would be reasonable for a nonincapacitating exposure.

If the authority having jurisdiction or the design professional is concerned with potential toxic fire effects, other than those addressed by the FED procedure as documented, the calculation procedure can be expanded by adding additional terms to the FED equation, with each term expressed as a ratio. The numerator of the ratio is the cumulative exposure to that fire effect, measured as an integral of the product of instantaneous exposure (concentration for toxic products) and time. The denominator of the ratio is the quantity of cumulative exposure for which FED equals the chosen threshold value (i.e., 0.8 or 0.3) based on that fire effect alone. A complete analysis of tenability requires consideration of tenability criteria for thermal effects (convected heat and radiated heat) and smoke obscuration, as well as those for smoke toxicity, and an example of the application of such criteria is shown in ASTM E 2280, *Standard Guide for Fire Hazard Assessment of the Effect of Upholstered Seating Furniture Within Patient Rooms of Health Care Facilities.*

For buildings where an unusually large fraction of the occupants are especially vulnerable, the calculation procedure for the smoke toxicity incapacitating criterion should be modified to use FED values lower than 0.8 or 0.3.

**Method 2.** For each design fire scenario and the design specifications, conditions, and assumptions, the design team can demonstrate that each room or area will be fully evacuated before the smoke and toxic gas layer in that room descends to a level lower than 6 ft (1830 mm) above the floor. The timing of such an evacuation means that no occupant is exposed to fire effects. Such an evacuation requires calculation of the locations, movement, and behavior of occupants, because fire effects and occupants are separated by moving the occupants. A level of 60 in. (1525 mm) is often used in calculations, but, at that level, a large fraction of the population would not be able to stand, walk, or run normally and still avoid inhalation of toxic gases. They would have to bend over or otherwise move their heads closer to the floor level.

**Method 3.** For each design fire scenario and the design specifications and assumptions, the design team can demonstrate that the smoke and toxic gas layer will not descend to a level lower than 6 ft (1830 mm) above the floor in any occupied room. The advantage of this procedure is that it conservatively ensures that no occupant is exposed to fire effects, regardless of where occupants are located or where they move. This eliminates the need for calculations regarding occupants, including those for their behavior, movement locations, pre-fire characteristics, and reactions to fire effects. This procedure is even more conservative and simpler than the procedure in Method 2, because it does not allow fire effects in occupied rooms to develop to a point where people could be affected at any time during the fire.

**Method 4.** For each design fire scenario and the design specifications and assumptions, the design team can demonstrate that no fire effects will reach any occupied room. The advantage of this procedure is that it eliminates the need for calculations regarding occupants, including those for their behavior, movement, locations, pre-fire characteristics, and reactions to fire effects. A further advantage is that it also eliminates the need for some of the modeling of fire effects, because it is not necessary to model the filling of rooms, only the spread of fire effects to those rooms. This procedure is even more conservative and simpler than the procedures in Methods 2 and 3, because it does not allow any fire effects in occupied rooms.

**A.5.3.1** This requirement applies both to systems and features required by the *Code* that reference applicable standards and to any additional systems or features included in the design at the discretion of the design team. The referenced standards are hereby expected to state maintenance, testing, and other requirements needed to provide positive assurance of an acceptable level of reliability. The referenced standards themselves might be prescriptive- or performance-based.

**A.5.4.1** The design specifications and other conditions form the input to evaluation of proposed designs *(see Section 5.6)*. Where a specification or condition is not known, a reasonable estimation is permitted. However, the design team must take steps to ensure that the estimation is valid during the life of the building. Any estimations need to be documented. *(See Section 5.8.)*

**A.5.4.4** Systems addressed by this requirement include automatic fire suppression systems and fire alarm systems. Performance issues that need to be documented might include response time indexes, discharge densities, and distribution patterns. Calculations should not include an unlimited supply of extinguishing agent if only a limited supply will be provided in the actual structure or building.

Emergency procedures addressed by this requirement might be of two types. The design team could include documentation from buildings that are operationally very similar, along with documented operational performance measures tied to the recruitment and training of emergency team personnel. Where such data are unavailable, or where the proposed design differs significantly from other buildings, the design could be based on detailed analyses of the decisions and tasks that need to be performed by emergency personnel, using plausible conservative assumptions about the occupant characteristics and training of those personnel.

**A.5.4.5.1** Examples of design features that might be incorporated to modify expected occupant characteristics include training, use of staff to assist with notification and movement, or type of notification appliance used.

**A.5.4.5.2** The four basic characteristics — sensibility, reactivity, mobility, and susceptibility — comprise a minimum, exhaustive set of mutually exclusive performance characteristics of people in buildings that can affect a fire safety system's ability to meet life safety objectives. The characteristics are briefly described as follows:

(1) Sensibility to physical cues, which is the ability to sense the sounding of an alarm and can also include discernment and discrimination of visual and olfactory cues in addition to auditory emanations from the fire itself
(2) Reactivity, which is the ability to interpret cues correctly and take appropriate action and can be a function of cognitive capacity, speed of instinctive reaction, or group dynamics; might need to consider reliability or likelihood of a wrong decision, as in situations where familiarity with the premises influences wayfinding
(3) Mobility (speed of movement), which is determined by individual capabilities, as well as crowding phenomena, such as arching at doorways
(4) Susceptibility to products of combustion, which includes metabolism, lung capacity, pulmonary disease, allergies, or other physical limitations that affect survivability in a fire environment

In application, as with the use of computer evacuation models, assumptions can address a larger number of factors that are components of the basic performance characteristics, including the following:

(1) Alertness — condition of being awake/asleep, can depend on time of day
(2) Responsiveness — ability to sense cues and react

(3) Commitment — degree to which occupant is committed to an activity underway before the alarm
(4) Focal point — point at which an occupant's attention is focused (e.g., to front of classroom, stage, or server in business environment)
(5) Physical and mental capabilities — influence on ability to sense, respond, and react to cues; might be related to age or disability
(6) Role — influence on whether occupant will lead or follow others
(7) Familiarity — influence of time spent in building or participation in emergency training
(8) Social affiliation — extent to which an occupant will act/react as an individual or as a member of a group
(9) Condition over the course of the fire — effects, both physiological and psychological, of the fire and its combustion products on each occupant

For a more detailed explanation of occupant characteristics, see the *SFPE Engineering Guide to Human Behavior in Fire.* Occupant characteristics that are discussed in the guide include the following:

(1) Population numbers and density
(2) Condition of being alone or with others
(3) Familiarity with the building
(4) Distribution and activities
(5) Alertness
(6) Physical and cognitive ability
(7) Social affiliation
(8) Role and responsibility
(9) Location
(10) Commitment
(11) Focal point
(12) Occupant condition
(13) Gender
(14) Culture
(15) Age

**A.5.4.5.4** The number of people expected to be contained in a room or area should be based on the occupant load factor specified in Table 7.3.1.2 or other approved sources.

**A.5.4.5.5** For example, in hospitals, staff characteristics such as number, location, quality, and frequency of training should be considered.

**A.5.4.7** Design proposals need to state explicitly any design specifications or estimations regarding building fire safety plans, inspection programs, or other ongoing programs whose performance is necessary for the building, when occupied and operational, to meet the stated goals and objectives. Programs of interest include any maintenance, training, labeling, or certification programs required to ensure operational status or reliability in building systems or features.

**A.5.4.9** The design elements required to be excluded by 5.4.9 include those regarding the interrelations between the performance of building elements and systems, occupant behavior, or emergency response actions that conflict with each other. For each fire scenario, care needs to be taken to ensure that conflicts in actions do not occur. Typical conflicts could include the following:

(1) Assuming a fire door will remain closed during the fire to contain smoke while this same door is used by occupants during egress from the area

(2) Assuming fire apparatus will arrive immediately from a distant location to provide water to fire department connections and similar situations

For example, an assumption that compartmentation blocking the passage of fire and smoke will be maintained at the door to a stairwell cannot be paired with an assumption that evacuation through that door will extend over many minutes.

**A.5.4.10** The provisions required by 5.4.10 to be documented include those that are in excess of basic requirements covered by referenced codes and standards, typical design requirements, and operating procedures. Such provisions include the following:

(1) More frequent periodic testing and maintenance to increase the reliability of fire protection systems
(2) Redundant systems to increase reliability
(3) On-site guard service to enhance detection of fires and aid in fire response procedures
(4) Staff training
(5) Availability and performance of emergency response personnel
(6) Other factors

**A.5.5** Design fire scenarios define the challenge a building is expected to withstand. Design fire scenarios capture and limit value judgments on the type and severity of the fire challenge to which a proposed fire safety system needs to respond. The system includes any and all aspects of the proposed design that are intended to mitigate the effects of a fire, such as egress system, automatic detection and suppression, barriers, staff training, and placement of manual extinguishers.

Design fire scenarios come from two sources: those that are specified in 5.5.3.1 through 5.5.3.8, and those that are developed by the design team based on the unique characteristics of the building as required by 5.5.2. In most, if not all, cases, more than one design fire scenario will be developed to meet the requirements of 5.5.2.

Once the set of design fire scenarios is established, both those specified by 5.5.3.1 through 5.5.3.8 and those that are developed as required by 5.5.2, they need to be quantified into a format that can be used for the evaluation of proposed designs. The *SFPE Engineering Guide to Performance-Based Fire Protection Analysis and Design of Buildings* outlines a process and identifies tools and references that can be used at each step of this process.

**A.5.5.2** The protection systems and features used to meet the challenge of the design fire scenario should be typical of, and consistent with, those used for other similar areas of the building. They should not be designed to be more effective in the building area addressed than in similar areas not included and that are, therefore, not explicitly evaluated.

**A.5.5.3** It is desirable to consider a wide variety of different fire scenarios to evaluate the complete life safety capabilities of the building or structure. Fire scenarios should not be limited to a single or a couple of worst-case fire scenarios.

The descriptive terms used to indicate the rate of fire growth for the scenarios are intended to be generic. Use of *t*-squared fires is not required for any scenario.

**A.5.5.3.1** An example of Design Fire Scenario 1 for a health care occupancy would involve a patient room with two occupied beds with a fire initially involving one bed and the room door open. This is a cursory example in that much of the explicitly required information indicated in 5.5.3.1 can be determined from the information provided in the example. Note that it is usually necessary to consider more than one scenario to capture the features and conditions typical of an occupancy.

**A.5.5.3.2** Design Fire Scenario 2 examples include a fire involving ignition of gasoline as an accelerant in a means of egress, clothing racks in corridors, renovation materials, or other fuel configurations that can cause an ultrafast fire. The means of egress chosen is the doorway with the largest egress capacity among doorways normally used in the ordinary operation of the building. The baseline occupant characteristics for the property are assumed. At ignition, doors are assumed to be open throughout the building.

**A.5.5.3.3** An example of Design Fire Scenario 3 is a fire in a storage room adjacent to the largest occupiable room in the building. The contents of the room of fire origin are specified to provide the largest fuel load and the most rapid growth in fire severity consistent with the normal use of the room. The adjacent occupiable room is assumed to be filled to capacity with occupants. Occupants are assumed to be somewhat impaired in whatever form is most consistent with the intended use of the building. At ignition, doors from both rooms are assumed to be open. Depending on the design, doorways connect the two rooms or they connect via a common hallway or corridor.

For purposes of this scenario, an occupiable room is a room that might contain people; that is, a location within a building where people are typically found.

**A.5.5.3.4** An example of Design Fire Scenario 4 is a fire originating in a concealed wall or ceiling space adjacent to a large, occupied function room. Ignition involves concealed combustibles, including wire or cable insulation and thermal or acoustical insulation. The adjacent function room is assumed to be occupied to capacity. The baseline occupant characteristics for the property are assumed. At ignition, doors are assumed to be open throughout the building.

**A.5.5.3.5** An example of Design Fire Scenario 5 is a cigarette fire in a trash can. The trash can is close enough to room contents to ignite more substantial fuel sources but is not close enough to any occupant to create an intimate-with-ignition situation. If the intended use of the property involves the potential for some occupants to be incapable of movement at any time, the room of origin is chosen as the type of room likely to have such occupants, filled to capacity with occupants in that condition. If the intended use of the property does not involve the potential for some occupants to be incapable of movement, the room of origin is chosen to be an assembly or function area characteristic of the use of the property, and the trash can is placed so that it is shielded by furniture from suppression systems. At ignition, doors are assumed to be open throughout the building.

**A.5.5.3.6** An example of Design Fire Scenario 6 is a fire originating in the largest fuel load of combustibles possible in normal operation in a function or assembly room, or in a process/manufacturing area, characteristic of the normal operation of the property. The configuration, type, and geometry of the combustibles are chosen so as to produce the most rapid and severe fire growth or smoke generation consistent with the normal operation of the property. The baseline occupant characteristics for the property are assumed. At ignition, doors are assumed to be closed throughout the building.

This scenario includes everything from a big couch fire in a small dwelling to a rack fire in combustible liquids stock in a big box retail store.

**A.5.5.3.7** An example of Design Fire Scenario 7 is an exposure fire. The initiating fire is the closest and most severe fire possible consistent with the placement and type of adjacent properties and the placement of plants and combustible adornments on the property. The baseline occupant characteristics for the property are assumed.

This category includes wildlands/urban interface fires and exterior wood shingle problems, where applicable.

**A.5.5.3.8** Design Fire Scenario 8 addresses a set of conditions with a typical fire originating in the building with any one passive or active fire protection system or feature being ineffective. Examples include unprotected openings between floors or between fire walls or fire barrier walls, failure of rated fire doors to close automatically, shutoff of sprinkler system water supply, nonoperative fire alarm system, inoperable smoke management system, or automatic smoke dampers blocked open. This scenario should represent a reasonable challenge to the other building features provided by the design and presumed to be available.

The concept of a fire originating in ordinary combustibles is intentionally selected for this scenario. This fire, although presenting a realistic challenge to the building and the associated building systems, does not represent the worst-case scenario or the most challenging fire for the building. Examples include the following:

(1) Fire originating in ordinary combustibles in the corridor of a patient wing of a hospital under the following conditions:
  (a) Staff is assumed not to close any patient room doors upon detection of fire.
  (b) The baseline occupant characteristics for the property are assumed, and the patient rooms off the corridor are assumed to be filled to capacity.
  (c) At ignition, doors to patient rooms are not equipped with self-closing devices and are assumed to be open throughout the smoke compartment.
(2) Fire originating in ordinary combustibles in a large assembly room or area in the interior of the building under the following conditions:
  (a) The automatic suppression systems are assumed to be out of operation.
  (b) The baseline occupant characteristics for the property are assumed, and the room of origin is assumed to be filled to capacity.
  (c) At ignition, doors are assumed to be closed throughout the building.
(3) Fire originating in ordinary combustibles in an unoccupied small function room adjacent to a large assembly room or area in the interior of the building under the following conditions:
  (a) The automatic detection systems are assumed to be out of operation.
  (b) The baseline occupant characteristics for the property are assumed, the room of origin is assumed to be unoccupied, and the assembly room is assumed to be filled to capacity.
  (c) At ignition, doors are assumed to be closed throughout the building.

**A.5.5.3.8(3)** The exemption is applied to each active or passive fire protection system individually and requires two different types of information to be developed by analysis and approved by the authority having jurisdiction. System reliability is to be analyzed and accepted. Design performance in the absence of the system is also to be analyzed and accepted, but acceptable performance does not require fully meeting the stated goals and objectives. It might not be possible to meet fully the goals and objectives if a key system is unavailable, and yet no system is totally reliable. The authority having jurisdiction will determine which level of performance, possibly short of the stated goals and objectives, is acceptable, given the very low probability (i.e., the system's unreliability probability) that the system will not be available.

**A.5.6** The *SFPE Engineering Guide to Performance-Based Fire Protection Analysis and Design of Buildings* outlines a process for evaluating whether trial designs meet the performance criteria during the design fire scenarios. Additional information on reviewing the evaluation of a performance-based design can be found in the *SFPE Enforcer's Guide to Performance-Based Design Review.*

The procedures described in Sections 5.2 and 5.4 identify required design fire scenarios among the design fire scenarios within which a proposed fire safety design is required to perform and the associated untenable conditions that are to be avoided in order to maintain life safety. Section 5.6 discusses methods that form the link from the scenarios and criteria to the goals and objectives.

Assessment methods are used to demonstrate that the proposed design will achieve the stated goals/objectives by providing information indicating that the performance criteria of Section 5.2 can be adequately met. Assessment methods are permitted to be either tests or modeling.

*Tests.* Test results can be directly used to assess a fire safety design when they accurately represent the scenarios developed by using Section 5.4 and provide output data matching the performance criteria in Section 5.2. Because the performance criteria for this *Code* are stated in terms of human exposure to lethal fire effects, no test will suffice. However, tests will be needed to produce data for use in models and other calculation methods.

*Standardized Tests.* Standardized tests are conducted on various systems and components to determine whether they meet some predetermined, typically prescriptive criteria. Results are given on a pass/fail basis — the test specimen either does or does not meet the pre-established criteria. The actual performance of the test specimen is not usually recorded.

*Scale.* Tests can be either small, intermediate, or full scale. Small-scale tests are used to test activation of detection and suppression devices and the flammability and toxicity of materials. Usually, the item to be tested is placed within the testing device or apparatus. Intermediate-scale tests can be used to determine the adequacy of system components — for example, doors and windows — as opposed to entire systems. The difference between small- and intermediate-scale tests is usually one of definition provided by those conducting the test. Full-scale tests are typically used to test building and structural components or entire systems. The difference between intermediate- and large-scale tests is also subject to the definition of those performing the test. Full-scale tests are intended to most closely depict performance of the test subject as installed in the field; that is, most closely represent real world performance.

Full-scale building evacuations can provide information on how the evacuation of a structure is likely to occur for an existing building with a given population without subjecting occupants to the real physical or psychological effects of a fire.

*Data Uses.* The data obtained from standardized tests have three uses for verification purposes. First, the test results can

be used instead of a model. This use is typically the role of full-scale test results. Second, the test results can be used as a basis for validating the model. The model predictions match well with the test results. Therefore, the model can be used in situations similar to the test scenario. Third, the test results can be used as input to models. This is typically the use of small-scale tests, specifically flammability tests.

*Start-Up Test.* Start-up test results can be used to demonstrate that the fire safety system performs as designed. The system design might be based on modeling. If the start-up test indicates a deficiency, the system needs to be adjusted and retested until it can be demonstrated that the design can meet the performance criteria. Typically, start-up tests apply only to the installation to which they are designed.

*Experimental Data.* Experimental data from nonstandardized tests can be used when the specified scenario and the experimental setup are similar. Typically, experimental data are applicable to a greater variety of scenarios than are standardized test results.

*Human and Organizational Performance Tests.* Certain tests determine whether inputs used to determine human performance criteria remain valid during the occupancy of a building. Tests of human and organizational performance might include any of the following:

(1) Measuring evacuation times during fire drills
(2) Querying emergency response team members to determine whether they know required procedures
(3) Conducting field tests to ensure that emergency response team members can execute tasks within predetermined times and accuracy limits

Design proposals should include descriptions of any tests needed to determine whether stated goals, objectives, and performance criteria are being met.

*Modeling.* Models can be used to predict the performance criteria for a given scenario. Because of the limitations on using only tests for this purpose, models are expected to be used in most, if not all, performance-based design assessments.

The effect of fire and its toxic products on the occupants can be modeled, as can the movement and behavior of occupants during the fire. The term *evacuation model* will be used to describe models that predict the location and movements of occupants, and the term *tenability model* will be used to describe models that predict the effects on occupants of specified levels of exposure to fire effects.

*Types of Fire Models.* Fire models are used to predict fire-related performance criteria. Fire models can be either probabilistic or deterministic. Several types of deterministic models are available: computational fluid dynamics (CFD or field) models, zone models, purpose-built models, and hand calculations. Probabilistic fire models are also available but are less likely to be used for this purpose.

Probabilistic fire models use the probabilities as well as the severity of various events as the basis of evaluation. Some probabilistic models incorporate deterministic models, but are not required to do so. Probabilistic models attempt to predict the likelihood or probability that events or severity associated with an unwanted fire will occur, or they predict the "expected loss," which can be thought of as the probability-weighted average severity across all possible scenarios. Probabilistic models can be manifested as fault or event trees or other system models that use frequency or probability data as input. These models tend to be manifested as computer software, but are not required to do so.

Furthermore, the discussion that follows under "Sources of Models" can also be applied to probabilistic models, although it concentrates on deterministic models.

CFD models can provide more accurate predictions than other deterministic models, because they divide a given space into many smaller-volume spaces. However, since they are still models, they are not absolute in their depiction of reality. In addition, they are much more expensive to use, because they are computationally intensive. Because of their expense, complexity, and intensive computational needs, CFD models require much greater scrutiny than do zone models.

It is much easier to assess the sensitivity of different parameters with zone models, because they generally run much faster and the output is much easier to interpret. Prediction of fire growth and spread has a large number of variables associated with it.

Purpose-built models (also known as stand-alone models) are similar to zone models in their ease of use. However, purpose-built models do not provide a comprehensive model. Instead, they predict the value of one variable of interest. For example, such a model can predict the conditions of a ceiling jet at a specified location under a ceiling, but a zone model would "transport" those conditions throughout the enclosure.

Purpose-built models might or might not be manifested as computer software. Models that are not in the form of software are referred to as hand calculations. Purpose-built models are, therefore, simple enough that the data management capabilities of a computer are not necessary. Many of the calculations are found in the *SFPE Handbook of Fire Protection Engineering*.

*Types of Evacuation Models.* Four categories of evacuation models can be considered: single-parameter estimation methods, movement models, behavioral simulation models, and tenability models.

*Single-parameter estimation methods* are generally used for simple estimates of movement time. They are usually based on equations derived from observations of movement in nonemergency situations. They can be hand calculations or simple computer models. Examples include calculation methods for flow times based on widths of exit paths and travel times based on travel distances. Sources for these methods include the *SFPE Handbook of Fire Protection Engineering* and the NFPA *Fire Protection Handbook*.

*Movement models* generally handle large numbers of people in a network flow similar to water in pipes or ball bearings in chutes. They tend to optimize occupant behavior, resulting in predicted evacuation times that can be unrealistic and far from conservative. However, they can be useful in an overall assessment of a design, especially in early evaluation stages where an unacceptable result with this sort of model indicates that the design has failed to achieve the life safety objectives.

*Behavioral simulation models* take into consideration more of the variables related to occupant movement and behavior. Occupants are treated as individuals and can have unique characteristics assigned to them, allowing a more realistic simulation of the design under consideration. However, given the limited availability of data for the development of these models, for their verification by their authors, or for input when using them, their predictive reliability is questionable.

*Tenability Models.* In general, tenability models will be needed only to automate calculations for the time-of-exposure effect equations referenced in A.5.2.2.

*Other Models.* Models can be used to describe combustion (as noted, most fire models only characterize fire effects), automatic system performance, and other elements of the calculation. There are few models in common use for these purposes, so they are not further described here.

*Sources of Models.* A compendia of computer fire models are found in the *SFPE Computer Software Directory* and in Olenick, S. and Carpenter, D., "An Updated International Survey of Computer Models for Fire and Smoke," *Journal of Fire Protection Engineering*, 13, 2, 2003, pp. 87–110. Within these references are models that were developed by the Building Fire Research Laboratory of the National Institute of Standards and Technology, which can be downloaded from the Internet at http://www.bfrl.nist.gov/864/fmabs.html. Evacuation models are discussed in the *SFPE Handbook of Fire Protection Engineering* and the NFPA *Fire Protection Handbook*.

*Validation.* Models undergo limited validation. Most can be considered demonstrated only for the experimental results they were based on or the limited set of scenarios to which the model developers compared the model's output, or a combination of both.

The Society of Fire Protection Engineers has a task group that independently evaluates computer models. In January 1998, they finished their first evaluation and had chosen a second model for evaluation. Until more models can be independently evaluated, the model user has to rely on the available documentation and previous experience for guidance regarding the appropriate use of a given model.

The design professional should present the proposal, and the authority having jurisdiction, when deciding whether to approve a proposal, should consider the strength of the evidence presented for the validity, accuracy, relevance, and precision of the proposed methods. An element in establishing the strength of scientific evidence is the extent of external review and acceptance of the evidence by peers of the authors of that evidence.

Models have limitations. Most are not user friendly, and experienced users are able to construct more reasonable models and better interpret output than are novices. For these reasons, the third-party review and equivalency provisions of 5.1.4 and 5.3.3 are provided. The intent is not to discourage the use of models, only to indicate that they should be used with caution by those who are well versed in their nuances.

*Input Data.* The first step in using a model is to develop the input data. The heat release rate curve specified by the user is the driving force of a fire effects model. If this curve is incorrectly defined, the subsequent results are not usable. In addition to the smoldering and growth phases that will be specified as part of the scenario definition, two additional phases are needed to complete the input heat release rate curve — steady burning and burnout.

Steady burning is characterized by its duration, which is a function of the total amount of fuel available to be burned. In determining the duration of this phase, the designer needs to consider how much fuel has been assumed to be consumed in the smoldering and growth phases and how much is assumed to be consumed in the burnout phase that follows. Depending on the assumptions made regarding the amount of fuel consumed during burnout, the time at which this phase starts is likely to be easy to determine.

The preceding discussion assumes that the burning objects are solid (e.g., tables and chairs). If liquid or gaseous fuels are involved, the shape of the curve will be different. For example, smoldering is not relevant for burning liquids or gases, and the growth period is very short, typically measured in seconds. Peak heat release rate can depend primarily on the rate of release, on the leak rate (gases and liquid sprays), or on the extent of spill (pooled liquids). The steady burning phase is once again dependent on the amount of fuel available to burn. Like the growth phase, the burnout phase is typically short (e.g., closing a valve), although it is conceivable that longer times might be appropriate, depending on the extinguishment scenario.

Material properties are usually needed for all fuel items, both initial and secondary, and the enclosure surfaces of involved rooms or spaces.

For all fires of consequence, it is reasonable to assume that the fire receives adequate ventilation. If there is insufficient oxygen, the fire will not be sustained. An overabundance of oxygen is only a concern in special cases (e.g., hermetically sealed spaces) where a fire might not occur due to dilution of the fuel (i.e., a flammable mixture is not produced). Therefore, given that the scenarios of interest will occur in nonhermetically sealed enclosures, it is reasonable to assume that adequate ventilation is available and that, if a fire starts, it will continue to burn until it either runs out of fuel or is extinguished by other means. The only variable that might need to be assumed is the total vent width.

Maximum fire extent is affected by two geometric aspects: burning object proximity to walls and overall enclosure dimensions.

The room dimensions affect the time required for a room to flashover. For a given amount and type of fuel, under the same ventilation conditions, a small room will flashover before a large room. In a large room with a small amount of fuel, a fire will behave as if it is burning outside — that is, adequate oxygen for burning and no concentration of heat exist. If the fuel package is unchanged but the dimensions of the room are decreased, the room will begin to have an affect on the fire, assuming adequate ventilation. The presence of the relatively smaller enclosure results in the buildup of a hot layer of smoke and other products of combustion under the ceiling. This buildup, in turn, feeds more heat back to the seat of the fire, which results in an increase in the pyrolysis rate of the fuel and, thus, increases the amount of heat energy released by the fire. The room enclosure surfaces themselves also contribute to this radiation feedback effect.

Probabilistic data are expressed as either a frequency (units of inverse time) or a probability (unitless, but applicable to a stated period of time). An example of the former is the expected number of failures per year, and the range of the latter is between zero and one, inclusive. Probabilities can be either objective or subjective. Subjective probabilities express a degree of belief that an event will occur. Objective probabilities are based on historical data and can be expressed as a reliability of an item, such as a component or a system.

**A.5.6.3.3** Procedures used to develop required input data need to preserve the intended conservatism of all scenarios and assumptions. Conservatism is only one means to address the uncertainty inherent in calculations and does not eliminate the need to consider safety factors, sensitivity analysis, and other methods of dealing with uncertainty. The *SFPE Engineering Guide to Performance-Based Fire Protection Analysis and Design of Buildings* outlines a process for identifying and treating uncertainty.

**A.5.6.4** An assessment method translates input data, which might include test specifications, parameters, or variables for modeling, or other data, into output data, which are measured against the performance criteria. Computer fire models should be evaluated for their predictive capability in accordance with ASTM E 1355, *Standard Guide for Evaluating the Predictive Capability of Deterministic Fire Models*.

**A.5.7** The assessment of precision required in 5.8.2 will require a sensitivity and uncertainty analysis, which can be translated into safety factors.

*Sensitivity Analysis.* The first run a model user makes should be labeled as the base case, using the nominal values of the various input parameters. However, the model user should not rely on a single run as the basis for any performance-based fire safety system design. Ideally, each variable or parameter that the model user made to develop the nominal input data should have multiple runs associated with it, as should combinations of key variables and parameters. Thus, a sensitivity analysis should be conducted that provides the model user with data that indicate how the effects of a real fire might vary and how the response of the proposed fire safety design might also vary.

The interpretation of a model's predictions can be a difficult exercise if the model user does not have knowledge of fire dynamics or human behavior.

*Reasonableness Check.* The model user should first try to determine whether the predictions actually make sense; that is, whether they do not upset intuition or preconceived expectations. Most likely, if the results do not pass this test, an input error has been committed.

Sometimes the predictions appear to be reasonable but are, in fact, incorrect. For example, a model can predict higher temperatures farther from the fire than closer to it. The values themselves might be reasonable; for example, they are not hotter than the fire, but they do not "flow" down the energy as expected.

A margin of safety can be developed using the results of the sensitivity analysis in conjunction with the performance criteria to provide the possible range of time during which a condition is estimated to occur.

Safety factors and margin of safety are two concepts used to quantify the amount of uncertainty in engineering analyses. Safety factors are used to provide a margin of safety and represent, or address, the gap in knowledge between the theoretically perfect model — reality — and the engineering models that can only partially represent reality.

Safety factors can be applied either to the predicted level of a physical condition or to the time at which the condition is predicted to occur. Thus, a physical or a temporal safety factor, or both, can be applied to any predicted condition. A predicted condition (i.e., a parameter's value) and the time at which it occurs are best represented as distributions. Ideally, a computer fire model predicts the expected or nominal value of the distribution. Safety factors are intended to represent the spread of the distributions.

Given the uncertainty associated with data acquisition and reduction, and the limitations of computer modeling, any condition predicted by a computer model can be thought of as an expected or nominal value within a broader range. For example, an upper layer temperature of 1110°F (600°C) is predicted at a given time. If the modeled scenario is then tested (i.e., full-scale experiment based on the computer model's input data), the actual temperature at that given time could be 1185°F or 1085°F (640°C or 585°C). Therefore, the temperature should be reported as 1110°F + 75°F/−25°F (600°C + 40°C/−15°C) or as a range of 1085°F to 1185°F (585°C to 640°C).

Ideally, predictions are reported as a nominal value, a percentage, or an absolute value. As an example, an upper layer temperature prediction could be reported as "1110°F (600°C), 55°F (30°C)," or "1110°F (600°C), 5 percent." In this case, the physical safety factor is 0.05 (i.e., the amount by which the nominal value should be degraded and enhanced). Given the state-of-the-art of computer fire modeling, this is a very low safety factor. Physical safety factors tend to be on the order of tens of percent. A safety factor of 50 percent is not unheard of.

Part of the problem in establishing safety factors is that it is difficult to state the percentage or range that is appropriate. These values can be obtained when the computer model predictions are compared to test data. However, using computer fire models in a design mode does not facilitate this comparison, due to the following:

(1) The room being analyzed has not been built yet.
(2) Test scenarios do not necessarily depict the intended design.

A sensitivity analysis should be performed, based on the assumptions that affect the condition of interest. A base case that uses all nominal values for input parameters should be developed. The input parameters should be varied over reasonable ranges, and the variation in predicted output should be noted. This output variation can then become the basis for physical safety factors.

The temporal safety factor addresses the issue of when a condition is predicted and is a function of the rate at which processes are expected to occur. If a condition is predicted to occur 2 minutes after the start of the fire, this prediction can be used as a nominal value. A process similar to that already described for physical safety factors can also be employed to develop temporal safety factors. In such a case, however, the rates (e.g., rates of heat release and toxic product generation) will be varied instead of absolute values (e.g., material properties).

The margin of safety can be thought of as a reflection of societal values and can be imposed by the authority having jurisdiction for that purpose. Because the time for which a condition is predicted will most likely be the focus of the authority having jurisdiction (e.g., the model predicts that occupants will have 5 minutes to safely evacuate), the margin of safety will be characterized by temporal aspects and tacitly applied to the physical margin of safety.

Escaping the harmful effects of fire (or mitigating them) is, effectively, a race against time. When assessing fire safety system designs based on computer model predictions, the choice of an acceptable time is important. When an authority having jurisdiction is faced with the predicted time of untenability, a decision needs to be made regarding whether sufficient time is available to ensure the safety of building occupants. The authority having jurisdiction is assessing the margin of safety. Is there sufficient time to get everyone out safely? If the authority having jurisdiction feels that the predicted egress time is too close to the time of untenability, the authority having jurisdiction can impose an additional period of time that the designer will have to incorporate into the system design. In other words, the authority having jurisdiction can impose a greater margin of safety than that originally proposed by the designer.

**A.5.8.1** The *SFPE Engineering Guide to Performance-Based Fire Protection Analysis and Design of Buildings* describes the documentation that should be provided for a performance-based design.

Proper documentation of a performance-based design is critical to design acceptance and construction. Proper documentation will also ensure that all parties involved understand the factors necessary for the implementation, maintenance, and continuity of the fire protection design. If attention to details is maintained in the documentation, there should be little dispute during approval, construction, start-up, and use.

Poor documentation could result in rejection of an otherwise good design, poor implementation of the design, inadequate system maintenance and reliability, and an incomplete record for future changes or for testing the design forensically.

**A.5.8.2** The sources, methodologies, and data used in performance-based designs should be based on technical references that are widely accepted and used by the appropriate professions and professional groups. This acceptance is often based on documents that are developed, reviewed, and validated under one of the following processes:

(1) Standards developed under an open consensus process conducted by recognized professional societies, codes or standards organizations, or governmental bodies
(2) Technical references that are subject to a peer review process and published in widely recognized peer-reviewed journals, conference reports, or other publications
(3) Resource publications, such as the *SFPE Handbook of Fire Protection Engineering*, which are widely recognized technical sources of information

The following factors are helpful in determining the acceptability of the individual method or source:

(1) Extent of general acceptance in the relevant professional community, including peer-reviewed publication, widespread citation in the technical literature, and adoption by or within a consensus document
(2) Extent of documentation of the method, including the analytical method itself, assumptions, scope, limitations, data sources, and data reduction methods
(3) Extent of validation and analysis of uncertainties, including comparison of the overall method with experimental data to estimate error rates, as well as analysis of the uncertainties of input data, uncertainties and limitations in the analytical method, and uncertainties in the associated performance criteria
(4) Extent to which the method is based on sound scientific principles
(5) Extent to which the proposed application is within the stated scope and limitations of the supporting information, including the range of applicability for which there is documented validation, and considering factors such as spatial dimensions, occupant characteristics, and ambient conditions, which can limit valid applications

In many cases, a method will be built from, and will include, numerous component analyses. Such component analyses should be evaluated using the same acceptability factors that are applied to the overall method, as outlined in items (1) through (5).

A method to address a specific fire safety issue, within documented limitations or validation regimes, might not exist. In such a case, sources and calculation methods can be used outside of their limitations, provided that the design team recognizes the limitations and addresses the resulting implications.

The technical references and methodologies to be used in a performance-based design should be closely evaluated by the design team and the authority having jurisdiction, and possibly by a third-party reviewer. The strength of the technical justification should be judged using criteria in items (1) through (5). This justification can be strengthened by the presence of data obtained from fire testing.

**A.5.8.11** Documentation for modeling should conform to ASTM E 1472, *Standard Guide for Documenting Computer Software for Fire Models*, although most, if not all, models were originally developed before this standard was promulgated. Information regarding the use of the model DETACT-QS can be found in the *SFPE Engineering Guide–the Evaluation of the Computer Fire Model DETACT-QS*.

**A.6.1.2.1 Assembly Occupancy.** Assembly occupancies might include the following:

(1) Armories
(2) Assembly halls
(3) Auditoriums
(4) Bowling lanes
(5) Club rooms
(6) College and university classrooms, 50 persons and over
(7) Conference rooms
(8) Courtrooms
(9) Dance halls
(10) Drinking establishments
(11) Exhibition halls
(12) Gymnasiums
(13) Libraries
(14) Mortuary chapels
(15) Motion picture theaters
(16) Museums
(17) Passenger stations and terminals of air, surface, underground, and marine public transportation facilities
(18) Places of religious worship
(19) Pool rooms
(20) Recreation piers
(21) Restaurants
(22) Skating rinks
(23) Special amusement buildings, regardless of occupant load
(24) Theaters

Assembly occupancies are characterized by the presence or potential presence of crowds with attendant panic hazard in case of fire or other emergency. They are generally or occasionally open to the public, and the occupants, who are present voluntarily, are not ordinarily subject to discipline or control. Such buildings are ordinarily not used for sleeping purposes. Special conference rooms, snack areas, and other areas incidental to, and under the control of, the management of other occupancies, such as offices, fall under the 50-person limitation.

Restaurants and drinking establishments with an occupant load of fewer than 50 persons should be classified as mercantile occupancies.

Occupancy of any room or space for assembly purposes by fewer than 50 persons in another occupancy, and incidental to such other occupancy, should be classified as part of the other occupancy and should be subject to the provisions applicable thereto.

For special amusement buildings, see 12.4.7 and 13.4.7.

**A.6.1.3.1 Educational Occupancy.** Educational occupancies include the following:

(1) Academies
(2) Kindergartens
(3) Schools

An educational occupancy is distinguished from an assembly occupancy in that the same occupants are regularly present.

**A.6.1.4.1 Day-Care Occupancy.** Day-care occupancies include the following:

(1) Adult day-care occupancies, except where part of a health care occupancy
(2) Child day-care occupancies
(3) Day-care homes
(4) Kindergarten classes that are incidental to a child day-care occupancy
(5) Nursery schools

In areas where public schools offer only half-day kindergarten programs, many child day-care occupancies offer state-approved kindergarten classes for children who need full-day care. As these classes are normally incidental to the day-care occupancy, the requirements of the day-care occupancy should be followed.

**A.6.1.5.1 Health Care Occupancy.** Health care occupancies include the following:

(1) Hospitals
(2) Limited care facilities
(3) Nursing homes

Occupants of health care occupancies typically have physical or mental illness, disease, or infirmity. They also include infants, convalescents, or infirm aged persons.

**A.6.1.6.1 Ambulatory Health Care Occupancy.** It is not the intent that occupants be considered to be incapable of self-preservation just because they are in a wheelchair or use assistive walking devices, such as a cane, a walker, or crutches. Rather, it is the intent to address emergency care centers that receive patients who have been rendered incapable of self-preservation due to the emergency, such as being rendered unconscious as a result of an accident or being unable to move due to sudden illness.

**A.6.1.7.1 Detention and Correctional Occupancy.** Detention and correctional occupancies include the following:

(1) Adult and juvenile substance abuse centers
(2) Adult and juvenile work camps
(3) Adult community residential centers
(4) Adult correctional institutions
(5) Adult local detention facilities
(6) Juvenile community residential centers
(7) Juvenile detention facilities
(8) Juvenile training schools

See A.22.1.1.1.6 and A.23.1.1.1.6.

**A.6.1.7.2** Chapters 22 and 23 address the residential housing areas of the detention and correctional occupancy as defined in 3.3.188.5. Examples of uses, other than residential housing, include gymnasiums or industries.

**A.6.1.8.1.1 One- and Two-Family Dwelling Unit.** The application statement of 24.1.1.1 limits each dwelling unit to being "occupied by members of a single family with not more than three outsiders." The *Code* does not define the term *family*. The definition of family is subject to federal, state, and local regulations and might not be restricted to a person or a couple (two people) and their children. The following examples aid in differentiating between a single-family dwelling and a lodging or rooming house:

(1) An individual or a couple (two people) who rent a house from a landlord and then sublease space for up to three individuals should be considered a family renting to a maximum of three outsiders, and the house should be regulated as a single-family dwelling in accordance with Chapter 24.

(2) A house rented from a landlord by an individual or a couple (two people) in which space is subleased to 4 or more individuals, but not more than 16, should be considered and regulated as a lodging or rooming house in accordance with Chapter 26.

(3) A residential building that is occupied by 4 or more individuals, but not more than 16, each renting from a landlord, without separate cooking facilities, should be considered and regulated as a lodging or rooming house in accordance with Chapter 26.

**A.6.1.8.1.3  Hotel.** So-called apartment hotels should be classified as hotels, because they are potentially subject to the same transient occupancy as hotels. Transients are those who occupy accommodations for less than 30 days.

**A.6.1.8.1.4  Dormitory.** Rooms within dormitories intended for the use of individuals for combined living and sleeping purposes are guest rooms or guest suites. Examples of dormitories include college dormitories, fraternity and sorority houses, and military barracks.

**A.6.1.9.1  Residential Board and Care Occupancy.** The following are examples of facilities classified as residential board and care occupancies:

(1) Group housing arrangement for physically or mentally handicapped persons who normally attend school in the community, attend worship in the community, or otherwise use community facilities

(2) Group housing arrangement for physically or mentally handicapped persons who are undergoing training in preparation for independent living, for paid employment, or for other normal community activities

(3) Group housing arrangement for the elderly that provides personal care services but that does not provide nursing care

(4) Facilities for social rehabilitation, alcoholism, drug abuse, or mental health problems that contain a group housing arrangement and that provide personal care services but do not provide acute care

(5) Assisted living facilities

(6) Other group housing arrangements that provide personal care services but not nursing care

**A.6.1.10.1  Mercantile Occupancy.** Mercantile occupancies include the following:

(1) Auction rooms
(2) Department stores
(3) Drugstores
(4) Restaurants with fewer than 50 persons
(5) Shopping centers
(6) Supermarkets

Office, storage, and service facilities incidental to the sale of merchandise and located in the same building should be considered part of the mercantile occupancy classification.

**A.6.1.11.1  Business Occupancy.** Business occupancies include the following:

(1) Air traffic control towers (ATCTs)
(2) City halls
(3) College and university instructional buildings, classrooms under 50 persons, and instructional laboratories
(4) Courthouses
(5) Dentists' offices
(6) Doctors' offices
(7) General offices

(8) Outpatient clinics (ambulatory)
(9) Town halls

Doctors' and dentists' offices are included, unless of such character as to be classified as ambulatory health care occupancies. *(See 3.3.188.1.)*

Birth centers should be classified as business occupancies if they are occupied by fewer than four patients, not including infants, at any one time; do not provide sleeping facilities for four or more occupants; and do not provide treatment procedures that render four or more patients, not including infants, incapable of self-preservation at any one time. For birth centers occupied by patients not meeting these parameters, see Chapter 18 or Chapter 19, as appropriate.

Service facilities common to city office buildings, such as newsstands, lunch counters serving fewer than 50 persons, barber shops, and beauty parlors are included in the business occupancy group.

City halls, town halls, and courthouses are included in this occupancy group, insofar as their principal function is the transaction of public business and the keeping of books and records. Insofar as they are used for assembly purposes, they are classified as assembly occupancies.

**A.6.1.12.1  Industrial Occupancy.** Industrial occupancies include the following:

(1) Drycleaning plants
(2) Factories of all kinds
(3) Food processing plants
(4) Gas plants
(5) Hangars (for servicing/maintenance)
(6) Laundries
(7) Power plants
(8) Pumping stations
(9) Refineries
(10) Sawmills
(11) Telephone exchanges

In evaluating the appropriate classification of laboratories, the authority having jurisdiction should treat each case individually, based on the extent and nature of the associated hazards. Some laboratories are classified as occupancies other than industrial; for example, a physical therapy laboratory or a computer laboratory.

**A.6.1.13.1  Storage Occupancy.** Storage occupancies include the following:

(1) Barns
(2) Bulk oil storage
(3) Cold storage
(4) Freight terminals
(5) Grain elevators
(6) Hangars (for storage only)
(7) Parking structures
(8) Truck and marine terminals
(9) Warehouses

Storage occupancies are characterized by the presence of relatively small numbers of persons in proportion to the area.

**A.6.1.14.1.3** Examples of uses that might be incidental to another occupancy include the following:

(1) Newsstand (mercantile) in an office building
(2) Giftshop (mercantile) in a hotel
(3) Small storage area (storage) in any occupancy

(4) Minor office space (business) in any occupancy

(5) Maintenance area (industrial) in any occupancy

**A.6.1.14.1.3(2)** Examples of uses that have occupant loads below the occupancy classification threshold levels include the following:

(1) Assembly use with fewer than 50 persons within a business occupancy

(2) Educational use with fewer than 6 persons within an apartment building.

**A.6.1.14.3.2** For example, a common path of travel that occurs wholly in a business tenant space, in a multiple occupancy building containing assembly and business occupancies, should not have to meet the assembly occupancy common path of travel limitation.

**A.6.2.1.3** Under the provision of 6.2.1.3, any violation of the requirements of Chapters 11 through 42 for separation or protection of hazardous operation or storage would inherently involve violation of the other sections of the *Code*, unless additional egress facilities appropriate to high hazard contents were provided.

**A.6.2.2.1** These classifications do not apply to the application of sprinkler protection classifications. *See* NFPA 13, *Standard for the Installation of Sprinkler Systems.* Depending on the use of the space, the area might require special hazard protection in accordance with Section 8.7.

**A.6.2.2.2** Chapter 42 recognizes storage of noncombustible materials as low hazard. In other occupancies, it is assumed that, even where the actual contents hazard is normally low, there is sufficient likelihood that some combustible materials or hazardous operations will be introduced in connection with building repair or maintenance, or some psychological factor might create conditions conducive to panic, so that the egress facilities cannot safely be reduced below those specified for ordinary hazard contents.

**A.6.2.2.3** Ordinary hazard classification represents the conditions found in most buildings and is the basis for the general requirements of this *Code*.

The fear of poisonous fumes or explosions is necessarily a relative matter to be determined on a judgment basis. All smoke contains some toxic fire gases but, under conditions of ordinary hazard, there should be no unduly dangerous exposure during the period necessary to escape from the fire area, assuming there are proper exits.

**A.6.2.2.4** High hazard contents include occupancies where flammable liquids are handled or used or are stored under conditions involving possible release of flammable vapors; where grain dust, wood flour or plastic dust, aluminum or magnesium dust, or other explosive dusts are produced; where hazardous chemicals or explosives are manufactured, stored, or handled; where materials are processed or handled under conditions producing flammable flyings; and other situations of similar hazard.

Chapters 40 and 42 include detailed provisions on high hazard contents.

**A.7.1.1** An installation of supplemental evacuation equipment is not recognized as a means of egress. Consequently, such equipment does not satisfy any requirement for minimum number of, capacity of, travel distance to, or remoteness of, means of egress.

**A.7.1.3.2.1(1)** In existing buildings, existing walls in good repair and consisting of lath and plaster, gypsum wallboard, or masonry units can usually provide satisfactory protection for the purposes of this requirement where a 1-hour fire resistance rating is required. Further evaluation might be needed where a 2-hour fire resistance rating is required. Additional guidelines can be found in Appendix D of NFPA 914, *Code for Fire Protection of Historic Structures*, and in the *SFPE Handbook of Fire Protection Engineering*.

**A.7.1.3.2.1(3)** In existing buildings, existing walls in good repair and consisting of lath and plaster, gypsum wallboard, or masonry units can usually provide satisfactory protection for the purposes of this requirement where a 1-hour fire resistance rating is required. Further evaluation might be needed where a 2-hour fire resistance rating is required. Additional guidelines can be found in Appendix D of NFPA 914, *Code for Fire Protection of Historic Structures*, and in the *SFPE Handbook of Fire Protection Engineering*.

**A.7.1.3.2.1(6)** It is not the intent to require the structural elements supporting outside stairs, or structural elements that penetrate within exterior walls or any other wall not required to have a fire resistance rating, to be protected by fire resistance–rated construction.

**A.7.1.3.2.1(9)** Means of egress from the level of exit discharge is permitted to pass through an exit stair enclosure or exit passageway serving other floors. Doors for convenience purposes and unrelated to egress also are permitted to provide access to and from exit stair enclosures and exit passageways, provided that such doors are from corridors or normally occupied spaces. It is also the intent of this provision to prohibit exit enclosure windows, other than approved vision panels in doors, that are not mounted in an exterior wall.

**A.7.1.3.2.1(10)(b)** Penetrations for electrical wiring are permitted where the wiring serves equipment permitted by the authority having jurisdiction to be located within the exit enclosure, such as security systems, public address systems, and fire department emergency communications devices.

**A.7.1.3.2.3** This provision prohibits the use of exit enclosures for storage or for installation of equipment not necessary for safety. Occupancy is prohibited other than for egress, refuge, and access. The intent is that the exit enclosure essentially be "sterile" with respect to fire safety hazards.

**A.7.1.4.1** See Chapters 12 through 42 for further limitations on interior wall and ceiling finish.

**A.7.1.4.2** See Chapters 12 through 42 for further limitations on interior floor finish.

**A.7.1.5** For the purpose of this requirement, projections include devices such as lighting equipment, emergency signaling equipment, environmental controls and equipment, security devices, signs, and decorations that are typically limited in area.

**A.7.1.6.4** The foreseeable slip conditions are those that are likely to be present at the location of the walking surface during the use of the building or area. A foreseeable condition of a swimming pool deck is that it is likely to be wet.

Regarding the slip resistance of treads, it should be recognized that, when walking up or down stairs, a person's foot exerts a smaller horizontal force against treads than is exerted when walking on level floors. Therefore, materials used for floors that are acceptable as slip resistant (as described by ASTM F 1637, *Standard Practice for Safe Walking Surfaces*) provide adequate slip resistance where used for stair treads. Such slip resistance includes the important leading edges of treads, the part of the tread that the foot first contacts during descent, which is the most critical direction of travel. If stair treads are wet, there is an increased danger of slipping, just as there is an increased danger of slipping on wet floors of similar materials. A small wash or drainage slope on exterior stair treads is, therefore, recommended to shed water. *(See Templer, J. A., The Staircase: Studies of Hazards, Falls, and Safer Design, Cambridge, MA: MIT Press, 1992.)*

**A.7.1.7.2** Aside from the problems created for persons who are mobility impaired, small changes of elevations in floors are best avoided because of the increased occurrence of missteps where the presence of single steps, a series of steps, or a ramp is not readily apparent. Although small changes of elevation pose significant fall risks in the case of individual movement, they are even more undesirable where crowds traverse the area.

A contrasting marking stripe on each stepping surface can be helpful at the nosing or leading edge so that the location of each step is readily apparent, especially when viewed in descent. Such stripes should be not less than 1 in. (25 mm), but should not exceed 2 in. (51 mm), in width. Other methods could include a relatively higher level of lighting, contrasting colors, contrasting textures, highly prominent handrails, warning signs, a combination thereof, or other similar means. The construction or application of marking stripes should be such that slip resistance is consistent over the walking surface and no tripping hazard is created *(see also A.7.2.2.3.3.2)*. Depending on the distractions of the surroundings, the familiarity of users with a particular small change of level, and especially the number of people that might be in a group traversing the change of level (thereby reducing visibility of the level changes), a strong argument can be made for the elimination of steps and ramps that might pose a risk of missteps.

**A.7.1.8** Elements of the means of egress that might require protection with guards include stairs, landings, escalators, moving walks, balconies, corridors, passageways, floor or roof openings, ramps, aisles, porches, and mezzanines.

Escalators and moving walks, other than previously approved existing escalators and moving walks, are prohibited from serving as components of the required means of egress. Building occupants using the escalator at the time of fire or similar emergency must traverse some portion of the escalator to gain access to a required egress route. For those building occupants using the escalator, such travel along the escalator is part of their means of egress. The requirement that guards be provided at the open side of means of egress that exceed 30 in. (760 mm) above the floor or grade below is meant to be applied to escalators and moving walks.

**A.7.1.10.1** A proper means of egress allows unobstructed travel at all times. Any type of barrier including, but not limited to, the accumulations of snow and ice in those climates subject to such accumulations is an impediment to free movement in the means of egress. Another example of an obstruction or impediment to full instant use of means of egress is any security device or system that emits any medium that could obscure a means of egress. It is, however, recognized that obstructions occur on a short-duration basis. In these instances, awareness training should be provided to ensure that blockages are kept to a minimum and procedures are established for the control and monitoring of the area affected.

**A.7.2.1.2.1** Figure A.7.2.1.2.1(a) and Figure A.7.2.1.2.1(b) illustrate the method of measuring clear width for doors.

In cases where a chapter requires a door width, for example, of not less than 36 in. (915 mm), this requirement can be met by a door leaf of the minimum specified width if the term *clear width* does not appear as part of the minimum width requirement. A pair of cross-corridor doors subject to such a requirement would be judged under the following criteria:

(1) Each door leaf is required to be not less than 36 in. (915 mm) in width.
(2) The pair of doors is required to provide sufficient, clear, unobstructed width (which will be less than the door leaf width measurement) to handle its assigned occupant load, based on a calculation using the appropriate egress capacity factor in Table 7.3.3.1.

Where swinging doors do not open at least 90 degrees, the clear width of the doorway should be measured between the face of the door and the stop.

It is not the intent to regulate projections above the 6 ft 8 in. (2030 mm) height.

**A.7.2.1.2.2** Figure A.7.2.1.2.2(a) and Figure A.7.2.1.2.2(b) illustrate the method of measuring egress capacity width for purposes of calculating door egress capacity.

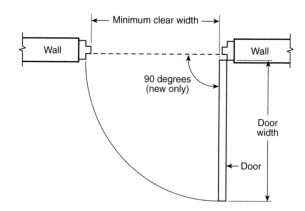

**FIGURE A.7.2.1.2.1(a)  Minimum Clear Width.**

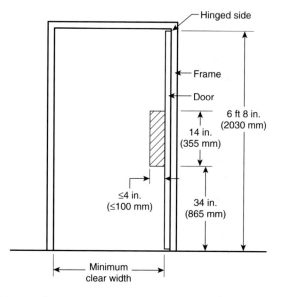

**FIGURE A.7.2.1.2.1(b)  Minimum Clear Width with Permitted Obstructions.**

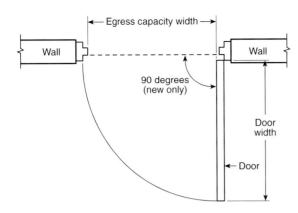

**FIGURE A.7.2.1.2.2(a)   Door Width — Egress Capacity.**

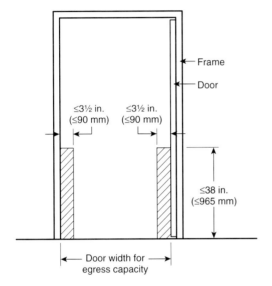

**FIGURE A.7.2.1.2.2(b)   Door Width — Egress Capacity with Permitted Obstructions.**

**A.7.2.1.2.3.2(9)**   The relative egress carrying capacity of door openings and stairs is based on the two-to-three ratio used in Table 7.3.3.1 to help balance the capacity of various egress elements and ensure that downstream egress facilities do not form a bottleneck or constriction to flow. For example, a stairway with a nominal width of 56 in. (1420 mm) should be served by an exit discharge door with a minimum width opening of 37 in. (940 mm) if only one discharge door is provided. It might be advantageous for two discharge doors to serve such a stairway, each with a more typical clear opening width of 32 in. (810 mm). This would facilitate access, into the exit, of fire fighters and other emergency responders without causing undue interference to evacuees attempting to transition from the stair to the exit discharge door.

**A.7.2.1.4.1**   Where doors are subject to two-way traffic, or where their opening can interfere with pedestrian traffic, an appropriately located vision panel can reduce the chance of accidents.

Swinging doors in horizontal- or vertical-rolling partitions should be permitted in a means of egress where the following criteria are met:

(1)  The door or doors comply with 7.2.1.4.
(2)  The partition in which the doors are mounted complies with the applicable fire protection rating and closes upon smoke detection or power failure at a speed not exceeding 9 in./s (230 mm/s) and not less than 6 in./s (150 mm/s).
(3)  The doors mounted in the partition are self-closing or automatic-closing in accordance with 7.2.1.8.

**A.7.2.1.4.3.1**   The requirements of 7.2.1.4.3 are not intended to apply to the swing of cross-corridor doors, such as smoke barrier doors and horizontal exits.

**A.7.2.1.5.2**   Some fire door assemblies are listed for use with fire pins or fusible links that render the door leaf release inoperative upon exposure to elevated temperature during a fire. The door leaf release mechanism is made inoperative where conditions in the vicinity of the door opening become untenable for human occupancy, and such door opening no longer provides a viable egress path.

**A.7.2.1.5.6(5)**   Separate power supplies might be provided to the electronic lock and the releasing hardware. In this case, it is critical that the lock be arranged to release upon loss of power to the releasing hardware to ensure occupants can egress in the event of a power failure.

**A.7.2.1.5.8**   It is intended that the re-entry provisions apply only to enclosed exit stairs, not to outside stairs. This arrangement makes it possible to leave the stairway at such floor if the fire renders the lower part of the stair unusable during egress or if the occupants seek refuge on another floor.

**A.7.2.1.5.10**   Examples of devices that might be arranged to release latches include knobs, levers, and bars. This requirement is permitted to be satisfied by the use of conventional types of hardware, whereby the door is released by turning a lever, knob, or handle or by pushing against a bar, but not by unfamiliar methods of operation, such as a blow to break glass. It is also within the intent of this requirement that switches integral to traditional doorknobs, lever handles, or bars, and that interrupt the power supply to an electromagnetic lock, be permitted, provided that they are affixed to the door leaf. The operating devices should be capable of being operated with one hand and should not require tight grasping, tight pinching, or twisting of the wrist to operate.

**A.7.2.1.5.10.3**   Examples of devices that, when used with a latch, can be arranged to require not more than one additional releasing operation include night latches, dead bolts, and security chains.

**A.7.2.1.5.12**   Examples of devices prohibited by this requirement include locks, padlocks, hasps, bars, chains, or combinations thereof.

**A.7.2.1.6**   None of the special locking arrangements addressed in 7.2.1.6 are intended to allow *credentialed egress, request to exit,* or similar provisions, where an occupant cannot leave the building without swiping a card through a reader. Where such an arrangement is desired to keep track of occupants, the swiping of cards needs to be procedural but not necessary for releasing the door lock or latch. Free egress needs to be available at all times. Another option to free egress is the use of a delayed-egress locking system.

**A.7.2.1.6.1.1(3)**   It is not the intent to require a direct physical or electrical connection between the door release device and the lock. It is the intent to allow door movement initiated by operating the door release device required in 7.2.1.5.10 as one option to initiate the irreversible process.

Several factors need to be considered in approving an increase in delay time from 15 seconds to 30 seconds. Some of the factors include occupancy, occupant density, ceiling height, fire hazards present, fire protection features provided, and the location of the delayed-egress locks. An example of a location where the increase in delay time might not be approved is at an exit stair discharge door.

**A.7.2.1.6.1.1(4)** In the event that the authority having jurisdiction has permitted increased operation time, the sign should reflect the appropriate time.

**A.7.2.1.6.2** It is not the intent to require doors that restrict access but that comply with 7.2.1.5.10 to comply with the access-controlled egress door provisions of 7.2.1.6.2. The term *access-controlled* was chosen when the requirements of 7.2.1.6.2 were first added to the *Code* to describe the function in which a door is electronically locked from the inside in a manner that restricts egress. It is not the *Code's* intent to prohibit methods of securing the door in a locked position from the outside with access control products, provided that the egress requirements of 7.2.1.6.2 are met.

**A.7.2.1.6.3(14)** It is not the intent to prohibit elevator lobby doors from being equipped with card access systems for gaining access, for example, to tenant spaces. It is the access-controlled egress door system described in 7.2.1.6.2 that is prohibited from being installed on the same door as the lock addressed by 7.2.1.6.3.

**A.7.2.1.8.1** Examples of doors designed to normally be kept closed include those to a stair enclosure or horizontal exit.

**A.7.2.1.9** Horizontal-sliding doors installed in accordance with 7.2.1.14 should not be considered powered doors subject to the provisions of 7.2.1.9.

Powered doors are divided into two categories — power assisted and power operated. Power-assisted doors that conform to ANSI/BHMA A156.19, *American National Standard for Power Assist and Low Energy Power Operated Doors*, use limited power to operate the door. They require fewer safeguards as compared to full power–operated doors. These door operators are for swinging doors only. Power-operated doors that conform to ANSI/BHMA A156.10, *American National Standard for Power Operated Pedestrian Doors*, require more power to operate the door and require additional safeguards to provide protection against personal injury. Power-operated doors can be swinging, sliding, or folding doors.

**A.7.2.1.9.1** An example of the type of door addressed by 7.2.1.9.1 is one actuated by a motion-sensing device upon the approach of a person.

**A.7.2.1.9.1.5** Although a single power-operated door leaf located within a two-leaf opening might alone not provide more than 30 in. (760 mm) of clear width in the emergency breakout mode, where both leaves are broken out to become side hinged, the required egress width is permitted to be provided by the width of the entire opening.

**A.7.2.1.15.1** Door assemblies (fire-rated and non-fire-rated) within the required means of egress (e.g., door assemblies that discharge from exit enclosures) require a higher level of care and maintenance throughout the life of their installations to ensure they perform as intended by the *Code*. Annual inspection and functional testing of these door assemblies is neces-

sary to verify that they are maintained in proper working condition. Panic hardware and fire exit hardware devices are specifically required to be used in assembly and educational occupancies. However, door leaves that are equipped with panic hardware or fire exit hardware, in areas not specifically required by the *Code* (e.g., stairwell entry doors and double-egress cross-corridor door assemblies not serving an assembly occupancy), should be subject to annual inspection and functional testing to ensure that the operating hardware functions correctly in accordance with 7.2.1.7, since the presence of panic hardware and fire exit hardware implies it is required by the *Code*.

Additionally, door assemblies that are electrically controlled egress doors in accordance with 7.2.1.5.5 and door assemblies that are equipped with special locking arrangements in accordance with 7.2.1.6 are outfitted with electrified hardware and access control devices that are susceptible to wear and abuse. Consequently, these door assemblies need to be inspected and tested on an annual basis, regardless of the occupant load being served.

In cases where the authority having jurisdiction determines there is a distinct hazard to building occupant safety, the inspection requirements of 7.2.1.15 should be applied to other exit access, exit, and exit discharge door assemblies.

**A.7.2.2.2.1.1(2)** It is the intent of 7.2.2.2.1.1(2) to permit the use of Table 7.2.2.2.1.1(b) in existing buildings, even where there is a change in occupancy per 4.6.11. Safety improvements should be made that are reasonable and feasible at minimal cost. Improvements include removal, repair, or replacement of step coverings, as described in A.7.2.2.3.5, particularly Figure A.7.2.2.3.5(e), and addition of functional handrails and guardrails in place of, or in conjunction with, other rails, as described in 7.2.2.4.

**A.7.2.2.2.1.2(B)** The stair width requirement of 7.2.2.2.1.2(B) is based on accumulating the occupant load on each story the stair serves.

The accumulating of occupant load is done for the purposes of the requirements of 7.2.2.2.1.2 only. The egress capacity requirements of Section 7.3 are NOT cumulative on a story-by-story basis.

If additional exits provide egress capacity, the occupant load served by such additional exits, up to the limit permitted for the egress capacity of such additional exits, is not added to the total occupant load considered for the minimum stair width requirements of 7.2.2.2.1.2.

If horizontal exits are provided on any of the stories, the total occupant load of all compartments on the story with the horizontal exits is used in the calculation of the minimum stair width requirements of 7.2.2.2.1.2. The number of stairs permitted through application of horizontal exit requirements in 7.2.4 is not affected by the minimum stair width requirements of 7.2.2.2.1.2.

The examples that follow illustrate applications of the minimum stair width requirement.

A stair in a building two stories in height above grade plane that has 2000 persons on the second story, among 10 equally sized stairs that serve the second story, would be considered to have an occupant load of 200 persons for the purposes of applying Table 7.2.2.2.1.2(B). The minimum width of such a stair would be 44 in. (1120 mm).

For a building with a relatively large floor area, a typical 44 in. (1120 mm) stair would not be required to be increased in width until it serves a building approximately 14 stories in height above grade plane, calculated as follows:

$$\frac{2000 \text{ persons}}{147 \text{ persons per floor for a } 44 \text{ in. (1120 mm) width stair}} \approx 14 \text{ stories}$$

For egress in the descending direction, only the stair width below the 14 stories with the total occupant load of 2000 persons per stair, or 4000 persons if served by two equally sized stairs, would need to be increased to 56 in. (1420 mm). If the building is 20 stories in height above grade plane, only the stairs on the lowest 7 stories would be required to have the 56 in. (1420 mm) width.

For a building 41 stories in height above grade plane with 200 persons on each story (or 8000 persons overall, not including the level of exit discharge), with two equally sized stairs, each stair would be considered to have an occupant load of 4000 persons for the purposes of applying Table 7.2.2.2.1.2(B). Only the portion of the stair serving 2000 persons would be required to have the wider width. If each story provides the same floor area for occupancy, the upper 20 stories would have 44 in. (1120 mm) stairs, and the lowest 20 stories would have the 56 in. (1420 mm) stairs, as a minimum.

**A.7.2.2.2.4** If properly designed and constructed, stairs with winders are not necessarily more dangerous than other stairs. Attention to the factors that follow helps to make winders generally more effective for egress and safety. Handrails should be continuous, without breaks at newel posts, from story to story. Handrails located at a greater than normal distance from the inner turn of winders can improve safety by constraining stair users to walk on the portion of the treads providing deeper treads, which should have not less than 11 in. (280 mm) of depth. Combinations of straight flights and winders are best arranged with winders located only below the straight flight. This arrangement is best because the winders provide larger tread dimensions over much of their width than do typical treads on straight flights. A descending person will, thus, be unlikely to experience a reduction of tread depth during descent, a condition of nonuniformity that is best avoided.

**A.7.2.2.3.3.2** The tripping hazard referred to in 7.2.2.3.3.2 occurs especially during descent, where the tread walking surface has projections such as strips of high-friction materials or lips from metal pan stairs that are not completely filled with concrete or other material. Tread nosings that project over adjacent treads can also be a tripping hazard. ICC/ANSI A117.1, *American National Standard for Accessible and Usable Buildings and Facilities,* illustrates projecting nosing configurations that minimize the hazard.

Where environmental conditions (such as illumination levels and directionality or a complex visual field that draws a person's attention away from stair treads) lead to a hazardous reduction in one's ability to perceive stair treads, they should be made of a material that allows ready discrimination of the number and position of treads. In all cases, the leading edges of all treads should be readily visible during both ascent and descent. A major factor in injury-producing stair accidents, and in the ability to use stairs efficiently in conditions such as egress, is the clarity of the stair treads as separate stepping surfaces.

**A.7.2.2.3.4** A small drainage slope for stair treads subject to wetting can improve tread slip resistance *(see also A.7.2.2.3.3.2).* A consistent slope to a side of the stair, where drainage is possible, might be preferable to a front-to-back slope of the treads. Providing a pitch of ⅛ in./ft to ¼ in./ft (10 mm/m to 21 mm/m) aids the shedding of water from a nominally horizontal surface.

**A.7.2.2.3.5** Figure A.7.2.2.3.5(a), Figure A.7.2.2.3.5(b), Figure A.7.2.2.3.5(c), and Figure A.7.2.2.3.5(d) illustrate the method for measuring riser height and tread depth. Stairs that are covered with resilient floor coverings might need additional tread depth beyond the minimum specified in the *Code.* Any horizontal projection of resilient covering materials beyond the tread nosing and riser, such as carpet and underlayment, can interfere with users' feet and thereby reduce usable tread depth. At the tread nosing, such resilient covering materials might not be capable of providing stable support for users' feet. Generally, effective tread depth is reduced by the uncompressed thickness of such resilient coverings, and might be further reduced over time if coverings are not well secured, and, consequently, might move forward at the nosings. *[See Figure A.7.2.2.3.5(e).]*

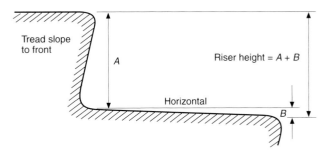

**FIGURE A.7.2.2.3.5(a)** Riser Measurement with Tread Slope to Front.

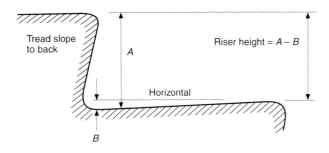

**FIGURE A.7.2.2.3.5(b)** Riser Measurement with Tread Slope to Back.

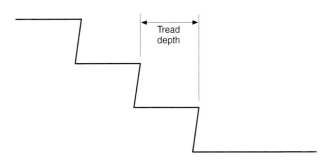

**FIGURE A.7.2.2.3.5(c)** Tread Depth.

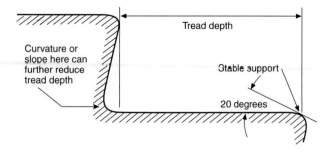

**FIGURE A.7.2.2.3.5(d)   Tread Measurement with Stable Support at Leading Edge.**

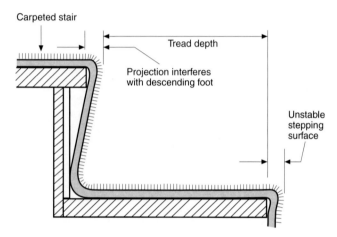

**FIGURE A.7.2.2.3.5(e)   Tread Measurement with Unstable Stepping Surface at Leading Edge.**

**A.7.2.2.3.6** A relatively common error in much of home stair construction and, more rarely, in other stair construction is a failure to make the landing nosing projection consistent with the projection of all other nosings in the stair flight. Such an error can easily occur if the stair flight is installed as a prefabricated unit where the top landing does not have a comparable nosing and the unit includes nosing projections. This heightens the risk of an overstepping misstep, at the second or third step down, by a person who is descending.

A fairly reliable test of step dimension uniformity is the crouch and sight test, in which the inspector crouches on the landing above a flight of stairs to confirm that all of the nosings, including the landing nosing, line up. Unless there is a rare matched variation in the height of a step riser and in the tread depth, both proportionally larger or smaller than other steps in the flight, such that the internosing slope or pitch is maintained consistent in the flight, the visual alignment of the nosings in the crouch and sight test will indicate dimensional uniformity. Thus, as a first task in any stair inspection, the crouch and sight test should be routinely performed. If the stair does not pass this visual test, careful measurements performed in accordance with 7.2.2.3.5 are essential. If the stair appears to pass this test, indicating that the internosing slope or pitch is consistent, a prudent second, quick test is to measure the internosing distances for each step to confirm their consistency.

Step dimensions or their uniformity should not be measured by simply laying a measuring tape or stick on the tread or against the riser. Such measurements could be misleading and erroneous relative to the criteria set out in 7.2.2.3.5, particularly if nosing projections are not uniform (as addressed in 7.2.2.3.6.5), if treads slope, or if the slopes vary within a stair flight.

**A.7.2.2.3.6.5** "Safety yellow" is the widely used, standard color (described in ANSI/NEMA Z535.1, *Standard for Safety Colors*) to be used for a "caution" function, as a solid color or in alternating, angled yellow-black bars or other geometric combination that draws attention beyond merely designating a nosing. Other nosings, not located above nonuniform risers, need only contrast with the remainder of the step and can be of any color providing contrast relative to the remainder of the tread. Note that similar specification of distinctive and contrasting nosing markings is called for in assembly seating aisle stairs (*see, respectively, 12.2.5.6.6(7) and 12.2.5.6.9.1*). The safety problems of exterior stairs in assembly aisles and adjacent to a sloping public way are similar, as each individual step has to be visually detected in a reliable fashion. In addition, the presence and location of steps with unavoidably nonuniform risers must be effectively communicated, especially when viewed in the descent direction. Widely varying light conditions further heighten the need for such markings.

**A.7.2.2.4.1.4** The intent of this provision is to place handrails for the required egress width only, regardless of the actual width. The required egress width is provided along the natural path of travel to and from the building. Examples of this requirement are shown in Figure A.7.2.2.4.1.4. The reduced intermediate handrail spacing of 60 in. (1525 mm), along with a handrail height within the permissible height limits, allows users to reach and grasp one handrail. Except as noted in 7.2.2.4.2 and 7.2.2.4.4, handrails are not required on stair landings.

**A.7.2.2.4.4** Figure A.7.2.2.4.4 illustrates some of the requirements of 7.2.2.4.4.

**A.7.2.2.4.4.4** Additional handrails, beyond those required by the *Code*, are permitted at heights other than those stipulated. For example, where children under the age of five are major users of a facility, an additional handrail at a height in the range of 28 in. to 32 in. (710 mm to 810 mm) might be useful. Generally, children prefer to use, and can effectively use, handrails that are located at shoulder to head height due to their developmental characteristics and their less developed balance and walking abilities. At age 3, head height ranges from 35 in. to 40 in. (890 mm to 1015 mm); shoulder height averages 29 in. (735 mm). At age 5, head height ranges from 39 in. to 46 in. (990 mm to 1170 mm); shoulder height ranges from 31 in. to 37 in. (785 mm to 940 mm).

**A.7.2.2.4.4.6(2)** Handrails should be designed so they can be grasped firmly with a comfortable grip and so the hand can be slid along the rail without encountering obstructions. The profile of the rail should comfortably match the hand grips. For example, a round profile, such as is provided by the simplest round tubing or pipe having an outside diameter of 1½ in. to 2 in. (38 mm to 51 mm), provides good graspability for adults. Factors such as the use of a handrail by small children and the wall-fixing details should be taken into account in assessing handrail graspability. The most functional, as well as the most preferred, handrail shape and size is circular with a 1½ in. (38 mm) outside diameter (according to research conducted using adults). Handrails used predominantly by children should be designed at the lower end of the permitted dimensional range.

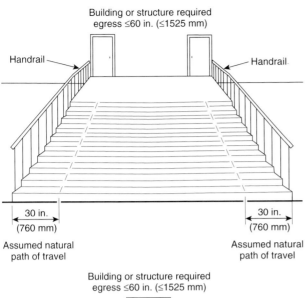

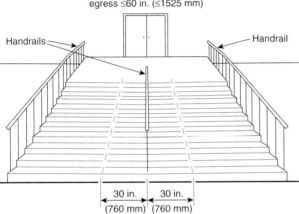

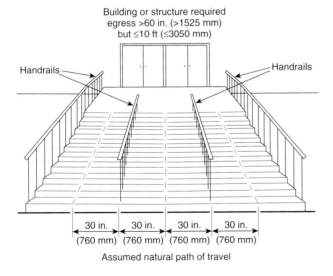

**FIGURE A.7.2.2.4.1.4   Assumed Natural Paths of Travel on Monumental Stairs with Various Handrail Locations.**

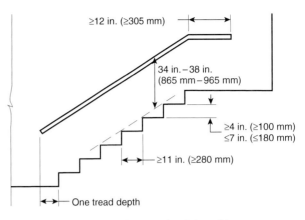

**ELEVATION VIEW (straight stair)**

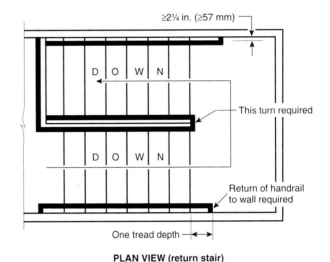

**PLAN VIEW (return stair)**

**FIGURE A.7.2.2.4.4   Handrail Details.**

Handrails are one of the most important components of a stair; therefore, design excesses such as oversized wood handrail sections should be avoided, unless there is a readily perceived and easily grasped handhold provided. In handrail design, it is useful to remember at all times the effectiveness of a simple round profile that allows some locking action by fingers as they curl around the handrail.

Perimeter dimension, referred to in 7.2.2.4.4.6(2), is the length of the shortest loop that wraps completely around the railing.

**A.7.2.2.4.5.2(3)**   This reduction in required height applies only to the stair, not to the landings.

**A.7.2.2.4.5.3**   Vertical intermediate rails are preferred to reduce climbability.

**A.7.2.2.5.2**   The purpose of this provision is to protect the exterior wall of a stairway from fires in other portions of the building. If the exterior wall of the stair is flush with the building exterior wall, the fire would need to travel around 180 degrees in order to impact the stair. This has not been a problem in existing buildings, so no protection is required. However, if the angle of exposure is less than 180 degrees, protection of either the stair wall or building wall is required.

Figure A.7.2.2.5.2(a), Figure A.7.2.2.5.2(b), and Figure A.7.2.2.5.2(c) illustrate the requirement, assuming nonrated glass on the exterior wall of the stair is used.

**A.7.2.2.5.3** An example of a use with the potential to interfere with egress is storage.

**A.7.2.2.5.4** Figure A.7.2.2.5.4 shows an example of a stairway marking sign.

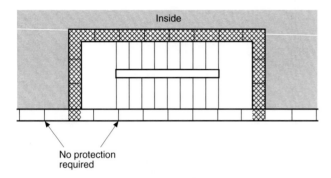

**FIGURE A.7.2.2.5.2(a)  Stairway with Nonrated Exterior Wall in Same Plane as Building Exterior Wall.**

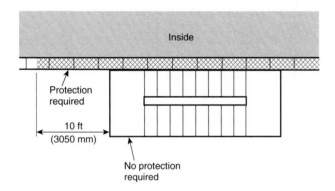

**FIGURE A.7.2.2.5.2(b)  Stairway with Unprotected Exterior Perimeter Protruding Past Building Exterior Wall.**

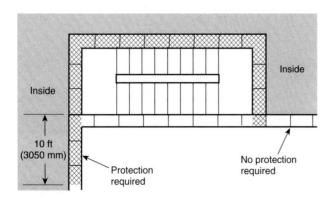

**FIGURE A.7.2.2.5.2(c)  Stairway with Nonrated Exterior Wall Exposed by Adjacent Exterior Wall of Building.**

**NORTH STAIR**

FLOOR

**5**

**SUB-BASEMENT TO 24TH FLOOR**
**NO ROOF ACCESS**
↓ **DOWN TO FIRST FLOOR**
**FOR EXIT DISCHARGE**

**FIGURE A.7.2.2.5.4  Example of a Stairway Marking Sign.**

**A.7.2.2.5.4.1(K)** It is not the intent to require a sign that reads ROOF ACCESS, as such message might be misinterpreted by building occupants as an alternative egress route. However signs that read ROOF ACCESS are not prohibited, as many such signs have been installed in existing buildings so as to make a requirement for removal impractical. Historically, the ROOF ACCESS sign has provided information for the fire department. Where there is no roof access, such information will be posted via a NO ROOF ACCESS sign. The absence of the NO ROOF ACCESS sign should be understood by the fire department to mean that roof access is possible.

**A.7.2.2.5.4.3** For stair nosing marking, surface-applied material, such as adhesive-backed tape and magnetic strips, should not be used, as it is not durable under the scuffing from users' feet and, in coming loose, it creates a tripping hazard. While a carefully applied and consistently maintained coating is acceptable, contrasting color or photoluminescent material integral with the nosings is preferable because of its permanence. See also 7.1.6.4 and 7.2.2.3.6 for slip resistance uniformity requirements, as well as prohibition of projections on the treads.

Guidance on the use of photoluminescent marking is provided by ASTM E 2030, *Guide for Recommended Uses of Photoluminesent (Phosphorescent) Safety Markings*. Additional marking, for example, at the side boundaries of the stair, should be applied in accordance with the guidance provided therein.

**A.7.2.2.5.4.4** Coatings and other applied markings, if used, should be durable for the expected usage, especially at end terminations of the marking and at changes in stair direction where usage is more extensive and hand forces are larger.

**A.7.2.2.5.5.5** Examples of obstacles addressed by 7.2.2.5.5.5 are standpipes, hose cabinets, and wall projections.

**A.7.2.2.5.5.7(1)** The marking stripe for door hardware should be of sufficient size to adequately mark the door hardware. This marking could be located behind, immediately adjacent to, or on the door handle or escutcheon.

**A.7.2.2.6.2** The guards that are required by 7.1.8 and detailed in 7.2.2.4.5 will usually meet this requirement where the stair is not more than 36 ft (11 m) above the finished ground level. Special architectural treatment, including application of such devices as metal or masonry screens and grilles, will usually be necessary to comply with the intent of this requirement for stairs over 36 ft (11 m) above the finished ground level.

**A.7.2.2.6.3.1** Where outside stairs are not required to be separated from interior portions of the building in accordance with 7.2.2.6.3.1(1) through (5), such stairs are considered exits and not exit access.

**A.7.2.2.6.5** See A.7.2.2.3.4.

**A.7.2.3.9.1** The design pressure differences required by 7.2.3.9.1 are based on specific gas temperatures and ceiling heights. The system is required to be approved, because anticipated conditions might be different from those on which the design pressure differences were calculated and, thus, different design pressure differences might be needed. For additional information on necessary minimum design pressure differences, including calculational techniques, or maximum pressure differences across doors to ensure reasonable operating forces, see NFPA 92, *Standard for Smoke Control Systems*.

**A.7.2.4.1.2** An example of one way to provide the required egress capacity from the upper floor of a department store building measuring 350 ft × 200 ft (107 m × 61 m), with an occupant load of 1166 per floor, would be to furnish eight 44 in. (1120 mm) stairs. *[See Figure A.7.2.4.1.2(a).]*

The building is assumed to be divided into two sections by a fire barrier meeting the requirements for a horizontal exit, one 130 ft × 200 ft (40 m × 61 m), and the other 220 ft × 200 ft (67 m × 61 m), with two pairs of 46 in. (1170 mm) double egress doors, with each door providing 44 in. (1120 mm) of clear egress width *[see Figure A.7.2.4.1.2(b)]*. The smaller section, considered separately, will require the equivalent of three 44 in. (1120 mm) exit stairs, and the larger section will require five such exits. The horizontal exits will serve as one of the three exits required for the smaller section, and two of the five exits required for the larger section. Therefore, only two 44 in. (1120 mm) exit stairs from the smaller section and three 44 in. (1120 mm) exit stairs from the larger section will be required if the exits can be arranged to meet the requirements for the 150 ft (46 m) travel distance permitted from any point in a nonsprinklered building. Thus, the total number of exit stairs required for the building will be five, as compared to eight if no horizontal exit had been provided.

Another option would be the use of two 56 in. (1420 mm) exit stairs from the larger section, which would reduce the total number of stairways required from the floor to four *[see Figure A.7.2.4.1.2(c)]*. However, if the building were further subdivided by a second fire wall meeting the requirements for a horizontal exit, no further reduction in stairways would be permitted in order to comply with the requirement that horizontal exits provide a maximum of one-half of egress capacity.

**A.7.2.4.3.10** Fusible link–actuated automatic-closing doors do not qualify for use in horizontal exits under these provisions, because smoke might pass through the opening before there is sufficient heat to release the hold-open device. Such doors are also objectionable because, once closed, they are difficult to open and would inhibit orderly egress.

**A.7.2.5.6.1** The guards required by 7.1.8 and detailed in 7.2.2.4.5 for the unenclosed sides of ramps will usually meet this requirement where the ramp is not more than 36 ft (11 m) above the finished ground level. Special architectural treatment, including application of such devices as metal or masonry screens and grilles, will usually be necessary to comply

with the intent of the requirements for ramps over 36 ft (11 m) above the finished ground level.

**A.7.2.5.6.2** Providing a pitch of ⅛ in./ft to ¼ in./ft (10 mm/m to 21 mm/m) will aid the shedding of water from a nominally horizontal surface.

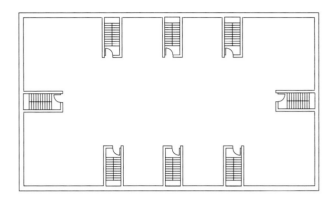

**FIGURE A.7.2.4.1.2(a)  Eight Exits, Required to Provide Necessary Egress Capacity, with None via Horizontal Exit.**

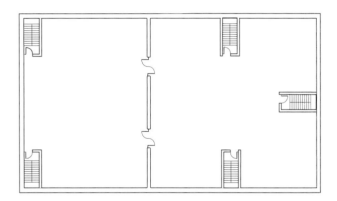

**FIGURE A.7.2.4.1.2(b)  Number of Stairs Reduced by Three Through Use of Two Horizontal Exits; Egress Capacity Not Reduced.**

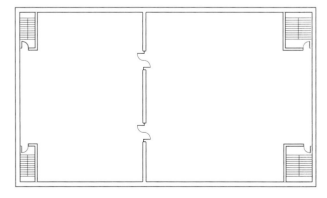

**FIGURE A.7.2.4.1.2(c)  Number of Stairs Further Reduced by Widening Stairs in Larger Compartment, But Not to Less than One-Half the Required Number and Capacity of Exits from That Compartment.**

**A.7.2.6** An exit passageway serves as a horizontal means of exit travel that is protected from fire in a manner similar to an enclosed interior exit stair. Where it is desired to offset exit stairs in a multistory building, an exit passageway can be used to preserve the continuity of the protected exit by connecting the bottom of one stair to the top of the stair that continues to the street floor. Probably the most important use of an exit passageway is to satisfy the requirement that at least 50 percent of the exit stairs discharge directly outside from multistory buildings (*see 7.7.2*). Thus, if it is impractical to locate the stair on an exterior wall, an exit passageway can be connected to the bottom of the stair to convey the occupants safely to an outside exit door. In buildings of extremely large area, such as shopping malls and some factories, the exit passageway can be used to advantage where the travel distance to reach an exit would otherwise be excessive.

**A.7.2.6.1** Examples of building elements that might be arranged as exit passageways include hallways, corridors, passages, tunnels, underfloor passageways, or overhead passageways.

**A.7.2.6.4.1(1)** Where an exit passageway serves occupants on the level of exit discharge as well as other floors, it should not be required that the occupant loads be added, thus increasing the width of the exit passageway. The situation is the same as that in which occupants from the level of exit discharge join occupants from upper floors for a few feet of horizontal travel through a stair enclosure.

**A.7.2.8.7** Swinging stairs, although superior to fire escape ladders, are generally unsatisfactory, even for emergency use. Although such stairs are permitted by this *Code*, they should not be used where it is reasonably possible to terminate the fire escape stair at the finished ground level.

**A.7.2.8.7.9** A latch is desirable for holding swinging stairs down after they have swung to the finished ground level.

**A.7.2.11** Special consideration should be given prior to the application of such devices where children, the elderly, or physically disabled persons use such devices. These devices present obstacles in ascent and descent that differ from those for stairs and ladders.

**A.7.2.12.2.3** The clear width of not less than 48 in. (1220 mm) is needed for a three-person carry of an occupied wheelchair up or down a stair. This procedure, as well as the more difficult two-person wheelchair carry or roll, requires training and experience. Safer, alternative stair descent measures for transporting a person who normally requires a wheelchair include evacuation chairs and self-braking stair descent devices. In addition to having such devices available where needed, and having persons trained and experienced in their use, it is important to have people trained and experienced in wheelchair transfer techniques.

In view of the logistical difficulties, as well as the dangers inherent in carrying occupied wheelchairs or otherwise transporting their occupants on stairs, the preferred means of egress from an area of refuge consists of facilities normally employed for ingress and egress by people using wheelchairs. Foremost among these options are elevators meeting the fire fighters' emergency operations requirements of ASME A17.1/CSA B44, *Safety Code for Elevators and Escalators.*

Stair descent devices using stair-bearing belted tracks, for example, provide a safer, more effective evacuation option than carrying an occupied wheelchair or other carried device down the stairs and on landings. The use of stair descent devices is recommended.

The design, manufacture, selection, maintenance, and operation of stairway descent devices should take into account the following recommendations and general guidance information:

(1) The minimum carrying capacity of the device should be 350 lb (159 kg) and should be rated for the maximum permitted slope of the stair served.
(2) For existing stairs with 8 in. (205 mm) riser height and 9 in. (230 mm) tread depth, the 350 lb (159 kg) carrying capacity specified in A.7.2.12.2.3(1) should be maintained at a slope or pitch of 42 degrees or 1.0 unit vertical for 1.1 units horizontal.
(3) The rated maximum carrying capacity and maximum stair pitch should be labeled on the device, and the device should only be operated within the labeled limits of load and stair slope or pitch.
(4) The maximum descent speed should be capable of being controlled during operation, without undue restraint by the operator(s), to 30 in./s (760 mm/s), measured along the slope of the stair.
(5) When operated according to the manufacturer's instructions and loaded to its maximum stated capacity, the device should come reliably to a complete stop within a distance of 12 in. (305 mm), measured along the landing or stair slope, on a stair with a slope or pitch within the device's maximum stated capability, and the following also should apply:
    (a) The device should have a brake.
    (b) On walking surfaces other than stairs, the device should maintain a parked position, without rolling, so that the operator can attend to other activities, including assisting the passenger to transfer from or to other mobility devices.
(6) Minimum stairway width and landing depth requirements, especially at landings where turns occur, should be disclosed by the manufacturer, and only devices appropriate to the building's stairways should be provided in the building.
(7) Slowing or delays when transitioning between stairway landings and stair flights should cause no more than minimal delay to the movement of pedestrians using the stairway in the vicinity of the device and, generally, should not significantly reduce the flow of evacuees using the stairway system.
(8) When descending stairs, the device should be easily operable by one person who is trained on its use, and the following also should apply:
    (a) Above-average weight or strength should not be required for proper operation.
    (b) Lifting or carrying of the device, when occupied, should normally not be required.
(9) Unless designed specifically for use on stairs with nonrectangular treads (e.g., winder treads), with operators trained for such use, the device should be operated only on stairways with straight flights having rectangular treads.
(10) On straight flights, the device should be designed to have supporting contact with at least two treads, except during the transition between landings and stair flights.
(11) The device should be equipped with restraining straps that securely hold the passenger, including the chest, waist, thighs, knees, ankles, and arms, to prevent injury, and the length and quantity of straps should be designed to accommodate a range of passenger sizes and weights up to the maximum capacity of the device.

(12) The seat or seat sling should have open sides and be positioned at an appropriate height to allow transfer with minimal operator assistance, and the specific procedure used for a particular transfer should be determined through discussion between the passenger and the operator(s).

(13) Setup of the device should be described in procedures posted on the device, should require no other tools or expertise beyond that of available operators, and should take approximately 10 seconds to set up from storage condition to readiness for transfer.

(14) In addition to descending stairs, the device should be able to travel across ramped or horizontal surfaces, such as stair landings and hallways, so that it can follow an entire egress route to the exterior of the building.

(15) The device's seating system should provide adequate support of the passenger to minimize the potential for discomfort or injury, recognizing that people with physical disabilities, who will be the main occupants of the device, are often unusually susceptible to pressure-related injuries and spasm, while being unable to perceive the warning signs of pain.

(16) If cabinets or storage covers are provided, they should include signage or labeling that clearly identifies the device and its use, and the device should be readily retrievable from storage without use of a key or special tool.

(17) The building evacuation plan should include the location of the devices, a process for verifying the availability of trained operators, and other critical information as is required for the building.

(18) The manufacturer of the device should provide comprehensive training materials with each device, and the following also should apply:

(a) All designated operators should be trained in accordance with these instructions.

(b) Evacuation drills that involve actual use of the device by the designated operators, including transfer and transport of building occupants with disabilities, should occur at least quarterly.

(19) The device should be inspected and tested annually in accordance with the manufacturer's recommendations, and preventive maintenance should be performed in accordance with the manufacturer's recommendations.

(20) Device capabilities differing significantly from those spelled out in the foregoing recommendations (e.g., carrying capacity, higher normal speed, and fail-safe braking systems) should be disclosed by the manufacturer in a manner readily known to operators whose training should take such differences into account.

(21) Limitations of the device based on stair nosing geometry and nature of stairway covering should be disclosed by the manufacturer in specifications, operating instructions, and labeling.

(22) Carrying handles, if installed on the device, should provide secure gripping surfaces and adequate structural and geometric design to facilitate carrying by two or more operators, and the following also should be considered:

(a) Carrying might be necessitated by damaged or otherwise irregular walking surfaces that do not facilitate rolling with the device's wheels or tracks.

(b) Carrying might also be necessitated by the existence of an ascending stair along the means of egress (e.g., in the exit discharge path.)

(23) Unless specialized operator training is undertaken, use of a stairway descent device on stairways with unusually large treads and on escalators should be attempted only under the following conditions:

(a) The device is designed for extra-long distance between the tread nosings [e.g., about 16 in. (405 mm) on escalators, as opposed to about 12 in. to 13 in. (305 mm to 330 mm) on typical exit stairways].

(b) Such specialized training might entail maintaining a downward force on the device operating handle.

**A.7.2.12.2.4** The use of elevators for egress, especially during an emergency such as a fire, is not an approach to be taken without considerable planning, ongoing effort, and a high degree of understanding by everyone involved with the evacuation of persons with mobility impairments. Due in part to the limited capacity of elevators, as well as to the conflicting demands for elevator use for fire-fighting activities, even elevators in accordance with 7.2.12.2.4 cannot be considered as satisfying any of the *Code's* requirements for egress capacity, number of means of egress, or travel distance to an exit.

**A.7.2.12.2.6** The instructions should include the following:

(1) Directions to find other means of egress
(2) Advice that persons able to use exit stairs do so as soon as possible, unless they are assisting others
(3) Information on planned availability of assistance in the use of stairs or supervised operation of elevators and how to summon such assistance
(4) Directions for use of the emergency communications system

To facilitate an adequate degree of understanding of the use of areas of refuge and of the associated assisted egress procedures, information should be provided to those using the facilities. The exact content of the information, its organization (e.g., as a set of instructions), and its format (e.g., either posted instructions in the area of refuge or information otherwise transmitted to facility users) should be determined on a case-by-case basis. The information should be tailored to the specific facility, its emergency plan, the intended audience, and the intended presentation format. Suggested information content addressing two situations follows.

*Refuge with Elevator Use.* An area of refuge provided in the elevator lobby serves as a staging area for persons unable to use stairs and needing assistance for their evacuation during an emergency. The elevator(s) will be taken out of automatic service and operated by emergency service personnel. Persons unable to evacuate down the exit stairs without assistance and needing transportation by elevator should make certain the elevator lobby doors are closed while they wait in the elevator lobby for assistance. The two-way communication system should be used if there is a delay of more than several minutes in the arrival of an elevator that will provide transportation to the level of exit discharge. Alternatively, another refuge area, and assistance with evacuation, is available in the designated exit stair.

*Refuge with Stair Use.* An area of refuge within the designated exit stair serves as a staging area for persons needing assistance for their evacuation during an emergency. Persons unable to use the stairs unassisted, or who wish to move down the stairs at a slower pace, should wait on the stair landing. The two-way communication system should be used if assistance is needed.

**A.7.2.12.3.1** Figure A.7.2.12.3.1 illustrates the application of the minimum space requirement to an area of refuge located within an exit stair enclosure. Note that each of the two required spaces is sufficient to allow the parking of a standard wheelchair. Preferably, such spaces should be provided adjacent to each other in a location where the presence of people taking temporary shelter in an area of refuge will be immediately apparent to rescue personnel and other evacuees.

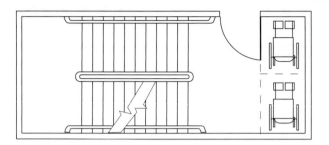

**FIGURE A.7.2.12.3.1 Exit Stair Used as an Area of Refuge.**

**A.7.2.12.3.2** The method of meeting the tenability performance criteria required of an area of refuge of less than 1000 ft$^2$ (93 m$^2$) can involve controlling the exposing fire (e.g., via automatic sprinkler protection), installing smoke-resisting doors in the smoke-resisting barriers *(see* NFPA 105, *Standard for Smoke Door Assemblies and Other Opening Protectives)*, providing smoke control to prevent or limit smoke migration through cracks or other leakage paths *(see* NFPA 92, *Standard for Smoke Control Systems)*, or providing other means or a combination of these means.

Calculations, if used, need to be based on established engineering relationships and equations. Such calculational procedures are described in NFPA 92 and the *SFPE Handbook of Fire Protection Engineering*. Tenable conditions are those that maintain the temperature of any smoke in the area of refuge at less than 200°F (93°C) if the smoke is more than 60 in. (1525 mm) above the floor, and at less than 120°F (49°C) if the smoke descends below the 60 in. (1525 mm) level in the area of refuge. Also, if the smoke descends below the 60 in. (1525 mm) level, tenable conditions require not less than 16 percent oxygen and not more than 30,000 ppm/min exposure to carbon monoxide. The exposing conditions used in the calculations should be in accordance with the following:

(1) The exposing space is sprinkler protected, and the following conditions also exist:
    (a) The temperature of the exposing smoke is 200°F (93°C).
    (b) The smoke layer extends to the floor.
    (c) The oxygen content is 16 percent.
    (d) The carbon monoxide concentration is 2000 ppm (0.2 percent).
(2) The exposing space is a nonsprinklered corridor finished with Class A interior wall and ceiling finish, and the following conditions also exist:
    (a) The temperature of the exposing smoke is 600°F (316°C).
    (b) The smoke layer extends to a level 24 in. (610 mm) above the floor.
    (c) The oxygen content is 3 percent.
    (d) The carbon monoxide concentration is 50,000 ppm (5 percent).
(3) The exposing space is either not a corridor or, if a corridor, the corridor is not finished with a Class A interior wall and ceiling finish, and the following conditions also exist:

    (a) The temperature of the exposing smoke is 1500°F (815°C).
    (b) The smoke layer extends to a level 24 in. (610 mm) above the floor.
    (c) The oxygen content is 3 percent.
    (d) The carbon monoxide concentration is 50,000 ppm (5 percent).

**A.7.2.12.3.4** Requirements for fire resistance ratings in excess of 1 hour, fire protection ratings in excess of 20 minutes, and prohibitions on duct penetrations appear in other *Code* sections. For example, if the barrier creating the area of refuge is also part of an exit stair enclosure that connects two or more stories, or is a horizontal exit, a minimum 2-hour fire resistance rating for the barrier and a minimum 1½-hour fire protection rating for opening protectives, such as doors, would be required for most occupancies.

For further information on door openings in smoke-resisting barriers, see NFPA 105, *Standard for Smoke Door Assemblies and Other Opening Protectives.*

Generally, by providing one barrier that subdivides a floor area, two areas of refuge can be created. This subdivision method and the possibility of creating areas of refuge within compartmented elevator lobbies or on enlarged stair landings of exit stair enclosures make less onerous any requirement for a story to have more than one accessible means of egress.

**A.7.2.13.1** It is the intent of 7.2.13.1 that elevators serving as a means of egress serve only independent towers or the tower portion of any integral structure. For elevators that are used as a component in the means of egress, the elevator lobbies, elevator shaft, and machine room need to be protected from the effects of fire.

**A.7.2.13.6** One or more of the following approaches can be used to restrict exposure of elevator equipment to water:

(1) A combination of sealed elevator lobby doors, sloped floors, floor drains, and sealed elevator shaft walls is used.
(2) The elevator is mounted on the building exterior that normally operates in the elements, and seals are used on the elevator lobby doors.
(3) The elevator shaft is separated from the building at each floor by an exterior elevator lobby designed to prevent water entry into the elevator shaft.

Information gained from ongoing research concerning waterflow and elevators could lead to the development of water-resistive or water-protected elevator equipment specifically for fire applications. Such equipment should be used only with the building elements (e.g., sealed elevator lobby doors, sloped floors, floor drains) for which it is developed. Further information is available from the NIST publication, *Feasibility of Fire Evacuation by Elevators at FAA Control Towers.*

**A.7.2.13.7** Cooling equipment dedicated to the elevator machine room can be used to minimize requirements for standby power.

**A.7.2.13.8** Communication between elevator lobbies and a central control point can be by telephone or intercom. Auditory alarms should be designed so that they do not interfere with people talking on communications systems.

**A.7.2.13.9** Smoke detection in the elevator lobby will result in a Phase I recall of the elevators. The elevators will then be automatically taken out of normal service and will be available to be operated by emergency service personnel.

**A.7.3.1.2** The normal occupant load is not necessarily a suitable criterion, because the greatest hazard can occur when an unusually large crowd is present, which is a condition often difficult for authorities having jurisdiction to control by regulatory measures. The principle of this *Code* is to provide means of egress for the maximum probable number of occupants, rather than to attempt to limit occupants to a number commensurate with available means of egress. However, limits of occupancy are specified in certain special cases for other reasons.

Suggested occupant load factors for components of large airport terminal buildings are given in Table A.7.3.1.2. However, the authority having jurisdiction might elect to use different occupant load factors, provided that egress requirements are satisfied.

**Table A.7.3.1.2 Airport Terminal Occupant Load Factors**

| Airport Terminal Area | ft² (gross) | m² (gross) |
|---|---|---|
| Concourse | 100 | 9.3 |
| Waiting areas | 15 | 1.4 |
| Baggage claim | 20 | 1.9 |
| Baggage handling | 300 | 27.9 |

The figure used in determining the occupancy load for mall shopping centers of varying sizes was arrived at empirically by surveying over 270 mall shopping centers, by studying mercantile occupancy parking requirements, and by observing the number of occupants per vehicle during peak seasons.

These studies show that, with an increase in shopping center size, there is a decrease in the number of occupants per square foot of gross leasable area.

This phenomenon is explained when one considers that, above a certain shopping center gross leasable area [approximately 600,000 ft² (56,000 m²)], there exists a multiplicity of the same types of stores. The purpose of duplicate types of stores is to increase the choices available to a customer for any given type of merchandise. Therefore, when shopping center size increases, the occupant load increases as well, but at a declining rate. In using Figure 7.3.1.2(a) or Figure 7.3.1.2(b), the occupant load factor is applied only to the gross leasable area that uses the mall as a means of egress.

**A.7.3.3** In egress capacity calculations, standard rounding should be used.

**A.7.3.3.2** The effective capacity of stairways has been shown by research to be proportional to the effective width of the stairway, which is the nominal width minus 12 in. (305 mm). This phenomenon, and the supporting research, were described in the chapter, "Movement of People," in the first, second, and third editions of the *SFPE Handbook of Fire Protection Engineering* and was also addressed in Appendix D of the 1985 edition of NFPA *101*, among several other publications. In 1988, this appendix was moved to form Chapter 2 of the 1988 edition of NFPA 101M, *Alternative Approaches to Life Safety*. (This document was later designated as NFPA 101A, *Guide on Alternative Approaches to Life Safety*, and this chapter remained in the document through the 1998 edition.) In essence, the effective width phenomenon recognizes that there is an edge or boundary effect at the sides of a circulation path. It has been best examined in relation to stairway width, where the edge effect was estimated to be 6 in. (150 mm) on each side, but a similar phenomenon occurs with other paths,

such as corridors and doors, although quantitative estimates of their edge effect are not as well established as they have been for stairways, at least those stairways studied in Canada during the late 1960s through the 1970s in office building evacuation drills and in crowd movement in a variety of buildings with assembly occupancy.

More recent studies have not been performed to determine how the edge effect might be changing (or has changed) with demographic changes to larger, heavier occupants moving more slowly, and thus swaying laterally, to maintain balance when walking. The impact of such demographic changes, which are significant and influential for evacuation flow and speed of movement on stairs, for example, has the effect of increasing the time of evacuation in a way that affects all stair widths, but will be most pronounced for nominal widths less than 56 in. (1422 mm).

Without taking into account occupant demographic changes in the last few decades that affect evacuation performance, especially on stairs, the formula for enhanced capacity of stairways wider than 44 in. (1120 mm) assumes that any portion of the nominal width greater than 44 in. (1120 mm) is as effective proportionally as the effective width of a nominal 44 in. (1120 mm) stair, that is, 32 in. (810 mm). Thus, the denominator (0.218) in the equation is simply the effective width of 32 in. (810 mm) divided by the capacity of 147 persons that is credited, by the 0.3 in. (7.6 mm) capacity factor in Table 7.3.3.1, to the corresponding nominal width, 44 in. (1120 mm).

The resulting permitted stairway capacities, based on occupant load of single stories (in accordance with 7.3.1.4), for several stairway widths are shown in Table A.7.3.3.2.

**Table A.7.3.3.2 Stairway Capacities**

| Permitted Capacity (no. of persons) | Nominal Width | | Clear Width Between Handrails[a] | | Effective Width | |
|---|---|---|---|---|---|---|
| | in. | mm | in. | mm | in. | mm |
| 120[b] | 36 | 915 | 28 | 710 | 24 | 610 |
| 147 | 44 | 1120 | 36 | 915 | 32 | 810 |
| 202 | 56 | 1420 | 48 | 1220 | 44 | 1120 |
| 257 | 68 | 1725 | 60 | 1525[c] | 56 | 1420 |

[a]A reasonable handrail incursion of only 4 in. (100 mm), into the nominal width, is assumed on each side of the stair, although 7.3.3.2 permits a maximum incursion of 4½ in. (114 mm) on each side.
[b]Other *Code* sections limit the occupant load for such stairs more severely, (e.g., 50 persons in 7.2.2.2.1.2). Such lower limits are partly justified by the relatively small effective width of such stairs, which, if taken into account by Table 7.3.3.1, would result in a correspondingly low effective capacity of only 110 persons (24 divided by 0.218), or a more realistic capacity factor of 0.327, applicable to nominal width.
[c]A clear width of 60 in. (1525 mm) is the maximum permitted by the handrail reachability criteria of 7.2.2.4.1.2. Although some prior editions of the *Code* permitted wider portions of stairs [up to 88 in. (2240 mm), between handrails], such wider portions are less effective for reasonably safe crowd flow and generally should not be used for major crowd movement. To achieve the maximum possible, reasonably safe egress capacity for such stairs, retrofit of an intermediate — not necessarily central — handrail is recommended; for example, with an intermediate handrail located 36 in. (915 mm) from the closest side handrail. In this case, the effective capacity would be 358 persons for the formerly permitted, now retrofitted, stair. This is based on a retrofitted, effective width of about 78 in. (1980 mm) [subtracting 2 in. (51 mm) from each usable side of a handrail and assuming a 2 in. (51 mm) wide, retrofitted intermediate handrail].

**A.7.3.4.1.1** The criteria of 7.3.4.1.1 provide for minimum widths for small spaces such as individual offices. The intent is that these reductions in required width apply to spaces formed by furniture and movable walls, so that accommodations can easily be made for mobility-impaired individuals. One side of a path could be a fixed wall, provided that the other side is movable. This does not exempt the door widths or widths of fixed-wall corridors, regardless of the number of people or length.

Figure A.7.3.4.1.1(a) and Figure A.7.3.4.1.1(b) present selected anthropometric data for adults. The male and female figures depicted in the figures are average, 50th percentile, in size. Some dimensions apply to very large, 97.5 percentile, adults (noted as 97.5 P).

**A.7.4** Section 7.4 requires a minimum number of means of egress, unless otherwise specified by an occupancy chapter in subsection ____.2.4, which addresses number of means of egress. Several occupancy chapters establish not only the minimum number of means of egress but also the minimum number of actual exits that must be provided on each floor. For example, for new educational occupancies, 14.2.4 requires access to two exits and further requires that both of the exits be provided on the floor. In contrast, for industrial occupancies, 40.2.4.1.1 requires access to two exits and further requires that at least one of the exits be located on the floor. Access to the other exit can involve traveling to another floor via an egress component such as an open stair, provided that such open

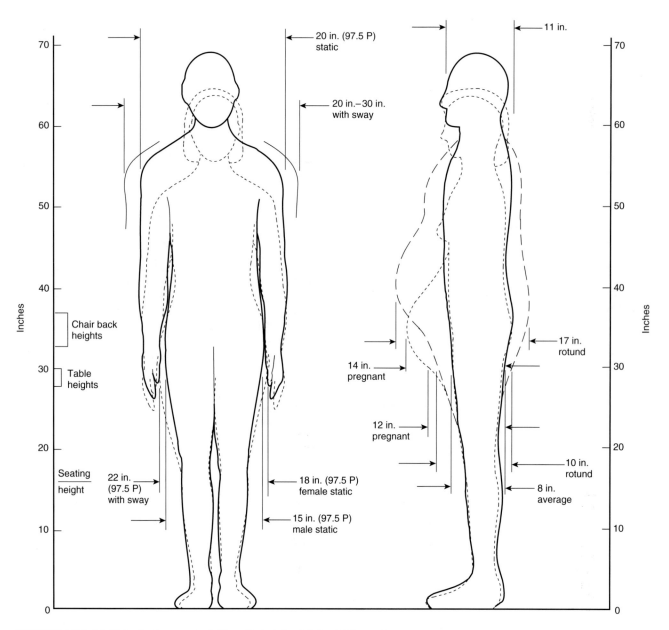

**FIGURE A.7.3.4.1.1(a)   Anthropometric Data (in in.) for Adults; Males and Females of Average, 50th Percentile, Size; Some Dimensions Apply to Very Large, 97.5 Percentile (97.5 P), Adults.**

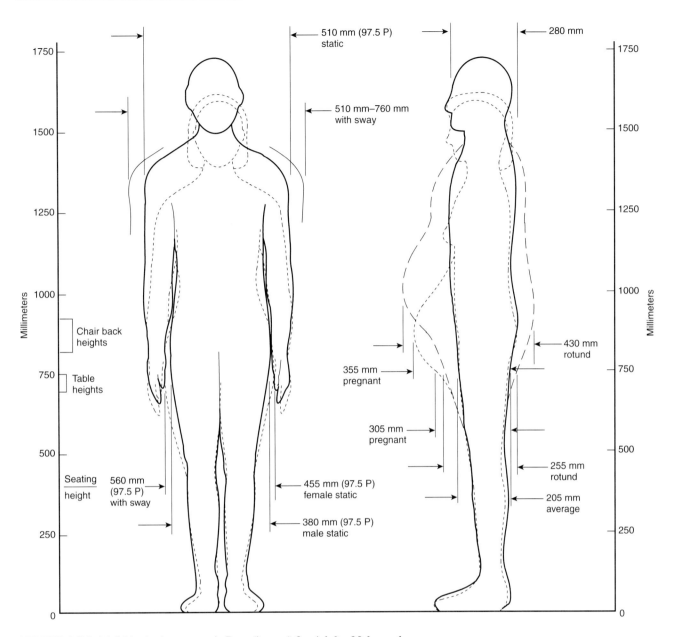

**FIGURE A.7.3.4.1.1(b)   Anthropometric Data (in mm) for Adults; Males and Females of Average, 50th Percentile, Size; Some Dimensions Apply to Very Large, 97.5 Percentile (97.5 P), Adults.**

stair is permitted by the occupancy chapter's provisions for the protection of vertical openings.

In most occupancy chapters, meeting the requirements for egress capacities and travel distances means the required minimum number of means of egress will automatically be met. However, in occupancies characterized by high occupant loads, such as assembly and mercantile occupancies, compliance with requirements for more than two exits per floor might require specific attention.

**A.7.5.1.1.1**   See A.7.5.1.5.

**A.7.5.1.3.2**   Figure A.7.5.1.3.2(a) through Figure A.7.5.1.3.2(e) illustrate the method of measurement intended by 7.5.1.3.2.

**A.7.5.1.3.4**   Figure A.7.5.1.3.4 illustrates the method of measuring exit separation distance along the line of travel within a minimum 1-hour fire resistance–rated corridor.

**A.7.5.1.4.2**   It is difficult in actual practice to construct scissor stairs so that products of combustion that have entered one stairway do not penetrate into the other. Their use as separate required exits is discouraged. The term *limited-combustible* is intentionally not included in 7.5.1.4.2. The user's attention is directed to the provisions for limited-combustible and noncombustible in 4.6.13 and 4.6.14, respectively.

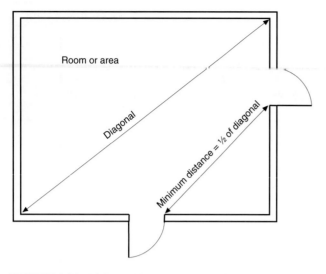

**FIGURE A.7.5.1.3.2(a)   Diagonal Rule for Exit Remoteness.**

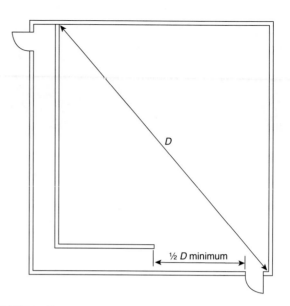

**FIGURE A.7.5.1.3.2(c)   Diagonal Rule for Exit and Access Remoteness.**

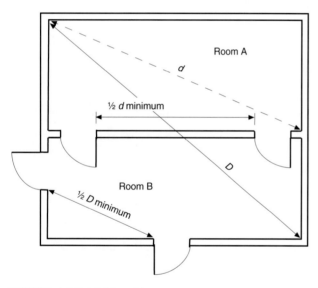

**FIGURE A.7.5.1.3.2(b)   Diagonal Rule for Exit and Exit Access Door Remoteness.**

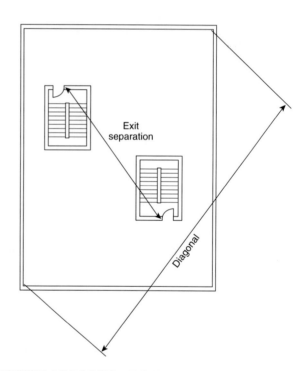

**FIGURE A.7.5.1.3.2(d)   Exit Separation and Diagonal Measurement of Area Served.**

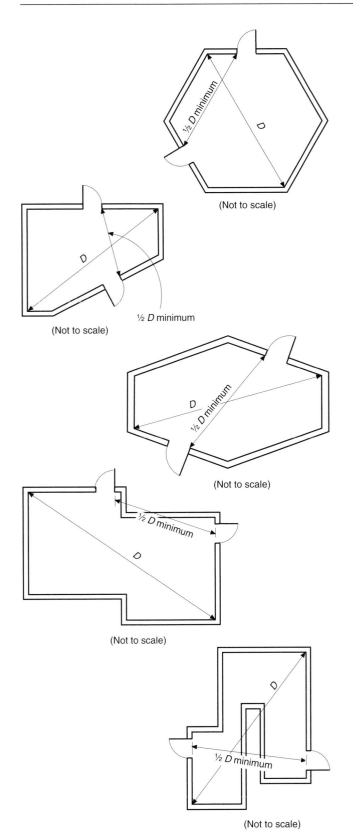

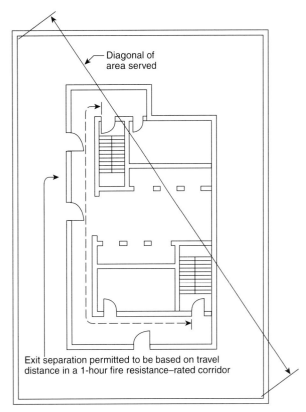

**FIGURE A.7.5.1.3.4   Exit Separation Measured Along Corridor Path.**

**FIGURE A.7.5.1.3.2(e)   Diagonal Measurement for Unusually Shaped Areas.**

**A.7.5.1.5** The terms *dead end* and *common path of travel* are commonly used interchangeably. Although the concepts of each are similar in practice, they are two different concepts.

A common path of travel exists where a space is arranged so that occupants within that space are able to travel in only one direction to reach any of the exits or to reach the point at which the occupants have the choice of two paths of travel to remote exits. Part (a) of Figure A.7.5.1.5 is an example of a common path of travel.

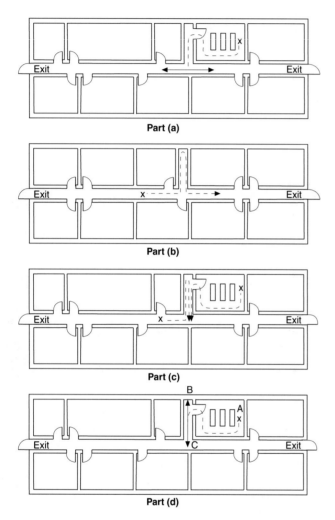

Part (a)

Part (b)

Part (c)

Part (d)

**FIGURE A.7.5.1.5  Common Paths of Travel and Dead-End Corridors.**

While a dead end is similar to a common path of travel, a dead end can exist where there is no path of travel from an occupied space but can also exist where an occupant enters a corridor thinking there is an exit at the end and, finding none, is forced to retrace his or her path to reach a choice of exits. Part (b) of Figure A.7.5.1.5 is an example of such a dead-end arrangement.

Combining the two concepts, part (c) of Figure A.7.5.1.5 is an example of a combined dead-end/common path of travel problem.

Common paths of travel and dead-end travel are measured using the same principles used to measure travel distance as described in Section 7.6. Starting in the room shown in part (d) of Figure A.7.5.1.5, measurement is made from the most remote point in the room, A, along the natural path of travel and through the doorway along the centerline of the corridor to point C, located at the centerline of the corridor, which then provides the choice of two different paths to remote exits; this is common path of travel. The space between point B and point C is a dead end. *(See 3.3.47 for the definition of common path of travel.)*

**A.7.5.2.1** It is not the intent that an area with equipment such as a beverage brewpot, microwave oven, and a toaster be considered a kitchen.

**A.7.5.2.2** Doors that lead through wall paneling, and that harmonize in appearance with the rest of the wall to avoid detracting from some desired aesthetic or decorative effect, are not acceptable, because casual occupants might not be aware of such means of egress even though it is visible.

**A.7.5.4.1** An accessible means of egress should comply with the accessible route requirements of ICC/ANSI A117.1, *American National Standard for Accessible and Usable Buildings and Facilities.*

**A.7.6** Table A.7.6 is a compilation of the requirements of the individual occupancy chapters (Chapters 12 through 42) for permissible length of common path of travel, dead-end corridors, and travel distance to not less than one of the required exits.

A dead end exists where an occupant enters a corridor thinking there is an exit at the end and, finding none, is forced to retrace the path traveled to reach a choice of egress travel paths. Although relatively short dead ends are permitted by this *Code*, it is better practice to eliminate them wherever possible, as they increase the danger of persons being trapped in case of fire. Compliance with the dead-end limits does not necessarily mean that the requirements for remoteness of exits have been met. Such lack of compliance is particularly true in small buildings or buildings with short public hallways. Adequate remoteness can be obtained in such cases by further reducing the length of dead ends. *(See also A.7.5.1.5.)*

**Table A.7.6 Common Path, Dead-End, and Travel Distance Limits (by occupancy)**

| Type of Occupancy | Common Path Limit | | | | Dead-End Limit | | | | Travel Distance Limit | | | |
|---|---|---|---|---|---|---|---|---|---|---|---|---|
| | Unsprinklered | | Sprinklered | | Unsprinklered | | Sprinklered | | Unsprinklered | | Sprinklered | |
| | ft | m | ft | m | ft | m | ft | m | ft | m | ft | m |
| **Assembly** | | | | | | | | | | | | |
| New | 20/75 | 6.1/23[a] | 20/75 | 6.1/23[a] | 20 | 6.1[b] | 20 | 6.1[b] | 200 | 61[c] | 250 | 76[c] |
| Existing | 20/75 | 6.1/23[a] | 20/75 | 6.1/23[a] | 20 | 6.1[b] | 20 | 6.1[b] | 200 | 61[c] | 250 | 76[c] |
| **Educational** | | | | | | | | | | | | |
| New | 75 | 23 | 100 | 30 | 20 | 6.1 | 50 | 15 | 150 | 46 | 200 | 61 |
| Existing | 75 | 23 | 100 | 30 | 20 | 6.1 | 50 | 15 | 150 | 46 | 200 | 61 |
| **Day Care** | | | | | | | | | | | | |
| New | 75 | 23 | 100 | 30 | 20 | 6.1 | 50 | 15 | 150 | 46[d] | 200 | 61[d] |
| Existing | 75 | 23 | 100 | 30 | 20 | 6.1 | 50 | 15 | 150 | 46[d] | 200 | 61[d] |
| **Health Care** | | | | | | | | | | | | |
| New | NA | NA | 100 | 30 | NA | NA | 30 | 9.1 | NA | NA | 200 | 61[d] |
| Existing | NR | NR | NR | NR | NR[e] | NR[e] | NR[e] | NR[e] | 150 | 46[d] | 200 | 61[d] |
| **Ambulatory Health Care** | | | | | | | | | | | | |
| New | 75 | 23[f] | 100 | 30[f] | 20 | 6.1 | 50 | 15 | 150 | 46 | 200 | 61 |
| Existing | 75 | 23[f] | 100 | 30[f] | 50 | 15 | 50 | 15 | 150 | 46 | 200 | 61 |
| **Detention and Correctional** | | | | | | | | | | | | |
| New — Use Condition II, III, IV | 50 | 15 | 100 | 30 | 50 | 15 | 50 | 15 | 150 | 46[d] | 200 | 61[d] |
| New — Use Condition V | 50 | 15 | 100 | 30 | 20 | 6.1 | 20 | 6.1 | 150 | 46[d] | 200 | 61[d] |
| Existing — Use Condition II, III, IV, V | 50 | 15[g] | 100 | 30[g] | NR | NR | NR | NR | 150 | 46[d] | 200 | 61[d] |
| **Residential** | | | | | | | | | | | | |
| One- and two-family dwellings | NR | NR | NR | NR | NR | NR | NR | NR | NR | NR | NR | NR |
| Lodging or rooming houses | NR | NR | NR | NR | NR | NR | NR | NR | NR | NR | NR | NR |
| Hotels and dormitories | | | | | | | | | | | | |
| New | 35 | 10.7[h,i] | 50 | 15[h,i] | 35 | 10.7 | 50 | 15 | 175 | 53[d,j] | 325 | 99[d,j] |
| Existing | 35 | 10.7[h] | 50 | 15[h] | 50 | 15 | 50 | 15 | 175 | 53[d,i] | 325 | 99[d,i] |
| Apartment buildings | | | | | | | | | | | | |
| New | 35 | 10.7[h] | 50 | 15[h] | 35 | 10.7 | 50 | 15 | 175 | 53[d,j] | 325 | 99[d,j] |
| Existing | 35 | 10.7[h] | 50 | 15[h] | 50 | 15 | 50 | 15 | 175 | 53[d,j] | 325 | 99[d,j] |
| Board and care | | | | | | | | | | | | |
| Small, new and existing | NR | NR | NR | NR | NR | NR | NR | NR | NR | NR | NR | NR |
| Large, new | NA | NA | 125 | 38[i] | NA | NA | 30 | 9.1 | NA | NA | 325 | 99[d,j] |
| Large, existing | 110 | 33 | 160 | 49 | 50 | 15 | 50 | 15 | 175 | 53[d,j] | 325 | 99[d,j] |
| **Mercantile** | | | | | | | | | | | | |
| Class A, B, C | | | | | | | | | | | | |
| New | 75 | 23 | 100 | 30 | 20 | 6.1 | 50 | 15 | 150 | 46 | 250 | 76 |
| Existing | 75 | 23 | 100 | 30 | 50 | 15 | 50 | 15 | 150 | 46 | 250 | 76 |
| Open air, new and existing | NR | NR | NR | NR | 0 | 0 | 0 | 0 | NR | NR | NR | NR |
| Mall | | | | | | | | | | | | |
| New | 75 | 23 | 100 | 30 | 20 | 6.1 | 50 | 15 | 150 | 46 | 400 | 120[k] |
| Existing | 75 | 23 | 100 | 30 | 50 | 15 | 50 | 15 | 150 | 46 | 400 | 120[k] |

*(continues)*

**Table A.7.6**  *Continued*

| Type of Occupancy | Common Path Limit | | | | Dead-End Limit | | | | Travel Distance Limit | | | |
| | Unsprinklered | | Sprinklered | | Unsprinklered | | Sprinklered | | Unsprinklered | | Sprinklered | |
| | ft | m | ft | m | ft | m | ft | m | ft | m | ft | m |
|---|---|---|---|---|---|---|---|---|---|---|---|---|
| **Business** | | | | | | | | | | | | |
| New | 75 | 23[l] | 100 | 30[l] | 20 | 6.1 | 50 | 15 | 200 | 61 | 300 | 91 |
| Existing | 75 | 23[l] | 100 | 30[l] | 50 | 15 | 50 | 15 | 200 | 61 | 300 | 91 |
| **Industrial** | | | | | | | | | | | | |
| General | 50 | 15 | 100 | 30 | 50 | 15 | 50 | 15 | 200 | 61[m] | 250 | 75[n] |
| Special purpose | 50 | 15 | 100 | 30 | 50 | 15 | 50 | 15 | 300 | 91 | 400 | 122 |
| High hazard | 0 | 0 | 0 | 0 | 0 | 0 | 0 | 0 | | | 75 | 23 |
| Aircraft servicing hangars, finished ground level floor | 50 | 15[o] | 100 | 30[o] | 50 | 15[o] | 50 | 15[o] | footnote m | footnote m | footnote m | footnote m |
| Aircraft servicing hangars, mezzanine floor | 50 | 15[o] | 75 | 23[o] | 50 | 15[o] | 50 | 15[o] | 75 | 23 | 75 | 23 |
| **Storage** | | | | | | | | | | | | |
| Low hazard | NR | NR | NR | NR | NR | NR | NR | NR | NR | NR | NR | NR |
| Ordinary hazard | 50 | 15 | 100 | 30 | 50 | 15 | 100 | 30 | 200 | 61 | 400 | 122 |
| High hazard | 0 | 0 | 0 | 0 | 0 | 0 | 0 | 0 | 75 | 23 | 100 | 30 |
| Parking structures, open[p] | 50 | 15 | 50 | 15 | 50 | 15 | 50 | 15 | 300 | 91 | 400 | 122 |
| Parking structures, enclosed | 50 | 15 | 50 | 15 | 50 | 15 | 50 | 15 | 150 | 46 | 200 | 60 |
| Aircraft storage hangars, finished ground level floor | 50 | 15[o] | 100 | 30[o] | 50 | 15[o] | 50 | 15[o] | footnote m | footnote m | footnote m | footnote m |
| Aircraft servicing hangars, mezzanine floor | 50 | 15[o] | 75 | 23[o] | 50 | 15[o] | 50 | 15[o] | 75 | 23 | 75 | 23 |
| Underground spaces in grain elevators | 50 | 15[o] | 100 | 30[o] | 50 | 15[o] | 100 | 30[o] | 200 | 61 | 400 | 122 |

NR: No requirement. NA: Not applicable.

[a]For common path serving >50 persons, 20 ft (6.1 m); for common path serving ≤50 persons, 75 ft (23 m).

[b]Dead-end corridors of 20 ft (6.1 m) permitted; dead-end aisles of 20 ft (6.1 m) permitted.

[c]See Chapters 12 and 13 for special considerations for smoke-protected assembly seating in arenas and stadia.

[d]This dimension is for the total travel distance, assuming incremental portions have fully utilized their permitted maximums. For travel distance within the room, and from the room exit access door to the exit, see the appropriate occupancy chapter.

[e]See 19.2.5.2.

[f]See business occupancies, Chapters 38 and 39.

[g]See Chapter 23 for special considerations for existing common paths.

[h]This dimension is from the room/corridor or suite/corridor exit access door to the exit; thus, it applies to corridor common path.

[i]See the appropriate occupancy chapter for requirements for second exit access based on room area.

[j]See the appropriate occupancy chapter for special travel distance considerations for exterior ways of exit access.

[k]See 36.4.4 and 37.4.4 for special travel distance considerations in covered malls considered to be pedestrian ways.

[l]See Chapters 38 and 39 for special common path considerations for single-tenant spaces.

[m]See Chapters 40 and 42 for special requirements on spacing of doors in aircraft hangars.

[n]See Chapter 40 for industrial occupancy special travel distance considerations.

[o]See Chapters 40 and 42 for special requirements if high hazard conditions exist.

[p]See 42.8.2.6.2 for special travel distance considerations in open parking structures.

**A.7.6.1** The natural exit access (path of travel) is influenced by the contents and occupancy of the building. Furniture, fixtures, machinery, or storage can serve to increase the length of travel. It is good practice in building design to recognize the influence of contents and occupancy by spacing exits for a completely open floor area at closer intervals than are required, thus reducing the hazard of excessive travel distances due to the introduction of furniture, fixtures, machinery, or storage and minimizing the possibility of violating the travel distance requirements of this *Code*.

**A.7.6.3** Examples of locations where open stairways might exist include between mezzanines or balconies and the floor below.

**A.7.7.1** An exit from the upper stories in which the direction of egress travel is generally downward should not be arranged so that it is necessary to change to travel in an upward direction at any point before discharging to the outside. A similar prohibition of reversal of the vertical component of travel should be applied to exits from stories below the floor of exit discharge. However, an exception is permitted in the case of stairs used in connection with overhead or underfloor exit passageways that serve the street floor only.

It is important that ample roadways be available from buildings in which there are large numbers of occupants so that exits will not be blocked by persons already outside. Two or more avenues of departure should be available for all but very small places. Location of a larger theater — for example, on a narrow dead-end street — might be prohibited by the authority having jurisdiction under this rule, unless some alternate way of travel to another street is available.

Exterior walking surfaces within the exit discharge are not required to be paved and often are provided by grass or similar surfaces. Where discharging exits into yards, across lawns, or onto similar surfaces, in addition to providing the required width to allow all occupants safe access to a public way, such access also is required to meet the following:

(1) Provisions of 7.1.7 with respect to changes in elevation
(2) Provisions of 7.2.2 for stairs, as applicable
(3) Provisions of 7.2.5 for ramps, as applicable
(4) Provisions of 7.1.10 with respect to maintaining the means of egress free of obstructions that would prevent its use, such as snow and the need for its removal in some climates

**A.7.7.3.4** Examples include partitions and gates. The design should not obstruct the normal movement of occupants to the exit discharge. Signs, graphics, or pictograms, including tactile types, might be permitted for existing exit enclosures where partitions or gates would obstruct the normal movement of occupants to the exit discharge.

**A.7.8.1.1** Illumination provided outside the building should be to either a public way or a distance away from the building that is considered safe, whichever is closest to the building being evacuated.

**A.7.8.1.2.3** A consideration for the approval of automatic, motion sensor–type lighting switches, controls, timers, or controllers is whether the equipment is listed as a fail-safe device for use in the means of egress.

**A.7.8.1.3** A desirable form of means of egress lighting is by lights recessed in walls about 12 in. (305 mm) above the floor. Such lights are not likely to be obscured by smoke.

**A.7.8.1.3(4)** Some processes, such as manufacturing or handling of photosensitive materials, cannot be performed in areas provided with the minimum specified lighting levels. The use of spaces with lighting levels below 1 ft-candle (10.8 lux) might necessitate additional safety measures, such as written emergency plans, training of new employees in emergency evacuation procedures, and periodic fire drills.

**A.7.8.1.4** An example of the failure of any single lighting unit is the burning out of an electric bulb.

**A.7.8.2.1** An example of a power source with reasonably ensured reliability is a public utility electric service.

**A.7.9.1.1** Emergency lighting outside the building should provide illumination to either a public way or a distance away from the building that is considered safe, whichever is closest to the building being evacuated.

**A.7.9.2.1** The illumination uniformity ratio is determined by the following formula:

$$\frac{\text{Maximum illumination at any point}}{\text{Minimum illumination at any point}}$$

**A.7.9.2.3** Where approved by the authority having jurisdiction, this requirement is permitted to be met by means such as the following.

(1) Two separate electric lighting systems with independent wiring, each adequate alone to provide the specified lighting, as follows:
   (a) One such system is permitted to be supplied from an outside source, such as a public utility service, and the other from an electric generator on the premises driven by an independent source of power.
   (b) Both sources of illumination should be in regular simultaneous operation whenever the building is occupied during periods of darkness.
(2) An electric circuit, or circuits, used only for means of egress illumination, with two independent electric sources arranged so that, on the failure of one, the other will automatically and immediately operate, as follows:
   (a) One such source is permitted to be a connection from a public utility, or similar outside power source, and the other an approved storage battery with suitable provision to keep it automatically charged.
   (b) The battery should be provided with automatic controls that, after operation of the battery due to failure of the primary power source or operation for the purpose of turning off the primary electric source for the lights, will shut off the battery after its specified period of operation and will automatically recharge and ready the battery for further service when the primary current source is turned on again.
(3) Electric battery–operated emergency lighting systems complying with the provisions of 7.9.2.3 and operating on a separate circuit and at a voltage different from that of the primary light can be used where permitted. (*See NFPA 70, National Electrical Code.*)

These requirements are not intended to prohibit the connection of a feeder serving exit lighting and similar emergency functions ahead of the service disconnecting means, but such provision does not constitute an acceptable alternate source of power. Such a connection furnishes only supplementary protection for emergency electrical functions, particularly where intended to allow the fire department to open the main disconnect without hampering exit activities. Provision should be made to alert the fire department that certain power and lighting is fed by an emergency generator and will continue operation after the service disconnect is opened.

Where emergency lighting is provided by automatic transfer between normal power service and an emergency generator, it is the intent to prohibit the installation, for any reason, of a single switch that can interrupt both energy sources.

**A.7.9.2.6**  Automobile-type lead storage batteries are not suitable by reason of their relatively short life when not subject to frequent discharge and recharge as occurs in automobile operation.

For proper selection and maintenance of appropriate batteries, see *NFPA 70, National Electrical Code.*

**A.7.9.3.1.1(2)**  Technical justification for extending test intervals past 30 days should be based on recorded event history (data) and should include evaluation of the following criteria:

(1) Number of egress lighting units
(2) Number of 30-second tests for analysis
(3) Re-evaluation period (confirm or adjust intervals)
(4) Number of fixtures found obstructed
(5) Number of fixtures found misaligned
(6) Fixtures found to be missing
(7) Fixtures found damaged
(8) Battery design
(9) Type of light source
(10) Fixture design (manufacturer)
(11) Number of light fixtures per exit path
(12) Existence of fire, smoke, and thermal barriers
(13) Evacuation capability
(14) Maximum egress time
(15) Hours of occupancy
(16) Number of recorded bulb failures
(17) Number of recorded fixture failures
(18) Single fixture reliability
(19) Repairs — mean time to repair
(20) Lighted egress path probability of success or failure — monthly upper tolerance limit
(21) Lighted egress path probability of success or failure — quarterly upper tolerance limit (estimated)

**A.7.10.1.2.1**  Where a main entrance also serves as an exit, it will usually be sufficiently obvious to occupants so that no exit sign is needed.

The character of the occupancy has a practical effect on the need for signs. In any assembly occupancy, hotel, department store, or other building subject to transient occupancy, the need for signs will be greater than in a building subject to permanent or semipermanent occupancy by the same people, such as an apartment house where the residents are presumed to be familiar with exit facilities by reason of regular use thereof. Even in a permanent residence–type building, however, there is a need for signs to identify exit facilities such as outside stairs that are not subject to regular use during the normal occupancy of the building.

There are many types of situations where the actual need for signs is debatable. In cases of doubt, however, it is desirable to be on the safe side by providing signs, particularly because posting signs does not ordinarily involve any material expense or inconvenience.

The requirement for the locations of exit signs visible from any direction of exit access is illustrated in Figure A.7.10.1.2.1.

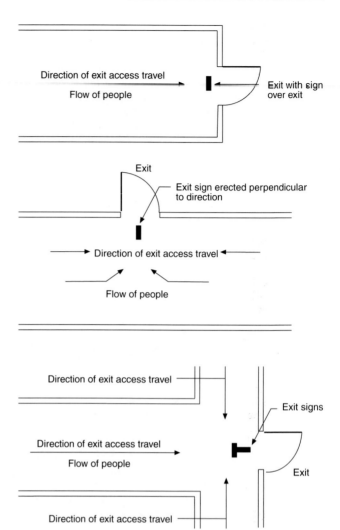

**FIGURE A.7.10.1.2.1  Location of Exit Signs.**

**A.7.10.1.2.2**  The direction of travel to the exit discharge within a stair enclosure with horizontal components in excess of the typical landings might need additional signage to be readily visible or obvious. Exit signs should be installed above doors through which the egress path leads. Directional exit signs should be installed where the horizontal egress path changes directions. The stairway marking signs required by 7.2.2.5.4, provided within the stair enclosure at each floor landing, indicate the vertical direction to exit discharge.

**A.7.10.1.5.2**  For externally illuminated signs in accordance with 7.10.6 and internally illuminated signs listed without a viewing distance, the rated viewing distance should be considered to be 100 ft (30 m). However, placing signs to meet the 100 ft (30 m) viewing distance in other than exit access corridors might create operating difficulties or encourage placement of a sign above the line of sight. To resolve the viewing distance versus placement issue, consideration should be given to proportionally increasing the level of illumination and the size of the exit legend to the viewing distance, if signs are placed at greater distances.

**A.7.10.1.6**  See A.7.10.3.

**A.7.10.1.7**  See 3.3.144.2 for the definition of the term *internally illuminated.*

**A.7.10.1.8**  In stores, for example, an otherwise adequate exit sign could be rendered inconspicuous by a high-intensity illuminated advertising sign located in the immediate vicinity.

Red is the traditional color for exit signs and is required by law in many places. However, at an early stage in the development of the *Code*, a provision made green the color for exit signs, following the concept of traffic lights in which green indicates safety and red is the signal to stop. During the period when green signs were specified by the *Code*, many such signs were installed, but the traditional red signs also remained. In 1949, the Fire Marshals Association of North America voted to request that red be restored as the required exit sign color, because it was found that the provision for green involved difficulties in law enactment that were out of proportion to the importance of safety. Accordingly, the 10th edition of the *Code* specified red where not otherwise required by law. The present text avoids any specific requirement for color, based on the assumption that either red or green will be used in most cases and that there are some situations in which a color other than red or green could actually provide better visibility.

**A.7.10.2.1**  A sign complying with 7.10.2 and indicating the direction of the nearest approved exit should be placed at the point of entrance to any escalator or moving walk. *(See A.7.10.3.)*

**A.7.10.3**  Where graphics are used, the symbols provided in NFPA 170, *Standard for Fire Safety and Emergency Symbols*, should be used. Such signs need to provide equal visibility and illumination and are to comply with the other requirements of Section 7.10.

**A.7.10.3.2**  Pictograms are permitted to be used in lieu of, or in addition to, signs with text.

**A.7.10.4**  It is not the intent of this paragraph to require emergency lighting but only to have the sign illuminated by emergency lighting if emergency lighting is required and provided.

It is not the intent to require that the entire stroke width and entire stroke height of all letters comprising the word EXIT be visible per the requirements of 7.10.6.3 under normal or emergency lighting operation, provided that the sign is visible and legible at a 100 ft (30 m) distance under all room illumination conditions.

**A.7.10.5.1**  See A.7.8.1.3(4).

**A.7.10.5.2**  It is the intent to prohibit the use of a freely accessible light switch to control the illumination of either an internally or externally illuminated exit sign.

**A.7.10.5.2.2**  The flashing repetition rate should be approximately one cycle per second, and the duration of the off-time should not exceed ¼ second per cycle. During on-time, the illumination levels need to be provided in accordance with 7.10.6.3. Flashing signs, when activated with the fire alarm system, might be of assistance.

**A.7.10.6.1**  Experience has shown that the word EXIT, or other appropriate wording, is plainly legible at 100 ft (30 m) if the letters are as large as specified in 7.10.6.1.

**A.7.10.6.2**  Figure A.7.10.6.2 shows examples of acceptable locations of directional indicators with regard to left and right orientation. Directional indicators are permitted to be placed under the horizontal stroke of the letter T, provided that spacing of not less than ⅜ in. (10 mm) is maintained from the horizontal and vertical strokes of the letter T.

EXIT>

<EXIT

<EXIT>

**FIGURE A.7.10.6.2  Directional Indicators.**

**A.7.10.6.3**  Colors providing a good contrast are red or green letters on matte white background. Glossy background and glossy letter colors should be avoided.

The average luminance of the letters and background is measured in footlamberts or candela per square meter. The contrast ratio is computed from these measurements by the following formula:

$$\text{Contrast} = \frac{L_g - L_e}{L_g}$$

Where $L_g$ is the greater luminance and $L_e$ is the lesser luminance, either the variable $L_g$ or $L_e$ is permitted to represent the letters, and the remaining variable will represent the background. The average luminance of the letters and background can be computed by measuring the luminance at the positions indicated in Figure A.7.10.6.3 by numbered circles.

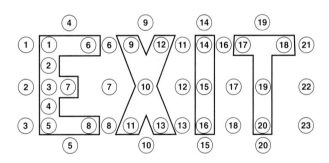

**FIGURE A.7.10.6.3  Measurement of Exit Sign Luminance.**

**A.7.10.7.2**  Photoluminescent signs need a specific minimum level of light on the face of the sign to ensure that the sign is charged for emergency operation and legibility in both the normal and emergency modes. Additionally, the type of light source (e.g., incandescent, fluorescent, halogen, metal halide) is important. Each light source produces different types of visible and invisible light (e.g., UV) that might affect the ability of some photoluminescent signs to charge and might also affect the amount of light output available during emergency mode. This type of sign would not be suitable where the illumination levels are permitted to decline. The charging light source should not be connected to automatic timers, because continuous illumination of the sign is needed; otherwise, the sign illumination would not be available, because it would be discharged.

**A.7.10.8.3** The likelihood of occupants mistaking passageways or stairways that lead to dead-end spaces for exit doors and becoming trapped governs the need for exit signs. Thus, such areas should be marked with a sign that reads as follows.

NO EXIT

Supplementary identification indicating the character of the area, such as TO BASEMENT, STOREROOM, LINEN CLOSET, or the like, is permitted to be provided.

**A.7.10.8.4(1)** These signs are to be used in place of signs that indicate that elevators are not to be used during fires. Examples of these signs include the following:

In the Event of Fire, This Elevator Will Be Used

by the Fire Department for Evacuation of People.

PROTECTED ELEVATOR—

USABLE IN EMERGENCIES

**A.7.10.8.4(2)** The wording of these signs should reflect human behavior in fires and the control specifics of the elevator system. Subparagraph 7.10.8.4 addresses signs, but provisions for notification of the vision impaired need to be considered. For information about human behavior with respect to elevator evacuation, see Groner and Levin, "Human Factor Considerations in the Potential for Using Elevators in Building Emergency Evacuation Plans"; Levin and Groner, "Human Behavior Aspects of Staging Areas for Fire Safety in GSA Buildings"; and Levin and Groner, "Human Factor Considerations for the Potential Use of Elevators for Fire Evacuation of FAA Air Traffic Control Towers." Some examples of messages on signs that could be displayed are shown in Table A.7.10.8.4(2).

**Table A.7.10.8.4(2) Elevator Status Messages**

| Elevator Status | Message |
|---|---|
| Normal use | Elevator in Service |
| Elevators recalled and waiting for fire service | Please Wait for Fire Department or Use Stairs |
| Elevator out of service | Elevator Out of Service |

**A.7.10.8.5** Egress paths with multiple turns can often be confusing with respect to which exit route will lead to the closest exit door. Floor evacuation diagrams can eliminate the guesswork by giving the occupant a point of reference by the YOU ARE HERE symbol. The entire floor plan should be shown with the primary and secondary exit routes, exit stairs, and elevators clearly identified. For further information, see ASTM E 2238, *Standard Guide for Evacuation Route Diagrams.*

**A.7.11.1** Seventy-five feet (23 m) can be traversed in approximately 10 seconds to 15 seconds, even when allowing for a momentary delay to decide which way to go, during which it can be assumed that the average individual can hold his or her breath.

**A.7.13.1** 29 CFR 1910.146 of the OSHA regulations describes the aspects of normally unoccupied areas. For example, hazardous atmosphere criteria are presented, and

asphyxiation risk due to an entrance becoming engulfed are addressed. The areas described by 29 CFR 1910.146, "Permitted Required Confined Spaces," would be considered hazardous if located within a building or structure regulated by NFPA *101.*

**A.7.13.2.1** Egress from normally unoccupied building service equipment support areas not exceeding 45,000 ft$^2$ (4180 m$^2$) is permitted to be by access panels or other hardware not complying with the door requirements of 7.2.1.

**A.7.14.1.1** The Phase I Emergency Recall Operation mandated by the firefighters' emergency operation provisions of ASME A17.1/CSA B 44, *Safety Code for Elevators and Escalators,* recalls elevators upon detection of smoke by smoke detectors installed in the following locations:

(1) At each floor served by the elevator in the lobby (landing) adjacent to the hoistway doors
(2) In the associated elevator machine room
(3) In the elevator hoistway where sprinklers are located in the hoistway

Where smoke from a fire remote from the elevator lobby (landing), elevator machine room, and elevator hoistway can be kept from reaching the elevator lobby (landing), elevator machine room, and elevator hoistway, the associated elevators can continue to operate in a fire emergency. The provisions of Section 7.14 address the features that need to be provided to make such elevator operation safe for evacuation.

**A.7.14.2.1** Building occupants have traditionally been taught not to use elevators in fire or similar emergencies. The emergency plan should include more than notification that the elevators can be used for emergency evacuation. The plan should include training to make occupants aware that the elevators will be available only for the period of time prior to elevator recall via smoke detection in the elevator lobby, elevator machine room, or elevator hoistway. Occupants should be prepared to use the exit stairs (which are required to be directly accessible from the elevator lobby by 7.14.8.3) where the elevator has been called out of service.

**A.7.14.3.2** The emergency voice/alarm communication system with the ability to provide voice directions on a selective basis to any building floor might be used to instruct occupants of the fire floor who are able to use stairs to relocate to a floor level below. The selective voice notification feature might be used to provide occupants of a given elevator lobby with a status report or supplemental instructions.

**A.7.14.3.3** An audible notification appliance will need to be positioned in the elevator lobby in order to meet the requirement of 7.14.3.4. The continued use of the occupant evacuation elevator system is predicated on elevator lobby doors that are closed to keep smoke from reaching the elevator lobby smoke detector that is arranged to initiate the Phase I Emergency Recall Operation.

**A.7.14.4.2** The presence of sprinklers in the elevator machine room would necessitate the installation of a shunt trip for automatically disconnecting the main line power for compliance with ASME A17.1/CSA B44, *Safety Code for Elevators and Escalators,* as it is unsafe to operate elevators while sprinkler water is being discharged in the elevator machine room. The presence of a shunt trip conflicts with the needs of the occupant evacuation elevator, as it disconnects the power without ensuring that the elevator is first returned to a safe floor so as to prevent trapping occupants. The provision of 7.14.4.2, prohibiting the sprinklering of elevator machine rooms, deviates from the requirements of NFPA 13, *Standard for the Installation of Sprinkler Systems*, which permits no such exemption. However, NFPA 13 permits a similar exemption for electrical equipment rooms where the room is dedicated to electrical equipment only; the equipment is installed in a 2-hour fire-rated enclosure, including protection for penetrations; and no combustible storage is stored in the room. Similar safeguards are imposed on the occupant evacuation elevator by 7.14.6.1 and 7.14.6.2.

**A.7.14.4.3** NFPA 13, *Standard for the Installation of Sprinkler Systems*, permits sprinklers to be omitted from the top of the elevator hoistway where the hoistway for passenger elevators is noncombustible and the car enclosure materials meet the requirements of ASME A17.1/CSA B44, *Safety Code for Elevators and Escalators*. The provision of 7.14.5.3 restricts occupant evacuation elevators to passenger elevators that are in noncombustible hoistways and for which the car enclosure materials meet the requirements of ASME A17.1/CSA B 44. (*See 7.14.5.3.*)

**A.7.14.5.2** Elevator shunt breakers are intended to disconnect the electric power to an elevator prior to sprinkler system waterflow impairing the functioning of the elevator. The provision of 7.14.4.2 prohibits the installation of sprinklers in the elevator machine room and at the top of the elevator hoistway, obviating the need for shunt breakers. The provision of 7.14.5.2 is not actually an exemption to the provisions of ASME A17.1/CSA B44, *Safety Code for Elevators and Escalators*, as ASME A17.1/CSA B44 requires the automatic main line power disconnect (shunt trip) only where sprinklers are located in the elevator machine room or in the hoistway more than 24 in. (610 mm) above the pit floor. The provision of 7.14.4.2 prohibits sprinklers in the elevator machine room. The provision of 7.14.4.3 prohibits sprinklers at the top of the hoistway and at other points in the hoistway more than 24 in. (610 mm) above the pit floor in recognition of the limitations on combustibility established by 7.14.5.3.

**A.7.14.6.1** The minimum 2-hour fire resistance–rated separation is based on the omission of sprinklers from the elevator machine room in accordance with 7.14.4.2.

**A.7.14.6.2** The requirement of 7.14.6.2 is consistent with that in ASME A17.1/CSA B44, *Safety Code for Elevators and Escalators*, which permits only machinery and equipment used in conjunction with the function or use of the elevator to be in the elevator machine room. An inspection program should be implemented to ensure that the elevator machine room is kept free of storage.

**A.7.14.7.2** Wiring or cables that provide control signals are exempt from the protection requirements of 7.14.7.2, provided that such wiring or cables, where exposed to fire, will not disable Phase II Emergency In-Car Operation once such emergency operation has been activated.

**A.7.14.8.2.1** Elevator lobbies provide a safe place for building occupants to await the elevators and extend the time available for such use by providing a barrier to smoke and heat that might threaten the elevator car or hoistway. Smoke detectors within the elevator lobbies are arranged to initiate a Phase I Emergency Recall Operation if the lobby is breached by smoke.

**A.7.14.8.6** The performance-based language of 7.14.8.6 permits alternate design options to prevent water from an operating sprinkler system from infiltrating the hoistway enclosure. For example, such approved means might include drains and sloping the floor. The objective of the water protection requirement is to limit water discharged from sprinklers operating on the floor of fire origin from entering the hoistway, as it might by flowing into the lobby and under the landing doors, interfering with safety controls normally located on the front of the elevator car. A small flow of water (of the order of the flow from a single sprinkler) should be able to be diverted by the landing doorway nose plate to the sides of the opening, where it can do little harm. The requirement is intended to protect from water from sprinklers outside the elevator lobby, since the activation of sprinklers in the lobby would be expected to be preceded by activation of the lobby smoke detector that recalls the elevators.

Water protection can be achieved in any of several ways. Mitigation features that should be effective in keeping the waterflow from a sprinkler out of the hoistway include the following:

(1) Raised lip in accordance with 7.1.6.2 and a floor drain
(2) Sloped floor and a floor drain
(3) Sealed sill plates and baseboards on both sides of the lobby partitions and along the perimeter of the hoistway shaft

**A.7.14.8.7** The elevator lobby doors addressed in 7.14.8.7 do not include the elevator hoistway doors. The elevator hoistway doors serving fire-rated hoistway enclosures in accordance with 8.6.5 must meet the criteria of Table 8.3.4.2.

**A.8.2.1.2** Table A.8.2.1.2 is from *NFPA 5000, Building Construction and Safety Code*, and is reproduced in this annex for the convenience of users of this *Code*.

**A.8.2.3.1** ASTM E 119, *Standard Test Methods for Fire Tests of Building Construction and Materials*, and ANSI/UL 263, *Standard for Fire Tests of Building Construction and Materials*, are considered nationally recognized methods of determining fire resistance and have been found to yield equivalent test methods.

**A.8.2.4.2** The intent of this provision is to allow the provisions of either ASCE/SFPE 29, *Standard Calculation Methods for Structural Fire Protection*, or ACI 216.1/TMS 0216.1, *Standard Method for Determining Fire Resistance of Concrete and Masonry Assemblies*, for the calculation for fire resistance of concrete or masonry elements or assemblies.

**A.8.3.1.1(4)** Walls in good condition with lath and plaster, or gypsum board of not less than ½ in. (13 mm) on each side, can be considered as providing a minimum ½-hour fire resistance rating. Additional information on archaic material assemblies can be found in Appendix I of NFPA 914, *Code for Fire Protection of Historic Structures*.

**A.8.3.1.2** To ensure that a fire barrier is continuous, it is necessary to seal completely all openings where the fire barrier abuts other fire barriers, the exterior walls, the floor below, and the floor or ceiling above. In 8.3.1.2(2), the fire resistance rating of the bottom of the interstitial space is provided by that membrane alone. Ceilings of rated floor/ceiling and roof/ceiling assemblies do not necessarily provide the required fire resistance.

**A.8.3.2.1.1** Fire resistance–rated glazing complying with 8.3.2, where not installed in a door, is considered a wall, not an opening protective.

**Table A.8.2.1.2 Fire Resistance Ratings for Type I Through Type V Construction (hours)**

| Construction Element | Type I | | Type II | | | Type III | | Type IV | Type V | |
|---|---|---|---|---|---|---|---|---|---|---|
| | 442 | 332 | 222 | 111 | 000 | 211 | 200 | 2HH | 111 | 000 |
| **Exterior Bearing Walls**[a] | | | | | | | | | | |
| Supporting more than one floor, columns, or other bearing walls | 4 | 3 | 2 | 1 | 0[b] | 2 | 2 | 2 | 1 | 0[b] |
| Supporting one floor only | 4 | 3 | 2 | 1 | 0[b] | 2 | 2 | 2 | 1 | 0[b] |
| Supporting a roof only | 4 | 3 | 1 | 1 | 0[b] | 2 | 2 | 2 | 1 | 0[b] |
| **Interior Bearing Walls** | | | | | | | | | | |
| Supporting more than one floor, columns, or other bearing walls | 4 | 3 | 2 | 1 | 0 | 1 | 0 | 2 | 1 | 0 |
| Supporting one floor only | 3 | 2 | 2 | 1 | 0 | 1 | 0 | 1 | 1 | 0 |
| Supporting roofs only | 3 | 2 | 1 | 1 | 0 | 1 | 0 | 1 | 1 | 0 |
| **Columns** | | | | | | | | | | |
| Supporting more than one floor, columns, or other bearing walls | 4 | 3 | 2 | 1 | 0 | 1 | 0 | H | 1 | 0 |
| Supporting one floor only | 3 | 2 | 2 | 1 | 0 | 1 | 0 | H | 1 | 0 |
| Supporting roofs only | 3 | 2 | 1 | 1 | 0 | 1 | 0 | H | 1 | 0 |
| **Beams, Girders, Trusses, and Arches** | | | | | | | | | | |
| Supporting more than one floor, columns, or other bearing walls | 4 | 3 | 2 | 1 | 0 | 1 | 0 | H | 1 | 0 |
| Supporting one floor only | 2 | 2 | 2 | 1 | 0 | 1 | 0 | H | 1 | 0 |
| Supporting roofs only | 2 | 2 | 1 | 1 | 0 | 1 | 0 | H | 1 | 0 |
| **Floor-Ceiling Assemblies** | 2 | 2 | 2 | 1 | 0 | 1 | 0 | H | 1 | 0 |
| **Roof-Ceiling Assemblies** | 2 | 1½ | 1 | 1 | 0 | 1 | 0 | H | 1 | 0 |
| **Interior Nonbearing Walls** | 0 | 0 | 0 | 0 | 0 | 0 | 0 | 0 | 0 | 0 |
| **Exterior Nonbearing Walls**[c] | 0[b] | 0[b] | 0[b] | 0[b] | 0[b] | 0[b] | 0[b] | 0[b] | 0[b] | 0[b] |

H: Heavy timber members *(see NFPA 5000 for requirements)*.
[a]See 7.3.2.1 of *NFPA 5000*.
[b]See Section 7.3 of *NFPA 5000*.
[c]See 7.2.3.2.12, 7.2.4.2.3, and 7.2.5.6.8 of *NFPA 5000*.
[*5000:* Table 7.2.1.1]

**A.8.3.3.2** Some door assemblies have been tested to meet the conditions of acceptance of ASTM E 119, *Standard Test Methods for Fire Tests of Building Construction and Materials*, or ANSI/UL 263, *Standard for Fire Tests of Building Construction and Materials*. Where such assemblies are used, the provisions of 8.3.2 should be applied instead of those of 8.3.3.2.

**A.8.3.3.2.3** In existing installations, it is important to be able to determine the fire protection rating of the fire door. However, steel door frames that are well set in the wall might be judged as acceptable even if the frame label is not legible.

**A.8.3.3.6** Some window assemblies have been tested to meet the conditions of acceptance of ASTM E 119, *Standard Test Methods for Fire Tests of Building Construction and Materials*, or ANSI/UL 263, *Standard for Fire Tests of Building Construction and*

*Materials*. Where such assemblies are used, the provisions of 8.3.2 should be applied instead of those of 8.3.3.6.

**A.8.3.4.2** Longer ratings might be required where opening protectives are provided for property protection as well as life safety. NFPA 80, *Standard for Fire Doors and Other Opening Protectives*, should be consulted for standard practice in the selection and installation of fire door assemblies and fire window assemblies.

**Table 8.3.4.2.** A vision panel in a fire door is not a fire window, and, thus, it is not the intent of the "NP" notations in the "Fire Window Assemblies" column of Table 8.3.4.2 to prohibit vision panels in fire doors.

**A.8.3.5.1** ASTM E 2174, *Standard Practice for On-Site Inspection of Installed Fire Stops*, provides guidance for the inspection of through-penetration fire stop systems tested in accordance with ASTM E 814, *Standard Test Method for Fire Tests of Through-Penetration Fire Stops*, and ANSI/UL 1479, *Standard for Fire Tests of Through-Penetration Firestops*.

**A.8.3.5.6.3(1)(c)** Criteria associated with fireblocking can be found in 8.14.2 of *NFPA 5000, Building Construction and Safety Code*.

**A.8.3.6.5** On-site inspection of firestopping is important in maintaining the integrity of any vertical or horizontal fire barrier. Two standard practice documents were developed with the ASTM process to allow inspections of through-penetration firestops, joints, and perimeter fire barrier systems. ASTM E 2393, *Standard Practice for On-Site Inspection of Installed Fire Resistive Joint Systems and Perimeter Fire Barriers*, provides guidance for the inspection of fire-resistive joints and perimeter fire barrier joint systems tested in accordance with the requirements of ASTM E 1966, *Standard Test Method for Fire-Resistive Joint Systems*, or with ANSI/UL 2079, *Standard for Tests for Fire Resistance of Building Joint Systems*. ASTM E 2393 contains a standardized report format, which would lead to greater consistency for inspections.

**A.8.3.6.7** The provisions of 8.3.6.7 are intended to restrict the interior vertical passage of flame and hot gases from one floor to another at the location where the floor intersects the exterior wall assembly. The requirements of 8.3.6.7 mandate sealing the opening between a floor and an exterior wall assembly to provide the same fire performance as that required for the floor. ASTM E 2307, *Standard Test Method for Determining Fire Resistance of Perimeter Fire Barrier Systems Using Intermediate-Scale, Multi-Story Test Apparatus*, is a test method for evaluating the performance of perimeter fire barrier systems. Some laboratories have tested and listed perimeter fire barrier systems essentially in accordance with the ASTM method. The ASTM test method evaluates the performance of perimeter fire barrier systems in terms of heat transfer and fire spread inside a building through the floor/exterior wall intersection. The current test method does not assess the ability of perimeter fire barrier systems to prevent the spread of fire from story to story via the exterior. However, some laboratories have included additional temperature measurement criteria in their evaluation of the exterior wall, and also evaluate vision glass breakage, as additional pass/fail criteria in an attempt to at least partially address this leapfrog effect.

**A.8.4.1** Although a smoke partition is intended to limit the free movement of smoke, it is not intended to provide an area that would be free of smoke.

**A.8.4.2(2)** An architectural, exposed, suspended-grid acoustical tile ceiling with penetrations for sprinklers, ducted HVAC supply and return-air diffusers, speakers, and recessed light fixtures is capable of limiting the transfer of smoke.

**A.8.4.3.4** Gasketing of doors should not be necessary, as the clearances in NFPA 80, *Standard for Fire Doors and Other Opening Protectives*, effectively achieve resistance to the passage of smoke if the door is relatively tight-fitting.

**A.8.4.6.2** An air-transfer opening, as defined in NFPA 90A, *Standard for the Installation of Air-Conditioning and Ventilating Systems*, is an opening designed to allow the movement of environmental air between two contiguous spaces.

**A.8.5.1** Wherever smoke barriers and doors therein require a degree of fire resistance, as specified by requirements in the various occupancy chapters (Chapters 12 through 42), the construction should be a fire barrier that has been specified to limit the spread of fire and restrict the movement of smoke.

Although a smoke barrier is intended to restrict the movement of smoke, it might not result in tenability throughout the adjacent smoke compartment. The adjacent smoke compartment should be safer than the area on the fire side, thus allowing building occupants to move to that area. Eventually, evacuation from the adjacent smoke compartment might be required.

**A.8.5.2** To ensure that a smoke barrier is continuous, it is necessary to seal completely all openings where the smoke barrier abuts other smoke barriers, fire barriers, exterior walls, the floor below, and the floor or ceiling above. It is not the intent to prohibit a smoke barrier from stopping at a fire barrier if the fire barrier meets the requirements of a smoke barrier (i.e., the fire barrier is a combination smoke barrier/fire barrier).

**A.8.5.4.1** For additional information on the installation of smoke control door assemblies, see NFPA 105, *Standard for Smoke Door Assemblies and Other Opening Protectives*.

**A.8.5.4.4** Where, because of operational necessity, it is desired to have smoke barrier doors that are usually open, such doors should be provided with hold-open devices that are activated to close the doors by means of the operation of smoke detectors and other alarm functions.

**A.8.6.2** Openings might include items such as stairways; hoistways for elevators, dumbwaiters, and inclined and vertical conveyors; shaftways used for light, ventilation, or building services; or expansion joints and seismic joints used to allow structural movements.

**A.8.6.5** The application of the 2-hour rule in buildings not divided into stories is permitted to be based on the number of levels of platforms or walkways served by the stairs.

**A.8.6.6(7)** Given that a mezzanine meeting the maximum one-third area criterion of 8.6.10.2.1 is not considered a story, it is permitted, therefore, to have 100 percent of its exit access within the communicating area run back through the story below.

**A.8.6.7** Where atriums are used, there is an added degree of safety to occupants because of the large volume of space into which smoke can be dissipated. However, there is a need to ensure that dangerous concentrations of smoke are promptly removed from the atrium, and the exhaust system needs careful design. For information about systems that can be used to provide smoke protection in these spaces, see the following:

(1) NFPA 92, *Standard for Smoke Control Systems*
(2) *Principles of Smoke Management*

**A.8.6.7(1)(c)** The intent of the requirement for closely spaced sprinklers is to wet the atrium glass wall to ensure that the surface of the glass is wet upon operation of the sprinklers, with a maximum spacing of sprinklers of 6 ft (1830 mm) on centers. Provided that it can be shown that the glass can be wet by the sprinklers using a given discharge rate, and that the 6 ft (1830 mm) spacing is not exceeded, the intent of the requirement is met. It is important that the entire glass area surface is wet. Due consideration should be given to the height of the glass panels and any horizontal members that might interfere with sprinkler wetting action.

**A.8.6.7(5)** See NFPA 92, *Standard for Smoke Control Systems*. The engineering analysis should include the following elements:

(1) Fire dynamics, including the following:
    (a) Fire size and location
    (b) Materials likely to be burning
    (c) Fire plume geometry
    (d) Fire plume or smoke layer impact on means of egress
    (e) Tenability conditions during the period of occupant egress
(2) Response and performance of building systems, including passive barriers, automatic detection and extinguishing, and smoke control
(3) Response time required for building occupants to reach building exits, including any time required to exit through the atrium as permitted by 8.6.7(2)

**A.8.6.7(6)** Activation of the ventilation system by manual fire alarms, extinguishing systems, and detection systems can cause unwanted operation of the system, and it is suggested that consideration be given to zoning of the activation functions so the ventilation system operates only when actually needed.

**A.8.6.9.1(4)** The intent of this requirement is to prohibit a communication of two compartments on the same floor via two convenience openings. This is represented in Figure A.8.6.9.1(4).

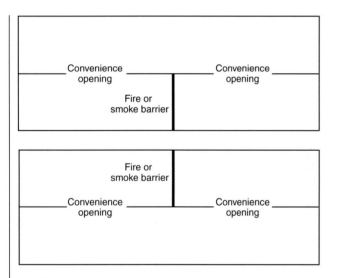

**FIGURE A.8.6.9.1(4) Examples of Convenience Openings That Communicate with Two Compartments on Same Floor in Violation of 8.6.9.1(4).**

**A.8.6.9.1(6)** This requirement prohibits means of egress down or up the convenience opening. It does not prohibit means of escape from running down or up the convenience opening within residential dwelling units.

**A.8.6.9.7(2)** The intent is to place a limitation on the size of the opening to which the protection applies. The total floor opening should not exceed twice the projected area of the escalator or moving walk at the floor. Also, the arrangement of the opening is not intended to circumvent the requirements of 8.6.7.

As with any opening through a floor, the openings around the outer perimeter of the escalators should be considered as vertical openings.

**A.8.6.11.2(2)** See NFPA 90A, *Standard for the Installation of Air-Conditioning and Ventilating Systems*.

**A.8.7.1.1** Areas requiring special hazard protection include, but are not limited to, areas such as those used for storage of combustibles or flammables, areas housing heat-producing appliances, or areas used for maintenance purposes.

**A.8.7.2** For details, see NFPA 68, *Standard on Explosion Protection by Deflagration Venting*.

**A.8.7.3.2** NFPA 58, *Liquefied Petroleum Gas Code*, permits portable butane-fueled appliances in restaurants and in attended commercial food catering operations where fueled by not in excess of two 10 oz (0.28 kg) LP-Gas capacity, nonrefillable butane containers having a water capacity not in excess of 1.08 lb (0.4 kg) per container. Containers are required to be directly connected to the appliance, and manifolding of containers is not permitted. Storage of cylinders is also limited to 24 containers, with an additional 24 permitted where protected by a 2-hour fire resistance–rated barrier.

**A.8.7.5** While the scope of NFPA 99, *Health Care Facilities Code*, is limited to health care occupancies, it is the intent that this requirement be applied to hyperbaric facilities used in all occupancies.

**A.9.4.1** Under certain conditions, elevators are recognized as means of egress.

The use of elevators for emergency evacuation purposes, where operated by trained emergency service personnel (e.g., building personnel, fire personnel), should be incorporated into the building evacuation program. Elevators are normally capable of manual, in-car fire fighter operation (Phase II) after elevator recall (Phase I). In addition, there usually are two or more shafts wherever there are more than three elevators, which further enhances the possibilities for elevator use during an emergency evacuation where operated by trained personnel.

In high-rise buildings, in towers, or in deep underground spaces where travel over considerable vertical distance on stairs can cause persons incapable of such physical effort to collapse before they reach the street exit, stairways are permitted to be used for initial escape from the immediate area of danger, and elevators are permitted to be used to complete the travel to the street.

It can be reasonably assumed that, in all buildings of sufficient height to indicate the need for elevators, elevators will be provided for normal use; for this reason, no requirements for mandatory installation of elevators are included in the *Code*.

For additional information on elevators, see ASME A17.1/CSA B44, *Safety Code for Elevators and Escalators*, and ASME A17.3, *Safety Code for Existing Elevators and Escalators*.

**A.9.4.5** Continued operation of solid-state elevator equipment is contingent on maintaining the ambient temperature in the range specified by the elevator manufacturer. If the machine room ventilation/air-conditioning is connected to the general building system, and that system is shut down during a fire, the fire department might lose the use of elevators due to excessive heat in the elevator machine room.

**A.9.6.1** The provisions of Section 9.6 cover the basic functions of a complete fire alarm system, including fire detection, alarm, and communications. These systems are primarily intended to provide the indication and warning of abnormal conditions, the summoning of appropriate aid, and the control of occupancy facilities to enhance protection of life.

Some of the provisions of Section 9.6 originated with *NFPA 72, National Fire Alarm and Signaling Code.* For purposes of this *Code,* some provisions of Section 9.6 are more stringent than those of *NFPA 72,* which should be consulted for additional details.

**A.9.6.1.5** Records of conducted maintenance and testing and a copy of the certificate of compliance should be maintained.

**A.9.6.1.6** A fire watch should at least involve some special action beyond normal staffing, such as assigning an additional security guard(s) to walk the areas affected. Such individuals should be specially trained in fire prevention and in occupant and fire department notification techniques, and they should understand the particular fire safety situation for public education purposes. *(Also see* NFPA 601, *Standard for Security Services in Fire Loss Prevention.)*

The term *out of service* in 9.6.1.6 is intended to imply that a significant portion of the fire alarm system is not in operation, such as an entire initiating device, signaling line, or notification appliance circuit. It is not the intent of the *Code* to require notification of the authority having jurisdiction, or evacuation of the portion of the building affected, for a single nonoperating device or appliance.

**A.9.6.1.8.1** The *Code* intends that only one smoke detector is required to be installed at the fire alarm control unit, the notification circuit power extenders, and the supervising station transmitting equipment, even when the area of the room would require more than one smoke detector if installed according to the spacing rules in *NFPA 72, National Fire Alarm and Signaling Code,* Chapter 5.

**A.9.6.2.5** It is not the intent of 9.6.2.5 to require manual fire alarm boxes to be attached to movable partitions or to equipment, nor is it the intent to require the installation of permanent structures for mounting purposes only.

**A.9.6.2.6** The manual fire alarm box required by 9.6.2.6 is intended to provide a means to manually activate the fire alarm system when the automatic fire detection system or waterflow devices are out of service due to maintenance or testing, or where human discovery of the fire precedes automatic sprinkler system or automatic detection system activation. Where the fire alarm system is connected to a monitoring facility, the manual fire alarm box required by 9.6.2.6 should be connected to a separate circuit that is not placed "on test" when the detection or sprinkler system is placed on test. The manual fire alarm box should be located in an area that is accessible to occupants of the building and should not be locked.

**A.9.6.2.7** Manual fire alarm boxes can include those with key-operated locks for detention areas or psychiatric hospitals, manual fire alarm boxes in areas where explosive vapors or dusts might be a hazard, or manual fire alarm boxes in areas with corrosive atmospheres. The appearance of manual fire alarm boxes for special uses often differs from those used in areas of normal occupancy. Manual fire alarm boxes, such as those with locks, that are located in areas where the general public has limited access might need to have signage advising persons to seek assistance from staff in the event a fire is noted.

**A.9.6.2.10.1.3** *NFPA 72, National Fire Alarm and Signaling Code,* mandates smoke alarms in all sleeping rooms, and interconnection of smoke alarms is required for both new and existing installations. Per, 9.6.2.10.1.2, the residential occupancy chapters determine whether smoke alarms are needed within sleeping rooms. Paragraph 9.6.2.10.1.3 limits the requirement for interconnection of smoke alarms to those in new construction. This *Code* does not intend to require compliant, existing smoke alarm installations to be interconnected. This *Code* is periodically revised to add retrospective requirements only where the need is clearly substantiated.

**A.9.6.2.10.3** A dwelling unit is that structure, area, room, or combination of rooms, including hotel rooms/suites, in which a family or individual lives. A dwelling unit includes living areas only and not common usage areas in multifamily buildings, such as corridors, lobbies, and basements.

**A.9.6.3.2.1** Elevator lobbies have been considered areas subject to unwanted alarms due to factors such as low ceilings and smoking. In the past several years, new features have become available to reduce this problem. These features are, however, not necessarily included in any specific installation.

**A.9.6.3.2.2** The concept addressed is that detectors used for releasing service, such as door or damper closing and fan shutdown, are not required to sound the building alarm.

**A.9.6.3.2.3** The concept addressed is that detectors used for releasing service, such as door or damper closing and fan shutdown, are not required to sound the building alarm.

**A.9.6.3.5.7** Visual notification appliances installed in large-volume spaces, such as arenas, stadiums, malls, and atriums, can be alternative devices that are not listed as visible notification appliances for fire alarm systems, provided that the notification objective of the visual signal is reasonably achieved. Examples of alternative devices include, but are not limited to, scoreboards, message boards, and other electronic devices that meet the performance objectives of visible fire alarm appliances in large-volume spaces.

It is the intent to permit the omission of visible notification appliances as identified in 9.6.3.5.7, provided that the adjacent areas that have not been specifically designated as exempt are provided with visible notification as required by 9.6.3.5.

**A.9.6.3.5.8** Documentation should be maintained with the as-built drawings so that inspection and testing personnel understand that the visible appliances have been exempted from certain areas and, therefore, can note the deviation on the acceptance test documentation and ongoing inspection reports. This will provide inspection and testing personnel with necessary details regarding the omission of visible notification appliances.

**A.9.6.3.6.2** To approve an evacuation plan to selectively notify building occupants, the authority having jurisdiction should consider several building parameters, including building compartmentation, detection and suppression system zones, occupant loads, and the number and arrangement of the means of egress.

In high-rise buildings, it is typical to evacuate the fire floor, the floor(s) above, and the floor immediately below. Other areas are then evacuated as the fire develops.

**A.9.6.3.9.2** The provisions of 9.6.3.9.2 offer an alternative to the emergency voice alarm and communications system provisions (live voice or recorded voice announcements) of *NFPA 72, National Fire Alarm and Signaling Code.* Occupancies such as large-venue assembly occupancies and mercantile mall buildings are occupancies in which the physical configuration (e.g., large-volume spaces), function, and human behavior (including elevated levels of occupant-generated noise) present challenges with respect to effective occupant notification by standard means in accordance with *NFPA 72.* Because the routine operation of these occupancies demands highly reliant, acoustically capable, and sufficiently audible public address systems, properly trained staff can be relied on to use these public address systems to effect occupant evacuation, relocation, or both.

As 9.6.3.9.2 specifically permits an alternative means of notification to that prescribed by *NFPA 72*, it does not mandate that the secondary power supply and the intelligibility and audibility facets of the public address system comply with *NFPA 72* or suggest that equivalency with the related provisions of *NFPA 72* is required. However, it is anticipated that, when approving the secondary power and audibility capabilities of public address systems, authorities having jurisdiction will ensure that these systems are conceptually comparable to the emergency voice alarm and communications system provisions of *NFPA 72*, such that a reliable and effective occupant notification system is provided.

**A.9.7.1.1** For a discussion of the effectiveness of automatic sprinklers, as well as a general discussion of automatic sprinklers, see the NFPA *Fire Protection Handbook.* Where partial sprinkler protection is permitted by another section of this *Code*, the limited area systems provisions of NFPA 13, *Standard for the Installation of Sprinkler Systems*, should apply.

**A.9.7.1.3** Properly designed automatic sprinkler systems provide the dual function of both automatic alarms and automatic extinguishment. Dual function is not provided in those cases where early detection of incipient fire and early notification of occupants are needed to initiate actions in behalf of life safety earlier than can be expected from heat-sensitive fire detectors.

**A.9.7.2.1** *NFPA 72, National Fire Alarm and Signaling Code*, provides details of standard practice in sprinkler supervision. Subject to the approval of the authority having jurisdiction, sprinkler supervision is also permitted to be provided by direct connection to municipal fire departments or, in the case of very large establishments, to a private headquarters providing similar functions. *NFPA 72* covers such matters. System components and parameters that are required to be monitored should include, but should not be limited to, control valves, water tank levels and temperatures, tank pressure, and air pressure on dry-pipe valves.

Where municipal fire alarm systems are involved, reference should also be made to NFPA 1221, *Standard for the Installation, Maintenance, and Use of Emergency Services Communications Systems*.

**A.9.7.4.1** For a description of standard types of extinguishers and their installation, maintenance, and use, see NFPA 10, *Standard for Portable Fire Extinguishers*. The labels of recognized testing laboratories on extinguishers provide evidence of tests indicating the reliability and suitability of the extinguisher for its intended use. Many unlabeled extinguishers are offered for sale that are substandard by reason of insufficient extinguishing capacity, questionable reliability, or ineffective extinguishing agents for fires in ordinary combustible materials or because they pose a personal hazard to the user.

**A.10.2** The requirements pertaining to interior finish are intended to restrict the spread of fire over the continuous surface forming the interior portions of a building.

Table A.10.2 shows the fire test methods and classification criteria that apply to different interior finish materials.

**Table A.10.2  Fire Testing of Interior Finish Materials**

| Material | Test Method | Acceptance Criterion | Application Requirement | Section |
|---|---|---|---|---|
| Interior wall and ceiling finish materials, except as shown in this table | ASTM E 84 or ANSI/UL 723 | Class A, in accordance with 10.2.3.4 (1) | As required by relevant sections | 10.2.3 |
| | ASTM E 84 or ANSI/UL 723 | Class B, in accordance with 10.2.3.4 (2) | As required by relevant sections | 10.2.3 |
| | ASTM E 84 or ANSI/UL 723 | Class C, in accordance with 10.2.3.4 (3) | As required by relevant sections | 10.2.3 |
| | NFPA 286 | In accordance with 10.2.3.7.2 | Permitted where Class A, B, or C is required by relevant sections | 10.2.3.2 |
| Materials having thickness <$\frac{1}{28}$ in. (<0.90 mm) applied directly to the surface of walls or ceilings | No testing required | | | 10.2.1.2 |
| Exposed portions of structural members complying with requirements for buildings of Type IV (2HH) construction in accordance with NFPA 220 | No testing required | | | 10.2.3.1 |
| Cellular or foamed plastics (exposed foamed plastics and foamed plastics used in conjunction with textile or vinyl facing or cover) | NFPA 286 | In accordance with 10.2.3.7.2 | Permitted where Class A, B, or C is required by relevant sections | 10.2.4.3.1.1(1) |
| | ANSI/UL 1715 | Pass | Permitted where Class A, B, or C is required by relevant sections | 10.2.4.3.1.1(2) |

**Table A.10.2** *Continued*

| Material | Test Method | Acceptance Criterion | Application Requirement | Section |
|---|---|---|---|---|
| | ANSI/UL 1040 | Pass | Permitted where Class A, B, or C is required by relevant sections | 10.2.4.3.1.1(3) |
| | FM 4880 | Pass | Permitted where Class A, B, or C is required by relevant sections | 10.2.4.3.1.1(4) |
| | Suitable large-scale fire test that substantiates combustibility characteristics for use intended under actual fire conditions | Pass | Permitted where Class A, B, or C is required by relevant sections | 10.2.4.3.1 |
| Textile wall coverings | NFPA 286 | In accordance with 10.2.3.7.2 | Permitted where Class A, B, or C is required by relevant sections | 10.2.4.1(6) |
| | NFPA 265, Method B | In accordance with 10.2.3.7.1 | Permitted on walls and partitions | 10.2.4.1(5) |
| | ASTM E 84 or ANSI/UL 723 | Class A, in accordance with 10.2.3.4 (1) | Permitted on walls, but also requires sprinklers per Section 9.7 | 10.2.4.1(1) |
| | ASTM E 84 or ANSI/UL 723 | Class A, in accordance with 10.2.3.4 (1) | Permitted on partitions not exceeding three-quarters of the floor-to-ceiling height or not exceeding 8 ft (2440 mm) in height, whichever is less | 10.2.4.1(2) |
| | ASTM E 84 or ANSI/UL 723 | Class A, in accordance with 10.2.3.4 (1) | Permitted to extend not more than 48 in. (1220 mm) above finished floor on ceiling-height walls and ceiling-height partitions | 10.2.4.1(3) |
| | ASTM E 84 or ANSI/UL 723 | Class A, in accordance with 10.2.3.4 (1) | Previously approved existing installations of textile material meeting the requirements of Class A permitted to be continued to be used | 10.2.4.1(4) |
| Expanded vinyl wall coverings | NFPA 286 | In accordance with 10.2.3.7.2 | Permitted where Class A, B, or C is required by relevant sections | 10.2.4.2(6) |
| | NFPA 265, Method B | In accordance with 10.2.3.7.1 | Permitted on walls and partitions | 10.2.4.2(5) |
| | ASTM E 84 or ANSI/UL 723 | Class A, in accordance with 10.2.3.4 (1) | Permitted on walls, but also requires sprinklers per Section 9.7 | 10.2.4.2(1) |
| | ASTM E 84 or ANSI/UL 723 | Class A, in accordance with 10.2.3.4 (1) | Permitted on partitions not exceeding three-quarters of the floor-to-ceiling height or not exceeding 8 ft (2440 mm) in height, whichever is less | 10.2.4.2(2) |
| | ASTM E 84 or ANSI/UL 723 | Class A, in accordance with 10.2.3.4 (1) | Permitted to extend not more than 48 in. (1220 mm) above finished floor on ceiling-height walls and ceiling-height partitions | 10.2.4.2(3) |
| | ASTM E 84 or ANSI/UL 723 | Class A, B, or C, in accordance with 10.2.3.4 | Existing installations of materials with appropriate wall finish classification for occupancy involved, and with classification in accordance with the provisions of 10.2.3.4 | 10.2.4.2(4) |

*(continues)*

**Table A.10.2** *Continued*

| Material | Test Method | Acceptance Criterion | Application Requirement | Section |
|---|---|---|---|---|
| Textile ceiling coverings | NFPA 286 | In accordance with 10.2.3.7.2 | Permitted where Class A, B, or C is required by relevant sections | 10.2.4.1(6) |
| | ASTM E 84 or ANSI/UL 723 | Class A, in accordance with 10.2.3.4 (1) | Permitted on walls, but also requires sprinklers per Section 9.7 | 10.2.4.1(1) |
| | ASTM E 84 or ANSI/UL 723 | Class A, in accordance with 10.2.3.4 (1) | Previously approved existing installations of textile material meeting the requirements of Class A permitted to be continued to be used | 10.2.4.1(4) |
| Expanded vinyl ceiling coverings | NFPA 286 | In accordance with 10.2.3.7.2 | Permitted where Class A, B, or C is required by relevant sections | 10.2.4.2(6) |
| | ASTM E 84 or ANSI/UL 723 | Class A, in accordance with 10.2.3.4 (1) | Permitted on walls, but also requires sprinklers per Section 9.7 | 10.2.4.2(1) |
| | ASTM E 84 or ANSI/UL 723 | Class A, B, or C, in accordance with 10.2.3.4 | Existing installations of materials with appropriate wall finish classification for occupancy involved, and with classification in accordance with the provisions of 10.2.3.4 | 10.2.4.2(4) |
| Interior trim, other than foamed plastic and other than wall base | ASTM E 84 or ANSI/UL 723 | Class C, in accordance with 10.2.3.4 | Interior wall and ceiling trim and incidental finish, other than wall base not in excess of 10 percent of the aggregate wall and ceiling areas of any room or space where interior wall and ceiling finish of Class A or Class B is required | 10.2.5.1 |
| | NFPA 286 | In accordance with 10.2.3.7.2 | Permitted where Class A, B, or C is required by relevant sections | 10.2.3.2 |
| Foamed plastic used as interior trim | ASTM E 84 or ANSI/UL 723 | Flame spread index ≤ 75 | (1) Minimum density of interior trim required to be 20 lb/ft$^3$ (320 kg/m$^3$) | 10.2.4.3.2 |
| | | | (2) Maximum thickness of interior trim required to be ½ in. (13 mm), and maximum width required to be 4 in. (100 mm) | 10.2.4.3.2 |
| | | | (3) Interior trim not permitted to constitute more than 10 percent of the wall or ceiling area of a room or space | 10.2.4.3.2 |
| | NFPA 286 | In accordance with 10.2.3.7.2 | Permitted where Class A, B, or C is required by relevant sections | 10.2.3.2 |

**Table A.10.2** *Continued*

| Material | Test Method | Acceptance Criterion | Application Requirement | Section |
|---|---|---|---|---|
| Fire-retardant coatings | NFPA 703 | Class A, B, or C, when tested by ASTM E 84 or ANSI/UL 723, in accordance with 10.2.3.4 | Required flame spread index or smoke developed index values of existing surfaces of walls, partitions, columns, and ceilings permitted to be secured by applying approved fire-retardant coatings to surfaces having higher flame spread index values than permitted; such treatments required to be tested, or listed and labeled for application to material to which they are applied | 10.2.6.1 |
| Carpet and carpetlike interior floor finishes | ASTM D 2859 | Pass | All areas | 10.2.7.1 |
| Floor coverings, other than carpet, judged to represent an unusual hazard (excluding traditional finish floors and floor coverings, such as wood flooring and resilient floor coverings) | NFPA 253 | Critical radiant flux $\geq 0.1$ W/cm$^2$ | All areas | 10.2.7.2 |
| Interior floor finish, other than carpet and carpetlike materials | NFPA 253 | Class I: Critical radiant flux $\geq 0.45$ W/cm$^2$, in accordance with 10.2.7.4 | As required by relevant sections | 10.2.7.3 |
| | NFPA 253 | Class II: Critical radiant flux $\geq 0.22$ W/cm$^2$, in accordance with 10.2.7.4 | As required by relevant sections | 10.2.7.3 |
| Wall base [interior floor trim material used at junction of wall and floor to provide a functional or decorative border, and not exceeding 6 in. (150 mm) in height] | NFPA 253 | Class II: Critical radiant flux $\geq 0.22$ W/cm$^2$, in accordance with 10.2.7.4 | All areas | 10.2.5.2 |
| | NFPA 253 | Class I: Critical radiant flux $\geq 0.45$ W/cm$^2$, in accordance with 10.2.7.4 | If interior floor finish is required to meet Class I critical radiant flux | 10.2.5.2 |
| Floor finish of traditional type, such as wood flooring and resilient floor coverings | No testing required | | | 10.2.2.2 |

**A.10.2.1** The presence of multiple paint layers has the potential for paint delamination and bubbling or blistering of paint. Testing (NFPA *Fire Technology*, August 1974, "Fire Tests of Building Interior Covering Systems," David Waksman and John Ferguson, Institute for Applied Technology, National Bureau of Standards) has shown that adding up to two layers of paint with a dry film thickness of about 0.007 in. (0.18 mm) will not change the fire properties of surface-covering systems. Testing has shown that the fire properties of the surface-covering systems are highly substrate dependent and that thin coatings generally take on the characteristics of the substrate. When exposed to fire, the delamination, bubbling, and blistering of paint can result in an accelerated rate of flame spread.

**A.10.2.1.4** Such partitions are intended to include washroom water closet partitions.

**A.10.2.2** Table A.10.2.2 provides a compilation of the interior finish requirements of the occupancy chapters (Chapters 12 through 42).

**A.10.2.2.2** This paragraph recognizes that traditional finish floors and floor coverings, such as wood flooring and resilient floor coverings, have not proved to present an unusual hazard.

**A.10.2.2.2(2)** Compliance with 16 CFR 1630, "Standard for the Surface Flammability of Carpets and Rugs" (FFI-70), is considered equivalent to compliance with ASTM D 2859, *Standard Test Method for Ignition Characteristics of Finished Textile Floor Covering Materials.*

**A.10.2.3** See A.10.2.4.1.

**Table A.10.2.2 Interior Finish Classification Limitations**

| Occupancy | Exits | Exit Access Corridors | Other Spaces |
|---|---|---|---|
| Assembly — New | | | |
| >300 occupant load | A | A or B | A or B |
| | I or II | I or II | NA |
| ≤300 occupant load | A | A or B | A, B, or C |
| | I or II | I or II | NA |
| Assembly — Existing | | | |
| >300 occupant load | A | A or B | A or B |
| ≤300 occupant load | A | A or B | A, B, or C |
| Educational — New | A | A or B | A or B; C on low partitions[†] |
| | I or II | I or II | NA |
| Educational — Existing | A | A or B | A, B, or C |
| Day-Care Centers — New | A | A | A or B |
| | I or II | I or II | NA |
| Day-Care Centers — Existing | A or B | A or B | A or B |
| Day-Care Homes — New | A or B | A or B | A, B, or C |
| | I or II | | NA |
| Day-Care Homes — Existing | A or B | A, B, or C | A, B, or C |
| Health Care — New | A | A | A |
| | NA | B on lower portion of corridor wall[†] | B in small individual rooms[†] |
| | I or II | I or II | NA |
| Health Care — Existing | A or B | A or B | A or B |
| Detention and Correctional — New (sprinklers mandatory) | A or B | A or B | A, B, or C |
| | I or II | I or II | NA |
| Detention and Correctional — Existing | A or B | A or B | A, B, or C |
| | I or II | I or II | NA |
| One- and Two-Family Dwellings and Lodging or Rooming Houses | A, B, or C | A, B, or C | A, B, or C |
| Hotels and Dormitories — New | A | A or B | A, B, or C |
| | I or II | I or II | NA |
| Hotels and Dormitories — Existing | A or B | A or B | A, B, or C |
| | I or II[†] | I or II[†] | NA |
| Apartment Buildings — New | A | A or B | A, B, or C |
| | I or II | I or II | NA |
| Apartment Buildings — Existing | A or B | A or B | A, B, or C |
| | I or II[†] | I or II[†] | NA |
| Residential Board and Care — (See Chapters 32 and 33.) | | | |
| Mercantile — New | A or B | A or B | A or B |
| | I or II | | NA |
| Mercantile — Existing | | | |
| Class A or Class B stores | A or B | A or B | Ceilings — A or B; walls — A, B, or C |
| Class C stores | A, B, or C | A, B, or C | A, B, or C |
| Business and Ambulatory Health Care — New | A or B | A or B | A, B, or C |
| | I or II | | NA |
| Business and Ambulatory Health Care — Existing | A or B | A or B | A, B, or C |
| Industrial | A or B | A, B, or C | A, B, or C |
| | I or II | I or II | NA |
| Storage | A or B | A, B, or C | A, B, or C |
| | I or II | | NA |

NA: Not applicable.

Notes:

(1) Class A interior wall and ceiling finish — flame spread index, 0–25 (new applications); smoke developed index, 0–450.

(2) Class B interior wall and ceiling finish — flame spread index, 26–75 (new applications); smoke developed index, 0–450.

(3) Class C interior wall and ceiling finish — flame spread index, 76–200 (new applications); smoke developed index, 0–450.

(4) Class I interior floor finish — critical radiant flux, not less than 0.45 W/cm².

(5) Class II interior floor finish — critical radiant flux, not more than 0.22 W/cm², but less than 0.45 W/cm².

(6) Automatic sprinklers — where a complete standard system of automatic sprinklers is installed, interior wall and ceiling finish with a flame spread rating not exceeding Class C is permitted to be used in any location where Class B is required, and Class B interior wall and ceiling finish is permitted to be used in any location where Class A is required; similarly, Class II interior floor finish is permitted to be used in any location where Class I is required, and no interior floor finish classification is required where Class II is required. These provisions do not apply to new detention and correctional occupancies.

(7) Exposed portions of structural members complying with the requirements for heavy timber construction are permitted.

†See corresponding chapters for details.

**A.10.2.3.4** It has been shown that the method of mounting interior finish materials might affect actual performance. Where materials are tested in intimate contact with a substrate to determine a classification, such materials should be installed in intimate contact with a similar substrate. Such details are especially important for "thermally thin" materials. For further information, see ASTM E 84, *Standard Test Method for Surface Burning Characteristics of Building Materials.*

Some interior wall and ceiling finish materials, such as fabrics not applied to a solid backing, do not lend themselves to a test made in accordance with ASTM E 84. In such cases, the large-scale test outlined in NFPA 701, *Standard Methods of Fire Tests for Flame Propagation of Textiles and Films,* is permitted to be used.

Prior to 1978, the test report described by ASTM E 84 included an evaluation of the fuel contribution as well as the flame spread index and the smoke developed index. However, it is now recognized that the measurement on which the fuel contribution is based does not provide a valid measure. Therefore, although the data are recorded during the test, the information is no longer normally reported. Classification of interior wall and ceiling finish thus relies only on the flame spread index and smoke developed index.

The 450 smoke developed index limit is based solely on obscuration. *(See A.10.2.4.1.)*

**A.10.2.3.7** The methodology specified in NFPA 265, *Standard Methods of Fire Tests for Evaluating Room Fire Growth Contribution of Textile or Expanded Vinyl Wall Coverings on Full Height Panels and Walls,* includes provisions for measuring smoke obscuration. Such measurement is considered desirable, but the basis for specific recommended values is not currently available. *(See A.10.2.4.1.)*

**A.10.2.4** Surface nonmetallic raceway products, as permitted by *NFPA 70, National Electrical Code,* are not interior finishes.

**A.10.2.4.1** Previous editions of the *Code* have regulated textile materials on walls and ceilings using ASTM E 84, *Standard Test Method for Surface Burning Characteristics of Building Materials,* or ANSI/UL 723, *Standard for Test for Surface Burning Characteristics of Building Materials.* Full-scale room/corner fire test research has shown that flame spread indices produced by ASTM E 84 or ANSI/UL 723 might not reliably predict all aspects of the fire behavior of textile wall and ceiling coverings.

NFPA 265, *Standard Methods of Fire Tests for Evaluating Room Fire Growth Contribution of Textile or Expanded Vinyl Wall Coverings on Full Height Panels and Walls,* and NFPA 286, *Standard Methods of Fire Tests for Evaluating Contribution of Wall and Ceiling Interior Finish to Room Fire Growth,* both known as room/corner tests, were developed for assessing the fire and smoke obscuration performance of textile wall coverings and interior wall and ceiling finish materials, respectively. As long as an interior wall or ceiling finish material is tested by NFPA 265 or NFPA 286, as appropriate, using a mounting system, substrate, and adhesive (if appropriate) that are representative of actual use, the room/corner test provides an adequate evaluation of a product's flammability and smoke obscuration behavior. Manufacturers, installers, and specifiers should be encouraged to use NFPA 265 or NFPA 286, as appropriate — but not both — because each of these standard fire tests has the ability to characterize actual product behavior, as opposed to data generated by tests using ASTM E 84 or ANSI/UL 723, which only allow comparisons of one product's performance with another. If a manufacturer or installer chooses to test a wall finish in accordance with NFPA 286, additional testing in accordance with ASTM E 84 or ANSI/UL 723 is not necessary.

The test results from ASTM E 84 or ANSI/UL 723 are suitable for classification purposes but should not be used as input into fire models, because they are not generated in units suitable for engineering calculations. Actual test results for heat, smoke, and combustion product release from NFPA 265, and from NFPA 286, are suitable for use as input into fire models for performance-based design.

**A.10.2.4.2** Expanded vinyl wall covering consists of a woven textile backing, an expanded vinyl base coat layer, and a non-expanded vinyl skin coat. The expanded base coat layer is a homogeneous vinyl layer that contains a blowing agent. During processing, the blowing agent decomposes, which causes this layer to expand by forming closed cells. The total thickness of the wall covering is approximately 0.055 in. to 0.070 in. (1.4 mm to 1.8 mm).

**A.10.2.4.3.1** Both NFPA 286, *Standard Methods of Fire Tests for Evaluating Contribution of Wall and Ceiling Interior Finish to Room Fire Growth,* and ANSI/UL 1715, *Standard for Fire Test of Interior Finish Material,* contain smoke obscuration criteria. ANSI/UL 1040, *Standard for Fire Test of Insulated Wall Construction,* and FM 4880, *Approval Standard for Class I Insulated Wall or Wall and Roof/Ceiling Panels; Plastic Interior Finish Materials; Plastic Exterior Building Panels; Wall/Ceiling Coating Systems; Interior or Exterior Finish Systems,* do not include smoke obscuration criteria. Smoke obscuration is an important component of the fire performance of cellular or foamed plastic materials.

**A.10.2.4.3.1.2** Both NFPA 286, *Standard Methods of Fire Tests for Evaluating Contribution of Wall and Ceiling Interior Finish to Room Fire Growth,* and ANSI/UL 1715, *Standard for Fire Test of Interior Finish Material,* contain smoke obscuration criteria. ANSI/UL 1040, *Standard for Fire Test of Insulated Wall Construction,* and FM 4880, *Approval Standard for Class I Insulated Wall or Wall and Roof/Ceiling Panels; Plastic Interior Finish Materials; Plastic Exterior Building Panels; Wall/Ceiling Coating Systems; Interior or Exterior Finish Systems,* do not. Smoke obscuration is an important component of the fire performance of cellular or foamed plastic materials.

**A.10.2.4.4** Light-transmitting plastics are used for a variety of purposes, including light diffusers, exterior wall panels, skylights, canopies, glazing, and the like. Previous editions of the *Code* have not addressed the use of light-transmitting plastics. Light-transmitting plastics will not normally be used in applications representative of interior finishes. Accordingly, ASTM E 84, *Standard Test Method for Surface Burning Characteristics of Building Materials,* or ANSI/UL 723, *Standard for Test for Surface Burning Characteristics of Building Materials,* can produce test results that might or might not apply.

Light-transmitting plastics are regulated by model building codes such as *NFPA 5000, Building Construction and Safety Code.* Model building codes provide adequate regulation for most applications of light-transmitting plastics. Where an authority having jurisdiction determines that a use is contemplated that differs from uses regulated by model building codes, light-transmitting plastics in such applications can be substantiated by fire tests that demonstrate the combustibility characteristics of the light-transmitting plastics for the use intended under actual fire conditions.

**A.10.2.6** Fire-retardant coatings need to be applied to surfaces properly prepared for the material, and application needs to be consistent with the product listing. Deterioration of coatings applied to interior finishes can occur due to repeated cleaning of the surface or painting over applied coatings.

**A.10.2.6.1** It is the intent of the *Code* to mandate interior wall and ceiling finish materials that obtain their fire performance and smoke developed characteristics in their original form. However, in renovations, particularly those involving historic buildings, and in changes of occupancy, the required fire performance or smoke developed characteristics of existing surfaces of walls, partitions, columns, and ceilings might have to be secured by applying approved fire-retardant coatings to surfaces having higher flame spread ratings than permitted. Such treatments should comply with the requirements of NFPA 703, *Standard for Fire Retardant–Treated Wood and Fire-Retardant Coatings for Building Materials.* When fire-retardant coatings are used, they need to be applied to surfaces properly prepared for the material, and application needs to be consistent with the product listing. Deterioration of coatings applied to interior finishes can occur due to repeated cleaning of the surface or painting over applied coatings, but permanency must be assured in some appropriate fashion. Fire-retardant coatings must possess the desired degree of permanency and be maintained so as to retain the effectiveness of the treatment under the service conditions encountered in actual use.

**A.10.2.7.2** The fire performance of some floor finishes has been tested, and traditional finish floors and floor coverings, such as wood flooring and resilient floor coverings, have not proved to present an unusual hazard.

**A.10.2.7.3** The flooring radiant panel provides a measure of a floor covering's tendency to spread flames where located in a corridor and exposed to the flame and hot gases from a room fire. The flooring radiant panel test method is to be used as a basis for estimating the fire performance of a floor covering installed in the building corridor. Floor coverings in open building spaces and in rooms within buildings merit no further regulation, provided that it can be shown that the floor covering is at least as resistant to spread of flame as a material that meets the U.S. federal flammability standard 16 CFR 1630, "Standard for the Surface Flammability of Carpets and Rugs" (FF 1-70). All carpeting sold in the U.S. since 1971 is required to meet this standard and, therefore, is not likely to become involved in a fire until a room reaches or approaches flashover. Therefore, no further regulations are necessary for carpet other than carpet in exitways and corridors.

It has not been found necessary or practical to regulate interior floor finishes on the basis of smoke development.

Full-scale fire tests and fire experience have shown floor coverings in open building spaces merit no regulation beyond the United States federally mandated DOC FF 1-70 "pill test." This is because floor coverings meeting the FF 1-70 regulation will not spread flame significantly until a room fire approaches flashover. At flashover, the spread of flame across a floor covering will have minimal impact on the already existing hazard. The minimum critical radiant flux of a floor covering that will pass the FF 1-70 regulation has been determined to be approximately $0.04 \text{ W/cm}^2$ (*see Annex B, Tu, King-Mon and Davis, Sanford, "Flame Spread of Carpet Systems Involved in Room Fires"*). The flooring radiant panel is only able to determine critical radiant flux values to $0.1 \text{ W/cm}^2$. This provision will prevent use of a noncomplying material, which might create a problem, especially when the *Code* is used outside the United States, where federal regulation FF 1-70 is not mandated.

**A.10.3.1** Testing per NFPA 701, *Standard Methods of Fire Tests for Flame Propagation of Textiles and Films,* applies to textiles and films used in a hanging configuration. If the textiles are to be applied to surfaces of buildings or backing materials as inte-

rior finishes for use in buildings, they should be treated as interior wall and ceiling finishes in accordance with Section 10.2 of this *Code,* and they should then be tested for flame spread index and smoke developed index values in accordance with ASTM E 84, *Standard Test Method for Surface Burning Characteristics of Building Materials,* or ANSI/UL 723, *Standard for Test for Surface Burning Characteristics of Building Materials,* or for flame spread and flashover in accordance with NFPA 265, *Standard Methods of Fire Tests for Evaluating Room Fire Growth Contribution of Textile or Expanded Vinyl Wall Coverings on Full Height Panels and Walls.* Films and other materials used as interior finish applied to surfaces of buildings should be tested for flame spread index and smoke developed index values in accordance with ASTM E 84 or ANSI/UL 723 or for heat and smoke release and flashover in accordance with NFPA 286, *Standard Methods of Fire Tests for Evaluating Contribution of Wall and Ceiling Interior Finish to Room Fire Growth.*

The test results from NFPA 701 are suitable for classification purposes but should not be used as input into fire models, because they are not generated in units suitable for engineering calculations.

**A.10.3.2.1** The Class I requirement associated with testing in accordance with NFPA 260, *Standard Methods of Tests and Classification System for Cigarette Ignition Resistance of Components of Upholstered Furniture,* or with ASTM E 1353, *Standard Test Methods for Cigarette Ignition Resistance of Components of Upholstered Furniture,* and the char length of not more than 1½ in. (38 mm) required with testing in accordance with NFPA 261, *Standard Method of Test for Determining Resistance of Mock-Up Upholstered Furniture Material Assemblies to Ignition by Smoldering Cigarettes,* or with ASTM E 1352, *Standard Test Method for Cigarette Ignition Resistance of Mock-Up Upholstered Furniture Assemblies,* are indicators that the furniture item or mattress is resistant to a cigarette ignition. A fire that smolders for an excessive period of time without flaming can reduce the tenability within the room or area of fire origin without developing the temperatures necessary to operate automatic sprinklers.

The test results from NFPA 260 or ASTM E 1353, and from NFPA 261 or ASTM E 1352, are suitable for classification purposes but should not be used as input into fire models, because they are not generated in units suitable for engineering calculations.

**A.10.3.2.2** The char length of not more than 2 in. (51 mm) required in 16 CFR 1632, "Standard for the Flammability of Mattresses and Mattress Pads" (FF 4-72), is an indicator that the mattress is resistant to a cigarette ignition. United States federal regulations require mattresses in this country to comply with 16 CFR 1632.

**A.10.3.3** The intent of the provisions of 10.3.3 is as follows:

(1) The peak heat release rate of not more than 80 kW by a single upholstered furniture item was chosen based on maintaining a tenable environment within the room of fire origin, and the sprinkler exception was developed because the sprinkler system helps to maintain tenable conditions, even if the single upholstered furniture item were to have a peak rate of heat release in excess of 80 kW.

(2) The total heat release of not more than 25 MJ by the single upholstered furniture item during the first 10 minutes of the test was established as an additional safeguard to protect against the adverse conditions that would be created by an upholstered furniture item that released its heat in other than the usual measured scenario, and the following should also be noted:

(a) During the test for measurement of rate of heat release, the instantaneous heat release value usually peaks quickly and then quickly falls off, so as to create a triangle-shaped curve.

(b) In the atypical case, if the heat release were to peak and remain steady at that elevated level, as opposed to quickly falling off, the 80 kW limit would not ensure safety.

(c) Only a sprinkler exception is permitted in lieu of the test because of the ability of the sprinkler system to control the fire.

Actual test results for heat, smoke, and combustion product release from ASTM E 1537, *Standard Test Method for Fire Testing of Upholstered Furniture*, might be suitable for use as input into fire models for performance-based design. Furthermore, California Technical Bulletin 133, "Flammability Test Procedure for Seating Furniture for Use in Public Occupancies," includes pass/fail criteria for a single upholstered furniture item of 80 kW peak heat release rate and 25 MJ total heat release over the first 10 minutes of the test.

**A.10.3.4**  The intent of the provisions of 10.3.4 is as follows:

(1) The peak heat release rate of not more than 100 kW by a single mattress was chosen based on maintaining a tenable environment within the room of fire origin, and the sprinkler exception was developed because the sprinkler system helps to maintain tenable conditions, even if the single mattress were to have a peak rate of heat release in excess of 100 kW.

(2) The total heat release of not more than 25 MJ by the single mattress during the first 10 minutes of the test was established as an additional safeguard to protect against the adverse conditions that would be created by a mattress that released its heat in other than the usual measured scenario, and the following should also be noted:

(a) During the test for measurement of rate of heat release, the instantaneous heat release value usually peaks quickly and then quickly falls off, so as to create a triangle-shaped curve.

(b) In the atypical case, if the heat release were to peak and remain steady at that elevated level, as opposed to quickly falling off, the 100 kW limit would not ensure safety.

(c) Only a sprinkler exception is permitted in lieu of the test because of the ability of the sprinkler system to control the fire.

Actual test results for heat, smoke, and combustion product release from ASTM E 1590, *Standard Test Method for Fire Testing of Mattresses*, might be suitable for use as input into fire models for performance-based design. Furthermore, California Technical Bulletin 129, "Flammability Test Procedure for Mattresses for Use in Public Buildings," includes pass/fail criteria for a single mattress of 100 kW peak heat release rate and 25 MJ total heat release over the first 10 minutes of test.

**A.10.3.5**  Christmas trees that are not effectively flame-retardant treated, ordinary crepe paper decorations, and pyroxylin plastic decorations might be classified as highly flammable.

**A.10.3.7**  Neither ANSI/UL 1975, *Standard for Fire Tests for Foamed Plastics Used for Decorative Purposes*, nor NFPA 289, *Standard Method of Fire Test for Individual Fuel Packages*, is intended for evaluating interior wall and ceiling finish materials.

Actual test results for heat, smoke, and combustion product release from ANSI/UL 1975 or from NFPA 289 might be suitable

for use as input into fire models intended for performance-based design.

**A.11.2.2**  Escape chutes, controlled descent devices, and elevators are permitted to provide escape routes in special structures; however, they should not be substituted for the provisions of this *Code*.

**A.11.2.2.4.1**  The grade level of open structures, which by their very nature contain an infinite number of means of egress, are exempt from the requirements for number of means of egress.

**A.11.3.1.3.1(2)**  The incidental accessory uses are intended to apply to small office spaces or lounge areas and similar uses that are used by tower employees.

**A.11.3.2.4**  The Washington Monument in Washington, DC, is an example of a tower where it would be impracticable to provide a second stairway.

**A.11.3.4.2(2)**  The incidental accessory uses are intended to apply to small office spaces or lounge areas and similar uses that are used by tower employees.

**A.11.3.4.4.6.2(2)**  Occupants of air traffic control towers might be required by administrative controls to remain in the facility when a fire occurs so they can perform orderly transfer of operations. Methods to limit compromising the means of egress might include a fire resistance–rated separation between discharge paths or smoke control in large spaces.

**A.11.5**  For further information on pier fire protection, see NFPA 307, *Standard for the Construction and Fire Protection of Marine Terminals, Piers, and Wharves*.

**A.11.6**  Fire safety information for manufactured home parks is found in NFPA 501A, *Standard for Fire Safety Criteria for Manufactured Home Installations, Sites, and Communities*.

**A.11.7.3.2**  It is not the intent that emergency access openings be readily openable from the exterior by the public but that they be easily opened with normal fire department equipment.

**A.11.8.3.1**  Where an occupancy chapter (Chapters 12 through 42) permits the omission of sprinklers in specific spaces, such as small bathrooms and closets in residential occupancies, the building is still considered to be protected throughout for the purposes of 11.8.3.1.

**A.11.8.4.1**  The need for voice communication can be based on a decision regarding staged or partial evacuation versus total evacuation of all floors. The determination of need is a function of occupancy classification and building height.

**A.11.8.6**  It is not the intent of the paragraph to require any of the equipment in the list, other than the telephone for fire department use, but only to provide the controls, panels, annunciators, and similar equipment at this location if the equipment is provided or required by another section of the *Code*.

**A.11.9.3.3.1**  The requirements of this paragraph can be considered as a Class 4, Type 60, system per NFPA 110, *Standard for Emergency and Standby Power Systems*.

**A.12.1.2**  Assembly occupancy requirements should be determined on a room-by-room basis, a floor-by-floor basis, and a total building basis. The requirements for each room should be based on the occupant load of that room, and the requirements for each floor should be based on the occupant load of that floor, but the requirements for the assembly building overall should be based on the total occupant load. Therefore, it is quite feasible to have several assembly occupancies with occupant loads of 300 or less grouped together in a single building. Such a building would be an assembly occupancy with an occupant load of over 1000.

**A.12.1.3.2** For example, an assembly room for the residents of a detention occupancy will not normally be subject to simultaneous occupancy.

**A.12.1.4.2** An understanding of the term *accessory room* might be useful to the enforcer of the *Code*, although the term is not used within the *Code*. An accessory room includes a dressing room, the property master's work and storage rooms, the carpenter's room, or similar rooms necessary for legitimate stage operations.

**A.12.1.7.1** The increase in occupant load above that calculated using occupant load factors from Table 7.3.1.2 is permitted if the provisions of 12.1.7.1 are followed. The owner or operator has the right to submit plans and to be permitted an increase in occupant load if the plans comply with the *Code*. The authority having jurisdiction is permitted to reject the plan for increase in occupant load if the plan is unrealistic, inaccurate, or otherwise does not properly reflect compliance with other *Code* requirements. It is not the intent of the provisions of 12.1.7.1 to prohibit an increase in occupant load solely on the basis of exceeding the limits calculated using occupant load factors from Table 7.3.1.2.

To assist in preventing serious overcrowding incidents in sports arenas, stadia, and similar occupancies, spectator standing room should not be permitted between the seating areas and the playing areas, except in horse race and dog track facilities.

Where a capacity or near-capacity audience is anticipated, all seating should be assigned with tickets showing the section, row, and seat number.

Where standing room is permitted, the capacity of the standing area should meet the following criteria:

(1) The capacity should be determined on the basis of 5 ft² (0.46 m²) per person.
(2) The capacity should be added to the seating capacity in determining egress requirements.
(3) The capacity should be located to the rear of the seating area.
(4) The capacity should be assigned standing-room-only tickets according to the area designated for the purpose.

The number of tickets sold, or otherwise distributed, should not exceed the aggregate number of seats plus the approved standing room numbers.

**A.12.2.2.3.1(1)** The seating plan and the means of egress should be reviewed each time the seating is substantially rearranged.

**A.12.2.3.2** The provisions of 12.2.3.2 should be applied within the audience seating chamber and to the room doors. The capacity of means of egress components encountered after leaving the audience seating chamber, such as concourses, lobbies, exit stair enclosures, and the exit discharge, should be calculated in accordance with Section 7.3.

**A.12.2.3.6.6** The original *Code* wording exempted sports arenas and railway stations. If an assembly occupancy was not similar to a sports arena or railway station, it was often judged ineligible to use the provision of 12.2.3.6.6. A list of exempted

assembly venues also raises the question of why other occupancies are not included and necessitates additions to the list. For example, an exhibit hall of very large size might have several main entrances/exits. A theater extending the width of a block cannot really have a main entrance/exit in one confined location. A restaurant might have a main entrance serving the parking lot and another main entrance for those entering from the street. The authority having jurisdiction needs to determine where such arrangements are acceptable.

**A.12.2.4** It is not the intent to require four means of egress from each level of an assembly occupancy building having a total occupant load of more than 1000 where, individually, the floors have occupant loads of less than 1000.

**A.12.2.5.4.2** This requirement and the associated requirement of 12.2.5.4.3 have the effect of prohibiting festival seating, unless it truly is a form of seating, such as lawn seating, where generous spaces are commonly maintained between individuals and small groups so that people can circulate freely at any time. Such lawn seating is characterized by densities of about one person per 15 ft² (1.4 m²). Both requirements prohibit uncontrolled crowd situations, such as in front of stages at rock music concerts where the number and density of people is uncontrolled by architectural or management features.

**A.12.2.5.4.3** This requirement is intended to facilitate rapid emergency access to individuals who are experiencing a medical emergency, especially in the case of cardiopulmonary difficulties, where there is a need for rapid medical attention from trained personnel. The requirement also addresses the need for security and law enforcement personnel to reach individuals whose behavior is endangering themselves and others.

**A.12.2.5.4.4** The catchment area served by an aisle accessway or aisle is the portion of the total space that is naturally served by the aisle accessway or aisle. Hence, the requirement for combining the required capacity where paths converge is, in effect, a restatement of the idea of a catchment area. The establishment of catchment areas should be based on a balanced use of all means of egress, with the number of persons in proportion to egress capacity.

**A.12.2.5.5** For purposes of the means of egress requirements of this *Code*, tablet-arm chair seating is not considered seating at tables. Dinner theater–style configurations are required to comply with the aisle accessway provisions applying to seating at tables and the aisle requirements of 12.2.5.6, if the aisles contain steps or are ramped. Generally, if aisles contain steps or are ramped, all of the *Code* requirements for aisles, stairs, and ramps are required to be met. *(Also see 7.1.7 and A.7.1.7.2.)*

**A.12.2.5.5.1** Seats having reclining backs are assumed to be in their most upright position when unoccupied.

**A.12.2.5.5.4** The system known as *continental seating* has one pair of egress doors provided for every five rows that is located close to the ends of the rows. In previous editions of the *Code*, such egress doors were required to provide a clear width of not less than 66 in. (1675 mm) discharging into a foyer, into a lobby, or to the exterior of the building. This continental seating arrangement can result in egress flow times (i.e., with nominal flow times of approximately 100 seconds, rather than 200 seconds) that are approximately one-half as long as those resulting where side aisles lead to more remote doors. Such superior egress flow time performance is desirable in some situations; however, special attention should be given either to a comparably good egress capacity for other parts of the egress system or to sufficient space to accommodate queuing outside the seating space.

**A.12.2.5.6.3**  It is the intent to permit handrails to project not more than 3½ in. (90 mm) into the clear width of aisles required by 12.2.5.6.3.

**A.12.2.5.6.4.1**  Technical information about the convenience and safety of ramps and stairs having gradients in the region of 1 in 8 clearly suggests that the goal should be slopes for ramps that are less steep and combinations of stair risers and treads that are, for example, superior to 4 in. (100 mm) risers and 32 in. (865 mm) treads. This goal should be kept in mind by designers in establishing the gradient of seating areas to be served by aisles.

**A.12.2.5.6.5(3)**  Tread depth is more important to stair safety than is riser height. Therefore, in cases where the seating area gradient is less than 5 in 11, it is recommended that the tread dimension be increased beyond 11 in. (280 mm), rather than reducing the riser height. Where the seating area gradient exceeds 8 in 11, it is recommended that the riser height be increased while maintaining a tread depth of not less than 11 in. (280 mm).

**A.12.2.5.6.8**  Failure to provide a handrail within a 30 in. (760 mm) horizontal distance of all required portions of the aisle stair width means that the egress capacity calculation is required to be modified as specified by 12.2.3.3(3). This modification might lead to an increase in the aisle width. Although this increase will compensate for reduced egress efficiency, it does not help individuals walking on such portions of stairs to recover from missteps, other than by possibly marginally reducing the crowding that might exacerbate the problem of falls. *(See also 7.2.2.4.)*

**A.12.2.5.6.9**  Certain tread cover materials such as plush carpets, which are often used in theaters, produce an inherently well-marked tread nosing under most lighting conditions. On the other hand, concrete treads have nosings with a sharp edge and, especially under outdoor lighting conditions, are difficult to discriminate. Therefore, concrete treads require an applied marking stripe. The slip resistance of such marking stripes should be similar to the rest of the treads, and no tripping hazard should be created; luminescent, self-luminous, and electroluminescent tread markings have the advantage of being apparent in reduced light or in the absence of light.

**A.12.2.5.7**  For purposes of the means of egress requirements of this *Code*, seating at counters or at other furnishings is considered to be the same as seating at tables.

**A.12.2.5.7.2**  Effectively, where the aisle accessway is bounded by movable seating, the 12 in. (305 mm) minimum width might be increased by about 15 in. to 30 in. (380 mm to 760 mm) as seating is pushed in toward tables. Moreover, it is such movement of chairs during normal and emergency egress situations that makes the zero-clearance allowance workable. The allowance also applies to booth seating where people sitting closest to the aisle normally move out ahead of people farthest from the aisle.

**A.12.2.5.7.3**  See A.12.2.5.8.3.

**A.12.2.5.7.4**  The minimum width requirement as a function of accessway length is as follows:

(1)  0 in. (0 mm) for the first 6 ft (1830 mm) of length toward the exit
(2)  12 in. (305 mm) for the next 6 ft (1830 mm); that is, up to 12 ft (3660 mm) of length
(3)  12 in. to 24 in. (305 mm to 610 mm) for lengths from 12 ft to 36 ft (3.7 m to 11 m), the maximum length to the closest aisle or egress doorway permitted by 12.2.5.7.5

Any additional width needed for seating is to be added to these widths, as described in 12.2.5.8.3.

**A.12.2.5.8.1**  See 7.1.7 and A.7.1.7.2 for special circulation safety precautions applicable where small elevation differences occur.

**A.12.2.5.8.2**  It is important to make facilities accessible to people using wheelchairs. See ICC/ANSI A117.1, *American National Standard for Accessible and Usable Buildings and Facilities,* which provides guidance on appropriate aisle widths.

**A.12.2.5.8.3**  Figure A.12.2.5.8.3 shows typical measurements involving seating and tables abutting an aisle. For purposes of the means of egress requirements of this *Code*, seating at counters or other furnishings is considered to be the same as seating at tables.

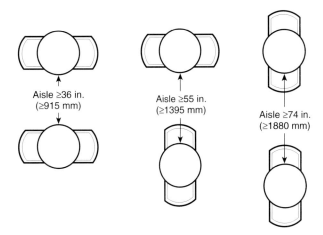

**FIGURE A.12.2.5.8.3  Seating at Tables Abutting an Aisle.**

**A.12.2.11.1.1**  This requirement includes provisions of guards and rails at the front of boxes, galleries, and balconies, and at aisle accessways adjacent to vomitories and orchestra pits.

**A.12.3.1(1)**  The allowance for unenclosed stairs or ramps presumes the balcony or mezzanine complies with the other provisions of the *Code*, such as travel distance to exits in accordance with 12.2.6 and number of exits in accordance with 12.2.4. For the purposes of this exception, a balcony with glazing that provides a visual awareness of the main assembly area is considered open.

**A.12.3.4.2.3**  The intent is to require detectors only in nonsprinklered hazardous areas that are unoccupied. When the building is occupied, the detectors in the unoccupied, unsprinklered hazardous areas will initiate occupant notification. If the building is unoccupied, the fire in the nonsprinklered hazardous area is not a life safety issue, and the detectors, upon activation, are not required to notify anyone. The signal from a detector is permitted to be sent to a control panel in an area that is occupied when the building is occupied, but that is unoccupied when the building is unoccupied, without the need for central station monitoring or the equivalent.

**A.12.3.4.3.5** Examples of devices that might be used to provide alternative visible means include scoreboards, message boards, and other electronic devices.

**A.12.3.5.3(1)** It is the intent to permit a single multipurpose room of less than 12,000 ft$^2$ (1115 m$^2$) to have certain small rooms as part of the single room. These rooms could be a kitchen, an office, an equipment room, and the like. It is also the intent that an addition could be made to an existing building, without requiring that the existing building be sprinklered, where both the new and existing buildings have independent means of egress and a fire-rated separation is provided to isolate one building from the other.

A school gymnasium with egress independent of, and separated from, the school would be included in this exception, as would a function hall attached to a church with a similar egress arrangement.

**A.12.4.1.1** Life safety evaluations are examples of performance-based approaches to life safety. In this respect, significant guidance in the form and process of life safety evaluations is provided by Chapter 5, keeping in mind the fire safety emphasis in Chapter 5. Performance criteria, scenarios, evaluation, safety factors, documentation, maintenance, and periodic assessment (including a warrant of fitness) all apply to the broader considerations in a life safety evaluation. A life safety evaluation deals not only with fire but also with storms, collapse, crowd behavior, and other related safety considerations for which a checklist is provided in A.12.4.1.3. Chapter 5 provides guidance, based on fire safety requirements, for establishing a documented case showing that products of combustion in all conceivable fire scenarios will not significantly endanger occupants using means of egress in the facility (e.g., due to fire detection, automatic suppression, smoke control, large-volume space, or management procedures). Moreover, means of egress facilities plus facility management capabilities should be adequate to cope with scenarios where certain egress routes are blocked for some reason.

In addition to making realistic assumptions about the capabilities of persons in the facility (e.g., an assembled crowd including many disabled persons or persons unfamiliar with the facility), the life safety evaluation should include a factor of safety of not less than 2.0 in all calculations relating to hazard development time and required egress time (the combination of flow time and other time needed to detect and assess an emergency condition, initiate egress, and move along the egress routes). The factor of safety takes into account the possibility that half of the egress routes might not be used (or be usable) in certain situations.

Regarding crowd behavior, the potential hazards created by larger masses of people and greater crowd densities (which can be problematic during ingress, occupancy, and egress) demand that technology be used by designers, managers, and authorities responsible for buildings to compensate for the relaxed egress capacity provisions of Table 12.4.2.3. In very large buildings for assembly use, the hazard of crowd crushes can exceed that of fire or structural failure. Therefore, the building designers, managers, event planners, security personnel, police authorities, and fire authorities, as well as the building construction authorities, should understand the potential problems and solutions, including coordination of their activities. For crowd behavior, this understanding includes factors of space, energy, time, and information, as well as specific crowd management techniques, such as metering. Published

guidance on these factors and techniques is found in the *SFPE Handbook of Fire Protection Engineering*, Section 3, Chapter 13, pp. 3-342–3-366 (Proulx, G., "Movement of People"), and the publications referenced therein.

Table 12.2.3.2 and Table 12.4.2.3 are based on a linear relationship between number of seats and nominal flow time, with not less than 200 seconds (3.3 minutes) for 2000 seats plus 1 second for every additional 50 seats up to 25,000. Beyond 25,000 total seats, the nominal flow time is limited to 660 seconds (11 minutes). Nominal flow time refers to the flow time for the most able group of patrons; some groups less familiar with the premises or less able groups might take longer to pass a point in the egress system. Although three or more digits are noted in the tables, the resulting calculations should be assumed to provide only two significant figures of precision.

**A.12.4.1.3** Factors to be considered in a life safety evaluation include the following:

(1) Nature of the events being accommodated, including the following:
  (a) Ingress, intra-event movement, and egress patterns
  (b) Ticketing and seating policies/practices
  (c) Event purpose (e.g., sports contest, religious meeting)
  (d) Emotional qualities (e.g., competitiveness) of event
  (e) Time of day when event is held
  (f) Time duration of single event
  (g) Time duration of attendees' occupancy of the building

(2) Occupant characteristics and behavior, including the following:
  (a) Homogeneity
  (b) Cohesiveness
  (c) Familiarity with building
  (d) Familiarity with similar events
  (e) Capability (as influenced by factors such as age, physical abilities)
  (f) Socioeconomic factors
  (g) Small minority involved with recreational violence
  (h) Emotional involvement with the event and other occupants
  (i) Use of alcohol or drugs
  (j) Food consumption
  (k) Washroom utilization

(3) Management, including the following:
  (a) Clear, contractual arrangements for facility operation/use as follows:
    i. Between facility owner and operator
    ii. Between facility operator and event promoter
    iii. Between event promoter and performer
    iv. Between event promoter and attendee
    v. With police forces
    vi. With private security services
    vii. With ushering services
  (b) Experience with the building
  (c) Experience with similar events and attendees
  (d) Thorough, up-to-date operations manual
  (e) Training of personnel
  (f) Supervision of personnel
  (g) Communications systems and utilization
  (h) Ratios of management and other personnel to attendees
  (i) Location/distribution of personnel
  (j) Central command location
  (k) Rapport between personnel and attendees

(l) Personnel support of attendee goals

(m) Respect of attendees for personnel due to the following:

    i. Dress (uniform) standards

    ii. Age and perceived experience

    iii. Personnel behavior, including interaction

    iv. Distinction between crowd management and control

    v. Management concern for facility quality (e.g., cleanliness)

    vi. Management concern for entire event experience of attendees (i.e., not just during occupancy of the building)

(4) Emergency management preparedness, including the following:

(a) Complete range of emergencies addressed in operations manual

(b) Power loss

(c) Fire

(d) Severe weather

(e) Earthquake

(f) Crowd incident

(g) Terrorism

(h) Hazardous materials

(i) Transportation accident (e.g., road, rail, air)

(j) Communications systems available

(k) Personnel and emergency forces ready to respond

(l) Attendees clearly informed of situation and proper behavior

(5) Building systems, including the following:

(a) Structural soundness

(b) Normal static loads

(c) Abnormal static loads (e.g., crowds, precipitation)

(d) Dynamic loads (e.g., crowd sway, impact, explosion, wind, earthquake)

(e) Stability of nonstructural components (e.g., lighting)

(f) Stability of movable (e.g., telescoping) structures

(g) Fire protection

(h) Fire prevention (e.g., maintenance, contents, housekeeping)

(i) Compartmentation

(j) Automatic detection and suppression of fire

(k) Smoke control

(l) Alarm and communications systems

(m) Fire department access routes and response capability

(n) Structural integrity

(o) Weather protection

(p) Wind

(q) Precipitation (attendees rush for shelter or hold up egress of others)

(r) Lightning protection

(s) Circulation systems

(t) Flowline or network analysis

(u) Waywinding and orientation

(v) Merging of paths (e.g., precedence behavior)

(w) Decision/branching points

(x) Route redundancies

(y) Counterflow, crossflow, and queuing situations

(z) Control possibilities, including metering

(aa) Flow capacity adequacy

(bb) System balance

(cc) Movement time performance

(dd) Flow times

(ee) Travel times

(ff) Queuing times

(gg) Route quality

(hh) Walking surfaces (e.g., traction, discontinuities)

(ii) Appropriate widths and boundary conditions

(jj) Handrails, guardrails, and other rails

(kk) Ramp slopes

(ll) Step geometries

(mm) Perceptual aspects (e.g., orientation, signage, marking, lighting, glare, distractions)

(nn) Route choices, especially for vertical travel

(oo) Resting/waiting areas

(pp) Levels of service (overall crowd movement quality)

(qq) Services

(rr) Washroom provision and distribution

(ss) Concessions

(tt) First aid and EMS facilities

(uu) General attendee services

A scenario-based approach to performance-based fire safety is addressed in Chapter 5. In addition to using such scenarios and, more generally, the attention to performance criteria, evaluation, safety factors, documentation, maintenance, and periodic assessment required when the Chapter 5 option is used, life safety evaluations should consider scenarios based on characteristics important in assembly occupancies. These characteristics include the following:

(1) Whether there is a local or mass awareness of an incident, event, or condition that might provoke egress

(2) Whether the incident, event, or condition stays localized or spreads

(3) Whether or not egress is desired by facility occupants

(4) Whether there is a localized start to any egress or mass start to egress

(5) Whether exits are available or not available

Examples of scenarios and sets of characteristics that might occur in a facility follow.

*Scenario 1.* Characteristics: mass start, egress desired (by management and attendees), exits not available, local awareness.

Normal egress at the end of an event occurs just as a severe weather condition induces evacuees at the exterior doors to retard or stop their egress. The backup that occurs in the egress system is not known to most evacuees, who continue to press forward, potentially resulting in a crowd crush.

*Scenario 2.* Characteristics: mass start, egress not desired (by management), exits possibly not available, mass awareness.

An earthquake occurs during an event. The attendees are relatively safe in the seating area. The means of egress outside the seating area are relatively unsafe and vulnerable to aftershock damage. Facility management discourages mass egress until the means of egress can be checked and cleared for use.

*Scenario 3.* Characteristics: local start, incident stays local, egress desired (by attendees and management), exits available, mass awareness.

A localized civil disturbance (e.g., firearms violence) provokes localized egress, which is seen by attendees, generally, who then decide to leave also.

*Scenario 4.* Characteristics: mass start, egress desired (by attendees), incident spreads, exits not available, mass awareness.

In an open-air facility unprotected from wind, precipitation, and lightning, sudden severe weather prompts egress to shelter, but not from the facility. The means of egress congest and block quickly as people in front stop once they are under shelter, while people behind them continue to press forward, potentially resulting in a crowd crush.

These scenarios illustrate some of the broader factors to be taken into account when assessing the capability of both building systems and management features on which reliance is placed in a range of situations, not just fire emergencies. Some scenarios also illustrate the conflicting motivations of management and attendees, based on differing perceptions of danger and differing knowledge of hazards, countermeasures, and capabilities. Mass egress might not be the most appropriate life safety strategy in some scenarios, such as Scenario 2.

Table A.12.4.1.3 summarizes the characteristics in the scenarios and provides a framework for developing other characteristics and scenarios that might be important for a particular facility, hazard, occupant type, event, or management.

**A.12.4.2** Outdoor facilities are not accepted as inherently smoke-protected but must meet the requirements of smoke-protected assembly seating in order to utilize the special requirements for means of egress.

**A.12.4.2.1(1)(b)** The engineering analysis should be part of the life safety evaluation required by 12.4.1.

**A.12.4.5.12** Prior editions of the *Code* required stages to be protected by a Class III standpipe system in accordance with NFPA 14, *Standard for the Installation of Standpipe and Hose Systems*. NFPA 14 requires that Class II and Class III standpipes be automatic — not manual — because they are intended to be used by building occupants. Automatic standpipe systems are required to provide not less than 500 gpm (1890 L/min) at 100 psi (689 kN/m²). This requirement often can be met only if a fire pump is installed. Installation of a fire pump presents an unreasonable burden for the system supplying the two hose outlets at the side of the stage. The revised wording of 12.4.5.12 offers some relief by permitting the hose outlets to be in accordance with NFPA 13, *Standard for the Installation of Sprinkler Systems*.

**A.12.4.7** Where a special amusement building is installed inside another building, such as within an exhibit hall, the special amusement building requirements apply only to the special amusement building. For example, the smoke detectors required by 12.4.7.4 are not required to be connected to the building's system. Where installed in an exhibit hall, such smoke detectors are also required to comply with the provisions applicable to an exhibit.

**A.12.4.7.1** The aggregate horizontal projections of a multilevel play structure are indicative of the number of children who might be within the structure and at risk from a fire or similar emergency. The word "aggregate" is used in recognition of the fact that the platforms and tubes that make up the multilevel play structure run above each other at various levels. In calculating the area of the projections, it is important to account for all areas that might be expected to be occupied within, on top of, or beneath the components of the structure when the structure is used for its intended function.

**A.12.4.7.2** See A.12.4.7.1.

**A.12.4.7.7.3** Consideration should be given to the provision of directional exit marking on or adjacent to the floor.

**A.12.4.10.2(2)** Delayed-egress locks on doors from the airport loading walkway into the airport terminal building might compromise life safety due to the limited period of time the airport loading walkway will provide protection for emergency egress. The requirement of 12.4.10.2(2) would not limit the use of access-controlled or delayed-egress hardware from the airport terminal building into the airport loading walkway.

**A.12.7.2.4(5)** NFPA 58, *Liquefied Petroleum Gas Code*, permits portable butane-fueled appliances in restaurants and in attended commercial food catering operations where fueled by not more than two 10 oz (0.3 L) LP-Gas capacity, nonrefillable butane containers that have a water capacity not exceeding 1.08 lb (0.5 kg) per container. The containers are required to be directly connected to the appliance, and manifolding of containers is not permitted. Storage of cylinders is also limited to 24 containers, with an additional 24 permitted where protected by a 2-hour fire resistance–rated barrier.

**A.12.7.3(3)(a)** Securely supported altar candles in churches that are well separated from any combustible material are permitted. On the other hand, lighted candles carried by children wearing cotton robes present a hazard too great to be permitted. There are many other situations of intermediate hazard where the authority having jurisdiction will have to exercise judgment.

**A.12.7.4.3** The phrase "unprotected materials containing foamed plastic" is meant to include foamed plastic items covered by "thermally thin" combustible fabrics or paint. *(See A.10.2.3.4.)*

**A.12.7.5.3.7.1(3)** See A.12.4.1.1.

**A.12.7.6** The training program in crowd management should develop a clear appreciation of factors of space, energy, time, and information, as well as specific crowd management techniques, such as metering. Published guidelines on these factors and techniques are found in the *SFPE Handbook of Fire Protection Engineering*, Section 3, Chapter 13.

**A.12.7.7** It is important that an adequate number of competent attendants is on duty at all times when the assembly occupancy is occupied.

**Table A.12.4.1.3 Life Safety Evaluation Scenario Characteristics Matrix**

| | | | | | Management | | Occupants | | | | | | |
| | | | | | Egress | Egress Not | Egress | Egress Not | | | | | |
| Scenario | Local Awareness | Mass Awareness | Incident Localized | Incident Spreads | Desired | Desired | Desired | Desired | Local Start | Mass Start | Exits Available | Exits Not Available | Other |
|---|---|---|---|---|---|---|---|---|---|---|---|---|---|
| 1 | X | — | — | — | X | — | X | — | — | X | — | X | — |
| 2 | — | X | — | — | — | X | — | — | — | X | — | X | — |
| 3 | — | X | X | — | X | — | X | — | X | — | X | — | — |
| 4 | — | X | — | X | — | — | X | — | — | X | — | X | — |

**A.12.7.7.3** It is not the intent of this provision to require an announcement in bowling alleys, cocktail lounges, restaurants, or places of worship.

**A.13.1.2** Assembly occupancy requirements should be determined on a room-by-room basis, a floor-by-floor basis, and a total building basis. The requirements for each room should be based on the occupant load of that room, and the requirements for each floor should be based on the occupant load of that floor, but the requirements for the assembly building overall should be based on the total occupant load. Therefore, it is quite feasible to have several assembly occupancies with occupant loads of 300 or less grouped together in a single building. Such a building would be an assembly occupancy with an occupant load of over 1000.

**A.13.1.3.2** For example, an assembly room for the residents of a detention occupancy will not normally be subject to simultaneous occupancy.

**A.13.1.4.2** An understanding of the term *accessory room* might be useful to the enforcer of the *Code*, although the term is not used within the *Code*. An accessory room includes a dressing room, the property master's work and storage rooms, the carpenter's room, or similar rooms necessary for legitimate stage operations.

**A.13.1.7.1** The increase in occupant load above that calculated using occupant load factors from Table 7.3.1.2 is permitted if the provisions of 13.1.7.1 are followed. The owner or operator has the right to submit plans and to be permitted an increase in occupant load if the plans comply with the *Code*. The authority having jurisdiction is permitted to reject the plan for increase in occupant load if the plan is unrealistic, inaccurate, or otherwise does not properly reflect compliance with other *Code* requirements. It is not the intent of the provisions of 13.1.7.1 to prohibit an increase in occupant load solely on the basis of exceeding the limits calculated using occupant load factors from Table 7.3.1.2.

Existing auditorium and arena structures might not be designed for the added occupant load beyond the fixed seating. The authority having jurisdiction should consider exit access and aisles before permitting additional occupant load in areas using seating such as festival seating or movable seating on the auditorium or arena floor area.

To assist in preventing serious overcrowding incidents in sports arenas, stadia, and similar occupancies, spectator standing room should not be permitted between the seating areas and the playing areas, except in horse race and dog track facilities.

Where a capacity or near-capacity audience is anticipated, all seating should be assigned with tickets showing the section, row, and seat number.

Where standing room is permitted, the capacity of the standing area should meet the following criteria:

(1) The capacity should be determined on the basis of 5 ft$^2$ (0.46 m$^2$) per person.
(2) The capacity should be added to the seating capacity in determining egress requirements.
(3) The capacity should be located to the rear of the seating area.
(4) The capacity should be assigned standing-room-only tickets according to the area designated for the purpose.

The number of tickets sold, or otherwise distributed, should not exceed the aggregate number of seats plus the approved standing room numbers.

**A.13.2.2.3.1(1)** The seating plan and the means of egress should be reviewed each time the seating is substantially rearranged.

**A.13.2.3.2** The provisions of 13.2.3.2 should be applied within the audience seating chamber and to the room doors. The capacity of means of egress components encountered after leaving the audience seating chamber, such as concourses, lobbies, exit stair enclosures, and the exit discharge, should be calculated in accordance with Section 7.3.

**A.13.2.3.6.6** The original *Code* wording exempted sports arenas and railway stations. If an assembly occupancy was not similar to a sports arena or railway station, it was often judged ineligible to use the provision of 13.2.3.6.6. A list of exempted assembly venues also raises the question of why other occupancies are not included and necessitates additions to the list. For example, an exhibit hall of very large size might have several main entrances/exits. A theater extending the width of a block cannot really have a main entrance/exit in one confined location. A restaurant might have a main entrance serving the parking lot and another main entrance for those entering from the street. The authority having jurisdiction needs to determine where such arrangements are acceptable.

**A.13.2.4** It is not the intent to require four means of egress from each level of an assembly occupancy building having a total occupant load of more than 1000 where, individually, the floors have occupant loads of less than 1000.

**A.13.2.5.4.2** This requirement and the associated requirement of 13.2.5.4.3 have the effect of prohibiting festival seating, unless it truly is a form of seating, such as lawn seating, where generous spaces are commonly maintained between individuals and small groups so that people can circulate freely at any time. Such lawn seating is characterized by densities of about one person per 15 ft$^2$ (1.4 m$^2$). Both requirements prohibit uncontrolled crowd situations, such as in front of stages at rock music concerts where the number and density of people is uncontrolled by architectural or management features.

**A.13.2.5.4.3** This requirement is intended to facilitate rapid emergency access to individuals who are experiencing a medical emergency, especially in the case of cardiopulmonary difficulties, where there is a need for rapid medical attention from trained personnel. The requirement also addresses the need for security and law enforcement personnel to reach individuals whose behavior is endangering themselves and others.

**A.13.2.5.4.4** The catchment area served by an aisle accessway or aisle is the portion of the total space that is naturally served by the aisle accessway or aisle. Hence, the requirement for combining the required capacity where paths converge is, in effect, a restatement of the idea of a catchment area. The establishment of catchment areas should be based on a balanced use of all means of egress, with the number of persons in proportion to egress capacity.

**A.13.2.5.5** For purposes of the means of egress requirements of this *Code*, tablet-arm chair seating is not considered seating at tables. Dinner theater–style configurations are required to comply with the aisle accessway provisions applying to seating at tables and the aisle requirements of 13.2.5.6, if the aisles contain steps or are ramped. Generally, if aisles contain steps or are ramped, all of the *Code* requirements for aisles, stairs, and ramps are required to be met. *(Also see 7.1.7 and A.7.1.7.2.)*

**A.13.2.5.5.1** Seats having reclining backs are assumed to be in their most upright position when unoccupied.

**A.13.2.5.5.4** The system known as *continental seating* has one pair of egress doors provided for every five rows that is located close to the ends of the rows. In previous editions of the *Code*, such egress doors were required to provide a clear width of not less than 66 in. (1675 mm) discharging into a foyer, into a lobby, or to the exterior of the building. This continental seating arrangement can result in egress flow times (i.e., with nominal flow times of approximately 100 seconds rather than 200 seconds) that are approximately one-half as long as those resulting where side aisles lead to more remote doors. Such superior egress flow time performance is desirable in some situations; however, special attention should be given either to a comparably good egress capacity for other parts of the egress system or to sufficient space to accommodate queuing outside the seating space.

**A.13.2.5.6.3** It is the intent to permit handrails to project not more than 3½ in. (90 mm) into the clear width of aisles required by 13.2.5.6.3.

**A.13.2.5.6.4.1** Technical information about the convenience and safety of ramps and stairs having gradients in the region of 1 in 8 clearly suggests that the goal should be slopes for ramps that are less steep and combinations of stair risers and treads that are, for example, superior to 4 in. (100 mm) risers and 32 in. (865 mm) treads. This goal should be kept in mind by designers in establishing the gradient of seating areas to be served by aisles.

**A.13.2.5.6.5(3)** Tread depth is more important to stair safety than is riser height. Therefore, in cases where the seating area gradient is less than 5 in 11, it is recommended that the tread dimension be increased beyond 11 in. (280 mm), rather than reducing the riser height. Where the seating area gradient exceeds 8 in 11, it is recommended that the riser height be increased while maintaining a tread depth of not less than 11 in. (280 mm).

**A.13.2.5.6.5(5)** Completely uniform tread dimensions are preferred over aisle stair designs where tread depths alternate between relatively small intermediate treads between seating platforms and relatively large treads at seating platforms. A larger tread that is level with the seating platform is not needed to facilitate easy access to, and egress from, a row of seating. If this arrangement is used, it is important to provide a tread depth that is better than minimum for the intermediate tread; hence, 13 in. (330 mm) is specified. Where nonuniformities exist due to construction tolerance, they should not exceed ³⁄₁₆ in. (4.8 mm) between adjacent treads.

**A.13.2.5.6.8** Failure to provide a handrail within a 30 in. (760 mm) horizontal distance of all required portions of the aisle stair width means that the egress capacity calculation is required to be modified as specified by 13.2.3.3(3). This modification might lead to an increase in the aisle width. Although this increase will compensate for reduced egress efficiency, it does not help individuals walking on such portions of stairs to recover from missteps, other than by possibly marginally reducing the crowding that might exacerbate the problem of falls. *(See also 7.2.2.4.)*

**A.13.2.5.6.9** Certain tread cover materials such as plush carpets, which are often used in theaters, produce an inherently well-marked tread nosing under most lighting conditions. On the other hand, concrete treads have nosings with a sharp edge and, especially under outdoor lighting conditions, are difficult to discriminate. Therefore, concrete treads require an applied marking stripe. The slip resistance of such marking stripes should be similar to the rest of the treads, and no tripping hazard should be created; luminescent, self-luminous, and electroluminescent tread markings have the advantage of being apparent in reduced light or in the absence of light.

**A.13.2.5.7** For purposes of the means of egress requirements of this *Code*, seating at counters or at other furnishings is considered to be the same as seating at tables.

**A.13.2.5.7.2** Effectively, where the aisle accessway is bounded by movable seating, the 12 in. (305 mm) minimum width might be increased by about 15 in. to 30 in. (380 mm to 760 mm) as seating is pushed in toward tables. Moreover, it is such movement of chairs during normal and emergency egress situations that makes the zero-clearance exception workable. The exception also applies to booth seating where people sitting closest to the aisle normally move out ahead of people farthest from the aisle.

**A.13.2.5.7.3** See A.13.2.5.8.3.

**A.13.2.5.7.4** The minimum width requirement as a function of accessway length is as follows:

(1) 0 in. (0 mm) for the first 6 ft (1830 mm) of length toward the exit
(2) 12 in. (305 mm) for the next 6 ft (1830 mm); that is, up to 12 ft (3660 mm) of length
(3) 12 in. to 24 in. (305 mm to 610 mm) for lengths from 12 ft to 36 ft (3.7 m to 11 m), the maximum length to the closest aisle or egress doorway permitted by 13.2.5.7.5

Any additional width needed for seating is to be added to these widths, as described in 13.2.5.8.3.

**A.13.2.5.8.1** See 7.1.7 and A.7.1.7.2 for special circulation safety precautions applicable where small elevation differences occur.

**A.13.2.5.8.2** It is important to make facilities accessible to people using wheelchairs. See ICC/ANSI A117.1, *American National Standard for Accessible and Usable Buildings and Facilities*, which provides guidance on appropriate aisle widths.

**A.13.2.5.8.3** Figure A.13.2.5.8.3 shows typical measurements involving seating and tables abutting an aisle. Note that, for purposes of the means of egress requirements of this *Code*, seating at counters or other furnishings is considered to be the same as seating at tables.

**A.13.3.1(1)** The allowance for unenclosed stairs or ramps presumes the balcony or mezzanine complies with the other provisions of the *Code*, such as travel distance to exits in accordance with 13.2.6 and number of exits in accordance with 13.2.4. For the purposes of this exception, a balcony with glazing that provides a visual awareness of the main assembly area is considered open.

**A.13.3.4.2.3** The intent is to require detectors only in nonsprinklered hazardous areas that are unoccupied. Where the building is occupied, the detectors in the unoccupied, unsprinklered hazardous areas will initiate occupant notification. If the building is unoccupied, the fire in the nonsprinklered hazardous area is not a life safety issue, and the detectors, upon activation, are not required to notify anyone. The signal from a detector is permitted to be sent to a control panel in an area that is occupied when the building is occupied, but that is unoccupied when the building is

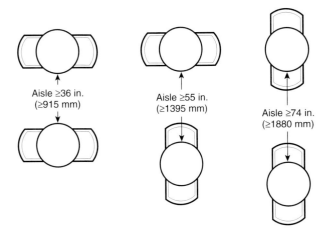

Aisle ≥36 in.
(≥915 mm)

Aisle ≥55 in.
(≥1395 mm)

Aisle ≥74 in.
(≥1880 mm)

**FIGURE A.13.2.5.8.3  Seating at Tables Abutting an Aisle.**

unoccupied, without the need for central station monitoring or the equivalent.

**A.13.4.1.1**  Life safety evaluations are examples of performance-based approaches to life safety. In this respect, significant guidance in the form and process of life safety evaluations is provided by Chapter 5, keeping in mind the fire safety emphasis in Chapter 5. Performance criteria, scenarios, evaluation, safety factors, documentation, maintenance, and periodic assessment (including a warrant of fitness) all apply to the broader considerations in a life safety evaluation. A life safety evaluation deals not only with fire but also with storms, collapse, crowd behavior, and other related safety considerations for which a checklist is provided in A.13.4.1.3. Chapter 5 provides guidance, based on fire safety requirements, for establishing a documented case showing that products of combustion in all conceivable fire scenarios will not significantly endanger occupants using means of egress in the facility (e.g., due to fire detection, automatic suppression, smoke control, large-volume space, or management procedures). Moreover, means of egress facilities plus facility management capabilities should be adequate to cope with scenarios where certain egress routes are blocked for some reason.

In addition to making realistic assumptions about the capabilities of persons in the facility (e.g., an assembled crowd including many disabled persons or persons unfamiliar with the facility), the life safety evaluation should include a factor of safety of not less than 2.0 in all calculations relating to hazard development time and required egress time (the combination of flow time and other time needed to detect and assess an emergency condition, initiate egress, and move along the egress routes). This factor of safety takes into account the possibility that half of the egress routes might not be used (or usable) in certain situations.

Regarding crowd behavior, the potential hazards created by larger masses of people and greater crowd densities (which can be problematic during ingress, occupancy, and egress) demand that technology be used by designers, managers, and authorities responsible for buildings to compensate for the relaxed egress capacity provisions of Table 13.4.2.3. In very large buildings for assembly use, the hazard of crowd crushes can exceed that of fire or structural failure. Therefore, the building designers, managers, event planners, security personnel, police authorities, and fire authorities, as well as the building construction authorities, should understand the potential problems and solutions, including coordination of their activi-

ties. For crowd behavior, this understanding includes factors of space, energy, time, and information, as well as specific crowd management techniques, such as metering. Published guidance on these factors and techniques is found in the *SFPE Handbook of Fire Protection Engineering*, Section 3, Chapter 13, pp. 3-342–3-366 (Proulx, G., "Movement of People"), and the publications referenced therein.

Table 13.2.3.2 and Table 13.4.2.3 are based on a linear relationship between number of seats and nominal flow time, with not less than 200 seconds (3.3 minutes) for 2000 seats plus 1 second for every additional 50 seats up to 25,000. Beyond 25,000 total seats, the nominal flow time is limited to 660 seconds (11 minutes). Nominal flow time refers to the flow time for the most able group of patrons; some groups less familiar with the premises or less able groups might take longer to pass a point in the egress system. Although three or more digits are noted in the tables, the resulting calculations should be assumed to provide only two significant figures of precision.

**A.13.4.1.3**  Factors to be considered in a life safety evaluation might include the following:

(1) Nature of the events being accommodated, including the following:
  (a) Ingress, intra-event movement, and egress patterns
  (b) Ticketing and seating policies/practices
  (c) Event purpose (e.g., sports contest, religious meeting)
  (d) Emotional qualities (e.g., competitiveness) of event
  (e) Time of day when event is held
  (f) Time duration of single event
  (g) Time duration of attendees' occupancy of the building

(2) Occupant characteristics and behavior, including the following:
  (a) Homogeneity
  (b) Cohesiveness
  (c) Familiarity with building
  (d) Familiarity with similar events
  (e) Capability (as influenced by factors such as age, physical abilities)
  (f) Socioeconomic factors
  (g) Small minority involved with recreational violence
  (h) Emotional involvement with the event and other occupants
  (i) Use of alcohol or drugs
  (j) Food consumption
  (k) Washroom utilization

(3) Management, including the following:
  (a) Clear, contractual arrangements for facility operation/ use as follows:
    i. Between facility owner and operator
    ii. Between facility operator and event promoter
    iii. Between event promoter and performer
    iv. Between event promoter and attendee
    v. With police forces
    vi. With private security services
    vii. With ushering services
  (b) Experience with the building
  (c) Experience with similar events and attendees
  (d) Thorough, up-to-date operations manual
  (e) Training of personnel
  (f) Supervision of personnel
  (g) Communications systems and utilization
  (h) Ratios of management and other personnel to attendees

(i) Location/distribution of personnel

(j) Central command location

(k) Rapport between personnel and attendees

(l) Personnel support of attendee goals

(m) Respect of attendees for personnel due to the following:

    i. Dress (uniform) standards

    ii. Age and perceived experience

    iii. Personnel behavior, including interaction

    iv. Distinction between crowd management and control

    v. Management concern for facility quality (e.g., cleanliness)

    vi. Management concern for entire event experience of attendees (i.e., not just during occupancy of the building)

(4) Emergency management preparedness, including the following:

(a) Complete range of emergencies addressed in operations manual

(b) Power loss

(c) Fire

(d) Severe weather

(e) Earthquake

(f) Crowd incident

(g) Terrorism

(h) Hazardous materials

(i) Transportation accident (e.g., road, rail, air)

(j) Communications systems available

(k) Personnel and emergency forces ready to respond

(l) Attendees clearly informed of situation and proper behavior

(5) Building systems, including the following:

(a) Structural soundness

(b) Normal static loads

(c) Abnormal static loads (e.g., crowds, precipitation)

(d) Dynamic loads (e.g., crowd sway, impact, explosion, wind, earthquake)

(e) Stability of nonstructural components (e.g., lighting)

(f) Stability of movable (e.g., telescoping) structures

(g) Fire protection

(h) Fire prevention (e.g., maintenance, contents, housekeeping)

(i) Compartmentation

(j) Automatic detection and suppression of fire

(k) Smoke control

(l) Alarm and communications systems

(m) Fire department access routes and response capability

(n) Structural integrity

(o) Weather protection

(p) Wind

(q) Precipitation (attendees rush for shelter or hold up egress of others)

(r) Lightning protection

(s) Circulation systems

(t) Flowline or network analysis

(u) Waywinding and orientation

(v) Merging of paths (e.g., precedence behavior)

(w) Decision/branching points

(x) Route redundancies

(y) Counterflow, crossflow, and queuing situations

(z) Control possibilities, including metering

(aa) Flow capacity adequacy

(bb) System balance

(cc) Movement time performance

(dd) Flow times

(ee) Travel times

(ff) Queuing times

(gg) Route quality

(hh) Walking surfaces (e.g., traction, discontinuities)

(ii) Appropriate widths and boundary conditions

(jj) Handrails, guardrails, and other rails

(kk) Ramp slopes

(ll) Step geometries

(mm) Perceptual aspects (e.g., orientation, signage, marking, lighting, glare, distractions)

(nn) Route choices, especially for vertical travel

(oo) Resting/waiting areas

(pp) Levels of service (overall crowd movement quality)

(qq) Services

(rr) Washroom provision and distribution

(ss) Concessions

(tt) First aid and EMS facilities

(uu) General attendee services

A scenario-based approach to performance-based fire safety is addressed in Chapter 5. In addition to utilizing such scenarios and, more generally, the attention to performance criteria, evaluation, safety factors, documentation, maintenance, and periodic assessment required when the Chapter 5 option is used, life safety evaluations should consider scenarios based on characteristics important in assembly occupancies. These characteristics include the following:

(1) Whether there is a local or mass awareness of an incident, event, or condition that might provoke egress

(2) Whether the incident, event, or condition stays localized or spreads

(3) Whether or not egress is desired by facility occupants

(4) Whether there is a localized start to any egress or mass start to egress

(5) Whether exits are available or not available

Examples of scenarios and sets of characteristics that might occur in a facility follow.

*Scenario 1.* Characteristics: mass start, egress desired (by management and attendees), exits not available, local awareness.

Normal egress at the end of an event occurs just as a severe weather condition induces evacuees at the exterior doors to retard or stop their egress. The backup that occurs in the egress system is not known to most evacuees, who continue to press forward, potentially resulting in a crowd crush.

*Scenario 2.* Characteristics: mass start, egress not desired (by management), exits possibly not available, mass awareness.

An earthquake occurs during an event. The attendees are relatively safe in the seating area. The means of egress outside the seating area are relatively unsafe and vulnerable to aftershock damage. Facility management discourages mass egress until the means of egress can be checked and cleared for use.

*Scenario 3.* Characteristics: local start, incident stays local, egress desired (by attendees and management), exits available, mass awareness.

A localized civil disturbance (e.g., firearms violence) provokes localized egress, which is seen by attendees, generally, who then decide to leave also.

*Scenario 4.* Characteristics: mass start, egress desired (by attendees), incident spreads, exits not available, mass awareness.

In an open-air facility unprotected from wind, precipitation, and lightning, sudden severe weather prompts egress to shelter but not from the facility. The means of egress congest and block quickly as people in front stop once they are under shelter, while people behind them continue to press forward, potentially resulting in a crowd crush.

These scenarios illustrate some of the broader factors to be taken into account when assessing the capability of both building systems and management features on which reliance is placed in a range of situations, not just fire emergencies. Some scenarios also illustrate the conflicting motivations of management and attendees based on differing perceptions of danger and differing knowledge of hazards, countermeasures, and capabilities. Mass egress might not be the most appropriate life safety strategy in some scenarios, such as Scenario 2.

Table A.13.4.1.3 summarizes the characteristics in the scenarios and provides a framework for developing other characteristics and scenarios that might be important for a particular facility, hazard, occupant type, event, or management.

**A.13.4.2** Outdoor facilities are not accepted as inherently smoke-protected but must meet the requirements of smoke-protected assembly seating in order to use the special requirements for means of egress.

**A.13.4.2.1(1)(b)** The engineering analysis should be part of the life safety evaluation required by 13.4.1.

**A.13.4.5.12** Prior editions of the *Code* required stages to be protected by a Class III standpipe system in accordance with NFPA 14, *Standard for the Installation of Standpipe and Hose Systems*. NFPA 14 requires that Class II and Class III standpipes be automatic — not manual — because they are intended to be used by building occupants. Automatic standpipe systems are required to provide not less than 500 gpm (1890 L/min) at 100 psi (689 kN/m²). This requirement often can be met only if a fire pump is installed. Installation of a fire pump presents an unreasonable burden for the system supplying the two hose outlets at the side of the stage. The revised wording of 13.4.5.12 offers some relief by permitting the hose outlets to be in accordance with NFPA 13, *Standard for the Installation of Sprinkler Systems*.

**A.13.4.7** Where a special amusement building is installed inside another building, such as within an exhibit hall, the special amusement building requirements apply only to the special amusement building. For example, the smoke detectors required by 13.4.7.4 are not required to be connected to the building's system. Where installed in an exhibit hall, such smoke detectors are also required to comply with the provisions applicable to an exhibit.

**A.13.4.7.1** The aggregate horizontal projections of a multilevel play structure are indicative of the number of children who might be within the structure and at risk from a fire or similar emergency. The word "aggregate" is used in recognition of the fact that the platforms and tubes that make up the multilevel play structure run above each other at various levels. In calculating the area of the projections, it is important to account for all areas that might be expected to be occupied within, on top of, or beneath the components of the structure when the structure is used for its intended function.

**A.13.4.7.2** See A.13.4.7.1.

**A.13.4.7.7.3** Consideration should be given to the provision of directional exit marking on or adjacent to the floor.

**A.13.4.10.2(2)** Delayed-egress locks on doors from the airport loading walkway into the airport terminal building might compromise life safety due to the limited period of time the airport loading walkway will provide protection for emergency egress. The requirement of 13.4.10.2(2) would not limit the use of access-controlled or delayed-egress hardware from the airport terminal building into the airport loading walkway.

**A.13.7.2.4(5)** NFPA 58, *Liquefied Petroleum Gas Code*, permits portable butane-fueled appliances in restaurants and in attended commercial food catering operations where fueled by not in excess of two 10 oz (0.3 L) LP-Gas capacity, nonrefillable butane containers that have a water capacity not exceeding 1.08 lb (0.5 kg) per container. The containers are required to be directly connected to the appliance, and manifolding of containers is not permitted. Storage of cylinders is also limited to 24 containers, with an additional 24 permitted where protected by a 2-hour fire resistance–rated barrier.

**A.13.7.3(3)(a)** Securely supported altar candles in churches that are well separated from any combustible material are permitted. On the other hand, lighted candles carried by children wearing cotton robes present a hazard too great to be permitted. There are many other situations of intermediate hazard where the authority having jurisdiction will have to exercise judgment.

**A.13.7.4.3** The phrase "unprotected materials containing foamed plastic" is meant to include foamed plastic items covered by "thermally thin" combustible fabrics or paint. *(See A.10.2.3.4.)*

**A.13.7.5.3.7.1(3)** See A.13.4.1.1.

**A.13.7.6** The training program in crowd management should develop a clear appreciation of factors of space, energy, time, and information, as well as specific crowd management techniques, such as metering. Published guidelines on these factors and techniques are found in the *SFPE Handbook of Fire Protection Engineering*, Section 3, Chapter 13.

**Table A.13.4.1.3 Life Safety Evaluation Scenario Characteristics Matrix**

| | | | | | Management | | Occupants | | | | | | |
| | Local Awareness | Mass Awareness | Incident Localized | Incident Spreads | Egress Desired | Egress Not Desired | Egress Desired | Egress Not Desired | Local Start | Mass Start | Exits Available | Exits Not Available | Other |
| Scenario | | | | | | | | | | | | | |
| 1 | X | — | — | — | X | — | X | — | — | X | — | X | — |
| 2 | — | X | — | — | — | X | — | — | — | X | — | X | — |
| 3 | — | X | X | — | X | — | X | — | X | — | X | — | — |
| 4 | — | X | — | X | — | — | X | — | — | X | — | X | — |

**A.13.7.7** It is important that an adequate number of competent attendants is on duty at all times when the assembly occupancy is occupied.

**A.13.7.7.3** It is not the intent of this provision to require an announcement in bowling alleys, cocktail lounges, restaurants, or places of worship.

**A.14.2.2.3** See A.7.2.2.4.4.4 regarding additional handrails on stairs that are used extensively by children 5 years of age or less.

**A.14.2.5.9** A corridor roofed over and enclosed on its long side and open to the atmosphere at the end is permitted to be considered an exterior corridor if either of the following criteria are met:

(1) Clear story openings for the corridor are provided on both sides of the corridor and above adjacent roofs or buildings, and such clear openings are not less than one-half the height of the corridor walls.
(2) The corridor roof has unobstructed openings to the sky not less than 50 percent of the area of the roof.

The openings detailed in A.14.2.5.9(1) and (2) are to be equally distributed, and, if louvers are installed, they are to be fixed open with a clear area based on the actual openings between louver vanes.

**A.14.2.11.1** It is highly desirable that all windows be of a type that can be readily opened from inside and that they are large enough and low enough for use by students, teachers, and fire fighters. Windows are permitted to serve as a supplementary means of emergency escape, particularly where ladders can be raised by fire fighters or others.

**A.14.3.3.2** The definition of interior wall finish is meant to include washroom water closet partitions.

**A.14.3.4.2.3.1** Occupied portions of the building should have access to a central point for manual activation of the evacuation signal.

**A.14.3.4.2.3.2** Occupied portions of the building should have access to a central point for manual activation of the evacuation signal.

**A.14.3.4.3.1.1** Use of the distinctive three-pulse temporal pattern fire alarm evacuation signal that is required by *NFPA 72, National Fire Alarm and Signaling Code*, will help educate students to recognize the need to evacuate when they are in other occupancies.

**A.14.3.5.1** It is the intent to permit use of the criteria of 8.2.1.3(1) to create separate buildings for purposes of limiting educational occupancy building area to not more than 12,000 ft$^2$ (1115 m$^2$).

**A.14.7.2.1** The requirements are, of necessity, general in scope, as it is recognized that they apply to all types of educational occupancies as well as conditions of occupancies, such as truant schools; schools for the mentally handicapped, vision impaired, hearing impaired, and speech impaired; and public schools. It is fully recognized that no one code can meet all the conditions of the various buildings involved, and it will be necessary for site administrators to issue supplements to these requirements, but all supplements should be consistent with these requirements.

**A.14.7.3.1** Particular attention should be given to keeping all doors unlocked; keeping doors that serve to protect the safety of paths of egress closed and under no conditions blocked open, such as doors on stairway enclosures; keeping outside stairs and fire escape stairs free from all obstructions and clear of snow and ice; and allowing no accumulation of snow or ice or materials of any kind outside exit doors that might prevent the opening of the door or interfere with rapid escape from the building.

Any condition likely to interfere with safe egress should be corrected immediately, if possible, or otherwise should be reported at once to the appropriate authorities.

**A.15.2.2.3** See A.7.2.2.4.4.4 regarding additional handrails on stairs that are used extensively by children 5 years of age or less.

**A.15.2.5.9** A corridor roofed over and enclosed on its long side and open to the atmosphere at the end is permitted to be considered an exterior corridor if either of the following criteria are met:

(1) Clear story openings for the corridor are provided on both sides of the corridor and above adjacent roofs or buildings, and such clear openings are not less than one-half the height of the corridor walls.
(2) The corridor roof has unobstructed openings to the sky not less than 50 percent of the area of the roof.

The openings detailed in A.15.2.5.9(1) are to be equally distributed, and, if louvers are installed, they are to be fixed open with a clear area based on the actual openings between louver vanes.

**A.15.2.11.1** It is highly desirable that all windows be of a type that can be readily opened from inside and that they are large enough and low enough for use by students, teachers, and fire fighters. Windows are permitted to serve as a supplementary means of emergency escape, particularly where ladders can be raised by fire fighters or others.

**A.15.3.4.2.3.1** Occupied portions of the building should have access to a central point for manual activation of the evacuation signal.

**A.15.3.4.2.3.2** Occupied portions of the building should have access to a central point for manual activation of the evacuation signal.

**A.15.3.4.3.1.1** The audible occupant notification signal for evacuation of an educational occupancy building should be the distinctive three-pulse temporal pattern fire alarm evacuation signal that is required of new systems by *NFPA 72, National Fire Alarm and Signaling Code*. The temporal pattern will help educate students to recognize the need to evacuate when they are in other occupancies. Existing fire alarm systems should be modified, as feasible, to sound the three-pulse temporal pattern.

**A.15.3.6(2)** This provision permits valve supervision in accordance with Section 9.7, rather than requiring that the entire automatic sprinkler system be electrically supervised. It is intended that the valve supervision be performed electrically, not by chaining and locking the valves in the open position.

**A.15.7.2.1** The requirements are, of necessity, general in scope, as it is recognized that they apply to all types of educational occupancies as well as conditions of occupancies, such as truant schools; schools for the mentally handicapped, vision impaired, hearing impaired, and speech impaired; and public schools. It is fully recognized that no one code can meet all the conditions of the various buildings involved, and it will be necessary for site administrators to issue supplements to these requirements, but all supplements should be consistent with these requirements.

**A.15.7.3.1** Particular attention should be given to keeping all doors unlocked; keeping doors that serve to protect the safety of paths of egress closed and under no conditions blocked open, such as doors on stairway enclosures; keeping outside stairs and fire escape stairs free from all obstructions and clear of snow and ice; and allowing no accumulation of snow or ice or materials of any kind outside exit doors that might prevent the opening of the door or interfere with rapid escape from the building.

Any condition likely to interfere with safe egress should be corrected immediately, if possible, or otherwise should be reported at once to the appropriate authorities.

**A.16.1.1** Day-care occupancies do not provide for the full-time maintenance of a client. Occupancies that provide a primary place of residence are addressed in other occupancy chapters. *(See Chapters 24 through 33.)*

The requirements of Chapter 16 are based on the need to adequately protect the occupants in case of fire. The requirements assume that adequate staffing will be available and are based on staffing similar to that outlined in Table A.16.1.1.

**Table A.16.1.1 Staffing**

| Staff-to-Client Ratio | Age (mo.) |
| --- | --- |
| 1:3 | 0–24 |
| 1:4 | 25–36 |
| 1:7 | 37–60 |
| 1:10 | 61–96 |
| 1:12 | ≥97 |
| 1:3 | Clients incapable of self-preservation |

If staff-to-client ratios fall below those suggested by Table A.16.1.1, it is the responsibility of the authority having jurisdiction to determine the additional safeguards beyond the requirements of Chapter 16 that are necessary. Typical additional provisions might include restricting the day-care occupancy to the level of exit discharge, requiring additional smoke detection, requiring automatic sprinkler protection, requiring better or additional means of egress, and requiring similar types of provisions, depending on the situation.

**A.16.1.2.3** A conversion from a day-care occupancy with more than 12 clients to a day-care home is not considered a change of occupancy. The resulting day-care home should be permitted to meet the requirements of Chapter 17 for existing day-care homes.

**A.16.2.2.4** The purpose of this requirement is to prevent arrangements whereby a client can be trapped in a space or area. It is intended that this provision be broadly interpreted by the authority having jurisdiction to include equipment such as refrigerators and freezers.

**A.16.2.2.3** See A.7.2.2.4.4.4 regarding additional handrails on stairs that are used extensively by children 5 years of age or less.

**A.16.3.2.1(2)(a)** It is not the intent to classify a room with a domestic-type clothes washer and a domestic-type clothes dryer as a laundry.

**A.16.6.1.4.2** A conversion from a day-care occupancy with more than 12 clients to a day-care home is not considered a change of occupancy. The resulting day-care home should be permitted to meet the requirements of Chapter 17 for existing day-care homes.

**A.16.7.1** The requirements are, of necessity, general in scope, because it is recognized that they apply to all types of day-care occupancies as well as conditions of occupancies, such as truant day-care occupancies; occupancies for the mentally handicapped, vision impaired, hearing impaired, and speech impaired; adult day-care; care of infants; and day-care occupancies. It is fully recognized that no one code can meet all the conditions of the various buildings involved, and it will be necessary for site administrators, through the written fire emergency response plan, to issue supplements to these requirements; however, all supplements should be consistent with these requirements. Additionally, it is recommended that fire safety be a part of the educational programs of the occupancy for clients.

Fire emergency response plans need to be written and made available to all employees, including temporary or substitute staff, so that all employees know what is expected of them during a fire emergency. The elements needed in the written plan should be identified in coordination with the authority having jurisdiction.

The facility fire emergency response plan might be a module of a facility disaster plan that covers other emergencies.

The proper safeguarding of clients during a fire emergency requires prompt and effective response by the facility employees in accordance with the fire emergency response plan. Duties covered under the plan should be assigned by position rather than by employee name. Such assignment ensures that, in the absence of an employee, the duties of the position will be performed by a substitute or temporary employee assigned to the position. Temporary or substitute employees should be instructed in advance regarding their duties under the plan for the position to which they are assigned.

Written fire emergency response plans should include, but should not be limited to, information for employees regarding methods and devices available for alerting occupants of a fire emergency. Employees should know how the fire department is to be alerted. Even where automatic systems are expected to alert the fire department, the written plan should provide for backup alerting procedures by staff. Other responses of employees to a fire emergency should include the following:

(1) Removal of clients in immediate danger to areas of safety, as set forth in the plan
(2) Methods of using building features to confine the fire and its byproducts to the room or area of origin
(3) Control of actions and behaviors of clients during removal or evacuation activities and at predetermined safe assembly areas

The written plan should state clearly the facility policy regarding the actions staff are to take or not take to extinguish a fire. It should also incorporate the emergency egress and relocation drill procedures set forth in 16.7.2.

For additional guidance on emergency plans, see *NFPA 1600, Standard on Disaster/Emergency Management and Business Continuity Programs*. This standard establishes a common set of criteria for disaster management, emergency management, and business continuity programs.

**A.16.7.2.1** The requirements are, of necessity, general in scope, because it is recognized that they apply to all types of day-care occupancies as well as conditions of occupancies, such as truant day-care occupancies; and day-care occupancies for the mentally handicapped, vision impaired, hearing impaired, and speech impaired. It is fully recognized that no one code can meet all the conditions of the various buildings involved, and it will be necessary for site administrators to issue supplements to these requirements, but all supplements should be consistent with these requirements.

**A.16.7.3.2** Particular attention should be given to keeping all doors unlocked; keeping doors that serve to protect the safety of paths of egress closed and under no conditions blocked open, such as doors on stairway enclosures; keeping outside stairs and fire escape stairs free from all obstructions and clear of snow and ice; and allowing no accumulation of snow or ice or materials of any kind outside exit doors that might prevent the opening of the door or interfere with rapid escape from the building.

**A.16.7.5** It is the intent that the requirement for adequate adult staff to be awake at all times when clients are present be applied to family day-care and group day-care homes that are operated at night, as well as day-care occupancies.

**A.17.1.1** Day-care occupancies do not provide for the full-time maintenance of a client. Occupancies that provide a primary place of residence are addressed in other occupancies. *(See Chapters 24 through 33.)*

The requirements of Chapter 17 are based on the need to adequately protect the occupants in case of fire. The requirements assume that adequate staffing will be available and are based on staffing similar to that outlined in Table A.17.1.1.

**Table A.17.1.1 Staffing**

| Staff-to-Client Ratio | Age (mo.) |
|---|---|
| 1:3 | 0–24 |
| 1:4 | 25–36 |
| 1:7 | 37–60 |
| 1:10 | 61–96 |
| 1:12 | ≥97 |
| 1:3 | Clients incapable of self-preservation |

If staff-to-client ratios fall below those suggested by Table A.17.1.1, it is the responsibility of the authority having jurisdiction to determine the additional safeguards beyond the requirements of Chapter 17 that are necessary. Typical additional provisions might include restricting the day-care occupancy to the level of exit discharge, requiring additional smoke detection, requiring automatic sprinkler protection, requiring better or additional means of egress, and requiring similar types of items, depending on the situation.

**A.17.1.2.3** A conversion from a day-care occupancy with more than 12 clients to a day-care home is not considered a change of occupancy. The resulting day-care home should be permitted to meet the requirements of Chapter 17 for existing day-care homes.

**A.17.2.2.2.4** The purpose of this requirement is to prevent arrangements where a client can be trapped in a space or area. It is intended that this provision be broadly interpreted by the authority having jurisdiction to include equipment such as refrigerators and freezers.

**A.17.2.2.3** See A.7.2.2.4.4.4 regarding additional handrails on stairs that are used extensively by children 5 years of age and under.

**A.17.3.2.1(2)(a)** It is not the intent to classify a room with a domestic-type clothes washer and a domestic-type clothes dryer as a laundry.

**A.17.6.1.1.2** Day-care homes do not provide for the full-time maintenance of a client. Day-care occupancies that provide a primary place of residence are addressed in other day-care occupancy chapters. *(See Chapters 24 through 33.)*

**A.17.6.1.4.2** A conversion from a day-care occupancy with more than 12 clients to a day-care home is not considered a change of occupancy. The resulting day-care home should be permitted to meet the requirements of Chapter 17 for existing day-care homes.

**A.17.7.1** The requirements are, of necessity, general in scope, because it is recognized that they apply to all types of day-care occupancies as well as conditions of occupancies, such as truant day-care occupancies; occupancies for the mentally handicapped, vision impaired, hearing impaired, and speech impaired; adult day-care; care of infants; and day-care occupancies. It is fully recognized that no one code can meet all the conditions of the various buildings involved, and it will be necessary for site administrators, through the written fire emergency response plan, to issue supplements to these requirements; however, all supplements should be consistent with these requirements. Additionally, it is recommended that fire safety be a part of the educational programs of the occupancy for clients.

Fire emergency response plans need to be written and made available to all employees, including temporary or substitute staff, so that all employees know what is expected of them during a fire emergency. The elements needed in the written plan should be identified in coordination with the authority having jurisdiction.

The facility fire emergency response plan might be a module of a facility disaster plan that covers other emergencies.

The proper safeguarding of clients during a fire emergency requires prompt and effective response by the facility employees in accordance with the fire emergency response plan. Duties covered under the plan should be assigned by position rather than by employee name. Such assignment ensures that, in the absence of an employee, the duties of the position will be performed by a substitute or temporary employee assigned to the position. Temporary or substitute employees should be instructed in advance regarding their duties under the plan for the position to which they are assigned.

Written fire emergency response plans should include, but should not be limited to, information for employees about methods and devices available for alerting occupants of a fire emergency. Employees should know how the fire department is to be alerted. Even where automatic systems are expected to alert the fire department, the written plan should provide for backup alerting procedures by staff. Other responses of employees to a fire emergency should include the following:

(1) Removal of clients in immediate danger to areas of safety, as set forth in the plan

(2) Methods of using building features to confine the fire and its byproducts to the room or area of origin

(3) Control of actions and behaviors of clients during removal or evacuation activities and at predetermined safe assembly areas

The written plan should state clearly the facility policy regarding the actions staff are to take or not take to extinguish a fire. It should also incorporate the emergency egress and relocation drill procedures set forth in 17.7.2.

For additional guidance on emergency plans, see *NFPA 1600, Standard on Disaster/Emergency Management and Business Continuity Programs*. This standard establishes a common set of criteria for disaster management, emergency management, and business continuity programs.

**A.17.7.2.1** The requirements are, of necessity, general in scope, because it is recognized that they apply to all types of day-care occupancies as well as conditions of occupancies, such as truant day-care occupancies; and day-care occupancies for the mentally handicapped, vision impaired, hearing impaired, and speech impaired. It is fully recognized that no one code can meet all the conditions of the various buildings involved, and it will be necessary for site administrators to issue supplements to these requirements, but all supplements should be consistent with these requirements.

**A.17.7.3.2** Particular attention should be given to keeping all doors unlocked; keeping doors that serve to protect the safety of paths of egress closed and under no conditions blocked open, such as doors on stairway enclosures; keeping outside stairs and fire escape stairs free from all obstructions and clear of snow and ice; and allowing no accumulation of snow or ice or materials of any kind outside exit doors that might prevent the opening of the door or interfere with rapid escape from the building.

**A.17.7.5** It is the intent that the requirement for adequate adult staff to be awake at all times when clients are present be applied to family day-care and group day-care homes that are operated at night, as well as day-care occupancies.

**A.18.1.1.1.1** In determining equivalency for conversions, modernizations, renovations, or unusual design concepts of hospitals or nursing homes, the authority having jurisdiction is permitted to accept evaluations based on the health care occupancies' fire safety evaluation system (FSES) of NFPA 101A, *Guide on Alternative Approaches to Life Safety*, utilizing the parameters for new construction.

**A.18.1.1.1.7** There are many reasons why doors in the means of egress in health care occupancies might need to be locked for the protection of the patients or the public. Examples of conditions that might justify door locking include dementia, mental health, infant care, pediatric care, or patients under court detention order requiring medical treatment in a health care facility. See 18.2.2.2.5 for details on door locking.

**A.18.1.1.1.10** The *Code* recognizes that certain functions necessary for the life safety of building occupants — such as the detection of fire and associated products of combustion, the closing of corridor doors, the operation of manual fire alarm devices, and the removal of patients from the room of fire origin — require the intervention of facility staff. It is not the intent of 18.1.1.1.10 to specify the levels or locations of staff necessary to meet this requirement.

**A.18.1.1.2** This objective is accomplished in the context of the physical facilities, the type of activities undertaken, the provisions for the capabilities of staff, and the needs of all occupants through requirements directed at the following:

(1) Prevention of ignition
(2) Detection of fire
(3) Control of fire development
(4) Confinement of the effects of fire
(5) Extinguishment of fire
(6) Provision of refuge or evacuation facilities, or both
(7) Staff reaction

**A.18.1.1.4.3.3** For the purpose of this requirement, a floor that is not divided by a smoke barrier is considered one smoke compartment. Where automatic sprinklers are retrofitted into existing nonsprinklered buildings, the construction alternatives for sprinklers provided in this *Code* are intended to apply to the renovated area.

**A.18.1.1.4.3.4** In minor rehabilitation, only the rehabilitation itself — not the entire smoke compartment or building — is required to be brought up to the requirements for new nonsprinklered facilities.

**A.18.1.3.3** Doctors' offices and treatment and diagnostic facilities that are intended solely for outpatient care and are physically separated from facilities for the treatment or care of inpatients, but that are otherwise associated with the management of an institution, might be classified as business occupancies rather than health care occupancies. Facilities that do not provide housing for patients on a 24-hour basis are required to be classified as other than health care occupancies per 18.1.1.1.7, except where services are provided routinely to four or more inpatients who are incapable of self-preservation.

**A.18.1.3.4.1** It is the intent that these requirements apply to mobile, transportable, and relocatable structures (in accordance with 1.3.2) where such structures are used to provide shared medical services on an extended or a temporary basis. Where properly separated from the health care occupancy and intended to provide services simultaneously for three or fewer health care patients who are litterborne, the level of protection for such structures should be based on the appropriate occupancy classification of other chapters of this *Code*. Mobile, transportable, or relocatable structures that are not separated from a contiguous health care occupancy, or that are intended to provide services simultaneously for four or more health care patients who are litterborne, should be classified and designed as health care occupancies.

**A.18.2.2** In planning egress, arrangements should be made to transfer patients from one section of a floor to another section of the same floor that is separated by a fire barrier or smoke barrier in such a manner that patients confined to their beds can be transferred in their beds. Where the building design will allow, the section of the corridor containing an entrance or elevator lobby should be separated from corridors leading from it by fire or smoke barriers. Such arrangement, where the lobby is centrally located, will, in effect, produce a smoke lock, placing a double barrier between the area to which patients might be taken and the area from which they need to be evacuated because of threatening smoke and fire.

**A.18.2.2.2.4(2)** Where delayed-egress locks complying with 7.2.1.6.1 are used, the provisions of 18.2.2.2.5 are not required.

**A.18.2.2.2.4(3)** Where access-controlled egress doors complying with 7.2.1.6.2 are used, the provisions of 18.2.2.2.5 are not required.

**A.18.2.2.2.5.1** Psychiatric units, Alzheimer units, and dementia units are examples of areas with patients who might have clinical needs that justify door locking. Forensic units and detention units are examples of areas with patients who might pose a security threat. Where Alzheimer or dementia patients in nursing homes are not housed in specialized units, the provisions of 18.2.2.2.5.1 should not apply. (See 18.2.2.2.5.2.)

**A.18.2.2.2.5.2** Pediatric units, maternity units, and emergency departments are examples of areas where patients might have special needs that justify door locking.

**A.18.2.2.2.5.2(3)** Where locked doors in accordance with 18.2.2.2.5.2 are proposed for an existing building that is not sprinklered throughout, the authority having jurisdiction might consider permitting the installation based on an analysis of the extent of sprinkler protection provided. Sprinklered areas should include, at a minimum, the secured compartment and compartments that the occupants of the secured compartment must travel through to egress the building.

**A.18.2.2.2.7** It is desirable to keep doors in exit passageways, stair enclosures, horizontal exits, smoke barriers, and required enclosures around hazardous areas closed at all times to impede the travel of smoke and fire gases. Functionally, however, this involves decreased efficiency and limits patient observation by the staff of a facility. To accommodate such needs, it is practical to presume that such doors will be kept open, even to the extent of employing wood chocks and other makeshift devices. Doors in exit passageways, horizontal exits, and smoke barriers should, therefore, be equipped with automatic hold-open devices activated by the methods described, regardless of whether the original installation of the doors was predicated on a policy of keeping them closed.

**A.18.2.3.4** It is not the intent that the required corridor width be maintained clear and unobstructed at all times. Projections into the required width are permitted by 7.3.2.2. It is not the intent that 18.2.3.4 supersede 7.3.2.2.

**A.18.2.3.4(1)** Occupant characteristics are an important factor to be evaluated in setting egress criteria. Egress components in nonpatient use areas, such as administrative office spaces, should be evaluated based on actual use. A clear corridor width of not less than 44 in. (1120 mm) is specified, assuming occupants in nonpatient areas will be mobile and capable of evacuation without assistance.

**A.18.2.3.4(2)** The intent of 18.2.3.4(2) is to permit limited noncontinuous projections along the corridor wall. These include hand-rub dispensing units complying with 18.3.2.6, nurse charting units, wall-mounted computers, telephones, artwork, bulletin boards, display case frames, cabinet frames, fire alarm boxes, and similar items. It is not the intent to permit the narrowing of the corridor by the walls themselves. The provision of 7.3.2.2 permits projections up to 4½ in. (114 mm) to be present at and below the 38 in. (965 mm) height specified in 18.2.3.4(2), and it is not the intent of 18.2.3.4(2) to prohibit such projections. Permitting projections above the 38 in. (965 mm) handrail height complies with the intent of the requirement, as such projections will not interfere with the movement of gurneys, beds, and wheelchairs. Projections below handrail height for limited items, such as fire extinguisher cabinets and recessed water coolers, also will not interfere with equipment movement.

**A.18.2.3.4(3)** Exit access should be arranged to avoid any obstructions to the convenient removal of nonambulatory persons carried on stretchers or on mattresses serving as stretchers.

**A.18.2.3.4(4)(c)** Wheeled equipment and carts in use include food service carts, housekeeping carts, medication carts, isolation carts, and similar items. Isolation carts should be permitted in the corridor only where patients require isolation precautions.

Unattended wheeled crash carts and other similar wheeled emergency equipment are permitted to be located in the corridor when "not in use," because they need to be immediately accessible during a clinical emergency. Note that "not in use" is not the same as "in storage." Storage is not permitted to be open to the corridor, unless it meets one of the provisions permitted in 18.3.6.1 and is not a hazardous area.

Wheeled portable patient lift or transport equipment needs to be readily available to clinical staff for moving, transferring, toileting, or relocating patients. These devices are used daily for safe handling of patients and to provide for worker safety. This equipment might not be defined as "in use" but needs to be convenient for the use of caregivers at all times.

**A.18.2.3.4(5)** The means for affixing the furniture can be achieved with removable brackets to allow cleaning and maintenance. Affixing the furniture to the floor or wall prevents the furniture from moving, so as to maintain a minimum 6 ft (1830 mm) corridor clear width. Affixing the furniture to the floor or wall also provides a sturdiness that allows occupants to safely transfer in and out.

**A.18.2.3.4(5)(f)** Examples of building service and fire protection equipment include fire extinguishers, manual fire alarm boxes, shutoff valves, and similar equipment.

**A.18.2.3.4(6)** The 8 ft (2440 mm) corridor width does not need to be maintained at the door or the open door leaf. A reduction for the frame and leaf is acceptable as long as the minimum clear width is provided at the door opening in the direction of egress travel. In situations where egress occurs only in one direction, it is permissible to have a single door leaf.

**A.18.2.3.5(1)** See A.18.2.3.4(1).

**A.18.2.3.5(2)** The intent of 18.2.3.5(2) is to permit limited noncontinuous projections along the corridor wall. These include hand-rub dispensing units complying with 18.3.2.6, nurse charting units, wall-mounted computers, telephones, artwork, bulletin boards, display case frames, cabinet frames, fire alarm boxes, and similar items. It is not the intent to permit the narrowing of the corridor by the walls themselves. The provision of 7.3.2.2 permits projections up to 4½ in. (114 mm) to be present at and below the 38 in. (965 mm) height specified in 18.2.3.5(2), and it is not the intent of 18.2.3.5(2) to prohibit such projections. Permitting projections above the 38 in. (965 mm) handrail height complies with the intent of the requirement, as such projections will not interfere with the movement of gurneys, beds, and wheelchairs. Projections below handrail height for limited items, such as fire extinguisher cabinets and recessed water coolers, also will not interfere with equipment movement.

**A.18.2.3.5(3)** See A.18.2.3.4(3).

**A.18.2.3.5(4)(c)** Wheeled equipment and carts in use include food service carts, housekeeping carts, medication carts, isolation carts, and similar items. Isolation carts should be permitted in the corridor only where patients require isolation precautions.

Unattended wheeled crash carts and other similar wheeled emergency equipment are permitted to be located in the corridor when "not in use," because they need to be immediately accessible during a clinical emergency. Note that "not in use," is not the same as "in storage." Storage is not permitted to be open to the corridor, unless it meets one of the provisions permitted in 18.3.6.1 and is not a hazardous area.

Wheeled portable patient lift or transport equipment needs to be readily available to clinical staff for moving, transferring, toileting, or relocating patients. These devices are used daily for safe handling of patients and to provide for worker safety. This equipment might not be defined as "in use" but needs to be convenient for the use of caregivers at all times.

**A.18.2.3.5(5)** The 6 ft 1830 mm) corridor width does not need to be maintained at the door or the open door leaf. A reduction for the frame and leaf is acceptable as long as the minimum clear width is provided at the door opening in the direction of egress travel. In situations where egress occurs only in one direction, it is permissible to have a single door leaf.

**A.18.2.4.4** An exit is not necessary for each individual smoke compartment if there is access to an exit through other smoke compartments without passing through the smoke compartment of fire origin.

**A.18.2.5.4** The term *intervening rooms or spaces* means rooms or spaces serving as a part of the required means of egress from another room.

**A.18.2.5.6.1** For the purposes of this paragraph, it is the intent that the term *habitable rooms* not include individual bathrooms, closets, and similar spaces, as well as briefly occupied work spaces, such as control rooms in radiology and small storage rooms in a pharmacy.

**A.18.2.5.7.1.2** Two or more contiguous suites with an aggregate area not exceeding the suite size limitations of 18.2.5.7.2.3 and 18.2.5.7.3.3 are permitted to be considered a single suite, so as not to require separation from each other.

**A.18.2.5.7.1.3(A)** The term *intervening room* means a room serving as a part of the required means of egress from another room.

**A.18.2.5.7.1.3(C)** Examples of suites that might be hazardous areas are medical records and pharmaceutical suites.

**A.18.2.5.7.2.1(A)** For the purposes of this paragraph, it is the intent that the term *habitable rooms* not include individual bathrooms, closets, and similar spaces, as well as briefly occupied work spaces, such as control rooms in radiology and small storage rooms in a pharmacy.

**A.18.2.5.7.2.1(D)(1)** The interior partitions or walls might extend full height to the ceiling, provided that they do not obscure visual supervision of the suite. Where they do obscure visual supervision, see 18.2.5.7.2.1(D)(2).

**A.18.2.5.7.2.2(B)** Where only one means of egress is required from the suite, it needs to be provided by a door opening directly to a corridor complying with 18.3.6.

**A.18.2.5.7.2.2(C)** Where the second exit access for a sleeping suite is through an adjacent suite, it is the intent that the 100 ft (30 m) travel distance limitation in the suite be applied only to the suite under consideration.

**A.18.2.5.7.2.3(C)(1)** The alternative of 18.2.5.7.2.1(D)(1)(b) is not to be applied, since 18.2.5.7.2.3(B)(2) requires total coverage automatic smoke detection for the suite that exceeds 5000 ft² (460 m²) but does not exceed 7500 ft² (700 m²).

**A.18.2.5.7.3.2(B)** Where only one means of egress is required from the suite, it needs to be provided by a door opening directly to a corridor complying with 18.3.6.

**A.18.2.5.7.3.2(C)** Where the second exit access for a non-sleeping suite is through an adjacent suite, it is the intent that the adjacent suite not be considered an intervening room.

**A.18.3.2.1** Provisions for the enclosure of rooms used for charging linen chutes and waste chutes or for rooms into which these chutes empty are provided in Section 9.5.

**A.18.3.2.2** The hazard level of a laboratory is considered severe if quantities of flammable, combustible, or hazardous materials are present that are capable of sustaining a fire of sufficient magnitude to breach a 1-hour fire separation. See the NFPA *Fire Protection Handbook* for guidance.

**A.18.3.2.5.2** This provision is intended to permit small appliances used for reheating, such as microwave ovens, hot plates, toasters, and nourishment centers to be exempt from the requirements for commercial cooking equipment and hazardous area protection.

**A.18.3.2.5.3** The intent of 18.3.2.5.3 is to limit the number of persons for whom meals are routinely prepared to not more than 30. Staff and feeding assistants are not included in this number.

**A.18.3.2.5.3(3)** The minimum airflow of 500 cfm (14,000 L/m) is intended to require the use of residential hood equipment at the higher end of equipment capacities. It is also intended to draw a sufficient amount of the cooking vapors into the grease baffle and filter system to reduce migration beyond the hood.

**A.18.3.2.5.3(6)** The intent of this provision is to limit cooking fuel to gas or electricity. The prohibition of solid fuels for cooking is not intended to prohibit charcoal grilling on grills located outside the facility.

**A.18.3.2.5.3(7)** Deep-fat frying is defined as a cooking method that involves fully immersing food in hot oil.

**A.18.3.2.5.3(9)** The intent of this requirement is that the fuel source for the cooktop or range is to be turned on only when staff is present or aware that the kitchen is being used. The timer function is meant to provide an additional safeguard if the staff forgets to deactivate the cooktop or range. If a cooking activity lasts longer than 120 minutes, the timer would be required to be manually reset.

**A.18.3.2.5.3(11)** The intent of requiring smoke alarms instead of smoke detectors is to prevent false alarms from initiating the building fire alarm system and notifying the fire department. Smoke alarms should be maintained a minimum of 20 ft (6.1 m) away from the cooktop or range as studies have shown this distance to be the threshold for significantly reducing false alarms caused by cooking. The intent of the interconnected smoke alarms, with silence feature, is that while the devices would alert staff members to a potential problem, if it is a false alarm, the staff members can use the silence feature instead of disabling the alarm. The referenced study indicates that nuisance alarms are reduced with photoelectric smoke alarms. Providing two, interconnected alarms provides a safety factor since they are not electrically supervised by the the the fire alarm system. *(Smoke Alarms – Pilot Study of Nuisance Alarms Associated with Cooking)*

**A.18.3.2.5.4** The provisions of 18.3.2.5.4 differ from those of 18.3.2.5.3, as they apply to cooking equipment that is separated from the corridor.

**A.18.3.2.5.5** The provision of 18.3.2.5.5 clarifies that protected commercial cooking equipment does not require an enclosure (separation) as a hazardous area in accordance with Section 8.7, as is required by 18.3.2.1.

**A.18.3.2.6** Extensive research, including fire modeling, has indicated that alcohol-based hand-rub solutions can be safely installed in corridors of health care facilities, provided that certain other precautions are taken. The total quantities of flammable liquids in any area should comply with the provisions of other recognized codes, including NFPA 1, *Fire Code*, and NFPA 30, *Flammable and Combustible Liquids Code*. In addition, special consideration should be given to the following:

(1) Obstructions created by the installation of hand-rub solution dispensers
(2) Location of dispensers with regard to adjacent combustible materials and potential sources of ignition, especially where dispensers are mounted on walls of combustible construction
(3) Requirements for other fire protection features, including complete automatic sprinkler protection, to be installed throughout the compartment
(4) Amount and location of the flammable solutions, both in use and in storage, particularly with respect to potential for leakage or failure of the dispenser

**A.18.3.3.2** The reductions in class of interior finish prescribed by 10.2.8.1 are permitted to be used.

**A.18.3.4.2** It is not the intent of this *Code* to require single-station smoke detectors that might be required by local codes to be connected to or to initiate the building fire alarm system.

**A.18.3.4.3.1(2)** It is the intent of this provision to permit a visible fire alarm signal instead of an audible signal to reduce interference between the fire alarm and medical equipment monitoring alarms.

**A.18.3.4.5.3** The requirement for smoke detectors in spaces open to the corridors eliminates the requirements of 18.3.6.1(1)(c), (2)(b), and (5)(b) for direct supervision by the facility staff of nursing homes.

**A.18.3.5.1** In areas where the replenishment of water supplies is not immediately available from on-site sources, alternate provisions for the water-fill rate requirements of NFPA 13, *Standard for the Installation of Sprinkler Systems*, and NFPA 22, *Standard for Water Tanks for Private Fire Protection*, that are acceptable to the authority having jurisdiction should be provided. Appropriate means for the replenishment of these supplies from other sources, such as fire department tankers, public safety organizations, or other independent contractors should be incorporated into the overall fire safety plan of the facility.

With automatic sprinkler protection required throughout new health care facilities and quick-response sprinklers required in smoke compartments containing patient sleeping rooms, a fire and its life-threatening byproducts can be reduced, thereby allowing the defend-in-place concept to continue. The difficulty in maintaining the proper integrity of life safety elements has been considered, and it has been judged that the probability of a sprinkler system operating as designed is equal to or greater than other life safety features.

**A.18.3.5.6** The requirements for use of quick-response sprinklers intend that quick-response sprinklers be the predominant type of sprinkler installed in the smoke compartment. It is recognized, however, that quick-response sprinklers might

not be approved for installation in all areas, such as those where NFPA 13, *Standard for the Installation of Sprinkler Systems*, requires sprinklers of the intermediate- or high-temperature classification. It is not the intent of the 18.3.5.6 requirements to prohibit the use of standard sprinklers in limited areas of a smoke compartment where intermediate- or high-temperature sprinklers are required.

Residential sprinklers are considered acceptable in patient sleeping rooms of all health care facilities, even though not specifically listed for this purpose in all cases.

Where the installation of quick-response sprinklers is impracticable in patient sleeping room areas, appropriate equivalent protection features acceptable to the authority having jurisdiction should be provided. It is recognized that the use of quick-response sprinklers might be limited in facilities housing certain types of patients or by the installation limitations of quick-response sprinklers.

**A.18.3.5.10** Although this exception is currently not recognized by NFPA 13, *Standard for the Installation of Sprinkler Systems*, a proposal for such exemption has been submitted for consideration by the Technical Committee on Sprinkler System Installation Criteria. This exception is limited to hospitals, as nursing homes and many limited care facilities might have more combustibles within the closets. The limited amount of clothing found in the small clothes closets in hospital patient rooms is typically far less than the amount of combustibles in casework cabinets that do not require sprinkler protection, such as nurse servers. In many hospitals, especially new hospitals, it is difficult to make a distinction between clothes closets and cabinet work. The exception is far more restrictive than similar exceptions for hotels and apartment buildings. NFPA 13 already permits the omission of sprinklers in wardrobes [*see 8.1.1(7) of NFPA 13*]. It is not the intent of 18.3.5.10 to affect the wardrobe provisions of NFPA 13. It is the intent that the sprinkler protection in the room covers the closet as if there were no door on the closet. (*See 8.5.3.2.3 of NFPA 13.*)

**A.18.3.5.11** For the proper operation of sprinkler systems, cubicle curtains and sprinkler locations need to be coordinated. Improperly designed systems might obstruct the sprinkler spray from reaching the fire or might shield the heat from the sprinkler. Many options are available to the designer including, but not limited to, hanging the cubicle curtains 18 in. (455 mm) below the sprinkler deflector; using a ½ in. (13 mm) diagonal mesh or a 70 percent open weave top panel that extends 18 in. (455 mm) below the sprinkler deflector; or designing the system to have a horizontal and minimum vertical distance that meets the requirements of NFPA 13, *Standard for the Installation of Sprinkler Systems*. The test data that form the basis of the NFPA 13 requirements are from fire tests with sprinkler discharge that penetrated a single privacy curtain.

**A.18.3.6.1(1)(a)** The presence of stored combustible materials in a room or space open to the corridor does not necessarily result in the room or space being classified as a hazardous area. In some circumstances, the amount and type of combustibles might result in the room or space being classified as a hazardous area by the authority having jurisdiction.

**A.18.3.6.1(3)** A typical nurses' station would normally contain one or more of the following with associated furniture and furnishings:

(1) Charting area
(2) Clerical area
(3) Nourishment station

(4) Storage of small amounts of medications, medical equipment and supplies, clerical supplies, and linens

(5) Patient monitoring and communication equipment

**A.18.3.6.2**  It is the intent of the *Code* that there be no required fire resistance or area limitations for vision panels in corridor walls and doors.

An architectural, exposed, suspended-grid acoustical tile ceiling with penetrating items, such as sprinkler piping and sprinklers; ducted HVAC supply and return-air diffusers; speakers; and recessed lighting fixtures, is capable of limiting the transfer of smoke.

**A.18.3.6.2.3**  While a corridor wall is required to form a barrier to limit the transfer of smoke, such a barrier is not required to be either a smoke barrier or a smoke partition — two terms for which specific *Code* definitions and requirements apply.

**A.18.3.6.3**  While it is recognized that closed doors serve to maintain tenable conditions in a corridor and adjacent patient rooms, such doors, which, under normal or fire conditions, are self-closing, might create a special hazard for the personal safety of a room occupant. Such closed doors might present a problem of delay in discovery, confining fire products beyond tenable conditions.

Because it is critical for responding staff members to be able to immediately identify the specific room involved, it is recommended that approved automatic smoke detection that is interconnected with the building fire alarm be considered for rooms having doors equipped with closing devices. Such detection is permitted to be located at any approved point within the room. When activated, the detector is required to provide a warning that indicates the specific room of involvement by activation of a fire alarm annunciator, nurse call system, or any other device acceptable to the authority having jurisdiction.

Where a nurse server penetrates a corridor wall, the access opening on the corridor side of the nurse server must be protected as is done for a corridor door.

**A.18.3.6.3.1**  Gasketing of doors should not be necessary to achieve resistance to the passage of smoke if the door is relatively tight-fitting.

**A.18.3.6.3.10**  Doors should not be blocked open by furniture, door stops, chocks, tie-backs, drop-down or plunger-type devices, or other devices that necessitate manual unlatching or releasing action to close. Examples of hold-open devices that release when the door is pushed or pulled are friction catches or magnetic catches.

**A.18.3.6.3.12**  It is not the intent of 18.3.6.3.12 to prohibit the application of push plates, hardware, or other attachments on corridor doors in health care occupancies.

**A.18.3.6.5.1**  It is not the intent of 18.3.6.5.1 to permit mail slots or pass-through openings in doors or walls of rooms designated as a hazardous area.

**A.18.3.7**  See A.18.2.2.

**A.18.3.7.3(2)**  Where the smoke control system design requires dampers so that the system will function effectively, it is not the intent of the provision to permit the damper to be omitted.

This provision is not intended to prevent the use of plenum returns where ducting is used to return air from a ceiling plenum through smoke barrier walls. Short stubs or jumper ducts

are not acceptable. Ducting is required to connect at both sides of the opening and to extend into adjacent spaces away from the wall. The intent is to prohibit open-air transfers at or near the smoke barrier walls.

**A.18.3.7.6**  Smoke barrier doors are intended to provide access to adjacent zones. The pair of cross-corridor doors are required to be opposite swinging. Access to both zones is required.

It is not the intent of 18.3.7.6 to prohibit the application of push plates, hardware, or other attachments on some barrier doors in health care occupancies.

**A.18.3.7.8**  Smoke barriers might include walls having door openings other than cross-corridor doors. There is no restriction in the *Code* regarding which doors or how many doors form part of a smoke barrier. For example, doors from the corridor to individual rooms are permitted to form part of a smoke barrier. Split astragals (i.e., astragals installed on both door leaves) are also considered astragals.

**A.18.3.7.9**  It is not the intent to require the frame to be a listed assembly.

**A.18.4.3.1**  For example, the provisions of 18.1.1.4.3.1(2) and 18.1.1.4.3.4 do not require the installation of sprinklers if the modification involves less than 50 percent of the area of the smoke compartment and less than 4500 ft$^2$ (420 m$^2$) of the area of the smoke compartment.

**A.18.5.2.2**  For both new and existing buildings, it is the intent to permit the installation and use of fireplace stoves and room heaters utilizing solid fuel as defined in NFPA 211, *Standard for Chimneys, Fireplaces, Vents, and Solid Fuel–Burning Appliances,* provided that all such devices are installed, maintained, and used in accordance with the appropriate provisions of that standard and all manufacturers' specifications. These requirements are not intended to permit freestanding solid fuel–burning appliances such as freestanding wood-burning stoves.

**A.18.5.2.3(2)(d)**  The glass front of a direct-vent fireplace can become extremely hot. Barriers such as screens or mesh installed over the direct-vent glass help reduce the risk of burn from touching the glass.

**A.18.5.2.3(2)(e)**  The intent of locating controls in a restricted location is to ensure staff is aware of use of the fireplace and to prevent unauthorized use. Examples of locked controls are a keyed switch or locating the switch in a staff-controlled location such as a staff station.

**A.18.7**  Health care occupants have, in large part, varied degrees of physical disability, and their removal to the outside, or even their disturbance caused by moving, is inexpedient or impractical in many cases, except as a last resort. Similarly, recognizing that there might be an operating necessity for the restraint of the mentally ill, often by use of barred windows and locked doors, fire exit drills are usually extremely disturbing, detrimental, and frequently impracticable.

In most cases, fire exit drills, as ordinarily practiced in other occupancies, cannot be conducted in health care occupancies. Fundamentally, superior construction, early discovery and extinguishment of incipient fires, and prompt notification need to be relied on to reduce the occasion for evacuation of buildings of this class to a minimum.

**A.18.7.1.4**  Many health care occupancies conduct fire drills without disturbing patients by choosing the location of the simulated emergency in advance and by closing the doors to patients' rooms or wards in the vicinity prior to initiation of the drill. The purpose of a fire drill is to test and evaluate the efficiency, knowledge, and response of institutional personnel in implementing the facility fire emergency plan. Its purpose is not to disturb or excite patients. Fire drills should be scheduled on a random basis to ensure that personnel in health care facilities are drilled not less than once in each 3-month period.

Drills should consider the ability to move patients to an adjacent smoke compartment. Relocation can be practiced using simulated patients or empty wheelchairs.

**A.18.7.2.1** Each facility has specific characteristics that vary sufficiently from other facilities to prevent the specification of a universal emergency procedure. The recommendations that follow, however, contain many of the elements that should be considered and adapted, as appropriate, to the individual facility.

Upon discovery of fire, personnel should immediately take the following action:

(1) If any person is involved in the fire, the discoverer should go to the aid of that person, calling aloud an established code phrase, which provides for both the immediate aid of any endangered person and the transmission of an alarm.
(2) Any person in the area, upon hearing the code called aloud, should activate the building fire alarm using the nearest manual fire alarm box.
(3) If a person is not involved in the fire, the discoverer should activate the building fire alarm using the nearest manual fire alarm box.
(4) Personnel, upon hearing the alarm signal, should immediately execute their duties as outlined in the facility fire safety plan.
(5) The telephone operator should determine the location of the fire as indicated by the audible signal.
(6) In a building equipped with an uncoded alarm system, a person on the floor of fire origin should be responsible for promptly notifying the facility telephone operator of the fire location.
(7) If the telephone operator receives a telephone alarm reporting a fire from a floor, the operator should regard that alarm in the same fashion as an alarm received over the fire alarm system and should immediately notify the fire department and alert all facility personnel of the place of fire and its origin.
(8) If the building fire alarm system is out of order, any person discovering a fire should immediately notify the telephone operator by telephone, and the operator should then transmit this information to the fire department and alert the building occupants.

**A.18.7.4** The most rigid discipline with regard to prohibition of smoking might not be nearly as effective in reducing incipient fires from surreptitious smoking as the open recognition of smoking, with provision of suitable facilities for smoking. Proper education and training of the staff and attendants in the ordinary fire hazards and their abatement is unquestionably essential. The problem is a broad one, varying with different types and arrangements of buildings; the effectiveness of rules of procedure, which need to be flexible, depends in large part on the management.

**A.18.7.5.1** In addition to the provisions of 10.3.1, which deal with ignition resistance, additional requirements with respect to the location of cubicle curtains relative to sprinkler placement are included in NFPA 13, *Standard for the Installation of Sprinkler Systems.*

**A.18.7.5.6(4)** The percentage of decorations should be measured against the area of any wall or ceiling, not the aggregate total of walls, ceilings, and doors. The door is considered part of the wall. The decorations must be located such that they do not interfere with the operation of any door, sprinkler, smoke detector, or any other life safety equipment. Other art might include hanging objects or three-dimensional items.

**A.18.7.5.7.1(3)** It is not the intent to permit collection receptacles with a capacity greater than 32 gal (121 L) to be positioned at or near a nurses' station based on the argument that such nurses' station is constantly attended. The large collection receptacle itself needs to be actively attended by staff. Staff might leave the large receptacle in the corridor outside a patient room while entering the room to collect soiled linen or trash, but staff is expected to return to the receptacle, move on to the next room, and repeat the collection function. Where staff is not actively collecting material for placement in the receptacle, the receptacle is to be moved to a room protected as a hazardous area.

**A.18.7.5.7.2** It is the intent that this provision permits recycling of bottles, cans, paper and similar clean items that do not contain grease, oil, flammable liquids, or significant plastic materials using larger containers or several adjacent containers and not require locating such containers in a room protected as a hazardous area. Containers for medical records awaiting shredding are often larger than 32 gal (121 L). These containers are not to be included in the calculations and limitations of 18.7.5.7.1. There is no limit on the number of these containers, as FM Approval Standard 6921, *Containers for Combustible Waste,* ensures that the fire will not spread outside of the container. FM approval standards are written for use with FM Approvals. The tests can be conducted by any approved laboratory. The portions of the standard referring to FM Approvals are not included in this reference.

**A.18.7.5.7.2(2)** See 18.7.5.7.1(3).

**A.19.1.1.1.1** In determining equivalency for existing hospitals or nursing homes, the authority having jurisdiction is permitted to accept evaluations based on the health care occupancies fire safety evaluation system (FSES) of NFPA 101A, *Guide on Alternative Approaches to Life Safety,* utilizing the parameters for existing buildings.

**A.19.1.1.1.7** There are many reasons why doors in the means of egress in health care occupancies might need to be locked for the protection of the patients or the public. Examples of conditions that might justify door locking include dementia, mental health, infant care, pediatric care, or patients under court detention order requiring medical treatment in a health care facility. See 19.2.2.2.5 for details on door locking.

**A.19.1.1.1.10** The *Code* recognizes that certain functions necessary for the life safety of building occupants — such as the detection of fire and associated products of combustion, the closing of corridor doors, the operation of manual fire alarm devices, and the removal of patients from the room of fire origin — require the intervention of facility staff. It is not the intent of 19.1.1.1.10 to specify the levels or locations of staff necessary to meet this requirement.

**A.19.1.1.2** This objective is accomplished in the context of the physical facilities, the type of activities undertaken, the provisions for the capabilities of staff, and the needs of all occupants through requirements directed at the following:

(1) Prevention of ignition
(2) Detection of fire
(3) Control of fire development
(4) Confinement of the effects of fire
(5) Extinguishment of fire
(6) Provision of refuge or evacuation facilities, or both
(7) Staff reaction

**A.19.1.1.4.3.3** For the purpose of this requirement, a floor that is not divided by a smoke barrier is considered one smoke compartment. Where automatic sprinklers are retrofitted into existing nonsprinklered buildings, the construction alternatives for sprinklers provided in this *Code* are intended to apply to the renovated area.

**A.19.1.1.4.3.4** In minor rehabilitation, only the rehabilitation itself is required to be brought up to the requirements for new nonsprinklered facilities, not the entire smoke compartment or building.

**A.19.1.3.3** Doctors' offices and treatment and diagnostic facilities that are intended solely for outpatient care and are physically separated from facilities for the treatment or care of inpatients, but that are otherwise associated with the management of an institution, might be classified as business occupancies rather than health care occupancies. Facilities that do not provide housing for patients on a 24-hour basis are required to be classified as other than health care occupancies per 19.1.1.1.7, except where services are provided routinely to four or more inpatients who are incapable of self-preservation.

**A.19.1.3.4.1** It is the intent of the *Code* that these requirements apply to mobile, transportable, and relocatable structures (in accordance with 1.3.2) when such structures are used to provide shared medical services on an extended or a temporary basis. Where properly separated from the health care occupancy and intended to provide services simultaneously for three or fewer health care patients who are litterborne, the level of protection for such structures should be based on the appropriate occupancy classification of other chapters of this *Code*. Mobile, transportable, or relocatable structures that are not separated from a contiguous health care occupancy, or that are intended to provide services simultaneously for four or more health care patients who are litterborne, should be classified and designed as health care occupancies.

**A.19.1.6.2** Unoccupied space, for the purposes of 19.1.6.2(3), is space not normally occupied by persons, fuel-fired equipment, or hazardous contents.

**A.19.2.2.2.4(2)** Where delayed-egress locks complying with 7.2.1.6.1 are used, the provisions of 19.2.2.2.5 are not required.

**A.19.2.2.2.4(3)** Where access-controlled egress doors complying with 7.2.1.6.2 are used, the provisions of 19.2.2.2.5 are not required.

**A.19.2.2.2.5.1** Psychiatric units, Alzheimer units, and dementia units are examples of areas with patients who might have clinical needs that justify door locking. Forensic units and detention units are examples of areas with patients who might pose a security threat. Where Alzheimer or dementia patients in nursing homes are not housed in specialized units, the provisions of 19.2.2.2.5.1 should not apply. *(See 19.2.2.2.5.2.)*

**A.19.2.2.2.5.2** Pediatric units, maternity units, and emergency departments are examples of areas where patients might have special needs that justify door locking.

**A.19.2.2.2.5.2(3)** Where locked doors in accordance with 19.2.2.2.5.2 are proposed for an existing building that is not sprinklered throughout, the authority having jurisdiction might consider permitting the installation based on an analysis of the extent of sprinkler protection provided. Sprinklered areas should include, at a minimum, the secured compartment and compartments that the occupants of the secured compartment must travel through to egress the building.

**A.19.2.2.2.7** It is desirable to keep doors in exit passageways, stair enclosures, horizontal exits, smoke barriers, and required enclosures around hazardous areas closed at all times to impede the travel of smoke and fire gases. Functionally, however, this involves decreased efficiency and limits patient supervision by the staff of a facility. To accommodate such needs, it is practical to presume that such doors will be kept open, even to the extent of employing wood chocks and other makeshift devices. Doors in exit passageways, horizontal exits, and smoke barriers should, therefore, be equipped with automatic hold-open devices actuated by the methods described, regardless of whether the original installation of the doors was predicated on a policy of keeping them closed.

**A.19.2.2.2.9** Doors to the enclosures of interior stair exits should be arranged to open from the stair side at not less than every third floor so that it will be possible to leave the stairway at such floor if fire renders the lower part of the stair unusable during egress or if occupants seek refuge on another floor.

**A.19.2.2.5.3** The waiver of the requirement for doors to swing in the direction of egress travel is based on the assumption that, in this occupancy, there is no possibility of a panic rush that might prevent the opening of doors that swing against egress travel.

A desirable arrangement, which is possible with corridors 8 ft (2440 mm) or more in width, is to have two 42 in. (1070 mm) doors, normally closed, each swinging with the egress travel (in opposite directions).

**A.19.2.3.4** It is not the intent that the required corridor width be maintained clear and unobstructed at all times. Projections into the required width are permitted by 7.3.2.2. It is not the intent that 19.2.3.4 supersede 7.3.2.2. Existing corridors more than 48 in. (1220 mm) in width are not permitted to be reduced in width, unless they exceed the width requirements of 18.2.3.4 or 18.2.3.5. *(See 4.6.7.4, 4.6.7.5, and 4.6.12.2.)*

**A.19.2.3.4(2)** The intent of 19.2.3.4(2) is to permit limited noncontinuous projections along the corridor wall. These include hand-rub dispensing units complying with 19.3.2.6, nurse charting units, wall-mounted computers, telephones, artwork, bulletin boards, display case frames, cabinet frames, fire alarm boxes, and similar items. It is not the intent to permit the narrowing of the corridor by the walls themselves. The provision of 7.3.2.2 permits projections up to 4½ in. (114 mm) to be present at and below the 38 in. (965 mm) handrail height, and it is not the intent of 19.2.3.4(2) to prohibit such projections.

**A.19.2.3.4(4)(c)** Wheeled equipment and carts in use include food service carts, housekeeping carts, medication carts, isolation carts, and similar items. Isolation carts should be permitted in the corridor only where patients require isolation precautions.

Unattended wheeled crash carts and other similar wheeled emergency equipment are permitted to be located in the corridor when "not in use," because they need to be immediately accessible during a clinical emergency. Note that "not in use" is not the same as "in storage." Storage is not permitted to be open to the corridor, unless it meets one of the provisions permitted in 19.3.6.1 and is not a hazardous area.

Wheeled portable patient lift or transport equipment needs to be readily available to clinical staff for moving, transferring, toileting, or relocating patients. These devices are used daily for safe handling of patients and to provide for worker safety. This equipment might not be defined as "in use" but needs to be convenient for the use of caregivers at all times.

**A.19.2.3.4(5)** The means for affixing the furniture can be achieved with removable brackets to allow cleaning and maintenance. Affixing the furniture to the floor or wall prevents the furniture from moving, so as to maintain a minimum 6 ft (1830 mm) corridor clear width. Affixing the furniture to the floor or wall also provides a sturdiness that allows occupants to safely transfer in and out.

**A.19.2.3.4(5)(f)** Examples of building service and fire protection equipment include fire extinguishers, manual fire alarm boxes, shutoff valves, and similar equipment.

**A.19.2.4.4** An exit is not necessary for each individual smoke compartment if there is access to an exit through other smoke compartments without passing through the smoke compartment of fire origin.

**A.19.2.5.2** Every exit or exit access should be arranged, if practical and feasible, so that no corridor has a dead end exceeding 30 ft (9.1 m).

**A.19.2.5.4** The term *intervening rooms or spaces* means rooms or spaces serving as a part of the required means of egress from another room.

**A.19.2.5.6.1** For the purposes of this paragraph, it is the intent that the term *habitable rooms* not include individual bathrooms, closets, and similar spaces, as well as briefly occupied work spaces, such as control rooms in radiology and small storage rooms in a pharmacy.

**A.19.2.5.7.1.2** Two or more contiguous suites with an aggregate area not exceeding the suite size limitation of 19.2.5.7.2.3 and 19.2.5.7.3.3 are permitted to be considered a single suite, so as not to require separation from each other. The intent of 19.2.5.7.1.2(2) is to continue to permit suites that have smoke-resisting walls separating them from the rest of the building, even though the walls might not have a fire resistance rating. This requirement includes walls that comply with 19.3.6.2.4, even though sprinkler protection is not provided.

**A.19.2.5.7.1.3(A)** The term *intervening room* means a room serving as a part of the required means of egress from another room.

**A.19.2.5.7.1.3(C)** Examples of suites that might be hazardous areas are medical records and pharmaceutical suites.

**A.19.2.5.7.1.3(D)** It is the intent that the provision of 19.2.5.7.1.3(D) apply only where the quantities of combustibles occupy an area exceeding 50 ft$^2$ (4.6 m$^2$) so as to be a hazardous contents area. Where quantities of combustibles occupy less than 50 ft$^2$ (4.6 m$^2$), there is no restriction on quantity.

**A.19.2.5.7.2.1(A)** For the purposes of this paragraph, it is the intent that the term *habitable rooms* not include individual bathrooms, closets, and similar spaces, as well as briefly occupied work spaces, such as control rooms in radiology and small storage rooms in a pharmacy.

**A.19.2.5.7.2.1(D)(1)** The interior partitions or walls might extend full height to the ceiling, provided that they do not obscure visual supervision of the suite. Where they do obscure visual supervision, see 19.2.5.7.2.1(D)(2).

**A.19.2.5.7.2.2(B)** Where only one means of egress is required from the suite, it needs to be provided by a door opening directly to a corridor complying with 19.3.6.

**A.19.2.5.7.2.2(C)** Where the second exit access for a sleeping suite is through an adjacent suite, it is the intent that the 100 ft (30 m) travel distance limitation in the suite be applied only to the suite under consideration.

**A.19.2.5.7.2.3(C)(1)** The alternative of 19.2.5.7.2.1(D)(1)(b) is not to be applied, since 19.2.5.7.2.3(C)(2) requires total coverage automatic smoke detection for the suite that exceeds 5000 ft$^2$ (460 m$^2$) but does not exceed 7500 ft$^2$ (700 m$^2$).

**A.19.2.5.7.3.2(B)** Where only one means of egress is required from the suite, it needs to be provided by a door opening directly to a corridor complying with 19.3.6.

**A.19.2.5.7.3.2(C)** Where the second exit access for a non-sleeping suite is through an adjacent suite, it is the intent that the adjacent suite not be considered an intervening room.

**A.19.3.2.1.2** Penetrations of hazardous area walls located above ceilings that comply with Section 8.4 are not required to be sealed to comply with 19.3.2.1.2.

**A.19.3.2.2** The hazard level of a laboratory is considered severe if quantities of flammable, combustible, or hazardous materials are present that are capable of sustaining a fire of sufficient magnitude to breach a 1-hour fire separation. See NFPA *Fire Protection Handbook* for guidance.

**A.19.3.2.5.2** This provision is intended to permit small appliances used for reheating, such as microwave ovens, hot plates, toasters, and nourishment centers, to be exempt from the requirements for commercial cooking equipment and hazardous area protection.

**A.19.3.2.5.3** The intent of 19.3.2.5.3 is to limit the number of persons for whom meals are routinely prepared to not more than 30. Staff and feeding assistants are not included in this number.

**A.19.3.2.5.3(3)** The minimum airflow of 500 cfm (14,000 L/m) is intended to require the use of residential hood equipment at the higher end of equipment capacities. It is also intended to draw a sufficient amount of the cooking vapors into the grease baffle and filter system to reduce migration beyond the hood.

**A.19.3.2.5.3(6)** The intent of this provision is to limit cooking fuel to gas or electricity. The prohibition of solid fuels for cooking is not intended to prohibit charcoal grilling on grills located outside the facility.

**A.19.3.2.5.3(7)** Deep-fat frying is defined as a cooking method that involves fully immersing food in hot oil.

**A.19.3.2.5.3(9)** The intent of this requirement is that the fuel source for the cooktop or range is to be turned on only when staff is present or aware that the kitchen is being used. The timer function is meant to provide an additional safeguard if the staff forgets to deactivate the cooktop or range. If a cooking activity lasts longer than 120 minutes, the timer would be required to be manually reset.

**A.19.3.2.5.3(11)** The intent of requiring smoke alarms instead of smoke detectors is to prevent false alarms from initiating the building fire alarm system and notifying the fire department. Smoke alarms should be maintained a minimum of 20 ft (6.1 m) away from the cooktop or range as studies have shown this distance to be the threshold for significantly reducing false alarms caused by cooking. The intent of the interconnected smoke alarms, with silence feature, is that while the devices would alert staff members to a potential problem, if it is a false alarm, the staff members can use the silence feature instead of disabling the alarm. The referenced study indicates that nuisance alarms are reduced with photoelectric smoke alarms. Providing two, interconnected alarms provides a safety factor since they are not electrically supervised by the the fire alarm system. *(Smoke Alarms – Pilot Study of Nuisance Alarms Associated with Cooking)*

**A.19.3.2.5.4** The provisions of 19.3.2.5.4 differ from those of 19.3.2.5.3, as they apply to cooking equipment that is separated from the corridor.

**A.19.3.2.5.5** The provision of 19.3.2.5.5 clarifies that protected commercial cooking equipment does not require an enclosure (separation) as a hazardous area in accordance with Section 8.7, as is required by 19.3.2.1.

**A.19.3.2.6** Extensive research, including fire modeling, has indicated that alcohol-based hand-rub solutions can be safely installed in corridors of health care facilities, provided that certain other precautions are taken. The total quantities of flammable liquids in any area should comply with the provisions of other recognized codes, including NFPA 1, *Fire Code*, and NFPA 30, *Flammable and Combustible Liquids Code*. In addition, special consideration should be given to the following:

(1) Obstructions created by the installation of hand-rub solution dispensers
(2) Location of dispensers with regard to adjacent combustible materials and potential sources of ignition, especially where dispensers are mounted on walls of combustible construction
(3) Requirements for other fire protection features, including complete automatic sprinkler protection, to be installed throughout the compartment
(4) Amount and location of the flammable solutions, both in use and in storage, particularly with respect to potential for leakage or failure of the dispenser

**A.19.3.3.2** The reduction in class of interior finish prescribed by 10.2.8.1 is permitted to be used.

**A.19.3.4.2** It is not the intent of this *Code* to require single-station smoke detectors, which might be required by local codes, to be connected to or to initiate the building fire alarm system.

**A.19.3.4.3.1(1)** It is the intent of this provision to permit a visible fire alarm signal instead of an audible signal to reduce interference between the fire alarm and medical equipment monitoring alarms.

**A.19.3.5.4** It is not the intent to require existing standard sprinklers in existing sprinkler systems to be replaced with listed quick-response or listed residential sprinklers. It is the intent that new sprinkler systems installed in existing buildings comply with the requirements of Chapter 18, including 18.3.5.6.

**A.19.3.5.7** It is intended that any valve that controls automatic sprinklers in the building or portions of the building, including sectional and floor control valves, be electrically supervised. Valves that control isolated sprinkler heads, such as in laundry

and trash chutes, are not required to be electrically supervised. Appropriate means should be provided to ensure that valves that are not electrically supervised remain open.

**A.19.3.5.8** The provisions of 19.3.5.8(6) and (7) are not intended to supplant NFPA 13, *Standard for the Installation of Sprinkler Systems*, which requires that residential sprinklers with more than a 10°F (5.6°C) difference in temperature rating not be mixed within a room. Currently there are no additional prohibitions in NFPA 13 on the mixing of sprinklers having different thermal response characteristics. Conversely, there are no design parameters to make practical the mixing of residential and other types of sprinklers.

Residential sprinklers are considered acceptable in patient sleeping rooms of all health care facilities, even though not specifically listed for this purpose in all cases.

**A.19.3.5.8(6)** It is not the intent of the *Code* to permit standard-response sprinklers to meet the criteria of 19.3.5.8 just because the sprinklers were installed before quick-response sprinklers were invented or listed. The intent of 19.3.5.8(6) is to permit older quick-response systems to be credited, even though there might be some standard-response sprinklers in existence due to the fact that quick-response sprinklers were unavailable for those specific locations at the time. For example, in the early days of quick-response sprinklers, there were no high-temperature quick-response sprinklers available.

**A.19.3.5.10** Although this exception is currently not recognized by NFPA 13, *Standard for the Installation of Sprinkler Systems*, a proposal for such exemption has been submitted for consideration by the Technical Committee on Sprinkler System Installation Criteria. This exception is limited to hospitals, as nursing homes and many limited care facilities might have more combustibles within the closets. The limited amount of clothing found in the small clothes closets in hospital patient rooms is typically far less than the amount of combustibles in casework cabinets that do not require sprinkler protection, such as nurse servers. In many hospitals, especially new hospitals, it is difficult to make a distinction between clothes closets and cabinet work. The exception is far more restrictive than similar exceptions for hotels and apartment buildings. NFPA 13 already permits the omission of sprinklers in wardrobes *[see 8.1.1(7) of NFPA 13]*. It is not the intent of 19.3.5.10 to affect the wardrobe provisions of NFPA 13. It is the intent that the sprinkler protection in the room covers the closet as if there were no door on the closet. *(See 8.5.3.2.3 of NFPA 13.)*

**A.19.3.5.11** For the proper operation of sprinkler systems, cubicle curtains and sprinkler locations need to be coordinated. Improperly designed systems might obstruct the sprinkler spray from reaching the fire or might shield the heat from the sprinkler. Many options are available to the designer including, but not limited to, hanging the cubicle curtains 18 in. (455 mm) below the sprinkler deflector; using ½ in. (13 mm) diagonal mesh or a 70 percent open weave top panel that extends 18 in. (455 mm) below the sprinkler deflector; or designing the system to have a horizontal and minimum vertical distance that meets the requirements of NFPA 13, *Standard for the Installation of Sprinkler Systems*. The test data that forms the basis of the NFPA 13 requirements is from fire tests with sprinkler discharge that penetrated a single privacy curtain.

**A.19.3.6.1(1)(a)** The presence of stored combustible materials in a room or space open to the corridor does not necessarily result in the room or space being classified as a hazardous area. In some circumstances, the amount and type of combustibles might result in the room or space being classified as a hazardous area by the authority having jurisdiction.

**A.19.3.6.1(3)** A typical nurses' station would normally contain one or more of the following with associated furniture and furnishings:

(1) Charting area
(2) Clerical area
(3) Nourishment station
(4) Storage of small amounts of medications, medical equipment and supplies, clerical supplies, and linens
(5) Patient monitoring and communication equipment

**A.19.3.6.1(7)(b)** A fully developed fire (flashover) occurs if the rate of heat release of the burning materials exceeds the capability of the space to absorb or vent that heat. The ability of common lining (wall, ceiling, and floor) materials to absorb heat is approximately 0.75 Btu/ft² (0.07 kJ/m²) of lining. The venting capability of open doors or windows is in excess of 20 Btu/ft² (1.95 kJ/m²) of opening. In a fire that has not reached flashover conditions, fire will spread from one furniture item to another only if the burning item is close to another furniture item. For example, if individual furniture items have a heat release rate of 500 Btu/s (525 kW) and are separated by 12 in. (305 mm) or more, the fire is not expected to spread from item to item, and flashover is unlikely to occur. *(See also the NFPA Fire Protection Handbook.)*

**A.19.3.6.1(8)** This provision permits waiting areas to be located across the corridor from each other, provided that neither area exceeds the 600 ft² (55.7 m²) limitation.

**A.19.3.6.2.2** The intent of the minimum ½-hour fire resistance rating for corridor partitions is to require a nominal fire rating, particularly where the fire rating of existing partitions cannot be documented. Examples of acceptable partition assemblies would include, but are not limited to, ½ in. (13 mm) gypsum board, wood lath and plaster, gypsum lath, or metal lath and plaster.

**A.19.3.6.2.3** The purpose of extending a corridor wall above a lay-in ceiling or through a concealed space is to provide a barrier to limit the passage of smoke. Such a barrier is not required to be either a smoke barrier or a smoke partition — two terms for which specific *Code* definitions and requirements apply. The intent of 19.3.6.2.3 is not to require light-tight barriers above lay-in ceilings or to require an absolute seal of the room from the corridor. Small holes, penetrations, or gaps around items such as ductwork, conduit, or telecommunication lines should not affect the ability of this barrier to limit the passage of smoke.

**A.19.3.6.2.4** An architectural, exposed, suspended-grid acoustical tile ceiling with penetrating items, such as sprinkler piping and sprinklers; ducted HVAC supply and return-air diffusers; speakers; and recessed lighting fixtures, is capable of limiting the transfer of smoke.

**A.19.3.6.2.6** Monolithic ceilings are continuous horizontal membranes composed of noncombustible or limited-combustible materials, such as plaster or gypsum board, with seams or cracks permanently sealed.

**A.19.3.6.3** Where a nurse server penetrates a corridor wall, the access opening on the corridor side of the nurse server must be protected as is done for a corridor door.

**A.19.3.6.3.1** Gasketing of doors should not be necessary to achieve resistance to the passage of smoke if the door is relatively tight-fitting.

**A.19.3.6.3.5** While it is recognized that closed doors serve to maintain tenable conditions in a corridor and adjacent patient rooms, such doors, which, under normal or fire conditions, are self-closing, might create a special hazard for the personal safety of a room occupant. Such closed doors might present a problem of delay in discovery, confining fire products beyond tenable conditions.

Because it is critical for responding staff members to be able to immediately identify the specific room involved, it is recommended that approved automatic smoke detection that is interconnected with the building fire alarm be considered for rooms having doors equipped with closing devices. Such detection is permitted to be located at any approved point within the room. When activated, the detector is required to provide a warning that indicates the specific room of involvement by activation of a fire alarm annunciator, nurse call system, or any other device acceptable to the authority having jurisdiction.

In existing buildings, use of the following options reasonably ensures that patient room doors will be closed and remain closed during a fire:

(1) Doors should have positive latches, and a suitable program that trains staff to close the doors in an emergency should be established.
(2) It is the intent of the *Code* that no new installations of roller latches be permitted; however, repair or replacement of roller latches is not considered a new installation.
(3) Doors protecting openings to patient sleeping or treatment rooms, or spaces having a similar combustible loading, might be held closed using a closer exerting a closing force of not less than 5 lbf (22 N) on the door latch stile.

**A.19.3.6.3.10** Doors should not be blocked open by furniture, door stops, chocks, tie-backs, drop-down or plunger-type devices, or other devices that necessitate manual unlatching or releasing action to close. Examples of hold-open devices that release when the door is pushed or pulled are friction catches or magnetic catches.

**A.19.3.6.3.12** It is not the intent of 19.3.6.3.12 to prohibit the application of push plates, hardware, or other attachments on corridor doors in health care occupancies.

**A.19.3.6.5.1** It is not the intent of 19.3.6.5.1 to permit mail slots or pass-through openings in doors or walls of rooms designated as a hazardous area.

**A.19.3.7.3(2)** Where the smoke control system design requires dampers in order that the system functions effectively, it is not the intent of the exception to permit the damper to be omitted.

This provision is not intended to prevent the use of plenum returns where ducting is used to return air from a ceiling plenum through smoke barrier walls. Short stubs or jumper ducts are not acceptable. Ducting is required to connect at both sides of the opening and to extend into adjacent spaces away from the wall. The intent is to prohibit open-air transfers at or near the smoke barrier walls.

**A.19.3.7.6.1** It is not the intent of 19.3.7.6.1 to prohibit the application of push plates, hardware, or other attachments on smoke barrier doors in health care occupancies.

**A.19.3.7.8** Smoke barriers might include walls having door openings other than cross-corridor doors. There is no restriction in the *Code* regarding which doors or how many doors form part of a smoke barrier. For example, doors from the corridor to individual rooms are permitted to form part of a smoke barrier.

**A.19.4.2.2** The provision of 19.4.2.2 is intended to prevent the phase-in period for the installation of sprinklers from being reset to 12 years upon adoption of the 2012 edition of the *Code* in jurisdictions where the 12-year period had already begun via the adoption of the 2009 edition.

**A.19.5.2.2** For both new and existing buildings, it is the intent to permit the installation and use of fireplace stoves and room heaters using solid fuel as defined in NFPA 211, *Standard for Chimneys, Fireplaces, Vents, and Solid Fuel–Burning Appliances,* provided that all such devices are installed, maintained, and used in accordance with the appropriate provisions of that standard and all manufacturers' specifications. These requirements are not intended to permit freestanding solid fuel–burning appliances such as freestanding wood-burning stoves.

**A.19.5.2.3(2)(d)** The glass front of a direct-vent fireplace can become extremely hot. Barriers such as screens or mesh installed over the direct-vent glass help reduce the risk of burn from touching the glass.

**A.19.5.2.3(2)(e)** The intent of locating controls in a restricted location is to ensure staff is aware of use of the fireplace and to prevent unauthorized use. Examples of locked controls are a keyed switch or locating the switch in a staff-controlled location such as a staff station.

**A.19.7** Health care occupants have, in large part, varied degrees of physical disability, and their removal to the outside, or even their disturbance caused by moving, is inexpedient or impractical in many cases, except as a last resort. Similarly, recognizing that there might be an operating necessity for the restraint of the mentally ill, often by use of barred windows and locked doors, fire exit drills are usually extremely disturbing, detrimental, and frequently impracticable.

In most cases, fire exit drills, as ordinarily practiced in other occupancies, cannot be conducted in health care occupancies. Fundamentally, superior construction, early discovery and extinguishment of incipient fires, and prompt notification need to be relied on to reduce the occasion for evacuation of buildings of this class to a minimum.

**A.19.7.1.4** Many health care occupancies conduct fire drills without disturbing patients by choosing the location of the simulated emergency in advance and by closing the doors to patients' rooms or wards in the vicinity prior to initiation of the drill. The purpose of a fire drill is to test and evaluate the efficiency, knowledge, and response of institutional personnel in implementing the facility fire emergency plan. Its purpose is not to disturb or excite patients. Fire drills should be scheduled on a random basis to ensure that personnel in health care facilities are drilled not less than once in each 3-month period.

Drills should consider the ability to move patients to an adjacent smoke compartment. Relocation can be practiced using simulated patients or empty wheelchairs.

**A.19.7.2.1** Each facility has specific characteristics that vary sufficiently from other facilities to prevent the specification of a universal emergency procedure. The recommendations that follow, however, contain many of the elements that should be considered and adapted, as appropriate, to the individual facility.

Upon discovery of fire, personnel should immediately take the following action:

(1) If any person is involved in the fire, the discoverer should go to the aid of that person, calling aloud an established code phrase, which provides for both the immediate aid of any endangered person and the transmission of an alarm.
(2) Any person in the area, upon hearing the code called aloud, should activate the building fire alarm using the nearest manual fire alarm box.
(3) If a person is not involved in the fire, the discoverer should activate the building fire alarm using the nearest manual fire alarm box.
(4) Personnel, upon hearing the alarm signal, should immediately execute their duties as outlined in the facility fire safety plan.
(5) The telephone operator should determine the location of the fire as indicated by the audible signal.
(6) In a building equipped with an uncoded alarm system, a person on the floor of fire origin should be responsible for promptly notifying the facility telephone operator of the fire location.
(7) If the telephone operator receives a telephone alarm reporting a fire from a floor, the operator should regard that alarm in the same fashion as an alarm received over the fire alarm system and should immediately notify the fire department and alert all facility personnel of the place of fire and its origin.
(8) If the building fire alarm system is out of order, any person discovering a fire should immediately notify the telephone operator by telephone, and the operator should then transmit this information to the fire department and alert the building occupants.

**A.19.7.4** The most rigid discipline with regard to prohibition of smoking might not be nearly as effective in reducing incipient fires from surreptitious smoking as the open recognition of smoking, with provision of suitable facilities for smoking. Proper education and training of the staff and attendants in the ordinary fire hazards and their abatement is unquestionably essential. The problem is a broad one, varying with different types and arrangements of buildings; the effectiveness of rules of procedure, which need to be flexible, depends in large part on the management.

**A.19.7.5.1** In addition to the provisions of 10.3.1, which deal with ignition resistance, additional requirements with respect to the location of cubicle curtains relative to sprinkler placement are included in NFPA 13, *Standard for the Installation of Sprinkler Systems.*

**A.19.7.5.6(4)** The percentage of decorations should be measured against the area of any wall or ceiling, not the aggregate total of walls, ceilings, and doors. The door is considered part of the wall. The decorations must be located such that they do not interfere with the operation of any door, sprinkler, smoke detector, or any other life safety equipment. Other art might include hanging objects or three-dimensional items.

**A.19.7.5.6(5)** When determining if the hazard for fire development or spread is present, consideration should be given to whether the building or area being evaluated is sprinklered.

**A.19.7.5.7.1(3)** It is not the intent to permit collection receptacles with a capacity greater than 32 gal (121 L) to be positioned at or near a nurses' station based on the argument that such nurses' station is constantly attended. The large collection receptacle itself needs to be actively attended by staff. Staff might leave the large receptacle in the corridor outside a patient room while entering the room to collect soiled linen or trash, but staff is expected to return to the receptacle, move on to the next room, and repeat the collection function. Where staff is not actively collecting material for placement in the receptacle, the receptacle is to be moved to a room protected as a hazardous area.

**A.19.7.5.7.2** It is the intent that this provision permits recycling of bottles, cans, paper, and similar clean items that do not contain grease, oil, flammable liquids, or significant plastic materials, using larger containers or several adjacent containers, and not require locating such containers in a room protected as a hazardous area. Containers for medical records awaiting shredding are often larger than 32 gal (121 L). These containers are not to be included in the calculations and limitations of 19.7.5.7.1. There is no limit on the number of these containers, as FM Approval Standard 6921, *Containers for Combustible Waste*, ensures that the fire will not spread outside of the container. FM approval standards are written for use with FM Approvals. The tests can be conducted by any approved laboratory. The portions of the standard referring to FM Approvals are not included in this reference.

**A.19.7.5.7.2(2)** See 19.7.5.7.1(3).

**A.19.7.7** A document that provides recognized engineering principles for the testing of smoke control systems is NFPA 92, *Standard for Smoke Control Systems*.

**A.20.1.1.1.8** The *Code* recognizes that certain functions necessary for the life safety of building occupants, such as the closing of corridor doors, the operation of manual fire alarm devices, and the removal of patients from the room of fire origin, require the intervention of facility staff. It is not the intent of 20.1.1.1.8 to specify the levels or locations of staff necessary to meet this requirement.

**A.20.1.1.2** This objective is accomplished in the context of the physical facilities, the type of activities undertaken, the provisions for the capabilities of staff, and the needs of all occupants through requirements directed at the following:

(1) Prevention of ignition
(2) Detection of fire
(3) Control of fire development
(4) Confinement of the effects of fire
(5) Extinguishment of fire
(6) Provision of refuge or evacuation facilities, or both
(7) Staff reaction

**A.20.1.3.2** Doctors' offices and treatment and diagnostic facilities that are intended solely for outpatient care and are physically separated from facilities for the treatment or care of inpatients, but are otherwise associated with the management of an institution, might be classified as business occupancies rather than health care occupancies.

**A.20.3.2.6** Extensive research, including fire modeling, has indicated that alcohol-based hand-rub solutions can be safely installed in corridors of health care facilities, provided that certain other precautions are taken. The total quantities of flammable liquids in any area should comply with the provisions of other recognized codes, including NFPA 1, *Fire Code*, and NFPA 30, *Flammable and Combustible Liquids Code*. In addition, special consideration should be given to the following:

(1) Obstructions created by the installation of hand-rub solution dispensers

(2) Location of dispensers with regard to adjacent combustible materials and potential sources of ignition, especially where dispensers are mounted on walls of combustible construction
(3) Requirements for other fire protection features, including complete automatic sprinkler protection, to be installed throughout the compartment
(4) Amount and location of the flammable solutions, both in use and in storage, particularly with respect to potential for leakage or failure of the dispenser

**A.20.3.7.9** Smoke barriers might include walls having door openings other than cross-corridor doors. There is no restriction in the *Code* regarding which doors or how many doors form part of a smoke barrier. For example, doors from the corridor to individual rooms are permitted to form part of a smoke barrier.

**A.20.3.7.13** Split astragals (i.e., astragals installed on both door leaves) are also considered astragals.

**A.20.7** Health care occupants have, in large part, varied degrees of physical disability, and their removal to the outside, or even their disturbance caused by moving, is inexpedient or impractical in many cases, except as a last resort. Similarly, recognizing that there might be an operating necessity for the restraint of the mentally ill, often by use of barred windows and locked doors, fire exit drills are usually extremely disturbing, detrimental, and frequently impracticable.

In most cases, fire exit drills, as ordinarily practiced in other occupancies, cannot be conducted in health care occupancies. Fundamentally, superior construction, early discovery and extinguishment of incipient fires, and prompt notification need to be relied on to reduce the occasion for evacuation of buildings of this class to a minimum.

**A.20.7.1.4** Many health care occupancies conduct fire drills without disturbing patients by choosing the location of the simulated emergency in advance and by closing the doors to patients' rooms or wards in the vicinity prior to the initiation of the drill. The purpose of a fire drill is to test and evaluate the efficiency, knowledge, and response of institutional personnel in implementing the facility fire emergency plan. Its purpose is not to disturb or excite patients. Fire drills should be scheduled on a random basis to ensure that personnel in health care facilities are drilled not less than once in each 3-month period.

Drills should consider the ability to move patients to an adjacent smoke compartment. Relocation can be practiced using simulated patients or empty wheelchairs.

**A.20.7.2.1** Each facility has specific characteristics that vary sufficiently from other facilities to prevent the specification of a universal emergency procedure. The recommendations that follow, however, contain many of the elements that should be considered and adapted, as appropriate, to the individual facility.

Upon discovery of fire, personnel should immediately take the following action:

(1) If any person is involved in the fire, the discoverer should go to the aid of that person, calling aloud an established code phrase, which provides for both the immediate aid of any endangered person and the transmission of an alarm.
(2) Any person in the area, upon hearing the code called aloud, should activate the building fire alarm using the nearest manual fire alarm box.

(3) If a person is not involved in the fire, the discoverer should activate the building fire alarm using the nearest manual fire alarm box.

(4) Personnel, upon hearing the alarm signal, should immediately execute their duties as outlined in the facility fire safety plan.

(5) The telephone operator should determine the location of the fire as indicated by the audible signal.

(6) In a building equipped with an uncoded alarm system, a person on the floor of fire origin should be responsible for promptly notifying the facility telephone operator of the fire location.

(7) If the telephone operator receives a telephone alarm reporting a fire from a floor, the operator should regard that alarm in the same fashion as an alarm received over the fire alarm system and should immediately notify the fire department and alert all facility personnel of the place of fire and its origin.

(8) If the building fire alarm system is out of order, any person discovering a fire should immediately notify the telephone operator by telephone, and the operator should then transmit this information to the fire department and alert the building occupants.

**A.20.7.4** The most rigid discipline with regard to prohibition of smoking might not be nearly as effective in reducing incipient fires from surreptitious smoking as the open recognition of smoking, with provision of suitable facilities for smoking. Proper education and training of the staff and attendants in the ordinary fire hazards and their abatement is unquestionably essential. The problem is a broad one, varying with different types and arrangements of buildings; the effectiveness of rules of procedure, which need to be flexible, depends in large part on the management.

**A.20.7.5.1** In addition to the provisions of 10.3.1, which deal with ignition resistance, additional requirements with respect to the location of cubicle curtains relative to sprinkler placement are included in NFPA 13, *Standard for the Installation of Sprinkler Systems.*

**A.20.7.7** A document that provides recognized engineering principles for the testing of smoke control systems is NFPA 92, *Standard for Smoke Control Systems.*

**A.21.1.1.1.8** The *Code* recognizes that certain functions necessary for the life safety of building occupants, such as the closing of corridor doors, the operation of manual fire alarm devices, and the removal of patients from the room of fire origin, require the intervention of facility staff. It is not the intent of 21.1.1.1.8 to specify the levels or locations of staff necessary to meet this requirement.

**A.21.1.1.2** This objective is accomplished in the context of the physical facilities, the type of activities undertaken, the provisions for the capabilities of staff, and the needs of all occupants through requirements directed at the following:

(1) Prevention of ignition
(2) Detection of fire
(3) Control of fire development
(4) Confinement of the effects of fire
(5) Extinguishment of fire
(6) Provision of refuge or evacuation facilities, or both
(7) Staff reaction

**A.21.1.3.2** Doctors' offices and treatment and diagnostic facilities that are intended solely for outpatient care and are physically separated from facilities for the treatment or care of inpatients, but that are otherwise associated with the management of an institution, might be classified as business occupancies rather than health care occupancies.

**A.21.2.2.6** The waiver of the requirement for doors to swing in the direction of egress travel is based on the assumption that, in this occupancy, there is little possibility of a panic rush that might prevent the opening of doors that swing against egress travel.

A desirable arrangement, which is possible with corridors 6 ft (1830 mm) or more in width, is to have two 32 in. (810 mm) doors, normally closed, each swinging with the egress travel (in opposite directions).

**A.21.3.2.6** Extensive research, including fire modeling, has indicated that alcohol-based hand-rub solutions can be safely installed in corridors of health care facilities, provided that certain other precautions are taken. The total quantities of flammable liquids in any area should comply with the provisions of other recognized codes, including NFPA 1, *Fire Code,* and NFPA 30, *Flammable and Combustible Liquids Code.* In addition, special consideration should be given to the following:

(1) Obstructions created by the installation of hand-rub solution dispensers
(2) Location of dispensers with regard to adjacent combustible materials and potential sources of ignition, especially where dispensers are mounted on walls of combustible construction
(3) Requirements for other fire protection features, including complete automatic sprinkler protection, to be installed throughout the compartment
(4) Amount and location of the flammable solutions, both in use and in storage, particularly with respect to potential for leakage or failure of the dispenser

**A.21.3.7.9** Smoke barriers might include walls having door openings other than cross-corridor doors. There is no restriction in the *Code* regarding which doors or how many doors form part of a smoke barrier. For example, doors from the corridor to individual rooms are permitted to form part of a smoke barrier.

**A.21.7** Health care occupants have, in large part, varied degrees of physical disability, and their removal to the outside, or even their disturbance caused by moving, is inexpedient or impractical in many cases, except as a last resort. Similarly, recognizing that there might be an operating necessity for the restraint of the mentally ill, often by use of barred windows and locked doors, fire exit drills are usually extremely disturbing, detrimental, and frequently impracticable.

In most cases, fire exit drills, as ordinarily practiced in other occupancies, cannot be conducted in health care occupancies. Fundamentally, superior construction, early discovery and extinguishment of incipient fires, and prompt notification need to be relied on to reduce the occasion for evacuation of buildings of this class to a minimum.

**A.21.7.1.4** Many health care occupancies conduct fire drills without disturbing patients by choosing the location of the simulated emergency in advance and by closing the doors to patients' rooms or wards in the vicinity prior to initiation of the drill. The purpose of a fire drill is to test and evaluate the efficiency, knowledge, and response of institutional personnel in implementing the facility fire emergency plan. Its purpose is not to disturb or excite patients. Fire drills should be scheduled on a random basis to ensure that personnel in health care facilities are drilled not less than once in each 3-month period.

Drills should consider the ability to move patients to an adjacent smoke compartment. Relocation can be practiced using simulated patients or empty wheelchairs.

**A.21.7.2.1** Each facility has specific characteristics that vary sufficiently from other facilities to prevent the specification of a universal emergency procedure. The recommendations that follow, however, contain many of the elements that should be considered and adapted, as appropriate, to the individual facility.

Upon discovery of fire, personnel should immediately take the following action:

(1) If any person is involved in the fire, the discoverer should go to the aid of that person, calling aloud an established code phrase, which provides for both the immediate aid of any endangered person and the transmission of an alarm.
(2) Any person in the area, upon hearing the code called aloud, should activate the building fire alarm using the nearest manual fire alarm box.
(3) If a person is not involved in the fire, the discoverer should activate the building fire alarm using the nearest manual fire alarm box.
(4) Personnel, upon hearing the alarm signal, should immediately execute their duties as outlined in the facility fire safety plan.
(5) The telephone operator should determine the location of the fire as indicated by the audible signal.
(6) In a building equipped with an uncoded alarm system, a person on the floor of fire origin should be responsible for promptly notifying the facility telephone operator of the fire location.
(7) If the telephone operator receives a telephone alarm reporting a fire from a floor, the operator should regard that alarm in the same fashion as an alarm received over the fire alarm system and should immediately notify the fire department and alert all facility personnel of the place of fire and its origin.
(8) If the building fire alarm system is out of order, any person discovering a fire should immediately notify the telephone operator by telephone, and the operator should then transmit this information to the fire department and alert the building occupants.

**A.21.7.4** The most rigid discipline with regard to prohibition of smoking might not be nearly as effective in reducing incipient fires from surreptitious smoking as the open recognition of smoking, with provision of suitable facilities for smoking. Proper education and training of the staff and attendants in the ordinary fire hazards and their abatement is unquestionably essential. The problem is a broad one, varying with different types and arrangements of buildings; the effectiveness of rules of procedure, which need to be flexible, depends in large part on the management.

**A.21.7.5.1** In addition to the provisions of 10.3.1, which deal with ignition resistance, additional requirements with respect to the location of cubicle curtains relative to sprinkler placement are included in NFPA 13, *Standard for the Installation of Sprinkler Systems.*

**A.21.7.7** A document that provides recognized engineering principles for the testing of smoke control systems is NFPA 92, *Standard for Smoke Control Systems.*

**A.22.1.1.1.4(2)** In determining equivalency for conversions, modernizations, renovations, or unusual design concepts of detention and correctional facilities, the authority having jurisdiction is permitted to accept evaluations based on the detention and correctional occupancies fire safety evaluation system (FSES) of NFPA 101A, *Guide on Alternative Approaches to Life Safety,* utilizing the parameters for new construction.

**A.22.1.1.1.6** It is not the intent to classify as detention and correctional occupancies the areas of health care occupancies in which doors are locked against patient egress where needed for the clinical needs of the patients. For example, a dementia treatment center can be adequately protected by the health care occupancies requirements of Chapter 18. *[See 18.1.1.1.7, 18.2.2.2.2, 18.2.2.2.4(1), and 18.2.2.2.6.]*

The one-resident threshold requirement of 22.1.1.1.6 is not meant to force a residential occupancy, where security is imposed on one or more occupants, to be reclassified as a detention and correctional occupancy.

**A.22.1.1.1.7** Lockups in which persons are detained with some degree of security imposed on them are common in many occupancies. Examples include the following:

(1) Immigration and naturalization facilities at border crossings
(2) Customs facilities at international airports
(3) Prisoner holding facilities at courthouses
(4) Local police department holding areas
(5) Security offices at sports stadia
(6) Security offices at shopping mall complexes

**A.22.1.2.1** Users and occupants of detention and correctional facilities at various times can be expected to include staff, visitors, and residents. The extent and nature of facility utilization vary according to the type of facility, its function, and its programs.

Figure A.22.1.2.1 illustrates the five use conditions.

**A.22.1.2.2** Prompt operation is intended to be accomplished in the period of time between detection of fire, either by the smoke detector(s) required by 22.3.4.4 or by other means, whichever occurs first, and the advent of intolerable conditions forcing emergency evacuation. Fire tests have indicated that the time available is a function of the volume and height of the space involved and the rate of fire development. In traditional one-story corridor arrangements, the time between detection by smoke detectors and the advent of lethal conditions down to head height can be as short as approximately 3 minutes. In addition, it should be expected that approximately 1 minute will be required to evacuate all the occupants of a threatened smoke compartment once the locks are released. In such a case, a prompt release time would be 2 minutes.

**A.22.1.2.3(2)** If the Use Condition I facility conforms to the requirements of residential occupancies under this *Code,* there are no staffing requirements. If the Use Condition I facility conforms to the requirements of Use Condition II facilities as permitted by this provision, staffing is required in accordance with 22.7.1.

**A.22.1.3** Detention and correctional facilities are a complex of structures, each serving a definite and usually different purpose. In many institutions, all, or almost all, the occupancy-type classifications found in this *Code* are represented. Means of egress and other features are governed by the type of occu-

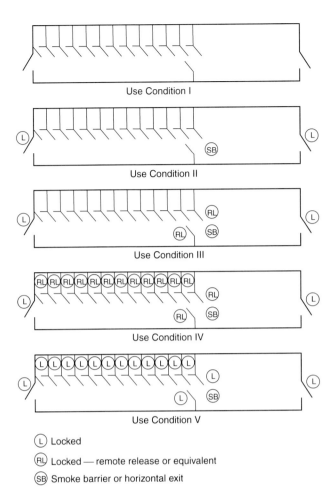

L Locked

RL Locked — remote release or equivalent

SB Smoke barrier or horizontal exit

**FIGURE A.22.1.2.1  Detention and Correctional Use Conditions.**

pancy classification and the hazard of occupancy, unless specific exemptions are made.

All buildings and structures are to be classified using Chapter 22 and Section 6.1 as a guide, subject to the ruling of the authority having jurisdiction where a question arises concerning the proper classification of any individual building or structure.

Use condition classification of the institution, as well as of individual areas within the complex, is always to be considered by the authority having jurisdiction.

**A.22.1.3.2.2**  Key-operated locking hardware of a lesser grade than institutional grade hardware might not be suitable for the heavy use to which such locks are expected to be subjected.

**A.22.2.2.5.2**  An exit is not necessary from each individual fire compartment or smoke compartment if there is access to an exit through other fire compartments or smoke compartments without passing through the fire compartment or smoke compartment of fire origin.

**A.22.2.11.4**  It might be necessary to provide a certain number of resident sleeping rooms with doors providing a clear width of not less than 32 in. (810 mm) *(see 7.2.1.2)* in order to comply with the requirements for the physically handicapped. Such sleeping

rooms should be located where there is a direct accessible route to the exterior or to an area of safe refuge. *(See 22.3.7.)*

**A.22.2.11.8**  A remote position is generally a control point where a number of doors can be unlocked simultaneously, either mechanically or electrically. In areas where there are a number of sleeping rooms, it is impractical for attendants to unlock doors individually. Doors in an exit should be unlocked prior to unlocking sleeping room doors. Sight and sound supervision of resident living areas can be by means of camera and communications systems.

This section of the *Code* does not intend to prohibit Use Condition V facilities, nor does it intend to limit Use Condition V facilities to 10 manually released locks.

**A.22.3.1(2)**  For purposes of providing control valves and waterflow devices, multilevel residential housing areas complying with this provision are considered to be one story.

**A.22.3.2.1**  Furnishings are usually the first items ignited in a detention and correctional environment. The type, quantity, and arrangement of furniture and other combustibles are important factors in determining how fast the fire will develop. Furnishings, including upholstered items and wood items, such as wardrobes, desks, and bookshelves, might provide sufficient fuel to result in room flashover, which is the full fire involvement of all combustibles within a room once sufficient heat has been built up within the room.

Combustible loading in any room opening onto a residential housing area should be limited to reduce the potential for room flashover. Rooms in which fuel loads are not controlled, thereby creating a potential for flashover, should be considered hazardous areas. Where fire-rated separation is provided, doors to such rooms, including sleeping rooms, should be self-closing.

It is strongly recommended that padded cells not be used due to their fire record. However, recognizing that they will be used in some cases, provisions for the protection of padded cells are provided. It is recognized that the minimum ¾-hour fire protection–rated fire door will be violated with the "plant on" of the padding, but a minimum ¾-hour fire protection–rated fire door should be the base of the assembly.

**A.22.3.4.2.2**  Where the fire alarm control unit is in an area that is not continuously occupied, automatic smoke detection is needed at the control unit. The provision of 22.3.4.2.2 exempts the smoke detection only at the notification appliance circuit power extenders and the supervising station transmitting equipment.

**A.22.3.4.3.1(2)**  The staff at the constantly attended location should have the capability to promptly initiate the general alarm function and contact the fire department or have direct communication with a control room or other location that can initiate the general alarm function and contact the fire department.

**A.22.3.4.4**  Examples of contiguous common spaces are galleries and corridors.

**A.22.3.4.4.3**  An open dormitory is a dormitory that is arranged to allow staff to observe the entire dormitory area at one time.

**A.22.3.5.4(1)**  Where access to portable fire extinguishers is locked, staff should be present on a 24-hour basis and should have keys readily available to unlock access to the extinguishers. Where supervision of sleeping areas is from a 24-hour attended staff location, portable fire extinguishers are permitted to be provided at the staff location in lieu of the sleeping area.

**A.22.3.5.4(2)** It is recognized that locating portable fire extinguishers at staff locations might only result in travel distances to extinguishers being in excess of those permitted by NFPA 10, *Standard for Portable Fire Extinguishers.*

**A.22.3.7.1(2)** A door to the outside, by itself, does not meet the intent of this provision if emergency operating procedures do not provide for the door to be unlocked when needed. In cases where use of the door is not ensured, a true smoke barrier per the base requirement of 22.3.7.1 would be needed.

**A.22.3.7.5** Structural fire resistance is defined as the ability of the assembly to stay in place and maintain structural integrity without consideration of heat transmission. Twelve-gauge steel plate suitably framed and stiffened meets this requirement.

**A.22.3.7.6(1)** As an example, a smoke barrier is permitted to consist of fire-rated glazing panels mounted in a security grille arrangement.

**A.22.3.8** The requirements in Table 22.3.8 for smoke-resistant separations include taking the necessary precautions to restrict the spread of smoke through the air-handling system. However, the intent is not that smoke dampers are required to be provided for each opening. Smoke dampers would be one acceptable method; however, other techniques, such as allowing the fans to continue to run with 100 percent supply and 100 percent exhaust, would be acceptable.

**A.22.4.4.3** This provision is intended to promote the use of horizontal exits in detention and correctional occupancies. Horizontal exits provide an especially effective egress system for an occupancy in which the occupants, due to security concerns, are not commonly released to the outside. This provision offers a *Code-*specified equivalent alternative to the requirement of 7.2.4.3.5 that horizontal exits are not to be penetrated by ducts in nonsprinklered buildings. The intended continuity of the fire resistance–rated and smoke-resisting barrier is maintained by requiring that duct penetrations of horizontal exits be protected by combination fire damper/smoke leakage–rated dampers that will close upon activation of a smoke detector and a heat-actuated mechanism before the barrier's ability to resist the passage of smoke and fire is compromised.

**A.22.4.4.6.2** It is not the intent of this requirement to restrict room face separations, which restrict visibility from the common space into individual sleeping rooms.

**A.22.4.4.6.4** The vertical separation between the lowest floor level and the uppermost floor level is not to exceed 13 ft (3960 mm). Figure A.22.4.4.6.4 illustrates how the height is to be determined.

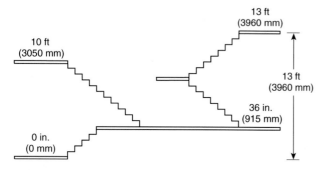

**FIGURE A.22.4.4.6.4  Vertical Height Measurement.**

**A.22.4.4.11** The requirements in Table 22.4.4.11 for smoke-resistant and fire-rated separations include taking the necessary precautions to restrict the spread of smoke through the air-handling system. However, the intent is that smoke dampers are required to be provided for each opening. Smoke dampers would be one acceptable method; however, other techniques, such as allowing the fans to continue to run with 100 percent supply and 100 percent exhaust, would be acceptable.

**A.22.4.4.12.2(2)** The automatic smoke venting should be in accordance with NFPA 204, *Standard for Smoke and Heat Venting,* for light hazard occupancies.

**A.22.4.4.13** Personal property provides combustible contents for fire development. Therefore, adequate controls are needed to limit the quantity and combustibility of fuels available to burn to reduce the probability of room flashover. The provisions of 22.4.4.13 will not, by themselves, prevent room flashover if personal property controls are not provided.

**A.22.4.4.13.2** Mattresses used in detention and correctional facilities should be evaluated with regard to the fire hazards of the environment. The potential for vandalism and excessive wear and tear also should be taken into account when evaluating the fire performance of the mattress. ASTM F 1870, *Standard Guide for Selection of Fire Test Methods for the Assessment of Upholstered Furnishings in Detention and Correctional Facilities,* provides guidance for this purpose.

**A.22.4.5.1.4(1)** The term *other physical restraints* is meant to include the use of personal restraint devices, such as handcuffs or shackles, where occupants are secured to the structure or furnishings to restrict movement.

**A.22.7.1.2** This requirement is permitted to be met by electronic or oral monitoring systems, visual monitoring, call signals, or other means.

**A.22.7.1.3** Periodic, coordinated training should be conducted and should involve detention and correctional facility personnel and personnel of the fire department legally committed to serving the facility.

**A.22.7.4** Personal property provides combustible contents for fire development. Therefore, adequate controls are needed to limit the quantity and combustibility of the fuels available to burn to reduce the probability of room flashover. The provisions of 22.7.4 will not, by themselves, prevent room flashover if personal property controls are not provided.

**A.23.1.1.1.4(2)** In determining equivalency for existing detention and correctional facilities, the authority having jurisdiction is permitted to accept evaluations based on the detention and correctional occupancies fire safety evaluation system (FSES) of NFPA 101A, *Guide on Alternative Approaches to Life Safety,* utilizing the parameters for existing buildings.

**A.23.1.1.1.6** It is not the intent to classify as detention and correctional occupancies the areas of health care occupancies in which doors are locked against patient egress where needed for the clinical needs of the patients. For example, a dementia treatment center can be adequately protected by the health care occupancies requirements of Chapter 19. [*See 19.1.1.1.7, 19.2.2.2.2, 19.2.2.2.4(1), and 19.2.2.2.6.*]

The one-resident threshold requirement of 23.1.1.1.6 is not meant to force a residential occupancy, where security is imposed on one or more occupants, to be reclassified as a detention and correctional occupancy.

**A.23.1.1.1.7** Lockups in which persons are detained with some degree of security imposed on them are common in many occupancies. Examples include the following:

(1) Immigration and naturalization facilities at border crossings
(2) Customs facilities at international airports
(3) Prisoner holding facilities at courthouses
(4) Local police department holding areas
(5) Security offices at sports stadia
(6) Security offices at shopping mall complexes

**A.23.1.2.1** Users and occupants of detention and correctional facilities at various times can be expected to include staff, visitors, and residents. The extent and nature of facility utilization will vary according to the type of facility, its function, and its programs.

Figure A.23.1.2.1 illustrates the five use conditions.

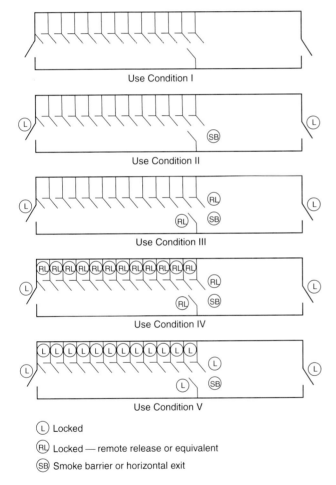

ⓛ Locked

ⓡⓛ Locked — remote release or equivalent

ⓢⓑ Smoke barrier or horizontal exit

**FIGURE A.23.1.2.1  Detention and Correctional Use Conditions.**

**A.23.1.2.2** Prompt operation is intended to be accomplished in the period of time between detection of fire, either by the smoke detector(s) required by 23.3.4.4 or by other means, whichever occurs first, and the advent of intolerable conditions forcing emergency evacuation. Fire tests have indicated that the time available is a function of the volume and height of the space involved and the rate of fire development. In traditional one-story corridor arrangements, the time between detection by smoke detectors and the advent of lethal conditions down to head height can be as short as approximately 3 minutes. In addition, it should be expected that approximately 1 minute will be required to evacuate all the occupants of a threatened smoke compartment once the locks are released. In such a case, a prompt release time would be 2 minutes.

**A.23.1.2.3(2)** If the Use Condition I facility conforms to the requirements of residential occupancies under this *Code*, there are no staffing requirements. If the Use Condition I facility conforms to the requirements of Use Condition II facilities as permitted by this exception, staffing is required in accordance with 23.7.1.

**A.23.1.3** Detention and correctional facilities are a complex of structures, each serving a definite and usually different purpose. In many institutions, all, or almost all, of the occupancy-type classifications found in this *Code* are represented. Means of egress and other features are governed by the type of occupancy classification and the hazard of occupancy, unless specific exemptions are made.

All buildings and structures are to be classified using Chapter 23 and Section 6.1 as a guide, subject to the ruling of the authority having jurisdiction where there is a question as to the proper classification of any individual building or structure.

Use condition classification of the institution, as well as of individual areas within the complex, is always to be considered by the authority having jurisdiction.

**A.23.1.3.2.1** Key-operated locking hardware should be of institutional grade. Lesser grade hardware might not be suitable for the heavy use to which such locks are expected to be subjected.

**A.23.2.2.5.2** An exit is not necessary from each individual fire compartment if there is access to an exit through other fire compartments without passing through the fire compartment of fire origin.

**A.23.2.2.5.3** This provision is intended to promote the use of horizontal exits in detention and correctional occupancies. Horizontal exits provide an especially effective egress system for an occupancy in which the occupants, due to security concerns, are not commonly released to the outside. This provision offers a *Code*-specified equivalent alternative to the requirement of 7.2.4.3.5 that horizontal exits are not to be penetrated by ducts. The intended continuity of the fire resistance–rated and smoke-resisting barrier is maintained by requiring that duct penetrations of horizontal exits be protected by combination fire damper/smoke leakage–rated dampers that close upon activation of a smoke detector and a heat-actuated mechanism before the barrier's ability to resist the passage of smoke and fire is compromised.

**A.23.2.4.2** Multilevel and multitiered residential housing areas meeting the requirements of 23.3.1.2 and 23.3.1.3 are considered one story. Therefore, two exits are not required from each level; only access to two exits is required.

**A.23.2.4.3** An exit is not necessary from each individual fire compartment and smoke compartment if there is access to an exit through other fire compartments or smoke compartments without passing through the fire compartment or smoke compartment of fire origin.

**A.23.2.5.2** Every exit or exit access should be arranged, if feasible, so that no corridor or aisle has a pocket or dead end exceeding 50 ft (15 m) for Use Conditions II, III, and IV and 20 ft (6100 mm) for Use Condition V.

**A.23.2.5.3(3)** In determining whether to approve the existing common path of travel that exceeds 50 ft (15 m), the authority having jurisdiction should ensure that the common path is not in excess of the travel distance permitted by 23.2.6.

**A.23.2.11.4** It might be necessary to provide a certain number of resident sleeping rooms with doors providing a clear width of not less than 32 in. (810 mm) *(see 7.2.1.2)* in order to comply with the requirements for the physically handicapped. Such sleeping rooms should be located where there is a direct accessible route to the exterior or to an area of safe refuge. *(See 23.3.7.)*

**A.23.2.11.8** A remote position is generally a control point where a number of doors can be unlocked simultaneously, either mechanically or electrically. In areas where there are a number of sleeping rooms, it is impractical for attendants to unlock doors individually. Doors in an exit should be unlocked prior to unlocking sleeping room doors. Sight and sound supervision of resident living areas can be by means of camera and communications systems.

This section of the *Code* does not intend to prohibit Use Condition V facilities, nor does it intend to limit Use Condition V facilities to 10 manually released locks.

**A.23.3.1.2.1** It is not the intent of this requirement to restrict room face separations, which restrict visibility from the common space into individual sleeping rooms.

**A.23.3.1.2.3** The vertical separation between the lowest floor level and the uppermost floor level is not to exceed 13 ft (3960 mm). Figure A.23.3.1.2.3 illustrates how the height is to be determined.

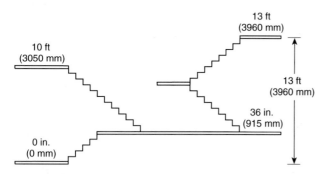

**FIGURE A.23.3.1.2.3 Vertical Height Measurement.**

**A.23.3.1.3** A recommended method of calculating the expected level of smoke in a smoke removal–equipped cell block follows.

This method for calculating the expected level of smoke has been developed from data experimentally produced in full-scale burnouts of test cells. The test cells were sized, loaded with fuel, and constructed to represent severe conditions of heavily fuel-loaded [approximately 6 lb/ft² (29 kg/m²)] cells as found in prison locations. The filling rate and temperature of the effluent gas and smoke have been calculated using the data from these tests and established formulae from plume dynamics.

The application of the method described in A.23.3.1.3 should be limited to situations where there is not less than 10 ft (3050 mm) from the floor level to the lowest acceptable level of smoke accumulation *(Z)*; the reservoir above the lowest acceptable level for *Z* is at least 20 percent of the *Z* dimension; the length of the cell block is not less than *Z*; and the fan is not less than 10 ft (3050 mm) higher than the floor of the highest cell.

The determination of smoke removal requirements is based on the dimensions of the cell opening. Where more than one cell opening is involved, the larger size on the level being calculated should be used.

The fan size, temperature rating, and operations means can be determined by the procedure that follows.

*Acceptable Smoke Level.* Determine the lowest acceptable level of smoke accumulation in accordance with 23.3.1.3. The vertical distance between that level and the floor level of the lowest open cell is the value of *Z* to be used in connection with Figure A.23.3.1.3(a).

*Characteristic Cell Opening.* Determine the opening of the cell face. Where there is more than one size of cell opening, use the largest. Match the actual opening to those shown in Figure A.23.3.1.3(b), and use the corresponding curve from Figure A.23.3.1.3(a). If there is no match between the size and shape of the opening and Figure A.23.3.1.3(a), interpolate between the curves. If the opening exceeds 6 ft × 6 ft (1.8 m × 1.8 m), use the curve for a 6 ft × 6 ft (1.8 m × 1.8 m) opening. This curve represents the maximum burning situation, and increasing the size of the opening will not increase the actual burning rate.

*Exhaust Fan Rate.* Determine the exhaust fan capacity needed to extract smoke at a rate that will maintain the smoke level at a point higher than *Z*. This is the rate shown on the baseline of Figure A.23.3.1.3(a) corresponding to the level of *Z* on the vertical axis for the solid line (ventilation rate) curve appropriate to the cell door size. This exhaust capability needs to be provided at a point higher than *Z*.

*Intake Air.* Provide intake air openings that either exist or are automatically provided at times of emergency smoke removal. These openings are to be located at or near the baseline of the cell block to allow for intake air at the rate to be vented by the fan. The openings provided shall be sufficient to avoid a friction load that can reduce the exhaust efficiency. Standard air-handling design criteria are used in making this calculation.

*Fan Temperature Rating.* Determine the potential temperature of gases that the fan might be required to handle by measuring the distance from the floor of the highest cell to the centerline of the fan, or fan ports if the fan is in a duct or similar arrangement. Determine the intersection of the new *Z* value with the appropriate ventilation rate curve (solid line) from Figure A.23.3.1.3(a). Estimate the temperature rise by interpolating along the appropriate ventilation rate curve and between the constant temperature rise curves (dashed lines) from Figure A.23.3.1.3(a). Provide all elements of the exhaust system that are to be above the acceptable smoke level with the capability to effectively operate with the indicated increase in temperature.

*Operation of Exhaust System.* Arrange the emergency exhaust system to initiate automatically on detection of smoke, on operation of a manual fire alarm system, or by direct manual operation. The capability to manually start the automatic exhaust system should be provided in a guard post in the cell block, at another control location, or both. Where appropriate, the emergency exhaust fans are permitted to be used for comfort ventilation as well as for serving their emergency purposes.

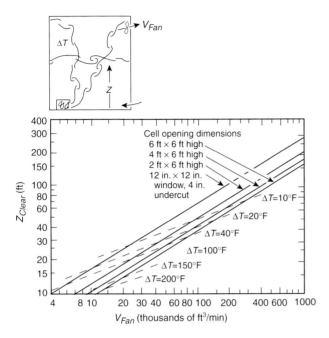

For SI units, 1 ft = 0.3048 m; 1 in. = 25.4 mm;
(°F – 32) ÷ 1.8 = °C; 1 ft³/min = 0.00047 m³/s

$\Delta T$ = Temperature of upper layer gases above ambient
$Z_{Clear}$ = Distance from cell floor to smoke layer
$V_{Fan}$ = Fan discharge capacity (as installed)

Solid lines: Ventilation rate curves
Dashed lines: Constant temperature rise curves

**FIGURE A.23.3.1.3(a)   Cell Block Smoke Control Ventilation Curves.**

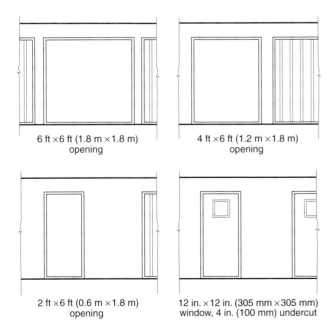

**FIGURE A.23.3.1.3(b)   Typical Cell Openings.**

**A.23.3.2.1**  It is strongly recommended that padded cells not be used due to their fire record. However, recognizing that they will be used in some cases, provisions for the protection of padded cells are provided. It is recognized that the minimum ¾-hour fire protection–rated fire door will be violated with the "plant on" of the padding, but a minimum ¾-hour fire protection–rated fire door should be the base of the assembly.

**A.23.3.4.2.2**  Where the fire alarm control unit is in an area that is not continuously occupied, automatic smoke detection is needed at the control unit. The provision of 23.3.4.2.2 exempts the smoke detection only at the notification appliance circuit power extenders and the supervising station transmitting equipment.

**A.23.3.4.3.1(2)**  The staff at the constantly attended location should have the capability to promptly initiate the general alarm function and contact the fire department or have direct communication with a control room or other location that can initiate the general alarm function and contact the fire department.

**A.23.3.4.4.3**  An open dormitory is a dormitory that is arranged to allow staff to observe the entire dormitory area at one time.

**A.23.3.5.2**  Where the openings in ceilings or partitions are ¼ in. (6.3 mm) or larger in the smallest dimension, where the thickness or depth of the material does not exceed the smallest dimension of the openings, and where such openings constitute not less than 70 percent of the area of the ceiling or partition material, the disruption of sprinkler spray patterns is permitted to be disregarded.

**A.23.3.5.4(1)**  Where access to portable fire extinguishers is locked, staff should be present on a 24-hour basis and should have keys readily available to unlock access to the extinguishers. Where supervision of sleeping areas is from a 24-hour attended staff location, portable fire extinguishers are permitted to be provided at the staff location in lieu of the sleeping area.

**A.23.3.5.4(2)**  It is recognized that locating portable fire extinguishers at staff locations might only result in travel distances to extinguishers being in excess of those permitted by NFPA 10, *Standard for Portable Fire Extinguishers.*

**A.23.3.7.1**  Consideration can be given for large open areas that might be permitted to function as smoke sinks as an alternative to the installation of more than one smoke barrier as required by 23.3.7.1. Vertical movement downward to an area of refuge might be permitted by the authority having jurisdiction in lieu of horizontal movement.

**A.23.3.7.1(2)**  A door to the outside, by itself, does not meet the intent of this provision if emergency operating procedures do not provide for the door to be unlocked when needed. In cases where use of the door is not ensured, a true smoke barrier per the base requirement of 23.3.7.1 would be needed.

**A.23.3.7.3(2)**  Consideration should be given to increasing the travel distance to a smoke barrier to coincide with existing range lengths and exits.

**A.23.3.7.5**  Structural fire resistance is defined as the ability of the assembly to stay in place and maintain structural integrity without consideration of heat transmission. Twelve-gauge steel plate suitably framed and stiffened meets this requirement.

**A.23.3.7.6(1)**  As an example, a smoke barrier is permitted to consist of fire-rated glazing panels mounted in a security grille arrangement.

**A.23.3.8** The requirements in Table 23.3.8 for smoke-resistant and fire-rated separations include taking the necessary precautions to restrict the spread of smoke through the air-handling system. However, the intent is not that smoke dampers are required to be provided for each opening. Smoke dampers would be one acceptable method; however, other techniques, such as allowing the fans to continue to run with 100 percent supply and 100 percent exhaust, would be acceptable.

**A.23.4.1.2(2)** The automatic smoke venting should be in accordance with NFPA 204, *Standard for Smoke and Heat Venting*, for light hazard occupancies.

**A.23.4.5.1.4(1)** The term *other physical restraints* is meant to include the use of personal restraint devices, such as handcuffs or shackles, where occupants are secured to the structure or furnishings to restrict movement.

**A.23.7.1.2** This requirement is permitted to be met by electronic or oral monitoring systems, visual monitoring, call signals, or other means.

**A.23.7.1.3** Periodic, coordinated training should be conducted and should involve detention and correctional facility personnel and personnel of the fire department legally committed to serving the facility.

**A.23.7.4** Personal property provides combustible contents for fire development. Therefore, adequate controls are needed to limit the quantity and combustibility of the fuels available to burn to reduce the probability of room flashover. The provisions of 23.7.4 will not, by themselves, prevent room flashover if personal property controls are not provided.

**A.23.7.4.3** Mattresses used in detention and correctional facilities should be evaluated with regard to the fire hazards of the environment. The potential for vandalism and excessive wear and tear also should be taken into account when evaluating the fire performance of the mattress. ASTM F 1870, *Standard Guide for Selection of Fire Test Methods for the Assessment of Upholstered Furnishings in Detention and Correctional Facilities*, provides guidance for this purpose.

**A.24.1.1.1** The *Code* specifies that, wherever there are three or more living units in a building, the building is considered an apartment building and is required to comply with either Chapter 30 or Chapter 31, as appropriate. A townhouse unit is considered to be an apartment building if there are three or more units in the building. The type of wall required between units in order to consider them as separate buildings is normally established by the authority having jurisdiction. If the units are separated by a wall of sufficient fire resistance and structural integrity to be considered as separate buildings, the provisions of Chapter 24 apply to each townhouse. Condominium status is a form of ownership, not occupancy; for example, there are condominium warehouses, condominium apartments, and condominium offices.

The provisions of 24.1.1.1 state that, in one- and two-family dwellings, each dwelling unit can be "occupied by members of a single family with not more than three outsiders." The *Code* does not define the term *family*. The definition of *family* is subject to federal, state, and local regulations and might not be restricted to a person or a couple (two people) and their children. The following examples aid in differentiating between a single-family dwelling and a lodging or rooming house:

(1) An individual or a couple (two people) who rent a house from a landlord and then sublease space for up to three individuals should be considered a family renting to a maximum of three outsiders, and the house should be regulated as a single-family dwelling in accordance with Chapter 24.

(2) A house rented from a landlord by an individual or a couple (two people) in which space is subleased to 4 or more individuals, but not more than 16, should be considered and regulated as a lodging or rooming house in accordance with Chapter 26.

(3) A residential building that is occupied by 4 or more individuals, but not more than 16, each renting from a landlord, without separate cooking facilities, should be considered and regulated as a lodging or rooming house in accordance with Chapter 26.

**A.24.2** The phrase "means of escape" indicates a way out of a residential unit that does not conform to the strict definition of means of egress but does meet the intent of the definition by providing an alternative way out of a building. (*See the definition of means of escape in 3.3.171.*)

**A.24.2.2.3.3** A window with dimensions of 20 in. × 24 in. (510 mm × 610 mm) has an opening of 3.3 ft$^2$ (0.31 m$^2$), which is less than the required 5.7 ft$^2$ (0.53 m$^2$). Therefore, either the height or width needs to exceed the minimum requirement to provide the required clear area (*see Figure A.24.2.2.3.3*). The current minimum width and height dimensions, as well as the minimum clear opening, became a requirement of this *Code* in the 1976 edition and were based on tests conducted to determine the minimum size of the wall opening required to allow a fire fighter wearing complete turnout gear and a self-contained breathing apparatus entry to the room from the exterior to effect search and rescue. Prior editions of the *Code* limited the width or height, or both, to not less than 22 in. (560 mm) and a clear opening of 5 ft$^2$ (0.47 m$^2$). For existing window frames and sash of steel construction, adherence to these dimensional criteria is essential to allow fire fighter entry. For existing window frames and sash of wood construction that can easily be removed prior to entry by fire fighters to achieve the 5 ft$^2$ (0.47 m$^2$) hole in the wall, the clear opening created by the occupant upon opening the window from the interior room side is required only to provide an opening measuring not less than 20 in. × 24 in. (510 mm × 610 mm) or 3.3 ft$^2$ (0.31 m$^2$).

**A.24.2.4.7** It is the intent of this requirement that security measures, where installed, do not prevent egress.

**A.24.3.4.1.1** Paragraph 11.5.1.3 of *NFPA 72, National Fire Alarm and Signaling Code*, contains related requirements. They specify that, where the interior floor area for a given level of a dwelling unit, excluding garage areas, is greater than 1000 ft$^2$ (93 m$^2$), smoke alarms are to be installed as follows:

(1) All points on the ceiling are to have a smoke alarm within a distance of 30 ft (9.1 m), measured along a path of travel, or to have one smoke alarm per 500 ft$^2$ (46.5 m$^2$) of floor area, which is calculated by dividing the total interior floor area per level by 500 ft$^2$ (46.5 m$^2$).

(2) Where dwelling units include great rooms or vaulted/cathedral ceilings extending over multiple floors, smoke alarms located on the upper floor that are intended to protect the aforementioned area are permitted to be considered as part of the lower floor(s) protection scheme used to meet the requirements of A.24.3.4.1.1(1).

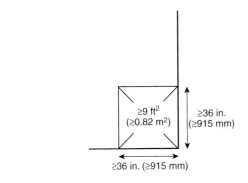

**PLAN VIEW**

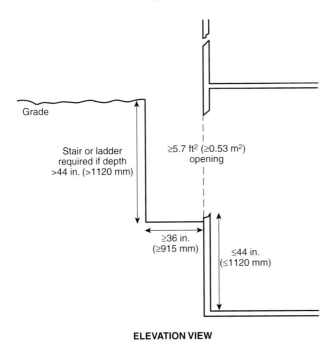

**ELEVATION VIEW**

**FIGURE A.24.2.2.3.3   Escape Window Utilizing a Window Well.**

**A.24.3.4.1.1(2)**   Paragraphs 11.5.1.1(2) and 11.5.1.2 of *NFPA 72, National Fire Alarm and Signaling Code*, contain related requirements. The requirement of 11.5.1.1(2) specifies that an alarm is to be installed outside of each separate dwelling unit sleeping area, within 21 ft (6.4 m) of any door to a sleeping room, with the distance measured along a path of travel. The requirement in 11.5.1.2 specifies that, where the area addressed in 11.5.1.1(2) is separated from the adjacent living areas by a door, a smoke alarm is to be installed in the area between the door and the sleeping rooms, and additional alarms are to be installed on the living area side of the door.

**A.24.3.4.2.2**   The placement requirements of NFPA 720, *Standard for the Installation of Carbon Monoxide (CO) Detection and Warning Equipment*, are modified specifically for one- and two-family dwellings as required by this *Code* and do not affect other regulations within a jurisdiction.

**A.24.3.5**   Automatic sprinklers are recognized as an excellent addition to homes to enhance life safety and property protection. Automatic sprinklers can be part of a comprehensive package of fire protection and can assist in the overall master planning of a community. Where all of the buildings within an area are sprinklered, including the single-family dwellings, the response times and personnel of local fire departments can be established at different levels than if the buildings were not sprinklered, saving considerable amounts of tax dollars. When whole developments are sprinklered, water mains, hydrant spacing, road widths, and building density can be altered to help alleviate the economic impact of the sprinklers.

**A.26.1.1.1**   Bed and breakfast occupancies with more than 3, but fewer than 17, occupants are considered lodging and rooming houses.

**A.26.2.3.5.1**   It is the intent of this requirement that security measures, where installed, do not prevent egress.

**A.26.3.1.2**   Such protection can be accomplished by separation by physical distance, arrangement of the stairs, protection of the openings exposing the stairs, or a combination thereof.

**A.26.3.4.3.1**   The proprietor is the owner or owner's agent with responsible charge.

**A.26.3.4.6.2**   The placement requirements of NFPA 720, *Standard for the Installation of Carbon Monoxide (CO) Detection and Warning Equipment*, are modified to accommodate lodging or rooming house occupancies that are part of multiple occupancy buildings (e.g., an on-call physicians' sleeping room in a hospital). The placement requirements of NFPA 720 are modified specifically for lodging or rooming houses as required by this *Code* and do not affect other regulations within a jurisdiction.

**A.26.3.4.7**   It is the intent that smoke detection be provided at each fire alarm control unit, regardless of the presence of sprinkler protection.

**A.26.3.6.3.3**   The decision to permit the use of the criteria from NFPA 13D, *Standard for the Installation of Sprinkler Systems in One- and Two-Family Dwellings and Manufactured Homes*, in these occupancies is based on the following:

(1) The desire to obtain a level of fire suppression and control that is approximately equivalent to that delivered by residential facilities protected by such systems *(see A.1.1 in NFPA 13D)*
(2) The fact that potential fire exposure and challenge to the suppression system in a small lodging and rooming occupancy is of the same nature and no more severe than that found in residences

**A.28.2.2.12.2**   The provision of 28.2.2.12.2 permits the entire floor to serve as an area of refuge where it is protected in accordance with 28.3.5. The provision is acceptable because supervised automatic sprinkler systems have built-in signals for monitoring features of the system, such as the opening and closing of water control valves. Such systems also monitor pump power supplies, water tank levels, and conditions that will impair the satisfactory operation of the sprinkler system. Because of these monitoring features, supervised automatic sprinkler systems have a high level of satisfactory performance and response to fire conditions.

**A.28.2.3.3**   The exemption contained in 28.2.3.3 applies to corridors within an individual room or suite and does not apply where a suite can be subdivided and rented separately.

**A.28.2.7.2** Where open stairways are permitted, they are considered as exit access to exits rather than as exits, and the requirements for travel distance to exits include the travel on such stairs. *(See 7.6.3.)*

**A.28.3.4.3.1** Visible signaling appliances might be governed by provisions of federal regulations in 28 CFR 36, Appendix A, "Americans with Disabilities Act Accessibility Guidelines for Buildings and Facilities," Section 4.28, Alarms.

**A.28.3.4.3.3** A quantity of such rooms and suites might be required to be equipped to accommodate hearing-impaired individuals based on the total number of rooms in a transient lodging facility. *(See 28 CFR 36, Appendix A, "Americans with Disabilities Act Accessibility Guidelines for Buildings and Facilities.")*

**A.28.3.4.5** Caution needs to be exercised in locating smoke alarms with regard to their proximity to bathrooms, cooking facilities, and HVAC outlets in order to prevent nuisance alarms.

**A.28.4.1.2** See 4.8.2.1(4).

**A.28.5.3.2** "Protected power supply" means a source of electrical energy of sufficient capacity to allow proper operation of the elevator and its associated control and communications systems. The power supply's point of origin, system of distribution, type and size of overcurrent protection, degree of isolation from other portions of the building electrical system, and degree of mechanical protection should be such that it is unlikely that the supply would be disrupted at any but the advanced stages of building fire involvement or by structural collapse.

A protected power supply might consist of, and should provide, not less than the level of reliability associated with an electrical distribution system with service equipment located and installed in accordance with 230.72(B) and 230.82(5) of *NFPA 70, National Electrical Code.* The distribution system is not to have any other connection to the building electrical distribution system. A protected power supply is not required to incorporate two sources of energy or automatic transfer capability from a normal to an emergency source; for example, an alternate set of service conductors.

The number and type of elevators to be connected to a protected power supply should be limited, or the characteristics of the protected power supply should be selected to ensure conformance to 230.95 of *NFPA 70,* without the provision of ground fault protection for the supply.

An elevator installation supplied by a protected power supply should comply with Article 620 of *NFPA 70* and ASME A17.1/CSA B44, *Safety Code for Elevators and Escalators.* The energy absorption means should always be connected on the load side of the disconnecting means. The energy absorption means should not consist of loads likely to become inoperative or disconnected under the conditions assumed to exist when the elevator is under the control of fire department personnel. Examples of such loads include light and power loads external to the elevator equipment room.

**A.28.7.1.1** Employers are obligated to determine the degree to which employees are to participate in emergency activities. Regulations of the U.S. Department of Labor (OSHA) govern these activities and provide options for employers, from total evacuation to aggressive structural fire fighting by employee brigades. *(For additional information, see 29 CFR 1910, Subparts E and L, "OSHA Regulations for Emergency Procedures and Fire Brigades.")*

**A.28.7.1.2** Emergencies should be assumed to have arisen at various locations in the occupancy in order to train employees in logical procedures.

**A.28.7.4.1** Floor diagrams should reflect the actual floor arrangement and should be oriented with the actual direction to the exits.

**A.28.7.4.2** Factors for developing the fire safety information include such items as construction type, suppression systems, alarm and detection systems, building layout, and building HVAC systems.

**A.29.2.2.8** Due to the nature of escalators, they are no longer acceptable as a component in a means of egress. However, because many escalators have been used for exit access and exit discharge in the past, they are permitted to continue to be considered in compliance. Very few escalators have ever been installed in a manner to qualify as an exit. For information on escalator protection and requirements, see previous editions of the *Code.*

**A.29.2.2.12.2** The provision of 29.2.2.12.2 permits the entire floor to serve as an area of refuge where it is protected in accordance with 29.3.5. The provision is acceptable because supervised automatic sprinkler systems have built-in signals for monitoring features of the system, such as the opening and closing of water control valves. Such systems also monitor pump power supplies, water tank levels, and conditions that will impair the satisfactory operation of the sprinkler system. Because of these monitoring features, supervised automatic sprinkler systems have a high level of satisfactory performance and response to fire conditions.

**A.29.2.7.2** Where open stairways or escalators are permitted, they are considered as exit access to exits rather than as exits, and the requirements for travel distance to exits include the travel on such stairs and escalators. *(See 7.6.3.)*

**A.29.3.4.3.6** The provision for immediate notification of the public fire department is intended to include, but is not limited to, all of the arrangements in 9.6.4.2. Other arrangements that depend on a clerk or other member of the staff to notify the fire department might also be permitted. In such cases, however, it is essential that a trained staff member and an immediately available means of calling the fire department are continuously available. If a telephone is to be used, it should not be of any type or arrangement that requires a coin or the unlocking of a device to contact the fire department.

**A.29.3.4.5** Caution needs to be exercised in locating smoke alarms with regard to their proximity to bathrooms, cooking facilities, and HVAC outlets in order to prevent nuisance alarms.

**A.29.3.5.3** Although not required by the *Code,* the use of residential sprinklers or quick-response sprinklers is encouraged for new installations of sprinkler systems within dwelling units, apartments, and guest rooms. Caution should be exercised, as the system needs to be designed for the sprinkler being used.

**A.29.4.1.2** See 4.8.2.1(4).

**A.29.7.1.1** Employers are obligated to determine the degree to which employees are to participate in emergency activities. Regulations of the U.S. Department of Labor (OSHA) govern these activities and provide options for employers, from total evacuation to aggressive structural fire fighting by employee brigades. *(For additional information, see 29 CFR 1910, Subparts E and L, "OSHA Regulations for Emergency Procedures and Fire Brigades.")*

**A.29.7.1.2** Emergencies should be assumed to have arisen at various locations in the occupancy in order to train employees in logical procedures.

**A.29.7.4.1** Floor diagrams should reflect the actual floor arrangement and should be oriented with the actual direction to the exits.

**A.29.7.4.2** Factors for developing the fire safety information include such items as construction type, suppression systems, alarm and detection systems, building layout, and building HVAC systems.

**A.30.2.2.2.2.1** It is the intent of this requirement that security measures, where installed, should not prevent egress.

**A.30.2.2.12.2** The provision of 30.2.2.12.2 permits the entire floor to serve as an area of refuge where it is protected in accordance with 31.3.5. The provision is acceptable because supervised automatic sprinkler systems have built-in signals for monitoring features of the system, such as the opening and closing of water control valves. Such systems also monitor pump power supplies, water tank levels, and conditions that will impair the satisfactory operation of the sprinkler system. Because of these monitoring features, supervised automatic sprinkler systems have a high level of satisfactory performance and response to fire conditions.

**A.30.3.4.5** Previous editions of the *Code* permitted the single-station smoke alarm required by 30.3.4.5 to be omitted from each apartment where a complete automatic smoke detection system was installed throughout the building. With such a system, when one detector is activated, an alarm is sounded throughout the building. Experience with complete smoke detection systems in apartment buildings has shown that numerous nuisance alarms are likely to occur. Where there is a problem with frequent nuisance alarms, occupants ignore the alarm, or the system is either disconnected or otherwise rendered inoperative.

**A.30.3.5.3** The 12 ft$^2$ (1.1 m$^2$) closet sprinkler exemption differs from requirements in NFPA 13, *Standard for the Installation of Sprinkler Systems*, because fire loss data supports the longstanding position of the *Code*, since the 1976 edition, to omit sprinklers from such closets. The provision is further supported by the lack of losses in buildings protected in accordance with NFPA 13D, *Standard for the Installation of Sprinkler Systems in One- and Two-Family Dwellings and Manufactured Homes*, and NFPA 13R, *Standard for the Installation of Sprinkler Systems in Residential Occupancies up to and Including Four Stories in Height*, which permit the omission of sprinklers from closets not exceeding 24 ft$^2$ (2.2 m$^2$).

**A.30.4.1.2** See 4.8.2.1(4).

**A.31.1** See Table A.31.1.

**A.31.2.2.8** Due to the nature of escalators, they are no longer acceptable as a component in a means of egress. However, because many escalators have been used for exit access and exit discharge in the past, they are permitted to continue to be considered in compliance. Very few escalators have ever been installed in a manner to qualify as an exit. For information on escalator protection and requirements, see previous editions of the *Code*.

**A.31.2.2.12.2** The provision of 31.2.2.12.2 permits the entire floor to serve as an area of refuge where it is protected in accordance with 31.3.5. The provision is acceptable because supervised automatic sprinkler systems have built-in signals for monitoring features of the system, such as the opening and closing of water control valves. Such systems also monitor pump power supplies, water tank levels, and conditions that

will impair the satisfactory operation of the sprinkler system. Because of these monitoring features, supervised automatic sprinkler systems have a high level of satisfactory performance and response to fire conditions.

**A.31.2.4.6** This single-exit exemption could be applied to an apartment building three stories in height with a basement.

**A.31.2.11.1** The provision of 31.2.11 recognizes the need to provide smoke control in existing buildings. Smokeproof enclosures can be accomplished without the use of a vestibule in accordance with 7.2.3.

**A.31.3.4.4.1** It is intended that a building compliant with Option 2 function as described in the paragraph that follows.

Occupants within a living unit become aware of a fire emergency, either through personal awareness or through being alerted by the smoke alarm(s) installed within the living unit. Other building occupants are alerted to the fire emergency by the building fire alarm system that is initiated by manual fire alarm boxes adjacent to the exits, heat detection within the living unit where the fire emergency exists, smoke detection in the common areas outside the living unit, or a combination thereof. The installation of system heat detectors versus smoke detectors within the living unit is intended to eliminate nuisance-type alarms and reduce occupant complacency from frequent false alarms. The installation of smoke detection within the living unit should only be contemplated after a careful analysis of the goals and with the approval of the authority having jurisdiction.

**A.31.3.4.5.1** *NFPA 101* provides adequate, balanced fire protection and takes into consideration the passive and active systems required in a given occupancy. The level of protection prescribed by *NFPA 72, National Fire Alarm and Signaling Code*, which includes smoke alarms in all sleeping rooms, without exception, does not necessarily take into consideration the complete protection package mandated by *NFPA 101*.

**A.31.3.5.2** Although not required by the *Code*, the use of residential sprinklers or quick-response sprinklers is encouraged for new installations of sprinkler systems within dwelling units, apartments, and guest rooms. Caution should be exercised, because the system needs to be designed for the sprinkler being used.

**A.31.3.5.10** For example, if an Option 3 sprinkler system were being used to justify use of Class C wall finish in an exit enclosure, the sprinkler system would need to be extended into the exit enclosure, even if the rest of the requirements for Option 3 did not require the sprinklers in the exit enclosure.

**A.31.3.5.11.3** This system might consist of a combination of any or all of the following systems:

(1) Partial automatic sprinkler protection
(2) Smoke detection alarms
(3) Smoke control
(4) Compartmentation or other approved systems, or both

**A.31.3.6.1** The intent is to recognize that existing partitions of sound wood lath and plaster, wire lath and plaster, or gypsum lath and plaster construction have demonstrated the ability to contain most room fires. Recent data on archaic construction methods have established the fire resistance rating of such construction at about 20 minutes. Such construction meets the intent of 31.3.6.1.

**A.31.4.1.2** See 4.8.2.1(4).

**Table A.31.1 Alternate Requirements for Existing Apartment Buildings According to Protection Provided**

| Feature | No Suppression or Detection System Option 1 | Complete Automatic Fire Detection Option 2 | Automatic Sprinkler Protection in Selected Areas Option 3 | Automatic Sprinkler Protection Throughout per NFPA 13 (with exceptions) Option 4 |
|---|---|---|---|---|
| **Exit Access** | | | | |
| Travel distance from apartment door to exit | 100 ft (30 m) | 150 ft (46 m) | 150 ft (46 m) | 200 ft (61 m) |
| Travel distance within apartment | 75 ft (23 m) | 125 ft (38 m) | 75 ft (23 m) | 125 ft (38 m) |
| Smoke barrier required *(See 31.3.7.)* | R | R | R | NR |
| Maximum single path corridor distance | 35 ft (10.7 m) | 35 ft (10.7 m) | 35 ft (10.7 m) | 35 ft (10.7 m) |
| Maximum dead end | 50 ft (15 m) | 50 ft (15 m) | 50 ft (15 m) | 50 ft (15 m) |
| *Corridor fire resistance* | | | | |
| Walls | ½ hr | ½ hr | ½ hr | ½ hr |
| Doors (fire protection rating) | 20 min. or 1¾ in. (44 mm) thick | 20 min. or 1¾ in. (44 mm) thick | Smoke resisting | Smoke resisting |
| **Interior Finish** | | | | |
| Lobbies and corridors | A or B | A or B | A or B | A, B, or C |
| Other spaces | A, B, or C | A, B, or C | A, B, or C | A, B, or C |
| Floors in corridors | I or II | I or II | NR | NR |
| **Exits** | | | | |
| *Wall fire resistance* | | | | |
| 1–3 stories† | 1 hr | 1 hr | 1 hr | 1 hr |
| >3 stories† | 2 hr | 2 hr | 2 hr | 1 hr |
| *Smokeproof enclosures* | | | | |
| Not high-rise | NR | NR | NR | NR |
| High-rise | R | R | R | NR |
| *Door fire resistance* | | | | |
| 1–3 stories† | 1 hr | 1 hr | 1 hr | 1 hr |
| >3 stories† | 1½ hr | 1½ hr | 1½ hr | 1 hr |
| *Interior finish* | | | | |
| Walls and ceilings | A or B | A or B | A or B | A, B, or C |
| Floors | I or II | I or II | I or II | NR |
| **Within Living Unit (Apartment)** | | | | |
| Escape windows, per Section 24.2 *(See 31.2.1.)* | R | R | R | NR |
| **Alarm System** | | | | |
| >3 stories or >11 units† | Manual initiation | Manual and auto initiation | Manual and auto initiation | Manual and auto initiation |
| >2 stories or >50 units† | Annunciator panel | Annunciator panel | Annunciator panel | Annunciator panel |

R: Required *(see Code for details and exemptions)*. NR: No requirements.
†Number of stories in height.

**A.32.1.6** The provisions of 8.3.1(4) address a ½-hour fire resistance rating. The information in A.8.3.1.1(4) addresses common materials used in barriers having a minimum ½-hour fire resistance rating.

**A.32.2.2.3.1(3)** A window with dimensions of 20 in. × 24 in. (510 mm × 610 mm) has an opening of 3.3 ft² (0.31 m²), which is less than the required 5.7 ft² (0.53 m²). Therefore, either the height or width needs to exceed the minimum requirement to provide the required clear area.

**A.32.2.2.6.3** Exterior stair protection can be accomplished through separation by physical distance, arrangement of the stairs, protection of the openings exposing the stairs, or other means acceptable to the authority having jurisdiction.

**A.32.2.3.2.1** Spaces containing approved, properly installed and maintained furnaces and heating equipment, furnace rooms, and cooking and laundry facilities should not be classified as hazardous areas solely on the basis of such equipment.

**A.32.2.3.5** All sprinkler systems installed in accordance with NFPA 13, *Standard for the Installation of Sprinkler Systems,* and NFPA 13R, *Standard for the Installation of Sprinkler Systems in Residential Occupancies up to and Including Four Stories in Height,* are required to be inspected, tested, and maintained in accordance with NFPA 25, *Standard for the Inspection, Testing, and Maintenance of Water-Based Fire Protection Systems.* However, systems installed in accordance with NFPA 13D, *Standard for the Installation of Sprinkler Systems in One- and Two-Family Dwellings and Manufactured Homes,* are historically exempt from applying NFPA 25. While there is a great deal of information in NFPA 25 that is not appropriate for NFPA 13D sprinkler systems, there are some basic concepts of inspection, testing, and maintenance that are critical to system performance and must be performed when an NFPA 13D sprinkler system is installed in a board and care occupancy. The frequencies mandated by this *Code* are slightly different from those required by NFPA 25. It is the intent of this *Code* to utilize the frequencies stated in Chapter 32, but to reference the purpose and the procedures for the inspections, tests, and maintenance from NFPA 25.

**A.32.2.3.5.1** Where any provision requires the use of an automatic sprinkler system in accordance with 32.2.3.5, the provision of 32.2.3.5.2 is not permitted to be used.

**A.32.2.3.5.2** Where a facility utilizing the provision of 32.2.3.5.2 is occupied by residents who can no longer comply with the 3-minute evacuation response, 33.1.8 requires the facility to comply with the requirements for new construction, including automatic sprinkler protection. *(See also A.33.1.8.)*

**A.32.2.3.5.3.2** The decision to permit the use of the criteria from NFPA 13D, *Standard for the Installation of Sprinkler Systems in One- and Two-Family Dwellings and Manufactured Homes,* in these occupancies is based on the following:

(1) The desire to obtain a level of fire suppression and control approximately equivalent to that delivered by residential facilities protected by such systems (*see A.1.1 in NFPA 13D*)

(2) The fact that potential fire exposure and challenge to the suppression system in a small board and care facility are of the same nature and are no more severe than those found in residences

Chapter 32 permits the use of NFPA 13D, and NFPA 13R, *Standard for the Installation of Sprinkler Systems in Residential Occupancies up to and Including Four Stories in Height,* outside of their scopes. This permission is based on a review of the occupancy and a recognition that the fires in board and care facilities are similar to those of other residential occupancies and that the level of protection is appropriate. The requirements of NFPA 13D and NFPA 13R have been supplemented with requirements for additional water supplies to compensate for the special needs of the board and care occupancy.

NFPA 13D contains additional requirements for a piping system serving both sprinkler and domestic needs.

**A.32.3.3.3** The provisions in 10.2.8 to permit modifications to interior finish requirements where automatic sprinklers are provided are permitted.

**A.32.3.3.4.6** Positive alarm sequence applies only to emergency forces notification. Occupant notification is required to occur immediately upon activation of the detection device or system.

**A.32.3.3.6** It is not the intent to prohibit furniture in corridors and spaces open to corridors, provided that the minimum required width is maintained. Storage is not permitted in corridors or spaces open to corridors.

**A.32.3.3.7.10** Where the smoke control system design requires dampers in order that the system functions effectively, it is not the intent of 32.3.3.7.10 to permit the damper to be omitted.

The term *fully ducted* means the supply and return-air systems are provided with continuous ducts from all air registers to the air-handling unit.

**A.32.3.3.7.13** Smoke barrier doors are intended to provide access to adjacent zones. The pair of cross-corridor doors are required to be opposite swinging. Access to both zones is required.

**A.32.3.3.7.17** Smoke barriers might include walls having door openings other than cross-corridor doors. There is no restriction in the *Code* regarding which doors or how many doors form part of a smoke barrier. For example, doors from the corridor to individual rooms are permitted to form part of a smoke barrier.

**A.32.3.3.7.18** It is not the intent to require the frame to be a listed assembly.

**A.32.3.3.8** The scope of NFPA 96, *Standard for Ventilation Control and Fire Protection of Commercial Cooking Operations,* is limited to appliances that produce grease-laden vapors and does not apply to domestic cooking equipment used for food warming or limited cooking.

**A.32.3.6.3.2** "Protected power supply" means a source of electrical energy of sufficient capacity to allow proper operation of the elevator and its associated control and communications systems. The power supply's point of origin, system of distribution, type and size of overcurrent protection, degree of isolation from other portions of the building electrical system, and degree of mechanical protection should be such that it is unlikely that the supply would be disrupted at any but the advanced stages of building fire involvement or by structural collapse.

A protected power supply might consist of, and should provide, not less than the level of reliability associated with an electrical distribution system with service equipment located and installed in accordance with 230.72(B) and 230.82(5) of *NFPA 70, National Electrical Code.* The distribution system is not to have any other connection to the building electrical distribution system. A protected power supply is not required to incorporate two sources of energy or automatic transfer capability from a normal to an emergency source; for example, an alternate set of service conductors.

The number and type of elevators to be connected to a protected power supply should be limited, or the characteristics of the protected power supply should be selected to ensure conformance with 230.95 of *NFPA 70,* without the provision of ground fault protection for the supply.

An elevator installation supplied by a protected power supply should comply with Article 620 of *NFPA 70,* except that the energy absorption means required by 620.91 of *NFPA 70* should always be connected on the load side of the disconnecting means. The energy absorption means should not consist of loads likely to become inoperative or disconnected under the conditions assumed to exist when the elevator is under the control of fire department personnel. Examples of such loads include light and power loads external to the elevator equipment room.

**A.32.4** Board and care occupancies in apartment buildings will usually be small facilities housing 16 or fewer residents. It is intended that the board and care occupancy conform to the requirements of Section 32.2 for small board and care facilities. In the unusual case where an apartment houses a large board and care facility, it would be reasonable for the authority having jurisdiction, using the requirement of 4.6.1, to apply the provisions of Section 32.3 to the apartment. In addition, the apartment building in which the facility is housed needs to comply with the requirements for apartment buildings in Chapters 30 and 31 and the additional criteria presented in Section 32.4.

**A.32.4.1.3** In determining equivalency for conversions, modernizations, renovations, or unusual design concepts, the authority having jurisdiction might permit evaluations based on the residential board and care occupancies fire safety evaluation system (FSES) of NFPA 101A, *Guide on Alternative Approaches to Life Safety.*

**A.32.7.4.1** Smoking regulations should include the following:

(1) Smoking should be prohibited in any room, compartment, or area where flammable or combustible liquids, combustible gases, or oxygen is used or stored and in any other hazardous location, and the following also should apply:

   (a) Such areas should be posted with signs that read NO SMOKING or the international symbol for no smoking.

   (b) In residential board and care facilities where smoking is totally prohibited and signs so indicating are placed at all major entrances, secondary signs with language that prohibits smoking are not required.

(2) Smoking by residents classified as not responsible with regard to their ability to safely use and dispose of smoking materials should be prohibited.

(3) Where a resident, as specified in A.32.7.4.1(2), is under direct supervision by staff or by a person approved by the administration, smoking might be permitted.

(4) Smoking materials should not be provided to residents or maintained by residents without the approval of the administration.

(5) Areas where smoking is permitted should be clearly identified.

(6) Ashtrays of noncombustible material and safe design should be provided and required to be used in all areas where smoking is permitted.

(7) Self-closing cover devices into which ashtrays can be emptied should be made available to all areas where smoking is permitted and should be required to be used.

**A.32.7.5** The requirements applicable to draperies/curtains, upholstered furniture, and mattresses apply only to new draperies/curtains, new upholstered furniture, and new mattresses. The term *new* means unused, normally via procurement from the marketplace, either by purchase or donation, of items not previously used. Many board and care facilities allow residents to bring into the board and care home upholstered furniture items from the resident's previous residence. Such items are not new and, thus, are not regulated. On the other hand, some of the larger board and care homes purchase contract furniture, as is done in hotels. Such new, unused furniture, whether purchased or received as a donation, is regulated by the requirements of 32.7.5.2. By federal law, mattresses manufactured and sold within the United States must pass testing per 16 CFR 1632, "Standard for the Flammability of Mattresses and Mattress Pads" (FF4-72).

**A.32.7.5.2** New upholstered furniture within board and care homes should be tested for rates of heat release in accordance with 10.3.3.

**A.32.7.5.3** New mattresses within board and care homes should be tested for rates of heat release in accordance with 10.3.4.

**A.33.1.1** The requirements of Chapter 33 are designed to accommodate typical changes in the capabilities of the resident, such as those due to accidents, temporary illness, cyclical variations in capabilities, and gradual aging. This approach is based on the assumption that the capabilities of the resident will be evaluated not less than annually, and for residents with geriatric problems or degenerative diseases, not less than every 6 months. Also, residents should be re-evaluated after each accident or illness that requires hospitalization.

The requirements of Chapter 33 were developed on the assumption that the occupants will normally evacuate the building in fire emergencies. During fire exit drills, all occupants should evacuate the building with staff assistance, as needed. Exceptions can be made in facilities with an evacuation capability rating of impractical. Managers of board and care homes with nursing home backgrounds sometimes are not aware of the differences between the requirements of 19.7.1 and 33.7.3.

**A.33.1.1.4** The provision of 33.1.1.4 was added after Chapter 32 was revised in its entirety to avoid potential conflicts between the two chapters. Occupancies meeting Chapter 32 requirements are deemed to comply with Chapter 33.

**A.33.1.6** The provisions of 8.3.1(4) address a ½-hour fire resistance rating. The information in A.8.3.1.1(4) addresses common materials used in barriers having a minimum ½-hour fire resistance rating.

**A.33.1.8** When the group evacuation capability changes to a level of greater risk, the owner/operator of the facility needs to take such action as is necessary, within a reasonable time frame, to restore the evacuation capability of the facility to that for which it was approved. If subsequent evaluations indicate that the original evacuation capability of the facility cannot or is not being maintained at the original level of risk, the facility would be considered as having changed the occupancy subclassification to one of greater risk, and the safeguards required for the level of greater risk would apply. If a facility improves its original evacuation capability to one of less risk, a re-evaluation and upgrading to the requirements for new construction are not needed.

**A.33.2.1.2.1.1** In determining equivalency for existing buildings, conversions, modernizations, renovations, or unusual design concepts, the authority having jurisdiction might permit evaluations based on the residential board and care occupancies fire safety evaluation system (FSES) of NFPA 101A, *Guide on Alternative Approaches to Life Safety.*

**A.33.2.2.3.1(3)** A window with dimensions of 20 in. × 24 in. (510 mm × 610 mm) has an opening of 3.3 ft$^2$ (0.31 m$^2$), which is less than the required 5.7 ft$^2$ (0.53 m$^2$). Therefore, either the height or width needs to exceed the minimum requirement to provide the required clear area.

**A.33.2.2.6.3** Exterior stair protection can be accomplished through separation by physical distance, arrangement of the stairs, protection of the openings exposing the stairs, or other means acceptable to the authority having jurisdiction.

**A.33.2.3.4.3** Most often, smoke alarms sounding an alarm at 85 dBA or greater, installed outside the bedroom area, will meet the intent of this requirement. Smoke alarms remotely located from the bedroom might not be loud enough to awaken the average person. In such cases, it is recommended that smoke alarms be interconnected so that the activation of any smoke alarm will cause all smoke alarms to activate.

*NFPA 101* provides adequate, balanced fire protection and takes into consideration the passive and active systems required in a given occupancy. The level of protection prescribed by *NFPA 72, National Fire Alarm and Signaling Code,* which includes smoke alarms in all sleeping rooms, without exception, does not necessarily take into consideration the complete protection package prescribed by *NFPA 101.*

**A.33.2.3.5** All sprinkler systems installed in accordance with NFPA 13, *Standard for the Installation of Sprinkler Systems,* and NFPA 13R, *Standard for the Installation of Sprinkler Systems in Residential Occupancies up to and Including Four Stories in Height,* are required to be inspected, tested, and maintained in accordance with NFPA 25, *Standard for the Inspection, Testing, and Maintenance of Water-Based Fire Protection Systems.* However, systems installed in accordance with NFPA 13D, *Standard for the Installation of Sprinkler Systems in One- and Two-Family Dwellings and Manufactured Homes,* are historically exempt from applying NFPA 25. While there is a great deal of information in NFPA 25 that is not appropriate for NFPA 13D sprinkler systems, there are some basic concepts of inspection, testing, and maintenance that are critical to system performance and must be performed when an NFPA 13D sprinkler system is installed in a board and care occupancy. The frequencies mandated by this *Code* are slightly different from those required by NFPA 25. It is the intent of this *Code* to utilize the frequencies stated in Chapter 32, but to reference the purpose and the procedures for the inspections, tests, and maintenance from NFPA 25.

**A.33.2.3.5.3.1** The decision to permit the use of the criteria from NFPA 13D, *Standard for the Installation of Sprinkler Systems in One- and Two-Family Dwellings and Manufactured Homes,* in these occupancies is based on the following:

(1) The desire to obtain a level of fire suppression and control approximately equivalent to that delivered by residential facilities protected by such systems *(see A.1.1 in NFPA 13D)*

(2) The fact that potential fire exposure and challenge to the suppression system in a small board and care facility are of the same nature and are no more severe than those found in residences

Chapter 33 permits the use of NFPA 13D and NFPA 13R, *Standard for the Installation of Sprinkler Systems in Residential Occupancies up to and Including Four Stories in Height,* outside of their scopes. This permission is based on a review of the occupancy and a recognition that the fires in board and care facilities are similar to those of other residential occupancies and that the level of protection is appropriate. In some circumstances, such as those for impractical evacuation capabilities, the requirements of NFPA 13D and NFPA 13R have been supplemented with requirements for additional water supplies to compensate for the special needs of the board and care occupancy.

**A.33.3.1.2.1.1** In determining equivalency for existing buildings, conversions, modernizations, renovations, or unusual design concepts, the authority having jurisdiction might permit evaluations based on the residential board and care occupancies fire safety evaluation system (FSES) of NFPA 101A, *Guide on Alternative Approaches to Life Safety.*

**A.33.3.1.2.2** In determining equivalency for existing buildings, the authority having jurisdiction might permit evaluations based on the health care occupancies fire safety evaluation system (FSES) of NFPA 101A, *Guide on Alternative Approaches to Life Safety,* substituting the mandatory safety requirements values of Table A.33.3.1.2.2 for those contained in NFPA 101A.

A residential board and care facility that selects the option to meet the requirements for limited care facilities in Chapter 19 is not considered a change in occupancy.

**Table A.33.3.1.2.2 Substitute Mandatory Safety Requirements Values**

| Zone Location | Containment $(S_a)$ | Extinguishment $(S_b)$ | People Movement $(S_c)$ |
|---|---|---|---|
| First floor | 5 | 6 | 3 |
| Above or below first floor | 9 | 8 | 5 |
| Over 75 ft (23 m) in height | 9 | 8 | 5 |

**A.33.3.3.4.6.1** See A.29.3.4.3.6.

**A.33.3.3.5.1** It is intended that this requirement apply to existing small facilities that are converted to large facilities.

Chapter 33 permits the use of NFPA 13D, *Standard for the Installation of Sprinkler Systems in One- and Two-Family Dwellings and Manufactured Homes,* and NFPA 13R, *Standard for the Installation of Sprinkler Systems in Residential Occupancies up to and Including Four Stories in Height,* outside of their scopes. This permission is based on a review of the occupancy and a recognition that the fires in board and care facilities are similar to those of other residential occupancies and that the level of protection is appropriate. In some circumstances, such as those for impractical evacuation capabilities, the requirements of NFPA 13D and NFPA 13R have been supplemented with requirements for additional water supplies to compensate for the special needs of the board and care occupancy.

**A.33.4** Board and care occupancies in apartment buildings will usually be small facilities housing 16 or fewer residents. It is intended that the board and care occupancy conform to the requirements of Section 33.2 for small board and care facilities. In the unusual case where an apartment houses a large board and care facility, it would be reasonable for the authority having jurisdiction, using the requirement of 4.6.1, to apply the provisions of Section 33.3 to the apartment. In addition, the apartment building in which the facility is housed needs to comply with the requirements for apartment buildings in Chapters 30 and 31 and the additional criteria presented in Section 33.4.

**A.33.4.1.3.1** In determining equivalency for existing buildings, conversions, modernizations, renovations, or unusual design concepts, the authority having jurisdiction might permit evaluations based on the residential board and care occupancies fire safety evaluation system (FSES) of NFPA 101A, *Guide on Alternative Approaches to Life Safety.*

**A.33.7.4.1** Smoking regulations should include the following:

(1) Smoking should be prohibited in any room, compartment, or area where flammable or combustible liquids, combustible gases, or oxygen is used or stored and in any other hazardous location, and the following also should apply:

    (a) Such areas should be posted with signs that read NO SMOKING or the international symbol for no smoking.

    (b) In residential board and care facilities where smoking is totally prohibited and signs so indicating are placed at all major entrances, secondary signs with language that prohibits smoking are not required.

(2) Smoking by residents classified as not responsible with regard to their ability to safely use and dispose of smoking materials should be prohibited.

(3) Where a resident, as specified in A.33.7.4.1(2), is under direct supervision by staff or by a person approved by the administration, smoking might be permitted.

(4) Smoking materials should not be provided to residents or maintained by residents without the approval of the administration.

(5) Areas where smoking is permitted should be clearly identified.

(6) Ashtrays of noncombustible material and safe design should be provided and required to be used in all areas where smoking is permitted.

(7) Self-closing cover devices into which ashtrays can be emptied should be made available to all areas where smoking is permitted and should be required to be used.

**A.33.7.5** The requirements applicable to draperies/curtains, upholstered furniture, and mattresses apply only to new draperies/curtains, new upholstered furniture, and new mattresses. The term *new* means unused, normally via procurement from the marketplace, either by purchase or donation, of items not previously used. Many board and care facilities allow residents to bring into the board and care home upholstered furniture items from the resident's previous residence. Such items are not new and, thus, are not regulated. On the other hand, some of the larger board and care homes purchase contract furniture, as is done in hotels. Such new, unused furniture, whether purchased or received as a donation, is regulated by the requirements of 33.7.5.2. By federal law, mattresses manufactured and sold within the United States must pass testing per 16 CFR 1632, "Standard for the Flammability of Mattresses and Mattress Pads" (FF 4-72).

**A.33.7.5.2** New upholstered furniture within board and care homes should be tested for rates of heat release in accordance with 10.3.3.

**A.33.7.5.3** New mattresses within board and care homes should be tested for rates of heat release in accordance with 10.3.4.

**A.36.1.3.2.2(4)** Means to prevent spilled fuel from accumulating and entering the mercantile occupancy building can be by curbs, scuppers, special drainage systems, sloping the floor away from the door openings, or floor elevation differences of not less than 4 in. (100 mm).

**A.36.2.2.2.2** The words "principal entrance/exit doors" describe doors that the authority having jurisdiction can reasonably expect to be unlocked in order for the facility to do business.

**A.36.2.2.7.2** Egress from a mall building should be designed as follows:

(1) The mall/pedestrian way has been assigned no occupant load, but it is required to be provided with means of egress sized to accommodate the total occupant load of the mall building based on the gross leasable area.

(2) The exits for the mall/pedestrian way are permitted to be provided by a combination of exterior exit doors and exit passageways.

(3) After completion of A.36.2.2.7.2(1), each tenant space is to be judged individually for occupant load and egress capacity, and the following also apply:

    (a) The step specified in A.36.2.2.7.2(3) normally sends a portion or all (per 36.4.4.3.4) of the tenant space's occupant load into the mall.

    (b) Any remaining occupants are sent through the back of the tenant space into an exit passageway that might serve multiple tenant spaces and the mall.

(4) The width of the exit passageway is required to be sized for the most restrictive of the following:

    (a) Width of not less than 66 in. (1675 mm) per 36.4.4.2.2(3)

    (b) Portion of the egress capacity from the largest single tenant space being served by the exit passageway

    (c) Portion of the egress capacity from the mall being provided by the exit passageway

The concepts used in A.36.2.2.7.2(4)(a) through (c) include the following:

(1) After proper egress capacity is provided for the mall/pedestrian way, each tenant space is then required to independently provide egress capacity for its occupants.

(2) The mall required exit passageway width and the tenant space required exit passageway width are not required to be added together.

(3) The required exit passageway width for a tenant space is not required to be added to that of other tenant spaces using the same exit passageway.

**A.36.2.5.10** To eliminate the obstruction to the means of egress of the interior exit access and the exterior exit discharge, it is the intent to provide adequate area for transit and parking of wheeled carts or buggies used by customers. This area includes corral areas adjacent to exits that are constructed to restrict the movement of wheeled carts or buggies therefrom.

**A.36.2.7.2** The basis for the exemption to the general rule on complete enclosure of exits up to their point of discharge to the outside of the building is that, with the specified safeguards, reasonable safety is maintained.

A stairway is not considered to discharge through the street floor area if it leads to the street through a fire resistance–rated enclosure (exit passageway) separating it from the main area, even though there are doors between the first-floor stairway landing and the main area.

The provisions of 36.2.7.2 should not be confused with those for open stairways, as permitted by 36.3.1(1).

**A.36.3.2.1** It is the intent to permit a suspended natural gas–fired unit heater that complies with the requirements of 9.2.2 to be installed and used in a mercantile occupancy without classifying the area in which it is located as hazardous.

**A.36.3.2.1.1** These areas can include, but are not limited to, areas used for general storage, boiler or furnace rooms, and maintenance shops that include woodworking and painting areas.

**A.36.3.2.2** The requirement for separating high hazard contents areas from other parts of the building is intended to isolate the hazard, and 8.2.3.3 is applicable.

**A.36.3.6.1** The intent of 36.3.6.1(2) and (3) is to permit spaces within single tenant spaces, or within buildings protected throughout by an approved, supervised automatic sprinkler system, to be open to the exit access corridor without separation.

**A.36.4.4.2.2(3)** The minimum requirement for terminating mall exit access in not less than 66 in. (1675 mm) of egress width relates to the minimum requirement for not less than one aisle in Class A mercantile occupancies with 30,000 ft² (2800 m²) or greater sales area to be 60 in. (1525 mm) in width.

**A.36.4.4.2.2(5)** Walls providing tenant separations are only required to extend to the underside of the ceiling assembly, regardless of the ceiling's fire-resistive rating. If a ceiling is not provided in either of the tenant spaces, then the wall should extend to the underside of the roof or floor above.

**A.36.4.4.2.2(6)** Fire experience in mall shopping centers indicates that the most likely place of fire origin is in the tenant space, where the combustible fire load is far greater than in the mall. Furthermore, any fires resulting from the comparatively low fire load in the mall are more likely to be detected and extinguished in their incipient stages. Early detection is likely due to the nature of the mall as a high-traffic pedestrian way. Such fires produce less smoke development in a greater volume of space than fires in the more confined adjacent tenant space. Smoke control systems that address fire experience in malls are necessary to ensure the integrity of the mall as a pedestrian way by maintaining it reasonably free of the products of combustion for a duration not less than that required to evacuate the area of the building that is affected by the fire. Secondary considerations should include the following:

(1) Confinement of the products of combustion to the area of origin
(2) Removal of the products of combustion, with a minimum of migration of such products of combustion from one tenant space to another
(3) Achievement of evacuation without the need for smoke control in one- and two-level mall buildings protected by automatic sprinklers

Systems, or combinations of systems, that can be engineered to address fires in malls of three or more levels include the following:

(1) Separate mechanical exhaust or control systems
(2) Mechanical exhaust or control systems in conjunction with heating, ventilating, and air-conditioning systems
(3) Automatically or manually released gravity roof vent devices, such as skylights, relief dampers, or smoke vents
(4) Combinations of items (1), (2), and (3) in this list, or any other engineered system designed to accomplish the purpose of this section

**A.36.4.4.3.5** It is not the intent of 36.4.4.3.5 to require that large tenant spaces be considered anchor stores. A tenant space not considered in determining the occupant load of the mall is required to be arranged so that all of its means of egress will be independent of the mall.

**A.36.4.4.4.3.2** It is the intent to permit the omission of visible alarm notification appliances from the mall or pedestrian way in mall buildings. It is anticipated that occupants with hearing impairments will receive cues from other building occupants and respond accordingly. Visible signals should be provided in public restrooms and other adjunct spaces in the mall subject to occupancy solely by persons with hearing impairments.

**A.36.4.4.6.2** Rooms opening onto the exit passageway are intended to include building service elevators, elevator machine rooms, electrical rooms, telephone rooms, janitor closets, restrooms, and similar normally unoccupied spaces not requiring hazardous area protection in accordance with Section 8.7.

**A.36.4.4.9** Fire experience in mall shopping centers indicates that the most likely place of fire origin is in the tenant space where the combustible fire load is far greater than in the mall.

Furthermore, any fires resulting from the comparatively low fire load in the mall are more likely to be detected and extinguished in their incipient stages. Early detection is likely due to the nature of the mall as a high-traffic pedestrian way. Such fires produce less smoke development in a greater volume of space than fires in the more confined adjacent tenant space.

Smoke control systems that address fire experience in malls are necessary in order to achieve the following:

(1) Ensure the integrity of the mall as a pedestrian way by maintaining it reasonably free of the products of combustion for a duration not less than that required to evacuate the building
(2) Confine the products of combustion to the area of origin
(3) Remove the products of combustion with a minimum of migration of such products of combustion from one tenant to another

Systems, or combinations of systems, that can be engineered to address fires in malls include the following:

(1) Separate mechanical exhaust or control systems
(2) Mechanical exhaust or control systems in conjunction with heating, ventilating, and air-conditioning systems
(3) Automatically or manually released gravity roof vent devices, such as skylights, relief dampers, or smoke vents
(4) Combinations of items (1), (2), and (3) in this list, or any other engineered system designed to accomplish the purpose of this section

**A.37.1.3.2.2(4)** Means to prevent spilled fuel from accumulating and entering the mercantile occupancy building can be by curbs, scuppers, special drainage systems, sloping the floor away from the door openings, or elevation differences of not less than 4 in. (100 mm).

**A.37.2.2.2.2** The words "principal entrance/exit doors" describe doors that the authority having jurisdiction can reasonably expect to be unlocked in order for the facility to do business.

**A.37.2.2.7.2** Egress from a mall building should be designed as follows:

(1) The mall/pedestrian way has been assigned no occupant load, but it is required to be provided with means of egress sized to accommodate the total occupant load of the mall building based on the gross leasable area.
(2) The exits for the mall/pedestrian way are permitted to be provided by a combination of exterior exit doors and exit passageways.

(3) After completion of A.37.2.2.7.2(1), each tenant space is to be judged individually for occupant load and egress capacity, and the following also apply:

    (a) The step specified in A.37.2.2.7.2(3) normally sends a portion or all (per 37.4.4.3.4) of the tenant space's occupant load into the mall.

    (b) Any remaining occupants are sent through the back of the tenant space into an exit passageway that might serve multiple tenant spaces and the mall.

(4) The width of the exit passageway is required to be sized for the most restrictive of the following:

    (a) Width of not less than 66 in. (1675 mm) per 37.4.4.2.2(3)

    (b) Portion of the egress capacity from the largest single tenant space being served by the exit passageway

    (c) Portion of the egress capacity from the mall being provided by the exit passageway

The concepts used in A.37.2.2.7.2(4)(a) through (c) include the following:

(1) After proper egress capacity is provided for the mall/pedestrian way, each tenant space is then required to independently provide egress capacity for its occupants.

(2) The mall required exit passageway width and the tenant space required exit passageway width are not required to be added together.

(3) The required exit passageway width for a tenant space is not required to be added to that of other tenant spaces using the same exit passageway.

**A.37.2.5.2** The purpose of 37.2.5.2 is to avoid pockets or dead ends of such size that they pose an undue danger of persons becoming trapped in case of fire.

It is recognized that dead ends exceeding the permitted limits exist and, in some cases, are impractical to eliminate. The authority having jurisdiction might permit such dead ends to continue to exist, taking into consideration any or all of the following:

(1) Tenant arrangement
(2) Automatic sprinkler protection
(3) Smoke detection
(4) Exit remoteness

**A.37.2.5.3** It is recognized that common paths of travel exceeding the permitted limits exist and, in some cases, are impractical to eliminate. The authority having jurisdiction might permit such paths of travel to continue to exist, taking into consideration any or all of the following:

(1) Tenant arrangement
(2) Automatic sprinkler protection
(3) Smoke detection
(4) Exit remoteness

**A.37.2.5.10** To eliminate the obstruction to the means of egress of the interior exit access and the exterior exit discharge, it is the intent to provide adequate area for transit and parking of wheeled carts or buggies used by customers. This area includes corral areas adjacent to exits that are constructed to restrict the movement of wheeled carts or buggies therefrom.

**A.37.2.7.2** The basis for the exemption to the general rule on complete enclosure of exits up to their point of discharge to the outside of the building is that, with the specified safeguards, reasonable safety is maintained.

A stairway is not considered to discharge through the street floor area if it leads to the street through a fire resistance–rated enclosure (exit passageway) separating it from the main area, even though there are doors between the first floor stairway landing and the main area.

The provisions of 37.2.7.2 should not be confused with those for open stairways, as permitted by 37.3.1(1) and (2).

**A.37.3.2.1** It is the intent to permit a suspended natural gas–fired unit heater that complies with the requirements of 9.2.2 to be installed and used in a mercantile occupancy without classifying the area in which it is located as hazardous.

**A.37.3.2.1.1** These areas can include, but are not limited to, areas used for general storage, boiler or furnace rooms, and maintenance shops that include woodworking and painting areas.

**A.37.3.2.2** The requirement for separating high hazard contents areas from other parts of the building is intended to isolate the hazard, and 8.2.3.3 is applicable.

**A.37.4.4.2.2(3)** The minimum requirement for terminating mall exit access in not less than 66 in. (1675 mm) of egress width relates to the minimum requirement for not less than one aisle in Class A mercantile occupancies with 30,000 ft$^2$ (2800 m$^2$) or greater sales area to be 60 in. (1525 mm) in width.

**A.37.4.4.2.2(6)** Fire experience in mall shopping centers indicates that the most likely place of fire origin is in the tenant space, where the combustible fire load is far greater than in the mall. Furthermore, any fires resulting from the comparatively low fire load in the mall are more likely to be detected and extinguished in their incipient stages. Early detection is likely due to the nature of the mall as a high-traffic pedestrian way. Such fires produce less smoke development in a greater volume of space than fires in the more confined adjacent tenant space. Smoke control systems that address fire experience in malls are necessary to ensure the integrity of the mall as a pedestrian way by maintaining it reasonably free of the products of combustion for a duration not less than that required to evacuate the area of the building that is affected by the fire. Secondary considerations should include the following:

(1) Confinement of the products of combustion to the area of origin
(2) Removal of the products of combustion, with a minimum of migration of such products of combustion from one tenant space to another
(3) Achievement of evacuation without the need for smoke control in one- and two-level mall buildings protected by automatic sprinklers

Systems, or combinations of systems, that can be engineered to address fires in malls of three or more levels include the following:

(1) Separate mechanical exhaust or control systems
(2) Mechanical exhaust or control systems in conjunction with heating, ventilating, and air-conditioning systems
(3) Automatically or manually released gravity roof vent devices, such as skylights, relief dampers, or smoke vents
(4) Combinations of items (1), (2), and (3) in this list, or any other engineered system designed to accomplish the purpose of this section

**A.37.4.4.3.5** It is not the intent of 37.4.4.3.5 to require that large tenant spaces be considered anchor stores. A tenant space not considered in determining the occupant load of the mall is required to be arranged so that all of its means of egress will be independent of the mall.

**A.37.4.4.6.2** Rooms opening onto the exit passageway are intended to include building service elevators, elevator machine rooms, electrical rooms, telephone rooms, janitor closets, restrooms, and similar normally unoccupied spaces not requiring hazardous area protection in accordance with Section 8.7.

**A.38.1.3.2.2(4)** Means to prevent spilled fuel from accumulating and entering the business occupancy building can be by curbs, scuppers, special drainage systems, sloping the floor away from the door openings, or elevation differences not less than 4 in. (100 mm).

**A.38.2.2.2.2** The words "principal entrance/exit doors" describe doors that the authority having jurisdiction can reasonably expect to be unlocked in order for the facility to do business.

**A.38.2.3.2** It is not the intent that this provision apply to non-corridor or nonpassageway areas of exit access, such as the spaces between rows of desks created by office layout or low-height partitions.

**A.38.3.2.1** It is not the intent of this provision that rooms inside individual tenant spaces that are used to store routine office supplies for that tenant be required to be either separated or sprinklered.

**A.38.3.2.2** The requirement for separating high hazard contents areas from other parts of the building is intended to isolate the hazard, and 8.2.3.3 is applicable.

**A.38.3.6.1** The intent of 38.3.6(1) through (3) is to permit spaces to be open to the exit access corridor without separation.

**A.38.3.6.1(1)** Where exits are available from an open floor area, such as open plan buildings, corridors are not required to be separated. An example of an open plan building is a building in which the work spaces and accesses to exits are delineated by the use of tables, desks, bookcases, or counters, or by partitions that are less than floor-to-ceiling height.

**A.38.3.6.1(2)** It is the intent of this provision that a single tenant be limited to an area occupied under a single management and work the same hours. The concept is that people under the same employ working the same hours would likely be familiar with their entire tenant space. It is not the intent to apply this provision simply because tenants are owned by the same organization. For example, in a government-owned office building, the offices of different federal agencies would be considered multiple tenants, because an employee normally works for one agency. The agencies might work various hours. Another example of multiple tenancy would be a classroom building of a university, because some classrooms might be in use at times when other classrooms are not being used.

**A.39.1.3.2.2(4)** Means to prevent spilled fuel from accumulating and entering the business occupancy building can be by curbs, scuppers, special drainage systems, sloping the floor away from the door openings, or elevation differences not less than 4 in. (100 mm).

**A.39.2.2.2.2** The words "principal entrance/exit doors" describe doors that the authority having jurisdiction can reasonably expect to be unlocked in order for the facility to do business.

**A.39.2.5.2** It is recognized that dead ends exceeding the permitted limits exist and, in some cases, are impractical to eliminate. The authority having jurisdiction might permit such dead ends to continue to exist, taking into consideration any or all of the following:

(1) Tenant arrangement
(2) Automatic sprinkler protection
(3) Smoke detection
(4) Exit remoteness

**A.39.2.5.3** It is recognized that common paths of travel exceeding the permitted limits exist and, in some cases, are impractical to eliminate. The authority having jurisdiction might permit such common paths of travel to continue to exist, taking into consideration any or all of the following:

(1) Tenant arrangement
(2) Automatic sprinkler protection
(3) Smoke detection
(4) Exit remoteness

**A.39.3.2.1** It is not the intent of this provision that rooms inside individual tenant spaces that are used to store routine office supplies for that tenant be required to be separated or sprinklered.

**A.39.3.2.2** The requirement for separating high hazard contents areas from other parts of the building is intended to isolate the hazard, and 8.2.3.3 is applicable.

**A.39.4.2.2** In some cases, appreciable cost might be involved in bringing an existing occupancy into compliance. Where this is true, it would be appropriate for the authority having jurisdiction to prescribe a schedule determined jointly with the facility, allowing suitable periods of time for the correction of the various deficiencies and giving due weight to the ability of the owner to secure the necessary funds.

**A.40.1.2.1.3** Additional information on the definition of high hazard industrial occupancy can be found in A.3.3.188.8.2.

**A.40.1.7** In most cases, the requirements for maximum travel distance to exits will be the determining factor, rather than number of occupants, because exits provided to satisfy travel distance requirements will be sufficient to provide egress capacity for all occupants, except in cases of an unusual arrangement of buildings or the high occupant load of a general manufacturing occupancy.

**A.40.2.1.2** Horizontal and vertical utility chases in large industrial buildings used for routing of piping, ducts, and wiring must provide a reasonable level of access for occasional maintenance workers but do not warrant compliance with the comprehensive egress requirements of Chapter 7. Minimum access in these cases is governed by electrical and mechanical codes; 40.2.5.2, Industrial Equipment Access; and the Occupational Safety and Health Administration (OSHA) for facilities in the United States. Utility chases governed by 40.2.1.2 might involve tunnels or large open spaces located above or below occupied floors; however, such spaces differ from mechanical equipment rooms, boiler rooms, and furnace rooms, based on the anticipated frequency of use by maintenance workers. Portions of utility chases where the anticipated presence of maintenance workers is routine are not intended to be included by this paragraph.

**A.40.2.2.5.2** The customary building code requirement for fire doors on both sides of an opening in a fire wall is permitted to be met by having an automatic-sliding fire door on one side and a self-closing fire door swinging out from the other side of the wall. This arrangement qualifies only as a horizontal exit from the sliding door side. For further information, see A.7.2.4.3.10.

**A.40.2.5.1.1** Ancillary facilities located within industrial occupancies might include administrative office, laboratory, control, and employee service facilities that are incidental to the predominant industrial function and are of such size that separate occupancy classification is not warranted.

**A.40.2.5.1.2** Occupants of ancillary facilities located within special-purpose industrial occupancies might be required by administrative controls to remain in the facility when a fire occurs in the predominant industrial area, so that they can perform an orderly shutdown of process equipment to control the spread of the fire and minimize damage to important equipment.

**A.40.2.9** The authority having jurisdiction should review the facility and designate the stairs, aisles, corridors, ramps, and passageways that should be required to be provided with emergency lighting. In large locker rooms or laboratories using hazardous chemicals, for example, the authority having jurisdiction should determine that emergency lighting is needed in the major aisles leading through those spaces.

**A.40.3.2** Emergency lighting should be considered where operations require lighting to perform orderly manual emergency operation or shutdown, maintain critical services, or provide safe start-up after a power failure.

**A.40.6** For further information on aircraft hangars, see NFPA 409, *Standard on Aircraft Hangars.*

**A.42.1.7** There is no occupant load factor specified for storage occupancies. Rather, the probable maximum number of persons present needs to be considered in determining the occupant load.

**A.42.2.1.2** Horizontal and vertical utility chases in large industrial buildings used for routing of piping, ducts, and wiring must provide a reasonable level of access for occasional maintenance workers but do not warrant compliance with the comprehensive egress requirements of Chapter 7. Minimum access in these cases is governed by the electrical and mechanical code; 40.2.5.2, Industrial Equipment Access; and the Occupational Safety and Health Administration (OSHA) for facilities in the United States. Utility chases governed by 42.2.1.2 might involve tunnels or large open spaces located above or below occupied floors; however, such spaces differ from mechanical equipment rooms, boiler rooms, and furnace rooms, based on the anticipated frequency of use by maintenance workers. Portions of utility chases where the anticipated presence of maintenance workers is routine are not intended to be included by this paragraph.

**A.42.2.2.5.2** The customary building code requirement for fire doors on both sides of an opening in a fire wall is permitted to be met by having an automatic-sliding fire door on one side and a self-closing fire door swinging out from the other side of the wall. This arrangement qualifies only as a horizontal exit from the sliding door side. For further information, see A.7.2.4.3.10.

**A.42.2.6** The travel distance to exits specified recognizes a low population density. Consideration should be given to locating areas that have a relatively high population, such as lunchrooms, meeting rooms, packaging areas, and offices, near the outside wall of the building to keep the travel distance to a minimum.

**A.42.6** For further information on aircraft hangars, see NFPA 409, *Standard on Aircraft Hangars.*

**A.42.7** For further information, see NFPA 61, *Standard for the Prevention of Fires and Dust Explosions in Agricultural and Food Processing Facilities.* The egress requirements for storage elevators are based on the possibility of fire and are not based on the possibility of grain dust explosions.

**A.42.8.3.1.1.3(3)** It is common practice to construct a parking structure that consists of open parking levels above grade plane meeting the opening requirements for an open parking structure, but that also has enclosed parking levels below grade plane that need to comply with the requirements for an enclosed parking structure. It is impractical to have enclosed ramps between the enclosed level(s) and the open level(s) of such a parking structure.

**A.43.2.2.1.4** It is not the intent that a corridor, aisle, or circulation space within a suite be considered as a corridor that is shared by more than one occupant space. The suite should be considered as only one occupant space. The following situations should be considered to involve more than one occupant space:

(1) Work affecting a corridor that is common to multiple guest rooms on a floor of a hotel occupancy
(2) Work affecting a corridor that is common to multiple living units on a floor of an apartment building occupancy
(3) Work affecting a corridor that is common to multiple tenants on a floor of a business occupancy

**A.43.2.2.2** Equipment or fixtures do not include manufacturing, production, or process equipment, but do include connections from building service to process equipment.

**A.43.4.2(2)** Some building codes have permitted an increase in egress capacity in buildings protected throughout by an approved automatic sprinkler system. The intent of 43.4.2(2) is that, during a renovation project, egress capacity is permitted to continue to be evaluated using the previously approved method.

**A.43.6.2.2** The provisions for marking of means of egress are those addressed in Section 7.10.

**A.43.7.2.1(2)** It is not the intent of 43.7.2.1(2) to supersede the provision of 32.2.3.5.2 that exempts automatic sprinklers from small board and care facility conversions serving eight or fewer residents when all occupants have the ability as a group to move reliably to a point of safety within 3 minutes.

**A.43.7.3** Table 43.7.3 groups all the residential occupancy classifications into the general category of residential. The category of residential includes one- and two-family dwellings, lodging or rooming houses, hotels and dormitories, and apartment buildings.

## Annex B   Supplemental Evacuation Equipment

*This annex is not a part of the requirements of this NFPA document but is included for informational purposes only. Information in this annex is intended to be adopted by the jurisdiction at the discretion of the adopting jurisdiction. Additionally, information in this annex is intended to be incorporated on a voluntary basis by building owners and developers who might have a desire to include supplemental evacuation equipment in their projects.*

*Although this annex is written in mandatory language, it is not intended to be enforced or applied unless specifically adopted by the jurisdiction or, if it is being applied on a voluntary basis, by the building owner or developer.*

Note: Traditionally, supplemental evacuation equipment has not been regulated or recognized by the *Code*. Until recently, such equipment was considered to include only items such as chain ladders and rope fire escape ladders for use in single-family homes. The criteria specified in Annex B also provides no regulation or recognition for the private installation and use of such equipment by an owner and family, while providing a framework of regulations for the use of controlled descent devices and platform rescue systems in commercial and residential multistory buildings. The broader term *supplemental evacuation equipment* provides for subsets of equipment to be added as further technologies develop.

### B.1  General.

### B.1.1  Definitions.

**B.1.1.1  Controlled Descent Device.** A system operating on the exterior of a building or structure that lowers one or two people per descent, each wearing a rescue harness, at a controlled rate from an upper level to the ground or other safe location.

**B.1.1.2  Platform Rescue System.** An enclosed platform, or set of enclosed platforms, moving vertically along guides or other means on the exterior of a building or structure, intended for the evacuation of multiple occupants from an upper level or levels to the ground or other safe location, that has the capability of transporting emergency responders to upper levels of a building.

**B.1.1.3  Supplemental Escape Device or System.** Dedicated equipment that supplements the means of egress or means of escape for exiting a building or structure.

Note: Supplemental escape devices and systems are not a substitute for the required means of egress or means of escape. If properly installed, maintained, and used, controlled descent devices and platform rescue systems might provide an added means of escape for the occupants where the required means of egress or means of escape is not usable or accessible, and where the event that has caused failure of the required system has not also impaired the functionality of the device or system itself.

**B.1.1.4  Supplemental Evacuation Equipment.** Devices or systems that are not a part of the required means of egress or escape, but that might enhance use of the means of egress or escape, or provide an alternate to the means of egress or escape.

### B.1.2  Reserved.

**B.2  Supplemental Escape Devices or Systems.** A supplemental escape device or system, other than that provided or installed for use by the owner and owner's family, and the installation of such device or system, shall comply with Section B.3 or Section B.4, as appropriate, and the following criteria:

Note: The provisions of Section B.2 are not intended to preclude the installation of supplemental escape devices or systems that do not meet these requirements where intended for personal use, such as by an owner and family.

It should be recognized that supplemental escape devices or systems addressed by these requirements are intended to be used only when all other means of egress are unusable and when remaining in place to await the restoration of the means of egress is considered untenable.

Generally, fire departments have the capability of providing external rescue of building occupants within reach of their portable ladders, aerial ladders, and aerial platform devices. Where a fire department responds to a building emergency and has the capability to provide timely assistance with external rescue, that assistance should be used instead of the supplemental escape devices or systems.

(1) Each supplemental escape device or system shall be of an approved type and shall comply with an approved product safety standard.

(2) The installation of escape devices or systems shall be approved.

  Note: Use of a supplemental escape device or system typically requires that a window or exterior door be opened. The window or door should be closed, except when it is in use for escape. Where the design of the building does not provide exterior doors or operable windows and a window must be broken to use the device or system, consideration should be given to the probable effect of that action, such as showering the emergency response personnel and equipment below with sharp pieces of glass. In such a situation, to obtain approval, it might be appropriate to require tempered safety glass on windows that must be broken to deploy the supplemental escape device and access the system.

(3) The supplemental escape device or system shall be installed, inspected, tested, maintained, and used in accordance with the manufacturer's instructions.

(4) The location of each supplemental escape system access point shall be identified with a readily visible sign complying with the following:

  (a) The sign shall be in plainly legible letters that read SUPPLEMENTAL ESCAPE DEVICE.

  (b) The minimum height of the lettering shall be ¾ in. (19 mm), with a stroke width of ⅛ in. (3 mm).

(5) Each sign required by B.2(4) shall comply with the following:

  (a) The sign shall include the following in plainly legible letters: "Use only when exits are not accessible and building evacuation is imperative, as directed by authorized building personnel or emergency responders."

  (b) The minimum height of the lettering shall be ½ in. (13 mm).

(6) A sign with instructions for use of the escape device or system shall be provided and shall comply with the following:

  (a) The sign shall be posted at the equipment and the equipment's access location.

  (b) The minimum height of lettering on the instructions shall be ½ in. (13 mm).

(c) Pictographs demonstrating use of the escape device or system shall be provided.

Note: Given the nature of the probable circumstances surrounding its deployment, the proper use of the supplemental escape device or system should be readily apparent to the user or trained operator.

(7) The signs and instructions specified in B.2(4), (5), and (6) shall be illuminated as follows:

(a) The signs shall be continuously illuminated while the building is occupied.

(b) The level of illumination provided shall be in accordance with 7.10.6.3, 7.10.7.2, or an approved equivalent.

(8) Where emergency lighting is required by Chapters 11 through 43, it shall be provided as follows:

(a) The illumination shall be in accordance with 7.9.1.

(b) The level of illumination required by 7.9.2.1 shall be provided to illuminate the supplemental escape device or system at its access location and the required signage.

(9) The supplemental escape device or system and its installation shall accommodate persons with various disabilities and of all ages.

Note: It is not the intent of B.2(9) that access ramps, doorways, controls, signage, and other features of the supplemental escape device or system meet all requirements for accessibility for persons with disabilities. The equipment is supplemental in nature and is not recognized as part of the required means of egress. A number of other occupants should be trained to assist persons with disabilities to access the equipment. In selecting the equipment and approving the installation, consideration should be given to how persons with mobility impairments will access the equipment.

Even when exit stairs are usable, elevators might not be able to be used. Use of a supplemental escape device or system to evacuate persons with mobility impairments might be desirable. Such circumstances should be considered and incorporated into the facility's evacuation plan, which should also identify the trained operators authorized to deploy the equipment for such use.

(10) The installation shall be approved such that use of the supplemental escape device or system shall not cause any harm or injury to the user, operator, or others who might be in the vicinity of the equipment when in use.

(11) Where an evacuation plan is required by Chapters 11 through 43, or by other regulation, an approved, written evacuation plan shall be provided as follows:

(a) The plan shall be in accordance with 4.8.2.

(b) The plan shall not rely on the use of supplemental escape devices and systems but shall accommodate the use of such a system by specifying the following:

　i. Role of the supplemental escape device or system in the overall plan

　ii. Role and authority of emergency response personnel with respect to the supplemental escape device or system

　iii. Person or persons authorized to direct the deployment of, and to operate, the escape device or system

　iv. Special considerations, if any, that affect the usability of the supplemental escape device or system

　v. Training required for operators and users

Note: An evacuation plan can be a highly effective tool in determining who should be evacuated under various scenarios and how that evacuation will be accomplished. Even where none is required, an evacuation plan is recommended to identify, among other things, those persons who are authorized to deploy supplemental escape devices and systems.

The more sophisticated the equipment and the greater the number of potential evacuees, the greater is the need to have a trained and authorized person decide which equipment to deploy and when it should be deployed, based on the circumstances at the time. Such a person would be the incident commander, typically the emergency response officer in charge, whether from a private brigade or public service.

Even where a building or facility is not required to have an approved evacuation plan by the *Code*, the supplemental escape device or system operating procedures should be integrated into the building evacuation and emergency procedures to the extent provided.

(12) User and operator training shall be provided in conjunction with the installation of the supplemental escape device or system, and periodically thereafter.

(13) Where an approved evacuation plan is required, training shall be provided in accordance with the approved plan.

(14) The supplemental escape device or system shall be inspected and tested in accordance with the manufacturer's instructions but not less frequently than annually, and the following also shall apply:

(a) Notification of testing shall be provided to building occupants or the authority having jurisdiction, as appropriate.

(b) Written records of the inspection and testing shall be maintained by the owner for a minimum of 1 year after the next scheduled inspection and testing.

Note: It is important that the supplemental escape device or system does not remain idle for many years in order to help ensure that it will be functional if it does need to be used. The manufacturer's instructions for the particular model of equipment involved should be followed.

(15) Supplemental escape devices and systems shall be listed, certified, or approved to operate as intended over the prevalent climatic conditions for the location in which they are installed.

**B.3 Platform Rescue Systems.** Where platform rescue systems are installed or provided, they shall comply with the following:

(1) The platform rescue system shall comply with ASTM E 2513, *Standard Specification for Multi-Story Building External Evacuation Platform Rescue Systems*, or an approved, equivalent product safety standard.

(2) The platform rescue system shall be deployed with trained operators to assist with evacuation of occupants.

(3) Where a fixed installation of electrical or other type power is required to operate the platform rescue system, a redundant source of power shall be provided.

(4) The installation shall be designed such that the vertical distance to be traversed by the platform rescue system shall not exceed the limit specified in the product's listing, certification, or approved installation.

(5) The platform access from within buildings shall be by ramps or stairs, and the following also shall apply:

    (a) Portable ramps and stairs shall be permitted.

    (b) The maximum slope of a ramp shall be as low as practical, but shall not be required to be less than 1 in 8.

    (c) The maximum riser height of stairs shall be 9 in. (230 mm).

    (d) The minimum tread depth of stairs shall be 9 in. (230 mm).

(6) The platform access opening shall be sized in accordance with the following:

    (a) For installations in new construction, the platform access opening shall be a minimum 32 in. (810 mm) in width and a minimum 48 in. (1220 mm) in height.

    (b) For installations in existing construction, the platform access opening shall be as large as practical but shall not be required to exceed 32 in. (810 mm) in width and 48 in. (1220 mm) in height.

(7) The platform access and egress shall not be by ladders.

(8) Rooftop operating equipment and systems shall be protected from accumulations of climatic ice or snow and fire suppression ice.

**B.4 Controlled Descent Devices.** Where controlled descent devices are installed or provided, they shall comply with the following:

(1) The controlled descent device shall comply with ASTM E 2484, *Standard Specification for Multi-Story Building External Evacuation Controlled Descent Devices*, or an approved, equivalent product safety standard.

(2) The installation shall be designed such that the vertical distance to be traversed by the controlled descent device shall not exceed the limit specified in the product's listing, certification, or approved installation.

(3) Where a fixed installation of electrical or other type power is required to operate the controlled descent device, a redundant source of power shall be provided.

(4) Rooftop operating equipment and systems shall be protected from accumulations of climatic ice or snow and fire suppression ice.

(5) Controlled descent device building access openings in new building installations shall be a minimum of 32 in. (810 mm) wide and 42 in. (1065 mm) high.

(6) Controlled descent device building access openings in existing buildings shall be a minimum of 20 in. (510 mm) wide and 24 in. (610 mm) high and shall provide a clear opening of not less than 5.7 ft$^2$ (0.53 m$^2$).

(7) The approved occupant load and weight limits shall be posted adjacent to the controlled descent device installation or building access opening in minimum $\frac{1}{2}$ in. (13 mm) letters, with a minimum $\frac{1}{16}$ in. (1.6 mm) stroke.

(8) The occupant load and weight limits shall not be exceeded in use.

## Annex C   Informational References

**C.1 Referenced Publications.** The documents or portions thereof listed in this annex are referenced within the informational sections of this code and are not part of the requirements of this document unless also listed in Chapter 2 for other reasons.

**C.1.1 NFPA Publications.** National Fire Protection Association, 1 Batterymarch Park, Quincy, MA 02169-7471.

NFPA 1, *Fire Code*, 2012 edition.

NFPA 10, *Standard for Portable Fire Extinguishers*, 2010 edition.

NFPA 13, *Standard for the Installation of Sprinkler Systems*, 2010 edition.

NFPA 13D, *Standard for the Installation of Sprinkler Systems in One- and Two-Family Dwellings and Manufactured Homes*, 2010 edition.

NFPA 13R, *Standard for the Installation of Sprinkler Systems in Residential Occupancies up to and Including Four Stories in Height*, 2010 edition.

NFPA 14, *Standard for the Installation of Standpipe and Hose Systems*, 2010 edition.

NFPA 22, *Standard for Water Tanks for Private Fire Protection*, 2008 edition.

NFPA 25, *Standard for the Inspection, Testing, and Maintenance of Water-Based Fire Protection Systems*, 2011 edition.

NFPA 30, *Flammable and Combustible Liquids Code*, 2012 edition.

NFPA 58, *Liquefied Petroleum Gas Code*, 2011 edition.

NFPA 61, *Standard for the Prevention of Fires and Dust Explosions in Agricultural and Food Processing Facilities*, 2008 edition.

NFPA 68, *Standard on Explosion Protection by Deflagration Venting*, 2007 edition.

*NFPA 70®, National Electrical Code®*, 2011 edition.

*NFPA 72®, National Fire Alarm and Signaling Code*, 2010 edition.

NFPA 80, *Standard for Fire Doors and Other Opening Protectives*, 2010 edition.

NFPA 90A, *Standard for the Installation of Air-Conditioning and Ventilating Systems*, 2012 edition.

NFPA 92, *Standard for Smoke Control Systems*, 2012 edition.

NFPA 96, *Standard for Ventilation Control and Fire Protection of Commercial Cooking Operations*, 2011 edition.

NFPA 99, *Health Care Facilities Code*, 2012 edition.

NFPA 101A, *Guide on Alternative Approaches to Life Safety*, 2010 edition.

NFPA 105, *Standard for Smoke Door Assemblies and Other Opening Protectives*, 2010 edition.

NFPA 110, *Standard for Emergency and Standby Power Systems*, 2010 edition.

NFPA 170, *Standard for Fire Safety and Emergency Symbols*, 2009 edition.

NFPA 204, *Standard for Smoke and Heat Venting*, 2012 edition.

NFPA 211, *Standard for Chimneys, Fireplaces, Vents, and Solid Fuel–Burning Appliances*, 2010 edition.

NFPA 220, *Standard on Types of Building Construction*, 2012 edition.

NFPA 241, *Standard for Safeguarding Construction, Alteration, and Demolition Operations*, 2009 edition.

NFPA 251, *Standard Methods of Tests of Fire Resistance of Building Construction and Materials*, 2006 edition.

NFPA 253, *Standard Method of Test for Critical Radiant Flux of Floor Covering Systems Using a Radiant Heat Energy Source*, 2011 edition.

NFPA 259, *Standard Test Method for Potential Heat of Building Materials*, 2008 edition.

NFPA 260, *Standard Methods of Tests and Classification System for Cigarette Ignition Resistance of Components of Upholstered Furniture*, 2009 edition.

NFPA 261, *Standard Method of Test for Determining Resistance of Mock-Up Upholstered Furniture Material Assemblies to Ignition by Smoldering Cigarettes*, 2009 edition.

NFPA 265, *Standard Methods of Fire Tests for Evaluating Room Fire Growth Contribution of Textile or Expanded Vinyl Wall Coverings on Full Height Panels and Walls*, 2011 edition.

NFPA 269, *Standard Test Method for Developing Toxic Potency Data for Use in Fire Hazard Modeling*, 2007 edition.

NFPA 286, *Standard Methods of Fire Tests for Evaluating Contribution of Wall and Ceiling Interior Finish to Room Fire Growth*, 2011 edition.

NFPA 289, *Standard Method of Fire Test for Individual Fuel Packages*, 2009 edition.

NFPA 307, *Standard for the Construction and Fire Protection of Marine Terminals, Piers, and Wharves*, 2011 edition.

NFPA 409, *Standard on Aircraft Hangars*, 2011 edition.

NFPA 501A, *Standard for Fire Safety Criteria for Manufactured Home Installations, Sites, and Communities*, 2009 edition.

NFPA 551, *Guide for the Evaluation of Fire Risk Assessments*, 2010 edition.

NFPA 601, *Standard for Security Services in Fire Loss Prevention*, 2010 edition.

NFPA 701, *Standard Methods of Fire Tests for Flame Propagation of Textiles and Films*, 2010 edition.

NFPA 703, *Standard for Fire Retardant–Treated Wood and Fire-Retardant Coatings for Building Materials*, 2012 edition.

NFPA 720, *Standard for the Installation of Carbon Monoxide (CO) Detection and Warning Equipment*, 2012 edition.

NFPA 914, *Code for Fire Protection of Historic Structures*, 2010 edition.

NFPA 1124, *Code for the Manufacture, Transportation, Storage, and Retail Sales of Fireworks and Pyrotechnic Articles*, 2006 edition.

NFPA 1221, *Standard for the Installation, Maintenance, and Use of Emergency Services Communications Systems*, 2010 edition.

NFPA 1600®, *Standard on Disaster/Emergency Management and Business Continuity Programs*, 2010 edition.

NFPA 5000®, *Building Construction and Safety Code®*, 2012 edition.

NFPA *Fire Protection Handbook*, 19th edition, 2003.

NFPA *Fire Protection Handbook*, 20th edition, 2008.

Waksman, D. and J. B. Ferguson. August 1974. Fire Tests of Building Interior Covering Systems. In *Fire Technology* 10:211 – 220.

*SFPE Handbook of Fire Protection Engineering*, 4th edition.

### C.1.2 Other Publications.

**C.1.2.1 ACI Publication.** American Concrete Institute, P.O. Box 9094, Farmington Hills, MI 48333. www.concrete.org

ACI 216.1/TMS 0216.1, *Code Requirements for Determining Fire Resistance of Concrete and Masonry Construction Assemblies*, 2008.

**C.1.2.2 ANSI Publications.** American National Standards Institute, Inc., 25 West 43rd Street, 4th floor, New York, NY 10036. www.ansi.org

ANSI/BHMA A156.10, *American National Standard for Power Operated Pedestrian Doors*, 1999.

ANSI/BHMA A156.19, *American National Standard for Power Assist and Low Energy Power Operated Doors*, 2002.

ICC/ANSI A117.1, *American National Standard for Accessible and Usable Buildings and Facilities*, 2009.

**C.1.2.3 ASCE Publications.** American Society of Civil Engineers, 1801 Alexander Bell Drive, Reston, VA 20191-4400. www.asce.org

ASCE/SFPE 29, *Standard Calculation Methods for Structural Fire Protection*, 2005.

**C.1.2.4 ASHRAE Publications.** American Society of Heating, Refrigerating and Air Conditioning Engineers, Inc., 1791 Tullie Circle, NE, Atlanta, GA 30329-2305. www.ashrae.org

ASHRAE *Handbook and Product Directory — Fundamentals*, 2001.

Klote, J.H., and Milke, J.A., *Principles of Smoke Management*, 2002.

**C.1.2.5 ASME Publications.** American Society of Mechanical Engineers, Three Park Avenue, New York, NY 10016-5990. www.asme.org

ASME A17.1/CSA B44, *Safety Code for Elevators and Escalators*, 2006.

ASME A17.3, *Safety Code for Existing Elevators and Escalators*, 2005.

**C.1.2.6 ASTM Publications.** ASTM International, 100 Barr Harbor Drive, P.O. Box C700, West Conshohocken, PA 19428-2959. www.astm.org

ASTM C 1629/C 1629M, *Standard Classification for Abuse-Resistant Nondecorated Interior Gypsum Panel Products and Fiber-Reinforced Cement Panels*, 2006.

ASTM D 2859, *Standard Test Method for Ignition Characteristics of Finished Textile Floor Covering Materials*, 2006.

ASTM E 84, *Standard Test Method for Surface Burning Characteristics of Building Materials*, 2010.

ASTM E 119, *Standard Test Methods for Fire Tests of Building Construction and Materials*, 2010a.

ASTM E 814, *Standard Test Method for Fire Tests of Through-Penetration Fire Stops*, 2010.

ASTM E 1352, *Standard Test Method for Cigarette Ignition Resistance of Mock-Up Upholstered Furniture Assemblies*, 2008a.

ASTM E 1353, *Standard Test Methods for Cigarette Ignition Resistance of Components of Upholstered Furniture*, 2008a.

ASTM E 1355, *Standard Guide for Evaluating the Predictive Capability of Deterministic Fire Models*, 2005a.

ASTM E 1472, *Standard Guide for Documenting Computer Software for Fire Models*, 2007.

ASTM E 1537, *Standard Test Method for Fire Testing of Upholstered Furniture*, 2007.

ASTM E 1590, *Standard Test Method for Fire Testing of Mattresses*, 2007.

ASTM E 1966, *Standard Test Method for Fire-Resistive Joint Systems*, 2007.

ASTM E 2030, *Standard Guide for Recommended Uses of Photoluminescent (Phosphorescent) Safety Markings*, 2009a.

ASTM E 2174, *Standard Practice for On-Site Inspection of Installed Fire Stops*, 2009.

ASTM E 2238, *Standard Guide for Evacuation Route Diagrams*, 2002.

ASTM E 2280, *Standard Guide for Fire Hazard Assessment of the Effect of Upholstered Seating Furniture Within Patient Rooms of Health Care Facilities*, 2003.

ASTM E 2307, *Standard Test Method for Determining Fire Resistance of Perimeter Fire Barrier Systems Using Intermediate-Scale, Multi-Story Test Apparatus*, 2010.

ASTM E 2393, *Standard Practice for On-Site Inspection of Installed Fire Resistive Joint Systems and Perimeter Fire Barriers*, 2010.

ASTM E 2484, *Standard Specification for Multi-Story Building External Evacuation Controlled Descent Devices*, 2008.

ASTM E 2513, *Standard Specification for Multi-Story Building External Evacuation Platform Rescue Systems*, 2007.

ASTM F 1637, *Standard Practice for Safe Walking Surfaces*, 2009.

ASTM F 1870, *Standard Guide for Selection of Fire Test Methods for the Assessment of Upholstered Furnishings in Detention and Correctional Facilities*, 2005.

**C.1.2.7 California Technical Bulletins.** State of California, Department of Consumer Affairs, Bureau of Home Furnishings and Thermal Insulation, 3485 Orange Grove Avenue, North Highlands, CA 95660-5595.

Technical Bulletin 129, "Flammability Test Procedure for Mattresses for Use in Public Buildings," October 1992.

Technical Bulletin 133, "Flammability Test Procedure for Seating Furniture for Use in Public Occupancies," January 1991.

**C.1.2.8 FMGR Publications.** FM Global Research, FM Global, 1301 Atwood Avenue, P.O. Box 7500, Johnston, RI 02919. www.fmglobal.com

FM 4880, *Approval Standard for Class I Insulated Wall or Wall and Roof/Ceiling Panels; Plastic Interior Finish Materials; Plastic Exterior Building Panels; Wall/Ceiling Coating Systems; Interior or Exterior Finish Systems*, 1994.

FM Approval Standard 6921, *Containers for Combustible Waste*, 2004.

**C.1.2.9 NEMA Publications.** National Electrical Manufacturers Association, 1300 North 17th Street, Suite 1847, Rosslyn, VA 22209.

ANSI/NEMA Z535.1, *Standard for Safety Colors*, 2006.

**C.1.2.10 NIST Publications.** National Institute of Standards and Technology, 100 Bureau Drive, Gaithersburg, MD 20899-1070. www.nist.gov

NISTIR 5445, *Feasibility of Fire Evacuation by Elevators at FAA Control Towers*, 1994.

**C.1.2.11 SFPE Publications.** Society of Fire Protection Engineers, 7315 Wisconsin Avenue, Suite 1225 W, Bethesda, MD 20814. www.sfpe.org

*SFPE Computer Software Directory.*

*SFPE Enforcer's Guide to Performance-Based Design Review.*

*SFPE Engineering Guide — Evaluation of the Computer Fire Model DETACT-QS.*

*SFPE Engineering Guide to Human Behavior in Fire.*

*SFPE Engineering Guide to Performance-Based Fire Protection Analysis and Design of Buildings*, 1998.

*SFPE Guidelines for Peer Review in the Fire Protection Design Process.*

**C.1.2.12 UL Publications.** Underwriters Laboratories Inc., 333 Pfingsten Road, Northbrook, IL 60062-2096. www.ul.com

UL *Fire Resistance Directory*, 2010.

ANSI/UL 263, *Standard for Fire Tests of Building Construction and Materials*, 2003, Revised 2007.

ANSI/UL 723, *Standard for Test for Surface Burning Characteristics of Building Materials*, 2008, Revised 2010.

ANSI/UL 1040, *Standard for Fire Test of Insulated Wall Construction*, 1996, Revised 2007.

ANSI/UL 1479, *Standard for Fire Tests of Through-Penetration Firestops*, 2003, Revised 2010.

ANSI/UL 1715, *Standard for Fire Test of Interior Finish Material*, 1997, Revised 2008.

UL 1975, *Standard for Fire Tests for Foamed Plastics Used for Decorative Purposes*, 2006.

ANSI/UL 2079, *Standard for Tests for Fire Resistance of Building Joint Systems*, 2004, Revised 2008.

**C.1.2.13 U.S. Government Publications.** U.S. Government Printing Office, Washington, DC 20402. www.access.gpo.gov/

Title 16, *Code of Federal Regulations*, Part 1630, "Standard for the Surface Flammability of Carpets and Rugs" (FF 1-70).

Title 16, *Code of Federal Regulations*, Part 1632, "Standard for the Flammability of Mattresses and Mattress Pads" (FF 4-72).

Title 28, *Code of Federal Regulations*, Part 36, Appendix A, "Americans with Disabilities Act Accessibility Guidelines for Buildings and Facilities."

Title 29, *Code of Federal Regulations*, Part 1910, Subparts E and L, "OSHA Regulations for Emergency Procedures and Fire Brigades."

Title 29, *Code of Federal Regulations*, Part 1910.146, "Permit-Required Confined Spaces."

Lee, A and Pineda, D. 2010, *Smoke Alarms – Pilot Study of Nuisance Alarms Associated with Cooking*, Bethesda, MD: US Consumer Product Safety Commission.

**C.1.2.14 Other Publications** *Australian Fire Engineering Guidelines*. 1996. Sydney, Australia: Fire Code Perform Centre, Ltd.

*British Standard Firesafety Engineering in Buildings*, DD240: Part 1. 1997. London, England: British Standards Institution.

Gann, R. G., V. Babrauskas, R. D. Peacock, and J. R. Hall. 1994. Fire conditions for smoke toxicity measurement. *Fire and Materials* 18(193): 193–99.

Kaplan, H. L., and G. E. Hartzell. 1984. Modeling of toxicological effects of fire gases: I. Incapacitation effects of narcotic fire gases. *Journal of Fire Sciences* 2:286–305.

Hirschler et al., "Carbon monoxide and human lethality: Fire and non-fire studies," Elsevier, 1993.

Olenick, S., and D. Carpenter. 2003. An updated international survey of computer models for fire and smoke. *Journal of Fire Protection Engineering* 3(2):87–110.

**C.2 Informational References.** The following documents or portions thereof are listed here as informational resources only. They are not a part of the requirements of this document.

Endsley, Bolte, and Jones. *Designing for Situation Awareness: An approach to user-centered design*. 2003. Boca Raton, FL: CRC Press, Taylor and Francis.

Freeman, J. R. 1889. Experiments relating to hydraulics of fire streams." Paper No. 426, *Transactions*, American Society of Civil Engineers, XXI:380–83.

Groner, N. E., and M. L. Levin. 1992. Human factor considerations in the potential for using elevators in building emergency evacuation plans, NIST-GCR-92-615. Gaithersburg, MD: National Institute of Standards and Technology.

Klote, J. H., B. M. Levin, and N. E. Groner. 1994. Feasibility of fire evacuations by elevators at FAA control towers, NISTIR 5445. Gaithersburg, MD: National Institute of Standards and Technology.

Klote, J. H., B. M. Levin, and N. E Groner. "Feasibility of Fire Evacuation by Elevators at FAA Control Towers," National Institute of Standards and Technology, NISTIR 5443, 1994.

Levin, B. M., and N. E. Groner. 1992. Human behavior aspects of staging areas for fire safety in GSA buildings, NIST-

GCR-92-606. Gaithersburg, MD: National Institute of Standards and Technology.

Levin, B. M., and N. E. Groner. 1994. Human factor considerations for the potential use of elevators for fire evacuation of FAA air traffic control towers, NIST-GCR-94-656. Gaithersburg, MD: National Institute of Standards and Technology.

Seigel, L. G. 1969. The protection of flames from burning buildings. *Fire Technology* 5(1):43–51.

Templer, J. A. 1992. *The Staircase: Studies of Hazards, Falls, and Safer Design.* Cambridge, MA: MIT Press.

Tu, K.-M., and S. Davis. 1976. Flame spread of carpet systems involved in room fires, NFSIR 76-1013. Washington, DC: Center for Fire Research, National Bureau of Standards.

### C.3 References for Extracts in Informational Sections.

NFPA 80, *Standard for Fire Doors and Other Opening Protectives,* 2010 edition.

*NFPA 5000®, Building Construction and Safety Code®,* 2012 edition.

# Index

## -F-

Tentative Interim Amendment

# NFPA® 101
## Life Safety Code®

### 2012 Edition

**Reference:** 18.3.2.5.3(11) – (12), 19.3.2.5.3(11) – (13), related 18.3.4 and 19.3.4 alarm system provisions, and associated advisory annex

**TIA 12-2**

*(SC 12-10-6/TIA Log #1075)*

Pursuant to Section 5 of the NFPA Regulations Governing Committee Projects, the National Fire Protection Association has issued the following Tentative Interim Amendment to NFPA 101, *Life Safety Code,* 2012 edition. The TIA was processed by the Technical Committee on Health Care Occupancies and the Correlating Committee on Safety to Life, and was issued by the Standards Council on October 30, 2012, with an effective date of November 19, 2012.

A Tentative Interim Amendment is tentative because it has not been processed through the entire standards-making procedures. It is interim because it is effective only between editions of the standard. A TIA automatically becomes a proposal of the proponent for the next edition of the standard; as such, it then is subject to all of the procedures of the standards-making process.

*1. Revise 18.3.2.5.3 (11) – (12), 19.3.2.5.3 (11) – (13), related 18.3.4 and 19.3.4 alarm system provisions, and associated advisory annex as follows:*

| Chapter 18  New Health Care Occupancies | Chapter 19  Existing Health Care Occupancies |
|---|---|
| **18.3.2.5.3*** Within a smoke compartment, where residential or commercial cooking equipment is used to prepare meals for 30 or fewer persons, one cooking facility shall be permitted to be open to the corridor, provided that all of the following conditions are met:<br>(1)    The portion of the health care facility served by the cooking facility is limited to 30 beds and is separated from other portions of the health care facility by a smoke barrier constructed in accordance with 18.3.7.3, 18.3.7.6, and 18.3.7.8.<br>(2)    The cooktop or range is equipped with a range hood of a width at least equal to the width of the cooking surface, with grease baffles or other grease-collecting and clean-out capability.<br>**(3)*** The hood systems have a minimum airflow of 500 cfm (14,000 L/min).<br>(4)    The hood systems that are not ducted to the exterior additionally have a charcoal filter to remove smoke and odor.<br>(5)    The cooktop or range complies with all of the following:<br>(a)    The cooktop or range is protected with a fire suppression system listed in accordance with UL 300, *Standard for Fire Testing of Fire Extinguishing Systems for Protection of Commercial Cooking Equipment,* or is tested and meets all requirements of UL 300A, *Extinguishing System Units for Residential Range Top* | **19.3.2.5.3*** Within a smoke compartment, where residential or commercial cooking equipment is used to prepare meals for 30 or fewer persons, one cooking facility shall be permitted to be open to the corridor, provided that all of the following conditions are met:<br>(1)    The portion of the health care facility served by the cooking facility is limited to 30 beds and is separated from other portions of the health care facility by a smoke barrier constructed in accordance with 19.3.7.3, 19.3.7.6, and 19.3.7.8.<br>(2)    The cooktop or range is equipped with a range hood of a width at least equal to the width of the cooking surface, with grease baffles or other grease-collecting and clean-out capability.<br>**(3)*** The hood systems have a minimum airflow of 500 cfm (14,000 L/min).<br>(4)    The hood systems that are not ducted to the exterior additionally have a charcoal filter to remove smoke and odor.<br>(5)    The cooktop or range complies with all of the following:<br>(a)    The cooktop or range is protected with a fire suppression system listed in accordance with UL 300, *Standard for Fire Testing of Fire Extinguishing Systems for Protection of Commercial Cooking Equipment,* or is tested and meets all requirements of UL 300A, *Extinguishing System Units for Residential Range Top* |

*Cooking Surfaces*, in accordance with the applicable testing document's scope.

(b)     A manual release of the extinguishing system is provided in accordance with NFPA 96, *Standard for Ventilation Control and Fire Protection of Commercial Cooking Operations*, Section 10.5.

(c)     An interlock is provided to turn off all sources of fuel and electrical power to the cooktop or range when the suppression system is activated.

**(6)***     The use of solid fuel for cooking is prohibited.

**(7)***     Deep-fat frying is prohibited

(8)     Portable fire extinguishers in accordance with NFPA 96 are located in all kitchen areas.

**(9)***     A switch meeting all of the following is provided:

(a)     A locked switch, or a switch located in a restricted location, is provided within the cooking facility that deactivates the cooktop or range.

(b)     The switch is used to deactivate the cooktop or range whenever the kitchen is not under staff supervision.

(c)     The switch is on a timer, not exceeding a 120-minute capacity, that automatically deactivates the cooktop or range, independent of staff action.

(10)  Procedures for the use, inspection, testing, and maintenance of the cooking equipment are in accordance with Chapter 11 of NFPA 96 and the manufacturer's instructions and are followed.

**(11)*** Not less than two AC-powered photoelectric smoke alarms with battery backup, interconnected in accordance with 9.6.2.10.3, and equipped with a silence feature are located not closer than 20 ft (6.1 m) and not further than 25 ft (7.6 m) from the cooktop or range.

(12)*  The smoke alarms required by 18.3.2.5.3(11) are permitted to be located outside the kitchen area where such placement is necessary for compliance with the 20-ft (7.6-m) minimum distance criterion.

(13)*  A single system smoke detector is permitted to be installed in lieu of the smoke alarms required in 18.3.2.5.3(11) provided the following criteria are met:

(a) The detector is located not closer than 20 ft (6.1 m) and not further than 25 ft (7.6 m) from the cooktop or range.

(b) The detector is permitted to initiate a local audible alarm signal only.

(c) The detector is not required to initiate a building-wide occupant notification signal.

(d) The detector is not required to notify the emergency forces.

(e) The local audible signal initiated by the detector is permitted to be silenced and reset by a button on the detector or by a switch installed within 10 ft (3.0 m) of the system smoke detector.

(14) System smoke detectors that are required to be installed in corridors or spaces open to the corridor by other sections of this chapter are not used to meet the requirements of 18.3.2.5.3(11) and are located not closer than 25 ft (7.6 m) to the cooktop or range.

*Cooking Surfaces*, in accordance with the applicable testing document's scope.

(b)     A manual release of the extinguishing system is provided in accordance with NFPA 96, *Standard for Ventilation Control and Fire Protection of Commercial Cooking Operations*, Section 10.5.

(c)     An interlock is provided to turn off all sources of fuel and electrical power to the cooktop or range when the suppression system is activated.

**(6)***     The use of solid fuel for cooking is prohibited.

**(7)***     Deep-fat frying is prohibited.

(8)     Portable fire extinguishers in accordance with NFPA 96 are located in all kitchen areas.

**(9)***     A switch meeting all of the following is provided:

(a)     A locked switch, or a switch located in a restricted location, is provided within the cooking facility that deactivates the cooktop or range.

(b)     The switch is used to deactivate the cooktop or range whenever the kitchen is not under staff supervision.

(c)     The switch is on a timer, not exceeding a 120-minute capacity, that automatically deactivates the cooktop or range, independent of staff action.

(10)  Procedures for the use, inspection, testing, and maintenance of the cooking equipment are in accordance with Chapter 11 of NFPA 96 and the manufacturer's instructions and are followed.

**(11)*** Not less than two AC-powered photoelectric smoke alarms with battery backup, interconnected in accordance with 9.6.2.10.3, and equipped with a silence feature are located not closer than 20 ft (6.1 m) and not further than 25 ft (7.6 m) from the cooktop or range.

(12)*  The smoke alarms required by 19.3.2.5.3(11) are permitted to be located outside the kitchen area where such placement is necessary for compliance with the 20-ft (7.6-m) minimum distance criterion.

(13)*  A single system smoke detector is permitted to be installed in lieu of the smoke alarms required in 19.3.2.5.3(11) provided the following criteria are met:

(a) The detector is located not closer than 20 ft (6.1 m) and not further than 25 ft (7.6 m) from the cooktop or range.

(b) The detector is permitted to initiate a local audible alarm signal only.

(c) The detector is not required to initiate a building-wide occupant notification signal.

(d) The detector is not required to notify the emergency forces.

(e) The local audible signal initiated by the detector is permitted to be silenced and reset by a button on the detector or by a switch installed within 10 ft (3.0 m) of the system smoke detector.

(14) System smoke detectors that are required to be installed in corridors or spaces open to the corridor by other sections of this chapter are not used to meet the requirements of 19.3.2.5.3(11) and are located not closer than 25 ft (7.6 m) to the cooktop or range.

(15) The smoke compartment is protected throughout by an approved, supervised automatic sprinkler system in accordance with Section 9.7.

**18.3.4.2.1** Initiation of the required fire alarm systems shall be by manual means in accordance with 9.6.2 and by means of any required sprinkler system waterflow alarms, detection devices, or detection systems, unless otherwise permitted by 18.3.4.2.2 and 18.3.4.2.3.

---

**18.3.4.2.3** The system smoke detector installed in accordance with 18.3.2.5.3(13) shall not be required to initiate the fire alarm system.

---

**18.3.4.3.1 Occupant Notification.** Occupant notification shall be accomplished automatically in accordance with 9.6.3, unless otherwise modified by the following:

(1) Paragraph 9.6.3.2.3 shall not be permitted to be used.

(2)*In lieu of audible alarm signals, visible alarm-indicating appliances shall be permitted to be used in critical care areas.

(3) The provision of 18.3.2.5.3(13)(c) shall be permitted to be used.

**18.3.4.3.2.1** Emergency forces notification shall be accomplished in accordance with 9.6.4, except that the provision of 18.3.2.5.3(13)(d) shall be permitted to be used.

---

**A.18.3.2.5.3**   The intent of 18.3.2.5.3 is to limit the number of persons for whom meals are routinely prepared to not more than 30. Staff and feeding assistants are not included in this number.
**A.18.3.2.5.3(3)**   The minimum airflow of 500 cfm (14,000 L/m) is intended to require the use of residential hood equipment at the higher end of equipment capacities. It is also intended to draw a sufficient amount of the cooking vapors into the grease baffle and filter system to reduce migration beyond the hood.
**A.18.3.2.5.3(6)**   The intent of this provision is to limit cooking fuel to gas or electricity. The prohibition of solid fuels for cooking is not intended to prohibit charcoal grilling on grills located outside the facility.
**A.18.3.2.5.3(7)**   Deep-fat frying is defined as a cooking method that involves fully immersing food in hot oil.
**A.18.3.2.5.3(9)**   The intent of this requirement is that the fuel source for the cooktop or range is to be turned on only when staff is present or aware that the kitchen is being used. The timer function is meant to provide an additional safeguard if the staff forgets to deactivate the cooktop or range. If a cooking activity lasts longer than 120 minutes, the timer would be required to be manually reset.

---

**19.3.4.2.1** Initiation of the required fire alarm systems shall be by manual means in accordance with 9.6.2 and by means of any required sprinkler system waterflow alarms, detection devices, or detection systems, unless otherwise permitted by 19.3.4.2.2 through 19.3.4.2.5.

---

**19.3.4.2.3** The system smoke detector installed in accordance with 19.3.2.5.3(13) shall not be required to initiate the fire alarm system.
**19.3.4.2.4** Fixed extinguishing systems protecting commercial cooking equipment in kitchens that are protected by a complete automatic sprinkler system shall not be required to initiate the fire alarm system.
**19.3.4.2.5** Detectors required by 19.7.5.3 and 19.7.5.5 shall not be required to initiate the fire alarm system.

---

**19.3.4.3.1 Occupant Notification.** Occupant notification shall be accomplished automatically in accordance with 9.6.3, unless otherwise modified by the following:
(1)*In lieu of audible alarm signals, visible alarm-indicating appliances shall be permitted to be used in critical care areas.
(2) Where visual devices have been installed in patient sleeping areas in place of an audible alarm, they shall be permitted where approved by the authority having jurisdiction.
(3) The provision of 19.3.2.5.3(13)(c) shall be permitted to be used.

**19.3.4.3.2.1** Emergency forces notification shall be accomplished in accordance with 9.6.4, except that the provision of 19.3.2.5.3(13)(d) shall be permitted to be used.

---

**A.19.3.2.5.3**   The intent of 19.3.2.5.3 is to limit the number of persons for whom meals are routinely prepared to not more than 30. Staff and feeding assistants are not included in this number.
**A.19.3.2.5.3(3)** The minimum airflow of 500 cfm (14,000 L/m) is intended to require the use of residential hood equipment at the higher end of equipment capacities. It is also intended to draw a sufficient amount of the cooking vapors into the grease baffle and filter system to reduce migration beyond the hood.
**A.19.3.2.5.3(6)**   The intent of this provision is to limit cooking fuel to gas or electricity. The prohibition of solid fuels for cooking is not intended to prohibit charcoal grilling on grills located outside the facility.
**A.19.3.2.5.3(7)**   Deep-fat frying is defined as a cooking method that involves fully immersing food in hot oil.
**A.19.3.2.5.3(9)**   The intent of this requirement is that the fuel source for the cooktop or range is to be turned on only when staff is present or aware that the kitchen is being used. The timer function is meant to provide an additional safeguard if the staff forgets to deactivate the cooktop or range. If a cooking activity lasts longer than 120 minutes, the timer would be required to be manually reset.

**A.18.3.2.5.3(11)** Protection of the cooktop or range is accomplished by the sprinklers that are required in the space and the required cooktop hood fire suppression system. The smoke alarms are intended to notify staff who might not be in the immediate area. Smoke alarms should be maintained a minimum of 20 ft (6.1 m) away from the cooktop or range as studies have shown this distance to be the threshold for significantly reducing nuisance alarms caused by cooking. The intent of the interconnected smoke alarms, with silence feature, is that while the devices would alert staff members to a potential problem, if it is a nuisance alarm, the staff members can use the silence feature instead of disabling the alarm. The referenced study indicates that nuisance alarms are reduced with photoelectric smoke alarms. Providing two, interconnected alarms provides a safety factor since they are not electrically supervised by the fire alarm system. *(Smoke Alarms – Pilot Study of Nuisance Alarms Associated with Cooking)*

**A.18.3.2.5.3(12)** The provision of 18.3.2.5.3(12) recognizes that it is more important to maintain the 20-ft (6.1-m) minimum spacing criterion between the smoke alarm and the cooktop or range, to minimize nuisance alarms, than to assure that the smoke alarm is located within the kitchen area itself.

**A.18.3.2.5.3(13)** The requirements of 18.3.2.5.3(13) are intended to allow the local staff to silence and reset the system smoke detector without the assistance of the engineering or maintenance personnel. This provision is not intended to require the system smoke detector to initiate a building-wide occupant alarm signal or to notify the emergency forces.

**A.19.3.2.5.3(11)** Protection of the cooktop or range is accomplished by the sprinklers that are required in the space and the required cooktop hood fire suppression system. The smoke alarms are intended to notify staff who might not be in the immediate area. Smoke alarms should be maintained a minimum of 20 ft (6.1 m) away from the cooktop or range as studies have shown this distance to be the threshold for significantly reducing nuisance alarms caused by cooking. The intent of the interconnected smoke alarms, with silence feature, is that while the devices would alert staff members to a potential problem, if it is a nuisance alarm, the staff members can use the silence feature instead of disabling the alarm. The referenced study indicates that nuisance alarms are reduced with photoelectric smoke alarms. Providing two, interconnected alarms provides a safety factor since they are not electrically supervised by the fire alarm system. *(Smoke Alarms – Pilot Study of Nuisance Alarms Associated with Cooking)*

**A.19.3.2.5.3(12)** The provision of 19.3.2.5.3(12) recognizes that it is more important to maintain the 20-ft (6.1-m) minimum spacing criterion between the smoke alarm and the cooktop or range, to minimize nuisance alarms, than to assure that the smoke alarm is located within the kitchen area itself.

**A.19.3.2.5.3(13)** The requirements of 19.3.2.5.3(13) are intended to allow the local staff to silence and reset the system smoke detector without the assistance of the engineering or maintenance personnel. This provision is not intended to require the system smoke detector to initiate a building-wide occupant alarm signal or to notify the emergency forces.

**Issue Date:** October 30, 2012

**Effective Date:** November 19, 2012

Tentative Interim Amendment

# NFPA 101®
## Life Safety Code®
### 2012 Edition

**Reference:** Table 17.1.6.1
**TIA 12-3**
*(SC 13-10-6/TIA Log #1113)*

Pursuant to Section 5 of the NFPA *Regulations Governing the Development of NFPA Standards*, the National Fire Protection Association has issued the following Tentative Interim Amendment to NFPA 101, *Life Safety Code®*, 2012 edition. The TIA was processed by the Technical Committee on Educational and Day-Care Occupancies and the Correlating Committee on Safety to Life, and was issued by the Standards Council on October 22, 2013, with an effective date of November 11, 2013.

A Tentative Interim Amendment is tentative because it has not been processed through the entire standards-making procedures. It is interim because it is effective only between editions of the standard. A TIA automatically becomes a public input of the proponent for the next edition of the standard; as such, it then is subject to all of the procedures of the standards-making process.

*1. Revise Table 17.1.6.1 to read as follows:*

**Table 17.1.6.1  Construction Type Limitations**

| Construction Type | Sprinklered[a] | Stories in Height[b] | | | | | |
|---|---|---|---|---|---|---|---|
| | | One Story Below[c] | 1 | 2 | 3–4 | >4 but Not High-Rise | High-Rise |
| I (442) | Yes | X | X | X | X | X | X |
| | No | X | X | X | X | X | NP |
| I (332) | Yes | X | X | X | X | X | X |
| | No | X | X | X | X | X | NP |
| II (222) | Yes | X | X | X | X | X | X |
| | No | X | X | X | X | X | NP |
| II (111) | Yes | X | X | X | X[d] | X[d] | NP |
| | No | X | X | X[d] | NP | NP | NP |
| II (000) | Yes | X | X | X | NP | NP | NP |
| | No | NP | X | NP | NP | NP | NP |
| III (211) | Yes | X | X | X | X[d] | NP | NP |
| | No | X | X | X[d] | NP | NP | NP |
| III (200) | Yes | NP | X | X | NP | NP | NP |
| | No | NP | X | NP | NP | NP | NP |
| IV (2HH) | Yes | X | X | X | NP | NP | NP |
| | No | X | X | X | NP | NP | NP |

**Table 17.1.6.1  Construction Type Limitations** *(continued)*

| Construction Type | Sprinklered[a] | Stories in Height[b] | | | | | |
|---|---|---|---|---|---|---|---|
| | | One Story Below[c] | 1 | 2 | 3–4 | >4 but Not High-Rise | High-Rise |
| V (111) | Yes | X | X | X | X[d] | NP | NP |
| | No | X | X | X[d] | NP | NP | NP |
| V (000) | Yes | NP | X | X | NP | NP | NP |
| | No | NP | X | NP | NP | NP | NP |

X: Permitted. NP: Not Permitted.

[a]Sprinklered throughout by an approved, supervised automatic sprinkler system in accordance with Section 9.7. *(See 17.3.5.)*

[b]See 4.6.3.

[c]One story below the level of exit discharge.

[d]Permitted only if clients capable of self-preservation.

**Issue Date:** October 22, 2013

**Effective Date:** November 11, 2013

Tentative Interim Amendment

# NFPA 101®
## Life Safety Code®
### 2012 Edition

**Reference:** 19.2.2.2.5.2
**TIA 12-4**
*(SC 13-10-7/TIA Log #1114)*

Pursuant to Section 5 of the NFPA *Regulations Governing the Development of NFPA Standards*, the National Fire Protection Association has issued the following Tentative Interim Amendment to NFPA 101, *Life Safety Code®*, 2012 edition. The TIA was processed by the Technical Committee on Health Care Occupancies and the Correlating Committee on Safety to Life, and was issued by the Standards Council on October 22, 2013, with an effective date of November 11, 2013.

A Tentative Interim Amendment is tentative because it has not been processed through the entire standards-making procedures. It is interim because it is effective only between editions of the standard. A TIA automatically becomes a public input of the proponent for the next edition of the standard; as such, it then is subject to all of the procedures of the standards-making process.

*1. Revise 19.2.2.2.5.2 to read as follows:*

**19.2.2.2.5.2\*** Door-locking arrangements shall be permitted where patient special needs require specialized protective measures for their safety, provided that all of the following are met:

(1)   Staff can readily unlock doors at all times in accordance with 19.2.2.2.6.

(2)   A total (complete) smoke detection system is provided throughout the locked space in accordance with 9.6.2.9, or locked doors can be remotely unlocked at an approved, constantly attended location within the locked space.

**(3)\***   The building is protected throughout by an approved, supervised automatic sprinkler system in accordance with 19.3.5.7.

(4)   The locks are electrical locks that fail safely so as to release upon loss of power to the device.

(5)   The locks release by independent activation of each of the following:

   (a)   Activation of the smoke detection system required by 19.2.2.2.5.2(2)

   (b)   Waterflow in the automatic sprinkler system required by 19.2.2.2.5.2(3)

**Issue Date:** October 22, 2013

**Effective Date:** November 11, 2013

# Sequence of Events Leading to Issuance of an NFPA Committee Document

## Step 1: Call for Proposals

- Proposed new Document or new edition of an existing Document is entered into one of two yearly revision cycles, and a Call for Proposals is published.

## Step 2: Report on Proposals (ROP)

- Committee meets to act on Proposals, to develop its own Proposals, and to prepare its Report.
- Committee votes by written ballot on Proposals. If two-thirds approve, Report goes forward. Lacking two-thirds approval, Report returns to Committee.
- Report on Proposals (ROP) is published for public review and comment.

## Step 3: Report on Comments (ROC)

- Committee meets to act on Public Comments to develop its own Comments, and to prepare its report.
- Committee votes by written ballot on Comments. If two-thirds approve, Report goes forward. Lacking two-thirds approval, Report returns to Committee.
- Report on Comments (ROC) is published for public review.

## Step 4: Technical Report Session

- *"Notices of intent to make a motion"* are filed, are reviewed, and valid motions are certified for presentation at the Technical Report Session. ("Consent Documents" that have no certified motions bypass the Technical Report Session and proceed to the Standards Council for issuance.)
- NFPA membership meets each June at the Annual Meeting Technical Report Session and acts on Technical Committee Reports (ROP and ROC) for Documents with "certified amending motions."
- Committee(s) vote on any amendments to Report approved at NFPA Annual Membership Meeting.

## Step 5: Standards Council Issuance

- Notification of intent to file an appeal to the Standards Council on Association action must be filed within 20 days of the NFPA Annual Membership Meeting.
- Standards Council decides, based on all evidence, whether or not to issue Document or to take other action, including hearing any appeals.

# Committee Membership Classifications

The following classifications apply to Technical Committee members and represent their principal interest in the activity of the committee.

M    *Manufacturer:* A representative of a maker or marketer of a product, assembly, or system, or portion thereof, that is affected by the standard.

U    *User:* A representative of an entity that is subject to the provisions of the standard or that voluntarily uses the standard.

I/M    *Installer/Maintainer:* A representative of an entity that is in the business of installing or maintaining a product, assembly, or system affected by the standard.

L    *Labor:* A labor representative or employee concerned with safety in the workplace.

R/T    *Applied Research/Testing Laboratory:* A representative of an independent testing laboratory or independent applied research organization that promulgates and/or enforces standards.

E    *Enforcing Authority:* A representative of an agency or an organization that promulgates and/or enforces standards.

I    *Insurance:* A representative of an insurance company, broker, agent, bureau, or inspection agency.

C    *Consumer:* A person who is, or represents, the ultimate purchaser of a product, system, or service affected by the standard, but who is not included in the *User* classification.

SE    *Special Expert:* A person not representing any of the previous classifications, but who has a special expertise in the scope of the standard or portion thereof.

NOTES:
1. "Standard" connotes code, standard, recommended practice, or guide.
2. A representative includes an employee.
3. While these classifications will be used by the Standards Council to achieve a balance for Technical Committees, the Standards Council may determine that new classifications of members or unique interests need representation in order to foster the best possible committee deliberations on any project. In this connection, the Standards Council may make appointments as it deems appropriate in the public interest, such as the classification of "Utilities" in the National Electrical Code Committee.
4. Representatives of subsidiaries of any group are generally considered to have the same classification as the parent organization.

# NFPA Document Proposal Form

For further information on the standards-making process, please contact the Codes and Standards Administration at 617-984-7249 or visit www.nfpa.org/codes.

For technical assistance, please call NFPA at 1-800-344-3555.

**FOR OFFICE USE ONLY**

Log #: _____

Date Rec'd: _____

**Please indicate in which format you wish to receive your ROP/ROC** ☐ electronic ☐ paper ☒ download
(Note: If choosing the download option, you must view the ROP/ROC from our website; no copy will be sent to you.)

**Date** April 1, 200X   **Name** John J. Doe   **Tel. No.** 716-555-1234

**Company** Air Canada Pilot's Association   **Email**

**Street Address** 123 Summer Street Lane   **City** Lewiston   **State** NY   **Zip** 14092

***If you wish to receive a hard copy, a street address MUST be provided. Deliveries cannot be made to PO boxes.*

**Please indicate organization represented (if any)** _____

**1. (a) NFPA Document Title** National Fuel Gas Code   **NFPA No. & Year** 54, 200X Edition

**(b) Section/Paragraph** 3.3

**2. Proposal Recommends (check one):** ☐ new text  ☒ revised text  ☐ deleted text

**3. Proposal (include proposed new or revised wording, or identification of wording to be deleted):** [Note: Proposed text should be in legislative format; i.e., use underscore to denote wording to be inserted (inserted wording) and strike-through to denote wording to be deleted (deleted wording).]

Revise definition of effective ground-fault current path to read:

3.3.78 Effective Ground-Fault Current Path. An intentionally constructed, permanent, low impedance electrically conductive path designed and intended to carry ~~underground~~ electric fault current ~~conditions~~ from the point of a ground fault on a wiring system to the electrical supply source.

**4. Statement of Problem and Substantiation for Proposal:** (Note: State the problem that would be resolved by your recommendation; give the specific reason for your Proposal, including copies of tests, research papers, fire experience, etc. If more than 200 words, it may be abstracted for publication.)

Change uses proper electrical terms.

**5. Copyright Assignment**

(a) ☐ I am the author of the text or other material (such as illustrations, graphs) proposed in the Proposal.

(b) ☒ Some or all of the text or other material proposed in this Proposal was not authored by me. Its source is as follows: (please identify which material and provide complete information on its source)

ABC Co.

*I hereby grant and assign to the NFPA all and full rights in copyright in this Proposal and understand that I acquire no rights in any publication of NFPA in which this Proposal in this or another similar or analogous form is used. Except to the extent that I do not have authority to make an assignment in materials that I have identified in (b) above, I hereby warrant that I am the author of this Proposal and that I have full power and authority to enter into this assignment.*

**Signature (Required)** *John J. Doe*

**PLEASE USE SEPARATE FORM FOR EACH PROPOSAL**

Mail to: Secretary, Standards Council · National Fire Protection Association
1 Batterymarch Park · Quincy, MA 02169-7471 OR
Fax to: (617) 770-3500 OR Email to: proposals_comments@nfpa.org

06/09-B

# NFPA Document Proposal Form

**NOTE:** All Proposals must be received by 5:00 pm EST/EDST on the published Proposal Closing Date.

For further information on the standards-making process, please contact the Codes and Standards Administration at 617-984-7249 or visit www.nfpa.org/codes.

For technical assistance, please call NFPA at 1-800-344-3555.

Please indicate in which format you wish to receive your ROP/ROC ☐ electronic ☐ paper ☐ download
(Note: If choosing the download option, you must view the ROP/ROC from our website; no copy will be sent to you.)

Date _____ Name _____ Tel. No. _____

Company _____ Email _____

Street Address _____ City _____ State _____ Zip _____

***If you wish to receive a hard copy, a street address MUST be provided. Deliveries cannot be made to PO boxes.**

Please indicate organization represented (if any) _____

1. (a) NFPA Document Title _____ NFPA No. & Year _____

   (b) Section/Paragraph _____

2. **Proposal Recommends (check one):**  ☐ new text  ☐ revised text  ☐ deleted text

3. **Proposal (include proposed new or revised wording, or identification of wording to be deleted):** [Note: Proposed text should be in legislative format; i.e., use underscore to denote wording to be inserted (inserted wording) and strike-through to denote wording to be deleted (deleted wording).]

4. **Statement of Problem and Substantiation for Proposal:** (Note: State the problem that would be resolved by your recommendation; give the specific reason for your Proposal, including copies of tests, research papers, fire experience, etc. If more than 200 words, it may be abstracted for publication.)

5. Copyright Assignment

   (a) ☐ I am the author of the text or other material (such as illustrations, graphs) proposed in the Proposal.

   (b) ☐ Some or all of the text or other material proposed in this Proposal was not authored by me. Its source is as follows: (please identify which material and provide complete information on its source)

*I hereby grant and assign to the NFPA all and full rights in copyright in this Proposal and understand that I acquire no rights in any publication of NFPA in which this Proposal in this or another similar or analogous form is used. Except to the extent that I do not have authority to make an assignment in materials that I have identified in (b) above, I hereby warrant that I am the author of this Proposal and that I have full power and authority to enter into this assignment.*

**Signature (Required)** _____

## PLEASE USE SEPARATE FORM FOR EACH PROPOSAL

Mail to: Secretary, Standards Council · National Fire Protection Association
1 Batterymarch Park · Quincy, MA 02169-7471 OR
Fax to: (617) 770-3500 OR Email to: proposals_comments@nfpa.org

06/09-C